Intermediate Algebra

Functions & Authentic Applications

Third Edition

Jay Lehmann

College of San Mateo

To Keri, who read with beginner's eyes
and counseled with words so wise.

Upper Saddle River, New Jersey 07458

Library of Congress Cataloging-in-Publication Data
Lehmann, Jay.
Intermediate algebra : functions and authentic applications / Jay Lehmann. -- 3rd ed.
p. cm.
Includes bibliographical references and index.
ISBN 0-13-195333-8
1. Algebra I. Title.
QA154.3.L44 2008
512.9--dc22

2006032589

Executive Editor: *Paul Murphy*
Editorial Director, Mathematics: *Christine Hoag*
Project Manager: *Mary Beckwith*
Senior Development Editor: *Karen Karlin*
Media Project Manager, Developmental Mathematics: *Audra J. Walsh*
Senior Managing Editor: *Linda Mihatov Behrens*
Executive Managing Editor: *Kathleen Schiaparelli*
Assistant Managing Editor: *Bayani Mendoza de Leon*
Manufacturing Buyer: *Maura Zaldivar*
Manufacturing Manager: *Alexis Heydt-Long*
Director of Marketing: *Patrice Jones*
Senior Marketing Manager: *Kate Valentine*
Marketing Assistant: *Jennifer de Leeuwerk*
Editor in Chief of Development: *Carol Trueheart*
Editorial Assistant/Print Supplements Editor: *Abigail Rethore*
Art Director/Interior Designer: *Maureen Eide*
Cover Designers: *Juan R. López/Kristine Carney*
Art Editor: *Thomas Benfatti*
Creative Director: *Juan R. López*
Director of Creative Services: *Paul Belfanti*
Manager, Cover Visual Research & Permissions: *Karen Sanatar*
Cover Image: *Aaron Graubart—Getty Images, Iconica*
Production Management/Composition: *ICC Macmillan, Inc.*
Art Studio: *Laserwords*

Pearson Prentice Hall
Pearson Education, Inc.
Upper Saddle River, New Jersey 07458

Printed in the United States of America

10 9 8 7 6 5 4 3 2

ISBN 0-13-195333-8

Pearson Education Ltd., *London*
Pearson Education Australia Pty. Limited, *Sydney*
Pearson Education Singapore, Pte. Ltd.
Pearson Education North Asia Ltd., *Hong Kong*
Pearson Education Canada Ltd., *Toronto*
Pearson Educacion de Mexico, S.A. de C.V.
Pearson Education – Japan, *Tokyo*
Pearson Education Malaysia, Pte. Ltd

Contents

PREFACE vii
TO THE STUDENT xii
INDEX OF APPLICATIONS xiv

1 LINEAR EQUATIONS AND LINEAR FUNCTIONS 1

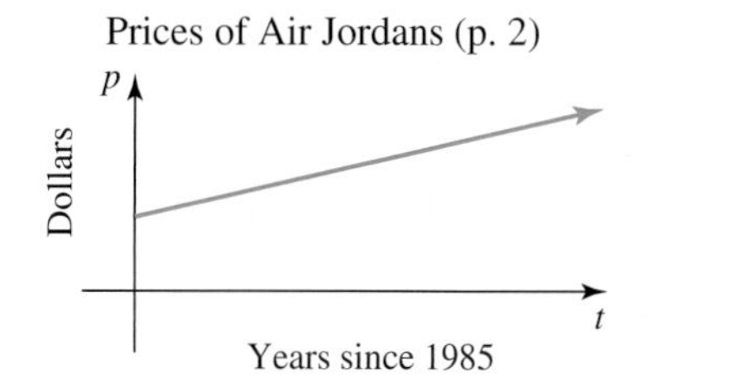

1.1 Using Qualitative Graphs to Describe Situations 1
1.2 Graphing Linear Equations 7
1.3 Slope of a Line 16
1.4 Meaning of Slope for Equations, Graphs, and Tables 25
1.5 Finding Linear Equations 34
1.6 Functions 41
CHAPTER SUMMARY 50
Key Points of Chapter 1 50
Chapter 1 Review Exercises 52
Chapter 1 Test 53

2 MODELING WITH LINEAR FUNCTIONS 55

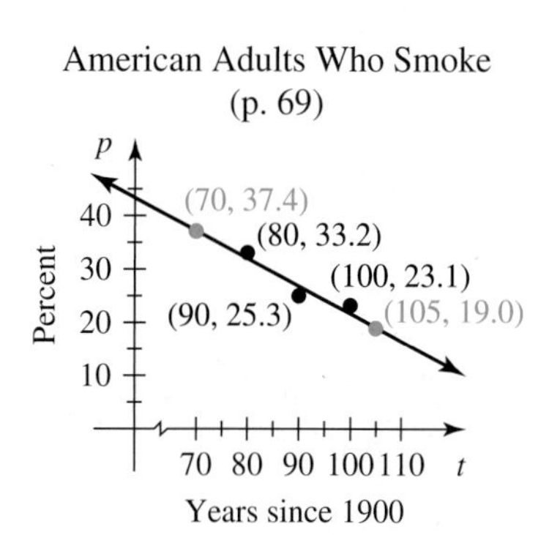

2.1 Using Lines to Model Data 55
2.2 Finding Equations of Linear Models 65
2.3 Function Notation and Making Predictions 74
2.4 Slope Is a Rate of Change 88
CHAPTER SUMMARY 98
Key Points of Chapter 2 98
Chapter 2 Review Exercises 100
Chapter 2 Test 101

3 SYSTEMS OF LINEAR EQUATIONS 103

U.S. Life Expectancies of Women and Men (p. 125)

Year of Birth	Women (years)	Men (years)
1980	77.4	70.0
1985	78.2	71.1
1990	78.8	71.8
1995	78.9	72.5
2000	79.5	74.1
2003	80.1	74.8

3.1 Using Graphs and Tables to Solve Systems 103
3.2 Using Substitution and Elimination to Solve Systems 113
3.3 Using Systems to Model Data 124
3.4 Value, Interest, and Mixture Problems 131
3.5 Using Linear Inequalities in One Variable to Make Predictions 143
CHAPTER SUMMARY 155
Key Points of Chapter 3 155
Chapter 3 Review Exercises 156
Chapter 3 Test 158
Cumulative Review of Chapters 1–3 159

Average Ticket Prices to Major League Baseball Games (p. 191)

Year	Average Ticket Price (dollars)
1950	1.54
1960	1.96
1970	2.72
1980	4.45
1991	8.84
2000	16.22
2005	21.17

Safe Exposure Times to Music at Rock Concerts (pp. 248–49)

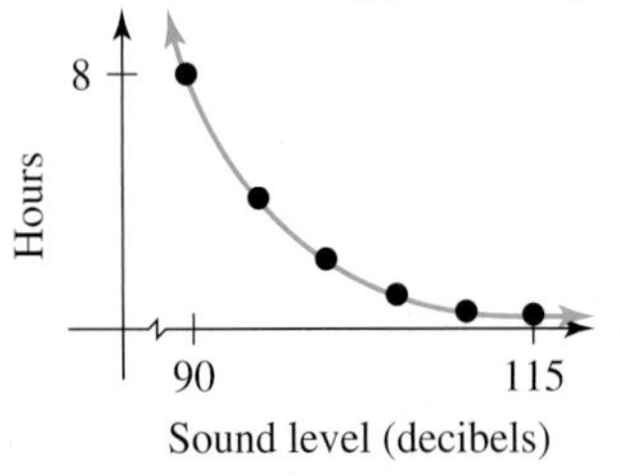

Percentages of Americans Who Think That the Environment Should Be Given Priority (p. 328)

Year	Percent
2000	69
2001	57
2002	54
2003	47
2004	49
2005	53

Numbers of Billionaires (p. 344)

Year	Number of Billionaires
2000	298
2001	266
2002	228
2003	262
2004	313

4 EXPONENTIAL FUNCTIONS 162

4.1 Properties of Exponents 162
4.2 Rational Exponents 175
4.3 Graphing Exponential Functions 183
4.4 Finding Equations of Exponential Functions 193
4.5 Using Exponential Functions to Model Data 200
CHAPTER SUMMARY 214
Key Points of Chapter 4 214
Chapter 4 Review Exercises 216
Chapter 4 Test 217

5 LOGARITHMIC FUNCTIONS 219

5.1 Inverse Functions 219
5.2 Logarithmic Functions 230
5.3 Properties of Logarithms 236
5.4 Using the Power Property with Exponential Models to Make Predictions 244
5.5 More Properties of Logarithms 253
5.6 Natural Logarithms 260
CHAPTER SUMMARY 267
Key Points of Chapter 5 267
Chapter 5 Review Exercises 269
Chapter 5 Test 270
Cumulative Review of Chapters 1–5 271

6 POLYNOMIAL FUNCTIONS 274

6.1 Adding and Subtracting Polynomial Expressions and Functions 274
6.2 Multiplying Polynomial Expressions and Functions 285
6.3 Factoring Trinomials of the Form $x^2 + bx + c$; Factoring Out the GCF 295
6.4 Factoring Polynomials 304
6.5 Factoring Special Binomials; A Factoring Strategy 311
6.6 Using Factoring to Solve Polynomial Equations 317
CHAPTER SUMMARY 331
Key Points of Chapter 6 331
Chapter 6 Review Exercises 333
Chapter 6 Test 334

7 QUADRATIC FUNCTIONS 336

7.1 Graphing Quadratic Functions in Vertex Form 336
7.2 Graphing Quadratic Functions in Standard Form 347
7.3 Using the Square Root Property to Solve Quadratic Equations 359
7.4 Solving Quadratic Equations by Completing the Square 371
7.5 Using the Quadratic Formula to Solve Quadratic Equations 377
7.6 Solving Systems of Linear Equations in Three Variables; Finding Quadratic Functions 388
7.7 Finding Quadratic Models 396
7.8 Modeling with Quadratic Functions 403

Preface

"The question of common sense is always, 'What is it good for?'—a question which would abolish the rose and be answered triumphantly by the cabbage."

—James Russell Lowell

These words seem to suggest that poet and editor James Russell Lowell (1819–1891) took Intermediate Algebra. How many times have your students asked, "What is it good for?" After years of responding "You'll find out in the next course," I began an ongoing quest to develop a more satisfying and substantial response to my students' query.

Curve-Fitting Approach Although there are many ways to center an Intermediate Algebra course around authentic applications, I chose a curve-fitting approach for several reasons. A curve-fitting approach

- allows great flexibility in choosing interesting, authentic, current situations to model.
- emphasizes concepts relating to functions in a natural, substantial way.
- deepens students' understanding of functions, because it requires students to describe functions graphically, numerically, symbolically, and verbally.
- unifies the many diverse topics of a typical Intermediate Algebra course.

To curve fit, students learn the following four-step modeling process:

1. Examine the data set to determine which type of model, if any, to use.
2. Find an equation of the model.
3. Verify that the model fits the data.
4. Use the model to make estimates and predictions.

This four-step process weaves together topics that are crucial to the course. Students must notice numerical patterns from data displayed in tables, recognize graphical patterns in scattergrams, find equations of functions, graph and evaluate functions, and solve equations.

Not only does curve fitting foster cohesiveness within chapters, but also it creates a parallel theme for each chapter that introduces and discusses a new function. This structure enhances students' abilities to observe similarities and differences among fundamental functions such as linear functions, exponential functions, and quadratic functions. The coverage of exponential functions directly follows that of linear functions so that students can see the dual nature of these two functions (by comparing the slope addition property with the base multiplier property). In addition, logarithms can then be studied relatively early in the course, when students have plenty of energy to learn about an unfamiliar function.

Some students find it hard to stay interested, because they have "seen it all before" in Elementary Algebra. To address this issue, content that will be new to most students is presented in Sections 1.1, 1.4, and 1.6 as well as in most of Chapters 2–11. Section 1.1 sets the tone that this course will be different, interesting, alive, and relevant, inviting students' creativity into the classroom.

In response to the requests of reviewers, I have made pedagogical improvements, added some topics, and reorganized content in the third edition.

NEW TO THE THIRD EDITION

Modeling

Most data tables have been updated or replaced. To establish a consistent theme, modeling discussions have been added to some sections that did not contain modeling in the second edition.

Modeling exercises have been added that require students to identify data described in paragraph form, to define variables, to find models, and to make predictions without prompts from the exercises.

New Sections

- The new Section 3.4, on value, interest, and mixture problems, employs a different set of problem-solving skills than curve-fitting applications warrant. Some of these problems require students to use functions to model situations and to reflect on the meaning of the parameters and graphs of such functions.
- The new Section 6.1 discusses how to add and subtract polynomials and polynomial functions. Sum functions and difference functions are used to model situations. (Product functions and quotient functions are now discussed in Sections 6.2 and 8.1, respectively.)

Revised or Expanded Sections

- The treatment of domains and ranges of functions has been greatly enhanced in Section 1.6.
- Sections 2.2, 2.3, 4.5, and 5.4 have been expanded to include modeling exercises in paragraph form.
- Section 2.4 now discusses rate of change before connecting this concept with the slope of a linear function.
- Section 6.4 has been expanded to include two methods of factoring trinomials of the form $ax^2 + bx + c$.
- Section 7.2 has been expanded to include two methods of graphing quadratic functions in standard form and now covers the topic of maximizing area.
- A discussion about deciding which method to use to solve a quadratic equation has been added to Section 7.5.
- Section 7.6 now covers solving systems of linear equations in three variables.

Reorganization

- Graphing quadratic functions has been moved from Chapter 6 to Chapter 7. Now Chapter 6 focuses on polynomial functions, and Chapter 7 focuses on quadratic functions.
- Factoring sums and differences of cubes has been moved from Chapter 11 to Section 6.5. Now all factoring techniques are discussed in Chapter 6.
- Finding complex-number solutions of quadratic equations has been moved from Chapter 11 to Sections 7.3, 7.4, and 7.5.

Enhancements to Homework Sets

NEW! Related Review exercises have been added to Chapters 4–11. These exercises relate current concepts to previously learned concepts.

NEW! Expressions, Equations, Functions, and Graphs exercises have been added throughout Chapters 4–11. These exercises are designed to help students gain a solid understanding of those core concepts, including how to distinguish among them.

Challenging exercises have been added to certain sections in response to some math departments' requests. For example, multivariate expressions have been added where appropriate.

Several sections now discuss **how to use graphs and tables to solve equations** in one variable.

The number of exercises in which students **evaluate functions** has been increased where appropriate.

Exercises that require students to **solve formulas** for a specified variable have been added where appropriate.

Some Pedagogical Enhancements

Warnings have been added to address students' common misunderstandings about key concepts.

Each chapter now begins with a description of an authentic situation that can be modeled by using the concepts discussed in the chapter.

CONTINUED FROM THE SECOND EDITION

Technology The text assumes that students have access to technology such as the TI-83 or TI-84 graphing calculator. Technology of this sort allows students to create scattergrams and check the fit of a model quickly and accurately. It also empowers students to verify their results from Homework exercises and efficiently explore mathematical concepts in the Group Explorations.

The text supports instructors in holding students accountable for all aspects of the course without the aid of technology, including finding equations of models. (Regression equations are included in the Answers section, because it can be difficult or impossible to anticipate which points a student will choose in trying to find a reasonable equation.)

Group Explorations Almost all sections of the text contain one or two explorations that support student investigation of a concept. Instructors use explorations as collaborative activities during class time or as part of homework assignments. Some explorations lead students to think about concepts introduced in the current section. Other, "Looking Ahead," explorations are directed-discovery activities that introduce key concepts to be discussed in the section that follows. The explorations empower students to become active explorers of mathematics and can open the door to the wonder and beauty of the subject.

Tips for Success Many sections close with tips that are intended to help students succeed in the course. A complete listing of these tips is included in the Index.

Additional Topics Chapter Topics typically taught in an Intermediate Algebra course that cannot be connected with a curve-fitting approach at the appropriate level are assembled in Chapter 11. Each section contains a Section Quiz feature. The union of these quizzes can be used as a set of review exercises for Chapter 11. Instructors who wish to "cut and paste" sections from Chapter 11 into earlier chapters can append these quizzes to the appropriate Chapter Review exercises.

Appendix A: Reviewing Prerequisite Material Appendix A can be used to remind students of important topics typically addressed in an Elementary Algebra course. Examples and exercises are included in each section.

Appendix B: Using a TI-83 or TI-84 Graphing Calculator Appendix B contains step-by-step instructions for using the TI-83 and TI-84 graphing calculators. A subset of this appendix can serve as a tutorial early in the course. In addition, when the text requires a new calculator skill, students are referred to the appropriate section in Appendix B.

RESOURCES FOR INSTRUCTORS

Instructor's Resource Manual This manual contains suggestions for course pacing and homework assignments. It also discusses how to incorporate technology and how to structure lab and project assignments.

The Instructor's Resource Manual contains section-by-section suggestions for lectures and for the explorations in the text. Lab assignments are included for most chapters. These labs are excellent for increasing students' understanding of concepts and the scientific method and are great for more in-depth writing assignments. The labs are also available as part of MyMathLab.

Instructor's Solutions Manual This manual includes complete solutions to the even-numbered exercises in the Homework sections of the text.

NEW MyMathLab Instructor Version MyMathLab is a series of text-specific, easily customizable online courses for Prentice Hall textbooks in mathematics and statistics. Powered by CourseCompass™ (Pearson Education's online teaching and learning environment) and MathXL® (our online homework, tutorial, and assessment system), MyMathLab gives you the tools you need to deliver all or a portion of your course online, whether your students are in a lab setting or working from home. MyMathLab provides a rich and flexible set of course materials, featuring free-response exercises that are algorithmically generated for unlimited practice and mastery. Students can also use online tools, such as animations and a multimedia textbook, to independently improve their understanding and performance. Instructors can use MyMathLab's homework and test managers to select and assign online exercises correlated directly to the textbook, and they can also create and assign their own online exercises and import TestGen tests for added flexibility. MyMathLab's online gradebook—designed specifically for mathematics and statistics—automatically tracks students' homework and test results and gives the instructor control over how to calculate final grades. Instructors can also add offline (paper-and-pencil) grades to the gradebook. MyMathLab is available to qualified adopters. For more information, visit our website at www.mymathlab.com, or contact your Prentice Hall sales representative.

NEW MathXL® Instructor Version MathXL® is a powerful online homework, tutorial, and assessment system that accompanies Prentice Hall textbooks in mathematics or statistics. With MathXL, instructors can create, edit, and assign online homework and tests by using algorithmically generated exercises correlated at the objective level to the textbook. They can also create and assign their own online exercises and import TestGen tests for added flexibility. All student work is tracked in MathXL's online gradebook. Students can take chapter tests in MathXL and receive personalized study plans based on their test results. The study plan diagnoses weaknesses and links students directly to tutorial exercises for the objectives they need to study and retest. MathXL is available to qualified adopters. For more information, visit our website at www.mathxl.com, or contact your Prentice Hall sales representative.

NEW InterAct Math Tutorial Website: www.interactmath.com Get practice and tutorial help online! This interactive tutorial website provides algorithmically generated practice exercises that correlate directly to the exercises in the textbook. Students can retry an exercise as many times as they like with new values each time for unlimited practice and mastery. Every exercise is accompanied by an interactive guided solution that provides helpful feedback for incorrect answers, and students can also view a worked-out sample problem that steps them through an exercise similar to the one they're working on.

TestGen TestGen enables instructors to build, edit, print, and administer tests by using a computerized bank of questions developed to cover the objectives of the text. TestGen is algorithmically based, allowing instructors to create multiple but equivalent versions of the same question or test with the click of a button. Instructors can also modify test bank questions or add new questions. Tests can be printed or administered online. The software is available on a dual-platform Windows/Macintosh CD-ROM.

RESOURCES FOR STUDENTS

Student Solutions Manual This manual contains the complete solutions to the odd-numbered exercises in the Homework sections of the text.

MathXL® Tutorials on CD This interactive tutorial CD-ROM provides algorithmically generated practice exercises that are correlated at the objective level to the exercises in the textbook. Every practice exercise is accompanied by an example and a guided solution designed to involve students in the solution process. The software provides helpful feedback for incorrect answers and can generate printed summaries of students' progress.

GETTING IN TOUCH

I would love to hear from you and would greatly appreciate receiving your comments or questions regarding this text. If you have any questions, please ask them, and I will respond.

Thank you for your interest in preserving the rose.

Jay Lehmann
MathnerdJay@aol.com

To the Student

You are about to embark on an exciting journey. In this course, you will learn not only more about algebra but also how to apply algebra to describe and make predictions about authentic situations. This text contains data that describe hundreds of situations. Most of the data have been collected from recent newspapers and Internet postings, so the information is current and of interest to the general public. I hope that includes you.

Working with authentic data will make mathematics more meaningful. While working with data about authentic situations, you will learn the meaning of mathematical concepts. As a result, the concepts will be easier to learn, because they will be connected to familiar contexts. And you will see that almost any situation can be viewed mathematically. That vision will help you understand the situation and make estimates and/or predictions.

Many of the problems you will explore in this course involve data collected in a scientific experiment, survey, or census. The practical way to deal with such data sets is to use technology. So, a graphing calculator or computer system is required.

Hands-on explorations are rewarding and fun. This text contains explorations with step-by-step instructions that will lead you to *discover* concepts, rather than hear or read about them. Because discovering a concept is exciting, it is more likely to leave a lasting impression on you. Also, as you progress through the explorations, your ability to make intuitive leaps will improve, as will your confidence in doing mathematics. Over the years, students have remarked to me time and time again that they never dreamed that learning math could be so much fun.

This text contains special features to help you succeed. Many sections contain a Tips for Success feature. These tips are meant to inspire you to try new strategies to help you succeed in this course and future courses. Some tips might remind you of strategies that you have used successfully in the past but have forgotten. If you browse through all of the tips early in the course, you can take advantage of as many of them as you wish. Then, as you progress through the text, you'll be reminded of your favorite strategies. A complete listing of Tips for Success is included in the Index.

Other special features that are designed to support you in this course include Warnings, which can help you avoid common misunderstandings; Key Points summaries, which can help you review and retain concepts and skills addressed in the chapter you have just read; Related Review exercises, which can help you understand current concepts in the context of previously learned concepts; and Expressions, Equations, Functions, and Graphs exercises, which can help you understand and distinguish among these four core concepts.

I have also included a review of key concepts from Elementary Algebra in Appendix A. Before or shortly after the course begins, consider reading this appendix and completing the exercises. If you need more review, refer to an Elementary Algebra text or ask your instructor or a tutor.

Feel free to contact me. It is my pleasure to read and respond to e-mails from students who are using my text. If you have any questions or comments about the text, feel free to contact me.

Jay Lehmann
MathnerdJay@aol.com

ACKNOWLEDGMENTS

Writing a modeling textbook is an endurance run that I could not have completed without the dedicated assistance of many people. First, I am greatly indebted to Keri, my wife, who yet again served as an irreplaceable sounding board for the multitude of decisions that went into creating this text. In particular, I credit her internal divining rod in selecting captivating data from a mound of data sets I have collected. And thanks to Dylan, our 8-year-old artist, who did rough drafts of many of the cartoons in the text.

I have received much support from the following instructors in my district: Robert Biagini-Komas, Ken Brown, Cheryl Gregory, Bob Hasson, and Rick Hough. Over the years, they have given much sound advice in responding to my countless e-mail inquiries.

I acknowledge several people at Prentice Hall. I am very grateful to Editorial Director Christine Hoag, who has shared in my vision for this text and has made significant investments to make that vision happen. The text has been greatly enhanced through the support of Executive Editor Paul Murphy, who has made a multitude of contributions, including assembling an incredible team to develop and produce this text. The team includes Project Manager Mary Beckwith, who has orchestrated the many aspects of this project, leading to a significantly better book. And I am deeply grateful to Senior Development Editor Karen Karlin, who has given incredible support in clarifying the text, updating hundreds of data sets, and responding to literally thousands of my queries.

I thank these reviewers, whose thoughtful, detailed comments helped me sculpt this text into its current form:

Joel Berman, *Valencia Community College*
Barbara Burke, *Hawaii Pacific University*
Paula Castagna, *Fresno City College*
Cathy Gardner, *Grand Valley State University*
Jane Mays, *Grand Valley State University*
Timothy Merzenich, *Chemeketa Community College*
Camille Moreno, *Cosumnes River College*
Charlie Naffziger, *Central Oregon Community College*
Donna Marie Norman, *Jefferson Community College*
Jody Rooney, *Jackson Community College*
James Ryan, *State Center Community College District, Clovis*
Ingrid Scott, *Montgomery College*
Janet E. Teeguarden, *Ivy Tech State College*
Lisa Winch, *Kalamazoo Valley Community College*

I also thank the reviewers of the second edition, whose comments contributed to this third edition as well:

Gwen Autin, *Southeastern Louisiana University*
Nancy Brien, *Middle Tennessee State University*
William P. Fox, *Francis Marion University*
Kathryn M. Gundersen, *Three Rivers Community College*
Diane Mathios, *De Anza College*
Jane E. Mays, *Grand Valley State University*
Scott McDaniel, *Middle Tennessee State University*
Jason L. Miner, *Santa Barbara City College*
Ernest Palmer, *Grand Valley State University*
Jody Rooney, *Jackson Community College*
John Szeto, *Southeastern Louisiana University*
Janet E. Teeguarden, *Ivy Tech State College*
Ollie Vignes, *Southeastern Louisiana University*

Index of Applications

academia
mathematical papers written by one or two authors, 153
acoustics
guitar string vibration, 492–93
loudness, 235
hearing loss and, 248–49
tuning fork frequency, 500
aeronautics
calcium loss in weightless environment, 249
agriculture
value of farmland, 293–94
air travel, 98
airline fuel prices, 401
charter flight, 407–8
climbing steepness, 23
coach and first-class ticket pricing, 141
domestic commercial airline boardings, 85
free tickets, 129
frequent-flier programs, 403–4, 432–33, 467
revenue maximization by airline, 407–8
taxi-out times, 342–43
unruly airline passengers, 70
archaeology
dating wood, 252
mummy of pharaoh, 252
art
names in glass sculpture, 560
painting frame, 330, 335
astronomy
period of planet, 518

biology
bacterial growth, 200–1, 248
calcium loss in weightless environment, 249
half-life of substances in bloodstream, 209, 252
business
advertising
annual billings, 248
cost of, 3
revenue from, 217
sales and, 494
television, 213–14
automobile consumers, 87
bond rating of, 328
chain letters, 577–78
computer use by, 405–6
consumer complaints about improper debt collection practices, 582–83
cost(s) to
of advertising, 3
of CD production, 469–71, 477
of employee, 294
of manufacturing, 476–77, 500
depreciation of cars, 152
employment
drug tests on employees, 401, 410
number of employees, 153
in oil company, 95
in passenger airlines, 269
export quotas on, 153
female bosses, 87
fiber-optic cable demand, 328–29
Ford's U.S. market shares, 85
fraudulent corporate reports, 70
gift cards, 251
health care benefits, 87
Home Depot stores, 402–3
hotel
extended-stay rooms, 334
new openings, 386
job offer comparison, 572, 577, 580
Kohl's stores, 217
lobbying expenditures, 101
mileage rates for, 101
music store, 132–33
oil production, 131
paid vacation days and holidays, 399–400
pharmaceutical industry's spending on government and politics, 560–61, 572–73
prices
of airline fuel, 401
of scrap iron and sheet metal, 369
for tickets, 140–41
profit, 91, 560
rental by
of conference room, 499
of trucks, 143, 150–51, 152, 158, 272
restaurant chain growth, 210
revenue(s)
from ads, 217
of American Express, 323–24
annual, 334
maximizing, 412, 419
from music downloads, 208
from ring tones, 208
salary(ies), 32, 40
annual increases, 558–59, 560, 564–65, 567, 571, 572, 573, 576, 577
average base pay of MBA graduate, 283, 467–68
of major league baseball players, 250
of professors, 77–79, 96, 294
sales
advertising budget and, 494
annual, 217, 265, 467
of bagged salads, 80–81
of beer, 402, 411
of blank audiocassettes, 250
of books, 209, 478, 583
of CDs, 140
at coffeehouses and doughnut shops, 420, 479
comparison between two companies, 161
of digital TVs and displays, 174–75
at duty-free shops, 101
of eight-track cartridges, 578
of energy and nutrition bars, 203–4
of fish oils, 209
at food-and-drink places, 573
on Internet, 329
of laserdiscs, 209
of light truck, 582
of Michael Jackson's albums, 210
of portable MP3 players, 209
of SUVs, 131
of tickets, 158, 272
of toys, 284
using debit cards, 335
of videocassette players, 249
of yogurt, 93
Sarbanes–Oxley Act, 266
shredder models, 95
Starbucks stores, 210, 249
TiVo subscribers, 208
Toyota market share, 131
unions, 95–96, 329, 420–21
videocassettes and DVDs bought by U.S. dealers, 406–7, 408–9, 499
weekly income, 81

cars. *See* motor vehicles
chemistry
half-life, 209, 217, 252, 270
radiation, 202–3
solutions
acid, 137–38, 142, 582
alcohol, 138–39, 142
antifreeze, 142, 159
pH of, 235
commodities
petroleum consumption, 157–58
communications. *See also* computers and computing
cell phone bill, 294
cell phone subscribers, 567
chain letters, 577–78
directory assistance calls, 111
fiber-optic cable demand, 328–29
phone charge, 96
rumor spreading, 207, 248, 567, 577
computers and computing
BlackBerry subscribers, 263–64
broadband cable and DSL subscribers, 472–73
CPU speeds, 211, 246
Google, 207–8
households with personal computers, 111, 128, 432, 467
households with webcams, 548
Internet news sources, 418
Internet sales, 329
Internet users, 63, 188, 429–30, 464–65
personal information into a pop-up, 419
prank e-mail, 213
spam e-mails, 531, 540
at work, 405–6
construction
auditorium seating, 572
of fencing, 357, 412
LEED-certified green buildings, 249
consumer behavior
complaints about improper debt collection practices, 582–83
contests. *See* games and contests
crime. *See also* law and law enforcement
crime indexes, 73–74
economy and, 74
fraudulent corporate reports, 70
keylogging programs in, 210
seizures of methamphetamine (meth) labs, 550
violent victimization rates by gender, 153

demographics. *See also* population
births
despite contraception, 549–50
outside marriage, 71, 84, 96
deaths
from heart disease, 335
infant mortality, 244–45
at intersections, 580
life expectancy, 597–98, 600
at birth, 72, 85
by gender, 125, 153
living alone, 479
marriages
married couples, 111
median ages at first, 411
median heights of boys, 554
rural population, 96–97
single-parent households headed by fathers, 548
distance. *See also* length
descent of hot-air balloon, 15, 32
height
apparent, 493
of sheer cliff, 549
to horizon, 524
across a lake, 608, 612
length of ladder needed to reach a window, 611
of lightning, 492
skidding, 510

stopping, 492
travel time and, 492
of trip to three cities, 612

economics
consumer confidence index, 329
crime and economy, 74
federal debt, 369–70
foreign aid, 329
GNP, 212–13
inflation, 253
percentages of Americans living below the poverty level, 64–65
transaction demand, 491
education
bachelor's degrees, 478–79
charges for supplies and field trips, 540–41
charter schools, 386–87
cost of, 95, 96
dentistry degrees, 62
distance learning, 508, 539
early decision, 251
enrollment
at college, 110–11, 128, 280–81, 478
in Intermediate Algebra, 73, 86–87
in journalism and mass communication programs, 561
rate of, 32–33
grade school mathematics test scores, 421
high school students taking college-level courses and tests, 182
hiring full-time and adjunct instructors, 140
history test scores, 87
master of divinity degree, 129
math tutor, 555, 557, 560
military aptitude test takers, 70
postsecondary institutions, 478
number of colleges and universities, 159
public school per-student expenditures, 87
salaries of professors, 77–79, 96, 294
satellite college, 140
student-to-faculty ratios, 357
tuition, 97
at Princeton University, 271
electronics
intensity of TV signal, 492
TV screen width, 611–12
energy
nuclear power plant capacities, 386
solar, 356
wind, 413
entertainment. *See also* leisure and recreation; music
American Idol viewership, 548
digital TVs and displays, 174–75
eight-track tape cartridge sales, 578
Halloween party attendance, 212
hour-long television shows, 418
movie attendance, 228–29
PlayStation ownership by households, 553
television ratings of World Series and of prime-time shows, 158
television screen width, 611–12
ticket prices, 133–35, 140–41
environment
Chernobyl nuclear accident, 202–3
LEED-certified green buildings, 249
protecting, 328
recycling, 96

finance. *See also* investment(s)
average annual expenditures, 356–57
base pay of MBA graduate, 283, 467–68
billionaires in United States, 344
car values, 40
charges for supplies and field trips, 540–41
checking accounts, 70
cost(s)
of conference room rental, 499
mean, per person, 476
of school photos, 495
shared, 476
of state corrections, 291–92
credit card offers, 66
credit scores, 228
debit cards, 335
depreciation of cars, 126–27, 130
federal debt, 212, 369–70
improper debt collection practices, 582–83
income
household, 477
personal, 101, 131, 159
inflation, 253
interest
compounded annually, 201, 208–9, 217, 244, 248, 269
from investment, 135–37, 141–42, 158, 161
interest-bearing accounts, 213
Internet banking, 102
Manhattan condominium prices, 95
online bill paying, 95
postage for first-class letter, 561
price(s)
of airline fuel, 401
of athletic shoes, 64
of gasoline, 356, 409, 477
of ski rental packages, 71–72, 85
of tickets, 133–35, 140–41
recreational expenditures, 583
salary(ies), 32, 40
of major league baseball players, 250
of professors, 77–79, 96, 294
Sarbanes–Oxley Act, 266
shared expenses, 440
spending on domestic travel, 369
tax returns filed electronically, 62–63
food and nutrition
bottled water consumption, 384, 471–72
chicken and red meat consumption, 110, 128, 283
demand for "lowfat" or "no fat" food products, 66–68
ice cream consumption rate, 88
milk consumption, 129, 153, 283
pizza weight, 493
soft drinks consumption, 129, 283
forestry
leaves on trees, 269
timber harvests, 249

games and contests
award plan choices, 578
poker, 249
gardens
dimensions of, 330
fencing around, 353–55, 357, 358, 417
mulching, 324–25, 330
sod coverage, 330
genealogy
ancestry, 567, 577
geology
carbon dating, 246
continental ice sheet, 252
tsunamis, 510, 548
volcanic eruption, 246–47
geometry
rectangle dimensions, 330
sphere volume, 492
government
classified federal documents, 70, 398
federal debt, 212, 369–70
federal tax code, 87
Food Stamp Program, 365, 432
Internal Revenue Service (IRS) standard mileage rate, 101
pharmaceutical industry's spending on, 560–61, 572–73
postage, 561
Sarbanes–Oxley Act, 266

health
AIDS cases, 420
amusement park injuries, 97
births
despite contraception, 549–50
multiple, of triplets or more, 218
Children's Health Insurance Program (CHIP), 87
diabetes, 72, 87
fitness club membership, 130–31
flu epidemic, 207
health care coverage, 87
hearing loss, 248–49
heart attack risk, 211
heart disease deaths, 335
house calls paid by Medicare, 396–97
obese adults, 269–70
polio cases, 249–50
seniors with severe memory impairment, 250–51
shingles, 480
smoking, 69, 79–80, 96
teenage, 400
syphilis rate, 273
weight-loss program promotions, 130

insurance
life, 211, 252
investment(s). *See also* finance
bond rating, 328
doubling of, 208
Dow Jones Industrial Average, 204–6
households owning stocks, 548
interest from, 135–37, 141–42, 158, 161, 420
compounded annually, 201, 208–9, 244, 248, 269

law and law enforcement. *See also* crime
cancer-related charges of discrimination, 344
costs of state corrections, 291–92
drug arrests, 544–46
ex-convicts re-arrested, 549
lawsuits filed against tobacco companies, 217
seizures of methamphetamine (meth) labs, 550
leisure and recreation. *See also* entertainment
amusement park injuries, 97
boating fatalities, 357
paid vacation days and holidays, 399–400
recreational expenditures, 583
theme and amusement park attendance, 249
ticket prices for Walt Disney World, 270–71
time-shares, 413
vacations in July, 401
visitors on snowmobiles at Yellowstone National Park, 131
visitors to Grand Canyon, 55–56, 57–58, 433
visitors to Ireland, 344
length
of striped bass, 70

media
ad-supported cable television, 131
cable channels per household, 208
newspapers
ad revenue, 217
circulations, 129–30
morning and evening, 479
purchases of, 357
trust in, 95
news sources, 418
TiVo subscribers, 208

medicine
cancer tests, 252
cancer treatment, 209
diabetes, 72, 87
flu treatment, 208
generic prescription drugs, 213
heart attack risk, 211
house calls paid by Medicare, 396–97
organ transplants, 402
polio cases, 249–50
radiation treatment, 459, 492
shingles, 480
stress fractures, 252
thallium-201 injection, 209
wrong-site surgery, 228

meteorology
atmospheric pressures, 96, 97–98

military
defense spending, 344
military aptitude test takers, 70
women in, 102

miscellaneous
adoptions, 161, 532, 540, 548–49
ball bearing radii and masses, 498–99
cremations, 228
firearm applications, 95
hanging cable, 266
household glass cleaner amount, 560
ISO paper-size system, 524–25
popular names, 567, 578
rug dimensions, 330
U.S. patent applications, 96
Valentine's Day celebration, 411

mixtures
acid solutions, 137–38, 142
alcohol solutions, 138–39, 142
antifreeze solution, 142, 159

motor vehicles. *See also* transportation; travel
acceleration times, 410
bicyclists hit and killed by, 161
Camaro production, 46–47
car values, 40
consumers under 50 years of age, 87
depreciation of cars, 126–27, 130, 152
ethanol production for, 248
Ford's U.S. market shares, 85
gasoline consumption, 32, 96, 477
gasoline mileage, 32
gasoline prices, 409
gasoline sale, 96
light truck sales, 582
skidding distance, 510
speeds of pickup truck, 273
SUV sales, 131
teenagers with driver's licenses, 102
temperatures inside, 509
Toyota market share, 131
traffic deaths at intersections, 580

music. *See also* entertainment
CD production, 469–71, 477
Michael Jackson's, 210
piano, 191
rock band profit, 560
sound recordings, 65

nutrition. *See* food and nutrition

physics
bounce heights, 481–82, 483–84, 488–89, 493
dropped objects, 567
escape velocity, 525
falling objects, 492, 493, 524, 549
flow rate, 517
force, 492
frequency, 500
illumination from light bulb, 494
intensity of signal, 492
pendulum motion, 493–94
pressure–volume relationship, 486–87, 489–90
projectile motion, 356, 387
vibration, 492–93
weight, 270, 488, 492

politics. *See also* government
lobbying, 101
pharmaceutical industry's spending on, 560–61, 572–73
presidential elections, 411–12
State of the Union address, 101
voters who voted, 541

population. *See also* demographics
bacterial, 200–201
of bald eagle, 174
of deer, 5
density of, 476
of Detroit, 85–86
of District of Columbia, 95
foreign born, 402, 412
most populous nations, 582
of Nevada, 97, 228
of Pacific Northwest salmon, 59, 60–61, 69–70
rural, 96–97
of U.S., 211–12, 401–2, 410
of world, 206, 210, 250

publishing
book sales, 478, 583
books in print, 272–73
poetry books, 369

pyrotechnics
fireworks, 353

ranching
fencing around ranch, 357, 412

rate
of car, 95, 98, 102
of gas consumption, 92, 96, 101
of pumping out flooded basement, 87
of train, 95

real estate
home prices, 158
householders who own a home, 410
Manhattan condominium prices, 95
value of farmland, 293–94

recreation. *See* leisure and recreation

seismology
Richter numbers, 233, 235
tsunamis, 510, 548

society
morals and values, 386

speed, 479
driving, 479, 480
land speed records, 480
of tsunami, 510, 548
typing, 495
wind, 509–10

sports
baseball, 55, 72, 84, 95, 356, 387, 417, 495, 549
home runs, 411
salaries in, 250
ticket prices, 191
basketball
ad spending, 102
1500-meter run, 130
football
advertising costs, 3
attendance at, 95
rushing yards, 70
Super Bowl advertising, 213–14
400-meter run, 125–26
men's, 73
women's, 72
ironman competition, 328
men's 400-meter run record times, 73
Olympics
speed skating, 110, 128
Winter, 182
runner's stride rate, 73
skiing, 23, 54
prices of rental packages, 71–72, 85
recommended ski lengths, 600
swimming, 491
200-meter run, 161

surveillance
red-light cameras, 550

surveys
Americans satisfied with U.S., 64

temperature
boiling point and elevation, 63–64
cooling time, 159, 264, 266
cricket chirps and, 86, 96
of Earth, 64
in enclosed vehicle, 509
heat index, 98
scales for measuring, 86, 96
windchills and, 65, 600

tests and testing
age versus mental functioning, 129
mathematics scores, 421

transportation. *See also* motor vehicles
collisions at highway-railroad crossings, 63
steepness of two roads, 17, 23

travel. *See also* air travel; motor vehicles; transportation
driving time, 440, 474–75, 479–80, 499, 500
spending on domestic, 369

waste disposal
dump capacities, 252–53

weather
heat index, 98
rainfall, 5
snow melt, 498
storms, 492
windchill, 65, 600
wind intensities, 509–10

weight
of Pacific albacore tuna, 70

work
union membership, 95–96
volunteer, 402, 409–10

Chapter 1

Linear Equations and Linear Functions

"I see a certain order in the universe[,] and math is one way of making it visible."
—*May Sarton,* As We Are Now, *1973*

Table 1 Camaro Production

Year	Camaro Production (thousands)
1996	67
1997	56
1998	48
1999	41
2000	42
2001	29

Do you have a favorite make of car? The production of the Chevrolet Camaro® has decreased steadily since 1996 (see Table 1). In an Exploration in Section 1.6, you will estimate the number of Camaros produced in 2002 and why it makes sense that that was the last year for Camaros to be manufactured.

In this course, we will use mathematics to describe many authentic situations, such as the Camaro data. We will use these descriptions to make estimates and predictions, as in estimating Camaro production in 2002. Later, you will estimate by how much the number of Internet users is increasing per year, you will predict when baseball players' minimum salary will be $600 thousand, and you will predict the average amount of money Americans will pay for gasoline in 2010.

A major objective of this text is to help you view the world in a mathematical manner. That viewpoint will allow you to recognize important patterns—patterns that will enable you to make estimates and predictions like the ones just mentioned.

In this chapter, we will discuss how to describe a line by using a *graph,* an *equation,* and a *table*. We will also discuss how to describe the steepness of a line. Finally, we will work with an important group of lines called *linear functions*. We will lay the groundwork so that in Chapter 2 we can use lines to describe authentic situations.

1.1 USING QUALITATIVE GRAPHS TO DESCRIBE SITUATIONS

Objectives

- Use qualitative graphs to describe situations.
- Identify independent variables and dependent variables.
- Know the meaning of an *intercept* of a curve.
- Identify increasing curves and decreasing curves.
- Describe a concept or procedure.

In this section, we will use qualitative graphs to describe authentic situations. A **qualitative graph** is a graph without scaling (tick marks and their numbers) on the axes.

Reading Qualitative Graphs

How can we use a qualitative graph to describe an authentic situation?

Example 1 Reading a Qualitative Graph

Since 1985, Michael Jordan has endorsed a successful line of shoes, called Air Jordan®. Let p be the retail price (in dollars) of Air Jordans and t be the number of years since

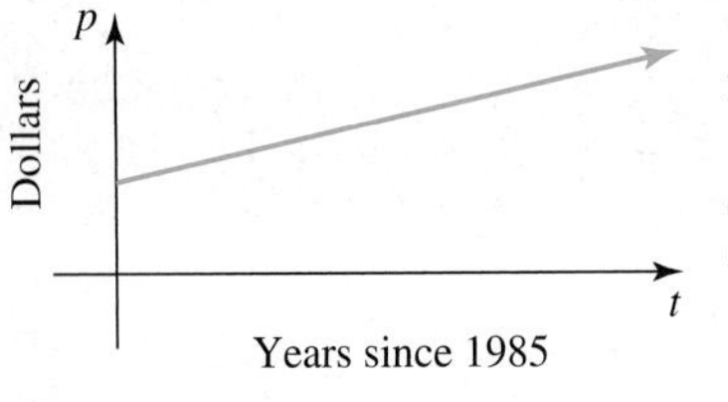

Figure 1 Retail price of Air Jordan shoes

1985. (For example, $t = 1$ represents the year 1986.) The qualitative graph displayed in Fig. 1 describes the prices of the shoes. What does the graph tell us?

Solution

The graph (or curve) tells us that the retail price of Air Jordans has increased steadily. ■

A curve is said to be *linear* if it forms a straight line. The curve in Fig. 1 is linear.

Independent and Dependent Variables

The retail price p of Air Jordans depends on the year. Due to inflation and to increasing popularity of the shoes, the price increases over time. Because p depends on t, we call p the *dependent variable*.

The year does *not* depend on the shoe's price. Raising or lowering the price has no effect on the passage of time. Time is independent of the price. Since t is independent of p, we call t the *independent variable*.

DEFINITION Independent and dependent variables

Assume that an authentic situation can be described by using the variables t and p and that p depends on t:

- We call t the **independent variable.**
- We call p the **dependent variable.**

Example 2 Identifying Independent and Dependent Variables

For each situation, identify the independent variable and the dependent variable.

1. You are waiting in line to go to a concert. Let T be the number of minutes you must wait, and let N be the number of people ahead of you when you first get in line.
2. Let n be the number of times a person can lift dumbbells that weigh w pounds.

Solution

1. The more people ahead of you when you first get in line, the more time you must wait. The wait time, T, depends on the number of people ahead of you, N. Thus, T is the dependent variable and N is the independent variable. (The number of people in line does *not* depend on your wait time.)
2. The heavier the dumbbells, the fewer times the person can lift them. The number of times the person can lift the dumbbells, n, depends on the dumbbells' weight, w. Thus, n is the dependent variable and w is the independent variable. (The weight of the dumbbells does *not* depend on the number of times the person can lift them.) ■

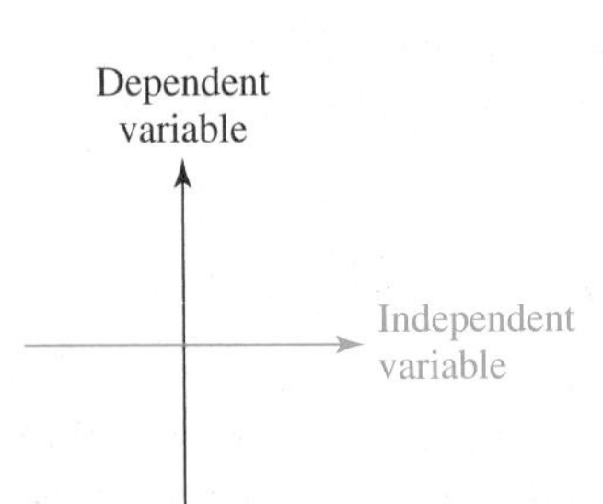

Figure 2 Match the vertical axis with the dependent variable and the horizontal axis with the independent variable

For graphs, we describe the values of the independent variable along the horizontal axis and the values of the dependent variable along the vertical axis (see Fig. 2). For example, in Fig. 1, we describe the values of the independent variable t along the horizontal axis, and we describe the values of the dependent variable p along the vertical axis.

Example 3 Reading a Qualitative Graph

Let A be the average age (in years) when men first marry, and let t be the number of years since 1900. In Fig. 3, the graph describes the relationship between the variables t and A. What does the graph tell us?

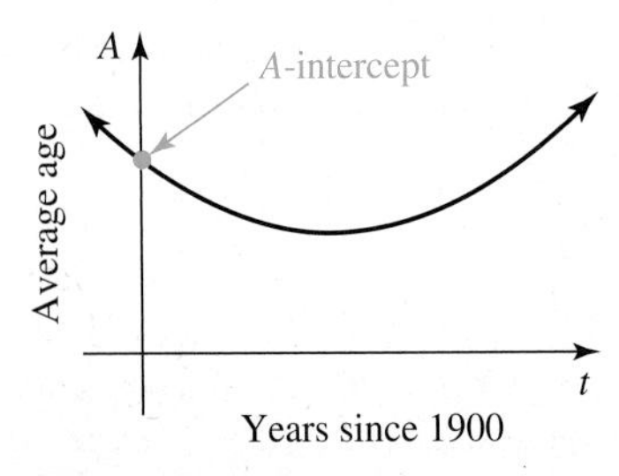

Figure 3 The average age when men first marry

Solution

The graph tells us that the average age when men first marry decreased each year for a while and then increased each year after that. ■

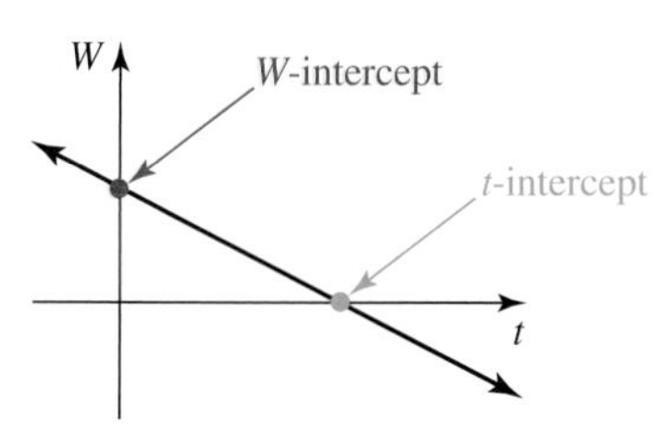

Figure 4 Intercepts of a line

We say that the curve sketched in Fig. 3 is a *parabola*.

In Fig. 3, note that the curve and the A-axis intersect. The point of intersection is an A-*intercept*. An **intercept** of a curve is any point where the curve and an axis (or axes) intersect. Two more examples of intercepts are shown in Fig. 4 for a linear curve.

Sketching Qualitative Graphs

In Examples 4–6, we sketch qualitative graphs that describe given situations.

Example 4 Sketching a Qualitative Graph

Let C be the cost (in dollars) of a 30-second ad during the Super Bowl at t years since 1987. For most years, the annual increase in cost is more than the previous annual increase in cost. Sketch a qualitative graph that describes the relationship between C and t.

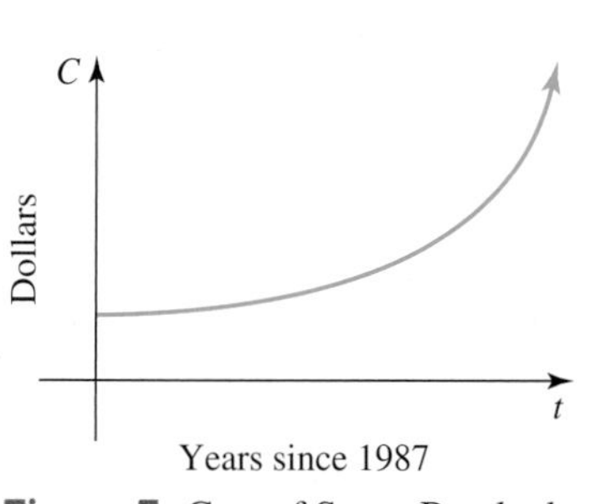

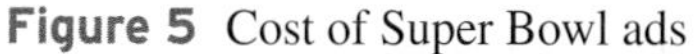

Figure 5 Cost of Super Bowl ads

Solution

Since the cost of an ad varies according to the year, C is the dependent variable and t is the independent variable (see Fig. 5). Because ads were not free in 1987 ($t = 0$), the C-intercept is above the origin. The costs are increasing, so we sketch a curve that goes upward from left to right. Since most increases are more than the previous increase, the curve should "bend" upward from left to right. ■

Some *exponential* curves have shapes similar to the shape of the curve sketched in Fig. 5.

If a curve goes upward from left to right, we say that the curve is **increasing** (see Fig. 6). For example, the cost curve in Fig. 5 is increasing. If a curve goes downward from left to right, we say that the curve is **decreasing** (see Fig. 7).

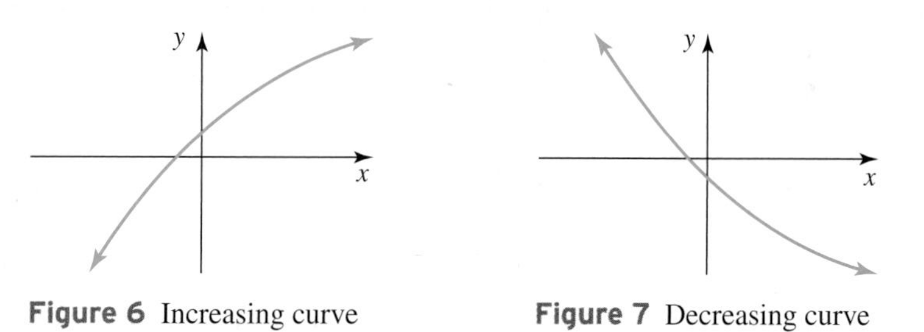

Figure 6 Increasing curve

Figure 7 Decreasing curve

In this chapter and future chapters, we will discuss the curves mentioned in this section more thoroughly and use them to make predictions—linear curves in this chapter and Chapter 2, exponential curves in Chapters 4 and 5, and quadratic curves in Chapters 6 and 7. In Chapter 3, you will make predictions with two linear curves.

Example 5 Sketching a Qualitative Graph

Hot coffee is poured into a cup at room temperature. Let F be the temperature (in degrees Fahrenheit) of the coffee at t minutes since the coffee was poured. Sketch a qualitative graph that describes the relationship between the variables t and F.

Solution

Note that F depends on t, so we let the vertical axis be the F-axis and the horizontal axis be the t-axis (see Fig. 8). Since the coffee cools with time, the curve should be decreasing. Further, the curve should show that the drop in temperature during any minute is less than the drop in temperature in the previous minute. (Why?)

The coffee's temperature will not go below room temperature, so the curve should eventually level off.

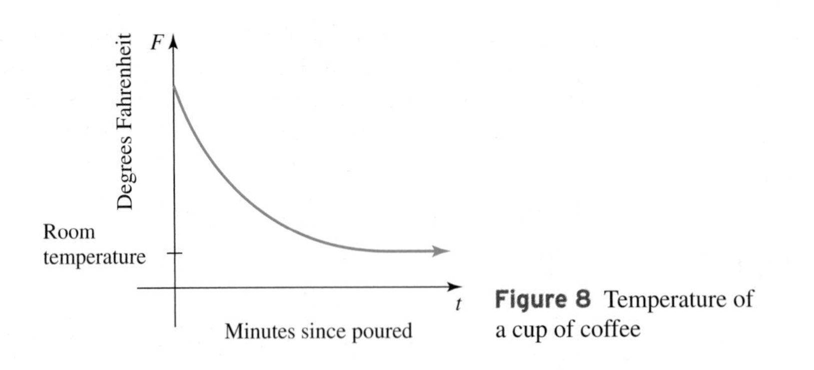

Figure 8 Temperature of a cup of coffee ■

In each situation we have discussed so far, the independent variable has represented time. Let's explore a situation in which the independent variable stands for something else.

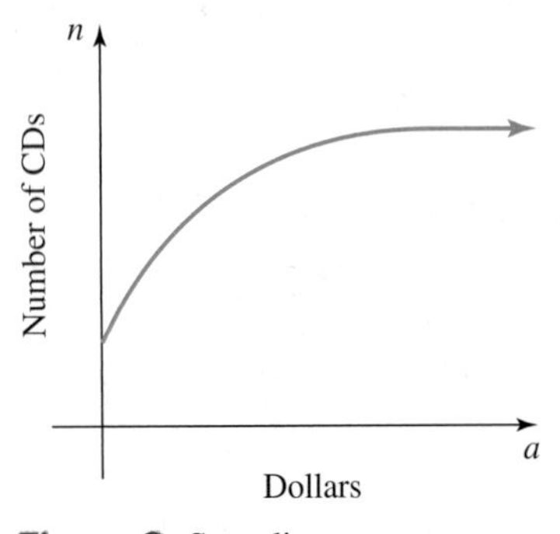

Figure 9 Spending on advertising, and CDs sold

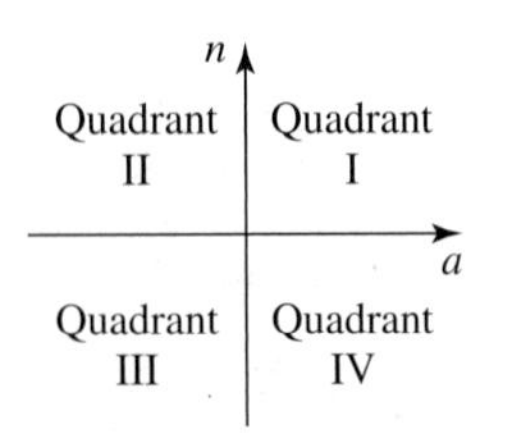

Figure 10 The four quadrants

Example 6 Sketching a Qualitative Graph

Suppose that the latest Radiohead CD is about to be released. Let n be the number of CDs that will be sold if a dollars are spent on advertising. Sketch a qualitative graph that describes the relationship between the variables a and n.

Solution

The number of CDs sold is in part determined by the amount of money spent on advertising, so n is on the vertical axis and a is on the horizontal axis (see Fig. 9). Because both n and a must be nonnegative (why?), the qualitative curve is in *quadrant I* (and one point of it is on the n-axis). The four quadrants are shown in Fig. 10.

Even if no money is spent on advertising, some CDs will be sold. So the n-intercept should be above the origin. The more money spent on advertising, the greater the sales, so the curve should be increasing. There are only so many people, however, who would buy the CD no matter how much advertising is done, so the curve should level off. ■

Describing a Concept or Procedure

In some homework exercises, you will describe in general a concept or procedure.

GUIDELINES ON WRITING A GOOD RESPONSE

- Create an example that illustrates the concept or outlines the procedure. Looking at examples or exercises may jump-start you into creating your own example.
- Using complete sentences and correct terminology, describe the key ideas or steps for your example. You can review the text for ideas, but write your description in your own words.
- Describe also the concept or the procedure in general, without referring to your example. It may help to reflect on several examples and what they all have in common.
- In some cases, it will be helpful to point out the similarities and the differences between the concept or the procedure you are describing and other concepts or procedures.
- Describe the benefits of knowing the concept or the procedure.
- If you have described the steps in a procedure, explain why it's permissible to follow these steps.
- Clarify any common misunderstandings about the concept, or discuss how to avoid making common mistakes when following the procedure.

Example 7 Responding to a General Question about a Concept

Describe the meaning of *independent variable* and *dependent variable.*

Solution

Assume that an authentic situation can be described by using the variables t and a and that a depends on t. Then t is the independent variable and a is the dependent variable.

For example, let a be the amount of money (in dollars) that a person is paid for working t hours at a gasoline station. Then t is the independent variable and a is the dependent variable, because the person's pay depends on the number of hours worked.

For graphs, we use the horizontal axis to describe values of the independent variable and the vertical axis to describe values of the dependent variable. ■

group exploration

Sketching a qualitative graph

A bathtub is filled with water, and then the plug is pulled out. Let V be the volume of water (in gallons) in the tub at t seconds after the plug is pulled out.

1. Which variable is the dependent variable? the independent variable? Explain.
2. Sketch a qualitative graph that describes the relationship between V and t. Explain.

Taking It One Step Further

3. Carefully describe an experiment you could run to verify the shape of your curve from Problem 2. In particular, explain how you could measure the volume of water at various times. Ask your instructor if you should run such an experiment.

HOMEWORK 1.1

1. The deer population in a forest is described during the years between 2000 and the present. Let p be the deer population in the forest and t be the number of years since 2000. Match each graph in Fig. 11 with each scenario. The population
   a. decreased steadily.
   b. increased steadily.
   c. remained steady.
   d. decreased for a while and then increased.

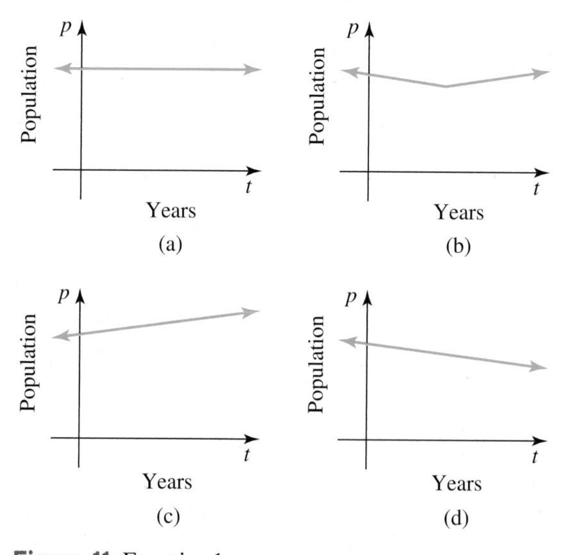

Figure 11 Exercise 1

2. Let A be the amount of rain (in inches) that has fallen in t hours. Match each graph in Fig. 12 with each scenario. The rain fell
   a. harder and harder.
   b. softly and then stopped. After a while, it began raining hard.
   c. hard and then stopped. After a while, it began raining softly.
   d. more and more softly.

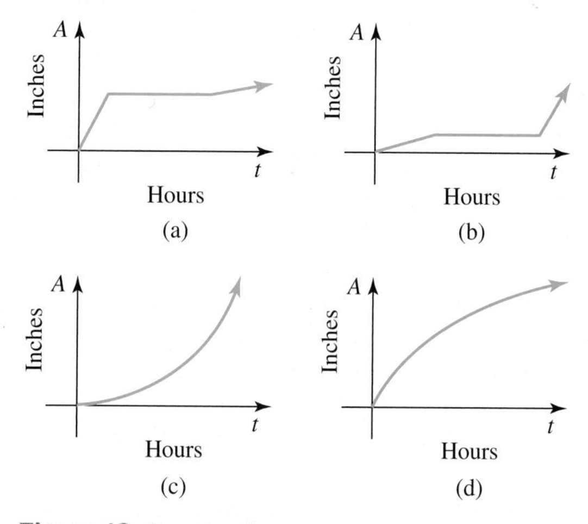

Figure 12 Exercise 2

For Exercises 3–12, identify the independent variable and the dependent variable.

3. Let T be the time (in minutes) it takes to grade N tests.
4. Let c be the total cost (in dollars) of n pencils.
5. Let F be the temperature (in degrees Fahrenheit) of an oven, and let T be the number of minutes it takes to cook a potato in the oven.

6. Let r be the rate (in gallons per hour) at which water is added to a swimming pool, and let t be the number of hours it takes to fill the pool.

7. Let L be the length of a song (in minutes), and let T be the number of seconds it takes to download the song.

8. Let a be the age (in years) of a car, and let c be the annual cost of repairs (in dollars) for the car.

9. Let P be the percentage of Americans who own a car and whose income is I dollars.

10. Let n be the number of people at a beach when the temperature is F degrees Fahrenheit.

11. Let r be the radius (in inches) of a plate, and let n be the number of ounces of spaghetti the plate can hold.

12. Let I be a person's annual income (in dollars), and let T be the federal taxes (in dollars) the person must pay.

For Exercises 13–36, sketch a graph that shows the relationship between the variables defined. Justify your graph. Various "correct" graphs are possible.

13. Let F be the temperature (in degrees Fahrenheit) at a specific outdoor location at t hours after 6 A.M. on a specific day. (So, $t = 0$ represents 6 A.M.)

14. A person went for a run. After a while, she stopped to rest. She then walked home. Let s be her speed (in feet per second) at t minutes after she began her run.

15. An airplane flew from New York to Chicago. Let h be the plane's altitude (in feet) at t seconds after takeoff.

16. Let h be the height (in feet) of a tennis ball at t seconds after it was dropped. (Allow for bounces.)

17. The number of people undergoing laser eye surgery has increased since 1996. Let n be the number of people who have undergone laser eye surgery during the year that is at t years since 1996.

18. The percentage of smokers in the United States has declined steadily since 1965. Let P be the percentage of smokers in the United States at t years since 1965.

19. The percentage of major firms that perform drug tests on employees and/or job applicants increased from 1987 to 1996 and decreased thereafter. Let p be the percentage of firms that perform drug tests at t years since 1987.

20. The percentage of the U.S. population that is foreign born decreased from 1950 to 1970 and increased thereafter. Let p be the percentage of the U.S. population that is foreign born at t years since 1950.

21. A commuter left home, drove toward her workplace, got gas, then continued driving to work. Let g be the amount of gas (in gallons) in the gas tank at t minutes after she left home.

22. At noon, a person began to breathe in. Let V be the volume of air (in liters) in this person's lungs at t seconds after noon.

23. Let h be the height (in feet) of a specific person at age a years.

24. Let A be the angle (in degrees) between the hour hand and the minute hand of a clock at t minutes from midnight.

25. Let d be the diameter (in inches) of a balloon after a person has blown into it n times.

26. Let n be the number of species in existence, and let d be the total amount of deforestation (in thousands of acres).

27. Let T be the time (in minutes) it takes to drive from home to school if there are an average of n cars per mile on the route.

28. Let s be the maximum speed (in miles per hour) that a specific speedboat can travel when going

a. downstream on a river with a current of c miles per hour.

b. upstream on a river with a current of c miles per hour.

29. Let S be the speed of a car (in miles per hour) driven on a level road when the end of the accelerator has been d inches from the floor of the car for several minutes.

30. Let p be the percentage of times a person who is playing darts at d feet from the target hits the bull's-eye.

31. Let T be the number of minutes it takes a person to make lasagna, and let n be the number of times the person has made it before.

32. Let T be the number of hours it takes a painting crew to paint a house, and let N be the number of people in the crew.

33. Let n be the number of people in the United States who would be willing to purchase a new Honda Civic® CX at a price of p dollars.

34. A person plans to drive from Rockville, Maryland, to Raleigh, North Carolina. Let T be the driving time (in hours) if the person drives at S miles per hour.

35. Let A be the area of a circle with radius r.

36. Air is blown into a balloon. The balloon is then tied so that no air can enter or leave it. If the balloon is squeezed, the air pressure inside increases. Let P be the air pressure (in pounds per square inch) inside the balloon when the balloon's volume is V cubic inches.

37. Write a scenario to match each graph in Fig. 13. Refer to the variables x and y in your description.

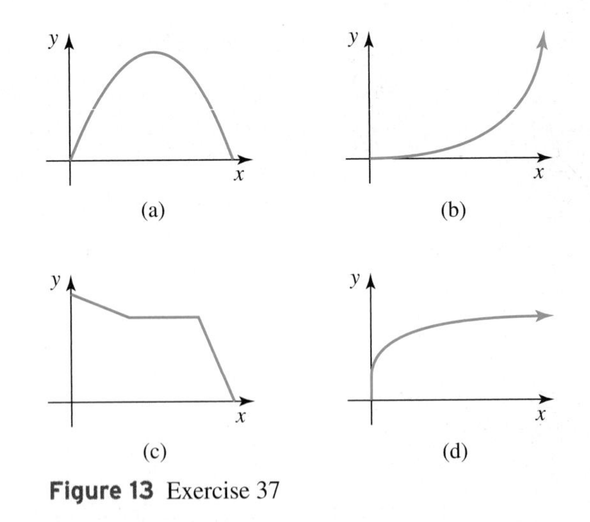

Figure 13 Exercise 37

38. Explain how to sketch a qualitative graph that describes a given situation. (See page 4 for guidelines on writing a good response.)

1.2 GRAPHING LINEAR EQUATIONS

Objectives

- Know the meaning of *solution, satisfy,* and *solution set.*
- Know the meaning of a *graph* of an equation.
- Know which equations have graphs that are lines.
- Sketch graphs of linear equations.
- Find intercepts of graphs of linear equations.
- Sketch graphs of vertical and horizontal lines.

In Section 1.1, we worked with many types of curves. For the rest of this chapter, we will focus on lines and various ways to represent them. When we are *not* describing an authentic situation, we will use the variables x and y to describe a line (or any other type of curve). We will treat x as the independent variable and y as the dependent variable.

Definition of Solution, Satisfy, and Solution Set

Consider the equation $y = 3x - 1$. Let's find y when $x = 2$:

$y = 3x - 1$	Original equation
$y = 3(2) - 1$	Substitute 2 for x.
$= 6 - 1$	Multiply before subtracting (see Section A.6).
$= 5$	Subtract.

So, $y = 5$ when $x = 2$, which we can represent by using the ordered pair $(2, 5)$. For an **ordered pair** (a, b), we write the value of the independent variable in the first (left) position and the value of the dependent variable in the second (right) position. The numbers a and b are called **coordinates.** For $(2, 5)$, the *x-coordinate* is 2 and the *y-coordinate* is 5.

The equation $y = 3x - 1$ becomes a true statement when we substitute 2 for x and 5 for y:

$y = 3x - 1$	Original equation
$5 \stackrel{?}{=} 3(2) - 1$	Substitute 2 for x and 5 for y.
$5 \stackrel{?}{=} 5$	Simplify.
true	

We say that $(2, 5)$ is a *solution* of the equation $y = 3x - 1$ and that $(2, 5)$ *satisfies* the equation $y = 3x - 1$.

A set is a container. Much like a garbage can contains garbage, a *solution set* contains solutions.

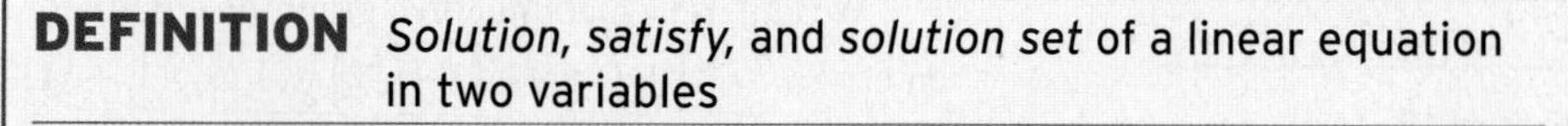

DEFINITION *Solution, satisfy,* and *solution set* of a linear equation in two variables

An ordered pair (a, b) is a **solution** of an equation in terms of x and y if the equation becomes a true statement when a is substituted for x and b is substituted for y. We say that (a, b) **satisfies** the equation.

The **solution** set of an equation is the set of all solutions of the equation.

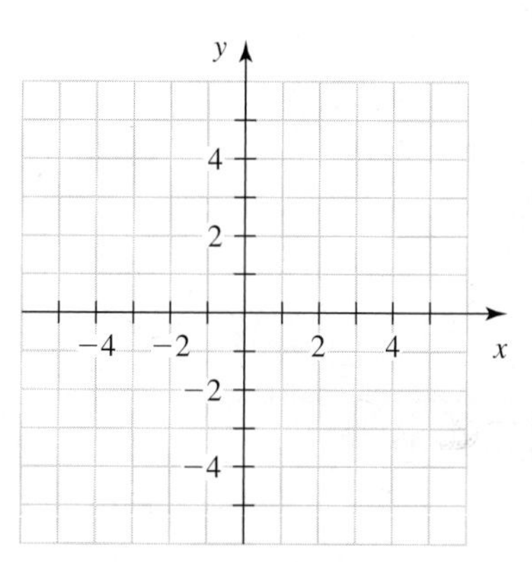

Figure 14 A coordinate system

A **coordinate system** is shown in Fig. 14. We can describe solutions of an equation by plotting points. For a review of plotting points, see Section A.1.

Example 1 Graphing an Equation

Find five solutions of the equation $y = -2x + 1$, and plot them.

Solution

We begin by arbitrarily choosing the values 0, 1, and 2 to substitute for x:

$$\begin{aligned} y &= -2(0) + 1 \\ &= 0 + 1 \\ &= 1 \end{aligned} \qquad \begin{aligned} y &= -2(1) + 1 \\ &= -2 + 1 \\ &= -1 \end{aligned} \qquad \begin{aligned} y &= -2(2) + 1 \\ &= -4 + 1 \\ &= -3 \end{aligned}$$

Solution: (0, 1) Solution: (1, −1) Solution: (2, −3)

For a review of performing operations with real numbers, see Section A.4.

The ordered pairs (−2, 5) and (−1, 3) are also solutions. We organize our findings in Table 2 and plot these five solutions in Fig. 15. ■

Table 2 Solutions of $y = -2x + 1$

x	y
−2	5
−1	3
0	1
1	−1
2	−3

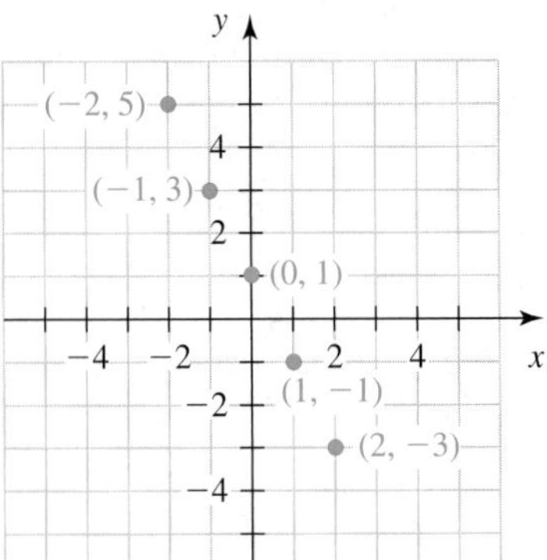

Figure 15 Five solutions of $y = -2x + 1$

Notice that a line contains the five points that we found in Example 1 (see Fig. 16). It turns out that every point on the line represents a solution of the equation $y = -2x + 1$. For example, the point (3, −5) lies on the line (see Fig. 17) and the ordered pair (3, −5) satisfies the equation $y = -2x + 1$. (Try it.)

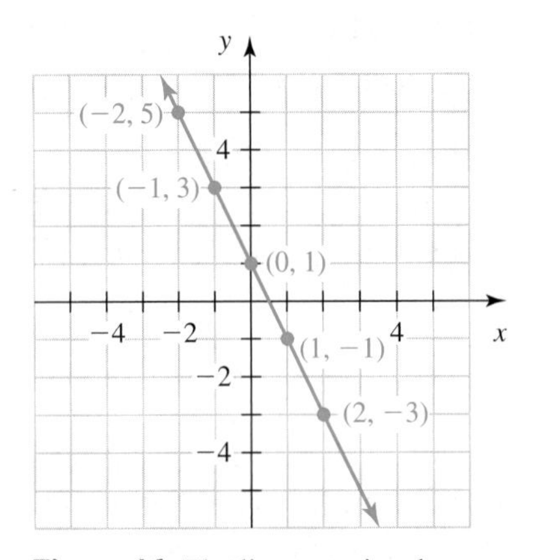

Figure 16 The line contains the points found in Example 1

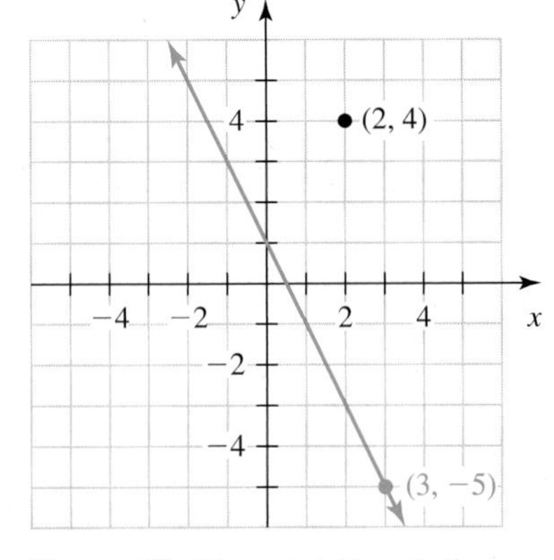

Figure 17 The point (3, −5) lies on the line, but (2, 4) does not

It also turns out that points that do not lie on the line represent ordered pairs that do not satisfy the equation. For example, the point (2, 4) does not lie on the line, and the ordered pair (2, 4) does not satisfy the equation $y = -2x + 1$:

$y = -2x + 1$	Original equation
$4 \stackrel{?}{=} -2(2) + 1$	Substitute 2 for x and 4 for y.
$4 \stackrel{?}{=} -4 + 1$	Multiply.
$4 \stackrel{?}{=} -3$	Add.
false	

We refer to the line in Fig. 16 as the *graph* of the equation $y = -2x + 1$.

We can use ZDecimal on a graphing calculator to verify our graph (see Fig. 18). To enter $y = -2x + 1$, press [(−)] **2** [X,T,Θ,n] [+] **1.** The key [−] is used for subtraction, and the key [(−)] is used for negative numbers as well as for taking opposites. For graphing calculator instructions, see Sections B.3, B.4, and B.6.

Plot1 Plot2 Plot3
\Y1 = -2X+1
\Y2 =
\Y3 =
\Y4 =
\Y5 =
\Y6 =
\Y7 =

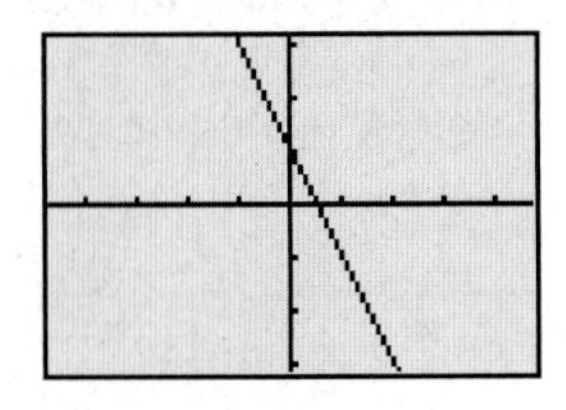

Figure 18 Use ZDecimal to verify our graph

> **DEFINITION** Graph
>
> The **graph** of an equation in two variables is the set of points that correspond to all solutions of the equation.

The graph of an equation in two variables is a visual description of the solutions of the equation. Every point on the graph represents a solution of the equation. Every point *not* on the graph represents an ordered pair that is *not* a solution.

Graphs of Linear Equations

We saw that the graph of the equation $y = -2x + 1$ is a line. Notice that the equation $y = -2x + 1$ has the form $y = mx + b$ (where $m = -2$ and $b = 1$).

Graphs of Equations That Can Be Put into $y = mx + b$ Form

If an equation can be put into the form

$$y = mx + b$$

where m and b are constants, then the graph of the equation is a line.

For example, the graphs of the equations

$$y = 3x - 7, \quad y = -\frac{5}{7}x - 2, \quad y = x + 9, \quad y = -2x, \quad \text{and} \quad y = 4$$

are lines. The equation $y = -2x$ is of the form $y = mx + b$, because we can write it as $y = -2x + 0$ (so, $m = -2$ and $b = 0$). The equation $y = 4$ is also of the form $y = mx + b$, because we can write it as $y = 0x + 4$ (so, $m = 0$ and $b = 4$).

Example 2 Graphing a Linear Equation

Sketch the graph of $4y - 8x + 12 = 0$.

Solution

First, we solve for y. (For a review of solving equations in two or more variables, see Section A.11.)

$$\begin{aligned} 4y - 8x + 12 &= 0 && \text{Original equation} \\ 4y - 8x + 12 + 8x &= 0 + 8x && \text{Add } 8x \text{ to both sides.} \\ 4y + 12 &= 8x && \text{Combine like terms (see Section A.9).} \\ 4y + 12 - 12 &= 8x - 12 && \text{Subtract 12 from both sides.} \\ 4y &= 8x - 12 && \text{Simplify.} \\ \frac{4y}{4} &= \frac{8x}{4} - \frac{12}{4} && \text{Divide both sides by 4.} \\ y &= 2x - 3 && \text{Simplify.} \end{aligned}$$

Table 3 Solutions of $y = 2x - 3$

x	y
0	$2(0) - 3 = -3$
1	$2(1) - 3 = -1$
2	$2(2) - 3 = 1$

Since the equation $y = 2x - 3$ is of the form $y = mx + b$, we know that the graph of the equation is a line. Although we can sketch a line from as few as two points, we plot a third point as a check. If the third point is not in line with the other two, then we know that we have computed or plotted at least one of the solutions incorrectly.

To begin, we calculate three solutions of $y = 2x - 3$ in Table 3. Then we plot the three corresponding points and sketch the line through them (see Fig. 19).

We can use ZStandard followed by ZSquare on a graphing calculator to view the graph as a partial check. This check will not reveal whether we isolated y correctly, because we entered $y = 2x - 3$ rather than the *original* equation $4y - 8x + 12 = 0$ (see Fig. 20). For graphing calculator instructions, see Sections B.3, B.4, and B.6.

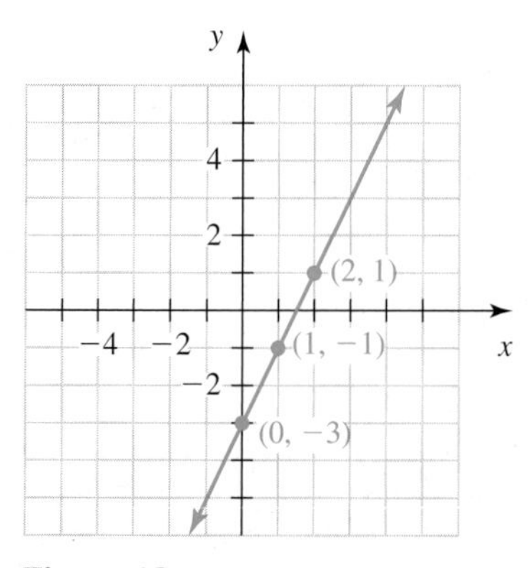

Figure 19 Graph of $y = 2x - 3$

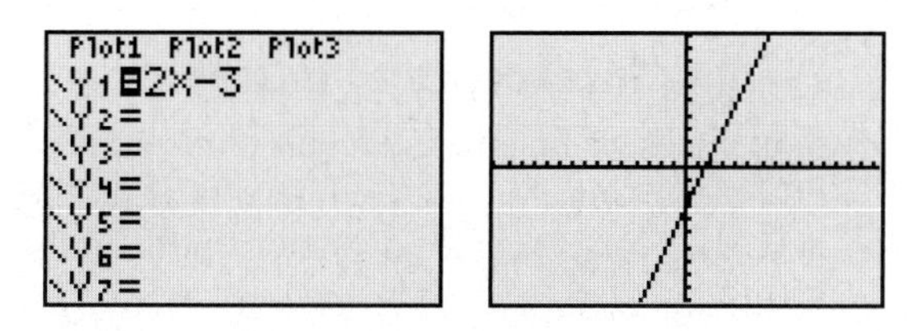

Figure 20 Graph of $y = 2x - 3$

Example 3 Using the Distributive Law to Help Graph a Linear Equation

Sketch the graph of $3(2y - 5) = 2x - 3 - 8x$.

Solution

First, we use the distributive law on the left-hand side of the equation and combine like terms on the right-hand side:

$$3(2y - 5) = 2x - 3 - 8x \quad \text{Original equation}$$

$$6y - 15 = -6x - 3 \quad \text{Distributive law; combine like terms (see Sections A.8 and A.9).}$$

$$6y - 15 + 15 = -6x - 3 + 15 \quad \text{Add 15 to both sides.}$$

$$6y = -6x + 12 \quad \text{Simplify.}$$

$$\frac{6y}{6} = \frac{-6x}{6} + \frac{12}{6} \quad \text{Divide both sides by 6.}$$

$$y = -x + 2 \quad \text{Simplify.}$$

Table 4 Solutions of $y = -x + 2$

x	y
0	$-(0) + 2 = 2$
1	$-(1) + 2 = 1$
2	$-(2) + 2 = 0$

Figure 21 Graph of $y = -x + 2$

Next, we calculate three solutions of $y = -x + 2$ in Table 4. Then we plot the solutions listed in Table 4 and sketch the line that contains them (see Fig. 21).

We can use ZStandard followed by ZSquare to view the graph of $y = -x + 2$ as a partial check (see Fig. 22). ■

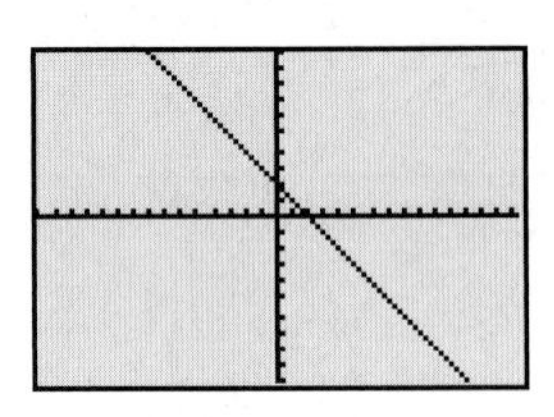

Figure 22 Graph of $y = -x + 2$

Example 4 Graphing an Equation That Contains Fractions

Sketch the graph of $y = \frac{1}{2}x - 1$.

Solution

Note that $\frac{1}{2}$ times an even number is an integer. So, in Table 5, we use even-number values of x to avoid fractional values of y. We plot the solutions and sketch a line that contains the points (see Fig. 23).

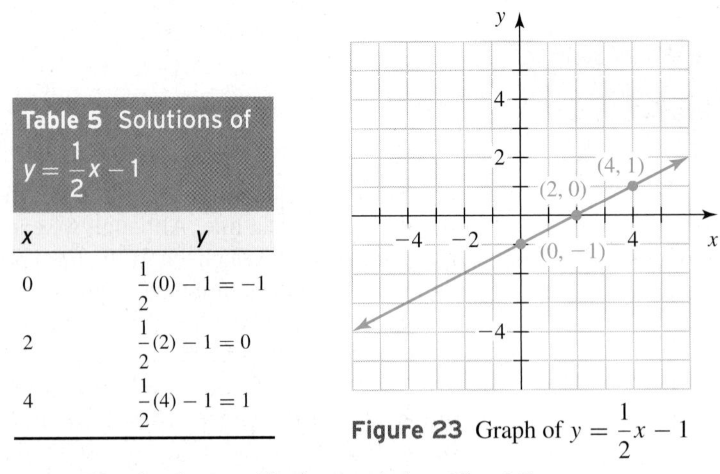

Table 5 Solutions of $y = \frac{1}{2}x - 1$

x	y
0	$\frac{1}{2}(0) - 1 = -1$
2	$\frac{1}{2}(2) - 1 = 0$
4	$\frac{1}{2}(4) - 1 = 1$

Figure 23 Graph of $y = \frac{1}{2}x - 1$

We can use ZDecimal to verify the graph (see Fig. 24). ■

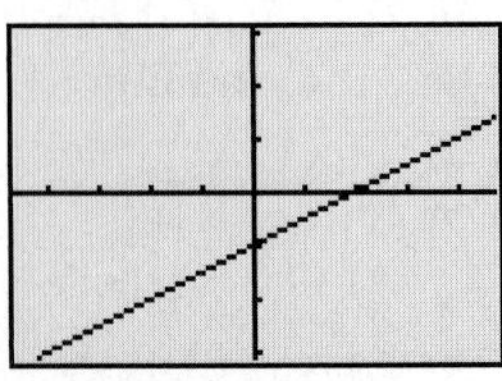

Figure 24 Verify the graph of $y = \frac{1}{2}x - 1$

Finding Intercepts of a Graph

Sometimes we find the intercepts of the graph of an equation to help us sketch the graph of the equation. Since an x-intercept is on the x-axis, we know that its y-coordinate is 0 (see Fig. 25). Since a y-intercept is on the y-axis, we know that its x-coordinate is 0.

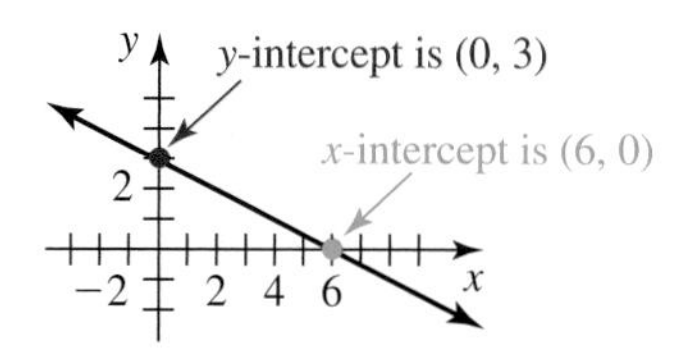

Figure 25 Intercepts of a line

Intercepts of the Graph of an Equation

For an equation containing the variables x and y,

- To find the x-coordinate of each x-intercept, substitute 0 for y and solve for x.
- To find the y-coordinate of each y-intercept, substitute 0 for x and solve for y.

In Example 5, we will find the intercepts of the graph of the equation $y = -2x + 4$ to help us graph that equation. Since the graph of $y = -2x + 4$ is a line, we say "the line $y = -2x + 4$" as shorthand for "the graph of the equation $y = -2x + 4$."

Example 5 Using Intercepts to Sketch a Graph

1. Find the x-intercept of $y = -2x + 4$.
2. Find the y-intercept of $y = -2x + 4$.
3. Sketch the graph of $y = -2x + 4$.

Table 6 Solutions of $y = -2x + 4$

x	y
0	4
1	2
2	0

Solution

1. To find the x-intercept, we substitute 0 for y and solve for x (for a review of solving linear equations in one variable, see Section A.10):

$$
\begin{aligned}
y &= -2x + 4 && \text{Original equation} \\
0 &= -2x + 4 && \text{Substitute 0 for } y. \\
0 + 2x &= -2x + 4 + 2x && \text{Add } 2x \text{ to both sides.} \\
2x &= 4 && \text{Simplify.} \\
x &= 2 && \text{Divide both sides by 2.}
\end{aligned}
$$

   The x-intercept is (2, 0).

2. To find the y-intercept, we substitute 0 for x and solve for y:

$$y = -2(0) + 4 = 4$$

   The y-intercept is (0, 4).

3. We list an additional solution of $y = -2x + 4$ in Table 6 and sketch the graph in Fig. 26.

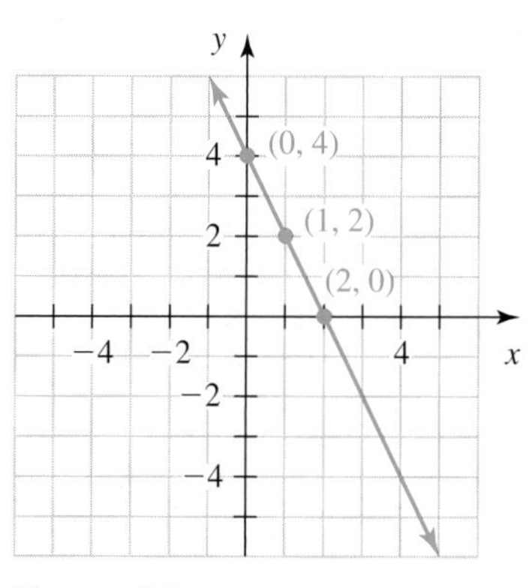

Figure 26 Graph of $y = -2x + 4$

   We use ZStandard followed by ZSquare to verify our graph (see Fig. 27). Then we use "zero" to verify the x-intercept (see Fig. 28) and TRACE to verify the y-intercept (see Fig. 29). For graphing calculator instructions, see Sections B.5 and B.21.

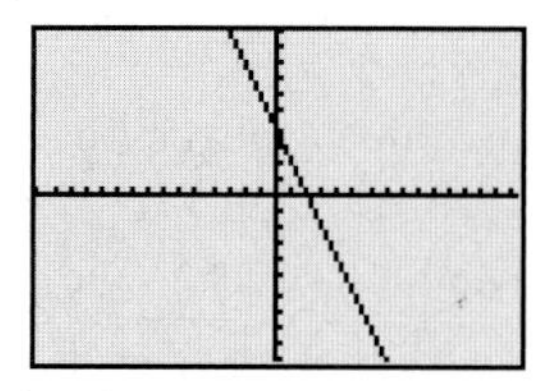

Figure 27 Verify the graph of $y = -2x + 4$

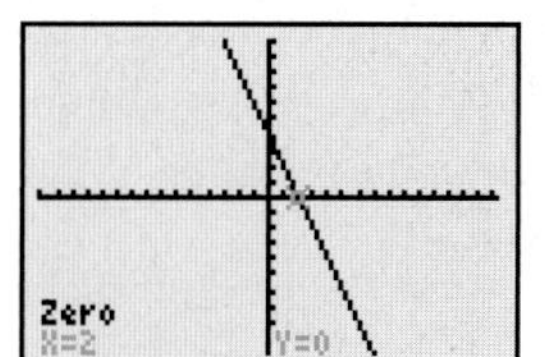

Figure 28 Verify the x-intercept

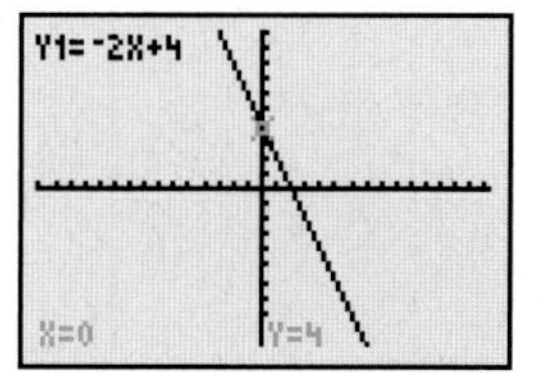

Figure 29 Verify the y-intercept

■

Vertical and Horizontal Lines

Which types of equations have graphs that are vertical or horizontal lines? We will begin to explore them in Example 6.

Example 6 Graphing a Vertical Line

Sketch the graph of $x = 3$.

Solution

Note that the values of x must be 3, but y can have any value. Some solutions of $x = 3$ are listed in Table 7. We see in Fig. 30 that the graph of $x = 3$ is a vertical line.

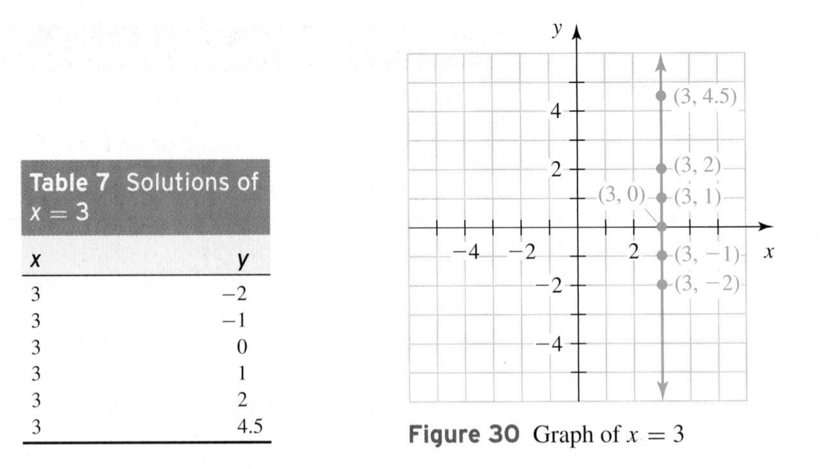

Table 7 Solutions of $x = 3$

x	y
3	−2
3	−1
3	0
3	1
3	2
3	4.5

Figure 30 Graph of $x = 3$ ■

In Example 6, we saw that the graph of the equation $x = 3$ is a vertical line. Any equation that can be put into the form $x = a$, where a is a constant, has a vertical line as its graph.

We cannot use a graphing calculator to graph an equation such as $x = 3$. We can only enter equations of the form $y = A$, where A is a constant or an expression in terms of x.

Table 8 Solutions of $y = -5$

x	y
−2	−5
−1	−5
0	−5
1	−5
2	−5
3.5	−5

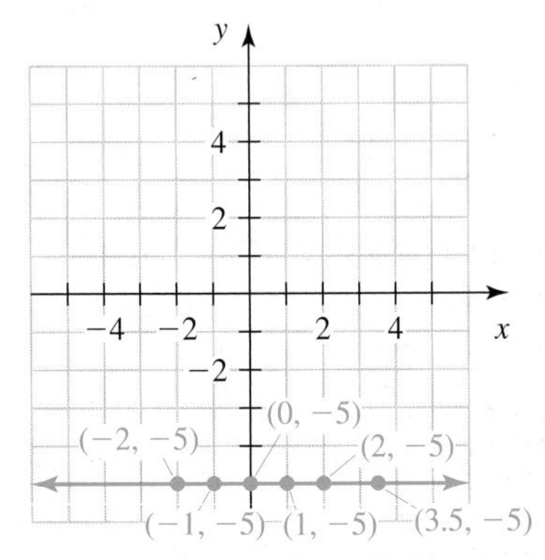

Figure 31 Graph of $y = -5$

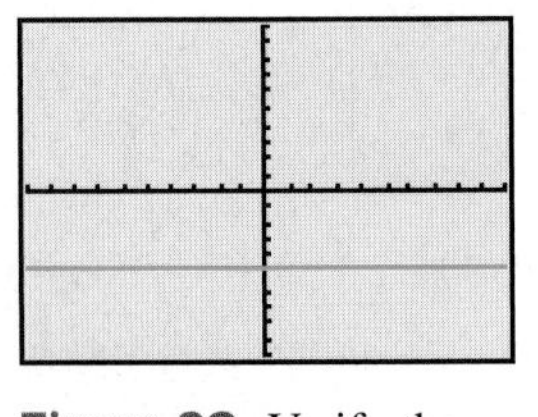

Figure 32 Verify the graph of $y = -5$

Example 7 Graphing a Horizontal Line

Sketch the graph of $y = -5$.

Solution

The value of y must be -5, but x can have any value. Some solutions of $y = -5$ are listed in Table 8. We see in Fig. 31 that the graph of $y = -5$ is a horizontal line.

We can use ZStandard to verify the graph (see Fig. 32). ■

In Example 7, we saw that the graph of the equation $y = -5$ is a horizontal line. Any equation that can be put into the form $y = b$, where b is a constant, has a horizontal line as its graph.

Equations of Vertical and Horizontal Lines

If a and b are constants, then

- An equation that can be put into the form $x = a$ has a vertical line as its graph (see Fig. 33).
- An equation that can be put into the form $y = b$ has a horizontal line as its graph (see Fig. 34).

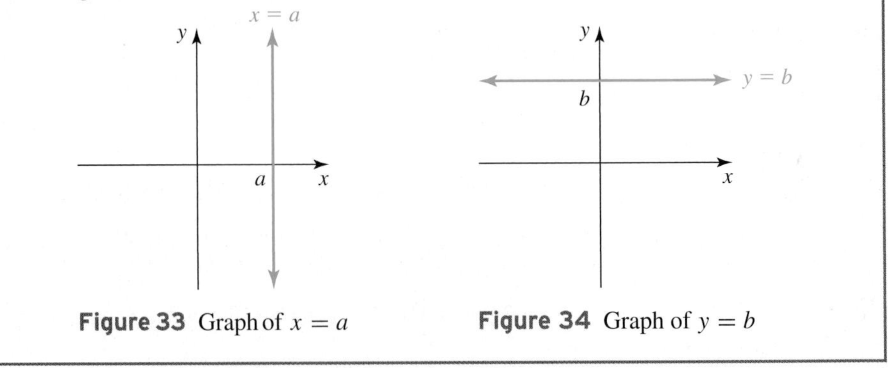

Figure 33 Graph of $x = a$

Figure 34 Graph of $y = b$

For example, the graphs of the equations $x = 4$ and $x = -7$ are vertical lines. The graphs of the equations $y = 6$ and $y = -2$ are horizontal lines.

Linear Equations in Two Variables

If an equation can be put into either form

$$y = mx + b \qquad \text{or} \qquad x = a$$

where m, a, and b are constants, then the graph of the equation is a line. We call such an equation a **linear equation in two variables.**

Any equation that can be put into the form $x = a$ has a vertical line as its graph. Any equation that can be put into the form $y = mx + b$ has a nonvertical line as its graph.

group exploration

Looking ahead: Graphical significance of *m* and *b*

1. Use ZDecimal to graph these equations of the form $y = mx$ in order, and describe what you observe:

$$y = 0.3x, \quad y = 0.7x, \quad y = x, \quad y = 2x, \quad \text{and} \quad y = 3x.$$

   Do the same with the equations

$$y = -0.3x, \quad y = -0.7x, \quad y = -x, \quad y = -2x, \quad \text{and} \quad y = -3x.$$

2. Use ZDecimal to graph these equations of the form $y = 2x + b$ in order, and describe what you observe:

$$y = 2x - 2, \quad y = 2x - 1, \quad y = 2x, \quad y = 2x + 1, \quad \text{and} \quad y = 2x + 2.$$

3. So far, you have graphed equations of the forms $y = mx$ (where $b = 0$) and $y = 2x + b$ (where $m = 2$). Now graph more equations of the form $y = mx + b$ until you are confident that you know the graphical significance of m and b for any values of m and b.

4. Make a guess about the graph of $y = mx + b$ in each situation. Test each guess by checking whether it is true for values of m and b other than the ones you have worked with so far.

   a. m is zero
   b. m is positive
   c. m is negative
   d. m is a large positive number
   e. m is a positive number near 0
   f. m is a negative number near 0
   g. $m < -10$ (for example, $m = -20$)
   h. b is equal to 5
   i. b is equal to -3
   j. b is equal to 0

TIPS FOR SUCCESS: Study Time

For each hour of class time, study for at least two hours outside class. If your math background is weak, you may need to spend more time studying.

One way to study is to do what you are doing now: Read the text. Class time is a great opportunity to be introduced to new concepts and to see how they fit together with previously learned ones. However, there is usually not enough time to address details as well as a textbook can. In this way, a textbook can serve as a supplement to what you learn in class.

HOMEWORK 1.2

FOR EXTRA HELP ▶ Student Solutions Manual | PH Math/Tutor Center | MathXL® | MyMathLab

*Graph the equation by hand. Verify your graph by using ZStandard followed by ZSquare on a graphing calculator. [**Graphing Calculator:** See Sections B.3, B.4, and B.6. Also, recall that the key* [−] *is used for subtraction and that the key* [(−)] *is used for negative numbers as well as for taking opposites.]*

1. $y = 4x + 3$ **2.** $y = 3x + 2$

3. $y = 3x - 10$ **4.** $y = 2x - 6$

5. $y = -2x + 3$ **6.** $y = -3x + 7$

7. $y = -3x - 1$ **8.** $y = -4x - 2$

9. $y = 2x$ **10.** $y = 4x$

11. $y = -3x$ **12.** $y = -2x$

13. $y = -x$ **14.** $y = x$

15. $9x - 3y = 0$ **16.** $0 = 4y - 20x$

17. $3y - 6x = 12$ **18.** $10x - 5y = 20$

19. $8x - 2y - 10 = 0$ **20.** $30x + 6y - 12 = 0$

21. $2y - 6x - 14 = -4$

22. $3y + 3x - 2 = 7$

23. $8y - 7x + 3 = -4x + 5y - 9$

24. $6y - 4x - 1 = 7y - 2x - 4$

25. $-3(y - 5) = 2(3x - 6)$

26. $2(y - 3) = 4(x + 1)$

27. $6x - 3(2y - 3) = y - 2(4x - 1)$

28. $5x - 2(3y - 1) = -2y - 3(x - 2)$

29. $y = \frac{1}{3}x$ **30.** $y = -\frac{1}{2}x$

31. $y = \frac{3}{4}x - 2$ **32.** $y = \frac{2}{5}x - 1$

33. $y = -\frac{1}{2}x + 1$ **34.** $y = -\frac{2}{3}x - 2$

35. Use a graphing calculator to sketch the graphs of equations of the form $y = mx + b$.
- **a.** Graph $y = -4.1x + 8.7$. Is the graph a line? (Here, $m = -4.1$ and $b = 8.7$.)
- **b.** Graph $y = 6$. Is the graph a line? (Here, $m = 0$ and $b = 6$.)
- **c.** Create and graph at least two more equations of the form $y = mx + b$. Are the graphs lines?

36. A student says that the graph of $y + x^2 = 5x + x^2 + 1$ is not a line, since the equation of a line does not have an x^2 term in it. Is the student correct? Explain.

37. a. Use a graphing calculator to graph each equation.
- **i.** $y = 2$
- **ii.** $y = -2$
- **iii.** $y = 5.4896$

b. Describe the graph of $y = b$, where b is a constant.

38. a. Graph each equation by hand.
- **i.** $x = 4$
- **ii.** $x = -5$
- **iii.** $x = 2.5$

b. Describe the graph of $x = a$, where a is a constant.

Graph the equation by hand.

39. $x = 6$ **40.** $x = 3$ **41.** $y = -4$

42. $y = -3$ **43.** $y = 0$ **44.** $x = 0$

*Use ZDecimal on a graphing calculator to graph the equations that follow. Use TRACE to find the coordinates of a point on the graph. Then verify that the ordered pair for that point satisfies the equation. [**Graphing Calculator:** See Sections B.6 and B.5.]*

45. $y = -3x + 1$ **46.** $2x - 3y = 6$

47. $0.83x = 4.98y - 2$

48. a. Use a graphing calculator to draw the graph of the line $y = -2.43x + 1.89$.

b. Use ZDecimal followed by TRACE to help you create a table of ordered-pair solutions of the equation. Include at least five ordered pairs.

Solve the equation. (To review solving equations in one variable, see Section A.10.)

49. $x - 1 + 2x = 3x - 9x + 17$

50. $3 - 5x - 2 = 4x + 9 - 7x$

51. $-2(3w + 5) = 3w - 4$

52. $4(2t - 1) = -5t + 7$

53. $4 - 6(2 - 3x) = 2x - (4 - 5x)$

54. $7 - 2(5 - 4x) = 3x - (8 - 3x)$

55. $4(r - 2) - 3(r - 1) = 2(r + 6)$

56. $-4(a + 6) + 5(a - 3) = 3(a - 1)$

57. $\frac{1}{2}x + \frac{1}{3} = \frac{5}{2}$

58. $\frac{1}{3}x - \frac{1}{4} = \frac{2}{3}$

59. $-\frac{5}{6}b + \frac{3}{4} = \frac{1}{2}b - \frac{2}{3}$

60. $-\frac{3}{4}w - \frac{5}{8} = \frac{3}{2}w + \frac{1}{4}$

Solve the equation. Round your result to the second decimal place.

61. $2.75x - 3.95 = -6.21x + 74.92$

62. $-6.54x + 87.35 = -4.66x - 99.03$

Solve for the specified variable. (To review solving equations in two or more variables, see Section A.11.)

63. $P = 2L + 2W$, for L **64.** $c = 3(x + y)$, for x

65. $ax + by = c$, for y **66.** $\frac{x}{a} + \frac{y}{a} = 1$, for y

Find all x-intercepts and y-intercepts.

67. $y = 2x + 10$ **68.** $y = -3x - 12$

69. $2x + 3y = 12$ **70.** $5x - 4y = 20$

71. $y = 3x$ **72.** $y = -2x$

73. $y = 3$ **74.** $x = -2$

Assuming that the graph of the equation has an x-intercept and a y-intercept, find both intercepts.

75. $y = mx + b$

76. $ax + by = c$

77. $\dfrac{x}{a} + \dfrac{y}{b} = 1$

78. $\dfrac{y - b}{m} = x$

79. The graph of an equation is sketched in Fig. 35. Describe five ordered-pair solutions of this equation by using a table.

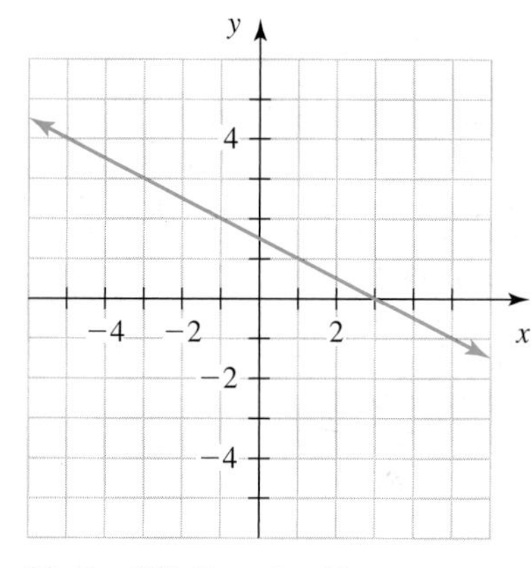

Figure 35 Exercise 79

80. The graph of an equation is sketched in Fig. 36. Describe five ordered-pair solutions of this equation by using a table.

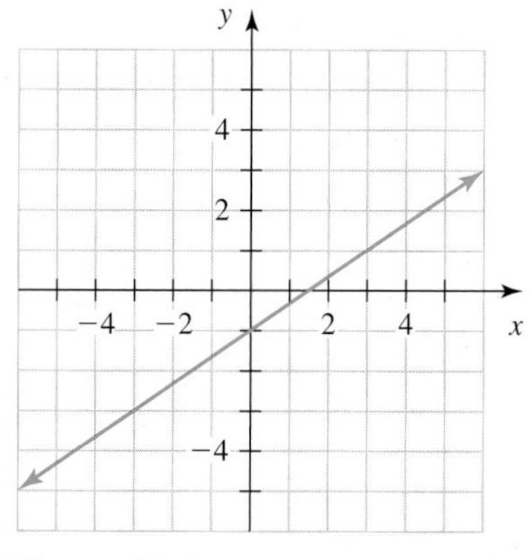

Figure 36 Exercise 80

81. Find an equation of the line sketched in Fig. 37.

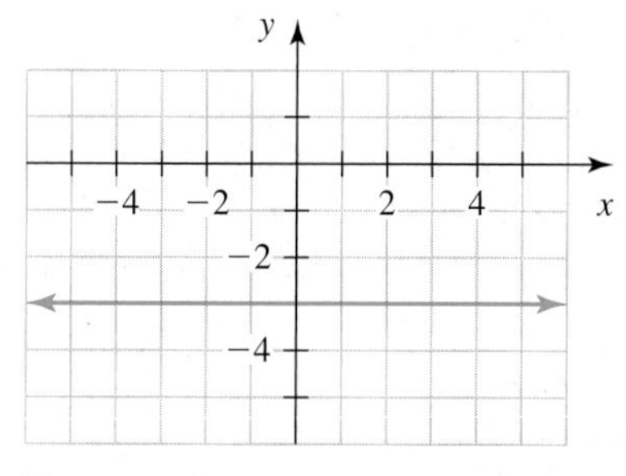

Figure 37 Exercise 81

82. Find an equation of the line sketched in Fig. 38.

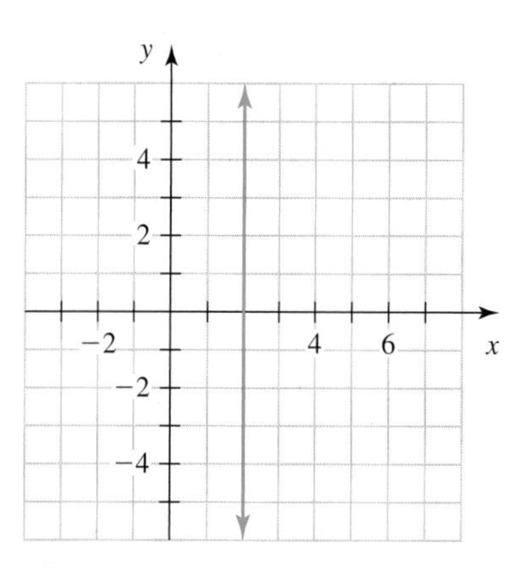

Figure 38 Exercise 82

For Exercises 83–92, refer to Fig. 39.

83. Estimate y when $x = 4$.

84. Estimate y when $x = 0$.

85. Estimate y when $x = -7$.

86. Estimate y when $x = -3$.

87. Estimate x when $y = 4$.

88. Estimate x when $y = 5$.

89. Estimate x when $y = 0$.

90. Estimate x when $y = -2$.

91. Estimate x when $y = -1.5$.

92. Estimate x when $y = 0.5$.

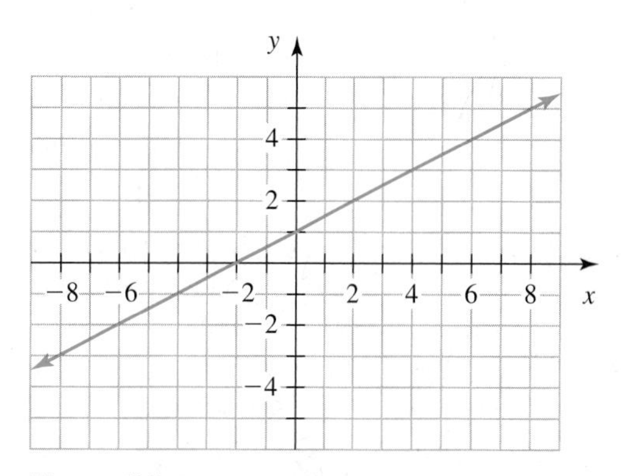

Figure 39 Exercises 83–92

93. A person lowers his hot-air balloon by gradually releasing air from it. Let x be the number of minutes that he has been releasing air from the balloon, and let y be the altitude of the balloon (in feet). Assume that the relationship between x and y is described by the equation $y = -200x + 800$.

a. Graph $y = -200x + 800$ by hand.

b. Find the y-intercept of the graph of $y = -200x + 800$. What does it mean in this situation?

c. Find the x-intercept of the graph of $y = -200x + 800$. What does it mean in this situation?

94. A person fills up her car's gas tank and then drives for a long time. Let x be the driving time (in hours) since she fueled up, and let y be the number of gallons of gas in the car's gas tank. Assume that the relationship between x and y is described by the equation $y = -2x + 10$.
a. Graph $y = -2x + 10$ by hand.
b. Find the y-intercept. What does it mean in this situation?
c. Find the x-intercept. What does it mean in this situation?

95. The graph of the equation $y = 2x + b$ contains the point $(7, 5)$. What is the constant b?

96. The graph of the equation $y = mx + 3$ contains the point $(2, 11)$. What is the constant m?

97. The graph of an equation is sketched in Fig. 40. Which of the points A, B, C, D, E, and F represent ordered pairs that satisfy the equation?

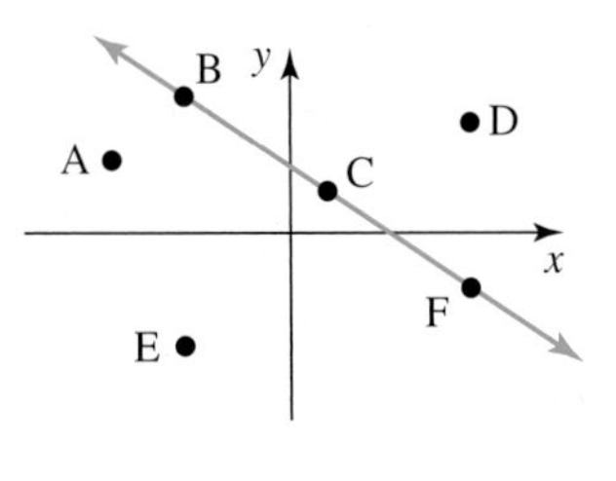

Figure 40 Exercise 97

98. The graphs of $y = ax + b$ and $y = cx + d$ are sketched in Fig. 41. For each part, decide which one or more of the points A, B, C, D, E, and F represent ordered pairs that
a. satisfy the equation $y = ax + b$.
b. satisfy the equation $y = cx + d$.
c. satisfy both equations.
d. do not satisfy either equation.

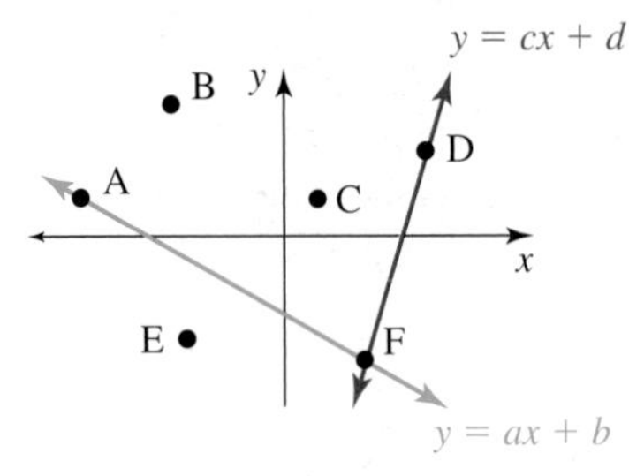

Figure 41 Exercise 98

99. Find the ordered pair(s) that satisfies both of the equations that follow. Explain. [**Hint:** Graph the equations on the same coordinate system.]

$$y = 2x + 1 \qquad y = -3x + 6$$

100. Explain why the y-coordinate of an x-intercept is 0. Explain why the x-coordinate of a y-intercept is 0.

101. Describe how to sketch a graph of a linear equation in two variables. Also, describe the meaning of a graph.

1.3 SLOPE OF A LINE

Objectives

- Compare the steepness of two objects.
- Know the meaning of, and how to calculate, the *slope* of a nonvertical line.
- Know the sign of the slope of an increasing or a decreasing line.
- Know the slopes of horizontal lines and vertical lines.
- Know the relationship between slopes of parallel lines.
- Know the relationship between slopes of perpendicular lines.

In this section, we will discuss how to measure the steepness of an object such as a ladder, road, or ski slope. Then we will focus on measuring the steepness of a nonvertical line.

Comparing the Steepness of Two Objects

How do we measure steepness? In this section, we will discuss the *slope* of a line. This important concept has numerous applications in engineering, medicine, surveying, physics, economics, mathematics, and many other fields.

Consider the sketch of two ladders leaning against a building in Fig. 42. Which ladder is steeper?

Ladder B is steeper than ladder A, even though both ladders reach a point at the same height on the building. To measure the steepness of each ladder, we compare the *vertical* distance from the base of the building to the ladder's top with the *horizontal* distance from the ladder's foot to the building. We calculate the ratio of vertical distance to horizontal distance for ladder A:

$$\text{Ladder A:} \quad \frac{\text{vertical distance}}{\text{horizontal distance}} = \frac{8 \text{ feet}}{4 \text{ feet}} = \frac{2}{1}$$

For ladder A, the vertical distance is 2 times the horizontal distance.

Figure 42 Ladder A and ladder B leaning against a building

Next, we calculate the ratio of vertical distance to horizontal distance for ladder B:

$$\text{Ladder B:} \quad \frac{\text{vertical distance}}{\text{horizontal distance}} = \frac{8 \text{ feet}}{2 \text{ feet}} = \frac{4}{1}$$

For ladder B, the vertical distance is 4 times the horizontal distance.

These calculations confirm that ladder B is steeper than ladder A in Fig. 42.

Comparing the Steepness of Two Objects

To compare the steepness of two objects such as two ramps, two roofs, or two ski slopes, compute the ratio

$$\frac{\text{vertical distance}}{\text{horizontal distance}}$$

for each object. The object with the larger ratio is the steeper object.

Example 1 Comparing the Steepness of Two Roads

Road A climbs steadily for 135 feet over a horizontal distance of 3900 feet. Road B climbs steadily for 120 feet over a horizontal distance of 3175 feet. Which road is steeper? Explain.

Solution

Figure 43 shows sketches of the two roads, but the horizontal distances and vertical distances are not drawn to scale.

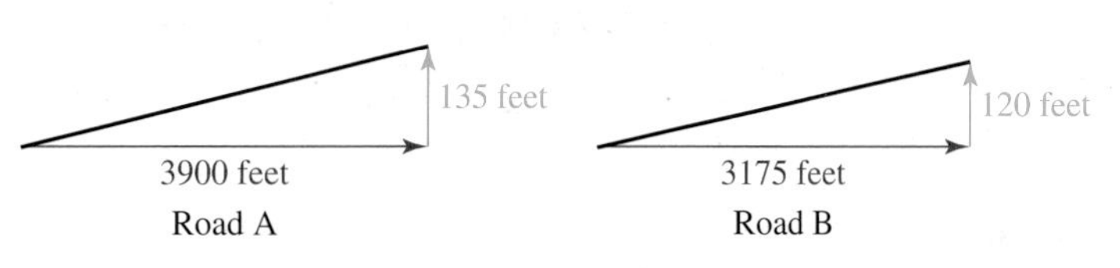

Figure 43 Roads A and B

Here, we calculate the approximate ratio of the vertical distance to the horizontal distance for each road:

$$\text{Road A:} \quad = \frac{\text{vertical distance}}{\text{horizontal distance}} = \frac{135 \text{ feet}}{3900 \text{ feet}} \approx \frac{0.035}{1}$$

$$\text{Road B:} \quad = \frac{\text{vertical distance}}{\text{horizontal distance}} = \frac{120 \text{ feet}}{3175 \text{ feet}} \approx \frac{0.038}{1}$$

Road B is a little steeper than road A, because road B's ratio of vertical distance to horizontal distance is greater than road A's. ■

The **grade** of a road is the ratio of the vertical distance to the horizontal distance, written as a percentage. To write a decimal number as a percentage, we move the decimal point two places to the right and insert the percent symbol. In Example 1, the grade of road A is about 3.5% and the grade of road B is about 3.8%.

Finding a Line's Slope

How do we calculate the steepness of a nonvertical line if we are given two points on the line? Let's use the subscript 1 to label x_1 and y_1 as the coordinates of the first point, (x_1, y_1). Likewise, we label x_2 and y_2 as the coordinates of the second point, (x_2, y_2). The horizontal change between point (x_1, y_1) and point (x_2, y_2), called the *run,* is the difference $x_2 - x_1$. The vertical change between these points, called the *rise,* is the difference $y_2 - y_1$ (see Fig. 44). The *slope* of the line is the ratio of the rise to the run. We use the letter m to represent the slope.

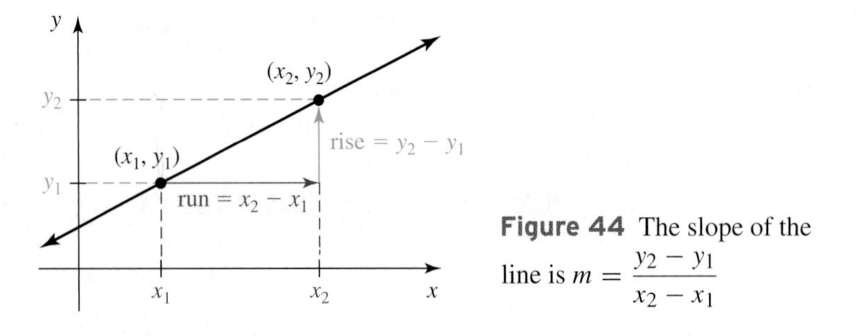

Figure 44 The slope of the line is $m = \dfrac{y_2 - y_1}{x_2 - x_1}$

DEFINITION Slope of a nonvertical line

Let (x_1, y_1) and (x_2, y_2) be two distinct points of a nonvertical line (see Fig. 44). The **slope** of the line is

$$m = \frac{\text{vertical change}}{\text{horizontal change}} = \frac{\text{rise}}{\text{run}} = \frac{y_2 - y_1}{x_2 - x_1}$$

In words: The slope of a nonvertical line is equal to the ratio of the rise to the run (in going from one point on the line to another point on the line).

A **formula** is an equation that contains two or more variables. We will refer to the equation $m = \dfrac{y_2 - y_1}{x_2 - x_1}$ as the **slope formula.**

Here, we list, in verbal and graphical forms, the directions (right, left, up, or down) associated with the signs of rises and runs:

Sign of rise or run	*Direction (verbal)*	*Direction (graphical)*
run is positive	goes to the right	
run is negative	goes to the left	
rise is positive	goes up	
rise is negative	goes down	

Example 2 Finding the Slope of a Line

Find the slope of the line that contains the points (1, 2) and (5, 4).

Solution

Using the slope formula, where $(x_1, y_1) = (1, 2)$ and $(x_2, y_2) = (5, 4)$, we have

$$m = \frac{4-2}{5-1} = \frac{2}{4} = \frac{1}{2}$$

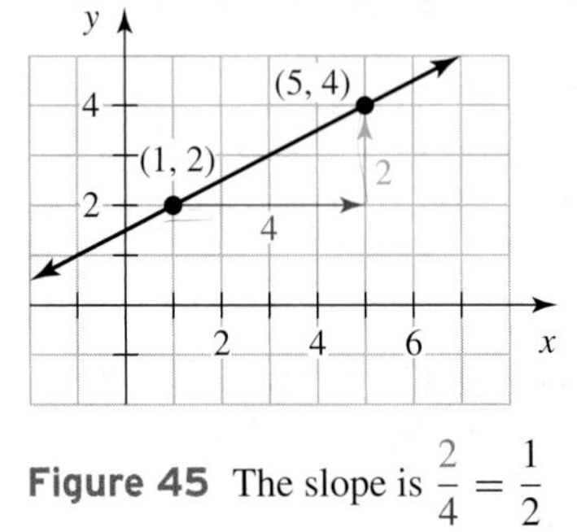

Figure 45 The slope is $\dfrac{2}{4} = \dfrac{1}{2}$

By plotting the points, we find that if the run is 4, then the rise is 2 (see Fig. 45). So, the slope is $m = \dfrac{\text{rise}}{\text{run}} = \dfrac{2}{4} = \dfrac{1}{2}$, which is our result from using the slope formula. ■

In Example 2, we calculated the slope of a line for $(x_1, y_1) = (1, 2)$ and $(x_2, y_2) = (5, 4)$. Here, we switch the roles of the two points to find the slope when $(x_1, y_1) = (5, 4)$ and $(x_2, y_2) = (1, 2)$ instead:

$$m = \frac{y_2 - y_1}{x_2 - x_1} = \frac{2-4}{1-5} = \frac{-2}{-4} = \frac{1}{2}$$

The result is the same as our result in Example 2. In general, when we use the slope formula with two points on a line, it doesn't matter which point we choose to be (x_1, y_1) and which we choose to be (x_2, y_2).

WARNING It is a common error to substitute into the slope formula incorrectly. Carefully consider why the middle and right-hand formulas are incorrect:

Correct	**Incorrect**	**Incorrect**
$m = \frac{y_2 - y_1}{x_2 - x_1}$	$m = \frac{y_2 - y_1}{x_1 - x_2}$	$m = \frac{x_2 - x_1}{y_2 - y_1}$

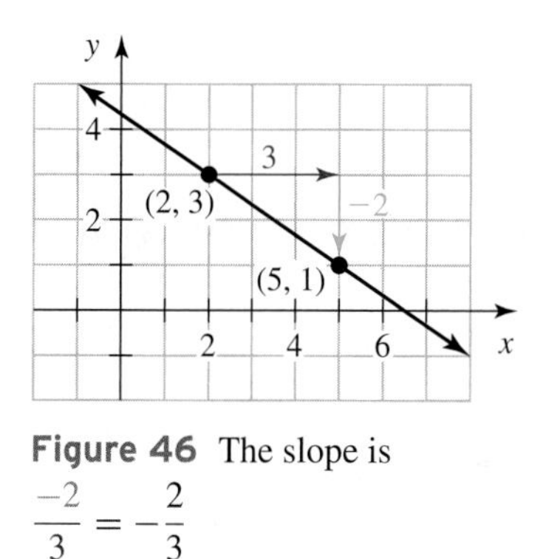

Figure 46 The slope is $\frac{-2}{3} = -\frac{2}{3}$

Example 3 Finding the Slope of a Line

Find the slope of the line that contains the points (2, 3) and (5, 1).

Solution

$$m = \frac{1-3}{5-2} = \frac{-2}{3} = -\frac{2}{3}$$

By plotting the points, we find that if the run is 3, then the rise is −2 (see Fig. 46). So, the slope is $\frac{-2}{3} = -\frac{2}{3}$, which is our result from using the slope formula. ■

Increasing and Decreasing Lines

Consider the increasing line in Fig. 47. Our work with the signs of the rise and run shows that the slope of the line is positive.

Now consider the decreasing line in Fig. 48. Our work with the signs of the rise and run shows that the slope of the line is negative.

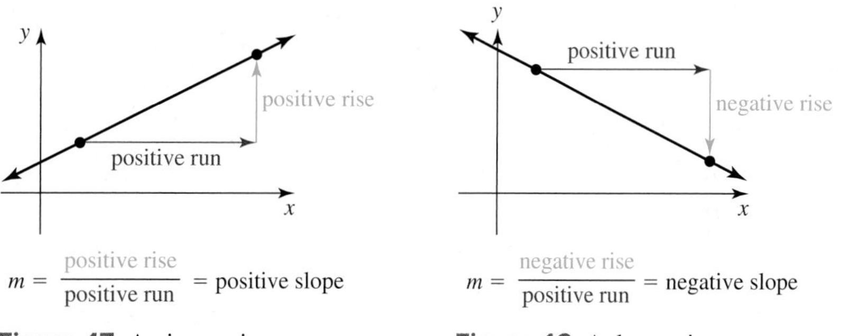

$m = \frac{\text{positive rise}}{\text{positive run}} = \text{positive slope}$

Figure 47 An increasing line has positive slope

$m = \frac{\text{negative rise}}{\text{positive run}} = \text{negative slope}$

Figure 48 A decreasing line has negative slope

> **Slopes of Increasing and Decreasing Lines**
>
> - An increasing line has positive slope (see Fig. 47).
> - A decreasing line has negative slope (see Fig. 48).

Example 4 Finding the Slope of a Line

Find the slope of the line that contains the points (−9, −4) and (12, −8).

Solution

$$m = \frac{-8-(-4)}{12-(-9)} = \frac{-8+4}{12+9} = \frac{-4}{21} = -\frac{4}{21}$$

Since the slope is negative, the line is decreasing. ■

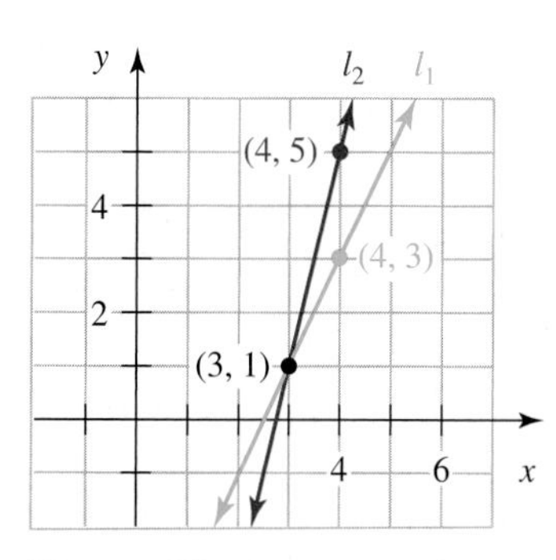

Figure 49 Find the slopes of the two lines

Example 5 Comparing the Slopes of Two Lines

Find the slopes of the two lines sketched in Fig. 49. Which line has the greater slope? Explain why this makes sense in terms of the steepness of a line.

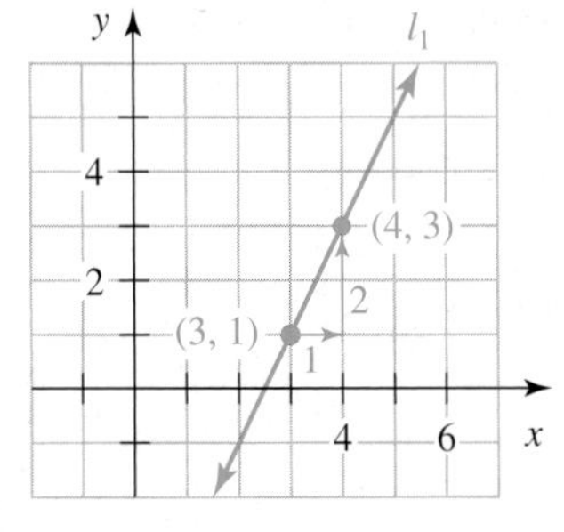

Figure 50 Line with lesser slope

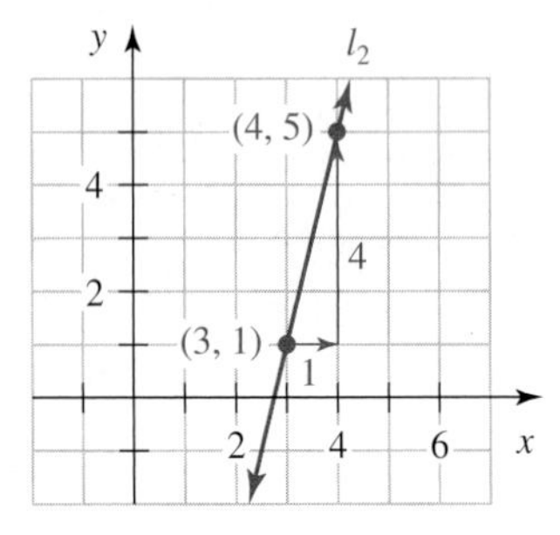

Figure 51 Line with greater slope

Solution

For line l_1 in Fig. 50, if the run is 1, the rise is 2. We calculate the slope of line l_1:

$$\text{Slope of line } l_1 = \frac{\text{rise}}{\text{run}} = \frac{2}{1} = 2$$

For line l_2 in Fig. 51, if the run is 1, the rise is 4. We calculate the slope of line l_2:

$$\text{Slope of line } l_2 = \frac{\text{rise}}{\text{run}} = \frac{4}{1} = 4$$

Note that the slope of line l_2 is greater than the slope of line l_1, which is what we would expect because line l_2 looks steeper than line l_1. ■

In general, **for two nonparallel increasing lines, the steeper line has the greater slope.**

Horizontal and Vertical Lines

What is the slope of a horizontal line or vertical line? We will explore this question in Examples 6 and 7.

Example 6 Investigating the Slope of a Horizontal Line

Find the slope of the line that contains the points (2, 3) and (6, 3).

Solution

We plot the points (2, 3) and (6, 3) and sketch the line that contains the points (see Fig. 52).

The slope formula gives

$$m = \frac{3-3}{6-2} = \frac{0}{4} = 0$$

So, the slope of the horizontal line is zero. It makes sense that the horizontal line has slope equal to zero, because such a line has "no steepness." ■

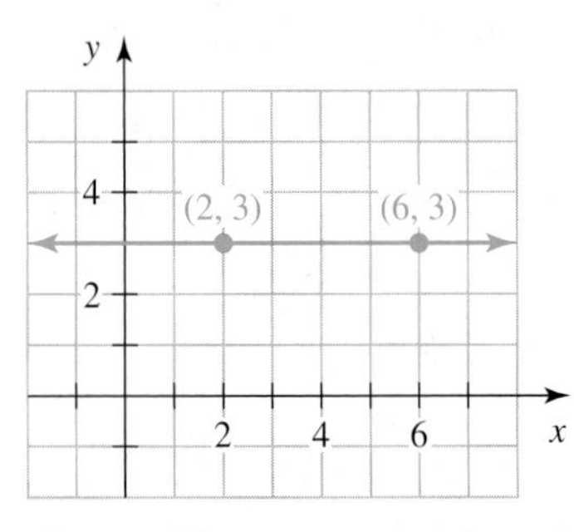

Figure 52 The horizontal line has slope equal to zero

Example 7 Investigating the Slope of a Vertical Line

Find the slope of the line that contains the points (4, 2) and (4, 5).

Solution

We plot the points (4, 2) and (4, 5) and sketch the line that contains the points (see Fig. 53).

The slope formula gives

$$m = \frac{5-2}{4-4} = \frac{3}{0}$$

Since division by zero is undefined, the slope of the vertical line is *undefined*. ■

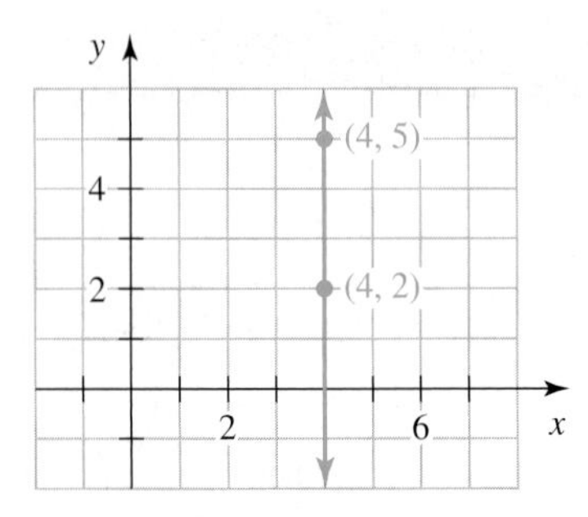

Figure 53 The vertical line has undefined slope

Slopes of Horizontal and Vertical Lines

- A horizontal line has slope equal to zero (see Fig. 54).
- A vertical line has undefined slope (see Fig. 55).

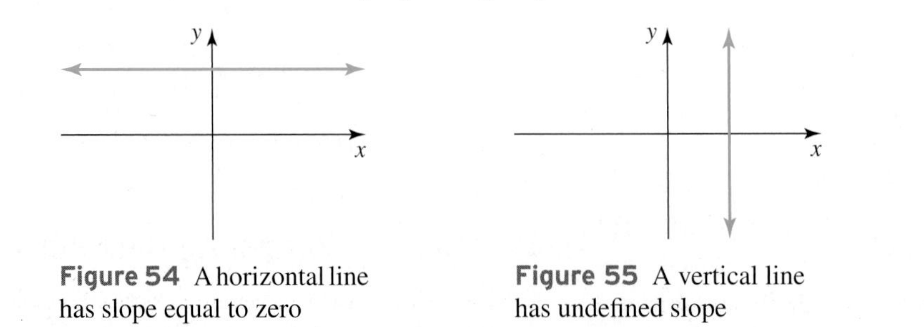

Figure 54 A horizontal line has slope equal to zero

Figure 55 A vertical line has undefined slope

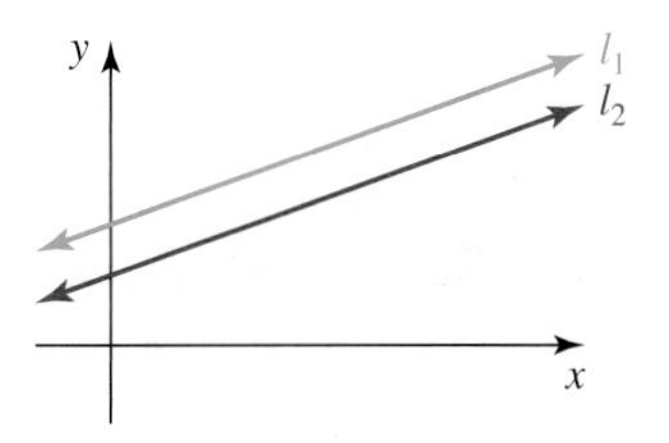

Figure 56 Two parallel lines

Parallel and Perpendicular Lines

Two lines are called **parallel** if they do not intersect (see Fig. 56). In Example 8, we compare the slopes of two parallel lines.

Example 8 Finding Slopes of Parallel Lines

Find the slopes of the parallel lines l_1 and l_2 sketched in Fig. 57.

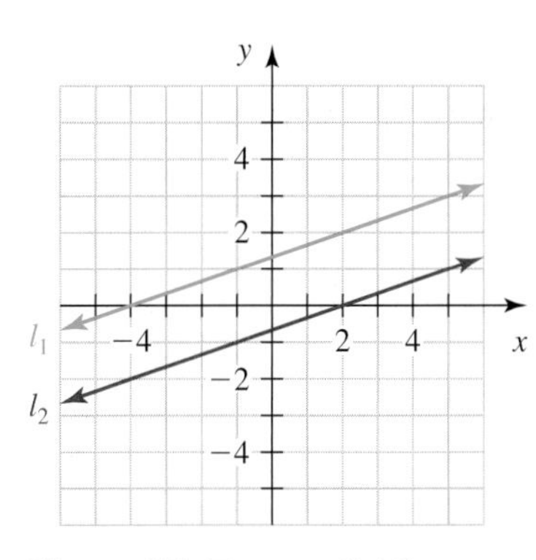

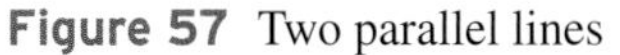

Figure 57 Two parallel lines

Solution

For both lines, if the run is 3, the rise is 1 (see Fig. 58). So, the slope of both lines is

$$m = \frac{\text{rise}}{\text{run}} = \frac{1}{3}$$

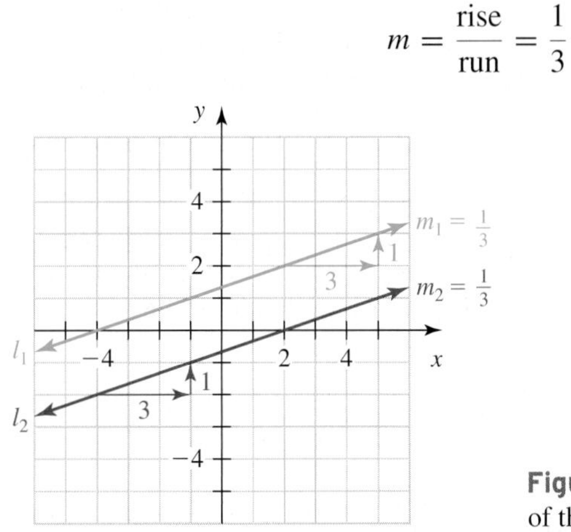

Figure 58 Calculate the slopes of the parallel lines ■

It makes sense that nonvertical parallel lines have equal slope, since parallel lines have the same steepness.

> **Slopes of Parallel Lines**
>
> If lines l_1 and l_2 are nonvertical parallel lines on the same coordinate system, then the slopes of the lines are equal:
>
> $$m_1 = m_2$$
>
> Also, if two distinct lines have equal slope, then the lines are parallel.

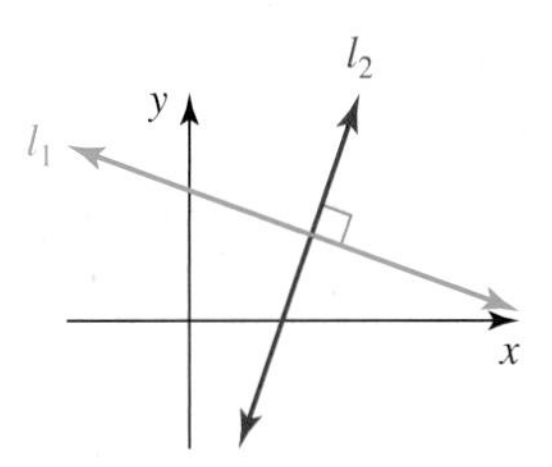

Figure 59 Two perpendicular lines

Two lines are called **perpendicular** if they intersect at a 90° angle (see Fig. 59). In Example 9, we compare the slopes of two perpendicular lines.

Example 9 Finding Slopes of Perpendicular Lines

Find the slopes of the perpendicular lines l_1 and l_2 in Fig. 60.

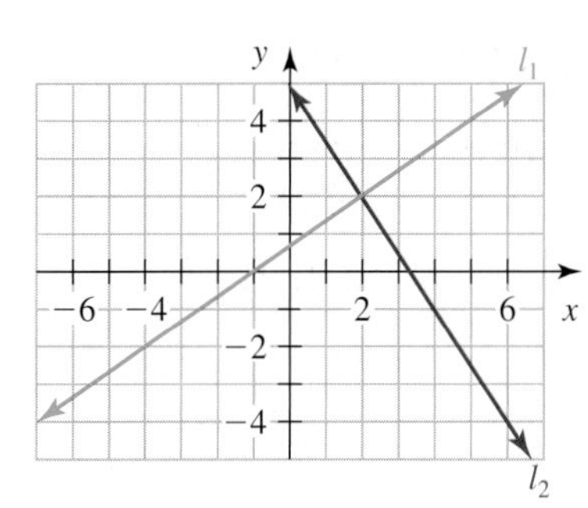

Figure 60 Two perpendicular lines

Solution

From Fig. 61, we see that the slope of line l_1 is $m_1 = \frac{2}{3}$ and that the slope of line l_2 is $m_2 = \frac{-3}{2} = -\frac{3}{2}$.

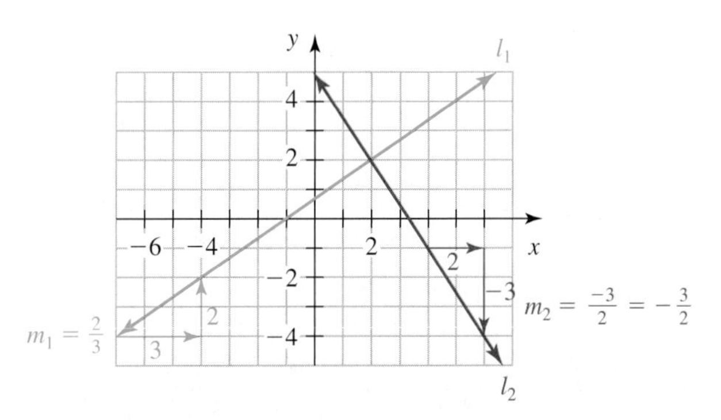

Figure 61 Two perpendicular lines ■

In Example 9, the slope $-\frac{3}{2}$ is the opposite of the reciprocal of the slope $\frac{2}{3}$.

Slopes of Perpendicular Lines

If lines l_1 and l_2 are nonvertical perpendicular lines, then the slope of one line is the opposite of the reciprocal of the slope of the other line:

$$m_2 = -\frac{1}{m_1}$$

Also, if the slope of one line is the opposite of the reciprocal of another line's slope, then the lines are perpendicular.

Example 10 Finding Slopes of Parallel and Perpendicular Lines

A line l_1 has slope $\frac{3}{7}$.

1. If line l_2 is parallel to line l_1, find the slope of line l_2.
2. If line l_3 is perpendicular to line l_1, find the slope of line l_3.

Solution

1. The slopes of lines l_2 and l_1 are equal, so line l_2 has slope $\frac{3}{7}$.
2. The slope of line l_3 is the opposite of the reciprocal of $\frac{3}{7}$, or $-\frac{7}{3}$. ■

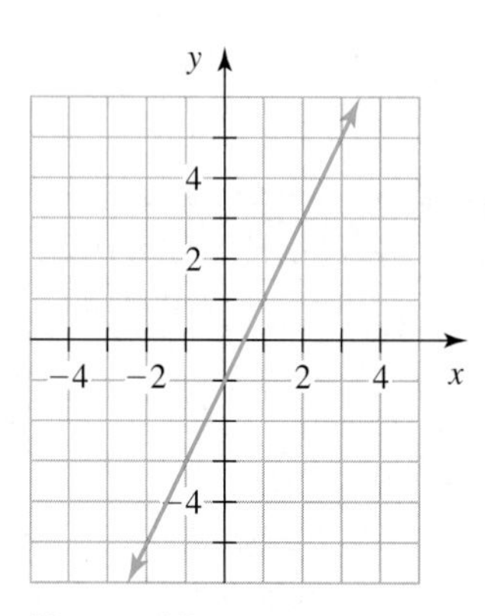

Figure 62 Use different pairs of points to calculate the slope

group exploration

For a line, rise over run is constant

1. A line is sketched in Fig. 62. Plot the points (−2, −5), (1, 1), and (3, 5). (Plotted correctly, these points will lie on the line.)
2. Using the points (−2, −5) and (1, 1), find the slope of the line.
3. Using the points (1, 1) and (3, 5), find the slope of the line.
4. Using the points (−2, −5) and (3, 5), find the slope of the line.
5. Using two other points of your choice, find the slope of the line.
6. What do you notice about the slopes you have calculated? Does it matter which two points on a line are used to find the slope of the line?

group exploration

Looking ahead: Slope addition property

1. Complete Table 9.

Table 9 Solutions to Three Equations

$y = 2x + 1$		$y = 3x - 5$		$y = -2x + 6$	
x	y	x	y	x	y
0		0		0	
1		1		1	
2		2		2	
3		3		3	
4		4		4	

2. In Table 9, the x-coordinates increase by 1 each time. For each equation, what do you notice about the y-coordinates? Compare what you notice with the coefficient of x in each equation.

3. Describe what the pattern from Problem 2 would be in general for any equation of the form $y = mx + b$.
4. Create an equation of the form $y = mx + b$, and check whether it behaves as you described in Problem 3.
5. Substitute 1 for x in the equation $y = mx + b$. Then substitute 2 for x. Then substitute 3. Explain why these results suggest that your description in Problem 3 is correct.

Helping students during office hours is part of an instructor's job. Your instructor wants you to succeed and hopes that you take advantage of all opportunities to learn.

Come prepared to office visits. For example, if you are having trouble with a concept, attempt some related exercises and bring your work so that your instructor can see where you are having difficulty. If you miss a class, read the material first, borrow class notes, and try completing assigned exercises before visiting your instructor to get the most out of the visit.

HOMEWORK 1.3

FOR EXTRA HELP ▶ Student Solutions Manual | PH Math/Tutor Center | MathXL® | MyMathLab

1. A portion of road A climbs steadily for 120 feet over a horizontal distance of 4000 feet. A portion of road B climbs steadily for 160 feet over a horizontal distance of 6500 feet. Which road is steeper? Explain.
2. Airplane A climbs steadily for 2500 feet over a horizontal distance of 8000 feet. Airplane B climbs steadily for 3100 feet over a horizontal distance of 9500 feet. Which plane is climbing more steeply? Explain.
3. Ski run A declines steadily for 90 yards over a horizontal distance of 300 yards. Ski run B declines steadily for 125 yards over a horizontal distance of 450 yards. Which run is steeper? Explain.
4. A ski run declines steadily from the top of a mountain to a chairlift. Then the run continues to decline with a different constant steepness, to end at a restaurant on the mountain. The horizontal distance for the entire run is 1300 yards over a vertical decline of 415 yards. The horizontal distance from the top of the mountain to the chairlift is 400 yards over a vertical decline of 100 yards. Find (the absolute value of) the "slopes" of each part of the run.

Find the slope of the line passing through the given points. State whether the line is increasing, decreasing, horizontal, or vertical.

5. $(2, 3)$ and $(5, 9)$
6. $(1, 8)$ and $(5, 4)$
7. $(-5, 7)$ and $(1, 3)$
8. $(2, 6)$ and $(8, -2)$
9. $(-4, 10)$ and $(2, -2)$
10. $(-3, 6)$ and $(1, -10)$
11. $(1, -2)$ and $(7, -4)$
12. $(2, -3)$ and $(6, -9)$
13. $(-5, -8)$ and $(4, -2)$
14. $(-1, -12)$ and $(7, -2)$
15. $(-4, -9)$ and $(-2, -1)$
16. $(-1, -6)$ and $(-2, -5)$
17. $(0, 0)$ and $(1, 1)$
18. $(0, 0)$ and $(100, 100)$
19. $(2, 6)$ and $(7, 6)$
20. $(-3, -1)$ and $(5, -1)$
21. $(-6, -2)$ and $(-6, 5)$
22. $(4, -1)$ and $(4, 8)$

Find the slope of the line passing through the given points. Round your result to the second decimal place. State whether the line is increasing, decreasing, horizontal, or vertical.

23. $(1.2, 5.4)$ and $(3.9, 2.6)$
24. $(-3.9, 2.2)$ and $(-5.1, -7.4)$
25. $(8.94, -17.94)$ and $(21.13, -2.34)$
26. $(-25.41, 82.78)$ and $(-11.26, -66.66)$
27. For each line sketched in Fig. 63, determine whether the line's slope is positive, negative, zero, or undefined.

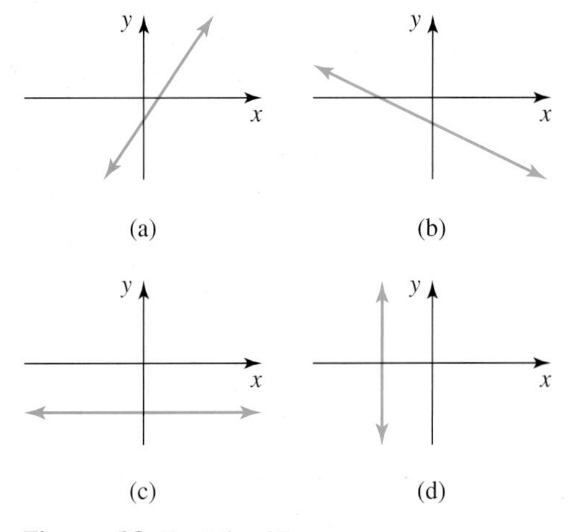

Figure 63 Exercise 27

28. Compute the slope of the line with x-intercept $(5, 0)$ and y-intercept $(0, -2)$.

29. Find the slope of the line sketched in Fig. 64.

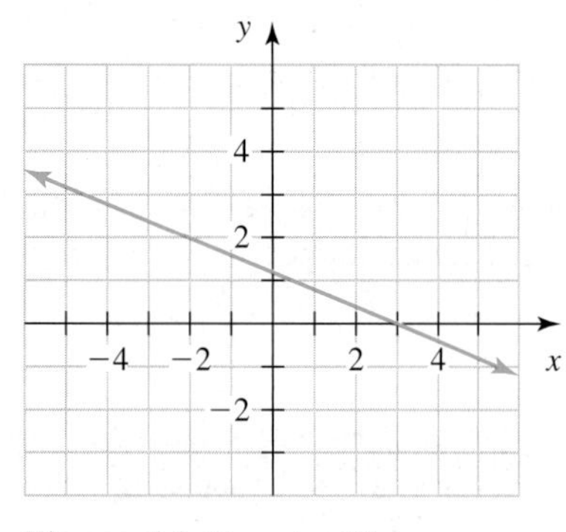

Figure 64 Exercise 29

30. Find the slope of the line sketched in Fig. 65.

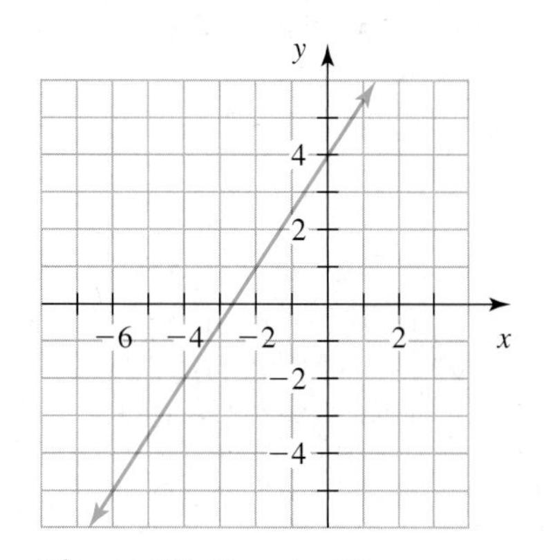

Figure 65 Exercise 30

Using the indicated slopes of lines l_1 and l_2, determine whether the lines are parallel, perpendicular, or neither. Assume that the lines are not the same.

31. $m_1 = 2, m_2 = 2$

32. $m_1 = 5, m_2 = -\frac{1}{5}$

33. $m_1 = 6, m_2 = -6$

34. $m_1 = 1, m_2 = -1$

35. $m_1 = \frac{2}{7}, m_2 = -\frac{7}{2}$

36. $m_1 = \frac{3}{5}, m_2 = \frac{5}{3}$

37. $m_1 = \frac{7}{4}, m_2 = \frac{4}{7}$

38. $m_1 = \frac{5}{8}, m_2 = -\frac{5}{8}$

39. $m_1 = 0$, m_2 is undefined

40. m_1 and m_2 are undefined

41. Are the lines sketched in Fig. 66 perpendicular? Explain.

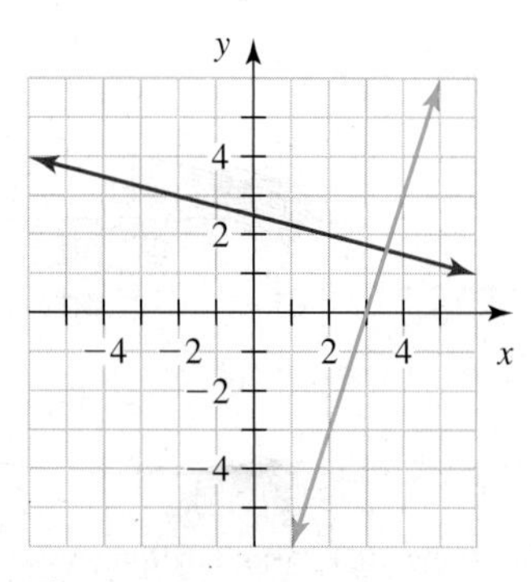

Figure 66 Exercise 41

42. Are the lines sketched in Fig. 67 parallel? Explain.

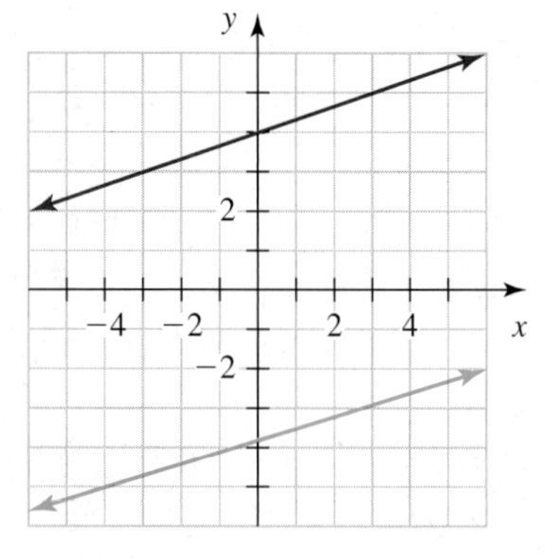

Figure 67 Exercise 42

Sketch a line that has the indicated slope.

43. $m = \frac{2}{5}$

44. $m = \frac{4}{3}$

45. $m = -\frac{4}{3}$ $\left[\textbf{Hint: } -\frac{4}{3} = \frac{-4}{3}\right]$

46. $m = -\frac{2}{3}$

47. $m = 3$ $\left[\textbf{Hint: } 3 = \frac{3}{1}\right]$

48. $m = -2$

49. $m = 0$

50. m is undefined

51. Sketch a line with slope 2 and another line with slope 3. Which line is steeper?

52. Sketch a line with slope 2 and another line with slope -2. Which line is steeper?

Sketch a line that meets the description. Find the slope of that line.

53. An increasing line that is nearly horizontal

54. A decreasing line that is nearly horizontal

55. A decreasing line that is nearly vertical

56. An increasing line that is nearly vertical

Sketch a line that meets the given description.

57. The slope is a large positive number.

58. The slope is a positive number near zero.

59. The slope is a negative number near zero.

60. The slope is less than -5.

61. A student tries to find the slope of the line that contains the points (2, 5) and (4, 8):

$$\frac{4-2}{8-5} = \frac{2}{3}$$

Describe any errors. Then find the slope correctly.

62. A student tries to find the slope of the line that contains the points (7, 1) and (2, 9):

$$\frac{9-1}{7-2} = \frac{8}{5}$$

Describe any errors. Then find the slope correctly.

63. A line contains the points (2, 7) and (3, 10). Find three more points that lie on the line.

64. A line contains the points $(-6, -4)$ and $(-3, 1)$. Find three more points that lie on the line.

65. a. Carefully graph the given equation by hand. Then find the slope of the line by using the ratio $\frac{\text{rise}}{\text{run}}$.

i. $y = 2x + 1$ **ii.** $y = 3x - 5$ **iii.** $y = -2x + 6$

b. Compare the slope of each line with the coefficient of x in the corresponding equation.

66. a. Use the expression $\frac{y_2 - y_1}{x_2 - x_1}$ to find the slope of the line that contains the points $(x_1, y_1) = (2, 3)$ and $(x_2, y_2) = (7, 5)$.

b. Use the expression $\frac{y_1 - y_2}{x_1 - x_2}$ to find the slope of the line that contains the points $(x_1, y_1) = (2, 3)$ and $(x_2, y_2) = (7, 5)$.

c. Compare your results from parts (a) and (b).

d. Show that $\frac{y_2 - y_1}{x_2 - x_1} = \frac{y_1 - y_2}{x_1 - x_2}$, where (x_1, y_1) and (x_2, y_2) are two distinct points of a nonvertical line. [**Hint:** $a - b = -(b - a)$]

e. When using two given points on a line to calculate the slope of the line, does it matter which point we choose to be first, (x_1, y_1), and which second, (x_2, y_2)? Explain.

67. Explore the relationship among three lines that pass through the origin (0, 0), where the slope of one of the lines is the reciprocal of the slope of one of the other lines.

a. By hand, carefully sketch the lines that pass through the origin (0, 0) and that have slopes 5, 1, and $\frac{1}{5}$.

b. Sketch the lines that pass through the origin (0, 0) and that have slopes $\frac{2}{5}$, 1, and $\frac{5}{2}$.

c. Sketch the lines that pass through the origin (0, 0) and that have slopes $\frac{3}{4}$, 1, and $\frac{4}{3}$.

d. What pattern do you notice from your graphs in parts (a)–(c)?

e. A line with slope m is sketched in Fig. 68. Sketch a line with slope $\frac{1}{m}$ that passes through the origin (0, 0). Assume that both axes are scaled the same.

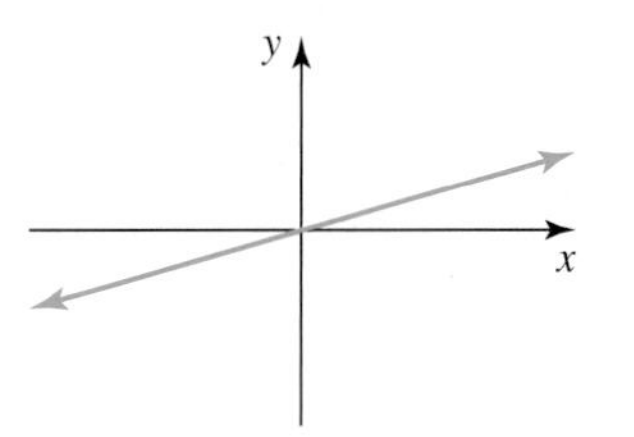

Figure 68 Exercise 67e

68. a. A square has vertices at (3, 1) and (3, 7). How many possible positions are there for the other two vertices? Find the coordinates for each possibility.

b. A parallelogram has vertices at (−7, −2), (3, 1), and (−4, 2). How many possible positions are there for the fourth vertex? Find the coordinates for each possibility. [**Hint:** Try drawing different line segments between the given vertices.]

69. Suppose that a line with slope $-\frac{2}{3}$ contains a point P. A point Q lies three units to the right and two units down from point P. A point S lies three units to the left and two units up from point P. Does the line contain point Q? point S? Explain.

70. Describe the meaning of the slope of a line. Sketch various types of lines and give the slope for each line. For each sketch, explain why the slope assignment makes sense. For example, you could sketch a horizontal line, state that the slope is zero, and explain why it makes sense that the slope of a horizontal line is zero in terms of rise and run.

1.4 MEANING OF SLOPE FOR EQUATIONS, GRAPHS, AND TABLES

"I love mathematics. Mathematics is the language of nature and force. Math is directly behind most technical innovations and has also been used to solve many of the mysteries of life."

—Tim P., student

Objectives

- Find the slope and y-intercept of a nonvertical line from an equation in *slope–intercept form.*
- Know the *vertical change property.*
- Use slope and y-intercept to sketch the graph of an equation of the form $y = mx + b$.
- Know the *slope addition property.*

In this section, we will discuss what slope means as regards equations, graphs, and tables. We also use these meanings to graph a linear equation with defined slope.

Finding Slope from a Linear Equation

How can we use an equation of a nonvertical line to find the line's slope?

Example 1 Finding the Slope of a Line

Find the slope of the line $y = 2x + 1$.

Solution

We use $x = 0, 1, 2, 3$ in Table 10 to list solutions, and we sketch the graph of the equation in Fig. 69.

Table 10 Solutions of $y = 2x + 1$

x	y
0	1
1	3
2	5
3	7

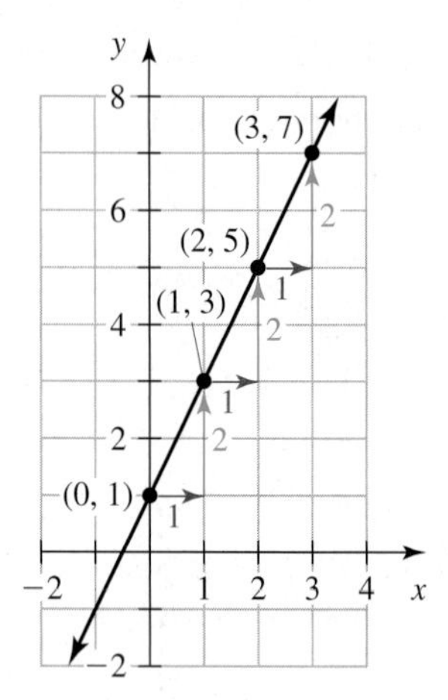

Figure 69 Graph of $y = 2x + 1$

If the run is 1, the rise is 2 (see Fig. 69). So, the slope is

$$m = \frac{\text{rise}}{\text{run}} = \frac{2}{1} = 2$$ ■

From Example 1, we can make three observations about the slope of the nonvertical line $y = 2x + 1$:

1. The coefficient of x is 2, which is the slope.
2. If the run is 1, then the rise is 2 (the slope). See Fig. 69.
3. As the value of x increases by 1, the value of y increases by 2 (the slope). See Table 10.

Example 2 Finding the Slope of a Line

Find the slope of the line $y = -3x + 8$.

Solution

Some solutions are listed in Table 11, and a graph is sketched in Fig. 70.

Table 11 Solutions of $y = -3x + 8$

x	y
0	8
1	5
2	2
3	−1

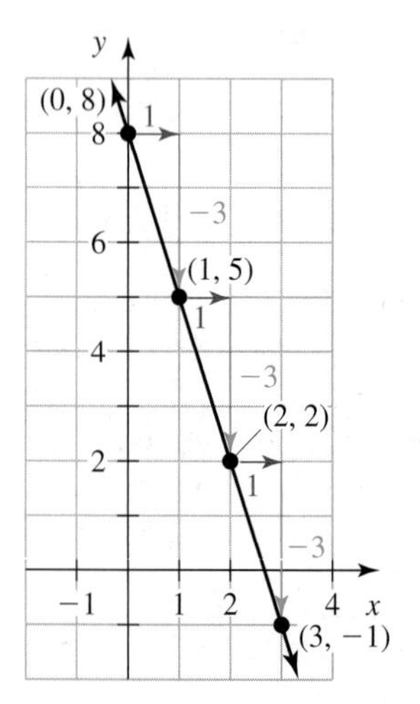

Figure 70 Graph of $y = -3x + 8$

The slope is $m = \frac{-3}{1} = -3$. ■

From Example 2, we can make three observations about the slope of the nonvertical line $y = -3x + 8$:

1. The coefficient of x is -3, which is the slope.
2. If the run is 1, then the rise is -3 (the slope). See Fig. 70.
3. As the value of x increases by 1, the value of y changes by -3 (the slope). See Table 11.

Our first observations made after both Examples 1 and 2 suggest a general property about slope.

> **Slope of a Linear Equation of the Form $y = mx + b$**
>
> For a linear equation of the form $y = mx + b$, m is the slope of the line.

Example 3 Identifying Parallel or Perpendicular Lines

Are the lines $y = \frac{5}{6}x + 3$ and $12y - 10x = 5$ parallel, perpendicular, or neither?

Solution

For the line $y = \frac{5}{6}x + 3$, the slope is $\frac{5}{6}$. For $12y - 10x = 5$, the slope is *not* -10. To find the slope, we begin by solving the equation $12y - 10x = 5$ for y:

$$12y - 10x = 5 \quad \text{Original equation}$$

$$12y - 10x + 10x = 5 + 10x \quad \text{Add } 10x \text{ to both sides.}$$

$$12y = 10x + 5 \quad \text{Combine like terms; rearrange terms.}$$

$$\frac{12y}{12} = \frac{10}{12}x + \frac{5}{12} \quad \text{Divide both sides by 12.}$$

$$y = \frac{5}{6}x + \frac{5}{12} \quad \text{Simplify.}$$

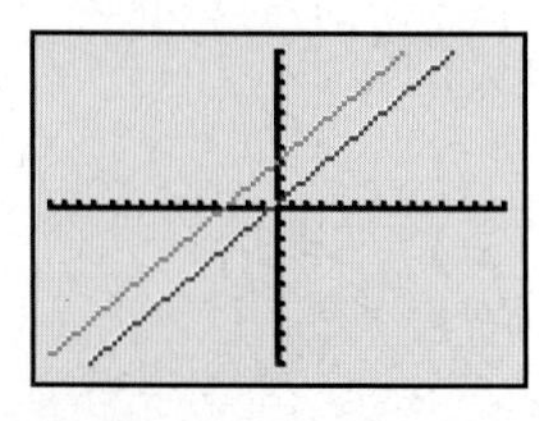

Figure 71 Graphs of the two parallel lines

For $y = \frac{5}{6}x + \frac{5}{12}$, the slope is $\frac{5}{6}$, the same as the slope of the line $y = \frac{5}{6}x + 3$. Therefore, the two lines are parallel. We use ZStandard followed by ZSquare to draw the lines on the same coordinate system (see Fig. 71). ■

Vertical Change Property

Our second observations made after both Examples 1 and 2 suggest a general property of a line $y = mx + b$.

Vertical Change Property

For a line $y = mx + b$, if the run is 1, then the rise is the slope m. (See Figs. 72 and 73.)

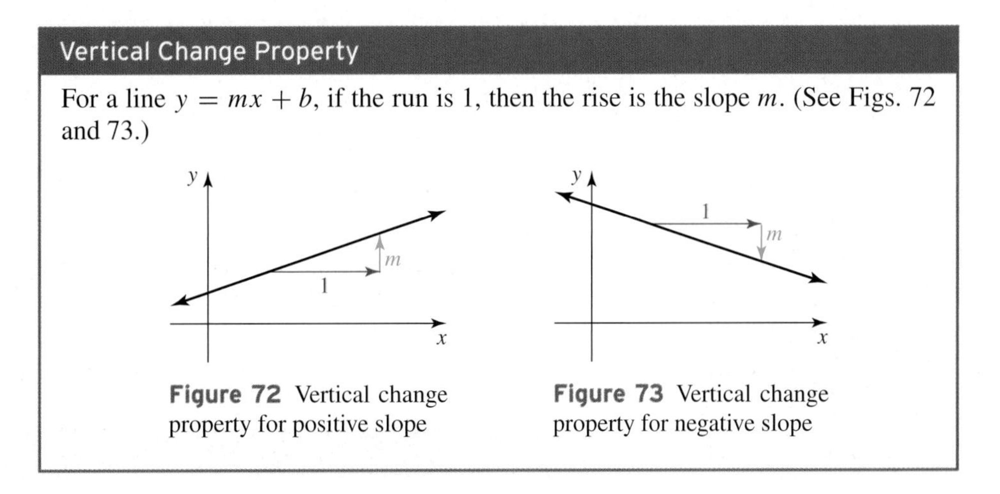

Figure 72 Vertical change property for positive slope

Figure 73 Vertical change property for negative slope

Finding the *y*-Intercept of a Linear Equation

When we sketch graphs of equations of the form $y = mx + b$, it is helpful to know the y-intercept. Substituting 0 for x in the equation $y = mx + b$ gives

$$y = m(0) + b = b$$

which shows that the y-intercept is $(0, b)$.

y-Intercept of a Linear Equation of the Form $y = mx + b$

For a linear equation of the form $y = mx + b$, the y-intercept is $(0, b)$.

For instance, in Example 1, the line $y = 2x+1$ has y-intercept $(0, 1)$. In Example 2, the line $y = -3x + 8$ has y-intercept $(0, 8)$. We say that both equations are in slope–intercept form.

DEFINITION Slope-intercept form

If an equation is of the form $y = mx + b$, we say that it is in **slope–intercept form.**

Graphing Linear Equations

In Example 4, we will graph an equation in slope–intercept form.

Example 4 Using Slope to Graph a Linear Equation

Sketch the graph of $y = 3x - 1$.

Solution

Note that the y-intercept is $(0, -1)$ and that the slope is $3 = \frac{3}{1} = \frac{\text{rise}}{\text{run}}$. To graph:

1. Plot the y-intercept, $(0, -1)$.
2. From $(0, -1)$, look 1 unit to the right and 3 units up to plot a second point, which we see by inspection is $(1, 2)$. See Fig. 74.
3. Sketch the line that contains these two points (see Fig. 75).

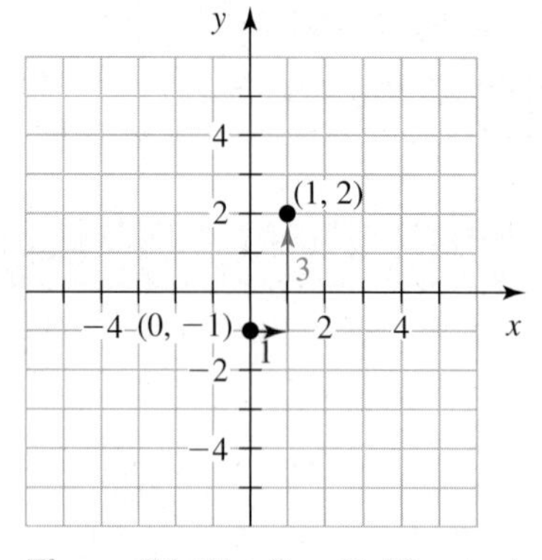

Figure 74 Plot $(0, -1)$. Then look 1 unit to the right and 3 units up, which gives $(1, 2)$

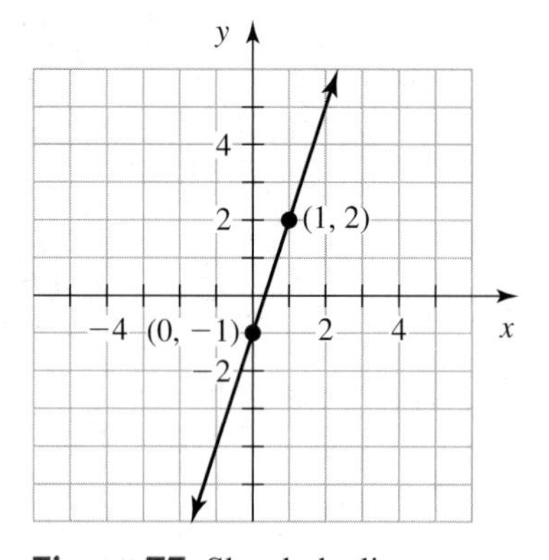

Figure 75 Sketch the line containing $(0, -1)$ and $(1, 2)$

We use ZDecimal to verify our graph (see Fig. 76). ■

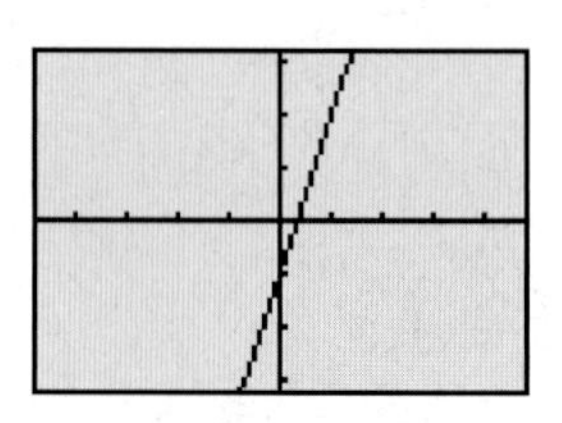

Figure 76 Use ZDecimal to verify our graph

Using Slope to Graph a Linear Equation of the Form $y = mx + b$

To sketch the graph of a linear equation of the form $y = mx + b$,

1. Plot the y-intercept $(0, b)$.
2. Use $m = \frac{\text{rise}}{\text{run}}$ to plot a second point. For example, if $m = -\frac{2}{5} = \frac{-2}{5}$, then look 5 units to the right (from the y-intercept) and 2 units down to plot another point.
3. Sketch the line that passes through the two plotted points.

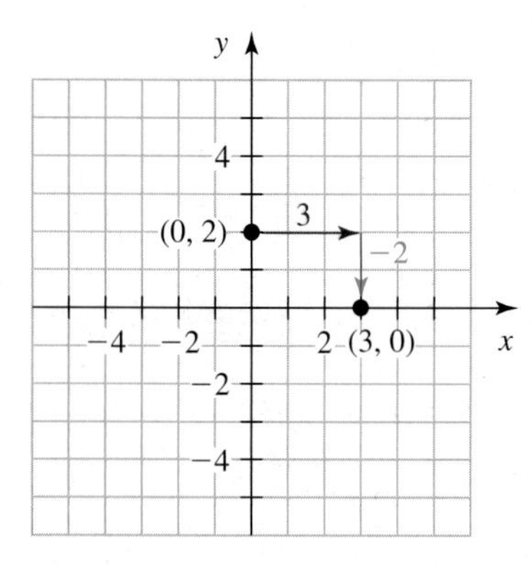

Figure 77 Plot $(0, 2)$. Then look 3 units to the right and 2 units down, which gives $(3, 0)$

Example 5 Using Slope to Graph a Linear Equation

Sketch the graph of $2x + 3y = 6$.

Solution

First, we write the equation in slope–intercept form:

$$
\begin{aligned}
2x + 3y &= 6 && \textit{Original equation} \\
2x + 3y - 2x &= 6 - 2x && \textit{Subtract 2x from both sides.} \\
3y &= -2x + 6 && \textit{Combine like terms; rearrange terms.} \\
\frac{3y}{3} &= \frac{-2x}{3} + \frac{6}{3} && \textit{Divide both sides by 3.} \\
y &= -\frac{2}{3}x + 2 && \textit{Simplify; } \frac{-a}{b} = -\frac{a}{b}
\end{aligned}
$$

The y-intercept is $(0, 2)$, and the slope is $-\frac{2}{3} = \frac{-2}{3} = \frac{\text{rise}}{\text{run}}$. To graph:

1. Plot the y-intercept, $(0, 2)$.
2. From $(0, 2)$, look 3 units to the right and 2 units down to plot a second point, which we see by inspection is $(3, 0)$. See Fig. 77.
3. Then sketch the line that contains these two points (see Fig. 78).

We can verify our result by checking that both $(0, 2)$ and $(3, 0)$ are solutions of $2x + 3y = 6$. ■

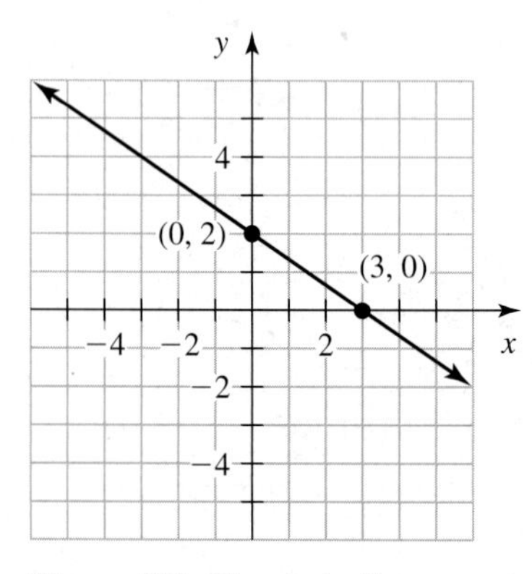

Figure 78 Sketch the line containing $(0, 2)$ and $(3, 0)$

Before we can use the y-intercept and the slope to graph a linear equation, we must solve for y to put the equation into the form $y = mx + b$.

WARNING It is a common error to think that the slope of the graph of an equation such as $2x + 3y = 6$ is 2, because 2 is the coefficient of x. We must first solve for y (and get $y = -\frac{2}{3}x + 2$) to determine the slope, which is $-\frac{2}{3}$ (see Example 5).

Example 6 Working with a General Linear Equation

1. Determine the slope and the y-intercept of the graph of $ax + by = c$, where a, b, and c are constants and b is nonzero.
2. Find the slope and the y-intercept of the graph of $3x + 7y = 5$.

Solution

1.

$$ax + by = c \qquad \text{Original equation}$$
$$ax + by - ax = c - ax \qquad \text{Subtract } ax \text{ from both sides.}$$
$$by = -ax + c \qquad \text{Combine like terms; rearrange terms.}$$
$$\frac{by}{b} = \frac{-ax}{b} + \frac{c}{b} \qquad \text{Divide both sides by } b.$$
$$y = -\frac{a}{b}x + \frac{c}{b} \qquad \text{Simplify; } \frac{-a}{b} = -\frac{a}{b}$$

The slope is $-\frac{a}{b}$, and the y-intercept is $\left(0, \frac{c}{b}\right)$.

2. We substitute 3 for a, 7 for b, and 5 for c in our results from Problem 1 to find that the slope is $-\frac{3}{7}$ and the y-intercept is $\left(0, \frac{5}{7}\right)$. ■

Slope Addition Property

Our third observations made after both Examples 1 and 2 suggest a general property of a table of solutions of an equation of the form $y = mx + b$.

Slope Addition Property

For a linear equation of the form $y = mx + b$, if the value of the independent variable increases by 1, then the value of the dependent variable changes by the slope m.

For example, consider the equation $y = -5x + 4$. We know that as the value of x increases by 1, the value of y changes by -5.

Example 7 Identifying Possible Linear Equations

Four sets of points are described in Table 12. For each set, decide whether there is a line that passes through every point. If so, find the slope of that line. If not, decide whether there is a line that comes close to every point.

Table 12 Four Sets of Points

Set 1		Set 2		Set 3		Set 4	
x	y	x	y	x	y	x	y
1	23	4	12	0	3	50	8
2	20	5	17	1	6	51	8
3	17	6	22	2	12	52	8
4	14	7	27	3	24	53	8
5	11	8	32	4	48	54	8

Solution

1. For set 1, when the value of x increases by 1, the value of y changes by -3. So, a line with slope -3 passes through every point.
2. For set 2, when the value of x increases by 1, the value of y changes by 5. Therefore, a line with slope 5 passes through every point.
3. For set 3, when the value of x increases by 1, the value of y does not change by the same value. So, a line does not pass through every point. Further, a line does

not come close to every point, because the value of y changes by such different amounts each time the value of x increases by 1.

4. For set 4, when the value of x increases by 1, the value of y changes by 0. Therefore, a line with slope 0 (the horizontal line $y = 8$) passes through every point. ■

group exploration

Drawing lines with various slopes

1. On a graphing calculator, graph a group of lines (a *family of lines*) to make a starburst like the one in Fig. 79. List the equations of your lines.
2. On a graphing calculator, graph a family of lines to make a starburst like the one in Fig. 80. List the equations of your lines.

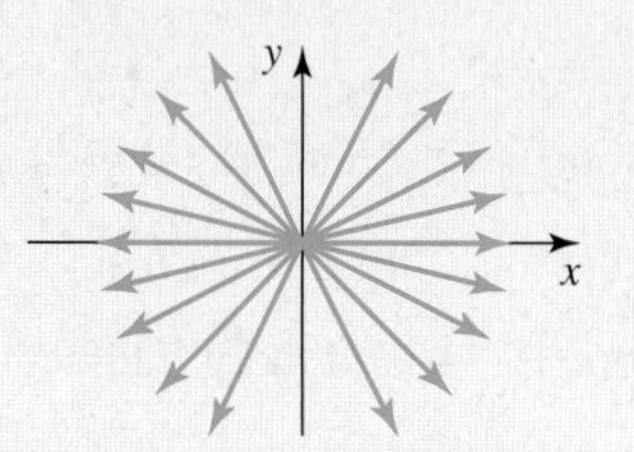

Figure 79 A starburst

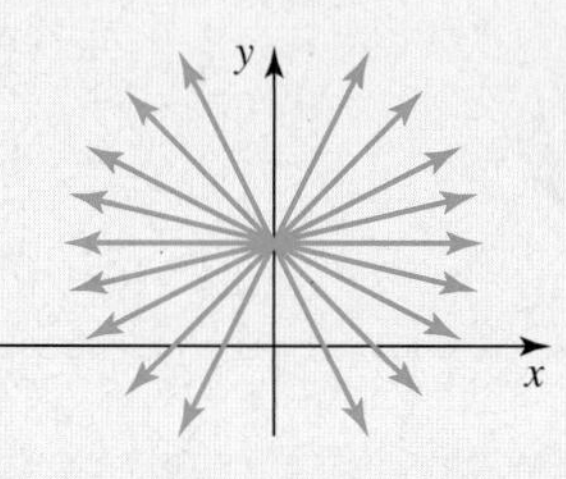

Figure 80 Another starburst

3. Summarize what you have learned about slope from this exploration, this section, and Section 1.3.

group exploration

Looking ahead: Finding an equation of a line

Some solutions of four linear equations are provided in Table 13. Find an equation of each of the four lines.

Table 13 Solutions of Four Linear Equations

Equation 1		Equation 2		Equation 3		Equation 4	
x	y	x	y	x	y	x	y
0	5	3	10	2	7	1	17
1	8	4	8	4	12	3	13
2	11	5	6	6	17	5	9
3	14	6	4	8	22	7	5

$y = 3x + 5$ | $y = -2x + 16$ | $y = \frac{5}{2}x + 2$ | $y = -2x + 19$

HOMEWORK 1.4

FOR EXTRA HELP ▶ Student Solutions Manual | PH Math/Tutor Center | MathXL® | MyMathLab

Determine the slope and y-intercept of the graph of the linear equation. Use the slope and y-intercept to graph the equation by hand. Verify your graph by using ZStandard followed by ZSquare on a graphing calculator.

1. $y = 6x + 1$

2. $y = 3x + 6$

3. $y = -2x + 7$

4. $y = -3x + 5$

5. $y = \frac{5}{4}x - 2$

6. $y = \frac{1}{2}x - 8$

7. $y = -\frac{3}{7}x + 2$

8. $y = -\frac{4}{3}x + 1$

9. $y = -\frac{5}{3}x - 1$

10. $y = -\frac{2}{5}x - 3$

11. $y + x = 5$

12. $2x + y = -1$

13. $-7x + 2y = 10$

14. $3y - 5x = 6$

15. $3(x - 2y) = 9$

16. $2(y - 3x) = 8$

17. $2x - 3y + 9 = 12$

18. $-5x - 15y + 23 = 3$

19. $4x - 5y + 3 = 2x - 2y - 3$

20. $3y - 6x + 2 = 7y - x - 6$

21. $1 - 3(y - 2x) = 7 + 3(x - 3y)$

22. $8 - 2(y - 3x) = 2 + 4(x - 2y)$

23. $y = 4x$ **24.** $y = -7x$

25. $y = -1.5x + 3$ **26.** $y = 0.25x - 2$

27. $y = x$ **28.** $y = -x$ **29.** $y = 4$

30. $y = 0$ **31.** $y + 2 = 0$ **32.** $y - 3 = 0$

Determine the slope and y-intercept of the graph of the given equation, where a, b, c, and d are nonzero constants.

33. $ax - by = c$ **34.** $a(x + y) = c$

35. $a(y + b) = x$ **36.** $ay = b(x + d)$

37. $\frac{x}{a} + \frac{y}{a} = 1$ **38.** $\frac{y + b}{a} = x$

39. Four sets of points are described in Table 14. For each set, decide whether there is a line that passes through every point. If so, find the slope of that line. If not, decide whether there is a line that comes close to every point.

Table 14 Four Sets of Points

Set 1		Set 2		Set 3		Set 4	
x	y	x	y	x	y	x	y
1	6	1	5.9	1	3	50	90
2	106	2	5.6	11	8	51	80
3	205	3	5.3	13	13	52	70
4	305	4	5.0	20	18	53	60
5	406	5	4.7	40	23	54	50
6	505	6	4.4	90	28	55	40

40. Four sets of points are described in Table 15. For each set, decide whether there is a line that passes through every point. If so, find the slope of that line. If not, decide whether there is a line that comes close to every point.

Table 15 Four Sets of Points

Set 1		Set 2		Set 3		Set 4	
x	y	x	y	x	y	x	y
0	50	3	2	1	8	5	1
1	47	5	5	2	8	5	9
2	44	7	8	3	8	5	10
3	41	9	11	4	8	5	40
4	38	11	14	5	8	5	46
5	35	13	17	6	8	5	99

41. Some values of four linear equations are provided in Table 16. Complete the table.

Table 16 Values of Four Linear Equations

Equation 1		Equation 2		Equation 3		Equation 4	
x	y	x	y	x	y	x	y
1	12	23	69	1		30	15
2	15	24	53	2		31	
3		25		3	35	32	
4		26		4		33	
5		27		5		34	
6		28		6	17	35	60

42. Some values of four linear equations are provided in Table 17. Complete the table.

Table 17 Values of Four Linear Equations

Equation 1		Equation 2		Equation 3		Equation 4	
x	y	x	y	x	y	x	y
0	16	0		1	36	10	80
1	23	1		2		11	
2		2		3		12	
3		3	6	4		13	
4		4		5		14	70
5		5	14	6	16	15	

43. Graphs of four equations are shown in Fig. 81. State the signs of the constants m and b for the $y = mx + b$ form of each equation.

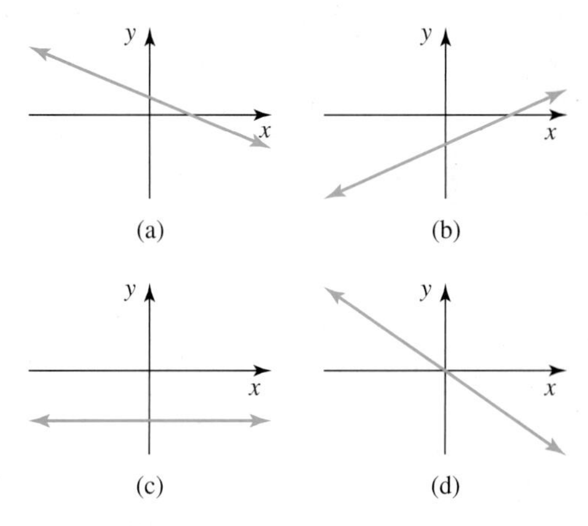

Figure 81 Exercise 43

44. On a graphing calculator, graph five equations whose graphs are five parallel lines like those in Fig. 82. List your equations.

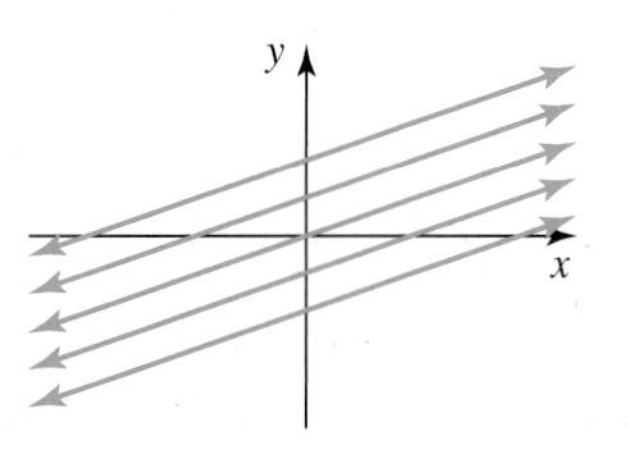

Figure 82 A family of parallel lines for Exercise 44

45. Find an equation of the line sketched in Fig. 83.

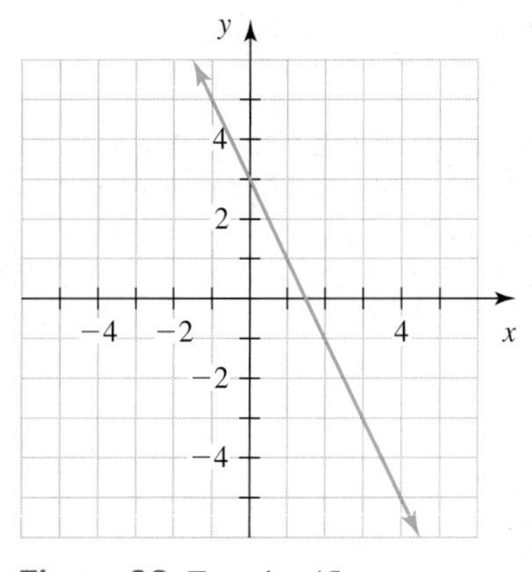

Figure 83 Exercise 45

46. Find an equation of the line sketched in Fig. 84.

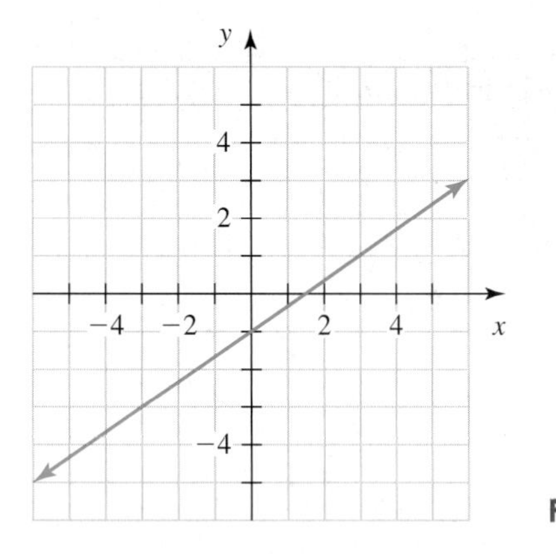

Figure 84 Exercise 46

Write an equation of

47. an increasing line that is nearly horizontal with y-intercept above the origin.

48. an increasing line that is nearly vertical with y-intercept below the origin.

49. a decreasing line that is nearly vertical with y-intercept below the origin.

50. a decreasing line that is nearly horizontal with y-intercept above the origin.

Determine whether the given pair of lines is parallel, perpendicular, or neither. Explain.

51. $y = 4x + 7$ and $y = 4x - 3$

52. $y = \frac{4}{9}x - 5$ and $y = -\frac{9}{4}x - 5$

53. $y = \frac{3}{8}x + 1$ and $y = \frac{8}{3}x + 4$

54. $y = \frac{2}{5}x$ and $y = \frac{2}{5}x - 3$

55. $2x + 3y = 6$ and $4x + 6y = 7$

56. $4x + y = 6$ and $x - 4y = 5$

57. $5x - 3y = 1$ and $3x + 5y = -2$

58. $8x - 4y = 1$ and $y = 8(x - 4) + 1$

59. $x = -3$ and $x = 1$

60. $y = 5$ and $y = -2$

61. $x = 0$ and $y = 0$

62. $x = -2$ and $y = -4$

63. Let x be the driving time (in hours) since a person has fueled up his car, and let y be the number of gallons of gas in the car's gas tank. Assume that the equation $y = -3x + 18$ describes the relationship between x and y.

a. Complete Table 18.

Table 18 Amounts of Gas in a Car's Gas Tank

Driving Time (hours) x	Amount of Gas (gallons) y
0	
1	
2	
3	
4	
5	
6	

b. By how much is the amount of gas decreasing each hour? Compare your result with the slope of the graph of $y = -3x + 18$. Discuss your observation in terms of the slope addition property.

c. If the person is driving at about 60 mph, what is the gas mileage of the car? [**Hint:** *Gas mileage* is the number of miles that the car can travel on 1 gallon of gas.]

64. A person lowers her hot-air balloon by gradually releasing air from it. Let x be the number of minutes that she has been releasing air from the balloon, and let y be the altitude of the balloon (in feet). Assume that the equation $y = -400x + 2400$ describes the relationship between x and y.

a. Complete Table 19.

Table 19 Altitudes of a Balloon

Time (minutes) x	Altitude (feet) y
0	
1	
2	
3	
4	
5	
6	

b. By how much is the altitude of the hot-air balloon decreasing each minute? Compare your result with the slope of the graph of $y = -400x + 2400$. Discuss your observation in terms of the slope addition property.

65. Let y be a person's salary (in thousands of dollars) after she has worked x years at a company. Assume that the equation $y = 2x + 26$ describes the relationship between x and y.

a. Complete Table 20.

Table 20 Salaries

Time at Company (years) x	Salary (thousands of dollars) y
0	
1	
2	
3	
4	

b. By how much does her salary increase each year? Compare your result with the slope of the graph of $y = 2x + 26$. Discuss your observation in terms of the slope addition property.

66. Let y be a college's enrollment (in thousands of students), and let x be the number of years that the college has been open. Assume that the equation $y = 0.5x + 6$ describes the relationship between and x and y.

a. Complete Table 21.

Table 21 Enrollments

Number of Years College Has Been Open x	Enrollment (thousands of students) y
0	
1	
2	
3	
4	

b. By how much does the college's enrollment increase each year? Compare your result with the slope of the graph of $y = 0.5x + 6$. Discuss your observation in terms of the slope addition property.

67. a. Use a graphing calculator to graph the line $y = x$ by using the ZDecimal window settings displayed in Fig. 85. [***Graphing Calculator:*** *See Section B.7.*]

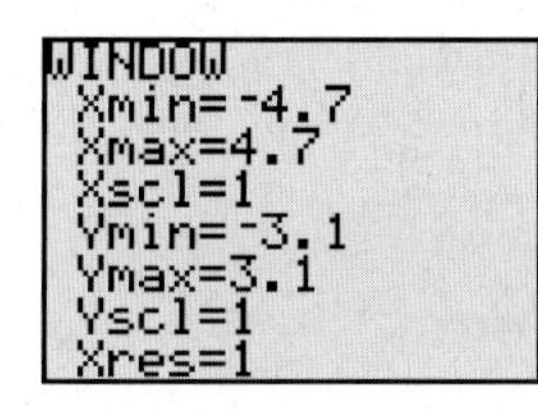

Figure 85 Window for Exercise 67a

b. Graph $y = x$ by using the window settings displayed in Fig. 86.

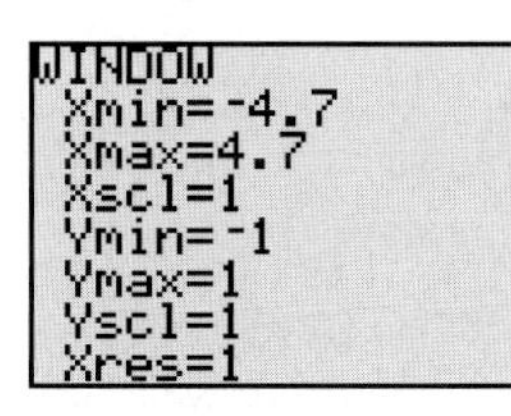

Figure 86 Window for Exercise 67b

c. Graph $y = x$ by using the window settings displayed in Fig. 87.

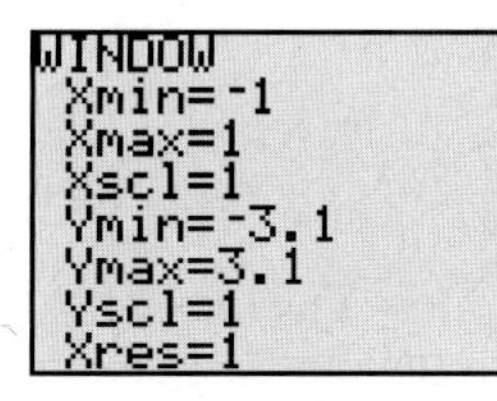

Figure 87 Window for Exercise 67c

d. Compare the results of parts (a) and (b). Explain the different views of $y = x$. Do the same for parts (a) and (c).

e. Can you make the graph of $y = x$ appear to be a decreasing line by adjusting the window settings? Can you make the graph of $y = x$ cross the y-axis at a point other than $(0, 0)$ (the origin) by changing the window settings? Describe all possible appearances of the graph of $y = x$ obtained by using various window settings.

68. a. Use a graphing calculator to graph $y = 0.0005x + 0.003$. You might observe that no graph appears to be drawn on your calculator screen. This does not mean that the calculator is broken! To locate the line, think about its slope and y-intercept. Then use TRACE to confirm your suspicion. Try changing your window settings so that you can see the graph clearly. Record a window setting that allows the graph to be seen.

b. Use a graphing calculator to graph $y = -3x + 20{,}000$. Record a window setting that allows you to see the graph.

c. Use a graphing calculator to graph $y = 4000x - 0.04$. Record a window setting that allows you to see the graph.

69. A line passes through the point $(0, -3)$ with slope 2. What is an equation of the line?

70. A line passes through the point $(0, 4)$ with slope -5. What is an equation of the line?

71. A line passes through the point $(3, 8)$ and has slope 5.

a. Sketch the line by hand.

b. Find an equation of the line. [**Hint:** Use part (a) to find b for $y = mx + b$.]

c. Use a graphing calculator to verify that your equation is correct.

72. A line passes through the point $(-2, 7)$ and has slope -6.

a. Sketch the line by hand.

b. Find an equation of the line. [**Hint:** Use part (a) to find b for $y = mx + b$.]

c. Use a graphing calculator to verify that your equation is correct.

73. A line passes through the point $(2, 6)$ and has slope $-\frac{3}{8}$.

a. Sketch the line by hand.

b. Estimate an equation of the line.

c. Use a graphing calculator to verify that your equation is correct.

74. A line passes through the point $(-3, 2)$ and has slope $\frac{1}{4}$.

a. Sketch the line by hand.

b. Estimate an equation of the line.

c. Use a graphing calculator to verify that your equation is correct.

75. a. Find the slope of each line: $y = 2$, $y = -5$, and $y = 3.72$.

b. Find the slope of the graph of any linear equation of the form $y = k$.

76. a. Find the slope of each line: $x = 3$, $x = -6$, and $x = -7.9$.

b. Find the slope of the graph of any equation of the form $x = k$.

77. A student says that the line $2x + 3y = 6$ has slope 2 because the coefficient of x is 2. Is the student correct? Explain.

78. A student says that the line $y = 3x - 5$ has slope $3x$. Is the student correct? Explain.

79. Explain why the slope addition property makes sense. Include a table of ordered pairs for a linear equation. (See page 4 for guidelines on writing a good response.)

80. The graph of a linear equation can be sketched by

- plotting points,
- using the slope and the y-intercept (sometimes), and
- using the x-intercept and the y-intercept (sometimes).

Discuss how to use each method to sketch the graph of a linear equation. For each method, describe the types of equations, if any, for which you would sketch graphs by that method. (See page 4 for guidelines on writing a good response.)

1.5 FINDING LINEAR EQUATIONS

Objectives

- Use the slope–intercept form to find an equation of a line.
- Use the *point–slope form* to find an equation of a line.

In this section, we will discuss two methods of finding an equation of a line: slope–intercept form, which we worked with in Section 1.4, and another form of an equation of a line, called *point–slope form*.

Method 1: Using Slope-Intercept Form

In Example 1, we will use the concept that a point that lies on a line satisfies an equation of the line.

Example 1 Using Slope and a Point to Find an Equation of a Line

Find an equation of the line that has slope $m = 3$ and contains the point (2, 5).

Solution

Recall from Section 1.4 that the equation for a nonvertical line can be put into the form $y = mx + b$. Since $m = 3$, we have

$$y = 3x + b$$

To find b, recall from Section 1.2 that every point on the graph of an equation represents a solution of that equation. In particular, the ordered pair (2, 5) should satisfy the equation $y = 3x + b$:

$$\begin{aligned} 5 &= 3(2) + b && \text{Substitute 2 for } x \text{ and 5 for } y. \\ 5 &= 6 + b && \text{Multiply.} \\ 5 - 6 &= 6 + b - 6 && \text{Subtract 6 from both sides.} \\ -1 &= b && \text{Simplify.} \end{aligned}$$

Now we substitute -1 for b in $y = 3x + b$:

$$y = 3x - 1$$

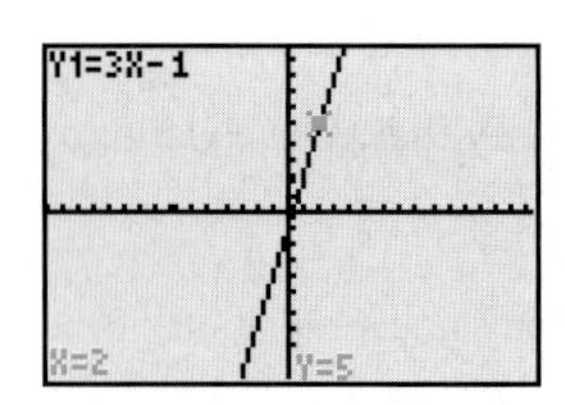

Figure 88 Check that the line contains (2, 5)

We can use TRACE on a graphing calculator to verify that the graph of $y = 3x - 1$ contains the point (2, 5). See Fig. 88. For graphing calculator instructions, see Section B.5. ■

In Example 1, we used a point and the slope of a line to find an equation of the line. We can also use two points to find an equation of a line.

Example 2 Using Two Points to Find an Equation of a Line

Find an equation of the line that contains the points $(-2, 6)$ and $(3, -4)$.

Solution

First, we find the slope of the line:

$$m = \frac{-4 - 6}{3 - (-2)} = \frac{-10}{3 + 2} = \frac{-10}{5} = -2$$

Thus, we have $y = -2x + b$. Since the line contains the point $(3, -4)$, we substitute 3 for x and -4 for y:

$$\begin{aligned} -4 &= -2(3) + b && \text{Substitute 3 for } x \text{ and } -4 \text{ for } y. \\ -4 &= -6 + b && \text{Multiply.} \\ -4 + 6 &= -6 + b + 6 && \text{Add 6 to both sides.} \\ 2 &= b && \text{Simplify.} \end{aligned}$$

So, the equation is $y = -2x + 2$. We can use a graphing calculator to check that the graph of $y = -2x + 2$ contains both $(-2, 6)$ and $(3, -4)$. See Fig. 89.

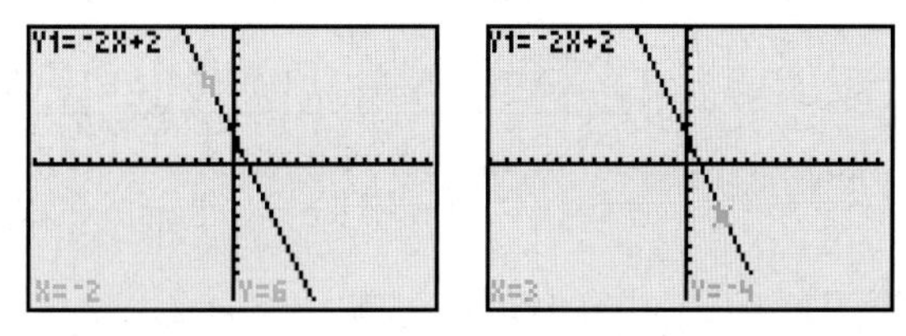

Figure 89 Check that the line contains both $(-2, 6)$ and $(3, -4)$

■

Finding a Linear Equation That Contains Two Given Points

To find an equation of the line that passes through two given points whose x-coordinates are different,

1. Use the slope formula, $m = \dfrac{y_2 - y_1}{x_2 - x_1}$, to find the slope of the line.
2. Substitute the m value you found in step 1 into the equation $y = mx + b$.
3. Substitute the coordinates of one of the given points into the equation you found in step 2, and solve for b.
4. Substitute the m value you found in step 1 and the b value you found in step 3 into the equation $y = mx + b$.
5. Use a graphing calculator to check that the graph of your equation contains the two given points.

In Example 3, we find an equation of a line whose slope is a fraction.

Example 3 Using Two Points to Find an Equation of a Line

Find an equation of the line that passes through the points $(-3, -5)$ and $(2, -1)$.

Solution

First, we find the slope of the line:

$$m = \frac{-1 - (-5)}{2 - (-3)} = \frac{-1 + 5}{2 + 3} = \frac{4}{5}$$

Thus, we have $y = \frac{4}{5}x + b$. Since the line contains the point $(2, -1)$, we substitute 2 for x and -1 for y:

$$-1 = \frac{4}{5}(2) + b \qquad \text{Substitute 2 for } x \text{ and } -1 \text{ for } y.$$

$$-1 = \frac{8}{5} + b \qquad \frac{4}{5}(2) = \frac{4}{5}\left(\frac{2}{1}\right) = \frac{8}{5}$$

$$5 \cdot (-1) = 5 \cdot \frac{8}{5} + 5 \cdot b \qquad \text{Multiply both sides by 5.}$$

$$-5 = 8 + 5b \qquad 5 \cdot \frac{8}{5} = \frac{5}{1} \cdot \frac{8}{5} = \frac{8}{1} = 8$$

$$-13 = 5b \qquad \text{Subtract 8 from both sides.}$$

$$-\frac{13}{5} = b \qquad \text{Divide both sides by 5; } \frac{-13}{5} = -\frac{13}{5}$$

So, the equation is $y = \frac{4}{5}x - \frac{13}{5}$. We can use a graphing calculator to check that the graph of $y = \frac{4}{5}x - \frac{13}{5}$ contains both $(-3, -5)$ and $(2, -1)$. See Fig. 90. ■

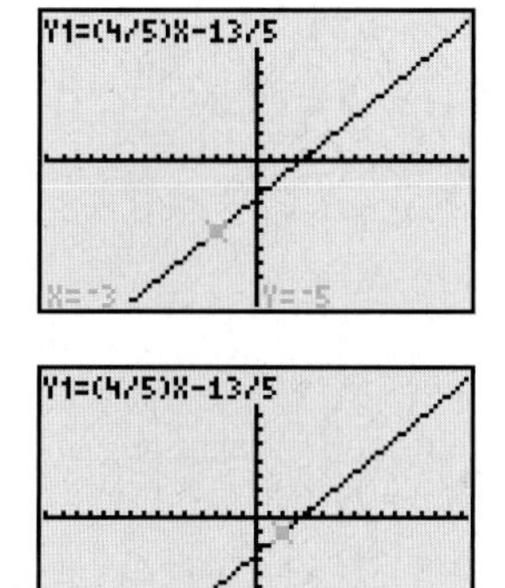

Figure 90 Check that the line contains both $(-3, -5)$ and $(2, -1)$

In Chapter 2, we will discuss how to find an approximate equation of a line to describe an authentic situation. Example 4 will help prepare us for that task.

Example 4 Finding an Approximate Equation of a Line

Find an approximate equation of the line that contains the points $(-6.81, 7.17)$ and $(-2.47, 4.65)$. Round the slope and the constant term to two decimal places.

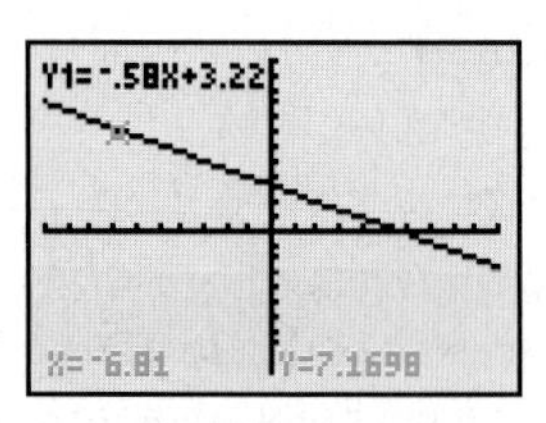

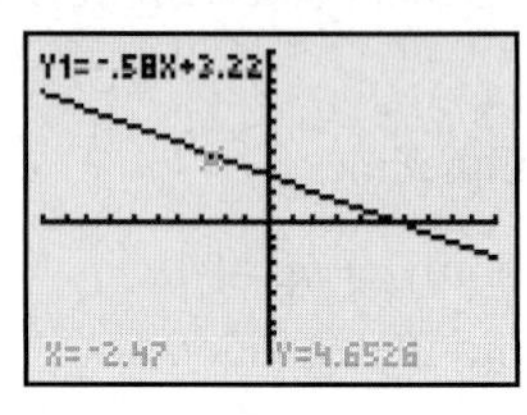

Figure 91 Check that the line comes very close to $(-6.81, 7.17)$ and $(-2.47, 4.65)$

Solution

First, we find the slope of the line:

$$m = \frac{4.65 - 7.17}{-2.47 - (-6.81)} = \frac{-2.52}{4.34} \approx -0.58$$

So, we have $y = -0.58x + b$. Since the line contains the point $(-6.81, 7.17)$, we substitute -6.81 for x and 7.17 for y:

$$\begin{aligned} 7.17 &= -0.58(-6.81) + b && \text{Substitute } -6.81 \text{ for } x \text{ and } 7.17 \text{ for } y. \\ 7.17 &= 3.9498 + b && \text{Multiply.} \\ 7.17 - 3.9498 &= 3.9498 + b - 3.9498 && \text{Subtract 3.9498 from both sides.} \\ 3.22 &\approx b && \text{Combine like terms.} \end{aligned}$$

The approximate equation is $y = -0.58x + 3.22$.

We use a graphing calculator to check that the line $y = -0.58x + 3.22$ comes very close to the points $(-6.81, 7.17)$ and $(-2.47, 4.65)$. See Fig. 91. ■

The line $y = -0.58x + 3.22$ does not contain the points $(-6.81, 7.17)$ and $(-2.47, 4.65)$, because we rounded the slope and the constant term to the second decimal place. However, the line does comes very close to these points. When finding an approximate equation, we will usually round numbers such as the slope or the constant term to the second decimal place.

In Example 5, we will find an equation of a line that contains a given point and is parallel to a given line. Recall from Section 1.3 that nonvertical parallel lines have equal slopes.

Example 5 Finding an Equation of a Line Parallel to a Given Line

Find an equation of a line l that contains the point $(5, 3)$ and is parallel to the line $y = 2x - 3$.

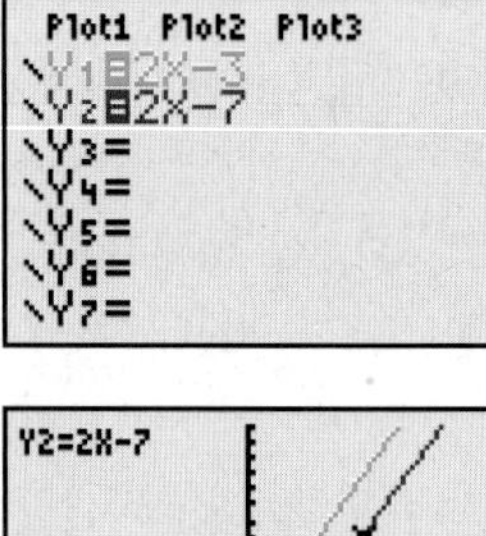

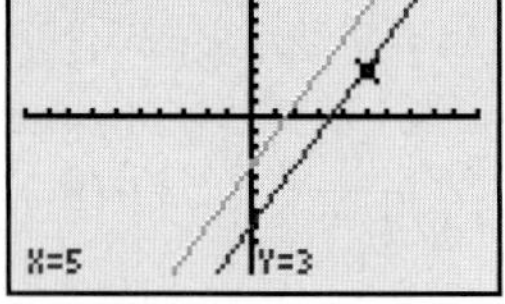

Figure 92 Check that the line contains $(5, 3)$ and is parallel to $y = 2x - 3$

Solution

For the line $y = 2x - 3$, the slope is 2. So, the slope of parallel line l is also 2. An equation of line l is $y = 2x + b$. To find b, we substitute the coordinates of $(5, 3)$ into the equation $y = 2x + b$:

$$\begin{aligned} 3 &= 2(5) + b && \text{Substitute 5 for } x \text{ and 3 for } y. \\ -7 &= b && \text{Multiply; subtract 10 from both sides.} \end{aligned}$$

An equation of l is $y = 2x - 7$. We use a graphing calculator to verify our equation (see Fig. 92). ■

Recall from Section 1.3 that if two nonvertical lines are perpendicular, then the slope of one line is the opposite of the reciprocal of the slope of the other line.

Example 6 Finding an Equation of a Line Perpendicular to a Given Line

Find an equation of the line l that contains the point $(2, 5)$ and is perpendicular to the line $-2x + 5y = 10$.

Solution

First, we isolate y in the equation $-2x + 5y = 10$:

$$-2x + 5y = 10 \qquad \text{Original equation: line perpendicular to line } l$$

$$-2x + 5y + 2x = 10 + 2x \qquad \text{Add } 2x \text{ to both sides.}$$

$$5y = 2x + 10 \qquad \text{Simplify.}$$

$$\frac{5y}{5} = \frac{2x}{5} + \frac{10}{5} \qquad \text{Divide both sides by 5.}$$

$$y = \frac{2}{5}x + 2 \qquad \text{Simplify.}$$

For the line $y = \frac{2}{5}x + 2$, the slope is $m = \frac{2}{5}$. The slope of the line l must be the opposite of the reciprocal of $\frac{2}{5}$, or $-\frac{5}{2}$. An equation of l is $y = -\frac{5}{2}x + b$. To find b, we substitute the coordinates of the given point (2, 5) into $y = -\frac{5}{2}x + b$:

$$5 = -\frac{5}{2}(2) + b \qquad \text{Substitute 2 for x and 5 for y.}$$

$$5 = -5 + b \qquad -\frac{5}{2}(2) = -\frac{5}{2}\left(\frac{2}{1}\right) = -\frac{5}{1} = -5$$

$$10 = b \qquad \text{Add 5 to both sides.}$$

An equation of l is $y = -\frac{5}{2}x + 10$. We use ZStandard followed by ZSquare to verify our work (see Fig. 93). ■

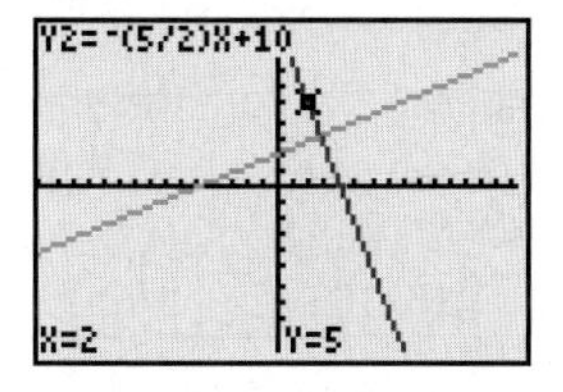

Figure 93 Check that the line contains (2, 5) and is perpendicular to $-2x + 5y = 10$

In Example 7, we will use a graphical approach to find the equation of a line.

Example 7 Finding an Equation of a Line Perpendicular to a Given Line

Find an equation of the line l that contains (4, 3) and is perpendicular to the line $x = 2$.

Solution

The graph of $x = 2$ is a vertical line (see Fig. 94). A line perpendicular to it must be horizontal, so there is an equation of l of the form $y = b$. To find b, we substitute the y-coordinate of the given point (4, 3) into $y = b$ and get $3 = b$. So, an equation of l is $y = 3$. ■

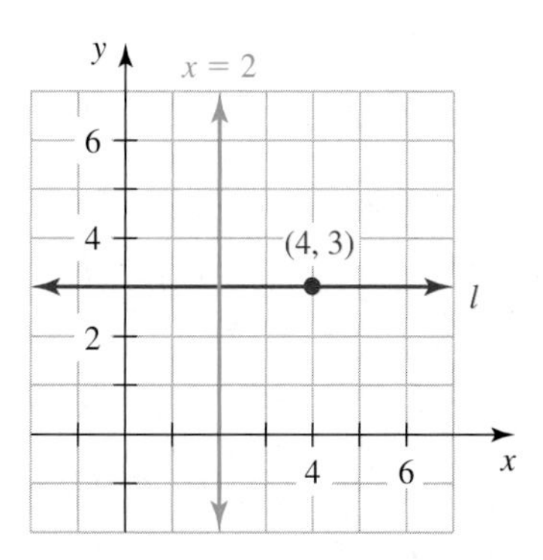

Figure 94 Find an equation of the line l

Method 2: Using Point-Slope Form

We can find an equation of a line by another method. Suppose that a nonvertical line has slope m and contains the point (x_1, y_1). Then, if (x, y) represents a different point on the line, the slope of the line is

$$\frac{y - y_1}{x - x_1} = m$$

Multiplying both sides of the equation by $x - x_1$ gives

$$\frac{y - y_1}{x - x_1} \cdot (x - x_1) = m(x - x_1)$$

$$y - y_1 = m(x - x_1)$$

We say that this linear equation is in **point–slope form.***

*Although we assumed that (x, y) is different from (x_1, y_1), note that (x_1, y_1) is a solution of the equation $y - y_1 = m(x - x_1)$: $y_1 - y_1 = m(x_1 - x_1)$, or $0 = 0$, a true statement.

Point-Slope Form

If a nonvertical line has slope m and contains the point (x_1, y_1), then an equation of the line is

$$y - y_1 = m(x - x_1)$$

Example 8 Using Point-Slope Form to Find an Equation of a Line

A line has slope $m = 2$ and contains the point $(3, -8)$. Find an equation of the line.

Solution

Substituting $x_1 = 3$, $y_1 = -8$, and $m = 2$ in the equation $y - y_1 = m(x - x_1)$ gives

$$y - (-8) = 2(x - 3) \quad \text{Substitute 3 for } x_1, -8 \text{ for } y_1, \text{ and 2 for } m.$$
$$y + 8 = 2x - 6 \quad \text{Simplify; distributive law.}$$
$$y + 8 - 8 = 2x - 6 - 8 \quad \text{Subtract 8 from both sides.}$$
$$y = 2x - 14 \quad \text{Simplify.}$$

We can use a graphing calculator to check that the graph of $y = 2x - 14$ contains the point $(3, -8)$. ■

Example 9 Using Point-Slope Form to Find an Equation of a Line

Use the point–slope form to find an equation of the line that contains the points $(-5, 2)$ and $(3, -1)$. Then write the equation in slope–intercept form.

Solution

We begin by finding the slope of the line:

$$m = \frac{-1 - 2}{3 - (-5)} = \frac{-3}{8} = -\frac{3}{8}$$

Then we substitute $x_1 = 3$, $y_1 = -1$, and $m = -\frac{3}{8}$ in the equation $y - y_1 = m(x - x_1)$:

$$y - (-1) = -\frac{3}{8}(x - 3) \quad \text{Substitute } x_1 = 3, y_1 = -1, \text{ and } m = -\frac{3}{8}.$$
$$y + 1 = -\frac{3}{8}x + \frac{9}{8} \quad \text{Simplify; distributive law}$$
$$y + 1 - 1 = -\frac{3}{8}x + \frac{9}{8} - 1 \quad \text{Subtract 1 from both sides.}$$
$$y = -\frac{3}{8}x + \frac{1}{8} \quad \text{Simplify; } \frac{9}{8} - 1 = \frac{9}{8} - \frac{8}{8} = \frac{1}{8}$$

■

So far, we have worked with equations, graphs, and tables. In Section 1.2, we sketched a graph by going from an equation to a table and then to a graph. In Section 1.4, we sketched a graph by going directly from an equation to a graph. In Exercises 71–74 at the end of this section, you will find an equation of a line by going from its graph to an equation. Many times throughout the rest of the course, we will use combinations of the six paths indicated in Fig. 95.

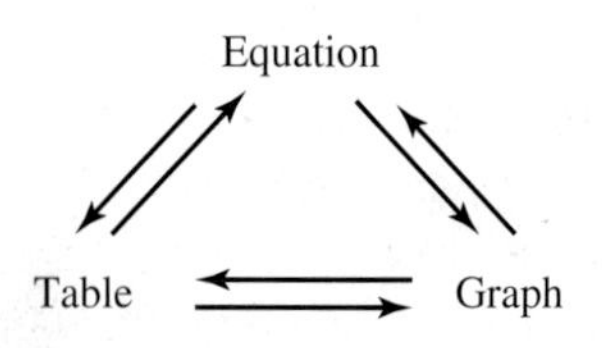

Figure 95 Six paths among equations, tables, and graphs

group exploration

Deciding which points to use to find an equation of a line

1. **a.** Use the method shown in Example 2 to find an equation of the line that contains the points $(1, 2)$ and $(3, 8)$.
   b. In part (a), you used either point $(1, 2)$ or point $(3, 8)$ to find the constant b for the equation $y = 3x + b$. Now use the other point to find b.
   c. Does it matter which point is used to find the constant b? Explain.

2. Imagine any line that is not parallel to either axis. Choose four points on the line. Name the points A, B, C, and D.
   a. Use points A and B to find an equation of the line. Write your equation in slope–intercept form.
   b. Use points C and D to find an equation of the line. Write your equation in slope–intercept form.
   c. Are the equations you found in parts (a) and (b) the same? Explain.

TIPS FOR SUCCESS: Affirmations

Do you ever tell yourself (or others) that you are not good at math? This is called *negative self-talk*. The more you talk that way, the more likely your subconscious will believe it—and you *will* do poorly in math.

You can counteract years of negative self-talk by telling yourself with conviction that you are good at math. It might seem strange to state that something is true that hasn't happened yet, but it works! Such statements are called *affirmations*.

There are four guiding principles for getting the most out of affirmations:

1. Say affirmations that imply that the desired event is currently happening. For example, say "I am good at algebra," not "I will be good at algebra."
2. Say affirmations that you are making progress toward the desired result. For example, say "I am good at algebra, and I continue to improve at it."
3. Say affirmations in the positive rather than in the negative. For example, say "I attend each class," not "I don't cut classes."
4. Say affirmations with conviction.

If you would like to learn more about affirmations, the book *Creative Visualization* (Bantam Books, 1985), by Shakti Gawain, is an excellent resource.

HOMEWORK 1.5

FOR EXTRA HELP ▶

Student Solutions Manual

PH Math/Tutor Center

MathXL®

MyMathLab

Find an equation of the line that has the given slope and contains the given point. Use a graphing calculator to verify that the graph of your equation passes through the given point. Determine whether your line is increasing, decreasing, horizontal, or vertical, and check that the sign of m agrees with your determination.

1. $m = 3, (5, 2)$

2. $m = 5, (-3, -1)$

3. $m = -2, (3, -9)$

4. $m = -4, (-2, -8)$

5. $m = \frac{3}{5}, (20, 7)$

6. $m = \frac{2}{3}, (6, -1)$

7. $m = -\frac{1}{6}, (2, -3)$

8. $m = -\frac{1}{4}, (-2, 3)$

9. $m = -\frac{5}{2}, (-3, -4)$

10. $m = \frac{3}{4}, (-5, 2)$

11. $m = 0, (1, 2)$

12. $m = 0, (-3, -4)$

13. m is undefined, $(3, 7)$

14. m is undefined, $(-5, 1)$

Find an approximate equation of the line that has the given slope and contains the given point. Round the constant term to two decimal places. Check that the given point approximately satisfies your equation.

15. $m = 1.6, (2.1, 3.8)$

16. $m = -2.7, (6.2, -4.9)$

17. $m = -3.24, (-5.28, 1.93)$

18. $m = 1.94, (-2.53, -3.77)$

Find an equation of the line that passes through the two given points. Verify your equation by using a graphing calculator.

19. $(2, 3)$ and $(4, 5)$

20. $(3, 5)$ and $(7, 1)$

21. $(-2, 6)$ and $(3, -4)$

22. $(-4, -6)$ and $(-2, 0)$

23. $(-8, -6)$ and $(-4, -14)$

24. $(-4, -1)$ and $(-2, -7)$

25. $(0, 0)$ and $(1, 1)$

26. $(0, 8)$ and $(4, 5)$

27. $(2, 1)$ and $(7, 5)$

28. $(3, 2)$ and $(5, 9)$

29. $(-4, 2)$ and $(2, -5)$

30. $(2, -1)$ and $(5, -3)$

31. $(-5, -7)$ and $(-3, -2)$

32. $(-5, -3)$ and $(-2, -4)$

33. $(2, 5)$ and $(4, 5)$

34. $(-5, -2)$ and $(1, -2)$

35. $(-3, -4)$ and $(-3, 6)$

36. $(4, -7)$ and $(4, -3)$

Find an approximate equation of the line that passes through the two given points. Round the slope and the constant term to two decimal places. Use a graphing calculator to verify your result.

37. $(-5.1, -3.9)$ and $(7.4, 2.2)$

38. $(-9.4, 7.1)$ and $(3.9, -2.3)$

39. $(-5.97, -6.24)$ and $(-1.25, -4.05)$

40. $(-7.13, -2.21)$ and $(-4.99, -7.78)$

Find an equation of the line that contains the given point and is parallel to the given line. Use a graphing calculator to verify your result.

41. $(4, 5)$, $y = 3x + 1$

42. $(1, 4)$, $y = 4x - 6$

43. $(-3, 8)$, $y = -2x + 7$

44. $(2, -3)$, $y = -x + 2$

45. $(4, 1)$, $y = \frac{1}{2}x - 3$

46. $(6, -3)$, $y = -\frac{2}{3}x - 1$

47. $(3, 4)$, $3x - 4y = 12$

48. $(4, -1)$, $5x + 2y = 10$

49. $(-3, -2)$, $6y - x = -7$

50. $(-1, -4)$, $3y + 5x = -11$

51. $(2, 3)$, $y = 6$

52. $(3, -1)$, $y = -4$

53. $(-5, 4)$, $x = 2$

54. $(-2, -5)$, $x = 1$

Find an equation of the line that contains the given point and is perpendicular to the given line. Use ZStandard followed by ZSquare with a graphing calculator to verify your result.

55. $(3, 8)$, $y = 2x + 5$

56. $(2, 1)$, $y = 5x - 4$

57. $(-1, 7)$, $y = -3x + 7$

58. $(-3, -2)$, $y = -6x - 13$

59. $(2, 7)$, $y = -\frac{2}{5}x + 3$

60. $(1, -2)$, $y = \frac{1}{3}x - 4$

61. $(10, 3)$, $4x - 5y = 7$

62. $(6, -1)$, $5x + 2y = -9$

63. $(-3, -1)$, $-2x + y = 5$

64. $(-1, 2)$, $-3x - 4y = 12$

65. $(2, 3)$, $x = 5$

66. $(-4, -2)$, $x = -1$

67. $(2, 8)$, $y = -3$

68. $(1, -1)$, $y = 7$

69. Let y be the value (in thousands of dollars) of a car when it is x years old. Some pairs of values of x and y are listed in Table 22.

Table 22 Values of a Car

Age (years) x	Value (thousands of dollars) y
0	19
1	17
2	15
3	13
4	11

Find an equation that describes the relationship between x and y.

70. Let y be a person's salary (in thousands of dollars) after he has worked at a company for x years. Some pairs of values of x and y are listed in Table 23.

Table 23 Salaries

Time at Company (years) x	Salary (thousands of dollars) y
0	25
1	28
2	31
3	34
4	37
5	40

Find an equation that describes the relationship between x and y.

71. Find an equation of the line sketched in Fig. 96. Check your equation with a graphing calculator.

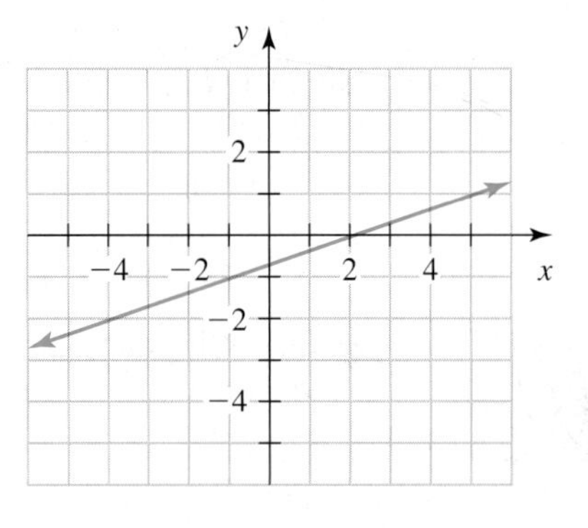

Figure 96 Exercise 71

72. Find an equation of the line sketched in Fig. 97. Check your equation with a graphing calculator.

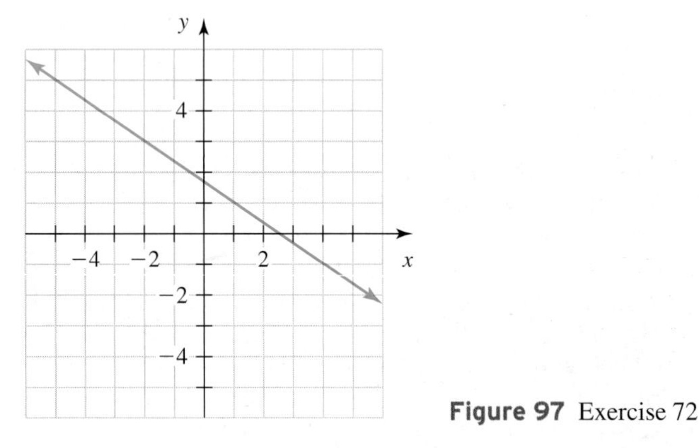

Figure 97 Exercise 72

73. Find an equation of the line sketched in Fig. 98. Check your equation with a graphing calculator.

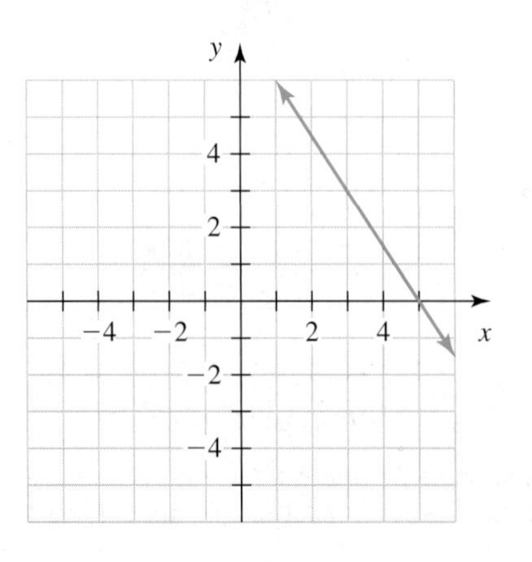

Figure 98 Exercise 73

74. Find an equation of the line sketched in Fig. 99. Check your equation with a graphing calculator.

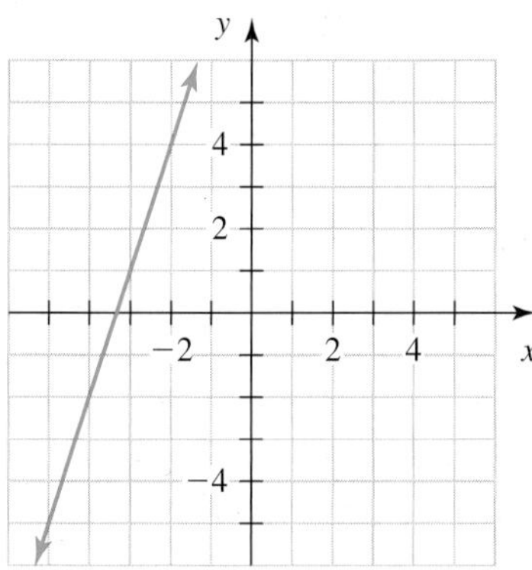

Figure 99 Exercise 74

75. Decide whether a line can have the indicated number of x-intercepts. If it is possible, find an equation of such a line. If it is not possible, explain why.

a. No x-intercepts **b.** Exactly one x-intercept
c. Exactly two x-intercepts
d. An infinite number of x-intercepts

76. Decide whether a line can have the indicated number of y-intercepts. If it is possible, find an equation of such a line. If it is not possible, explain why.

a. No y-intercepts **b.** Exactly one y-intercept
c. Exactly two y-intercepts
d. An infinite number of y-intercepts

77. Is there a line that contains all of the given points? If so, find an equation of it. If not, find an equation of a line that contains most of the given points.

$(-4, 15), (-1, 9), (3, 1), (4, -1), (9, -11)$

78. Is there a line that contains all of the given points? If so, find an equation of it. If not, find an equation of a line that contains most of the given points.

$(-3, 7), (-1, 5), (1, 1), (3, -3), (4, -5)$

79. Create a table of seven pairs of values of x and y for which

a. each point lies on the line $y = 3x - 6$.
b. each point lies close to, but not on, the line $y = 3x - 6$.
c. the points do not lie close to the line $y = 3x - 6$, but all of them lie close to another line. In addition to creating the table, provide an equation of the other line.

80. Suppose that a set of points all lie 0.5 unit above the line $y = -4x + 3$. Find an equation of the line that passes through the points of the set.

81. a. Find an equation of a line with slope -4.

b. Find an equation of a line with y-intercept $\left(0, \frac{3}{7}\right)$. Verify your result with a graphing calculator.

c. Find an equation of a line that contains the point $(-2, 8)$. Verify your result with a graphing calculator.

d. Determine whether there is a line that has slope -4, has y-intercept $\left(0, \frac{3}{7}\right)$, and contains the point $(-2, 8)$. Explain.

82. Find equations of two perpendicular lines that intersect at the point $(3, 1)$.

83. A student thinks that if a line has slope 2 and contains the point $(3, 5)$, then the equation of the line is $y = 2x + 5$, because the slope is 2 (the coefficient of x) and the y-coordinate of $(3, 5)$ is 5 (the constant term). What would you tell the student?

84. A student tries to find an equation of the line that contains the points $(1, 5)$ and $(3, 9)$. The student believes that an equation of the line is $y = 4x + 1$. The student then checks whether $(1, 5)$ satisfies $y = 4x + 1$:

$$y = 4x + 1$$
$$5 \stackrel{?}{=} 4(1) + 1$$
$$5 \stackrel{?}{=} 5$$
$$\text{true}$$

The student concludes that $y = 4x + 1$ is an equation of the line. Find any errors. Then find an equation correctly.

85. Describe how to find an equation of a line that contains two given points. How can you verify that the graph of the equation contains the two points?

Table 24 A Relationship Described by a Table

x	y
3	2
4	1
5	3
5	4

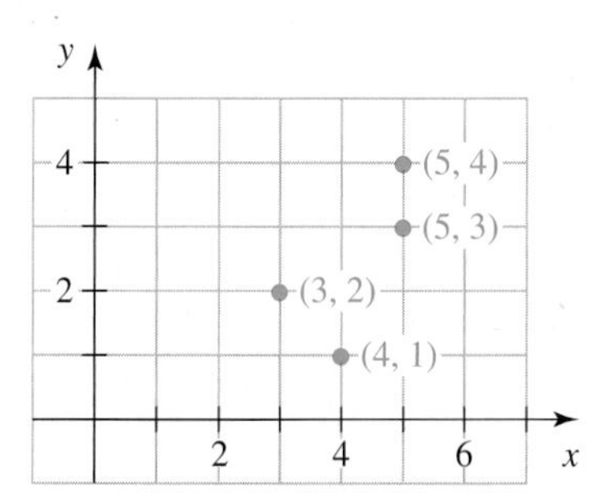

Figure 100 The relationship of Table 24 described by a graph

1.6 FUNCTIONS

Objectives

- Know the meanings of *relation, domain, range,* and *function.*
- Identify functions by using the *vertical line test.*
- Know the definition of a *linear function.*
- Know the Rule of Four for functions.
- Use the graph of a function to find the function's domain and range.

Throughout this chapter, we have described relationships between two variables. In this section, we will discuss how to describe some of these relationships by using an extremely important concept called a *function.*

Relation, Domain, Range, and Function

In this chapter we have used graphs, tables, and equations to describe the relationship between two variables. For example, Table 24 describes a relationship between the variables x and y. This relationship is also described graphically in Fig. 100.

We call the set of ordered pairs listed in Table 24 a *relation*. This relation consists of the ordered pairs (3, 2), (4, 1), (5, 3), and (5, 4). The *domain* of the relation is the set of all values of x (the independent variable)—in this case, 3, 4, and 5. The *range* of the relation is the set of all values of y (the dependent variable)—here, 1, 2, 3, and 4.

DEFINITION Relation, domain, and range

A **relation** is a set of ordered pairs. The **domain** of a relation is the set of all values of the independent variable, and the **range** of the relation is the set of all values of the dependent variable.

We can think of a relation as a machine in which values of x are "inputs" and values of y are "outputs." In general, each member of the domain is an **input,** and each member of the range is an **output.**

For the relation described in Table 24, we can think of the values of x as being sent to the values of y (see Fig. 101).

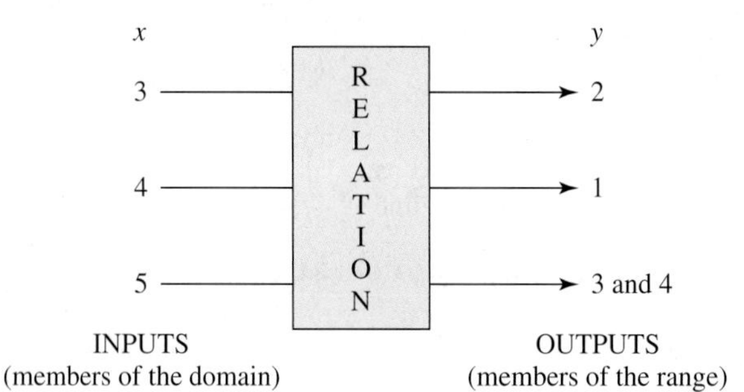

Figure 101 Think of a relation as an input–output machine

Note that the input $x = 5$ is sent to *two* outputs: $y = 3$ and $y = 4$. In a special type of relation called a *function,* each input is sent to exactly *one* output. The relation described in Table 24 is not a function.

DEFINITION Function

A **function** is a relation in which each input leads to exactly one output.

The equation $y = x + 2$ describes a relation consisting of an infinite number of ordered pairs. We will determine whether the relation is a function in Example 1.

Example 1 Deciding whether an Equation Describes a Function

Is the relation $y = x + 2$ a function? Find the domain and range of the relation.

Solution

Let's consider some input–output pairs (in Fig. 102).

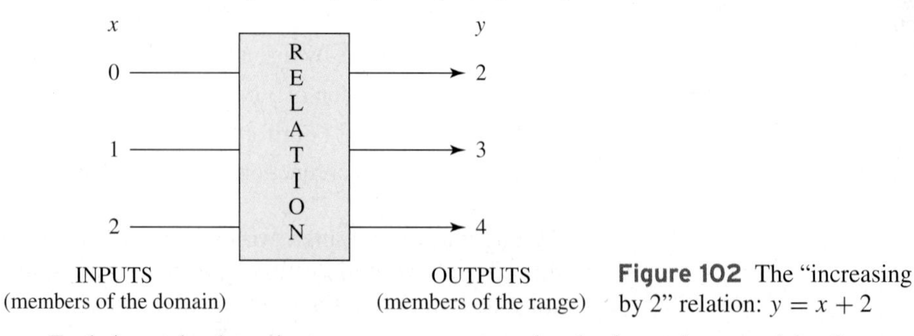

Figure 102 The "increasing by 2" relation: $y = x + 2$

Each input leads to just *one* output—namely, the input increased by 2—so the relation $y = x + 2$ is a function.

The domain of the relation $y = x + 2$ is the set of all real numbers, since we can add 2 to *any* real number. The range of $y = x + 2$ is also the set of real numbers, since any real number is the output of the number that is 2 units less than it. ■

Example 2 Deciding whether an Equation Describes a Function

Is the relation $y = \pm x$ a function?

Solution

If $x = 1$, then $y = \pm 1$. So, the input $x = 1$ leads to *two* outputs: $y = -1$ and $y = 1$. Therefore, the relation $y = \pm x$ is not a function. ■

Example 3 Deciding whether an Equation Describes a Function

Is the relation $y^2 = x$ a function?

Solution

Let's consider the input $x = 4$. We substitute 4 for x and solve for y:

$$y^2 = 4 \qquad \text{Substitute 4 for } x.$$
$$y = -2 \quad \text{or} \quad y = 2 \qquad (-2)^2 = 4, 2^2 = 4$$

The input $x = 4$ leads to *two* outputs: $y = -2$ and $y = 2$. So, the relation $y^2 = x$ is not a function. ■

Table 25 Input-Output Pairs of a Relation

x (input)	y (output)
0	2
1	3
1	5
2	7
3	10

Example 4 Deciding whether a Table Describes a Function

Is the relation described by Table 25 a function?

Solution

The input $x = 1$ leads to *two* outputs: $y = 3$ and $y = 5$. So the relation is not a function. ■

Example 5 Deciding whether a Graph Describes a Function

Is the relation described by the graph in Fig. 103 a function?

Solution

See Fig. 104. The input $x = 3$ leads to *two* outputs: $y = -4$ and $y = 4$. So, the relation is not a function.

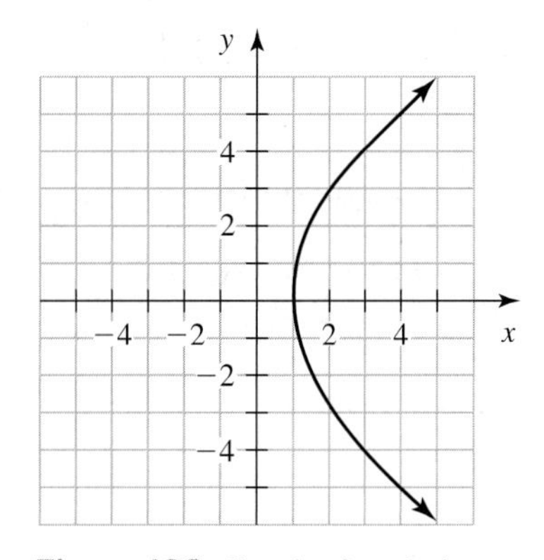

Figure 103 Graph of a relation

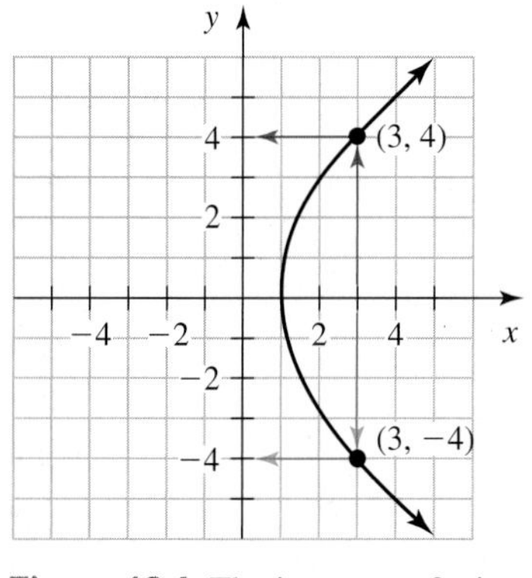

Figure 104 The input $x = 3$ gives two outputs: $y = -4$ and $y = 4$ ■

Vertical Line Test

Notice that the relation described in Example 5 is not a function because some vertical lines would intersect the graph more than once.

Vertical Line Test

A relation is a function if and only if every vertical line intersects the graph of the relation at no more than one point. We call this requirement the **vertical line test.**

Example 6 Deciding whether a Graph Describes a Function

Determine whether the graph represents a function.

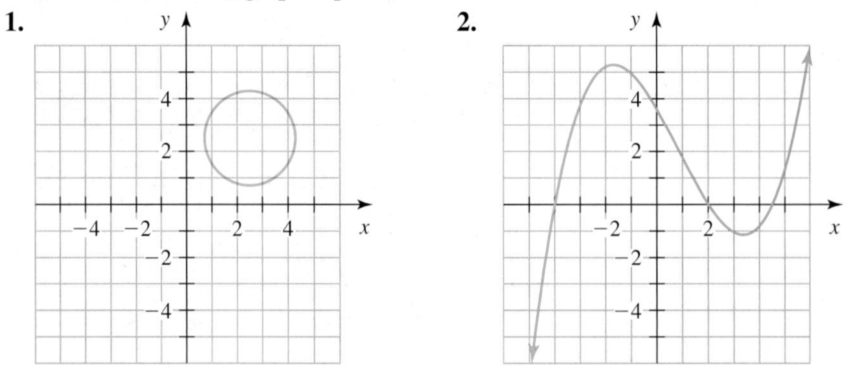

Solution

1. Since the vertical line sketched in Fig. 105 intersects the circle more than once, the relation is not a function.
2. Each vertical line sketched in Fig. 106 intersects the curve at one point. In fact, *any* vertical line would intersect this curve at just one point. So, the relation is a function.

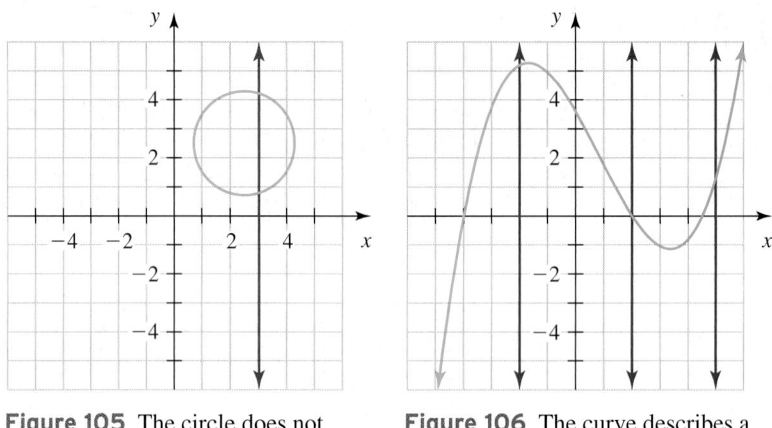

Figure 105 The circle does not describe a function

Figure 106 The curve describes a function

Example 7 Deciding whether an Equation Describes a Function

Is the relation $y = 2x + 1$ a function?

Solution

We begin by sketching the graph of $y = 2x + 1$ in Fig. 107. Note that each vertical line would intersect the line $y = 2x + 1$ at just one point. So, the relation $y = 2x + 1$ is a function.

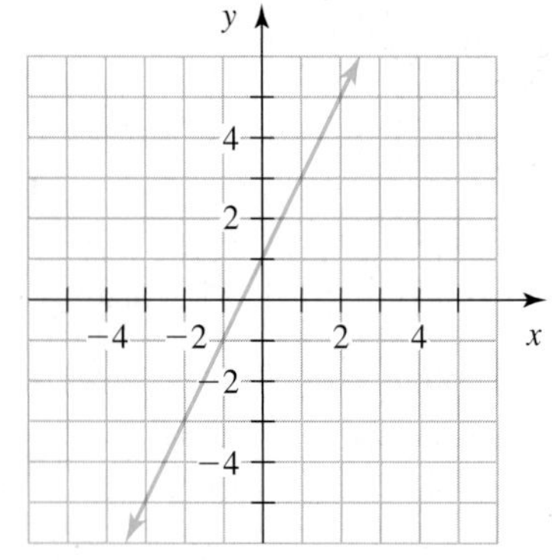

Figure 107 Graph of $y = 2x + 1$

Linear Functions

In Example 7, we saw that the line $y = 2x + 1$ is a function. In fact, any nonvertical line is a function, since it passes the vertical line test.

DEFINITION Linear function

A **linear function** is a relation whose equation can be put into the form

$$y = mx + b$$

where m and b are constants.

In this chapter, we have made many observations about linear equations. Since a linear function can be described by a linear equation, these observations tell us about linear functions. Let's summarize what we know about a linear function $y = mx + b$:

1. The graph of the function is a nonvertical line.
2. The constant m is the slope of the line, a measure of the line's steepness.
3. If $m > 0$, the graph of the function is an increasing line.
4. If $m < 0$, the graph of the function is a decreasing line.
5. If $m = 0$, the graph of the function is a horizontal line.
6. If an input increases by 1, then the corresponding output changes by the slope m.
7. If the run is 1, the rise is the slope m.
8. The y-intercept of the line is $(0, b)$.

Finally, since a linear equation of the form $y = mx + b$ is a *function,* we know that each input leads to exactly one output.

Table 26 Input-Output Pairs for $y = 2x + 1$

x	y
0	1
1	3
2	5
3	7
4	9

Rule of Four for Functions

We can describe functions in four ways. For instance, in Example 7 we described the function $y = 2x + 1$ by using (1) the equation and (2) a graph (see Fig. 107). We can also describe some of the input–output pairs for the same function by using (3) a table (see Table 26). Finally, we can describe the function (4) verbally: In this case, for each input–output pair, the output is 1 more than twice the input.

Rule of Four for Functions

We can describe some or all of the input–output pairs of a function by means of

1. an equation,
2. a graph,
3. a table, or
4. words.

These four ways to describe input–output pairs of a function are known as the **Rule of Four** for functions.

Example 8 Describing a Function by Using the Rule of Four

1. Is the relation $y = -2x - 1$ a function?
2. List some input–output pairs of $y = -2x - 1$ by using a table.
3. Describe the input–output pairs of $y = -2x - 1$ by using a graph.
4. Describe the input–output pairs of $y = -2x - 1$ by using words.

Table 27 Input-Output Pairs of $y = -2x - 1$

x	y
-2	$-2(-2) - 1 = 3$
-1	$-2(-1) - 1 = 1$
0	$-2(0) - 1 = -1$
1	$-2(1) - 1 = -3$
2	$-2(2) - 1 = -5$

Solution

1. Since $y = -2x - 1$ is of the form $y = mx + b$, it is a (linear) function.
2. We list five input–output pairs in Table 27.
3. We graph $y = -2x - 1$ in Fig. 108.
4. For each input–output pair, the output is 1 less than -2 times the input. ■

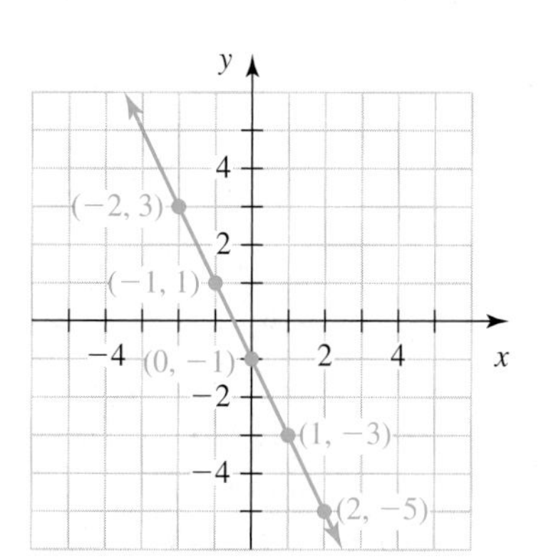

Figure 108 Graph of $y = -2x - 1$

Using a Graph to Find the Domain and Range of a Function

To describe the domain or range of a function, it is sometimes helpful to use the *inequality symbols* $\leq$ and $\geq$. The symbol $\leq$ means "is less than or equal to"; the symbol $\geq$ means "is greater than or equal to." For example, the inequality $x \leq 4$ means that all values of x are less than or equal to 4. And the inequality $y \geq 7$ means that all values of y are greater than or equal to 7.

The inequality $5 \leq x$ means that 5 is *less* than or equal to all values of x. Notice that it is more natural to say that all values of x are *greater* than or equal to 5, which is true.

The inequality $2 \leq x \leq 6$ means that $2 \leq x$ *and* $x \leq 6$: All values of x are *both* greater than or equal to 2 *and* less than or equal to 6. In other words, all values of x are between 2 and 6, inclusive.

Example 9 Finding the Domain and Range

Use the graph of the function to determine the function's domain and range.

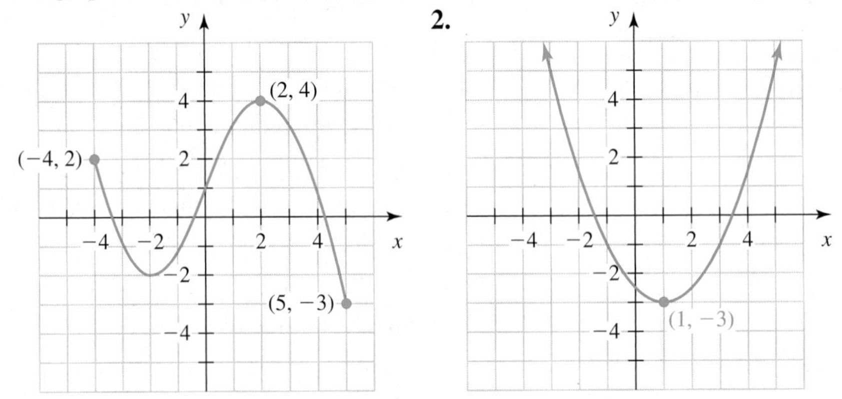

Solution

1. The domain is the set of all x-coordinates of points in the graph. Since there are no breaks in the graph, and since the leftmost point is $(-4, 2)$ and the rightmost point is $(5, -3)$, the domain is $-4 \leq x \leq 5$.

   The range is the set of all y-coordinates of points in the graph. Since the lowest point is $(5, -3)$ and the highest point is $(2, 4)$, the range is $-3 \leq y \leq 4$.
2. The graph extends to the left and right indefinitely without breaks, so every real number is an x-coordinate of some point in the graph. The domain is the set of all real numbers.

   The output -3 is the smallest number in the range, because $(1, -3)$ is the lowest point in the graph. The graph also extends upward indefinitely without breaks, so every number larger than -3 is also in the range. The range is $y \geq -3$. ■

group exploration

Vertical line test

Table 28 A Relation Described by a Table

x	y
2	1
2	5
2	7

Table 29 A Relation Described by a Table

x	y
4	2
4	3
4	6

1. Consider the relation described by Table 28. Is the relation a function? Explain. Now plot the points on a coordinate system. What do you notice about them?
2. Consider the relation described by Table 29. Is the relation a function? Explain. Now plot the points on a coordinate system. What do you notice about them?
3. Describe the graph of a relation that is not a function.
4. Determine whether each graph in Fig. 109 is the graph of a function. Explain.

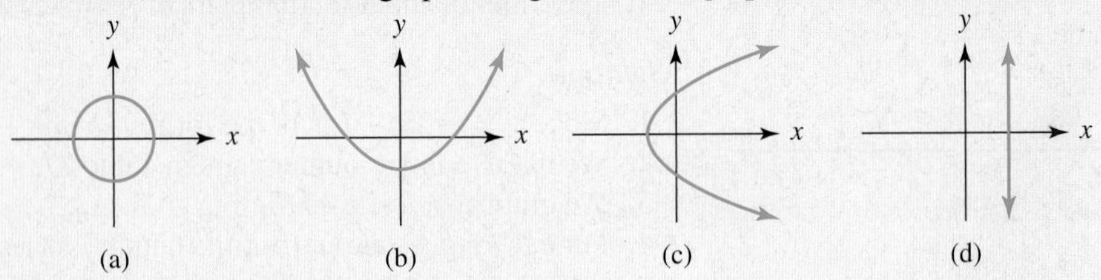

Figure 109 Which graphs describe functions?

group exploration

Looking ahead: Linear modeling

Table 30 Camaro Production

Year	Camaro Production (thousands)
1996	67
1997	56
1998	48
1999	41
2000	42
2001	29

Source: *BlueOvalNews, Independent Voice of the Ford Community.*

The production of the Chevrolet Camaro has decreased since 1996 (see Table 30).

1. Let C be Camaro production (in thousands), and let t be the number of years since 1995. For example, $t = 1$ represents 1996, because 1996 is 1 year since 1995. So, the 1996 production of 67 thousand Camaros is represented by $t = 1$ and $C = 67$. The information in Table 30 can be summarized with a table of values for t and C. Create such a table by filling in the missing entries in Table 31.

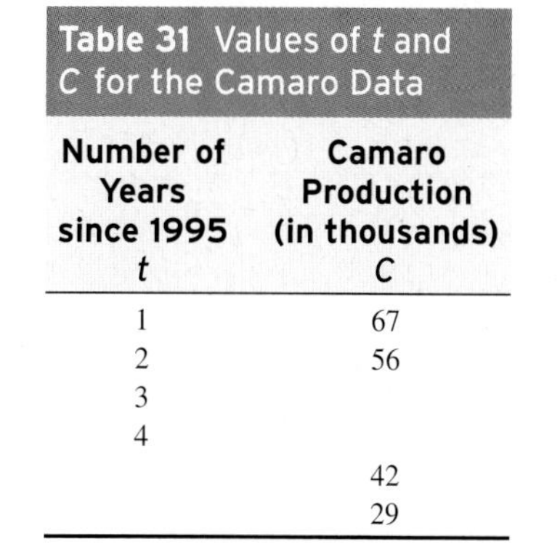

Table 31 Values of t and C for the Camaro Data

Number of Years since 1995 t	Camaro Production (in thousands) C
1	67
2	56
3	
4	
	42
	29

2. Plot the points (t, C) that you listed in Table 31 on a coordinate system like the one in Fig. 110.

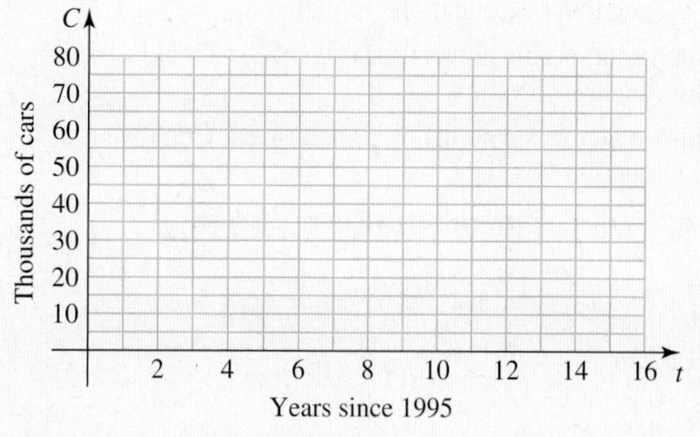

Figure 110 Plot the data points

3. When examining your graph, what do you notice about the arrangement of your plotted points?
4. Sketch a line that comes close to the six data points.
5. Use the line to estimate when 10 thousand Camaros were produced.
6. Use the line to estimate the number of Camaros produced in 2002.
7. Find the t-intercept of the line. What does the t-intercept mean in this situation? Will this prediction happen for certain? Explain.
8. Explain why it is not surprising that 2002 was the last year for Camaros to be manufactured.

TIPS FOR SUCCESS: Practice Exams

When studying for an exam (or a quiz), try creating your own exam to take for practice. Select several Homework exercises from each section on which you will be tested. Choose a variety of exercises that address concepts that your instructor has emphasized. Include many exercises that are moderately difficult and some that are challenging. Completing such a practice test will help you reflect on important concepts and pin down what types of problems you need to study more.

Work on the practice exam for a predetermined time period. Doing so will help you get used to a timed exam, build your confidence, and lower your anxiety for the real exam.

If you are studying with another student, each of you can create a test and then take each other's test. Or create a test together and each take it separately.

HOMEWORK 1.6

FOR EXTRA HELP ▶

Student Solutions Manual

PH Math/Tutor Center

MathXL® MyMathLab

1. Some ordered pairs of four relations are listed in Table 32. Which of these relations could be functions? Explain.

Table 32 Which Relations Might Be Functions? (Exercise 1)

Relation 1		Relation 2		Relation 3		Relation 4	
x	y	x	y	x	y	x	y
1	1	3	27	0	4	5	10
2	3	4	24	1	4	6	20
3	5	5	21	2	4	7	30
3	7	6	18	3	4	8	40
4	9	7	15	4	4	8	50

2. Some ordered pairs of four relations are listed in Table 33.
   a. Which of the relations could be functions? Explain.
   b. Which could be linear functions? Explain.

Table 33 Which Relations Might Be Functions? (Exercise 2)

Relation 1		Relation 2		Relation 3		Relation 4	
x	y	x	y	x	y	x	y
1	3	5	27	0	50	3	11
2	4	5	24	1	45	4	13
3	5	5	21	2	40	5	17
3	6	5	18	3	35	6	25
4	7	5	15	4	30	7	40

3. For a certain relation, an input leads to two different outputs. Could the relation be a function? Explain.

4. For a certain relation, two different inputs lead to the same output. Could the relation be a function? Explain.

5. A relation's graph contains the points (2, 3) and (5, 3). Could the relation be a function? Explain.

6. A relation's graph contains the points (4, 5) and (4, 9). Could the relation be a function? Explain.

Determine whether the graph represents a function. Explain.

7. **8.** **9.** **10.** **11.** **12.** **13.** **14.**

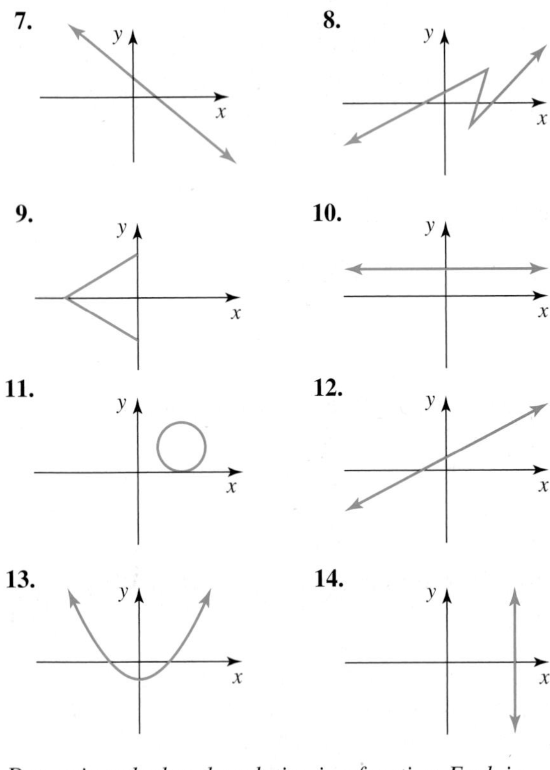

Determine whether the relation is a function. Explain.

15. $y = 5x - 1$

16. $y = -3x + 8$

17. $2x - 5y = 10$

18. $4x + 3y = 24$

19. $y = 4$

20. $y = -1$

21. $x = -3$

22. $x = 0$

23. $7x - 2y = 21 + 3(y - 5x)$

24. $2x + 5y = 9 - 4(x + 2y)$

25. Is a nonvertical line the graph of a function? Explain.

26. Is a vertical line the graph of a function? Explain.

27. Is a circle the graph of a function? Explain.

28. Is a semicircle that is the "upper half" of a circle the graph of a function? Explain.

29. Describe the Rule of Four as applied to the function $y = 3x - 2$:

a. Describe five input–output pairs by using a table.
b. Describe the input–output pairs by using a graph.
c. Describe the input–output pairs by using words.

30. Describe the Rule of Four as applied to the function $y = \frac{1}{2}x + 2$:

a. Describe five input–output pairs by using a table.
b. Describe the input–output pairs by using a graph.
c. Describe the input–output pairs by using words.

Use the graph of the function to determine the function's domain and range.

31.

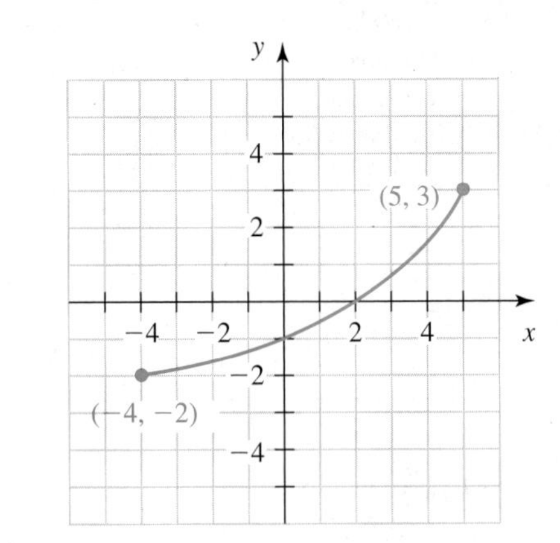

32.

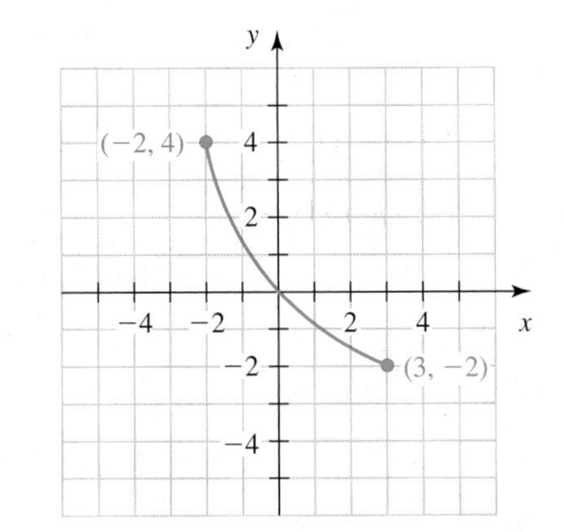

33.

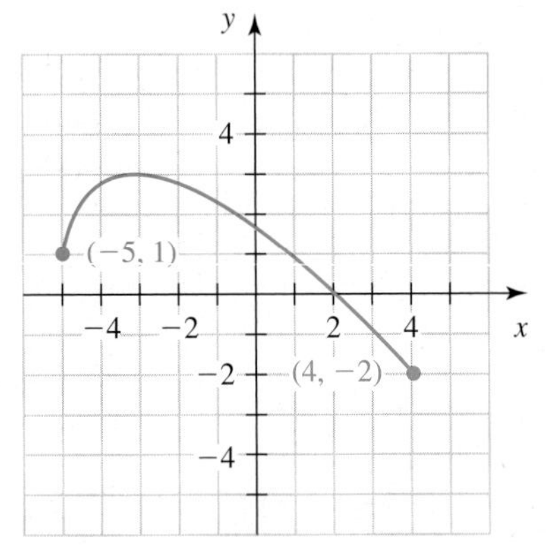

34.

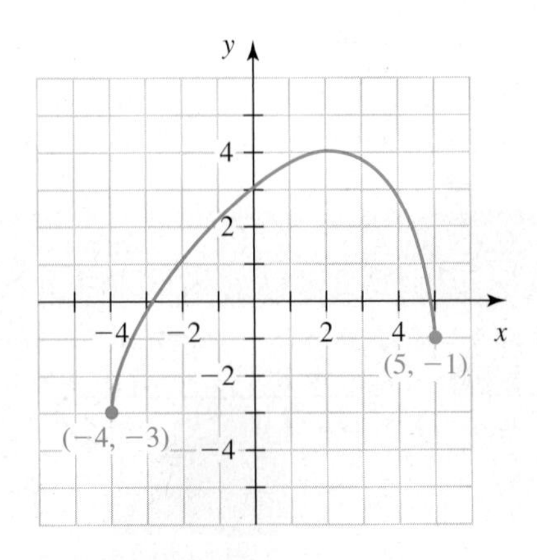

35.

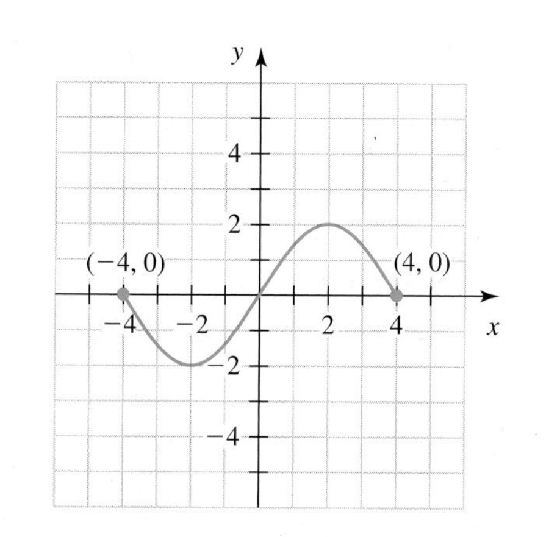

36.

37.

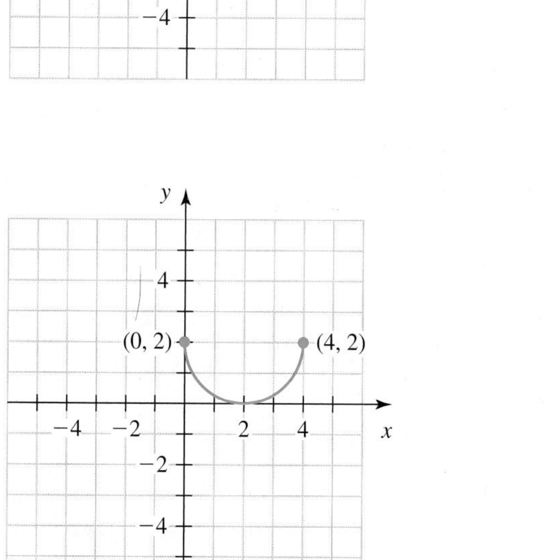

38.

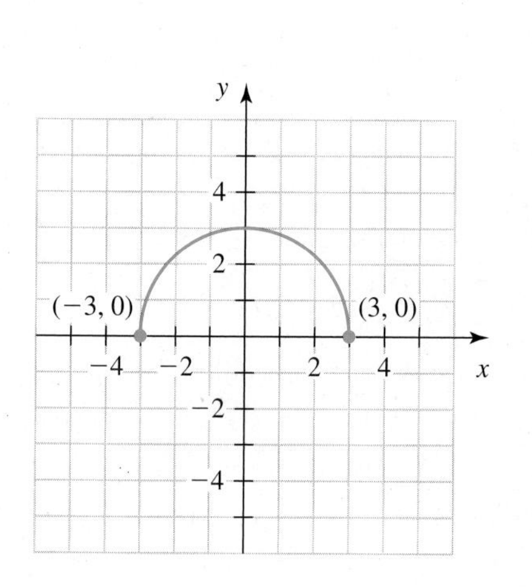

39.

40.

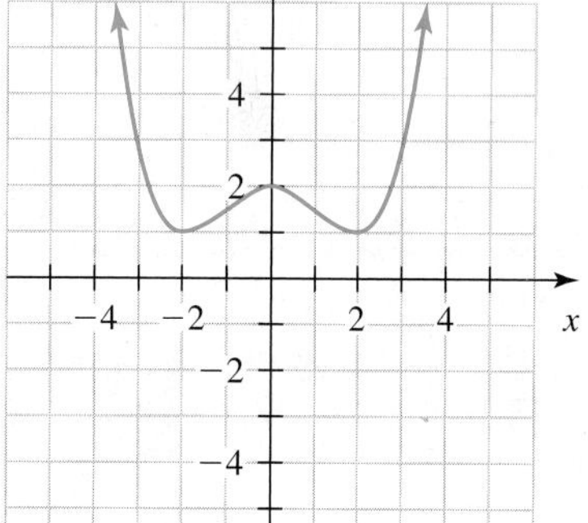

41.

42.

43. Describe the input–output pairs of a function (different from those in this section) by using an equation, a graph, and words. Describe also five input–output pairs of the function by using a table. Explain why your relation is a function.

44. Sketch the graph of a relation (different from those in this section) that is not a function. Next, create a table that lists five ordered pairs of the relation. Explain why your relation is not a function.

45. Sketch the graph of a relation for which the input $x = 2$ gives exactly two outputs and the input $x = 6$ gives exactly one output. Is the relation a function? Explain.

46. Sketch the graph of a relation for which the input $x = -4$ gives exactly three outputs and the input $x = 5$ gives exactly one output. Is the relation a function? Explain.

47. Sketch the graph of a function whose domain is $-3 \le x \le 5$ and whose range is $-2 \le y \le 4$.

48. Sketch the graph of a function whose domain is $-2 \le x \le 4$ and whose range is $-1 \le y \le 5$.

49. Sketch the graph of a function whose domain is the set of all real numbers and whose range is $y \ge 2$.

50. Sketch the graph of a function whose domain is the set of all real numbers and whose range is $y \le 3$.

Decide whether the relation is a function. Explain.

51. $y = \sqrt{x}$ [**Hint:** Sketch a graph.]

52. $y = x^4$ [**Hint:** Sketch a graph.]

53. $y^4 = x$ [**Hint:** Substitute 16 for x; then solve for y.]

54. $y^3 = x$ [**Hint:** Substitute 0, 1, and 8 for x, and solve for y after each substitution.]

55. A student tries to determine whether the relation $y = x^2$ is a function. She finds that both inputs $x = -3$ and $x = 3$ give the same output, $y = 9$. The student concludes that the relation is not a function. Is her conclusion correct? Explain.

56. Explain how you can determine whether a relation is a function.

CHAPTER SUMMARY

Key Points OF CHAPTER 1

Qualitative graph (Section 1.1)
A **qualitative graph** is a graph without scaling on the axes.

Independent and dependent variables (Section 1.1)
Assume that an authentic situation can be described by using the variables t and p and that p depends on t:

- We call t the **independent variable.**
- We call p the **dependent variable.**

Axes of a graph (Section 1.1)
For graphs, we describe the values of the independent variable along the horizontal axis and the values of the dependent variable along the vertical axis.

Intercept (Section 1.1)
An **intercept** of a curve is any point where the curve and an axis (or axes) intersect.

Increasing curve (Section 1.1)
If a curve goes upward from left to right, the curve is an **increasing curve.**

Decreasing curve (Section 1.1)
If a curve goes downward from left to right, the curve is a **decreasing curve.**

Solution, satisfy, and solution set of a linear equation in two variables (Section 1.2)
An ordered pair (a, b) is a **solution** of an equation in terms of x and y if the equation becomes a true statement when a is substituted for x and b is substituted for y. We say that (a, b) **satisfies** the equation. The **solution set** of an equation is the set of all solutions of the equation.

Graph (Section 1.2)
The **graph** of an equation in two variables is the set of points that correspond to all solutions of the equation.

Intercepts of the graph of an equation (Section 1.2)
For an equation containing the variables x and y,

- To find the x-coordinate of each x-intercept, substitute 0 for y and solve for x.
- To find the y-coordinate of each y-intercept, substitute 0 for x and solve for y.

Equations of vertical and horizontal lines (Section 1.2)
If a and b are constants, then

- An equation that can be put into the form $x = a$ has a vertical line as its graph.
- An equation that can be put into the form $y = b$ has a horizontal line as its graph.

Linear equations in two variables (Section 1.2)
If an equation can be put into either form $y = mx + b$ or $x = a$, where m, a, and b are constants, then the graph of the equation is a line. We call such an equation a **linear equation in two variables.**

Comparing the steepness of two objects (Section 1.3)
To compare the steepness of two objects, compute the ratio

$$\frac{\text{vertical distance}}{\text{horizontal distance}}$$

for each object. The object with the larger ratio is the steeper object.

Slope of a nonvertical line (Section 1.3)
Let (x_1, y_1) and (x_2, y_2) be two distinct points of a nonvertical line. The **slope** of the line is

$$m = \frac{\text{vertical change}}{\text{horizontal change}} = \frac{\text{rise}}{\text{run}} = \frac{y_2 - y_1}{x_2 - x_1}$$

Slopes of increasing and decreasing lines (Section 1.3)	An increasing line has positive slope. A decreasing line has negative slope.
Comparing the steepness of two lines (Section 1.3)	For two nonparallel increasing lines, the steeper line has the greater slope.
Slopes of horizontal and vertical lines (Section 1.3)	A horizontal line has slope equal to zero. A vertical line has undefined slope.
Slopes of parallel lines (Section 1.3)	If lines l_1 and l_2 are nonvertical parallel lines on the same coordinate system, then the slopes of the lines are equal: $m_1 = m_2$. Also, if two distinct lines have equal slope, then the lines are parallel.
Slopes of perpendicular lines (Section 1.3)	If lines l_1 and l_2 are nonvertical perpendicular lines, then the slope of one line is the opposite of the reciprocal of the slope of the other line: $m_2 = -\frac{1}{m_1}$. Also, if the slope of one line is the opposite of the reciprocal of another line's slope, then the lines are perpendicular.
Slope and y-intercept of a linear equation of the form $y = mx + b$; slope–intercept form (Section 1.4)	For a linear equation of the form $y = mx + b$, m is the slope of the line, and the y-intercept is $(0, b)$. We say that the equation is in **slope–intercept form.**
Vertical change property (Section 1.4)	For a line $y = mx + b$, if the run is 1, then the rise is the slope m.
Using slope to graph a linear equation of the form $y = mx + b$ (Section 1.4)	To sketch the graph of a linear equation of the form $y = mx + b$, **1.** Plot the y-intercept $(0, b)$. **2.** Use $m = \frac{\text{rise}}{\text{run}}$ to plot a second point. **3.** Sketch the line that passes through the two plotted points.
Solve for y first (Section 1.4)	Before we can use the y-intercept and the slope to graph a linear equation, we must solve for y to put the equation into the form $y = mx + b$.
Slope addition property (Section 1.4)	For a linear equation of the form $y = mx + b$, if the value of the independent variable increases by 1, then the value of the dependent variable changes by the slope m.
Finding an equation of a line that contains two given points (Section 1.5)	To find an equation of the line that passes through two given points whose x- coordinates are different, **1.** Use the slope formula, $m = \frac{y_2 - y_1}{x_2 - x_1}$, to find the slope of the line. **2.** Substitute the m value you found in step 1 into the equation $y = mx + b$. **3.** Substitute the coordinates of one of the given points into the equation you found in step 2, and solve for b. **4.** Substitute the m value you found in step 1 and the b value you found in step 3 into the equation $y = mx + b$. **5.** Use a graphing calculator to check that the graph of your equation contains the two given points.
Point–slope form (Section 1.5)	If a nonvertical line has slope m and contains the point (x_1, y_1), then an equation of the line is $y - y_1 = m(x - x_1)$. We say that such an equation is in **point–slope form.**
Relation, domain, and range (Section 1.6)	A **relation** is a set of ordered pairs. The **domain** of a relation is the set of all values of the independent variable, and the **range** of the relation is the set of all values of the dependent variable.
Input and output (Section 1.6)	Each member of the domain is an **input,** and each member of the range is an **output.**
Function (Section 1.6)	A **function** is a relation in which each input leads to exactly one output.
Vertical line test (Section 1.6)	A relation is a function if and only if each vertical line intersects the graph of the relation at no more than one point.
Linear function (Section 1.6)	A **linear function** is a relation whose equation can be put into the form $y = mx + b$, where m and b are constants.
Rule of Four for functions (Section 1.6)	We can describe some or all of the input–output pairs of a function by means of (1) an equation, (2) a graph, (3) a table, or (4) words. These four ways to describe input–output pairs of a function are known as the **Rule of Four** for functions.

CHAPTER 1 REVIEW EXERCISES

1. Let n be the number of e-mails that a person receives, and let t be the amount of time (in minutes) it takes to read and reply to them. Identify the independent variable and the dependent variable.

Sketch a qualitative graph that shows the relationship between the variables defined in each exercise. Justify your sketch.

2. Let L be the length (in inches) of a candle at t minutes after it is lit.
3. Let T be the number of seconds it takes to cook a marshmallow that is d inches from a campfire.

Solve.

4. $3(2x - 4) - 2 = 5x - (3 - 4x)$
5. $\frac{2}{3}w - \frac{1}{2} = \frac{5}{6}w + \frac{4}{3}$
6. $a(x - c) = d$ for x.
7. Find the x-intercept and y-intercept of the graph of $3x - 5y = 17$.
8. Find the x-intercept and y-intercept of the graph of $ax + b = cy$.

For Exercises 9–12, refer to Fig. 111.

9. Estimate y when $x = 5$.
10. Estimate y when $x = 0$.
11. Estimate x when $y = 2$.
12. Estimate x when $y = 0$.

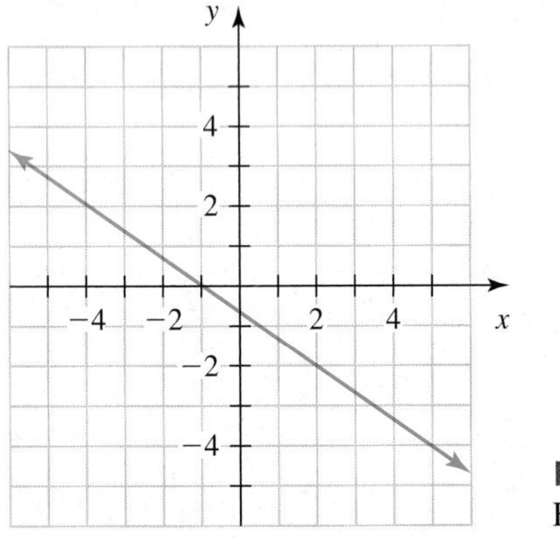

Figure 111 Exercises 9–12

Find the slope of the line that passes through the given points. State whether the line is increasing, decreasing, horizontal, or vertical.

13. $(-3, -2)$ and $(2, -5)$
14. $(-9, -7)$ and $(-1, -3)$
15. $(4, -1)$ and $(4, 3)$
16. Find the slope of the line that passes through the points $(-5.27, 2.99)$ and $(3.54, -8.48)$. Round your result to the second decimal place. State whether the line is increasing, decreasing, horizontal, or vertical.

Graph the equation by hand.

17. $y = -3x + 10$
18. $y + 2x = 0$
19. $y = 7$
20. $3x - 2y = 12$
21. $-3(y + 2) = 2x + 9$
22. $3x - 2(2y - 1) = 8x - 3(x + 2)$
23. Determine the slope and y-intercept of the graph of $a(x - y) = c$.

Determine whether the given lines are parallel, perpendicular, or neither. Explain.

24. $2x + 5y = 7$ and $y = \frac{2}{5}x + 7$
25. $3x - 8y = 7$ and $-6x + 16y = 5$

Find an equation of the line that has the given slope and contains the given point.

26. $m = -4$, $(-3, 7)$
27. $m = -\frac{2}{3}$, $(5, -4)$

Find an equation of the line that passes through the given points.

28. $(-3, -2)$ and $(2, 6)$
29. $(-4, 6)$ and $(2, -2)$
30. $(3, -2)$ and $(3, 5)$
31. Find an approximate equation of the line that passes through the points $(-3.62, -8.79)$ and $(2.51, -6.38)$. Round the slope and the constant term to two decimal places.
32. Find equations of each line sketched in Fig. 112.

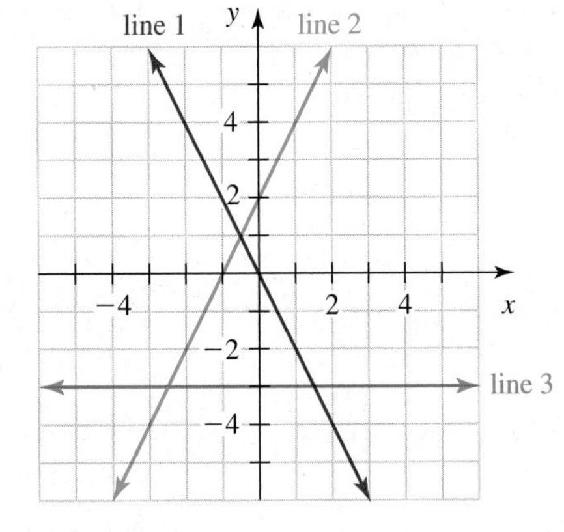

Figure 112 Exercise 32

33. Some values of a linear equation are provided in Table 34. Complete the table.

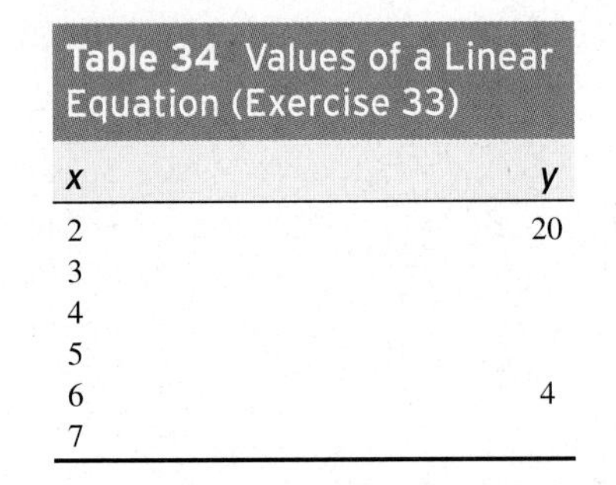
Table 34 Values of a Linear Equation (Exercise 33)

x	y
2	20
3	
4	
5	
6	4
7	

34. Graphs of the functions $y = 0.5x + 6$ and $y = -1.5x + 3$ are shown in Fig. 113. Find an equation of line 1, also sketched in Fig. 113. Assume that line 1 and the line $y = 0.5x + 6$ are parallel.

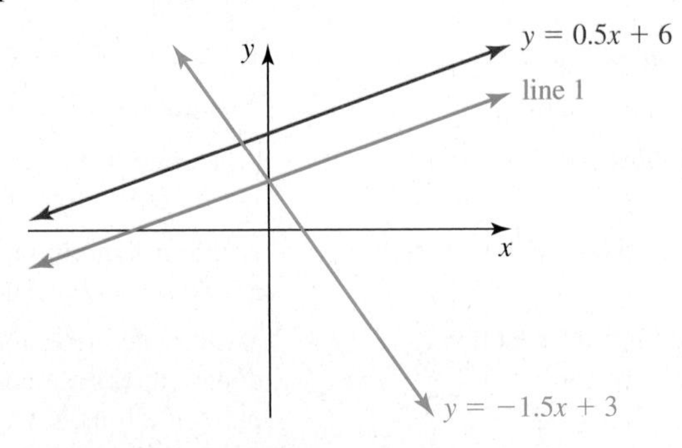

Figure 113 Exercise 34

35. The graphs of $y = ax + b$ and $y = cx + d$ are sketched in Fig. 114.

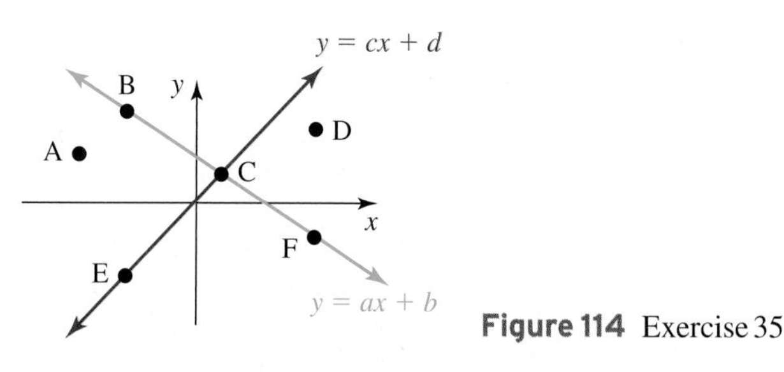

Figure 114 Exercise 35

For each part, decide which one or more of the points A, B, C, D, E, and F represent ordered pairs that

a. satisfy the equation $y = ax + b$.
b. satisfy the equation $y = cx + d$.
c. satisfy both equations.
d. do not satisfy either equation.

36. Find an equation of the line that contains the point $(-2, 5)$ and is parallel to the line $3x - y = 6$.

37. Decide whether a line can have an infinite number of x-intercepts. If it can, find an equation of such a line. If it cannot, explain why.

38. Some ordered pairs for four relations are listed in Table 35. Which of these relations could be functions? Explain.

Table 35 Which Relations Might Be Functions? (Exercise 38)

Relation 1		Relation 2		Relation 3		Relation 4	
x	y	x	y	x	y	x	y
1	12	3	27	0	7	2	1
2	15	4	24	1	7	2	2
3	18	4	21	2	7	2	3
4	21	5	18	3	7	2	4
5	24	6	15	4	7	2	5

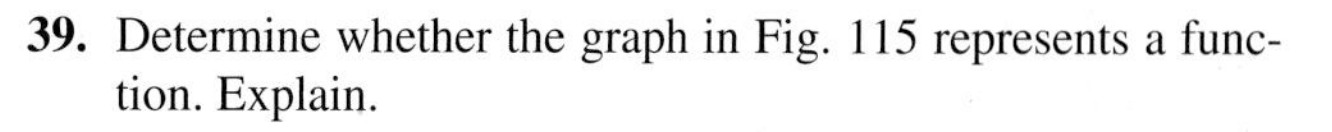
39. Determine whether the graph in Fig. 115 represents a function. Explain.

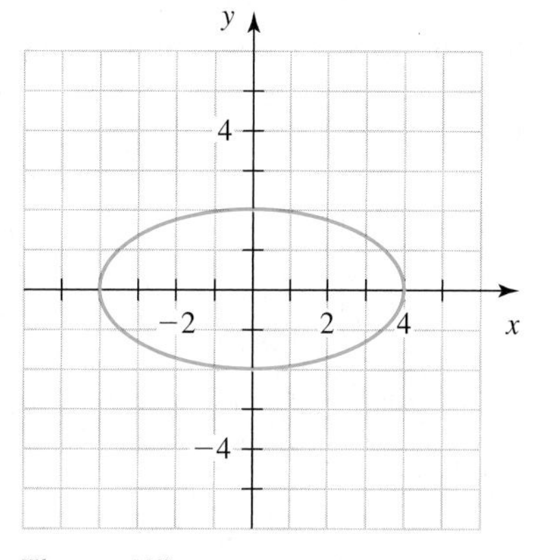

Figure 115 Exercise 39

Determine whether each relation is a function. Explain.

40. $5x - 6y = 3$

41. $x = 9$

42. $y^2 = x$

43. Use the graph of the function in Fig. 116 to determine the function's domain and range.

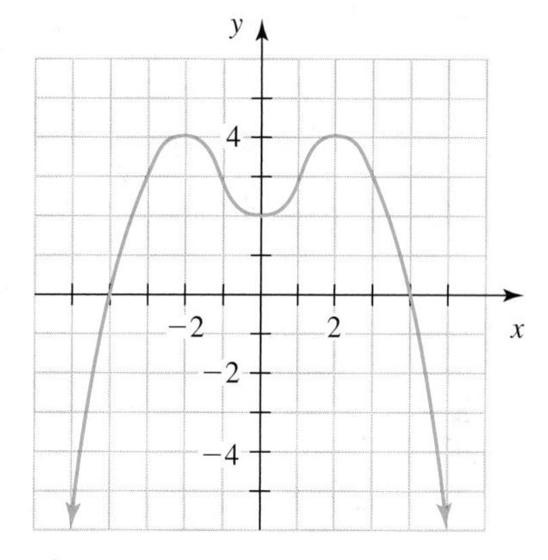

Figure 116 Exercise 43

CHAPTER 1 TEST

1. Let w be the weight (in pounds) of a gold bar, and let v be its value (in dollars). Identify the independent variable and the dependent variable.

2. A student eats breakfast at home. She then begins walking to school. After a while, she jogs. After jogging for some time, she runs the rest of the way at a faster pace. Let d be how far she is from home (in yards) at t minutes since she started eating breakfast. Sketch a qualitative graph that describes the relationship between t and d.

3. Write a scenario to match the graph in Fig. 117. Refer to the variables x and y in your description.

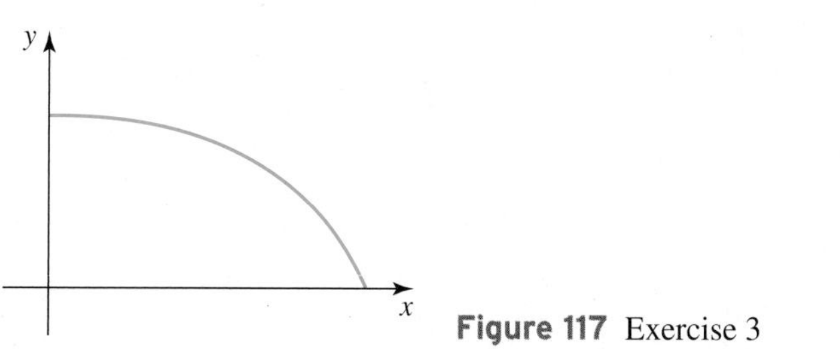

Figure 117 Exercise 3

4. Solve the equation $5 - 3(4x - 2) = 8 - (7x + 1)$.

5. Find equations of each of the three lines sketched in Fig. 118.

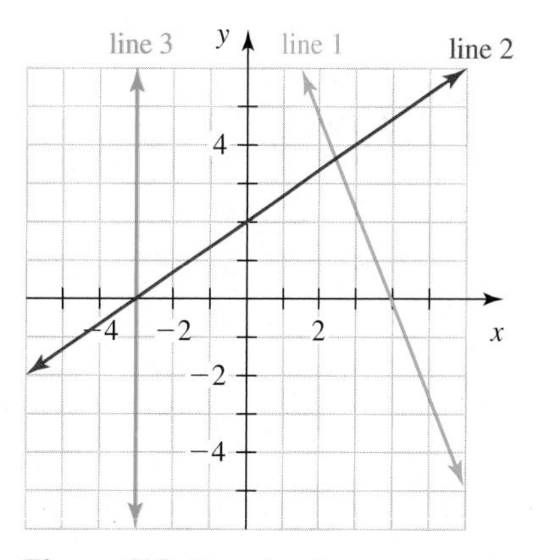

Figure 118 Exercise 5

6. Graphs of the equations $y = mx + b$ and $y = kx + c$ are sketched in Fig. 119.

a. Which is greater, m or k? Explain.

b. Which is greater, b or c? Explain.

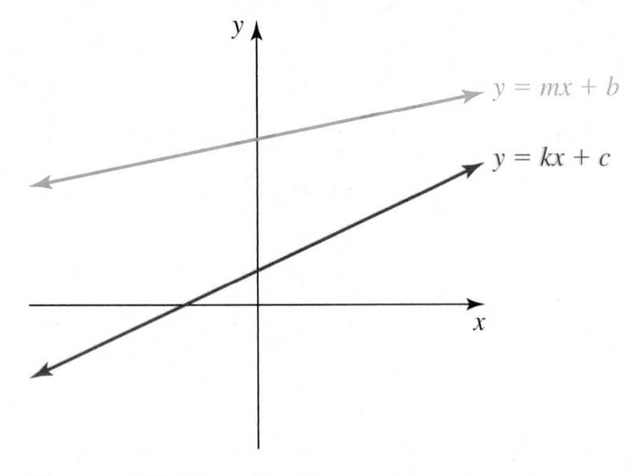

Figure 119 Exercise 6

7. Ski run A declines steadily for 85 yards over a horizontal distance of 270 yards. Ski run B declines steadily for 140 yards over a horizontal distance of 475 yards. Which run is steeper? Explain.

8. Some values of a linear equation are provided in Table 36. Complete the table.

Table 36 Values of a Linear Equation (Exercise 8)

x	y
4	
5	
6	33
7	
8	
9	45

Graph the equation by hand.

9. $y = -\frac{1}{5}x + 4$

10. $2(2x - y) = 2x + 9 + y$

11. Find the slope of the line that contains the points $(-3, 2)$ and $(5, -8)$.

12. A line contains the points $(2, 8)$ and $(5, 6)$. Find three more points that lie on the line.

13. Find an equation of the line with slope $-\frac{3}{7}$ that contains the point $(-2, 5)$.

14. Find an equation of the line that contains the points $(-3, 7)$ and $(2, -5)$.

15. Is there a line that contains all of the given points? If so, find an equation of the line. If not, find an equation that contains most of the points.

$$(-2, 5), (0, 2), (2, -3), (3, -5), (5, -9)$$

16. Find the equation of a line that contains the point $(4, -1)$ and is perpendicular to the line $3x - 5y = 20$.

17. Find the x-intercept and y-intercept of the graph of $2y + 5 = 4(x - 1) + 3$.

18. Describe the Rule of Four as applied to the function $y = 2x - 4$:

a. Describe five input–output pairs by using a table.

b. Describe the input–output pairs by using a graph.

c. Describe the input–output pairs by using words.

19. Sketch the graph of a relation that is *not* a function. Explain.

20. Determine whether the relation described by $y = \pm\sqrt{x}$ is a function. Explain.

21. Determine whether the relation described by $y = -2x + 5$ is a function. Explain.

22. Use the graph of the relation in Fig. 120 to determine the relation's domain and range. Determine whether the relation is a function.

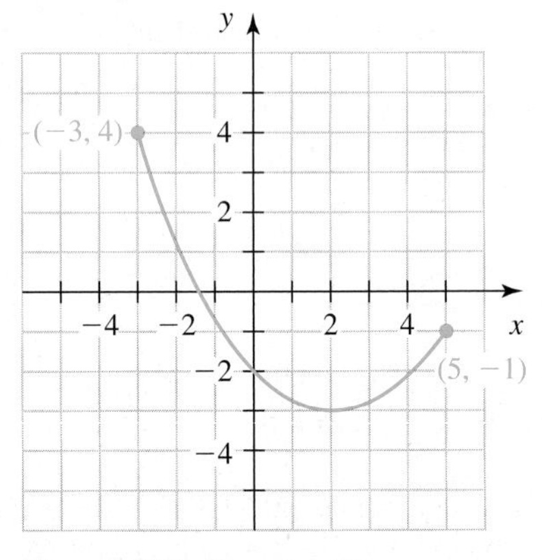

Figure 120 Exercise 22

Chapter 2

Modeling with Linear Functions

"I have learned throughout my life as a composer chiefly through my mistakes and pursuits of false assumptions, not by my exposure to founts of wisdom and knowledge."
—*Igor Stravinsky*

Table 1 Percentages of Americans Whose Favorite Sport Is Baseball

Year	Percent
1994	19
1995	16
1997	14
1998	16
2000	13
2002	12
2004	10

Source: *The Gallup Organization*

What is your favorite sport? Baseball has been dubbed "America's favorite pastime." However, the popularity of baseball is waning. Table 1 shows, for various years, the percentage of Americans whose favorite sport is baseball. In Exercise 81 of Homework 2.3, you will predict when only 6% of Americans will choose baseball as their favorite sport.

In Section 1.1, we used qualitative graphs to describe situations. In this chapter, we will discuss how to use linear functions to describe authentic situations. We will use graphs and equations of these functions to make estimates and predictions, such as predicting when 6% of Americans will choose baseball as their favorite sport. We will also discuss the meaning of the slope of a linear function used to describe an authentic situation.

2.1 USING LINES TO MODEL DATA

Objectives

- Know the meaning of *scattergram, approximately linearly related, model,* and *linear model.*
- Use a linear model to make estimates and predictions.
- Find intercepts of a linear model.
- Know the meaning of *interpolate, extrapolate,* and *model breakdown.*

In this section, we use graphs of linear functions to make estimates and predictions about authentic situations.

Scattergrams

We begin by creating a special type of graph called a *scattergram.*

Example 1 Using a Graph to Describe an Authentic Situation

The Grand Canyon is a beautiful landmark, yet the difficulty of finding a parking spot can detract from visitors' enjoyment. The numbers of Grand Canyon visitors are listed in Table 2 for various years. Describe the data with a graph.

Solution

Let v be the number (in millions) of visitors in the year that is t years since 1960. For example, $t = 10$ represents 1970, since 1970 is 10 years after 1960. Then we can describe the data with a table of values for t and v (see Table 3).

Table 2 Visitors to the Grand Canyon

Year	Number of Visitors (millions)
1960	1.2
1970	2.3
1980	2.6
1990	3.8
2000	4.8

Source: *National Park Service*

Table 3 Visitors to the Grand Canyon

Number of Years Since 1960 t	Number of Visitors (millions) v
0	1.2
10	2.3
20	2.6
30	3.8
40	4.8

Recall from Section 1.2 that, for an ordered pair (a, b), we write the value of the independent variable in the first (left) position and the value of the dependent variable in the second (right) position. Since v depends on t, we write ordered pairs in the form (t, v). For example, the ordered pair $(10, 2.3)$ indicates that when $t = 10$, $v = 2.3$—or that in 1970 there were 2.3 million visitors.

Next, we plot the (t, v) data points shown in Fig. 1. Recall from Section 1.1 that the values of the independent variable are described by the horizontal axis and the values of the dependent variable are described by the vertical axis. So, we let the horizontal axis be the t-axis and the vertical axis be the v-axis. We write the units "Years since 1960" and "Millions of visitors" on the appropriate axes.

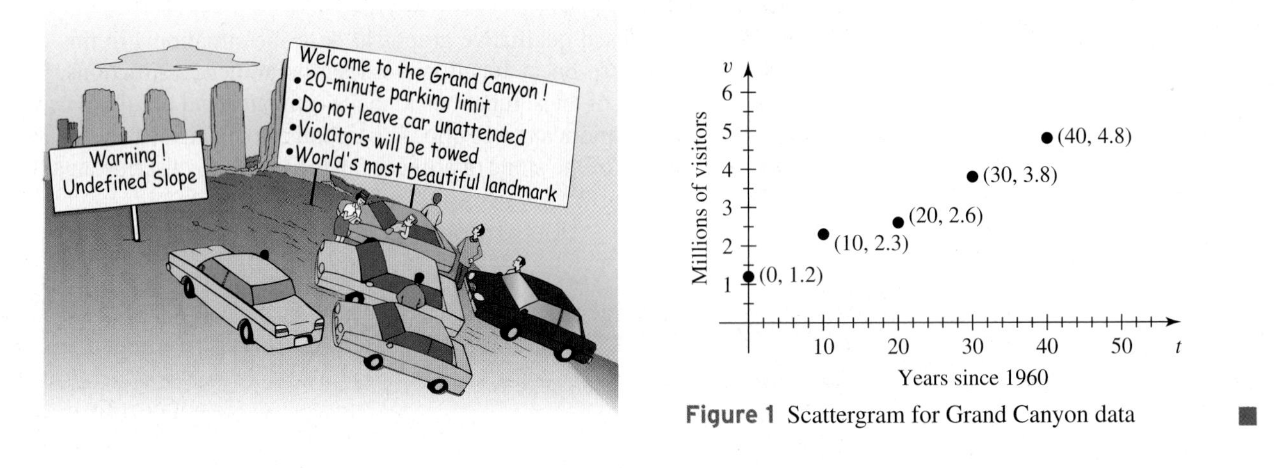

Figure 1 Scattergram for Grand Canyon data ■

A graph of plotted ordered pairs, such as the graph in Fig. 1, is called a **scattergram. A scattergram should have scaling on both axes and labels indicating the variable names and scale units.**

We can sketch a line that comes close to (or on) the data points of Fig. 1 (see Fig. 2).

There are many lines that come close to (or on) the data points (see Fig. 3). Each of the three lines shown in Fig. 3 does a reasonable job of describing the data.

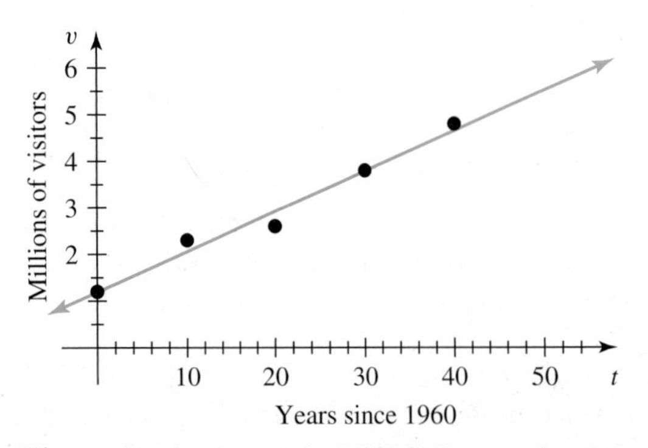

Figure 2 Linear model of Grand Canyon data

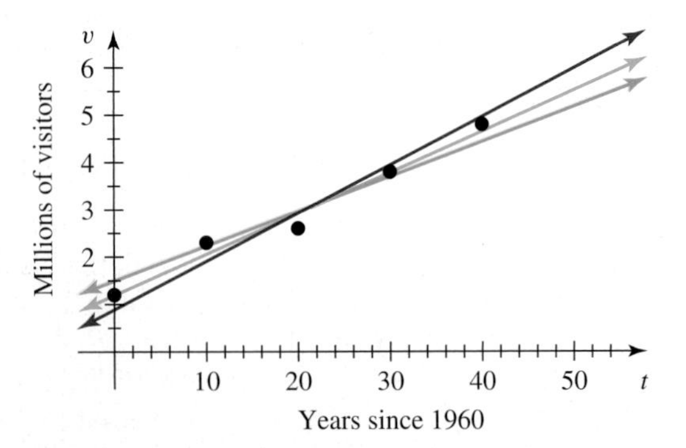

Figure 3 A few of the many lines that come close to (or on) the data points

Linear Models

In Fig. 2 we sketched a line that describes the number of visitors to the Grand Canyon. However, this description is not exact. For example, the line does not describe exactly what happened in the years 1970, 1980, or 2000, because the line does not contain any of these data points. However, the line does come very close to the data points, so it suggests very good approximations for these years.

If the points in a scattergram of data lie close to (or on) a line, then we say that the relevant variables are **approximately linearly related.** For the Grand Canyon situation, variables t and v are approximately linearly related.

Because the line in Fig. 2 is nonvertical, it is the graph of a linear function. The process of choosing a linear function to represent the relationship between the number of visitors in a year and the number of years since 1960 is an example of *modeling*.

DEFINITION Model

A **model** is a mathematical description of an authentic situation. We say that the description *models* the situation.

We call the Grand Canyon linear function a *linear model.* In later chapters, we will discuss other types of models. The term "model" is being used in much the same way as it is used in "airplane model." Just as an airplane designer can use the behavior of an airplane model in a wind tunnel to predict the behavior of an actual airplane, a linear model can be used to predict what might happen in a situation in which two variables are approximately linearly related.

DEFINITION Linear model

A **linear model** is a linear function, or its graph, that describes the relationship between two quantities for an authentic situation.

Every linear model is a linear function. So, depending on what we choose to emphasize, we can refer to the Grand Canyon model as a "model" or a "function."

However, not every linear function is a linear model. Models are used only to describe situations. Functions are used both to describe situations *and* to describe certain *mathematical* relationships between two variables. For example, if the equation $y = 2x$ is not being used to describe a situation, then it is a function, not a model.

Using a Linear Model to Make Estimates and Predictions

Since all of the Grand Canyon data points lie close to (or on) our linear model, it seems reasonable that data points for the years between 1960 and 2000 that are not given in Table 3 might also lie close to the line. Similarly, it is reasonable that data points for at least a few years before 1960 or for at least a few years after 2000 might also lie near the line.

Example 2 Using a Linear Model to Make a Prediction and an Estimate

1. Use the linear model shown in Fig. 2 to predict the number of visitors in 2010.
2. Use the linear model to estimate in what year there were 4 million visitors.

Solution

1. The year 2010 corresponds to $t = 50$, because $2010 - 1960 = 50$. To estimate the number of visitors, we locate the point on the linear model where the t-coordinate is 50. We see that the corresponding v-coordinate is about 5.6 (see Fig. 4). So, according to the model, there will be 5.6 million visitors in 2010.

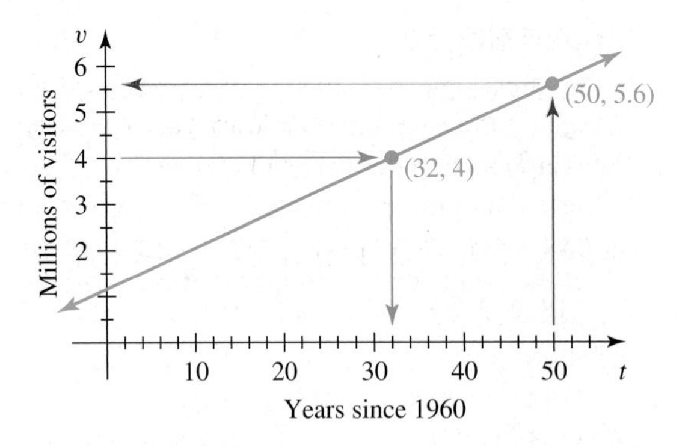

Figure 4 Using the Grand Canyon model to make a prediction and an estimate

2. To find the year in which there were 4 million visitors, we locate the point on the linear model where the v-coordinate is 4. We see that the corresponding t-coordinate is about 32. So, *according to the linear model,* there were 4 million visitors in $1960 + 32 = 1992$. ■

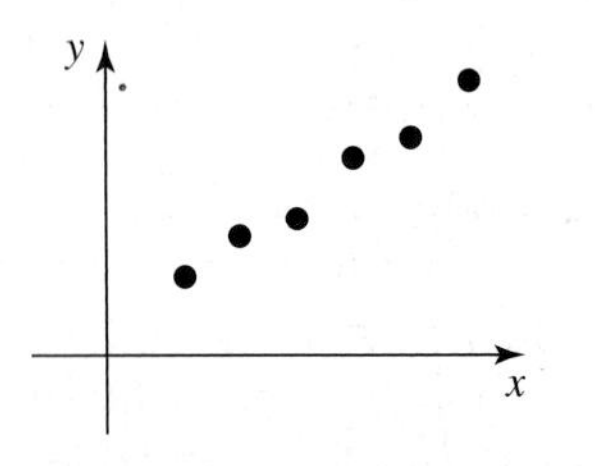

Figure 5 Scattergram for situation 1

When to Use a Linear Function to Model Data

How do we determine whether an authentic situation can be described well by a linear model?

Example 3 Deciding Whether to Use a Linear Function to Model Data

Consider the scattergrams of data shown in Figs. 5, 6, and 7 for situations 1, 2, and 3, respectively. For each situation, determine whether a linear function would model it well.

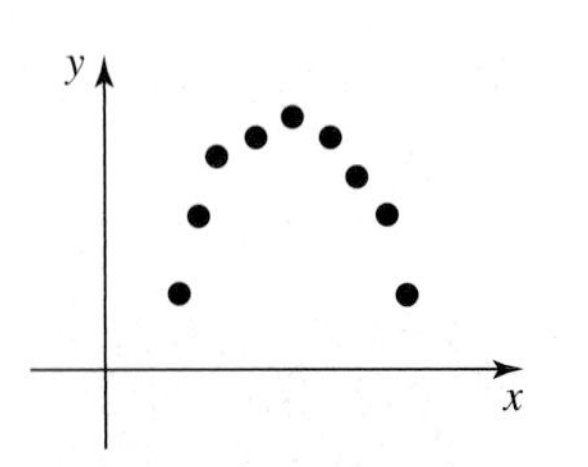

Figure 6 Scattergram for situation 2

Solution

It appears that the data points for situation 1 lie close to a line, so a linear model would describe situation 1 well. The data points for situation 2 do not lie close to any one line; a linear model would not describe situation 2. (In Chapters 6 and 7, we will discuss a type of nonlinear model that would describe situation 2 well.) The data points for situation 3 do not lie near a line; a linear model would not describe situation 3. ■

We create a scattergram of data to determine whether the relevant variables are approximately linearly related. If they are, we draw a line that comes close to the data points and use the line to make estimates and predictions.

WARNING

It is a common error to try to find a line that contains the greatest number of data points. Our goal is to find a line that comes close to *all* of the data points. For example, even though model 1 in Fig. 8 does not contain any of the data points shown, it fits the complete set of data points much better than does model 2, which contains three data points.

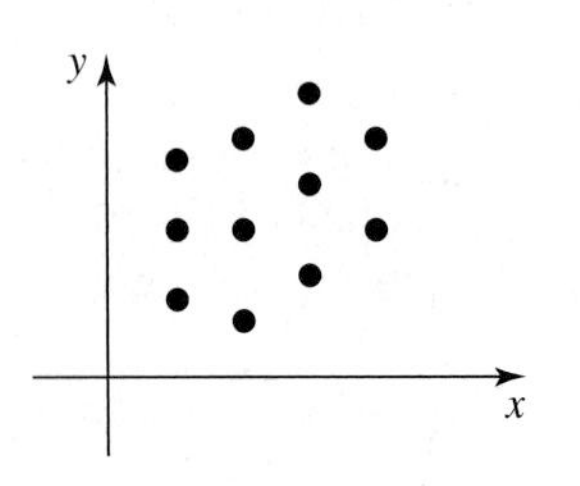

Figure 7 Scattergram for situation 3

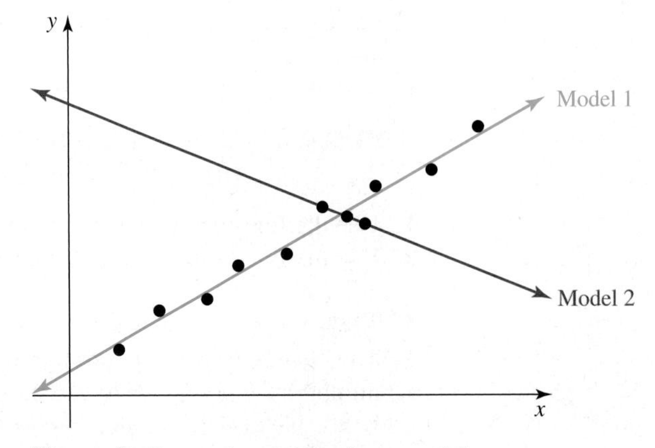

Figure 8 Comparing the fit of two models

Intercepts of a Model and Model Breakdown

In Example 4, we will find the intercepts of a linear model. By doing so, we will see that even if a model describes *known* data well, the model may not describe the situation well for all values of the independent variable.

Example 4 Intercepts of a Model; Model Breakdown

The wild Pacific Northwest salmon populations are listed in Table 4 for various years.

Table 4 Wild Pacific Northwest Salmon Populations

Year*	Population (millions)
1960	10.02
1965	10.00
1970	7.61
1975	3.15
1980	4.59
1985	3.11
1990	2.22

Source: *Golden Gate Anglers' Club*
*For convenience, we assume throughout this text that, unless otherwise stated, all yearly data were collected on December 31 of the given year.

1. Let P be the salmon population (in millions) at t years since 1950. Find a linear model that describes the relationship between t and P.
2. Find the P-intercept of the model. What does it mean in this situation?
3. Use the model to predict when the salmon will become extinct.

Solution

1. We describe the data in terms of t and P in Table 5. Next, we sketch a scattergram (see Fig. 9). It appears that t and P are approximately linearly related, so we sketch a line that comes close to the data points.

Table 5 Values of t and P for the Wild Pacific Northwest Salmon Data

Number of Years Since 1950 t	Salmon Population (millions) P
10	10.02
15	10.00
20	7.61
25	3.15
30	4.59
35	3.11
40	2.22

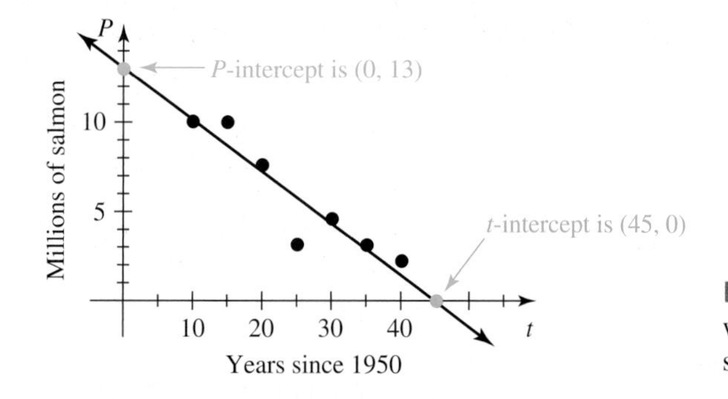

Figure 9 Intercepts of the wild Pacific Northwest salmon model

2. The P-intercept is (0, 13), or $P = 13$ when $t = 0$ (the year 1950). According to the model, there were 13 million salmon in 1950.
3. The t-intercept is (45, 0), or $P = 0$ when $t = 45$. According to the model, the salmon became extinct in $1950 + 45 = 1995$. Fortunately, this did not happen. There are wild Pacific Northwest salmon alive today, so our model gives a false prediction. ■

To draw the salmon model in Fig. 9, we used a scattergram consisting of data points representing various years from 1960 to 1990. In Fig. 10, we draw that portion of the model in blue, and we draw the rest of the model in red. When we use the blue portion of the model to make estimates, we are performing *interpolation*. When we use the red portions of the model, we are performing *extrapolation*.

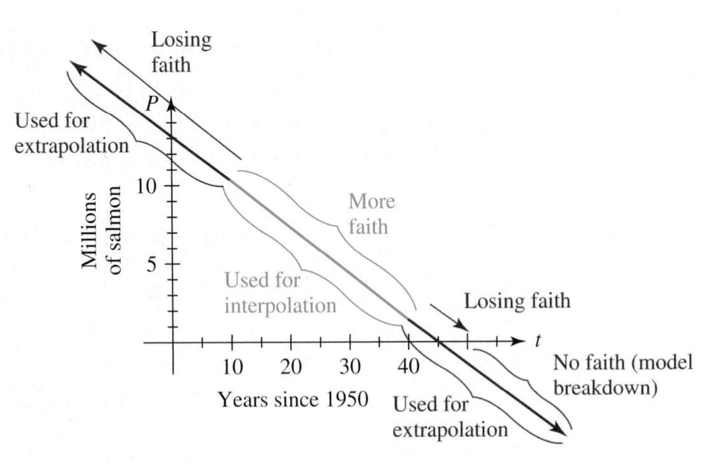

Figure 10 Interpolation versus extrapolation

> **DEFINITION** Interpolation, extrapolation
>
> For a situation that can be modeled by a function whose independent variable is t: We perform **interpolation** when we use a part of the model whose t-coordinates are between the t-coordinates of two data points. We perform **extrapolation** when we use a part of the model whose t-coordinates are not between the t-coordinates of any two data points.

Although we could get large errors from interpolating with the salmon model, we have more faith in our results from interpolating than from extrapolating. That's because the blue portion of the model comes close to several known data points, whereas we have no idea whether the red portion of the model comes close to *any* data points.

When we extrapolate, our faith declines more and more as we stray farther and farther from the blue portion of the salmon model. In fact, we have no faith in the portion of the model from $1950 + 45 = 1995$ on, because the model predicts nonpositive salmon populations for these years. We say that *model breakdown* has occurred from 1995 on.

> **DEFINITION** Model breakdown
>
> When a model gives a prediction that does not make sense or an estimate that is not a good approximation, we say that **model breakdown** has occurred.

When model breakdown occurs, it is time to modify our model or possibly rethink our modeling process. A different model might give more reasonable predictions. It could be helpful to gather more data to check our choice of model.

Example 5 Modifying a Model

In 2002, there were 3 million wild Pacific Northwest salmon. For each of the scenarios that follow, use the data for 2002 and the data in Table 6 to sketch a model. Let P be the wild Pacific Northwest salmon population (in millions) at t years since 1950.

Table 6 Wild Pacific Northwest Salmon Populations

Year	Population (millions)
1960	10.02
1965	10.00
1970	7.61
1975	3.15
1980	4.59
1985	3.11
1990	2.22

1. The salmon population levels off at 10 million.
2. The salmon become extinct.

Solution

1. We represent the 3 million salmon in 2002 by the ordered pair (52, 3), because $2002 - 1950 = 52$. We plot this data point in addition to the data points in our original scattergram and then sketch an appropriate model (see Fig. 11).
2. See Fig. 12.

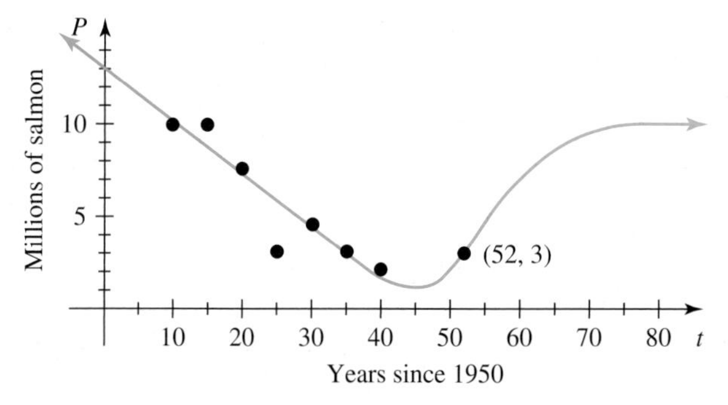

Figure 11 Population levels off at 10 million

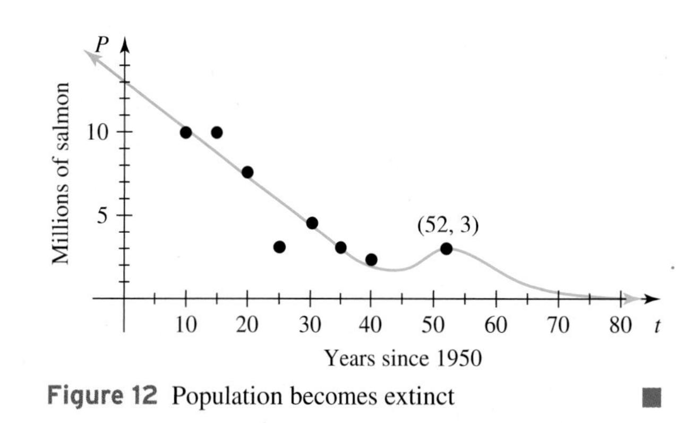

Figure 12 Population becomes extinct

group exploration

How defining the independent variable affects a model

Here you will compare two salmon models that have different definitions for the independent variable t.

Table 7 Wild Pacific Northwest Salmon Populations

Year	Population (millions)
1960	10.02
1965	10.00
1970	7.61
1975	3.15
1980	4.59
1985	3.11
1990	2.22

1. Let P be the wild Pacific Northwest salmon population (in millions) at t years *since 1960* (see Table 7). *Carefully* sketch a scattergram of the salmon data.
2. For the scattergram in Fig. 9 (page 59), the variable t is defined to be the number of years *since 1950*. Carefully sketch a line that comes close to the points in your scattergram from Problem 1 in much the same way as the line in Fig. 9 comes close to the points in Fig. 9.
3. Find each result both by using your linear model from Problem 2 and by using the linear model in Fig. 9.
   - **a.** Find the P-intercept of the linear model.
   - **b.** Find the t-intercept.
   - **c.** Predict the salmon population in 1988.
   - **d.** Predict when there were 9 million salmon.
   - **e.** Find the slope of the linear model.
4. For each part of Problem 3, which of your results are the same? Explain why this makes sense in terms of the two definitions of t.
5. For each part of Problem 3, which of your results are different? Explain why this makes sense in terms of the two definitions of t.

group exploration

Identifying types of modeling errors

Here you will explore possible causes of error for predictions based on a linear model for the Pacific salmon data (see Table 8).

Table 8 Wild Pacific Northwest Salmon Populations

Year	Population (millions)
1960	10.02
1965	10.00
1970	7.61
1975	3.15
1980	4.59
1985	3.11
1990	2.22

1. Let P be the wild Pacific Northwest salmon population (in millions) at t years since 1950. Sketch a scattergram of the salmon data.
2. Sketch a line that comes close to the data points.
3. Use your linear model to estimate the salmon population in 1975. What is the actual number of salmon? Calculate the error in your estimate for 1975. (The *error* is the difference between the estimated value and the actual value.) Why is the error so great?
4. Use your linear model to predict the salmon population in the year 2020. Is this an accurate prediction? If not, why is the error so great?

5. Take another look at your sketch from Problem 2. Is the t-axis perfectly horizontal and the P-axis perfectly vertical? Are the scalings of both axes precise? Is your line straight? How might these considerations relate to the accuracy of a prediction? Explain.

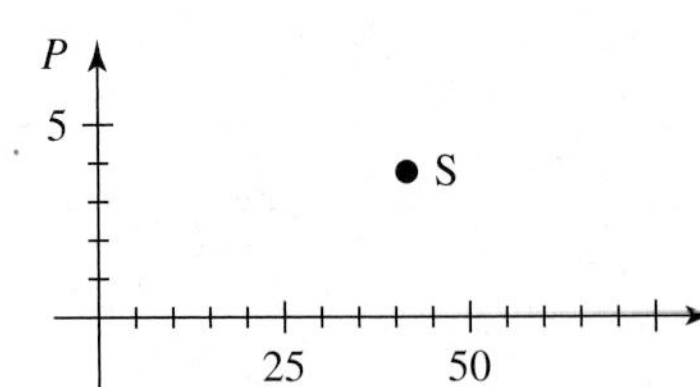

Figure 13 Problem 6

6. What are the coordinates of point S plotted in Fig. 13? Do you think you have found the correct first decimal place (tenths place) for these coordinates? How about the second decimal place?
7. Problems 3–6 of this exploration suggest several possible causes of error for predictions based on a linear model. Describe the possible causes of error.

TIPS FOR SUCCESS: Get in Touch with Classmates

Exchange phone numbers and e-mail addresses with some classmates so that, if you must miss class, then you can contact someone to find out what you missed and what homework was assigned.

HOMEWORK 2.1

FOR EXTRA HELP ▶

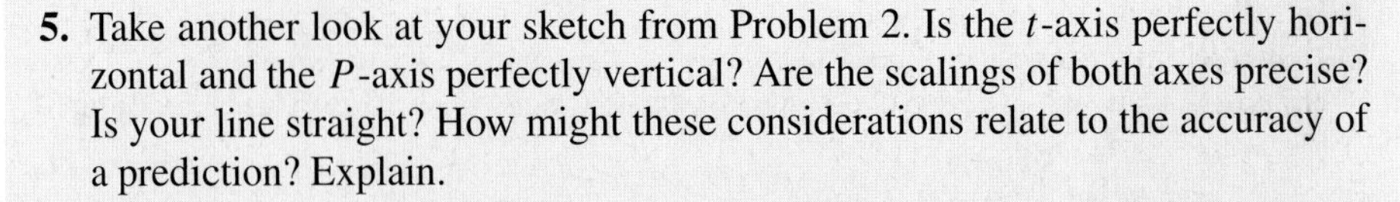

1. The percentages of dentistry degrees earned by women are shown in Table 9 for various years.

Table 9 Percentage of Dentistry Degrees Earned by Women

Year	Percent
1970	1
1980	13
1990	31
2000	40
2002	39

Source: *U.S. National Center for Education Statistics*

a. Let p be the percentage of dentistry degrees earned by women at t years since 1970. For example, $t = 0$ represents 1970 and $t = 10$ represents 1980. Create a scattergram of the data (see Fig. 14).

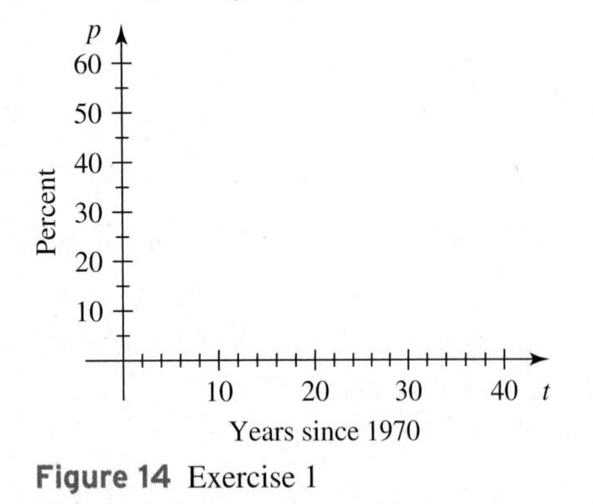

Figure 14 Exercise 1

b. Draw a line that comes close to the points in your scattergram.
c. Predict the percentage of dentistry degrees that will be earned by women in 2010. [**Note:** When responding to exercises in this text, always use sentences, not phrases.]
d. Estimate when women earned 20% of dentistry degrees.

2. The percentages of Americans who sent their tax returns to the Internal Revenue Service (IRS) electronically are shown in Table 10 for various years.

Table 10 Percentages of Tax Returns Sent Electronically

Year	Percent
1998	19
2000	27
2002	36
2004	45
2006	55

Source: *IRS*

a. Let p be the percentage of Americans who sent their tax returns to the IRS electronically at t years since 1995. For example, $t = 0$ represents 1995 and $t = 2$ represents 1997. Create a scattergram of the data (see Fig. 15).

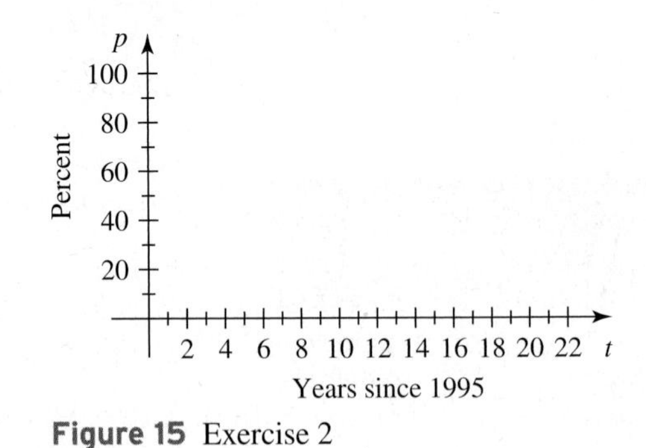

Figure 15 Exercise 2

b. Draw a line that comes close to the points in your scattergram.
c. Estimate when half of Americans sent returns electronically. [**Note:** When responding to exercises in this text, always use sentences, not phrases.]

d. By a 1998 law, the IRS must motivate 80% of taxpayers to send their returns electronically by 2007. Does your model predict that this goal will be reached by then? Explain.

e. In 2005, an independent oversight board recommended giving the IRS until 2011 to reach the 80% goal. Will this goal be reached by then? Explain.

3. Due to improved technology and public service campaigns, the number of collisions at highway–railroad crossings per year has declined since 1986 (see Table 11).

Table 11 Numbers of Collisions at Highway-Railroad Crossings

Year	Number of Collisions (thousands)
1986	6.5
1990	5.8
1994	5.0
1998	3.5
2002	3.1
2003	3.0

Source: *Federal Highway Association*

Let n be the number of collisions (in thousands) for the year that is t years since 1980.

a. Create a scattergram of the data. Extend your t-axis and n-axis so that you can make estimates about numbers of collisions before 1986 and predictions about future numbers of collisions. [**Note:** Include scaling on both axes and labels indicating the variable names and the scale units.]

b. Draw a line that comes close to the data points in your scattergram.

c. Use your linear model to estimate the number of collisions in 1997. Did you perform interpolation or extrapolation? Explain.

d. Use your linear model to predict in which year there will be 1.0 thousand collisions. Did you perform interpolation or extrapolation? Explain.

e. Find the n-intercept of your linear model. What does it mean in this situation?

f. Find the t-intercept. What does it mean in this situation? [**Note:** If model breakdown occurs, say so, say where, and explain why.]

4. Repeat Exercise 3, but let n be the number of collisions (in thousands) for the year that is t years *since 1986*. Which of your responses for this exercise are the same as those for Exercise 3? Explain why it makes sense that these responses are the same. Explain why it makes sense that the other responses are different.

5. The numbers of Internet users in the United States are shown in Table 12 for various years.

Table 12 Numbers of Internet Users in the United States

Year	Number of Users (millions)
1996	39
1997	60
1998	84
1999	105
2000	122
2001	143
2002	166
2003	183
2004	207

Source: *Jupiter MMXI*

a. Let n be the number of Internet users (in millions) in the United States at t years since 1990. Create a scattergram of the data.

b. Sketch a line that comes close to the data points.

c. Find the t-intercept of your linear model. What does it mean in this situation? Did you perform interpolation or extrapolation? Explain.

d. Estimate by how much the number of Internet users is increasing per year.

e. Use your linear model to predict when everyone in the United States will be an Internet user. Assume that the U.S. population is 301 million. Did you perform interpolation or extrapolation? Explain.

6. The temperature at which water boils (the *boiling point*) depends on elevation: The higher the elevation, the lower is the boiling point. At sea level, water boils at 212°F; at an elevation of 10,000 meters, water boils at about 151°F. Boiling points are listed in Table 13 for various elevations.

Table 13 Boiling Points of Water

Elevation (in thousands of meters)	Boiling Point (°F)
0	212
1	205
2	200
5	181
10	151
15	123

a. Let B be the boiling point (in degrees Fahrenheit) at an elevation of E thousand meters. Create a scattergram of the data.

b. Sketch a line that comes close to the data points.

c. Mount Everest, the highest mountain in the world, reaches 8850 meters at its peak. What is the boiling point of water at the peak? Did you perform interpolation or extrapolation? Explain.

d. We say that water is *lukewarm* if its temperature is close to body temperature (about 98.6°F). At what elevation would boiling water feel lukewarm? Did you perform interpolation or extrapolation? Explain.

e. The cooking time required to make a hard-boiled egg depends on the water temperature. Let T be the amount of time (in minutes) it takes to cook an egg in boiling water at an elevation of E thousand meters. Sketch a *qualitative* graph that describes the relationship between E and T.

7. In Example 1 of Section 1.1, we sketched a qualitative graph that described the relationship between the prices of Air Jordan shoes and years since 1985. Now you will use the data in Table 14 to make some predictions. (Assume that all prices quoted are suggested retail prices.)

Table 14 Prices of Air Jordans

Name of Shoe	Year	Price per Pair (dollars)
Air Jordan I	1985	85.00
Air Jordan III	1988	99.50
Air Jordan V	1990	113.50
Air Jordan VIII	1993	135.00
Air Jordan XVII	2002	200.00

Source: *Consumer Reports*

a. Let p be the price (in dollars) of Air Jordans at t years since 1985. Create a scattergram of the data.

b. Sketch a line that comes close to the data points.

c. Use your line to estimate the price of the Air Jordan XIV in 1999. The actual price was \$150. By how much did the model overestimate the price?

d. Use your line to estimate the price of the Air Jordan XVI in 2001. The actual price was \$160. By how much did the model overestimate the price?

e. Michael Jordan was retired when both the Air Jordan XIV and Air Jordan XVI were released. He returned from retirement before the Air Jordan XVII was released. Give a possible explanation for why your estimates in parts (c) and (d) were overestimates.

f. On the same axes that you used to sketch your linear model in part (b), sketch a curve that might have described the price of the shoes if Michael Jordan had never returned from the retirement described in part (e). [**Hint:** Your new curve will not be a linear model.]

8. Most scientists agree that Earth's average surface temperature has increased by about 1°F over the past 120 years and that the warming is due to carbon emissions from the burning of fossil fuels (see Table 15).

Table 15 Carbon Emissions from Fossil-Fuel Burning

Year	Carbon Emissions (billions of tons)
1950	1.6
1960	2.6
1970	4.1
1980	5.3
1990	6.1

Source: *UN, BP, DOE, IEA*

Let c be the carbon emissions (in billions of tons) at t years since 1950.

a. Sketch a scattergram of the carbon-emission data.

b. Sketch a line that comes close to the data points.

c. Use your line to estimate the carbon emissions in 2000.

d. The actual amount of carbon emissions in 2000 was 6.3 billion tons. Is this amount less than, equal to, or greater than your estimate in part (c)? What does this comparison suggest may be starting to happen? On the same set of axes that you used in part (b), sketch a curve that describes what might be starting to happen. [**Hint:** Your new curve will not be a linear model.]

9. The percentages of Americans who are satisfied with the way things are in the United States are shown in Table 16 for various years.

Table 16 Percentages of Americans Who Are Satisfied

Year	Percent
1992	21
1993	28
1994	33
1995	32
1996	39
1997	49
1998	60
1999	59

Source: *The Gallup Organization*

Let p be the percentage of Americans at t years since 1990 who are satisfied with the way things are.

a. Create a scattergram of the data in Table 16.

b. Draw a line that comes close to the data points.

c. Use your line to estimate the percentage of Americans who were satisfied in 2004.

d. Data for the years 2000–2004 are shown in Table 17. Create a scattergram of the data for the years 1992–2004.

Table 17 Percentages of Americans Who Are Satisfied

Year	Percent
2000	60
2001	56
2002	52
2003	46
2004	43

Source: *The Gallup Organization*

e. Compute the error in the estimation for 2004 that you made in part (c). (The *error* is the difference between the estimated percentage and the actual percentage.) Explain why the error in your estimate is so large.

10. The percentages of Americans living below the poverty level are shown in Table 18 for various years.

Table 18 Percentages of Americans Living Below the Poverty Level

Year	Percent
1993	15.1
1994	14.5
1995	13.8
1996	13.7
1997	13.3
1998	12.7
1999	11.9
2000	11.3

Source: *U.S. Census Bureau*

Let p be the percentage of Americans living below the poverty level at t years since 1990.

a. Create a scattergram of the data in Table 18.

b. Draw a line that comes close to the data points.

c. Use your line to estimate the percentage in 2004.

d. Data for the years 2001–2004 are shown in Table 19. Create a scattergram of the data for the years 1993–2004.

Table 19 Percentages of Americans Living Below the Poverty Level

Year	Percent
2001	11.7
2002	12.1
2003	12.5
2004	12.7

Source: *U.S. Census Bureau*

e. Compute the error in the estimation for 2004 that you made in part (c). (The *error* is the difference between the estimated percentage and the actual percentage.) Explain why the absolute value of the error in your estimate is so large.

11. Table 20 describes the difference in the percentages of sound recording sales between pop and rap/hip-hop recordings for various years. Pop recordings had greater sales in the early 1990s, but the opposite was true by the late 1990s.

Table 20 Percentage Differences in Sound Recordings

Year	Difference in Percentages
1990	5.2
1994	2.4
1996	0.4
1999	−0.5
2000	−1.9

Source: *Recording Industry Association of America, Inc.*

Let D be the difference in the percentages at t years since 1990.

a. Without graphing, estimate the coordinates of the t-intercept for a line that comes close to the data points. What does that point mean in this situation? If you don't see how to estimate the coordinates, create a scattergram of the data first.

b. Without graphing, estimate the coordinates of the D-intercept for a line that comes close to the data points. What does that point mean in this situation? If you don't see how to estimate the coordinates, create a scattergram of the data first.

12. The *windchill* (or *windchill factor*) is a measure of how cold you feel as a result of being exposed to wind. Table 21 provides some data on windchills for various temperatures when the wind speed is 10 mph.

Table 21 Windchills for a 10-mph Wind

Temperature (°F)	Windchill (°F)
−15	−35
−10	−28
−5	−22
5	−10
10	−4
15	3
20	9
25	15

Source: *National Weather Service Forecast Office*

Let w be the windchill (in degrees Fahrenheit) corresponding to a temperature of t degrees Fahrenheit when the wind speed is 10 mph.

a. Without graphing, estimate the coordinates of the t-intercept for a line that comes close to the data points. What does that point mean in this situation? If you don't see how to estimate the coordinates, create a scattergram of the data first.

b. Without graphing, estimate the coordinates of the w-intercept for a line that comes close to the data points. What does that point mean in this situation? If you don't see how to estimate the coordinates, create a scattergram of the data first.

13. Describe the meaning of a linear function and a linear model. Is a linear function necessarily a model? Is a linear model necessarily a function? (See page 4 for guidelines on writing a good response.)

14. When using a line to model a situation, do we usually have more faith in a result obtained by interpolation or extrapolation? Explain. (See page 4 for guidelines on writing a good response.)

15. Which is more desirable, finding a linear model whose graph contains several, but not all, data points or finding a linear model whose graph does not contain any data points but comes close to all data points? Include some sketches of scattergrams and linear models.

16. Describe how to find a linear model for a situation and how to use the model to make estimates and predictions.

2.2 FINDING EQUATIONS OF LINEAR MODELS

Objectives

- Find an equation of a linear model by using data described in words.
- Find an equation of a linear model by using data displayed in a table.

In Section 2.1, we used graphs of linear functions to model data. In this section, we use *equations* of linear functions to model data.

Finding an Equation of a Linear Model by Using Data Described in Words

In Example 1, we will use data described in words to find an equation of a model.

Example 1 Finding an Equation of a Linear Model

The average number of credit card offers a household receives in one month has increased approximately linearly from 5.1 offers in 2002 to 5.9 offers in 2005 (Source: *Synovate*). Let n be the average number of credit card offers a household receives in one month at t years since 2000. Find an equation of a linear model.

Solution

Known values of t and n are shown in Table 22.

Table 22 Known Values of t and n

Years Since 2000 t	Number of Offers per Month n
2	5.1
5	5.9

A linear function can be put into the form $y = mx + b$, where y depends on x. Since t and n are approximately linearly related and n depends on t, we will find an equation of the form $n = mt + b$.

First, we use the data points (2, 5.1) and (5, 5.9) to find the slope of the model:

$$m = \frac{5.9 - 5.1}{5 - 2} = \frac{0.8}{3} \approx 0.27$$

So, we can substitute 0.27 for m in the equation $n = mt + b$:

$$n = 0.27t + b$$

Next, we can find the constant b by substituting the coordinates of the point (2, 5.1) into the equation $n = 0.27t + b$ and then solving for b:

$$5.1 = 0.27(2) + b \quad \text{Substitute 2 for } t \text{ and 5.1 for } n.$$
$$5.1 = 0.54 + b \quad \text{Multiply.}$$
$$5.1 - 0.54 = 0.54 + b - 0.54 \quad \text{Subtract 0.54 from both sides.}$$
$$4.56 = b \quad \text{Combine like terms.}$$

Now we can substitute 4.56 for b in the equation $n = 0.27t + b$:

$$n = 0.27t + 4.56$$

We verify our equation by using TRACE on a graphing calculator to check that our line approximately contains the points (2, 5.1) and (5, 5.9). See Fig. 16.

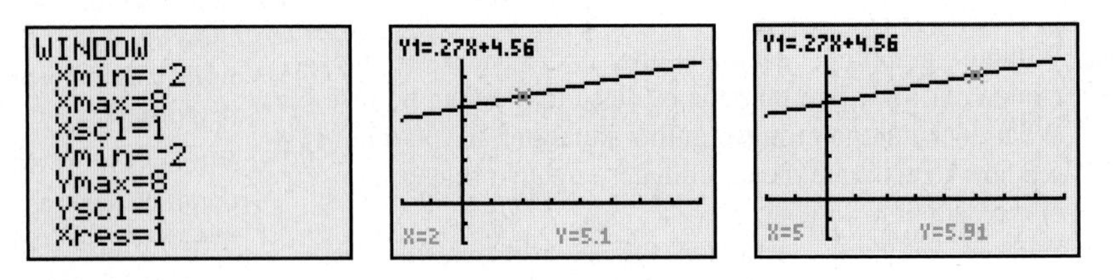

Figure 16 Checking that the model approximately contains both (2, 5.1) and (5, 5.9)

■

Finding an Equation of a Linear Model by Using Data Displayed in a Table

In Example 2, we will use data shown in a table to find an equation of a linear model.

Example 2 Finding an Equation of a Linear Model

During the early 1990s, there was great consumer demand for food products claiming to be "low fat" or "no fat." Since then, this demand has declined greatly. Table 23 shows the percentages of new products that claimed to be "low fat" or "no fat" from 1996 to 2001.

Table 23 Percentages of New Products Claiming to Be "Low Fat" or "No Fat"

Year	Percent
1996	29
1997	26
1998	22
1999	17
2000	16
2001	11

Source: *Productscan Online*

Let p be the percentage of new food products claiming to be "low fat" or "no fat" at t years since 1995. Find an equation of a line that comes close to the points in the scattergram of the data.

Solution

We begin by viewing the positions of the points in the scattergram (see Fig. 17). To save time and improve accuracy in plotting points, we can use a graphing calculator to view a scattergram of the data (see Fig. 18). For graphing calculator instructions, see Section B.8.

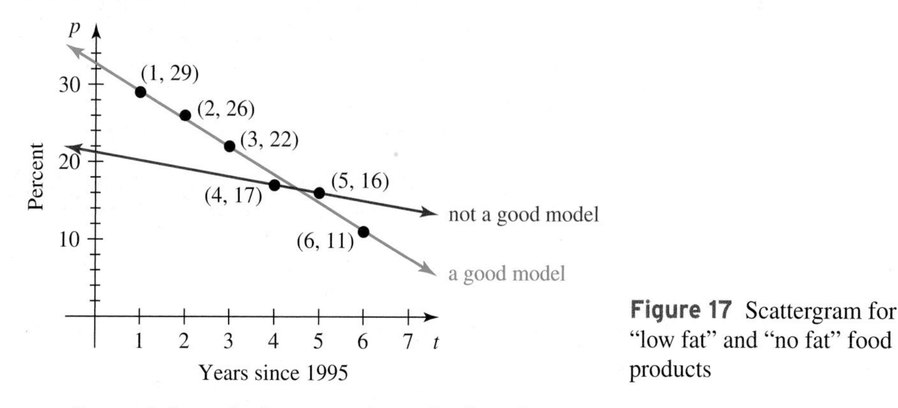

Figure 17 Scattergram for "low fat" and "no fat" food products

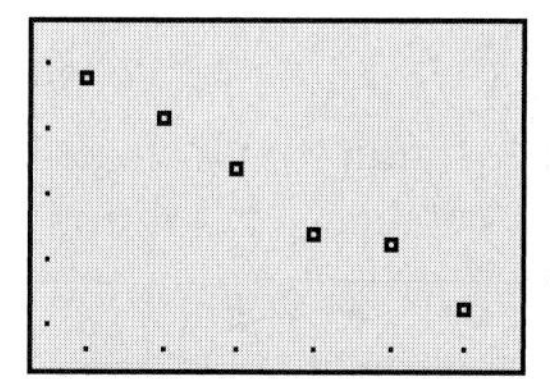

Figure 18 Graphing calculator scattergram

Our task is to find an equation of a line that comes close to the data points. It is not necessary to use two *data* points to find an equation, although it is often convenient and satisfactory to do so.

The red line that contains points (4, 17) and (5, 16) does *not* come close to the other data points (see Fig. 17). However, the green line that passes through points (1, 29) and (3, 22) appears to come close to the rest of the points. We will find the equation of this line.

For an equation of the form $p = mt + b$, we first use the points (1, 29) and (3, 22) to find m (see Fig. 19):

$$m = \frac{22 - 29}{3 - 1} = \frac{-7}{2} = -3.5$$

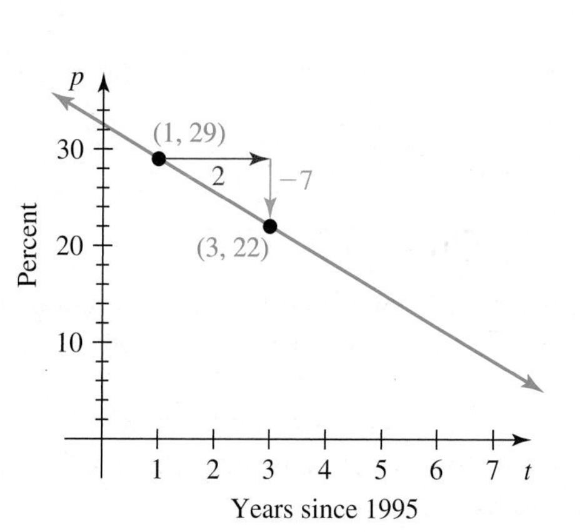

Figure 19 Two points of the "low-fat" scattergram

Then we substitute -3.5 for m in the equation $p = mt + b$:

$$p = -3.5t + b$$

To find the constant b, we substitute the coordinates of the point $(1, 29)$ into the equation $p = -3.5t + b$ and then solve for b:

$$\begin{aligned} 29 &= -3.5(1) + b && \text{Substitute 1 for } t \text{ and 29 for } p. \\ 29 &= -3.5 + b && \text{Multiply.} \\ 29 + 3.5 &= -3.5 + b + 3.5 && \text{Add 3.5 to both sides.} \\ 32.5 &= b && \text{Simplify.} \end{aligned}$$

Now we substitute 32.5 for b in the equation $p = -3.5t + b$:

$$p = -3.5t + 32.5$$

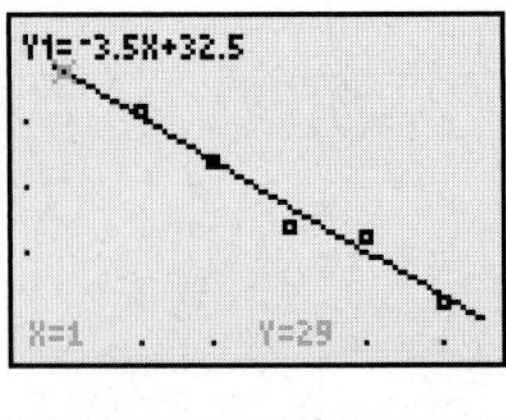

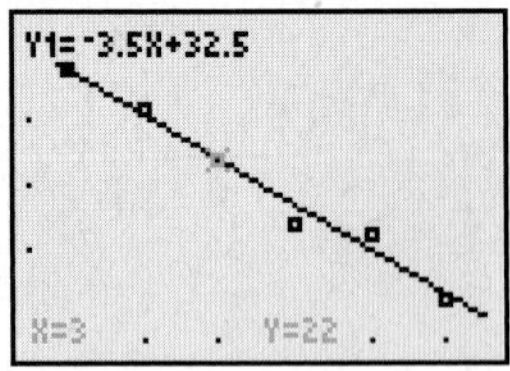

Figure 20 Checking that the model contains both $(1, 29)$ and $(3, 22)$

We can check the correctness of our equation by using a graphing calculator to verify that our line contains the points $(1, 29)$ and $(3, 22)$. See Fig. 20. For graphing calculator instructions, see Sections B.10 and B.11. ■

WARNING It is a common error to skip creating a scattergram when we find an equation of a model. However, we benefit in many ways by viewing a scattergram of data. First, we can determine whether the data are approximately linearly related. Second, if the data are approximately linearly related, viewing a scattergram helps us choose two good points with which to find an equation of a linear model. Third, by graphing the model with the scattergram, we can assess whether the model fits the data reasonably well.

Example 2 took four steps to find an equation of a linear model.

Finding an Equation of a Linear Model

To find an equation of a linear model, given some data,

1. Create a scattergram of the data.
2. Determine whether there is a line that comes close to the data points. If so, choose two points (not necessarily data points) that you can use to find the equation of a linear model.
3. Find an equation of the line you identified in step 2.
4. Use a graphing calculator to verify that the graph of your equation comes close to the points of the scattergram.

What should you do if you discover that a model does not fit a data set well? A good first step is to check for graphing or calculation errors. If your work appears to be correct, then try using different points to derive your equation, or increase or decrease the slope m and/or the constant term b of your equation $y = mx + b$ until the fit is good.

In Example 2, we used the linear equation $p = -3.5t + 32.5$ to describe the linear "low-fat" model. Depending on what aspect of the model we want to emphasize, we can refer to the model as the "low-fat model," "low-fat function," or "equation $p = -3.5t + 32.5$."

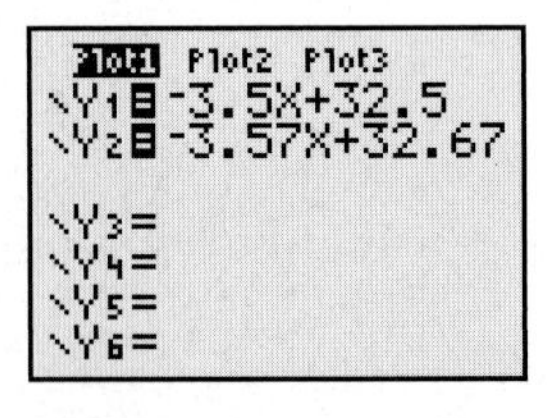

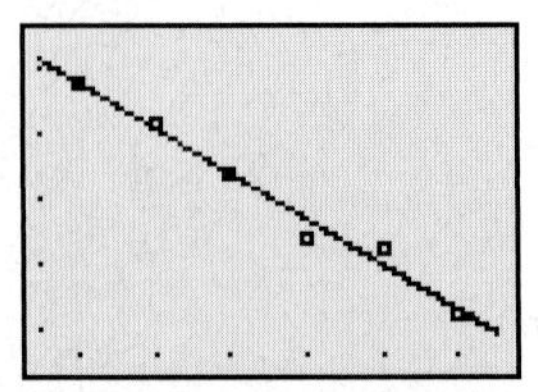

Figure 21 Comparing two "low-fat" models' graphs that are so close together that there appears to be only one line

In Example 2, we used two points to find the model $p = -3.5t + 32.5$. There is another way to find a model. Most graphing calculators have a built-in **linear regression** feature for finding an equation of a linear model. Linear regression gives the equation $p = -3.57t + 32.67$. In Fig. 21, we see that both models fit the data well. For graphing calculator instructions, see Section B.16.

A linear equation found by linear regression is called a **linear regression equation,** and the function described by the equation is called a **linear regression function.** The graph is called a **regression line.** (You can learn more about linear regression in a statistics course.)

Now you have two ways to find the equation of a linear model. No matter which method you use, the objective is the same: Find an equation of a line that comes close to the data points.

Table 24 Percentages of American Adults Who Smoke

Year	Percent Who Smoke
1970	37.4
1980	33.2
1990	25.3
2000	23.1
2005	19.0

Source: *National Center for Health Statistics*

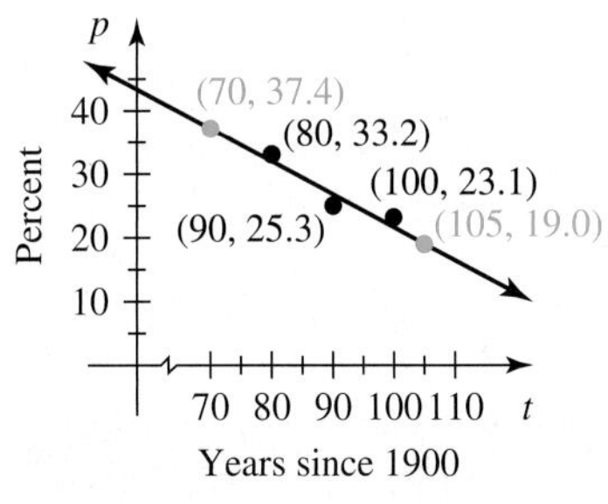

Figure 22 Smoking scattergram and linear model

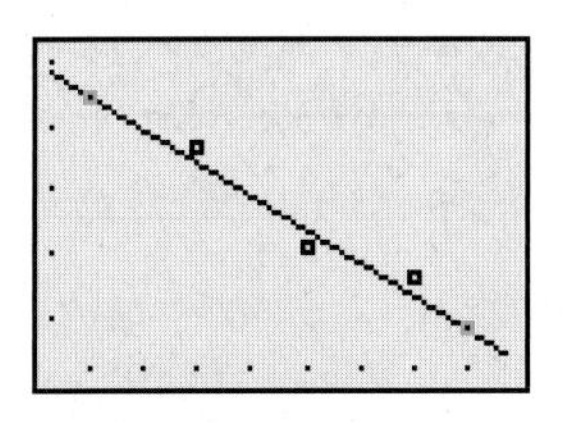

Figure 23 Verifying the smoking model

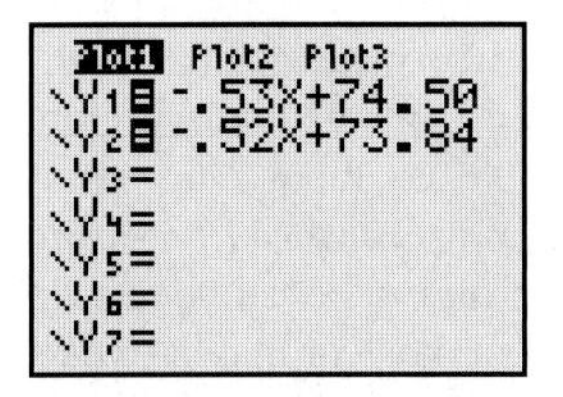

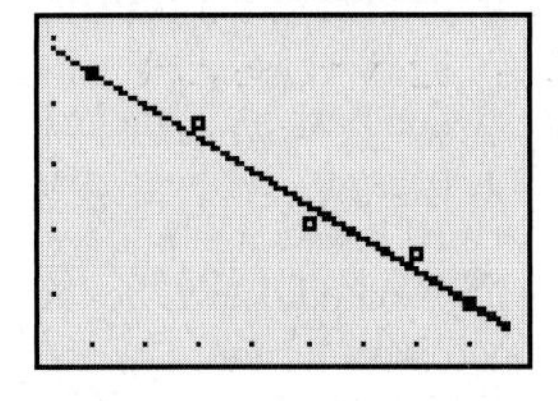

Figure 24 Comparing two smoking models' graphs that are so close together that there appears to be only one line

Example 3 Finding the Equation of a Linear Model

Cigarette smoking has been on the decline for the past several decades (see Table 24). Let p be the percentage of American adults who smoke at t years since 1900.

1. Use two well-chosen points to find an equation of a model that describes the relationship between t and p.
2. Find the linear regression equation and line by using a graphing calculator. Compare this model with the one you found in Problem 1.

Solution

1. We see from the scattergram in Fig. 22 that a line containing the points (70, 37.4) and (105, 19.0) comes close to the rest of the data points. The zigzag lines on the t-axis near the origin indicate that part of the scale is missing. This is done to show a clearer view of the data points.

   For an equation of the form $p = mt + b$, we first use the points (70, 37.4) and (105, 19.0) to find m:

   $$m = \frac{19.0 - 37.4}{105 - 70} = -0.53$$

   So, the equation has the form

   $$p = -0.53t + b$$

   To find b, we use the point (70, 37.4) and substitute 70 for t and 37.4 for p in the equation $p = -0.53t + b$:

   $$37.4 = -0.53(70) + b \qquad \text{Substitute 70 for } t \text{ and 37.4 for } p.$$
   $$37.4 = -37.10 + b \qquad \text{Multiply.}$$
   $$37.4 + 37.1 = -37.10 + b + 37.1 \qquad \text{Add 37.1 to both sides.}$$
   $$74.50 = b \qquad \text{Simplify.}$$

   So, the equation is $p = -0.53t + 74.50$.

   We can use a graphing calculator to verify that the linear model contains the points (70, 37.4) and (105, 19.0) and comes close to the other data points (see Fig. 23).
2. The regression equation is $p = -0.52t + 73.84$; it is "close" to the equation $p = -0.53t + 74.50$. Further, both models appear to fit the data well (see Fig. 24).

■

When the graph of a function is increasing, we say that the function is **increasing.** When the graph of a function is decreasing, we say that the function is **decreasing.** So, the smoking function $p = -0.53t + 74.50$ is decreasing.

group exploration

Choosing "good points" to find a model

The wild Pacific Northwest salmon data are listed in Table 25. The table includes a first column that indicates a name for each data point. For example, point D refers to the point (25, 3.15).

Table 25 Values of t and P for Wild Pacific Northwest Salmon

Name of Point	Years Since 1950 t	Population (millions) P
A	10	10.02
B	15	10.00
C	20	7.61
D	25	3.15
E	30	4.59
F	35	3.11
G	40	2.22

1. Find an equation of the line that contains points A and B.
2. Use a graphing calculator to verify that the graph of your equation passes through both points A and B. Does the line come close to the other data points?
3. If you had used points A and F, you would have found the equation $P = -0.28t + 12.78$. Compare its graph with the graph you drew in Problem 2. Explain why the graphs look so different.
4. List all pairs of points that yield equations that you think would be good linear models. (You do not have to find the equations.)
5. Several pairs of points from a scattergram yield equations that could serve as models of the data. Discuss how to choose two such data points to find an equation that comes reasonably close to all the data points.
6. It is not necessary to use data points to find an equation of a linear model. While viewing the salmon scattergram, use the arrow keys on a graphing calculator to identify two nondata points that you feel would yield an equation of a line that is close to the data points. For graphing calculator instructions, see Section B.23. Find an equation of the line that contains these two points. Then, use a graphing calculator to verify that the graph of your equation comes close to the data points.

TIPS FOR SUCCESS: Verify Your Work

Use a graphing calculator to verify your work. In this section, for example, you can use a graphing calculator to check your equations. Checking your work increases your chances of catching errors and thus will likely improve your performance on homework assignments, quizzes, and tests.

HOMEWORK 2.2

FOR EXTRA HELP ▶

Student Solutions Manual | PH Math/Tutor Center | MathXL® | MyMathLab

1. The number of classified documents has increased approximately linearly from 8.7 million documents in 2001 to 15.6 million documents in 2004 (Source: *OpenTheGovernment.Org*). Let n be the number of documents (in millions) labeled as classified in the year that is t years since 2000. Find an equation of a linear model to describe the data.
2. The number of unruly airline passengers per year has increased approximately linearly from 184 passengers in 1996 to 303 passengers in 2004 (Source: *Federal Aviation Administration*). Let n be the number of unruly passengers in the year that is t years since 1990. Find an equation of a linear model to describe the data.
3. The number of rushing yards per year by Shaun Alexander of the Seattle Seahawks has increased approximately linearly from 1175 yards in 2002 to 1880 yards in 2005 (Source: *NFL.com*). Let r be Alexander's number of rushing yards in the year that is t years since 2000. Find an equation of a linear model to describe the data.
4. Due to a crackdown on fraudulent corporate reports, the number of restatements of facts in corporate reports has increased approximately linearly from 217 restatements in 1999 to 410 restatements in 2004 (Source: *Huron Consulting Group*). Let n be the number of restatements in the year that is t years since 1990. Find an equation of a linear model to describe the data.
5. The average minimum balance required for non-interest-bearing checking accounts decreased approximately linearly from \$440.00 in 2001 to \$206.33 in 2004 (Source: *Bankrate.com*). Let b be the average minimum balance (in dollars) for non-interest-bearing checking at t years since 2000. Find an equation of a linear model to describe the data.
6. The number of high school students who have taken the military's aptitude test has decreased approximately linearly from about 857.6 thousand students in 1999 to about 722.4 thousand students in 2004 (Source: *Military Entrance Processing Command*). Let n be the number of high school students (in thousands) who have taken the test in the year that is t years since 1990. Find an equation of a linear model to describe the data.
7. The average length of a 6-year-old striped bass is 27 inches, and that of a 20-year-old striped bass is 54.5 inches (Source: *daybreakfishing.com*). A striped bass's age and average length are approximately linearly related. Let L be the average length (in inches) of a striped bass at age a years. Find an equation of a linear model to describe the data.
8. For Pacific albacore tuna, the weight and mercury concentration are approximately linearly related. A 4-kilogram tuna has an average mercury concentration of 0.10 part per million. A 10-kilogram tuna has an average mercury concentration of 0.19 part per million. Let c be the average mercury concentration (in parts per million) of a Pacific albacore tuna that weighs w kilograms. Find an equation of a linear model to describe the data.

Consider the scattergram of data and the graph of the model $y = mx + b$ in the indicated figure. Sketch the graph of a linear *model that describes the data better. Then explain how you would adjust the values of m and b of the original model so that it would describe the data better.*

9. See Fig. 25.

Figure 25 Exercise 9

10. See Fig. 26.

Figure 26 Exercise 10

11. Find an equation of a line that comes close to the points listed in Table 26. Then use a graphing calculator to check that your line comes close to the points. [***Graphing Calculator:*** See Sections B.8 and B.10.]

Table 26 Finding a Linear Model

x	y
3	5
4	7
5	10
6	12
7	15

12. Find an equation of a line that comes close to the points listed in Table 27. Then use a graphing calculator to check that your line comes close to the points. [***Graphing Calculator:*** See Sections B.8 and B.10.]

Table 27 Finding a Linear Model

x	y
1	18
4	14
5	12
7	8
10	5

13. Three students are to find a linear model of the data in Table 28. Student A uses points (1, 5.9) and (2, 6.4), student B uses points (3, 9.0) and (4, 11.0), and student C uses points (5, 12.1) and (6, 15.5). Which student seems to have made the best choice of points? Explain.

Table 28 Three Students' Model Data

x	y
1	5.9
2	6.4
3	9.0
4	11.0
5	12.1
6	15.5
7	16.5

14. Three students are to find a linear model of the data in Table 29. Student A uses points (3, 13.8) and (4, 10.1), student B uses points (6, 7.8) and (7, 4.3), and student C uses points (5, 9.1) and (8, 3.1). Which student seems to have made the best choice of points? Explain.

Table 29 Three Students' Model Data

x	y
3	13.8
4	10.1
5	9.1
6	7.8
7	4.3
8	3.1
9	1.1

15. The percentages of births outside marriage in the United States are shown in Table 30 for various years.

Table 30 Births Outside Marriage

Year	Percent of Births Outside Marriage
1970	10.7
1975	14.3
1980	18.4
1985	22.0
1990	28.0
1995	32.2
2000	33.2
2002	34.0

Source: *National Center for Health Statistics*

Let p be the percentage of births outside marriage in the United States at t years since 1900.

a. Use a graphing calculator to draw a scattergram of the data.

b. Find an equation of a linear model to describe the data.

c. Draw your line and the scattergram in the same viewing window. Verify that the line passes through your two chosen points and that it comes close to all of the data points.

16. Repeat Exercise 15, but let p be the percentage of births outside marriage in the United States at t years *since 1970*. Compare the slope of your model with the slope of the model that you found in Exercise 15. Compare the p-intercepts. Explain why your comparisons make sense.

17. The prices of ski rental packages from Gold Medal Sports® are shown in Table 31 for various numbers of days.

Table 31 Prices of Ski Rental Packages

Number of Days	Price of Package (dollars)
1	15.00
2	30.00
3	45.00
4	56.00
5	70.00
6	78.00

Source: *Gold Medal Sports*

Let p be the price (in dollars) of a ski rental package for n days.

a. Use a graphing calculator to draw a scattergram of the data.

b. Find an equation of a linear model to describe the data.

c. Draw your line and the scattergram in the same viewing window. Verify that the line passes through the two points you chose in finding the equation in part (b) and that it comes close to all of the data points.

18. The life expectancies at birth of Americans are shown in Table 32 for various years.

Table 32 Life Expectancies at Birth

Year of Birth	Life Expectancy (years)
1980	73.7
1985	74.7
1990	75.4
1995	75.8
2000	77.0
2003	77.6

Source: *U.S. Census Bureau*

a. Let L be the life expectancy at birth (in years) of an American born t years after 1980. Use a graphing calculator to draw a scattergram of the data.

b. Find an equation of a linear model to describe the data.

c. Draw your line and the scattergram in the same viewing window. Verify that the line passes through the two points you chose in finding the equation in part (b) and that it comes close to all of the data points.

19. Table 33 shows, for various years, the percentage of Americans whose favorite sport is baseball.

Table 33 Percentages of Americans Whose Favorite Sport Is Baseball

Year	Percent
1994	19
1995	16
1997	14
1998	16
2000	13
2002	12
2004	10

Source: *The Gallup Organization*

Let p be the percentage of Americans whose favorite sport is baseball at t years since 1990.

a. Use a graphing calculator to draw a scattergram of the data.

b. Find an equation of a linear model to describe the data.

c. Draw your line and the scattergram in the same viewing window. Verify that the line passes through the two points you chose in finding the equation in part (b) and that it comes close to all of the data points.

20. The percentages of Americans who have been diagnosed with diabetes are shown in Table 34 for various age groups.

Table 34 Percentages of Americans Diagnosed with Diabetes, by Age Group

Age Group (years)	Age Used to Represent Age Group (years)	Percent
35–39	37	2
40–44	42	4
45–49	47	5
50–54	52	8
55–59	57	10
60–64	62	13
65–69	67	14

Source: *National Health Interview Survey*

Let p be the percentage of Americans at age a years who have been diagnosed with diabetes at some point in their lives.

a. Use a graphing calculator to draw a scattergram of the data.

b. Find an equation of a linear model to describe the data.

c. Draw your line and the scattergram in the same viewing window. Verify that the line passes through the two points you chose in finding the equation in part (b) and that it comes close to all of the data points.

d. Find the regression equation to describe the data.

e. Use a graphing calculator to graph the equations you found in parts (b) and (d) in the same viewing window. Compare the graphs.

21. Table 35 lists world record times for the women's 400-meter run. Let r be the record time (in seconds) at t years since 1900.

Table 35 Women's 400-Meter Run Record Times

Year	Runner	Country	Record Time (seconds)
1957	Marlene Mathews	Australia	57.0
1959	Maria Itkina	USSR	53.4
1962	Shin Geum Dan	North Korea	51.9
1969	Nicole Duclos	France	51.7
1972	Monika Zehrt	E. Germany	51.0
1976	Irena Szewinska	Poland	49.29
1979	Marita Koch	E. Germany	48.60
1983	Jarmila Kratochvílová	Chad	47.99
1985	Marita Koch	E. Germany	47.60

Source: *International Association of Athletics Federations*

a. Use a graphing calculator to draw a scattergram of the data.

b. Find an equation of a linear model to describe the data.

c. Draw your line and the scattergram in the same viewing window. Verify that the line passes through the two points you chose in finding the equation in part (b) and that it comes close to all of the data points.

22. Table 36 lists world record times for the men's 400-meter run.

Table 36 Men's 400-Meter Run Record Times

Year	Runner	Country	Record Time (seconds)
1900	Maxie Long	USA	47.8
1916	Ted Meredith	USA	47.4
1928	Emerson Spencer	USA	47.0
1932	Bill Carr	USA	46.2
1941	Graver Klemmer	USA	46.0
1950	George Rhoden	Jamaica	45.8
1960	Carl Kaufmann	Germany	44.9
1968	Lee Evans	USA	43.86
1988	Harry Reynolds	USA	43.29
1999	Michael Johnson	USA	43.18

Source: *International Association of Athletics Federations*

a. Let r be the record time (in seconds) at t years since 1900. Use a graphing calculator to draw a scattergram of the data.

b. Find an equation of a linear model to describe the data.

c. Draw your line and the scattergram in the same viewing window. Verify that the line passes through the two points you chose in finding the equation in part (b) and that it comes close to all of the data points.

23. In Exercises 21 and 22, you found equations for the women's and men's 400-meter run record times. Equations that model the data well are

$$r = -0.27t + 70.45 \quad \text{women's model}$$
$$r = -0.053t + 48.08 \quad \text{men's model}$$

where r represents the record time (in seconds) at t years since 1900.

a. Graph both models by hand for the years from 1900 to 2050. (If you are able to use a graphing calculator to do this exercise, you may do so.)

b. Do the models predict that the women's record time will ever equal the men's record time? If so, what is that record time, and when will the record be set?

c. Do the models predict that the women's record time will ever be less than the men's record time? If so, in what years?

24. A runner's *stride rate* is the number of steps per second. The average stride rates of the top female and male runners are shown in Table 37 for various speeds.

Table 37 Top Female and Male Runners' Speeds and Average Stride Rates

Speed (feet per second)	Average Stride Rate (number of steps per second)	
	Women	Men
15.86	3.05	2.92
16.88	3.12	2.98
17.50	3.17	3.03
18.62	3.25	3.11
19.97	3.36	3.22
21.06	3.46	3.31
22.11	3.55	3.41

Source: Biomechanical comparison of male and female runners, *R. C. Nelson et al., 1977*

a. Let r be the average stride rate of a woman running at s feet per second. Find the regression equation to describe the data.

b. Let r be the average stride rate of a man running at s feet per second. Find the regression equation to describe the data.

c. Which of your two models has the r-intercept with the larger r-coordinate? What does this tell you about the graphs of the models?

d. Which model has the larger slope? What does this tell you about the graphs of the models?

e. Explain why your work in parts (c) and (d) suggests that the graphs of your two models do not intersect in quadrant I (where s and r are positive). What does that mean in this situation?

25. To enroll in Intermediate Algebra, a student at College of San Mateo (CSM) must score at least 21 points (out of 50) on a placement test. Using four semesters of data, the CSM Mathematics Department computed the percentages of students who succeeded in Intermediate Algebra (grade of A, B, or C) for various groups of scores on the placement test (see Table 38).

Table 38 Percentages of Intermediate Algebra Students Who Succeeded

Placement Score Group	Score Used to Represent Score Group	Percentage Who Succeeded in Intermediate Algebra
21–25	23	34
26–30	28	47
31–35	33	55
36–40	38	71
41–45	43	84
46–50	48	*

Source: *College of San Mateo Mathematics Department*
*There were not enough students in this group to give useful data.

a. Let p be the percentage of Intermediate Algebra students succeeding in the course who scored x points on the placement test. Use a graphing calculator to draw a scattergram of the data.

b. Find an equation of a line that you think comes close to the points in the scattergram.

c. Draw your line and the scattergram in the same viewing window. Verify that the line contains the two points you chose in finding the equation in part (b) and that it comes close to all of the data points.

26. "Crime index" refers to the number of incidents of crime. The number of burglaries, aggravated assaults, and all types of crime per 100,000 Americans are shown in Table 39 for various years.

Table 39 Crime Indexes

Year	Crime Index (number of incidents per 100,000 people)		
	Burglary	Aggravated Assault	All Types of Crime
1993	1099	440	5484
1994	1042	428	5374
1995	987	418	5276
1996	945	391	5087
1997	919	382	4930
1998	863	361	4619
1999	770	336	4267
2000	728	324	4124
2001	741	319	4161
2002	746	310	4119
2003	741	295	4063

Source: *FBI*

a. Let A be the crime index of aggravated assaults for the year that is t years since 1990. Can the data be modeled well by a linear model? If yes, find such a model. If no, explain why not.

b. Let B be the crime index of burglaries for the year that is t years since 1990. Can the data be modeled well by a linear model? If yes, find such a model. If no, explain why not.

c. Let C be the crime index of all types of crime for the year that is t years since 1990. Can the data be modeled well by a linear model? If yes, find such a model. If no, explain why not.

d. Economist Steven D. Levitt believes that crime and the economy are not related. The U.S. economy performed well from 1993 to 2000 and did poorly from 2000 to 2003.

i. Explain why the aggravated assault data support Levitt's theory.

ii. Explain why the burglary data are less supportive (than the aggravated assault data) of Levitt's theory.

iii. Explain why the data for all types of crime are less supportive (than the aggravated assault data) of Levitt's theory.

27. Explain how to find an equation of a linear model for a given situation. Also, explain how to verify that the linear function models the situation reasonably well.

28. A student comes up with a shortcut for modeling a situation described by a table that contains several rows of data. Instead of creating a scattergram of the data, the student chooses two data points at random and uses them to find an equation of a line. Give at least two examples to illustrate what can go wrong with this shortcut.

2.3 FUNCTION NOTATION AND MAKING PREDICTIONS

"Mathematics is a door to the world. If you ignore it, you injure your chances at becoming successful. But if you embrace it, you hold the power to do anything you want and to be able to do it successfully. This is why I like math."

—Shaun O., student

Objectives

- Use *function notation*.
- Find inputs and outputs of a function.
- Use a model to make estimates and predictions.
- Find intercepts of a model.
- Use data described in words to make estimates and predictions.
- Know the meaning of *domain* and *range* of a model.

In this section, we will discuss how to name a function. We will also find inputs and outputs of a function and use a model to make estimates and predictions. Finally, we will consider how such inputs and outputs relate to the *domain* and *range* of a model.

Function Notation

Rather than use an equation, table, graph, or words to refer to a function, it would be easier to name the function. For example, to use "f" as the name of the linear function $y = 2x + 1$, we use "$f(x)$" (read "f of x") to represent y:

$$y = f(x)$$

We refer to "$f(x)$" as *function notation*. To use function notation to write the equation of this function, we substitute $f(x)$ for y in the equation $y = 2x + 1$:

$$f(x) = 2x + 1$$

WARNING The notation "$f(x)$" does *not* mean f times x. It is another variable name for y.

Recall from Section 1.6 that we can think of a function as a machine that sends inputs to outputs. Here we substitute 4 for x in the equation $y = 2x + 1$: $y = 2(4) + 1 = 9$. So, the input $x = 4$ leads to the output $y = 9$. Now we substitute 4 for x in the equation $f(x) = 2x + 1$:

$$\begin{aligned} f(x) &= 2x + 1 && \text{Equation of } f \\ f(4) &= 2(4) + 1 && \text{Substitute 4 for } x. \\ &= 9 && \text{Simplify.} \end{aligned}$$

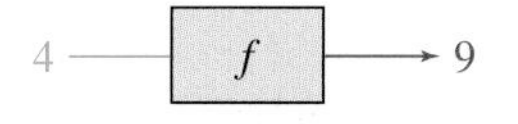

Figure 27 A function "machine"

The equation $f(4) = 9$ means that the input $x = 4$ leads to the output $y = 9$. Figure 27 shows the "machine" f sending the input 4 to the output 9.

Notice that $f(4) = 9$ is of the form

$$f(\text{input}) = \text{output}$$

This is true for any function f.

The number $f(4)$ is the value of y when x is 4. To find $f(4)$, we say that we **evaluate** the function f at $x = 4$.

Example 1 Evaluating a Function

Evaluate $f(x) = -4x + 2$ at 5.

Solution

$$f(x) = -4x + 2 \quad \text{Equation of } f$$
$$f(5) = -4(5) + 2 \quad \text{Substitute 5 for } x.$$
$$= -18 \quad \text{Simplify.}$$

■

We can also use "g" to name the function $y = -4x + 2$:

$$g(x) = -4x + 2$$

The most commonly used symbols to name functions are f, g, and h.

Example 2 Evaluating Functions

For $f(x) = 2x^2 - 3x$, $g(x) = \dfrac{4x - 2}{5x - 1}$, and $h(x) = 3x - 5$, find the following (to review exponentiation and order of operations, see Sections A.5 and A.6):

1. $f(-2)$ **2.** $g(3)$ **3.** $h(a)$ **4.** $h(a - 2)$

Solution

1.

$$f(-2) = 2(-2)^2 - 3(-2) \quad \text{Evaluate } f \text{ at } -2.$$
$$= 2(4) - 3(-2) \quad \text{Perform exponentiation first: } (-2)^2 = (-2)(-2) = 4$$
$$= 8 + 6 \quad \text{Multiply.}$$
$$= 14 \quad \text{Add.}$$

2.

$$g(3) = \frac{4 \cdot 3 - 2}{5 \cdot 3 - 1} \quad \text{Evaluate } g \text{ at 3.}$$
$$= \frac{12 - 2}{15 - 1} \quad \text{Multiply first.}$$
$$= \frac{10}{14} \quad \text{Subtract.}$$
$$= \frac{5}{7} \quad \text{Simplify.}$$

3. To find $h(a)$, we substitute a for x in the equation $h(x) = 3x - 5$:

$$h(a) = 3a - 5 \quad \text{Evaluate } h \text{ at } a.$$

4.

$$h(a - 2) = 3(a - 2) - 5 \quad \text{Evaluate } h \text{ at } a - 2.$$
$$= 3a - 6 - 5 \quad \text{Distributive law}$$
$$= 3a - 11 \quad \text{Combine like terms.}$$

■

So far, we have used equations to evaluate functions. Next, we will use a table to find an output and an input of a function.

Table 40 Input-Output Pairs of g

x	$g(x)$
3	12
4	9
5	8
6	9
7	12

Example 3 Using a Table to Find an Output and an Input

Some input–output pairs of a function g are shown in Table 40.

1. Find $g(7)$.

2. Find x when $g(x) = 9$.

Solution

1. From Table 40, we see that the input $x = 7$ leads to the output $y = 12$. So, $g(7) = 12$.
2. To find x when $g(x) = 9$, we need to find all inputs in the table that lead to the output $y = 9$. From Table 40, we see that both inputs $x = 4$ and $x = 6$ lead to the output $y = 9$. So, the values of x are 4 and 6. ■

In Example 4, we will use an equation to find an output and an input of a function.

Example 4 Using an Equation to Find an Output and an Input

Let $f(x) = \frac{3}{2}x - 1$.

1. Find $f(4)$.
2. Find x when $f(x) = -4$.

Solution

1.

$$
\begin{aligned}
f(4) &= \frac{3}{2}(4) - 1 && \text{Substitute 4 for } x. \\
&= 6 - 1 && \frac{3}{2}(4) = \frac{3}{2} \cdot \frac{4}{1} = 6 \\
&= 5 && \text{Subtract.}
\end{aligned}
$$

TABLE SETUP
TblStart=0
ΔTbl=1
Indpnt: Auto Ask
Depend: Auto Ask

Figure 28 Putting table in "Ask" mode

2. We substitute -4 for $f(x)$ in $f(x) = \frac{3}{2}x - 1$ and solve for x:

$$
\begin{aligned}
-4 &= \frac{3}{2}x - 1 && \text{Substitute } -4 \text{ for } f(x). \\
2(-4) &= 2 \cdot \frac{3}{2}x - 2 \cdot 1 && \text{Multiply both sides by LCD, 2.} \\
-8 &= 3x - 2 && \text{Multiply; simplify.} \\
-6 &= 3x && \text{Add 2 to both sides.} \\
-2 &= x && \text{Divide both sides by 3.}
\end{aligned}
$$

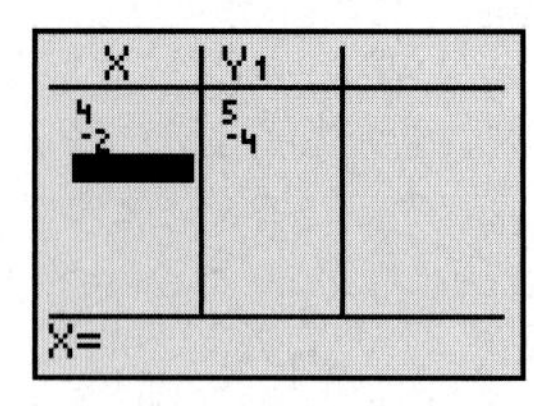

Figure 29 Verify the work

We can verify our work in Problems 1 and 2 by putting a graphing calculator table in Ask mode (see Figs. 28 and 29). For graphing calculator instructions, see Section B.15. ■

WARNING Example 4 asks for both a value of y (in Problem 1) and a value of x (in Problem 2). Be sure you know which value you need to find. When you are asked for $f(x)$, what you are looking for is a value of y, not a value of x.

For $f(x) = \frac{3}{2}x - 1$, we found in Problem 1 of Example 4 that $f(4) = 5$. Since $y = 5$ when $x = 4$, we know that the ordered pair $(4, 5)$ is a solution of $f(x) = \frac{3}{2}x - 1$ and that the point $(4, 5)$ is on the graph of f (see Fig. 30). We use blue arrows to show that the input $x = 4$ leads to the output $y = 5$.

In Problem 2 of Example 4, we found that $x = -2$ when $f(x) = -4$. So, the point $(-2, -4)$ is on the graph of f. We use red arrows in Fig. 30 to show that the output $y = -4$ originates from the input $x = -2$.

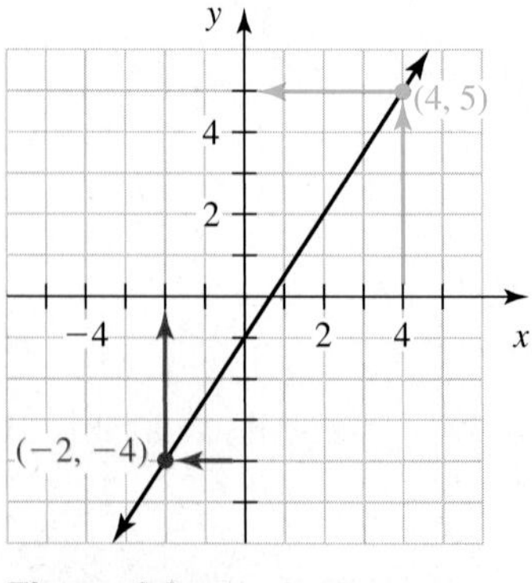

Figure 30 Graph of $f(x) = \frac{3}{2}x - 1$

Example 5 Using a Graph to Find Values of x or $f(x)$

A graph of a function f is sketched in Fig. 31.

1. Find $f(4)$.
2. Find $f(0)$.
3. Find x when $f(x) = -2$.
4. Find x when $f(x) = 0$.

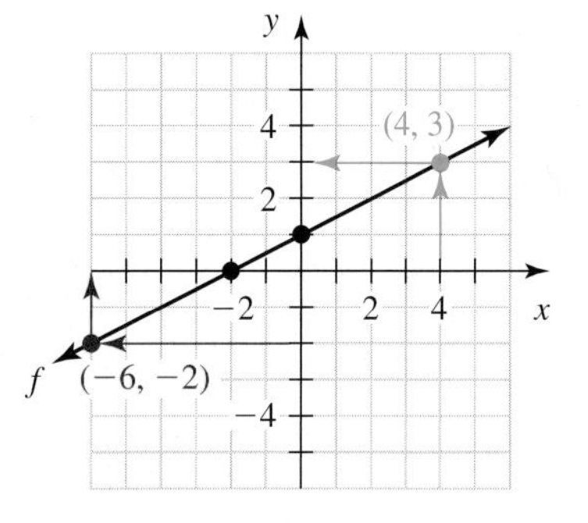

Figure 31 Problems 1–4 of Example 5

Solution

1. Recall that $y = f(x)$. The notation $f(4)$ refers to $f(x)$ when $x = 4$. So, we want the value of y when $x = 4$. The blue arrows in Fig. 31 show that the input $x = 4$ leads to the output $y = 3$. Hence, $f(4) = 3$.
2. To find $f(0)$, we want the value of y when $x = 0$. The line contains the point $(0, 1)$, so $f(0) = 1$.
3. We have $y = f(x) = -2$. Thus, $y = -2$. So, we want the value of x when $y = -2$. The red arrows in Fig. 31 show that the output $y = -2$ originates from the input $x = -6$. Hence, $x = -6$.
4. We have $y = f(x) = 0$. Thus, $y = 0$. The line contains the point $(-2, 0)$, so $x = -2$. ■

Using Function Notation with Models

Recall from Section 1.2 that when we are *not* describing an authentic situation, we treat x as the independent variable and y as the dependent variable. Here we label the independent variable, dependent variable, and function name of the equation $y = f(x)$:

dependent variable ↓ independent variable ↓

$$y = f(x)$$

↑ function name

We follow this same format for a function f that *is* a model.

DEFINITION Function notation

The dependent variable of a function f can be represented by the expression formed by writing the independent variable name within the parentheses of $f(\ )$:

$$\text{dependent variable} = f(\text{independent variable})$$

We call this representation **function notation.**

For instance, in Example 1 of Section 2.2, we found the model $n = 0.27t + 4.56$, where n is the average number of credit card offers a household receives in one month at t years since 2000. Since t is the independent variable and n is the dependent variable, we can use the function name f to write $n = f(t)$. To use function notation to write the equation of the model, we substitute $f(t)$ for n in the equation $n = 0.27t + 4.56$:

$$f(t) = 0.27t + 4.56$$

In Example 6, we will use function notation in making predictions.

Table 41 Average Salaries of Professors

Year	Average Salary (thousands of dollars)
1975	16.6
1980	22.1
1985	31.2
1990	41.9
1995	49.1
2000	57.7
2004	65.0

Source: *American Association of University Professors*

Example 6 Using an Equation of a Linear Model to Make Predictions

Table 41 shows the average salaries of professors at four-year public colleges and universities.

Let s be the professors' average salary (in thousands of dollars) at t years since 1900. A possible model is

$$s = 1.71t - 113.12$$

1. Verify that the function $s = 1.71t - 113.12$ models the data well.
2. Rewrite the equation $s = 1.71t - 113.12$ with the function name f.
3. Predict the average salary in 2011.
4. Predict when the average salary will be \$80,000.

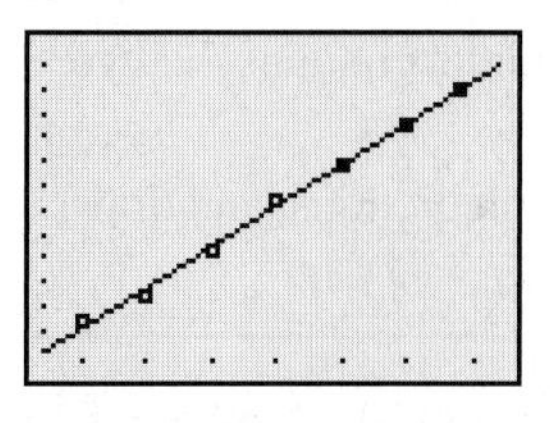

Figure 32 Check how well $s = 1.71t - 113.12$ models the data

Solution

1. We draw the graph of the model and the scattergram of the data in the same viewing window (see Fig. 32). The function appears to model the data quite well.

2. Here, t is the independent variable and s is the dependent variable. Since the function name is f, we can write $s = f(t)$. Then we substitute $f(t)$ for s in the equation $s = 1.71t - 113.12$:

$$f(t) = 1.71t - 113.12$$

3. We represent the year 2011 by $t = 111$. To find the average salary, we substitute 111 for t in the equation $f(t) = 1.71t - 113.12$:

$$\begin{aligned} f(111) &= 1.71(111) - 113.12 && \text{Substitute 111 for } t. \\ &= 76.69 && \text{Simplify.} \end{aligned}$$

The model predicts that the average salary will be \$76,690 in 2011.

4. We can represent the salary \$80,000 by $s = 80$. Since $s = f(t)$, we can write $f(t) = 80$. To find the year, we substitute 80 for $f(t)$ in the equation $f(t) = 1.71t - 113.12$ and solve for t:

$$\begin{aligned} 80 &= 1.71t - 113.12 && \text{Substitute 80 for } f(t). \\ 80 + 113.12 &= 1.71t - 113.12 + 113.12 && \text{Add 113.12 to both sides.} \\ 193.12 &= 1.71t && \text{Combine like terms.} \\ \frac{193.12}{1.71} &= \frac{1.71t}{1.71} && \text{Divide both sides by 1.71.} \\ 112.94 &\approx t && \text{Simplify.} \end{aligned}$$

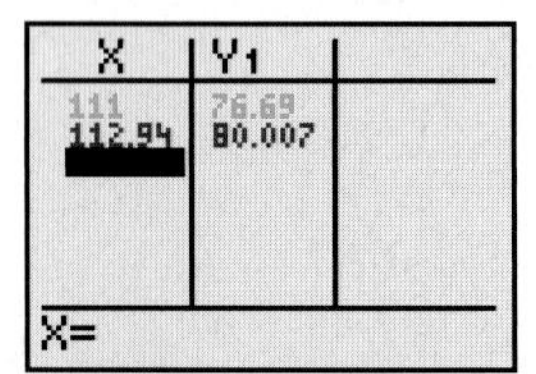

Figure 33 Verify the predictions

According to the model, the average salary will be \$80,000 in $1900 + 113 = 2013$. We can verify our work in both Problems 3 and 4 by using a graphing calculator table (see Fig. 33).

Or we can graphically verify our work by using TRACE (see Fig. 34). ■

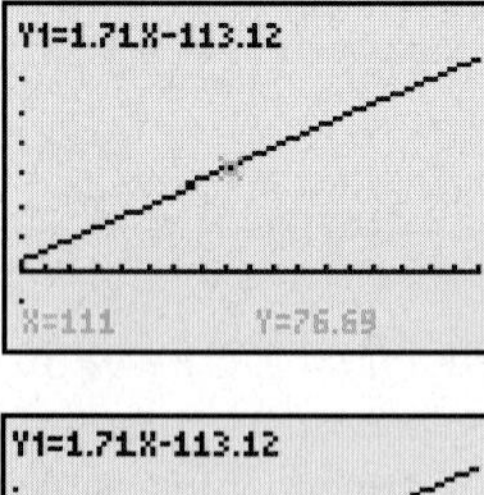

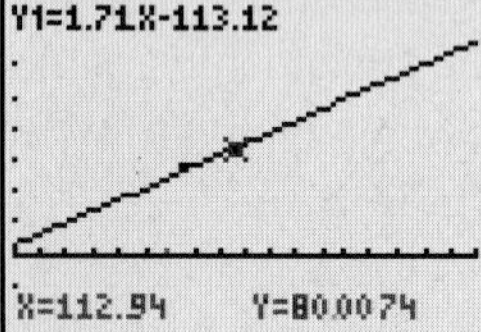

Figure 34 Verify predictions by using TRACE

Here we summarize how to use an equation of a model to make predictions (or estimates).

Using an Equation of a Linear Model to Make Predictions

- When making a prediction about the dependent variable of a linear model, substitute a chosen value for the independent variable in the model. Then solve for the dependent variable.
- When making a prediction about the independent variable of a linear model, substitute a chosen value for the dependent variable in the model. Then solve for the independent variable.

In Section 2.2 and this section, we discussed how to find linear models and how to use these models to make estimates and predictions. Here is a summary of this process.

Four-Step Modeling Process

To find a linear model and make estimates and predictions,

1. Create a scattergram of the data to determine whether there is a nonvertical line that comes close to the data points. If so, choose two points (not necessarily data points) that you can use to find the equation of a linear model.
2. Find an equation of your model.
3. Verify your equation by checking that the graph of your model contains the two chosen points and comes close to all of the data points.
4. Use the equation of your model to make estimates, make predictions, and draw conclusions.

In Example 6, we used f to name the function $f(t) = 1.71t - 113.12$, where $f(t)$ represents the average salary of professors at four-year *public* colleges and universities. When we use more than one function to model situations, naming the functions helps us distinguish among them. For example, we can also use a linear function to model the average salaries of professors at four-year *private* colleges and universities. A good model is $s = 2.22t - 155.06$ where s is the professors' average salary (in thousands of dollars) at t years since 1900. We can distinguish this function from f by using g as its name:

$$g(t) = 2.22t - 155.06$$

Finding Intercepts

In Example 7, we use a model to make a prediction and an estimate, find the intercepts of the model, and interpret the meaning of the intercepts.

Example 7 Using Function Notation; Finding Intercepts

Table 42 American Adults Who Smoke

Year	Percent
1970	37.4
1980	33.2
1990	25.3
2000	23.1
2005	19.0

Source: *National Center for Health Statistics*

In Example 3 of Section 2.2, we found the equation $p = -0.53t + 74.50$, where p is the percentage of American adults who smoke at t years since 1900 (see Table 42).

1. Rewrite the equation $p = -0.53t + 74.50$ with the function name g.
2. Find $g(110)$. What does the result mean in this situation?
3. Find the value of t when $g(t) = 30$. What does it mean in this situation?
4. Find the p-intercept of the model. What does it mean in this situation?
5. Find the t-intercept. What does it mean in this situation?

Solution

1. To use the name g, substitute $g(t)$ for p in the equation $p = -0.53t + 74.50$:

$$g(t) = -0.53t + 74.50$$

2. To find $g(110)$, we substitute 110 for t in the equation $g(t) = -0.53t + 74.50$:

$$\begin{aligned} g(t) &= -0.53t + 74.50 && \text{Equation of } g \\ g(110) &= -0.53(110) + 74.50 && \text{Substitute 110 for } t. \\ &= 16.2 && \text{Simplify.} \end{aligned}$$

So, $p = 16.2$ when $t = 110$. According to the model, 16.2% of American adults will smoke in 2010.

3. We substitute 30 for $g(t)$ in the equation $g(t) = -0.53t + 74.50$ and solve for t:

$$\begin{aligned} g(t) &= -0.53t + 74.50 && \text{Equation of } g \\ 30 &= -0.53t + 74.50 && \text{Substitute 30 for } g(t). \\ 30 - 74.50 &= -0.53t + 74.50 - 74.50 && \text{Subtract 74.50 from both sides.} \\ -44.5 &= -0.53t && \text{Combine like terms.} \\ \frac{-44.5}{-0.53} &= \frac{-0.53t}{-0.53} && \text{Divide both sides by } -0.53. \\ 83.96 &\approx t && \text{Simplify.} \end{aligned}$$

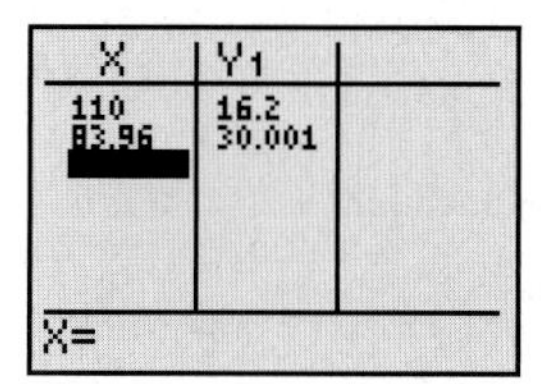

Figure 35 Verify the work

The model estimates that 30% of Americans smoked in $1900 + 83.96 \approx 1984$. We can verify our work in Problems 2 and 3 by using a graphing calculator table (see Fig. 35).

4. Since the model $g(t) = -0.53t + 74.50$ is in slope–intercept form, the p-intercept is (0, 74.50). So, the model estimates that 74.5% of American adults smoked in 1900. Research would show that this estimate is too high; model breakdown has occurred.
5. To find the t-intercept, we substitute 0 for $g(t)$ and solve for t:

$$\begin{aligned} 0 &= -0.53t + 74.50 && \text{Substitute 0 for } g(t). \\ 0 + 0.53t &= -0.53t + 74.50 + 0.53t && \text{Add 0.53t to both sides.} \\ 0.53t &= 74.50 && \text{Combine like terms.} \\ \frac{0.53t}{0.53} &= \frac{74.50}{0.53} && \text{Divide both sides by 0.53.} \\ t &\approx 140.57 && \text{Simplify.} \end{aligned}$$

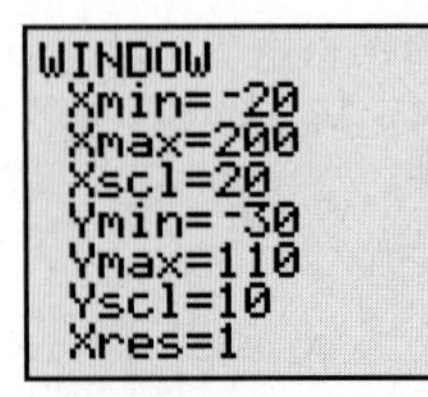

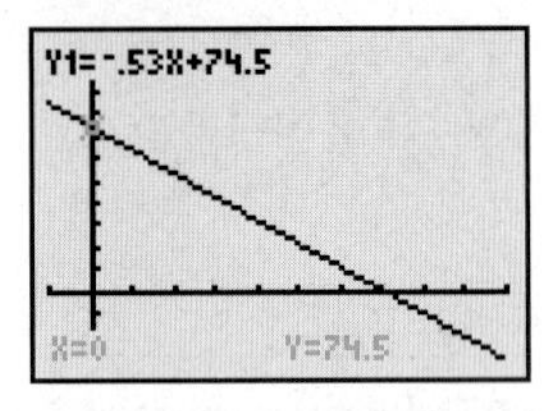

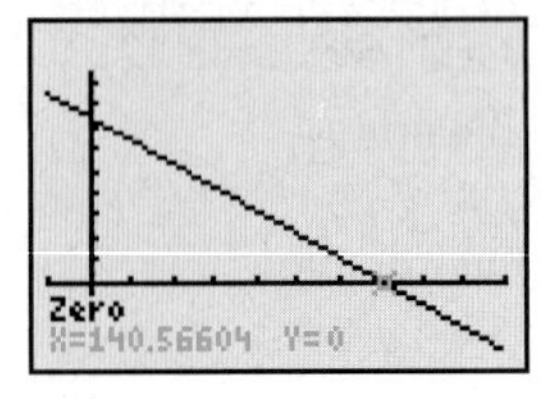

Figure 36 Verify the intercepts

The t-intercept is (140.57, 0). So, the model predicts that no American adults will smoke in $1900 + 140.57 \approx 2041$. However, common sense suggests that this event probably won't occur.

We can use TRACE to verify the p-intercept and the "zero" option to verify the t-intercept (see Fig. 36). ■

Intercepts of a Model

If a function of the form $p = mt + b$, where $m \neq 0$, is used to model a situation, then

- The p-intercept is $(0, b)$.
- To find the t-coordinate of the t-intercept, substitute 0 for p in the model's equation and solve for t.

Using Data Described in Words to Make Predictions

In most application problems in this text so far, we have been provided variable names and their definitions. In Example 8, a key step will be to create variable names and define the variables.

Example 8 Making a Prediction

Sales of bagged salads increased approximately linearly from \$0.9 billion in 1996 to \$2.7 billion in 2004 (Source: *Produce Marketing Association, ACNielsen*). Predict in which year the sales will be \$4 billion.

Solution

Let s be the sales (in billions of dollars) of bagged salads in the year that is t years since 1990. Known values of t and s are shown in Table 43.

Table 43 Known Values of t and s

Years Since 1990 t	Sales (billions of dollars) s
6	0.9
14	2.7

Since the variables t and s are approximately linearly related, we want an equation of the form $s = mt + b$. First, we use the values in Table 43 to find the slope of the model:

$$m = \frac{2.7 - 0.9}{14 - 6} \approx 0.23$$

So, we can substitute 0.23 for m in the equation $s = mt + b$:

$$s = 0.23t + b$$

To find b, we use the point (6, 0.9) and substitute 6 for t and 0.9 for s in the equation $s = 0.23t + b$ and then solve for b:

$$\begin{aligned} 0.9 &= 0.23(6) + b && \text{Substitute 6 for } t \text{ and 0.9 for } s. \\ 0.9 &= 1.38 + b && \text{Multiply.} \\ 0.9 - 1.38 &= 1.38 + b - 1.38 && \text{Subtract 1.38 from both sides.} \\ -0.48 &= b && \text{Combine like terms.} \end{aligned}$$

Then we substitute -0.48 for b in the equation $s = 0.23t + b$:

$$s = 0.23t - 0.48$$

Finally, to predict when sales will be \$4 billion, we substitute 4 for s in the equation $s = 0.23t - 0.48$ and solve for t:

$$\begin{aligned} 4 &= 0.23t - 0.48 && \text{Substitute 4 for } s. \\ 4 + 0.48 &= 0.23t - 0.48 + 0.48 && \text{Add 0.48 to both sides.} \\ 4.48 &= 0.23t && \text{Combine like terms.} \\ 19.48 &\approx t && \text{Divide both sides by 0.23.} \end{aligned}$$

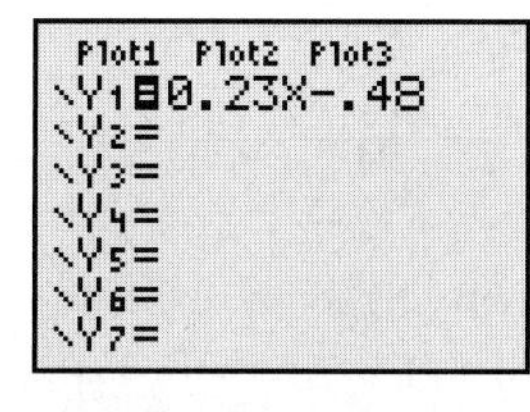

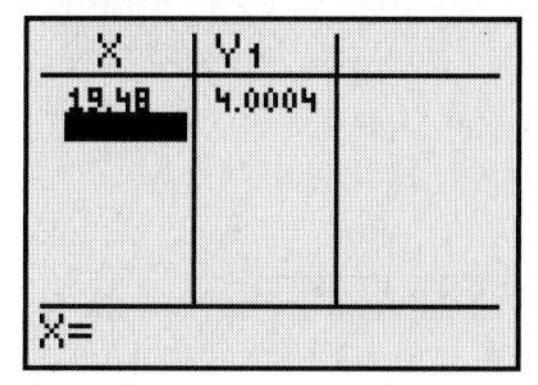

Figure 37 Verify the work

The model predicts that sales will be \$4 billion in $1990 + 19 = 2009$. We can verify our work by using a graphing calculator table (see Fig. 37). ■

In Example 8, we defined t to be the number of years since *1990*. If we had defined t to be the number of years since *1950* (or any other year), we would have obtained the same prediction of 2009, although the equation of our model would have been different. **If an exercise does not state a year from which to begin counting, choose any year.**

Domain and Range of a Model

Recall from Section 1.6 that the domain of a function is the set of all inputs and that the range of a function is the set of all outputs. For the **domain** and **range** of a model, we consider input–output pairs only when both the input and the output make sense in the situation. The domain of the model is the set of all such inputs, and the range of the model is the set of all such outputs.

Example 9 Finding the Domain and Range of a Model

A store is open from 9 A.M. to 5 P.M., Mondays through Saturdays. Let $I = f(t)$ be an employee's weekly income (in dollars) from working t hours each week at \$10 per hour.

1. Find an equation of the model f.
2. Find the domain and range of the model f.

Solution

1. The employee's weekly income (in dollars) is equal to the pay per hour times the number of hours worked per week:

$$f(t) = 10t$$

2. To find the domain and range of the model f, we consider input–output pairs only when both the input and the output make sense in this situation. Time is the input. Since the store is open 8 hours a day, 6 days a week, the employee can work up to 48 hours each week. So, the domain is the set of numbers between 0 and 48, inclusive: $0 \le t \le 48$.

   Income is the output. Since the number of hours worked is between 0 and 48 hours, inclusive, and the pay is \$10 per hour, the range is the set of numbers between 0 and $10(48) = 480$, inclusive: $0 \le f(t) \le 480$.

   In Fig. 38, we illustrate the inputs 22, 35, and 48 being sent to the outputs 220, 350, and 480, respectively. We also label the part of the t-axis that represents the domain and the part of the I-axis that represents the range.

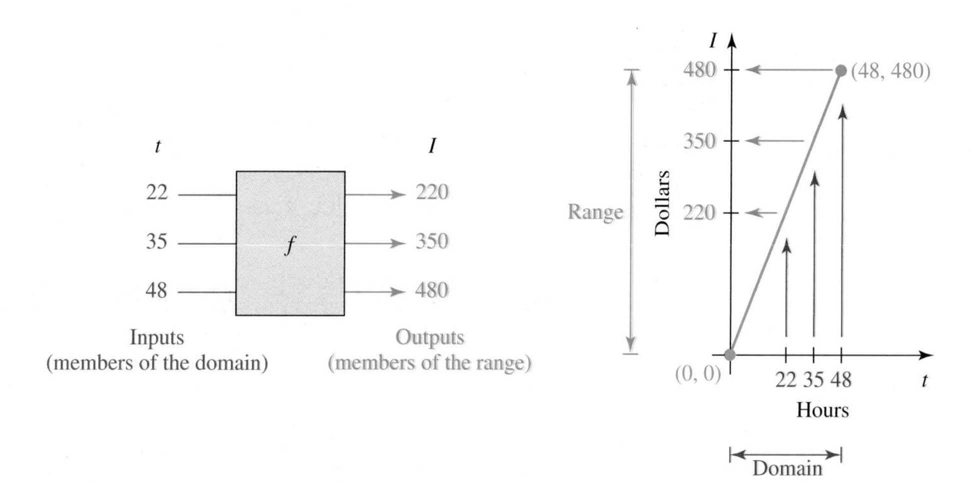

Figure 38 Domain and range of the employee income model ■

group exploration

Formula for slope

1. Let $f(x) = 2x + 1$. Find each of the following and compare all three results. [**Hint for part (b):** First find $f(5)$ and $f(3)$. Then subtract. Finally, divide.]

   a. the slope of the graph of f **b.** $\frac{f(5) - f(3)}{5 - 3}$ **c.** $\frac{f(7) - f(4)}{7 - 4}$

2. Let $g(x) = 3x + 5$. Find each of the following and compare all three results.

   a. the slope of the graph of g **b.** $\frac{g(3) - g(1)}{3 - 1}$ **c.** $\frac{g(4) - g(0)}{4 - 0}$

3. Let f be a function of the form $f(x) = mx + b$. Describe $\frac{f(c) - f(d)}{c - d}$, where $c \neq d$. Explain.

group exploration

Looking ahead: Significance of the slope and the dependent variable's intercept of a model

1. A small airplane is traveling at a constant speed of 100 miles per hour. Let d be the distance (in miles) the airplane can travel in t hours.
   a. Complete Table 44.
   b. Find an equation of a linear model.
   c. Compare the slope of your model with the speed of the airplane.
   d. What is the d-intercept? What does it mean in this situation?
2. In 2005, a company was worth \$10 million. Each year, its value increases by \$2 million. Let V be the company's value (in millions of dollars) at t years since 2005.
   a. Complete Table 45.
   b. Find an equation of a linear model.
   c. Compare the slope of your model with the rate at which the company's value is increasing.
   d. What is the V-intercept? What does it mean in this situation?
3. A person is in a hot-air balloon at an altitude of 1600 feet. The person begins gradually letting air out of the balloon, and the balloon descends at a rate of 200 feet per minute. Let H be the balloon's altitude (in feet) after air has been released for t minutes.
   a. Complete Table 46.
   b. Find an equation of a linear model.
   c. Compare the slope of your model with the rate at which the balloon is descending.
   d. What is the H-intercept? What does it mean in this situation?
4. In general, what is the meaning of slope in terms of an authentic situation? What is the meaning of the dependent variable's intercept?

Table 44 Distances Traveled by an Airplane

Time (hours) t	Distance (miles) d
0	
1	
2	
3	
4	

Table 45 Values of a Company

Year Since 2005 t	Value (millions of dollars) V
0	
1	
2	
3	
4	

Table 46 Altitudes of a Balloon

Time (minutes) t	Altitude (feet) H
0	
1	
2	
3	
4	

TIPS FOR SUCCESS: Math Journal

Do you tend to make the same mistakes repeatedly throughout a math course? If so, it might help to keep a journal in which you list errors you have made on assignments, quizzes, and tests. For each error you list, include the correct solution as well as a description of the concept needed to solve the problem correctly. Review this journal from time to time to help you avoid making these errors.

HOMEWORK 2.3

FOR EXTRA HELP ▶

Student Solutions Manual · PH Math/Tutor Center · MathXL® · MyMathLab

Evaluate $f(x) = 6x - 4$ at the given value of x.

1. $f(5)$ **2.** $f(-2)$ **3.** $f\left(\frac{2}{3}\right)$

4. $f\left(\frac{5}{2}\right)$ **5.** $f(a+2)$ **6.** $f(a-3)$

Evaluate $g(x) = 2x^2 - 5x$ at the given value of x. (To review exponentiation and order of operations, see Sections A.5 and A.6.)

7. $g(2)$ **8.** $g(3)$ **9.** $g(-3)$ **10.** $g(-2)$

Evaluate $h(x) = \frac{3x-4}{5x+2}$ at the given value of x.

11. $h(2)$ **12.** $h(-4)$

13. $h(a-3)$ **14.** $h(3a)$

For $f(x) = -2x + 7$, $g(x) = -3x^2 + 2x$, and $h(x) = -4$, find the following.

15. $g(-2)$ = −16 **16.** $g(-1)$ **17.** $f(5)$

18. $f(-4)$ **19.** $h(7)$ **20.** $h(-9)$

Evaluate $f(x) = -4x - 7$ at the given value of x.

21. $f(5a)$ **22.** $f(-3a)$

23. $f\left(\frac{a}{2}\right)$ **24.** $f\left(\frac{3a}{2}\right)$

25. $f(a+4)$ **26.** $f(a-4)$

27. $f(a+h)$ **28.** $f(a-h)$

For $f(x) = -3x + 7$, find the value of x that leads to the given value of $f(x)$.

29. $f(x) = 6$ **30.** $f(x) = 0$

31. $f(x) = \frac{5}{2}$ **32.** $f(x) = -\frac{4}{3}$

33. $f(x) = a$ **34.** $f(x) = a + 2$

For Exercises 35–38, let $f(x) = -5.95x + 183.22$. Round any results to the second decimal place.

35. Find $f(10.91)$. **36.** Find $f(17.28)$.

37. Find x when $f(x) = 99.34$.

38. Find x when $f(x) = 72.06$.

39. A student tries to find x when $f(x) = 5$ for $f(x) = x + 2$:

$$f(5) = 5 + 2 = 7$$

Describe any errors. Then find x correctly.

40. A student tries to find $g(-5)$, where $g(x) = x^2$:

$$g(-5) = -5^2 = -25$$

Describe any errors. Then find $g(-5)$ correctly.

41. **a.** For $f(x) = 4x$, find $f(3)$, $f(5)$, and $f(8)$. Is the equation $f(3+5) = f(3) + f(5)$ a true statement?
b. For $f(x) = x^2$, find $f(2)$, $f(3)$, and $f(5)$. Is the equation $f(2+3) = f(2) + f(3)$ a true statement?
c. For $f(x) = \sqrt{x}$, find $f(9)$, $f(16)$, and $f(25)$. Is $f(9+16) = f(9) + f(16)$ a true statement?
d. Is $f(a+b) = f(a) + f(b)$ a true statement for every function f?

42. **a.** For $f(x) = 3x + 2$, find $f(5) - f(4)$. Compare your result with the slope of the graph of f. [**Hint:** Find $f(5)$ and $f(4)$. Then subtract.]
b. For $f(x) = 2x + 5$, find $f(7) - f(6)$. Compare your result with the slope of the graph of f.
c. For $f(x) = 4x + 1$, find $f(3) - f(2)$. Compare your result with the slope of the graph of f.
d. For $f(x) = mx + b$, find $f(a+1) - f(a)$. Compare your result with the slope of the graph of f, and discuss what this result means. [**Hint:** Consider the slope addition property.]

For Exercises 43–46, refer to Table 47.

43. Find $f(2)$. **44.** Find $f(4)$.

45. Find x when $f(x) = 2$. **46.** Find x when $f(x) = 4$.

Table 47 Values of f (Exercises 43–46)

x	$f(x)$
0	0
1	2
2	4
3	2
4	0

For Exercises 47–58, refer to Fig. 39.

47. Estimate $f(-6)$. **48.** Estimate $f(0)$.

49. Estimate $f(2.5)$. **50.** Estimate $f\left(-\frac{11}{2}\right)$.

51. Estimate x when $f(x) = 0$.

52. Estimate x when $f(x) = 1$.

53. Estimate x when $f(x) = 3$.

54. Estimate x when $f(x) = 3.5$.

55. Estimate x when $f(x) = \frac{1}{2}$.

56. Estimate x when $f(x) = \frac{5}{2}$.

57. Find the domain of f. **58.** Find the range of f.

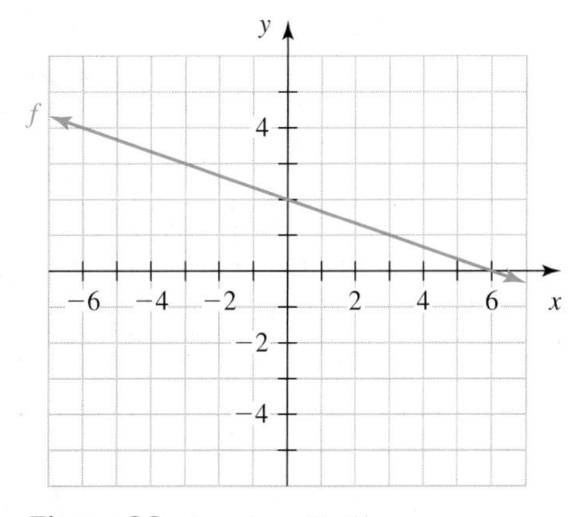

Figure 39 Exercises 47–58

For Exercises 59–62, refer to Fig. 40.

59. Find $g(-2)$.

60. Find x when $g(x) = 3$.

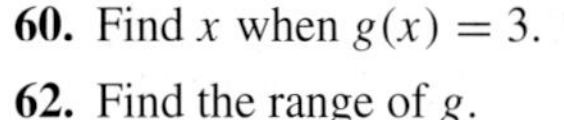

61. Find the domain of g.

62. Find the range of g.

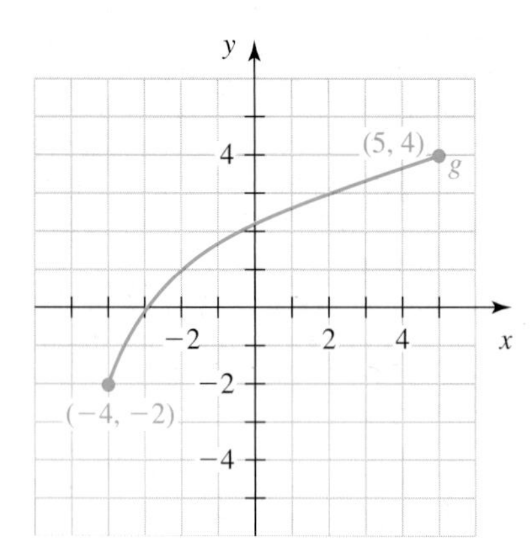

Figure 40
Exercises 59–62

For Exercises 63–66, refer to Fig. 41.

63. Find $h(1)$.

64. Find x when $h(x) = -1$.

65. Find the domain of h.

66. Find the range of h.

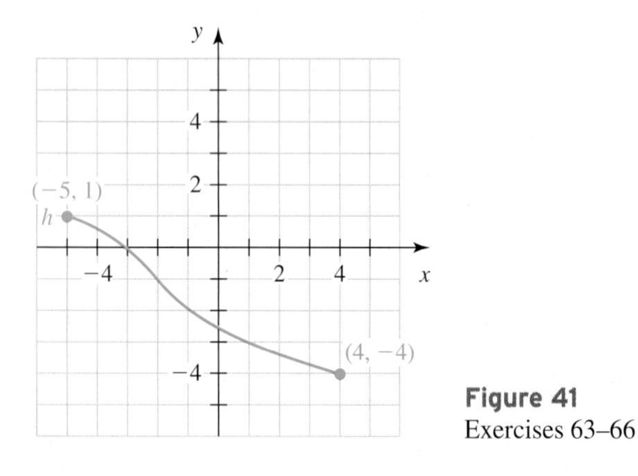

Figure 41
Exercises 63–66

Find all x-intercepts and y-intercepts. If no intercept exists, say so.

67. $f(x) = 5x - 8$

68. $f(x) = 4x + 2$

69. $f(x) = 3x$

70. $f(x) = -7x$

71. $f(x) = 5$

72. $f(x) = -2$

73. $f(x) = \frac{1}{2}x - 3$

74. $f(x) = -\frac{3}{4}x + \frac{1}{2}$

Find the approximate x-intercept and approximate y-intercept of the graph of the equation. Round the coordinates to the second decimal place.

75. $f(x) = 2.58x - 45.21$

76. $f(x) = -4.29x + 37.58$

77. Describe the Rule of Four as applied to the function $g(x) = -3x + 4$:

a. Describe five input–output pairs of g by using a table.
b. Describe the input–output pairs of g by using a graph.
c. Describe the input–output pairs of g by using words.

78. Describe the Rule of Four as applied to the function $f(x) = -\frac{3}{5}x - 1$:

a. Describe five input–output pairs of f by using a table.
b. Describe the input–output pairs of f by using a graph.
c. Describe the input–output pairs of f by using words.

79. In Exercise 15 of Homework 2.2, you found an equation close to $p = 0.77t - 42.90$ that models the percentage p of births outside marriage in the United States at t years since 1900 (see Table 48).

Table 48 Births Outside Marriage

Year	Percent of Births Outside Marriage
1970	10.7
1975	14.3
1980	18.4
1985	22.0
1990	28.0
1995	32.2
2000	33.2
2002	34.0

Source: *National Center for Health Statistics*

a. Rewrite the equation $p = 0.77t - 42.90$ with the function name f.
b. Find $f(110)$. What does your result mean in this situation?
c. Find the value of t so that $f(t) = 44$. What does your result mean in this situation?
d. According to the model, in what year will all births be outside marriage?
e. Estimate the percentage of births outside marriage in 1997. The actual percentage was 32.4%. What is the error in your estimate? (The error is the difference between the estimated value and the actual value.)

80. In Exercise 16 of Homework 2.2, you found an equation close to $p = 0.77t + 10.93$ that models the percentage p of births outside marriage in the United States at t years *since 1970*. Repeat Exercise 79, but use the model $p = 0.77t + 10.93$. In what instances are your answers the same? When are they different?

81. In Exercise 19 of Homework 2.2, you found an equation close to $p = -0.77t + 20.86$, where p is the percentage of Americans whose favorite sport is baseball at t years since 1990 (see Table 49).

Table 49 Percentages of Americans Whose Favorite Sport Is Baseball

Year	Percent
1994	19
1995	16
1997	14
1998	16
2000	13
2002	12
2004	10

Source: *The Gallup Organization*

a. Rewrite the equation $p = -0.77t + 20.86$ with the function name f.
b. Find $f(6)$. What does it mean in this situation?
c. Find t when $f(t) = 6$. What does it mean in this situation?
d. Find the p-intercept of the model. What does it mean in ths situation?
e. Find the t-intercept. What does it mean in this situation?

82. The U.S. market shares (percentage of U.S. vehicle sales) of Ford® are shown in Table 50 for various years.

Table 50 Ford's U.S. Market Shares

Year	Market Share (%)
1998	26
2000	24
2002	21
2004	19
2005	18

Source: *Ward's AutoInfoBank*

a. Let $s = f(t)$ be Ford's U.S. market share (in percent) at t years since 1995. Find an equation of f. Does the model fit the data well?
b. Find $f(15)$. What does it mean in this situation?
c. Find t when $f(t) = 15$. What does it mean in this situation?
d. Find the t-intercept. What does it mean in this situation?
e. Find the s-intercept. What does it mean in this situation?

83. In Exercise 17 of Homework 2.2, you found an equation close to $p = 12.74n + 4.40$, where p is the price (in dollars) of a Gold Medal Sports ski rental package for n days (see Table 51).

Table 51 Prices of Ski Rental Packages

Number of Days	Price of Package (dollars)
1	15.00
2	30.00
3	45.00
4	56.00
5	70.00
6	78.00

Source: *Gold Medal Sports*

a. Rewrite the equation $p = 12.74n + 4.40$ with the function name f.
b. Find the p-intercept of the model. What does it mean in this situation?
c. Use f to estimate the price of renting skis for 7 days.
d. Use a graphing calculator table to find $f(1)$, $f(2)$, $f(3), \ldots,$ and $f(6)$. Then estimate by how much p increases each time n is increased by 1. Compare these results with the slope of the graph of f. What does that mean in this situation?
e. Let $g(n) = \dfrac{f(n)}{n}$. Find $g(1), g(2), g(3), \ldots,$ and $g(6)$. Which of these results is the least? What does that mean in this situation? [**Hint:** To find $g(2)$, first find $f(2)$. Then divide the result by 2 to get $g(2) = \dfrac{f(2)}{2}$.]
f. For ski rental packages for over 6 days, Gold Medal Sports charges a *daily* fee of \$13.00 *per day*. Compare this result with your results in parts (d) and (e).

84. In Exercise 18 of Homework 2.2, you found an equation close to $L = 0.16t + 73.73$, where L is the life expectancy at birth (in years) of an American born t years after 1980 (see Table 52).

Table 52 Life Expectancies at Birth

Year of Birth	Life Expectancy (years)
1980	73.7
1985	74.7
1990	75.4
1995	75.8
2000	77.0
2003	77.6

Source: *U.S. Census Bureau*

a. Use the linear model to predict the life expectancy of an American born in 2011.
b. Use the linear model to predict the birth year in which the life expectancy of an American will be 81 years.
c. Use the linear model to estimate what your (or someone else's) life expectancy was at birth. Given the fact that you have made it to your current age, do you think that your life expectancy is now less than, the same as, or more than it was at your time of birth?
d. Find the t-intercept. What does it mean in this situation?
e. Sketch by hand a *qualitative* graph to describe human life expectancy at birth during the existence of humans on Earth. [**Hint:** Your curve should be nonlinear.]

85. The number of commercial airline boardings on domestic flights increased steadily during the 1990s (see Table 53).

Table 53 Numbers of Commercial Airline Boardings on Domestic Flights

Year	Number of Boardings (millions)
1991	452
1993	487
1995	547
1997	599
1999	635
2000	666

Source: *Bureau of Transportation Statistics*

a. Let $f(t)$ be the number of commercial airline boardings on domestic flights (in millions) for the year that is t years since 1990. Use a graphing calculator to draw a scattergram of the data.
b. Find an equation of f. Does your model fit the data well?
c. Use your model f to estimate the number of boardings in 2001. The actual number was 622 million. What is the error in your estimate? (The error is the difference between the estimated value and the actual value.)
d. The number of boardings in 2001 was low due to the terrorist attacks on September 11, 2001. By making the following assumptions, estimate the amount of money airlines lost in 2001.
- All trips were round trips.
- The average number of boardings for a round trip was four (two flights out, two back).
- The average round-trip fare was \$340.

86. In 1950, 1.8 million people lived in Detroit. However, Detroit's population has declined since then; it slipped below the 1 million mark in the 2000 census. As a result, Detroit lost over \$100 million in federal funds and income tax.

During the 1990s, to retain these monies, Detroit began restoring abandoned buildings to attract additional residents and began trying to locate all of its homeless so that everyone would be counted in the 2000 census. Detroit's populations are listed in Table 54 for various years.

Table 54 Detroit's Populations

Year	Populations (millions)
1950	1.85
1960	1.67
1970	1.51
1980	1.20
1990	1.03

Source: *U.S. Census Bureau*

a. Let $P = f(t)$ be Detroit's population (in millions) at t years since 1950. Use a graphing calculator to draw a scattergram of the data.

b. Find an equation of f. Does the graph of your equation come close to the points in the scattergram?

c. Use your model f to estimate the population in 2000. Assume that your result would have been Detroit's population but for the city's attempts to increase the count. The actual count was 951,270. Estimate by how much Detroit increased its population.

d. Find the linear model g that *contains* the data points in Table 55, where $g(t)$ is Detroit's population (in millions) at t years since 1950.

Table 55 Detroit's Populations

Year	Population (millions)
1990	1,027,974
2000	951,270

Source: *FAIR—Detroit Immigration Factsheet*

e. Find $f(60)$ and $g(60)$. If Detroit continues to attract new residents and locate its homeless, which of these results will likely be the better predictor of the population in 2010? Explain.

87. The United States and Great Britain use the Fahrenheit temperature scale, but most countries use the Celsius temperature scale. The Celsius reading 0°C is the temperature at which water freezes, and 100°C is the temperature at which water boils (at sea level). Table 56 shows equivalent Celsius and Fahrenheit temperatures.

Table 56 Equivalent Temperature Readings

Celsius Reading (°C)	Fahrenheit Reading (°F)
0	32
20	68
40	104
60	140
80	176
100	212

a. Let $F = f(C)$ be the Fahrenheit reading corresponding to a Celsius reading of C degrees. Find an equation of f.

b. If the temperature is 25°C, what is the Fahrenheit reading?

c. If the temperature is 40°F, what is the Celsius reading?

88. The rate at which a cricket chirps depends on the temperature of the surrounding air. You can estimate the air temperature by counting chirps! Some data are provided in Table 57.

Table 57 Rates of Cricket Chirping

Temperature (°F)	Rate (number of chirps per minute)
50	43
60	86
70	129
80	172
90	215

Source: Eric Sloane's Weather Book, *Eric Sloane, 2005*

a. Let $g(F)$ be the number of chirps per minute a cricket makes when the temperature is F degrees Fahrenheit. Find an equation of g. Verify that the graph of your equation comes close to the points in the scattergram of the data.

b. Find $g(73)$. What does it mean in this situation?

c. Find the value of F, where $g(F) = 100$. What does your result mean in this situation?

d. What are the possible air temperatures at a field where the crickets are not chirping?

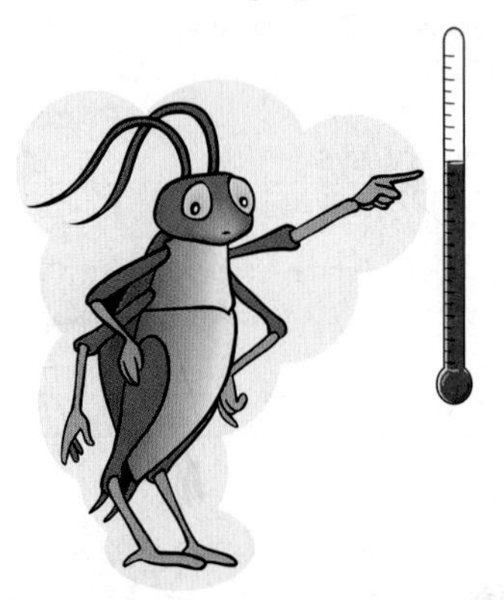

89. In Exercise 25 of Homework 2.2, you found an equation close to $p = 2.48x - 23.64$, where p is the percentage of Intermediate Algebra students at College of San Mateo (CSM) succeeding in the course (grade of A, B, or C) who scored x points on the placement test (see Table 58).

Table 58 Percentages of Intermediate Algebra Students Who Succeeded

Placement Score Group	Score Used to Represent Score Group	Percent Who Succeeded in Intermediate Algebra
21–25	23	34
26–30	28	47
31–35	33	55
36–40	38	71
41–45	43	84
46–50	48	*

Source: *College of San Mateo Mathematics Department*
*There were not enough students in this group to give useful data.

a. Rewrite the equation $p = 2.48x - 23.64$ with the function name f.
b. Students who score below 21 points (out of 50) on the placement test cannot enroll in Intermediate Algebra. Use the model f to estimate how high the cutoff score would have to be to ensure that all students succeed in the course.
c. Use the model f to estimate for which scores no students would succeed in the course.
d. If, in one semester, 145 students scored in the 16–20-point range on the placement test, predict how many of these students would have succeeded in the course if they had been allowed to enroll in it. Would you advise CSM to lower the placement score cutoff to 16? Explain.
e. Table 59 shows the numbers of students in various placement score groups for one semester. For students who scored at least 21 points on the placement test that semester, estimate how many succeeded in the course.

Table 59 Placement Test Scores for One Semester

Placement Score Group	Number of Students
21–25	94
26–30	44
31–35	19
36–40	12
41–45	9
46–50	4

90. In Exercise 20 of Homework 2.2, you found an equation close to $p = 0.42a - 13.91$, where p is the percentage of Americans at age a years who have been diagnosed with diabetes at some point in their lives (see Table 60).

Table 60 Percentages of Americans Diagnosed with Diabetes, by Age Group

Age Group (years)	Age Used to Represent Age Group (years)	Percent
35–39	37	2
40–44	42	4
45–49	47	5
50–54	52	8
55–59	57	10
60–64	62	13
65–69	67	14

Source: *National Health Interview Survey*

a. Rewrite the equation $p = 0.42a - 13.91$ with the function name f.
b. Estimate the percentage of 40-year-old Americans who have been diagnosed with diabetes.
c. Estimate at what age 7% of Americans have been diagnosed with diabetes.
d. Find the a-intercept of the model. What does it mean in this situation?
e. The chance of any one person being diagnosed increases as the person grows older. However, 13% of all Americans over the age of 70 have been diagnosed at some point in their lives—less than the percentage for ages 65–69 years. How is this possible?

91. Public school per-student expenditures increased approximately linearly from \$5.2 thousand in 1981 to \$9.0 thousand in 2002 (Source: *U.S. Department of Education*). Predict the per-student expenditure in 2011.

92. The number of words in the federal tax code increased approximately linearly from 1.4 million words in 1955 to 8.9 million words in 2000 (Source: *The Tax Foundation*). Predict the number of words in the federal tax code in 2010.

93. The percentage of male workers who prefer a female boss over a male boss increased approximately linearly from 4% in 1975 to 13% in 2002 (Source: *Bureau of Labor Statistics*). Predict when 16% of male workers will prefer a female boss.

94. The percentage of female workers who prefer a female boss over a male boss increased approximately linearly from 10% in 1975 to 23% in 2002 (Source: *Bureau of Labor Statistics*). Predict when 27% of female workers will prefer a female boss.

95. The percentage of large or medium-sized companies paying 100% of their employees' health care premiums decreased approximately linearly from 33% in 1999 to 17% in 2004 (Source: *Hay Group*).

a. Predict when no large or medium-sized companies will pay 100% of their employees' health care premiums.
b. Estimate the percentage of large or medium-sized companies that paid 100% of their employees' health care premiums in 2006.

96. The percentage of U.S. automobile consumers who are under 50 years of age decreased approximately linearly from 56.2% in 1990 to 50.1% in 2005 (Source: *CNW Marketing Research*).

a. Predict when that percentage will be 47%.
b. Predict the percentage in 2010.

97. The average score on the National Assessment of Educational Progress test in U.S. history was 195 points for fourth-graders who studied history about 45 minutes per week. The average score was 211 points for fourth-graders who studied history about 150 minutes per week. There is an approximate linear relationship between the number of hours that fourth-graders study history per week and the average score on the test (Source: *U.S. Department of Education*). Estimate the average score for fourth-graders who study history about 200 minutes per week.

98. In Mississippi, a child is eligible for the Children's Health Insurance Program (CHIP) if the child's family meets an income limit. For a family of four, family income must be no more than \$3067. For a family of six, family income must be no more than \$4114. There is a linear relationship between family size and the income limit (Source: *CHIP*). What is the income limit of a family of seven?

99. A basement is flooded with 640 cubic feet of water. It takes 4 hours to pump out the water. Let $f(t)$ be the number of cubic feet of water that remains in the basement after t hours of pumping.

a. Find a linear equation of f. [**Hint:** You are given information about two points that can be used to find an equation.]
b. Graph f by hand. Use a graphing calculator to verify your graph.
c. What are the domain and range of the model? Explain.

100. It takes a person 5 minutes to eat all 12 ounces of ice cream in a cup. Let $f(t)$ be the number of ounces of ice cream remaining in the cup t minutes after the person began eating the ice cream.

a. Find a linear equation of f. [**Hint:** You are given information about two points that can be used to find an equation.]

b. Graph f by hand. Use a graphing calculator to verify your graph.

c. What are the domain and range of the model? Explain.

101. For a function f, assume that $f(3) = 5$. Name an input and an output of f. Also, find three possible equations of f.

102. Describe the four-step modeling process in your own words.

2.4 SLOPE IS A RATE OF CHANGE

Objectives

- Calculate the *rate of change* of a quantity.
- Understand why slope is a rate of change.
- Use the rate of change to help find a linear model.
- Perform a *unit analysis* of a linear model.

How quickly does a quantity change in relation to another quantity? For example, how quickly has Nevada's population increased, or how quickly has the number of hotel fires declined?

Calculating Rate of Change

The **ratio** of a to b is the fraction $\frac{a}{b}$. A **unit ratio** is a ratio written as $\frac{a}{b}$ with $b = 1$.

Suppose that sea level increased *steadily* by 12 inches in the past 4 hours as it approaches high tide. We can compute how much sea level changed *per hour* by finding the unit ratio of the change in sea level (12 inches) to the change in time (4 hours):

$$\frac{12 \text{ inches}}{4 \text{ hours}} = \frac{3 \text{ inches}}{1 \text{ hour}}$$

So, sea level increased by 3 inches per hour. This is an example of a *rate of change*. We say that the rate of change of sea level with respect to time is 3 inches per hour. The rate of change is a *constant* because sea level increased *steadily*.

Here are some other examples of rates of change:

- The number of members of a club increases by five people per month.
- The value of a stock decreases by \$2 per week.
- The cost of a gallon of gasoline increases by 10¢ per month.

Suppose that sea level increases by 5 inches in 1 hour but by 3 inches in the next hour. We can find the *average rate of change* of sea level with respect to time by finding the unit ratio of the *total* change in sea level (8 inches) to the *total* change in time (2 hours):

$$\frac{8 \text{ inches}}{2 \text{ hours}} = \frac{4 \text{ inches}}{1 \text{ hour}}$$

So, the average rate of change is 4 inches per hour.

Formula for Rate of Change and Average Rate of Change

Suppose that a quantity y changes steadily from y_1 to y_2 as a quantity x changes steadily from x_1 to x_2. Then the **rate of change** of y with respect to x is the ratio of the change in y to the change in x:

$$\frac{\text{change in } y}{\text{change in } x} = \frac{y_2 - y_1}{x_2 - x_1}$$

If either quantity does not change steadily, then this formula is the **average rate of change** of y with respect to x.

Example 1 Finding Rates of Change

1. The number of fires in U.S. hotels declined approximately steadily from 7100 fires in 1990 to 4200 fires in 2002 (Source: *U.S. Fire Administration*). Find the average rate of change of the number of hotel fires per year between 1990 and 2002.
2. In San Bruno, California, the average value of a two-bedroom home is \$543 thousand, and the average value of a five-bedroom home is \$793 thousand (Source: *Green Banker*). Find the average rate of change of the average value of a home with respect to the number of bedrooms.

Solution

1.
$$\frac{\text{change in number of fires}}{\text{change in time}} = \frac{4200 \text{ fires} - 7100 \text{ fires}}{\text{year } 2002 - \text{year } 1990}$$
Change in a quantity is ending amount minus beginning amount.
$$= \frac{-2900 \text{ fires}}{12 \text{ years}}$$
Subtract.
$$\approx \frac{-241.67 \text{ fires}}{1 \text{ year}}$$
Find unit ratio.

The average rate of change of the number of fires per year was about -241.67 fires per year. So, on average, the number of fires declined yearly by about 242 fires.

2. To be consistent in finding the signs of the changes, we assume that the number of bedrooms increases from two to five and that the average value increases from \$543 thousand to \$793 thousand:

$$\frac{\text{change in average value}}{\text{change in number of bedrooms}} = \frac{793 \text{ thousand dollars} - 543 \text{ thousand dollars}}{5 \text{ bedrooms} - 2 \text{ bedrooms}}$$
Change in a quantity is ending amount minus beginning amount.
$$= \frac{250 \text{ thousand dollars}}{3 \text{ bedrooms}}$$
Subtract.
$$\approx \frac{83.33 \text{ thousand dollars}}{1 \text{ bedroom}}$$
Find unit ratio.

The average rate of change of the average value with respect to the number of bedrooms is about \$83.33 thousand per bedroom. So, the average value increases by about \$83.33 thousand per bedroom. ■

Our work in Example 1 shows a connection between the sign of a rate of change and whether the changing quantity is increasing or decreasing. In Problem 2, the average rate of change was *positive,* because the average value of a home *increases* (as the number of bedrooms increases). In Problem 1, the average rate of change was *negative,* because the number of hotel fires *decreased* (as time increased).

Increasing and Decreasing Quantities

Suppose that a quantity p depends on a quantity t:

- If p increases steadily as t increases steadily, then the rate of change of p with respect to t is positive.
- If p decreases steadily as t increases steadily, then the rate of change of p with respect to t is negative.

Slope Is a Rate of Change

The expression

$$\frac{y_2 - y_1}{x_2 - x_1}$$

that we have been using to calculate rate of change is the same expression that we use to calculate the slope of a line. This means that slope is a rate of change. We will explore this important concept in Example 2.

Table 61 Times and Distances

Time (hours) t	Distance (miles) d
0	0
1	60
2	120
3	180
4	240
5	300

Example 2 Comparing Slope with a Rate of Change

Suppose that a student drives at a constant rate. Let d be the distance (in miles) that the student can drive in t hours. Some values of t and d are shown in Table 61.

1. Create a scattergram. Then draw a linear model.
2. Find the slope of the linear model.
3. Find the rate of change of distance per hour for each given period. Compare each result with the slope of the linear model.
   a. From $t = 2$ to $t = 3$
   b. From $t = 0$ to $t = 4$

Solution

1. We draw a scattergram and then draw a line that contains the data points (see Fig. 42).

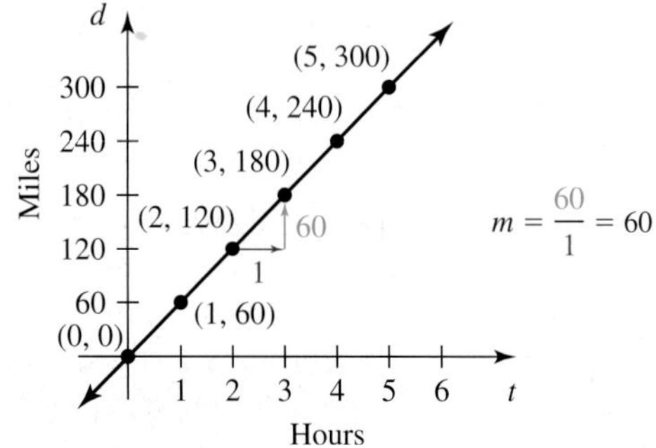

Figure 42 Car model and scattergram

2. The slope formula is $m = \dfrac{y_2 - y_1}{x_2 - x_1}$. So, with the variables t and d, we have

$$\frac{d_2 - d_1}{t_2 - t_1}$$

We arbitrarily use the points (2, 120) and (3, 180) to calculate the slope:

$$\frac{180 - 120}{3 - 2} = \frac{60}{1} = 60$$

So, the slope is 60. This checks with the calculation shown in Fig. 42.

3. a. Here we calculate the rate of change of distance per hour from $t = 2$ to $t = 3$:

$$\frac{\text{change in distance}}{\text{change in time}} = \frac{180 \text{ miles} - 120 \text{ miles}}{3 \text{ hours} - 2 \text{ hours}} \quad \text{Change in a quantity is ending amount minus beginning amount.}$$

$$= \frac{60 \text{ miles}}{1 \text{ hour}} \quad \text{Subtract.}$$

$$= 60 \text{ miles per hour} \quad \text{Divide.}$$

The rate of change (60 miles per hour) is equal to the slope (60).

b. Here we calculate the rate of change of distance per hour from $t = 0$ to $t = 4$:

$$\frac{\text{change in distance}}{\text{change in time}} = \frac{240 \text{ miles} - 0 \text{ miles}}{4 \text{ hours} - 0 \text{ hours}} \quad \text{Change in a quantity is ending amount minus beginning amount.}$$

$$= \frac{240 \text{ miles}}{4 \text{ hours}} \quad \text{Subtract.}$$

$$= \frac{60 \text{ miles}}{1 \text{ hour}} \quad \text{Find unit ratio.}$$

$$= 60 \text{ miles per hour} \quad \text{Divide.}$$

The rate of change (60 miles per hour) is equal to the slope (60). ■

In Example 2, we found that the time t and the distance d are linearly related. We also found that the slope is equal to the rate of change of distance per hour.

Slope Is a Rate of Change

If there is a linear relationship between quantities t and p, and if p depends on t, then the slope of the linear model is equal to the rate of change of p with respect to t.

In Problem 3 of Example 2, we calculated the same rate of change (60 miles per hour) for two different time periods. In fact, the rate of change is 60 miles per hour for *any* time period within the first five hours. This makes sense because the rate of change is equal to the slope of the line (60), which is a constant.

Constant Rate of Change

Suppose that a quantity p depends on a quantity t:

- If there is a linear relationship between t and p, then the rate of change of p with respect to t is constant.
- If the rate of change of p with respect to t is constant, then there is a linear relationship between t and p.

Finding an Equation of a Linear Model

We can use what we have learned about rate of change to help us find an equation of a linear model.

Example 3 Finding a Model

A company's profit was \$10 million in 2005 and has increased by \$3 million per year. Let p be the profit (in millions of dollars) in the year that is t years since 2005.

1. Is there a linear relationship between t and p? Explain.
2. Find the p-intercept of a linear model.
3. Find the slope of the linear model.
4. Find an equation of the linear model.

Solution

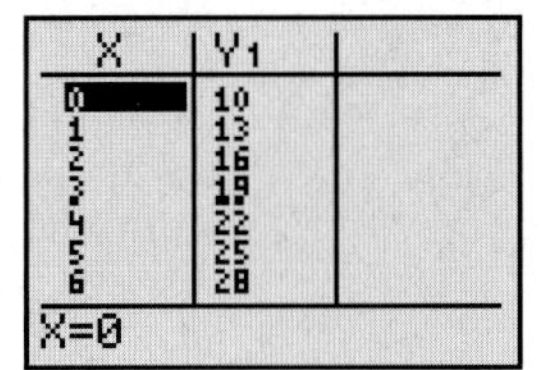

Figure 43 Verify the work

1. Since the rate of change of profit is a *constant* \$3 million per year, the variables t and p are linearly related.
2. The profit was \$10 million in 2005. Because 2005 is 0 years since 2005, we can represent this relationship by the ordered pair (0, 10). So, the p-intercept of the linear model is (0, 10).
3. The rate of change of profit per year is \$3 million per year. So, the slope of the linear model is 3.
4. An equation of the linear model can be written in the form $p = mt + b$. Since the slope is 3 and the p-intercept is (0, 10), we have $p = 3t + 10$.

   We can use a graphing calculator table to verify our equation (see Fig. 43). The ordered pair (0, 10) shown in the first row of the table means that the profit was \$10 million in 2005, which checks. Also, as the input increases by 1, the output increases by 3. This means that the profit increases by \$3 million per year, which also checks. ■

Unit Analysis of a Linear Model

In Example 3, we found the linear model $p = 3t + 10$. We can perform a *unit analysis* of the model by determining the units of the expressions on both sides of the equation:

$$\underbrace{p}_{\text{millions of dollars}} = \underbrace{3}_{\frac{\text{millions of dollars}}{\text{year}}} \cdot \underbrace{t}_{\text{years}} + \underbrace{10}_{\text{millions of dollars}}$$

We can use the fact that $\frac{\text{years}}{\text{year}} = 1$ to simplify the units of the expression on the right-hand side of the equation:

$$\frac{\text{millions of dollars}}{\text{year}} \cdot \text{years} + \begin{matrix}\text{millions}\\\text{of dollars}\end{matrix} = \begin{matrix}\text{millions}\\\text{of dollars}\end{matrix} + \begin{matrix}\text{millions}\\\text{of dollars}\end{matrix}$$
$$= \text{millions of dollars}$$

So, the units of the expressions on both sides of the equation are millions of dollars, which suggests that the equation is correct.

DEFINITION Unit analysis

We perform a **unit analysis** of a model's equation by determining the units of the expressions on both sides of the equation. The simplified units of the expressions on both sides of the equation should be the same.

We can perform a unit analysis of a model's equation to help verify the equation.

Example 4 Finding a Model

A driver fills her car's 12-gallon gasoline tank and drives at a constant speed. The car consumes 0.04 gallon per mile. Let G be the number of gallons of gasoline remaining in the tank after she has driven d miles since filling up.

1. Is there a linear relationship between d and G? Explain.
2. Find the G-intercept of a linear model.
3. Find the slope of the linear model.
4. Find an equation of the linear model.
5. Perform a unit analysis of the equation.

Solution

1. Since the rate of change of gallons remaining per mile is a *constant* -0.04 gallon per mile, the variables d and G are linearly related.
2. When the tank was filled, it contained 12 gallons of gasoline. We can represent this by the ordered pair (0, 12), which is the G-intercept.
3. The rate of change of gasoline remaining in the tank with respect to distance traveled is -0.04 gallon per mile. So, the slope of the linear model is -0.04.
4. An equation of the linear model can be written in the form $G = md + b$. Since the slope is -0.04 and the G-intercept is (0, 12), we have $G = -0.04d + 12$.
5. Here is a unit analysis of the equation $G = -0.04d + 12$:

$$\underbrace{G}_{\text{gallons}} = \underbrace{-0.04}_{\frac{\text{gallons}}{\text{mile}}} \cdot \underbrace{d}_{\text{miles}} + \underbrace{12}_{\text{gallons}}$$

We can use the fact that $\frac{\text{miles}}{\text{mile}} = 1$ to simplify the units of the expression on the right-hand side of the equation:

$$\frac{\text{gallons}}{\text{mile}} \cdot \text{miles} + \text{gallons} = \text{gallons} + \text{gallons}$$
$$= \text{gallons}$$

So, the units on both sides of the equation are gallons, which suggests that the equation is correct. ■

Two Variables That Are Approximately Linearly Related

In both Examples 3 and 4, we worked with two variables that are linearly related. We will now explore the meaning of the slope of a linear model in which two variables are *approximately* linearly related.

Example 5 Analyzing a Model

Table 62 Yogurt Sales

Year	Sales (billions of dollars)
2001	2.3
2002	2.5
2003	2.7
2004	2.8
2005	3.0

Source: *ACNielsen*

Yogurt sales (in billions of dollars) in the United States are shown in Table 62 for various years.

Let s be yogurt sales (in billions of dollars) in the year that is t years since 2000. A model of the situation is

$$s = 0.17t + 2.15$$

1. Use a graphing calculator to draw a scattergram and the model in the same viewing window. Check whether the line comes close to the data points.
2. What is the slope of the model? What does it mean in this situation?
3. Find the rates of change in sales from one year to the next. Compare the rates of change with the result in Problem 2.
4. Predict the sales in 2010.

Solution

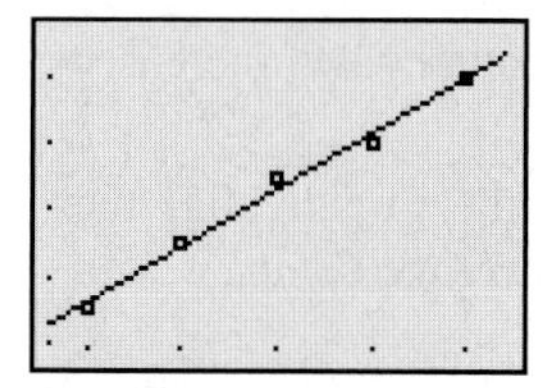

Figure 44 Yogurt scattergram and model

1. We draw the scattergram and the model in the same viewing window (see Fig. 44). For graphing calculator instructions on drawing scattergrams and models, see Sections B.8 and B.10.

   The line comes close to the data points, so the model is a reasonable one.
2. The slope is 0.17, because $s = 0.17t + 2.15$ is of the form $y = mx + b$ and $m = 0.17$. According to the model, sales are increasing by 0.17 billion dollars per year.
3. The rates of change of sales are shown in Table 63. All of the rates of change are fairly close to 0.17 billion dollars per year.

Table 63 Rates of Change of Yogurt Sales

Year	Sales (billions of dollars)	Rate of Change of Sales from Previous Year (billions of dollars per year)
2001	2.3	—
2002	2.5	$(2.5 - 2.3) \div (2002 - 2001) = 0.20$
2003	2.7	$(2.7 - 2.5) \div (2003 - 2002) = 0.20$
2004	2.8	$(2.8 - 2.7) \div (2004 - 2003) = 0.10$
2005	3.0	$(3.0 - 2.8) \div (2005 - 2004) = 0.20$

4. We substitute the input 10 for t in the equation $s = 0.17t + 2.15$:

$$s = 0.17(10) + 2.15 = 3.85$$

According to the model, yogurt sales will be \$3.85 billion in 2010. ■

In Example 5, we found that the slope of the yogurt model is 0.17, which means that, according to the model, yogurt sales increased by 0.17 billion dollars per year. In reality, yogurt sales did not increase by 0.17 billion dollars in any of the years between 2001 and 2005, inclusive. However, 0.17 billion dollars per year *is* a reasonable estimate of the *average* yearly increase.

Slope Is an Average Rate of Change

If two quantities t and p are approximately linearly related, and if p depends on t, then the slope of a reasonable linear model is approximately equal to the average rate of change of p with respect to t.

WARNING It is a common error to be vague in describing the meaning of the slope of a model. For example, a description such as

The slope means that it is increasing.

neither specifies the quantity that is increasing nor the rate of increase. The following statement includes the missing information:

The slope of 0.17 means that sales increased by 0.17 billion dollars per year.

group exploration

Slope of a linear model for approximately linearly related data

Table 64 Numbers of U.S. Satellite Television Subscribers

Year	Number of Subscribers (millions)
1998	9.3
1999	11.8
2000	14.5
2001	16.9
2002	19.0
2003	20.9
2004	23.4

Source: *Satellite Broadcasting and Communications Association*

The numbers of U.S. satellite television subscribers are listed in Table 64 for various years. Let $s = f(t)$ be the number of U.S. satellite television subscribers (in millions) at t years since 1995.

1. Use a graphing calculator to draw a scattergram of the data. Does it appear that the data can be modeled well by a linear function?
2. Use the data points (4, 11.8) and (9, 23.4) to find a linear model.
3. What is the slope of your model? According to the model, by how much will the number of subscribers increase each year?
4. Refer to Table 64 to find by how much the number of subscribers actually increased each year. List the increases in the number of subscribers from 1998 to 1999, from 1999 to 2000, and so on. How do the actual sales increases compare with the slope of your model?
5. Find the average of the increases in the number of subscribers by dividing the sum of the six numbers by the number 6. How does this average compare with the slope of your model?

group exploration

Looking ahead: Using a system of equations to model a situation

Table 65 Annual Fuel Consumption

	Annual Fuel Consumption (billions of gallons)	
Year	**Cars**	**Light Trucks**
1996	69.2	47.4
1998	71.7	50.5
2000	73.1	52.9
2002	75.5	55.2
2003	74.6	56.3

Source: *U.S. Federal Highway Administration*

Fuel consumption by cars and light trucks (vans, pickups, and SUVs) is shown in Table 65 for various years. Annual fuel consumption (in billions of gallons) $C(t)$ and $L(t)$ for cars and light trucks, respectively, are modeled by the system

$$F = C(t) = 0.84t + 64.63$$
$$F = L(t) = 1.25t + 40.16$$

where t is the number of years since 1990.

1. Find the F-intercepts of the graphs of C and L. What do they mean in this situation?
2. Find the slopes of the graphs of C and L. What do they mean in this situation?
3. Your responses to Problems 1 and 2 should suggest an event that will happen in the future. Describe that event.
4. On a graphing calculator, use Zoom Out twice and then "intersect" to estimate the coordinates of the point where the graphs of C and L intersect. (For graphing calculator instructions, see Sections B.6 and B.18.) What does it mean in terms of fuel consumption that the graphs intersect at this point? State the year and fuel consumption for this event.

TIPS FOR SUCCESS: Study in a Test Environment

Do you feel that you understand your homework assignments yet perform poorly on quizzes and tests? If so, you may not be studying enough to be ready to solve problems *in a test environment*. For example, although it is a good idea to refer to your lecture notes when you are stumped on a homework exercise, you must continue to solve similar exercises until you can solve them *without* referring to your lecture notes (unless your instructor uses open-notebook tests). The same idea applies to getting help from someone, referring to examples in the text, looking up answers in the back of the text, or any other form of support. Spend the last part of your study time completing exercises without such support. Try making up a practice quiz or test for yourself to do in a given amount of time.

HOMEWORK 2.4

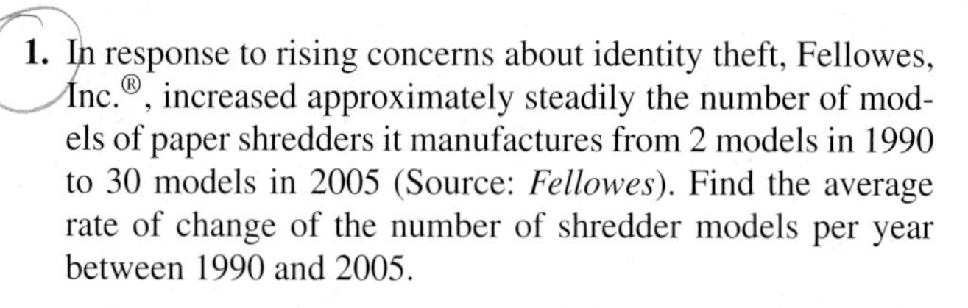

1. In response to rising concerns about identity theft, Fellowes, Inc.®, increased approximately steadily the number of models of paper shredders it manufactures from 2 models in 1990 to 30 models in 2005 (Source: *Fellowes*). Find the average rate of change of the number of shredder models per year between 1990 and 2005.

2. The average attendance at University of Southern California home football games increased approximately linearly from 59.0 thousand in 2001 to 90.8 thousand in 2005 (Source: *USC*). Find the average rate of change per year of the attendance.

3. Oil company employment declined approximately linearly from 1.6 million in 1982 to 0.5 million in 2004 (Source: *Wood Mackenzie*). Find the average rate of change of employment per year.

4. The District of Columbia's population decreased approximately linearly from 571,045 people in 2000 to 553,523 people in 2004 (Source: *U.S. Census Bureau*). Find the average rate of change of population in the District of Columbia per year.

5. The percentage of Americans who have trust in newspapers declined approximately linearly from 37% in 2000 to 28% in 2005 (Source: *The Gallup Organization*). Find the average rate of change of the percentage of Americans who have trust in newspapers per year.

6. The percentage of firearm applications that were rejected declined approximately linearly from 2.4% in 1999 to 1.6% in 2003 (Source: *FBI*). Find the average rate of change of this percentage per year.

7. In-district students at Triton College pay $504 for 9 credit hours (units) of classes and $672 for 12 credit hours of classes (Source: *Triton College*). Find the average rate of change of the total cost of classes with respect to the number of credit hours of classes.

8. In Manhattan, the average price of a one-bedroom condominium is $496,773, and that of a three-bedroom condominium is $1,360,571 (Source: *Condo Sales*). Find the average rate of change of price with respect to the number of bedrooms.

9. An Intermediate Algebra student drives a car at 70 miles per hour. Let d be the distance (in miles) that the student travels in t hours.
   a. Is there a linear relationship between t and d? Explain. If the relationship is linear, find the slope and describe what it means in this situation.
   b. Find an equation of the model.

10. A train is moving at 40 miles per hour. Let d be the distance (in miles) the train travels in t hours.
    a. Create a table of values of t and d. Then make a scattergram.
    b. Is there a linear relationship between t and d? Explain. If the relationship is linear, find the slope and describe what it means in this situation.

11. The number of U.S. households that paid bills online was 18.9 million in 2003 and has increased by about 6.7 million per year (Source: *Jupiter Research*). Let n be the number of households (in millions) that paid bills online at t years since 2003.
    a. Is there an approximate linear relationship between t and n? Explain. If the relationship is approximately linear, find the slope and describe what it means in this situation.
    b. What is the n-intercept of the model? What does it mean in this situation?
    c. Find an equation of the model.
    d. Perform a unit analysis of the equation you found in part (c).
    e. Predict the number of households that will pay bills online in 2011.

12. Retail sales of skin care products were $4.7 billion in 2002 and have increased by about $0.23 billion per year (Source: *NPD Group*). Let s be the retail sales (in billions of dollars) in the year that is t years since 2002.
    a. Is there an approximate linear relationship between t and s? Explain. If the relationship is approximately linear, find the slope and describe what it means in this situation.
    b. What is the s-intercept of the model? What does it mean in this situation?
    c. Find an equation of the model.
    d. Perform a unit analysis of the equation you found in part (c).
    e. Predict in which year retail sales of skin care products will be $7 billion.

13. Major league baseball games lasted an average of 2 hours and 46 minutes in 2003. Each year since then, the average time has decreased by about 3.8 minutes. Let $g(t)$ be the average time (in minutes) of major league baseball games in the year that is t years since 2003.
    a. What is the slope of the graph of g? What does it mean in this situation?
    b. Find an equation of g.
    c. Predict the average time of a baseball game in 2010.
    d. Predict when a baseball game will last $2\frac{1}{2}$ hours.
    e. Find the t-intercept of the model. What does it mean in this situation?

14. The percentage of all U.S. workers who are in a union was 13% in 2004 and has decreased by about 0.29 percentage point per year (Source: *U.S. Census Bureau*). Let $f(t)$ be the percentage of workers who are in a union at t years since 2004.
    a. What is the slope of the graph of f? What does it mean in this situation?
    b. Find an equation of f.
    c. Predict when 11% of all workers will be in a union.
    d. Predict the percentage of workers who will be in a union in 2010.
    e. Find the t-intercept of the model. What does it mean in this situation?

15. For fall 2005, full-time undergraduate residents paid \$1585.50 for tuition at Southeastern Louisiana University. Students could rent textbooks for \$25 per course. Let $f(n)$ be the total one-semester cost (in dollars) of tuition and textbook rental for n courses.

a. What is the slope of the graph of f? What does it mean in this situation?
b. Find an equation of f.
c. Perform a unit analysis of your equation of f.
d. Find $f(4)$. What does it mean in this situation?
e. Find n when $f(n) = 1710.50$. What does it mean in this situation?

16. For spring 2006, students taking up to 15 credit hours (units) at Clackamas Community College paid \$56 per credit hour for tuition, \$4 per credit hour for a general fee, and a \$15 billing fee per semester. Let $h(c)$ be the total one-semester cost (in dollars) of tuition and fees for a student who is taking c credit hours.

a. What is the slope of the graph of h? What does it mean in this situation?
b. Find an equation of h.
c. Perform a unit analysis of your equation of h.
d. Find $h(9)$. What does it mean in this situation?
e. Find c when $h(c) = 735$. What does it mean in this situation?

17. A person drives her Honda CR-V® on a road trip. At the start of the trip, she fills up the 15.3-gallon tank with gasoline. During the trip, the car uses about 0.05 gallon of gas per mile. Let $g(x)$ be the number of gallons of gasoline remaining in the tank (which has not been refilled) after she has driven x miles.

a. What is the slope of the graph of g? What does it mean in this situation?
b. Find an equation of g.
c. Find the x-intercept of the model. What does it mean in this situation?
d. What are the domain and range of g?
e. If the driver will refuel the car when 1 gallon of gasoline remains in the tank, how far can she drive the car before refueling?

18. Atmospheric pressure at sea level is 1 atmosphere (atm). Under water, pressure increases by approximately 0.0303 atm for every 1-foot increase in water depth. For example, water pressure in the ocean is 1.0303 atm at a depth of 1 foot. Let $f(d)$ be the water pressure (in atm) at a water depth of d feet.

a. What is the slope of the graph of f? What does it mean in this situation?
b. Find an equation of f.
c. Perform a unit analysis of your equation of f.
d. How deep must you dive for the water pressure to be twice the pressure at sea level?
e. Crater Lake, in Oregon, is the deepest lake in the United States. Find the water pressure at 1943 feet, the lake's greatest depth.

19. The percentage of materials that were recycled in the United States was 30% in 2001 and has increased by 0.86 percentage point per year (Source: *Franklin Associates, Ltd.*). Predict when 40% of materials in the United States will be recycled.

20. The number of U.S. patent applications was 344 thousand applications in 2001 and has increased by 31 thousand applications per year (Source: *U.S. Patent and Trademark Office*). Predict when there will be 650 thousand patent applications.

21. MCI charges \$1.25 per minute, plus a \$1.75 per-call surcharge, for state-to-state calls under a certain calling plan. If a person is charged \$48 for a state-to-state call, how long was the call?

22. Gasoline sells for \$2.95 per gallon at a certain gas station. A half-gallon of milk sells for \$1.39. If a person pays a total of \$39.74 for gasoline and a half-gallon of milk, how many gallons of gasoline did he buy?

23. In Example 6 of Section 2.3, we made predictions with the model $s = 1.71t - 113.12$, where s is the average salary (in thousands of dollars) of professors at four-year public colleges and universities at t years since 1900. What is the slope of this model? What does it mean in this situation?

24. In Exercise 15 of Homework 2.2, you found an equation close to $p = 0.77t - 42.90$ that models the percentage p of births outside marriage in the United States at t years since 1900. What is the slope of this model? What does it mean in this situation?

25. In Exercise 21 of Section 2.2, you found an equation close to $r = -0.27t + 70.45$, where r is the record time (in seconds) for the women's 400-meter run at t years since 1900. What is the slope of this model? What does it mean in this situation?

26. In Example 3 of Section 2.2, we found the equation

$$p = -0.53t + 74.50$$

where p is the percentage of adult Americans who smoke at t years since 1900. What is the slope of this model? What does it mean in this situation?

27. In Exercise 87 of Homework 2.3, you found the equation $f(C) = 1.8C + 32$, where $f(C)$ is the Fahrenheit reading corresponding to a Celsius reading of C degrees. What is the slope of this model? What does it mean in this situation?

28. In Exercise 88 of Homework 2.3, you found the equation $g(F) = 4.3F - 172$, where $g(F)$ is the number of chirps per minute a cricket makes when the temperature is F degrees Fahrenheit. What is the slope of this model? What does it mean in this situation?

29. The percentage of the world's population that lives in rural areas has decreased steadily for the past 50 years (see Table 66).

Table 66 Percentages of World Population Living in Rural Areas

Year	Percent
1950	70
1960	66
1970	63
1980	61
1990	57
2000	52
2003	52

Source: *Food and Agricultural Organization of the United Nations*

a. Let $f(t)$ be the percentage of world population that lives in rural areas at t years since 1950. Find an equation of f.

b. Find the slope of the model. What does it mean in this situation?

c. Estimate the *number* of people who lived in rural areas in 2006, when world population was 6.6 billion.

d. Predict when exactly half of the world population will live in rural areas.

e. Find the t-intercept of the model. What does it mean in this situation?

30. Nevada was the fastest-growing U.S. state for 19 consecutive years (see Table 67).

Table 67 Nevada's Population

Year	Population (millions)
1986	1.0
1990	1.2
1995	1.5
2000	2.0
2005	2.4

Source: *U.S. Census Bureau*

a. Let $f(t)$ be Nevada's population (in millions) at t years since 1985. Find an equation of f.

b. What is the slope of the model? What does it mean in this situation?

c. Predict Nevada's population in 2012.

d. Find the t-intercept of the model. What does it mean in this situation?

e. Predict when Nevada's population will reach Kansas's current population of 2.7 million.

31. Average tuitions at four-year colleges are listed in Table 68 for various years.

Table 68 Average Tuitions at Four-Year Colleges

Year	Public Tuition (dollars)	Private Tuition (dollars)
1984	2074	9202
1989	2395	12,146
1994	3188	13,844
1999	3632	16,454
2004	4694	19,710

Source: *The College Board*

a. Let $f(t)$ be the average tuition (in dollars) for public colleges at t years since 1980. Find an equation of f.

b. Let $g(t)$ be the average tuition (in dollars) for private colleges at t years since 1980. Find an equation of g.

c. What is the slope of the graph of f? The slope of the graph of g? What do these slopes tell you about this situation?

d. A student intends to earn a bachelor's degree by attending a college for four years, starting in 2010. Compare the approximate total four-year cost of attending a public college with that of attending a private college.

32. Amusement park injuries from mobile rides (such as roller coasters) and inflatable rides (such as inner-tube and raft rides) are shown in Table 69 for various years.

Table 69 Amusement Park Injuries

	Injuries (thousands)	
Year	Mobile Rides	Inflatable Rides
1997	3.0	1.3
1998	3.2	1.6
1999	3.2	2.2
2000	4.3	2.0
2001	2.8	2.3
2002	3.0	3.6
2003	3.0	4.3
2004	2.5	4.9

Source: *U.S. Consumer Product Safety Commission, NEISS*

a. Let $M(t)$ be the number of injuries (in thousands) from mobile rides in the year that is t years since 1990. Can the data be modeled well by using a linear model? If yes, find such a model. If no, explain why not.

b. Use the data for 1997 and 2003 in Table 69 to find the average rate of change of injuries from mobile rides from 1997 to 2003. Explain how your result relates to the result that you found in part (a).

c. Let $I(t)$ be the number of injuries (in thousands) from inflatable rides in the year that is t years since 1990. Can the data be modeled well by a linear model? If yes, find such a model. If no, explain why not.

d. Estimate the rate of change of the number of injuries per year from inflatable rides.

e. Predict in which year there will be 8000 injuries from inflatable rides. How many injuries is that per state, on average?

33. Table 70 shows various altitudes and the corresponding atmospheric pressures.

Table 70 Relationship Between Altitude and Pressure

Altitude (thousands of feet above sea level)	Pressure (inches of mercury)
0	29.92
1	28.86
2	27.82
3	26.82
4	25.84
5	24.89
6	23.98

Source: *Abbess Instruments*

a. Let $f(a)$ be the pressure (in inches of mercury) at altitude a (in thousands of feet above sea level). Find an equation of f.

b. What is the slope of the model? What does it mean in this situation?

c. Find the average rate of change of pressure with respect to altitude for each of the following changes in altitude. Compare each result with the result that you found in part (b):

i. From 1 thousand feet to 4 thousand feet

ii. From 2 thousand feet to 5 thousand feet

iii. From sea level to 6 thousand feet

d. Mount Elbert, in Colorado, is the second-highest mountain in the continental United States. Estimate the pressure at its peak, which is at 14,440 feet.

34. The heat index is the temperature the average person "feels" at a given humidity and air temperature. Heat indexes for various humidities when the air temperature is 75°F are shown in Table 71.

Table 71 Heat Indexes at Air Temperature of 75°F

Relative Humidity (percent)	Heat Index (degrees Fahrenheit)
0	69
20	72
40	74
60	76
80	78
100	80

Source: *National Weather Service*

a. Let $f(p)$ be the heat index (in degrees Fahrenheit) when the relative humidity is p percent (at air temperature of 75°F). Find an equation of f.

b. What is the slope of the model? What does it mean in this situation?

c. Find the average rate of change of the heat index with respect to relative humidity (at air temperature of 75°F) for each of the following changes in relative humidity. Compare each result with the result that you found in part (b):

i. From 0% humidity to 100% humidity

ii. From 20% humidity to 80% humidity

iii. From 40% humidity to 60% humidity

d. If the heat index is 77.6°F and the air temperature is 75°F, what is the relative humidity?

35. An airplane flies at a speed of 500 miles per hour for 3 hours. It then runs into strong headwinds and travels at a speed of 400 miles per hour for 2 more hours. Let $f(t)$ be the distance (in miles) traveled in t hours.

a. Complete a table consisting of values of t and $f(t)$. Use 0, 1, 2, 3, 4, and 5 for t. Assume that it takes no time for the airplane to decelerate from 500 miles per hour to 400 miles per hour.

b. Graph the function f by hand.

c. Discuss the assumption that it takes no time for the airplane to decelerate from 500 miles per hour to 400 miles per hour. Is that possible? Explain.

d. Discuss the connections between the speeds of the airplane and the graph of f.

36. At noon, a math instructor drives at 50 miles per hour for 2 hours. During the next minute, she accelerates to 70 miles per hour, then travels at that speed for 3 more hours. (That is, she is on the road for a total of 5 hours and 1 minute.) Let d be the distance traveled (in miles) after t hours have elapsed.

a. Complete Table 72.

Table 72 Accelerating from 50 mph to 70 mph in One Minute

Time of Day	t (in hours)	d (in miles)
12:00 P.M.	0	
1:00 P.M.	1	
2:00 P.M.	2	
2:01 P.M.	2.017	
3:01 P.M.	3.017	
4:01 P.M.	4.017	
5:01 P.M.	5.017	

b. Sketch a curve that describes the relationship between t and d.

For each scenario, sketch a qualitative graph that relates the distance d from home (in miles) to the number of minutes t that have elapsed.

37. A student drives at a constant speed. After a while, his favorite song plays on the radio, and he drives the rest of the way at a faster constant speed.

38. A commuter drives toward work at a constant speed. She notices a police car and slows down to a slower constant speed. After a while, the police car exits, and the commuter quickly resumes driving at her original constant speed.

39. A student drives toward school at a constant speed. After a while, she gets a flat tire, which takes some time to fix. She then continues to drive toward school, but at a faster constant speed to make up for lost time.

40. A student drives toward school at a constant speed. After a while, he realizes that he left his graphing calculator at home. He turns around and heads home at a faster constant speed in hopes of getting to class on time. After getting his calculator, he drives toward school at an even faster constant speed.

41. Explain the statement "Slope is a rate of change." Give an example other than one of those given in the text.

42. Give an example to illustrate that if the rate of change of one quantity with respect to another quantity is constant, then there is a linear relationship between the two quantities.

Chapter Summary

Key Points OF CHAPTER 2

Scattergram (Section 2.1)	A graph of plotted ordered pairs is called a **scattergram.**
Creating a complete scattergram (Section 2.1)	A scattergram should have scaling on both axes and labels indicating the variable names and scale units.
Approximately linearly related (Section 2.1)	If the points in a scattergram of data lie close to (or on) a line, then we say that the relevant variables are **approximately linearly related.**

Model (Section 2.1)

A **model** is a mathematical description of an authentic situation.

Linear model (Section 2.1)

A **linear model** is a linear function, or its graph, that describes the relationship between two quantities for an authentic situation.

Creating scattergrams and making estimates and predictions (Section 2.1)

We create a scattergram of data to determine whether the relevant variables are approximately linearly related. If they are, we draw a line that comes close to the data points and use the line to make estimates and predictions.

Interpolation and extrapolation (Section 2.1)

For a situation that can be modeled by a function whose independent variable is t: We perform **interpolation** when we use a part of the model whose t-coordinates are between the t-coordinates of two data points. We perform **extrapolation** when we use a part of the model whose t-coordinates are not between the t-coordinates of any two data points.

Model breakdown (Section 2.1)

When a model gives a prediction that does not make sense or an estimate that is not a good approximation, we say that **model breakdown** has occurred.

Finding an equation of a linear model (Section 2.2)

To find an equation of a linear model, given some data,

1. Create a scattergram of the data.
2. Determine whether there is a line that comes close to the data points. If so, choose two points (not necessarily data points) that you can use to find the equation of a linear model.
3. Find an equation of the line you identified in step 2.
4. Use a graphing calculator to verify that the graph of your equation comes close to the points of the scattergram.

Function notation (Section 2.3)

The dependent variable of a function f can be represented by the expression formed by writing the independent variable name within the parentheses of $f(\ )$:

$$\text{dependent variable} = f(\text{independent variable})$$

We call this representation **function notation.**

Four-step modeling process (Section 2.3)

To find a linear model and make estimates and predictions:

1. Create a scattergram of the data to determine whether there is a nonvertical line that comes close to the data points. If so, choose two points (not necessarily data points) that you can use to find the equation of a linear model.
2. Find an equation of your model.
3. Verify your equation by checking that the graph of your model contains the two chosen points and comes close to all of the data points.
4. Use the equation of your model to make estimates, make predictions, and draw conclusions.

Making a prediction about the dependent variable (Section 2.3)

When making a prediction about the dependent variable of a linear model, substitute a chosen value for the independent variable in the model. Then solve for the dependent variable.

Making a prediction about the independent variable (Section 2.3)

When making a prediction about the independent variable of a linear model, substitute a chosen value for the dependent variable in the model. Then solve for the independent variable.

Intercepts of a model (Section 2.3)

If a function of the form $p = mt + b$, where $m \neq 0$, is used to model a situation, then

- The p-intercept is $(0, b)$.
- To find the t-coordinate of the t-intercept, substitute 0 for p in the model's equation and solve for t.

Domain and range of a model (Section 2.3)

For the **domain** and **range** of a model, we consider input–output pairs only when both the input and the output make sense in the situation. The domain of the model is the set of all such inputs, and the range of the model is the set of all such outputs.

Rate of change and average rate of change (Section 2.4)

Suppose that a quantity y changes steadily from y_1 to y_2 as a quantity x changes steadily from x_1 to x_2. Then the **rate of change** of y with respect to x is the ratio of the change in y to the change in x:

$$\frac{\text{change in } y}{\text{change in } x} = \frac{y_2 - y_1}{x_2 - x_1}$$

If either quantity does not change steadily, then this formula is the **average rate of change** of y with respect to x.

Increasing and decreasing quantities (Section 2.4)	Suppose that a quantity p depends on a quantity t: • If p increases steadily as t increases steadily, then the rate of change of p with respect to t is positive. • If p decreases steadily as t increases steadily, then the rate of change of p with respect to t is negative.
Slope is a rate of change (Section 2.4)	If there is a linear relationship between quantities t and p, and if p depends on t, then the slope of the linear model is equal to the rate of change of p with respect to t.
Constant rate of change (Section 2.4)	Suppose that a quantity p depends on a quantity t: • If there is a linear relationship between t and p, then the rate of change of p with respect to t is constant. • If the rate of change of p with respect to t is constant, then there is a linear relationship between t and p.
Unit analysis (Section 2.4)	We perform a **unit analysis** of a model's equation by determining the units of the expressions on both sides of the equation. The simplified units of the expressions on both sides of the equation should be the same.
Slope is an average rate of change (Section 2.4)	If two quantities t and p are approximately linearly related, and if p depends on t, then the slope of a reasonable linear model is approximately equal to the average rate of change of p with respect to t.

CHAPTER 2 REVIEW EXERCISES

For $f(x) = 3x^2 - 7$, $g(x) = \dfrac{2x+5}{3x+6}$, *and* $h(x) = -10x - 3$, *find the following.*

1. $f(3)$ **2.** $f(-3)$ **3.** $g(2)$

4. $h\left(\dfrac{3}{5}\right)$ **5.** $h(a+3)$

For $f(x) = 2x + 3$, *find the value of* x *that corresponds to the given value of* $f(x)$.

6. $f(x) = -6$ **7.** $f(x) = \dfrac{2}{3}$

8. $f(x) = a + 7$

For Exercises 9–16, refer to Fig. 45.

9. Estimate $f(2)$. **10.** Estimate $f(0)$.

11. Estimate $f(-3)$.

12. Estimate x when $f(x) = 3$.

13. Estimate x when $f(x) = 0$.

14. Estimate x when $f(x) = -1$.

15. Find the domain of f. **16.** Find the range of f.

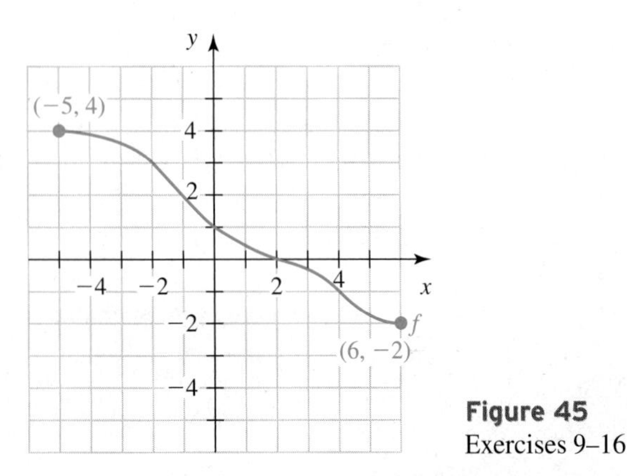

Figure 45 Exercises 9–16

For Exercises 17–20, refer to Table 73.

17. Find $f(0)$. **18.** Find $f(2)$.

19. Find x when $f(x) = 0$. **20.** Find x when $f(x) = 2$.

Table 73 Values of f (Exercises 17–20)

x	$f(x)$
0	1
1	2
2	4
3	3
4	0

Find all x-intercepts and y-intercepts of the graph of the function.

21. $f(x) = -7x + 3$ **22.** $f(x) = 4$

23. $f(x) = -\dfrac{4}{7}x + 2$

24. Find the approximate x-intercept and the approximate y-intercept of the graph of $2.56x - 9.41y = 78.25$. Round the coordinates to the second decimal place.

25. Copy the graphs of the data points and the model $y = mx + b$ in Fig. 46. Sketch the graph of a linear model that describes the data better. Then explain how you would adjust the slope and the y-intercept of the original model to describe the data better.

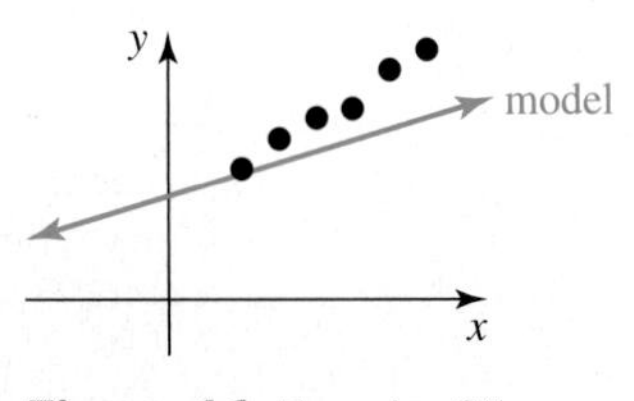

Figure 46 Exercise 25

26. A student's car has a 13-gallon gasoline tank. The car uses 1.8 gallons per hour when driven at 65 miles per hour. After filling up the gas tank, the student begins driving at 65 miles per hour and continues to drive at this speed until he runs out of gas. Let $A = f(t)$ be the amount of gasoline (in gallons) in the tank at t hours after the student filled the tank.
   a. Find an equation of f.
   b. What is the slope of the model? What does it mean in this situation?
   c. What is the A-intercept of the model? What does it mean in this situation?
   d. Perform a unit analysis of the equation you found in part (a).
   e. What is the t-intercept of the model? What does it mean in this situation?
   f. What are the domain and range of the model? Explain.

27. Average U.S. personal income in 2004 was \$32,937. Each year, the income increases by about \$962 (Source: *Bureau of Economic Analysis*). Let $g(t)$ be the average personal income (in dollars) at t years since 2004.
   a. Find the slope of the graph of g. What does it mean in this situation?
   b. Find an equation of g.
   c. Find $g(6)$. What does it mean in this situation?
   d. Find t, where $g(t) = 40{,}000$. What does it mean in this situation?

28. The number of times that President George W. Bush used the word "economy" in his State of the Union address increased approximately linearly from 4 times in 2002 to 16 times in 2006 (Source: The New York Times). Find the rate of change of the number of times he used the word "economy" in State of the Union addresses per year.

29. Sales at duty-free shops worldwide have increased approximately linearly from \$2.0 billion in 1980 to \$25.0 billion in 2005 (Source: *Generation Group*). Predict when the sales will be \$31 billion.

30. The Internal Revenue Service (IRS) standard mileage rate is a way of computing an automobile expense deduction on a tax return. Mileage rates for businesses are provided in Table 74 for various years. Let $M = f(t)$ be the standard mileage rate (in cents per mile) at t years since 1990.
   a. Find an equation of f.
   b. What is the slope? What does it mean in this situation?
   c. Find the M-intercept. What does it mean in this situation?
   d. Predict when the standard mileage rate will be 45 cents per mile.

Table 74 IRS Standard Mileage Rates for Businesses

Year	Standard Mileage Rate (cents per mile)
1999	31.4
2000	32.5
2001	34.5
2002	36.5
2003	36.0
2004	37.5

Source: *IRS*

   e. If a person will drive 12,500 miles on business trips in 2012, predict how much money she will be able to deduct for driving expenses.
   f. For the first three months of 1999, the mileage rate was 32.5 cents per mile. For the rest of that year, it was 31 cents per mile. Explain why the estimate of 31.4 cents was used in Table 74. [**Hint:** Find the average of the mileage rates for the 4 three-month periods of 1999.]

31. *Lobbying* means conducting activities aimed at influencing public officials. Lobbying expenditures by Fortune 500 technology companies are shown in Table 75 for various years.

Table 75 Lobbying Expenditures by Fortune 500 Tech Companies

Year	Lobbying Expenditures (millions of dollars)
1998	22.1
1999	23.0
2000	28.3
2001	31.7
2002	36.2
2003	38.9

Sources: *USA Today; Senate Office of Public Records*

Let $f(t)$ be the lobbying expenditure (in millions of dollars) by Fortune 500 tech companies at t years since 1990.
   a. Find an equation of f.
   b. What is the slope? What does it mean in this situation?
   c. Find the t-intercept. What does it mean in this situation?
   d. Find $f(20)$. What does it mean in this situation?
   e. Find t when $f(t) = 20$. What does it mean in this situation?
   f. Use the model to predict in which year Fortune 500 tech companies will spend \$72 million on lobbying.

CHAPTER 2 TEST

For Exercises 1–10, refer to Fig. 47.

1. Estimate $f(-3)$.
2. Estimate $f(3)$.
3. Estimate $f(0)$.
4. Estimate $f(-5)$.
5. Estimate x when $f(x) = -3$.
6. Estimate x when $f(x) = -2$.
7. Estimate x when $f(x) = 0$.
8. Estimate x when $f(x) = 0.5$.
9. Find the domain of f.
10. Find the range of f.

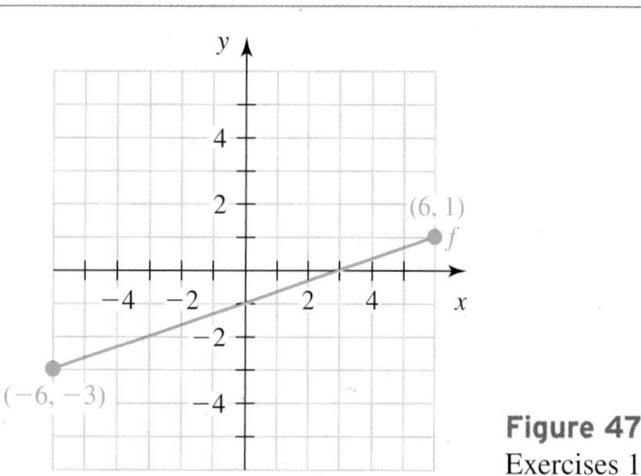

Figure 47 Exercises 1–10

Let $f(x) = -4x + 7$.

11. Find $f(-3)$.

12. Find $f(a - 5)$.

13. Find x when $f(x) = 2$.

14. Find x when $f(x) = a$.

Find all x-intercepts and y-intercepts.

15. $f(x) = 3x - 7$

16. $g(x) = -2x$

17. $k(x) = \frac{1}{3}x - 8$

18. The percentages of U.S. teenagers with driver's licenses are shown in Table 76 for various ages.

Table 76 Percentages of U.S. Teens with Driver's Licenses

Age (years)	Percent
16	56.8
17	61.3
18	72.0
19	79.4
20	81.8

Source: *National Longitudinal Survey of Youth*

Let $p = f(a)$ be the percentage of teenagers at age a years who have driver's licenses.

a. Find an equation of f.

b. What is the slope of the model? What does it mean in this situation?

c. Find the a-intercept. What does it mean in this situation?

d. Estimate the percentage of 21-year-old adults who have driver's licenses.

e. Estimate at what age all adults have driver's licenses.

19. The portion of the U.S. military who are women has increased over the past four decades (see Table 77).

Table 77 Percentages of the Military Who Are Women

Year	Percent
1965	1.2
1975	4.6
1985	9.8
1995	12.9
2003	15.1

Source: *U.S. Department of Defense*

a. Let $f(t)$ be the percentage of the military who are women at t years since 1900. Find an equation of f.

b. What is the slope of your model? What does it mean in this situation?

c. Find $f(100)$. What does it mean in this situation?

d. Find the value of t when $f(t) = 100$. What does it mean in this situation?

e. For what values of t does model breakdown occur for certain? Explain.

20. The total annual ad spending (in millions of dollars) for the NCAA basketball tournament (known as "March Madness") has increased approximately linearly from $311.2 million in 2001 to $467.7 million in 2005 (Source: *TNS Media Intelligence*). Predict the total annual ad spending for March Madness in 2010.

21. In 2001, 41% of community banks offered Internet banking. Each year since, the percentage has increased by about 6.8 percentage points. Let $p = f(t)$ be the percentage of community banks offering Internet banking at t years since 2001.

a. Is there an approximate linear relationship between t and p? Explain. If the relationship is approximately linear, find the slope and describe what it means in this situation.

b. What is the p-intercept? What does it mean in this situation?

c. Find an equation of the model.

d. Predict the percentage of community banks that will offer Internet banking in 2008.

e. Predict when all of the community banks will offer Internet banking.

22. A student drives at a constant rate from home to a movie theater. The movie she wanted to see is sold out. After staying at the theater for a few minutes, she drives home, but at a slower constant rate. Sketch a qualitative graph that relates the distance d from home (in miles) to the number of hours t that have elapsed.

Chapter 3

Systems of Linear Equations

"Mathematics . . . possesses not only truth, but supreme beauty—a beauty cold and austere, like that of sculpture. . . ."

—*Lord Bertrand Russell, "A Free Man's Worship,"* Mysticism and Logic

Table 1 Percentages of Airline Tickets That Were Free

	Percent of Tickets That Were Free	
Year	**Continental**	**American**
1999	6.0	10.3
2000	6.4	9.8
2001	6.2	9.5
2002	6.6	9.6
2003	7.3	9.2
2004	7.3	8.7
2005	7.4	8.6

Source: *Back Aviation Solutions*

Have you ever flown for free thanks to an airline's frequent-flier program? The percentages of Continental Airlines® tickets that were free and the percentages of American Airlines® tickets that were free are shown in Table 1 for various years. In Exercise 6 of Section 3.3, you will predict when the percentage of Continental tickets that are free will be equal to the percentage of American tickets that are free.

In Chapter 2, we used a single linear model to describe an authentic situation. In this chapter, we will use two linear models to describe authentic situations. In particular, we will predict when two quantities will be equal, such as the percentages of Continental and American tickets that are free. We will also predict when one quantity will be less than or greater than another quantity.

3.1 USING GRAPHS AND TABLES TO SOLVE SYSTEMS

Objectives

- Use graphing to make estimates and predictions about situations that can be modeled by using two linear functions.
- Know the meaning of *solution* and *solution set* of a *system of linear equations in two variables.*
- Use a graphical approach to solve systems of linear equations.
- Know the three types of linear systems of two equations.
- Use tables to solve a system of linear equations.

In this section, we will use graphs and tables to work with two or more linear equations in two variables.

Using Two Linear Models to Make a Prediction

We can use graphing to make estimates and predictions about some authentic situations that can be modeled by using two linear functions.

Example 1 Using Two Models to Make a Prediction

In the United States, life expectancies of women have been longer than life expectancies of men for many years (see Table 2). The life expectancies (in years) $W(t)$ and $M(t)$ of women and men, respectively, are modeled by the system

$$L = W(t) = 0.105t + 77.53$$
$$L = M(t) = 0.203t + 69.91$$

where t is the number of years since 1980. Use graphs of W and M to predict when life expectancies of women and men will be equal.

Table 2 U.S. Life Expectancies of Women and Men

Year of Birth	Women (years)	Men (years)
1980	77.4	70.0
1985	78.2	71.1
1990	78.8	71.8
1995	78.9	72.5
2000	79.5	74.1
2003	80.1	74.8

Source: *U.S. Census Bureau*

Solution

We begin by sketching graphs of W and M on the same coordinate system (see Fig. 1).

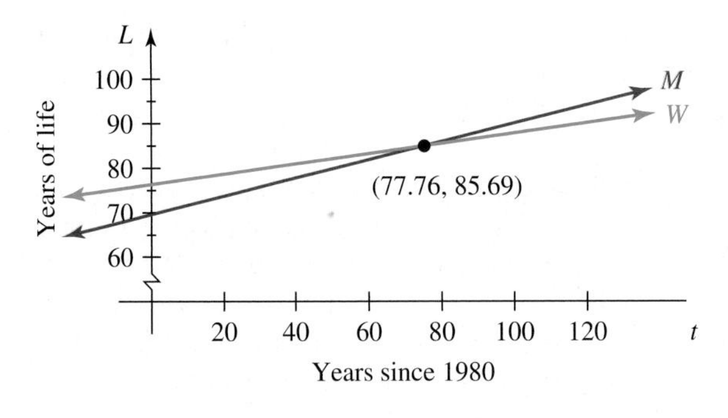

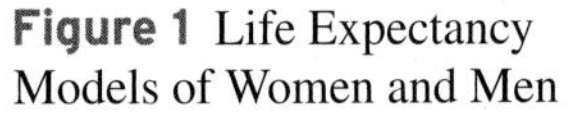
Figure 1 Life Expectancy Models of Women and Men

The intersection point is approximately (77.76, 85.69). So, the models predict that the life expectancy of both women and men will be about 86 years in 2058. We are not very confident about this prediction, however, because it is for so far into the future.

We verify our work by using "intersect" on a graphing calculator (see Figs. 2–4). For graphing calculator instructions, see Sections B.7 and B.18. ■

Figure 2 Enter the functions

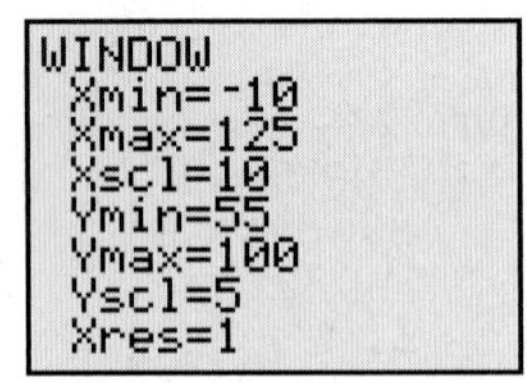

Figure 3 Set up the window

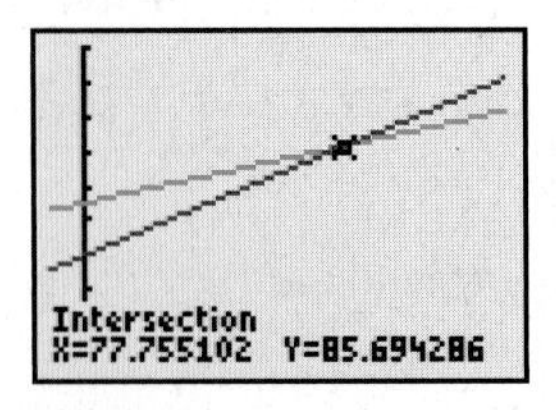

Figure 4 Find the intersection point

> **Intersection Point of the Graphs of Two Models**
>
> If the independent variable of two models represents time, then an intersection point of the graphs of the two models indicates a time when the quantities represented by the dependent variables were or will be equal.

Systems of Two Linear Equations

The two equations in Example 1 are an example of a system of linear equations in two variables. A **system of linear equations in two variables,** or a **linear system** for short, consists of two or more linear equations in two variables. Here is another example of a system of two linear equations in two variables:

$$y = 2x + 1$$
$$y = -3x + 6$$

We will work with such systems throughout this chapter.

Recall from Section 1.2 that every point on the graph of an equation represents a solution of the equation and that every point *not* on the graph represents an ordered pair that is *not* a solution. Knowing the meaning of a graph will help us greatly in this section.

Example 2 Finding Ordered Pairs That Satisfy Both of Two Given Equations

Find all ordered pairs that satisfy both of the equations

$$y = 2x + 1$$
$$y = -3x + 6$$

Solution

To begin, we graph each equation on the same coordinate system (see Fig. 5).

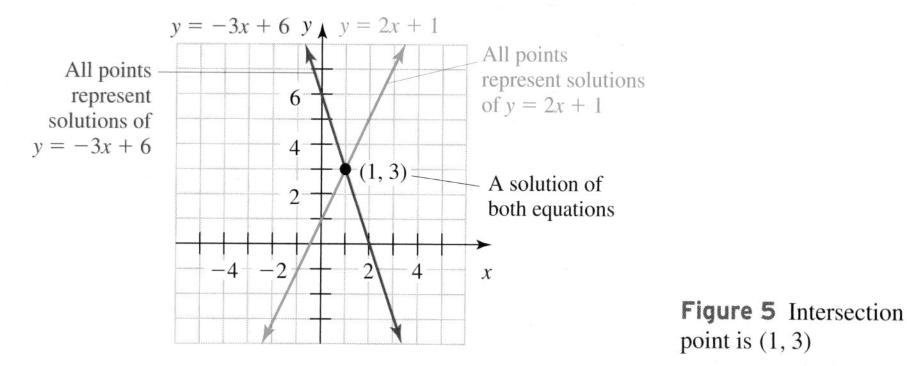

Figure 5 Intersection point is (1, 3)

For an ordered pair to be a solution of *both* equations, it must represent a point that lies on *both* lines. The intersection point (1, 3) is the only point that lies on both lines. So, the ordered pair (1, 3) is the only ordered pair that satisfies both equations.

We can verify that (1, 3) satisfies both equations:

$y = 2x + 1$	$y = -3x + 6$
$3 \stackrel{?}{=} 2(1) + 1$	$3 \stackrel{?}{=} -3(1) + 6$
$3 \stackrel{?}{=} 3$	$3 \stackrel{?}{=} 3$
true	true

■

SOLUTION SET OF A SYSTEM

In Example 2, we worked with the system

$$y = 2x + 1$$
$$y = -3x + 6$$

We found that the only point whose coordinates satisfy both equations is the intersection point (1, 3). We call the set containing only (1, 3) the *solution set of the system.*

> **DEFINITION** Solution of a system
>
> We say that an ordered pair (a, b) is a **solution** of a system of two equations in two variables if it satisfies both equations. The **solution set** of a system is the set of all solutions of the system. We **solve** a system by finding its solution set.

In general, **the solution set of a system of two linear equations can be found by locating any intersection point(s) of the graphs of the two equations.**

Example 3 Solving a System of Two Linear Equations by Graphing

Solve the system

$$y = 2x + 4$$
$$y = -x + 1$$

Solution

The graphs of the equations are sketched in Fig. 6. The intersection point is $(-1, 2)$. So, the solution is the ordered pair $(-1, 2)$. We can verify that $(-1, 2)$

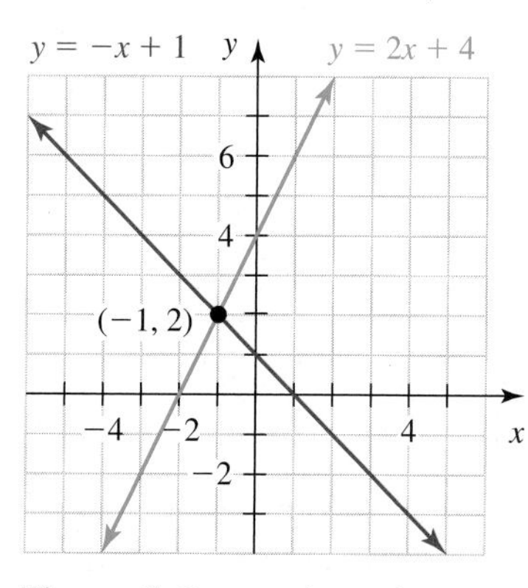

Figure 6 Intersection point is $(-1, 2)$

satisfies both equations:

$$\begin{array}{ll} y = 2x + 4 & y = -x + 1 \\ 2 \stackrel{?}{=} 2(-1) + 4 & 2 \stackrel{?}{=} -(-1) + 1 \\ 2 \stackrel{?}{=} 2 & 2 \stackrel{?}{=} 2 \\ \text{true} & \text{true} \end{array}$$

We can also verify our work by using "intersect" on a graphing calculator (see Fig. 7). ■

WARNING After solving a system of two linear equations, it is a common error to check that a result satisfies only one of the two equations. It is important to check that your result satisfies *both* equations.

Plot1 Plot2 Plot3
\Y1=2X+4
\Y2=-X+1
\Y3=
\Y4=
\Y5=
\Y6=
\Y7=

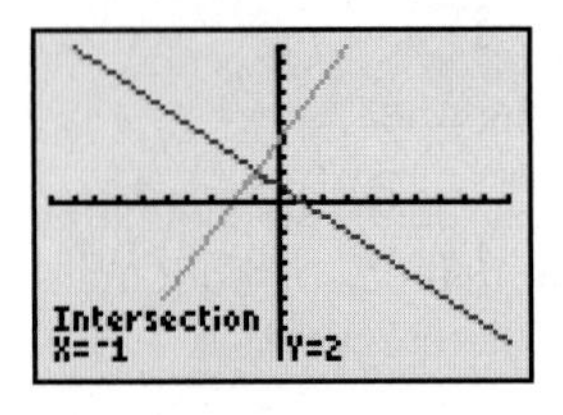

Figure 7 Verify that the intersection point is $(-1, 2)$

Example 4 Solving a System of Two Linear Equations by Graphing

Solve the system

$$\frac{1}{2}x + \frac{5}{4}y = \frac{5}{2} \qquad \text{Equation (1)}$$
$$y = 3x - 4 \qquad \text{Equation (2)}$$

Solution

First, we multiply both sides of equation (1) by the LCD, 4, to "clear the equation of fractions" (see Section A.11 for a review of solving an equation with fractions):

$$\begin{array}{rll} \frac{1}{2}x + \frac{5}{4}y &= \frac{5}{2} & \text{Equation (1)} \\ 4\left(\frac{1}{2}x + \frac{5}{4}y\right) &= 4 \cdot \frac{5}{2} & \text{Multiply both sides by LCD, 4.} \\ 4 \cdot \frac{1}{2}x + 4 \cdot \frac{5}{4}y &= 4 \cdot \frac{5}{2} & \text{Distributive law} \\ 2x + 5y &= 10 & \text{Simplify.} \\ 5y &= -2x + 10 & \text{Subtract } 2x \text{ from both sides.} \\ y &= -\frac{2}{5}x + 2 & \text{Divide both sides by 5.} \end{array}$$

Next, we sketch a graph of the equations $y = -\frac{2}{5}x + 2$ and $y = 3x - 4$ (see Fig. 8).

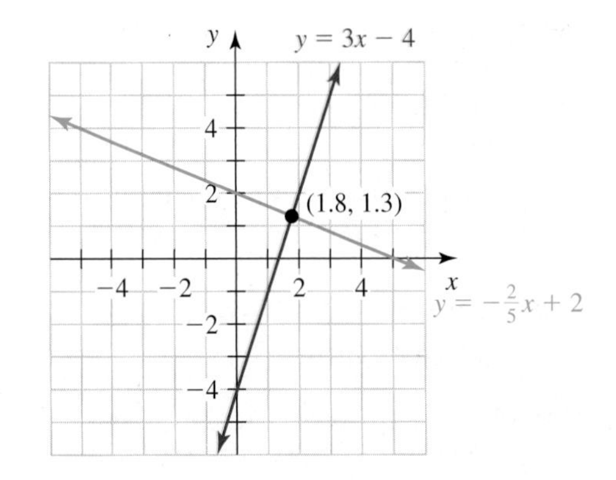

Figure 8 Approximate solution is (1.8, 1.3)

We can estimate the solution to be approximately (1.8, 1.3). However, we can get a better estimate by using "intersect" on a graphing calculator (see Fig. 9).

It turns out that the ordered pair (1.7647059, 1.2941176) found by "intersect" is accurate to seven decimal places. For ease in calculator entry, we round the

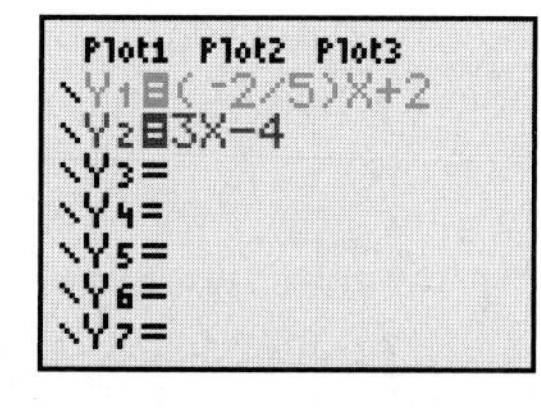

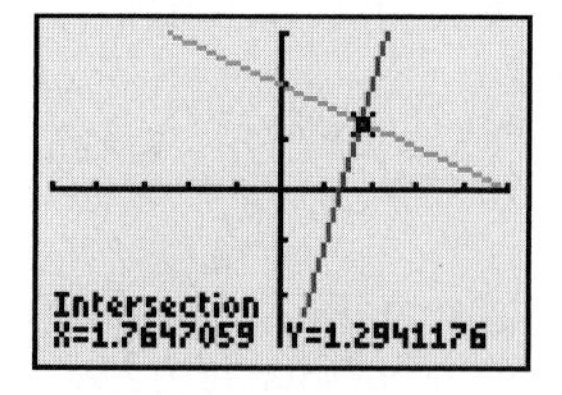

Figure 9 Using "intersect" to find the intersection point

coordinates to the second decimal place and check that (1.76, 1.29) satisfies both equations approximately:

$$\frac{1}{2}x + \frac{5}{4}y = \frac{5}{2} \qquad\qquad y = 3x - 4$$

$$\frac{1}{2}(1.76) + \frac{5}{4}(1.29) \stackrel{?}{=} \frac{5}{2} \qquad\qquad 1.29 \stackrel{?}{=} 3(1.76) - 4$$

$$2.4925 \approx 2.5 \qquad\qquad 1.29 \approx 1.28$$

Because (1.76, 1.29) satisfies both equations approximately, we know that (1.76, 1.29) is a good approximation of the *exact* solution, which we will learn to find in Section 3.2. ■

THREE TYPES OF LINEAR SYSTEMS

Each of the systems in Examples 2–4 has one solution. Not all systems have exactly one solution, however.

Example 5 Solving an Inconsistent System

Solve the system

$$y = \frac{1}{2}x + 1$$
$$y = \frac{1}{2}x - 2$$

Solution

Since the two lines have equal slopes, these lines are parallel (see Fig. 10). Parallel lines do not intersect, so there is no ordered pair that satisfies both equations. The solution set is the empty set.

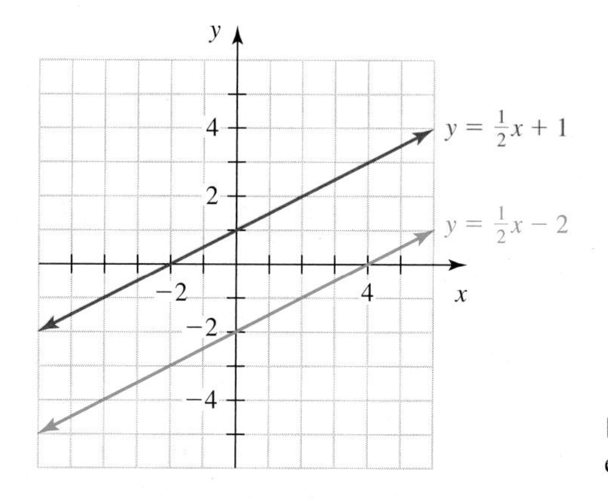

Figure 10 The solution is the empty set ■

A linear system whose solution set is the empty set is called an **inconsistent system.**

Example 6 Solving a Dependent System

Solve the system

$$y = 2x + 1 \qquad \text{Equation (1)}$$
$$6x - 3y = -3 \qquad \text{Equation (2)}$$

Solution

We write equation (2) in slope–intercept form:

$$6x - 3y = -3 \qquad \text{Equation (2)}$$
$$2x - y = -1 \qquad \text{Divide both sides by 3.}$$
$$-y = -2x - 1 \qquad \text{Subtract } 2x \text{ from both sides.}$$
$$y = 2x + 1 \qquad \text{Multiply both sides by } -1.$$

So, the graphs of $6x - 3y = -3$ and $y = 2x + 1$ are the same line. The solution set of the system is the set of the infinite number of ordered pairs that correspond to points that lie on the line $y = 2x + 1$ and on the (same) line $6x - 3y = -3$. ■

A linear system that has an infinite number of solutions is called a **dependent system.**

In Examples 4, 5, and 6, we have seen three types of systems. We now describe these three types.

Types of Linear Systems

There are three types of linear systems of two equations:

1. *One-solution system:* The lines intersect in one point. The solution set of the system contains only the ordered pair that corresponds to that point. See Fig. 11.
2. *Inconsistent system:* The lines are parallel. The solution set of the system is the empty set. See Fig. 12.
3. *Dependent system:* The lines are identical. The solution set of the system is the set of the infinite number of solutions represented by points on the same line. See Fig. 13.

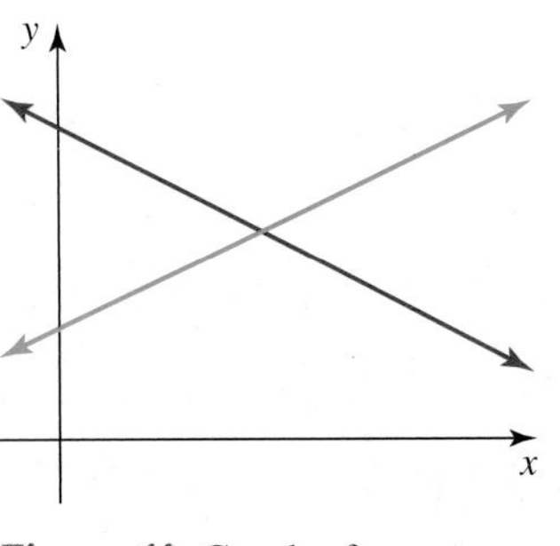

Figure 11 Graph of a system with one solution

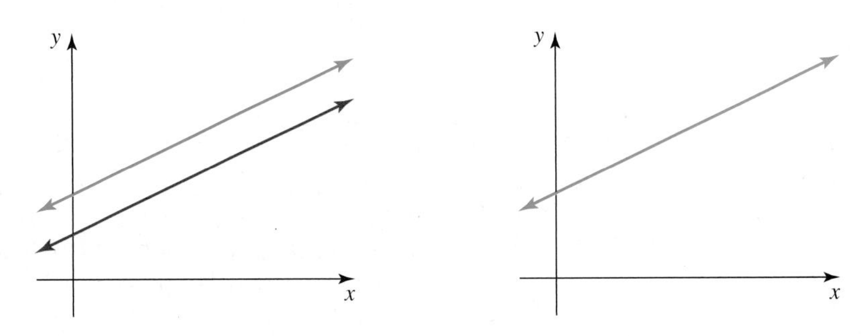

Figure 12 Graph of an inconsistent system

Figure 13 Graph of a dependent system

Solving a System From a Table of Solutions of Equations

We can use tables instead of graphs to solve a system.

Example 7 Using a Table to Solve a System

Use a table of solutions to solve the following system of two equations:

$$y = 2x - 3$$
$$y = -3x + 7$$

Solution

Some solutions of the two equations are shown in Table 3.

Table 3 Some Solutions of $y = 2x - 3$ and $y = -3x + 7$

x	0	1	2	3	4
$y = 2x - 3$	−3	−1	1	3	5
$y = -3x + 7$	7	4	1	−2	−5

Since the ordered pair (2, 1) is a solution of both equations, it is a solution of the system of equations.

The lines $y = 2x - 3$ and $y = -3x + 7$ have different slopes, so there is only one intersection point. Thus, the ordered pair (2, 1) is the *only* solution of the system. ■

In Example 7, we used a table of solutions of two linear equations to help us find the solution of the linear system. **If an ordered pair is listed in a table as a solution of both of two linear equations, then that ordered pair is a solution of the system.**

group exploration

Comparing the three types of systems

1. Is the system

$$y = -2x + 3$$
$$y = -2x + 5$$

a dependent system, an inconsistent system, or a one-solution system? Explain.

2. Consider the system

$$y = 3x - 5$$
$$y = mx + b$$

where m and b are constants.

   a. Find values of m and b such that the given system is inconsistent. What is the solution set of your system? Use a graphing calculator to verify your work.
   b. Find values of m and b such that the given system is dependent. What is the solution set of your system?
   c. Find values of m and b such that the given system is a one-solution system. Use "intersect" on a graphing calculator to find the solution.

3. Now consider this general system of two linear equations:

$$y = m_1x + b_1$$
$$y = m_2x + b_2$$

Discuss dependent systems, inconsistent systems, and one-solution systems in terms of m_1, m_2, b_1, and b_2.

TIPS FOR SUCCESS: Visualize

To help prepare mentally and physically for competition, many exceptional athletes often visualize themselves performing well at their event throughout their training. For example, a runner training for the 100-meter-dash might imagine getting set in the starting blocks, taking off right after the gunshot, being in front of the other runners, and so on, up until the moment of breaking the tape at the finish line.

In an experiment, three groups of basketball players were used to test the effectiveness of visualization. The first group warmed up by shooting baskets before a game. The second group visualized shooting baskets but did not shoot any baskets during the warm-up. The third group did not warm up or visualize before the game. The visualization group outperformed not only the group that did not warm up, but also the group that warmed up by shooting baskets!

Visualize doing all the things you feel you need to do to succeed in this course. If you do this regularly, you will have better follow-through with what you intend to do. You will also feel more confident about succeeding.

HOMEWORK 3.1

FOR EXTRA HELP ▶ Student Solutions Manual | PH Math/Tutor Center | MathXL® | MyMathLab

Find the solution set of the system by graphing the equations by hand. If the system is inconsistent or dependent, say so. If your result is one ordered pair, check that it satisfies both equations.

1. $y = 2x + 2$
   $y = -3x + 7$

2. $y = x - 5$
   $y = -2x + 4$

3. $y = -\frac{1}{2}x + 3$
   $y = \frac{1}{3}x + 8$

4. $y = -\frac{1}{4}x + 3$
   $y = \frac{1}{2}x$

5. $y = 3(x - 1)$
   $y = -2x + 7$

6. $y = 2(x - 3)$
   $y = -3x + 4$

7. $x + 4y = 20$
   $2x - 4y = -8$

8. $2x - 3y = -6$
   $x + 3y = -3$

9. $5(y - 2) = 21 - 2(x + 3)$
   $y = 3(x - 1) + 8$

10. $3(y - 2) = 2(3x + 1) + 7$
    $y = 7 - 4(x + 2)$

11. $y = -2x + 3$
$4y - 12 = -8x$

12. $y = 3x - 7$
$9x - 3y = 21$

13. $4x - 6y = 24$
$6x - 9y = 18$

14. $20x - 8y = 16$
$-15x + 6y = 18$

15. $\frac{1}{2}x - \frac{1}{2}y = 1$
$\frac{1}{4}x + \frac{1}{2}y = 2$

16. $\frac{1}{3}x + \frac{1}{2}y = -3$
$\frac{1}{2}x - \frac{1}{3}y = -\frac{7}{3}$

*Use "intersect" on a graphing calculator to solve the system. Round the coordinates of solutions to the second decimal place. If the system is inconsistent or dependent, say so. If your result is one ordered pair, check that it satisfies both equations approximately. [**Graphing Calculator:** See Section B.18.]*

17. $y = -2.51x - 6.49$
$y = 1.74x - 1.75$

18. $y = 5.437x - 2.136$
$y = -2.752x + 3.984$

19. $2x + 5y = 7$
$3x - 4y = -13$

20. $4x - 3y = 13$
$2y + 5x = 4$

21. $y = 2x - 1$
$2(2x - y) = 2$

22. $y = 2x + 1$
$3y - 1 = 2(3x + 1)$

23. $y = 5x + 2$
$0.2y - x = 1$

24. $y = -2x + 4$
$0.5y + x = 3$

25. $\frac{1}{2}x - \frac{1}{2}y = 1$
$\frac{1}{3}x + \frac{2}{3}y = 2$

26. $\frac{1}{3}y + x = -2$
$-\frac{1}{4}x + \frac{1}{2}y = 2$

27. The winning times for the Olympic 500-meter speed skating event have generally decreased since 1972 (see Table 4).

Table 4 Olympic 500-Meter Speed Skating Times

Year	Winning Time (seconds)	
	Women	**Men**
1972	43.33	39.44
1976	42.76	39.17
1980	41.78	38.03
1984	41.02	38.19
1988	39.10	36.45
1992	40.33	37.14
1994	39.25	36.33
1998	38.21	35.59
2002	37.375	34.615
2006	38.285	34.880

Source: The Universal Almanac

The winning times (in seconds) $W(t)$ and $M(t)$ for women and men, respectively, are modeled by the system

$$w = W(t) = -0.172t + 43.44$$
$$w = M(t) = -0.147t + 39.80$$

where t is the number of years since 1970.

a. Use the equations of the models to estimate the winning time for women and the winning time for men in 2006. Find the errors in your estimates.

b. Compare the slopes of the two models. What does your comparison tell you about this situation?

c. Explain why your work in parts (a) and (b) suggests that there may be a time when the women's winning time will be equal to the men's winning time.

d. Use "intersect" on a graphing calculator to predict when the women's winning time will be equal to the men's winning time. What will be that winning time?

28. Annual U.S. consumption of chicken and red meat (in pounds per person) is described for various years in Table 5.

Table 5 Annual U.S. Per-Person Consumption of Chicken and Red Meat

Year	Annual Consumption (pounds per person)	
	Chicken	**Red Meat**
1970	27.4	131.9
1975	26.3	125.8
1980	32.7	126.4
1985	36.4	124.9
1990	42.4	112.2
1995	48.2	113.6
2000	54.2	113.7
2003	57.5	111.9

Source: *U.S. Department of Agriculture*

Let $C(t)$ be the annual consumption of chicken and $R(t)$ be the annual consumption of red meat, both in pounds per person, in the year that is t years since 1900. The consumption of each can be modeled by the system

$$C(t) = 0.99t - 45.68$$
$$R(t) = -0.62t + 174.40$$

a. Compare $C(110)$ with $R(110)$. What does your comparison tell you about this situation?

b. Use "intersect" on a graphing calculator to predict when the consumption of chicken will equal that of red meat. What will be that consumption? How confident are you?

29. Table 6 lists women's and men's total enrollments at U.S. community colleges, professional schools, universities, and colleges (both publicly and privately controlled) for various years.

Table 6 College Enrollments

Year	Enrollment (millions)	
	Women	**Men**
1985	6.4	5.8
1990	7.5	6.3
1995	7.9	6.3
2000	8.6	6.7
2003	9.3	7.3

Source: *U.S. Census Bureau*

Let $W(t)$ and $M(t)$ be the enrollments (in millions) of women and men, respectively, both at t years since 1980.

a. Find equations of W and M.

b. Use "intersect" on a graphing calculator to estimate when women's and men's enrollments were approximately equal. What was that enrollment?

c. Predict the total enrollment of women and men in 2011.

30. The number of U.S. households with personal computers (PCs) and the total number of U.S. households are shown in Table 7 for various years.

Table 7 U.S. Households with Personal Computers

Year	Households with PCs (millions)	Total Households (millions)
1995	31.36	99.0
1997	39.56	101.0
1999	50.00	103.6
2001	60.81	108.2
2003	68.78	111.3

Sources: *U.S. Department of Commerce; Media Metrix; U.S. Census Bureau*

a. Let $C(t)$ be the number (in millions) of U.S. households with PCs and $H(t)$ be the total number (in millions) of U.S. households, both at t years since 1990. Find equations of C and H.

b. Use "intersect" on a graphing calculator to predict when all U.S. households will have PCs.

c. Use the models to predict the *percentage* of households that will have PCs in 2010.

31. The numbers of landline and wireless calls placed to directory assistance (411) are shown in Table 8 for various years.

Table 8 Numbers of Landline and Wireless 411 Calls

Year	Number of 411 Calls (billions)	
	Landline	Wireless
2001	4.6	1.3
2002	4.5	1.5
2003	4.4	1.7
2004	4.3	1.8
2005	4.2	2.0

Source: *Pierz Group*

a. Let $L(t)$ be the number of landline 411 calls and $W(t)$ be the number of wireless 411 calls, both in billions, in the year that is t years since 2000. Find equations of L and W.

b. Use "intersect" on a graphing calculator to predict in which year the number of landline 411 calls will equal the number of wireless 411 calls. What will be that number of calls?

c. In 2006, the average cost of a typical 411 call was \$1.25 for landline calls and \$1.50 for wireless calls. Estimate the total money collected (*total revenue*) from landline and wireless 411 calls in that year.

32. The percentages of women and men in the United States who are married are shown in Table 9 for various years.

Table 9 Percentages of Women and Men Who Are Married

Year	Percent of Women Who Are Married	Percent of Men Who Are Married
1990	59.7	64.3
1995	59.2	62.7
1997	57.9	61.5
2000	57.6	61.5
2004	57.1	60.3

Source: *U.S. Census Bureau*

a. Let $W(t)$ be the percentage of women who are married and $M(t)$ be the percentage of men who are married, both at t years since 1990. Find equations of W and M.

b. Use "intersect" on a graphing calculator to predict when the percentage of women who are married will equal the percentage of men who are married. What will be that percentage?

c. From 1990 to 2004, the percentage of men who were married was greater than the percentage of women who were married. Use this fact to determine whether there were more women or more men in the United States in that time period. Explain.

33. The graphs of $y = ax + b$ and $y = cx + d$ are sketched in Fig. 14.

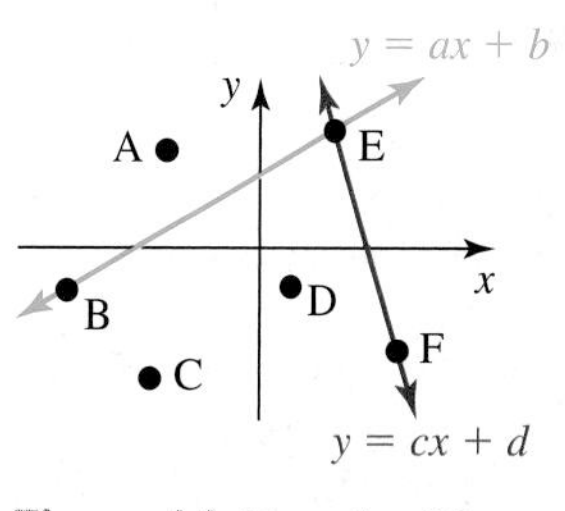

Figure 14 Exercise 33

For each part, decide which one or more of the points A, B, C, D, E, or F represent ordered pairs that

a. satisfy the equation $y = ax + b$.

b. satisfy the equation $y = cx + d$.

c. satisfy both equations.

d. do not satisfy either equation.

34. Consider the system

$$y = 3x - 7$$
$$y = -2x + 3$$

Find an ordered pair that

a. satisfies $y = 3x - 7$ but does not satisfy $y = -2x + 3$.

b. satisfies $y = -2x + 3$ but does not satisfy $y = 3x - 7$.

c. satisfies both equations.

d. does not satisfy either equation.

35. Figure 15 shows the graphs of two linear equations. To the first decimal place, estimate the coordinates of the solution of the system.

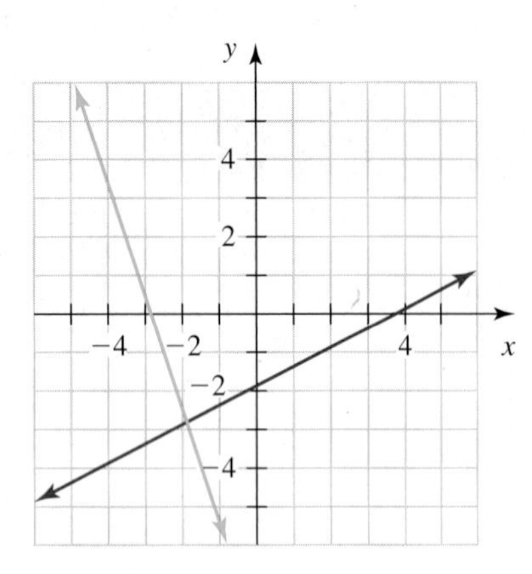

Figure 15 Exercise 35

36. Figure 16 shows the graphs of two linear equations. To the first decimal place, estimate the coordinates of the solution of the system.

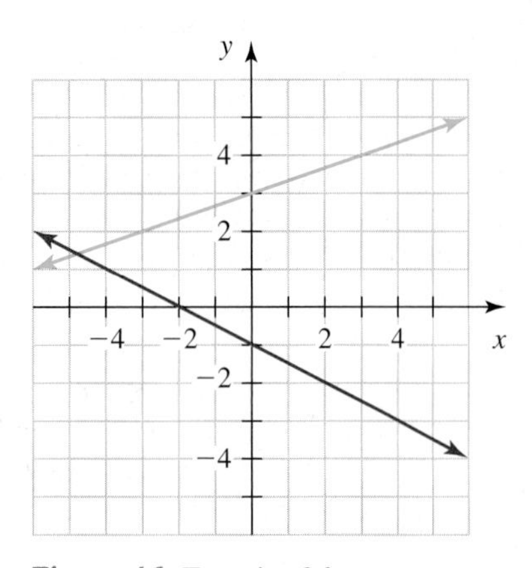

Figure 16 Exercise 36

37. Figure 17 shows the graphs of two linear equations. To the nearest whole number, estimate the coordinates of the solution of the system. Explain. [**Hint:** Use the slope of each line.]

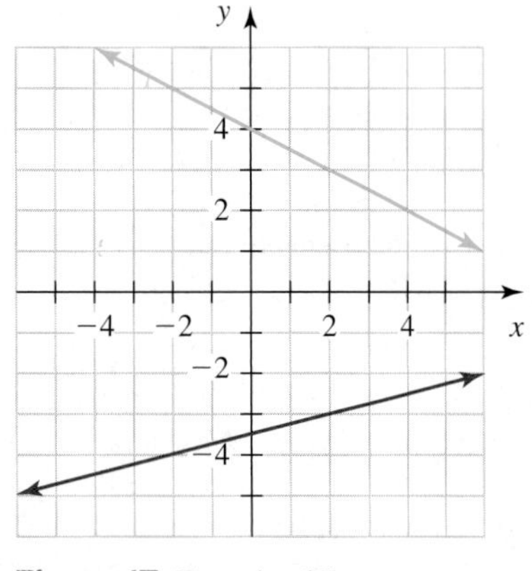

Figure 17 Exercise 37

38. Figure 18 shows the graphs of two linear equations. To the nearest whole number, estimate the coordinates of the solution of the system. Explain. [**Hint:** Use the slope of each line.]

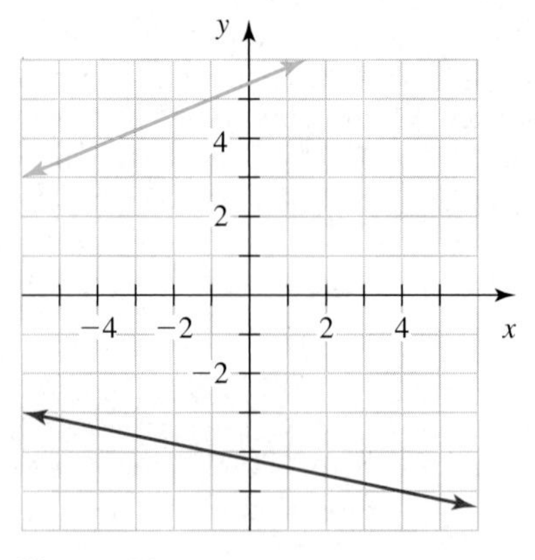

Figure 18 Exercise 38

39. Use Table 10 to solve the system

$$y = -5x + 11$$
$$y = 7x - 25$$

Table 10 Some Solutions of $y = -5x + 11$ and $y = 7x - 25$ (Exercise 39)

x	0	1	2	3	4
$y = -5x + 11$	11	6	1	−4	−9
$y = 7x - 25$	−25	−18	−11	−4	3

40. Use Table 11 to solve the system

$$y = 4x - 1$$
$$y = -3x + 6$$

Table 11 Some Solutions of $y = 4x - 1$ and $y = -3x + 6$ (Exercise 40)

x	0	1	2	3	4
$y = 4x - 1$	−1	3	7	11	15
$y = -3x + 6$	6	3	0	−3	−6

41. Some values of linear functions f and g are listed in Table 12. Estimate the solution of a system of two equations that describe f and g.

Table 12 Values of Functions f and g (Exercise 41)

x	0	1	2	3	4	5	6	7	8
$f(x)$	30	27	24	21	18	15	12	9	6
$g(x)$	2	7	12	17	22	27	32	37	42

42. Some values of linear functions f and g are listed in Table 13. Estimate the solution of a system of two equations that describe f and g.

Table 13 Values of Functions f and g (Exercise 42)

x	0	1	2	3	4	5	6	7	8
$f(x)$	99	95	91	87	83	79	75	71	67
$g(x)$	3	5	7	9	11	13	15	17	19

For Exercises 43–48, refer to Fig. 19.

43. Find $f(-4)$.

44. Find $g(-4)$.

45. Find x where $f(x) = 3$.

46. Find x where $g(x) = -1$.

47. Find x where $f(x) = g(x)$.

48. Estimate x where $g(x) = 0$.

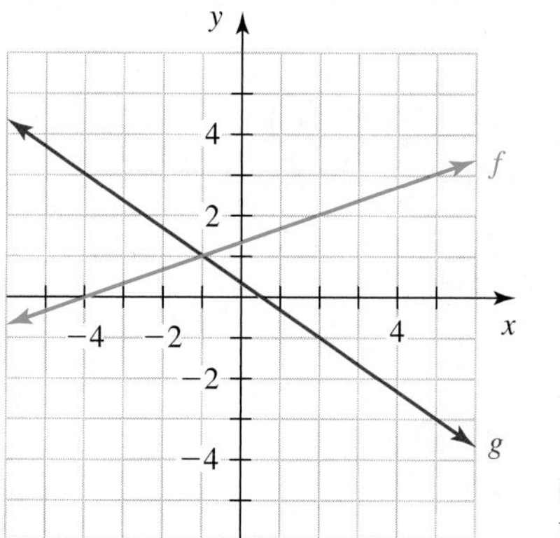

Figure 19 Exercises 43–48

49. Create a system of two linear equations as indicated. Verify your system graphically.
a. The solution of the system is (2, 1).
b. The system is inconsistent.
c. The system is dependent.

50. Solve the system. [**Hint:** Sketch the graphs on the same coordinate system.]

$$y = 2$$
$$x = -3$$

51. Find all ordered pairs that satisfy all three equations:

$$y = x + 3$$
$$y = -2x + 9$$
$$y = 3x - 1$$

52. Find all ordered pairs that satisfy all three equations:

$$y = 2x - 5$$
$$y = 0.6x + 1$$
$$y = -1.2x + 5$$

53. Create a system of three linear equations whose solution is $(-4, 3)$. Verify your result by checking that $(-4, 3)$ satisfies all three equations. Also, verify your result graphically.

54. A system of linear equations has $(-2, 3)$ and $(4, 1)$ as solutions.
a. Find a third solution.
b. How many solutions are there?

55. A student tries to solve the system

$$y = 3x - 1$$
$$y = -2x + 9$$

After graphing the equations, she believes that the solution is $(1, 2)$. She then checks whether $(1, 2)$ satisfies $y = 3x - 1$:

$$y = 3x - 1$$
$$2 \stackrel{?}{=} 3(1) - 1$$
$$2 \stackrel{?}{=} 2$$
true

The student concludes that (1, 2) is the solution of the system. Describe any errors. Then solve the system correctly.

56. Explain why any solutions of a system of two linear equations correspond to the intersection points of the graphs of the two equations. (See page 4 for guidelines on writing a good response.)

57. Describe the three types of systems of two linear equations and how to solve these systems. Also, explain how to verify your work. (See page 4 for guidelines on writing a good response.)

3.2 USING SUBSTITUTION AND ELIMINATION TO SOLVE SYSTEMS

Objectives

- Use substitution to solve a system of two linear equations.
- Use elimination to solve a system of two linear equations.
- Solve inconsistent and dependent systems by substitution or elimination.
- Use graphs and tables to solve an equation in one variable.

In Section 3.1, we used graphs or tables of linear functions to solve linear systems. In this section, we use *equations* of linear functions to solve such systems.

Using Substitution to Solve Systems

In Example 1, we solve a system by a technique called *substitution.*

Example 1 Solving a System by Substitution

Solve the system

$$y = x - 1 \quad \text{Equation (1)}$$
$$3x + 2y = 13 \quad \text{Equation (2)}$$

Solution

From equation (1), we know that the value of y is equal to the value of $x - 1$. So, we substitute $x - 1$ for y in equation (2):

$$3x + 2y = 13 \quad \text{Equation (2)}$$
$$3x + 2(x - 1) = 13 \quad \text{Substitute } x - 1 \text{ for } y.$$

By making this substitution, we now have an equation in terms of *one* variable. Next, we solve that equation for x:

$$3x + 2(x - 1) = 13 \quad \text{Equation in one variable}$$
$$3x + 2x - 2 = 13 \quad \text{Distributive law}$$
$$5x - 2 = 13 \quad \text{Combine like terms.}$$
$$5x = 15 \quad \text{Add 2 to both sides.}$$
$$x = 3 \quad \text{Divide both sides by 5.}$$

This means that the x-coordinate of the solution is 3. To find the y-coordinate, we substitute 3 for x in either of the original equations and solve for y:

$$y = x - 1 \quad \text{Equation (1)}$$
$$y = 3 - 1 \quad \text{Substitute 3 for } x.$$
$$y = 2 \quad \text{Subtract.}$$

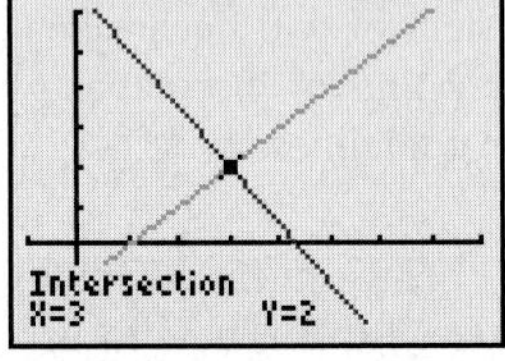

Figure 20 Verify that the intersection point is (3, 2)

So, the solution is (3, 2). We can check that (3, 2) satisfies both of the system's equations:

$$y = x - 1 \qquad\qquad 3x + 2y = 13$$
$$2 \stackrel{?}{=} 3 - 1 \qquad\qquad 3(3) + 2(2) \stackrel{?}{=} 13$$
$$2 \stackrel{?}{=} 2 \qquad\qquad 9 + 4 \stackrel{?}{=} 13$$

true true

Instead, we can check that (3, 2) is the solution by graphing the two equations and checking that (3, 2) is the intersection point of the two lines (see Fig. 20). To do so on a graphing calculator, we must first solve $3x + 2y = 13$ for y:

$$y = -\frac{3}{2}x + \frac{13}{2}$$

■

Using Substitution to Solve a Linear System

To use **substitution** to solve a system of two linear equations,

1. Isolate a variable on one side of either equation.
2. Substitute the expression for the variable found in step 1 into the other equation.
3. Solve the equation in one variable found in step 2.
4. Substitute the solution found in step 3 into one of the original equations, and solve for the other variable.

Example 2 Solving a System by Substitution

Solve the system

$$x - 3y = 4 \quad \text{Equation (1)}$$
$$2x - 5y = 5 \quad \text{Equation (2)}$$

Solution

We begin by solving for one of the variables in one of the equations. We can avoid fractions by choosing to solve equation (1) for x:

$$x - 3y = 4 \quad \text{Equation (1)}$$
$$x = 3y + 4 \quad \text{Add } 3y \text{ to both sides.}$$

Next, we substitute $3y + 4$ for x in equation (2) and solve for y:

$$\begin{aligned} 2x - 5y &= 5 && \text{Equation (2)} \\ 2(3y+4) - 5y &= 5 && \text{Substitute } 3y+4 \text{ for } x. \\ 6y + 8 - 5y &= 5 && \text{Distributive law} \\ y + 8 &= 5 && \text{Combine like terms.} \\ y &= -3 && \text{Subtract 8 from both sides.} \end{aligned}$$

Finally, we substitute -3 for y in the equation $x = 3y + 4$ and solve for x:

$$\begin{aligned} x &= 3(-3) + 4 && \text{Substitute } -3 \text{ for } y. \\ x &= -5 && \text{Simplify.} \end{aligned}$$

The solution is $(-5, -3)$. We could verify our work by checking that $(-5, -3)$ satisfies *both* of the original equations. ■

Using Elimination to Solve Systems

In addition to solving systems by graphing or by substitution, we can solve systems by a method called *elimination*. To solve systems by this method, we will need to use the following property.

Adding Left Sides and Adding Right Sides of Two Equations

If $a = b$ and $c = d$, then

$$a + c = b + d$$

In words: The sum of the left sides of two equations is equal to the sum of the right sides.

For example, if we add the left sides and add the right sides of the equations $2 = 2$ and $3 = 3$, we obtain the true statement $2 + 3 = 2 + 3$. We can apply this property to some pairs of equations to eliminate a variable; that is the key step in the method of elimination.

Example 3 Solving a System by Elimination

Solve the system

$$\begin{aligned} 4x - 5y &= 3 && \text{Equation (1)} \\ 3x + 5y &= 11 && \text{Equation (2)} \end{aligned}$$

Solution

We begin by adding the left sides and adding the right sides of the two equations:

$$\begin{aligned} 4x - 5y &= 3 && \text{Equation (1)} \\ 3x + 5y &= 11 && \text{Equation (2)} \\ \hline 7x + 0 &= 14 && \text{Add left sides and add right sides; combine like terms.} \end{aligned}$$

Having "eliminated" the variable y, we are left with an equation in *one* variable. Next, we solve that equation for x:

$$\begin{aligned} 7x + 0 &= 14 \\ 7x &= 14 && a + 0 = a \\ x &= 2 && \text{Divide both sides by 7.} \end{aligned}$$

Then, we substitute 2 for x in either of the original equations and solve for y:

$$
\begin{aligned}
4x - 5y &= 3 && \text{Equation (1)} \\
4(2) - 5y &= 3 && \text{Substitute 2 for } x. \\
8 - 5y &= 3 && \text{Multiply.} \\
-5y &= -5 && \text{Subtract 8 from both sides.} \\
y &= 1 && \text{Divide both sides by } -5.
\end{aligned}
$$

The solution is (2, 1). We could check that (2, 1) satisfies *both* of the original equations. ■

Example 4 Solving a System by Elimination

Solve the system

$$
\begin{aligned}
3x + 2y &= 18 && \text{Equation (1)} \\
6x - 5y &= 9 && \text{Equation (2)}
\end{aligned}
$$

Solution

If we add the left sides and add the right sides of the equations as they are now, neither variable would be eliminated. Therefore, we first multiply both sides of equation (1) by -2, yielding the system

$$
\begin{aligned}
-6x - 4y &= -36 && \text{Multiply both sides of equation (1) by } -2. \\
6x - 5y &= 9 && \text{Equation (2)}
\end{aligned}
$$

Now that the coefficients of the x terms are equal in absolute value (see Section A.3) and opposite in sign, we add the left sides and add the right sides of the equations and solve for y:

$$
\begin{aligned}
-6x - 4y &= -36 \\
6x - 5y &= 9 \\
\hline
0 - 9y &= -27 && \text{Add left sides and add right sides; combine like terms.} \\
-9y &= -27 && 0 + a = a \\
y &= 3 && \text{Divide both sides by } -9.
\end{aligned}
$$

We substitute 3 for y in equation (1) and solve for x:

$$
\begin{aligned}
3x + 2y &= 18 && \text{Equation (1)} \\
3x + 2(3) &= 18 && \text{Substitute 3 for } y. \\
3x + 6 &= 18 && \text{Multiply.} \\
3x &= 12 && \text{Subtract 6 from both sides.} \\
x &= 4 && \text{Divide both sides by 3.}
\end{aligned}
$$

The solution is (4, 3). ■

Using Elimination to Solve a Linear System

To use **elimination** to solve a system of two linear equations,

1. If needed, multiply both sides of one equation by a number (and, if necessary, multiply both sides of the other equation by another number) to get the coefficients of one variable to be equal in absolute value and opposite in sign.
2. Add the left sides and add the right sides of the equations to eliminate one of the variables.
3. Solve the equation in one variable found in step 2.
4. Substitute the solution found in step 3 into one of the original equations, and solve for the other variable.

Example 5 Solving a System by Elimination

Solve the system

$$4x - 3y = -3 \quad \text{Equation (1)}$$
$$5x + 2y = 25 \quad \text{Equation (2)}$$

Solution

To eliminate the y terms, we multiply both sides of equation (1) by 2 and multiply both sides of equation (2) by 3. That yields the system

$$8x - 6y = -6 \quad \text{Multiply both sides of equation (1) by 2.}$$
$$15x + 6y = 75 \quad \text{Multiply both sides of equation (2) by 3.}$$

The coefficients of the y terms are now equal in absolute value and opposite in sign. Next, we add the left sides and add the right sides of the equations and solve for x:

$$\begin{aligned} 8x - 6y &= -6 \\ 15x + 6y &= 75 \\ \hline 23x + 0 &= 69 && \text{Add left sides and add right sides; combine like terms.} \\ 23x &= 69 && a + 0 = a \\ x &= 3 && \text{Divide both sides by 23.} \end{aligned}$$

Substituting 3 for x in equation (1) gives

$$\begin{aligned} 4x - 3y &= -3 && \text{Equation (1)} \\ 4(3) - 3y &= -3 && \text{Substitute 3 for x.} \\ 12 - 3y &= -3 && \text{Multiply.} \\ -3y &= -15 && \text{Subtract 12 from both sides.} \\ y &= 5 && \text{Divide both sides by } -3. \end{aligned}$$

The solution is (3, 5). ■

In Example 5, we eliminated y by getting the coefficients of the y terms to be equal in absolute value and opposite in sign. Note that this process is similar to finding a least common multiple.

Example 6 Using Elimination to Solve a System with Fractions

Solve the system

$$\frac{1}{3}x - \frac{1}{2}y = \frac{1}{6} \quad \text{Equation (1)}$$
$$\frac{1}{5}x + \frac{1}{4}y = \frac{13}{20} \quad \text{Equation (2)}$$

Solution

First, we clear the fractions in equation (1) by multiplying both sides by the LCD, 6, and clear the fractions in equation (2) by multiplying both sides by the LCD, 20:

$$2x - 3y = 1 \quad \text{Multiply both sides of equation (1) by 6.}$$
$$4x + 5y = 13 \quad \text{Multiply both sides of equation (2) by 20.}$$

To eliminate the variable x, we multiply both sides of $2x - 3y = 1$ by -2:

$$\begin{aligned} -4x + 6y &= -2 && \text{Multiply both sides of } 2x - 3y = 1 \text{ by } -2. \\ 4x + 5y &= 13 \\ \hline 0 + 11y &= 11 && \text{Add left sides and add right sides; combine like terms.} \\ y &= 1 && \text{Divide both sides by 11.} \end{aligned}$$

To find the value of x, we can substitute 1 for y in any of the equations that have both x and y. We substitute 1 for y in the equation $2x - 3y = 1$ and solve for x:

$$2x - 3(1) = 1 \quad \text{Substitute 1 for } y \text{ in } 2x - 3y = 1.$$
$$2x = 4 \quad \text{Add 3 to both sides.}$$
$$x = 2 \quad \text{Divide both sides by 2.}$$

The solution is (2, 1). ■

In this section and Section 3.1, we have solved many systems by graphing, substitution, and elimination. **Any linear system of two equations can be solved by graphing, substitution, or elimination. All three methods will give the same result.**

Solving Inconsistent and Dependent Systems

Each system in Examples 1–6 has one solution. What happens in using substitution or elimination when we solve an inconsistent system (empty-set solution) or a dependent system (infinitely many solutions)?

Example 7 Using Substitution to Solve an Inconsistent System

Consider the linear system

$$y = 2x + 1 \quad \text{Equation (1)}$$
$$y = 2x + 3 \quad \text{Equation (2)}$$

The graphs of the equations are parallel lines (why?), so the system is inconsistent and the solution set is the empty set. What happens when we solve this system by substitution?

Solution

We substitute $2x + 1$ for y in equation (2) and solve for x:

$$y = 2x + 3 \quad \text{Equation (2)}$$
$$2x + 1 = 2x + 3 \quad \text{Substitute } 2x + 1 \text{ for } y.$$
$$1 = 3 \quad \text{Subtract } 2x \text{ from both sides.}$$
$$\text{false}$$

We get the *false* statement $1 = 3$. ■

Inconsistent System of Two Equations

If the result of applying substitution or elimination to a linear system of two equations is a false statement, the system is inconsistent—that is, the solution set is the empty set. (no solution)

Example 8 Applying Substitution to a Dependent System

In Example 6 of Section 3.1, we found that the system

$$y = 2x + 1 \quad \text{Equation (1)}$$
$$6x - 3y = -3 \quad \text{Equation (2)}$$

is dependent and that the solution set is the infinite set of solutions of the equation $y = 2x + 1$. What happens when we solve this system by substitution?

Solution

We substitute $2x + 1$ for y in equation (2) and solve for x:

$$6x - 3y = -3 \quad \text{Equation (2)}$$
$$6x - 3(2x + 1) = -3 \quad \text{Substitute } 2x + 1 \text{ for } y.$$
$$6x - 6x - 3 = -3 \quad \text{Distributive law}$$
$$-3 = -3 \quad \text{Combine like terms.}$$
$$\text{true}$$

We get the *true* statement $-3 = -3$. ■

Dependent System of Two Linear Equations

If the result of applying substitution or elimination to a linear system of two equations is a true statement (one that can be put into the form $a = a$), then the system is dependent—that is, the solution is the set of ordered pairs represented by every point on the (same) line.

Using Graphing to Solve an Equation in One Variable

Some equations in *one* variable that would be difficult or impossible to solve by performing the same operations on both sides can be solved easily by graphing. To see how, consider the system of two equations in *two* variables

$$y = 2x - 4 \quad \text{Equation (1)}$$
$$y = x - 1 \quad \text{Equation (2)}$$

We can use substitution to solve this system. To begin, we substitute $2x - 4$ for y in equation (2):

$$2x - 4 = x - 1 \quad \text{Substitute } 2x - 4 \text{ for } y \text{ in equation (2).}$$
$$2x - x = -1 + 4 \quad \text{Subtract } x \text{ from both sides; add 4 to both sides.}$$
$$x = 3 \quad \text{Combine like terms.}$$

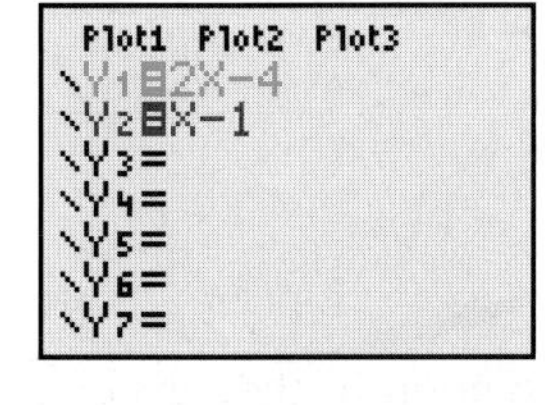

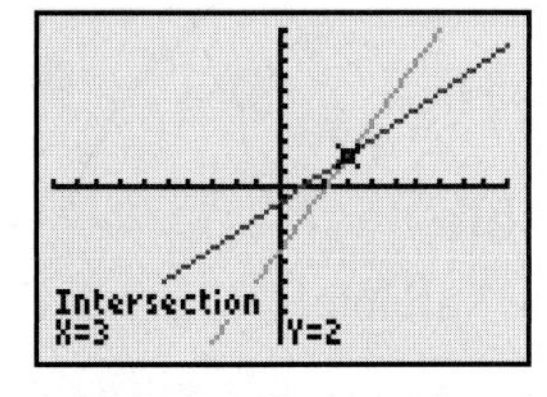

Figure 21 Solve the system

Then we substitute 3 for x in equation (1):

$$y = 2(3) - 4 = 2$$

The solution is (3, 2).

Now consider solving $2x - 4 = x - 1$, which is an equation in one variable. Our earlier work shows that the solution is 3. So, the solution 3 of $2x - 4 = x - 1$ is equal to the x-coordinate of the solution (3, 2) of the system

$$y = 2x - 4$$
$$y = x - 1$$

One way to find the solution 3 of the equation $2x - 4 = x - 1$ is to use "intersect" on a graphing calculator to find the x-coordinate of the intersection point (3, 2) of the lines $y = 2x - 4$ and $y = x - 1$ (see Fig. 21).

Using Graphing to Solve an Equation in One Variable

To use graphing to solve an equation $A = B$ in one variable, x, where A and B are expressions,

1. Use graphing to solve the system

$$y = A$$
$$y = B$$

2. The x-coordinates of any solutions of the system are the solutions of the equation $A = B$.

Example 9 Solving an Equation in One Variable by Graphing

Solve the equation by referring to the graphs shown in Fig. 22.

1. $\frac{1}{4}x + 2 = -\frac{3}{2}x - 5$

2. $\frac{1}{4}x + 2 = 3$

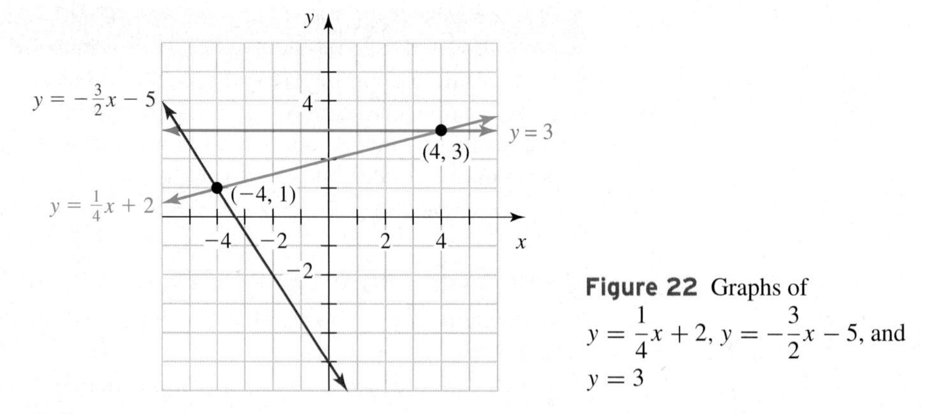

Figure 22 Graphs of $y = \frac{1}{4}x + 2$, $y = -\frac{3}{2}x - 5$, and $y = 3$

Solution

1. From Fig. 22, we see that the solution of the system

$$y = \frac{1}{4}x + 2$$
$$y = -\frac{3}{2}x - 5$$

is the ordered pair $(-4, 1)$. So, the x-coordinate -4 is the solution of the equation $\frac{1}{4}x + 2 = -\frac{3}{2}x - 5$.

2. From Fig. 22, we see that the solution of the system

$$y = \frac{1}{4}x + 2$$
$$y = 3$$

is the ordered pair $(4, 3)$. So, the x-coordinate 4 is the solution of the equation $\frac{1}{4}x + 2 = 3$. ■

Using Tables to Solve an Equation in One Variable

We can use tables rather than graphs to solve an equation in one variable.

Example 10 Solving an Equation in One Variable by Using Tables

Use a table to solve $3x - 8 = -7x + 2$.

Solution

Recall from Section 3.1 that we can use a table to solve a system such as

$$y = 3x - 8$$
$$y = -7x + 2$$

The solution is $(1, -5)$, which has an x-coordinate of 1 (see Fig. 23). For the equation $3x - 8 = -7x + 2$, the solution is 1. ■

Plot1 Plot2 Plot3
\Y1=3X−8
\Y2=−7X+2
\Y3=
\Y4=
\Y5=
\Y6=
\Y7=

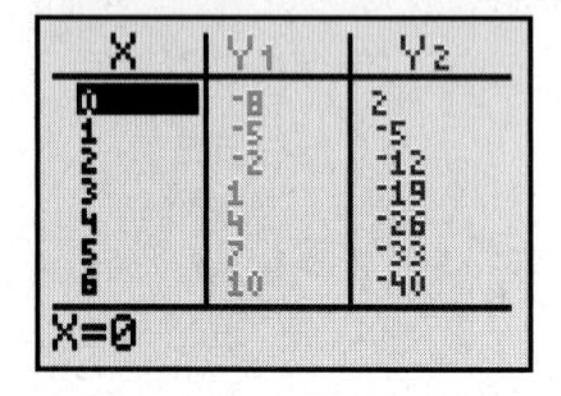

X	Y1	Y2
0	-8	2
1	-5	-5
2	-2	-12
3	1	-19
4	4	-26
5	7	-33
6	10	-40

X=0

Figure 23 Solve the system

Using a table to solve an equation in one variable is similar to using graphing to solve such an equation.

> **Using Tables to Solve an Equation in One Variable**
>
> To use a table to solve an equation $A = B$ in one variable, x, where A and B are expressions,
>
> 1. Use a table to solve the system
>
> $$y = A$$
> $$y = B$$
>
> 2. The x-coordinates of any solutions of the system are the solutions of the equation $A = B$.

group exploration

Comparing techniques of solving systems

Consider the system

$$2x + y = 4$$
$$x = 5 - 2y$$

1. Use substitution to solve the system.
2. Use elimination to solve the system.
3. Use graphing to solve the system.
4. Compare your results from Problems 1, 2, and 3.
5. Give an example of a one-solution system that is easiest to solve by substitution. Also, give an example of such a system that is easiest to solve by elimination. Finally, give an example of such a system that is easiest to solve by graphing. Explain. Solve your three systems.

group exploration

Using graphing to solve an equation in one variable

1. **a.** Solve

$$y = 2x + 1$$
$$y = 7 - x$$

by substitution. Use "intersect" on a graphing calculator to verify your result.

b. Solve $2x + 1 = 7 - x$. Check that your solution satisfies the equation.

c. Compare your solution to the equation in part (b) with the x-coordinate of the solution to the system in part (a).

2. **a.** Solve

$$y = 3x - 5$$
$$y = x + 3$$

by substitution. Use "intersect" to verify your result.

b. Solve $3x - 5 = x + 3$. Check that your solution satisfies the equation.

c. Compare your solution of the equation in part (b) with the x-coordinate of the solution of the system in part (a).

3. Solve $3x - 7 = 8 - 2x$ by a symbolic method. Explain how you can verify your solution by using "intersect."
4. It is difficult to find the exact solution of the equation $x^3 = 5 - x$. Use "intersect" to solve the equation, with the result rounded to the second decimal place. To enter x^3, press [X,T,Θ,n] [^] 3.

TIPS FOR SUCCESS: Cross-checks

If you finish a quiz or an exam early, verify your answers with cross-checks. For example, suppose you determine by elimination that the solution of the system

$$2x + 3y = 9$$
$$4x + 5y = 17$$

is (3, 1). There are several ways to verify your answer. You could check that (3, 1) satisfies both equations. You could graph each equation and check that the intersection point is (3, 1). Or you could solve the system by substitution.

HOMEWORK 3.2

FOR EXTRA HELP ▸ Student Solutions Manual · PH Math/Tutor Center · MathXL® · MyMathLab

Solve the system by substitution. Verify your solution graphically or by checking that it satisfies both equations in the system.

1. $y = x - 5$; $x + y = 9$
2. $y = 2x - 1$; $x + y = 5$
3. $2x - 3y = -1$; $x = 4y + 7$
4. $4x + 3y = -14$; $x = 2x + 8$
5. $3x - 5y - 29 = 0$; $y = 2(x - 5)$
6. $y = 3(x - 2)$; $7x - 3y - 10 = 0$
7. $y = 99x$; $y = 100x$
8. $y = x$; $y = -x$
9. $y = 4x + 5$; $y = 2x - 1$
10. $y = 11 - 3x$; $y = 2x + 1$
11. $y = 0.2x + 0.6$; $2y - 3x = -4$
12. $x = 2 - y$; $0.6x + 0.3y = 0.3$
13. $4x + 3y = 2$; $2x - y = -4$
14. $3x - y = 10$; $2x + 5y = -16$
15. $y = \frac{1}{2}x - 5$; $2x + 3y = -1$
16. $4x - 7y = 3$; $x = \frac{2}{3}y + 4$

Solve the system by elimination. Verify your solution graphically or by checking that it satisfies both equations of the system.

17. $-x + 3y = -25$; $x - 5y = 39$
18. $3x - 5y = 26$; $-3x - 2y = 2$
19. $3x - 4y = -6$; $5x - 4y = -2$
20. $2x + 3y = -2$; $x + 3y = 2$
21. $2x + y = 2$; $5x - 2y = -13$
22. $3x - 4y = 18$; $4x - y = 11$
23. $3x - 2y = 7$; $-6x - 5y = 4$
24. $3x + 2y = 3$; $9x - 8y = -33$
25. $3x + 5y = 3$; $7x - 2y = -34$
26. $-3x - 5y = -22$; $4x - 7y = -39$
27. $8x - 9y = -43$; $12x + 15y = 21$
28. $6x + 5y = 13$; $9x + 4y = 2$
29. $4x - 7y = -29$; $-5x - 2y = 4$
30. $-8x - 3y = -1$; $6x - 5y = 37$
31. $3y = 2x - 6$; $5x - 4y = 1$
32. $4x = 7y - 25$; $3x + 5y = 12$
33. $0.9x + 0.4y = 1.9$; $0.3x - 0.2y = 1.3$
34. $0.2x - 0.5y = 0.2$; $0.8x + 1.5y = -6.2$
35. $3(2x - 1) + 4(y - 3) = 1$; $4(x + 5) - 2(4y + 1) = 18$
36. $2(x - 3) - 3(y + 1) = -5$; $-4(x - 2) + 5(y + 3) = 13$
37. $\frac{1}{5}x + \frac{3}{2}y = 7$; $\frac{2}{5}x - \frac{9}{2}y = -16$
38. $-\frac{1}{2}x - \frac{1}{3}y = -1$; $-\frac{3}{2}x + \frac{2}{3}y = -8$
39. $\frac{2}{3}x + \frac{1}{2}y = \frac{1}{6}$; $\frac{1}{2}x + \frac{5}{4}y = \frac{11}{4}$ (see ex. 6)
40. $\frac{1}{4}x + \frac{5}{2}y = 2$; $\frac{5}{6}x - \frac{1}{3}y = -2$

Solve the system by either elimination or substitution. If the system is inconsistent or dependent, say so. For a one-solution system, verify the solution graphically or check that the solution satisfies both equations of the system.

41. $y = 2x + 5$; $6x - 3y = -3$
42. $y = 5x - 4$; $10x - 2y = 12$
43. $13x + 10y = -7$; $17x - 15y = 47$
44. $3x - 5y = 10$; $7x + 2y = 37$
45. $4x - 5y = 3$; $-12x + 15y = -9$
46. $2x - y = -4$; $-8x + 4y = 16$
47. $4x - 3y = 1$; $-20x + 15y = -3$
48. $-4x + 8y = 2$; $6x - 12y = -5$
49. $y = -2x + 4$; $y = -4x + 10$
50. $y = 3x + 7$; $y = -5x - 1$
51. $2(x + 3) - (y + 5) = -6$; $5(x - 2) + 3(y - 4) = -34$
52. $-(x - 6) + 6(y + 1) = 58$; $3(x + 1) - 4(y - 2) = -15$
53. $y = \frac{1}{2}x + 3$; $2y - x = 6$
54. $y = -\frac{1}{3}x + 4$; $x + 3y = 12$
55. $\frac{5}{6}x + \frac{1}{4}y = 3$; $-\frac{1}{3}x + \frac{5}{2}y = 4$
56. $\frac{1}{2}x - \frac{3}{4}y = -4$; $\frac{2}{3}x - \frac{1}{3}y = -4$
57. $\frac{x + 2y}{3} - \frac{x - y}{2} = \frac{13}{6}$; $\frac{x + 3y}{2} + \frac{x + y}{4} = \frac{17}{4}$
58. $\frac{2x + y}{3} - \frac{3x - y}{6} = 1$; $\frac{x + y}{3} + \frac{2x - y}{4} = \frac{31}{12}$

Use elimination or substitution to solve the system. Round the coordinates of solutions to the second decimal place. Verify your solution graphically or by checking that it satisfies both equations approximately.

59. $y = 2.58x - 8.31$
$y = -3.25x + 7.86$

60. $y = -4.25x - 2.19$
$y = 3.65x + 9.38$

61. $y = -0.77x + 4.84$
$y = -2.31x - 1.49$

62. $y = -3.38x - 8.57$
$y = -0.35x + 4.28$

Solve the system of equations three times, once by each of the three methods: elimination, substitution, and graphing. Decide which method you prefer for the system.

63. $3x + y = 11$
$y = -2x + 9$

64. $5x - 2y = 8$
$3x + 4y = 10$

65. Consider the system

$$2x + 4y = 10$$
$$3x - 7y = 2$$

a. Solve the system by eliminating the x terms.
b. Solve the system by eliminating the y terms.
c. Compare your results in parts (a) and (b).

66. Consider the system

$$-4x + 3y = 15$$
$$5x - 4y = -19$$

a. Solve the system by eliminating the x terms.
b. Solve the system by eliminating the y terms.
c. Compare your results in parts (a) and (b).

For Exercises 67–72, solve the given equation or system by referring to the graphs shown in Fig. 24.

67. $\frac{1}{2}x + \frac{5}{2} = 2x + 7$

68. $\frac{1}{2}x + \frac{5}{2} = 4$

69. $\frac{1}{2}x + \frac{5}{2} = 3$

70. $2x + 7 = 3$

71. $2x + 7 = -3$

72. $y = \frac{1}{2}x + \frac{5}{2}$
$y = 2x + 7$

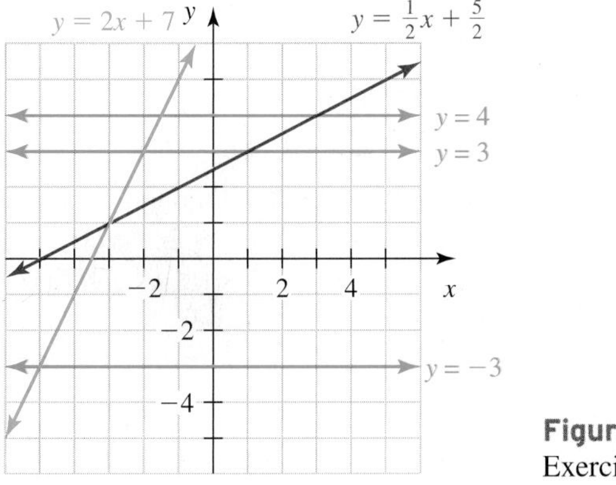

Figure 24 Exercises 67–72

For Exercises 73–78, solve the given equation or system by referring to the graphs shown in Fig. 25.

73. $\frac{1}{3}x + \frac{5}{3} = x - 1$

74. $\frac{1}{3}x + \frac{5}{3} = -3x - 5$

75. $\frac{1}{3}x + \frac{5}{3} = 2$

76. $\frac{1}{3}x + \frac{5}{3} = 0$

77. $y = \frac{1}{3}x + \frac{5}{3}$
$y = -3x - 5$

78. $y = \frac{1}{3}x + \frac{5}{3}$
$y = x - 1$

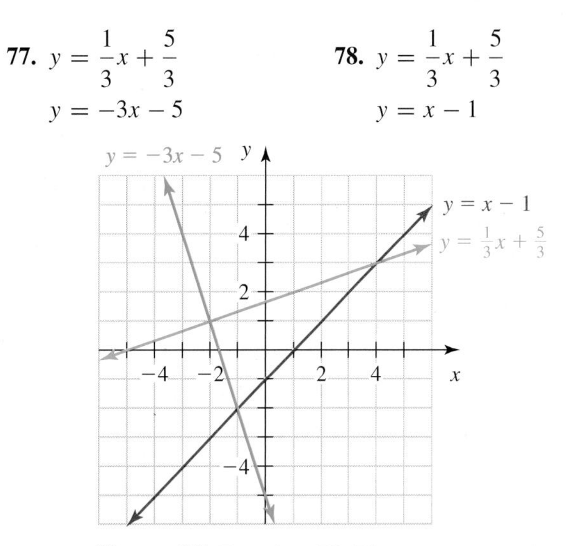

Figure 25 Exercises 73–78

Use "intersect" on a graphing calculator to solve the equation. Round the solution to the second decimal place.

79. $5x - 4 = -2x + 7$

80. $-3x - 8 = 4x + 1$

81. $-0.39x - 4.98 = 1.04x - 1.52$

82. $0.48x - 5.37 = 1.32x - 8.14$

83. $\frac{5}{6}x + \frac{9}{2} = -\frac{2}{7}x + \frac{7}{3}$

84. $\frac{4}{9}x + \frac{9}{2} = -\frac{1}{5}x - \frac{5}{3}$

For Exercises 85–90, solve the given equation or system by referring to the solutions of the functions shown in Table 14.

85. $\frac{1}{2}x + \frac{7}{2} = \frac{4}{5}x + 2$

86. $\frac{1}{2}x + \frac{7}{2} = \frac{11}{10}x + \frac{17}{10}$

87. $\frac{11}{10}x + \frac{17}{10} = 5$

88. $\frac{1}{2}x + \frac{7}{2} = 4$

89. $y = \frac{4}{5}x + 2$
$y = \frac{11}{10}x + \frac{17}{10}$

90. $y = \frac{1}{2}x + \frac{7}{2}$
$y = \frac{4}{5}x + 2$

Table 14 Some Solutions of Three Functions (Exercises 85–90)

x	0	1	2	3	4	5	6
$y = \frac{1}{2}x + \frac{7}{2}$	3.5	4	4.5	5	5.5	6	6.5
$y = \frac{4}{5}x + 2$	2	2.8	3.6	4.4	5.2	6	6.8
$y = \frac{11}{10}x + \frac{17}{10}$	1.7	2.8	3.9	5	6.1	7.2	8.3

91. Some values of linear functions f and g are listed in Table 15. Find the solution of a system of two equations that describes the functions f and g. [**Hint:** Find equations of f and g.]

Table 15 Values of Functions f and g (Exercise 91)

x	0	1	2	3	4	5	6	7	8
$f(x)$	3	7	11	15	19	23	27	31	35
$g(x)$	50	44	38	32	26	20	14	8	2

92. Some values of linear functions f and g are listed in Table 16. Find the solution of a system of two equations that describes the functions f and g. [**Hint:** Find equations of f and g.]

Table 16 Values of Functions f and g (Exercise 92)

x	0	1	2	3	4	5	6	7	8
$f(x)$	201	204	207	210	213	216	219	222	225
$g(x)$	6	11	16	21	26	31	36	41	46

93. A student decides to solve the system

$$y = 2x + 3$$
$$y = 2.01x + 1$$

by graphing the equations (see Fig. 26). He decides that the solution is the empty set. Describe any errors. Then solve the system correctly.

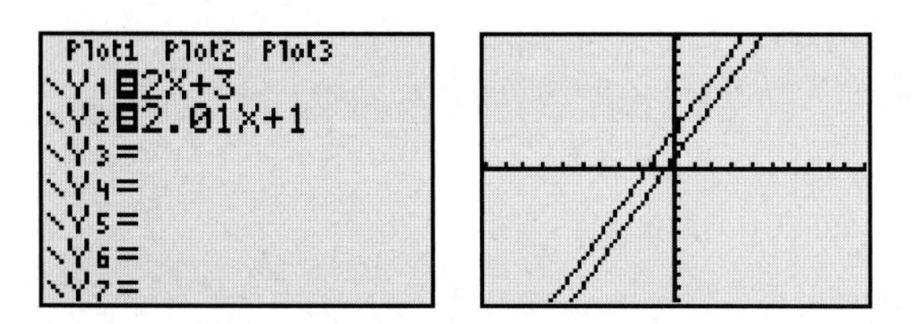

Figure 26 Graphs of two lines (Exercise 93)

94. To solve the system

$$x = 3$$
$$y = 4$$

a student adds the left sides and adds the right sides to get $x + y = 7$. She thinks that the solution set is the set of points on the line $y = -x + 7$. Is the student correct? Explain.

95. Find the coordinates of points A, B, C, D, E, and F as shown in Fig. 27. The equations of lines l_1–l_4 are provided, but no attempt has been made to sketch the lines accurately, except for showing the intersection points. Verify your results graphically.

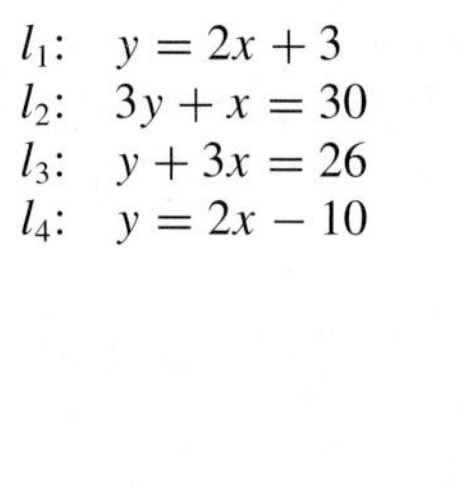

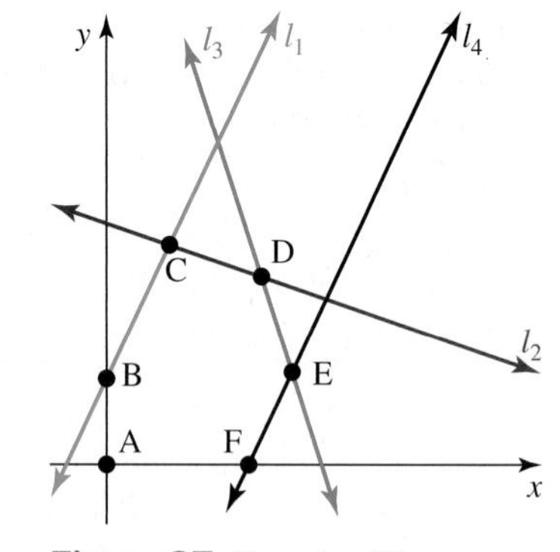

Figure 27 Exercise 95

96. Determine the constants a and b so that (2, 3) is the solution of the system. Verify your result by checking that (2, 3) satisfies each of your equations, and verify your result graphically.

$$7x - 4y = a$$
$$5x + 2y = b$$

97. a. Let a, b, c, d, k, and p be constants such that the two equations form a one-solution system. Solve the system.

$$ax + by = c$$
$$kx + py = d$$

b. Use your result from part (a) to solve this system:

$$3x + 5y = 2$$
$$4x + 3y = 4$$

98. Explain why the solution set of a system of two linear equations either is empty, consists of exactly one solution, or consists of infinitely many solutions. Why cannot such a solution set have exactly two solutions?

99. Describe how to solve a system by elimination. Include in your discussion the result of solving a system by elimination if the system is a one-solution system, an inconsistent system, or a dependent system. Finally, describe how to solve a system by substitution.

3.3 USING SYSTEMS TO MODEL DATA

"I like math because it's interesting. I feel mentally challenged whenever I see a math problem. Math is very crucial to the development of our society, and it applies to virtually every aspect of our lives. Many fields require the knowledge of math, such as computer science, architecture, engineering, and so on."

—Steven W., student

Objectives

- Use substitution and elimination to make predictions about situations described by a table of data.
- Use substitution and elimination to make predictions about situations described by rates of change.

In Section 3.1, we used a graphical approach to find the intersection point of the graphs of two linear models. In this section, we discuss how to use substitution and elimination to find such an intersection point.

Using a Table of Data to Find a System for Modeling

In Example 1, we will use two linear models from Section 3.1 and work with a table of data.

Example 1 Solving a System to Make a Prediction

In Example 1 of Section 3.1, we modeled life expectancies (in years) $W(t)$ and $M(t)$ of U.S. women and men, respectively, by the system

$$L = W(t) = 0.105t + 77.53$$
$$L = M(t) = 0.203t + 69.91$$

where t is the number of years since 1980 (see Table 17).

Use a symbolic method to predict when the life expectancies of women and men will be equal.

Table 17 U.S. Life Expectancies of Women and Men

Year of Birth	Women (years)	Men (years)
1980	77.4	70.0
1985	78.2	71.1
1990	78.8	71.8
1995	78.9	72.5
2000	79.5	74.1
2003	80.1	74.8

Source: *U.S. Census Bureau*

Solution

In Example 1 of Section 3.1, we found that the intersection point of the graphs of the two models represents the event when the life expectancies of women and men will be equal. We can find this intersection point by substitution (or elimination). To solve by substitution, we substitute $0.105t + 77.53$ for L in the equation $L = 0.203t + 69.91$ and solve for t:

$$0.105t + 77.53 = 0.203t + 69.91 \quad \text{Substitute } 0.105t + 77.53 \text{ for } L.$$
$$0.105t - 0.203t = 69.91 - 77.53 \quad \text{Subtract } 0.203t \text{ and } 77.53 \text{ from both sides.}$$
$$-0.098t = -7.62 \quad \text{Combine like terms.}$$
$$t \approx 77.76 \quad \text{Divide both sides by } -0.098.$$

Next, we substitute 77.76 for t in the equation $L = 0.105t + 77.53$:

$$L = 0.105(77.76) + 77.53 \approx 85.69$$

So, the approximate solution to the system is (77.76, 85.69), the same result that we found in Example 1 of Section 3.1. According to the models, the life expectancies of women and men will be equal in 2058. ■

In Example 1, we used substitution to predict when the life expectancy of women will equal the life expectancy of men. Graphically, this event is described by the intersection point of the graphs of the two life-expectancy models. In general, **we can find an intersection point of the graphs of two linear models by graphing, substitution, or elimination.**

Example 2 Solving a System to Make a Prediction

In Exercises 21 and 22 of Homework 2.2, you modeled world record times for the 400-meter run (see Table 18).

Table 18 400-Meter Run Record Times

Women		Men	
Year	Record Time (seconds)	Year	Record Time (seconds)
1957	57.0	1900	47.8
1959	53.4	1916	47.4
1962	51.9	1928	47.0
1969	51.7	1932	46.2
1972	51.0	1941	46.0
1976	49.29	1950	45.8
1979	48.60	1960	44.9
1983	47.99	1968	43.86
1985	47.60	1988	43.29
		1999	43.18

Source: *International Association of Athletics Federations*

The record times (in seconds) $W(t)$ and $M(t)$ for women and men, respectively, are modeled by the system

$$r = W(t) = -0.27t + 70.45$$
$$r = M(t) = -0.053t + 48.08$$

where t is the number of years since 1900. Predict when the women's record time and the men's record time will be equal.

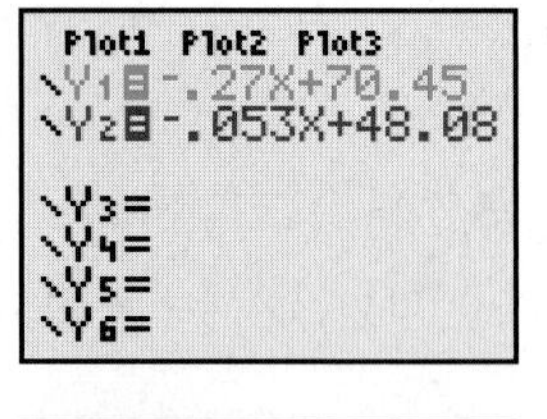

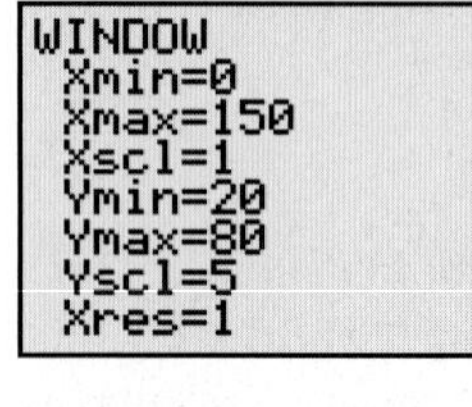

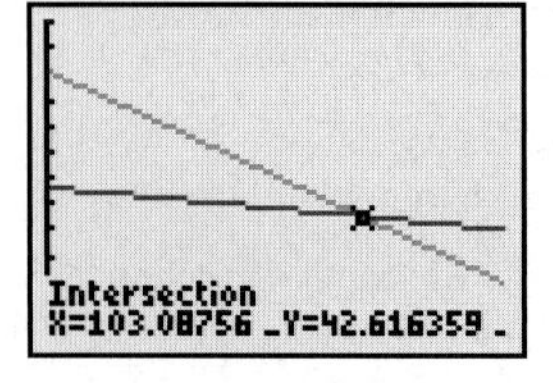

Figure 28 Verify that the intersection point is approximately $(103.09, 42.62)$

Solution

We solve the system

$$r = -0.27t + 70.45$$
$$r = -0.053t + 48.08$$

by substitution. We do so by substituting $-0.27t + 70.45$ for r in the equation $r = -0.053t + 48.08$:

$$-0.27t + 70.45 = -0.053t + 48.08 \quad \text{Substitute } -0.27t + 70.45 \text{ for } r.$$
$$-0.27t + 0.053t = 48.08 - 70.45 \quad \text{Add } 0.053t \text{ to both sides; subtract } 70.45 \text{ from both sides.}$$
$$-0.217t = -22.37 \quad \text{Combine like terms.}$$
$$t \approx 103.09 \quad \text{Divide both sides by } -0.217.$$

Next, we substitute 103.09 for t in the equation $r = -0.27t + 70.45$ and solve for r:

$$r = -0.27(103.09) + 70.45 \approx 42.62$$

So, according to the models, the world record times for both women and men will be 42.62 seconds in 2003. Model breakdown has occurred, because in 2006 the women's record time was still 47.60 seconds and the men's record time was still 43.18 seconds.

We can verify our result by using "intersect" on a graphing calculator (see Fig. 28). ■

Using Rate of Change to Find a System for Modeling

Recall from Section 2.4 that if the rate of change of the dependent variable with respect to the independent variable is constant, then there is a linear relationship between the variables. And for the linear model that describes the relationship, the slope is equal to the constant rate of change. We use these ideas in Example 3.

Example 3 Solving a System to Make a Prediction

In 2005, a 2004 Cadillac DeVille® cost about \$26,440, and a 2004 Acura Integra RSX® cost about \$18,243. The DeVille depreciates by \$3548 each year, and the Integra depreciates by \$1333 each year (Source: *Edmunds Automobile Buyer's Guide*). When will these 2004 cars have the same value?

Solution

Let $V = D(t)$ be the value (in dollars) of a 2004 DeVille and $V = I(t)$ be the value (in dollars) of a 2004 Integra at t years since 2005.

Because a 2004 Deville's value decreases by a *constant* \$3548 each year, the function D is linear and its slope is -3548. The V-intercept is $(0, 26{,}440)$, since the car is worth \$26,440 at year $t = 0$. So, an equation of D is

$$V = D(t) = -3548t + 26{,}440$$

Plot1 Plot2 Plot3
\Y1 = -3548X+26440
\Y2 = -1333X+18243
\Y3=
\Y4=
\Y5=

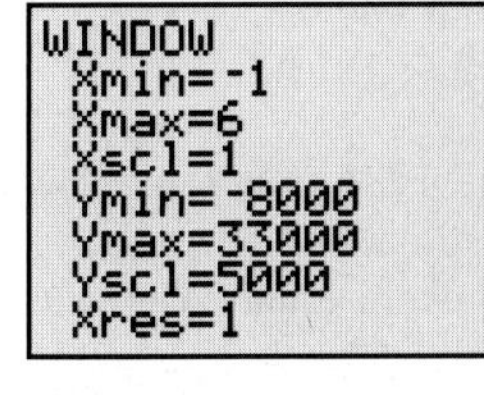

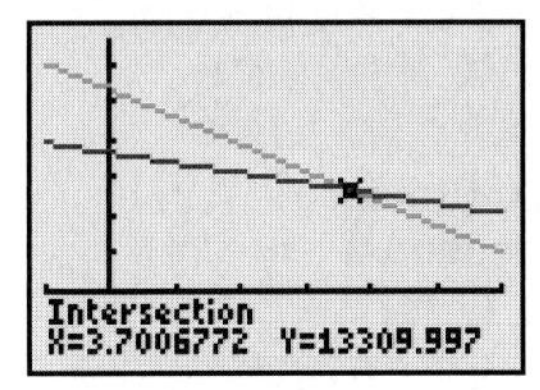

Figure 29 Verify the work

Similar work in finding the equation of the function I gives

$$V = I(t) = -1333t + 18{,}243$$

Next, we substitute $-3548t + 26{,}440$ for V in the equation $V = -1333t + 18{,}243$:

$$-3548t + 26{,}440 = -1333t + 18{,}243$$

Then we solve for t:

$$-3548t + 1333t = 18{,}243 - 26{,}440 \quad \text{Add } 1333t \text{ to both sides; subtract } 26{,}440 \text{ from both sides.}$$

$$-2215t = -8197 \quad \text{Combine like terms.}$$

$$t \approx 3.70 \quad \text{Divide both sides by } -2215.$$

We conclude that the cars will have the same value in approximately 4 years (in 2009).

Next, we find the common value of the cars:

$$D(3.70) = -3548(3.70) + 26{,}440 \approx 13{,}312$$

So, both cars will be worth about \$13,312 in 2009, according to the models. Using "intersect" on a graphing calculator gives the result that both cars will be worth about \$13,310 in 2009 (see Fig. 29). Our result of \$13,312 is \$2 more than the calculator result because we evaluated D at a rounded value of t. ■

group exploration

Looking ahead: Total value

1. A cup of coffee costs \$3. Find the total cost of
   a. 2 cups of coffee
   b. 4 cups of coffee
   c. 5 cups of coffee
   d. n cups of coffee
2. Let T be the total value of n objects that each have value v. Write an equation in terms of T, n, and v.
3. A pen costs \$2. A notebook costs \$4. Find the total cost of
   a. 3 pens and 2 notebooks
   b. 5 pens and 3 notebooks
   c. p pens and n notebooks
4. A 7000-seat theater has tickets for sale at \$30 and \$45. Let x be the number of \$30 tickets and y be the number of \$45 tickets.
   a. What is the meaning of the equation $30x + 45y = 232{,}500$?
   b. What is the meaning of the equation $x + y = 7000$?
   c. Solve the system. What does your result mean in this situation?

$$30x + 45y = 232{,}500$$
$$x + y = 7000$$

TIPS FOR SUCCESS: Study with a Classmate

It can be helpful to meet with a friend from class and discuss what happened in class that day. Not only can you ask questions *of* each other, but also you will learn just by explaining concepts *to* each other. Explaining a concept to someone else forces you to clarify your own understanding of the concept.

HOMEWORK 3.3

FOR EXTRA HELP ▶

Student Solutions Manual | PH Math/Tutor Center | MathXL® | MyMathLab

1. In Exercise 27 of Homework 3.1, the winning times (in seconds) $W(t)$ and $M(t)$ for women and men, respectively, in Olympic 500-meter speed skating are modeled by the system

$$y = W(t) = -0.172t + 43.44$$
$$y = M(t) = -0.147t + 39.80$$

where t is the number of years since 1970 (see Table 19). Use substitution or elimination to predict when the women's winning time will be equal to the men's winning time. What will be that winning time?

Table 19 Olympic 500-Meter Speed Skating Times

Year	Winning Time (seconds)	
	Women	Men
1972	43.33	39.44
1976	42.76	39.17
1980	41.78	38.03
1984	41.02	38.19
1988	39.10	36.45
1992	40.33	37.14
1994	39.25	36.33
1998	38.21	35.59
2002	37.375	34.615
2006	38.285	34.880

Source: The Universal Almanac

2. In Exercise 28 of Homework 3.1, the annual U.S. consumption (in pounds per person) $C(t)$ and $R(t)$ of chicken and red meat, respectively, are modeled by the system

$$y = C(t) = 0.99t - 45.68$$
$$y = R(t) = -0.62t + 174.40$$

where t is the number of years since 1900 (see Table 20).

Table 20 Annual U.S. Per-Person Consumption of Chicken and Red Meat

Year	Annual Consumption (pounds per person)	
	Chicken	Red Meat
1970	27.4	131.9
1975	26.3	125.8
1980	32.7	126.4
1985	36.4	124.9
1990	42.4	112.2
1995	48.2	113.6
2000	54.2	113.7
2003	57.5	111.9

Source: *U.S. Department of Agriculture*

Use substitution or elimination to predict when annual consumption of chicken will equal annual consumption of red meat. What will be that consumption? How confident are you in this prediction?

3. In Exercise 29 of Homework 3.1, the enrollments (in millions) $W(t)$ and $M(t)$ for women and men, respectively, at U.S. colleges are modeled by the system

$$E = W(t) = 0.15t + 5.76$$
$$E = M(t) = 0.072t + 5.43$$

where t is the number of years since 1980 (see Table 21). Use substitution or elimination to estimate when women's and men's enrollments were approximately equal. What was that enrollment?

Table 21 College Enrollments

Year	Enrollment (millions)	
	Women	Men
1985	6.4	5.8
1990	7.5	6.3
1995	7.9	6.3
2000	8.6	6.7
2003	9.3	7.3

Source: *U.S. Census Bureau*

4. In Exercise 30 of Homework 3.1, the number (in millions) of U.S. households with PCs $C(t)$ and the total number (in millions) of U.S. households $H(t)$ are modeled by the system

$$y = C(t) = 4.80t + 6.86$$
$$y = H(t) = 1.59t + 90.31$$

where t is the number of years since 1990 (see Table 22).

Table 22 U.S. Households with Personal Computers

Year	Households with PCs (millions)	Total Households (millions)
1995	31.36	99.0
1997	39.56	101.0
1999	50.00	103.6
2001	60.81	108.2
2003	68.78	111.3

Sources: *U.S. Department of Commerce; Media Metrix; U.S. Census Bureau*

a. Find the slopes of C and H. What do they mean in this situation?

b. Since the number of U.S. households is increasing, it is not surprising that the number of households with PCs is also increasing. Explain why your results in part (a) show that this is not the only reason the number of households with PCs is increasing.

c. Predict the *percentage* of U.S. households that will have PCs in 2011.

d. Use substitution or elimination to predict when all U.S. households will have PCs.

5. The annual U.S. per-person consumption of milk and soft drinks is shown in Table 23 for various years.

Table 23 Annual U.S. Per-Person Consumption of Milk and Soft Drinks

	Annual Consumption (gallons per person)	
Year	**Milk**	**Soft Drinks**
1950	36.4	10.8
1960	32.6	13.4
1970	29.8	24.3
1980	26.5	35.1
1990	24.3	46.2
2000	22.6	49.3

Source: *USDA/Economic Research Service*

a. Let $M(t)$ be the annual consumption of milk and $S(t)$ be the annual consumption of soft drinks, both in gallons per person, in the year that is t years since 1950. Find equations of M and S.

b. Use substitution or elimination to estimate when the annual consumption of milk was equal to the annual consumption of soft drinks. What was that consumption?

c. Use "intersect" on a graphing calculator to verify your work.

6. The percentages of Continental Airlines tickets that were free and the percentages of American Airlines tickets that were free are shown in Table 24 for various years.

Table 24 Percentages of Airline Tickets That Were Free

	Percent of Tickets That Were Free	
Year	**Continental**	**American**
1999	6.0	10.3
2000	6.4	9.8
2001	6.2	9.5
2002	6.6	9.6
2003	7.3	9.2
2004	7.3	8.7
2005	7.4	8.6

Source: *Back Aviation Solutions*

a. Let $c(t)$ be the percentage of Continental tickets that are free and $a(t)$ be the percentage of American tickets that are free, both at t years since 1990. Find equations of c and a.

b. Predict when the percentage of Continental tickets that are free will be equal to the percentage of American tickets that are free. What will be that percentage?

c. Use "intersect" on a graphing calculator to verify your work.

7. Scores on tests that evaluate general knowledge and vocabulary and scores on tests that evaluate memory and information-processing speed are shown in Table 25 for various ages. (The higher the score, the better the mental function. A score of 0 is average.)

a. Let $K(a)$ be the score on general knowledge and vocabulary tests and $M(a)$ be the score on memory and information-processing speed tests, both at age a years. Find equations of K and M.

b. Use substitution or elimination to estimate at what age a person's score on general knowledge and vocabulary tests

Table 25 Age Versus Mental Functioning

Age Group	Age Used to Represent Age Group	General Knowledge and Vocabulary Score	Memory and Information-Processing Speed Score
20–30	25	−0.4	1.0
30–40	35	−0.3	0.7
40–50	45	0	0.3
60–70	65	0.2	−0.2
70–80	75	0.3	−0.4

Source: *Denise Park, University of Illinois, Champaign–Urbana*

is equal to the person's score on memory and information-processing speed tests.

c. Use "intersect" on a graphing calculator to verify your work.

8. The percentages of students studying for a master of divinity degree who are women versus those who are men are shown in Table 26 for various years.

Table 26 Students Studying for a Master of Divinity Degree

	Percent	
Year	**Women**	**Men**
1973	5	95
1983	17	83
1993	25	75
1996	28	72
1998	30	70
2000	31	69
2002	32	68

Source: USA Today

a. Let $W(t)$ and $M(t)$ be the percentages of divinity majors who are women and men, respectively, at t years since 1970. Find equations of W and M.

b. Compare the slopes of your models. What do the slopes mean in this situation?

c. Use substitution or elimination to predict when the number of women studying for a master of divinity degree will equal the number of men studying for such a degree.

d. Using only the equation of W, predict when the number of women studying for a master of divinity degree will equal the number of men studying for such a degree. Compare your result with that in part (c).

9. The *Denver Post* and the *Rocky Mountain News* are competing newspapers in Denver, Colorado. Their circulations are listed in Table 27 for various years.

Table 27 Newspaper Circulations

Year	*Denver Post* Circulation (thousands)	*Rocky Mountain News* Circulation (thousands)
1992	255	355
1993	270	350
1994	285	345
1995	295	340

Source: *Audit Bureau of Circulations*

Let $C = D(t)$ be the circulation (in thousands) of the *Denver Post* and $C = R(t)$ be the circulation (in thousands) of the *Rocky Mountain News,* where t is the number of years since 1990. Reasonable equations of D and R are

$$C = D(t) = 13.5t + 229$$
$$C = R(t) = -5t + 365$$

a. Use substitution or elimination to estimate when the two newspapers had equal circulation.
b. The newspaper with greater circulation can usually charge more for advertisements, thus increasing its revenue. On March 1–2, 1997, both newspapers included articles claiming that the other newspaper had made false reports of its circulation to the audit bureau. Use your result in part (a) to explain why it is not surprising that the competition between the two newspapers heated up in 1997.
c. Throughout the years, both newspapers increased their circulation by giving away "bonus issues" free to those who subscribed to less than a full week. During the circulation wars from 1997 to 2000, both newspapers increased their distribution of bonus issues dramatically. In 2000, the combined circulation of the two newspapers was 826 thousand. Estimate the combined increase in number of bonus issues from the two newspapers due to the circulation wars. [**Hint:** Begin by finding $D(10)$ and $R(10)$.]
d. Use the models D and R to estimate the combined circulation of the two newspapers in 2001.
e. In January 2001, the two newspapers joined their revenue streams under a single entity, the Denver Newspaper Agency (DNA). In 2001, the combined circulation was 638 thousand. Is your estimate in part (d) an underestimate or an overestimate? Give a possible explanation for this in terms of the circulation wars, DNA, and bonus issues.

10. World record times for the 1500-meter run are listed in Table 28.

Table 28 1500-Meter Run Record Times

Women		Men	
Year	**Record Time (seconds)**	**Year**	**Record Time (seconds)**
1927	318	1926	231
1936	287	1941	227
1946	277	1955	220
1957	269	1980	211
1962	259	1995	207
1993	230	1998	206

Source: *International Association of Athletics Federations*

a. Let $W(t)$ and $M(t)$ be the record times (in seconds) of women and men, respectively, at t years since 1900. Find the regression equations of W and of M.
b. Use substitution or elimination to predict when the women's record time will equal the men's record time. Verify your result by using "intersect" on a graphing calculator.
c. Now find the regression equation of the women's record times, excluding the record set in 1927. Also, find the regression equation of the men's times, excluding the record set in 1926. Use these equations to predict when the women's record time will equal the men's record time.
d. Explain why your result in part (b) is so different from that in part (c).

11. In 2005, the price of a 2004 Ford Taurus® was about $12,281, with a depreciation of about $1725 per year. The price of a 2004 Ford Focus® was about $10,952, with a depreciation of about $1424 per year (Source: *Edmunds Automobile Buyer's Guide*).
a. Let $V = T(t)$ be the value (in dollars) of a 2004 Ford Taurus and $V = E(t)$ be the value (in dollars) of a 2004 Ford Focus, both at t years since 2005. Find equations of T and E.
b. Use substitution or elimination to predict when the cars will have the same value. What is that value?
c. Use a graphical method to verify your work in part (b).

12. In 2005, the price of a 2004 Honda Accord® was about $18,329, with a depreciation of about $2256 per year (Source: *Edmunds Automobile Buyer's Guide*). A student had $500 in 2005 and saves $1700 each year. Assume that the student does not earn interest on his savings.
a. Let $H(t)$ be the value (in dollars) of a 2004 Honda Accord and $S(t)$ be the student's total savings (in dollars), both at t years since 2005. Find equations of H and S.
b. Use substitution or elimination to predict when the student will be able to buy a 2004 Honda Accord in cash.
c. Use a graphical method to verify your work in part (b).
d. Now assume that the student earns interest on his savings. Is your answer in part (b) an underestimate or an overestimate? Explain.

13. Jenny Craig® offers a promotion for which members pay a start-up fee of $19 (to lose 19 pounds) and a weekly fee of $72 for Jenny Craig food (Source: *Jenny Craig*). Members meet with a food counselor free of charge.

Weight Watchers® offers a promotion for which members do not pay a start-up fee but pay a fee of $17 per week to meet with a food counselor. Members purchase their own food, which usually costs about $60 per week.

a. Let $J(t)$ be the per-person cost (in dollars) of the Jenny Craig program (including food) for t weeks. Let $W(t)$ be the per-person cost (in dollars) of the Weight Watchers program *plus food* for t weeks. Find equations of J and W.
b. Perform a unit analysis of the equations of J and W.
c. Use substitution or elimination to estimate how many weeks it will take for the total cost at Jenny Craig to equal the total cost at Weight Watchers (plus the cost of food). What is that total cost?
d. Use a graphical method to verify your work in part (b).

14. Fitness USA® offers membership for a flat fee of $35 plus a monthly fee of $26. Gold's Gym® offers membership for a flat fee of $14 plus a monthly fee of $29.95, with a year's commitment (Sources: *Fitness USA; Gold's Gym*). Let $f(t)$ be the total cost (in dollars) of joining Fitness USA and $g(t)$ be the total cost (in dollars) of joining Gold's Gym, both for t months.

a. Find equations of f and g.
b. Perform a unit analysis of the equations of f and g.
c. Use substitution or elimination to estimate when the total cost at each health club will be equal.
d. Use a graphing calculator table or graph to verify your work in part (b).

15. The sales of large and midsize SUVs were 281 thousand cars in 2002 and have since decreased by 7 thousand cars per year. The sales of small SUVs and crossover vehicles were 140 thousand in 2002 and have since increased by 33 thousand cars per year (Source: *Ward's AutoInfoBank*). Estimate when the sales of large and midsize SUVs were equal to the sales of small SUVs and crossover vehicles.

16. The number of visitors on snowmobiles at Yellowstone National Park was 84.5 thousand visitors in 2001 and has since decreased by 15.1 thousand visitors per year. The number of park visitors traveling by snowcoach—a van on treads—was 11.7 thousand visitors in 2001 and has since increased by 12.3 thousand visitors per year (Source: *National Park Service*). Predict when the annual number of snowmobile visitors will equal the annual number of snowcoach visitors.

17. Personal income (in 1998 dollars) in New Mexico increased approximately linearly from \$20,656 in 1998 to \$22,629 in 2003. Personal income (in 1998 dollars) in Ohio increased approximately linearly from \$26,017 in 1998 to \$26,530 in 2003 (Source: *Bureau of Economic Analysis*).

a. Let $f(t)$ be personal income (in 1998 dollars) in New Mexico and $g(t)$ be personal income (in 1998 dollars) in Ohio, both in the year that is t years since 1998. Find equations of f and g.
b. Use substitution or elimination to predict when personal income in New Mexico will equal personal income in Ohio.
c. Use a graphing calculator table or graph to verify your work in part (b).

18. Oil production in the United States decreased approximately linearly from 11.2 million barrels per day in 1985 to 8.7 million barrels per day in 2003. Oil consumption in the United States increased approximately linearly from 15.7 million barrels per day in 1985 to 20.0 million barrels per day in 2003 (Source: *U.S. Department of Energy*).

a. Let $c(t)$ be U.S. oil consumption and $p(t)$ be U.S. oil production, both in million barrels per day, at t years since 1980. Find equations of c and p.
b. Use substitution or elimination to estimate when the United States produced as much oil as it consumed.
c. Use a graphing calculator table or graph to verify your work in part (b).

19. Toyota's® U.S. market share increased approximately linearly from 10% in 1999 to 12% in 2004. Ford's® U.S. market share decreased approximately linearly from 25% in 1999 to 19% in 2004 (Source: *Autodata*). Predict when Toyota's market share will be equal to Ford's market share. What is that market share?

20. The percentage of households with a television tuned in to ad-supported cable increased approximately linearly from 24.6% in 1994 to 46.5% in 2004. The percentage of households with a television tuned in to the major networks decreased approximately linearly from 48.5% in 1994 to 26.2% in 2004 (Source: *Nielsen*). Estimate when the percentages were equal for cable and for the major networks.

21. To solve the system

$$y = f(t) = m_1 t + b_1$$
$$y = g(t) = m_2 t + b_2$$

where m_1, b_1, m_2, and b_2 are constants, a student eliminates y and finds a noninteger value of t. He rounds the value of t *up* to an integer I.

a. He is confused, because $f(I)$ is not equal to $g(I)$. Draw a graph to illustrate what happened.
b. If m_1 is larger than m_2, which will be larger, $f(I)$ or $g(I)$? Draw a graph to illustrate this.

22. Describe how you can find a system of linear equations to model a situation. Also, explain how you can use the system to make an estimate or prediction about the situation.

3.4 VALUE, INTEREST, AND MIXTURE PROBLEMS

Objectives

- Know a five-step problem-solving method.
- Use a system of two linear equations or a linear function to solve value, interest, and mixture problems.

In this section, we will solve problems that involve the (dollar) value of an object, the interest from an investment, or the percentage of a substance in a mixture. First, we will use a five-step problem-solving method that involves solving a linear system. Later, we will use a system of equations to find a function that we can use to analyze many aspects of a given situation.

Using a Five-Step Problem-Solving Method

We begin by discussing a method we can use to solve some problems involving two quantities.

Five-Step Problem-Solving Method

To solve some problems in which we want to find two quantities, it is useful to perform the following five steps:

- ***Step 1: Define each variable.*** For each quantity that we are trying to find, we usually define a variable to be that unknown quantity.
- ***Step 2: Write a system of two equations.*** We find a system of two equations by using the variables from step 1. We can usually write both equations either by translating into mathematics the information stated in the problem or by making a substitution into a formula.
- ***Step 3: Solve the system.*** We solve the system of equations from step 2.
- ***Step 4: Describe each result.*** We use a complete sentence to describe the found quantities.
- ***Step 5: Check.*** We reread the problem and check that the quantities we found agree with the given information.

Value Problems

How do we know that four dimes are worth 40 cents? We find the total value of the dimes by multiplying the value of one dime (10 cents) by the number of dimes (4): $10 \cdot 4 = 40$ cents.

Total-Value Formula

If n objects each have value v, then their total value T is given by

$$T = vn$$

In words: The total value is equal to the value of one object times the number of objects.

When some objects are sold, we refer to the total money collected as the **revenue** from selling the objects.

Example 1 Solving a Value Problem

A music store charges \$5 for a six-string pack of electric-guitar strings and \$20 for a four-string pack of electric-bass strings. If the store sells 35 packs of strings for a total revenue of \$295, how many packs of each type of string were sold?

Solution

Step 1: Define each variable. Let x be the number of packs of guitar strings sold and y be the number of packs of bass strings sold.

Step 2: Write a system of two equations. The revenue from the guitar strings is equal to the price per pack times the number of packs sold: $5x$. The revenue from the bass strings is equal to the price per pack times the number of packs sold: $20y$. We add the revenue from the guitar strings and the revenue from the bass strings to find a formula of the total revenue T (in dollars):

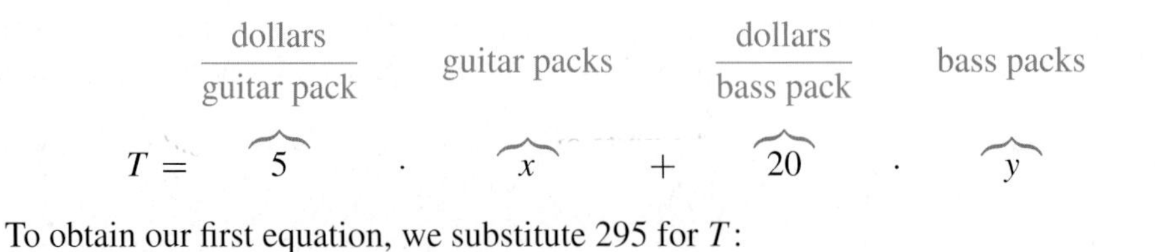

$$T = \overbrace{5}^{\frac{\text{dollars}}{\text{guitar pack}}} \cdot \overbrace{x}^{\text{guitar packs}} + \overbrace{20}^{\frac{\text{dollars}}{\text{bass pack}}} \cdot \overbrace{y}^{\text{bass packs}}$$

To obtain our first equation, we substitute 295 for T:

$$295 = 5x + 20y$$

Since the store sells 35 packs of strings, our second equation is

$$x + y = 35$$

The system is

$$5x + 20y = 295 \quad \text{Equation (1)}$$
$$x + y = 35 \quad \text{Equation (2)}$$

Step 3: Solve the system. We can use elimination to solve the system. To eliminate the x terms, we multiply both sides of equation (2) by -5, yielding the system

$$5x + 20y = 295 \quad \text{Equation (1)}$$
$$-5x - 5y = -175 \quad \text{Multiply both sides of equation (2) by } -5.$$

Then we add the left sides and add the right sides of the equations and solve for y:

$$5x + 20y = 295$$
$$-5x - 5y = -175$$
$$0 + 15y = 120 \quad \text{Add left sides and right sides; combine like terms.}$$
$$y = 8 \quad \text{Divide both sides by 15.}$$

Next, we substitute 8 for y in equation (2) and solve for x:

$$x + y = 35 \quad \text{Equation (2)}$$
$$x + 8 = 35 \quad \text{Substitute 8 for } y.$$
$$x = 27 \quad \text{Subtract 8 from both sides.}$$

Step 4: Describe each result. The store sold 27 packs of guitar strings and 8 packs of bass strings.

Step 5: Check. First, we find the sum $27 + 8 = 35$, which is equal to the total number of packs of strings sold. Next, we find the total revenue from selling 27 packs of guitar strings and 8 packs of bass strings: $5 \cdot 27 + 20 \cdot 8 = 295$, which checks. ■

Example 2 Solving a Value Problem

The American Analog Set will play at an auditorium that has 400 balcony seats and 1600 main-level seats. If tickets for balcony seats will cost \$15 less than tickets for main-level seats, what should the price be for each type of ticket so that the total revenue from a sellout performance will be \$70,000?

Solution

Step 1: Define each variable. Let b be the price for balcony seats and m be the price for main-level seats, both in dollars.

Step 2: Write a system of two equations. Since tickets for balcony seats will cost \$15 less than tickets for main-level seats, our first equation is

$$\overbrace{b}^{\text{balcony ticket price}} \overbrace{=}^{\text{is}} \overbrace{m - 15}^{\text{\$15 less than main-level ticket price}}$$

Since the total revenue is \$70,000, our second equation is

$$\overbrace{b}^{\frac{\text{dollars}}{\text{balcony ticket}}} \cdot \overbrace{400}^{\text{balcony tickets}} + \overbrace{m}^{\frac{\text{dollars}}{\text{main-level ticket}}} \cdot \overbrace{1600}^{\text{main-level tickets}} = \overbrace{70{,}000}^{\text{total revenue in dollars}}$$

The units of the expressions on both sides of the equation are dollars, which suggests that our work is correct. The system is

$$b = m - 15 \quad \text{Equation (1)}$$
$$400b + 1600m = 70{,}000 \quad \text{Equation (2)}$$

Step 3: Solve the system. We can use substitution to solve the system. We substitute $m - 15$ for b in equation (2) and solve for m:

$$
\begin{aligned}
400b + 1600m &= 70{,}000 && \text{Equation (2)} \\
400(m - 15) + 1600m &= 70{,}000 && \text{Substitute } m - 15 \text{ for } b. \\
400m - 6000 + 1600m &= 70{,}000 && \text{Distributive law.} \\
2000m - 6000 &= 70{,}000 && \text{Combine like terms.} \\
2000m &= 76{,}000 && \text{Add 6000 to both sides.} \\
m &= 38 && \text{Divide both sides by 2000.}
\end{aligned}
$$

Then we substitute 38 for m in equation (1) and solve for b:

$$b = 38 - 15 = 23$$

Step 4: Describe each result. Tickets for balcony seats should be priced at \$23 each, and tickets for main-level seats should be priced at \$38 each.

Step 5: Check. First we find the difference in the ticket prices: $38 - 23 = 15$ dollars, which checks. Then we compute the total revenue from selling 400 of the \$23 tickets and 1600 of the \$38 tickets: $23 \cdot 400 + 38 \cdot 1600 = 70{,}000$ dollars, which checks. ■

In Examples 1 and 2, we analyzed *one* aspect of a situation by working with a linear system. **If we want to analyze many aspects of a certain situation, it can help to use a system of equations to find a linear function. We can then use the function to analyze the situation in various ways.**

Example 3 Using a Function to Model a Value Situation

A 10,000-seat amphitheater will sell general-seat tickets at \$45 and reserved-seat tickets at \$65 for a Foo Fighters concert. Let x and y be the number of tickets that will sell for \$45 and \$65, respectively. Assume that the show will sell out.

1. Let $T = f(x)$ be the total revenue (in dollars) from selling the \$45 and \$65 tickets. Find an equation of f.
2. Use a graphing calculator to sketch a graph of f for $0 \le x \le 10{,}000$. What is the slope? What does it mean in this situation?
3. Find $f(8500)$. What does it mean in this situation?
4. Find $f(11{,}000)$. What does it mean in this situation?
5. The total cost of the production is \$350,000. How many of each type of ticket must be sold to make a profit of \$150,000?

Solution

1. We add the revenues from the general tickets and the reserved tickets to find an equation of the total revenue T:

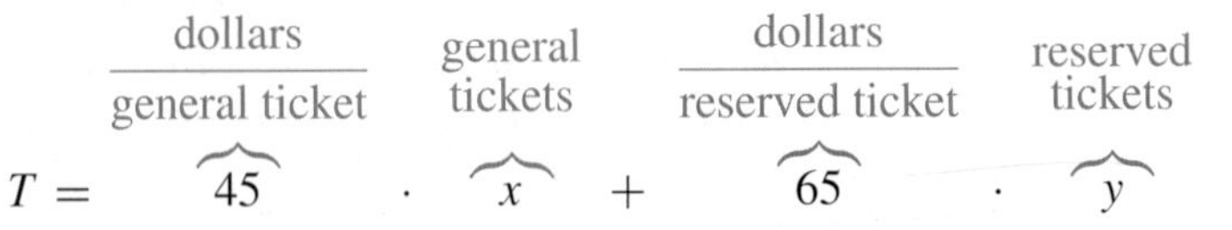

So far, we have described T in terms of x and y. Next, we describe T in terms of just x. The total number of tickets sold for a sellout performance is 10,000:

$$x + y = 10{,}000$$

Now we get y alone on one side of the equation:

$$y = 10{,}000 - x$$

Next, we substitute $10{,}000 - x$ for y in the equation $T = 45x + 65y$:

$$
\begin{aligned}
T &= 45x + 65(10{,}000 - x) && \text{Substitute } 10{,}000 - x \text{ for } y. \\
&= 45x + 650{,}000 - 65x && \text{Distributive law} \\
&= -20x + 650{,}000 && \text{Combine like terms.}
\end{aligned}
$$

So, an equation of f is

$$f(x) = -20x + 650{,}000$$

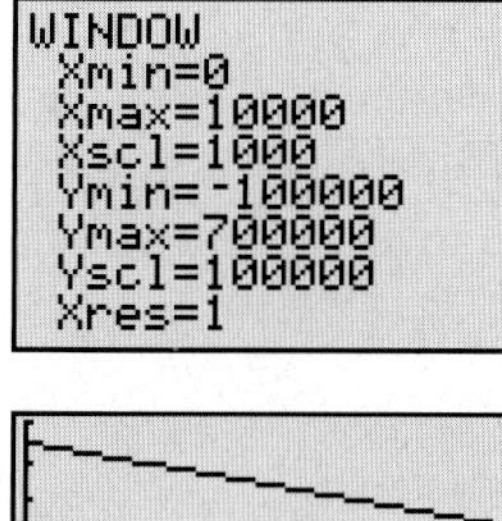

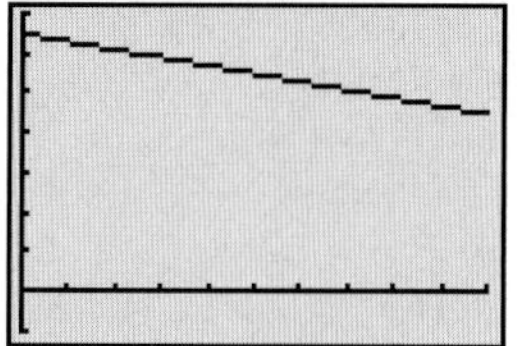

Figure 30 Graphing the revenue model

2. We draw a sketch of f in Fig. 30. The graph of f is a decreasing line with slope -20. This means that if one more ticket is sold for \$45 (and one less ticket is sold for \$65), the revenue will decrease by \$20.
3. Here, $f(8500) = -20(8500) + 650{,}000 = 480{,}000$. This means that if 8500 tickets sell for \$45 (and 1500 tickets sell for \$65), the total revenue will be \$480,000.
4. Here, $f(11{,}000) = -20(11{,}000) + 650{,}000 = 430{,}000$. This means that if 11,000 tickets sell for \$45, the total revenue will be \$430,000. Since there are only 10,000 seats, model breakdown has occurred.
5. To make a profit of \$150,000, the revenue would need to be $350{,}000 + 150{,}000 = 500{,}000$ dollars. We substitute 500,000 for T in the equation $T = -20x + 650{,}000$ and solve for x:

$$500{,}000 = -20x + 650{,}000 \quad \text{Substitute 500,000 for } T.$$
$$-150{,}000 = -20x \quad \text{Subtract 650,000 from both sides.}$$
$$7500 = x \quad \text{Divide both sides by } -20.$$

So, 7500 \$45 tickets and $10{,}000 - 7500 = 2500$ \$65 tickets would need to be sold for the profit to be \$150,000. ■

Interest Problems

In this section, we will model situations that involve money. We begin by considering interest from investments.

Money deposited in an account such as a savings account, CD, or mutual fund is called the **principal.** A person invests money in hopes of later getting back the principal plus additional money called the **interest,** which is a percentage of the principal (see Fig. 31). The **annual simple interest rate** is the percentage of the principal that equals the interest earned per year. So, if we invest \$100 and earn \$5 per year, then the annual simple interest rate is 5%.

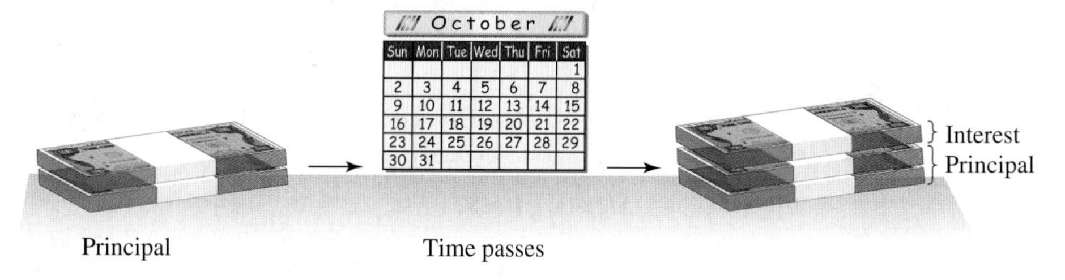

Figure 31 Invest the principal; time passes; get back the principal plus interest

Example 4 Interest from an Investment

How much interest will a person earn by investing \$3200 in an account at 4% simple annual interest for one year?

Solution

We find 4% of 3200:

$$0.04(3200) = 128$$

The person will earn \$128 in interest. ■

Some people invest in a variety of accounts, including lower-risk accounts and certain higher-risk accounts. These investors might earn large amounts of interest from the higher-risk accounts but will have a safety net of principal and interest from the lower-risk accounts.

Example 5 Solving an Interest Problem

A person plans to invest twice as much money in an Elfun Trusts account at 2.7% annual interest as in a Vanguard Morgan Growth account at 5.5% annual interest. Both interest rates are 5-year averages. How much will the person have to invest in each account to earn a total of \$218 in one year?

Solution

Step 1: Define each variable. Let x be the money (in dollars) invested at 2.7% annual interest and y be the money (in dollars) invested at 5.5% annual interest.

Step 2: Write a system of two equations. The person is investing twice as much in the 2.7% account as in the 5.5% account, so our first equation is

$$x = 2y$$

Since the total interest is \$218, our second equation is

$$\overbrace{0.027x}^{\text{interest from 2.7\% account}} + \overbrace{0.055y}^{\text{interest from 5.5\% account}} = \overbrace{218}^{\text{total interest}}$$

The system is

$$\begin{aligned} x &= 2y && \text{Equation (1)} \\ 0.027x + 0.055y &= 218 && \text{Equation (2)} \end{aligned}$$

Step 3: Solve the system. We can use substitution to solve the system. We substitute $2y$ for x in equation (2) and then solve for y:

$$\begin{aligned} 0.027(2y) + 0.055y &= 218 && \text{Substitute } 2y \text{ for } x. \\ 0.054y + 0.055y &= 218 && \text{Multiply.} \\ 0.109y &= 218 && \text{Combine like terms.} \\ y &= 2000 && \text{Divide both sides by 0.109.} \end{aligned}$$

Then we substitute 2000 for y in equation (1) and solve for x:

$$x = 2y = 2(2000) = 4000$$

Step 4: Describe each result. The person should invest \$4000 at 2.7% annual interest and \$2000 at 5.5% annual interest.

Step 5: Check. First, we note that 4000 is twice 2000, which checks. Next, we calculate the total interest earned from investing \$4000 at 2.7% and \$2000 at 5.5%:

$$0.027(4000) + 0.055(2000) = 218$$

which checks, too. ■

Example 6 Using a Function to Model a Situation Involving Interest

A person plans to invest a total of \$6000 in a Gabelli ABC mutual fund that has a 3-year average annual interest of 6% and in a Presidential Bank Internet CD account at 2.25% annual interest. Let x and y be the money (in dollars) invested in the mutual fund and the CD, respectively.

1. Let $I = f(x)$ be the total interest (in dollars) earned from investing the \$6000 for one year. Find an equation of f.
2. Use a graphing calculator to draw a graph of f for $0 \le x \le 6000$. What is the slope of f? What does it mean in this situation?
3. Use a graphing calculator to create a table of values of f. Explain how such a table could help the person decide how much money to invest in each account.
4. How much money should be invested in each account to earn \$300 in one year?

Solution

1. The interest earned from investing x dollars in an account earning 6% annual interest is $0.06x$, and the interest earned from investing y dollars in an account

earning 2.25% annual interest is $0.0225y$ dollars. We add the two interest earnings to find the total interest earned:

$$\overbrace{I}^{\text{total interest}} = \overbrace{0.06x}^{\substack{\text{interest from} \\ 6\% \text{ account}}} + \overbrace{0.0225y}^{\substack{\text{interest from} \\ 2.25\% \text{ account}}}$$

Next, we describe I in terms of just x. The person plans to invest \$6000:

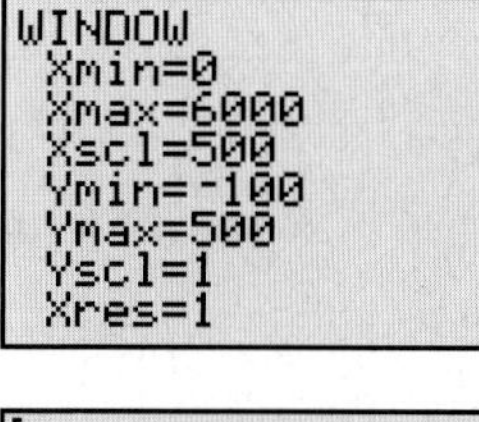

$$x + y = 6000$$

Then, we isolate y:

$$y = 6000 - x$$

Now we substitute $6000 - x$ for y in the equation $I = 0.06x + 0.0225y$:

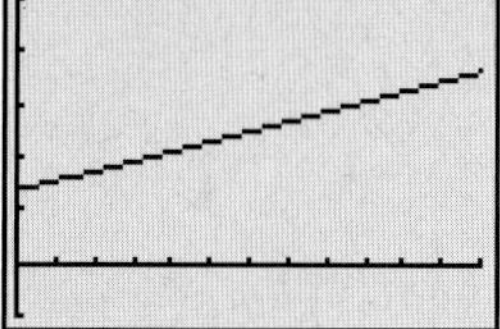

Figure 32 The revenue model

$$\begin{aligned} I &= 0.06x + 0.0225(6000 - x) && \text{Substitute } 6000 - x \text{ for } y. \\ &= 0.06x + 135 - 0.0225x && \text{Distributive law} \\ &= 0.0375x + 135 && \text{Combine like terms.} \end{aligned}$$

So, our equation of f is

$$f(x) = 0.0375x + 135$$

2. We draw a graph of f in Fig. 32. The graph of f is an increasing line with slope 0.0375. This means that if one more dollar is invested at 6% (and one less dollar is invested at 2.25%), the total interest will increase by 3.75 cents.
3. We create a table in Fig. 33. The person may know that she wants to invest some of the \$6000 in the safe Presidential CD account and the rest in the riskier Gabelli mutual fund, but she may not be clear as to exactly how much risk she is willing to take. By seeing some possible interest earnings, she may get a clearer idea of how much money to invest in each account.

X	Y1	
0	135	
1000	172.5	
2000	210	
3000	247.5	
4000	285	
5000	322.5	
6000	360	

X=0

Figure 33 Table of values of f

4. We substitute 300 for I in the equation $I = 0.0375x + 135$ and solve for x:

$$\begin{aligned} 300 &= 0.0375x + 135 && \text{Substitute 300 for } I. \\ 165 &= 0.0375x && \text{Subtract 135 from both sides.} \\ 4400 &= x && \text{Divide both sides by 0.0375.} \end{aligned}$$

The person should invest \$4400 in the Gabelli mutual fund account and $6000 - 4400 = 1600$ dollars in the Presidential CD account. ■

Mixture Problems

Regardless of whether we are chemists or cooks, pharmacists or mechanics, we often have to mix different substances (typically liquids). Suppose that 2 ounces of lime juice is mixed with 8 ounces of water to make 10 ounces of unsweetened limeade. Note that $\frac{2}{10} = 0.20 = 20\%$ of the limeade is lime juice. We call the limeade a *20% lime-juice solution*.

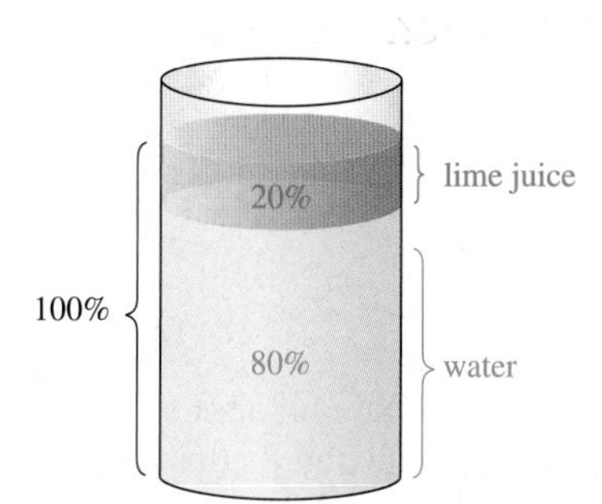

Figure 34 A 20% lime-juice solution

Then, the remaining $\frac{8}{10} = 0.80 = 80\%$ of the limeade is water. The percentage of the solution that is lime juice plus the percentage of the solution that is water is equal to 100% (all) of the solution: $20\% + 80\% = 100\%$. See Fig. 34.

So, for a 10% lime-juice solution, 10% of the solution would be lime juice and $100\% - 10\% = 90\%$ of the solution would be water. In general, **for an $x\%$ solution of two substances that are mixed, $x\%$ of the solution is one substance and $(100 - x)\%$ is the other substance.**

Example 7 Solving a Mixture Problem

A chemist needs 5 quarts of a 17% acid solution, but he has only a 15% acid solution and a 25% acid solution. How many quarts of the 15% acid solution should he mix with the 25% acid solution to make 5 quarts of a 17% acid solution?

Solution

Step 1: Define each variable. Let x be the number of quarts of 15% acid solution and y be the number of quarts of 25% acid solution.

Step 2: Write a system of two equations. Since he wants 5 quarts of the total mixture, our first equation is

$$x + y = 5$$

The total amount of pure acid doesn't change, regardless of how it is distributed in the two solutions. We find our second equation from the fact that the sum of the amounts of pure acid in both the 15% acid solution and the 25% acid solution is equal to the amount of pure acid in the desired mixture:

$$\overbrace{0.15x}^{\text{pure acid in 15\% solution}} + \overbrace{0.25y}^{\text{pure acid in 25\% solution}} = \overbrace{0.17(5)}^{\text{pure acid in mixture}}$$

The system is

$$x + y = 5 \qquad \text{Equation (1)}$$
$$0.15x + 0.25y = 0.85 \qquad \text{Equation (2)}$$

Step 3: Solve the system. We can solve the system by substitution. First, we solve equation (1) for y:

$$x + y = 5 \qquad \text{Equation (1)}$$
$$y = 5 - x \qquad \text{Subtract } x \text{ from both sides.}$$

Then we substitute $5 - x$ for y in equation (2) and solve for x:

$$0.15x + 0.25(5 - x) = 0.85 \qquad \text{Substitute } 5 - x \text{ for } y.$$
$$0.15x + 1.25 - 0.25x = 0.85 \qquad \text{Distributive law}$$
$$-0.10x + 1.25 = 0.85 \qquad \text{Combine like terms.}$$
$$-0.10x = -0.40 \qquad \text{Subtract 1.25 from both sides.}$$
$$x = 4 \qquad \text{Divide both sides by } -0.10.$$

Next, we substitute 4 for x in the equation $y = 5 - x$ and solve for y:

$$y = 5 - 4 = 1$$

Step 4: Describe each result. Four quarts of the 15% acid solution and 1 quart of the 25% acid solution are required.

Step 5: Check. First, we compute the total amount (in quarts) of pure acid in 4 quarts of 15% acid solution and 1 quart of 25% acid solution:

$$0.15(4) + 0.25(1) = 0.85$$

Next, we compute the amount (in quarts) of pure acid in 5 quarts of 17% acid solution:

$$0.17(5) = 0.85$$

Since the two results are equal, this checks. Also, $4 + 1 = 5$, which checks with the chemist wanting 5 quarts of the 17% solution. ■

Example 8 Solving a Mixture Problem

A chemist needs 8 cups of a 15% alcohol solution but has only a 20% alcohol solution. How much 20% solution and water should she mix to form the desired 8 cups of 15% solution?

Solution

Step 1: Define each variable. Let x be the number of cups of the 20% solution and y be the number of cups of water.

Step 2: Write a system of two equations. Since she wants 8 cups of the total mixture, our first equation is

$$x + y = 8$$

There is no alcohol in pure water, so we find our second equation from the fact that the amount of pure alcohol in the 20% alcohol solution is equal to the amount of pure alcohol in the desired mixture:

$$\overbrace{0.20x}^{\text{amount of pure alcohol in 20\% solution}} = \overbrace{0.15(8)}^{\text{amount of pure alcohol in mixture}}$$

The system is

$$\begin{aligned} x + y &= 8 && \text{Equation (1)} \\ 0.20x &= 1.2 && \text{Equation (2)} \end{aligned}$$

Step 3: Solve the system. We begin by solving equation (2) for x:

$$\begin{aligned} 0.20x &= 1.2 && \text{Equation (2)} \\ x &= 6 && \text{Divide both sides by 0.20.} \end{aligned}$$

Next, we substitute 6 for x in equation (1):

$$\begin{aligned} x + y &= 8 && \text{Equation (1)} \\ 6 + y &= 8 && \text{Substitute 6 for } x. \\ y &= 2 && \text{Subtract 6 from both sides.} \end{aligned}$$

Step 4: Describe each result. The chemist needs to mix 6 cups of the 20% alcohol solution with 2 cups of water.

Step 5: Check. First, we find $6 + 2 = 8$, which checks with the chemist wanting 8 cups of the 15% solution. Next, we compute the amount of pure alcohol in 6 cups of the 20% solution: $0.20(6) = 1.2$ cups. Finally, we compute the amount of pure alcohol in the 15% solution: $0.15(8) = 1.2$ cups. The computed amounts of pure alcohol in the 20% solution and the 15% solution are equal, so this checks. ■

group exploration

Looking ahead: Connection between a system of linear equations and a linear inequality in one variable

Recall from Example 2 of Section 3.3 that the equations

$$\begin{aligned} r = W(t) &= -0.27t + 70.45 && \text{Women's records} \\ r = M(t) &= -0.053t + 48.08 && \text{Men's records} \end{aligned}$$

model 400-meter run record times (in seconds) of women and men, respectively, at t years since 1900.

1. Use a graphing calculator to draw the graphs of W and M on the same coordinate system, and find the intersection point. Then copy the graphs on a piece of paper and mark the point of intersection.
2. Now find values of t for which the models predict that the women's record time will be less than the men's record time. Which years do these values of t represent? On your graph, shade the part of the t-axis that represents those years.
3. Try a *numerical* verification by using a graphing calculator. To do this, enter the functions as displayed in Fig. 35. Use the table setup displayed in Fig. 36, but replace the rectangle to the right of "TblStart=" with the t value for which the men's and women's time will be equal. Then display the tables and use the arrow keys to verify your result from Problem 2. (See Section B.14 for calculator instructions.)
4. Which of these two inequalities states that the women's record time is less than the men's record time?

$$W(t) < M(t) \qquad\qquad M(t) < W(t)$$

Plot1 Plot2 Plot3
\Y1 = -.27X+70.45
\Y2 = -.053X+48.08
\Y3 =
\Y4 =
\Y5 =
\Y6 =

Figure 35 Entered functions

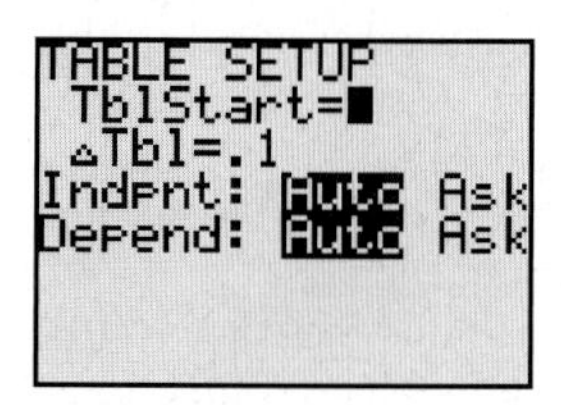

Figure 36 Table setup

5. Substitute $-0.27t + 70.45$ for $W(t)$ and $-0.053t + 48.08$ for $M(t)$ in the inequality that you chose in Problem 4. An inequality such as this is called a *linear inequality in one variable*. This inequality becomes a true statement when any of the values for t that you found in Problem 2 are substituted for t in the inequality. Explain.

TIPS FOR SUCCESS: Take Notes

It is always a good idea to take notes during classroom activities. Not only will you have something to refer to later, when you do the homework, but also you will have something to help you prepare for tests. In addition, the process of taking notes makes you even more involved with the material, which means that you will probably increase both your understanding and retention of it.

HOMEWORK 3.4

FOR EXTRA HELP ▶ Student Solutions Manual | PH Math/Tutor Center  MathXL® | MyMathLab

1. A 5000-seat theater has tickets for sale at \$27 and \$40. How many tickets should be sold at each price for a sellout performance to generate a total revenue of \$150,600?

2. A 9000-seat amphitheater has tickets for sale at \$35 and \$62. How many tickets should be sold at each price for a sellout performance to generate a total revenue of \$382,500?

3. On Amazon.com®, the CD *Plans* sells for \$12.96, and the EP *Forbidden Love,* both by Death Cab for Cutie, sells for \$9.99. If total sales of both CDs in one month are 836 CDs and total revenue of both CDs is \$10,475.19, how many of each CD were sold?

4. On Amazon.com, the CD *Good News for People Who Love Bad News* sells for \$8.97, and the CD *The Lonesome Crowded West,* both by Modest Mouse, sells for \$12.98. If total sales of both CDs in one month are 422 CDs and total revenue of both CDs is \$3917.67, how many of each CD were sold?

5. An auditorium has 500 balcony seats and 1800 main-level seats. If tickets for balcony seats will cost \$15 less than tickets for main-level seats, what should the prices be for each type of ticket so that the total revenue from a sellout performance will be \$84,500?

6. An auditorium has 360 balcony seats and 1200 main-level seats. If tickets for balcony seats will cost \$10 less than tickets for main-level seats, what should the prices be for each type of ticket so that the total revenue from a sellout performance will be \$33,840?

7. A board of trustees of a college is considering opening a satellite college. Researchers estimate that there will be three times as many full-time students as part-time students and that full-time students will take an average of 14 units and part-time students will take an average of 3 units each semester. The charge for tuition will be \$13 per unit. How many of each type of student would need to attend the new college for the total revenue to be \$877,500 each semester?

8. A new college may open in the future. Full-time instructors would teach 15 units each semester. Researchers estimate that adjunct (part-time) instructors teach an average of 6 units each semester and that there is student demand for a total of 3300 units of courses. There would be enough office space for 250 instructors. To keep costs low by hiring as many adjunct instructors as possible, yet have office space for all instructors, how many full-time instructors and adjunct instructors should be hired?

9. A 20,000-seat amphitheater will sell tickets at \$50 and \$75 for a Green Day concert. Let x and y be the number of tickets that will sell for \$50 and \$75, respectively. Assume that the show will sell out.

a. Let $R = f(x)$ be the total revenue (in dollars) from selling the \$50 and \$75 tickets. Find an equation of f.

b. Use a graphing calculator to draw a graph of f for $0 \le x \le 20{,}000$. What is the slope? What does it mean in this situation?

c. Find $f(16{,}000)$. What does it mean in this situation?

d. The total cost of the production is \$475,000. How many of each ticket must be sold to make a profit of \$600,000?

10. An 8000-seat amphitheater will sell tickets at \$30 and \$55 for a Garbage concert. Let x and y be the number of tickets that will sell for \$30 and \$55, respectively. Assume that the show will sell out.

a. Let $R = f(x)$ be the total revenue (in dollars) from selling the \$30 and \$55 tickets. Find an equation of f.

b. Use a graphing calculator to draw a graph of f for $0 \le x \le 8000$. What is the slope? What does it mean in this situation?

c. Find $f(6500)$. What does it mean in this situation?

d. How many of each ticket must be sold for the revenue to be \$250,000?

11. A 12,000-seat amphitheater will sell tickets at \$45 and \$70 for a Cake concert. Let x and y be the number of tickets that will sell for \$45 and \$70, respectively. Assume that the show will sell out.

a. Let $R = f(x)$ be the total revenue (in dollars) from selling the \$45 and \$70 tickets. Find an equation of f.

b. Use a graphing calculator table to find the values $f(0)$, $f(2000)$, $f(4000)$, ..., $f(12{,}000)$. What do they mean in this situation?

c. Describe the various possible total revenues from selling the tickets.
d. How many of each ticket must be sold for the revenue to be $602,500?

12. A 25,000-seat amphitheater will sell tickets at $55 and $85 for a U2 concert. Let x and y be the number of tickets that will sell for $55 and $85, respectively. Assume that the show will sell out.
a. Let $R = f(x)$ be the total revenue (in dollars) from selling the $55 and $85 tickets. Find an equation of f.
b. Use a graphing calculator table to find the values $f(0)$, $f(5000)$, $f(10{,}000), \ldots, f(25{,}000)$. What do they mean in this situation?
c. Describe the various possible total revenues from selling the tickets.
d. How many of each ticket must be sold for the revenue to be $1,510,000?

13. For flights between Los Angeles and San Francisco, United Airlines® usually uses a 737 airplane that has 8 first-class seats and 126 coach seats. The average price of round-trip first-class tickets is $242 more than the average price of round-trip coach tickets. Assume that a round-trip flight is sold out. Let x and y be the prices (in dollars) of coach and first-class tickets, respectively.
a. Let $R = f(x)$ be the total revenue (in dollars) from selling the tickets. Find an equation of f.
b. Find the slope of the graph of f. What does it mean in this situation?
c. If United Airlines wants the total revenue from the tickets for the round trip to be $14,130, what should be the average selling prices of both types of tickets?

14. An orchestra will perform in an auditorium with 1400 main-level seats and 300 balcony seats. Tickets for the balcony seats will cost $12 less than tickets for the main-level seats. Assume that there is a sellout performance. Let x and y be the prices (in dollars) for the main-level seats and the balcony seats, respectively.
a. Let $R = f(x)$ be the total revenue (in dollars) from ticket sales. Find an equation of f.
b. Find $f(55)$. What does it mean in this situation?
c. Find the value of x when $f(x) = 101{,}800$. What does it mean in this situation?

For Exercises 15–22, all interest rates are 5-year averages.

15. A person plans to invest $15,000. She will invest in both an American Funds New Perspective F account at 9% annual interest and an Oppenheimer Global Y account at 11% annual interest. How much should she invest in each account so that the total interest in one year will be $1410?

16. A person plans to invest $9000. He will invest in both a Winslow Green Growth account at 17% annual interest and a Columbia Small Cap Growth I Z account at 10% annual interest. How much should he invest in each account so that the total interest in one year will be $1040?

17. A person plans to invest $8500. He will invest in both a GMO Growth III account at 2.3% annual interest and a Gartmore Destinations Mod Agg Svc account at 6.6% annual interest. How much should he invest in each account so that the total interest in one year will be $303?

18. A person plans to invest $12,500. She will invest in both a Hartford Small Company Y account at 12.7% annual interest and a CMG Small Cap account at 10.8% annual interest. How much should she invest in each account so that the total interest in one year will be $1426?

19. A person plans to invest twice as much money in a Lord Abbett Developing Growth B account at 8.2% annual interest as in a Bridgeway Micro-Cap Limited account at 21.5% annual interest. How much should the person invest in each account to earn a total of $758 in one year?

20. A person plans to invest twice as much money in an MTP Small-Cap Growth Inst I account at 8% annual interest as in a Times Square Small Cap Growth Premier account at 13% annual interest. How much should the person invest in each account to earn a total of $1885 in one year?

21. A person plans to invest three times as much in a Dreyfus Premier Worldwide Growth R account at 4.5% interest as in an Oppenheimer Global Opportunities Y account at 14.9% annual interest. How much should the person invest in each account to earn a total of $426 in one year?

22. A person plans to invest three times as much in a Turner Small Cap Growth account at 10.4% interest as in a Merrill Lynch Small Cap Growth I account at 12.2% annual interest. How much should the person invest in each account to earn a total of $217 in one year?

23. A person plans to invest a total of $10,000 in a Charter One Bank CD at 2.87% annual interest and in a Dodge & Cox Balanced mutual fund that has a 3-year average annual interest rate of 8.10%. Let x and y be the money (in dollars) invested in the CD and mutual fund, respectively.
a. Let $I = f(x)$ be the total interest (in dollars) earned from investing the $10,000 for one year. Find an equation of f.
b. Use a graphing calculator to draw a graph of f for $0 \le x \le 10{,}000$. What is the slope? What does it mean in this situation?
c. How much money should be invested in each account to earn a total of $400 in one year?

24. A person plans to invest a total of $5000 in a Citizens Bank CD at 2% annual interest and in a Calamos Market Neutral A mutual fund that has a 3-year average annual interest rate of 9.45%. Let x and y be the money (in dollars) invested in the CD and the mutual fund, respectively.
a. Let $I = f(x)$ be the total interest (in dollars) earned from investing the $5000 for one year. Find an equation of f.
b. Use a graphing calculator to draw a graph of f for $0 \le x \le 5000$. What is the slope? What does it mean in this situation?
c. How much money should be invested in each account to earn a total of $350 in one year?

25. A person plans to invest \$9000 in an ING Direct CD at 2.5% annual interest and in a Thompson Plumb Balanced mutual fund that has a 3-year average annual interest rate of 9.45%. Let x and y be the money (in dollars) invested in the CD and the mutual fund, respectively.

a. Let $I = f(x)$ be the total interest (in dollars) earned from investing the \$9000 for one year. Find an equation of f.

b. Find $f(500)$. What does it mean in this situation?

c. Find the value of x when $f(x) = 500$. What does it mean in this situation?

d. Find $f(10{,}000)$. What does it mean in this situation?

26. A person plans to invest a total of \$12,000 in a Savings Bank of Manchester CD at 1.92% annual interest and in a Vanguard Wellesley Income mutual fund that has a 3-year average annual interest rate of 8%. Let x and y be the money (in dollars) invested in the CD and the mutual fund, respectively.

a. Let $I = f(x)$ be the total interest (in dollars) earned from investing the \$12,000 for one year. Find an equation of f.

b. Find $f(800)$. What does it mean in this situation?

c. Find the value of x when $f(x) = 800$. What does it mean in this situation?

d. Find $f(15{,}000)$. What does it mean in this situation?

27. A person plans to invest a total of \$8000 in a LaSalle Bank CD at 1.5% annual interest and in a Bridgeway Aggressive Investors 1 mutual fund that has a 3-year average annual interest rate of 11.6%. Let x and y be the money (in dollars) invested in the CD and the mutual fund, respectively.

a. Let $I = f(x)$ be the total interest (in dollars) earned from investing the \$8000 for one year. Find an equation of f.

b. A minimum principal of \$2500 is required for the LaSalle Bank CD account. Describe the various possible total interest earnings from investing the \$8000 in the accounts for one year.

c. How much of the \$8000 should be invested in each account so that the total interest earned in one year is \$400?

28. A person plans to invest a total of \$7000 in a Security Bank USA CD at 2.55% annual interest and in an Artisan Mid Cap mutual fund that has a 3-year average annual interest rate of 8%. Let x and y be the money (in dollars) invested in the CD and the mutual fund, respectively.

a. Let $I = f(x)$ be the total interest (in dollars) earned from investing the \$7000 for one year. Find an equation of f.

b. A minimum principal of \$500 is required for the Security Bank USA CD account. Describe the various possible total interest earnings from investing the \$7000 in the accounts for one year.

c. How much of the \$7000 should be invested in each account so that the total interest earned in one year is \$300?

29. A person plans to invest a total of \$6000 in a Nexity Bank CD at 2.85% annual interest and in an FMI Focus mutual fund that has a 3-year average annual interest rate of 9%. Let x and y be the money (in dollars) invested in the CD and the mutual fund, respectively.

a. Let $I = f(x)$ be the total interest (in dollars) earned from investing the \$6000 for one year. Find an equation of f.

b. Find the I-intercept of the model. What does it mean in this situation?

c. Find the x-intercept. What does it mean in this situation?

d. What is the slope? What does it mean in this situation?

30. A person plans to invest a total of \$14,000 in a North Middlesex Savings Bank CD at 2.27% annual interest and in a Calamos Growth A mutual fund that has a 3-year average annual interest rate of 15%. Let x and y be the money (in dollars) invested in the CD and the mutual fund, respectively.

a. Let $I = f(x)$ be the total interest (in dollars) earned from investing the \$14,000 for one year. Find an equation of f.

b. Find the I-intercept of the model. What does it mean in this situation?

c. Find the x-intercept. What does it mean in this situation?

d. What is the slope? What does it mean in this situation?

31. A chemist wants to mix a 10% alcohol solution and a 30% alcohol solution to make a 22% alcohol solution. How many ounces of each solution must be mixed to make 10 ounces of the 22% solution?

32. A chemist wants to mix a 10% alcohol solution and a 20% alcohol solution to make a 12% alcohol solution. How many cups of each solution must be mixed to make 5 cups of the 12% solution?

33. How many gallons each of a 5% antifreeze solution and a 20% antifreeze solution must be mixed to make 3 gallons of a 15% antifreeze solution?

34. How many liters each of a 15% antifreeze solution and a 30% antifreeze solution must be mixed to make 6 liters of a 20% antifreeze solution?

35. How many cups each of a 10% acid solution and a 25% acid solution must be mixed to make 6 cups of a 15% solution?

36. How many gallons each of a 10% acid solution and a 30% acid solution must be mixed to make 4 gallons of a 25% solution?

37. A chemist needs 5 liters of a 20% alcohol solution but has only a 25% alcohol solution. How many liters each of the 25% solution and water should she mix to make the desired 5 liters of 20% solution?

38. A chemist needs 9 ounces of a 10% alcohol solution but has only a 15% alcohol solution. How many ounces each of the 15% solution and water should he mix to make the desired 9 ounces of 10% solution?

For many worked-out examples, we wrote a system of two equations that helped us find quantities for a situation. In the following exercises, you will work backward. For example, given the system

$$\begin{aligned} 24x + 35y &= 53{,}500 \\ x + y &= 2000 \end{aligned}$$

we can relate the following to it:

A 2000-seat theater has tickets for sale at \$24 and \$35. How many tickets should be sold at each price for a sellout performance to generate a total revenue of \$53,500?

Describe an authentic situation for the given equation. Find the unknown quantities for your situation.

39. $$\begin{aligned} x &= y - 30 \\ 9500x + 2500y &= 579{,}000 \end{aligned}$$

40. $$\begin{aligned} x + y &= 9000 \\ 0.04x + 0.07y &= 420 \end{aligned}$$

41. $$\begin{aligned} x &= 3y \\ 0.05x + 0.12y &= 135 \end{aligned}$$

42. $$\begin{aligned} x + y &= 10 \\ 0.15x + 0.35y &= 0.27(10) \end{aligned}$$

3.5 USING LINEAR INEQUALITIES IN ONE VARIABLE TO MAKE PREDICTIONS

Objectives

- Know properties of inequalities.
- Know the meaning of *satisfy, solution,* and *solution set* for a *linear inequality in one variable*.
- Solve a linear inequality in one variable, and graph the solution set.
- Solve a three-part inequality in one variable, and graph the solution set.
- Use linear inequalities to make estimates and predictions about authentic situations.

So far in this chapter, we have estimated when one quantity will be equal to another quantity. In this section, we investigate when one quantity is less than (or greater than) another quantity.

Using Models to Compare Quantities

In Example 1, we compare the one-day cost of renting a pickup truck from two companies.

Example 1 Using Models to Compare Two Quantities

One Budget® office rents pickup trucks for \$39.95 per day plus \$0.19 per mile. One U-Haul® location charges \$19.95 per day plus \$0.49 per mile (Sources: *Budget; U-Haul*).

1. Find models that describe the one-day cost of renting a pickup truck from the companies.
2. Use graphs of your models to estimate for which mileages Budget offers the lower price.

Solution

1. Let $B(d)$ be the one-day cost (in dollars) of driving a Budget pickup truck d miles. Let $U(d)$ be the one-day cost (in dollars) of driving a U-Haul pickup truck d miles. Equations of B and U are

$$C = B(d) = 0.19d + 39.95$$
$$C = U(d) = 0.49d + 19.95$$

2. First, we sketch a graph of B and U in the same coordinate system (see Fig. 37). Note that the graph of B is below the graph of U for $d > 66.7$. Since the height of a point represents a price, we see that Budget offers the lower price for mileages over approximately 66.7 miles.

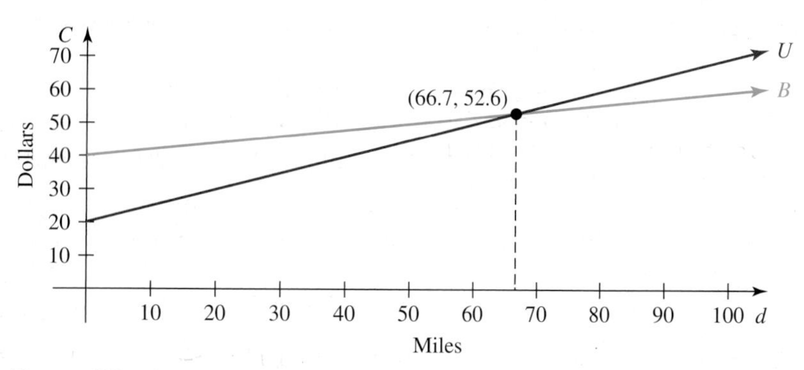

Figure 37 Budget and U-Haul models

■

In Example 1, we used graphing to estimate the mileages for which Budget's prices are lower than U-Haul's prices. Now we explore how to use inequalities to arrive at such an estimation. We begin by examining properties of inequalities.

Addition Property of Inequalities

In Section 1.6, we discussed the meaning of the inequality symbols $\leq$ (is less than or equal to) and $\geq$ (is greater than or equal to). In this section, we will also work with the symbols $<$ (is less than) and $>$ (is greater than). For example, the *inequality* $2 < 5$ means that 2 is less than 5, which is true. The inequality $9 > 5$ means that 9 is greater than 5, which is true.

What happens if we add 2 to both sides of the inequality $4 < 7$?

$$\begin{aligned} 4 &< 7 && \text{Original inequality} \\ 4+2 &\overset{?}{<} 7+2 && \text{Add 2 to both sides.} \\ 6 &\overset{?}{<} 9 && \text{Simplify.} \end{aligned}$$

true

What happens if we add -2 to both sides of the inequality $4 < 7$?

$$\begin{aligned} 4 &< 7 && \text{Original inequality} \\ 4+(-2) &\overset{?}{<} 7+(-2) && \text{Add } -2 \text{ to both sides.} \\ 2 &\overset{?}{<} 5 && \text{Simplify.} \end{aligned}$$

true

These examples suggest the following property.

Addition Property of Inequalities

If $a < b$, then $a + c < b + c$.

Similar properties hold for $\leq$, $>$, and $\geq$.

Similar properties hold for subtraction as well, since subtracting a number is the same as adding the opposite of the number.

We can use a number line to illustrate that if $a < b$, then $a + c < b + c$. From Figs. 38 and 39 we see that if a lies to the left of b (that is, $a < b$), then $a + c$ lies to the left of $b + c$ (that is, $a + c < b + c$). In Fig. 38, c is negative; in Fig. 39, c is positive.

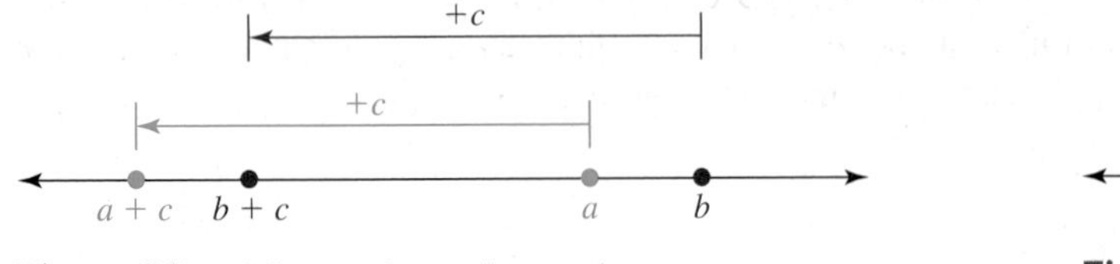

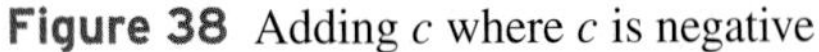

Figure 38 Adding c where c is negative

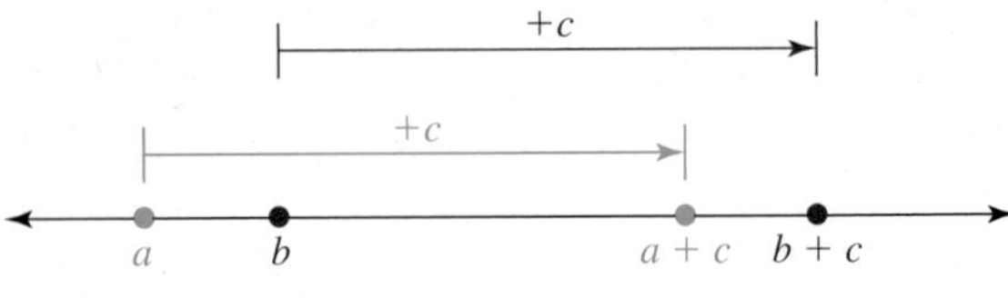

Figure 39 Adding c where c is positive

Multiplication Property of Inequalities

What if we multiply both sides of the inequality $4 < 7$ by 2?

$$\begin{aligned} 4 &< 7 && \text{Original inequality} \\ 4(2) &\overset{?}{<} 7(2) && \text{Multiply both sides by 2.} \\ 8 &\overset{?}{<} 14 && \text{Simplify.} \end{aligned}$$

true

Finally, what happens if we multiply both sides of $4 < 7$ by -2?

$$4 < 7 \quad \text{Original inequality}$$
$$4(-2) \overset{?}{<} 7(-2) \quad \text{Multiply both sides by } -2.$$
$$-8 \overset{?}{<} -14 \quad \text{Simplify.}$$
$$\text{false}$$

The result is the false statement $-8 < -14$. We can get a *true* statement if we *reverse the inequality symbol* when we multiply both sides of $4 < 7$ by -2:

$$4 < 7 \quad \text{Original inequality}$$
$$4(-2) \overset{?}{>} 7(-2) \quad \text{Reverse inequality symbol.}$$
$$-8 \overset{?}{>} -14 \quad \text{Simplify.}$$
$$\text{true}$$

So, when we multiply both sides of an inequality by a *negative* number, we *reverse* the inequality symbol.

Multiplication Property of Inequalities

- For a *positive* number c: If $a < b$, then $ac < bc$.
- For a *negative* number c: If $a < b$, then $ac > bc$.

Similar properties hold for $\leq$, $>$, and $\geq$.

In words: When we multiply both sides of an inequality by a positive number, we keep the inequality symbol. When we multiply by a negative number, we reverse the inequality symbol.

Similar rules apply for division as well, since dividing by a nonzero number is the same as multiplying by its reciprocal. Therefore, **when we multiply or divide both sides of an inequality by a negative number, we reverse the inequality symbol.**

Consider multiplying both sides of $a < b$ by -1:

$$a < b \quad \text{Original inequality}$$
$$-1a > -1b \quad \text{Multiply both sides by } -1\text{; reverse inequality symbol.}$$
$$-a > -b \quad -1a = -a$$

So, if $a < b$, then $-a > -b$. We can use a number line to illustrate this fact. To plot the point for $-a$, we move the point for a to the other side of the origin so that the points for $-a$ and a are the same distance from the origin (see Fig. 40).

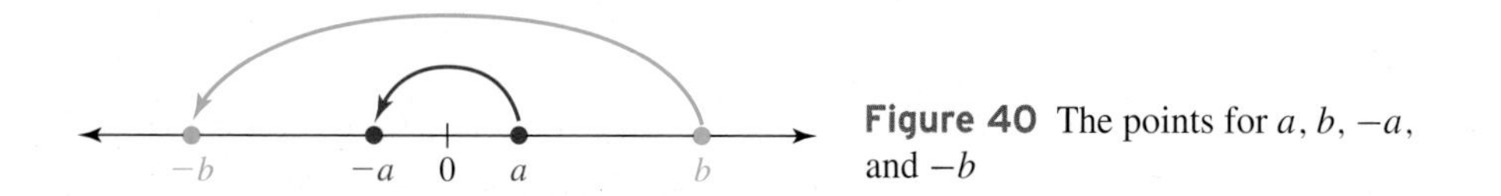

Figure 40 The points for a, b, $-a$, and $-b$

From Fig. 40, we see that if a lies to the *left* of b ($a < b$), then $-a$ lies to the *right* of $-b$ (that is, $-a > -b$).

Solving Linear Inequalities in One Variable

Here are some examples of *linear inequalities in one variable:*

$$4x - 6 < 8 \qquad 5x \leq 13 \qquad 2(x - 4) + 7 > -3 \qquad 5x \geq 2x - 6$$

DEFINITION Linear inequality in one variable

A **linear inequality in one variable** is an inequality that can be put into one of the forms

$$mx + b < 0, \qquad mx + b \leq 0, \qquad mx + b > 0, \qquad mx + b \geq 0$$

where m and b are constants and $m \neq 0$.

We say that a number **satisfies** an inequality in one variable if the inequality becomes a true statement after we have substituted the number for the variable. We call such a number a *solution* of the inequality.

Example 2 Finding Solutions of an Inequality

1. Does the number 2 satisfy the inequality $3x - 5 < 7$?
2. Does the number 6 satisfy the inequality $3x - 5 < 7$?
3. Find all solutions of the inequality $3x - 5 < 7$.

Solution

1. We substitute 2 for x in the inequality $3x - 5 < 7$:

$$3(2) - 5 \stackrel{?}{<} 7 \qquad \text{Substitute 2 for } x.$$

$$6 - 5 \stackrel{?}{<} 7 \qquad \text{Multiply.}$$

$$1 \stackrel{?}{<} 7 \qquad \text{Subtract.}$$

true

So, 2 satisfies the inequality $3x - 5 < 7$.

2. Here, we substitute 6 for x in the inequality $3x - 5 < 7$:

$$3(6) - 5 \stackrel{?}{<} 7 \qquad \text{Substitute 6 for } x.$$

$$18 - 5 \stackrel{?}{<} 7 \qquad \text{Multiply.}$$

$$13 \stackrel{?}{<} 7 \qquad \text{Subtract.}$$

false

So, 6 does not satisfy the inequality $3x - 5 < 7$.

3. To find all solutions of the inequality, we use properties of inequalities:

$$3x - 5 < 7 \qquad \text{Original inequality}$$

$$3x - 5 + 5 < 7 + 5 \qquad \text{Add 5 to both sides.}$$

$$3x < 12 \qquad \text{Combine like terms.}$$

$$\frac{3x}{3} < \frac{12}{3} \qquad \text{Divide both sides by 3.}$$

$$x < 4 \qquad \text{Simplify.}$$

All numbers less than 4 satisfy the inequality $3x - 5 < 7$. The solutions of the inequality are all numbers less than 4. ■

DEFINITION Solution of an inequality in one variable

We say that a number is a **solution** of an inequality in one variable if it satisfies the inequality. The **solution set** of an inequality is the set of all solutions of the inequality. We **solve** an inequality by finding its solution set.

To solve a linear inequality in one variable, we apply properties of inequalities to get the variable alone on one side of the inequality.

Example 3 Solving a Linear Inequality

Solve the inequality $-2x \geq 10$.

Solution

We divide both sides of the inequality by -2, a negative number:

$$-2x \geq 10 \quad \text{Original inequality}$$

$$\frac{-2x}{-2} \leq \frac{10}{-2} \quad \text{Divide both sides by } -2\text{; reverse inequality symbol.}$$

$$x \leq -5 \quad \text{Simplify.}$$

Since we divided by a negative number, we reversed the direction of the inequality. The solution set is the set of all numbers less than or equal to -5. ■

WARNING It is a common error to forget to reverse an inequality symbol when you multiply or divide both sides of an inequality by a negative number. For instance, in Example 3, it is important that we reversed the inequality symbol $\geq$ when we divided both sides of the inequality $-2x \geq 10$ by -2.

In Example 3, we found that the solution set of $-2x \geq 10$ is the set of all real numbers less than or equal to -5. We can represent these solutions graphically on a number line by shading the part of the number line that lies to the left of -5 (see Fig. 41). We draw a filled-in circle at -5 to indicate that -5 is a solution, too.

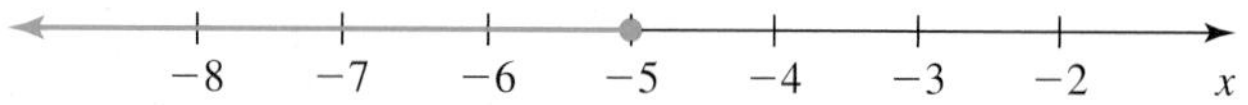

$-8 \quad -7 \quad -6 \quad -5 \quad -4 \quad -3 \quad -2 \quad x$

Figure 41 Graph of $x \leq -5$

If the solution set of an inequality is the set of numbers where $x < -5$, we shade the part of the number line that lies to the left of -5 but draw an *open* circle at -5 to indicate that -5 is *not* a solution (see Fig. 42).

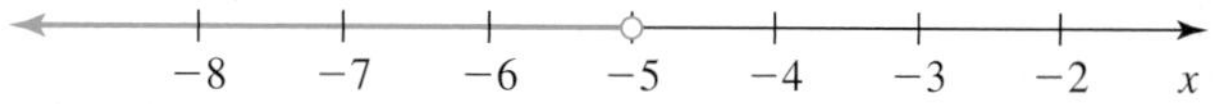

$-8 \quad -7 \quad -6 \quad -5 \quad -4 \quad -3 \quad -2 \quad x$

Figure 42 Graph of $x < -5$

More examples of inequalities, with matching graphs, are given in Fig. 43.

In Words	Inequality	Graph	Interval Notation
Numbers less than 2	$x < 2$	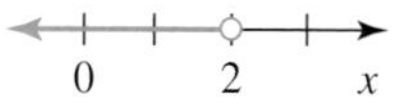 0 2 x	$(-\infty, 2)$
Numbers less than or equal to 2	$x \leq 2$	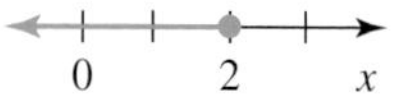 0 2 x	$(-\infty, 2]$
Numbers greater than 2	$x > 2$	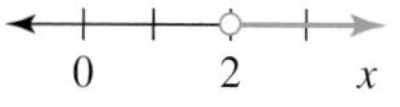 0 2 x	$(2, \infty)$
Numbers greater than or equal to 2	$x \geq 2$	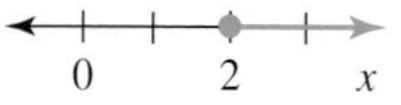 0 2 x	$[2, \infty)$

Figure 43 Words, inequalities, graphs, and interval notation

We can use **interval notation** to describe the solution set of an inequality. For example, we describe the numbers greater than 2 by $(2, \infty)$. We describe the numbers greater than or equal to 2 by $[2, \infty)$. We describe the set of real numbers by $(-\infty, \infty)$. More examples of interval notation are shown in Fig. 43.

Each of the four sets of numbers described in Fig. 43 is an interval. An **interval** is a set of real numbers represented by the number line or by an unbroken portion of it.

Example 4 Solving a Linear Inequality

Solve $-3(4x-5)-1 \le 17-6x$. Describe the solution set as an inequality, in a graph, and in interval notation.

Solution

$-3(4x-5)-1 \le 17-6x$	Original inequality
$-12x+15-1 \le 17-6x$	Distributive law
$-12x+14 \le 17-6x$	Combine like terms.
$-12x+14+6x \le 17-6x+6x$	Add 6x to both sides.
$-6x+14 \le 17$	Combine like terms.
$-6x+14-14 \le 17-14$	Subtract 14 from both sides.
$-6x \le 3$	Combine like terms.
$\dfrac{-6x}{-6} \ge \dfrac{3}{-6}$	Divide both sides by -6; reverse inequality symbol.
$x \ge -\dfrac{1}{2}$	Simplify.

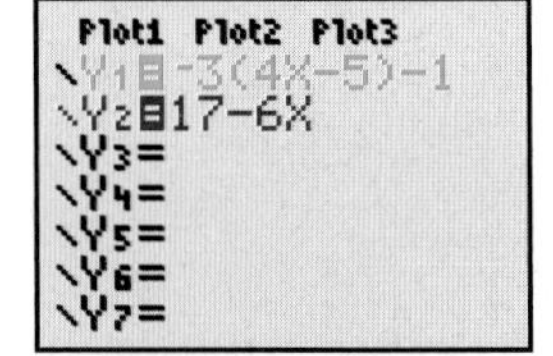

We can graph the solution set on a number line (see Fig. 44), or we can describe the solution set in interval notation as $\left[-\frac{1}{2}, \infty\right)$.

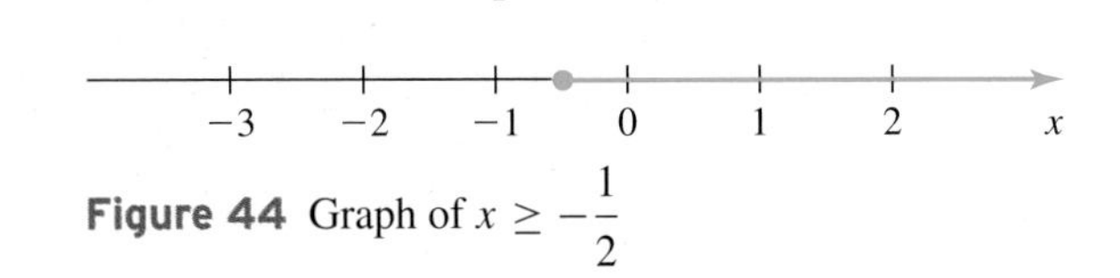

Figure 44 Graph of $x \ge -\frac{1}{2}$

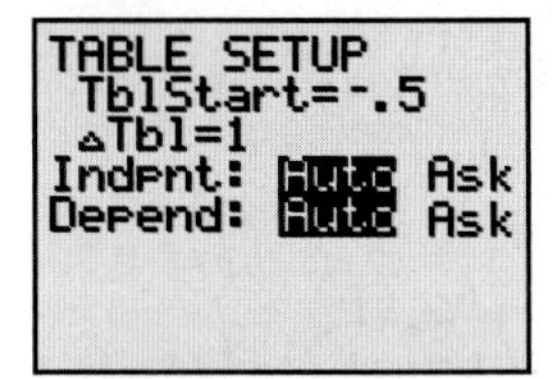

To verify our result, we check that, for inputs greater than or equal to $-\frac{1}{2}$, the outputs of $y=-3(4x-5)-1$ are less than or equal to the outputs of $y=17-6x$ (see Fig. 45). We do this by setting up the table so that x begins at -0.5 and increases by 1. Then, we scroll up two rows so that we can view both values of x that are less than -0.5 and values greater than -0.5. For graphing calculator instructions, see Section B.14. ■

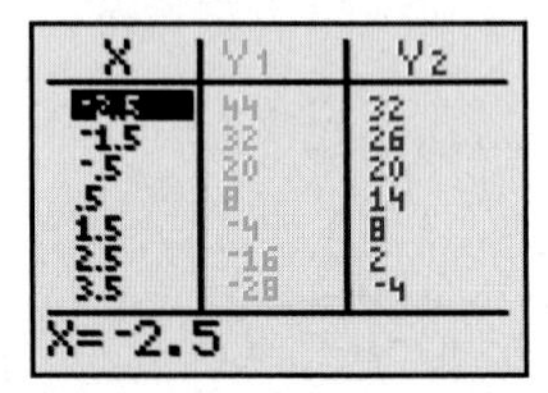

Figure 45 Verify the result

Example 5 Solving a Linear Inequality

Solve $\dfrac{3x+5}{4}-\dfrac{2x-7}{6}<\dfrac{11}{3}$. Describe the solution set as an inequality, in interval notation, and in a graph.

Solution

First, we multiply both sides of the inequality by the LCD, 12:

$12\left(\dfrac{3x+5}{4}-\dfrac{2x-7}{6}\right) < 12\cdot\dfrac{11}{3}$	Multiply both sides by LCD, 12.
$12\cdot\dfrac{3x+5}{4}-12\cdot\dfrac{2x-7}{6} < 12\cdot\dfrac{11}{3}$	Distributive law
$3(3x+5)-2(2x-7) < 44$	Simplify.
$9x+15-4x+14 < 44$	Distributive law
$5x+29 < 44$	Combine like terms.
$5x+29-29 < 44-29$	Subtract 29 from both sides.
$5x < 15$	Combine like terms.
$\dfrac{5x}{5} < \dfrac{15}{5}$	Divide both sides by 5.
$x < 3$	Simplify.

We graph the solution set, $(-\infty, 3)$, in Fig. 46.

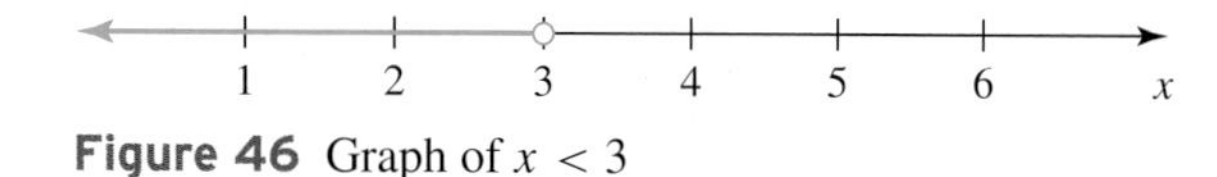

Figure 46 Graph of $x < 3$

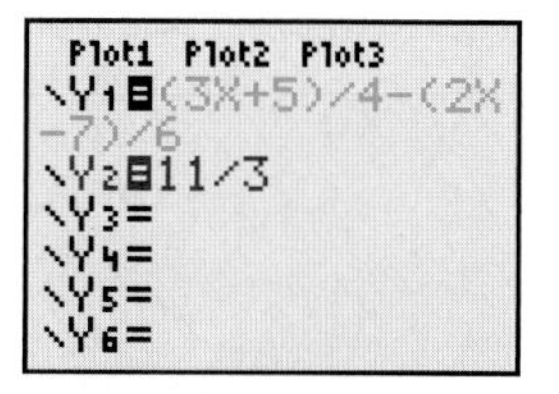

Figure 47 Verify the work

To verify our result, we check that, for values of x less than 3, the graph of the equation $y = \dfrac{3x+5}{4} - \dfrac{2x-7}{6}$ is below the horizontal line $y = \dfrac{11}{3}$. See Fig. 47. ■

Three-Part Inequalities

Now we will work with *three-part inequalities in one variable,* such as $3 \le x \le 7$. Recall from Section 1.6 that $3 \le x \le 7$ means that the values of x are *both* greater than or equal to 3 *and* less than or equal to 7. In other words, all values of x are between 3 and 7, inclusive. To graph the solutions, we shade the part of the number line that lies between 3 and 7 (see Fig. 48). We draw filled-in circles at 3 and 7 to indicate that 3 and 7 are solutions, too.

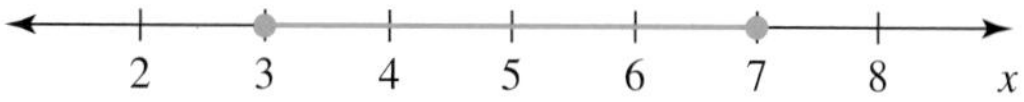

Figure 48 Graph of $3 \le x \le 7$

We describe the numbers between 3 and 7, inclusive, in interval notation by $[3, 7]$. More examples of three-part inequalities, with matching graphs and interval notation, are shown in Fig. 49.

In Words	Inequality	Graph	Interval Notation
Numbers between 1 and 3	$1 < x < 3$	 1 3 x	$(1, 3)$
Numbers between 1 and 3, inclusive	$1 \le x \le 3$	 1 3 x	$[1, 3]$
Numbers between 1 and 3, as well as 1	$1 \le x < 3$	 1 3 x	$[1, 3)$
Numbers between 1 and 3, as well as 3	$1 < x \le 3$	 1 3 x	$(1, 3]$

Figure 49 Words, inequalities, graphs, and interval notations

WARNING We use notation such as $(3, 7)$ in two ways. When we work with one variable, the *interval* $(3, 7)$ is the set of numbers between 3 and 7. When we work with two variables, such as x and y, the *ordered pair* $(3, 7)$ means that $x = 3$ and $y = 7$.

Example 6 Solving a Three-Part Inequality

Solve $-5 < 2x - 1 < 7$.

Solution

We can get x alone in the "middle part" of the inequality by applying the same operations to all three parts of the inequality:

$$-5 < 2x - 1 < 7 \quad \text{Original inequality}$$
$$-5 + 1 < 2x - 1 + 1 < 7 + 1 \quad \text{Add 1 to all three parts.}$$
$$-4 < 2x < 8 \quad \text{Combine like terms.}$$
$$\frac{-4}{2} < \frac{2x}{2} < \frac{8}{2} \quad \text{Divide all three parts by 2.}$$
$$-2 < x < 4 \quad \text{Simplify.}$$

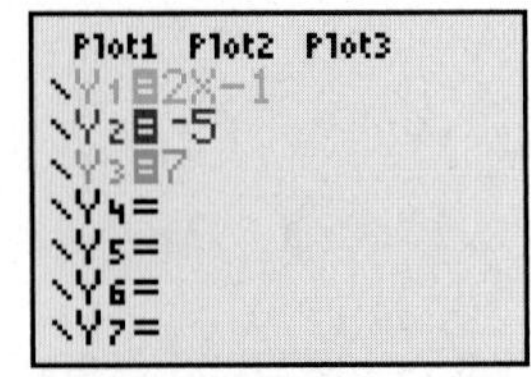

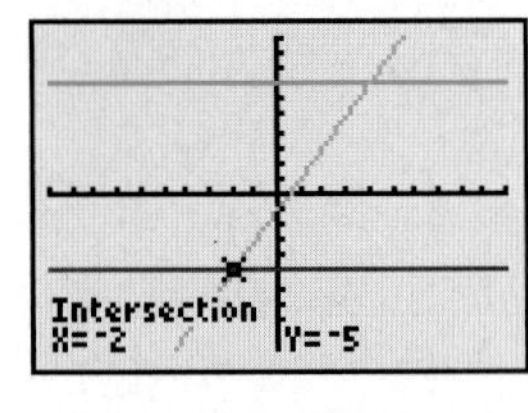

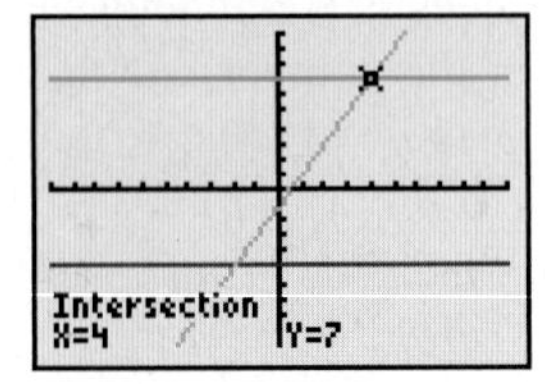

Figure 51 Verify the work

So, the solution set is the set of numbers between -2 and 4. We can graph the solution set on a number line (see Fig. 50), or we can describe the solution set in interval notation as $(-2, 4)$.

$-4\ -3\ -2\ -1\ 0\ 1\ 2\ 3\ 4\ 5\ 6\ x$

Figure 50 Graph of $-2 < x < 4$

To verify our result, we check that, for values of x between -2 and 4, the graph of $y = 2x - 1$ is between the horizontal lines $y = -5$ and $y = 7$ (see Fig. 51). ■

Example 7 Solving a Three-Part Inequality

Solve $\frac{1}{2} \le 5 - \frac{3}{2}w \le 4$.

Solution

$$\frac{1}{2} \le 5 - \frac{3}{2}w \le 4 \qquad \text{Original inequality}$$

$$2 \cdot \frac{1}{2} \le 2 \cdot 5 - 2 \cdot \frac{3}{2}w \le 2 \cdot 4 \qquad \text{Multiply all three parts by LCD, 2.}$$

$$1 \le 10 - 3w \le 8 \qquad \text{Simplify.}$$

$$1 - 10 \le 10 - 3w - 10 \le 8 - 10 \qquad \text{Subtract 10 from all three parts.}$$

$$-9 \le -3w \le -2 \qquad \text{Combine like terms.}$$

$$\frac{-9}{-3} \ge \frac{-3w}{-3} \ge \frac{-2}{-3} \qquad \text{Divide all three parts by } -3\text{; reverse inequality symbols.}$$

$$3 \ge w \ge \frac{2}{3} \qquad \text{Simplify.}$$

$$\frac{2}{3} \le w \le 3 \qquad \text{Write in form } a \le w \le b.$$

So, the solution set is the set of numbers between $\frac{2}{3}$ and 3, inclusive. We can graph the solution set on a number line (see Fig. 52), or we can describe the solution set in interval notation as $\left[\frac{2}{3}, 3\right]$.

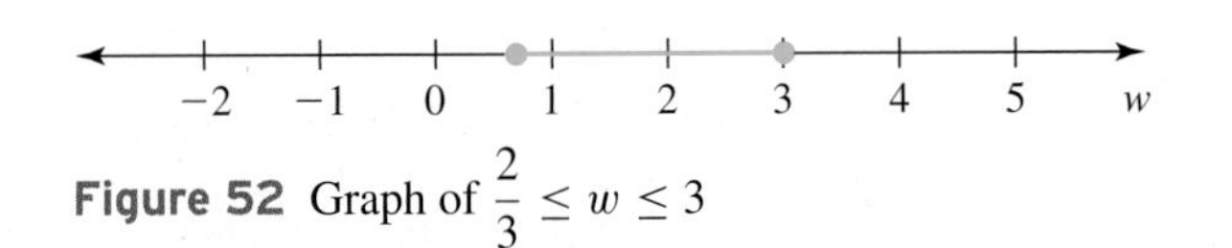

Figure 52 Graph of $\frac{2}{3} \le w \le 3$ ■

Using Linear Inequalities to Make Estimates and Predictions

In Example 1, we used graphing to estimate for what mileages Budget offers a lower price than U-Haul. In Example 8, we solve an inequality to make the estimate.

Example 8 Using Models to Compare Two Quantities

In Example 1, we modeled the one-day pickup truck costs (in dollars) $B(d)$ and $U(d)$ at Budget and U-Haul, respectively, by the system

$$C = B(d) = 0.19d + 39.95$$

$$C = U(d) = 0.49d + 19.95$$

where d is the number of miles driven. Use inequalities to estimate for which mileages Budget offers the lower price.

Solution

Budget offers the lower price when

$$B(d) < U(d)$$

We substitute $0.19d + 39.95$ for $B(d)$ and $0.49d + 19.95$ for $U(d)$ to get a linear inequality in one variable:

$$0.19d + 39.95 < 0.49d + 19.95$$

Then, we solve the inequality by isolating d on the left side of the inequality:

$0.19d + 39.95 < 0.49d + 19.95$	Original inequality
$0.19d + 39.95 - 0.49d < 0.49d + 19.95 - 0.49d$	Subtract $0.49d$ from both sides.
$-0.30d + 39.95 < 19.95$	Combine like terms.
$-0.30d + 39.95 - 39.95 < 19.95 - 39.95$	Subtract 39.95 from both sides.
$-0.30d < -20$	Combine like terms.
$\dfrac{-0.30d}{-0.30} > \dfrac{-20}{-0.30}$	Divide both sides by -0.30; reverse inequality symbol.
$d > 66.\overline{6}$	$66.\overline{6}$ represents $66.666\ldots$.

Plot1 Plot2 Plot3
\Y1=.19X+39.95
\Y2=.49X+19.95
\Y3=
\Y4=
\Y5=
\Y6=
\Y7=

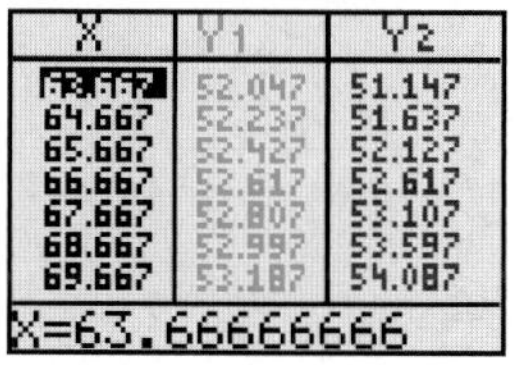

X	Y1	Y2
63.667	52.047	51.147
64.667	52.237	51.637
65.667	52.427	52.127
66.667	52.617	52.617
67.667	52.807	53.107
68.667	52.997	53.597
69.667	53.187	54.087

X=63.66666666

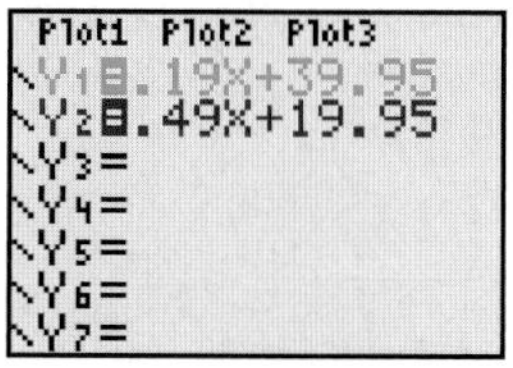

Figure 53 Verify the result

Budget offers the lower price if the truck is driven over $66.\overline{6}$ miles, about the same result that we found in Example 1.

To verify our result, we check that, for inputs greater than $66.\overline{6}$, the outputs of $y = 0.19x + 39.95$ are less than the outputs of $y = 0.49x + 19.95$ (see Fig. 53). We do this by setting up the table so that x begins at about $66.\overline{6}$ and increases by 1. Then, we scroll up three rows so that we can view values of x that are less than $66.\overline{6}$ and values that are greater than $66.\overline{6}$.

From Fig. 53, we see that Budget's costs, in column Y_1, are less than U-Haul's costs, in column Y_2, for distances above $66.\overline{6}$ miles. ■

group exploration

Meaning of the solution set of an inequality

We solve the inequality $-3x + 7 < 1$:

$$-3x + 7 < 1$$
$$-3x + 7 - 7 < 1 - 7$$
$$-3x < -6$$
$$\frac{-3x}{-3} > \frac{-6}{-3}$$
$$x > 2$$

1. Choose a number greater than 2. Check that your number satisfies the inequality $-3x + 7 < 1$.
2. Choose two more numbers greater than 2. Check that both of these numbers satisfy the inequality $-3x + 7 < 1$.
3. Choose three numbers that are *not* greater than 2. Show that each of these numbers does *not* satisfy the inequality $-3x + 7 < 1$.
4. Explain what it means when we write $x > 2$ as the last step in solving the inequality $-3x + 7 < 1$.

TIPS FOR SUCCESS: Form a Study Team

You can prepare for an exam by meeting with a study team. It usually works best if you form a team with students who are at your level of ability in this course. This will allow everyone to make contributions and will make everyone feel more comfortable asking questions about troublesome concepts.

Also, spend some time studying alone. This will ensure that you understand the concepts and can solve the relevant problems without help from your study team.

HOMEWORK 3.5

FOR EXTRA HELP ▶ Student Solutions Manual | PH Math/Tutor Center | MathXL® | MyMathLab

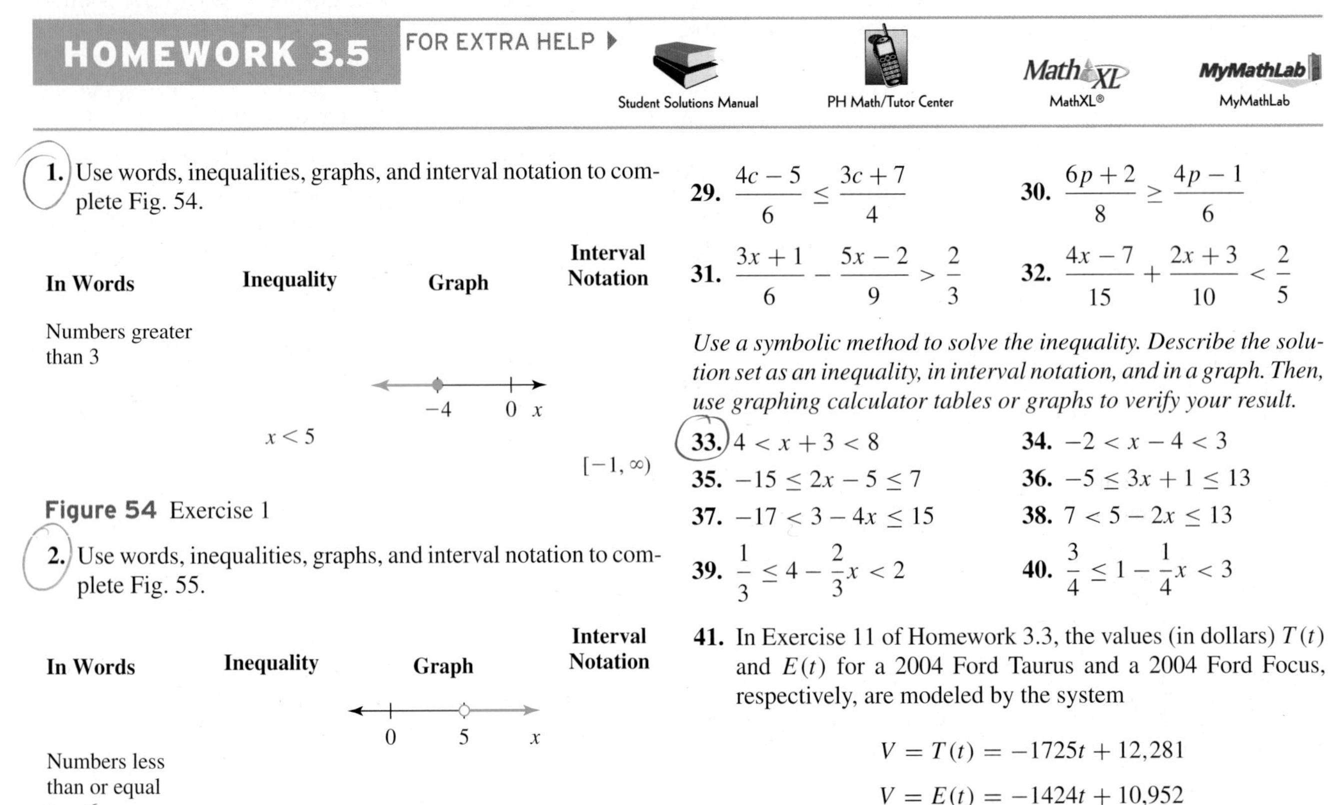

1. Use words, inequalities, graphs, and interval notation to complete Fig. 54.

In Words	Inequality	Graph	Interval Notation
Numbers greater than 3			

	$x < 5$		
			$[-1, \infty)$

Figure 54 Exercise 1

2. Use words, inequalities, graphs, and interval notation to complete Fig. 55.

In Words	Inequality	Graph	Interval Notation
		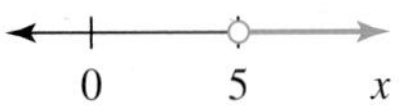	
Numbers less than or equal to -6			
			$[-\infty, 4)$
	$x \geq -3$		

Figure 55 Exercise 2

Use a symbolic method to solve the inequality. Describe the solution set as an inequality, in interval notation, and in a graph. Then, use graphing calculator tables or graphs to verify your result.

3. $x + 2 \geq 5$ **4.** $x - 3 < 8$

5. $-4x \geq 12$ **6.** $-7x > 21$

7. $2w + 7 < 11$ **8.** $15 - 4t \leq 0$

9. $9x < 4 + 5x$ **10.** $-5x > -15 - 2x$

11. $2.1x - 7.4 \leq 4.36$ **12.** $6.6 - 5.2x > 20.64$

13. $2b - 3 > 7b + 22$ **14.** $6p - 2 \geq 4p - 14$

15. $3 - 2(x - 4) > 4x + 1$

16. $-5(2x + 4) + 1 \leq 3x - 3$

17. $6.2a + 61.31 < 5(3.1 - 2.7a) + 0.5$

18. $13.5 - 4.1w \leq 2(w + 6.4) - 5.9w$

19. $7(x + 1) - 8(x - 2) \leq 0$

20. $-3(2x + 1) - 2(x + 4) > 0$

21. $5r - 4(2r - 6) - 1 \geq 3(3r - 1) + r$

22. $3p - 6(4p + 3) + 4 < 5(2p - 7) - 4p$

23. $-\frac{2}{3}x > 4$ **24.** $-\frac{1}{4}x \leq 2$

25. $\frac{2}{3} - \frac{3}{4}t \leq \frac{5}{2}$ **26.** $\frac{5}{8} + \frac{1}{6}b > \frac{2}{3}$

27. $-\frac{1}{2}x - \frac{5}{6} \geq \frac{1}{3} + \frac{3}{2}x$ **28.** $-\frac{3}{4}x + \frac{5}{2} < \frac{7}{8} - \frac{5}{2}x$

29. $\frac{4c - 5}{6} \leq \frac{3c + 7}{4}$ **30.** $\frac{6p + 2}{8} \geq \frac{4p - 1}{6}$

31. $\frac{3x + 1}{6} - \frac{5x - 2}{9} > \frac{2}{3}$ **32.** $\frac{4x - 7}{15} + \frac{2x + 3}{10} < \frac{2}{5}$

Use a symbolic method to solve the inequality. Describe the solution set as an inequality, in interval notation, and in a graph. Then, use graphing calculator tables or graphs to verify your result.

33. $4 < x + 3 < 8$ **34.** $-2 < x - 4 < 3$

35. $-15 \leq 2x - 5 \leq 7$ **36.** $-5 \leq 3x + 1 \leq 13$

37. $-17 < 3 - 4x \leq 15$ **38.** $7 < 5 - 2x \leq 13$

39. $\frac{1}{3} \leq 4 - \frac{2}{3}x < 2$ **40.** $\frac{3}{4} \leq 1 - \frac{1}{4}x < 3$

41. In Exercise 11 of Homework 3.3, the values (in dollars) $T(t)$ and $E(t)$ for a 2004 Ford Taurus and a 2004 Ford Focus, respectively, are modeled by the system

$$V = T(t) = -1725t + 12{,}281$$
$$V = E(t) = -1424t + 10{,}952$$

where t is the number of years since 2005 (Source: Edmunds Automobile Buyer's Guide). For what years will the value of the 2004 Taurus be less than the value of the 2004 Focus?

42. In Example 3 of Section 3.3, the values (in dollars) $D(t)$ and $I(t)$ for a 2004 Cadillac DeVille and a 2004 Acura Integra RSX, respectively, are modeled by the system

$$V = D(t) = -3548t + 26{,}440$$
$$V = I(t) = -1333t + 18{,}243$$

where t is the number of years since 2005 (Source: Edmunds Automobile Buyer's Guide). For what years will the value of the 2004 DeVille be less than the value of the 2004 Integra?

43. One U-Haul office rents 10-foot trucks for a one-day fee of \$19.95 plus \$0.69 per mile. One Penske® office charges a one-day fee of \$29.95 plus \$0.39 per mile (Sources: *U-Haul; Penske*).

a. Let $U(d)$ be U-Haul's charge (in dollars) and $P(d)$ be Penske's charge (in dollars), both for driving d miles in one day. Find equations of U and P.

b. For how many miles driven is the one-day charge at U-Haul less than the charge at Penske?

44. One Penske office rents 15-foot trucks for a one-day fee of \$39.95 plus \$0.29 per mile. One Budget® office charges a one-day fee of \$59.95 plus \$0.19 per mile (Source: *Penske; Budget*).

a. Let $P(d)$ be Penske's charge (in dollars) for driving d miles in one day. Let $B(d)$ be Budget's charge (in dollars) for driving d miles in one day. Find equations of P and B.

b. For how many miles driven is the one-day charge at Penske less than the charge at Budget?

45. The number of employees at Northwest Airlines® was 50.1 thousand employees in 2001 and has since decreased by about 4.2 thousand employees per year. The number of employees at Continental Airlines was 42.1 thousand employees in 2001 and has since decreased by about 2.4 thousand employees per year (Source: *U.S. Department of Transportation*). Predict in which years there will be more employees at Continental than at Northwest.

46. Export quotas (limits) have been placed on various types of caviar to avoid the depletion of sturgeon. The quota on sevruga caviar was 72 tons in 2001 and has since decreased by about 10.3 tons per year. The quota on osetra caviar was 33 tons in 2001 and has since decreased by about 2.0 tons per year (Source: *Fish & Fisheries*). Estimate in which years the quota for sevruga caviar has been less than the quota for osetra caviar.

47. The violent victimization rates for both women and men in the United States are shown in Table 29 for various years.

Table 29 Violent Victimization Rates by Gender

	Number of Crimes (per 1000 persons age 12+)	
Year	**Women**	**Men**
1995	38.1	55.7
1997	33.0	45.8
1999	28.8	37.0
2001	23.0	27.3
2002	20.8	25.5

Source: *National Crime Victimization Survey*

a. Let $W(t)$ be the violent victimization rate for women (number of female victims per 1000 women) and $M(t)$ be the violent victimization rate for men (number of male victims per 1000 men), both at t years since 1990. Find equations of W and M.

b. Estimate when the violent victimization rate for women was less than the victimization rate for men.

48. The annual U.S. per-person consumption of whole milk and lower fat milk is shown in Table 30 for various years.

Table 30 Annual U.S. Per-Person Consumption of Whole Milk and Lower Fat Milk

	Annual Consumption (gallons per person)	
Year	**Whole Milk**	**Lower Fat Milk**
1955	33.5	2.9
1965	28.8	3.7
1975	21.7	8.1
1985	14.3	12.2
1995	9.1	15.3
2000	8.1	14.5

Source: *USDA/Economic Research Service*

a. Let $W(t)$ be the annual consumption of whole milk and $L(t)$ be the annual consumption of lower fat milk, both in gallons per person, in the year that is t years since 1950. Find equations of W and L.

b. Estimate whether there was greater consumption of whole milk or lower fat milk in 2005. By how much?

c. Estimate in which years the annual consumption of whole milk was greater than the annual consumption of lower fat milk.

49. In Example 1 of Section 3.1, birth-year life expectancies (in years) $W(t)$ and $M(t)$ for women and men in the United States, respectively, are modeled by the system

$$L = W(t) = 0.105t + 77.53$$
$$L = M(t) = 0.203t + 69.91$$

where t is the number of years since 1980 (see Table 31).

Table 31 U.S. Life Expectancies of Women and Men

Year of Birth	Women (years)	Men (years)
1980	77.4	70.0
1985	78.2	71.1
1990	78.8	71.8
1995	78.9	72.5
2000	79.5	74.1
2003	80.1	74.8

Source: *U.S. Census Bureau*

a. Of those born in 2011, how much longer are women likely to live than men, on average?

b. Use a symbolic method to predict the birth years for which men will have a longer life expectancy than women. Verify your result by a graphical method.

c. A woman born in 1980 wants to choose a man to marry so that she does not outlive him.

i. According to the linear models, should the woman marry a younger or older man? Explain your reasoning.

ii. What are acceptable birth years for potential husbands? [**Hint:** There is a way to do this by using an inequality. If you draw a blank, do it *by trial and error*.]

50. As their work becomes more specialized, there is a greater need for mathematicians to work together so that their combined knowledge spans the topic of study (Source: *The Future of Scientific Communication*). Table 32 shows the percentages of mathematical papers written by one or two authors for various years.

Table 32 Mathematical Papers Written by One or Two Authors

	Percent	
Year	**One Author**	**Two Authors**
1970	81	18
1975	77	20
1980	69	24
1985	66	26
1990	62	29
1995	57	32
1999	50	34

Source: *Patrick Ion,* Mathematical Reviews

Let $f(t)$ be the percentage of papers written by one author and $g(t)$ be the percentage of papers written by two authors.

a. Find equations of f and g.

b. In 2010, what percentage of papers will be written by one author? two authors? three or more authors?

c. Predict the years when more papers will be written by two authors than by one author.

51. A student tries to solve the inequality $3x + 7 > 1$:

$$\begin{aligned} 3x + 7 &> 1 \\ 3x + 7 - 7 &> 1 - 7 \\ 3x &> -6 \\ \frac{3x}{3} &< \frac{-6}{3} \\ x &< -2 \end{aligned}$$

Describe any errors. Then solve the inequality correctly.

52. A student tries to solve the inequality $-10x + 3 \le -17$:

$$\begin{aligned} -10x + 3 &\le -17 \\ -10x + 3 - 3 &\le -17 - 3 \\ -10x &\le -20 \\ \frac{-10x}{-10} &\le \frac{-20}{-10} \\ x &\le 2 \end{aligned}$$

Describe any errors. Then solve the inequality correctly.

53. a. List three numbers that satisfy the inequality $3(x - 2) + 1 \ge 7 - 4x$.

b. List three numbers that do not satisfy the inequality $3(x - 2) + 1 \ge 7 - 4x$.

54. Solve the inequality $2x + 1 > 2x + 1$.

55. Find values of m and c so that the solution set to the inequality $mx < c$ is the set of numbers where $x > 2$.

56. a. Is the following statement true? Explain.

If $a < b$, then $a - c < b - c$.

b. Is the following statement true? Explain.

If $a < b$ and $c \ne 0$, then $\frac{a}{c} < \frac{b}{c}$.

57. a. Solve $x + 1 = -2x + 10$.

b. Solve $x + 1 < -2x + 10$.

c. Solve $x + 1 > -2x + 10$.

d. Graph the solutions in parts (a), (b), and (c) on the same number line. Use three colors to identify the different solutions. Make observations about the solutions.

e. Use a graphing calculator to graph $f(x) = x + 1$ and $g(x) = -2x + 10$ in the same viewing window. Explain how the observations you made in part (d) are related to these graphs.

58. a. Solve $2x - 1 = -3x + 9$.

b. Solve $2x - 1 < -3x + 9$.

c. Solve $2x - 1 > -3x + 9$.

d. Graph the solutions in parts (a), (b), and (c) on the same number line. Use three colors to identify the different solutions. Make observations about the solutions.

e. Use a graphing calculator to graph $f(x) = 2x - 1$ and $g(x) = -3x + 9$ in the same viewing window. Explain how the observations you made in part (d) are related to these graphs.

For Exercises 59–62, refer to Fig. 56. Is the statement true? Explain.

59. $f(-4) > g(-4)$

60. $f(2) \ge g(2)$

61. $f(-1) < g(-1)$

62. $f(-1) \le g(-1)$

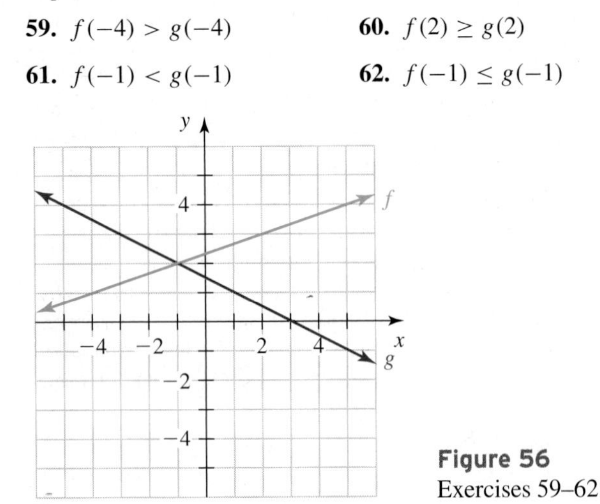

Figure 56 Exercises 59–62

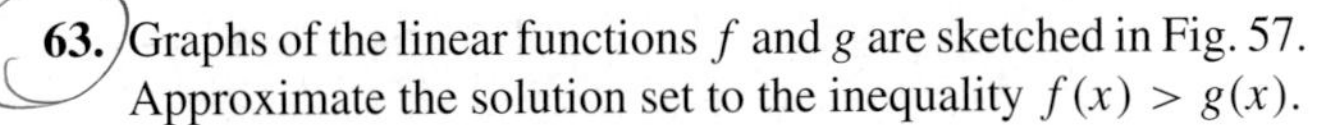

63. Graphs of the linear functions f and g are sketched in Fig. 57. Approximate the solution set to the inequality $f(x) > g(x)$.

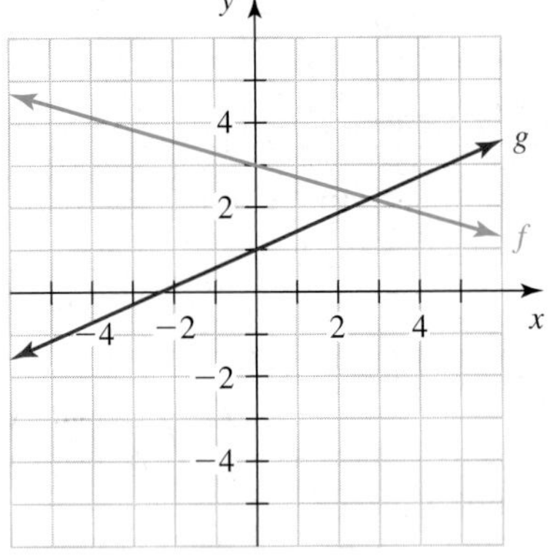

Figure 57 Exercise 63

64. Graphs of the linear functions f and g are sketched in Fig. 58. Determine which is the graph of f and which is the graph of g if the solution set to the inequality $f(x) \le g(x)$ is the set of numbers where $x \ge 2$.

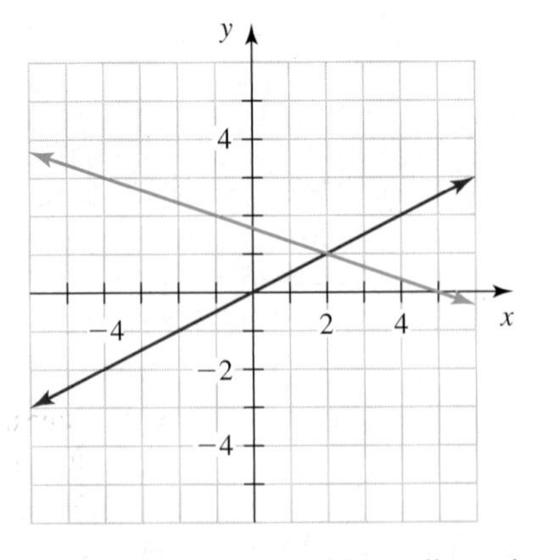

Figure 58 Exercise 64

65. Describe how to solve a linear inequality in one variable. Include a description of when and why you need to reverse the inequality symbol. Finally, explain what you have accomplished by solving an inequality.

CHAPTER SUMMARY

Key Points OF CHAPTER 3

Intersection point of the graphs of two models (Section 3.1)

If the independent variable of two models represents time, then an intersection point of the graphs of the two models indicates a time when the quantities represented by the dependent variables were or will be equal.

Solution of a system (Section 3.1)

We say that an ordered pair (a, b) is a **solution** of a system of two equations in two variables if it satisfies both equations. The **solution set** of a system is the set of all solutions of the system. We **solve** a system by finding its solution set.

Solving a system (Section 3.1)

The solution set of a system of two linear equations can be found by locating any intersection point(s) of the graphs of the two equations.

Types of linear systems (Section 3.1)

There are three types of linear systems of two equations:

1. *One-solution system:* The lines intersect in one point. The solution set of the system contains only the ordered pair that corresponds to that point.
2. **Inconsistent system:** The lines are parallel. The solution set of the system is the empty set.
3. **Dependent system:** The lines are identical. The solution set of the system is the set of the infinite number of solutions represented by points on the same line.

Using a table to solve a system (Section 3.1)

If an ordered pair is listed in a table as a solution of both of two linear equations, then that ordered pair is a solution of the system.

Using substitution to solve a linear system (Section 3.2)

To use **substitution** to solve a system of two linear equations,

1. Isolate a variable on one side of either equation.
2. Substitute the expression for the variable found in step 1 into the other equation.
3. Solve the equation in one variable found in step 2.
4. Substitute the solution found in step 3 into one of the original equations, and solve for the other variable.

Adding left sides and adding right sides of two equations (Section 3.2)

If $a = b$ and $c = d$, then $a + c = b + d$.

Using elimination to solve a linear system (Section 3.2)

To use **elimination** to solve a system of two linear equations,

1. If needed, multiply both sides of one equation by a number (and, if necessary, multiply both sides of the other equation by another number) to get the coefficients of one variable to be equal in absolute value and opposite in sign.
2. Add the left sides and add the right sides of the equations to eliminate one of the variables.
3. Solve the equation in one variable found in step 2.
4. Substitute the solution found in step 3 into one of the original equations, and solve for the other variable.

Methods of solving a system (Section 3.2)

Any linear system of two equations can be solved by graphing, substitution, or elimination. All three methods will give the same result.

Inconsistent and dependent systems (Section 3.2)

If the result of applying substitution or elimination to a linear system of two equations is

- a false statement, the system is inconsistent—that is, the solution set is the empty set.
- a true statement (one that can be put into the form $a = a$), then the system is dependent—that is, the solution is the set of ordered pairs represented by every point on the (same) line.

Using graphing or tables to solve an equation in one variable (Section 3.2)

To use graphing or a table to solve an equation $A = B$ in one variable, x, where A and B are expressions,

1. Use graphing or a table to solve the system

$$\begin{aligned} y &= A \\ y &= B \end{aligned}$$

2. The x-coordinates of any solutions of the system are the solutions of the equation $A = B$.

Intersection point of the graphs of two linear models (Section 3.3)	We can find an intersection point of the graphs of two linear models by graphing, substitution, or elimination.
Five-step problem-solving method (Section 3.4)	To solve some problems in which we want to find two quantities, it is useful to perform the following five steps: • ***Step 1: Define each variable.*** • ***Step 2: Write a system of two equations.*** • ***Step 3: Solve the system.*** • ***Step 4: Describe each result.*** • ***Step 5: Check.***
Total-value formula (Section 3.4)	If n objects each have value v, then their total value T is given by $T = vn$.
Using a function to analyze a situation (Section 3.4)	If we want to analyze many aspects of a certain situation, it can help to use a system of equations to find a linear function. We can then use the function to analyze the situation in various ways.
Annual simple interest rate (Section 3.4)	The **annual simple interest rate** is the percentage of the **principal** that equals the **interest** earned per year.
Percentage of solution (Section 3.4)	For an $x\%$ solution of two substances that are mixed, $x\%$ of the solution is one substance and $(100 - x)\%$ is the other substance.
Addition property of inequalities (Section 3.5)	If $a < b$, then $a + c < b + c$. Similar properties hold for $\leq$, $>$, and $\geq$.
Multiplication property of inequalities (Section 3.5)	• For a *positive* number c: If $a < b$, then $ac < bc$. • For a *negative* number c: If $a < b$, then $ac > bc$. Similar properties hold for $\leq$, $>$, and $\geq$.
Reversing the inequality symbol (Section 3.5)	When we multiply or divide both sides of an inequality by a negative number, we reverse the inequality symbol.
Linear inequality in one variable (Section 3.5)	A **linear inequality in one variable** is an inequality that can be put into one of the forms $mx + b < 0, \quad mx + b \leq 0, \quad mx + b > 0, \quad mx + b \geq 0$ where m and b are constants and $m \neq 0$.
Solution of an inequality in one variable (Section 3.5)	We say that a number is a **solution** of an inequality in one variable if it satisfies the inequality. The **solution set** of an inequality is the set of all solutions of the inequality. We **solve** an inequality by finding its solution set.

CHAPTER 3 REVIEW EXERCISES

Find the solution set of the system by graphing the equations by hand.

1. $y = -\frac{3}{2}x + 1$
$y = \frac{1}{4}x - 6$

2. $3x - 5y = -1$
$y = -2(x - 4)$

Solve the system by either elimination or substitution. If the system is inconsistent or dependent, say so.

3. $4x - 5y = -22$
$3x + 2y = -5$

4. $3x - 7y = 5$
$6x - 14y = -1$

5. $-4x - 5y = 3$
$10y = -8x - 6$

6. $y = 4.2x - 7.9$
$y = -2.8x + 0.5$

7. $y = 4.9x$
$-3.2y = x$

8. $3x - 5y = 21$
$y = \frac{1}{2}x - 4$

9. $\frac{3}{5}x - \frac{2}{3}y = 4$
$-\frac{6}{5}x + \frac{8}{3}y = -4$

10. $2(3x - 4) + 3(2y - 1) = -5$
$-3(2x + 1) + 4(y + 3) = -7$

11. Solve the system of equations three times, once by each of the three methods: elimination, substitution, and graphing:

$$2x - 5y = 15$$
$$y = -2x + 9$$

12. Create a system of two linear equations as indicated.
a. The system is dependent.
b. The system is inconsistent.
c. The solution of the system is (4, 6).

For Exercises 13 and 14, solve the given equation by referring to the graphs of $y = \frac{1}{2}x - \frac{5}{2}$ *and* $y = -\frac{2}{3}x - \frac{4}{3}$ *shown in Fig. 59.*

13. $\frac{1}{2}x - \frac{5}{2} = -\frac{2}{3}x - \frac{4}{3}$

14. $-\frac{2}{3}x - \frac{4}{3} = 2$

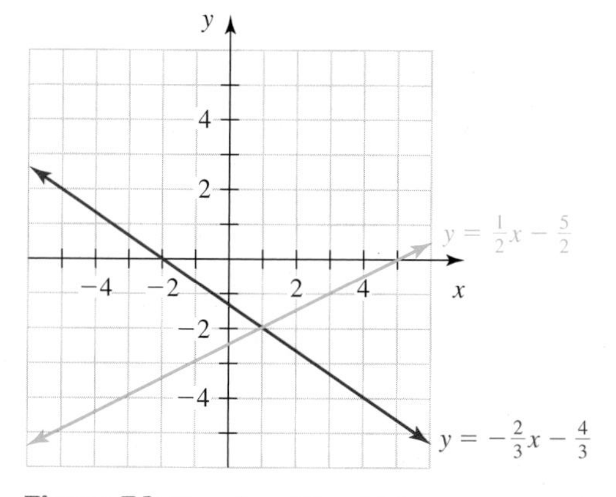

Figure 59 Exercises 13 and 14

15. Some values of functions f and g are given in Table 33. Determine two approximate solutions to the system of equations that describe the functions f and g.

Table 33 Some Values of Functions f and g (Exercise 15)

x	−4	−3	−2	−1	0	1	2	3	4
$f(x)$	2	5	8	11	14	17	20	23	26
$g(x)$	31	24	19	16	15	16	19	24	31

16. Determine the constants a and b such that (5, 3) is the solution of the system. Verify your result by checking that (5, 3) satisfies each of your equations. Also, verify your result graphically.

$$2x + 3y = a$$
$$6x - 4y = b$$

17. Find the coordinates of the points A, B, C, D, E, and F as shown in Fig. 60. The equations of the sketched lines are provided, but no attempt has been made to sketch the lines accurately except for showing the intersection points.

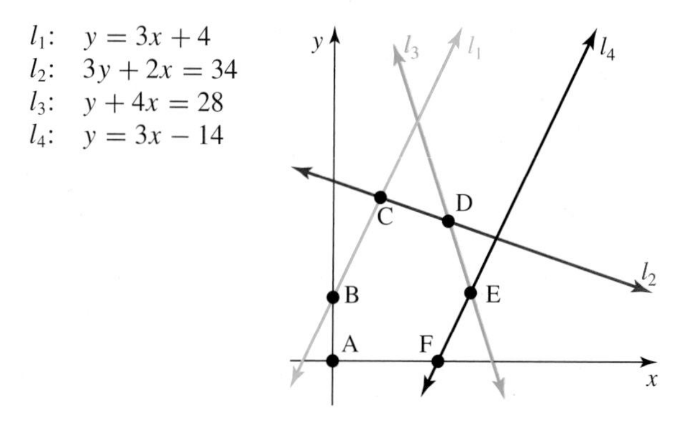

Figure 60 Exercise 17

18. Find all ordered pairs that satisfy all three equations:

$$y = -x + 2$$
$$y = -2x + 7$$
$$y = 3x - 6$$

Use a symbolic method to solve the inequality. Describe the solution set as an inequality, in interval notation, and in a graph.

19. $3x - 8 \leq 13$

20. $29.19 - 3.6a \geq 3.9(a + 2.1)$

21. $-5(2x + 3) \geq 2(3x - 4)$

22. $\frac{2x - 1}{4} - \frac{4x + 3}{6} > \frac{5}{3}$

23. $1 \leq 2x + 5 < 11$

24. **a.** List three numbers that satisfy the inequality $7 - 2(3x + 5) < 4x + 1$.

b. List three numbers that do not satisfy the inequality $7 - 2(3x + 5) < 4x + 1$.

25. A student tries to solve the inequality $5 - 3x \leq 11$:

$$5 - 3x \leq 11$$
$$5 - 3x - 5 \leq 11 - 5$$
$$-3x \leq 6$$
$$\frac{-3x}{-3} \leq \frac{6}{-3}$$
$$x \leq -2$$

Describe any errors. Then solve the inequality correctly.

For Exercises 26–29, refer to Fig. 61.

26. Find $f(4)$.

27. Find x where $g(x) = 0$.

28. Find x where $f(x) = g(x)$.

29. Find x where $f(x) > g(x)$.

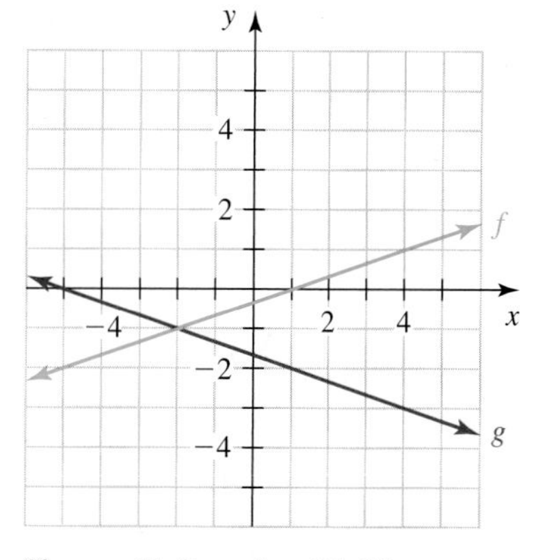

Figure 61 Exercises 26–29

30. Find values of a, b, and c such that the solution set to the inequality $ax + b > c$ is the set of numbers where $x < 5$. Use a symbolic method to verify your result.

31. Petroleum consumption by North America and petroleum consumption by Asia and Oceania (combined) are shown in Table 34 for various years. (One quadrillion Btu is 1,000,000,000,000,000 British thermal units.)

Table 34 Petroleum Consumptions

Year	Petroleum Consumption (quadrillion Btu) North America	Asia and Oceania
1985	37.0	22.3
1990	40.5	28.5
1995	41.7	37.1
2000	46.5	43.3
2003	47.3	45.8

Source: *Energy Information Administration*

Let $N(t)$ be the petroleum consumption by North America and $A(t)$ be the petroleum consumption by Asia and Oceania (combined), both in quadrillion Btu, where t is the number of years since 1980.

a. Find equations of N and A.
b. Compare the slopes of the two models. What do the slopes mean in this situation?
c. Estimate when petroleum consumption in North America was the same as in Asia and Oceania. What was that consumption?
d. Find values of t where $N(t) < A(t)$. What do they mean in this situation?

32. The television ratings of the baseball World Series and the average television ratings of prime-time shows on the major networks are shown in Table 35 for various years.

Table 35 Television Ratings of World Series and of Prime-Time Shows

Year	Rating of World Series	Average Rating of Prime-Time Shows on Major Networks
1973	30.7	19.2
1983	23.3	16.7
1993	17.3	12.5
2003	12.8	7.3

Sources: *Fox Sports; Nielsen Media Research*

a. Let $w(t)$ be the rating of the World Series and $p(t)$ be the average rating of prime-time shows on the major networks, both at t years since 1970. Find equations of w and p.
b. Use substitution or elimination to predict when the rating of the World Series will equal the average rating of prime-time shows on the major networks. What will that rating be?

33. One Rent-A-Wreck® office rents 16-foot trucks for a one-day fee of \$75 plus \$0.22 per mile. One U-Haul office charges a one-day fee of \$29.95 plus \$0.69 per mile.

a. Let $R(d)$ be Rent-A-Wreck's charge (in dollars) and $U(d)$ be U-Haul's charge (in dollars), both for driving d miles in one day. Find equations of R and U.
b. For how many miles driven is the one-day charge at Rent-A-Wreck equal to the charge at U-Haul? What is that charge?
c. For how many miles driven is the one-day charge at Rent-A-Wreck less than the charge at U-Haul?

34. In 2000, the average price of a home in a community was \$250,000, and it has increased by about \$9000 each year. A family had \$12,000 on January 1, 2000, and planned to save \$230 each month. Predict how long it should take the family to pay a 10% down payment on an average-priced house in the community.

35. A 20,000-seat theater has tickets for sale at \$55 and \$70. How many tickets should be sold at each price for a sellout performance to generate a total revenue of \$1,197,500?

36. A person plans to invest \$8000. She will invest in both a Hartford Global Leaders Y account at 6.8% annual interest and a Mutual Discovery Z account at 13.0% annual interest. Both interest rates are 5-year averages. Let x and y be the money (in dollars) invested in the 6.8% account and the 13.0% account, respectively.

a. Let $I = f(x)$ be the total interest (in dollars) earned from investing the \$8000 for one year. Find an equation of f.
b. Find $f(575)$. What does it mean in this situation?
c. Find the value of x when $f(x) = 575$. What does it mean in this situation?

CHAPTER 3 TEST

Solve the system by either elimination or substitution. If the system is inconsistent or dependent, say so.

1. $y = 3x - 1$
$3x - 2y = -1$

2. $2x - 5y = 3$
$6x = 15y + 9$

3. $4x - 6y = 5$
$6x - 9y = -2$

4. $\frac{2}{5}x - \frac{3}{4}y = 8$
$\frac{3}{5}x + \frac{1}{4}y = 1$

5. $-4(x + 2) + 3(2y - 1) = 21$
$5(3x - 2) - (4y + 3) = -59$

6. Create a system of two linear equations that has (5, 2) as its *only* solution.

7. Solve the system of equations three times, once by each of the three methods: elimination, substitution, and graphing.

$$4x - 3y = 9$$
$$y = 2x - 5$$

8. Consider the solution set to the system

$$y = 5x - 13$$
$$y = mx + b$$

where m and b are constants. If the system's solution set is the empty set, what can you say about m? About b?

Solve the inequality. Describe the solution set as an inequality, in interval notation, and in a graph.

9. $2 - 10x \geq 3x + 14$

10. $3(x + 4) + 1 < 5(x - 2)$

11. $2.6(t - 3.1) > 4.7t - 9.74$

12. $-\frac{5}{3}w + \frac{1}{6} \le \frac{7}{4}w$

For Exercises 13–16, refer to Fig. 62.

13. Find $f(5)$.

14. Find x where $g(x) = 3$.

15. Find x where $f(x) = g(x)$.

16. Find x where $f(x) \le g(x)$.

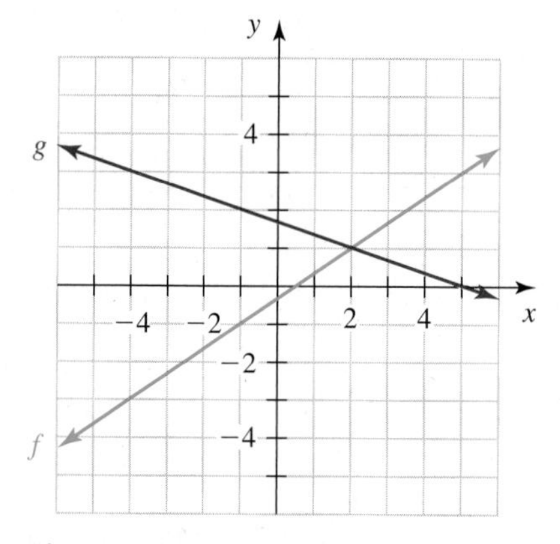

Figure 62 Exercises 13–16

17. Use "intersect" on a graphing calculator to solve the equation $\frac{2}{3}x - \frac{5}{2} = \frac{7}{2}x + \frac{4}{3}$. Round the solution to the second decimal place.

18. The graphs of f and g are sketched in Fig. 63. Solve the inequality $f(x) < g(x)$.

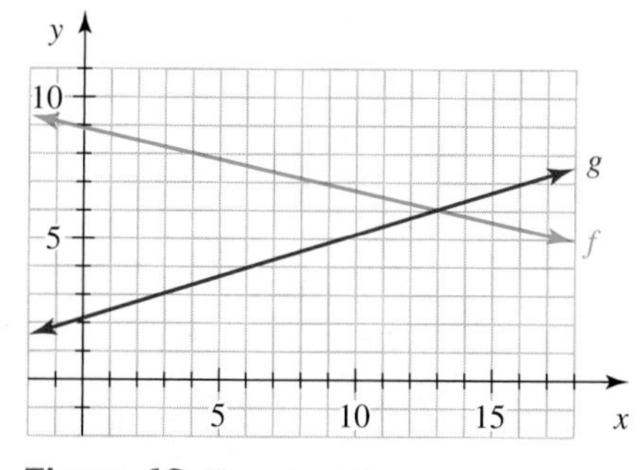

Figure 63 Exercise 18

19. a. Determine three numbers that satisfy the inequality $3x - 11 < 7 - 6x$.

b. Determine three numbers that do not satisfy the inequality $3x - 11 < 7 - 6x$.

20. The number of two-year colleges and of four-year colleges and universities are listed in Table 36 for various years.

Table 36 Numbers of Two-Year Colleges and Four-Year Colleges and Universities

Year	Two-Year	Four-Year
1970	886	1639
1980	1195	1957
1990	1408	2127
2000	1721	2363
2003	1844	2324

Source: *National Center for Education Statistics*

Let $f(t)$ be the number of two-year colleges and $g(t)$ be the number of four-year colleges and universities at t years since 1970.

a. Find equations of f and g.

b. Predict the total number of two-year colleges and four-year colleges and universities in 2011.

c. Use f and g to predict when there will be the same number of two-year colleges as four-year colleges and universities. What will that number be?

d. Explain in terms of the slopes of the two models why your result in part (c) is so far in the future.

21. Personal income (in 1998 dollars) in Maine increased approximately linearly from \$23,596 in 1998 to \$25,544 in 2003. Personal income (in 1998 dollars) in Georgia increased approximately linearly from \$25,279 in 1998 to \$26,086 in 2003 (Source: *Bureau of Economic Analysis*).

a. Let $M(t)$ be personal income (in 1998 dollars) in Maine and $G(t)$ be personal income (in 1998 dollars) in Georgia, both in the year that is t years since 1998. Find equations of M and G.

b. Predict when personal income in Maine will equal personal income in Georgia. What will that income be?

c. For which years was personal income in Maine less than personal income in Georgia?

22. How many gallons of a 10% antifreeze solution and a 20% antifreeze solution must be mixed to make 10 gallons of a 16% antifreeze solution?

CUMULATIVE REVIEW OF CHAPTERS 1-3

1. A person places a pitcher of hot tea in the refrigerator. Let T be the temperature (in degrees Fahrenheit) of the tea at t minutes after the tea was put into the refrigerator. Sketch a qualitative graph that describes the relationship between t and T.

Graph the equation.

2. $5x - 3y = 15$

3. $3(x - 4) = -2(y + 5) + 4$

4. Find the slope of the line that contains the points $(-4, 2)$ and $(3, -1)$.

5. Find an equation of a line with slope $-\frac{3}{5}$ that contains the point $(2, -3)$. Write your result in slope–intercept form.

6. Find an equation of a line that contains the points $(-5, -2)$ and $(-2, 3)$.

7. Find an equation of the line that contains the point $(-5, 3)$ and is perpendicular to the line $2x - 5y = 20$.

8. Find three points that lie between the lines $y = 2x + 3$ and $y = 2x + 3.1$.

For Exercises 9–11, refer to Table 37, which provides some values of four linear functions.

9. Complete Table 37.

10. Find $f(5)$.

11. Find x where $g(x) = 30$.

Table 37 Values of Four Linear Functions (Exercises 9–11)

Equation 1		Equation 2		Equation 3		Equation 4	
x	$f(x)$	x	$g(x)$	x	$h(x)$	x	$k(x)$
0	97	4		1	23	10	−28
1		5		2		11	
2		6	4	3		12	
3	58	7		4		13	
4		8		5		14	−16
5		9	43	6	−22	15	

For Exercises 12–16, let $f(x) = -\frac{3}{2}x + 7$.

12. Find $f(-4)$.

13. Find x where $f(x) = \frac{5}{3}$.

14. Find the x-intercept of the graph of f.

15. Find the y-intercept of the graph of f.

16. Sketch the graph of f.

For Exercises 17–22, refer to Fig. 64.

17. Find $g(3)$.

18. Find x where $f(x) = 1$.

19. Find the y-intercept of the graph of f.

20. Find an equation of f.

21. Find x where $f(x) = g(x)$.

22. Find x where $f(x) \leq g(x)$.

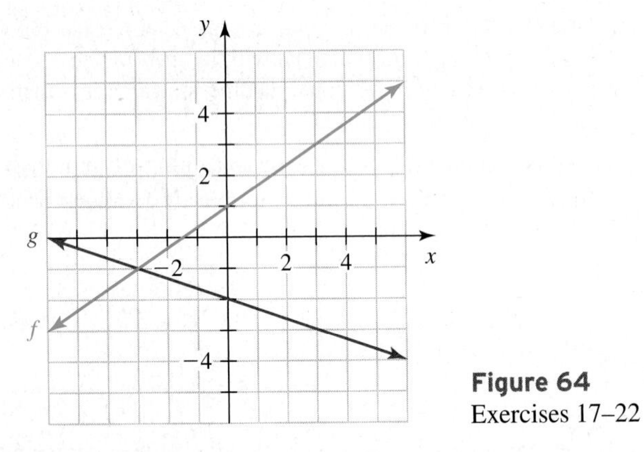

Figure 64 Exercises 17–22

Solve the equation.

23. $-5x - 3(2x + 4) = 8 - 2x$

24. $\frac{7}{8} - \frac{1}{4}b = \frac{1}{2} + \frac{3}{8}b$

25. Solve for b:

$$\frac{b}{c} - d = \frac{k}{c}$$

26. Find the x-intercept and y-intercept of the equation $5x - 3y + 2 = 0$.

For Exercises 27–29, refer to the relation graphed in Fig. 65.

27. Find the domain of the relation.

28. Find the range of the relation.

29. Is the relation a function? Explain.

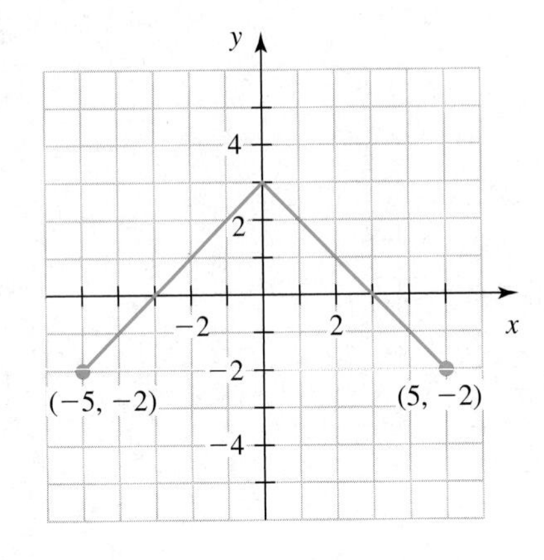

Figure 65 Exercises 27–29

Solve the system.

30. $2x + 4y = -8$
$5x - 3y = 19$

31. $3x - 7y = 14$
$y = \frac{3}{7}x - 2$

For Exercises 32–34, solve the given equation or system by referring to the graphs shown in Fig. 66.

32. $-\frac{1}{4}x - \frac{3}{2} = 2x + 3$

33. $-\frac{1}{4}x - \frac{3}{2} = -3$

34. $y = -\frac{5}{2}x + 3$
$y = -\frac{1}{4}x - \frac{3}{2}$

$y = -\frac{5}{2}x + 3$, $y = 2x + 3$, $y = -\frac{1}{4}x - \frac{3}{2}$

Figure 66 Exercises 32–34

35. Solve $-2(4x + 5) \geq 3(x - 7) + 1$. Describe the solution set as an inequality, in interval notation, and in a graph.

36.
a. Solve $-3x + 6 = 0$.
b. Solve $-3x + 6 < 0$. Graph the solution set on a number line.

c. Graph $y = -3x + 6$.
d. Solve the system:

$$y = -3x + 6$$
$$y = 4x - 1$$

37. Give an example of each of the following, and describe the solution set to each of your examples [**Hint:** In some cases, it will be helpful to describe the solution set by using a graph.]
a. equation in one variable
b. equation in two variables
c. system of two equations in two variables
d. inequality in one variable

38. The numbers of Chinese children adopted by American families are shown in Table 38 for various years.

Table 38 Numbers of Chinese Children Adopted by American Families

Year	Number of Adoptions (thousands)
1995	2.0
1997	3.9
1999	4.4
2001	4.8
2003	7.0
2005	7.9

Source: *U.S. Department of State*

Let $f(t)$ be the number of Chinese children (in thousands) adopted by American families in the year that is t years since 1990.
a. Find an equation of f.
b. Find $f(10)$. What does it mean in this situation?
c. Find t when $f(t) = 10$. What does the result mean in this situation?
d. Find the slope of the model. What does it mean in this situation?
e. China relaxed its adoption laws in 1991. Estimate the number of Chinese children adopted by American families in that year.

39. The number of bicyclists younger than 16 who were hit and killed by motor vehicles was 142 children in 2003 and has declined by about 17 children each year (Source: *Insurance Institute for Highway Safety*).
a. Let $n = f(t)$ be the number of bicyclists younger than 16 who were hit and killed by motor vehicles in the year that is t years since 2003. Find an equation of f.
b. What is the slope of the model? What does it mean in this situation?
c. Find the n-intercept. What does it mean in this situation?
d. Find the t-intercept. What does it mean in this situation?
e. For which years in the future is there model breakdown for certain? Explain.

40. Company A sold 9.5 million dollars of sports equipment in 2005, and its sales have increased by 1.3 million dollars each year. Company B sold 5.2 million dollars of sports equipment in 2005, and its sales have increased by 1.8 million dollars each year.
a. Let $A(t)$ and $B(t)$ be the sales (in millions of dollars) by company A and company B, respectively, at t years since 2005. Find equations of A and B.
b. Sketch the graphs of A and B by hand on the same coordinate system. According to your sketch, when will sales at the companies be equal? What will those sales be?
c. Now use substitution or elimination to predict when sales at the companies will be equal. What will those sales be? Compare your results with your results in part (b).
d. Solve the inequality $A(t) < B(t)$. What does your result mean in this situation?

41. World record times for the 200-meter run are listed in Table 39.

Table 39 200-Meter Run Record Times

Women		Men	
Year	Record Time (seconds)	Year	Record Time (seconds)
1973	22.38	1951	20.6
1974	22.21	1963	20.3
1978	22.06	1967	20.14
1984	21.71	1979	19.72
1988	21.34	1996	19.32

Source: *International Association of Athletics Federations*

a. Let $r = W(t)$ and $r = M(t)$ be the record times (in seconds) for women and men, respectively, at t years since 1900. Find the equations of W and of M.
b. Find $W(111)$ and $M(111)$. What do they mean in this situation?
c. Compare the slopes of the two models. What does your comparison tell you about the situation?
d. Explain why your results from parts (b) and (c) suggest that there may be a time when the women's record time will be equal to the men's record time.
e. Predict when the women's record time will equal the men's record time. What will that time be?
f. Solve $W(t) > M(t)$. What does your result mean in this situation?
g. Find the t-intercepts of the two models. What do your results mean in this situation?
h. On the same axes, sketch qualitative graphs that describe the relationship between t and r for hundreds of years into the past and future.

42. A person plans to invest twice as much in a UBS Global Equity Y account at 7.2% annual interest as in a Fidelity Worldwide account at 9.4% interest. Both interest rates are 5-year averages. How much will the person have to invest in each account to earn a total of $595 in one year?

Chapter 4

Exponential Functions

"There is not much difference between the delight a novice experiences in cracking a clever brain teaser and the delight a mathematician experiences in mastering a more advanced problem. Both look on beauty bare—that clean, sharply defined, mysterious, entrancing order that underlies all structure."

—Martin Gardner

Table 1 Numbers of Starbucks Stores

Year	Number of Stores
1991	116
1993	272
1995	676
1997	1412
1999	2135
2001	4709
2003	7225

Source: *Starbucks Corporation*

Is there a Starbucks® store nearby? The number of these coffeehouses worldwide has increased greatly since 1991 (see Table 1). In Exercise 33 of Homework 4.5, you will predict the number of stores in 2010 and describe the increase in the number of stores over time.

In Chapters 1–3, we worked with linear expressions and equations. In this chapter, we will work with a new type of expression and equation. In addition to working with linear functions, we will now use *exponential functions* to model some authentic situations. We will use these functions to make predictions, such as predicting the number of Starbucks stores in 2010.

4.1 PROPERTIES OF EXPONENTS

Objectives

- Know the meaning of *exponent,* zero exponent, and negative exponent.
- Know properties of exponents.
- Simplify expressions involving exponents.
- Know the meaning of *exponential function.*
- Use *scientific notation.*

In this section, we will simplify expressions involving *exponents*.

Definition of an Exponent

If n is a counting number (Section A.2), what is the meaning of b^n? The notation b^3 stands for $b \cdot b \cdot b$. So, $4^3 = 4 \cdot 4 \cdot 4 = 64$. The notation b^4 stands for $b \cdot b \cdot b \cdot b$. So, $2^4 = 2 \cdot 2 \cdot 2 \cdot 2 = 16$.

DEFINITION Exponent

For any counting number n,

$$b^n = \underbrace{b \cdot b \cdot b \cdot \ldots \cdot b}_{n \text{ factors of } b}$$

We refer to b^n as the **power,** the ***n*th power of *b*,** or ***b* raised to the *n*th power.** We call b the **base** and n the **exponent.**

The expression 3^5 is a power. It is the 5th power of 3, or 3 raised to the 5th power. For 3^5, the base is 3 and the exponent is 5. Here, we label the base and the exponent of 3^5 and calculate the power:

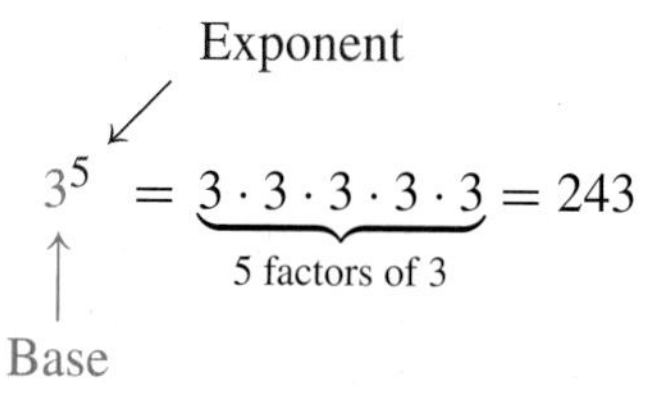

When we calculate a power, we say that we have performed an **exponentiation.**

Notice that the notation b^1 stands for one factor of b, so $b^1 = b$.

Two powers of b have specific names. We refer to b^2 as the **square of *b*** or ***b* squared.** We refer to b^3 as the **cube of *b*** or ***b* cubed.**

For an expression of the form $-b^n$, we compute b^n before finding the opposite. For example,

$$-2^4 = -\left(2^4\right) = -(2 \cdot 2 \cdot 2 \cdot 2) = -16$$

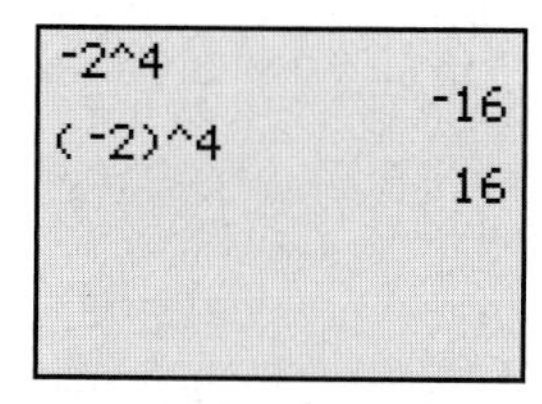

Figure 1 Compute -2^4 and $(-2)^4$

For -2^4, the base is 2, *not* -2. If we want the base to be -2, we must enclose -2 in parentheses:

$$(-2)^4 = (-2)(-2)(-2)(-2) = 16$$

We can use a graphing calculator to check both computations (see Fig. 1). To find -2^4, press [(−)] **2** [∧] **4** [ENTER].

Properties of Exponents

In this section and Section 4.2, we discuss five properties of exponents.

> **Properties of Exponents**
>
> If m and n are counting numbers, then
>
> - $b^m b^n = b^{m+n}$ — *Product property for exponents*
> - $\dfrac{b^m}{b^n} = b^{m-n}$, $b \neq 0$ and $m > n$ — *Quotient property for exponents*
> - $(bc)^n = b^n c^n$ — *Raising a product to a power*
> - $\left(\dfrac{b}{c}\right)^n = \dfrac{b^n}{c^n}$, $c \neq 0$ — *Raising a quotient to a power*
> - $\left(b^m\right)^n = b^{mn}$ — *Raising a power to a power*

In Example 1 (and the first Exploration), we will investigate why these properties make sense.

Example 1 Meaning of Exponential Properties

1. Show that $b^2 b^3 = b^5$.
2. Show that $b^m b^n = b^{m+n}$, where m and n are counting numbers.
3. Show that $\left(\dfrac{b}{c}\right)^n = \dfrac{b^n}{c^n}$, where n is a counting number and $c \neq 0$.

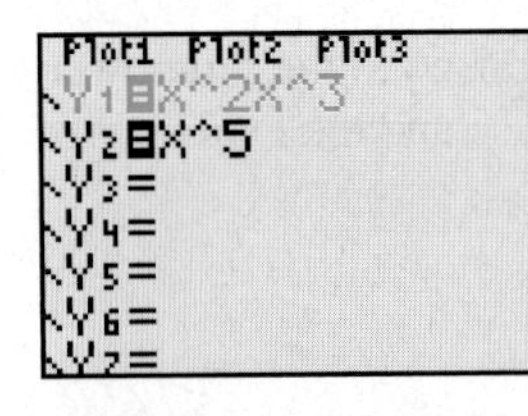

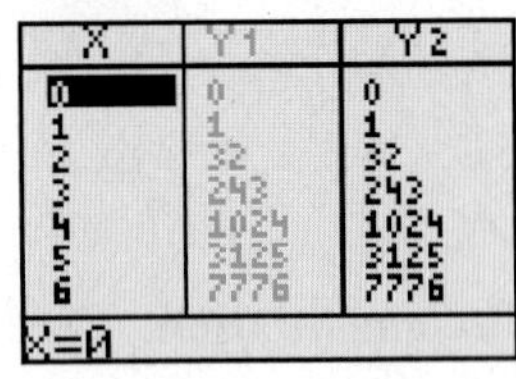

Figure 2 Comparing tables for $y = x^2x^3$ and $y = x^5$

Solution

1. By writing b^2b^3 without exponents, we see that

$$\begin{aligned} b^2b^3 &= (bb)(bbb) && \text{Write without exponents.} \\ &= bbbbb && \text{Remove parentheses.} \\ &= b^5 && \text{Write with an exponent.} \end{aligned}$$

We can verify that this result is correct for various constant bases by examining graphing calculator tables for both $y = x^2x^3$ and $y = x^5$ (see Fig. 2). For graphing calculator instructions, see Section B.14.

2. We write $b^m b^n$ without exponents:

$$b^m b^n = (\underbrace{bbb\cdots b}_{m \text{ factors}})(\underbrace{bbb\cdots b}_{n \text{ factors}}) = \underbrace{bbb\cdots b}_{m+n \text{ factors}} = b^{m+n}$$

3. We write $\left(\frac{b}{c}\right)^n$, where $c \neq 0$, without exponents:

$$\left(\frac{b}{c}\right)^n = \underbrace{\left(\frac{b}{c}\right)\left(\frac{b}{c}\right)\left(\frac{b}{c}\right)\cdots\left(\frac{b}{c}\right)}_{n \text{ factors}} = \frac{\overbrace{bbb\cdots b}^{n \text{ factors}}}{\underbrace{ccc\cdots c}_{n \text{ factors}}} = \frac{b^n}{c^n}$$

■

Simplifying Expressions Involving Exponents

We can use properties of exponents to simplify expressions involving exponents.

> **Simplifying Expressions Involving Exponents**
>
> An expression involving exponents is simplified if
>
> 1. It includes no parentheses.
> 2. Each variable or constant appears as a base as few times as possible. For example, we write $x^2x^4 = x^6$.
> 3. Each numerical expression (such as 7^2) has been calculated, and each numerical fraction has been simplified.
> 4. Each exponent is positive.

Example 2 Simplifying Expressions Involving Exponents

Simplify.

1. $(2b^2c^3)^5$
2. $(3b^3c^4)(2b^6c^2)$
3. $\dfrac{3b^7c^6}{12b^2c^5}$
4. $\left(\dfrac{24b^7c^8}{16b^2c^5d^3}\right)^4$

Solution

1. $$\begin{aligned} (2b^2c^3)^5 &= 2^5(b^2)^5(c^3)^5 && \text{Raise factors to nth power: } (bc)^n = b^nc^n \\ &= 32b^{10}c^{15} && \text{Multiply exponents: } (b^m)^n = b^{mn} \end{aligned}$$

2. $$\begin{aligned} (3b^3c^4)(2b^6c^2) &= (3\cdot 2)(b^3b^6)(c^4c^2) && \text{Rearrange factors.} \\ &= 6b^9c^6 && \text{Add exponents: } b^mb^n = b^{m+n} \end{aligned}$$

3. $$\begin{aligned} \frac{3b^7c^6}{12b^2c^5} &= \frac{b^{7-2}c^{6-5}}{4} && \text{Subtract exponents: } \frac{b^m}{b^n} = b^{m-n} \\ &= \frac{b^5c}{4} && \text{Subtract.} \end{aligned}$$

4. $\left(\dfrac{24b^7c^8}{16b^2c^5d^3}\right)^4 = \left(\dfrac{3b^5c^3}{2d^3}\right)^4$ Subtract exponents: $\dfrac{b^m}{b^n} = b^{m-n}$

$= \dfrac{(3b^5c^3)^4}{(2d^3)^4}$ Raise numerator and denominator to nth power: $\left(\dfrac{b}{c}\right)^n = \dfrac{b^n}{c^n}$

$= \dfrac{3^4(b^5)^4(c^3)^4}{2^4(d^3)^4}$ Raise factors to nth power: $(bc)^n = b^nc^n$

$= \dfrac{81b^{20}c^{12}}{16d^{12}}$ Multiply exponents: $(b^m)^n = b^{mn}$ ■

WARNING The expressions $3b^2$ and $(3b)^2$ are *not* equivalent expressions:

$$3b^2 = 3b \cdot b$$
$$(3b)^2 = (3b)(3b) = 9b \cdot b$$

For $3b^2$, the base is the variable b. For $(3b)^2$, the base is the product $3b$.

Here, we show a typical error and the correct way to find the power $(3b)^2$:

$$(3b)^2 = 3b^2 \quad \text{Incorrect}$$
$$(3b)^2 = 3^2b^2 = 9b^2 \quad \text{Correct}$$

Since the base $3b$ is a product, we need to distribute the exponent 2 to *both* factors 3 and b.

In general, when finding a power of the form $(bc)^n$, don't forget to distribute the exponent n to both factors b and c.

Zero as an Exponent

What is the meaning of b^0? If the property $\dfrac{b^m}{b^n} = b^{m-n}$ is to be true for $m = n$, then

$$1 = \frac{b^n}{b^n} = b^{n-n} = b^0, \quad b \neq 0$$

So, a reasonable definition of b^0 is 1.

DEFINITION Zero exponent

For $b \neq 0$,

$$\mathbf{b^0 = 1}$$

For example, $7^0 = 1$, $(-3)^0 = 1$, and $(ab)^0 = 1$, where $ab \neq 0$.

Negative Exponents

If n is an integer (Section A.2), what is the meaning of b^n? In particular, what is the meaning of a negative integer exponent? If the property $\dfrac{b^m}{b^n} = b^{m-n}$ is to be true for $m = 0$, then

$$\frac{1}{b^n} = \frac{b^0}{b^n} = b^{0-n} = b^{-n}, \quad b \neq 0$$

So, we should define b^{-n} to be $\dfrac{1}{b^n}$.

DEFINITION Negative integer exponent

If $b \neq 0$ and n is a counting number, then

$$b^{-n} = \frac{1}{b^n}$$

In words: To find b^{-n}, take its reciprocal and switch the sign of the exponent.

For example, $3^{-2} = \frac{1}{3^2} = \frac{1}{9}$ and $b^{-5} = \frac{1}{b^5}$.

Next, we write $\frac{1}{b^{-n}}$ in another form, where $b \neq 0$ and n is a counting number:

$$\begin{aligned}
\frac{1}{b^{-n}} &= 1 \div b^{-n} && \frac{a}{b} = a \div b \\
&= 1 \div \frac{1}{b^n} && \text{Write power so exponent is positive: } b^{-n} = \frac{1}{b^n} \\
&= 1 \cdot \frac{b^n}{1} && \text{Multiply by reciprocal of } \frac{1}{b^n}\text{, which is } \frac{b^n}{1}. \\
&= b^n && \text{Simplify.}
\end{aligned}$$

So, $\frac{1}{b^{-n}} = b^n$.

Negative Exponent in a Denominator

If $b \neq 0$ and n is a counting number, then

$$\frac{1}{b^{-n}} = b^n$$

In words: To find $\frac{1}{b^{-n}}$, take its reciprocal and switch the sign of the exponent.

For example, $\frac{1}{2^{-4}} = 2^4 = 16$ and $\frac{1}{b^{-8}} = b^8$.

Simplifying More Expressions Involving Exponents

As we have seen, simplifying an expression involving exponents includes writing the expression so that each exponent is positive.

Example 3 Simplifying Expressions Involving Exponents

Simplify.

1. $9b^{-7}$ **2.** $\frac{5}{b^{-3}}$ **3.** $3^{-1} + 4^{-1}$

Solution

1. $9b^{-7} = 9 \cdot \frac{1}{b^7} = \frac{9}{b^7}$

2. $\frac{5}{b^{-3}} = 5 \cdot \frac{1}{b^{-3}} = 5b^3$

3. $3^{-1} + 4^{-1} = \frac{1}{3} + \frac{1}{4} = \frac{4}{12} + \frac{3}{12} = \frac{7}{12}$ ■

It turns out that the five properties discussed at the start of this section are also true for all negative-integer exponents and the zero exponent.

Properties of Integer Exponents

If m and n are integers, $b \neq 0$, and $c \neq 0$, then

- $b^m b^n = b^{m+n}$ — Product property for exponents
- $\dfrac{b^m}{b^n} = b^{m-n}$ — Quotient property for exponents
- $(bc)^n = b^n c^n$ — Raising a product to a power
- $\left(\dfrac{b}{c}\right)^n = \dfrac{b^n}{c^n}$ — Raising a quotient to a power
- $\left(b^m\right)^n = b^{mn}$ — Raising a power to a power

Example 4 Simplifying Expressions Involving Exponents

Simplify.

1. $2^{-1003}2^{1000}$ **2.** $\dfrac{b^{-6}}{b^{-4}}$ **3.** $\dfrac{35b^{-9}c^3}{25b^{-7}c^{-5}}$

Solution

1.

$$\begin{aligned} 2^{-1003}2^{1000} &= 2^{-1003+1000} && \text{Add exponents: } b^m b^n = b^{m+n} \\ &= 2^{-3} && \text{Simplify.} \\ &= \frac{1}{2^3} && \text{Write powers so exponents are positive: } b^{-n} = \frac{1}{b^n} \\ &= \frac{1}{8} && \text{Simplify.} \end{aligned}$$

2.

$$\begin{aligned} \frac{b^{-6}}{b^{-4}} &= b^{-6-(-4)} && \text{Subtract exponents: } \frac{b^m}{b^n} = b^{m-n} \\ &= b^{-6+4} && a - b = a + (-b) \\ &= b^{-2} && \text{Simplify.} \\ &= \frac{1}{b^2} && \text{Write powers so exponents are positive: } b^{-n} = \frac{1}{b^n} \end{aligned}$$

3.

$$\begin{aligned} \frac{35b^{-9}c^3}{25b^{-7}c^{-5}} &= \frac{7b^{-9-(-7)}c^{3-(-5)}}{5} && \text{Subtract exponents: } \frac{b^m}{b^n} = b^{m-n} \\ &= \frac{7b^{-2}c^8}{5} && \text{Simplify.} \\ &= \frac{7c^8}{5b^2} && \text{Write powers so exponents are positive: } b^{-n} = \frac{1}{b^n} \end{aligned}$$

■

In the first step of Problem 2 of Example 4, we found that

$$\frac{b^{-6}}{b^{-4}} = b^{-6-(-4)}$$

WARNING We need a subtraction symbol *and* a negative symbol in the expression on the right-hand side. It is a common error to omit writing one of these two symbols in such problems.

Example 5 Simplifying Expressions Involving Exponents

Simplify.

1. $\dfrac{(3bc^5)^2}{(2b^{-2}c^2)^3}$

2. $\left(\dfrac{18b^{-4}c^7}{6b^{-3}c^2}\right)^{-4}$

Solution

1.
$$\frac{(3bc^5)^2}{(2b^{-2}c^2)^3} = \frac{3^2b^2(c^5)^2}{2^3(b^{-2})^3(c^2)^3} \qquad \text{Raise factors to a power: } (bc)^n = b^nc^n$$
$$= \frac{9b^2c^{10}}{8b^{-6}c^6} \qquad \text{Multiply exponents: } (b^m)^n = b^{mn}$$
$$= \frac{9b^{2-(-6)}c^{10-6}}{8} \qquad \text{Subtract exponents: } \frac{b^m}{b^n} = b^{m-n}$$
$$= \frac{9b^8c^4}{8} \qquad \text{Simplify.}$$

2.
$$\left(\frac{18b^{-4}c^7}{6b^{-3}c^2}\right)^{-4} = \left(3b^{-4-(-3)}c^{7-2}\right)^{-4} \qquad \text{Subtract exponents: } \frac{b^m}{b^n} = b^{m-n}$$
$$= \left(3b^{-1}c^5\right)^{-4} \qquad \text{Simplify.}$$
$$= 3^{-4}\left(b^{-1}\right)^{-4}\left(c^5\right)^{-4} \qquad \text{Raise factors to } n\text{th power: } (bc)^n = b^nc^n$$
$$= 3^{-4}b^4c^{-20} \qquad \text{Multiply exponents: } (b^m)^n = b^{mn}$$
$$= \frac{b^4}{3^4c^{20}} \qquad \text{Write powers so exponents are positive: } b^{-n} = \frac{1}{b^n}$$
$$= \frac{b^4}{81c^{20}} \qquad \text{Simplify.}$$

■

Definition of an Exponential Function

In this chapter and Chapter 5, we will work with *exponential functions*. Here are some examples of such functions:

$$f(x) = 2(3)^x, \qquad g(x) = -7\left(\frac{1}{2}\right)^x, \qquad h(x) = 5^x$$

Notice that, in exponential functions, the variable appears as an exponent.

> **DEFINITION** Exponential function
>
> An **exponential function** is a function whose equation can be put into the form
>
> $$f(x) = ab^x$$
>
> where $a \neq 0$, $b > 0$, and $b \neq 1$. The constant b is called the **base.**

Example 6 Evaluating Exponential Functions

For $f(x) = 3(2)^x$ and $g(x) = 5^x$, find the following.

1. $f(3)$ **2.** $f(-4)$ **3.** $g(a+3)$ **4.** $g(2a)$

Solution

1. $f(3) = 3(2)^3 = 3 \cdot 8 = 24$

2. $f(-4) = 3(2)^{-4} = \dfrac{3}{2^4} = \dfrac{3}{16}$

3. $g(a+3) = 5^{a+3}$ — Substitute $a+3$ for x in 5^x.

$= 5^a \cdot 5^3$ — Write as product: $b^{m+n} = b^m b^n$

$= 125(5)^a$ — $5^3 = 125$; rearrange factors: $ab = ba$

4. $g(2a) = 5^{2a}$ — Substitute $2a$ for x in 5^x.

$= (5^2)^a$ — $b^{mn} = (b^m)^n$

$= 25^a$ — $5^2 = 25$ ■

WARNING It is a common error to confuse exponential functions such as $E(x) = 2^x$ with linear functions such as $L(x) = 2x$. For the *exponential* function $E(x) = 2^x$, the variable x is an *exponent*. For the *linear* function $L(x) = 2x^1$, the variable x is a *base*.

Scientific Notation

Now we will discuss how to use exponents to describe numbers in *scientific notation*. This will enable us to describe compactly a number whose absolute value is very large or very small. For example, Earth is approximately 4,500,000,000 years old. We write 4,500,000,000 in scientific notation:

$$4.5 \times 10^9$$

The symbol "×" stands for multiplication.

As another example, light can travel 1 mile in 0.00000537 second. We write 0.00000537 in scientific notation:

$$5.37 \times 10^{-6}$$

DEFINITION Scientific notation

A number is written in **scientific notation** if it has the form $N \times 10^k$, where k is an integer and either $-10 < N \le -1$ or $1 \le N < 10$.

Here are more examples of numbers in scientific notation:

5.2×10^{17} $\quad 3.638 \times 10^9$ $\quad -5.86 \times 10^{-12}$ $\quad 2.13 \times 10^{-84}$

In Example 7, we will convert some numbers from scientific notation to standard decimal notation.

Example 7 Converting to Standard Decimal Notation

Simplify.

1. 5×10^3 **2.** 5×10^{-3}

Solution

1. $5 \times 10^3 = 5 \times 1000 = 5000$

We simplify $5 \times 10^3 = 5.0 \times 10^3$ by *multiplying* 5.0 by 10 three times, hence moving the decimal point three places to the *right:*

$$5.0 \times 10^3 = 5000.0 = 5000$$

three places to the right

2. $5 \times 10^{-3} = 5 \times \frac{1}{10^3}$ — Write powers so exponents are positive: $b^{-n} = \frac{1}{b^n}$

$= \frac{5}{1} \times \frac{1}{1000}$ — $a = \frac{a}{1}$; simplify.

$= \frac{5}{1000}$ — Multiply.

$= 0.005$ — $\frac{5}{1000}$ is 5 thousandths.

We simplify 5.0×10^{-3} by *dividing* 5.0 by 10 three times, hence moving the decimal point three places to the *left:*

$$5.0 \times 10^{-3} = 0.005$$

three places to the left

The problems in Example 7 suggest the way to convert a number from scientific notation $N \times 10^k$ to standard decimal notation.

Converting from Scientific Notation to Standard Decimal Notation

To write the scientific notation $N \times 10^k$ in standard decimal notation, we move the decimal point of the number N as follows:

- If k is *positive,* we multiply N by 10 k times; hence, we move the decimal point k places to the *right.*
- If k is *negative,* we divide N by 10 k times; hence, we move the decimal k places to the *left.*

Example 8 Converting to Standard Decimal Notation

Write the number in standard decimal notation.

1. 3.462×10^5 **2.** 7.38×10^{-4}

Solution

1. We *multiply* 3.462 by 10 five times; hence, we move the decimal point of 3.462 five places to the *right:*

$$3.462 \times 10^5 = 346{,}200.0$$

five places to the right

2. We *divide* 7.38 by 10 four times; hence, we move the decimal point of 7.38 four places to the *left:*

$$7.38 \times 10^{-4} = 0.000738$$

four places to the left

In Examples 7 and 8, we converted numbers from scientific notation to standard decimal notation. In Example 9, we will investigate the way to convert numbers from standard decimal notation to scientific notation.

Example 9 Converting to Scientific Notation

Write the number in scientific notation.

1. 6,257,000,000 **2.** 0.00000721

Solution

1. In scientific notation, we would have

$$6.257 \times 10^k$$

We must move the decimal point of 6.257 nine places to the right to get 6,257,000,000. So, $k = 9$ and the scientific notation is

$$6.257 \times 10^9$$

2. In scientific notation, we would have

$$7.21 \times 10^k$$

We must move the decimal point of 7.21 six places to the left to get 0.00000721. So, $k = -6$ and the scientific notation is

$$7.21 \times 10^{-6}$$

■

The problems in Example 9 suggest the way to convert a number from standard decimal notation to scientific notation.

Converting from Standard Decimal Notation to Scientific Notation

To write a number in scientific notation, count the number of places k that the decimal point must be moved so that the new number N meets the condition $-10 < N \leq -1$ or $1 \leq N < 10$:

- If the decimal point is moved to the left, then the scientific notation is written as $N \times 10^k$.
- If the decimal point is moved to the right, then the scientific notation is written as $N \times 10^{-k}$.

Example 10 Converting to Scientific Notation

Write the number in scientific notation.

1. 92,900,000 (the average distance in miles between Earth and the Sun)
2. 0.0024 (the average weight, in grams, of a grain of sand)

Solution

1. For 92,900,000, the decimal point must be moved seven places to the left so that the new number is between 1 and 10. Therefore, the scientific notation is 9.29×10^7.
2. For 0.0024, the decimal point must be moved three places to the right so that the new number is between 1 and 10. Therefore, the scientific notation is 2.4×10^{-3}.

■

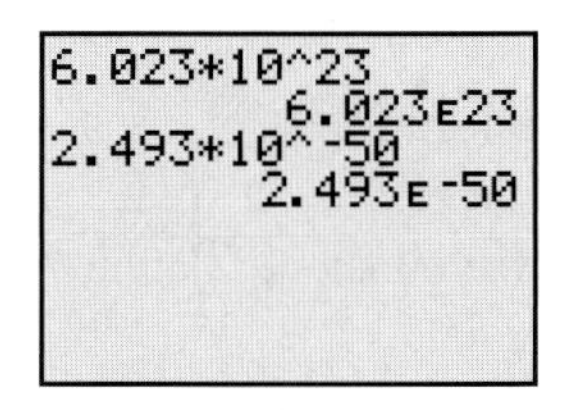

Figure 3 The numbers 6.023×10^{23} and 2.493×10^{-50}

Calculators express numbers in scientific notation so that the numbers "fit" on the screen. To represent 6.023×10^{23}, most calculators use the notation 6.023 E 23, where E stands for exponent (of 10). Calculators represent 2.493×10^{-50} by 2.493 E −50 (see Fig. 3).

group exploration

Properties of exponents

1. In Example 1, we showed that the statement

$$b^2b^3 = b^5$$

makes sense by first writing the expression b^2b^3 without exponents. For each part, show that the given statement makes sense by first writing an expression without exponents.

a. $(bc)^4 = b^4c^4$

b. $\frac{b^7}{b^3} = b^4, \quad b \neq 0$

c. $(b^3)^4 = b^{12}$

2. In Example 1, we also showed that the general statements

$$b^mb^n = b^{m+n} \quad \text{and} \quad \left(\frac{b}{c}\right)^n = \frac{b^n}{c^n}, \quad c \neq 0$$

make sense for counting numbers m and n. For each part, show that the general statement makes sense. Assume that m and n are counting numbers.

a. $(bc)^n = b^n c^n$

b. $\dfrac{b^m}{b^n} = b^{m-n}$, where $m > n$ and $b \neq 0$

c. $\left(b^m\right)^n = b^{mn}$

3. Choose values of b, c, and counting number n to show that the statement $(b+c)^n = b^n + c^n$ is false, in general.

group exploration

Looking ahead: Definition of $b^{1/n}$

Throughout this exploration, assume that $\left(b^m\right)^n = b^{mn}$ for rational numbers m and n.

1. First, you will explore the meaning of $b^{1/2}$, where b is nonnegative.

a. For now, do not use a calculator. You will explore how you should define $9^{1/2}$. You can determine a reasonable value of $9^{1/2}$ by first finding the *square* of the value:

$$\left(9^{1/2}\right)^2 = 9^{\frac{1}{2}\cdot 2} = 9^1 = 9$$

What would be a good meaning of $9^{1/2}$? [**Hint:** Can you think of a positive number whose square equals 9?]

b. What would be a good meaning of $16^{1/2}$? Of $25^{1/2}$?

c. Now use a graphing calculator to find $9^{1/2}$, $16^{1/2}$, and $25^{1/2}$. For example, to find $9^{1/2}$, press **9** [∧] [(] **1** [÷] **2** [)] [ENTER]. Is the calculator interpreting $b^{1/2}$ as you would expect?

d. What would be a good meaning of $b^{1/2}$, where b is nonnegative?

2. Now you will explore the meaning of $b^{1/3}$.

a. For now, do not use a calculator. You will explore how you should define $8^{1/3}$. You can first find the *cube* of the value:

$$\left(8^{1/3}\right)^3 = 8^{\frac{1}{3}\cdot 3} = 8^1 = 8$$

What would be a good meaning of $8^{1/3}$? Explain.

b. What would be a good meaning of $27^{1/3}$? Of $64^{1/3}$?

c. Use a graphing calculator to find $8^{1/3}$, $27^{1/3}$, and $64^{1/3}$. Is the calculator interpreting $b^{1/3}$ as you would expect?

d. What would be a good meaning of $b^{1/3}$?

3. What would be a good meaning of $b^{1/n}$, where n is a counting number and b is nonnegative?

TIPS FOR SUCCESS: Make Changes

If you have not had passing scores on tests and quizzes during the first part of this course, it is time to determine what the problem is, what changes you should make, and whether you can commit to making those changes.

Sometimes students must change the way they study. For example, Rosie did poorly on exams and quizzes for the first third of the course. It was not clear why she was not passing, since she had good attendance, was actively involved in classroom work, and was doing the homework assignments. Suddenly Rosie started getting A's on every quiz and test. What had happened? Rosie said, "I figured out that, to do well, it was not enough practice to do just the exercises you assigned. Now I do a lot of extra exercises from each section."

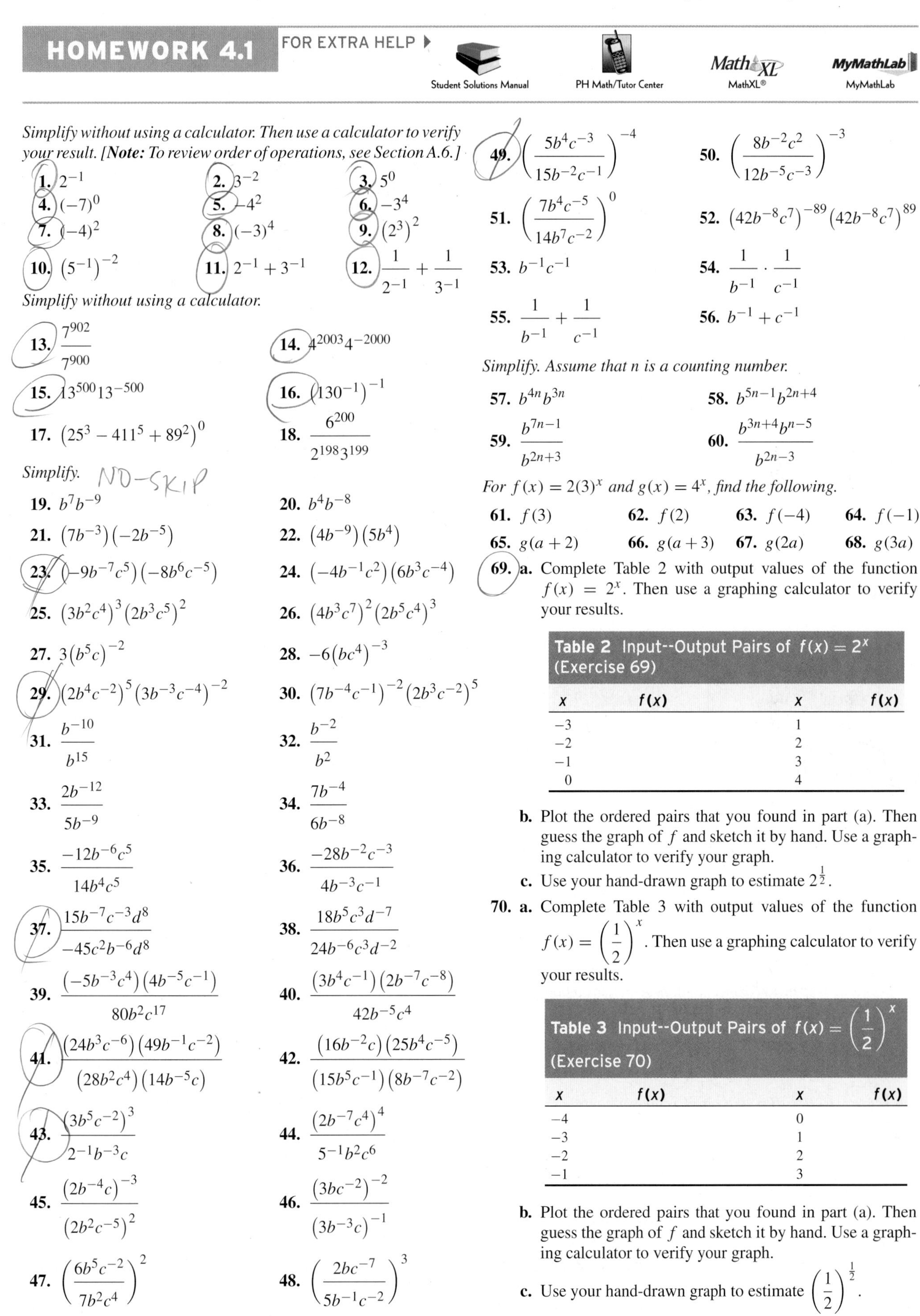

HOMEWORK 4.1

FOR EXTRA HELP ▸ Student Solutions Manual | PH Math/Tutor Center | MathXL® | MyMathLab

*Simplify without using a calculator. Then use a calculator to verify your result. [**Note:** To review order of operations, see Section A.6.]*

1. 2^{-1}
2. 3^{-2}
3. 5^{0}
4. $(-7)^{0}$
5. -4^{2}
6. -3^{4}
7. $(-4)^{2}$
8. $(-3)^{4}$
9. $(2^{3})^{2}$
10. $(5^{-1})^{-2}$
11. $2^{-1}+3^{-1}$
12. $\dfrac{1}{2^{-1}}+\dfrac{1}{3^{-1}}$

Simplify without using a calculator.

13. $\dfrac{7^{902}}{7^{900}}$
14. $4^{2003}4^{-2000}$
15. $13^{500}13^{-500}$
16. $(130^{-1})^{-1}$
17. $(25^{3}-411^{5}+89^{2})^{0}$
18. $\dfrac{6^{200}}{2^{198}3^{199}}$

Simplify.

19. $b^{7}b^{-9}$
20. $b^{4}b^{-8}$
21. $(7b^{-3})(-2b^{-5})$
22. $(4b^{-9})(5b^{4})$
23. $(-9b^{-7}c^{5})(-8b^{6}c^{-5})$
24. $(-4b^{-1}c^{2})(6b^{3}c^{-4})$
25. $(3b^{2}c^{4})^{3}(2b^{3}c^{5})^{2}$
26. $(4b^{3}c^{7})^{2}(2b^{5}c^{4})^{3}$
27. $3(b^{5}c)^{-2}$
28. $-6(bc^{4})^{-3}$
29. $(2b^{4}c^{-2})^{5}(3b^{-3}c^{-4})^{-2}$
30. $(7b^{-4}c^{-1})^{-2}(2b^{3}c^{-2})^{5}$
31. $\dfrac{b^{-10}}{b^{15}}$
32. $\dfrac{b^{-2}}{b^{2}}$
33. $\dfrac{2b^{-12}}{5b^{-9}}$
34. $\dfrac{7b^{-4}}{6b^{-8}}$
35. $\dfrac{-12b^{-6}c^{5}}{14b^{4}c^{5}}$
36. $\dfrac{-28b^{-2}c^{-3}}{4b^{-3}c^{-1}}$
37. $\dfrac{15b^{-7}c^{-3}d^{8}}{-45c^{2}b^{-6}d^{8}}$
38. $\dfrac{18b^{5}c^{3}d^{-7}}{24b^{-6}c^{3}d^{-2}}$
39. $\dfrac{(-5b^{-3}c^{4})(4b^{-5}c^{-1})}{80b^{2}c^{17}}$
40. $\dfrac{(3b^{4}c^{-1})(2b^{-7}c^{-8})}{42b^{-5}c^{4}}$
41. $\dfrac{(24b^{3}c^{-6})(49b^{-1}c^{-2})}{(28b^{2}c^{4})(14b^{-5}c)}$
42. $\dfrac{(16b^{-2}c)(25b^{4}c^{-5})}{(15b^{5}c^{-1})(8b^{-7}c^{-2})}$
43. $\dfrac{(3b^{5}c^{-2})^{3}}{2^{-1}b^{-3}c}$
44. $\dfrac{(2b^{-7}c^{4})^{4}}{5^{-1}b^{2}c^{6}}$
45. $\dfrac{(2b^{-4}c)^{-3}}{(2b^{2}c^{-5})^{2}}$
46. $\dfrac{(3bc^{-2})^{-2}}{(3b^{-3}c)^{-1}}$
47. $\left(\dfrac{6b^{5}c^{-2}}{7b^{2}c^{4}}\right)^{2}$
48. $\left(\dfrac{2bc^{-7}}{5b^{-1}c^{-2}}\right)^{3}$
49. $\left(\dfrac{5b^{4}c^{-3}}{15b^{-2}c^{-1}}\right)^{-4}$
50. $\left(\dfrac{8b^{-2}c^{2}}{12b^{-5}c^{-3}}\right)^{-3}$
51. $\left(\dfrac{7b^{4}c^{-5}}{14b^{7}c^{-2}}\right)^{0}$
52. $(42b^{-8}c^{7})^{-89}(42b^{-8}c^{7})^{89}$
53. $b^{-1}c^{-1}$
54. $\dfrac{1}{b^{-1}}\cdot\dfrac{1}{c^{-1}}$
55. $\dfrac{1}{b^{-1}}+\dfrac{1}{c^{-1}}$
56. $b^{-1}+c^{-1}$

Simplify. Assume that n is a counting number.

57. $b^{4n}b^{3n}$
58. $b^{5n-1}b^{2n+4}$
59. $\dfrac{b^{7n-1}}{b^{2n+3}}$
60. $\dfrac{b^{3n+4}b^{n-5}}{b^{2n-3}}$

For $f(x)=2(3)^{x}$ and $g(x)=4^{x}$, find the following.

61. $f(3)$
62. $f(2)$
63. $f(-4)$
64. $f(-1)$
65. $g(a+2)$
66. $g(a+3)$
67. $g(2a)$
68. $g(3a)$
69. **a.** Complete Table 2 with output values of the function $f(x)=2^{x}$. Then use a graphing calculator to verify your results.

Table 2 Input--Output Pairs of $f(x)=2^{x}$ (Exercise 69)

x	$f(x)$	x	$f(x)$
−3		1	
−2		2	
−1		3	
0		4	

b. Plot the ordered pairs that you found in part (a). Then guess the graph of f and sketch it by hand. Use a graphing calculator to verify your graph.

c. Use your hand-drawn graph to estimate $2^{\frac{1}{2}}$.

70. **a.** Complete Table 3 with output values of the function $f(x)=\left(\dfrac{1}{2}\right)^{x}$. Then use a graphing calculator to verify your results.

Table 3 Input--Output Pairs of $f(x)=\left(\dfrac{1}{2}\right)^{x}$ (Exercise 70)

x	$f(x)$	x	$f(x)$
−4		0	
−3		1	
−2		2	
−1		3	

b. Plot the ordered pairs that you found in part (a). Then guess the graph of f and sketch it by hand. Use a graphing calculator to verify your graph.

c. Use your hand-drawn graph to estimate $\left(\dfrac{1}{2}\right)^{\frac{1}{2}}$.

Write the number in standard decimal form.

71. 3.965×10^2

72. 8.23172×10^3

73. 2.39×10^{-1}

74. 7.46×10^{-3}

75. 5.2×10^2

76. 7.74×10^6

77. 9.113×10^{-5}

78. 7.3558×10^{-2}

79. -6.52×10^{-4}

80. -3.006×10^{-3}

81. 9×10^5

82. 4×10^3

83. -8×10^0

84. -6.1×10^0

Write the number in scientific notation.

85. 54,260,000

86. 173,229

87. 23,587

88. 6,541,883

89. 0.00098

90. 0.08156

91. 0.0000346

92. 0.00000387

93. −42,215

94. −647,000

95. −0.00244

96. −0.000013

For Exercises 97 and 98, numbers are displayed in a graphing calculator table's version of scientific notation. Write each number in the Y_1 column in standard decimal form.

97. See Fig. 4.

98. See. Fig. 5.

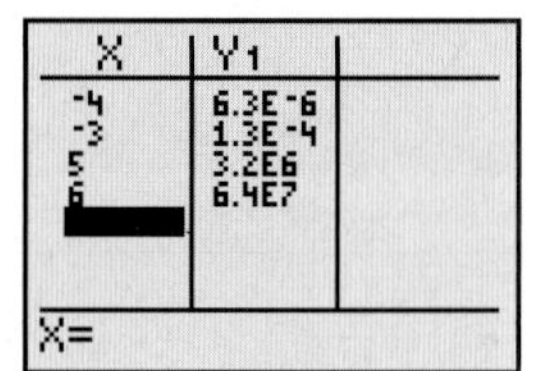

Figure 4 Exercise 97

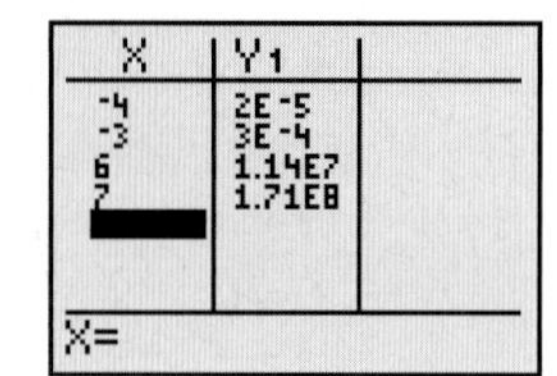

Figure 5 Exercise 98

For Exercises 99–102, the given sentence contains a number written in scientific notation. Write the number in standard decimal form.

99. The first evidence of life on Earth dates back to 3.6×10^9 years ago.

100. The Moon has an average distance from Earth of approximately 2.389×10^5 miles.

101. The hydrogen ion concentration in human blood is about 6.3×10^{-8} mole per liter.

102. The faintest sound that humans can hear has an intensity of about 10^{-12} watt per square meter.

For Exercises 103–106, the given sentence contains a number (other than a date) written in standard decimal form. Write the number in scientific notation.

103. The tanker *Exxon Valdez* spilled about 10,080,000 gallons of oil in Prince William Sound, Alaska, in 1989.

104. The average distance from Earth to Alpha Centauri is about 25,000,000,000,000 miles.

105. The wavelength of violet light is about 0.00000047 meter.

106. One second is about 0.0000000317 year.

107. The numbers of bald eagle pairs in the continental United States are shown in Table 4 for various years.

Table 4 Numbers of Bald Eagle Pairs

Year	Number of Bald Eagle Pairs (thousands)
1963	0.4
1974	0.8
1981	1.2
1986	1.9
1990	3.0
1995	4.7
2000	6.5
2005	7.7

Source: *U.S. Fish and Wildlife Service*

Let n be the number of bald eagle pairs (in thousands) in the continental United States at t years since 1960. The situation can be described by the linear function $n = 0.18t - 1.63$ and the exponential function $n = 0.29(1.078)^t$.

a. Use a graphing calculator to draw the graphs of the two functions and, in the same viewing window, the scattergram of the data. Which function describes the situation better?

b. Use the exponential model to predict the number of bald eagle pairs in 2012.

c. Use the linear model to predict the number of bald eagle pairs in 2012. Explain why your result is so much smaller than your result from part (b). [**Hint:** Zoom Out at least once.]

108. The sales of digital TV sets and displays are shown in Table 5 for various years.

Table 5 Sales of Digital TV Sets and Displays

Year	Number of Digital TV Sets and Displays Sold (millions)
2000	0.6
2001	1.4
2002	2.5
2003	4.0
2004	7.5
2005	15.0

Source: *Consumer Electronics Association*

Let n be the sales of digital TVs and displays (in millions) in the year that is t years since 2000. The situation can be described by the linear function $n = 2.62t - 1.39$ and the exponential function $n = 0.67(1.85)^t$.

a. Use a graphing calculator to draw the graphs of the two functions and, in the same viewing window, the scattergram of the data. Which function describes the situation better?

b. Use the exponential model to predict the sales of digital TVs and displays in 2011.

c. Use the linear model to predict the sales of digital TVs and displays in 2011. Explain why your result is so much smaller than your result from part (b). [**Hint:** Zoom Out at least once.]

109. Two students try to simplify $(5b^2)^{-1}$:

Student A

$$(5b^2)^{-1} = -5b^{-2} = \frac{-5}{b^2}$$

Student B

$$(5b^2)^{-1} = 5^{-1}(b^2)^{-1} = 5^{-1}b^{-2} = \frac{1}{5b^2}$$

Did either student simplify the expression correctly? Describe any errors.

110. Two students try to simplify an expression:

Student 1

$$\frac{7b^8}{b^{-3}} = 7b^{8-(-3)} = 7b^{11}$$

Student 2

$$\frac{7b^8}{b^{-3}} = 7b^{8-3} = 7b^5$$

Did either student simplify the expression correctly? Describe any errors.

111. A student tries to simplify $\dfrac{3b^{-2}c^4}{d^7}$:

$$\frac{3b^{-2}c^4}{d^7} = \frac{c^4}{3b^2d^7}$$

Describe any errors. Then simplify the expression correctly.

112. A student tries to simplify $(7x^4)^5$:

$$(7x^4)^5 = 7(x^4)^5 = 7x^{20}$$

Describe any errors. Then simplify the expression correctly.

113. It is common to confuse expressions such as $2^2, 2^{-1}, 2(-1)$, $\left(\frac{1}{2}\right)^2, \left(\frac{1}{2}\right)^{-1}, -2^2, (-2)^2$, and $\frac{1}{2}$. List these numbers from least to greatest. Are there any "ties"?

114. a. Simplify $\left(\dfrac{b}{c}\right)^{-2}$.

b. Simplify $\left(\dfrac{b}{c}\right)^{-n}$.

c. Use your result from part (b) to simplify $\left(\dfrac{b}{c}\right)^{-5}$ in one step.

115. Explore "0^0":

a. Simplify 5^0, 4^0, 3^0, 2^0, and 1^0. On the basis of these values, what would be a reasonable value of 0^0?

b. Simplify 0^5, 0^4, 0^3, 0^2, and 0^1. On the basis of these values, what would be a reasonable value of 0^0?

c. Why is it a good idea to leave 0^0 meaningless?

116. Simplify each expression.

a. b^{-1} **b.** $(b^{-1})^{-1}$

c. $((b^{-1})^{-1})^{-1}$ **d.** $(((b^{-1})^{-1})^{-1})^{-1}$

e. $\underbrace{((((b^{-1})^{-1})^{-1})\cdots)^{-1}}_{n \text{ exponents}}$

117. It is a common error to confuse the properties $b^m b^n = b^{m+n}$ and $(b^m)^n = b^{mn}$. Explain why each property makes sense, and compare the properties. Give examples to illustrate your comparison. (See page 4 for guidelines on writing a good response.)

118. Describe what it means to use exponential properties to simplify an expression. Include several examples in your description. (See page 4 for guidelines on writing a good response.)

Related Review

For $f(x) = 2x$ and $g(x) = 2^x$, find the following.

119. $f(3)$ **120.** $f(-3)$ **121.** $g(3)$ **122.** $g(-3)$

Expressions, Equations, Functions, and Graphs

Perform the indicated instruction. Then use words such as linear, exponential, function, one variable, *and* two variables *to describe the expression, equation, or system. For instance, to describe $2x = 10$, you could say "$2x = 10$ is a linear equation in one variable."*

123. Solve:

$$y = 3x + 1$$
$$y = 2x - 4$$

124. Simplify $5(3x + 1) - 4(2x - 4)$.

125. Solve $3x + 1 = 2x - 4$.

126. Graph $f(x) = 3x + 1$ by hand.

4.2 RATIONAL EXPONENTS

Objectives

- Know definitions of *rational exponents.*
- Simplify expressions that have rational exponents.

In Section 4.1, we worked with integer exponents. In this section, we work with exponents that are rational numbers (Section A.2).

Definitions of Rational Exponents

How should we define $b^{1/n}$, where n is a counting number? If the exponential property $(b^m)^n = b^{mn}$ is to be true for $m = \frac{1}{2}$ and $n = 2$, then

$$\left(9^{\frac{1}{2}}\right)^2 = 9^{\frac{1}{2} \cdot 2} = 9^1 = 9$$

Since $(-3)^2 = 9$ and $3^2 = 9$, the statement suggests that a good meaning of $9^{1/2}$ is -3 or 3. We define $9^{1/2} = 3$. We call the nonnegative number 3 the *principal second root,* or **principal square root,** of 9, written $\sqrt{9}$.

Similarly, if the property $(b^m)^n = b^{mn}$ is to be true for $m = \frac{1}{3}$ and $n = 3$, then

$$\left(8^{\frac{1}{3}}\right)^3 = 8^{\frac{1}{3} \cdot 3} = 8^1 = 8$$

Since $2^3 = 8$, the statement suggests that a good meaning of $8^{1/3}$ is 2. The number 2 is called the *third root,* or **cube root,** of 8, written $\sqrt[3]{8}$.

For $(-8)^{1/3}$, a good meaning is -2, since $(-2)^3 = -8$. We do not assign a real-number value to $(-9)^{1/2}$, since no real number squared is equal to -9.

DEFINITION $b^{1/n}$

For the counting number n, where $n \neq 1$,

- If n is odd, then $\boldsymbol{b^{1/n}}$ is the number whose nth power is b, and we call $b^{1/n}$ the ***n*th root of *b*.**
- If n is even and $b \geq 0$, then $\boldsymbol{b^{1/n}}$ is the nonnegative number whose nth power is b, and we call $b^{1/n}$ the **principal *n*th root of *b*.**
- If n is even and $b < 0$, then $b^{1/n}$ is not a real number.

$\boldsymbol{b^{1/n}}$ may be represented by $\sqrt[n]{b}$.

Example 1 Simplifying Expressions Involving Rational Exponents

Simplify.

1. $25^{1/2}$ **2.** $64^{1/3}$ **3.** $(-64)^{1/3}$

4. $16^{1/4}$ **5.** $-16^{1/4}$ **6.** $(-16)^{1/4}$

Solution

1. $25^{1/2} = 5$, since $5^2 = 25$.

2. $64^{1/3} = 4$, since $4^3 = 64$.

3. $(-64)^{1/3} = -4$, since $(-4)^3 = -64$.

4. $16^{1/4} = 2$, since $2^4 = 16$.

5. $-16^{1/4} = -(16^{1/4}) = -2$.

6. $(-16)^{1/4}$ is not a real number, since the fourth power of any real number is nonnegative.

Graphing calculator checks for Problems 1, 2, and 3 are shown in Fig. 6. For example, to find $25^{1/2}$, press **25** [∧] [(] **1** [÷] **2** [)] [ENTER]. If an exponent involves an operation, you must use parentheses. ■

```
25^(1/2)
                5
64^(1/3)
                4
(-64)^(1/3)
               -4
```

Figure 6 Checks for Problems 1, 2, and 3

What would be a reasonable definition of $b^{m/n}$? If the properties of exponents we discussed in Section 4.1 are to hold true for rational exponents, we have

$$8^{\frac{2}{3}} = 8^{\frac{1}{3} \cdot 2} = \left(8^{\frac{1}{3}}\right)^2 = 2^2 = 4 \quad \text{or} \quad 8^{\frac{2}{3}} = 8^{2 \cdot \frac{1}{3}} = (8^2)^{\frac{1}{3}} = 64^{\frac{1}{3}} = 4$$

Likewise,

$$32^{\frac{3}{5}} = 32^{\frac{1}{5}\cdot 3} = \left(32^{\frac{1}{5}}\right)^3 = 2^3 = 8 \quad \text{or} \quad 32^{\frac{3}{5}} = 32^{3\cdot\frac{1}{5}} = \left(32^3\right)^{\frac{1}{5}} = 32{,}768^{\frac{1}{5}} = 8$$

Also,

$$32^{-\frac{3}{5}} = \frac{1}{32^{\frac{3}{5}}} = \frac{1}{8}$$

DEFINITION Rational exponent

For the counting numbers m and n, where $n \neq 1$ and b is any real number for which $b^{1/n}$ is a real number,

- $b^{m/n} = \left(b^{1/n}\right)^m = \left(b^m\right)^{1/n}$
- $b^{-m/n} = \dfrac{1}{b^{m/n}}, \quad b \neq 0$

A power of the form $b^{m/n}$ or $b^{-m/n}$ is said to have a **rational exponent.**

Example 2 Simplifying Expressions Involving Rational Exponents

Simplify.

1. $25^{3/2}$ **2.** $(-27)^{2/3}$ **3.** $32^{-2/5}$ **4.** $(-8)^{-5/3}$

Solution

1. $25^{3/2} = \left(25^{1/2}\right)^3 = 5^3 = 125$

2. $(-27)^{2/3} = \left((-27)^{1/3}\right)^2 = (-3)^2 = 9$

3. $32^{-2/5} = \dfrac{1}{32^{2/5}} = \dfrac{1}{\left(32^{1/5}\right)^2} = \dfrac{1}{2^2} = \dfrac{1}{4}$

4. $(-8)^{-5/3} = \dfrac{1}{(-8)^{5/3}} = \dfrac{1}{\left((-8)^{1/3}\right)^5} = \dfrac{1}{(-2)^5} = \dfrac{1}{-32} = -\dfrac{1}{32}$

```
25^(3/2)
                125
(-27)^(2/3)
                  9
32^(-2/5)
                .25
```

Figure 7 Checks for Problems 1, 2, and 3

Graphing calculator checks for Problems 1, 2, and 3 are shown in Fig. 7. ■

Example 3 Evaluating an Exponential Function

For $f(x) = 64^x$, $g(x) = 3(16)^x$, and $h(x) = -5(9)^x$, find the following.

1. $f\left(\dfrac{2}{3}\right)$ **2.** $g\left(\dfrac{3}{4}\right)$ **3.** $h\left(-\dfrac{1}{2}\right)$

Solution

1. $f\left(\dfrac{2}{3}\right) = 64^{2/3} = \left(64^{1/3}\right)^2 = 4^2 = 16$

2. $g\left(\dfrac{3}{4}\right) = 3(16)^{3/4} = 3\left(16^{1/4}\right)^3 = 3(2)^3 = 3 \cdot 8 = 24$

3. $h\left(-\dfrac{1}{2}\right) = -5(9)^{-1/2} = \dfrac{-5}{9^{1/2}} = -\dfrac{5}{3}$ ■

Properties of Rational Exponents

The properties of exponents that we discussed in Section 4.1 are valid for *rational* exponents.

Properties of Rational Exponents

If m and n are rational numbers and b and c are any real numbers for which b^m, b^n, and c^n are real numbers, then

- $b^m b^n = b^{m+n}$ — Product property for exponents
- $\dfrac{b^m}{b^n} = b^{m-n}, b \neq 0$ — Quotient property for exponents
- $(bc)^n = b^n c^n$ — Raising a product to a power
- $\left(\dfrac{b}{c}\right)^n = \dfrac{b^n}{c^n}, \quad c \neq 0$ — Raising a quotient to a power
- $\left(b^m\right)^n = b^{mn}$ — Raising a power to a power

We can use properties of exponents to help us simplify expressions involving rational exponents.

Example 4 Simplifying Expressions Involving Rational Exponents

Simplify. Assume that b is positive.

1. $\left(4b^6\right)^{3/2}$ **2.** $\dfrac{b^{2/7}}{b^{-3/7}}$

Solution

1.
$$\begin{aligned}
\left(4b^6\right)^{3/2} &= 4^{3/2}\left(b^6\right)^{3/2} && \text{Raise factors to } n\text{th power: } (bc)^n = b^n c^n \\
&= \left(4^{1/2}\right)^3 b^{\frac{6}{1}\cdot\frac{3}{2}} && b^{m/n} = \left(b^{1/n}\right)^m\text{; multiply exponents: } (b^m)^n = b^{mn} \\
&= 2^3 b^9 && 4^{\frac{1}{2}} = 2\text{; multiply.} \\
&= 8b^9 && \text{Simplify.}
\end{aligned}$$

2.
$$\begin{aligned}
\frac{b^{2/7}}{b^{-3/7}} &= b^{\frac{2}{7}-\left(-\frac{3}{7}\right)} && \text{Subtract exponents: } \frac{b^m}{b^n} = b^{m-n} \\
&= b^{\frac{2}{7}+\frac{3}{7}} && \text{Simplify.} \\
&= b^{5/7} && \text{Add.}
\end{aligned}$$

■

Example 5 Simplifying Expressions Involving Rational Exponents

Simplify. Assume that b is positive.

1. $b^{2/3}b^{1/2}$ **2.** $\left(\dfrac{32b^2}{b^{12}}\right)^{2/5}$

Solution

1.
$$\begin{aligned}
b^{2/3}b^{1/2} &= b^{\frac{2}{3}+\frac{1}{2}} && \text{Add exponents: } b^m b^n = b^{m+n} \\
&= b^{\frac{4}{6}+\frac{3}{6}} && \text{Find common denominator.} \\
&= b^{7/6} && \text{Add.}
\end{aligned}$$

2.
$$\left(\frac{32b^2}{b^{12}}\right)^{2/5} = \left(32b^{2-12}\right)^{2/5}$$ Subtract exponents: $\frac{b^m}{b^n} = b^{m-n}$

$$= \left(32b^{-10}\right)^{2/5}$$ Subtract.

$$= \left(\frac{32}{b^{10}}\right)^{2/5}$$ Write powers so exponents are positive: $b^{-n} = \frac{1}{b^n}$

$$= \frac{32^{2/5}}{\left(b^{10}\right)^{2/5}}$$ Raise numerator and denominator to nth power: $\left(\frac{b}{c}\right)^n = \frac{b^n}{c^n}$

$$= \frac{\left(32^{1/5}\right)^2}{b^{10\cdot\frac{2}{5}}}$$ $b^{m/n} = \left(b^{1/n}\right)^m$; multiply exponents: $\left(b^m\right)^n = b^{mn}$

$$= \frac{2^2}{b^4}$$ $32^{1/5} = 2$; multiply.

$$= \frac{4}{b^4}$$ Simplify. ■

Example 6 Simplifying an Expression Involving Rational Exponents

Simplify $\frac{\left(81b^6c^{20}\right)^{1/2}}{\left(27b^{12}c^9\right)^{2/3}}$. Assume that b and c are positive.

Solution

$$\frac{\left(81b^6c^{20}\right)^{1/2}}{\left(27b^{12}c^9\right)^{2/3}} = \frac{81^{1/2}\left(b^6\right)^{1/2}\left(c^{20}\right)^{1/2}}{27^{2/3}\left(b^{12}\right)^{2/3}\left(c^9\right)^{2/3}}$$ Raise factors to a power: $(bc)^n = b^nc^n$

$$= \frac{9b^{6\cdot\frac{1}{2}}c^{20\cdot\frac{1}{2}}}{\left(27^{1/3}\right)^2b^{12\cdot\frac{2}{3}}c^{9\cdot\frac{2}{3}}}$$ $81^{1/2} = 9$; $b^{m/n} = \left(b^{1/n}\right)^m$; multiply exponents: $\left(b^m\right)^n = b^{mn}$

$$= \frac{9b^3c^{10}}{3^2b^8c^6}$$ $27^{1/3} = 3$; multiply.

$$= \frac{9b^{-5}c^4}{9}$$ Subtract exponents: $\frac{b^m}{b^n} = b^{m-n}$

$$= \frac{c^4}{b^5}$$ Write powers so exponents are positive: $b^{-n} = \frac{1}{b^n}$ ■

group exploration

Looking ahead: Graphical significance of *a* and *b* for $y = ab^x$

1. Use ZDecimal to graph these equations of the form $y = b^x$ in order, and describe what you observe:

$$y = 1.2^x, \quad y = 1.5^x, \quad y = 2^x, \quad \text{and} \quad y = 5^x$$

If you want a better view, set Ymin = 0. To change window settings, see Section B.7.

Do the same with the equations

$$y = 0.3^x, \quad y = 0.5^x, \quad y = 0.7^x, \quad \text{and} \quad y = 0.9^x$$

2. Use ZStandard to graph these equations of the form $y = a(1.1)^x$ in order, and describe what you observe:

$$y = 2(1.1)^x, \quad y = 3(1.1)^x, \quad y = 4(1.1)^x, \quad \text{and} \quad y = 5(1.1)^x$$

If you want a better view, set Ymin = 0.

Use ZStandard to do the same with the equations

$$y = -2(1.1)^x, \quad y = -3(1.1)^x, \quad y = -4(1.1)^x, \quad \text{and} \quad y = -5(1.1)^x$$

If you want a better view, set Ymax = 0.

3. So far, you have sketched the graphs of equations of only the forms $y = b^x$ (where $a = 1$) and $y = a(1.1)^x$ (where $b = 1.1$). Graph more equations of the form $y = ab^x$, until you are confident that you know the graphical significance of the constants a and b, for any possible combination of values of a and b. If you have any new insights into the graphical significance of a and b, describe those insights.

4. Describe the graph of $y = ab^x$ in the following situations.

a. a is positive
b. a is negative
c. $b > 1$
d. $0 < b < 1$
e. $b = 1$
f. b is negative

5. Describe the connection between the y-intercept of $y = ab^x$ and the values of a and b.

group exploration

Looking ahead: Numerical significance of *a* and *b* for $f(x) = ab^x$

In this exploration, you will investigate the nature of exponential functions of the form $f(x) = ab^x$.

1. Use a graphing calculator to create a table of ordered pairs for $f(x) = 2(3)^x$, $g(x) = 64\left(\frac{1}{2}\right)^x$, and a third exponential function of your choice. (See Figs. 8 and 9.) Use the following values for the x-coordinates: 0, 1, 2, ..., 6.

Figure 8 Enter the three functions

Figure 9 Table setup

2. a. What connection do you notice between the y-coordinates of each function and the base b of the function $y = ab^x$?

b. Test the connection you described in part (a) by choosing yet another exponential function, and check whether it behaves as you think it should.

c. For $f(x) = ab^x$, we have $f(0) = a$, $f(1) = ab$, $f(2) = abb$, and $f(3) = abbb$. Explain why these results suggest that your response to part (a) is correct.

3. **a.** What connection do you notice between the y-coordinates of each function and the coefficient a of the function $y = ab^x$?
   b. Test the connection you described in part (a) by choosing yet another exponential function, and check whether it behaves as you think it should.
   c. Use pencil and paper to find $f(0)$, where $f(x) = ab^x$. Explain why your result shows that your response to part (a) is correct.

TIPS FOR SUCCESS: Complete Exercises Without Help

If you work an exercise by referring to a similar example in your notebook or in the text, try the exercise again without that help. If you need to refer to your source of help to solve the exercise a second time, try the exercise a third time without help. When you complete the exercise without help, reflect on which concepts you used to work the exercise, where you had difficulty, and what key idea opened the door of understanding for you. You can use a similar strategy in getting help from another student, an instructor, or a tutor.

If this sounds like a lot of work, it is! But this work is well worth it. Although it is important to complete each assignment, it is also important to learn as much as possible while progressing through it.

HOMEWORK 4.2

FOR EXTRA HELP ▶ Student Solutions Manual | PH Math/Tutor Center | MathXL® | MyMathLab

*Simplify without using a calculator. Then use a graphing calculator to verify your result. [**Graphing Calculator:** Instructions for $x^{m/n}$: Press* X, T, Θ, n ^ (**m** ÷ **n**) *.]*

1. $16^{1/2}$ **2.** $27^{1/3}$ **3.** $1000^{1/3}$
4. $32^{1/5}$ **5.** $49^{1/2}$ **6.** $81^{1/4}$
7. $125^{1/3}$ **8.** $64^{1/6}$ **9.** $8^{4/3}$
10. $16^{3/4}$ **11.** $9^{3/2}$ **12.** $64^{2/3}$
13. $32^{2/5}$ **14.** $27^{4/3}$ **15.** $4^{5/2}$
16. $81^{3/4}$ **17.** $27^{-1/3}$ **18.** $16^{-1/4}$
19. $-36^{-1/2}$ **20.** $-32^{-1/5}$ **21.** $4^{-5/2}$
22. $9^{-3/2}$ **23.** $(-27)^{-4/3}$ **24.** $(-32)^{-3/5}$

Simplify without using a calculator. Then use a graphing calculator to verify your result.

25. $2^{1/4}2^{3/4}$ **26.** $3^{7/5}3^{3/5}$ **27.** $(3^{1/2}2^{3/2})^2$
28. $(2^{2/3}5^{1/3})^3$ **29.** $\dfrac{7^{1/3}}{7^{-5/3}}$ **30.** $\dfrac{5^{4/3}}{5^{1/3}}$

For $f(x) = 81^x$, $g(x) = 4(27)^x$, and $h(x) = -2(4)^x$, find the following.

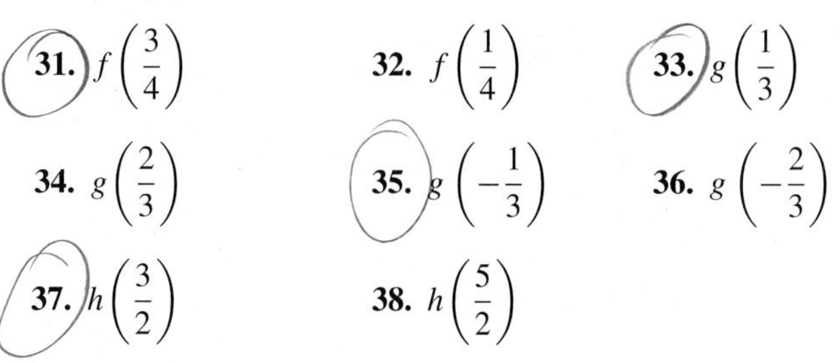

31. $f\left(\frac{3}{4}\right)$ **32.** $f\left(\frac{1}{4}\right)$ **33.** $g\left(\frac{1}{3}\right)$
34. $g\left(\frac{2}{3}\right)$ **35.** $g\left(-\frac{1}{3}\right)$ **36.** $g\left(-\frac{2}{3}\right)$
37. $h\left(\frac{3}{2}\right)$ **38.** $h\left(\frac{5}{2}\right)$

39. Without using a calculator, complete Table 6 with values of the function $f(x) = 16^x$. Then use a graphing calculator to verify your results.

Table 6 Values of the Function $f(x) = 16^x$

x	$f(x)$	x	$f(x)$
$-\frac{3}{4}$		$\frac{1}{4}$	
$-\frac{1}{2}$		$\frac{1}{2}$	
$-\frac{1}{4}$		$\frac{3}{4}$	
0		1	

40. Without using a calculator, complete Table 7 with values of the function $f(x) = 64^x$. Then use a graphing calculator to verify your results.

Table 7 Values of the Function $f(x) = 64^x$

x	$f(x)$	x	$f(x)$
$-\frac{5}{6}$		$\frac{1}{6}$	
$-\frac{2}{3}$		$\frac{1}{3}$	
$-\frac{1}{2}$		$\frac{1}{2}$	
$-\frac{1}{3}$		$\frac{2}{3}$	
$-\frac{1}{6}$		$\frac{5}{6}$	
0		1	

SKIP

Simplify. Assume that b and c are positive.

41. $b^{7/6}b^{5/6}$

42. $b^{1/5}b^{3/5}$

43. $b^{3/5}b^{-13/5}$

44. $b^{2/7}b^{-6/7}$

45. $\left(16b^{8}\right)^{1/4}$

46. $\left(27b^{27}\right)^{1/3}$

47. $4\left(25b^{8}c^{14}\right)^{-1/2}$

48. $-\left(8b^{-6}c^{12}\right)^{2/3}$

49. $\left(b^{3/5}c^{-1/4}\right)\left(b^{2/5}c^{-7/4}\right)$

50. $\left(b^{-4/3}c^{1/2}\right)\left(b^{-2/3}c^{-3/2}\right)$

51. $(5bcd)^{1/5}(5bcd)^{4/5}$

52. $\left(6bc^{2}\right)^{5/7}\left(6bc^{2}\right)^{2/7}$

53. $\left[\left(3b^{5}\right)^{3}\left(3b^{9}c^{8}\right)\right]^{1/4}$

54. $\left[\left(4b^{3}\right)^{2}\left(b^{2}c^{12}\right)\right]^{1/4}$

55. $\dfrac{b^{-2/5}c^{11/8}}{b^{18/5}c^{-5/8}}$

56. $\dfrac{b^{3/4}c^{1/2}}{b^{-1/4}c^{-1/2}}$

57. $\left(\dfrac{9b^{3}c^{-2}}{25b^{-5}c^{4}}\right)^{-1/2}$

58. $\left(\dfrac{16b^{12}c^{2}}{2b^{-3}c^{-4}}\right)^{-1/3}$

59. $32^{1/5}b^{3/7}b^{2/5}$

60. $16^{1/4}b^{1/4}b^{1/3}$

61. $\dfrac{b^{5/6}}{b^{1/4}}$

62. $\dfrac{b^{-2/3}}{b^{1/7}}$

63. $\dfrac{\left(9b^{5}\right)^{3/2}}{\left(27b^{4}\right)^{2/3}}$

64. $\dfrac{\left(32b^{3}\right)^{3/5}}{\left(16b^{3}\right)^{3/2}}$

65. $\left(\dfrac{8b^{2/3}}{2b^{4/5}}\right)^{3/2}$

66. $\left(\dfrac{27b^{1/3}c^{3/4}}{8b^{-2/3}c^{1/2}}\right)^{4/3}$

67. $\dfrac{\left(8bc^{3}\right)^{1/3}}{\left(81b^{-5}c^{3}\right)^{3/4}}$

68. $\dfrac{\left(1000b^{-7}c^{8}\right)^{2/3}}{\left(32b^{15}c^{4}\right)^{3/5}}$

69. $b^{2/5}\left(b^{8/5}+b^{3/5}\right)$

70. $c^{1/3}\left(c^{8/3}-c^{5/3}\right)$

71. The numbers of countries that have participated in the Winter Olympics are shown in Table 8 for various years.

Table 8 Numbers of Countries Participating in Winter Olympics

Year	Number of Countries
1924	16
1948	28
1968	37
1988	57
2006	85

Source: The Complete Book of the Winter Olympics

Let n be the number of countries participating in the Winter Olympics at t years since 1900. The situation can be described by the linear function $n = 0.81t - 9.41$ and the exponential function $n = 10.1(1.02)^{t}$.

a. Use a graphing calculator to draw the graphs of the two functions and, in the same viewing window, the scattergram of the data. Which function describes the situation better?

b. Use the exponential model to predict the number of countries that will participate in the 2010 Winter Olympics.

c. Use the exponential model and "intersect" on a graphing calculator to estimate when 42 countries participated in the Winter Olympics. [**Hint:** Graph the model and the horizontal line $n = 42$.]

72. The numbers of high school students who take college-level courses and test for credit are shown in Table 9 for various years.

Table 9 Numbers of High School Students Taking College-Level Courses and Testing for Credit

Year	Number of Students Who Test for Credit (thousands)
1963	21.8
1970	55.4
1977	82.7
1984	177.4
1991	359.1
1998	635.2
2005	1200.0

Source: *The College Board*

Let n be the number (in thousands) of high school students who take college-level courses and test for credit in the year that is t years since 1960. The situation can be described by the linear function $n = 25.36t - 246.99$ and the exponential function $n = 18.36(1.098)^{t}$.

a. Use a graphing calculator to draw the graphs of the two functions and, in the same viewing window, the scattergram of the data. Which function describes the situation better?

b. Use the exponential model to predict the number of students who will take college-level courses and test for credit in 2011.

c. Use the exponential model and "intersect" on a graphing calculator to estimate in which year 1000 high school students took college-level courses and tested for credit. [**Hint:** Graph the model and the horizontal line $n = 1000$.]

73. We can represent $\sqrt{5}$ by $5^{1/2}$. Explain.

74. To use a graphing calculator to find that $16^{1/2} = 4$, we press 16 [^] [(] 1 [÷] 2 [)]. If we omit the parentheses, we get the incorrect result 8. Explain why.

75. To convert from the scientific notation $N \times 10^{k}$ to standard decimal notation, we move the decimal point of the number N to the right by k places if k is positive. Explain.

76. List the exponent definitions and properties that are discussed in this section and Section 4.1. Explain how you can recognize which definition or property will help you simplify a given expression.

Related Review

For $f(x) = 8x$ and $g(x) = 8^x$, find the following.

77. $f\left(\frac{1}{3}\right)$ **78.** $f\left(\frac{4}{3}\right)$ **79.** $g\left(\frac{1}{3}\right)$

80. $g\left(\frac{4}{3}\right)$ **81.** $f\left(-\frac{1}{3}\right)$ **82.** $f\left(-\frac{2}{3}\right)$

83. $g\left(-\frac{1}{3}\right)$ **84.** $g\left(-\frac{2}{3}\right)$

Expressions, Equations, Functions, and Graphs

Perform the indicated instruction. Then use words such as linear, exponential, function, one variable, *and* two variables *to describe the expression, equation, or system. For instance, to describe $f(x) = 7(4)^x$, you could say "$f(x) = 7(4)^x$ is an exponential function."*

85. Graph $f(x) = \frac{3}{2}x - 4$ by hand.

86. Solve:

$$y = \frac{3}{2}x - 4$$
$$y = -\frac{1}{4}x + 3$$

87. Let $f(x) = \frac{3}{2}x - 4$. Find x when $f(x) = 5$.

88. Solve $\frac{3}{2}x - 4 = -\frac{1}{4}x + 3$.

4.3 GRAPHING EXPONENTIAL FUNCTIONS

Objectives

- Sketch the graph of an exponential function.
- Know the graphical significance of a and b for a function of the form $f(x) = ab^x$.
- Know the *base multiplier property*, the *increasing or decreasing property*, and the *reflection property*.

Recall from Section 4.1 that an exponential function is a function whose equation can be put into the form $f(x) = ab^x$, where $a \neq 0$, $b > 0$, and $b \neq 1$. In this section, we discuss how to use the values of a and b to help us graph an exponential function.

Graphing Exponential Functions

When graphing a certain type of function for the first time, we often begin by finding outputs for integer inputs near zero.

Example 1 Graphing an Exponential Function with $b > 1$

Graph $f(x) = 2^x$ by hand.

Solution

First, we list input–output pairs of the function f in Table 10. Note that as the value of x increases by 1, the value of y is multiplied by 2 (the base).

Next, we plot the solutions from Table 10 in Fig. 10 and sketch an increasing curve that contains the plotted points. The graph shows that as the value of x increases by 1, the value of y is doubled.

Table 10 Input--Output Pairs of $f(x) = 2^x$

x	$f(x)$
−3	$2^{-3} = \frac{1}{2^3} = \frac{1}{8}$
−2	$2^{-2} = \frac{1}{2^2} = \frac{1}{4}$
−1	$2^{-1} = \frac{1}{2^1} = \frac{1}{2}$
0	$2^0 = 1$
1	$2^1 = 2$
2	$2^2 = 4$
3	$2^3 = 8$

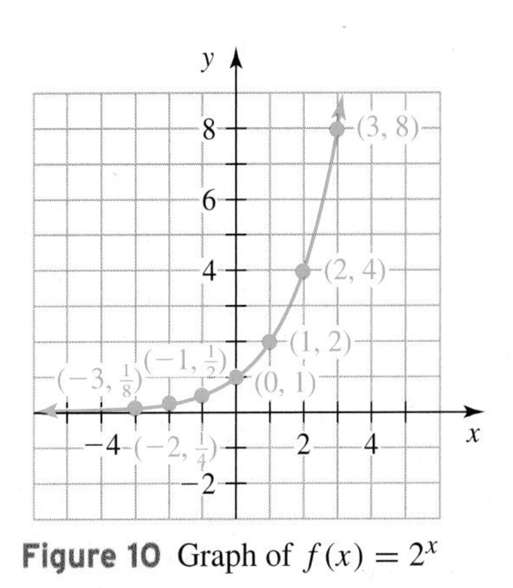

Figure 10 Graph of $f(x) = 2^x$

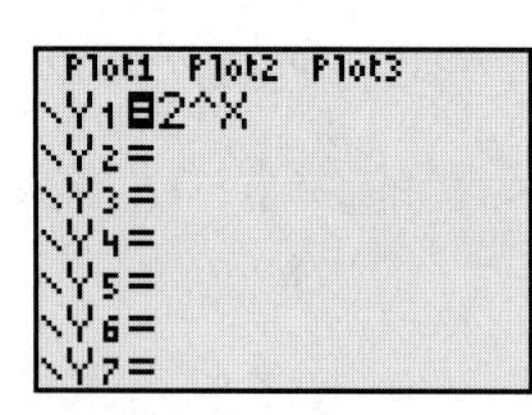

Figure 11 Enter the function

We can set up a window to verify our graph (see Figs. 11–13). For graphing calculator instructions, see Section B. 7. ■

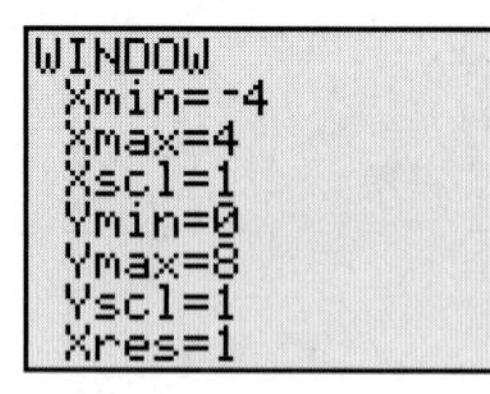

Figure 12 Set up the window

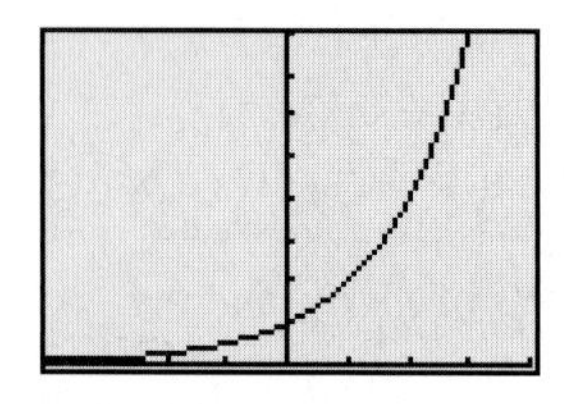

Figure 13 Graph the function

The smooth curve sketched in Fig. 10 implies that 2^x has meaning for *any* real-number exponent x. This is indeed true. In general, for $b > 0$, b^x has meaning for any real-number exponent x. The exponents are defined so that the graph of any exponential function is a smooth graph. Also, the exponential properties that we have discussed for rational exponents apply to real-number exponents as well. We can use a calculator to find *real-number powers* of numbers.

Recall from Section 1.2 that every point on the graph of an equation represents a solution of the equation. Every point *not* on the graph represents an ordered pair that is *not* a solution. The graph of an exponential function is called an **exponential curve.**

Example 2 Graphing an Exponential Function with $0 < b < 1$

Graph $g(x) = 4\left(\frac{1}{2}\right)^x$ by hand.

Solution

Input–output pairs of g are listed in Table 11. For example,

$$g(-1) = 4\left(\frac{1}{2}\right)^{-1} = 4\left(\frac{1}{2^{-1}}\right) = 4(2^1) = 8$$

So, $(-1, 8)$ is an input–output pair. Note that as the value of x increases by 1, the value of y is multiplied by $\frac{1}{2}$.

We plot the found points in Fig. 14 and sketch a decreasing exponential curve that contains the plotted points. The graph shows that as the value of x increases by 1, the value of y is halved.

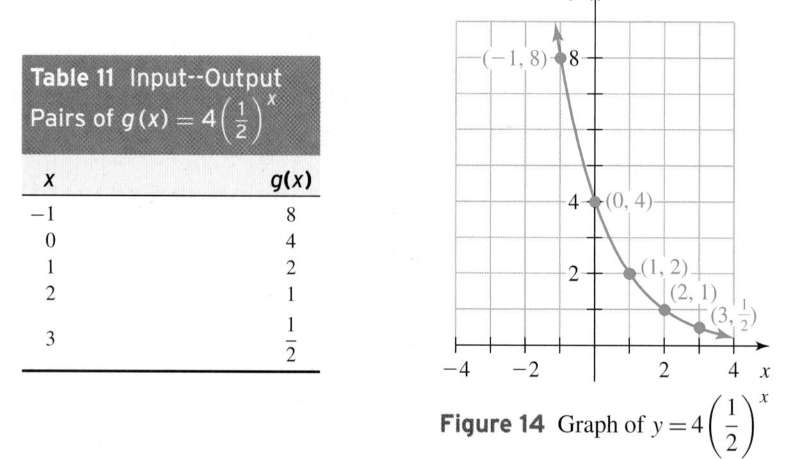

Table 11 Input--Output Pairs of $g(x) = 4\left(\frac{1}{2}\right)^x$

x	$g(x)$
-1	8
0	4
1	2
2	1
3	$\frac{1}{2}$

Figure 14 Graph of $y = 4\left(\frac{1}{2}\right)^x$

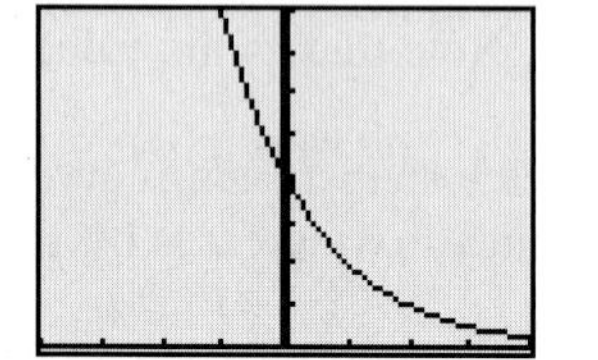

Figure 15 Graph of

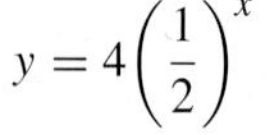

$y = 4\left(\frac{1}{2}\right)^x$

We can use a graphing calculator to verify our graph (see Fig. 15). ■

Base Multiplier Property; Increasing or Decreasing Property

Examples 1 and 2 suggest the *base multiplier property* and the *increasing or decreasing property*.

> **Base Multiplier Property**
>
> For an exponential function of the form $y = ab^x$, if the value of the independent variable increases by 1, the value of the dependent variable is multiplied by b.

We have seen two examples of this property in Examples 1 and 2. Here are two more examples of the base multiplier property:

1. For the function $f(x) = 2(3)^x$, as the value of x increases by 1, the value of y is multiplied by 3.

2. For the function $f(x) = 5\left(\frac{3}{4}\right)^x$, as the value of x increases by 1, the value of y is multiplied by $\frac{3}{4}$.

To prove the base multiplier property for the exponential function $f(x) = ab^x$, we compare outputs for the inputs k and $k + 1$, which differ by 1:

$$f(k) = ab^k \qquad \begin{aligned} f(k+1) &= ab^{k+1} \\ &= ab^k b^1 \\ &= f(k)b \end{aligned}$$

Since $f(k + 1) = f(k)b$, we conclude that if the value of the independent variable increases by 1, the value of the dependent variable is multiplied by b, which is what we set out to show.

For the increasing or decreasing property, we note in Example 1 that the base b is greater than 1 and the graph is increasing. In Example 2, the positive base is less than 1 and the graph is decreasing. For $f(x) = ab^x$ with $a > 0$, in general, we have the property that each multiplication by a base greater than 1 gives a larger value of y, whereas each multiplication by a positive base less than 1 gives a smaller value of y.

Increasing or Decreasing Property

Let $f(x) = ab^x$, where $a > 0$. Then

- If $b > 1$, then the function f is increasing. We say that the function **grows exponentially** (see Fig. 16).
- If $0 < b < 1$, then the function f is decreasing. We say that the function **decays exponentially** (see Fig. 17).

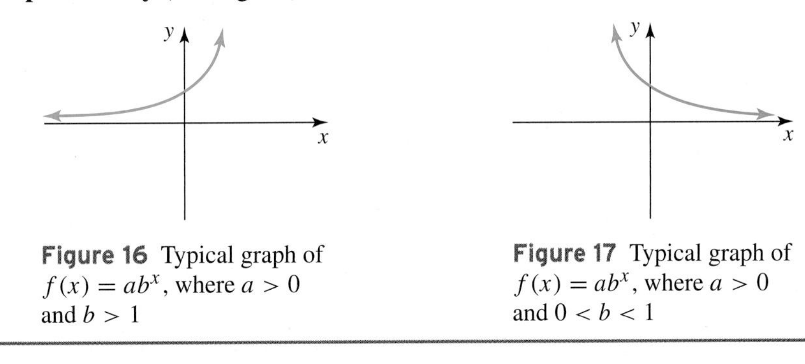

Figure 16 Typical graph of $f(x) = ab^x$, where $a > 0$ and $b > 1$

Figure 17 Typical graph of $f(x) = ab^x$, where $a > 0$ and $0 < b < 1$

Intercepts

When we sketch the graph of an exponential function, it is helpful to plot the y-intercept first. Substituting 0 for x in the general equation $y = ab^x$ gives

$$y = ab^0 = a(1) = a$$

So, the y-intercept is $(0, a)$.

y-Intercept of an Exponential Function

For an exponential function of the form

$$y = ab^x,$$

the y-intercept is $(0, a)$.

For the function $y = 5(8)^x$, the y-intercept is $(0, 5)$. For the function $y = 4\left(\frac{1}{7}\right)^x$, the y-intercept is $(0, 4)$.

WARNING For an exponential function of the form $y = b^x$ (rather than $y = ab^x$), the y-intercept is *not* $(0, b)$. By writing $y = b^x = 1b^x$, we see that the y-intercept is $(0, 1)$. For example, for $y = 2^x$, the y-intercept is $(0, 1)$. See Example 1.

Example 3 Intercepts and Graph of an Exponential Function

Let $f(x) = 6\left(\frac{1}{2}\right)^x$.

1. Find the y-intercept of f.
2. Find the x-intercept of f.
3. Graph f by hand.

Table 12 Input–Output Pairs of $f(x) = 6\left(\frac{1}{2}\right)^x$

x	$f(x)$
0	6
1	3
2	$\frac{3}{2}$
3	$\frac{3}{4}$
4	$\frac{3}{8}$

Solution

1. Since $f(x) = 6\left(\frac{1}{2}\right)^x$ is of the form $f(x) = ab^x$, we know that the y-intercept is $(0, a)$, or $(0, 6)$.
2. By the base multiplier property, we know that as the value of x increases by 1, the value of y is multiplied by $\frac{1}{2}$ (see Table 12).

   When we halve a number, it becomes smaller. But no number of halvings will give a result that is zero. So, as x grows large, y will become extremely close to, but never equal, 0. Likewise, the graph of f gets arbitrarily close to, but never reaches, the x-axis (see Fig. 18). In this case, we call the x-axis a **horizontal asymptote.** We conclude that the function f has no x-intercepts.
3. We plot five solutions from Table 12 and sketch a decreasing exponential curve that contains the five points (see Fig. 18). If we had not already found a table of solutions, we could have plotted the y-intercept and plotted additional solutions by increasing the value of x by 1 and going half as high for the value of y each time.

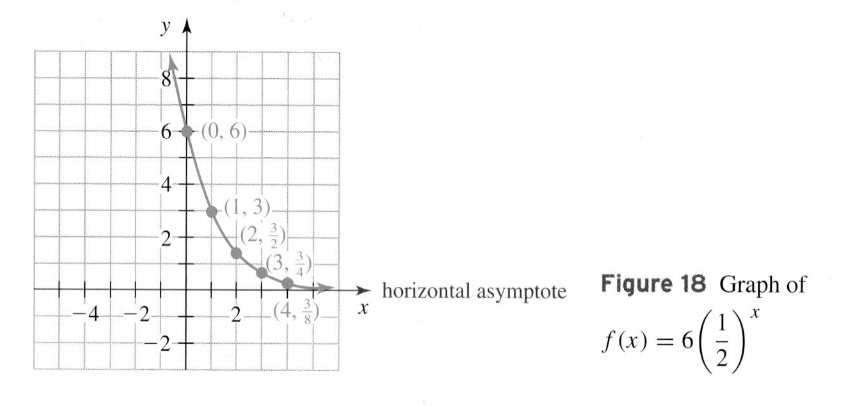

Figure 18 Graph of $f(x) = 6\left(\frac{1}{2}\right)^x$

As a check, we note that according to the increasing or decreasing property, the function f is decreasing, since the base, $\frac{1}{2}$, is between 0 and 1. For a more thorough check, we can use a graphing calculator to verify our graph (see Fig. 19). ■

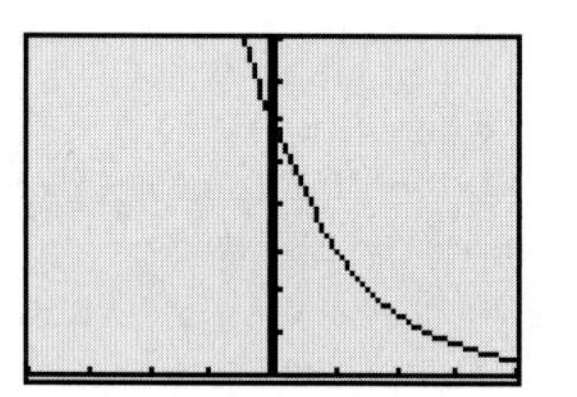

Figure 19 Graph of $y = 6\left(\frac{1}{2}\right)^x$

In Fig. 20, the graphs of both exponential functions get closer and closer to, but never reach, the x-axis. For both functions, the x-axis is a horizontal asymptote.

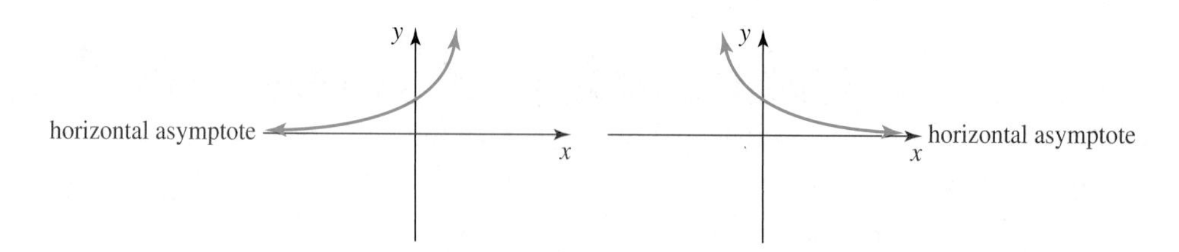

Figure 20 For both exponential functions, the x-axis is a horizontal asymptote

Reflection Property

In Example 4, we will graph two related exponential functions that will help us understand the *reflection property.*

Table 13 Input--Output Pairs of $f(x) = 5(3)^x$ and $g(x) = -5(3)^x$

x	$f(x)$	$g(x)$
0	5	−5
1	15	−15
2	45	−45
3	135	−135
4	405	−405

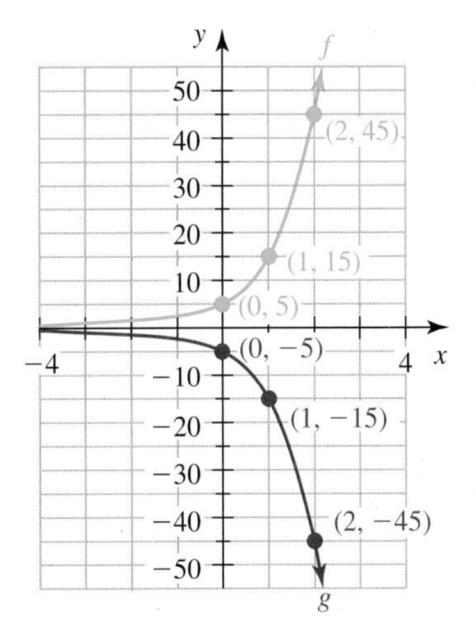

Figure 21 Graphs of $f(x) = 5(3)^x$ and $g(x) = -5(3)^x$

Example 4 Graphs of Functions of the Form $y = ab^x$ and $y = -ab^x$

1. Sketch and compare the graphs of $f(x) = 5(3)^x$ and $g(x) = -5(3)^x$.
2. Find the domain and range of f.
3. Find the domain and range of g.

Solution

1. Input–output pairs of f and g are listed in Table 13 and plotted in Fig. 21.
   In Table 13, we see that for each value of x, the outputs of g are the opposites of the outputs of f. Because of this, the graph of g is the **reflection,** or mirror image, of the graph of f, with the mirror along the x-axis. We can find the graph of g by *reflecting* the graph of f *across the x-axis.*
2. The expression $5(3)^x$ is defined for any real number x. So, the domain of f is the set of all real numbers. From Fig. 21, we see that the range of f (the set of all outputs of f) is the set of all positive real numbers.
3. The expression $-5(3)^x$ is defined for any real number x. So, the domain of g is the set of all real numbers. From Fig. 21, we see that the range is the set of all negative real numbers. ■

Reflection Property

The graphs of $f(x) = -ab^x$ and $g(x) = ab^x$ are reflections of each other across the x-axis.

We illustrate the reflection property and summarize four types of exponential curves in Figs. 22 and 23. **For all exponential functions, the x-axis is a horizontal asymptote.**

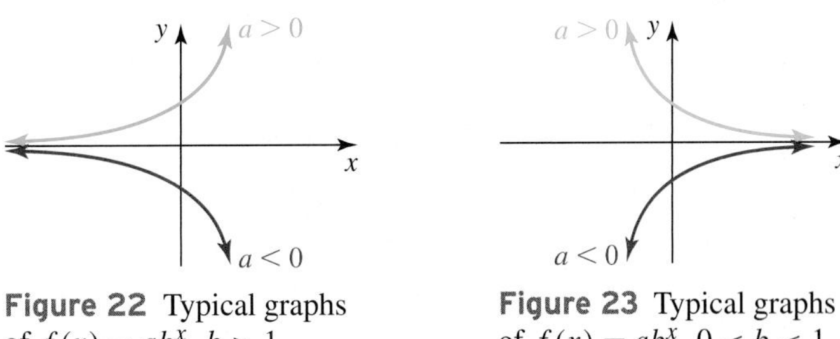

Figure 22 Typical graphs of $f(x) = ab^x$, $b > 1$

Figure 23 Typical graphs of $f(x) = ab^x$, $0 < b < 1$

Recall that for $b > 0$, b^x has meaning for any real-number exponent x. So, **the domain of any exponential function $f(x) = ab^x$ is the set of real numbers.**

Further, Figs. 22 and 23 show that $f(x) = ab^x$ has positive outputs if $a > 0$ and negative outputs if $a < 0$. Therefore, **the range of an exponential function $f(x) = ab^x$ is the set of all positive real numbers if $a > 0$, and the range is the set of all negative real numbers if $a < 0$.**

In Example 5, we use the graph of an exponential function f to find input or output values of f.

Example 5 Finding Values of a Function from Its Graph

The graph of an exponential function f is shown in Fig. 24.

1. Find $f(2)$.
2. Find x when $f(x) = 2$.
3. Find x when $f(x) = 0$.

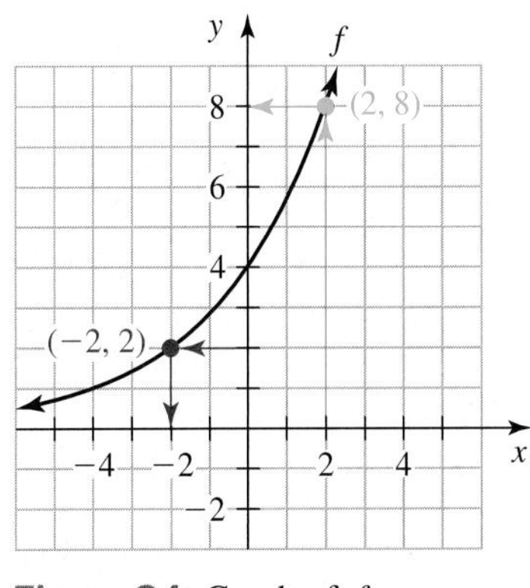

Figure 24 Graph of f

Solution

1. The blue arrows in Fig. 24 show that the input $x = 2$ leads to the output $y = 8$. We conclude that $f(2) = 8$.
2. The red arrows in Fig. 24 show that the output $y = 2$ originates from the input $x = -2$. We conclude that $x = -2$ when $f(x) = 2$.
3. Recall that the graph of an exponential function gets close to, but never reaches, the x-axis. So, there is no value of x where $f(x) = 0$. ■

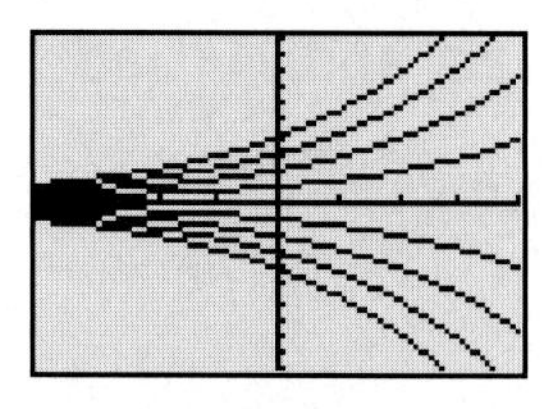

Figure 25 A family of exponential curves

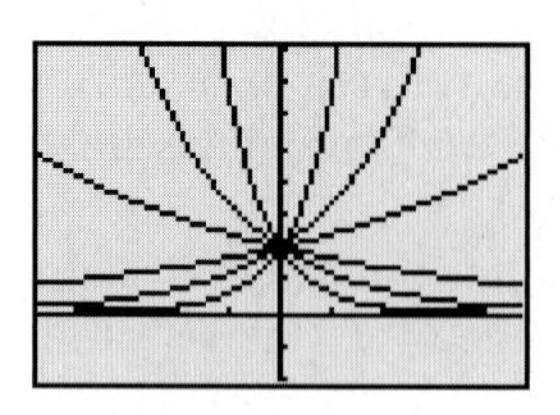

Figure 26 A family of exponential curves passing through (0, 2)

group exploration

Drawing families of exponential curves

For each problem, use a graphing calculator to graph a family of curves.

1. List the equations of a family of exponential curves like the ones shown in Fig. 25.
2. List the equations of a family of exponential curves like the ones shown in Fig. 26. All of these curves pass through the point (0, 2).
3. Summarize what you have learned from this exploration and this section about the coefficient a and the base b in functions of the form $f(x) = ab^x$.

group exploration

Looking ahead: Using trial and error to find a model

In Exercise 5 of Homework 2.1, you found that the number of Internet users in the United States can be modeled by a linear function. Here, you will model the number of Internet users *worldwide* (see Table 14).

Let $n = f(t)$ be the number of users (in millions) at t years since 1995.

1. Use a graphing calculator to draw a scattergram of the data. Would it be better to model the data with a linear or an exponential function? Explain.
2. Imagine an exponential function $f(t) = ab^t$ whose graph comes close to the data points in your scattergram. What is the n-intercept? What does this tell you about the value of a or b? Explain.
3. Guess a reasonable value of b for your function $f(t) = ab^t$. [**Hint:** The base multiplier property may help.]
4. Substitute your values of a and b from Problems 2 and 3 into the equation $f(t) = ab^t$.
5. Graph f and the scattergram in the same viewing window to see how well your model fits the data.
6. Now find better values of a and b through trial and error. When you are satisfied with your values of a and b, write the equation of f that you have found.

Table 14 Numbers of Internet Users Worldwide

Year	Number of Users (millions)
1995	34
1996	54
1997	90
1998	149
1999	230
2000	311
2001	457
2002	581
2003	727
2004	934

Source: Computer Industry Almanac

TIPS FOR SUCCESS: Desire and Faith

To accomplish anything worthwhile, including succeeding in this course, requires substantial effort and faith that you will succeed:

> "The secret of making something work in your lives is, first of all, the deep desire to make it work. Then the faith and belief that it can work. Then to hold that clear definite vision in your consciousness and see it working out step by step without one thought of doubt or disbelief."
>
> —Eileen Caddy, *Footprints on the Path* (1991)

Your deep desire to succeed in this course might be to earn a degree so that you can earn more money, to learn algebra for the love of learning, or to experience setting a goal and reaching it. Your faith and belief can come from knowing that you, your instructor, and your college will do everything possible to ensure your success. To hold your vision of success "without one thought of doubt or disbelief" is a tall order, but the more you look for ways to succeed rather than feel discouraged, the better are your chances of success.

HOMEWORK 4.3

Graph the given function by hand. Then use a graphing calculator to verify your graph.

1. $y = 3^x$
2. $y = 4^x$
3. $y = 10^x$
4. $y = 5^x$
5. $y = 3(2)^x$
6. $y = 2(3)^x$
7. $y = 6(3)^x$
8. $y = 3(5)^x$
9. $y = 15\left(\frac{1}{3}\right)^x$
10. $y = 20\left(\frac{1}{4}\right)^x$
11. $y = 12\left(\frac{1}{2}\right)^x$
12. $y = 6\left(\frac{2}{3}\right)^x$

Graph both functions by hand on the same coordinate system. Then use a graphing calculator to verify your graphs.

13. $f(x) = 2^x,\ g(x) = -2^x$
14. $f(x) = 3^x,\ g(x) = -3^x$
15. $f(x) = 4(3)^x,\ g(x) = -4(3)^x$
16. $f(x) = 2(10)^x,\ g(x) = -2(10)^x$
17. $f(x) = 8\left(\frac{1}{2}\right)^x,\ g(x) = -8\left(\frac{1}{2}\right)^x$
18. $f(x) = 6\left(\frac{1}{3}\right)^x,\ g(x) = -6\left(\frac{1}{3}\right)^x$

Graph the function by hand. Then use a graphing calculator to verify your graph. Find the domain and range of the function.

19. $f(x) = 5(2)^x$
20. $f(x) = 9\left(\frac{1}{3}\right)^x$
21. $f(x) = -8\left(\frac{1}{4}\right)^x$
22. $f(x) = -3(3)^x$

23. Graphs of four functions of the form $y = ab^x$ are shown in Fig. 27. Describe the constants a and b of each function.

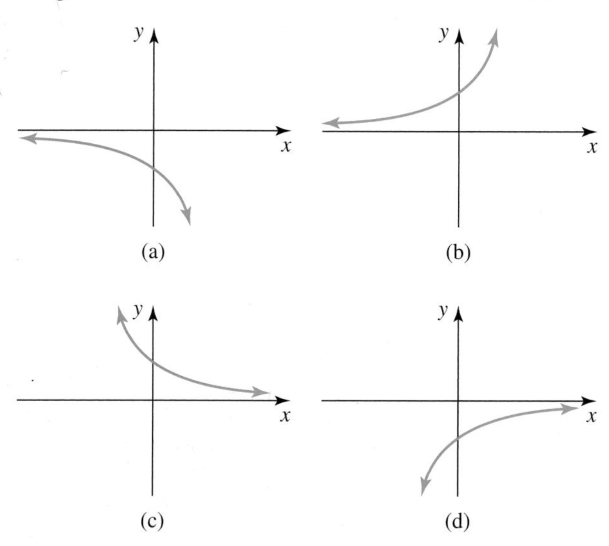

Figure 27 Exercise 23

24. The graphs of functions $f(x) = ab^x$ and $g(x) = cd^x$ are shown in Fig. 28.
    a. Which coefficient is greater, a or c? Explain.
    b. Which base is greater, b or d? Explain.

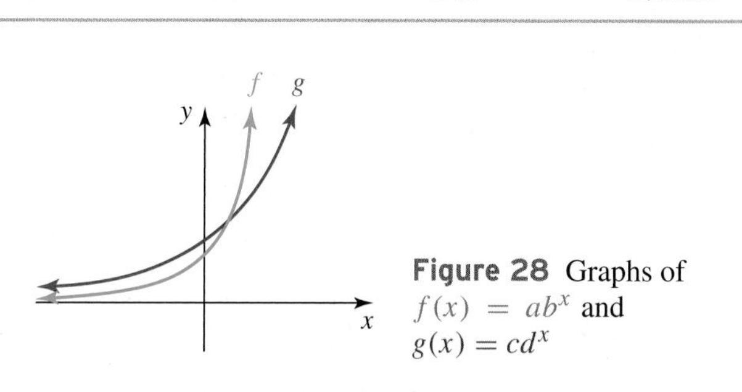

Figure 28 Graphs of $f(x) = ab^x$ and $g(x) = cd^x$

25. Use a graphing calculator to graph a family of exponential curves similar to the family graphed in Fig. 29. List the equations of that family.

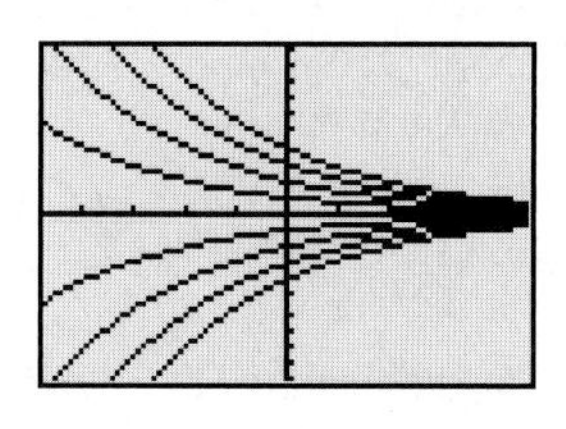

Figure 29 A family of exponential curves

26. Use a graphing calculator to graph a family of exponential curves similar to the family graphed in Fig. 30. All of these curves pass through the point $(0, -2)$. List the equations of that family.

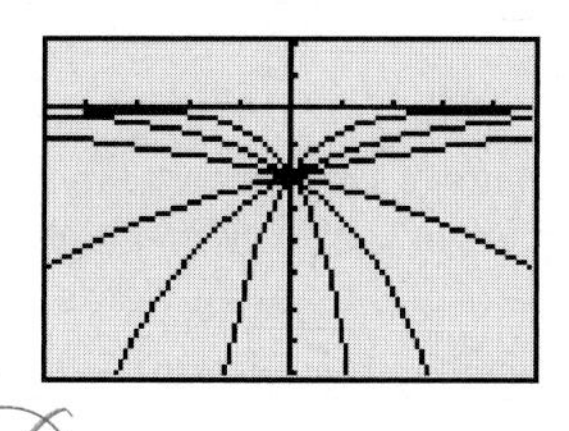

Figure 30 A family of exponential curves passing through $(0, -2)$

27. Use a graphing calculator to draw a graph similar to the one in Fig. 31. Use an equation of the form $f(x) = ab^x$, where a and b are constants that you specify. What equation works?

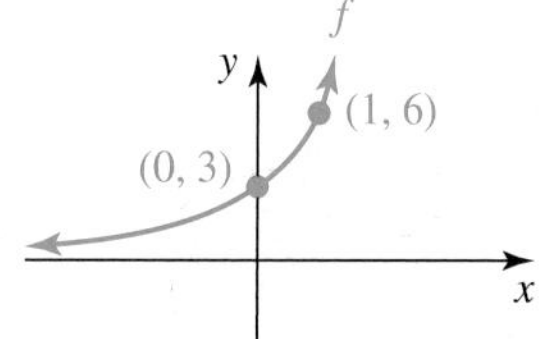

Figure 31 Exercise 27

28. Use a graphing calculator to draw a graph similar to the one in Fig. 32. Use an equation of the form $g(x) = ab^x$, where a and b are constants that you specify. What equation works? Use trial and error.

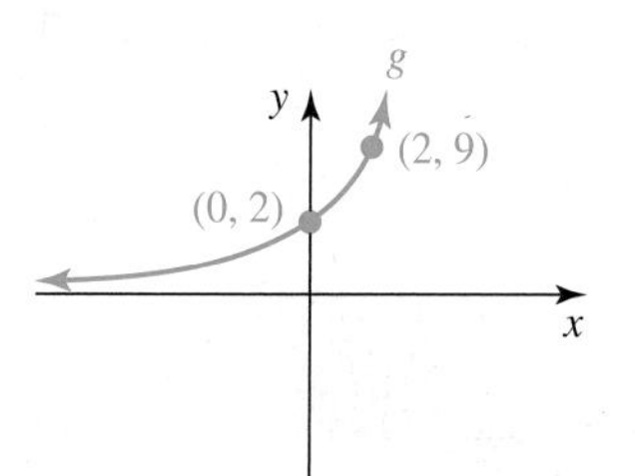

Figure 32 Exercise 28

29. Find equations of exponential functions that could correspond to the graphs shown in Fig. 33.

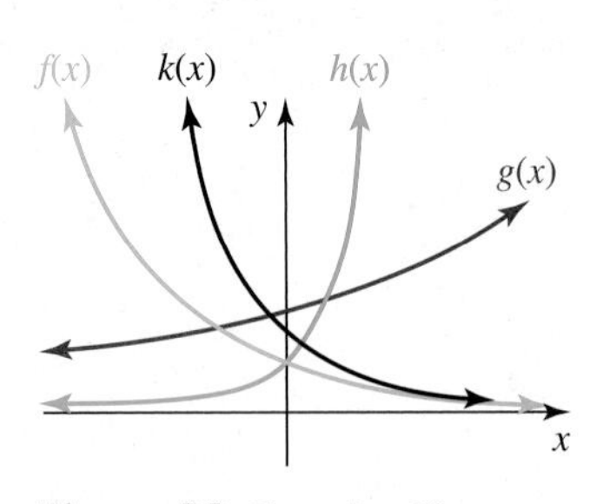

Figure 33 Exercise 29

30. a. Simplify and compare both expressions.

i. $4(3)^2$ and 12^2

ii. $4^2 \cdot 3^2$ and 12^2

b. Build a graphing calculator table that shows the same input values for both functions. Explain in terms of order of operations or exponential properties why the tables are the same or different.

i. $f(x) = 4(3)^x$ and $g(x) = 12^x$

ii. $f(x) = 4^x \cdot 3^x$ and $g(x) = 12^x$

31. Describe the Rule of Four as applied to the function $f(x) = 4(2)^x$:

a. Describe five input–output pairs of f by using a table.

b. Describe the input–output pairs of f by using a graph.

c. Describe the input–output pairs of f by using words.

32. Describe the Rule of Four as applied to the function $g(x) = 16\left(\frac{1}{2}\right)^x$:

a. Describe five input–output pairs of g by using a table.

b. Describe the input–output pairs of g by using a graph.

c. Describe the input–output pairs of g by using words.

33. Input–output pairs of four exponential functions are listed in Table 15. Complete the table.

Table 15 Complete the Table (Exercise 33)

x	$f(x)$	$g(x)$	$h(x)$	$k(x)$
0	162	3	2	800
1	54	12	10	400
2	18	48		
3	6			
4				

34. Input–output pairs of four exponential functions are listed in Table 16. Complete the table.

Table 16 Complete the Table (Exercise 34)

x	$f(x)$	$g(x)$	$h(x)$	$k(x)$
0	3	64	2	100
1	6	32	6	10
2	12	16		
3	24			
4				

35. Input–output pairs of four exponential functions are listed in Table 17. Complete the table.

Table 17 Complete the Table (Exercise 35)

x	$f(x)$	$g(x)$	$h(x)$	$k(x)$
0	5			
1		80	54	
2	20			
3		20		192
4			2	768

36. Input–output pairs of four exponential functions are listed in Table 18. Complete the table.

Table 18 Complete the Table (Exercise 36)

x	$f(x)$	$g(x)$	$h(x)$	$k(x)$
0			3	400
1		3		
2	25			
3		147		
4	1		30,000	25

For Exercises 37–44, refer to Fig. 34.

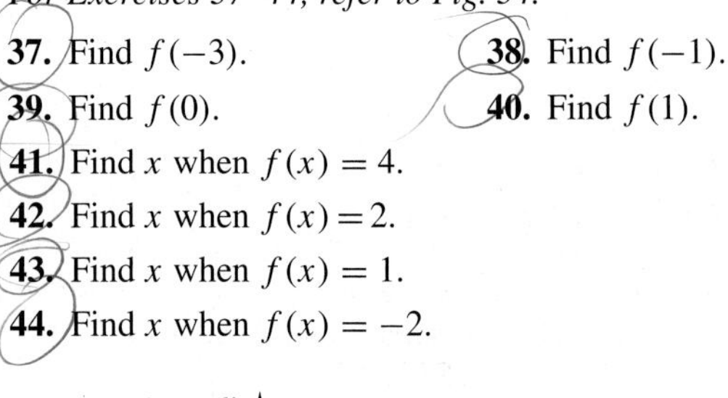

37. Find $f(-3)$.

38. Find $f(-1)$.

39. Find $f(0)$.

40. Find $f(1)$.

41. Find x when $f(x) = 4$.

42. Find x when $f(x) = 2$.

43. Find x when $f(x) = 1$.

44. Find x when $f(x) = -2$.

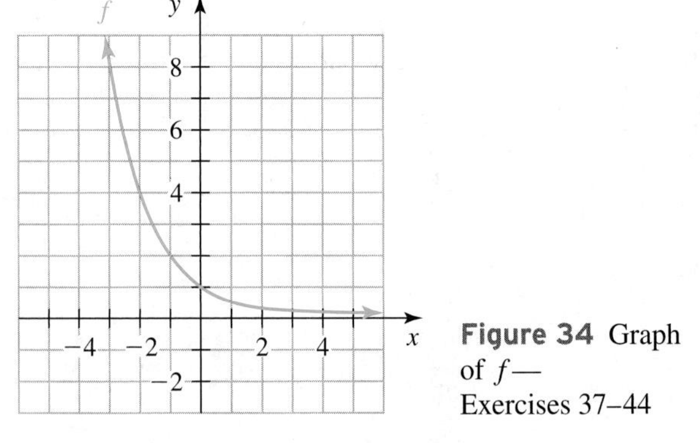

Figure 34 Graph of f—Exercises 37–44

For Exercises 45–52, refer to Table 19.

45. Find $f(3)$.

46. Find $f(6)$.

47. Find $f(5)$.

48. Find $f(0)$.

49. Find x when $f(x) = 3$.

50. Find x when $f(x) = 6$.

51. Find x when $f(x) = 24$.

52. Find x when $f(x) = 96$.

Table 19 Some Values of an Exponential Function f

x	$f(x)$
0	3
1	6
2	12
3	24
4	48
5	96
6	192

53. The average ticket prices to major league baseball games are shown in Table 20 for various years.

Table 20 Average Ticket Prices to Major League Baseball Games

Year	Average Ticket Price (dollars)
1950	1.54
1960	1.96
1970	2.72
1980	4.45
1991	8.84
2000	16.22
2005	21.17

Sources: The Sporting News and the Sporting News Baseball Dope Book, 1950–85; *Team Marketing Report, 1991–2004*

a. Let $f(t)$ be the average ticket price (in dollars) to major league baseball games for the year that is t years since 1950. Use a graphing calculator to draw a scattergram of the data. Is it better to use a linear or an exponential function to model the data? Explain.

b. Draw the graph of the function $f(t) = 1.22(1.051)^t$ and the scattergram in the same viewing window. Does the graph of f come close to the data points?

c. Use f to predict the average ticket price in 2010.

d. The most expensive average ticket price in 2005 was \$44.56, at Fenway Park, home of the Boston Red Sox. Use TRACE and Zoom Out on a graphing calculator to predict when the average ticket price to *all* major league baseball games will reach \$44.56.

54. If you place your hand on a piano and play a note, you will feel the piano vibrate. The number of vibrations per second (hertz) of a note is called its *frequency*. If you strike the piano keys from left to right, the frequencies of the notes increase. We use some of the letters of the alphabet, sometimes in conjunction with the "sharp" symbol ♯, to refer to these notes (see Fig. 35). The frequencies of 13 notes in a row are listed in Table 21.

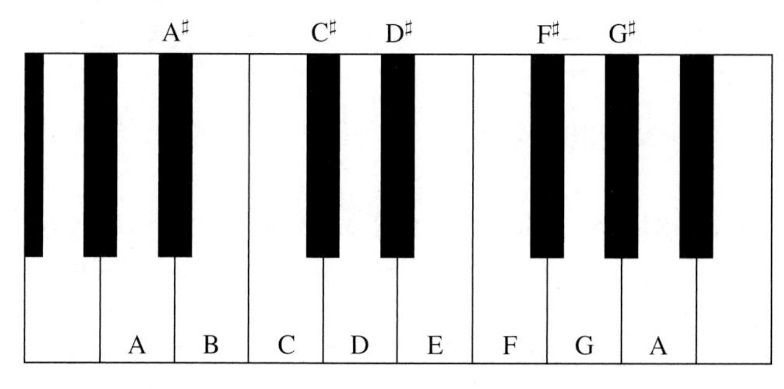

Figure 35 Notes on a piano

a. Let $f(n)$ be the frequency (in hertz) of the note that is n notes above the note A (the one with frequency 220.0 hertz). Use a graphing calculator to draw a scattergram of the data listed in Table 21. Will a linear function or an exponential function model the data better?

b. Draw the graph of the function $f(n) = 220(2)^{n/12}$ and the scattergram in the same viewing window. Does the graph come close to the data points? [***Graphing Calculator:*** For the exponential expression $220(2)^{x/12}$, press **220** [(] **2** [)] [^] [(] [X, T,Θ,*n*] [÷] **12** [)].]

Table 21 Frequencies of Notes on a Piano

Note	Number of Notes Above A	Frequency (in hertz)
A	0	220.0
A♯	1	233.1
B	2	246.9
C	3	261.6
C♯	4	277.2
D	5	293.7
D♯	6	311.1
E	7	329.6
F	8	349.2
F♯	9	370.0
G	10	392.0
G♯	11	415.3
A	12	440.0

Source: Math and Music *by Garland and Kahn*

c. Estimate the frequency of the note D that is 17 notes above the note A (the one with frequency 220.0 hertz).

d. Use TRACE to find which note has a frequency of 523.25 hertz.

e. Use a graphing calculator table to find $f(0)$, $f(12)$, $f(24)$, $f(36)$, and $f(48)$. What pattern do you notice? Describe this pattern in terms of the situation.

Find the x- and y-intercepts of the function.

55. $y = 7^x$

56. $y = 8(4)^x$

57. $y = 3\left(\frac{1}{5}\right)^x$

58. $y = -9\left(\frac{2}{3}\right)^x$

59. In this exercise, you will compare the function $f(x) = 100(2)^x$ with the function $g(x) = 5(3)^x$.

a. Find the y-intercept of each function.

b. What does the base multiplier property tell you about each function?

c. Based on your comments in parts (a) and (b), which function's outputs will eventually be much greater than the other's outputs? Explain.

d. Use a graphing calculator table to verify your comments to parts (a)–(c). To do this, enter the functions and set up a table as indicated in Figs. 36 and 37, respectively.

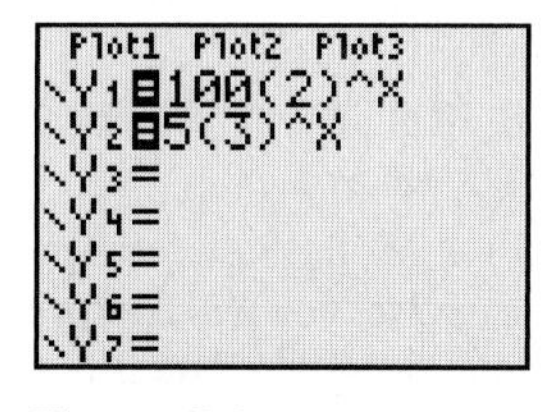

Figure 36 Enter the functions

Figure 37 Set up the table

60. What are the x-intercepts and y-intercepts of a function of the form $y = ab^x$, where $b > 0$?

Let $f(x) = 2^x + 3^x$.

61. Find $f(2)$.

62. Find $f(0)$.

63. Find $f(-2)$.

64. Find $f(-1)$.

Let $f(x) = 3^x$.

65. Find x when $f(x) = 3$.

66. Find x when $f(x) = 9$.

67. Find x when $f(x) = 1$.

68. Find x when $f(x) = \frac{1}{3}$.

*Use graphing calculator tables to compare each pair of functions f and g. What do you observe? Use exponential properties to show why this is so. [**Graphing Calculator:** For 2^{3x}, press* 2 [^] [(] 3 [X, T, Θ, n] [)]. *Recall that if an exponent involves an operation, you must use parentheses.]*

69. $f(x) = 2^{3x}$, $g(x) = 8^x$

70. $f(x) = 2^{-x}$, $g(x) = \left(\frac{1}{2}\right)^x$

71. $f(x) = 2^{x+3}$, $g(x) = 8(2)^x$

72. $f(x) = 3^x 3^x$, $g(x) = 3^{2x}$

73. $f(x) = \frac{6^x}{3^x}$, $g(x) = 2^x$

74. $f(x) = 2^0$, $g(x) = 3^0$

75. $f(x) = \frac{3^{2x}}{3^x}$, $g(x) = 3^x$

76. $f(x) = 2^x 3^x$, $g(x) = 6^x$

77. $f(x) = x^{1/2}$, $g(x) = \sqrt{x}$ [**Graphing Calculator:** For $\sqrt{x}$, press [2nd] [x^2] [X,T,Θ,n] [)].]

78. $f(x) = 5^{x/3}$, $g(x) = (5^{1/3})^x$

79. $f(x) = 2^x$, $g(x) = 8^{x/3}$

80. $f(x) = 25^{x/2} \cdot 5^x$, $g(x) = 25^x$

81. Is the statement true for $f(x) = 2^x$?
- **a.** $f(3 + 4) = f(3) + f(4)$
- **b.** $f(x + y) = f(x) + f(y)$

82. Is the statement true for $g(x) = 3^x$?
- **a.** $g(2 + 5) = g(2) \cdot g(5)$
- **b.** $g(4 + 6) = g(4) \cdot g(6)$
- **c.** $g(2 + 4) = g(2) \cdot g(4)$
- **d.** $g(x + y) = g(x) + g(y)$

83. Let $f(x) = ab^x$, where $a > 0$. Explain why f is increasing if $b > 1$ and f is decreasing if $0 < b < 1$.

84. In an exponential function $f(x) = b^x$, the base b is a positive number not equal to 1. In this exercise, you will explore what happens if we try to define a function whose base is negative. Consider $f(x) = (-4)^x$.
- **a.** Explain why $f\left(\frac{1}{2}\right)$ is undefined.
- **b.** Explain why $f\left(\frac{1}{4}\right)$ is undefined.
- **c.** List three more values of x that result in undefined outputs.

85. The graphs of the exponential functions $f(x) = -ab^x$ and $g(x) = ab^x$ are reflections of each other across the x-axis. Explain why this makes sense.

86. Explain how to sketch the graph of a function of the form $f(x) = ab^x$, where $b > 0$. Include the effect of a value of a or b on the graph.

Related Review

Graph the given function by hand. Then use a graphing calculator to verify your graph.

87. $y = 4 + 2x$

88. $y = 4(2)^x$

89. $y = -4 + 2x$

90. $y = -4(2)^x$

Find all x-intercepts and y-intercepts.

91. $y = 8 + 4x$

92. $y = 8(4)^x$

93. Some input–output pairs of the functions f, g, h, and k are provided in Table 22. For each function, determine whether the given values suggest that the function is linear, exponential, or neither.

Table 22 Identifying Functions (Exercise 93)

x	$f(x)$	$g(x)$	$h(x)$	$k(x)$
0	13	4	48	5
1	9	12	24	55
2	5	36	12	555
3	1	108	6	5555
4	−3	324	3	55555

94. The graphs of the equation $y = \frac{1}{3}x + 2$ and an equation of the form $y = ab^x$ are shown in Fig. 38. Find the values of a and b.

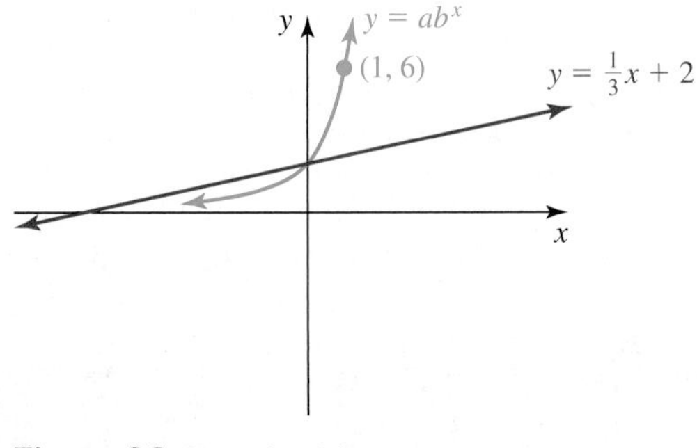

Figure 38 Exercise 94

Expressions, Equations, Functions, and Graphs

Perform the indicated instruction. Then use words such as linear, exponential, function, one variable, *and* two variables *to describe the expression, equation, or system. For instance, to describe $3x + 5 = 7$, you could say "$3x + 5 = 7$ is a linear equation in one variable."*

95. Graph $f(x) = 6\left(\frac{1}{2}\right)^x$ by hand.

96. Graph $y = -3x + 4$ by hand.

97. Find $f(-2)$, where $f(x) = 6\left(\frac{1}{2}\right)^x$.

98. Solve:

$$y = -3x + 4$$
$$2x - 5y = 31$$

4.4 FINDING EQUATIONS OF EXPONENTIAL FUNCTIONS

"Never stop learning; knowledge doubles every fourteen months."
—Anthony J. D'Angelo, *The College Blue Book*

Objectives

- Use the base multiplier property to find an exponential equation.
- Solve an equation of the form $ab^n = k$ for the base b.
- Use two points to find an exponential equation.

In this section, we will discuss two ways to find an equation of an exponential function.

Using the Base Multiplier Property to Find Exponential Functions

One way to find an equation of an exponential function of the form $y = ab^x$ is to use the base multiplier property, which we discussed in Section 4.3. That is, if the value of the independent variable increases by 1, then the value of the dependent variable is multiplied by the base b.

Table 23 Solutions of an Exponential Equation

x	$f(x)$
0	3
1	6
2	12
3	24
4	48

Example 1 Finding an Equation of an Exponential Curve

An exponential curve contains the points listed in Table 23. Find an equation of the curve.

Solution

For $f(x) = ab^x$, recall from Section 4.3 that the y-intercept is $(0, a)$. From Table 23, we see that the y-intercept is $(0, 3)$, so $a = 3$. As the value of x increases by 1, the value of y is multiplied by 2. By the base multiplier property, we know that $b = 2$. Therefore, an equation of the curve is

$$f(x) = 3(2)^x$$

We check our result with a graphing calculator table (see Fig. 39). For graphing calculator instructions, see Section B.13.

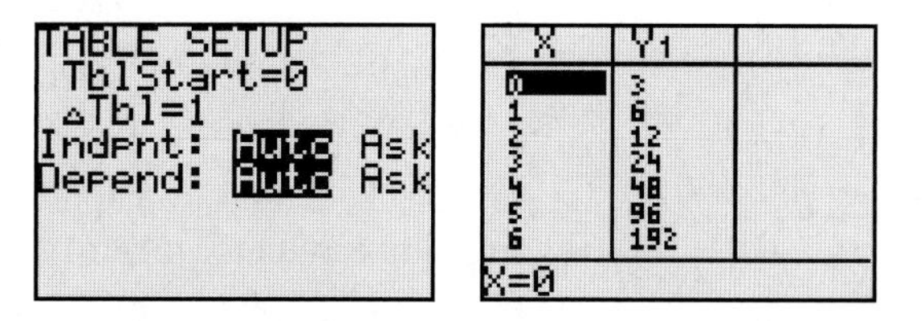

Figure 39 Verify the exponential equation $f(x) = 3(2)^x$ ■

For Example 2, it will be helpful to review the slope addition property from Section 1.4: For a linear function of the form $y = mx + b$, if the value of the independent variable increases by 1, then the value of the dependent variable changes by the slope m.

Example 2 Linear versus Exponential Functions

1. Find a possible equation of a function whose input–output pairs are listed in Table 24.
2. Find a possible equation of a function whose input–output pairs are listed in Table 25.

Table 24 Input--Output Pairs for f

x	$f(x)$
0	162
1	54
2	18
3	6
4	2

Solution

1. As the value of x increases by 1 throughout Table 24, the value of y is multiplied by $\frac{1}{3}$. This suggests that there is an exponential function $f(x) = a\left(\frac{1}{3}\right)^x$ that contains the points in Table 24. Since the y-intercept is $(0, 162)$, we have $f(x) = 162\left(\frac{1}{3}\right)^x$.

Table 25 Input--Output Pairs for g

x	$g(x)$
0	50
1	46
2	42
3	38
4	34

2. As the value of x increases by 1 throughout Table 25, the value of y changes by adding -4. This suggests that there is a linear function $g(x) = -4x + b$ that contains the points in Table 25. Since the y-intercept is $(0, 50)$, we have $g(x) = -4x + 50$. ■

Solving Equations of the Form $ab^n = k$ for b

So far, we have discussed how to find an equation of an exponential curve that contains points whose x-coordinates are *consecutive* integers. Later in this section, we will discuss how to find an equation of an exponential curve that contains two given points, such as $(2, 5)$ and $(5, 63)$, whose x-coordinates are *not* consecutive integers. To use this method, we first need to discuss how to solve equations of the form $ab^n = k$ for the base b.

Example 3 One-Variable Equations Involving Exponents

Find all real-number solutions.

1. $b^2 = 25$ **2.** $b^3 = 8$ **3.** $2b^4 = 32$
4. $10b^5 = 90$ **5.** $b^6 = -28$

Solution

1. $b^2 = 25$ Original equation
$b = -5$ or $b = 5$ $(-5)^2 = 25$ and $5^2 = 25$

So, the solutions are -5 and 5. We can use the notation ± 5 to stand for the numbers -5 and 5.

2. $b^3 = 8$ Original equation
$b = 2$ $2^3 = 8$

3. $2b^4 = 32$ Original equation
$b^4 = 16$ Divide both sides by 2.
$b = \pm 2$ $(-2)^4 = 16$ and $2^4 = 16$

We can check that both -2 and 2 satisfy the equation $2b^4 = 32$.

4. $10b^5 = 90$ Original equation
$b^5 = 9$ Divide both sides by 10.
$b = 9^{1/5}$ $9^{1/5}$ is the number whose 5th power is 9.
$b \approx 1.55$ $1.55^5 \approx 9$

We can check that 1.55 approximately satisfies the equation $10b^5 = 90$ by using a graphing calculator to verify that $10(1.55)^5 \approx 90$ (see Fig. 40).

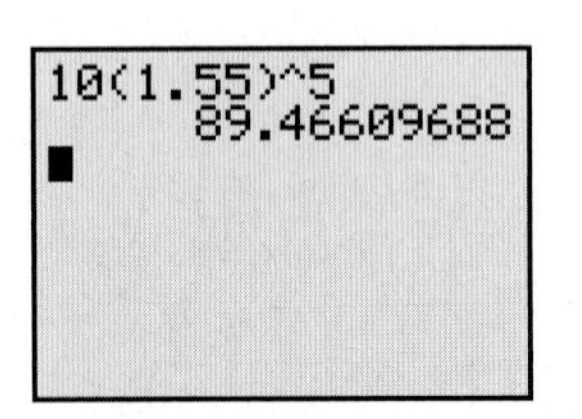

Figure 40 Checking that 1.55 approximately satisfies $10b^5 = 90$

5. The equation $b^6 = -28$ has no real-number solutions, since an even-numbered exponent gives a positive number. ■

The problems in Example 3 suggest how to solve equations of the form $b^n = k$ for b.

Solving Equations of the Form $b^n = k$ for b

To solve an equation of the form $b^n = k$ for b,

1. If n is odd, the real-number solution is $k^{1/n}$.
2. If n is even and $k \geq 0$, the real-number solutions are $\pm k^{1/n}$.
3. If n is even and $k < 0$, there is no real-number solution.

Example 4 One-Variable Equations Involving Exponents

Find all real-number solutions. Round any results to the second decimal place.

1. $5.42b^6 - 3.19 = 43.74$ **2.** $\dfrac{b^9}{b^4} = \dfrac{70}{3}$

Solution

1.
$$\begin{aligned} 5.42b^6 - 3.19 &= 43.74 && \text{Original equation} \\ 5.42b^6 &= 43.74 + 3.19 && \text{Add 3.19 to both sides.} \\ 5.42b^6 &= 46.93 && \text{Add.} \\ b^6 &= \frac{46.93}{5.42} && \text{Divide both sides by 5.42.} \\ b &= \pm\left(\frac{46.93}{5.42}\right)^{1/6} && \text{The solutions of } b^6 = k \text{ are } \pm k^{1/6} \text{ if } k \ge 0. \\ b &\approx \pm 1.43 && \text{Compute.} \end{aligned}$$

2.
$$\begin{aligned} \frac{b^9}{b^4} &= \frac{70}{3} && \text{Original equation} \\ b^5 &= \frac{70}{3} && \text{Subtract exponents: } \frac{b^m}{b^n} = b^{m-n} \\ b &= \left(\frac{70}{3}\right)^{1/5} && \text{The solution of } b^5 = k \text{ is } k^{1/5}. \\ b &\approx 1.88 && \text{Compute.} \end{aligned}$$

■

Using Two Points to Find Equations of Exponential Functions

Now that we have discussed how to solve equations of the form $ab^n = k$ for b, we can discuss a second way to find an equation of an exponential function.

Example 5 Finding an Equation of an Exponential Curve

Find an approximate equation $y = ab^x$ of the exponential curve that contains the points $(0, 3)$ and $(4, 70)$. Round the value of b to two decimal places.

Solution

Since the y-intercept is $(0, 3)$, we know the equation has the form $y = 3b^x$. Next, we substitute $(4, 70)$ in the equation $y = 3b^x$ and solve for b:

$$\begin{aligned} 70 &= 3b^4 && \text{Substitute 4 for } x \text{ and 70 for } y. \\ 3b^4 &= 70 && \text{If } c = d, \text{ then } d = c. \\ b^4 &= \frac{70}{3} && \text{Divide both sides by 3.} \\ b &= \pm\left(\frac{70}{3}\right)^{1/4} && \text{The solutions of } b^4 = k \text{ are } \pm k^{1/4} \text{ if } k \ge 0. \\ b &\approx 2.20 && \text{Compute; base of an exponential function is positive.} \end{aligned}$$

So, our equation is $y = 3(2.20)^x$; its graph contains the given point $(0, 3)$. Since we rounded the value b, the graph of the equation comes close to, but does not pass through, the given point $(4, 70)$.

We use a graphing calculator to verify our work (see Fig. 41). ■

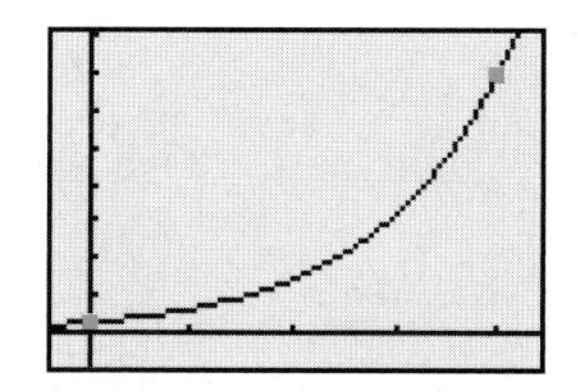

Figure 41 Verify the work

In Example 6, we will find an equation of a curve that approximates the exponential curve containing two given points. Neither point will be the y-intercept. To do this, we will use the following property.

Dividing Left Sides and Right Sides of Two Equations

If $a = b$ and $c = d$, then

$$\frac{a}{c} = \frac{b}{d}$$

In words: The quotient of the left sides of two equations is equal to the quotient of the right sides.

For example, if we divide the left sides and divide the right sides of the equations $2 = 2$ and $3 = 3$, we obtain the true statement $\frac{2}{3} = \frac{2}{3}$.

Example 6 Finding an Equation of an Exponential Curve

Find an approximate equation $y = ab^x$ of the exponential curve that contains (2, 5) and (5, 63). Round the values of a and b to two decimal places.

Solution

Since both of the ordered pairs (2, 5) and (5, 63) must satisfy the equation $y = ab^x$, we have the following system of equations:

$$5 = ab^2 \quad \text{Substitute 2 for } x \text{ and 5 for } y.$$
$$63 = ab^5 \quad \text{Substitute 5 for } x \text{ and 63 for } y.$$

It will be slightly easier to solve this system if we switch the equations to list the equation with the greater exponent of b first:

$$63 = ab^5$$
$$5 = ab^2$$

We divide the left sides and divide the right sides of the two equations to get the following result for nonzero a and b:

$$\frac{63}{5} = \frac{ab^5}{ab^2}$$

By then applying the properties $\frac{b^m}{b^n} = b^{m-n}$ and $\frac{a}{a} = 1$, where a and b are nonzero, to the right-hand side of the equation, we have an equation in terms of b (and not a):

$$\frac{63}{5} = b^3$$

We can now solve for b by finding the cube root of $\frac{63}{5}$:

$$b^3 = \frac{63}{5} \quad \text{If } c = d, \text{ then } d = c.$$
$$b = \left(\frac{63}{5}\right)^{1/3} \quad \text{The solution of } b^3 = k \text{ is } k^{1/3}.$$
$$\approx 2.33 \quad \text{Compute.}$$

So, we can substitute 2.33 for the constant b in the equation $y = ab^x$:

$$y \approx a(2.33)^x$$

To find a, we substitute the coordinates of the given point (2, 5) into $y = a(2.33)^x$:

$$5 = a(2.33)^2 \quad \text{Substitute 2 for } x \text{ and 5 for } y.$$
$$\frac{5}{(2.33)^2} = a \quad \text{Divide both sides by } 2.33^2.$$
$$a \approx 0.92 \quad \text{Compute.}$$

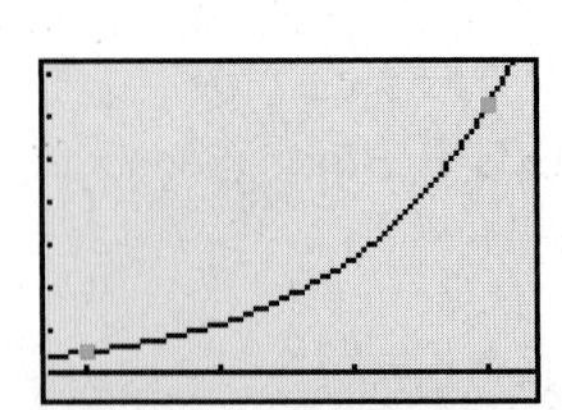

Figure 42 Check that the curve approximately contains (2, 5) and (5, 63)

So, an equation that approximates the exponential curve that passes through (2, 5) and (5, 63) is $y = 0.92(2.33)^x$.

We use a graphing calculator to verify our work (see Fig. 42). ■

In summary, **we can find an equation of an exponential function by using the base multiplier property or by using two points. Both methods give the same result.**

group exploration

Comparing three ways to find exponential equations

An exponential curve contains the points listed in Table 26.

1. Use the point (0, 5) and one other point in Table 26 to find an equation of the curve (see Example 5).
2. Use two points in Table 26 other than (0, 5) to find an equation of the curve (see Example 6).
3. Use the base multiplier property to find an equation of the curve. [**Hint:** First find $f(1)$ by recognizing a pattern.]
4. Compare your equations from Problems 1, 2, and 3.
5. An exponential curve contains the points listed in Table 27. Which method would you use to find an equation $y = ab^x$ that approximates the exponential curve? Explain. Also, find the equation. Round the value of b to two decimal places.

Table 26 Solutions of an Exponential Equation

x	$f(x)$
0	5
2	20
4	80
6	320
8	1280

Table 27 Solutions of an Exponential Equation

x	$f(x)$
0	400
3	200
6	100
9	50
12	25

TIPS FOR SUCCESS: Ask Questions

When you have a question during class time, do you ask it? Many students are reluctant to ask questions. If you tend to shy away from asking questions, keep in mind that the main idea of school is for you to learn through open communication with your instructor and other students.

If you are confused about a concept, it's likely that other students in your class are confused, too. If you ask your question, everyone else who is confused will be grateful that you asked. Most instructors want students to ask questions. It helps an instructor know when students understand the material and when they are having trouble.

HOMEWORK 4.4

FOR EXTRA HELP ▶

Student Solutions Manual PH Math/Tutor Center MathXL® MyMathLab

1. Some values of functions f, g, h, and k are provided in Table 28. Find a possible equation of each function. Verify your results with a graphing calculator table.

Table 28 Values of Four Functions (Exercise 1)

x	$f(x)$	$g(x)$	$h(x)$	$k(x)$
0	4	36	5	250
1	8	12	50	50
2	16	4	500	10
3	32	$\frac{4}{3}$	5000	2
4	64	$\frac{4}{9}$	50,000	$\frac{2}{5}$

2. Some values of functions f, g, h, and k are provided in Table 29. Find a possible equation of each function. Verify your results with a graphing calculator table.

Table 29 Values of Four Equations (Exercise 2)

x	$f(x)$	$g(x)$	$h(x)$	$k(x)$
0	80	4	3	3700
1	40	12	15	370
2	20	36	75	37
3	10	108	375	3.7
4	5	324	1875	0.37

3. Some values of functions f, g, h, and k are provided in Table 30. Find a possible equation of each function. Verify your results with a graphing calculator table. [**Hint:** Use linear or exponential equations.]

Table 30 Values of Four Functions (Exercise 3)

x	$f(x)$	$g(x)$	$h(x)$	$k(x)$
0	100	100	2	2
1	50	50	6	6
2	25	0	10	18
3	12.5	−50	14	54
4	6.25	−100	18	162

4. Some values of functions f, g, h, and k are provided in Table 31. Find a possible equation of each function. Verify your results with a graphing calculator table. [**Hint:** Use linear or exponential equations.]

Table 31 Values of Four Functions (Exercise 4)

x	$f(x)$	$g(x)$	$h(x)$	$k(x)$
0	3	19	2	512
1	12	13	9	128
2	48	7	16	32
3	192	1	23	8
4	768	−5	30	2

Find all real-number solutions. Round your result(s) to the second decimal place. Verify that your results satisfy the equation.

5. $b^2 = 16$

6. $b^4 = 81$

7. $b^3 = 27$

8. $b^5 = 100{,}000$

9. $3b^5 = 96$

10. $5b^2 = 45$

11. $35b^4 = 15$

12. $44b^3 = 12$

13. $3.6b^3 = 42.5$

14. $1.7b^4 = 86.4$

15. $32.7b^6 + 8.1 = 392.8$

16. $2.1b^5 - 8.2 = 237.5$

17. $\frac{1}{4}b^3 - \frac{1}{2} = \frac{9}{4}$

18. $\frac{1}{6}b^4 + \frac{5}{3} = \frac{11}{2}$

19. $\frac{b^6}{b^2} = 81$

20. $\frac{b^{10}}{b^3} = 2187$

21. $\frac{b^8}{b^3} = \frac{79}{5}$

22. $\frac{b^9}{b^6} = \frac{2}{9}$

Solve the equation for b, where n and m − n are odd, $a \neq 0$, and $p \neq 0$.

23. $b^n + k = p$

24. $ab^n + k = p$

25. $\frac{b^n}{a} + k = p$

26. $\frac{b^m}{b^n} = p$

Find an approximate equation $y = ab^x$ of the exponential curve that contains the given pair of points. Round the value of b to two decimal places. Verify your result with a graphing calculator.

27. (0, 4) and (1, 8)

28. (0, 5) and (1, 15)

29. (0, 3) and (5, 100)

30. (0, 8) and (4, 79)

31. (0, 87) and (6, 14)

32. (0, 256) and (7, 23)

33. (0, 7.4) and (3, 1.3)

34. (0, 2.1) and (5, 9.7)

35. (0, 5.5) and (2, 73.9)

36. (0, 97.2) and (4, 17.1)

37. (0, 39.18) and (15, 3.66)

38. (0, 12.94) and (20, 357.03)

39. Find an equation of the exponential curve sketched in Fig. 43. [**Hint:** Choose two points whose coordinates appear to be integers.]

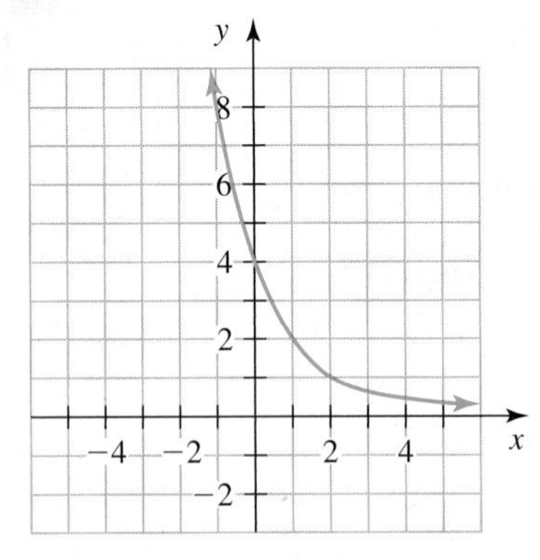

Figure 43 Graph of f—Exercise 39

40. Find an equation of the exponential curve sketched in Fig. 44.

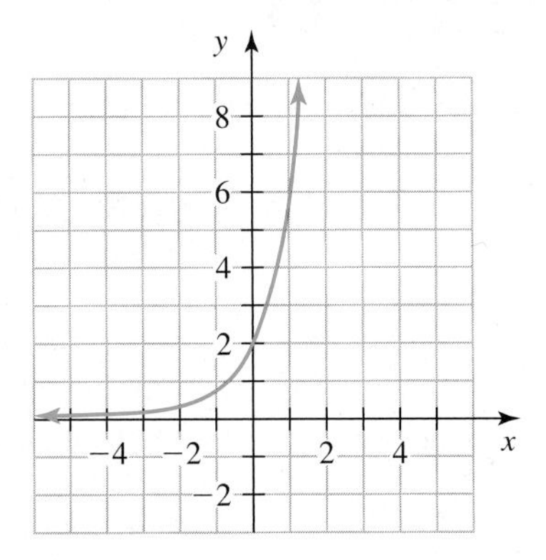

Figure 44 Graph of f—Exercise 40

Find an approximate equation $y = ab^x$ of the exponential curve that contains the given pair of points. Round the values of a and b to two decimal places. Verify your result with a graphing calculator.

41. (1, 4) and (2, 12)

42. (2, 5) and (3, 10)

43. (3, 4) and (5, 9)

44. (2, 7) and (5, 1)

45. (10, 329) and (30, 26)

46. (11, 8) and (17, 492)

47. (5, 8.1) and (9, 2.4)

48. (1, 3.5) and (5, 1.3)

49. (2, 73.8) and (7, 13.2)

50. (4, 6.3) and (10, 250.8)

51. (13, 24.71) and (21, 897.35)

52. (8, 39.43) and (12, 6.52)

53. Find an equation of the exponential curve sketched in Fig. 45.

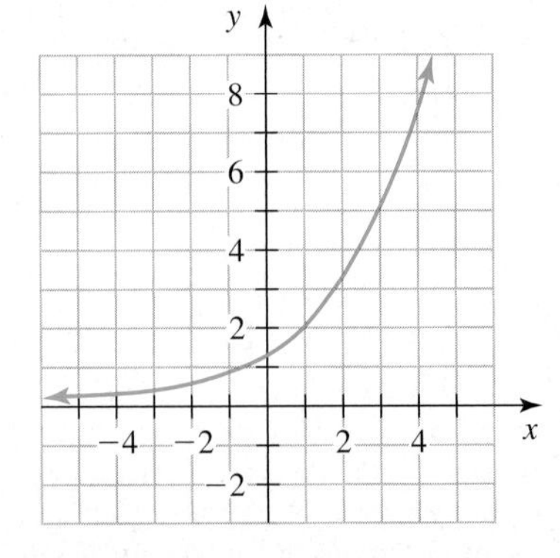

Figure 45 Graph of f—Exercise 53

54. Find an equation of the exponential curve sketched in Fig. 46.

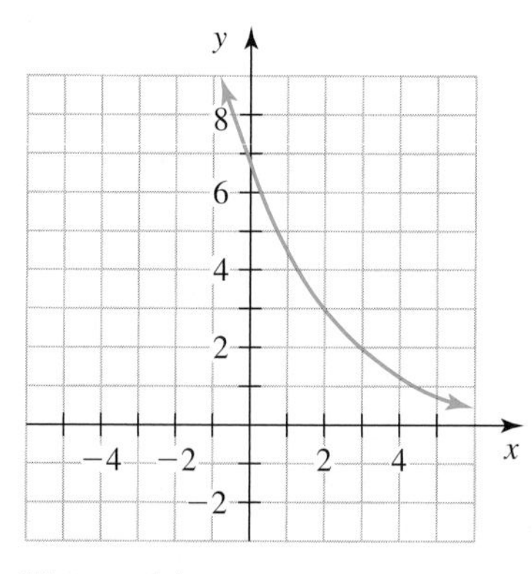

Figure 46 Graph of f—Exercise 54

55. Solve the system. [**Hint:** Think graphically.]

$$y = 6(4)^x$$
$$y = 6\left(\frac{1}{3}\right)^x$$

56. Solve the system. [**Hint:** Think graphically.]

$$y = 7(3)^x$$
$$y = 4(3)^x$$

57. a. Is there an exponential curve that contains the given point? If so, find an equation of the curve. If not, explain.

i. (0, 2)

ii. (2, 0)

b. Is there an exponential curve that contains the points (2, −1) and (3, 1)? If so, find an equation of the curve. If not, explain.

58. a. Is there an exponential curve that passes through the given points? If so, find an equation of the curve. If not, explain.

i. (0, 5) and (1, 5)

ii. (0, 3) and (7, 3)

b. Is there an exponential curve that passes through two given points that have the same y-coordinate? Explain.

59. Describe the base multiplier property and explain why it makes sense. Include an example.

60. Describe how to find the equation of an exponential curve that contains two given points. Include both the case in which one of the points is the y-intercept and the case in which neither of the points is the y-intercept.

Related Review

Find all real-number solutions or simplify, whichever is appropriate. Round your solution(s) to the second decimal place.

61. $\dfrac{b^7}{b^2}$

62. $\dfrac{b^8}{b^4} = \dfrac{65}{3}$

63. $\dfrac{b^7}{b^2} = 76$

64. $\dfrac{b^8}{b^4}$

65. $\dfrac{8b^3}{6b^{-1}}$

66. $\dfrac{10b^{-7}}{15b^{-2}}$

67. $\dfrac{8b^3}{6b^{-1}} = \dfrac{3}{7}$

68. $\dfrac{10b^{-7}}{15b^{-2}} = \dfrac{4}{7}$

Let L be a linear function and E be an exponential function. Assume that the graphs of both L and E contain the pair of given points. Find equations of L and E. Use a graphing calculator to draw the graphs of L and E in the same viewing window.

69. (0, 2) and (1, 6)

70. (2, 8) and (5, 2)

The graph of a function contains the given pair of points. Could the function be linear, exponential, either linear or exponential, or neither? Explain.

71. (5, 3) and (7, 6)

72. (2, 6) and (4, 6)

73. In this exercise, you will compare the linear function $L(x) = 2x + 100$ with the exponential function $E(x) = 3(2)^x$.

a. Find the y-intercept of both functions.

b. For functions L and E, describe what happens to the value of y as the value of x increases by 1.

c. Based on your responses to parts (a) and (b), which function's outputs will eventually dominate the other's outputs? Explain.

d. Use a graphing calculator table to verify your responses to parts (a)–(c). To do this, enter the functions and set up a table as indicated in Figs. 47 and 48, respectively.

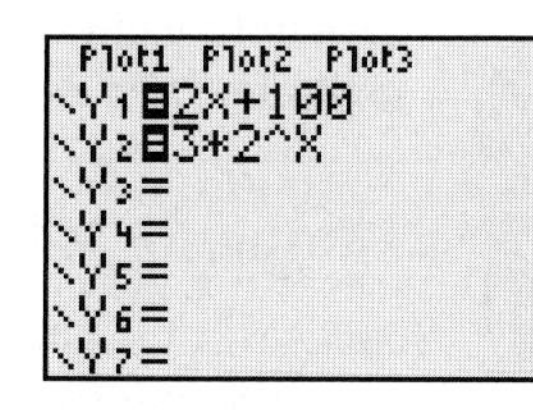

Figure 47 Enter the functions

Figure 48 Set up a table

74. Is it possible for a linear function and an exponential function to have the indicated number of intersection points? If so, give equations of the two functions. If not, explain. [**Hint:** First sketch some graphs.]

a. 3 intersection points

b. 2 intersection points

c. 1 intersection point

d. 0 intersection points

Expressions, Equations, Functions, and Graphs

Perform the indicated instruction. Then use words such as linear, exponential, function, one variable, *and* two variables *to describe the expression, equation, or system.*

75. Graph $f(x) = 3(2)^x$ by hand.

76. Find all real-number solutions of $2b^6 = 259$. Round any result(s) to the fourth decimal place.

77. Find $f(-3)$, where $f(x) = 3(2)^x$.

78. Simplify $\dfrac{8b^{-3}c^6}{12b^2c^3}$.

4.5 USING EXPONENTIAL FUNCTIONS TO MODEL DATA

Objectives

- Find an equation of an *exponential model* by using the base multiplier property.
- Model a *half-life* situation.
- Find an equation of an exponential model by using data described in words.
- Find an equation of an exponential model by using data displayed in a table.
- For a model $f(t) = ab^t$, know the meaning of the coefficient a and the base b in terms of the situation being modeled.
- Make estimates and predictions by using an exponential model.

In Section 4.4, we found equations of exponential functions. In this section, we use this skill to model authentic situations.

Using the Base Multiplier Property to Find a Model

We can use the base multiplier property to find an exponential model.

DEFINITION Exponential model, exponentially related, approximately exponentially related

An **exponential model** is an exponential function, or its graph, that describes the relationship between two quantities for an authentic situation. If all of the data points for a situation lie on an exponential curve, then we say that the independent and dependent variables are **exponentially related.** If no exponential curve contains all of the data points, but an exponential curve comes close to all of the data points (and perhaps contains some of them), then we say that the variables are **approximately exponentially related.**

Example 1 Modeling with an Exponential Function

Suppose that a peach has 3 million bacteria on it at noon on Monday and that one bacterium divides into two bacteria every hour, on average (see Fig. 49).

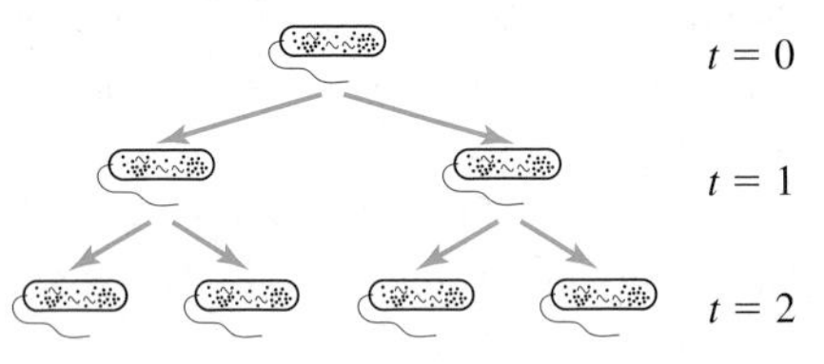

Figure 49 A result of bacteria dividing every hour

Let $B = f(t)$ be the number of bacteria (in millions) on the peach at t hours after noon on Monday.

1. Find an equation of f.
2. Predict the number of bacteria on the peach at noon on Tuesday.

Table 32 Values of a Bacteria Model

t (hours)	$B = f(t)$ (millions)
0	3
1	6
2	12
3	24
4	48

Solution

1. We complete a table of values of f based on the assumption that one bacterium divides into two bacteria every hour (see Table 32).

   As the value of t increases by 1, the value of B changes by greater and greater amounts, so it would *not* be appropriate to model the data by using a linear function. Note, though, that as the value of t increases by 1, the value of B is multiplied by 2, so we *can* model the situation by using an *exponential* model of the form $f(t) = a(2)^t$. The B-intercept is $(0, 3)$, so $f(t) = 3(2)^t$.

   We use a graphing calculator table and graph to verify our work (see Figs. 50 and 51).

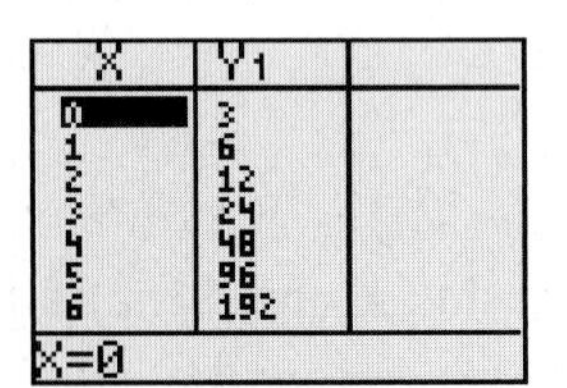

X	Y1
0	3
1	6
2	12
3	24
4	48
5	96
6	192

X=0

Figure 50 Table for bacteria model

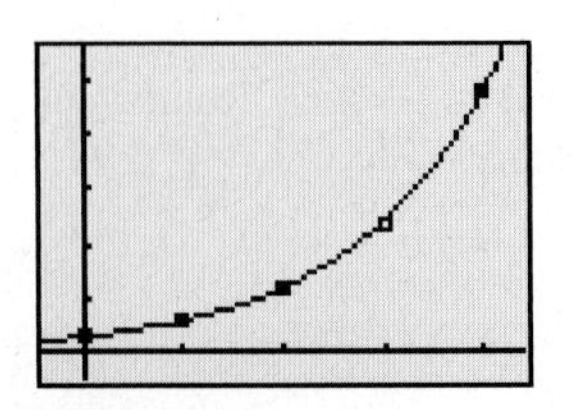

Figure 51 Graph of bacteria model

2. We use $t = 24$ to represent noon on Tuesday. We substitute 24 for t in our equation of f:

$$f(24) = 3(2)^{24} = 50{,}331{,}648$$

According to the model, there would be 50,331,648 million bacteria. To omit writing "million," we must add six zeros to 50,331,648—that is, 50,331,648,000,000. There would be about 50 trillion bacteria at noon on Tuesday. ■

For an exponential function $y = ab^t$, the y-intercept is $(0, a)$. So, **if $y = ab^t$ is an exponential model where y is a quantity at time t, then the coefficient a is the value of that quantity present at time $t = 0$.** For example, the bacteria model $f(t) = 3(2)^t$ has coefficient 3, which represents the 3 million bacteria that were present at time $t = 0$ (noon on Monday).

We can use an exponential function to model the value of an investment that earns r percent interest compounded annually. The term **r percent interest compounded annually** means that the interest earned each year equals r percent of the principal and any interest earned in previous years (all of which becomes part of the investment).

Example 2 Modeling with an Exponential Function

A person invests \$5000 in an account that earns 6% interest compounded annually.

1. Let $V = f(t)$ be the value (in dollars) of the account at t years after the money is invested. Find an equation of f.
2. What will be the value after 10 years?

Solution

Table 33 Values of a Compound Interest (6%) Account

t	$V = f(t)$
0	5000.00
1	$5000.00(1.06) = 5300.00$
2	$5300.00(1.06) = 5618.00$
3	$5618.00(1.06) = 5955.08$
4	$5955.08(1.06) \approx 6312.38$

1. Each year, the investment value is equal to the previous year's value (100% of it) plus 6% of the previous year's value. So, the value is equal to 106% of the previous year's value. For example, after one year, the value will be 106% of \$5000, or $1.06(5000) = 5300$ dollars. After two years, the value will be $1.06(5300) = 5618$ dollars. See Table 33.

   As the value of t increases by 1, the value of V is multiplied by 1.06. So, f is the exponential function $f(t) = a(1.06)^t$. Since the value of the account at the start is \$5000, we have $a = 5000$. So, $f(t) = 5000(1.06)^t$.
2. To find the value in 10 years, we substitute 10 for t:

$$f(10) = 5000(1.06)^{10} \approx 8954.24$$

The value will be \$8954.24 in 10 years. ■

In Example 2, we used the function $f(t) = 5000(1.06)^t$ to model the value of the 6% compounded-interest account. Note that subtracting 1 from the base 1.06 gives the interest rate in decimal form:

$$b - 1 = 1.06 - 1 = 0.06 = \text{interest rate (in decimal form)}$$

Half-life Applications

If a quantity decays exponentially, we can describe how quickly it decays by its *half-life*.

DEFINITION Half-life

If a quantity decays exponentially, the **half-life** is the amount of time it takes for that quantity to be reduced to half (see Fig. 52).

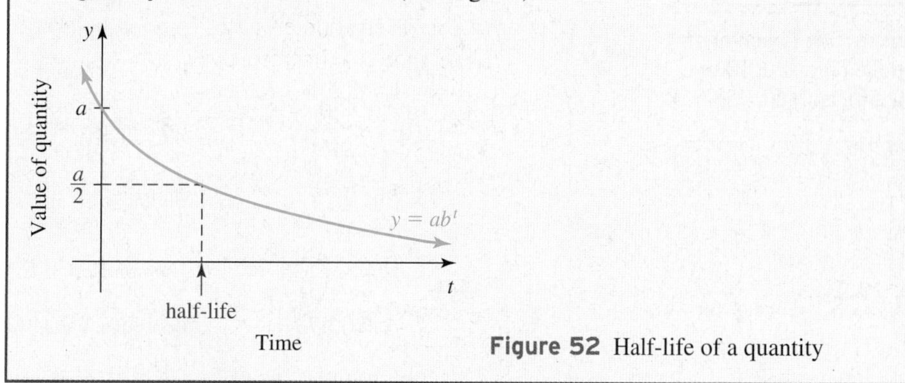

Figure 52 Half-life of a quantity

For example, the half-life of the radioactive element hydrogen-3 is 12.3 years, which means that every 12.3 years the number of hydrogen-3 atoms is reduced to half. The half-lives of some radioactive elements are much different from that of hydrogen-3. For instance, polonium-214 has a half-life of 0.164 millisecond, and uranium-238 has a half-life of 4.5 billion years!

Example 3 Modeling with an Exponential Function

The world's worst nuclear accident occurred in Chernobyl, Ukraine, on April 26, 1986. Immediately afterward, 28 people died from acute radiation sickness. So far, about 25,000 people have died from exposure to radiation, mostly due to the release of the radioactive element cesium-137 (Source: *Medicine Worldwide*).

Cesium-137 has a half-life of 30 years. Let $P = f(t)$ be the percent of the cesium-137 that remains at t years since 1986.

1. Find an equation of f.
2. Describe the meaning of the base of f.
3. What percent of the cesium-137 will remain in 2010?

Table 34 Percentages of Cesium-137 That Remain

Year t	Percent P
0	$100 = 100\left(\frac{1}{2}\right)^0$
30	$100 \cdot \frac{1}{2} = 100 \cdot \left(\frac{1}{2}\right)^1$
60	$100 \cdot \frac{1}{2} \cdot \frac{1}{2} = 100\left(\frac{1}{2}\right)^2$
90	$100 \cdot \frac{1}{2} \cdot \frac{1}{2} \cdot \frac{1}{2} = 100\left(\frac{1}{2}\right)^3$
t	$100\left(\frac{1}{2}\right)^{t/30}$

Solution

1. We discuss two methods of finding an equation of f.

Method 1 At time $t = 0$, 100% of the cesium-137 is present. At time $t = 30$, there will be $\frac{1}{2}(100) = 50$ percent. At time $t = 60$, there will be $\frac{1}{2} \cdot \frac{1}{2}(100) = 25$ percent. We organize these results, and one more calculation, in Table 34.

From Table 34, we see that the situation can be modeled well with an exponential function. Each exponent in the second column of the table is equal to the value of t in the first column, divided by 30. Thus, the equation of f is

$$f(t) = 100\left(\frac{1}{2}\right)^{t/30}$$

We can use a graphing calculator table and graph to verify our equation (see Figs. 53 and 54). We can write this equation in the form $f(t) = ab^t$:

$$f(t) = 100\left(\frac{1}{2}\right)^{t/30} = 100\left(\frac{1}{2}\right)^{\frac{1}{30}\cdot t} = 100\left(\left(\frac{1}{2}\right)^{\frac{1}{30}}\right)^t$$

Since $\left(\frac{1}{2}\right)^{1/30} \approx 0.977$, we can write

$$f(t) = 100(0.977)^t$$

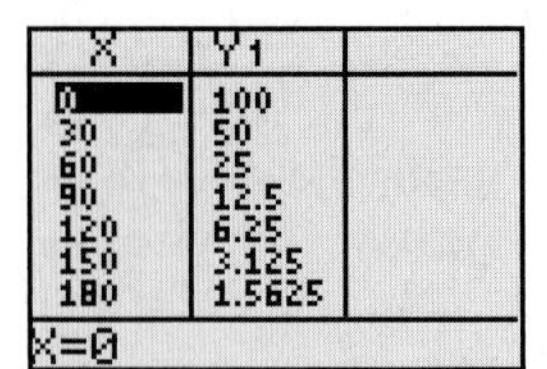

X	Y1
0	100
30	50
60	25
90	12.5
120	6.25
150	3.125
180	1.5625

X=0

Figure 53 Table for f

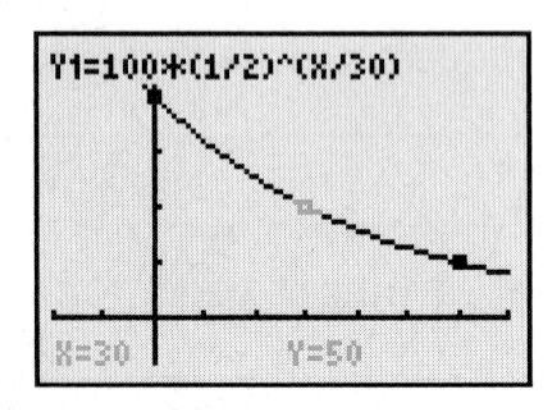

Figure 54 Graph of f and the points (0, 100), (30, 50), and (60, 25)

Method 2 Instead of recognizing a pattern from a table, we can find an equation of f by using the points (0, 100) and (30, 50). Since the P-intercept is (0, 100), we have

$$P = f(t) = 100b^t$$

To find b, we substitute the coordinates of (30, 50) into the equation $f(t) = 100b^t$:

$$\begin{aligned}
50 &= 100b^{30} && \text{Substitute 30 for } t \text{ and 50 for } f(t).\\
100b^{30} &= 50 && \text{If } c = d, \text{ then } d = c.\\
b^{30} &= \frac{50}{100} && \text{Divide both sides by 100.}\\
b^{30} &= \frac{1}{2} && \text{Simplify.}\\
b &= \pm\left(\frac{1}{2}\right)^{1/30} && \text{The solution of } b^{30} = k \text{ is } \pm k^{1/30}.\\
&\approx 0.977 && \text{Compute; base of an exponential function is positive.}
\end{aligned}$$

So, an equation of f is $f(t) = 100(0.977)^t$, the same equation we found earlier.

2. The base of f is 0.977. Each year, 97.7% of the previous year's cesium-137 is present. In other words, the cesium-137 decays by 2.3% each year.
3. Since $2010 - 1986 = 24$, we substitute 24 for t in the equation $f(t) = 100(0.977)^t$:

$$f(24) = 100(0.977)^{24} \approx 57.21$$

In 2010, about 57.2% of the cesium-137 will remain. ■

Meaning of the Base of an Exponential Model

If $f(t) = ab^t$, where $a > 0$, models a quantity at time t, then the percent rate of change is constant. In particular,

- If $b > 1$, then the quantity grows exponentially at a rate of $b - 1$ percent (in decimal form) per unit of time.
- If $0 < b < 1$, then the quantity decays exponentially at a rate of $1 - b$ percent (in decimal form) per unit of time.

In Example 2, we found the compound-interest model $f(t) = 5000(1.06)^t$. Since $1.06 - 1 = 0.06$, the base 1.06 indicates that the value grows exponentially by 6% per year.

In Example 3, we found the cesium-137 model $f(t) = 100(0.977)^t$. Since $1 - 0.977 = 0.023$, the base 0.977 indicates that the cesium-137 decays exponentially by 2.3% per year.

Finding a Model by Using Data Described in Words

Instead of using the base multiplier property, we can sometimes find an equation of a model by using two data points described in words.

Example 4 Modeling with an Exponential Function

Sales of energy and nutrition bars have grown approximately exponentially from \$0.2 billion in 1997 to \$1.2 billion in 2004 (Source: *Frost & Sullivan*). Predict the sales in 2011.

Solution

Table 35 Known Values of t and s

Years Since 1997 t	Sales (billions of dollars) s
0	0.2
7	1.2

Let s be the sales (in billions of dollars) of energy and nutrition bars in the year that is t years since 1997. Known values of t and s are shown in Table 35.

Since the variables t and s are approximately exponentially related, we want an equation of the form $s = ab^t$. From Table 35, we see that the s-intercept is (0, 0.2). So, the equation is of the form

$$s = 0.2b^t$$

To find b, we substitute the coordinates of (7, 1.2) into the equation $s = 0.2b^t$ and then solve for b:

$$\begin{aligned} 1.2 &= 0.2b^7 && \text{Substitute 7 for } t \text{ and 1.2 for } s. \\ 0.2b^7 &= 1.2 && \text{If } c = d, \text{ then } d = c. \\ b^7 &= 6 && \text{Divide both sides by 0.2.} \\ b &= 6^{1/7} && \text{The solution of } b^7 = k \text{ is } k^{1/7}. \\ b &\approx 1.292 && \text{Compute.} \end{aligned}$$

Then we substitute 1.292 for b in the equation $s = 0.2b^t$:

$$s = 0.2(1.292)^t$$

Finally, to predict the sales in 2011, we substitute $2011 - 1997 = 14$ for t in the equation $s = 0.2(1.292)^t$ and solve for s:

$$s = 0.2(1.292)^{14} \approx 7.22$$

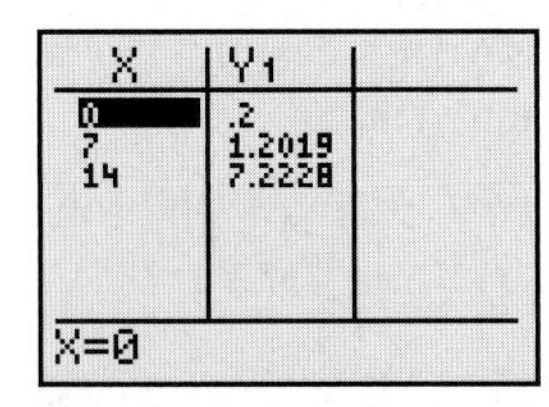

Figure 55 Verify the work

The model predicts that the sales will be \$7.22 billion in 2011. We use a graphing calculator table (see Fig. 55) to check that each of the three ordered pairs (0, 0.2), (7, 1.2), and (14, 7.22) approximately satisfies the equation $s = 0.2(1.292)^t$. ■

Finding an Exponential Model by Using Data Displayed in a Table

In Example 5, we will find an equation of an exponential model by using data shown in a table.

Example 5 Modeling with an Exponential Function

The Dow Jones Industrial Average is a measure of the strength of companies on the New York Stock Exchange (see Table 36).

Table 36 Dow Jones Industrial Averages*

Year	Dow Jones Average	Year	Dow Jones Average
1980	839	1990	2753
1982	875	1992	3169
1984	1259	1994	3754
1986	1547	1996	5117
1988	1939		

Source: *Market Watch, Inc.*
*For the start of each given year.

Let $A = f(t)$ be the Dow Jones Average at t years since 1980.

1. Find an equation of f.
2. Estimate the percent rate of growth of the Dow Jones Average.

Solution

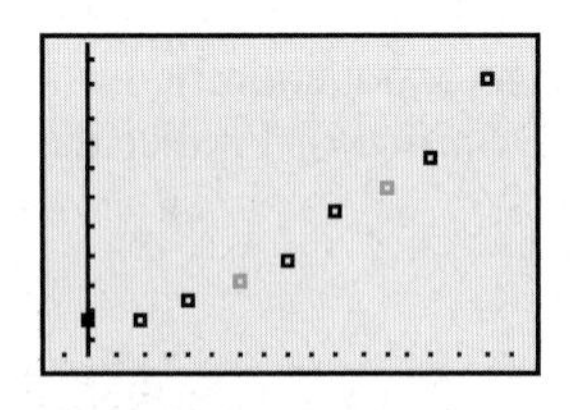

Figure 56 Dow Jones scattergram

1. We use a graphing calculator to view a scattergram of the data (see Fig. 56). Since the points "bend upward," the scattergram suggests that we can model the data better with an exponential function than with a linear function. If we imagine an exponential curve that contains the points (6, 1547) and (12, 3169), it appears that the curve might come close to the other data points. To find an equation of this curve, we substitute the coordinates of the points (6, 1547) and (12, 3169) into the equation $f(t) = ab^t$:

$$3169 = ab^{12}$$
$$1547 = ab^6$$

Next, we divide the two left sides and divide the two right sides and solve for b:

$$\frac{3169}{1547} = \frac{ab^{12}}{ab^6}, \quad \text{where } a \neq 0 \text{ and } b \neq 0$$ Divide left sides and divide right sides.

$$\frac{3169}{1547} = b^6$$ Simplify; subtract exponents: $\frac{b^m}{b^n} = b^{m-n}$

$$b = \pm\left(\frac{3169}{1547}\right)^{1/6}$$ The solutions of $b^6 = k$ are $\pm k^{1/6}$.

$$\approx 1.127$$ Compute; base of an exponential function is positive.

So, an equation is $f(t) = a(1.127)^t$. To find a, we substitute the coordinates of (6, 1547) into the equation:

$$1547 = a(1.127)^6$$ Substitute 6 for t and 1547 for $f(t)$.

$$a = \frac{1547}{1.127^6}$$ Divide both sides by 1.127^6.

$$\approx 755.00$$ Compute.

Figure 57 Check how well the model fits the data

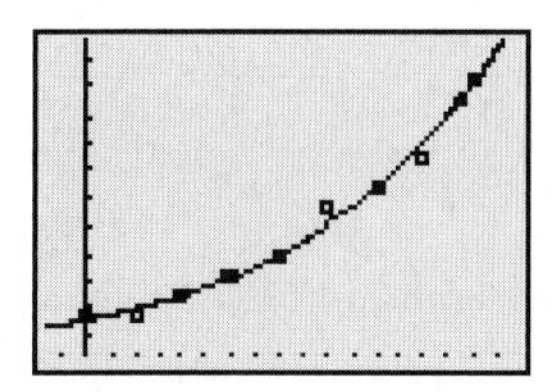

Figure 58 Compare the fit of models f and r

The equation is $f(t) = 755.00(1.127)^t$. The graph in Fig. 57 shows that our exponential model fits the data quite well.

2. The base of $f(t) = 755.00(1.127)^t$ is 1.127. So, the percent rate of growth of the Dow Jones Average between 1980 and 1996 was about 12.7%. ■

We used two points to find the exponential model $f(t) = 755.00(1.127)^t$ in Example 5. We can also find an exponential model by using a graphing calculator's *exponential regression*. Exponential regression gives the model $r(t) = 776.79(1.124)^t$. In Fig. 58, we see that the graphs of the models f and r are so similar that when we use ZoomStat, there appears to be just one curve. For graphing calculator instructions, see Section B.16.

We call the equation $r(t) = 776.79(1.124)^t$ the **exponential regression equation** for the given data. We refer to its graph as an **exponential regression curve.**

The coefficients of both models are approximately equal, as are the bases. In particular, the base 1.127 of f estimates that the Dow Jones Average grows exponentially by 12.7% each year. The base 1.124 of r estimates that the Dow Jones Average grows exponentially by 12.4% each year.

Example 6 Using a Model to Analyze a Situation

The exponential regression equation for the Dow Jones Industrial Average $r(t)$ is

$$r(t) = 776.79(1.124)^t$$

where t is the number of years since 1980.

Table 37 Dow Jones Industrial Averages

Year	Dow Jones Average
1998	7908
2000	11,497
2002	10,022
2004	10,454
2006	10,718

Source: *Market Watch, Inc.*

1. The Dow Jones Averages for more recent years than those in Example 5 are shown in Table 37. Calculate the error in using the model r to estimate the Dow Jones Averages for the years shown in Table 37.
2. Use a graphing calculator to graph the model r and the scattergram of the data in the same viewing window.
3. Based on your work in Problems 1 and 2, describe how well the function r models this situation.

Solution

1. First, we use our model to estimate the Average in 1998 by evaluating r at 18:

$$r(18) = 776.79(1.124)^{18} \approx 6369.38$$

Our estimate is 6369. Next, we compute the difference of our estimate, 6369, and the true value, 7908:

$$6369 - 7908 = -1539$$

So, our model underestimates the Average in 1998 by 1539. We perform similar calculations for 2000, 2002, 2004, and 2006 and list these calculations in Table 38.

Table 38 Estimates of Dow Jones Averages

Year	Model Estimate	Dow Jones Average	Difference of Estimate and Average
1998	6369	7908	−1539
2000	8047	11,497	−3450
2002	10,166	10,022	144
2004	12,844	10,454	2390
2006	16,227	10,718	5509

2. We graph the model and the scattergram of the data in Fig. 59.

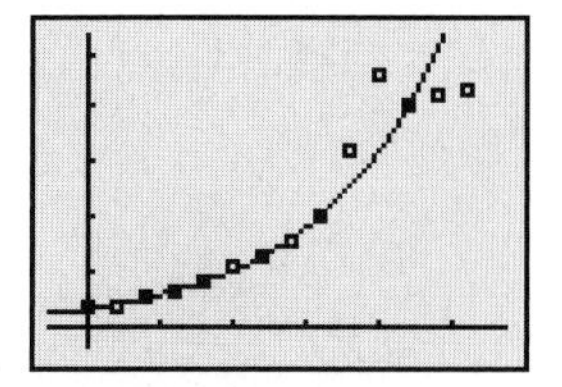

Figure 59 The graph of r and the scattergram for 1980–2006

3. From Table 38, we see that the model r underestimates the Average in 1998 and 2000 and overestimates the Average in 2002, 2004, and 2006. We can draw the same conclusion from Fig. 59, although it is difficult to tell that the model overestimates the Average in 2002, because the error in that year is small compared with the scale of the vertical axis.

Although the model describes the Averages quite well from 1980 to 1996, it does a poor job of describing most of the Averages from 1998 to 2006. However, throughout history, the Averages have swung wildly up and down in the short run, yet they have grown approximately exponentially in the long run. It is possible that in the long run the Averages will grow by about 12.4% per year, as suggested by the base of the model $r(t) = 776.79(1.124)^t$. ■

At this point in the course, we have a choice between modeling some data with a linear function or an exponential function. For some authentic situations, we will know which type of model to use based on our knowledge of the situation. For other situations, we can decide by viewing a scattergram of the data. Here, we review the four-step modeling process.

Four-Step Modeling Process

To find a model and then make estimates and predictions,

1. Create a scattergram of the data. Decide whether a line or an exponential curve comes close to the points.
2. Find an equation of your function.
3. Verify your equation by checking that the graph comes close to all of the data points.
4. Use your equation of the model to draw conclusions, make estimates, and/or make predictions.

If you discover that a model does not fit a data set well, a good first step is to check for any graphing or calculation errors. If your work appears to be correct, then try using different points to find your equation. Another option is to increase or decrease one or both of the constants in your equation until the fit is good.

group exploration

Comparing a linear model with an exponential model

In 1950, world population was 2.5 billion. In 1987, it was 5.0 billion (Source: *U.S. Census Bureau*).

1. First, assume that world population is growing exponentially. Let $E(t)$ be the world's population (in billions) at t years since 1950. Find an equation of E.
2. Now assume that world population is growing linearly. Let $L(t)$ be the world's population (in billions) at t years since 1950. Find an equation of L.
3. Use your equations of E and L to make two predictions of the world's population for each of the following years.

   a. 2010 **b.** 2050 **c.** 2150
4. Use the window settings shown in Fig. 60 to compare the graphs of E and L.
5. Will there be much difference in the world's population if it grows exponentially or linearly in the short run? in the long run? Explain.

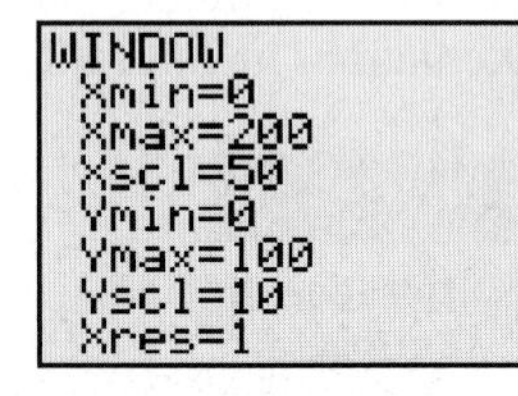

Figure 60 Compare the graphs of the exponential and linear models

group exploration

Looking ahead: Inverse function

A company's profit in 2005 was \$5 million. Each year, the profit increases by \$2 million. Let $p = f(t)$ be the profit (in millions of dollars) for the year that is t years after 2005.

1. Find an equation of f. Write the equation in function notation. Also, write the equation with the variables p and t.
2. Use an equation of f to predict when the company will have a profit of \$21 million.
3. In Problem 1, you found an equation with p and t. Solve this equation for t.
4. Substitute 21 for p in the equation that you found in Problem 3, and solve for t.
5. Compare your results for Problems 2 and 4.
6. Enter the equation you found in Problem 3 in a graphing calculator to help you complete Table 39.
7. In Problems 2 and 4, you used two different methods to find when the company will have a profit of \$21 million. If the company wants to know when it might attain 15 different profit levels, which method would be best to use? Explain.

Table 39 From Profit to Year

Profit (millions of dollars) p	Years Since 2005 t
21	
22	
23	
24	
25	

TIPS FOR SUCCESS: Show What You Know

Even if you don't know how to do one step of a problem, show your instructor that you understand the other parts of the problem. Check with your instructor first, but you may earn partial credit if you pick a number (which would probably be incorrect) to be the result for the one step and then show what you would do with that number in the remaining steps of the solution.

For example, suppose you want to find an equation of the line that passes through the points (4, 7) and (5, 9), but you have forgotten how to find slope. You could write,

I've drawn a blank on finding the slope. However, if the slope is 3, then

$$\begin{aligned} y &= 3x + b \\ 7 &= 3(4) + b \\ 7 &= 12 + b \\ -5 &= b \end{aligned}$$

Therefore, $y = 3x - 5$.

You could point out that you know that your result is incorrect, because the graph of $y = 3x - 5$ does not pass through the point (5, 9). Also, seeing your result (with a graph) may jog your memory about finding the slope and allow you to go back and do the problem correctly.

HOMEWORK 4.5

FOR EXTRA HELP ▶

Student Solutions Manual

PH Math/Tutor Center

MathXL®

MyMathLab

1. Suppose a rumor is spreading that the U.S. government plans to give each college student \$2000. Assume that 40 people have heard the rumor as of today and that each day the total number of people who have heard it triples. Let $f(t)$ be the total number of people who have heard the rumor at t days since today.
   a. Find an equation of f.
   b. How many people will have heard the rumor 10 days from now?
   c. How many people will have heard the rumor 15 days from now? Has model breakdown occurred? [**Hint:** The U.S. population was about 299 million in 2006.]
2. Suppose a flu epidemic has broken out at your school. Assume that on February 10, a total of 20 people have the flu and that each day the total number of people who have gotten the flu triples. Let $f(t)$ be the total number of people who have gotten the flu by the day that is t days since February 10.
   a. Find an equation of f.
   b. Find $f(4)$. What does it mean in this situation?
   c. Find $f(10)$. What does it mean in this situation?
3. The name of the search engine Google™ is a play on the word "googol," which refers to the number 1 followed by one hundred zeros.
   a. In 2004, Google's index contained 8 billion web pages. According to founders Larry Page and Sergey Brin, the size of the index doubles every year. Let $f(t)$ be the number of web pages (in billions) in Google's index at t years since 2004. Find an equation of f.

b. Predict the number of web pages that Google's index will contain in 2010.

c. If the web pages in Google's index in 2010 were printed and stacked in one pile, what would be the height (in miles) of that pile? [**Hint:** Take a stack of 500 pages of computer paper to be 2 inches tall.]

4. One treatment of Roche's Tamiflu® is usually 10 pills taken over 5 days. Tamiflu production was 5.5 million treatments in 2002 and has approximately doubled each year since then (Source: *Roche*).

a. Let $n = g(t)$ be the number of treatments (in millions) produced in the year that is t years since 2002. Find an equation of g.

b. What is the n-intercept of the model? What does it mean in this situation?

c. Predict the number of treatments that will be produced in 2007.

d. Roche plans to produce 300 million treatments in 2007. If that happens, will production continue to double each year from 2002 to 2007? Explain.

5. The revenue from ring tones was \$91 million in 2003 and has grown by about 114% per year since then (Source: *Jupiter Research*). That is, each year the revenue is about 2.14 times the previous year's revenue.

a. Let $r = h(t)$ be the revenue (in millions of dollars) from ring tones in the year that is t years since 2003. Find an equation of h.

b. What is the r-intercept of the model? What does it mean in this situation?

c. What is the base b of $h(t) = ab^t$? What does it mean in this situation?

d. Predict the revenue from ring tones in 2009.

e. Jupiter Research predicts that revenue from ring tones will be \$724 million in 2009. If that happens, will each year's revenue continue to be 2.14 times the revenue of the previous year? Explain.

6. The revenue from music downloads was \$0.3 billion in 2004 and has grown by about 86% per year since then (Source: *Forrester Research*). That is, each year the revenue is about 1.86 times the previous year's revenue.

a. Let $r = f(t)$ be the revenue (in billions of dollars) in the year that is t years since 2004. Find an equation of f.

b. What is the r-intercept of the model? What does it mean in this situation?

c. What is the base b of $f(t) = ab^t$? What does it mean in this situation?

d. Predict the revenue in 2010.

7. a. About 194 thousand TiVo® subscribers got TiVo through Direct TV® in 2001, and that number has grown by about 60% per year since then (Source: *TiVo*). That is, each year there are about 1.6 times as many subscribers as in the preceding year. Let $D(t)$ be the number (in thousands) of such subscribers at t years since 2001. Find an equation of D.

b. About 15 thousand people were stand-alone TiVo subscribers in 2001, and that number has grown by about 190% per year since then (Source: *TiVo*). That is, each year there are about 2.9 times as many subscribers as in the preceding year. Let $S(t)$ be the number (in thousands) of such subscribers at t years since 2001. Find an equation of S.

c. Use "intersect" on a graphing calculator to find the intersection point of the graphs of D and S. What does it mean in this situation? [**Hint:** Use the window settings shown in Fig. 61.]

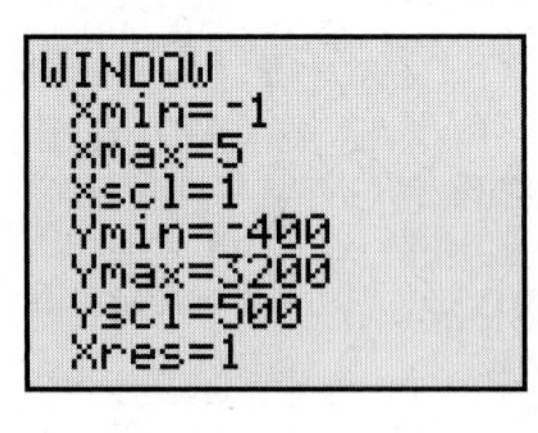

Figure 61 Exercise 7c

8. The average number of cable channels per household was 88 channels in 2002 and has grown by about 15% per year since then (Source: *National Cable & Telecommunications Association*). That is, each year since then, there have been about 1.15 times the number of channels as in the previous year.

a. Let $f(t)$ be the average number of cable channels at t years since 2002. Find an equation of f.

b. Estimate the average number of cable channels per household in 2006.

c. Currently, people must subscribe to bundles of channels rather than individual channels. In 2002, subscribers watched only 19% of the channels to which they subscribed, on average. The Federal Communications Commission has urged Congress to eliminate bundling by 2008. If that happens, predict the average number of channels per household in 2008.

9. Someone invests \$3000 in an account at 8% interest compounded annually. Let $f(t)$ be the value (in dollars) of the account at t years after she has invested the \$3000.

a. Find an equation of f.

b. What is the base b of your model $f(t) = ab^t$? What does it mean in this situation?

c. What is the coefficient a of your model $f(t) = ab^t$? What does it mean in this situation?

d. What will be the account's value in 15 years?

10. A person invests \$7000 at 10% interest compounded annually. Let $f(t)$ be the value (in dollars) of the account at t years after he has deposited the \$7000.

a. Find an equation of f.

b. What is the base b of your model $f(t) = ab^t$? What does it mean in this situation?

c. What is the coefficient a of $f(t) = ab^t$? What does it mean in this situation?

d. What will be the account's value in 10 years? Explain why the value has more than doubled, even though the investment earned 10% for 10 years.

11. A person invests \$4000 in stocks today, and their value doubles every 6 years. Let $f(t)$ be the value (in dollars) of the investment at t years from now.

a. Find an equation of f.

b. Find the value of the investment 20 years from now.

12. Someone invests \$2800 in stocks today, and their value doubles every 11 years. Let $g(t)$ be the value (in dollars) of the investment at t years from now.

a. Find an equation of g.

b. Find the value of the investment 25 years from now.

13. A person invests \$5000 at 6% interest compounded annually for 3 years and then invests the balance (the \$5000 plus the interest earned) in an account at 8% interest for 5 years. Find the value of the investment after the 8 years.

14. A person invests \$3500 at 3% interest compounded annually for 4 years and then invests the balance (the \$3500 plus the interest earned) in an account at 7% interest for 9 years. Find the value of the investment after the 13 years have elapsed.

15. Suppose a college bookstore sells 984 copies of a new (not used) textbook in 2003. Each year, the new-textbook sales are half of the previous year's sales (due to used-textbook sales) from 2003 to 2006. Let $s = g(t)$ be the sales of copies of the new textbook in the year that is t years since 2003.
a. Find an equation of g.
b. What is the s-intercept of the model? What does it mean in this situation?
c. Find $g(3)$. What does it mean in this situation?
d. What is the half-life of new-textbook sales?

16. Laserdisc player sales were \$108 million in 1995. Sales approximately halved each year from 1995 to 1999 (Source: *Consumer Electronics Association*). Let $s = f(t)$ be the sales (in millions of dollars) for the year that is t years since 1995.
a. Find an equation of f.
b. What is the s-intercept of the model? What does it mean in this situation?
c. Find $f(4)$. What does it mean in this situation?
d. What is the half-life of laserdisc player sales?

17. A storage tank contains radium-226, a radioactive element with a half-life of 1600 years. Let $f(t)$ be the percentage of radium-226 that remains at t years since the element was placed in the tank.
a. Find an equation of f.
b. What percentage of the radium-226 will remain after 100 years?
c. What percentage of the radium-226 will remain after 3200 years? Explain how you can find this result without using the equation of f.

18. A storage tank contains californium-251, a radioactive element with a half-life of about 900 years. Let $g(t)$ be the percentage of californium-251 that remains at t years since the element was placed in the tank.
a. Find an equation of g.
b. What percentage of the californium-251 will remain after 600 years?
c. What percentage of the californium-251 will remain after 3600 years? Explain how you can find this result without using the equation of g.

19. A thyroid cancer patient ingests a single dose of radioactive iodine-131 to kill the cancer cells. Iodine-131 has an *effective half-life* of 7.56 days—some is lost to radioactive decay, and some is removed through urination. Let $f(t)$ be the percentage of the iodine-131 that remains in the patient's body at t days since he ingested the iodine-131.
a. Find an equation of f.
b. For 3 days after ingesting the iodine-131, the patient must stay at least 1 meter away from other people, because the radiation he emits could be harmful. What percentage of the iodine-131 will remain in his body after 3 days?
c. The patient can safely spend a lot of time near a child when at most 5% of the iodine-131 remains. Use "intersect" on a graphing calculator to estimate when that time will be. [**Hint:** Use ZStandard followed by Zoom Out.]

20. A physician injects a patient with thallium-201 to determine how well blood flows to the patient's heart muscle. Thallium-201 has an *effective half-life* of 2.3 days—some is lost to exponential decay, and some is removed through the digestive and urinary tracts. Let $f(t)$ be the percentage of the thallium-201 that remains in the patient's body at t days since she was injected.
a. Find an equation of f.
b. What percentage of thallium-201 will remain after 5 days?
c. Use "intersect" on a graphing calculator to estimate when only 10% of the thallium-201 will remain. [**Hint:** Use ZStandard followed by Zoom Out.]

21. The half-life of caffeine in a person's bloodstream is about 6 hours. If a person's bloodstream contains 80 milligrams of caffeine, how much of that caffeine will remain after 14 hours?

22. The half-life of aspirin in a person's bloodstream is about 15 minutes. If a person's bloodstream contains 200 milligrams of aspirin, how much of that aspirin will remain after 40 minutes?

23. A storage tank contains a radioactive element. Let $p = f(t)$ be the percentage of the element that remains at t years since the element was placed in the tank. A graph of f is shown in Fig. 62.
a. What is the half-life of the element?
b. What percentage of the element will remain in the tank after 40 years?

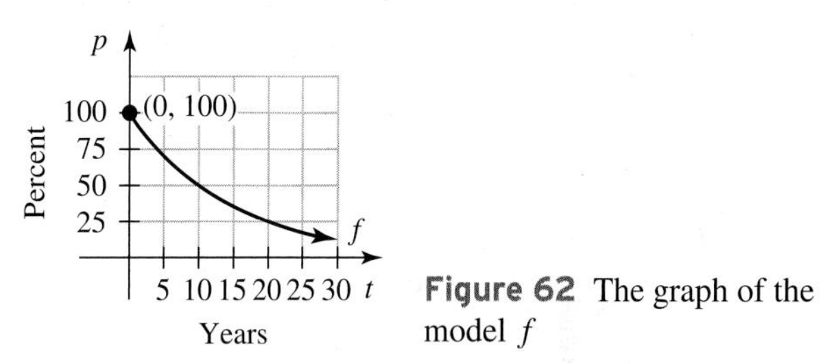

Figure 62 The graph of the model f

24. A storage tank contains a radioactive element. Let $p = g(t)$ be the percentage of the element that remains at t years since the element was placed in the tank. Some values of g are shown in Table 40.

Table 40 Percentages of a Radioactive Element

Year t	Percent $g(t)$
0	100.0
1	79.4
2	63.0
3	50.0
4	39.7
5	31.5
6	25.0

a. What is the half-life of the element?
b. What percentage of the element will remain in the tank after 12 years?

25. Sales of portable MP3 players grew approximately exponentially from \$0.08 billion in 2000 to \$1.2 billion in 2004 (Source: *Consumer Electronics Association*). Predict the sales of MP3 players in 2011.

26. Sales of fish oils grew approximately exponentially from \$30 million in 1995 to \$310 million in 2005 (Source: Nutrition Business Journal). Predict the sales of fish oils in 2012.

27. The number of Quiznos® restaurants grew approximately exponentially from one restaurant in 1981 to 3000 restaurants in 2004 (Source: *Quiznos*). Predict the number of restaurants in 2010.

28. A growing number of thieves are using keylogging programs to steal passwords and other personal information from Internet users. The number of keylogging programs reported grew approximately exponentially from 0.2 thousand programs in 2001 to 6.2 thousand programs in 2005 (Source: *iDefense*). Predict the number of keylogging programs that will be reported in 2011.

Consider the scattergram of data and the graph of the model $f(t) = ab^t$ in the figure. Sketch the graph of an exponential model that describes the data better; then explain how you would adjust the values of a and b of the original model to describe the data better.

29. See Fig. 63.

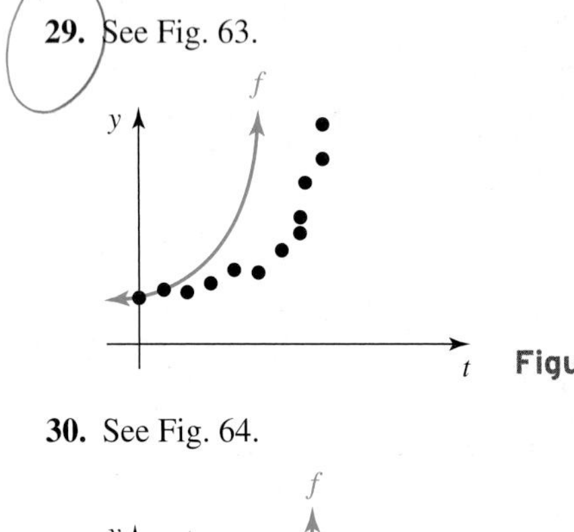

Figure 63 Exercise 29

30. See Fig. 64.

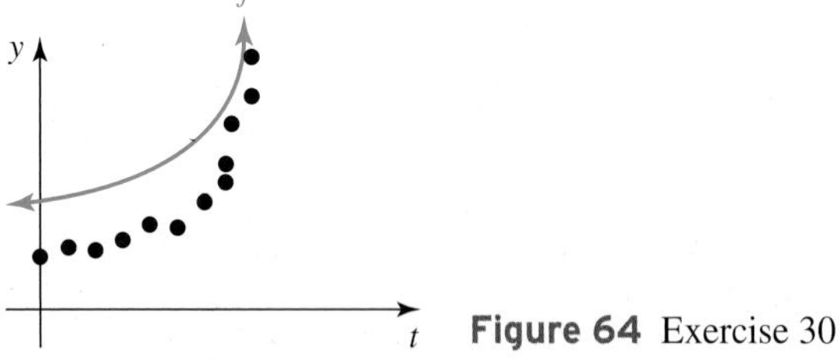

Figure 64 Exercise 30

31. World population is provided in Table 41 for various years.

Table 41 World Population

Year	Population (billions)
1930	2.070
1940	2.295
1950	2.500
1960	3.050
1970	3.700
1980	4.454
1990	5.279
2000	6.080
2006	6.522

Source: *U.S. Census Bureau*

Let $p = f(t)$ be the world's population (in billions) at t years since 1900.

a. Use a graphing calculator to draw a scattergram to describe the data. Is it better to use a linear or an exponential function to model the data? Explain.

b. Find an equation of f.

c. What is the p-intercept of the model? What does it mean in this situation?

d. Find $f(110)$. What does it mean in this situation?

32. Sales of some of Michael Jackson's albums are shown in Table 42.

Table 42 Sales of Michael Jackson's Albums

Year	Title	Sales (millions of albums)
1982	*Thriller*	26.0
1987	*Bad*	8.0
1991	*Dangerous*	5.8
1995	*History*	2.5
1997	*Blood on the Dance Floor*	1.3
2001	*Greatest Hits Vol. 1*	0.8
2003	*Number Ones*	0.9

Sources: *Nielsen Soundscan; RIAA; Sony Music*

Let $g(t)$ be the sales (in millions of albums) of the album released at t years since 1982.

a. Use a graphing calculator to draw a scattergram to describe the data. Is it better to use a linear or an exponential function to model the data? Explain.

b. Find an equation of g.

c. Estimate the sales of *Dangerous*. Have you performed interpolation or extrapolation? Explain. Find the error in your estimate. [**Hint:** The error is the difference of your estimate and the true value.]

d. Estimate the sales of *The Ultimate Collection,* which was released in 2004. Have you performed interpolation or extrapolation? Explain.

e. Estimate the revenue (total money collected) from *The Essential Michael Jackson,* which was released in 2005. Assume that the average price of the album is \$16.97.

33. The number of Starbucks stores worldwide has increased substantially since 1991 (see Table 43).

Table 43 Numbers of Starbucks Stores Worldwide

Year	Number of Stores
1991	116
1993	272
1995	676
1997	1412
1999	2135
2001	4709
2003	7225

Source: *Starbucks Corporation*

Let $n = f(t)$ be the number of Starbucks stores worldwide at t years since 1990.

a. Find an equation of f.

b. What is the percentage rate of growth of Starbucks stores?

c. Estimate the number of stores in 2010.

d. Use "intersect" on a graphing calculator to estimate when there were 18,000 stores. [**Hint:** Use Zoom Out and graph $n = 18{,}000$.]

34. The speed of a computer depends (in part) on its chip speed. The speeds of various chips, in megahertz (MHz), and the years they went on the market are listed in Table 44.

Table 44 Chip Speeds

Type of Chip	Year	Speed (MHz)
4004	1971	0.1
8080	1974	2
80186	1982	12
386	1985	16
486	1992	50
Pentium I	1995	120
Pentium II	1997	300
Pentium III	2000	1000
Pentium IV	2004	3200

Source: *i-probe*

Let $f(t)$ be the speed (in MHz) of a chip introduced into the market at t years since 1971.

a. Find an equation of f.

b. If f is linear, find the slope. If f is exponential, find the base b of $f(t) = ab^t$. What does it mean in this situation?

c. Suppose that you purchase a computer today. How many times faster will it be than a computer manufactured 10 years ago? [**Hint:** Find the ratio of the two chip speeds.]

d. Although it is true that the speed of a computer depends on its chip speed, explain why the ratio of the two chip speeds in part (c) may not equal the ratio of the computer speeds. [**Hint:** What does "depend" mean in this context?]

35. The University of Michigan offers a $500,000 life insurance policy. Monthly rates for nonsmoking faculty are shown in Table 45 for various ages.

Table 45 University of Michigan Life Insurance Monthly Rates

Age Group (years)	Age Used to Represent Age Group (years)	Monthly Rate (dollars)
30–34	32	15.00
35–39	37	18.50
40–44	42	26.00
45–49	47	46.00
50–54	52	75.50
55–59	57	118.00
60–64	62	195.50
65–69	67	326.60

Source: *University of Michigan*

Let $g(t)$ be the monthly rate (in dollars) for a nonsmoking faculty member at t years of age.

a. Find an equation of g.

b. What is the coefficient a of your model $g(t) = ab^t$? What does it mean in this situation?

c. What is the base b of your model $g(t) = ab^t$? What does it mean in this situation?

d. Find $g(47)$. What does it mean in this situation?

e. For many life insurance policies, monthly rates for women are different from monthly rates for men. Assume that these rates depend on life expectancy only. Given that the life expectancy of women is higher than that of men, would women or men pay higher monthly rates? Explain.

36. A person's heart attack risk can be estimated by using *Framingham point scores,* which are based on such factors as age, cholesterol level, blood pressure, and smoking habits. Men's risks of having a heart attack in the next 10 years are shown in Table 46 for various scores.

Table 46 Risks of Having a Heart Attack

Framingham Point Scores	Risk (percent)
0	1
5	2
10	6
15	20
17	30

Sources: The Journal of the American Medical Association; *Framingham Heart Study*

Let $f(s)$ be a man's risk of having a heart attack in the next 10 years if his score is s points.

a. Find an equation of f.

b. A 47-year-old man with high cholesterol has high blood pressure but does not smoke. His score is 11 points. What is the risk that he will have a heart attack in the next 10 years?

c. Another 47-year-old man has the same cholesterol level and blood pressure as the man described in part (b). However, this man's score is 5 points higher, because he smokes. What is the risk that he will have a heart attack in the next 10 years?

d. What is the coefficient a of your model $f(s) = ab^s$? What does it mean in this situation?

e. What is the base b of your model $f(s) = ab^s$? What does it mean in this situation?

37. From 1790 to 1860, U.S. population grew rapidly (see Table 47).

Table 47 U.S. Population

Year	Population (millions)	Population Ratio (current to previous)
1790	3.9	[leave blank]
1800	5.3	1.36
1810	7.2	
1820	9.6	
1830	12.9	
1840	17.1	
1850	23.2	
1860	31.4	

Source: *U.S. Census Bureau*

a. Complete the third column of Table 47. The first entry is 1.36, since $\frac{1800 \text{ population}}{1790 \text{ population}} = \frac{5.3}{3.9} \approx 1.36$.

b. What do you observe about the ratios in the third column?

c. Based on your observation in part (b), is it better to use an exponential or a linear function to model the data? Explain.

d. Let $f(t)$ be the U.S. population (in millions) at t years since 1790. Find an equation of an exponential function that models the data from 1790 to 1860.

e. Complete the third column of Table 48.

Table 48 U.S. Population

Year	Population (millions)	Population Ratio (current to previous)
1860	31.4	[leave blank]
1870	39.8	
1880	50.2	
1890	62.9	
1900	76.0	

Source: *U.S. Census Bureau*

f. Is it likely that f gives reasonable population estimates after 1860? Explain.

g. Use f to estimate the population in 2006. The actual population was 298.5 million. What is the error in your estimate?

38. The amounts of the federal debt are listed in Table 49 for various years.

Table 49 Federal Debt Amounts

Year	Federal Debt (billions of dollars)
1960	290
1965	321
1970	389
1975	577
1980	930
1985	1946
1990	3233
1995	4974
2000	5674
2005	7933

Source: *Bureau of the Public Debt*

a. Let $D = f(t)$ be the federal debt (in billions of dollars) at t years since 1960. Find an equation of f.

b. Predict the federal debt in 2010.

c. If the federal debt were paid off in 2010 by each U.S. citizen contributing an equal amount of money, how much would each person have to pay? Assume that the population will be about 309 million in 2010.

d. Predict the federal debt in 2050.

e. If the federal debt were paid off in 2050 by each U.S. citizen contributing an equal amount of money, how much would each person have to pay? Assume that the population will be about 394 million in 2050. Explain why some people want to reduce or eliminate the debt now rather than later.

39. Percentages of adults surveyed who plan to attend a Halloween party this year are shown in Table 50 for various age groups.

Table 50 Percentages of Adults Who Plan to Attend a Halloween Party

Age Group (years)	Age Used to Represent Age Group (years)	Percent
18–24	21.0	44
25–34	29.5	34
35–44	39.5	25
45–54	49.5	14
55–64	59.5	10
65 or Over	70.0	6

Source: *International Mass Retail Association*

Let p be the percentage of adults at age t years who plan to attend a Halloween party this year.

a. Find an equation of a model f by hand, and find an equation of a regression model r by using a graphing calculator. Compare the graphs of f and r.

b. Find $r(26)$. What does it mean in this situation?

c. Use r to estimate the number of 42-year-old adults who plan to attend a Halloween party. Assume that there are about 4.6 million 42-year-old adults.

d. What is the p-intercept of the model r? What does it mean in this situation?

e. What is the base b of the model $r(t) = ab^t$? What does it mean in this situation?

40. Table 51 compares the economic strength of a country with the percentage of the population that was involved in producing agriculture. *Gross national product* (*GNP*) is a measure of the amount of goods and services a country produces.

Table 51 Percentage of Population in Agriculture versus GNP

Country	Percent of Population in Agriculture	GNP per Person (dollars)*
United States	1	14,100
Great Britain	1	9300
West Germany†	3	11,400
Canada	5	12,300
Australia	6	11,400
France	8	10,500
Japan	9	10,100
Italy	10	6400
Argentina	12	2050
Soviet Union‡	16	6700
Mexico	33	2700
South Korea	36	1950
Brazil	37	1900
Nigeria	49	700
Egypt	49	600
Pakistan	51	400
Indonesia	55	500
China	56	350
India	61	300
Bangladesh	83	200

Source: Population and Resources in a Changing World, *Kingsley Davis et al.*

*GNP for 1983.

†West Germany reunified with East Germany in 1990.

‡The Soviet Union dissolved into 15 separate states in 1991.

a. Let $f(p)$ be the GNP per person (in dollars) for a country in which p percent of the population was involved in agriculture. Use a graphing calculator to find the regression equation of f.

b. Use a graphing calculator to sketch a scattergram of the data and your model on the same coordinate system. Does the model fit the data well?

c. What is the base b of your function $f(p) = ab^p$? What does it mean in this situation? Explain why this makes sense in terms of productivity.

d. Which country's data point is farthest from the regression curve? What does the position of the point in relation to the other data points and the regression curve suggest about this situation?

41. Describe how to use the base multiplier property to find an exponential model for a given situation.

42. Explain how to find an exponential model for a situation described by a table of data. Also, explain how to use the model to make an estimate or prediction for the situation.

Related Review

43. In this exercise, you will explore two types of interest-bearing accounts.

a. Suppose that \$800 is deposited into a savings account that earns 3% interest compounded annually. Let $C(t)$ be the value (in dollars) of the account at t years after \$800 has been deposited. Find an equation of C.

b. Now suppose that \$800 is deposited into a savings account that earns 3% simple interest. Recall from Section 3.4 that the term *simple interest* means that the interest earned each year is 3% of the \$800 only. Let $S(t)$ be the value (in dollars) of the account at t years after \$800 has been deposited. Find an equation of S. [**Hint:** Each year, the balance increases by $800(0.03) = 24$ dollars.]

c. Find $C(1)$, $C(2)$, $S(1)$, and $S(2)$. Explain in terms of the situation why it makes sense that $C(1)$ is equal to $S(1)$ but that $C(2)$ is not equal to $S(2)$.

d. Compare $C(20)$ with $S(20)$. What does your comparison mean in this situation?

44. On Monday, 20 people receive a prank e-mail warning them of an impending gasoline shortage. On Tuesday, 40 more people receive the e-mail.

a. Assume that the number of people receiving the e-mail is growing exponentially. Let $E(t)$ be the number of people who receive the e-mail on the day that is t days since Monday. Find an equation of E.

b. Now assume that the number of people receiving the e-mail is growing linearly. Let $L(t)$ be the number of people who receive the e-mail on the day that is t days since Monday. Find an equation of L.

c. Compare $E(7)$ with $L(7)$. What does your comparison mean in this situation?

d. Compare $E(28)$ with $L(28)$. What does your comparison mean in this situation?

e. Use a graphing calculator to draw the graphs of E and L on the same coordinate system. Compare the graphs.

f. Is there much difference in the number of people who will receive the e-mail if it grows exponentially or linearly? Explain.

45. The percentage of prescription drugs that are generic has greatly increased since 1984 (see Table 52).

Table 52 Percentages of Prescription Drugs That Are Generic

Year	Percent
1984	19
1989	32
1994	36
1999	50
2004	56

Source: *IMS Health*

a. Let $f(t)$ be the percentage of prescription drugs that are generic at t years since 1980. Use a graphing calculator to draw a scattergram of the data. Is it better to use a linear or an exponential function to model the data? Explain.

b. Find an equation of f.

c. Predict the percentage of prescription drugs that will be generic in 2011.

d. Predict when all prescription drugs will be generic.

46. The most expensive slot for television advertising occurs during the Super Bowl. In 2004, a 30-second Super Bowl ad slot cost \$2.3 million—about \$76,667 per second! The costs of 30-second ad slots have increased greatly even when adjusted for inflation (see Table 53).

Table 53 Costs of Television Ad Slots During the Super Bowl

Super Bowl	Year	Cost* for 30 Seconds (millions of dollars)
V	1971	0.3
X	1976	0.3
XV	1981	0.6
XX	1986	0.8
XXV	1991	1.0
XXX	1996	1.2
XXXIV	2000	2.0
XL	2006	2.4

Source: *NFL Research*
*Adjusted for inflation

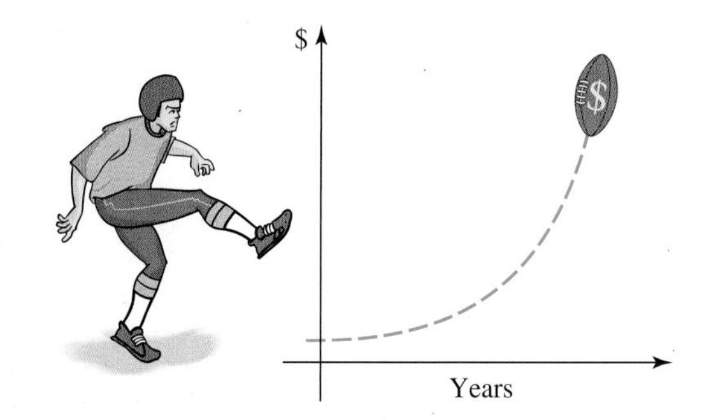

a. A linear model of the cost (in millions of dollars) of a 30-second ad slot is $L(t) = 0.061t - 0.64$, and an *exponential model* is $E(t) = 0.14(1.065)^t$, where t is the number of years since 1960. Draw the graph of L, the graph of E, and the scattergram of the data in the same viewing window. Describe how close each graph comes to the points in the scattergram. Also, use Zoom Out and decide which model appears to make better estimates for the years before 1971.

b. Use L and E to estimate the cost of a 30-second ad slot during Super Bowl I in 1967. The actual cost (adjusted for inflation) was $200,000. Which function does a better job of modeling the situation for 1967?

c. Find the slope of L and the base b of $E(t) = ab^t$, and describe what they mean in this situation.

d. In Super Bowl XXXVI in 2002, a 30-second ad slot cost $1.9 million, the first decline in cost. Use E to help estimate the total loss in revenue from ad slots during Super Bowl XXXVI, likely due to a poor economy. [**Hint:** First find $E(42)$.]

Expressions, Equations, Functions, and Graphs

Give an example of the following. Then solve, simplify, or graph, as appropriate.

47. exponential equation in one variable

48. linear function

49. exponential function

50. system of two linear equations in two variables

51. expression involving exponents

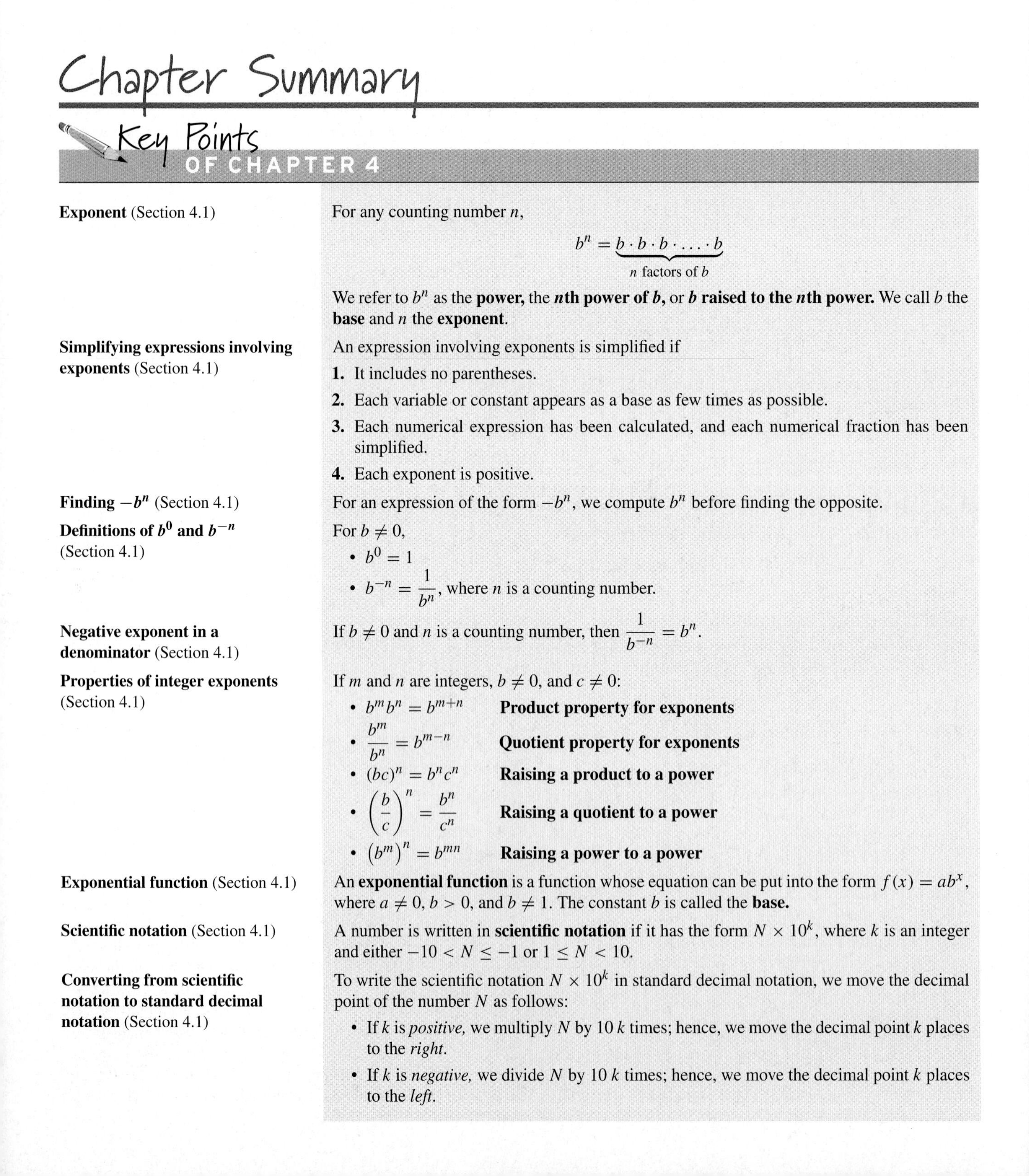

Chapter Summary

Key Points OF CHAPTER 4

Exponent (Section 4.1)

For any counting number n,

$$b^n = \underbrace{b \cdot b \cdot b \cdot \ldots \cdot b}_{n \text{ factors of } b}$$

We refer to b^n as the **power,** the **nth power of b,** or **b raised to the nth power.** We call b the **base** and n the **exponent**.

Simplifying expressions involving exponents (Section 4.1)

An expression involving exponents is simplified if

1. It includes no parentheses.
2. Each variable or constant appears as a base as few times as possible.
3. Each numerical expression has been calculated, and each numerical fraction has been simplified.
4. Each exponent is positive.

Finding $-b^n$ (Section 4.1)

For an expression of the form $-b^n$, we compute b^n before finding the opposite.

Definitions of b^0 and b^{-n} (Section 4.1)

For $b \neq 0$,

- $b^0 = 1$
- $b^{-n} = \dfrac{1}{b^n}$, where n is a counting number.

Negative exponent in a denominator (Section 4.1)

If $b \neq 0$ and n is a counting number, then $\dfrac{1}{b^{-n}} = b^n$.

Properties of integer exponents (Section 4.1)

If m and n are integers, $b \neq 0$, and $c \neq 0$:

- $b^m b^n = b^{m+n}$ **Product property for exponents**
- $\dfrac{b^m}{b^n} = b^{m-n}$ **Quotient property for exponents**
- $(bc)^n = b^n c^n$ **Raising a product to a power**
- $\left(\dfrac{b}{c}\right)^n = \dfrac{b^n}{c^n}$ **Raising a quotient to a power**
- $\left(b^m\right)^n = b^{mn}$ **Raising a power to a power**

Exponential function (Section 4.1)

An **exponential function** is a function whose equation can be put into the form $f(x) = ab^x$, where $a \neq 0$, $b > 0$, and $b \neq 1$. The constant b is called the **base.**

Scientific notation (Section 4.1)

A number is written in **scientific notation** if it has the form $N \times 10^k$, where k is an integer and either $-10 < N \leq -1$ or $1 \leq N < 10$.

Converting from scientific notation to standard decimal notation (Section 4.1)

To write the scientific notation $N \times 10^k$ in standard decimal notation, we move the decimal point of the number N as follows:

- If k is *positive,* we multiply N by 10 k times; hence, we move the decimal point k places to the *right.*
- If k is *negative,* we divide N by 10 k times; hence, we move the decimal point k places to the *left.*

Converting from standard decimal notation to scientific notation (Section 4.1)

To write a number in scientific notation, count the number of places k that the decimal point must be moved so that the new number N meets the condition $-10 < N \leq 1$ or $1 \leq N < 10$:

- If the decimal point is moved to the left, then the scientific notation is written as $N \times 10^k$.
- If the decimal point is moved to the right, then the scientific notation is written as $N \times 10^{-k}$.

Definition of $b^{1/n}$ (Section 4.2)

For the counting number n, where $n \neq 1$,

- If n is odd, then $\boldsymbol{b^{1/n}}$ is the number whose nth power is b, and we call $b^{1/n}$ the ***n*th root of *b***.
- If n is even and $b \geq 0$, then $b^{1/n}$ is the nonnegative number whose nth power is b, and we call $b^{1/n}$ the **principal *n*th root of *b***.
- If n is even and $b < 0$, then $b^{1/n}$ is not a real number.

$b^{1/n}$ may be represented by $\sqrt[n]{b}$.

Rational exponent (Section 4.2)

For the counting numbers m and n, where $n \neq 1$ and b is any real number for which $b^{1/n}$ is a real number,

- $\boldsymbol{b^{m/n}} = \left(b^{1/n}\right)^m = \left(b^m\right)^{1/n}$
- $\boldsymbol{b^{-m/n}} = \dfrac{1}{b^{m/n}}, b \neq 0$

Properties of rational exponents (Section 4.2)

If m and n are rational numbers and b and c are any real numbers for which b^m, b^n, and c^n are real numbers, then

- $b^m b^n = b^{m+n}$ **Product property for exponents**
- $\dfrac{b^m}{b^n} = b^{m-n}, b \neq 0$ **Quotient property for exponents**
- $(bc)^n = b^n c^n$ **Raising a product to a power**
- $\left(\dfrac{b}{c}\right)^n = \dfrac{b^n}{c^n}, c \neq 0$ **Raising a quotient to a power**
- $\left(b^m\right)^n = b^{mn}$ **Raising a power to a power**

Base multiplier property (Section 4.3)

For an exponential function of the form $y = ab^x$, if the value of the independent variable increases by 1, the value of the dependent variable is multiplied by b.

Increasing or decreasing property (Section 4.3)

Let $f(x) = ab^x$, where $a > 0$. Then

- If $b > 1$, then the function f is increasing. We say that the function **grows exponentially.**
- If $0 < b < 1$, then the function f is decreasing. We say that the function **decays exponentially.**

***y*-Intercept of an exponential function** (Section 4.3)

For an exponential function of the form $y = ab^x$, the y-intercept is $(0, a)$.

Reflection property (Section 4.3)

The graphs of $f(x) = -ab^x$ and $g(x) = ab^x$ are **reflections** of each other across the x-axis.

Horizontal asymptote (Section 4.3)

For all exponential functions, the x-axis is a horizontal asymptote.

Domain (Section 4.3)

The domain of any exponential function $f(x) = ab^x$ is the set of real numbers.

Range (Section 4.3)

The range of an exponential function $f(x) = ab^x$ is the set of all positive real numbers if $a > 0$, and the range is the set of all negative real numbers if $a < 0$.

Solving $b^n = k$ for b (Section 4.4)

To solve an equation of the form $b^n = k$ for b,

- If n is odd, the solution is $k^{1/n}$.
- If n is even and $k \geq 0$, the solutions are $\pm k^{1/n}$.
- If n is even and $k < 0$, there is no real-number solution.

Dividing left sides and right sides of two equations (Section 4.4)

If $a = b$ and $c = d$, then $\dfrac{a}{c} = \dfrac{b}{d}$.

Finding an equation (Section 4.4)

We can find an equation of an exponential function by using the base multiplier property or by using two points. Both methods give the same result.

Exponential model, exponentially related, approximately exponentially related (Section 4.5)

An **exponential model** is an exponential function, or its graph, that describes the relationship between two quantities for an authentic situation. If all of the data points for a situation lie on an exponential curve, then we say that the independent and dependent variables are **exponentially related.** If no exponential curve contains all of the data points, but an exponential curve comes close to all of the data points (and perhaps contains some of them), then we say that the variables are **approximately exponentially related.**

Quantity present at time $t = 0$ (Section 4.5)	If $y = ab^t$ is an exponential model where y is a quantity at time t, then the coefficient a is the value of that quantity present at time $t = 0$.
***r* percent interest compounded annually** (Section 4.5)	The term ***r* percent interest compounded annually** means that the interest earned each year equals r percent of the principal and any interest earned in previous years (all of which becomes part of the investment).
Half-life (Section 4.5)	If a quantity decays exponentially, the **half-life** is the amount of time it takes for that quantity to be reduced to half.
Meaning of the base *b* (Section 4.5)	If $f(t) = ab^t$, where $a > 0$, models a quantity at time t, then the percent rate of change is constant. In particular, • If $b > 1$, then the quantity grows exponentially at a rate of $b - 1$ percent (in decimal form) per unit of time. • If $0 < b < 1$, then the quantity decays exponentially at a rate of $1 - b$ percent (in decimal form) per unit of time.
Four-step modeling process (Section 4.5)	To find a model and then make estimates and predictions, **1.** Create a scattergram of the data. Decide whether a line or an exponential curve comes close to the points. **2.** Find an equation of your function. **3.** Verify your equation by checking that the graph comes close to all of the data points. **4.** Use your equation of the model to draw conclusions, make estimates, and/or make predictions.

CHAPTER 4 REVIEW EXERCISES

Simplify. Assume that b and c are positive.

1. $\dfrac{2^{-400}}{2^{-405}}$ **2.** $\dfrac{4b^{-3}c^{12}}{16b^{-4}c^3}$

3. $(2b^{-5}c^{-2})^3(3b^4c^{-6})^{-2}$ **4.** $\dfrac{(20b^{-2}c^{-9})(27b^5c^3)}{(18b^3c^{-1})(30b^{-1}c^{-4})}$

5. $32^{4/5}$ **6.** $16^{-3/4}$ **7.** $\dfrac{b^{-1/3}}{b^{4/3}}$ **8.** $\dfrac{(16b^8c^{-4})^{1/4}}{(25b^{-6}c^4)^{3/2}}$

9. $\left(\dfrac{32b^2c^5}{2b^{-6}c^1}\right)^{1/4}$ **10.** $(8^{2/3}b^{-1/3}c^{3/4})(64^{-1/3}b^{1/2}c^{-5/2})$

Simplify. Assume that n is a counting number.

11. $b^{2n-1}b^{4n+3}$ **12.** $\dfrac{b^{n/2}}{b^{n/3}}$

13. Use properties of exponents to show why $3^{2x} = 9^x$.

For $f(x) = 3(5)^x$ and $g(x) = 6^x$, find the following.

14. $f(-2)$ **15.** $g(a + 2)$

For $f(x) = 49^x$ and $g(x) = 2(81)^x$, find the following.

16. $f\left(\dfrac{1}{2}\right)$ **17.** $g\left(-\dfrac{3}{4}\right)$

Write the number in standard decimal form.

18. 4.4487×10^7 **19.** 3.85×10^{-5}

Write the number in scientific notation.

20. 54,000,000 **21.** -0.00897

22. Graph $f(x) = 2(3)^x$ by hand.

Graph the function by hand. Find its domain and range.

23. $h(x) = -3(2)^x$ **24.** $g(x) = 12\left(\dfrac{1}{2}\right)^x$

Find all real-number solutions of the equation. Round any result(s) to the second decimal place.

25. $3.9b^7 = 283.5$ **26.** $5b^4 - 13 = 67$

27. $\dfrac{1}{3}b^2 - \dfrac{1}{5} = \dfrac{2}{3}$

28. Some values of the functions f, g, h, and k are provided in Table 54. For each function, determine whether the given values suggest that the function is linear, exponential, or neither. If the function could be linear or exponential, find a possible equation for it.

Table 54 Values of Four Functions (Exercises 28--30)

x	$f(x)$	$g(x)$	$h(x)$	$k(x)$
1	30	5	2	96
2	26	15	3	48
3	22	45	6	24
4	18	135	11	12
5	14	405	18	6

For Exercises 29 and 30, refer to Table 54.

29. Find $f(4)$. **30.** Find x when $k(x) = 6$.

Find an approximate equation $y = ab^x$ of the exponential curve that contains the given pair of points. Round the values of a and b to two decimal places.

31. $(0, 2)$ and $(5, 3)$ **32.** $(3, 30)$ and $(9, 7)$

33. Consider the scattergram of data and the graph of the model $f(x) = ab^t$ in Fig. 65. Sketch the graph of an exponential model that describes the data better; then explain how you would adjust the values of a and b of the original model to describe the data better.

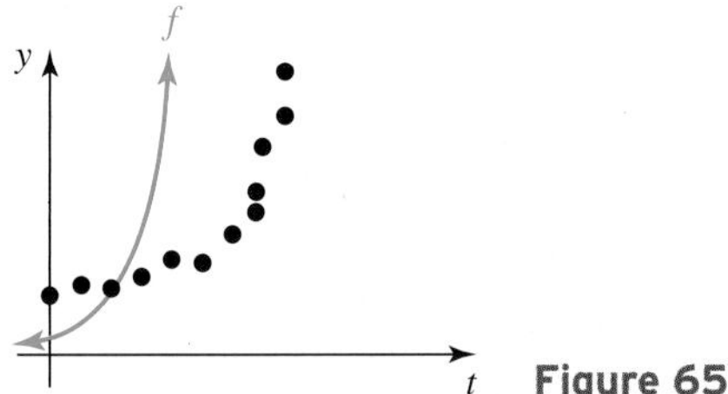

Figure 65 Exercise 33

34. Suppose that \$2000 is deposited into an account that earns 7% interest compounded annually. Let $f(t)$ be the value (in dollars) of the account at t years after the \$2000 is deposited.
a. Find an equation of f.
b. Find the value of the account after 5 years.

35. A corporation's annual total sales have doubled every year. The total sales in 2005 were \$17 thousand.
a. Let $g(t)$ be the total sales (in thousands of dollars) for the year that is t years since 2005. Find an equation of g.
b. Predict the corporation's total sales in 2011.

36. A storage tank contains carbon-14, a radioactive element with a half-life of 5730 years. Let $f(t)$ be the percentage of carbon-14 that remains at t years since the element was placed in the tank.
a. Find an equation of f.
b. Predict the percentage of carbon-14 remaining after 100 years.

37. The ad revenue from Latino newspapers in the United States has grown exponentially from \$121 million in 1990 to \$996 million in 2005 (Source: *Latino Print Network*). Predict the ad revenue from Latino newspapers in 2011.

38. Table 55 shows the number of Kohl's® stores for various years.

Table 55 Numbers of Kohl's Stores

Year	Number of Stores
1992	79
1996	150
1999	259
2002	457
2005	675

Source: *Kohl's*

a. Let $f(t)$ be the number of stores at t years since 1990. Use a graphing calculator to draw a scattergram of the data. Is it better to use a linear or an exponential function to model the data? Explain.
b. Find an equation of f.
c. What is the coefficient a of your model $f(t) = ab^t$? What does it mean in this situation?
d. What is the base b of $f(t) = ab^t$? What does it mean in this situation?
e. Find $f(12)$. What does it mean in this situation?

39. The number of lawsuits filed against tobacco companies has been increasing (see Table 56).

Table 56 Numbers of Lawsuits Filed Against Tobacco Companies

Year	Number of Lawsuits
1993	49
1994	73
1995	200
1996	352
1997	733

Source: *R.J. Reynolds Tobacco Co.*

a. Let $f(t)$ be the number of lawsuits filed during the year that is t years since 1993. Find an equation of f.
b. Predict the number of lawsuits that will be filed during 2008.
c. Some experts believe that many of the cases filed will disappear or fail. Suppose that only 10% of the lawsuits filed go to court and that the tobacco industry loses half of these cases and pays about \$10 million for each lost case.* Predict how much money the tobacco industry will pay for lost cases during 2008.
d. Before a settlement required the tobacco industry to pay \$206 billion, tobacco industry executives said that they would much rather fight cases in court than be required to pay money due to such a settlement. If part (c) shows what would have happened without a settlement, would the strategy suggested by the tobacco industry executives have been good for their industry? Explain.

*Some verdicts are for substantially more than \$10 million.

CHAPTER 4 TEST

Simplify.

1. $32^{2/5}$

2. $-8^{-4/3}$

Simplify. Assume that b and c are positive.

3. $\left(2b^3c^8\right)^3$

4. $\left(\dfrac{4b^{-3}c}{25b^5c^{-9}}\right)^0$

5. $\dfrac{b^{1/2}}{b^{1/3}}$

6. $\dfrac{25b^{-9}c^{-8}}{35b^{-10}c^{-3}}$

7. $\left(\dfrac{6b\left(b^3c^{-2}\right)}{3b^2c^5}\right)^2$

8. $\dfrac{\left(25b^8c^{-6}\right)^{3/2}}{\left(7b^{-2}\right)\left(2c^3\right)^{-1}}$

9. Use properties of exponents to show that $8^{x/3}2^{x+3} = 8(4)^x$.

For $f(x) = 4^x$, find the following.

10. $f(-2)$

11. $f\left(-\frac{3}{2}\right)$

Graph the function by hand. Find its domain and range.

12. $f(x) = -5(2)^x$

13. $f(x) = 18\left(\frac{1}{3}\right)^x$

14. For each graph in Fig. 66, find an equation of an exponential function that could fit the graph.

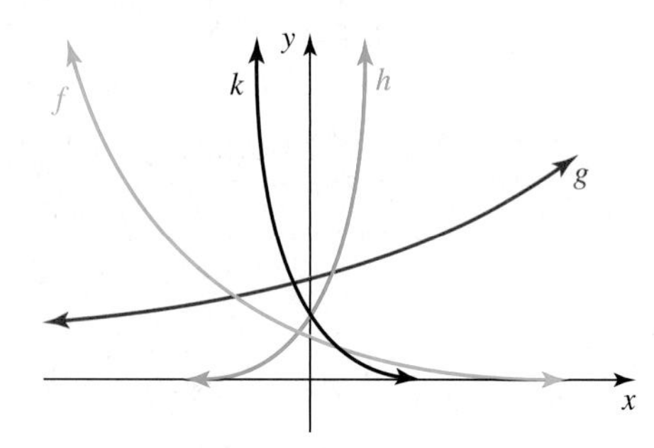

Figure 66 Exercise 14

15. Some values of a function f are provided in Table 57. Find an equation of f in terms of t.

Table 57 Values of a Function f (Exercise 15)

t	$f(t)$
0	160
1	80
2	40
3	20
4	10
5	5

16. Find all real-number solutions of $3b^6 + 5 = 84$. Round any result(s) to the second decimal place.

Find an approximate equation $y = ab^x$ of an exponential curve that contains the given pair of points. Round the values of a and b to two decimal places.

17. (0, 70) and (6, 20)

18. (4, 9) and (7, 50)

For Exercises 19–21, refer to Fig. 67.

19. Find $f(0)$.

20. Find x when $f(x) = 3$.

21. Find an equation of f.

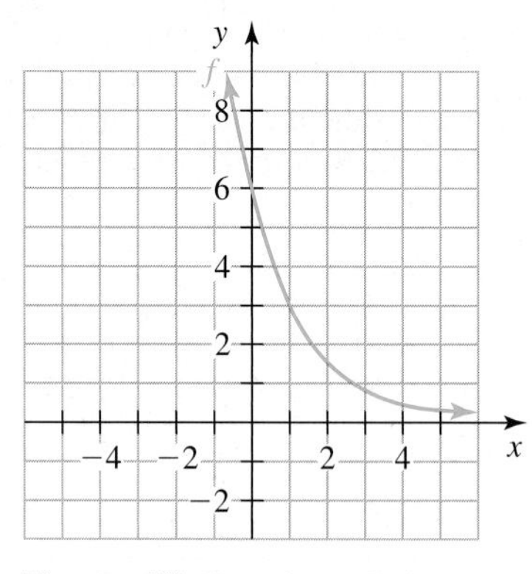

Figure 67 Exercises 19–21

22. On March 1, there are 400 leaves on a tree. For a while, the total number of leaves triples each week. Let $f(t)$ be the total number of leaves on the tree at t weeks since March 1.

a. Find an equation of f.

b. Find $f(6)$. What does it mean in this situation?

c. Find $f(52)$. What does it mean in this situation?

23. Due in part to fertility treatments, there has been a substantial increase in the number of multiple births of triplets or more in the United States (see Table 58).

Table 58 Numbers of Multiple Births of Triplets or More

Year	Number of Multiple Births of Triplets or More
1971	1000
1975	1100
1980	1337
1985	1925
1990	3028
1995	4973
2000	7325
2003	7663

Source: *National Center for Health Statistics*

a. Let $f(t)$ be the number of multiple births of triplets or more in the United States in the year that is t years since 1970. Find an equation of f.

b. Estimate the percent growth in multiple births of triplets or more each year.

c. What is the coefficient a of your model $f(t) = ab^t$? What does it mean in this situation?

d. Predict the number of multiple births of triplets or more in 2010.

Chapter 5

Logarithmic Functions

"Never regard study as a duty, but as the enviable opportunity to learn to know the liberating influence of beauty in the realm of the spirit for your own personal joy and to the profit of the community to which your later work belongs."

—*Albert Einstein*

Table 1 Numbers of LEED-Certified Green Buildings

Year	Number of Buildings with LEED Certification
2000	5
2001	14
2002	39
2003	80
2004	142

Source: *United States Green Building Council*

Have you heard of "green" buildings? They are buildings designed to meet certain environmental standards. The numbers of buildings with Leadership in Energy and Environmental Design (LEED) certification for being green are shown in Table 1 for various years. In Exercise 18 of Homework 5.4, you will predict when there will be an average of 100 LEED-certified green buildings per state.

In Chapter 4, we worked with exponential functions. In this chapter, we will work with *logarithmic functions,* which are closely related to exponential functions. We will also discuss how to simplify *logarithmic expressions* and solve *logarithmic equations in one variable*. In Section 4.5, we made predictions for the dependent variable of an exponential model—but not for the independent variable. In Section 5.4, we will be able to make predictions for the independent variable, such as when there will be an average of 100 LEED-certified green buildings per state.

5.1 INVERSE FUNCTIONS

Objectives

- Know the meaning of *inverse of a function, invertible function,* and *one-to-one function.*
- Know the *reflection property of inverse functions.*
- Graph the inverse of a function.
- Find an equation of the inverse of a model.
- Find an equation of the inverse of a function that is not a model.

In this section, we will study a type of function that has a special relationship to a given function. It is called the *inverse* of a function. We will begin the section with an example that will allow us to develop a definition of the inverse of a function. Then we will describe inverses of functions by using first tables, then graphs, and finally equations.

Definition of an Inverse of a Function

Although the United States and several other countries use the Fahrenheit scale (in °F) to measure temperature, most countries use the Celsius scale (in °C). A comparison of the two scales is shown in Table 2.

If an American visiting Europe hears that the local temperature will reach 20°C, it would be helpful to be able to convert 20°C to 68°F. There is a function g that converts

Table 2 Celsius and Fahrenheit Equivalent Readings

Celsius (°C)	Fahrenheit (°F)
0	32
20	68
40	104
60	140
80	176
100	212

Celsius inputs to Fahrenheit outputs (see Fig. 1). We let C be the Celsius temperature and F be the Fahrenheit temperature.

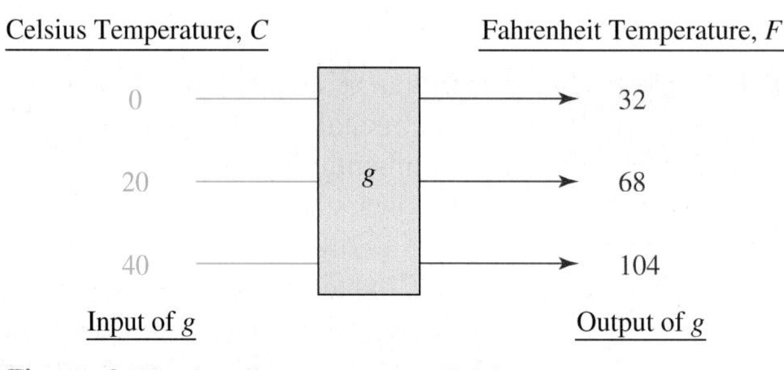

Figure 1 The function g converts Celsius temperatures to Fahrenheit temperatures

If a European visiting the United States hears that the local temperature will reach 68°F, it would be helpful to be able to convert 68°F to 20°C. If we reverse the arrows of Fig. 1, we have an input–output diagram of a new relation (see Fig. 2). For each Fahrenheit temperature, there is exactly one Celsius temperature, so the new relation is also a function. We call this function the *inverse* of g and show it as "g^{-1}" (read "g-inverse").

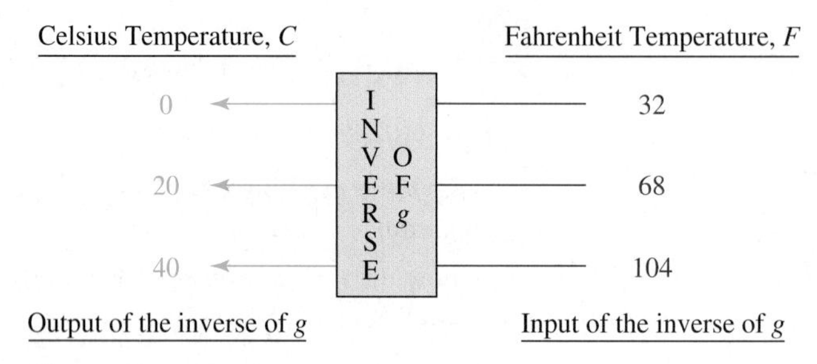

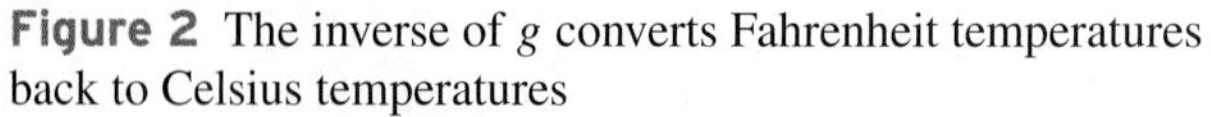

Figure 2 The inverse of g converts Fahrenheit temperatures back to Celsius temperatures

There are two key observations we can make about g^{-1}:

1. g^{-1} sends outputs of g to inputs of g. For example, g sends the input 0 to the output 32 and g^{-1} sends 32 to 0 (see Figs. 1 and 2). Using symbols, we write

$$g(0) = 32 \qquad \text{and} \qquad g^{-1}(32) = 0$$

We say that these two statements are **equivalent,** which means that one statement implies the other and vice versa.

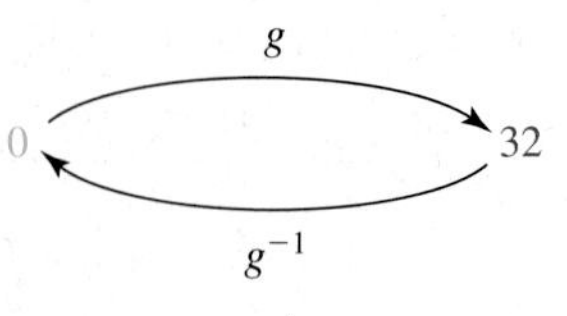

Figure 3 g^{-1} undoes g

2. g^{-1} undoes g. For example, g sends 0 to 32 and g^{-1} undoes this action by sending 32 *back* to 0 (see Fig. 3).

The **inverse of a function** f is a relation that sends b to a if $f(a) = b$. The inverse of a function is not necessarily a function. If the inverse of a function f is also a function, we say that f is **invertible** and use "f^{-1}" as a name for the inverse of f. We say that f^{-1} is the **inverse function of f.**

Property of an Inverse Function

For an invertible function f, the following statements are equivalent:

$$f(a) = b \qquad \text{and} \qquad f^{-1}(b) = a$$

In words: If f sends a to b, then f^{-1} sends b to a. If f^{-1} sends b to a, then f sends a to b.

Example 1 Evaluating an Inverse Function

Let f be an invertible function where $f(2) = 5$. Find $f^{-1}(5)$.

Solution

Since f sends 2 to 5, we know that f^{-1} sends 5 back to 2. So, $f^{-1}(5) = 2$. ■

Example 2 Evaluating f and f^{-1}

Table 3 Input-Output Values of f

x	$f(x)$
0	1
1	3
2	9
3	27
4	81

Some values of an invertible function f are shown in Table 3.

1. Find $f(3)$. **2.** Find $f^{-1}(9)$.

Solution

1. $f(3) = 27$.

2. Since f sends 2 to 9, we conclude that f^{-1} sends 9 back to 2. Therefore, $f^{-1}(9) = 2$. ■

WARNING The -1 in "$f^{-1}(x)$" is *not* an exponent. It is part of the function notation "f^{-1}"—which stands for the inverse of the function f. Here, we simplify 3^{-1} and use the values of f shown in Table 3 to find $f^{-1}(3)$:

$$3^{-1} = \frac{1}{3} \quad \text{"}-1\text{" is an exponent: Use } b^{-n} = \frac{1}{b^n}$$

$$f^{-1}(3) = 1 \quad \text{"}f^{-1}\text{" stands for the inverse of } f\text{; } f \text{ sends 1 to 3, so } f^{-1} \text{ sends 3 back to 1.}$$

Example 3 Evaluating f and f^{-1}

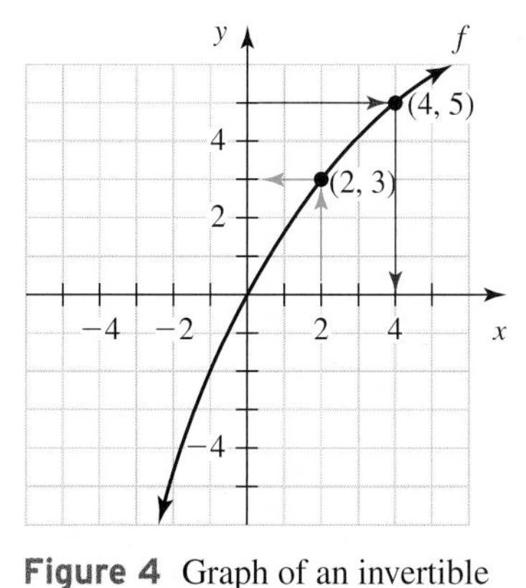

Figure 4 Graph of an invertible function f

The graph of an invertible function f is shown in Fig. 4.

1. Find $f(2)$. **2.** Find $f^{-1}(5)$.

Solution

1. The blue arrows in Fig. 4 show that f sends 2 to 3. So, $f(2) = 3$.

2. The function f sends 4 to 5. So, f^{-1} sends 5 back to 4 (see the red arrows). Therefore, $f^{-1}(5) = 4$. ■

All linear functions with nonzero slope and all exponential functions are invertible. (We will see why at end of this section.) Therefore, we can use the notation f^{-1} whenever we describe the inverse of either of these two types of functions.

Example 4 Finding Input-Output Values of an Inverse Function

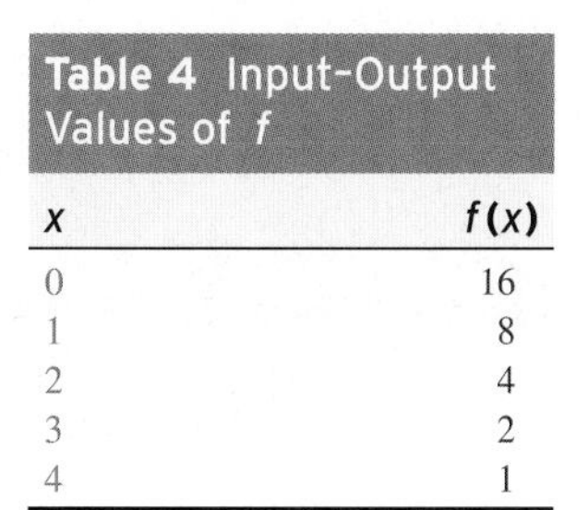

Table 4 Input-Output Values of f

x	$f(x)$
0	16
1	8
2	4
3	2
4	1

Let $f(x) = 16\left(\frac{1}{2}\right)^x$.

1. Find five input–output values of f^{-1}. **2.** Find $f^{-1}(8)$.

Solution

1. We begin by finding input–output values of f (see Table 4). Since f^{-1} sends outputs of f to inputs of f, we conclude that f^{-1} sends 16 to 0, 8 to 1, 4 to 2,

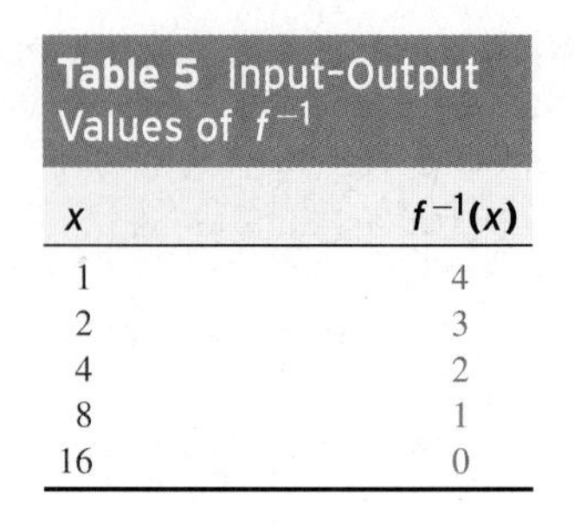

Table 5 Input-Output Values of f^{-1}

x	$f^{-1}(x)$
1	4
2	3
4	2
8	1
16	0

2 to 3, and 1 to 4. We list these results from the smallest to the largest input in Table 5.

2. From Table 5, we see that f^{-1} sends the input 8 to the output 1, so $f^{-1}(8) = 1$. ■

Let's return to the function g that converts Celsius temperatures to Fahrenheit temperatures. Consider the input–output diagrams of g and g^{-1} in Figs. 5 and 6. We've noted that to find the inverse of g, we reverse the arrow for g. But also notice that if we reverse the arrow for g^{-1}, we get the arrow for g. This suggests that the inverse of g^{-1} is g, which is true. So, g and g^{-1} are inverses of each other.

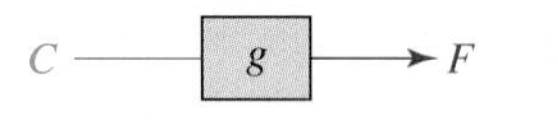

Figure 5 g sends values of C to values of F

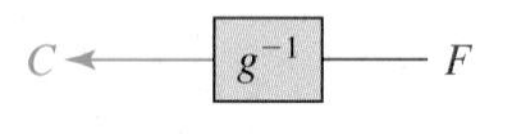

Figure 6 g^{-1} sends values of F to values of C

f and f^{-1} Are Inverses of Each Other

If f is an invertible function, then

- f^{-1} is invertible, and
- f and f^{-1} are inverses of each other.

Graphing Inverse Functions

Next, we discuss how to graph the inverse of an invertible function.

Example 5 Comparing the Graphs of a Function and Its Inverse

Sketch the graphs of $f(x) = 2^x$, f^{-1}, and $y = x$ on the same set of axes.

Solution

We list some input–output values of f and f^{-1} in Tables 6 and 7.

Table 6 Input-Output Pairs for f

x	$f(x)$
−3	$\frac{1}{8}$
−2	$\frac{1}{4}$
−1	$\frac{1}{2}$
0	1
1	2
2	4

Table 7 Input-Output Pairs for f^{-1}

x	$f^{-1}(x)$
$\frac{1}{8}$	−3
$\frac{1}{4}$	−2
$\frac{1}{2}$	−1
1	0
2	1
4	2

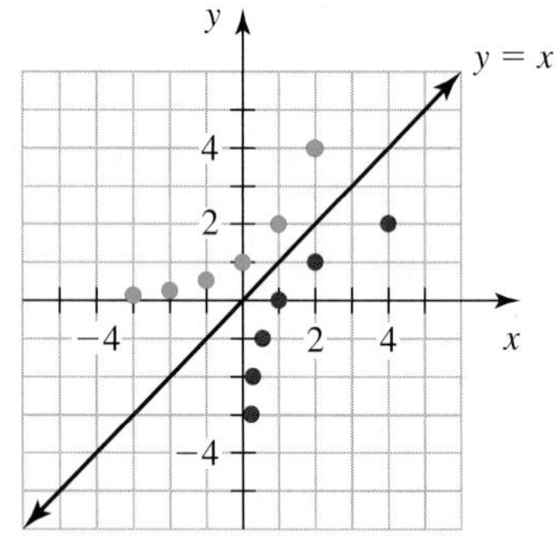

Figure 7 Some points of the graph of f (in blue), some points of the graph of f^{-1} (in red), and the line $y = x$

From Tables 6 and 7, we see that if a point (a, b) is on the graph of f, then the point (b, a) is on the graph of f^{-1}. This observation will lead us to a key step in graphing inverses in future problems.

Next, we plot the input–output pairs of f with blue dots, plot the input–output pairs of f^{-1} with red dots, and sketch the line $y = x$ (see Fig. 7).

If we were to draw the blue dots in wet ink and fold the paper along the line $y = x$, the ink would make dots where the red dots are. We say that the red dots are the reflection of the blue dots across the line $y = x$.

In Fig. 8, we sketch a graph of $f(x) = 2^x$ with a blue exponential curve. By reflecting all the blue points on the graph of f across the line $y = x$, we obtain the red graph of f^{-1}. The graph of f^{-1} is the reflection of the graph of $f(x) = 2^x$ across the line $y = x$. ■

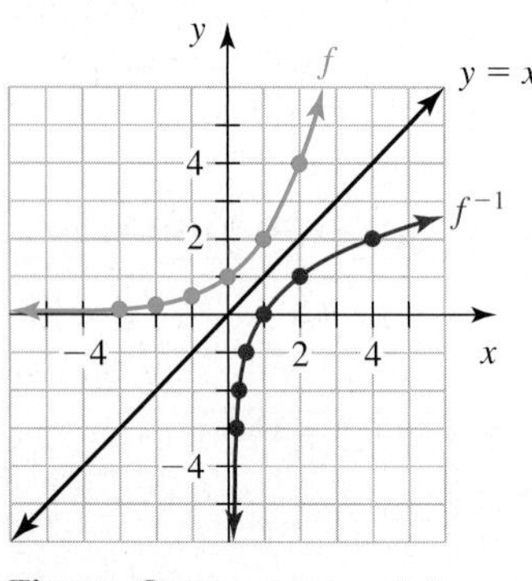

Figure 8 The graphs of f, f^{-1}, and $y = x$

Reflection Property of Inverse Functions

For an invertible function f, the graph of f^{-1} is the reflection of the graph of f across the line $y = x$.

In Example 5, we saw that if a point (a, b) is on the graph of f, then the point (b, a) is on the graph of the inverse of f.

Graphing an Inverse Function

For an invertible function f, we sketch the graph of f^{-1} by the following steps:

1. Sketch the graph of f.
2. Choose several points that lie on the graph of f.
3. For each point (a, b) chosen in step 2, plot the point (b, a).
4. Sketch the curve that contains the points plotted in step 3.

Example 6 Graphing an Inverse Function

Let $f(x) = \frac{1}{3}x - 1$. Sketch the graph of f, f^{-1}, and $y = x$ on the same set of axes.

Solution

We apply the four steps to graph the inverse function:

Step 1. Sketch the graph of f: See Fig. 9.
Step 2. Choose several points that lie on the graph of f: $(-6, -3)$, $(-3, -2)$, $(0, -1)$, $(3, 0)$, and $(6, 1)$.
Step 3. For each point (a, b) chosen in step 2, plot the point (b, a): We plot $(-3, -6)$, $(-2, -3)$, $(-1, 0)$, $(0, 3)$, and $(1, 6)$ in Fig. 10.
Step 4. Sketch the curve that contains the points plotted in step 3: The points from step 3 lie on a line. See Fig. 10.

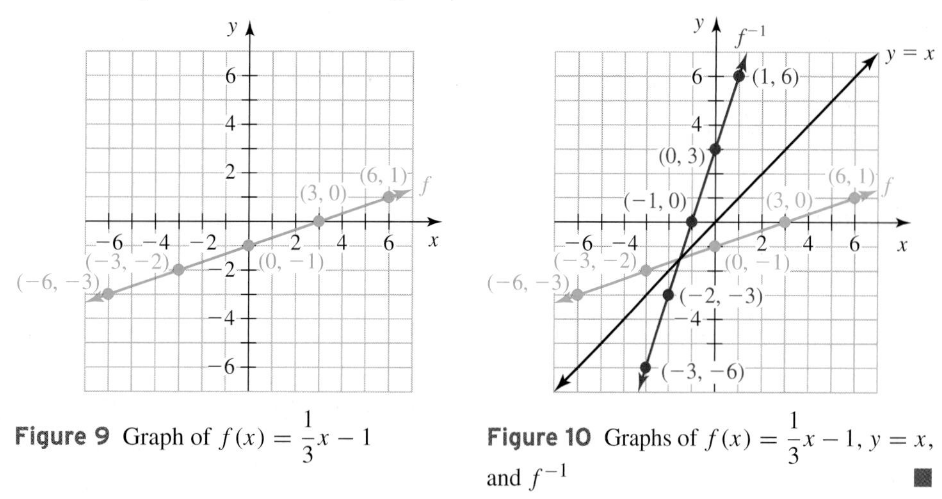

Figure 9 Graph of $f(x) = \frac{1}{3}x - 1$

Figure 10 Graphs of $f(x) = \frac{1}{3}x - 1$, $y = x$, and f^{-1} ■

Finding an Equation of the Inverse of a Model

So far, we have described inverse functions by using tables and graphs. We now describe them by using equations. To begin, we will use a three-step process to find an equation of the inverse of a model.

Table 8 Revenues from Sales of Digital Cameras

Year	Sales (millions of dollars)
2000	1.8
2001	2.0
2002	2.8
2003	3.9
2004	4.5

Source: *Consumer Electronics Association*

Example 7 Finding an Equation of the Inverse of a Model

Revenues from digital camera sales in the United States are shown in Table 8 for various years. Let $r = f(t)$ be the revenue (in millions of dollars) from digital camera sales in the year that is t years since 2000. A reasonable model is

$$f(t) = 0.73t + 1.54$$

1. Find an equation of f^{-1}.
2. Find $f(10)$. What does it mean in this situation?

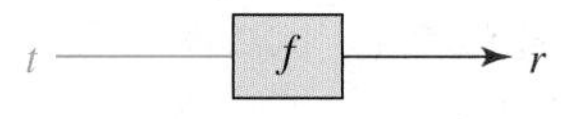

Figure 11 f sends values of t to values of r

r — f^{-1} → t

Figure 12 f^{-1} sends values of r to values of t

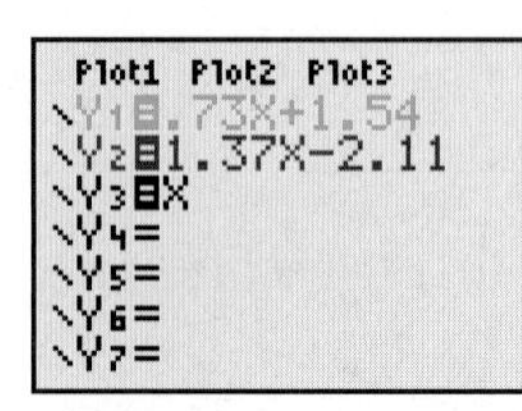

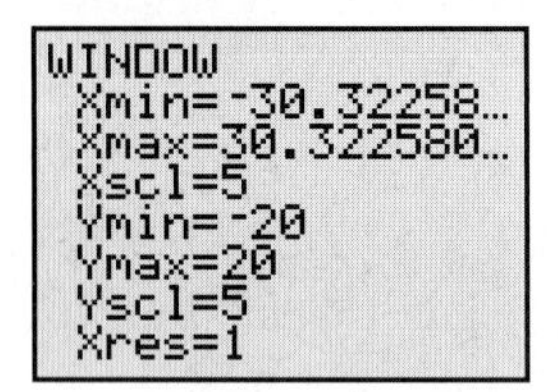

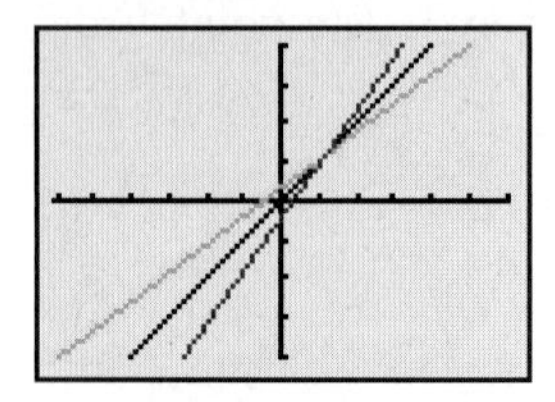

Figure 13 Check that the graph of f^{-1} is the reflection of f across the line $y = x$

3. Find $f^{-1}(10)$. What does it mean in this situation?
4. What is the slope of f? What does it mean in this situation?
5. What is the slope of f^{-1}? What does it mean in this situation?

Solution

1. Since f sends values of t to values of r, f^{-1} sends values of r to values of t (see Figs. 11 and 12).

   To find an equation of f^{-1}, we want to write t in terms of r. Here are three steps to follow to find an equation of f^{-1}:

   Step 1. We replace $f(t)$ with r: $r = 0.73t + 1.54$.

   Step 2. We solve the equation for t:

   $$r = 0.73t + 1.54 \qquad \text{Equation from step 1.}$$
   $$r - 1.54 = 0.73t \qquad \text{Subtract 1.54 from both sides.}$$
   $$\frac{r}{0.73} - \frac{1.54}{0.73} = t \qquad \text{Divide both sides by 0.73.}$$

   An approximate equation is $t = 1.37r - 2.11$.

   Step 3. Since f^{-1} sends values of r to values of t, we have $f^{-1}(r) = t$. So, we can substitute $f^{-1}(r)$ for t in the equation $t = 1.37r - 2.11$:

   $$f^{-1}(r) = 1.37r - 2.11$$

   To verify our work, we use ZStandard followed by ZSquare to check that the graph of f^{-1} is the reflection of the graph of f across the line $y = x$ (see Fig. 13).

2. $f(10) = 0.73(10) + 1.54 = 8.84$. Since f sends values of t to values of r, this means that $r = 8.84$ when $t = 10$. According to the model f, the revenue will be about \$8.8 million in 2010.
3. $f^{-1}(10) = 1.37(10) - 2.11 = 11.59$. Since f^{-1} sends values of r to values of t, this means that $t = 11.59$ when $r = 10$. According to the model f^{-1}, the revenue will be \$10 million in 2012.
4. The slope of $f(t) = 0.73t + 1.54$ is 0.73. This means that the rate of change of r with respect to t is 0.73. According to the model f, the revenue increases by \$0.73 million each year.
5. The slope of $f^{-1}(r) = 1.37r - 2.11$ is 1.37. This means that the rate of change of t with respect to r is 1.37. According to the model f^{-1}, 1.37 years pass each time the revenue increases by \$1 million. ■

In Example 7, we performed three steps to find an equation of the inverse of a model.

Three-Step Process for Finding the Inverse of a Model

To find the inverse of an invertible *model* f, where $p = f(t)$,

1. Replace $f(t)$ with p.
2. Solve for t.
3. Replace t with $f^{-1}(p)$.

Finding an Equation of the Inverse of a Function That Is Not a Model

We can take three similar steps, followed by one more step, to find an equation of the inverse of a function that is *not* a model.

Example 8 Finding the Inverse of a Function That Is Not a Model

Find the inverse of $f(x) = 2x - 3$.

Solution

Step 1. Substitute y for $f(x)$: $y = 2x - 3$

Step 2. Solve for x:

$$\begin{aligned} y &= 2x - 3 && \text{Equation from step 1} \\ y + 3 &= 2x && \text{Add 3 to both sides.} \\ 2x &= y + 3 && \text{If } a = b \text{, then } b = a. \\ \frac{2x}{2} &= \frac{y}{2} + \frac{3}{2} && \text{Divide both sides by 2.} \\ x &= \frac{1}{2}y + \frac{3}{2} && \text{Simplify.} \end{aligned}$$

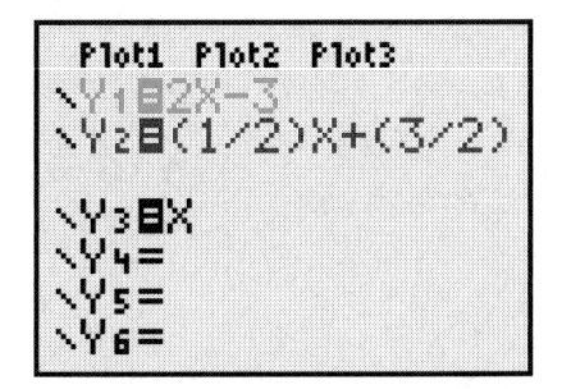

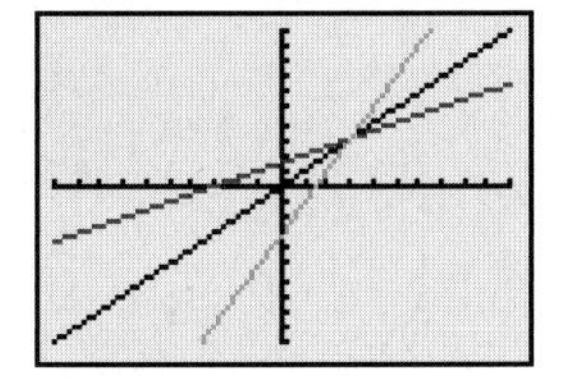

Figure 14 Check that the graph of f^{-1} is the reflection of f across the line $y = x$

Step 3. Replace x with $f^{-1}(y)$: $f^{-1}(y) = \frac{1}{2}y + \frac{3}{2}$.

Step 4. When a function is not a model, we usually want the input variable to be x. So, we rewrite the equation $f^{-1}(y) = \frac{1}{2}y + \frac{3}{2}$ in terms of x:

$$f^{-1}(x) = \frac{1}{2}x + \frac{3}{2}$$

To verify our work, we use ZStandard followed by ZSquare to check that the graph of f^{-1} is the reflection of the graph of f across the line $y = x$ (see Fig. 14). ■

When finding the inverse of a function f, we must keep in mind whether f is a model. If the function is not a model, we perform all four steps as shown in Example 8. However, if the function is a model, we perform only the first three steps, so that the variables retain their original meaning.

Four-Step Process for Finding the Inverse of a Function That Is Not a Model

Let f be an invertible function that is *not* a model. To find the inverse of f, where $y = f(x)$,

1. Replace $f(x)$ with y.
2. Solve for x.
3. Replace x with $f^{-1}(y)$.
4. Write the equation of f^{-1} in terms of x.

So far, we have used tables and graphs to describe the inverse of an exponential function, such as $f(x) = 2^x$. Can we find an equation of the inverse of $f(x) = 2^x$? To find such an equation, we would have to solve the equation $y = 2^x$ for x. We cannot do this by using familiar operations such as adding or subtracting. Nonetheless, the inverse of an exponential function is a powerful tool that we will continue to explore throughout the rest of this chapter.

One-to-One Functions

We said earlier in this section that the inverse of a function is not necessarily a function. Consider $f(x) = x^2$. Note that $f(-2) = 4$ and $f(2) = 4$. So, f sends both of the inputs -2 and 2 to the output 4 (see Fig. 15). Therefore, the inverse of f sends the input 4 to the *two* outputs -2 and 2 (see Fig. 16). Thus, the inverse of f is not a function. Note that the inverse of f is not a function because an output of f originates from more than one input of f.

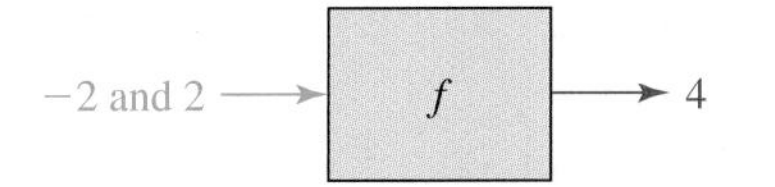

Figure 15 f sends both -2 and 2 to 4

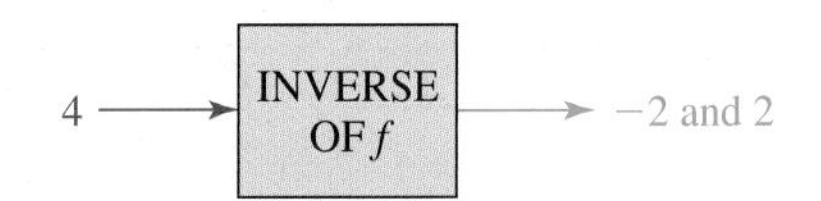

Figure 16 The inverse of f sends 4 to both -2 and 2

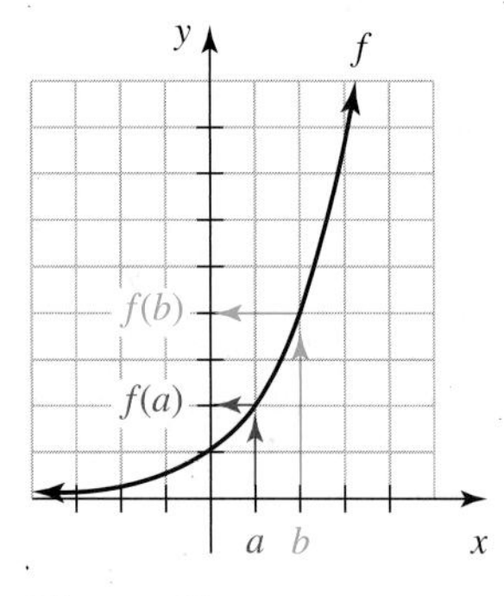

Figure 17 Graph of an increasing exponential function

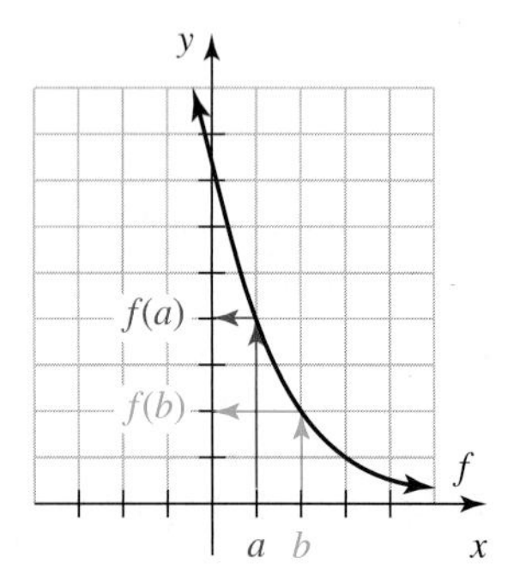

Figure 18 Graph of a decreasing exponential function

If each output of a function originates from exactly one input, we say that the function is **one-to-one. A one-to-one function is invertible.** All exponential functions are one-to-one and, hence, invertible. To see why, note that if f is an increasing exponential function, then $a < b$ implies that $f(a) < f(b)$ (see Fig. 17). If f is a decreasing exponential function, then $a < b$ implies that $f(a) > f(b)$. See Fig. 18. Both implications show that if two inputs a and b are different, then their outputs $f(a)$ and $f(b)$ are different, too. It follows that each output originates from exactly one input.

Using similar approaches, we could show that all linear functions with nonzero slope are one-to-one and, hence, invertible. Or we could show that such a function is invertible by finding the equation of the inverse of the function; you are asked to do this in Exercise 85.

group exploration

Looking ahead: A logarithm is an exponent

1. **a.** Solve $2^x = 16$.
   b. Solve $2^x = 32$.
   c. Approximate the solution of $2^x = 24$ by using trial and error. Your result should be correct up to four decimal places. [**Hint:** Parts (a) and (b) should suggest a reasonable first guess. Then you can use calculator tables to speed up the trial-and-error process.]
2. Approximate the solution of $3^x = 15$ by using trial and error. We call the solution $\log_3(15)$, where *log* is shorthand for *logarithm*.
3. Approximate the solution of $10^x = 500$. We call the solution $\log_{10}(500)$.

TIPS FOR SUCCESS: Review Material

At various times throughout this course, you can improve your understanding of algebra by reviewing material that you have learned so far. Re-solve problems, redo explorations, and consider again key points from previous chapters.

HOMEWORK 5.1

1. Let f be an invertible function where $f(4) = 7$. Find $f^{-1}(7)$.

2. Let g be an invertible function where $g^{-1}(3) = 2$. Find $g(2)$.

Some values of an invertible function f are given in Table 9. For Exercises 3–8, refer to this table.

3. Find $f(4)$.

4. Find $f(2)$.

5. Find $f^{-1}(4)$.

6. Find $f^{-1}(2)$.

7. Use a table to describe five input–output values of f^{-1}.

8. Find x when $f(x) = 2$.

Table 9 Values of f (Exercises 3-8)

x	f(x)
2	10
3	8
4	6
5	4
6	2

Some values of an invertible function g are given in Table 10. For Exercises 9–14, refer to this table.

9. Find $g(2)$.

10. Find $g(6)$.

11. Find $g^{-1}(2)$.

12. Find $g^{-1}(6)$.

13. Use a table to describe six input–output values of g^{-1}.

14. Find x when $g(x) = 6$.

Table 10 Values of g (Exercises 9-14)

x	g(x)
1	2
2	6
3	18
4	54
5	162
6	486

15. Complete Table 11 by using the table of values of f to complete the table of values of f^{-1}.

Table 11 Finding Values of f^{-1} (Exercise 15)

x	$f(x)$	x	$f^{-1}(x)$
1	34	4	
2	28	10	
3	22	16	
4	16	22	
5	10	28	
6	4	34	

16. Complete Table 12 by using the table of values of f to complete the table of values of f^{-1}.

Table 12 Finding Values of f^{-1} (Exercise 16)

x	$f(x)$	x	$f^{-1}(x)$
2	5	2	
3	2	3	
4	3	4	
5	6	5	
6	4	6	

Let $f(x) = 3(2)^x$.

17. Use a table to describe five input–output values of f^{-1}.

18. Find $f(2)$.

19. Find $f(3)$.

20. Find $f^{-1}(24)$.

21. Find $f^{-1}(3)$.

22. Find x when $f(x) = 6$.

Graph the given function, its inverse, and $y = x$ by hand on the same set of axes. Label each graph as f, f^{-1}, or $y = x$.

23. $f(x) = 3^x$

24. $f(x) = 4^x$

25. $f(x) = 3(2)^x$

26. $f(x) = 2(3)^x$

27. $f(x) = 2x$

28. $f(x) = 3x$

29. $f(x) = 3x - 2$

30. $f(x) = 2x - 5$

31. $f(x) = \frac{1}{2}x + 1$

32. $f(x) = \frac{2}{3}x - 1$

33. $f(x) = 4\left(\frac{1}{2}\right)^x$

34. $f(x) = 6\left(\frac{1}{3}\right)^x$

35. $f(x) = \left(\frac{1}{3}\right)^x$

36. $f(x) = 2\left(\frac{1}{2}\right)^x$

For Exercises 37–42, refer to Fig. 19, which shows the graph of an invertible function g.

37. Find $g(2)$.

38. Find $g(-6)$.

39. Find $g^{-1}(2)$.

40. Find $g^{-1}(-1)$.

41. Find $g^{-1}(0)$.

42. Graph g^{-1} by hand.

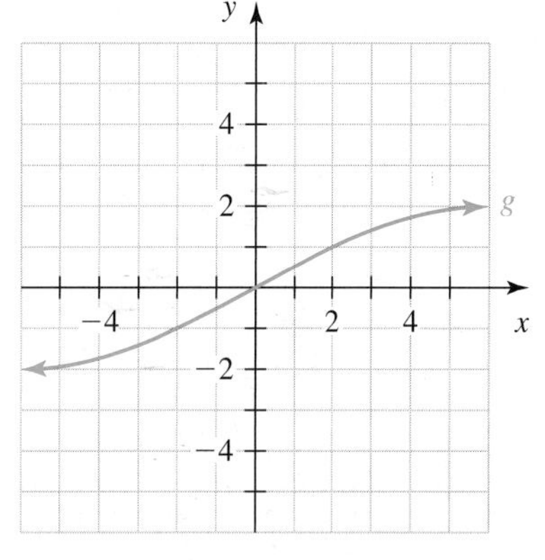

Figure 19 Exercises 37–42

For Exercises 43–48, refer to Fig. 20, which shows the graph of an invertible function f.

43. Find $f(2)$.

44. Find $f(-1)$.

45. Find $f^{-1}(4)$.

46. Find $f^{-1}(0)$.

47. Graph f^{-1} by hand.

48. Find $f^{-1}(1)$.

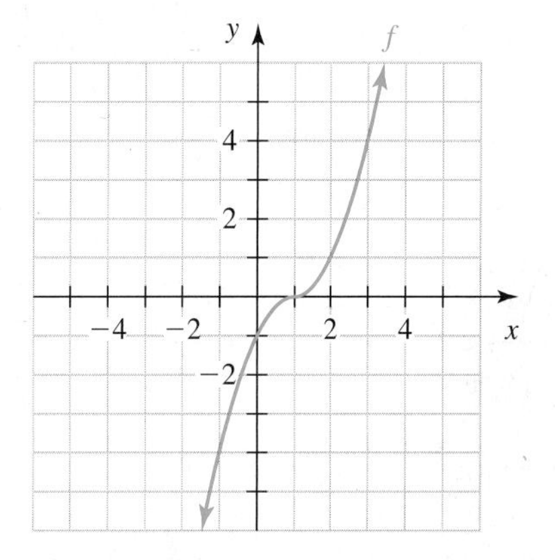

Figure 20 Exercises 43–48

49. In Exercise 15 of Homework 2.2, you found an equation close to $p = f(t) = 0.77t - 42.90$ that models the percentage p of births outside marriage at t years since 1900 (see Table 13).

Table 13 Births Outside Marriage

Year	Percent
1970	10.7
1975	14.3
1980	18.4
1985	22.0
1990	28.0
1995	32.2
2000	33.2
2002	34.0

Source: *National Center for Health Statistics*

a. Find an equation of the inverse function of f.

b. Find $f(100)$. What does it mean in this situation?

c. Find $f^{-1}(100)$. What does it mean in this situation?

d. What is the slope of f^{-1}? What does it mean in this situation?

50. In Exercise 30 of Homework 2.4, you found an equation close to $p = f(t) = 0.075t + 0.85$ that models Nevada's population p (in millions) at t years since 1985 (see Table 14).

Table 14 Nevada's Population

Year	Population (millions)
1986	1.0
1990	1.2
1995	1.5
2000	2.0
2005	2.4

Source: *U.S. Census Bureau*

a. Find an equation of f^{-1}.
b. Find $f(3)$. What does it mean in this situation?
c. Find $f^{-1}(3)$. What does it mean in this situation?
d. What is the slope of f^{-1}? What does it mean in this situation?

51. The number of reported cases of wrong-site surgery, in which a surgeon operated on the wrong limb, organ, or patient, are shown in Table 15 for various years.

Table 15 Reported Cases of Wrong-Site Surgery

Year	Number of Reported Cases
2000	50
2001	58
2002	65
2003	70
2004	70
2005	84

Source: *Joint Commission on Accreditation of Healthcare Organizations*

Let $n = f(t)$ be the number of cases of reported wrong-site surgery in the year that is t years since 2000.
a. Use a graphing calculator to draw a scattergram of the data. Can the data be modeled better by using a linear or an exponential function? Explain.
b. Find an equation of f.
c. Find an equation of f^{-1}.
d. Use f to predict in which year there will be 115 reported cases of wrong-site surgery.
e. Now use f^{-1} to predict in which year there will be 115 reported cases of wrong-site surgery.
f. Compare your results in parts (d) and (e).

52. The number of cremations in the United States has increased steadily since 1980 (see Table 16).

Table 16 Percentages of Bodies That Are Cremated

Year	Percent
1980	9.7
1985	13.9
1990	17.1
1995	21.1
2000	26.2
2004	29.6

Source: *Cremation Association of North America Victimization Survey*

Let $p = f(t)$ be the percentage of bodies that are cremated at t years since 1980.
a. Use a graphing calculator to draw a scattergram. Is it better to use a linear or an exponential function to model the situation?
b. Find an equation of f.
c. Find an equation of f^{-1}.
d. Use f to predict when half of people who die will be cremated.
e. Now use f^{-1} to predict when half of people who die will be cremated.
f. Compare your results in parts (d) and (e).

53. Credit scores measure financial responsibility. Credit scores of Americans are shown in Table 17 for various age groups.

Table 17 Credit Scores of Americans

Age Group (years)	Age Used to Represent Age Group (years)	Average Credit Score (points)
18–29	23.5	637
30–39	34.5	654
40–49	44.5	675
50–59	54.5	697
60–69	64.5	722
70 or more	75	747

Source: *Experian*

Let $c = f(a)$ be the average credit score (in points) of adults at a years of age.
a. Use a graphing calculator to draw a scattergram of the data. Can the data be modeled better by using a linear or an exponential function? Explain.
b. Find a linear equation of f.
c. Find an equation of f^{-1}.
d. The average credit score of all adults in 677 points. Use f^{-1} to estimate the (single) age of adults whose average credit score is 677 points.
e. The highest possible score is 830 points. Use f^{-1} to estimate the age of adults whose average credit score is 830 points. (The oldest American ever was Sarah Knauss, who died at age 119 years.)
f. What is the slope of f^{-1}? What does it mean in this situation?

54. The percentages of Americans who went to the movies at least once in the past year are shown in Table 18 for various age groups.

Table 18 Percentages of Americans Who Go to the Movies

Age Group (years)	Age Used to Represent Age Group (years)	Percent
18–24	21.0	88
25–34	29.5	79
35–44	39.5	73
45–54	49.5	65
55–64	59.5	46
65–74	69.5	38
over 74	80	28

Source: *U.S. National Endowment for the Arts*

Let $p = f(a)$ be the percentage of Americans at age a years who go to the movies.

a. Use a graphing calculator to draw a scattergram of the data. Can the data be modeled better by using a linear or an exponential function? Explain.
b. Find an equation of f.
c. Find an equation of f^{-1}.
d. Use f^{-1} to estimate at what age half of Americans go to the movies.
e. What is the slope of f^{-1}? What does it mean in this situation?

Find the inverse of the given function. Use a graphing calculator to verify your work by graphing f, f^{-1}, and $y = x$ on the same set of axes.

55. $f(x) = x + 8$

56. $f(x) = x - 6$

57. $f(x) = -4x$

58. $f(x) = 5x$

59. $f(x) = \frac{x}{7}$

60. $f(x) = -\frac{x}{2}$

61. $f(x) = -6x - 2$

62. $f(x) = 3x - 8$

63. $f(x) = 0.4x - 7.9$

64. $f(x) = -6.25x + 12.5$

65. $f(x) = \frac{7}{3}x + 1$

66. $f(x) = \frac{8}{5}x + 4$

67. $f(x) = -\frac{5}{6}x - 3$

68. $f(x) = -\frac{2}{5}x - 8$

69. $f(x) = \frac{6x - 2}{5}$

70. $f(x) = \frac{x - 7}{4}$

71. $f(x) = 7 - 8(x + 1)$

72. $f(x) = 2(x - 1) + 5$

73. $f(x) = x$

74. $f(x) = -x$

75. $f(x) = x^3$

76. $f(x) = x^5$

77. Let $f(x) = 5x - 9$.
a. Find an equation of f^{-1}.
b. Find $f(4)$.
c. Find $f^{-1}(4)$.

78. Let $g(x) = \frac{3}{5}x - 1$.
a. Find an equation of g^{-1}.
b. Find $g(20)$.
c. Find $g^{-1}(20)$.

79. Describe the Rule of Four as applied to the inverse of the function $f(x) = 3x - 5$:
a. Describe the input–output pairs of f^{-1} by using an equation.
b. Describe five input–output pairs of f^{-1} by using a table.
c. Describe the input–output pairs of f^{-1} by using a graph.
d. Describe the input–output pairs of f^{-1} by using words.

80. Describe the Rule of Four as applied to the inverse of the function $g(x) = \frac{4}{5}x + 2$:
a. Describe the input–output pairs of g^{-1} by using an equation.
b. Describe five input–output pairs of g^{-1} by using a table.
c. Describe the input–output pairs of g^{-1} by using a graph.
d. Describe the input–output pairs of g^{-1} by using words.

81. Explain why it makes sense that the function $g(x) = x - 5$ is the inverse of the function $f(x) = x + 5$.

82. Explain why it makes sense that the function $g(x) = \frac{x}{3}$ is the inverse of the function $f(x) = 3x$.

83. Is the function $f(x) = 3$ one-to-one? Is it invertible? Explain.

84. Is the function $f(x) = x^4$ one-to-one? Is it invertible? Explain.

85. In this exercise, you will show that all linear functions with nonzero slope are invertible.
a. Let $f(x) = mx + b$, where $m \neq 0$. Find an equation of the inverse of f.
b. Explain why your work in part (a) shows that the inverse of f is a function.

86. Explain how to find the inverse of an invertible linear function. Also, explain the meaning of an inverse function. (See page 4 for guidelines on writing a good response.)

Related Review

87. Let $f(x) = 2x - 3$ and $g(x) = \frac{1}{2}x + 3$.
a. Find any points of intersection of the graphs of f and g.
b. Find an equation of f^{-1}.
c. Find an equation of g^{-1}.
d. Find any points of intersection of the graphs of f^{-1} and g^{-1}.
e. What do you observe about your results from parts (a) and (d)? Explain why this makes sense in terms of a property of an inverse function.

88. Let $f(x) = 3x + 5$.
a. Solve the equation $3x + 5 = 11$.
b. Find an equation of f^{-1}.
c. Find $f^{-1}(11)$.
d. What do you observe about your results from parts (a) and (c)? Explain why this makes sense in terms of a property of an inverse function.
e. Use f^{-1} to solve each of the following equations.
i. $3x + 5 = 17$
ii. $3x + 5 = 2$
iii. $3x + 5 = -6$

Expressions, Equations, Functions, and Graphs

Perform the indicated instruction. Then use words such as linear, exponential, function, one variable, *and* two variables *to describe the expression, equation, or system.*

89. Solve:
$$3x - 5y = 10$$
$$7x + 4y = 39$$

90. Simplify $\left(\frac{25b^7c^{-3}}{49b^3c^{-9}}\right)^{1/2}$.

91. Graph $3x - 5y = 10$ by hand.

92. Find all real-number solutions of $6b^5 = 349$. Round any results to the second decimal place.

5.2 LOGARITHMIC FUNCTIONS

Objectives

- Know the meaning of *logarithm* and *logarithmic function*.
- Find logarithms and evaluate logarithmic functions.
- Know properties of logarithmic functions.
- Graph logarithmic functions.
- Use logarithms to model authentic situations.

In Section 5.1, we discussed the inverse of a function. In this section, we will focus on the inverse of an exponential function, which can be used to measure the energy released by earthquakes and the noise level of sounds.

Table 19 Input-Output Pairs of $f(x) = 2^x$

x	$f(x)$
0	2^0
1	2^1
2	2^2
3	2^3
4	2^4

Table 20 Input-Output Pairs of f^{-1}

x	$f^{-1}(x)$
2^0	0
2^1	1
2^2	2
2^3	3
2^4	4

Definition of a Logarithm

Consider the function $f(x) = 2^x$. Input–output values of f and f^{-1} are shown in Tables 19 and 20.

From Table 20, we see that $f^{-1}(2^3) = 3$. The output 3 is the *exponent* of the input 2^3. In fact, all of the outputs in Table 20 are *exponents* of the corresponding inputs. We know that $f^{-1}(2^6) = 6$, since 6 is the exponent of 2^6. Likewise, $f^{-1}(2^7) = 7$.

For $f(x) = 2^x$, the function f^{-1} is so useful that we give it a specific name and call it $\log_2$ (read "logarithm, base 2"). So, we write $\log_2(2^3) = 3$, $\log_2(2^6) = 6$, and so on.

To find $\log_2(32)$, we first write 32 as a power of 2:

$$\log_2(32) = \log_2(2^5) = 5$$

So, $\log_2(32) = 5$ because $2^5 = 32$. Likewise, $\log_2(16) = 4$ because $2^4 = 16$. In general,

$$\log_2(a) = k \quad \text{if} \quad 2^k = a$$

We define $\log_3$ in a similar way:

$$\log_3(a) = k \quad \text{if} \quad 3^k = a$$

For example, $\log_3(9) = 2$, since $3^2 = 9$. The logarithm 2 is the *exponent* on the base 3 that gives 9.

> **DEFINITION** Logarithm
>
> For $b > 0$, $b \neq 1$, and $a > 0$,
>
> the **logarithm $\log_b(a)$** is the number k such that $b^k = a$.
>
> In words: $\log_b(a)$ is the exponent on the base b that gives a. We call b the **base** of the logarithm.

In short, a logarithm is an exponent.

Example 1 Finding Logarithms

Find the logarithm.

1. $\log_7(49)$ **2.** $\log_8(64)$ **3.** $\log_3(81)$
4. $\log_5(125)$ **5.** $\log_4(64)$ **6.** $\log_{10}(100{,}000)$

Solution

1. $\log_7(49) = 2$, since $7^2 = 49$.
2. $\log_8(64) = 2$, since $8^2 = 64$.
3. $\log_3(81) = 4$, since $3^4 = 81$.
4. $\log_5(125) = 3$, since $5^3 = 125$.
5. $\log_4(64) = 3$, since $4^3 = 64$.
6. $\log_{10}(100{,}000) = 5$, since $10^5 = 100{,}000$.

■

DEFINITION Common logarithm

A **common logarithm** is a logarithm with base 10. We write $\log(a)$ to represent $\log_{10}(a)$.

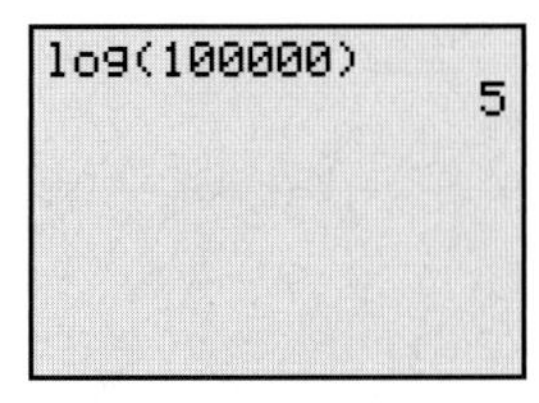

Figure 21 Use a graphing calculator to find that $\log(100{,}000) = 5$

For example, $\log(100{,}000)$ represents $\log_{10}(100{,}000)$. We can use the "log" key on a calculator to find base-10 logarithms. In Fig. 21, we find that $\log(100{,}000) = 5$.

It is important to keep in mind that a logarithm is an exponent. In Example 2, we use properties of exponents to find more logarithms.

Example 2 Finding Logarithms

Find the logarithm.

1. $\log_7(7)$ **2.** $\log_4(1)$ **3.** $\log_3\left(\frac{1}{9}\right)$

4. $\log_6(\sqrt{6})$ **5.** $\log_b(b^5)$ **6.** $\log(0.001)$

Solution

1. $\log_7(7) = 1$, since $7^1 = 7$.
2. $\log_4(1) = 0$, since $4^0 = 1$.
3. $\log_3\left(\frac{1}{9}\right) = -2$, since $3^{-2} = \frac{1}{3^2} = \frac{1}{9}$.
4. $\log_6\left(\sqrt{6}\right) = \frac{1}{2}$, since $6^{\frac{1}{2}} = \sqrt{6}$.
5. $\log_b\left(b^5\right) = 5$, since $b^5 = b^5$.
6. Remember that "$\log(a)$" is shorthand for $\log_{10}(a)$, so $\log(0.001) = -3$, since $10^{-3} = \frac{1}{10^3} = \frac{1}{1000} = 0.001$. ■

Properties of Logarithms

What are the general properties of logarithms? In the discussion that follows, we assume that $b > 0$ and $b \neq 1$.

In Example 2, we found that $\log_7(7) = 1$. In general, $\log_b(b) = 1$, since $b^1 = b$. We also found that $\log_4(1) = 0$. In general, $\log_b(1) = 0$, since $b^0 = 1$.

Properties of Logarithms

For $b > 0$ and $b \neq 1$,

- $\log_b(b) = 1$
- $\log_b(1) = 0$

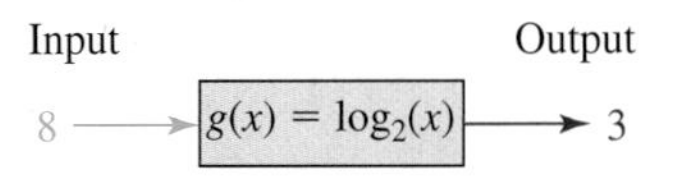

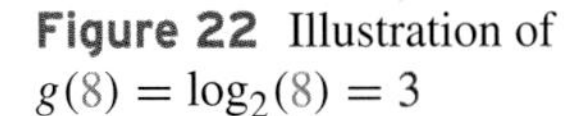

Figure 22 Illustration of $g(8) = \log_2(8) = 3$

Recall that $\log_2$ is the name of a function. When we write $\log_2(8) = 3$, we mean that the function $\log_2$ sends the input 8 to the output 3 (see Fig. 22). What follows is the general definition of a logarithmic function.

DEFINITION Logarithmic function

A **logarithmic function, base *b*,** is a function that can be put into the form

$$f(x) = \log_b(x)$$

where $b > 0$ and $b \neq 1$.

The nonpositive numbers are not in the domain of $\log_2$. Consider $\log_2(0)$. No exponent on the base 2 gives 0. Now consider $\log_2(-8)$. No exponent on the base 2 gives a negative number. In general, **the domain of a logarithmic function $\log_b$ is the set of all positive real numbers.**

The function $\log_2$ is the inverse of $f(x) = 2^x$. From our work in Section 5.1, we can conclude that $f(x) = 2^x$ is also the inverse of $\log_2$.

Logarithmic and Exponential Functions Are Inverses of Each Other

- For an exponential function $f(x) = b^x$, $f^{-1}(x) = \log_b(x)$.
- For a logarithmic function $g(x) = \log_b(x)$, $g^{-1}(x) = b^x$.

In words: $g(x) = \log_b(x)$ and $f(x) = b^x$ are inverse functions of each other.

Example 3 Finding an Inverse Function

Find the inverse of the function.

1. $f(x) = 4^x$ **2.** $h(x) = \log_9(x)$

Solution

1. $f^{-1}(x) = \log_4(x)$

2. $h^{-1}(x) = 9^x$ ■

Example 4 Evaluating f and f^{-1}

Let $f(x) = 3^x$.

1. Find $f(4)$. **2.** Find $f^{-1}(9)$.

Solution

1. $f(4) = 3^4 = 81$

2. $f^{-1}(9) = \log_3(9) = 2$ ■

Graphing a Logarithmic Function

Because a logarithmic function is the inverse of an exponential function, we can use the four-step graphing method of Section 5.1 to help us graph a logarithmic function.

Example 5 Graphing a Logarithmic Function

Sketch the graph of $y = \log_3(x)$.

Solution

The inverse of $f(x) = 3^x$ is $f^{-1}(x) = \log_3(x)$. So, we can apply the four-step method to graph the inverse of f:

Step 1. Sketch the graph of f: See Fig. 23.

Step 2. Choose several points on the graph of f: $\left(-1, \frac{1}{3}\right)$, $(0, 1)$, $(1, 3)$, and $(2, 9)$.

Step 3. For each point (a, b) chosen in step 2, plot point (b, a): We plot $\left(\frac{1}{3}, -1\right)$, $(1, 0)$, $(3, 1)$, and $(9, 2)$ in Fig. 24.

Step 4. Sketch the curve that contains the points plotted in step 3: See Fig. 24. The red curve is the graph of $y = \log_3(x)$.

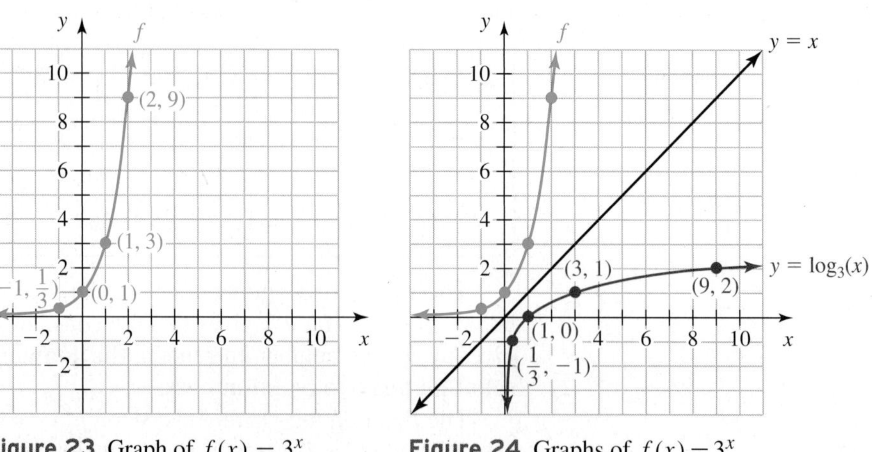

Figure 23 Graph of $f(x) = 3^x$

Figure 24 Graphs of $f(x) = 3^x$, $y = \log_3(x)$, and $y = x$ ■

To graph a logarithmic function $y = \log_b(x)$, we use the four-step graphing method from Section 5.1 to sketch the inverse of $f(x) = b^x$.

Using Logarithms to Model Authentic Situations

Scientists often use logarithms to rescale measurements of objects or phenomena when the measurements tend to be very small (e.g., 3.2×10^{-8}) or very large (e.g., 7.9×10^{13}). For example, scientists use logarithms for measurements of amplitudes of earthquakes, noise levels of sounds, and pH values of solutions.

The energy released by an earthquake is sometimes measured on the *Richter scale*. The *Richter number*, R, of an earthquake is given by

$$R = \log\left(\frac{A}{A_0}\right)$$

where A is the amplitude (maximum value) of a seismic wave and A_0, called the *reference amplitude,* is the amplitude of the smallest seismic wave that a seismograph can detect.

Example 6 Richter Numbers

In 1906, an earthquake in San Francisco had an amplitude 2×10^8 times the reference amplitude A_0. In 1989, an earthquake in San Francisco had an amplitude 8×10^6 times A_0.

1. Find the Richter number of both earthquakes.
2. Find the ratio of the amplitudes of the 1906 and 1989 earthquakes.

Solution

1. The Richter number of the 1906 earthquake is

$$\begin{aligned} R &= \log\left(\frac{2 \times 10^8 A_0}{A_0}\right) && \text{Substitute } 2 \times 10^8 A_0 \text{ for } A. \\ &= \log(2 \times 10^8) && \text{Simplify.} \\ &\approx 8.3 && \text{Compute.} \end{aligned}$$

The Richter number of the 1989 earthquake is

$$\begin{aligned} R &= \log\left(\frac{8 \times 10^6 A_0}{A_0}\right) && \text{Substitute } 8 \times 10^6 A_0 \text{ for } A. \\ &= \log(8 \times 10^6) && \text{Simplify.} \\ &\approx 6.9 && \text{Compute.} \end{aligned}$$

2. The ratio of the amplitudes of the 1906 earthquake and the 1989 earthquake is

$$\frac{2 \times 10^8 A_0}{8 \times 10^6 A_0} = 25$$

So, the 1906 earthquake had an amplitude 25 times greater than that of the 1989 earthquake. ■

group exploration

Looking ahead: Power property for logarithms

1. Use a calculator to compare $\log(3^2)$ with $2\log(3)$.
2. Use a calculator to compare $\log(7^4)$ with $4\log(7)$.
3. Use a graphing calculator table to compare values of $f(x) = \log(x^3)$ and $g(x) = 3\log(x)$. Also, compare the graphs of f and g in the same viewing window.
4. Use a graphing calculator table to compare values of $f(x) = \log(x^5)$ and $g(x) = 5\log(x)$. Also, compare the graphs of f and g in the same viewing window.
5. What do Problems 1–4 suggest about $\log(x^p)$? Test your observation.

HOMEWORK 5.2

FOR EXTRA HELP ▸ Student Solutions Manual · PH Math/Tutor Center · MathXL® · MyMathLab

Find the logarithm.

1. $\log_9(81)$
2. $\log_6(36)$
3. $\log_3(27)$
4. $\log_5(625)$
5. $\log_4(256)$
6. $\log_3(243)$
7. $\log_6(216)$
8. $\log_2(64)$
9. $\log(100)$
10. $\log(1000)$
11. $\log_4\left(\frac{1}{4}\right)$
12. $\log_3\left(\frac{1}{3}\right)$
13. $\log_2\left(\frac{1}{8}\right)$
14. $\log_3\left(\frac{1}{81}\right)$
15. $\log\left(\frac{1}{10{,}000}\right)$
16. $\log\left(\frac{1}{100}\right)$
17. $\log_5(1)$
18. $\log_8(1)$
19. $\log_9(9)$
20. $\log_4(4)$
21. $\log_9(3)$
22. $\log_{36}(6)$
23. $\log_8(2)$
24. $\log_{32}(2)$
25. $\log_7(\sqrt{7})$
26. $\log_2(\sqrt{2})$
27. $\log_5(\sqrt[4]{5})$
28. $\log_7(\sqrt[3]{7})$
29. $\log_2(\log_2(16))$
30. $\log_2(\log_3(81))$
31. $\log(\log(10))$
32. $\log_3(\log_3(27))$
33. $\log_b(b)$
34. $\log_b(1)$
35. $\log_b(b^4)$
36. $\log_b(b^6)$
37. $\log_b\left(\frac{1}{b^5}\right)$
38. $\log_b\left(\frac{1}{b^3}\right)$
39. $\log_b(\sqrt{b})$
40. $\log_b(\sqrt[5]{b})$
41. $\log_b(\log_b(b))$
42. $\log_2(\log_b(\sqrt{b}))$

Find the inverse of the given function.

43. $f(x) = 3^x$
44. $g(x) = 8^x$
45. $h(x) = 10^x$
46. $g(x) = \left(\frac{1}{3}\right)^x$
47. $f(x) = \log_5(x)$
48. $g(x) = \log_4(x)$
49. $h(x) = \log(x)$
50. $f(x) = \log_{\frac{1}{2}}(x)$

Let $f(x) = 2^x$.

51. Find $f(2)$.
52. Find $f(4)$.
53. Find $f^{-1}(2)$.
54. Find $f^{-1}(4)$.

Let $g(x) = \log_3(x)$.

55. Find $g(3)$.
56. Find $g(81)$.
57. Find $g^{-1}(3)$.
58. Find $g^{-1}(2)$.

For Exercises 59–62, refer to the values of the function $f(x) = 3^x$ listed in Table 21.

59. Find $f(1)$.
60. Find $f(3)$.
61. Find $f^{-1}(1)$. Also, write your result as a logarithm.
62. Find $f^{-1}(3)$. Also, write your result as a logarithm.

Table 21 Values of $f(x) = 3^x$ (Exercises 59–62)

x	$f(x)$
0	1
1	3
2	9
3	27
4	81

Graph the function by hand.

63. $y = \log_2(x)$
64. $y = \log_4(x)$
65. $y = \log(x)$
66. $y = \log_6(x)$
67. $y = \log_{\frac{1}{2}}(x)$
68. $y = \log_{\frac{1}{3}}(x)$

Match the function with its graph in Fig. 25.

69. $f(x) = b^x, 0 < b < 1$

70. $g(x) = b^x, b > 1$

71. $h(x) = \log_b(x), 0 < b < 1$

72. $k(x) = \log_b(x), b > 1$

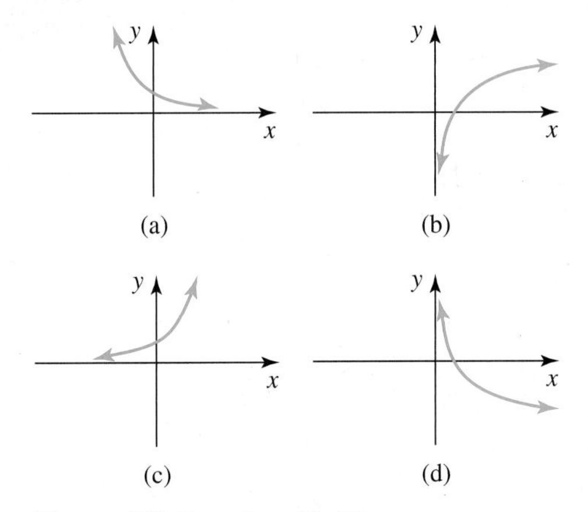

Figure 25 Exercises 69–72

73. Describe the Rule of Four as applied to the logarithmic function $f(x) = \log_5(x)$:
a. Describe five input–output pairs of f by using a table.
b. Describe the input–output pairs of f by using a graph.
c. Describe the input–output pairs of f by using words.

74. Describe the Rule of Four as applied to the logarithmic function $g(x) = \log_{\frac{1}{4}}(x)$:
a. Describe five input–output pairs of g by using a table.
b. Describe the input–output pairs of g by using a graph.
c. Describe the input–output pairs of g by using words.

75. a. Solve the equation $5^x = 25$.
b. Find $\log_5(25)$.
c. Explain why the results of parts (a) and (b) are the same.

76. a. Find log(100).
b. Complete Table 22.

Table 22 Values of log(x) (Exercise 76)

x	log(x)
0.001	
0.01	
0.1	
1	
10	
100	
1000	

c. Examine the entries in Table 22 and describe all patterns that you observe.

77. In 2004, an earthquake in the Indian Ocean had an amplitude 1.6×10^9 times the reference amplitude A_0. In 1985, an earthquake in Mexico City had an amplitude 6.3×10^7 times A_0.
a. Find the Richter number of the Indian Ocean earthquake.
b. Find the Richter number of the Mexico City earthquake.
c. Find the ratio of the amplitudes of the Indian Ocean and Mexico City earthquakes.

78. In 1920, an earthquake in Gansu, China, had an amplitude 4.0×10^8 times the reference amplitude A_0. In 1980, an earthquake in Naples, Italy, had an amplitude 1.6×10^7 times A_0.
a. Find the Richter number of the Gansu earthquake.
b. Find the Richter number of the Naples earthquake.
c. Find the ratio of the amplitudes of the Gansu and Naples earthquakes.

79. The loudness of sound can be measured on a *decibel scale.* The sound level L (in decibels) of a sound is given by

$$L = 10 \log\left(\frac{I}{I_0}\right)$$

where I is the intensity of the sound (in watts per square meter, W/m^2) and $I_0 = 10^{-12}$ W/m^2. The constant I_0 is the approximate intensity of the softest sound a human can hear. Find the decibel values of the sounds listed in Table 23.

Table 23 Examples of Sound Intensities

Sound	Intensity of Sound (W/m^2)
Faintest sound heard by humans	10^{-12}
Whisper	10^{-10}
Inside a running car	10^{-8}
Conversation	10^{-6}
Noisy street corner	10^{-4}
Soft-rock concert	10^{-2}
Threshold of pain	1

80. The acidity or alkalinity of a solution is measured on a *pH scale.* The pH of a solution is given by

$$\text{pH} = -\log(H^+)$$

where H^+ is the hydrogen ion concentration (in moles per liter) of the solution. Distilled water has a pH of 7. Acidic solutions have a pH less than 7, and basic (alkaline) solutions have a pH greater than 7. Most solutions have a pH between 1 and 14. Find the pH of each solution listed in Table 24, and determine whether the solution is acidic or basic. Round your results to the first decimal place.

Table 24 Hydrogen Ion Concentrations of Some Solutions

Solution	Hydrogen Ion Concentration (moles per liter)
Vinegar	1.6×10^{-3}
Human blood	6.3×10^{-8}
Shampoo	7.4×10^{-10}
Orange juice	6.3×10^{-4}
Hydrochloric acid	2.5×10^{-2}

81. Give an example of an exponential function f of the form $f(x) = b^x$. List five input–output pairs of f. Then list five input–output pairs of f^{-1}. What is another name for f^{-1}?

82. Explain how to find a logarithm. Also, explain why a logarithmic function is the inverse of a function.

Related Review

83. In this exercise, you will compare the logarithmic function $f(x) = \log_2(x)$, the linear function $g(x) = 2x$, and the exponential function $h(x) = 2^x$.

a. Complete Table 25.

Table 25 Complete the Table (Exercise 83)

x	$f(x)$	$g(x)$	$h(x)$
1			
2			
4			
8			
16			

b. As the inputs in Table 25 increase, for which function are the outputs growing at the fastest rate? the next fastest rate?

84. a. Sketch the graphs of the functions $f(x) = 3(2)^x$ and $g(x) = -3(2)^x$ on the same coordinate system. Are the graphs reflections of each other across the x-axis?

b. Sketch the graphs of the functions $f(x) = \frac{1}{3}x$ and $g(x) = -\frac{1}{3}x$ on the same coordinate system. Are the graphs reflections of each other across the x-axis?

c. Considering your observations in parts (a) and (b), graph the functions $f(x) = \log_2(x)$ and $g(x) = -\log_2(x)$ by hand on the same coordinate system.

Expressions, Equations, Functions, and Graphs

Perform the indicated instruction. Then use words such as linear, exponential, logarithmic, function, one variable, *and* two variables *to describe the expression, equation, or system.*

85. Find an approximate equation $y = ab^x$ of an exponential curve that contains the points (3, 5) and (7, 89). Round a and b to the second decimal place.

86. Graph $f(x) = -\frac{4}{3}x - 1$ by hand.

87. Simplify $4b^{2/3}c^{-5/4}\left(2b^{-1/5}c^{3/4}\right)$.

88. Find $f^{-1}(2)$ where $f(x) = -\frac{4}{3}x - 1$.

5.3 PROPERTIES OF LOGARITHMS

Objectives

- Convert equations in *logarithmic form* to *exponential form* and vice versa.
- Know the *power property for logarithms.*
- Use properties of logarithms to solve exponential and logarithmic equations.
- Solve equations in one variable by graphing.

In this section, we will discuss some properties of logarithms and how to use these properties to solve exponential and logarithmic equations in one variable.

Exponential/Logarithmic Forms Property

In Section 5.2, we discussed how to find logarithms. For example,

$$\log_2(8) = 3, \qquad \text{since} \qquad 2^3 = 8$$

We say that the equation $\log_2(8) = 3$ is in *logarithmic form* and the equation $2^3 = 8$ is in *exponential form*. The forms $\log_2(8) = 3$ and $2^3 = 8$ are equivalent.

Here are more examples of equations in equivalent logarithmic and exponential forms:

Logarithmic form	*Exponential form*
$\log_2(16) = 4$	$2^4 = 16$
$\log_3(9) = 2$	$3^2 = 9$
$\log_5(125) = 3$	$5^3 = 125$
$\log(100{,}000) = 5$	$10^5 = 100{,}000$

Exponential/Logarithmic Forms Property

For $a > 0$, $b > 0$, and $b \neq 1$, the equations

$$\log_b(a) = c \quad \text{and} \quad b^c = a$$

are equivalent.

The equation $\log_b(a) = c$ is in **logarithmic form** and the equation $b^c = a$ is in **exponential form.** Either form can replace the other when you solve a problem.

Example 1 Solving Equations in Logarithmic Form

Solve for x.

1. $\log_4(x) = 3$ **2.** $\log(3x - 2) = 2$

Solution

1. We write $\log_4(x) = 3$ in exponential form and solve for x:

$$4^3 = x \quad \text{Write in exponential form.}$$
$$64 = x \quad \text{Simplify.}$$

2. We write $\log(3x - 2) = 2$ in exponential form and solve for x:

$$10^2 = 3x - 2 \quad \text{Write in exponential form.}$$
$$100 = 3x - 2 \quad \text{Simplify.}$$
$$102 = 3x \quad \text{Add 2 to both sides.}$$
$$34 = x \quad \text{Divide both sides by 3.}$$

A logarithmic equation in one variable is an equation in one variable that contains one or more logarithms. Here are some examples of logarithmic equations in one variable:

$$\log_3(x) = 4 \quad \log_b(87) = 6 \quad 3\log_2(t) - 7 = 8 \quad \log_5(x^4) + \log_5(3x) = 3$$

Example 2 Solving Logarithmic Equations in One Variable

Solve for x.

1. $6\log_9(t) + 1 = 4$ **2.** $\log_3(x^4) = 2$

Solution

1. To begin, we get $\log_9(t)$ alone on the left side of the equation:

$$6\log_9(t) + 1 = 4 \quad \text{Original equation}$$
$$6\log_9(t) = 3 \quad \text{Subtract 1 from both sides.}$$
$$\log_9(t) = \frac{1}{2} \quad \text{Divide both sides by 6.}$$
$$9^{\frac{1}{2}} = t \quad \text{Write in exponential form.}$$
$$3 = t \quad \text{Simplify.}$$

2. We write $\log_3(x^4) = 2$ in exponential form and solve for x:

$$3^2 = x^4 \quad \text{Write in exponential form.}$$
$$x^4 = 9 \quad \text{If } c = d\text{, then } d = c\text{; simplify.}$$
$$x = \pm 9^{1/4} \quad \text{The solution of } x^4 = k \text{ is } \pm k^{1/4} \text{ if } k \geq 0.$$
$$x \approx \pm 1.7321 \quad \text{Compute.}$$

In most cases throughout this chapter, we will round approximate solutions to the fourth decimal place.

In Example 3, we will solve some more logarithmic equations—but now for the *base* of a logarithm. A key step will still be to write an equation in logarithmic form in exponential form instead.

Example 3 Solving for the Base of a Logarithm

Solve for b.

1. $\log_b(81) = 4$ **2.** $\log_b(67) = 5$

Solution

1. We write $\log_b(81) = 4$ in exponential form and solve for b:

$$b^4 = 81 \qquad \text{Write in exponential form.}$$
$$b = \pm 81^{1/4} \qquad \text{The solution of } b^4 = k \text{ is } \pm k^{1/4}.$$
$$b = 3 \qquad \text{Simplify; the base of a logarithm is positive.}$$

2. We write $\log_b(67) = 5$ in exponential form and solve for b:

$$b^5 = 67 \qquad \text{Write in exponential form.}$$
$$b = 67^{1/5} \qquad \text{The solution of } b^5 = k \text{ is } k^{1/5}.$$
$$b \approx 2.3185 \qquad \text{Compute.}$$

■

In summary, **for an equation of the form $\log_b(x) = k$, we can solve for b or x by writing the equation in exponential form.**

Power Property for Logarithms

An exponential equation in one variable is an equation in one variable in which an exponent contains a variable. Here are some examples of exponential equations in one variable:

$$3^x = 50 \quad 5(2)^x = 71 \quad 4(7)^x + 5 = 785 \quad 4^{3x-2} = 391$$

An important property called the **power property for logarithms** will help us solve exponential equations.

> **Power Property for Logarithms**
>
> For $x > 0$, $b > 0$, and $b \neq 1$,
>
> $$\log_b\left(x^p\right) = p \log_b(x)$$
>
> In words: A logarithm of a power of x is the exponent times the logarithm of x.

For example, $\log_3\left(x^5\right) = 5\log_3(x)$. Also, $\log_2\left(x^7\right) = 7\log_2(x)$.

A proof of the power property for logarithms follows: Let $k = \log_b\left(x^p\right)$. The exponential form of this equation is $b^k = x^p$. Taking the exponent $\frac{1}{p}$ on both sides and simplifying gives

$$\left(b^k\right)^{1/p} = \left(x^p\right)^{1/p} \qquad \text{Raise both sides to the power } \frac{1}{p}.$$
$$b^{k\left(\frac{1}{p}\right)} = x^{p\left(\frac{1}{p}\right)} \qquad \text{Multiply exponents: } \left(b^m\right)^n = b^{mn}$$
$$b^{k/p} = x \qquad \text{Simplify.}$$

The logarithmic form of this equation is

$$\log_b(x) = \frac{k}{p}$$

Multiplying both sides by p and simplifying gives

$$p\log_b(x) = p\left(\frac{k}{p}\right) \qquad \text{Multiply both sides by } p.$$
$$p\log_b(x) = k \qquad \text{Simplify.}$$
$$k = p\log_b(x) \qquad \text{If } a = b, \text{ then } b = a.$$

But $k = \log_b\left(x^p\right)$ as well, so by substitution, we have

$$\log_b\left(x^p\right) = p\log_b(x)$$

This statement is what we set out to prove.

Next, we describe the **logarithm property of equality,** which also is helpful in solving exponential equations.

Logarithm Property of Equality

For positive real numbers a, b, and c, where $b \neq 1$, the equations

$$a = c \quad \text{and} \quad \log_b(a) = \log_b(c)$$

are equivalent.

If we have the equation $a = c$ and then write $\log(a) = \log(b)$, we say that we "take the log of both sides" of the equation $a = c$.

To solve an equation in exponential form, such as $3^x = 17$, we first take the log of both sides and then apply the power property for logarithms.

Example 4 Solving an Exponential Equation

Solve the equation $2^x = 12$.

Solution

$2^x = 12$	Original equation
$\log(2^x) = \log(12)$	Take the log of both sides.
$x\log(2) = \log(12)$	Power property: $\log_b(x^p) = p\log_b(x)$
$x = \dfrac{\log(12)}{\log(2)}$	Divide both sides by $\log(2)$.
$x \approx 3.5850$	Compute.

We check that 3.5850 approximately satisfies the equation $2^x = 12$:

$$2^{3.5850} \approx 12.0003 \approx 12$$

■

WARNING When we compute a quotient of logarithms on a graphing calculator, it pays to watch the use of parentheses. Here we show a correct and an incorrect way to compute $\dfrac{\log(12)}{\log(2)}$:

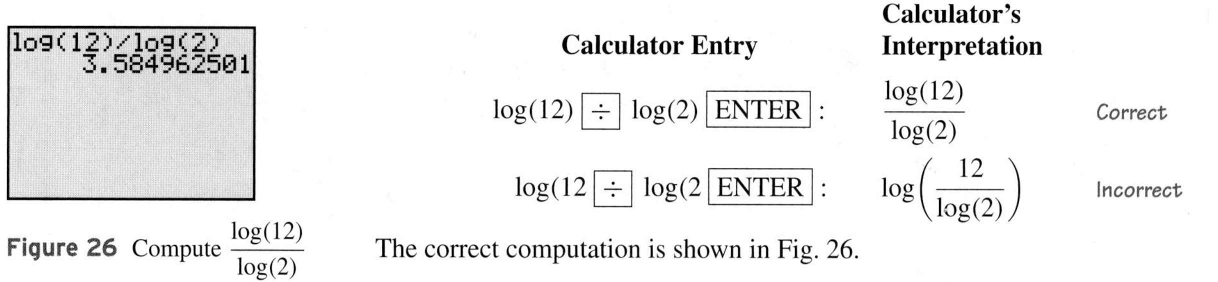

Figure 26 Compute $\dfrac{\log(12)}{\log(2)}$

Calculator Entry	Calculator's Interpretation	
log(12) [÷] log(2) [ENTER] :	$\dfrac{\log(12)}{\log(2)}$	Correct
log(12 [÷] log(2 [ENTER] :	$\log\left(\dfrac{12}{\log(2)}\right)$	Incorrect

The correct computation is shown in Fig. 26.

Example 5 Solving an Exponential Equation

Solve $3(4)^x = 71$.

Solution

$3(4)^x = 71$	Original equation
$4^x = \dfrac{71}{3}$	Divide both sides by 3.
$\log(4^x) = \log\left(\dfrac{71}{3}\right)$	Take the log of both sides.
$x\log(4) = \log\left(\dfrac{71}{3}\right)$	Power property: $\log_b(x^p) = p\log_b(x)$
$x = \dfrac{\log\left(\dfrac{71}{3}\right)}{\log(4)}$	Divide both sides by $\log(4)$.
$x \approx 2.2824$	Compute.

We check that 2.2824 approximately satisfies the equation $3(4)^x = 71$:

$$3(4)^{2.2824} \approx 71.0008 \approx 71$$ ■

WARNING Since $3(4)^x \neq (3 \cdot 4)^x$, we *cannot* begin to solve $3(4)^x = 71$ in Example 5 by saying

$$3(4)^x = 71$$
$$\log\left[3(4)^x\right] = \log(71)$$
$$x\log(3 \cdot 4) = \log(71) \quad \text{Cannot do this, since } 3(4)^x \neq (3 \cdot 4)^x.$$

That is why we began by dividing both sides of $3(4)^x = 71$ by 3.

In general,

$$\log_b\left(ax^p\right) \neq p\log_b(ax)$$

To solve some equations of the form $ab^x = c$ for x, we divide both sides of the equation by a, and then take the log of both sides. Next, we use the **power property for logarithms.**

Example 6 Solving an Exponential Equation

Solve $5^{3w-1} = 17$.

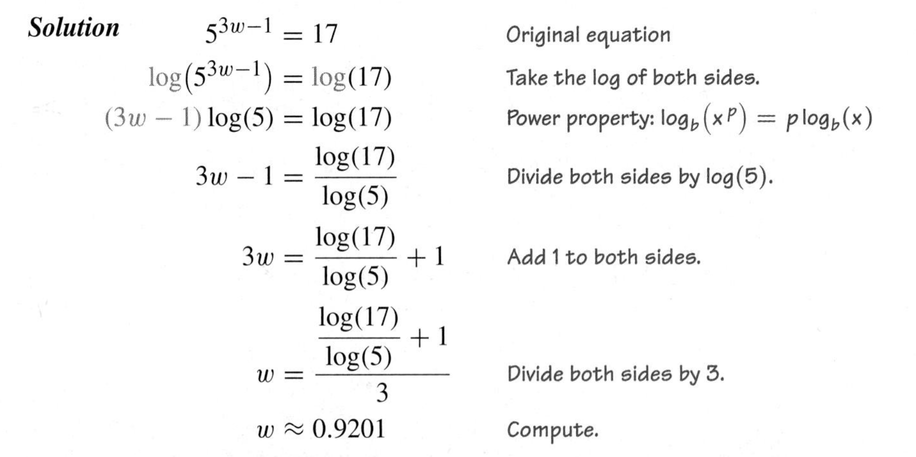

Solution

$$5^{3w-1} = 17 \quad \text{Original equation}$$
$$\log\left(5^{3w-1}\right) = \log(17) \quad \text{Take the log of both sides.}$$
$$(3w-1)\log(5) = \log(17) \quad \text{Power property: } \log_b\left(x^p\right) = p\log_b(x)$$
$$3w - 1 = \frac{\log(17)}{\log(5)} \quad \text{Divide both sides by } \log(5).$$
$$3w = \frac{\log(17)}{\log(5)} + 1 \quad \text{Add 1 to both sides.}$$
$$w = \frac{\frac{\log(17)}{\log(5)} + 1}{3} \quad \text{Divide both sides by 3.}$$
$$w \approx 0.9201 \quad \text{Compute.}$$

We use a graphing calculator table to check that, for the function $y = 5^{3x-1}$, the input 0.9201 leads approximately to the output 17 (see Fig. 27). ■

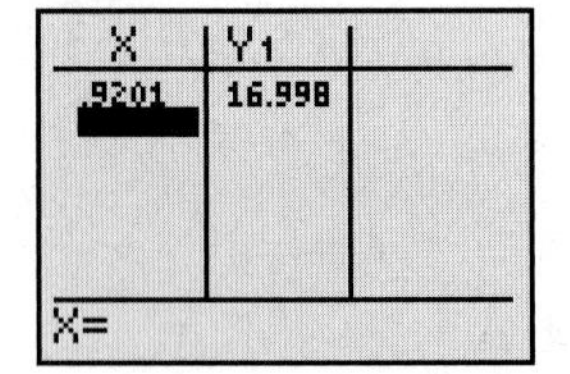

Figure 27 Verify the work

Example 7 Solving an Exponential Equation

Solve $7(2)^x - 4 = 20 + 3(2)^x$.

Solution

$$7(2)^x - 4 = 20 + 3(2)^x \quad \text{Original equation}$$
$$7(2)^x - 3(2)^x = 24 \quad \text{Subtract } 3(2)^x \text{ from both sides; add 4 to both sides.}$$
$$4(2)^x = 24 \quad 7(2)^x - 3(2)^x = (7-3)(2)^x = 4(2)^x$$
$$2^x = 6 \quad \text{Divide both sides by 4.}$$
$$\log\left(2^x\right) = \log(6) \quad \text{Take the log of both sides.}$$
$$x\log(2) = \log(6) \quad \text{Power property: } \log_b(x^p) = p\log_b(x)$$
$$x = \frac{\log(6)}{\log(2)} \quad \text{Divide both sides by } \log(2).$$
$$x \approx 2.5850 \quad \text{Compute.}$$

An approximate solution is 2.5850. Recall from Section 3.2 that we can use "intersect" on a graphing calculator to solve an equation in one variable. In Fig. 28, we graph the equations $y = 7(2)^x - 4$ and $y = 20 + 3(2)^x$ and find an approximate intersection point (2.5850, 38), which has x-coordinate 2.5850. So, an approximate solution of the original equation is 2.5850, which checks. ■

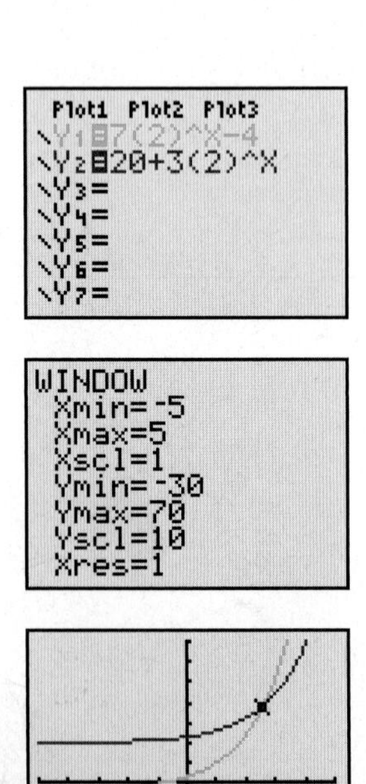

Figure 28 Verify the work

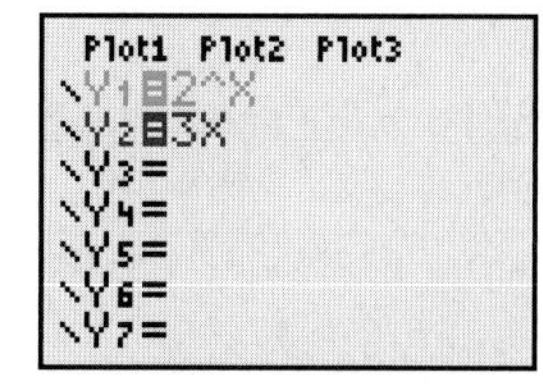

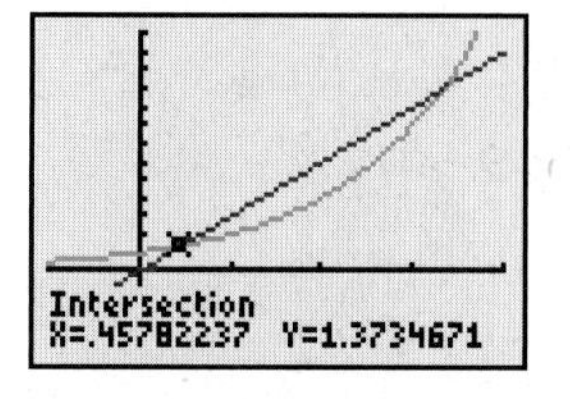

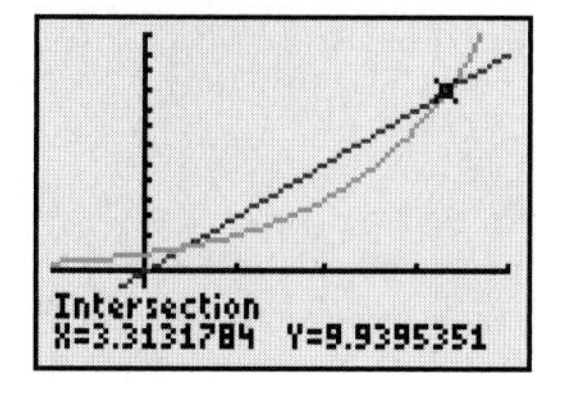

Figure 29 Solve the system

Solving Equations in One Variable by Using Graphs

Recall from Section 3.2 that some equations that are impossible to solve by performing operations on both sides can be solved by graphing. We will work with one such equation in Example 8.

Example 8 Using Graphing to Solve an Equation in One Variable

Use graphing to solve $2^x = 3x$.

Solution

We use "intersect" on a graphing calculator to find the solutions of the system

$$y = 2^x$$
$$y = 3x$$

See Fig. 29.

The approximate solutions of the system are (0.4578, 1.3735) and (3.3132, 9.9395). The x-coordinates of these ordered pairs, 0.4578 and 3.3132, are the approximate solutions of the equation $2^x = 3x$. ■

group exploration

Comparing the power property with other statements

Consider the following equations, where $x > 0$, $a > 0$, $b > 0$, and $b \neq 1$:

$$\log_b(x^p) = p\log_b(x)$$
$$\log_b[a(x^p)] = p\log_b(ax)$$
$$\log_b[(ax)^p] = p\log_b(ax)$$

1. Which, if any, of these equations are true in general? Explain why in terms of the power property for logarithms.
2. Show that the other equation or equations are false by using the substitutions $b = 10$, $a = 10$, $x = 10$, and $p = 2$.
3. Three students tried to solve $5(4)^x = 30$. Which students, if any, solved the equation correctly? Describe any errors and where they occurred.

Student 1's work	***Student 2's work***	***Student 3's work***
$5(4)^x = 30$	$5(4)^x = 30$	$5(4)^x = 30$
$4^x = 6$	$\log[5(4)^x] = \log(30)$	$20^x = 30$
$\log(4^x) = \log(6)$	$x\log[5(4)] = \log(30)$	$\log(20^x) = \log(30)$
$x\log(4) = \log(6)$	$x\log(20) = \log(30)$	$x\log(20) = \log(30)$
$x = \dfrac{\log(6)}{\log(4)}$	$x = \dfrac{\log(30)}{\log(20)}$	$x = \dfrac{\log(30)}{\log(20)}$
$x \approx 1.2925$	$x \approx 1.1353$	$x \approx 1.1353$

TIPS FOR SUCCESS: Take a Break

Have you ever had trouble solving a problem but returned to the problem hours later and found it easy to solve? By taking a break, you can return to the problem with a different perspective and renewed energy. You've also given your subconscious mind a chance to reflect on the problem while you take your break. You can strategically take advantage of this phenomenon by allocating time at two different points in your day to complete your homework assignment.

HOMEWORK 5.3

Write the equation in exponential form. Assume that all constants are positive and not equal to 1.

1. $\log_3(243) = 5$

2. $\log_2(32) = 5$

3. $\log(100) = 2$

4. $\log(10{,}000) = 4$

5. $\log_b(a) = c$

6. $\log_r(s) = t$

7. $\log(m) = n$

8. $\log(y) = z$

Write the equation in logarithmic form. Assume that all constants are positive and not equal to 1.

9. $5^3 = 125$

10. $2^5 = 32$

11. $10^3 = 1000$

12. $10^5 = 100{,}000$

13. $y^w = x$

14. $r^s = t$

15. $10^p = q$

16. $10^x = y$

Solve.

17. $\log_4(x) = 2$

18. $\log_2(x) = 3$

19. $\log(x) = -2$

20. $\log(x) = -3$

21. $\log_4(x) = 0$

22. $\log_9(x) = 0$

23. $\log_{27}(t) = \dfrac{4}{3}$

24. $\log_{16}(p) = \dfrac{3}{4}$

25. $2\log_8(2x-5) = 4$

26. $3\log_5(4x+1) = 9$

27. $4\log_{81}(x) - 3 = -2$

28. $3\log_8(x) + 5 = 6$

29. $\log_2(\log_3(y)) = 3$

30. $\log_3(\log_2(p)) = -1$

Solve. Round any solutions to the fourth decimal place.

31. $\log_6(x^3) = 2$

32. $\log_4(x^5) = 3$

Solve for b. Round any approximate solutions to the fourth decimal place.

33. $\log_b(49) = 2$

34. $\log_b(16) = 2$

35. $\log_b(8) = 3$

36. $\log_b(125) = 3$

37. $\log_b(16) = 5$

38. $\log_b(95) = 9$

Solve. Round any approximate solutions to the fourth decimal place. Verify your result by checking that it satisfies or approximately satisfies the original equation.

39. $4^x = 9$

40. $10^x = 50$

41. $5(4)^x = 80$

42. $3(2)^x = 17$

43. $3.83(2.18)^t = 170.91$

44. $1.73(4.09)^w = 526.44$

45. $8 + 5(2)^x = 79$

46. $20 = -3 + 4(2)^x$

47. $2^{4x+5} = 17$

48. $8 = 3^{7x-1}$

49. $6(3)^x - 7 = 85 + 4(3)^x$

50. $5^x + 7 = 50 - 3(5)^x$

51. $4^{3p} \cdot 4^{2p-1} = 100$

52. $3^{2r} \cdot 3^{r-4} = 97$

53. $3^x = -8$

54. $1^x = 13$

Solve. Round any approximate solutions to the fourth decimal place.

55. $\log_4(x) = 3$

56. $\log_3(x) = 4$

57. $3(4)^t + 15 = 406$

58. $2(6)^w - 17 = 3$

59. $\log_b(73) = 5$

60. $\log_b(19) = 4$

61. $3\log_{27}(y-1) = 2$

62. $2\log_4(3r+2) = 5$

63. $3(2)^{4x-2} = 83$

64. $5(4)^{2x+1} = 974$

For Exercises 65–70, estimate any solutions of the equation or system by referring to the graphs shown in Fig. 30.

65. $2^x = 4\left(\dfrac{1}{2}\right)^x$

66. $2^x = 4 - x$

67. $4\left(\dfrac{1}{2}\right)^x = 4 - x$

68. $2^x = 5$

69. $4\left(\dfrac{1}{2}\right)^x = 1$

70. $y = 2^x$
$y = 4\left(\dfrac{1}{2}\right)^x$

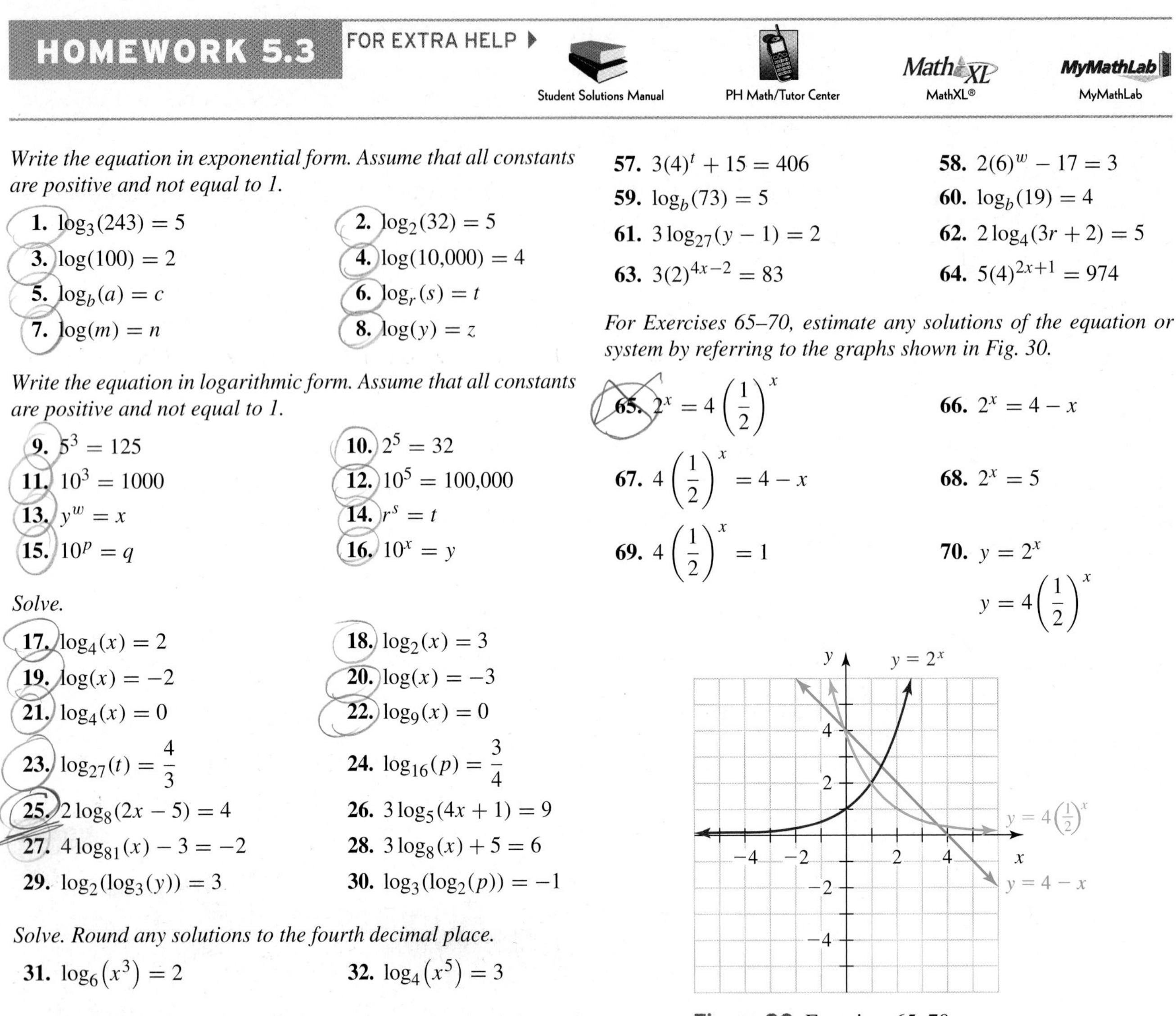

Figure 30 Exercises 65–70

Use "intersect" on a graphing calculator to solve the equation. Round any solutions to the fourth decimal place.

71. $3^x = 5 - x$

72. $2^x = 5 - 2x$

73. $7\left(\dfrac{1}{2}\right)^x = 2x$

74. $8\left(\dfrac{1}{3}\right)^x = x$

75. $\log(x+1) = 3 - \dfrac{2}{5}x$

76. $6 - x = 3\log(x+5)$

For Exercises 77–82, solve the given equation or system by referring to the values of $y = 3^{x-1}$, $y = 12\left(\frac{1}{2}\right)^x$, and $y = x - \frac{3}{2}$ shown in Table 26.

77. $3^{x-1} = 12\left(\dfrac{1}{2}\right)^x$

78. $12\left(\dfrac{1}{2}\right)^x = x - \dfrac{3}{2}$

79. $3^{x-1} = 81$

80. $12\left(\dfrac{1}{2}\right)^x = 6$

81. $y = 12\left(\frac{1}{2}\right)^x$
$y = x - \frac{3}{2}$

82. $y = 3^{x-1}$
$y = 12\left(\frac{1}{2}\right)^x$

Table 26 Some Values of Three Functions

x	1	2	3	4	5	6
$y = 3^{x-1}$	1	3	9	27	81	243
$y = 12\left(\frac{1}{2}\right)^x$	6	3	1.5	0.75	0.375	0.1875
$y = x - \frac{3}{2}$	−0.5	0.5	1.5	2.5	3.5	4.5

83. A student tries to solve $3(8^x) = 7$:

$$3(8)^x = 7$$
$$\log\left[3(8)^x\right] = \log(7)$$
$$x\log[3(8)] = \log(7)$$
$$x\log(24) = \log(7)$$
$$x = \frac{\log(7)}{\log(24)}$$
$$x \approx 0.6123$$

Describe any errors. Then solve the equation correctly.

84. A student tries to solve $2(3)^x = 10$:

$$2(3)^x = 10$$
$$6^x = 10$$
$$\log\left(6^x\right) = \log(10)$$
$$x\log(6) = 1$$
$$x = \frac{1}{\log(6)}$$
$$x \approx 1.2851$$

Describe any errors. Then solve the equation correctly.

In Exercises 85–88, solve for the specified variable. Assume that $b > 0$, that $b \neq 1$, and that the constants have values for which the equation has exactly one real-number solution.

85. $ab^x = c$, for x

86. $ab^x + d = c$, for x

87. $ab^{kx} + d = c$, for x

88. $ab^{kx-p} + d = c$, for x

Let $f(x) = 4^x$. Round any approximate results to the fourth decimal place.

89. Find $f(4)$.

90. Find $f^{-1}(16)$.

91. Find x when $f(x) = 3$.

92. Find x when $f^{-1}(x) = 3$.

Let $g(x) = \log_2(x)$. Round any approximate results to the fourth decimal place.

93. Find $g(8)$.

94. Find $g^{-1}(4)$.

95. Find a when $g(a) = 5$.

96. Find a when $g^{-1}(a) = 5$.

97. Determine whether the statement is true. Explain.

a. $\frac{\log_2(4)}{\log_2(16)} = \frac{4}{16}$

b. $\frac{\log_3(1)}{\log_3(27)} = \frac{1}{27}$

c. $\frac{\log(1000)}{\log(10{,}000)} = \frac{1000}{10{,}000}$

d. $\frac{\log_b(c)}{\log_b(d)} = \frac{c}{d}$, where b, c, and d are positive, $b \neq 1$, and $d \neq 1$.

98. Determine whether the statement is true. Explain. [**Hint:** For parts (a), (b), and (c), use the "log" key on a calculator.]

a. $\log\left(4 \cdot 3^2\right) = 2\log(4 \cdot 3)$

b. $\log\left(5 \cdot 2^3\right) = 3\log(5 \cdot 2)$

c. $\log\left(6 \cdot 10^4\right) = 4\log(6 \cdot 10)$

d. $\log\left(ax^p\right) = p\log(ax)$, where a and x are positive.

99. a. Use ZDecimal to graph $f(x) = \log\left(x^3\right) - 3\log(x) + 1$. Describe the graph in words.

b. Explain why the graph of f is in neither quadrant II nor quadrant III.

c. Use properties of logarithms to write the right-hand side of the equation of f as a constant. Use your result to explain why the graph that you found in part (a) makes sense.

100. The incorrect work that follows shows that the logarithm of any positive number is 0. Describe any errors. Assume that $b > 0$, $b \neq 1$, and $k > 0$:

$$\log_b(k) = \log_b(k \cdot 1) = \log_b\left(k \cdot 5^0\right) = 0 \cdot \log_b(5k) = 0$$

101. Describe the power property for logarithms. Give an example. Does the power property imply that $x^p = px$? Explain.

102. Describe how to use the power property for logarithms to solve an exponential equation.

Related Review

Solve. Find the exact solution if the equation is linear. For other types of equations, round the solution(s) to the fourth decimal place.

103. $5(3p - 7) - 9p = -4p + 23$

104. $99 - 2(3)^x = 12$

105. $5b^6 - 88 = 56$

106. $\log_b(75) = 4$

107. $\frac{3}{8}r = \frac{5}{6}r - \frac{2}{3}$

108. $\log_2(3x + 7) = 4$

Expressions, Equations, Functions, and Graphs

Perform the indicated instruction. Then use words such as linear, exponential, logarithmic, function, one variable, *and* two variables *to describe the expression, equation, or system.*

109. Solve $\log_2(x) = -5$.

110. Find an equation of a line that contains the points $(-2, -4)$ and $(3, 5)$.

111. Graph $y = \log_2(x)$ by hand.

112. Find $f^{-1}(-4)$, where $f(x) = 3x - 7$.

5.4 USING THE POWER PROPERTY WITH EXPONENTIAL MODELS TO MAKE PREDICTIONS

"The beginning of knowledge is the discovery of something we do not understand."

—Frank Herbert

Objectives

- Use the power property for logarithms with exponential models to make predictions.
- Use the half-life of carbon-14 to date archeological artifacts and fossils.

In Section 4.5, we used exponential functions to model data. In this section, we will use exponential functions with the power property for logarithms to make predictions about the independent variable of the function.

Example 1 Using the Power Property to Make a Prediction

A person invests \$7000 in a bank account with a yearly interest rate of 6% compounded annually. When will the balance be \$10,000?

Solution

Let $B = f(t)$ be the balance (in thousands of dollars) after t years or any fraction thereof. From our work with compound-interest accounts in Section 4.5, we know that we can model the situation well by using an exponential model of the form $f(t) = ab^t$ with y-intercept $(0, a)$. The B-intercept is $(0, 7)$, for \$7000 when $t = 0$, so $a = 7$ and $f(t) = 7b^t$. By the end of each year, the account has increased by 6% of the previous year's balance, so $b = 1.06$. Thus,

$$f(t) = 7(1.06)^t$$

To find when the balance is \$10,000 ($B = 10$), we substitute 10 for $f(t)$ and solve for t:

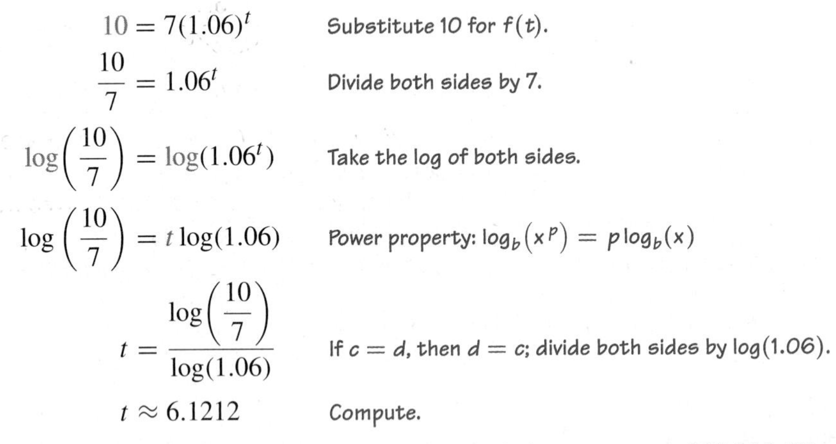

$$10 = 7(1.06)^t \quad \text{Substitute 10 for } f(t).$$

$$\frac{10}{7} = 1.06^t \quad \text{Divide both sides by 7.}$$

$$\log\left(\frac{10}{7}\right) = \log(1.06^t) \quad \text{Take the log of both sides.}$$

$$\log\left(\frac{10}{7}\right) = t\log(1.06) \quad \text{Power property: } \log_b(x^p) = p\log_b(x)$$

$$t = \frac{\log\left(\frac{10}{7}\right)}{\log(1.06)} \quad \text{If } c = d \text{, then } d = c \text{; divide both sides by } \log(1.06).$$

$$t \approx 6.1212 \quad \text{Compute.}$$

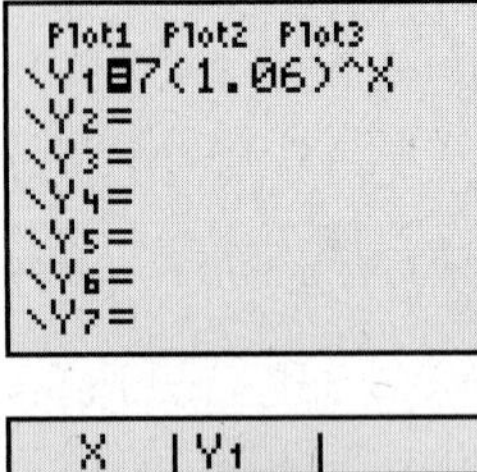

Figure 31 Verify the work

So, it will take about 6 years and 45 days for the balance to reach \$10,000. We use a graphing calculator table to check that, for the function $y = 7(1.06)^x$, the input 6.1212 leads approximately to the output 10 (see Fig. 31). ■

To make a prediction about the independent variable t of an exponential model of the form $f(t) = ab^t$, we substitute a value for $f(t)$ and divide both sides of the equation by the coefficient a. Next, we take the log of both sides of the equation and use the power property to help solve for t.

Example 2 Using the Power Property to Make a Prediction

The infant mortality rate is the number of deaths of infants under one year old per 1000 births.* In 1915, the rate was almost 100 deaths per 1000 infants, or 1 death

*The rate does not include fetal deaths.

per 10 infants. The infant mortality rate has decreased substantially since then (see Table 27).

Table 27 Infant Mortality Rates

Year	Rate (number of deaths per 1000 infants)
1915	99.9
1920	85.8
1930	64.6
1940	47.0
1950	29.2
1960	26.0
1970	20.0
1980	12.6
1990	9.2
2000	6.9
2005	6.5

Source: *National Center for Health Statistics*

1. Let $I = f(t)$ be the infant mortality rate (number of deaths per 1000 infants) at t years since 1900. Find an equation of f.
2. What is the percent rate of decay for infant mortality rates?
3. Find $f^{-1}(5)$. What does the result mean in this situation?

Solution

1. The scattergram in Fig. 32 shows that the points "bend" and that an exponential function will model the data better than a linear function.

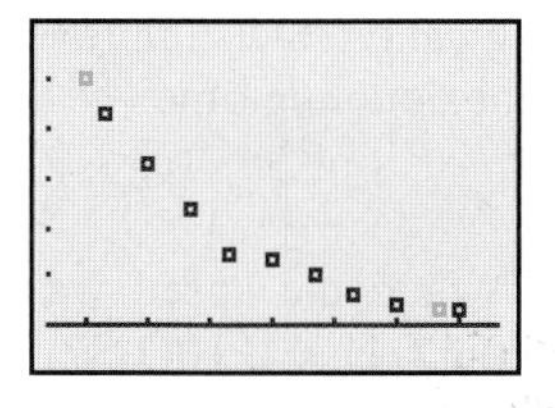

Figure 32 Scattergram of the data

We use the points $(15, 99.9)$ and $(100, 6.9)$ to find an equation of the form $I = ab^t$. We substitute the coordinates of the two chosen points into $f(t) = ab^t$:

$$6.9 = ab^{100}$$
$$99.9 = ab^{15}$$

We divide the left sides and divide the right sides of the equations and solve for b:

$$\frac{6.9}{99.9} = \frac{ab^{100}}{ab^{15}} \quad \text{Divide left sides and divide right sides.}$$

$$\frac{6.9}{99.9} = b^{85} \quad \frac{a}{a} = 1\text{; subtract exponents: } \frac{b^m}{b^n} = b^{m-n}$$

$$b = \left(\frac{6.9}{99.9}\right)^{1/85} \quad \text{The solution of } b^{85} = k \text{ is } k^{1/85}.$$

$$b \approx 0.969 \quad \text{Compute.}$$

So, $f(t) = a(0.969)^t$. We substitute the coordinates of $(100, 6.9)$ into the equation $f(t) = a(0.969)^t$ and solve for a:

$$6.9 = a(0.969)^{100} \quad \text{Substitute 100 for } t \text{ and 6.9 for } f(t).$$

$$a = \frac{6.9}{0.969^{100}} \quad \text{Divide both sides by } 0.969^{100}.$$

$$a \approx 160.87 \quad \text{Compute.}$$

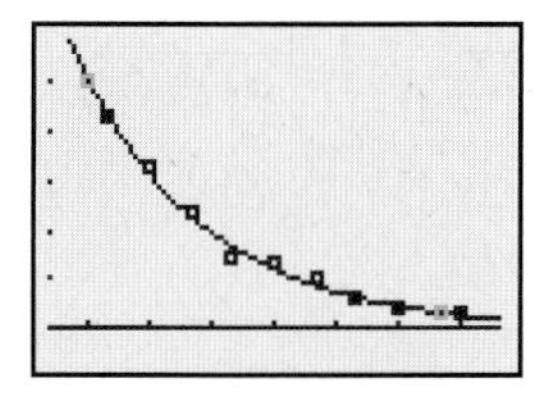

Figure 33 Verifying the model

Thus, $f(t) = 160.87(0.969)^t$. In Fig. 33, the model appears to fit the data very well.

2. The base b is 0.969. Since $1 - 0.969 = 0.031$, we conclude that the model estimates that the infant mortality rate has decayed by 3.1% per year.
3. Since f sends values of t to values of I, f^{-1} sends values of I to values of t (see Figs. 34 and 35).

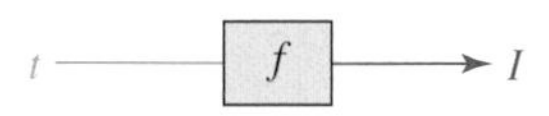

Figure 34 f sends values of t to values of I

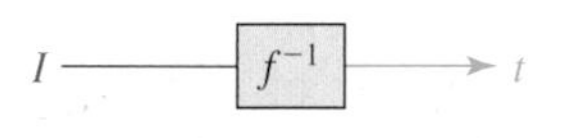

Figure 35 f^{-1} sends values of I to values of t

Therefore, $f^{-1}(5)$ represents the year (since 1900) when the infant mortality rate will be 5 deaths per 1000 infants. To find the year, we substitute 5 for $f(t)$ in the equation $f(t) = 160.87(0.969)^t$ and solve for t:

$$5 = 160.87(0.969)^t \quad \text{Substitute 5 for } f(t).$$

$$\frac{5}{160.87} = 0.969^t \quad \text{Divide both sides by 160.87.}$$

$$\log\left(\frac{5}{160.87}\right) = \log(0.969^t) \quad \text{Take the log of both sides.}$$

$$\log\left(\frac{5}{160.87}\right) = t\log(0.969) \quad \text{Power property: } \log_b(x^p) = p\log_b(x)$$

$$t = \frac{\log\left(\frac{5}{160.87}\right)}{\log(0.969)} \quad \text{Divide both sides by } \log(0.969).$$

$$t \approx 110.23 \quad \text{Compute.}$$

The model predicts that the infant mortality rate will be 5 deaths per 1000 infants in 2010. ■

Before we perform more modeling, it will be helpful to know a fact related to the base multiplier property (Section 4.3). For an exponential function $y = f(t) = ab^t$, we know by the base multiplier property that, as the value of t increases by 1, the value of y is multiplied by the base b. Also, if the value of y is multiplied by M in going from $t = 0$ to $t = k$, then the value of y will continue to be multiplied by M each time t is increased by k. To see this, recall that $f(0) = a$ and note that

$$f(k) = ab^k = aM, \quad \text{so} \quad b^k = M$$

We can use the fact that $b^k = M$ to show that if t is increased by k a total of n times, then y is multiplied by M a total of n times:

$$f(nk) = ab^{nk} = a(b^k)^n = a(M)^n$$

We use this idea in Example 3.

Example 3 Using the Power Property to Estimate Doubling Time

In Exercise 34 of Section 4.5, on the speeds of computer chips, you may have found that $f(t) = 0.10(1.37)^t$, where $f(t)$ is the chip speed (in MHz) at t years since 1971 (see Table 28). A rule of thumb for estimating how quickly technological products improve is that they double in speed every 2 years.* Use f to estimate the doubling time.

Table 28 Chip Speeds

Year	Chip Speed (MHz)
1971	0.1
1974	2
1982	12
1985	16
1992	50
1995	120
1997	300
2000	1000
2004	3200

Source: *i-probe*

Solution

In 1971, the chip speed was 0.10 MHz. We find the year when the speed was $2(0.10) = 0.20$ MHz by substituting 0.20 for $f(t)$ in the equation $f(t) = 0.10(1.37)^t$ and solving for t:

$$0.20 = 0.10(1.37)^t \quad \text{Substitute 0.20 for } f(t).$$

$$2 = 1.37^t \quad \text{Divide both sides by 0.10.}$$

$$\log(2) = \log(1.37^t) \quad \text{Take the log of both sides.}$$

$$\log(2) = t\log(1.37) \quad \text{Power property: } \log_b(x^p) = p\log_b(x)$$

$$t = \frac{\log(2)}{\log(1.37)} \quad \text{Divide both sides by } \log(1.37).$$

$$t \approx 2.20 \quad \text{Compute.}$$

According to the exponential function f, it took 2.20 years (about 2 years and 2 months) to double the 1971 chip speed. By the discussion preceding this example, we can conclude that the chip speed will double *every* 2.20 years. So, according to the model, the doubling time is about two months more than the rule-of-thumb estimate of 2 years. ■

Recall from Section 4.5 that the half-life of an element is the amount of time it takes for the number of atoms to be reduced to half. All organisms are, in part, composed of the elements carbon-12 and carbon-14. Carbon-14 is radioactive. After an animal or plant dies, its carbon-14 decays exponentially with a half-life of 5730 years. However, the amount of carbon-12 remains constant. Scientists know the ratio of carbon-14 to carbon-12 in *living* organisms. Hence, scientists can determine how long ago an organism lived by measuring the decreased ratio of carbon-14 to carbon-12 in a bone, piece of wood, or other artifact that was once part of the organism.

Example 4 Using the Power Property to Make an Estimate

A violent volcanic eruption and subsequent collapse of the former Mount Mazama created Crater Lake, the deepest lake in the United States. Scientists found a charcoal sample from a tree that burned in the eruption. If only 39.40% of the carbon-14 remains in the sample, when did Crater Lake form?

*In 1965, Gordon Moore used modeling to show that the number of transistors per integrated circuit doubled about every 2 years. He predicted that this rule of thumb (now referred to as "Moore's law") would model the situation well until at least 1975. In fact, the model still accurately describes the situation.

Solution

Let $P = f(t)$ be the percentage of carbon-14 that remains at t years after the sample formed. Since the percentage is halved every 5730 years, we will find an exponential equation of the form

$$f(t) = ab^t$$

At time $t = 0$, 100% (all) of the carbon-14 remained, so the P-intercept is (0, 100). Therefore, $a = 100$ and $f(t) = 100b^t$. At time $t = 5730$, $\frac{1}{2}(100) = 50\%$ of the carbon-14 remained. So, we substitute the coordinates of the point (5730, 50) into the equation $f(t) = 100b^t$ and solve for b:

$$\begin{aligned} 50 &= 100b^{5730} && \text{Substitute 5730 for } t \text{ and 50 for } f(t). \\ 0.5 &= b^{5730} && \text{Divide both sides by 100.} \\ b &= \pm 0.5^{1/5730} && \text{The solutions of } b^{5730} = k \text{ are } \pm k^{1/5730}. \\ b &\approx 0.999879 && \text{Compute; the base of an exponential function is positive.} \end{aligned}$$

The equation is $f(t) = 100(0.999879)^t$. We use more digits than usual for the base, as even a small change in it would greatly affect estimates well into the past (or future).

To estimate the age of the sample, we substitute 39.40 for $f(t)$ and solve for t:

$$\begin{aligned} 39.40 &= 100(0.999879)^t && \text{Substitute 39.40 for } f(t). \\ 0.3940 &= 0.999879^t && \text{Divide both sides by 100.} \\ \log(0.3940) &= \log(0.999879^t) && \text{Take the log of both sides.} \\ \log(0.3940) &= t\log(0.999879) && \text{Power property: } \log_b(x^p) = p\log_b(x) \\ t &= \frac{\log(0.3940)}{\log(0.999879)} && \text{Divide both sides by } \log(0.999879). \\ t &\approx 7697 && \text{Compute.} \end{aligned}$$

So, the age of Crater Lake (and the sample) is approximately 7697 years. ■

group exploration

Finding an equation of the inverse for an exponential model

In Example 1, we found the model $B = f(t) = 7(1.06)^t$, where B is the balance (in thousands of dollars) of an account at 6% interest compounded annually and the starting balance is \$7000. In this exploration, you will find the equation of the inverse function f^{-1}.

1. Substitute B for $f(t)$ in the equation $f(t) = 7(1.06)^t$.
2. Solve your equation for t.
3. Replace t with $f^{-1}(B)$ in your equation. You now have an equation of f^{-1}.
4. Use your equation of f^{-1} to find $f^{-1}(12)$. What does it mean in this situation?
5. Use a graphing calculator to help you complete Table 29.
6. How could the information in your completed table help an investor?

Table 29 Values of f^{-1}

B	$f^{-1}(B)$
7	
8	
9	
10	
11	
12	
13	
14	

group exploration

Looking ahead: Product and quotient properties for logarithms

1. **a.** Use a calculator to compare $\log(2 \cdot 3)$ with $\log(2) + \log(3)$.
   b. Use a calculator to compare $\log(7 \cdot 2)$ with $\log(7) + \log(2)$.
   c. Use a calculator to compare $\log(4 \cdot 6)$ with $\log(4) + \log(6)$.
   d. What do parts (a)–(c) of this exploration suggest about $\log(xy)$? Check whether your observation is true for other values of x and y. (Your observation is referred to as the *product property for logarithms.*)

2. Determine how $\log\left(\frac{x}{y}\right)$ can be expressed in terms of two logarithms. [**Hint:** Choose specific values for x and y. Then compute $\log\left(\frac{x}{y}\right)$, $\log(x)$, and $\log(y)$, and compare the values.] Check whether your observation is true for other values of x and y. (Your observation is referred to as the *quotient property* for logarithms.)

HOMEWORK 5.4

FOR EXTRA HELP ▶ Student Solutions Manual | PH Math/Tutor Center | MathXL® | MyMathLab

1. A person invests \$2000 in an account at 5% interest compounded annually. Let $V = f(t)$ be the value (in dollars) of the account after t years or any fraction thereof.
   a. Find an equation of f.
   b. What is the V-intercept? What does it mean in this situation?
   c. What will be the value of the investment in 5 years?
   d. When will the value of the investment be \$3000?
2. A person invests \$12,500 in an account at 8% interest compounded annually. Let $V = f(t)$ be the value (in dollars) of the account after t years or any fraction thereof.
   a. Find an equation of f.
   b. What is the V-intercept? What does it mean in this situation?
   c. What will be the value of the investment in 4 years?
   d. When will the value of the investment be \$20,000?
3. A person invests \$9300 in an account at 6% interest compounded annually. When will the value of the investment be \$13,700?
4. A person invests \$4500 in an account at 3% interest compounded annually. When will the value of the investment be \$5900?
5. A person invests \$6000 in an account at 10% interest compounded annually. When will the value of the investment be doubled? Explain why it will take less than 10 years, even though the rate is 10%.
6. A person invests \$4000 in an account at 5% interest compounded annually. When will the value of the investment be doubled? Explain why it will take less than 20 years, even though the rate is 5%.
7. The U.S. annual production of ethanol, used as fuel for automobiles, was 0.18 billion gallons in 1980 and has grown by about 13% per year since then (Source: *Department of Energy*). Predict when annual ethanol production will reach 7 billion gallons.
8. The advertising agency Crispin Porter & Bogusky® has produced creative ads and websites for companies such as Burger King®, The Gap®, and Volkswagen®. The agency's annual billings were \$0.145 million in 2000 and have grown by about 39% per year since then (Source: *Crispin Porter & Bogusky*). Predict when the annual billings will be \$5.5 billion.
9. Suppose that a rumor is spreading in the United States that the airlines will soon be giving away free promotional tickets for U.S. flights. Assume that 30 people have heard the rumor as of today and that each day the total number of people who have heard the rumor triples. Let $f(t)$ be the total number of people who have heard the rumor at t days since today.
   a. Find an equation of f.
   b. What is the total number of Americans who will have heard the rumor 8 days from now?
   c. Predict when all Americans will have heard the rumor. Assume that the U.S. population is 299 million.
10. There are 4 million bacteria on a peach at noon on Tuesday. Assume that a bacterium divides into two bacteria every hour, on average. Let $f(t)$ be the number of bacteria (in millions) on the peach at t hours after Tuesday noon.
    a. Find an equation of f.
    b. Find $f(24)$. What does it mean in this situation?
    c. Find $f^{-1}(8000)$. What does it mean in this situation?
11. According to the U.S. Occupational Safety and Health Administration standard, an average person can listen to 8 hours of sound per day at a sound level of 90 decibels without experiencing hearing loss. Recall from Exercise 79 of Homework 5.2 that a decibel is a unit for measurement of sound intensity. For each increase of 5 decibels, the exposure time must be cut in half. For example, an average person can listen to 4 hours of sound per day at a sound level of 95 decibels without experiencing hearing loss. (One overexposure may result in temporary, but probably not permanent, hearing loss.)

    Some examples of sound being made at various decibel levels are listed in Table 30.

Table 30 Examples of Decibel Levels

Sound Level (decibels)	Example
0	Faintest sound heard by humans
20	Whisper
40	Inside a running car
60	Conversation
80	Noisy street corner
100	Soft-rock concert
120	Threshold of pain

Source: Math and Music, *by Garland and Kahn*

a. Let $T = f(d)$ be the number of hours of safe exposure time in one day to a sound at a level d decibels *above 90 decibels*. Find an equation of f.
b. Many rock bands play at about 114 decibels. Use your equation of f to predict how long they can play at concerts so that their fans do not experience hearing loss. Based on your result, do you think that these types of fans experience hearing loss? Assume that most people are not wearing earplugs.
c. Many rock concerts last about 3 hours. At what sound level should the bands play so that fans who attend a lot of concerts will not experience hearing loss?

12. A round trip to Mars would involve a total of 12 months of flight time and 19 months of exploration on the planet, until Earth and Mars would be in the correct position for the return flight. In a weightless environment, astronauts lose about 1.5% of the calcium in their bodies per month (Source: *NASA*). Estimate the percentage of calcium in the body just after an astronaut would return from Mars. Assume that there would be no calcium loss (or gain) while the astronaut was on Mars.

13. First prize for the World Series of Poker's main event has grown approximately exponentially from \$0.21 million in 1975 to \$5 million in 2004 (Source: *Harrah's Entertainment*). Predict when the first prize will be \$10 million.

14. The number of people who have attended theme and amusement parks has grown approximately exponentially from 78.7 million in 1996 to 92.4 million in 2002 (Source: *Travel Industry Association of America*). Predict in which year 115 million people will attend theme and amusement parks.

15. The timber harvests in the Tongass National Forest in Alaska have decayed approximately exponentially from 471 million board feet in 1990 to 48 million board feet in 2003 (Source: *U.S. Forest Service*). Predict when the timber harvest will be 10 million board feet.

16. The revenue from sales of videocassette players has decayed approximately exponentially from \$14 million in 2000 to \$2 million in 2004 (Source: *Consumer Electronics Association*). Predict when the sales will be \$0.2 million.

17. In Exercise 33 of Homework 4.5, you found an equation close to $f(t) = 100.84(1.41)^t$, where $f(t)$ is the number of Starbucks stores worldwide at t years since 1990 (see Table 31).
a. Estimate when the first Starbucks store opened.
b. The first Starbucks store opened in 1971 in Seattle, Washington. Do you think the number of stores grew exponentially from 1971 to 2003? Explain.

Table 31 Numbers of Starbucks Stores Worldwide

Year	Number of Stores
1991	116
1993	272
1995	676
1997	1412
1999	2135
2001	4709
2003	7225

Source: *Starbucks Corporation*

c. What is the base b of the model $f(t) = ab^t$? What does it mean in this situation?
d. Predict when there will be an average of 500 Starbucks stores per country. (There are 192 countries.)
e. Predict when there will be one Starbucks store for every person in the world. Assume that world population will be 8.6 billion then.

18. The numbers of LEED-certified green buildings in the United States are shown in Table 32 for various years.

Table 32 Numbers of LEED-Certified Green Buildings

Year	Number of Buildings with LEED Certification
2000	5
2001	14
2002	39
2003	80
2004	142

Source: *United States Green Building Council*

Let $f(t)$ be the number of LEED-certified green buildings in the United States at t years since 2000.
a. Use a graphing calculator to draw a scattergram of the data. Is it better to model the data by using a linear or an exponential function? Explain.
b. Find an equation of f.
c. What is the base b of your model $f(t) = ab^t$? What does it mean in this situation?
d. Estimate the number of LEED-certified green buildings in 2005.
e. Predict when there will be an average of 100 LEED-certified green buildings per state.

19. The numbers of polio cases in the world are shown in Table 33 for various years.

Table 33 Numbers of Polio Cases Worldwide

Year	Number of Polio Cases (thousands)
1992	138
1994	73
1996	33
1998	11
2000	4
2002	1.9
2004	1.2

Source: *World Health Organization*

Let $f(t)$ be the number of polio cases (in thousands) in the year that is t years since 1990.

a. Use a graphing calculator to draw a scattergram of the data. Is it better to model the data by using a linear or an exponential function? Explain.

b. Find an equation of f.

c. In 1988, the World Health Assembly passed a resolution to eradicate polio by the year 2000. When the goal was not met, the organization reset the eradication date to December 31, 2005. Use f to estimate the number of cases in 2006. Does this suggest that the World Health Assembly reached its goal? Explain.

d. Predict in which year there will be 1 case of polio.

e. Find the approximate half-life of the number of polio cases. [**Hint:** Use f to estimate the number of cases in 1990. Then use f to estimate when there was half that number of cases.]

20. The revenues from sales of blank audiocassettes are shown in Table 34 for various years.

Table 34 Annual Revenues from Sales of Blank Audiocassettes

Year	Revenue (millions of dollars)
2000	162
2001	129
2002	98
2003	77
2004	66

Source: *Consumer Electronics Association*

Let $f(t)$ be the revenue (in millions of dollars) from sales of blank audiocassettes in the year that is t years since 2000.

a. Use a graphing calculator to draw a scattergram of the data. Is it better to model the data by using a linear or an exponential function? Explain.

b. Find an equation of f.

c. Predict the revenue from sales of blank audiocassettes in 2010.

d. Predict when the revenue will be $1 million.

e. Find the approximate half-life of the revenue from sales of blank audiocassettes. [**Hint:** Use f to estimate the revenue in 2000. Then use f to estimate when the revenue was half that amount.]

21. In Exercise 31 of Homework 4.5, you may have found an equation close to $f(t) = 1.2(1.0162)^t$, where $p = f(t)$ models world population (in billions) at t years since 1900 (see Table 35).

Table 35 World Population

Year	World Population (billions)
1930	2.070
1940	2.295
1950	2.500
1960	3.050
1970	3.700
1980	4.454
1990	5.279
2000	6.080
2006	6.522

Source: *U.S. Census Bureau*

a. The United Nations predicts that world population will reach 9.3 billion in 2075. Use f to predict when world population will reach 9.3 billion.

b. Use a graphing calculator to draw a scattergram to describe the part of the data in Table 35 *from 1970 to 2006*. Is it better to use a linear or an exponential function to model the data? Explain. Find an equation for such a function. Use the function notation "g" for this function.

c. Use g to predict when world population will reach 9.3 billion. Is that year before or after the predicted year you found in part (a)? Why does this make sense?

d. The United Nations describes a possible post-2075 scenario in which world population will reach 9.3 billion in 2075 and then decline to 9.1 billion in 2100. Draw by hand a scattergram of all of the data shown in Table 35, and plot points for the scenario's predictions for 2075 and 2100. Graph the functions f and g by hand. Then sketch a model that describes the scenario better than the functions f and g do.

22. The minimum and average salaries for major league baseball players are shown in Table 36 for various years.

Table 36 Salaries for Major League Baseball Players

Year	Minimum Salary (thousands of dollars)	Average Salary (thousands of dollars)
1970	12	29
1975	16	45
1980	30	144
1985	60	372
1990	100	579
1995	109	1071
2000	200	1998
2005	316	2633

Source: *baseball-reference.com*

Let $M(t)$ be the minimum salary and $A(t)$ be the average salary, both in thousands of dollars, at t years since 1970.

a. Find regressions equations of M and A.

b. Predict the minimum salary and the average salary in 2010.

c. What is the percentage rate of growth of minimum salaries? of average salaries?

d. In Exercise 53 of Homework 4.3, you may have worked with the model $f(t) = 1.22(1.051)^t$, where $f(t)$ is the average ticket price (in dollars) to major league baseball games at t years since 1950. What is the percentage rate of growth of average ticket prices? Can the growth in minimum salaries or average salaries be accounted for by the growth in average ticket prices alone? Explain.

e. Predict when the minimum salary will be $600 thousand.

f. In Exercise 13 of Homework 2.4, you may have worked with the model $g(t) = -3.8t + 166$, where $g(t)$ is the average length (in minutes) of a major league baseball game at t years since 2003. Use g to help you predict the pay per hour of the 162 regular-season games in 2012 for baseball players who are paid the minimum salary.

23. The percentages of seniors with severe memory impairment (based on memory tests) are shown in Table 37 for various age groups.

Let $p = f(t)$ be the percentage of seniors at age t years with severe memory impairment.

Table 37 Percentages of Seniors with Severe Memory Impairment

Age Group (years)	Age Used to Represent Age Group (years)	Percent
65–69	67	1.1
70–74	72	2.5
75–79	77	4.5
80–84	82	6.4
over 84	88	12.9

Source: *Federal Interagency Forum on Aging-Related Statistics*

a. Find an equation of f.
b. What is the base b of your model $f(t) = ab^t$? What does it mean in this situation?
c. Estimate what percentage of 70-year-old seniors have severe memory impairment.
d. Estimate at what age 10% of seniors have severe memory impairment.
e. In Exercise 7 of Section 3.3, you may have found that there is a linear relationship between an adult's age and an adult's score on a test measuring memory and information-processing speed. Would that *linear* relationship necessarily conflict with an *exponential* relationship between a senior's age and the percentage of seniors with severe memory impairment? Explain.

24. Saks Fifth Avenue® offered a promotional sale in which customers could receive a gift card. The values of the gift cards are shown in Table 38 for various expenditures.

Table 38 Saks Fifth Avenue Gift Card Values

Expenditure Group (dollars)	Expenditure Used to Represent Expenditure Group (dollars)	Gift Card Value (dollars)
250–499	375	25
500–999	750	50
1000–1999	1500	100
2000–2999	2500	200
3000 or more	3500	450

Source: *Saks Fifth Avenue*

Let $v = f(s)$ be the value (in dollars) of a gift card that a customer who spends s dollars will receive.
a. Find an equation of f.
b. What is the coefficient a of your model $f(s) = ab^s$? What does it mean in this situation?
c. What is the base b of the your model $f(s) = ab^s$? What does it mean in this situation?
d. Customer A spends \$2000, customer B spends \$2500, and customer C spends \$2999. According to your model f, what are the values of the gift cards that these customers will receive? Compare these values with the actual values of the gift cards they will receive.
e. Use your model to estimate for what expenditure a customer would receive a \$700 gift card.
f. If a new promotion is to include a gift card for \$700, as well as gift cards for the values shown in Table 38, determine reasonable expenditure groups for \$450 and \$700 gift cards.

25. In a study of 10 of the most selective U.S. colleges and universities, researchers found that a student applicant has a better chance of being accepted to a college through early decision (students apply early and colleges decide early) than by regular decision. Table 39 shows a comparison of SAT scores (out of 1600) and acceptance rates by both systems.

Table 39 Percentages of Applicants Accepted by Early Decision and Regular Decision

SAT Score Group	Score Used to Represent SAT Score Group	Percent: Early Decision	Percent: Regular Decision
1100–1190	1145	21	10
1200–1290	1245	35	17
1300–1390	1345	52	31
1400–1490	1445	70	48
1500–1600	1550	93	72

Source: *Professor Christopher Avery, Kennedy School of Government, Harvard University*

For students who score s points, let $E(s)$ and $R(s)$ be the percentages of early-decision and regular-decision applicants, respectively, who are accepted.
a. Find regression equations of E and R.
b. What percentage of early-decision applicants who score 1425 points get accepted? What about regular-decision applicants who score 1425 points?
c. For what score do half of early-decision applicants get accepted? What about regular-decision applicants?
d. The study concluded that students who apply for early decision have the equivalent of 100 points added to their SAT score, compared with students applying for regular decision. What do your results from part (c) suggest the equivalent number of added points to be?
e. Use "intersect" on a graphing calculator to find the intersection point of the graphs of E and R. What does it mean in this situation? [Challenge: Use substitution and the property $\frac{b^t}{c^t} = \left(\frac{b}{c}\right)^t$, where b and c are positive, to find the intersection point.]

26. New York Life offers a $250,000 life insurance policy. Quarterly rates for women and men are shown in Table 40 for various ages.

Table 40 New York Life Quarterly Rates for a $250,000 Policy

Age (years)	Quarterly Rate (dollars)	
	Women	Men
35	25.00	28.75
40	33.75	35.75
45	51.25	57.50
50	70.00	87.50
55	104.50	145.00
60	145.75	230.75
64	220.00	355.00

Source: *New York Life*

Let $W(t)$ and $M(t)$ be the quarterly rates (in dollars) for women and men, respectively, both at t years of age.

a. Find regression equations of W and M.

b. For a $250,000 policy, how much would a 52-year-old woman pay per quarter? How much would a 52-year-old man pay per quarter?

c. At what age would a woman pay $120 per quarter for a $250,000 policy? At what age would a man pay that much for a $250,000 policy?

d. Due to Montana insurance regulations, both genders must pay the same quarterly rates. So, New York Life uses the male rates for all residents of Montana. Estimate how much more a 62-year-old woman would pay per quarter for a $250,000 policy if she lived in Montana rather than in some other state.

e. Use "intersect" on a graphing calculator to find the intersection point of the graphs of W and M. What does it mean in this situation? [Challenge: Use substitution and the property $\frac{b^t}{c^t} = \left(\frac{b}{c}\right)^t$, where b and c are positive, to find the intersection point.]

27. Physicians use gallium citrate-67 to detect certain types of cancer, including lymphoma. Gallium citrate-67 has an effective half-life of 3.25 days—some is lost to radioactive decay, and some is removed through the digestive and urinary tracts. A patient who is breast-feeding is injected with the radioactive element.

a. Let $f(t)$ be the percentage of the gallium citrate-67 that remains in the patient's body at t days since she was injected. Find an equation of f.

b. A scan of the gallium citrate-67 is performed 2 days after the injection. What percentage of the element remains?

c. The patient can resume breast-feeding when only 0.39% of the gallium citrate-67 remains. When can she resume breast-feeding?

28. Physicians use technetium-99m to locate stress fractures in bones. Technetium-99m has an effective half-life of 5.3 hours—some is lost to radioactive decay, and some is removed through urination. A patient with a possible stress fracture in his foot is injected with the radioactive element.

a. Let $f(t)$ be the percentage of the technetium-99m that remains in the patient's body at t hours since he was injected. Find an equation of f.

b. What percentage of the technetium-99m will remain after 1 day?

c. When will only 1% of the technetium-99m remain?

29. Scientists used a sample of spruce wood from the Two Creeks Forest Bed in Wisconsin to date an advance of the continental ice sheet into the United States during the last Ice Age. If 24.46% of the carbon-14 remains in the sample, when did the ice sheet advance? (Assume that this advance killed the tree.) The half-life of carbon-14 is 5730 years.

30. A mummy was on display at a museum in Niagara Falls until it was sold in 1999. A few years later, researchers identified the mummy as the ancient Egyptian pharaoh Rameses I. The mummy was eventually returned to Egypt. If 69.57% of the carbon-14 in the mummy remains, estimate how long ago Rameses I lived. The half-life of carbon-14 is 5730 years.

31. An archeologist discovers a tool made of wood.

a. If 50% of the wood's carbon-14 remains, how old is the wood? Explain how you can find this result without using an equation. The half-life of carbon-14 is 5730 years.

b. If 25% of the wood's carbon-14 remains, how old is the wood? Explain how you can find this result without using an equation.

c. If 10% of the wood's carbon-14 remains, how old is the wood? First, guess an approximate age without solving an equation. Explain how you decided on your estimate. Then, use an equation to find the age.

32. A person drinks a cup of coffee. Assume that the caffeine enters his bloodstream immediately and that there was no caffeine in his bloodstream before he drank the coffee. The half-life of caffeine in a person's bloodstream is about 6 hours. A cup of coffee contains about 240 milligrams of caffeine.

a. Let $f(t)$ be the number of milligrams of caffeine in the person's bloodstream at t hours after he drank the coffee. Find an equation of f.

b. The person drinks the coffee at 8 A.M. and goes to bed at 11 P.M. Use f to predict the amount of caffeine in his bloodstream when he goes to bed.

c. The person drinks another cup of coffee 24 hours after the first cup. How much caffeine will be in his bloodstream from these 2 cups of coffee just after he drank the second cup? Explain how you can find this result without using an equation.

d. Now assume that the person drinks the cup of coffee at 8 A.M. and then drinks a cup of coffee every morning at 8 A.M. from then on. Sketch a qualitative graph that describes the relationship between caffeine in his bloodstream and time. Describe any assumptions that you make.

33. A storage tank contains a liquid radioactive element with a half-life of 100 years. It will be relatively safe for the contents to leak from the tank when 0.01% of the radioactive element remains. How long must the tank remain intact for this storage procedure to be safe?

34. Describe how you can use the power property for logarithms to make estimates and predictions.

Related Review

35. The combined dump capacities of the three largest U.S. waste companies are shown in Table 41 for various years.

Let c be the combined dump capacity (in billions of tons) of the three largest U.S. waste companies at t years since 2000.

Table 41 Combined Dump Capacities of the Three Largest U.S. Waste Companies

Year	Combined Dump Capacity (billions of tons)
2000	5.34
2001	5.63
2002	5.77
2003	6.39
2004	6.62

Source: *Securities and Exchange Commission*

a. Use the points (0, 5.34) and (4, 6.62) to find a linear model L to describe the situation. Then use the same pair of points to find an exponential model E to describe the situation.

b. Is the graph of E similar to the graph of L for values of t between 0 and 4, inclusive? If so, does this lead you to believe that the relationship between t and c is linear or exponential? Explain.

c. What is the slope of the linear model? What does it mean in this situation?

d. What is the base b of the exponential model $E(t) = ab^t$? What does it mean in this situation?

36. Due to inflation, an item that cost \$10,000 in 1980 cost \$24,126 in 2005. Costs comparable to \$10,000 in 1980 are shown in Table 42 for various years.

Let $c = f(t)$ be the cost (in dollars) in the year that is t years since 1980 that is comparable to \$10,000 in 1980.

a. Use a graphing calculator to draw a scattergram of the data. Is it better to model the data by using a linear or an exponential function? Explain.

b. Find an equation of f.

Table 42 Costs Comparable to \$10,000 in 1980

Year	Comparable Cost (dollars)
1980	10,000
1985	13,058
1990	15,862
1995	18,495
2000	20,898
2005	24,126

Source: *Bureau of Labor Statistics*

c. If your model is linear, find the slope. If your model is exponential, find the base b of $f(t) = ab^t$. What does your result mean in this situation?

d. What is the c-intercept? What does it mean in this situation?

e. Use f to predict when the cost of \$30,000 would be comparable to the cost of \$10,000 in 1980.

Expressions, Equations, Functions, and Graphs

Perform the indicated instruction. Then use words such as linear, exponential, logarithmic, function, one variable, *and* two variables *to describe the expression, equation, or system.*

37. Simplify $\dfrac{-25b^{3/8}}{40b^{2/5}}$.

38. Find the inverse of $f(x) = \dfrac{2}{5}x - 3$.

39. Solve $4(6)^x - 31 = 180$. Round any solutions to the fourth decimal place.

40. Graph $f(x) = \dfrac{2}{5}x - 3$ by hand.

5.5 MORE PROPERTIES OF LOGARITHMS

Objectives

- Know the *product, quotient,* and *change of base properties* for logarithms.
- Use properties of logarithms to simplify expressions and solve equations.
- Use a calculator to evaluate a logarithm with a base other than 10.

In Section 5.3, we studied some properties of logarithms. In this section, we will discuss three more.

Product Property for Logarithms

We can use the **product property for logarithms** to add two logarithms that have the same base.

Product Property for Logarithms

For $x > 0$, $y > 0$, $b > 0$, and $b \neq 1$,

$$\log_b(x) + \log_b(y) = \log_b(xy)$$

In words: The sum of logarithms is the logarithm of the product of their inputs.

For example, for $x > 0$, $\log_3(5) + \log_3(x) = \log_3(5x)$. A proof of the product property for logarithms follows.

Let $m = \log_b(x)$ and $n = \log_b(y)$. Writing both equations in exponential form, we have

$$x = b^m$$
$$y = b^n$$

Multiplying the left sides and multiplying the right sides yields

$$xy = (b^m)(b^n) \quad \text{Multiply left sides and multiply right sides.}$$
$$= b^{m+n} \quad \text{Add exponents: } b^m b^n = b^{m+n}$$

Writing $xy = b^{m+n}$ in logarithmic form gives

$$m + n = \log_b(xy)$$

Substituting $\log_b(x)$ for m and $\log_b(y)$ for n yields

$$\log_b(x) + \log_b(y) = \log_b(xy)$$

This statement is what we set out to prove.

Example 1 Using the Product Property for Logarithms

Simplify. Write the sum of logarithms as a single logarithm.

1. $\log_b(2x) + \log_b(x)$ **2.** $3\log_b(x^2) + 2\log_b(6x)$

Solution

1. $\log_b(2x) + \log_b(x) = \log_b(2x \cdot x)$ Product property: $\log_b(x) + \log_b(y) = \log_b(xy)$

$= \log_b(2x^2)$ Add exponents: $b^m b^n = b^{m+n}$

2. $3\log_b(x^2) + 2\log_b(6x) = \log_b(x^2)^3 + \log_b(6x)^2$ Power property: $p\log_b(x) = \log_b(x^p)$

$= \log_b\left[(x^2)^3 \cdot (6x)^2\right]$ Product property: $\log_b(x) + \log_b(y) = \log_b(xy)$

$= \log_b\left[x^6 \cdot 36x^2\right]$ Multiply exponents; raise factors to 2nd power.

$= \log_b(36x^8)$ Add exponents: $b^m b^n = b^{m+n}$

■

WARNING For us to apply the product property for logarithms, the coefficient of each logarithm must be 1. So, in Problem 2 of Example 1, we first applied the power property to get coefficients of 1:

$$3\log_b(x^2) + 2\log_b(6x) = \log_b(x^2)^3 + \log_b(6x)^2$$

Then, we applied the product property.

Quotient Property

We use the product property to simplify the sum of two logarithms with the same base. We use the **quotient property for logarithms** to simplify the difference of two logarithms with the same base.

Quotient Property for Logarithms

For $x > 0$, $y > 0$, $b > 0$, and $b \neq 1$,

$$\log_b(x) - \log_b(y) = \log_b\left(\frac{x}{y}\right)$$

In words: The difference of two logarithms is the logarithm of the quotient of their inputs.

For example, for $x > 0$, $\log_4(x) - \log_4(7) = \log_4\left(\frac{x}{7}\right)$. You are asked to prove the quotient property in Exercise 50.

Example 2 Product and Quotient Properties

Simplify. Write the result as a single logarithm with a coefficient of 1.

1. $\log_b(6w^7) - \log_b(w^2)$
2. $2\log_b(3p) + 3\log_b(p^2) - 4\log_b(2p)$

Solution

1. $\log_b(6w^7) - \log_b(w^2) = \log_b\left(\frac{6w^7}{w^2}\right)$ Quotient property: $\log_b(x) - \log_b(y) = \log_b\left(\frac{x}{y}\right)$

$= \log_b(6w^5)$ Subtract exponents: $\frac{b^m}{b^n} = b^{m-n}$

2. $2\log_b(3p) + 3\log_b(p^2) - 4\log_b(2p)$

$= \log_b(3p)^2 + \log_b(p^2)^3 - \log_b(2p)^4$ Power property: $p\log_b(x) = \log_b(x^p)$

$= \log_b\left[(3p)^2(p^2)^3\right] - \log_b(2p)^4$ Product property: $\log_b(x) + \log_b(y) = \log_b(xy)$

$= \log_b\frac{(3p)^2(p^2)^3}{(2p)^4}$ Quotient property: $\log_b(x) - \log_b(y) = \log_b\left(\frac{x}{y}\right)$

$= \log_b\frac{9p^2\cdot p^6}{16p^4}$ Raise factors to a power; multiply exponents.

$= \log_b\frac{9p^8}{16p^4}$ Add exponents: $b^m b^n = b^{m+n}$

$= \log_b\frac{9p^4}{16}$ Subtract exponents: $\frac{b^m}{b^n} = b^{m-n}$ ■

Solving Logarithmic Equations

We can use the power, product, and quotient properties to solve logarithmic equations.

Example 3 Solving a Logarithmic Equation

Solve $2\log_5(3x) + 4\log_5(2x) = 3$.

Solution

$2\log_5(3x) + 4\log_5(2x) = 3$ Original equation

$\log_5(3x)^2 + \log_5(2x)^4 = 3$ Power property: $p\log_b(x) = \log_b(x^p)$

$\log_5\left[(3x)^2(2x)^4\right] = 3$ Product property: $\log_b(x) + \log_b(y) = \log_b(xy)$

$\log_5\left[9x^2(16x^4)\right] = 3$ Raise factors to a power: $(bc)^n = b^n c^n$

$\log_5(144x^6) = 3$ Add exponents: $b^m b^n = b^{m+n}$

$5^3 = 144x^6$ Write in exponential form.

$x^6 = \frac{125}{144}$ Divide both sides by 144.

Although there is a negative sixth root of $\frac{125}{144}$, the original equation contains $4\log_5(2x)$, and the domain of a logarithmic function is the set of *positive* numbers. So, $2x$ must be positive; hence, x must be positive:

$$x = \left(\frac{125}{144}\right)^{1/6}$$

$$x \approx 0.9767$$

■

Example 4 Solving a Logarithmic Equation

Solve $5\log_7(t^3) - 2\log_7(3t) = 2$

Solution

$5\log_7(t^3) - 2\log_7(3t) = 2$	Original equation
$\log_7(t^3)^5 - \log_7(3t)^2 = 2$	Power property: $p\log_b(x) = \log_b(x^p)$
$\log_7\dfrac{(t^3)^5}{(3t)^2} = 2$	Quotient property: $\log_b(x) - \log_b(y) = \log_b\left(\dfrac{x}{y}\right)$
$\log_7\dfrac{t^{15}}{9t^2} = 2$	Multiply exponents; raise factors to 2nd power.
$\log_7\dfrac{t^{13}}{9} = 2$	Subtract exponents: $\dfrac{b^m}{b^n} = b^{m-n}$
$7^2 = \dfrac{t^{13}}{9}$	Write in exponential form.
$t^{13} = 441$	Multiply both sides by 9.
$t = 441^{1/13}$	The solution of $b^{13} = k$ is $k^{1/13}$.
$t \approx 1.5974$	Compute. ■

We solved the equations in Examples 3 and 4 by first applying the power property so that the coefficient of each logarithm was 1. Next, we combined the logarithms on one side of the equation by using the product property or the quotient property. Then, we solved the equation by using techniques discussed in Section 5.3.

Change-of-Base Property

The "log" key on a calculator finds logarithms, base 10. We use the **change-of-base property** to find logarithms for bases other than 10.

> **Change-of-Base Property**
>
> For $a > 0, b > 0, a \neq 1, b \neq 1$, and $x > 0$,
>
> $$\log_b(x) = \frac{\log_a(x)}{\log_a(b)}$$

For example, $\log_3(5) = \dfrac{\log_2(5)}{\log_2(3)}$. Also, $\log_3(5) = \dfrac{\log_4(5)}{\log_4(3)}$ and $\log_3(5) = \dfrac{\log(5)}{\log(3)}$. We are free to write a logarithm in terms of any new base, including base 10.

To prove the change-of-base property, we let $k = \log_b(x)$. In exponential form, we have

$$b^k = x$$

Next, we take $\log_a$ of both sides and solve for k:

$\log_a(b^k) = \log_a(x)$	Take the $\log_a$ of both sides.
$k\log_a(b) = \log_a(x)$	Power property: $\log_a(b^k) = k\log_a(b)$
$k = \dfrac{\log_a(x)}{\log_a(b)}$	Divide both sides by $\log_a(b)$.

But $k = \log_b(x)$, so, by substitution, we have

$$\log_b(x) = \frac{\log_a(x)}{\log_a(b)}$$

which is what we set out to prove.

To find a logarithm to a base other than 10, we use the change-of-base property to convert to $\log_{10}$; then we can use the log key on a calculator.

Example 5 Converting to $\log_{10}$

Find $\log_2(12)$.

Solution

We can use the change-of-base property to write $\log_2(12)$ in terms of base 10:

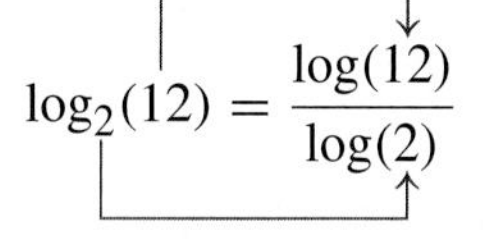

$$\log_2(12) = \frac{\log(12)}{\log(2)}$$

Figure 36 Compute $\frac{\log(12)}{\log(2)}$

Using the log key on a calculator, we compute that $\frac{\log(12)}{\log(2)} \approx 3.5850$ (see Fig. 36). So, $\log_2(12) \approx 3.5850$. ■

Example 6 Using the Change-of-Base Property

Write $\frac{\log_7(x)}{\log_7(4)}$ as a single logarithm.

Solution

By the change-of-base property, we have $\frac{\log_7(x)}{\log_7(4)} = \log_4(x)$. ■

In Section 5.2, we sketched the graph of a logarithmic function. From now on, we can use a graphing calculator to verify such a graph by converting the logarithmic function to $\log_{10}$.

Example 7 Using a Graphing Calculator to Graph a Logarithmic Function

Use a graphing calculator to draw the graph of $y = \log_3(x)$.

Solution

By the change-of-base property, we have $y = \log_3(x) = \frac{\log(x)}{\log(3)}$. Using the log key on a graphing calculator, we enter the function and draw the graph (see Fig. 37). This graph verifies the graph that we sketched by hand in Example 5 of Section 5.2. ■

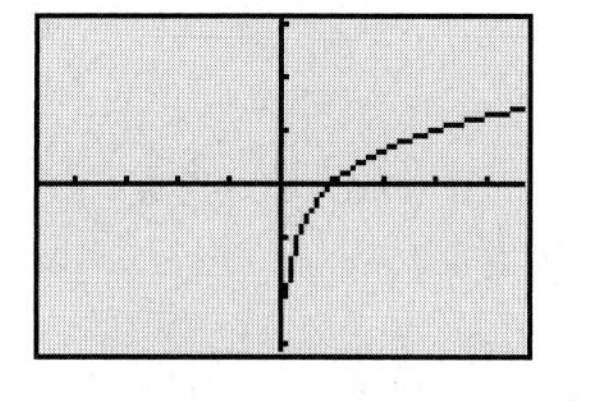

Figure 37 Graph of $y = \frac{\log(x)}{\log(3)}$

Recall from Sections 3.2 and 5.3 that some equations in one variable that are impossible to solve by performing operations on both sides can be solved by graphing. We will work with one such equation in Example 8.

Example 8 Using Graphing to Solve an Equation in One Variable

Use graphing to solve $\log_2(x+1) + \log_3(x+2) = 6 - x$.

Solution

We use the change-of-base property on the left side of the equation to write

$$\frac{\log(x+1)}{\log(2)} + \frac{\log(x+2)}{\log(3)} = 6 - x$$

Then we use "intersect" on a graphing calculator to solve the system

$$y = \frac{\log(x+1)}{\log(2)} + \frac{\log(x+2)}{\log(3)}$$
$$y = 6 - x$$

See Fig. 38.

The approximate solution of the system is (2.7024, 3.2976). The x-coordinate, 2.7024, is the approximate solution of the equation $\log_2(x+1) + \log_3(x+2) = 6 - x$. ■

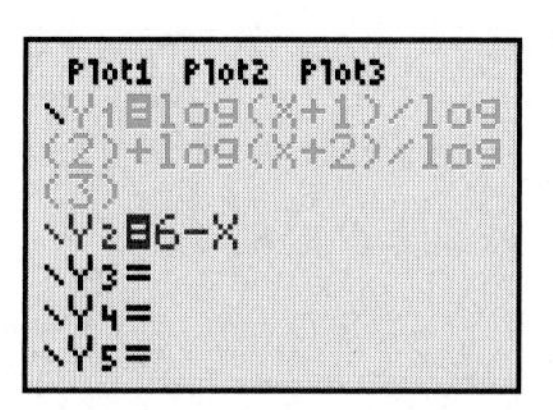

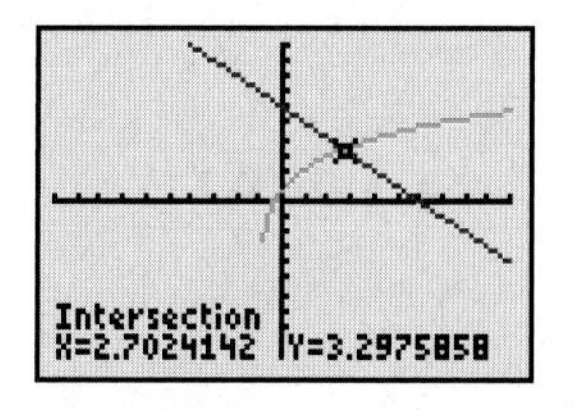

Figure 38 Solve the system

Comparing Properties of Logarithms

How do the properties of logarithms compare? Throughout this discussion, we assume that x, y, a, and b are positive and that a and b are not equal to 1.

The quotient property for logarithms tells us that a difference of logarithms is equal to a logarithm of a quotient:

$$\log_b(x) - \log_b(y) = \log_b\left(\frac{x}{y}\right)$$

The change-of-base property tells us that a logarithm is equal to a quotient of logarithms (with a "new" base):

$$\log_b(x) = \frac{\log_a(x)}{\log_a(b)}$$

WARNING It is a common error to confuse the quotient property and the change-of-base property for logarithms. In general,

$$\log_b(x) - \log_b(y) \neq \frac{\log_b(x)}{\log_b(y)}$$

and

$$\log_b\left(\frac{x}{y}\right) \neq \frac{\log_b(x)}{\log_b(y)}$$

group exploration

Function of a sum

1. Substitute 2 for x and 3 for y and use a calculator to help you decide whether the resulting statement is true or false.
   a. $\log(x+y) = \log(x) + \log(y)$
   b. $2^{x+y} = 2^x + 2^y$
   c. $(x+y)^2 = x^2 + y^2$
   d. $\sqrt{x+y} = \sqrt{x} + \sqrt{y}$
2. All of the statements in Problem 1 are of the form $f(x+y) = f(x) + f(y)$. Is the statement $f(x+y) = f(x) + f(y)$ true for every function f? Explain.
3. According to the distributive law, $a(x+y) = ax + ay$. Explain why this statement is true for all values of a but the statement $f(x+y) = f(x) + f(y)$ is not true for all functions f.
4. Give an example of a function f such that the statement $f(x+y) = f(x) + f(y)$ *is* true.

HOMEWORK 5.5

FOR EXTRA HELP ▶ Student Solutions Manual | PH Math/Tutor Center | MathXL® | MyMathLab

Simplify. Write your result as a single logarithm with a coefficient of 1.

1. $\log_b(x) + \log_b(3x)$
2. $\log_b(5x) + \log_b(x)$
3. $\log_b(8x) - \log_b(2)$
4. $\log_b(24x) - \log_b(6)$
5. $4\log_b(t) + \log_b(5t)$
6. $\log_b(7w) + 3\log_b(w)$
7. $\log_b(3x^2) - 5\log_b(x)$
8. $7\log_b(x) - \log_b(6x^4)$
9. $2\log_b(3x) + 3\log_b(x^3)$
10. $4\log_b(x^2) + 5\log_b(x)$
11. $3\log_b(2m) + 5\log_b(m^2) - \log_b(3m)$
12. $2\log_b(3k) + 4\log_b(k^3) - \log_b(5k)$

Solve. Round any solutions to the fourth decimal place.

13. $\log_5(6x) + \log_5(x) = 2$
14. $\log_3(x) + \log_3(6x) = 3$
15. $\log_2(9x) - \log_2(3) = 5$
16. $\log_4(12x) - \log_4(6) = 3$
17. $\log_7(w^2) + 2\log_7(3w) = 2$
18. $4\log_3(2r) + \log_3(r^3) = 4$
19. $2\log(x^5) - \log(x^7) = 1$
20. $\log(x^9) - 3\log(x^2) = 3$

21. $3\log(x^2)+4\log(2x)=2$

22. $2\log(2x)+3\log(x^4)=4$

23. $3\log_5(p^4)-5\log_5(2p)=3$

24. $4\log_2(k^2)-3\log_2(2k)=4$

Evaluate. Round your result to the fourth decimal place.

25. $\log_3(7)$

26. $\log_2(11)$

27. $\log_9(3.58)$

28. $\log_{12}(2.88)$

29. $\log_8\left(\frac{1}{70}\right)$

30. $\log_{\frac{1}{2}}(7)$

Solve by using "intersect" on a graphing calculator. Round any solutions to the fourth decimal place.

31. $\log(x+5)+\log(x+2)=3-x$

32. $\log(x+1)+\log(x+4)=8-2x$

33. $\log_5(x+3)+\log_2(x+4)=-2x+9$

34. $\log_3(x+2)+\log_4(x+1)=-x+7$

35. $\log_2(x+4)+\log_3(x+5)=2^x+1$

36. $\log_3(x+6)+\log_2(x+8)=3^x+2$

Write the expression as a single logarithm.

37. $\dfrac{\log_2(x)}{\log_2(7)}$

38. $\dfrac{\log_4(x)}{\log_4(5)}$

39. $\dfrac{\log_b(r)}{\log_b(s)}$

40. $\dfrac{\log_b(p)}{\log_b(q)}$

41. Three students try to solve the equation $3(2^x)=7$:

Student 1's work

$$3(2)^x=7$$
$$2^x=\frac{7}{3}$$
$$\log(2^x)=\log\left(\frac{7}{3}\right)$$
$$x\log(2)=\log\left(\frac{7}{3}\right)$$
$$x=\frac{\log\left(\frac{7}{3}\right)}{\log(2)}$$

Student 2's work

$$3(2)^x=7$$
$$2^x=\frac{7}{3}$$
$$x=\log_2\left(\frac{7}{3}\right)$$

Student 3's work

$$3(2)^x=7$$
$$\log\left[3(2)^x\right]=\log(7)$$
$$\log(3)+\log(2^x)=\log(7)$$
$$\log(2^x)=\log(7)-\log(3)$$
$$x\log(2)=\log(7)-\log(3)$$
$$x=\frac{\log(7)-\log(3)}{\log(2)}$$

Which student(s) solved the equation correctly?

42. A student tries to write the expression $3\log_2(x)+\log_2(x^2)$ as a single logarithm:

$$3\log_2(x)+\log_2(x^2)=3\log_2(x\cdot x^2)$$
$$=3\log_2(x^3)$$

Describe any errors. Then write the expression correctly as a single logarithm.

Let $g(x)=\log_{12}(x)$. Find each output. Round your result to the fourth decimal place.

43. $g(17)$

44. $g(50)$

45. $g(8)$

46. $g(5)$

47. Which of the following expressions are equal?

$\log_b(b^2)$ $\quad\log_b(b^6)-\log_b(b^4)$ $\quad\log_b(b^6)$ $\quad 2$ $\quad\log_b\left(\frac{b^6}{b^4}\right)$ $\quad\dfrac{\log_b(b^6)}{\log_b(b^4)}$

48. a. Use ZDecimal to graph $f(x)=\log(100x)-\log(x)$. Describe the graph in words.

b. Explain why the graph of f is in neither quadrant II nor quadrant III.

c. Use properties of logarithms to write the right-hand side of the equation of f as a constant. Use your result to explain why the graph that you found in part (a) makes sense.

49. Clearly, $\log_b(x)-\log_b(x)=0$. Use a property of logarithms to write $\log_b(x)-\log_b(x)$ in another form to show that $\log_b(1)=0$. Assume that $b>0$, $x>0$, and $b\neq 1$.

50. Prove the quotient property for logarithms. [**Hint:** Try to find a creative way to use the product property followed by the power property, with the expression $\log_b\left(\frac{x}{y}\right)$.]

51. a. Simplify $\log_2(x^3)+\log_2(x^5)$.

b. Solve $\log_2(x^3)+\log_2(x^5)=7$.

c. Compare the process of simplifying an expression with solving an equation.

d. Explain how simplifying an expression can help when you are solving an equation.

52. List the properties for logarithms discussed in this section and in Section 5.3. Explain how each property can be used. Give examples to illustrate your points.

Related Review

Simplify by writing your result as a single logarithm with a coefficient of 1, or solve, as appropriate. Round any solutions to the fourth decimal place.

53. $\log_2(x^4)+\log_2(x^3)$

54. $\log_2(t^9)-\log_2(t^5)=5$

55. $\log_2(x^4)+\log_2(x^3)=4$

56. $\log_2(t^9)-\log_2(t^5)$

57. $2\log_9(x^3)-3\log_9(2x)=2$

58. $3\log_5(3x)+2\log_5(x^2)$

59. $2\log_9(x^3)-3\log_9(2x)$

60. $3\log_5(3x)+2\log_5(x^2)=3$

Simplify. If an expression contains two logarithms, write your result as a single logarithm with a coefficient of 1.

61. $(16b^{16}c^{-7})^{1/4}(27b^{27}c^5)^{1/3}$

62. $\dfrac{(25b^9c^{-6})^{1/2}}{(81b^3c^{-8})^{1/4}}$

63. $3\log_b(2x^5) + 2\log_b(3x^4)$

64. $4\log_b(3x^2) - 5\log_b(2x^7)$

Solve the system.

65. $y = 3x - 7$
$y = -2x + 3$

66. $y = \frac{1}{2}x - 4$
$y = -\frac{2}{3}x + 3$

67. $y = \log_2(4x^2) - 3$
$y = \log_2(x) + 2$

68. $y = 2 + \log_3(3x^2)$
$y = 8 - \log_3(9x)$

Expressions, Equations, Functions, and Graphs

Perform the indicated instruction. Then use words such as linear, exponential, logarithmic, function, one variable, *and* two variables *to describe the expression, equation, or system.*

69. Solve: $2x - 3y = 6$
$y = \frac{2}{3}x - 2$

70. Solve $\log_3(2m - 1) = 5$.

71. Graph $2x - 3y = 6$ by hand.

72. Find $f(5)$, where $f(x) = \log_3(2x - 1)$.

5.6 NATURAL LOGARITHMS

Objectives

- Know the meaning of a *natural logarithm.*
- Evaluate natural logarithms.
- Use properties of natural logarithms to simplify expressions and solve equations.
- Use exponential models with base e to make estimates and predictions.

In Chapter 4 and in this chapter, we have worked with exponential and logarithmic functions with various values of the base. In statistics and calculus, it is helpful to use a special constant called "e" as the base for these two types of functions. In this section, we will describe this constant and use it as the base for logarithmic and exponential functions.

Definition of Natural Logarithm

In this section, we discuss a logarithm with a special base called "e," where e is an irrational number:

$$e \approx 2.718281828459045\ldots$$

To the nearest ten-thousandth, $e = 2.7183$.

Many equations for useful models contain e. For example, the equation for one type of "bell curve" is

$$f(x) = \frac{e^{-0.5x^2}}{\sqrt{2\pi}}$$

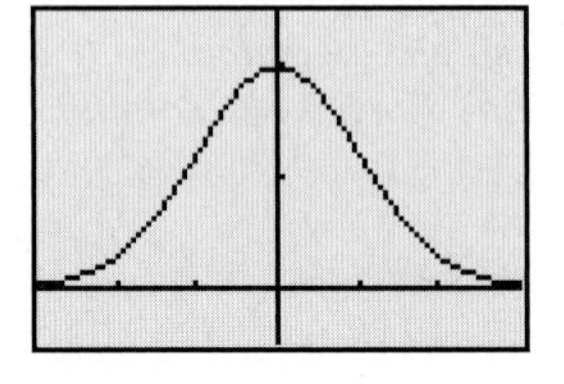

Figure 39 Graph of

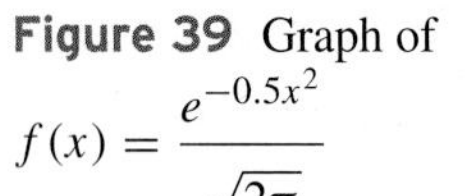

$f(x) = \frac{e^{-0.5x^2}}{\sqrt{2\pi}}$

The graph of f has the shape of a bell (see Fig. 39). Bell curves can be used to model heights of women (or men), IQs, widths of trunks of redwood trees, and many other quantities.

DEFINITION Natural logarithm

A **natural logarithm** is a logarithm with base e. We write $\ln(a)$ to represent $\log_e(a)$.

Throughout the following discussion, assume that $a > 0$, $b > 0$, and $b \neq 1$.

Recall from Section 5.2 that $\log_b(a)$ is the exponent on the base b that gives a. So, **for $a > 0$, $\ln(a)$ is the exponent on the base e that gives a.**

Recall from Section 5.3 that $\log_b(a) = c$ and $b^c = a$ are equivalent forms. In terms of base e, this means that $\ln(a) = c$ and $e^c = a$ are equivalent forms.

Exponential/Natural Logarithmic Forms Property

For $a > 0$, the equations

$$\ln(a) = c \quad \text{and} \quad e^c = a$$

are equivalent.

The equation $\ln(a) = c$ is in logarithmic form, and the equation $e^c = a$ is in exponential form. Either form can replace the other when you solve a problem.

The key on most graphing calculators that is labeled "ln" or "LN" will give you the natural logarithm of a number.

Example 1 Finding a Natural Logarithm

Figure 40 Computing ln(50)

Use a calculator to find ln(50).

Solution

By pressing [LN] 50 [)] [ENTER], we get $\ln(50) \approx 3.9120$ (see Fig. 40). This means that $e^{3.9120} \approx 50$. We check that $e^{3.9120} \approx 50$ by pressing [2nd] [LN] 3.9120 [)] [ENTER]. ■

We can find the natural logarithm of powers of e without using a calculator.

Example 2 Finding a Natural Logarithm

Find $\ln(e^5)$.

Solution

$$\begin{aligned} \ln(e^5) &= \log_e(e^5) && \text{Definition of natural logarithm} \\ &= 5 && \text{Simplify.} \end{aligned}$$ ■

From Example 2, we see that for the function ln, the input e^5 leads to the output 5. So, the *positive* real number e^5 is in the domain of ln. Recall from Section 5.2 that for *any* logarithmic function, the domain is the set of all positive real numbers.

Solving Logarithmic and Exponential Equations

In Example 3, we solve two logarithmic equations.

Example 3 Solving Logarithmic Equations

Figure 41 Computing e^4

Solve the equation.

1. $\ln(x) = 4$ **2.** $3\ln(4x) - 2 = 5$

Solution

1. We write $\ln(x) = 4$ in the exponential form $e^4 = x$. The approximate solution is 54.5982 (see Fig. 41).
2. $$\begin{aligned} 3\ln(4x) - 2 &= 5 && \text{Original equation} \\ 3\ln(4x) &= 7 && \text{Add 2 to both sides.} \\ \ln(4x) &= \frac{7}{3} && \text{Divide both sides by 3.} \\ e^{7/3} &= 4x && \text{Write in exponential form.} \\ \frac{e^{7/3}}{4} &= x && \text{Divide both sides by 4.} \\ x &\approx 2.5781 && \text{Compute.} \end{aligned}$$

We check that 2.5781 approximately satisfies the equation $3\ln(4x) - 2 = 5$:

$$3\ln[4(2.5781)] - 2 \approx 5.00004 \approx 5$$ ■

Example 4 Solving an Exponential Equation

Solve $5e^{x-1} = 100$.

Solution

$5e^{x-1} = 100$	Original equation
$e^{x-1} = 20$	Divide both sides by 5.
$\ln(20) = x - 1$	Write in logarithmic form.
$\ln(20) + 1 = x$	Add 1 to both sides.
$x \approx 3.9957$	Compute.

We check that 3.9957 approximately satisfies the equation $5e^{x-1} = 100$:

$$5e^{3.9957-1} \approx 99.9968 \approx 100$$

■

How do the properties for $\log_b(x)$ correspond to the properties for $\ln(x)$? Assume that $x > 0$, $y > 0$, $b > 0$, and $b \neq 1$. The following properties apply:

Properties of Logarithms	**Properties of Natural Logarithms**	
$\log_b(1) = 0$	$\ln(1) = 0$	The natural logarithm of 1 is 0.
$\log_b(b) = 1$	$\ln(e) = 1$	The natural logarithm of e is 1.
$\log_b(x^p) = p\log_b(x)$	$\ln(x^p) = p\ln(x)$	Power property
$\log_b(x) + \log_b(y) = \log_b(xy)$	$\ln(x) + \ln(y) = \ln(xy)$	Product property
$\log_b(x) - \log_b(y) = \log_b\left(\frac{x}{y}\right)$	$\ln(x) - \ln(y) = \ln\left(\frac{x}{y}\right)$	Quotient property

We can use the power property for natural logarithms to solve exponential equations.

Example 5 Solving an Equation

Solve $2(5)^t + 3 = 63$.

Solution

$2(5)^t + 3 = 63$	Original equation
$2(5)^t = 60$	Subtract 3 from both sides.
$5^t = 30$	Divide both sides by 2.
$\ln(5^t) = \ln(30)$	Take the natural logarithm of both sides.
$t\ln(5) = \ln(30)$	Power property: $\ln(x^p) = p\ln(x)$
$t = \dfrac{\ln(30)}{\ln(5)}$	Divide both sides by $\ln(5)$.
$t \approx 2.1133$	Compute.

We check that 2.1133 approximately satisfies the equation $2(5)^t + 3 = 63$:

$$2(5)^{2.1133} + 3 \approx 63.0017 \approx 63$$

■

In Example 5, we used ln to solve an exponential equation. In Section 5.3, we used log to solve exponential equations. It does not matter whether we use ln or log to solve an exponential equation such as $2(5)^t + 3 = 63$. Both methods require about the same amount of work and give the same result.

In Example 6, we use the power and quotient properties for logarithms to simplify a logarithmic expression.

Example 6 Power and Quotient Properties

Write $5\ln(x^3) - 3\ln(2x)$ as a single logarithm with a coefficient of 1. Simplify the result.

Solution

$$
\begin{aligned}
5\ln(x^3) - 3\ln(2x) &= \ln(x^3)^5 - \ln(2x)^3 && \text{Power property: } p\ln(x) = \ln(x^p) \\
&= \ln\frac{(x^3)^5}{(2x)^3} && \text{Quotient property: } \ln(x) - \ln(y) = \ln\left(\frac{x}{y}\right) \\
&= \ln\frac{x^{15}}{8x^3} && \text{Multiply exponents; raise factors to 3rd power.} \\
&= \ln\frac{x^{12}}{8} && \text{Subtract exponents: } \frac{b^m}{b^n} = b^{m-n}
\end{aligned}
$$

We verify our work by comparing tables for the functions $y = 5\ln(x^3) - 3\ln(2x)$ and $y = \ln\frac{x^{12}}{8}$ (see Fig. 42). ■

Figure 42 Verify the work

Example 7 Solving an Equation

Solve $3\ln(4x) + \ln(5x) = 7$.

Solution

$$
\begin{aligned}
3\ln(4x) + \ln(5x) &= 7 && \text{Original equation} \\
\ln(4x)^3 + \ln(5x) &= 7 && \text{Power property: } p\ln(x) = \ln(x^p) \\
\ln\left[(4x)^3(5x)\right] &= 7 && \text{Product property: } \ln(x) + \ln(y) = \ln(xy) \\
\ln\left[64x^3(5x)\right] &= 7 && \text{Raise factors to nth power: } (bc)^n = b^n c^n \\
\ln\left(320x^4\right) &= 7 && \text{Add exponents: } b^m b^n = b^{m+n} \\
e^7 &= 320x^4 && \text{Write in exponential form.} \\
x^4 &= \frac{e^7}{320} && \text{Divide both sides by 320.}
\end{aligned}
$$

Although there is a negative fourth root of $\frac{e^7}{320}$, the original equation contains $3\ln(4x)$, and the domain of a (natural) logarithm function is the set of *positive* numbers. So, $4x$ must be positive. Hence, x must be positive:

$$x = \left(\frac{e^7}{320}\right)^{1/4}$$

$$x \approx 1.3606$$

We use a graphing calculator table to check that for the function $y = 3\ln(4x) + \ln(5x)$, the input 1.3606 leads approximately to the output 7 (see Fig. 43). ■

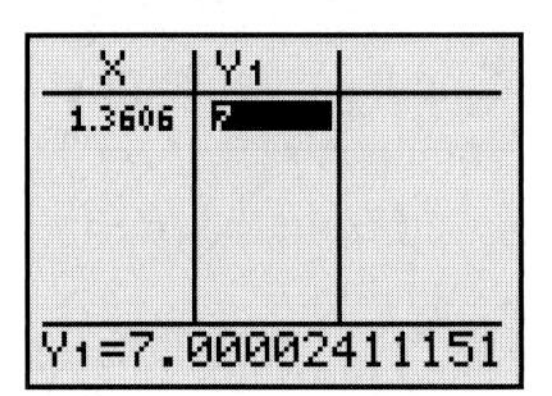

Figure 43 Verify the work

Exponential Models with Base e

So far, we have worked with exponential models of the form $f(t) = ab^t$. In calculus, it is often helpful to write such models in the form $f(t) = ae^{kt}$, where a and k are constants.

Example 8 Using a Model with Base e to Make a Prediction

The BlackBerry® is a wireless handheld device. The numbers of BlackBerry subscribers are shown in Table 43 for various years.

Table 43 Numbers of BlackBerry Subscribers

Year	Number of Subscribers (millions)
2000	0.03
2001	0.17
2002	0.32
2003	0.53
2004	1.07
2005	2.51

Source: *Research in Motion*

Let $f(t)$ be the number of BlackBerry subscribers (in millions) at t years since 2000. A possible equation of f is

$$f(t) = 0.049e^{0.8t}$$

1. Verify that f models the situation well.
2. Predict when there will be 100 million subscribers.

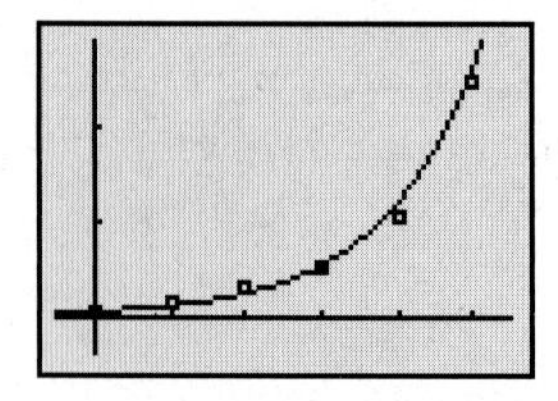

Figure 44 Verify the fit

Solution

1. We draw the graph of f and the scattergram of the data in the same viewing window (see Fig. 44). It appears that f is a reasonable model.
2. To predict when there will be 100 million subscribers, we substitute 100 for $f(t)$ and solve for t:

$$
\begin{aligned}
100 &= 0.049e^{0.8t} && \text{Substitute 100 for } f(t). \\
\frac{100}{0.049} &= e^{0.8t} && \text{Divide both sides by 0.049.} \\
\ln\left(\frac{100}{0.049}\right) &= \ln\left(e^{0.8t}\right) && \text{Take the natural logarithm of both sides.} \\
\ln\left(\frac{100}{0.049}\right) &= 0.8t && \ln\left(e^{a}\right) = a \\
t &= \frac{\ln\left(\frac{100}{0.049}\right)}{0.8} && \text{Divide both sides by 0.8.} \\
t &\approx 9.53 && \text{Compute.}
\end{aligned}
$$

The model predicts that there will be 100 million subscribers in 2010. ■

group exploration

Newton's law of cooling

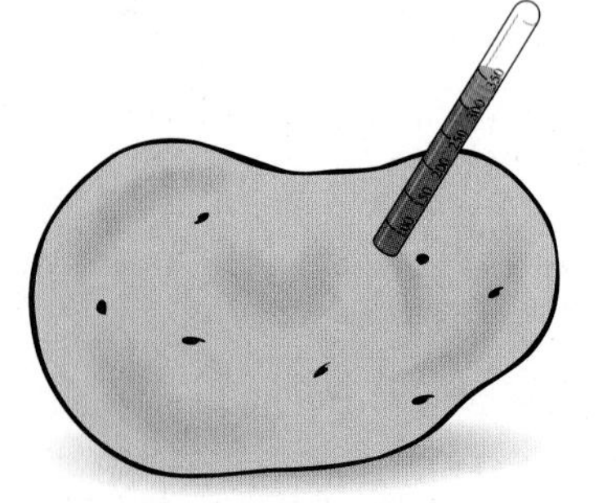

A hot potato is taken out of an oven and allowed to cool to room temperature. Let p be the temperature (in degrees Fahrenheit) of the potato at t minutes after it is removed from the oven.

1. Sketch a qualitative graph that describes the relationship between t and p.
2. Newton's law of cooling states that

$$p - r = ae^{-kt}$$

   where r is room temperature (in degrees Fahrenheit) and a and k are constants. The room temperature is 70°F. Substitute $r = 70$ into the equation.
3. The temperature of the potato was 350°F when it was removed from the oven. Find the value of the constant a, and substitute it into your equation.
4. The temperature of the potato was 200°F 5 minutes later. Find the value of k, and substitute it into your equation.
5. Isolate p on one side of your equation. Then use a graphing calculator to draw the graph of your equation. Compare your graph with your sketch in Problem 1.
6. What will be the temperature of the potato after a long time? Explain.
7. Estimate at what temperature a potato can be comfortably eaten. How long will it take for the potato to reach that temperature?

TIPS FOR SUCCESS: Calm Down During a Test

If you get flustered during a test, close your eyes, take a couple of deep breaths, and think about something pleasant or nothing at all for a moment. This short break might give you some perspective and help you relax.

Use a calculator to find the natural logarithm. Round your result to the fourth decimal place.

1. $\ln(54.8)$ **2.** $\ln(37.28)$ **3.** $\ln\left(\frac{1}{2}\right)$ **4.** $\ln\left(\frac{5}{8}\right)$

Find the natural logarithm. Verify your result by using a graphing calculator.

5. $\ln(e^4)$ **6.** $\ln(e^6)$ **7.** $\ln(e)$ **8.** $\ln(1)$

9. $\ln\left(\frac{1}{e}\right)$ **10.** $\ln\left(\frac{1}{e^2}\right)$ **11.** $\frac{1}{2}\ln(e^6)$ **12.** $\ln\left(\sqrt{e}\right)$

Solve the equation. Round any solutions to the fourth decimal place. Use graphing calculator tables or graphs to verify your result.

13. $\ln(x) = 2$ **14.** $\ln(x) = 5$

15. $\ln(p + 5) = 3$ **16.** $\ln(t - 4) = 5$

17. $7e^x = 44$ **18.** $3e^x = 85$

19. $5\ln(3x) + 2 = 7$ **20.** $2\ln(5x) - 3 = 1$

21. $4e^{3m-1} = 68$ **22.** $7e^{2p+10} = 100$

23. $e^{3x-5} \cdot e^{2x} = 135$ **24.** $e^{2x-3} \cdot e^x = 83$

25. $3.1^x = 49.8$ **26.** $2.4^x = 63.5$

27. $3(6)^x - 1 = 97$ **28.** $5(2)^x + 3 = 264$

29. $5e^x - 20 = 2e^x + 67$

30. $3e^x - 12 = e^x + 44$

Simplify. Write the expression as a single logarithm with a coefficient of 1. Use graphing calculator tables or graphs to verify your result.

31. $\ln(4x) + \ln(3x^4)$ **32.** $\ln(8x^2) + \ln(4x)$

33. $\ln(25x^4) - \ln(5x^3)$ **34.** $\ln(6x^3) - \ln(2x^4)$

35. $2\ln(w^4) + 3\ln(2w)$ **36.** $4\ln(3r) + 5\ln(r^3)$

37. $3\ln(3x) - 2\ln(x^2)$

38. $2\ln(x^3) - 3\ln(2x)$

39. $3\ln(2k) + 4\ln(k^2) - \ln(k^7)$

40. $5\ln(p^2) + 2\ln(3p) - \ln(p^9)$

Solve the equation. Round any solutions to the fourth decimal place. Check your result.

41. $\ln(3x) + \ln(x) = 4$

42. $\ln(2x) + \ln(5x) = 6$

43. $\ln(4x^5) - 2\ln(x^2) = 5$

44. $2\ln(x^3) - \ln(8x^5) = 1$

45. $2\ln(3x) + 2\ln(x^3) = 8$

46. $4\ln(2x) + 3\ln(x^3) = 9$

47. $5\ln(m^3) - 3\ln(2m) = 7$

48. $3\ln(4y) - 4\ln(y^3) = 5$

Solve by using "intersect" on a graphing calculator. Round any solutions to the fourth decimal place.

49. $e^x = 5 - x$

50. $2e^x = 9 - 2x$

51. $3\ln(x + 2) = -2x + 6$

52. $2\ln(x + 1) = -x + 7$

53. $3\ln(x + 3) = 0.7x + 2$

54. $4\ln(x + 5) = x + 5$

For Exercises 55–58, let $f(x) = 4\ln(x)$.

55. Find $f(e^5)$.

56. Find $f\left(\frac{1}{e^3}\right)$.

57. Find x when $f(x) = -8$.

58. Find x when $f(x) = 2$.

59. Explain in your own words why $\ln(e) = 1$.

60. Explain in your own words why $\ln(1) = 0$.

61. Assume that the equation $ae^{bx} = c$ has a solution for x, where $a \neq 0$ and $b \neq 0$. Solve for x.

62. Assume that the equation $ae^{bx+d} + k = c$ has a solution for x, where $a \neq 0$ and $b \neq 0$. Solve for x.

63. Which expressions are equal? Assume that $x > 0$ and $x \neq 1$.

$3\ln(x)$ $\ln(x^7) - \ln(x^4)$ $2\ln(x)\ln(x)$ $\frac{\ln(x^7)}{\ln(x^4)}$

$\ln(x^3)$ $\ln(3x)$

64. a. Solve $e^x = 30$ by writing the equation in logarithmic form.

b. Solve $e^x = 30$ by taking the natural logarithm of both sides of the equation.

c. Compare your results in parts (a) and (b).

65. Annual sales at Urban Outfitters® are shown in Table 44 for various years.

Table 44 Annual Sales at Urban Outfitters

Year	Annual Sales (billions of dollars)
2002	0.3
2003	0.4
2004	0.5
2005	0.8
2006	1.1

Source: *Urban Outfitters*

Let $f(t)$ be the annual sales at Urban Outfitters (in billions of dollars) at t years since 2000. An equation of f is

$$f(t) = 0.15e^{0.33t}$$

a. Verify that f models the situation well.

b. Predict the annual sales in 2010.

c. Predict when annual sales will reach $6 billion.

66. The Sarbanes–Oxley Act of 2002 has motivated many companies to restate their earnings in their financial reports (see Table 45).

Table 45 Numbers of Earnings Restatements

Year	Number of Restatements
1997	130
1999	205
2001	260
2003	535
2004	650

Source: *Glass, Lewis, & Company*

Let $f(t)$ be the number of earnings restatements in the year that is t years since 1990. A possible equation of f is

$$f(t) = 24.67e^{0.231t}$$

a. Verify that f models the situation well.
b. Predict the number of earnings restatements in 2010.
c. Predict in which year there will be 3000 restatements.

67. A person buys a cup of coffee and drinks it in the store. The coffee's temperature y (in degrees Fahrenheit) is given by

$$y = 70 + 137e^{-0.06t}$$

where t is the number of minutes since he bought the coffee.
a. What was the temperature of the coffee when he bought it?
b. If the person begins drinking the coffee when it reaches 180°F, how much time must he wait after buying it?
c. Use a graphing calculator table or graph to estimate the room temperature of the *store*.

68. A person makes a cup of tea. The tea's temperature y (in degrees Fahrenheit) is given by

$$y = 68 + 132e^{-0.05t}$$

where t is the number of minutes since she made the tea.
a. What was the temperature of the tea when she made it?
b. If the person waits 5 minutes to begin drinking the tea, what is the temperature of the tea then?
c. The tea is *lukewarm* at a temperature of about 98.6°F. If the person lets the tea sit until it is lukewarm, how much time has gone by since she made it?

69. A cable hangs between two poles that are 20 feet apart (see Fig. 45). The height of the cable (in feet) is given by

$$h(x) = 10\left(e^{0.03x} + e^{-0.03x}\right), \qquad -10 \le x \le 10$$

where x is the horizontal position (in feet) as indicated in Fig. 45.

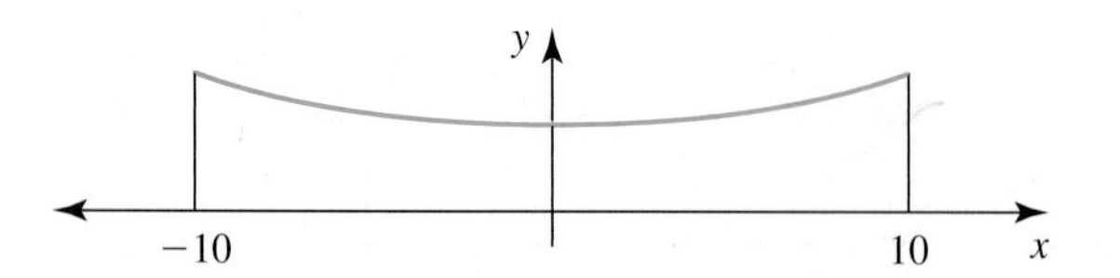

Figure 45 Exercise 69

a. Find the height of the cable at either pole.
b. Find $h(6)$. What does it mean in this situation?
c. How high is the cable where it is closest to the ground?

70. A cable hangs between two poles that are 30 feet apart (see Fig. 46). The height of the cable (in feet) is given by

$$h(x) = 20\left(e^{0.05x} + e^{-0.05x}\right), \qquad -15 \le x \le 15$$

where x is the horizontal position (in feet) as indicated in Fig. 46.

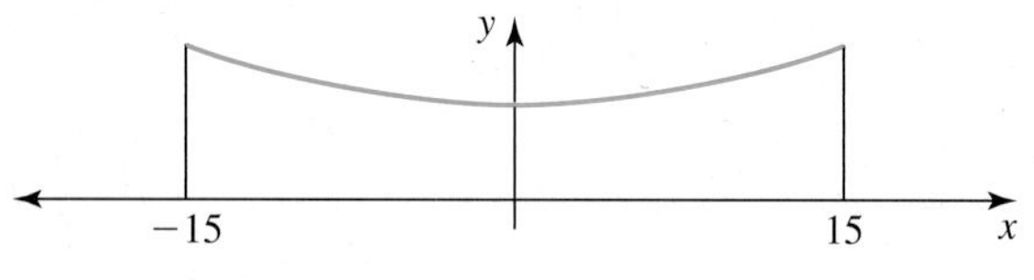

Figure 46 Exercise 70

a. Find $h(-8)$. What does it mean in this situation?
b. How high is the cable where it is closest to the ground?
c. Explain why there is model breakdown for $x < -15$ and for $x > 15$.

71. a. **i.** To solve $3^x = 58$, begin by taking the natural logarithm of both sides.
ii. To solve $3^x = 58$, begin by taking the common log (base 10) of both sides.
iii. Compare your results in parts (i) and (ii).
b. Create an equation of the form $b^x = c$, where b and c are positive and $b \ne 1$.
i. To solve your equation, begin by taking the natural logarithm of both sides.
ii. To solve your equation, begin by taking the common log (base 10) of both sides.
iii. Compare your results in parts (i) and (ii).
c. Summarize the main point of this exercise.

72. Explain how to use the power property for natural logarithms to solve an exponential equation in one variable.

Related Review

Simplify by writing your result as a single logarithm with a coefficient of 1, or solve, as appropriate. Round any solutions to the fourth decimal place.

73. $\ln(x^8) - \ln(x^3)$

74. $2\ln(4x) + 3\ln(x^2) = 7$

75. $\ln(x^8) - \ln(x^3) = 4$

76. $2\ln(4x) + 3\ln(x^2)$

Solve. Find the exact solution if the equation is linear. For any other type of equation, round the solution(s) to the fourth decimal place.

77. $3e^x - 5 = 7$

78. $4\log_2(x^2) - 2\log_2(3x) = 5$

79. $7 - 3(2t - 4) = 5t + 6$

80. $\log(w - 17) = 2$

81. $\dfrac{b^7}{b^3} = 16$

82. $3\ln(2x) + 4\ln(x^2) = 7$

Expressions, Equations, Functions, and Graphs

Give an example of the following. Then solve, simplify, or graph, as appropriate.

83. expression involving exponents
84. system of two linear equations in two variables
85. logarithmic equation in one variable
86. sum of two natural logarithmic expressions
87. linear function
88. exponential function

Chapter Summary

Key Points OF CHAPTER 5

Inverse of a function f (Section 5.1)

The **inverse of a function f** is a relation that sends b to a if $f(a) = b$.

Invertible function (Section 5.1)

If the inverse of a function f is also a function, we say that f is **invertible** and use "f^{-1}" as a name for the inverse of f.

Property of an inverse function (Section 5.1)

For an invertible function f, the following statements are equivalent: $f(a) = b$ and $f^{-1}(b) = a$.

f and f^{-1} are inverses of each other (Section 5.1)

If f is an invertible function, then

- f^{-1} is invertible, and
- f and f^{-1} are inverses of each other.

Reflection property of inverse functions (Section 5.1)

For an invertible function f, the graph of f^{-1} is the reflection of the graph of f across the line $y = x$.

Graphing an inverse function (Section 5.1)

For an invertible function f, we sketch the graph of f^{-1} by the following steps:

1. Sketch the graph of f.
2. Choose several points that lie on the graph of f.
3. For each point (a, b) chosen from step 2, plot the point (b, a).
4. Sketch the curve that contains the points plotted in step 3.

Three-step process for finding the inverse of a model (Section 5.1)

To find the inverse of an invertible *model* f, where $p = f(t)$,

1. Replace $f(t)$ with p.
2. Solve for t.
3. Replace t with $f^{-1}(p)$.

Four-step process for finding the inverse of a function that is not a model (Section 5.1)

Let f be an invertible function that is *not* a model. To find the inverse of f, where $y = f(x)$,

1. Replace $f(x)$ with y.
2. Solve for x.
3. Replace x with $f^{-1}(y)$.
4. Write the equation of f^{-1} in terms of x.

One-to-one function (Section 5.1)

If each output of a function originates from exactly one input, we say that the function is **one-to-one.**

Definition of a logarithm (Section 5.2)

For $b > 0$, $b \neq 1$, and $a > 0$, the **logarithm $\log_b(a)$** is the number k such that $b^k = a$. In words: $\log_b(a)$ is the exponent on the base b that gives a. We call b the **base** of the logarithm.

Common logarithm (Section 5.2)

A **common logarithm** is a logarithm with base 10. We write $\log(a)$ to represent $\log_{10}(a)$.

Properties of logarithms (Section 5.2)

For $b > 0$ and $b \neq 1$,

- $\log_b(b) = 1$
- $\log_b(1) = 0$

Logarithmic function (Section 5.2)

A **logarithmic function, base b**, is a function that can be put into the form $f(x) = \log_b(x)$, where $b > 0$ and $b \neq 1$.

Domain of a logarithmic function (Section 5.2)

The domain of a logarithmic function $\log_b$ is the set of all positive real numbers.

Logarithmic and exponential functions are inverses of each other (Section 5.2)

For an exponential function $f(x) = b^x$, $f^{-1}(x) = \log_b(x)$.

For a logarithmic function $g(x) = \log_b(x)$, $g^{-1}(x) = b^x$.

Graphing $y = \log_b(x)$ (Section 5.2)

To graph a logarithmic function $y = \log_b(x)$, use the four-step graphing method from Section 5.1 to sketch the inverse of $f(x) = b^x$.

Richter number (Section 5.2)	The *Richter number, R,* of an earthquake is given by $$R = \log\left(\frac{A}{A_0}\right)$$ where A is the amplitude (maximum value) of a seismic wave and A_0, called the *reference amplitude,* is the amplitude of the smallest seismic wave that a seismograph can detect.
Exponential/logarithmic forms property (Section 5.3)	For $a > 0$, $b > 0$, and $b \neq 1$, $\log_b(a) = c$ and $b^c = a$ are equivalent.
Logarithmic equation in one variable (Section 5.3)	A **logarithmic equation in one variable** is an equation in one variable that contains one or more logarithms.
Solving $\log_b(x) = k$ (Section 5.3)	For an equation of the form $\log_b(x) = k$, we can solve for b or x by writing the equation in exponential form.
Exponential equation in one variable (Section 5.3)	An **exponential equation in one variable** is an equation in one variable in which an exponent contains a variable.
Power property for logarithms (Section 5.3)	For $x > 0$, $b > 0$, and $b \neq 1$, $\log_b\left(x^p\right) = p\log_b(x)$.
Logarithm property of equality (Section 5.3)	For positive real numbers a, b, and c, where $b \neq 1$, the equations $a = c$ and $\log_b(a) = \log_b(c)$ are equivalent.
Solving $ab^x = c$ for x (Section 5.3)	To solve some equations of the form $ab^x = c$ for x, we divide both sides of the equation by a and then take the log of both sides. Next, we use the power property for logarithms.
Making a prediction (Section 5.4)	To make a prediction for the independent variable t of an exponential model of the form $f(t) = ab^t$, we substitute a value for $f(t)$, then divide both sides of the equation by the coefficient a. Next, we take the log of both sides of the equation, then use the power property to help solve for t.
Properties of logarithms (Section 5.5)	For $x > 0$, $y > 0$, $a > 0$, $b > 0$, $a \neq 1$, and $b \neq 1$: • $\log_b(x) + \log_b(y) = \log_b(xy)$ — Product property for logarithms • $\log_b(x) - \log_b(y) = \log_b\left(\frac{x}{y}\right)$ — Quotient property for logarithms • $\log_b(x) = \frac{\log_a(x)}{\log_a(b)}$ — Change-of-base property
Finding $\log_b(a)$ where $b \neq 10$ (Section 5.5)	To find a logarithm to a base other than 10, we use the change-of-base property for logarithms to convert to $\log_{10}$; then we can use the log key on a calculator.
Approximation of e (Section 5.6)	To the nearest ten-thousandth, $e = 2.7183$.
Definition of a natural logarithm (Section 5.6)	A **natural logarithm** is a logarithm with base e. We write $\ln(a)$ to represent $\log_e(a)$.
Meaning of $\ln(a)$ (Section 5.6)	For $a > 0$, $\ln(a)$ is the exponent on the base e that gives a.
Exponential/natural logarithmic forms property (Section 5.6)	For $a > 0$, the equations $\ln(a) = c$ and $e^c = a$ are equivalent.
Properties of natural logarithms (Section 5.6)	$\ln(1) = 0$ $\ln(e) = 1$ For $x > 0$ and $y > 0$, • $\ln\left(x^p\right) = p\ln(x)$ — Power property for natural logarithms • $\ln(x) + \ln(y) = \ln(xy)$ — Product property for natural logarithms • $\ln(x) - \ln(y) = \ln\left(\frac{x}{y}\right)$ — Quotient property for natural logarithms

CHAPTER 5 REVIEW EXERCISES

For Exercises 1 and 2, refer to Table 46.

1. Find $f(2)$. **2.** Find $f^{-1}(2)$.

Table 46 Values of f (Exercises 1 and 2)

x	f(x)
0	1
1	2
2	4
3	3
4	0

Graph f, f^{-1}, and $y = x$ by hand on the same set of axes.

3. $f(x) = 2x - 3$ **4.** $g(x) = 3^x$

5. The numbers of full-time-equivalent employees for all passenger airlines are shown in Table 47 for various years.

Table 47 Employment for All Passenger Airlines

Year	Number of Full-Time-Equivalent Employees (thousands)
2002	470
2003	465
2004	437
2005	432
2006	405

Source: *Bureau of Transportation Statistics*

Let $n = f(t)$ be the number (in thousands) of full-time-equivalent employees for all passenger airlines at t years since 2000.

a. Find an equation of f.

b. Find an equation of f^{-1}.

c. Find $f(11)$. What does it mean in this situation?

d. Find $f^{-1}(300)$. What does it mean in this situation?

Find the inverse of the function.

6. $f(x) = 3x$ **7.** $g(x) = \dfrac{4x - 7}{8}$

Evaluate the logarithmic function at the given value. Round approximate results to the fourth decimal place.

8. $\log_5(25)$ **9.** $\log(100{,}000)$

10. $\log_3\left(\dfrac{1}{9}\right)$ **11.** $\ln\left(\dfrac{1}{e^3}\right)$ **12.** $\log_4\left(\sqrt[3]{4}\right)$

13. $\log_3(7)$ **14.** $\ln(5)$ **15.** $\log_b\left(b^7\right)$

Find the inverse of the function.

16. $h(x) = 3^x$ **17.** $h(x) = \log(x)$

18. Sketch the graph of $y = \log_4(x)$ by hand.

19. Write the equation $d^t = k$ in logarithmic form.

20. Write the equation $\log_y(w) = r$ in exponential form.

Solve. Round any approximate solutions to the fourth decimal place.

21. $6(2)^x = 30$ **22.** $\log_3(x) = -4$

23. $4.3(9.8)^x - 3.3 = 8.2$ **24.** $\log_b(83) = 6$

25. $5\log_{32}(m) - 3 = -1$ **26.** $5(4)^{3r-7} = 40$

27. $2^{4t} \cdot 2^{3t-5} = 94$

For Exercises 28–30, find any solutions of the equation or system by referring to the graphs shown in Fig. 47.

28. $\log_2(x) = -\dfrac{3}{4}x + 5$ **29.** $2^x - 3 = -2$

30. $y = \log_2(x)$

$y = -\dfrac{3}{4}x + 5$

Figure 47 Exercises 28–30

Let $f(x) = 3^x$.

31. Find $f(4)$. **32.** Find $f^{-1}(25)$.

33. Find x when $f(x) = 6$. **34.** Find x when $f^{-1}(x) = 6$.

35. Suppose that \$8000 is deposited into an account that earns 5% interest compounded annually. Let $f(t)$ be the value (in dollars) of the account after t years or any fraction thereof.

a. Find a formula of f.

b. Find the value of the account after 9 years.

c. After how many years will the value of the account double?

36. On April 1, a tree has 30 leaves. Each week, the number of leaves quadruples (increases by four times). Let $f(t)$ be the number of leaves on the tree at t weeks since April 1.

a. Find an equation of f.

b. Predict the number of leaves at 5 weeks after April 1.

c. Predict when there will be 100,000 leaves.

37. The percentages of American adults who are obese are shown in Table 48 for various years.

Table 48 Percentages of American Adults Who Are Obese

Year	Percent
1990	12
1992	13
1994	14
1996	17
1998	18
2000	20
2002	22
2004	25

Source: *National Center for Chronic Disease Prevention and Health Promotion*

Let $p = f(t)$ be the percentage of adults who are obese at t years since 1900.

a. Use a graphing calculator to draw a scattergram of the data. Can the data be modeled better by using a linear or an exponential function? Explain.

b. Find an equation of f.

c. Find $f(100)$. What does it mean in this situation?

d. Find $f^{-1}(100)$. What does it mean in this situation?

e. Find t when $f(t) = 100$. What does it mean in this situation?

38. A student ran an experiment to investigate the relationship between the length of a rubber band and the weight applied to one end of it. He hooked one end of a rubber band to a horizontal pole supported between two chairs. He attached a bag to the other end of the rubber band. Then, he recorded the lengths of the rubber band with various numbers of tape cassettes in the bag (see Table 49).

Table 49 Lengths of a Rubber Band Stretched by Tape Cassettes

Number of Cassettes	Length of Rubber Band (inches)
0	10.00
1	12.00
2	15.38
3	19.50
4	28.31
5	33.50
6	45.45
7	64.15

Source: *Michael S.*

a. Let $f(n)$ be the length (in inches) of the rubber band stretched by n cassettes. Find an equation of f.

b. What is the base b of your function $f(n) = ab^n$? What does it mean in this situation?

c. What is the coefficient a of your function $f(n) = ab^n$. What does it mean in this situation?

d. Estimate the length of the rubber band if it were stretched by eight cassettes. Describe two scenarios in which model breakdown might occur for your estimate.

e. Estimate the number of cassettes needed to stretch the rubber band to 139 inches. If model breakdown occurs for your estimate from part (d), does that imply that model breakdown occurs for the estimate you made in this part? Explain.

39. A storage tank contains cobalt-60, which has a half-life of 5.3 years. Predict when 15% of the cobalt-60 will remain.

Simplify. Write your result as a single logarithm with a coefficient of 1.

40. $\log_b(p) + \log_b(6p) - \log_b(2p)$

41. $3\log_b(2x) + 2\log_b(3x)$

42. $4\log_b(x^2) - 2\log_b(x^5)$

43. $\dfrac{\log_b(w)}{\log_b(y)}$

44. Which of the following expressions are equal?

$$\log_b(b^5) - \log_b(b^2) \qquad 3 \qquad \log_b(b^5)$$

$$\log_b(b^3) \qquad \frac{\log_b(b^5)}{\log_b(b^2)} \qquad \log_b\left(\frac{b^5}{b^2}\right)$$

Solve. Round any solutions to the fourth decimal place.

45. $2\log_9(3w) + 3\log_9(w^2) = 5$

46. $5\log_6(2x) - 3\log_6(4x) = 2$

Simplify. Write your result as a single logarithm with a coefficient of 1.

47. $3\ln(4x) + 2\ln(2x)$

48. $\ln(2m^7) - 4\ln(m^3) + 3\ln(m^2)$

Solve. Round any solutions to the fourth decimal place.

49. $4e^x = 75$

50. $-3\ln(p) + 7 = 1$

51. $3\ln(t^5) - 5\ln(2t) = 7$

CHAPTER 5 TEST

1. Graph $f(x) = 3x - 6$, f^{-1}, and $y = x$ by hand on the same set of axes.

2. The graph of a function f is sketched in Fig. 48. Graph f^{-1} by hand.

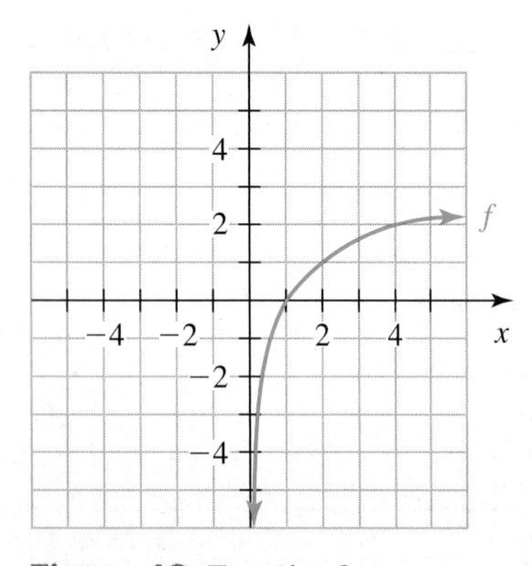

Figure 48 Exercise 2

3. The prices of an adult one-day ticket to Walt Disney World® are shown in Table 50 for various years.

Table 50 Prices of an Adult One-Day Ticket to Walt Disney World

Year	Ticket Price (dollars)
1995	39
1997	43
1999	47
2001	51
2003	55
2004	58

Source: *The Walt Disney Company*

Let $p = f(t)$ be the price (in dollars) of an adult one-day ticket at t years since 1990.

a. Find an equation of f.

b. Find an equation of f^{-1}.

c. Use f^{-1} to predict when the price of an adult one-day ticket will be \$70.

d. What is the slope of f^{-1}? What does it mean in this situation?

4. Find the inverse of the function $g(x) = 2x - 9$.

Evaluate. Round approximate results to the fourth decimal place.

5. $\log_2(16)$

6. $\log_4\left(\frac{1}{64}\right)$

7. $\log_7(10)$

8. $\log(0.1)$

9. $\log_b(\sqrt{b})$

10. $\ln\left(\frac{1}{e^2}\right)$

Find the inverse of the function.

11. $h(x) = 4^x$

12. $f(x) = \log_5(x)$

13. Write $s^t = w$ in logarithmic form.

14. Write $\log_c(a) = d$ in exponential form.

Solve. Round any solutions to the fourth decimal place.

15. $\log_b(50) = 4$

16. $6(2)^x - 9 = 23$

17. $\log_4(7p + 5) = -\frac{3}{2}$

18. Use "intersect" on a graphing calculator to solve the equation $4^x - 8 = -\frac{1}{2}x + 3$. Round any solutions to the second decimal place.

19. The tuition rates at Princeton University are shown in Table 51 for the academic years ending in the indicated year. Let $r = f(t)$ be the tuition rate (in dollars per year) at t years since 1950.

a. Find an equation of f.

b. What is the r-intercept of f? What does it mean in this situation?

c. What is the percentage rate of growth of the tuition rate?

d. Predict the tuition rate in 2011.

e. Predict when the tuition rate at Princeton will be \$60,000 per year.

Table 51 Tuition Rates at Princeton

Year	Tuition Rates (dollars per year)
1950	600
1960	1450
1970	2350
1980	5585
1990	14,390
2000	25,430
2005	29,910
2006	31,450

Source: *Princeton University*

20. Scientists wanted to date a sample of cloth wrappings of a mummified bull from a pyramid in Dashur, Egypt. If 78.04% of the carbon-14 remains in the sample, estimate the age of the mummy.

Simplify. Write your result as a single logarithm with a coefficient of 1.

21. $\log_b(x^3) + \log_b(5x)$

22. $3\log_b(4p^2) - 2\log_b(8w^5) + \log_b(2p^4)$

Solve. Round any solutions to the fourth decimal place.

23. $\log_3(x) + \log_3(2x) = 5$

24. $2\log_4(x^4) - 3\log_4(3x) = 3$

25. Simplify $2\ln(5w) + 3\ln(w^6)$. Write your result as a single logarithm with a coefficient of 1.

Solve. Round any solutions to the fourth decimal place.

26. $2e^{3x-1} = 54$

27. $7\ln(t - 2) - 1 = 4$

CUMULATIVE REVIEW OF CHAPTERS 1–5

Solve. Round approximate results to the fourth decimal place.

1. $2(4)^{5x-1} = 17$

2. $\log_3(x - 5) = 4$

3. $3b^7 - 18 = 7$

4. $8 + 2e^x = 15$

5. $4\log_5(3x^2) + 3\log_5(6x^4) = 3$

6. $7 - 3(4w - 2) = 2(3w + 5) - 4(2w + 1)$

For Exercises 7 and 8, estimate any solutions of the equation by referring to the graphs of $f(x) = 3^x$, $g(x) = 9\left(\frac{1}{3}\right)^x$, and $h(x) = x - 1$ shown in Fig. 49.

7. $3^x = 9\left(\frac{1}{3}\right)^x$

8. $9\left(\frac{1}{3}\right)^x = x - 1$

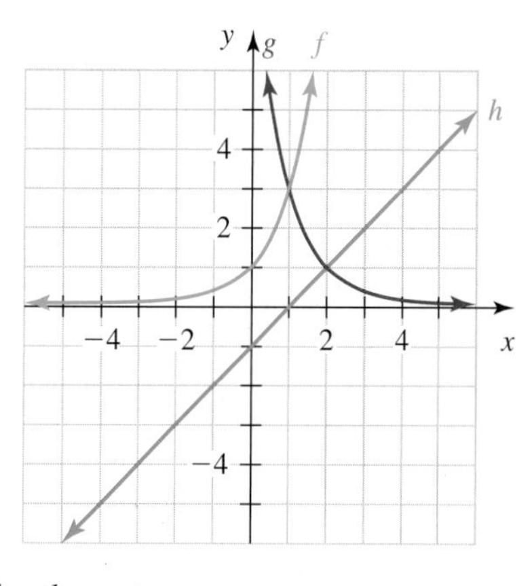

Figure 49 Exercises 7 and 8

Solve the system.

9. $x = 2y - 5$
$4x - 5y = -14$

10. $3(2 - 4x) = -10 - 2y$
$2x - 3y = -8$

11. Solve the inequality $8x - 3 \geq -3(4x - 5)$. Describe the solution set as an inequality, in interval notation, and in a graph.

Simplify.

12. $(4b^{-3}c^2)^3(5b^{-7}c^{-1})^2$ **13.** $\dfrac{8b^{1/3}c^{-1/2}}{6b^{-1/2}c^{3/4}}$

Simplify. Write your result as a single logarithm with a coefficient of 1.

14. $4\log_b(x^7) - 2\log_b(7x)$

15. $3\ln(p^6) + 4\ln(p^2)$

For Exercises 16–20, refer to Table 52.

16. Find an equation of f.

17. Find an equation of g.

18. Find the slope of h.

19. Find $k(5)$.

20. Find $f^{-1}(5)$.

Table 52 Values of Four Functions (Exercises 16–20)

x	$f(x)$	x	$g(x)$	x	$h(x)$	x	$k(x)$
0	5	0	25	4	83	3	160
1	15	1	28	5	76	4	80
2	45	2	31	6	69	5	40
3	135	3	34	7	62	6	20
4	405	4	37	8	55	7	10
5	1215	5	40	9	48	8	5

Graph the function by hand.

21. $y = 8\left(\dfrac{1}{2}\right)^x$ **22.** $y = -\dfrac{2}{5}x + 3$

23. $2(2x - y) + 2y = 3(4 + y)$ **24.** $y = \log_2(x)$

25. Find an equation of a line that contains the points $(-4, 7)$ and $(5, -3)$.

26. Find an equation of an exponential curve that contains the points $(3, 85)$ and $(7, 13)$.

Let $f(x) = 2(3)^x$. Round approximate results to the fourth decimal place.

27. Find $f(-4)$. **28.** Graph f by hand.

29. Graph f^{-1} by hand. **30.** Find $f^{-1}(35)$.

Find the logarithm.

31. $\log_3\left(\dfrac{1}{81}\right)$

32. $\log_b(\sqrt[7]{b})$

33. Find $\log_8(73)$. Round your result to the fourth decimal place.

For Exercises 34–37, refer to Fig. 50.

34. Estimate $f(-1)$.

35. Find an equation of f.

36. Graph f^{-1} by hand.

37. Estimate $f^{-1}(2)$.

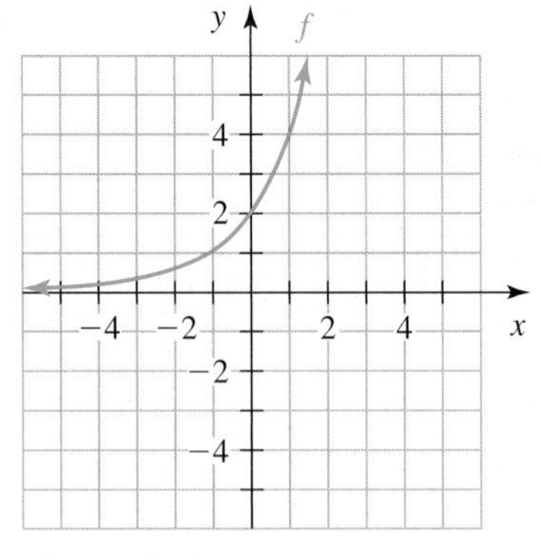

Figure 50 Exercises 34–37

Find the inverse of the function.

38. $f(x) = \dfrac{2}{7}x - 3$ **39.** $g(x) = 8^x$

40. Sketch a graph of a relation that is not a function. Next, create a table of five ordered pairs of the relation. Explain why your relation is not a function.

41. Compare the function $f(x) = 3x + 2$ with the function $g(x) = 2(3)^x$.

a. Find the y-intercept of the graphs of both functions.

b. For both f and g, describe what happens to the value of y as the value of x increases by 1.

c. For a large input value of x, which function will have a greater output value of y? Explain.

d. Graph f and g by hand on the same coordinate system.

42. Let f be the linear function and g be the exponential function whose graphs contain the points $(4, 3)$ and $(7, 8)$.

a. Find an equation of f.

b. Find an equation of g.

c. Use a graphing calculator to draw the graphs of f and g in the same viewing window.

43. Let $f(x) = 3x$ and $g(x) = 3^x$.

a. Find $f(2)$ and $g(2)$.

b. Find an equation of f^{-1} and an equation of g^{-1}.

c. Find $f^{-1}(81)$ and $g^{-1}(81)$.

44. One U-Haul office rents pickup trucks for a one-day fee of \$19.95 plus \$0.69 per mile. One Budget office charges a one-day fee of \$29.95 plus \$0.45 per mile (Sources: *U-Haul; Budget*).

a. Let $U(x)$ be U-Haul's charge and $B(x)$ be Budget's charge, both in dollars, for driving x miles in one day. Find equations of U and B.

b. Find the slopes of U and B. What do they mean in this situation?

c. For how many miles driven is the one-day charge at U-Haul equal to the one-day charge at Budget?

d. Solve the inequality $U(x) < B(x)$. What does your result mean in this situation?

45. A 15,000-seat amphitheater has tickets for sale at \$43 and \$60. How many tickets should be sold at each price for a sellout performance to generate a total revenue of \$721,500?

46. In 2003, 2.51 million books were in print, and that number has grown by 10% per year since then (Source: *Andrew Grabois, R. R. Bowker*).

a. Let $n = f(t)$ be the number of books (in millions) in print at t years since 2003. Find an equation of f.

b. What is the n-intercept of the model? What does it mean in this situation?
c. What is the base b of your function $f(t) = ab^t$? What does it mean in this situation?
d. Predict when there will be 5 million books in print.

47. During the 1990s, the syphilis rate in the United States decreased rapidly (see Table 53).

Table 53 Syphilis Rates in the United States

Year	Syphilis Rate (number of cases per 100,000 people)	Year	Syphilis Rate (number of cases per 100,000 people)
1990	20.3	1996	4.2
1991	17.0	1997	3.1
1992	13.3	1998	2.5
1993	10.2	1999	2.4
1994	7.8	2000	2.1
1995	6.2		

Source: *Centers for Disease Control and Prevention*

Let $s = f(t)$ be the syphilis rate (number of cases per 100,000 people) at t years since 1990.
a. Find an equation of f.
b. What is the s-intercept of the model? What does it mean in this situation?
c. What is the percent rate of decay? What does it mean in this situation?
d. Use f to predict when there will be 1 syphilis case per 100,000 people.
e. Find $f^{-1}(0.1)$. What does it mean in this situation?
f. Federal health officials had developed a plan designed to eradicate syphilis from the United States by 2004. Use f to estimate the syphilis rate in 2004. The actual rate was 2.7 syphilis cases per 100,000 people in 2004. Although the eradication goal was not met, is the rate heading in the right direction?

48. The average maximum speeds of a pickup truck are shown in Table 54 for various years.

Table 54 Average Maximum Speeds of a Pickup Truck

Year	Average Maximum Speed (miles per hour)
1985	109
1990	117
1995	120
2000	128
2005	136

Source: *Ward's AutoInfoBank*

Let $s = f(t)$ be the average maximum speed (in miles per hour) of a pickup truck at t years since 1980.
a. Use a graphing calculator to draw a scattergram of the data. Is it better to model the data by using a linear or an exponential function? Explain.
b. Find an equation of f.
c. If your model is linear, find the slope. If your model is exponential, find the base b of $f(t) = ab^t$. What does your result mean in this situation?
d. What is the t-intercept? What does it mean in this situation?
e. Find $f(30)$. What does it mean in this situation?
f. Find an equation of f^{-1}.
g. Find $f^{-1}(145)$. What does it mean in this situation?

Chapter 6

Polynomial Functions

"No other field can offer, to such an extent as mathematics, the joy of discovery, which is perhaps the greatest human joy."
—*Rósza Péter, "Mathematics Is Beautiful,"* Mathematics Intelligencer *12 (Winter 1990): 62*

Table 1 MBA Average Base Pay Amounts and Average Signing Bonuses

	Average Pay/Bonus (thousands of dollars)	
Year	**Base Pay**	**Signing Bonus**
2001	88.6	18.9
2002	82.0	13.0
2003	80.9	8.9
2004	88.8	10.6
2005	93.8	12.8

Source: *Graduate Management Admission Council*

Did you know that many recent graduates with a Master of Business Administration (MBA) degree get a signing bonus when they are hired? The average base pay and average signing bonus have changed significantly since 2001 (see Table 1). In Exercise 77 of Homework 6.1, you will predict the average total money from base pay and signing bonus for a certain year in the future.

So far, we have studied linear, exponential, and logarithmic functions. In this chapter, we will work with *polynomial expressions* and *polynomial functions*. We will perform operations with polynomial expressions and solve *polynomial equations* in one variable. We will also use polynomial functions to model authentic situations, such as the number of participants in the Ironman World Championship. We will use these functions to make predictions, such as the average total money MBA graduates will earn from base pay and a signing bonus for a certain year in the future.

6.1 ADDING AND SUBTRACTING POLYNOMIAL EXPRESSIONS AND FUNCTIONS

Objectives

- Know the meaning of *term, monomial, polynomial, degree, coefficient, like terms, polynomial function, quadratic function, parabola, vertex,* and *cubic function.*
- Add and subtract polynomials.
- Evaluate polynomial functions.
- Know typical graphs of quadratic functions and cubic functions.
- Find *sum functions* and *difference functions.*
- Use sum functions and difference functions to describe authentic situations.

In this section, we will discuss how to add and subtract expressions called *polynomials*. We will discuss how to combine two *polynomial functions* to form a "new" function and how to use some of these new functions to model authentic situations.

Polynomials

A **term** is a constant, a variable, or a product of a constant and one or more variables raised to powers. Here are some examples of terms: $3x^6$, x, -3, $x^{1/2}$, and $-2x^3y^{-5}$.

A **monomial** is a constant, a variable, or a product of a constant and one or more variables raised to *counting number* powers. Here are some examples of monomials:

$$3x^6 \qquad x \qquad -3 \qquad -5x^7y^4$$

A **polynomial,** or **polynomial expression,** is a monomial or a sum of monomials. Here are some examples of polynomials:

$$5x^3-2x^2+7x-4 \qquad 4x^5y^2-x^2 \qquad 4x+1 \qquad 5 \qquad x \qquad -2x^3$$

The polynomial $4x^3 - 8x^2 + 6x - 9$ is a *polynomial in one variable*. It has four terms: $4x^3$, $-8x^2$, $6x$, and -9. We usually write polynomials in one variable so that the exponents of the terms decrease from left to right, which is called **descending order.** If a polynomial contains more than one variable, we usually write the polynomial so that the exponents of one of the variables decrease from left to right.

The **degree of a term** in one variable is the exponent on the variable. For example, the degree of the term $7x^4$ is 4. The degree of a term in two or more variables is the sum of the exponents on the variables. For example, the term $2x^3y^5$ has degree $3 + 5 = 8$. The **degree of a polynomial** is the highest degree of any nonzero term of the polynomial. For example, the polynomial $4x^5 - 9x^3 + 1$ has degree 5. A constant polynomial, such as 7, has degree 0.

Polynomials with degrees 1, 2, or 3 have special names:

Degree	Name	Examples
1	linear (1st-degree) polynomial	$4x + 9$, $-6x$
2	quadratic (2nd-degree) polynomial	$7x^2 - 3x + 8$, $x^2 + 5$
3	cubic (3rd-degree) polynomial	$-5x^3 - 4x^2 + 8x + 2$, $8x^3 + 7x$

Example 1 Describing Polynomials

Use words such as *linear, quadratic, cubic, polynomial, degree, one variable,* and *two variables* to describe the expression.

1. $-3x^2 + 8x - 4$ **2.** $5x^3 - 2x^2 + 9x + 1$ **3.** $6a^5b^2 - 9a^3b^3 - 2ab^4$

Solution

1. The term $-3x^2$ has degree 2, which is larger than the degrees of the other terms. So, $-3x^2 + 8x - 4$ is a quadratic (2nd-degree) polynomial in one variable.

2. The term $5x^3$ has degree 3, which is larger than the degrees of the other terms. So, $5x^3 - 2x^2 + 9x + 1$ is a cubic (3rd-degree) polynomial in one variable.

3. The term $6a^5b^2$ has degree 7 (the sum of the exponents of the variables), which is larger than the degrees of the other terms. So, $6a^5b^2 - 9a^3b^3 - 2ab^4$ is a 7th-degree polynomial in two variables. ■

Combining Like Terms

The **coefficient** of a term is the constant factor of the term. For the term $-7x^3$, the coefficient is -7. For the term x^2, the coefficient is 1, because $x^2 = 1x^2$. The coefficients of a polynomial are the coefficients of the variable terms. For example, the coefficients of the polynomial $3x^3 + 6x^2 - 4x - 9$ are 3, 6, -4, and -9. The **leading coefficient** of a polynomial is the coefficient of the term with the largest degree. For $3x^3 + 6x^2 - 4x - 9$, the leading coefficient is 3.

Like terms are either constant terms or variable terms that contain the same variable(s) raised to exactly the same power(s). For example, the terms $2x^3y^5$ and $7x^3y^5$ are like terms, because both terms have an x with the exponent 3 and a y with the exponent 5. The terms $6x^5$ and $8x^2$ are **unlike terms** (not like terms), because the exponents of x are different.

We can combine like terms by using the distributive law. For example,

$$2x^3 + 4x^3 = (2 + 4)x^3 = 6x^3$$

Note that we can find $2x^3 + 4x^3$ in one step by adding the coefficients:

$$2x^3 + 4x^3 = 6x^3$$
$$\downarrow \quad \downarrow \quad \uparrow$$
$$2 + 4 = 6$$

Similarly, we can find $7x^2 - 3x^2$ by adding the coefficients:

$$7x^2 - 3x^2 = 4x^2$$
$$\downarrow \quad \downarrow \quad \uparrow$$
$$7 + (-3) = 4$$

However, we can't add the coefficients in $2x^4 + 5x^3$, because $2x^4$ and $5x^3$ are not like terms. There is no helpful way to use the distributive law for unlike terms.

Combining Like Terms

To combine like terms, add the coefficients of the terms.

Example 2 Combining Like Terms

Combine like terms when possible.

1. $5x^3 - 4x^2 + 2x^3 - x^2$
2. $3p^3t^2 + p^2t - 8p^3t^2 + 7p^2t$

Solution

1. We rearrange the terms so that the terms with x^3 are adjacent and the terms with x^2 are adjacent:

$$\begin{aligned} 5x^3 - 4x^2 + 2x^3 - x^2 &= 5x^3 + 2x^3 - 4x^2 - x^2 && \text{Rearrange terms.} \\ &= 7x^3 - 5x^2 && \text{Combine like terms.} \end{aligned}$$

2. $3p^3t^2 + p^2t - 8p^3t^2 + 7p^2t$

$$\begin{aligned} &= 3p^3t^2 - 8p^3t^2 + p^2t + 7p^2t && \text{Rearrange terms.} \\ &= -5p^3t^2 + 8p^2t && \text{Combine like terms.} \end{aligned}$$

■

Addition of Polynomials

Now that we know how to combine like terms, we can add polynomials.

Adding Polynomials

To add polynomials, combine like terms.

Example 3 Adding Polynomials

Find the sum $(5a^2 - 7ab + 3b^2) + (2a^2 - 4ab - 9b^2)$.

Solution

$$\begin{aligned} &(5a^2 - 7ab + 3b^2) + (2a^2 - 4ab - 9b^2) \\ &\quad = 5a^2 + 2a^2 - 7ab - 4ab + 3b^2 - 9b^2 && \text{Rearrange terms.} \\ &\quad = 7a^2 - 11ab - 6b^2 && \text{Combine like terms.} \end{aligned}$$

■

Subtraction of Polynomials

We can use the fact that $a - b = a - 1b$ to help us subtract two polynomials.

Example 4 Subtracting Polynomials

Find the difference $(7x^2 - 2x + 5) - (6x^2 - 4x + 1)$.

Solution

To begin, we write $(7x^2 - 2x + 5) - (6x^2 - 4x + 1)$ as $(7x^2 - 2x + 5) - 1(6x^2 - 4x + 1)$ and distribute the -1:

$$\begin{aligned} &(7x^2 - 2x + 5) - (6x^2 - 4x + 1) \\ &\quad = (7x^2 - 2x + 5) - 1(6x^2 - 4x + 1) && a - b = a - 1b \\ &\quad = 7x^2 - 2x + 5 - 6x^2 + 4x - 1 && \text{Distributive law} \\ &\quad = 7x^2 - 6x^2 - 2x + 4x + 5 - 1 && \text{Rearrange terms.} \\ &\quad = x^2 + 2x + 4 && \text{Combine like terms.} \end{aligned}$$

We use a graphing calculator table to verify the work (see Fig. 1). ■

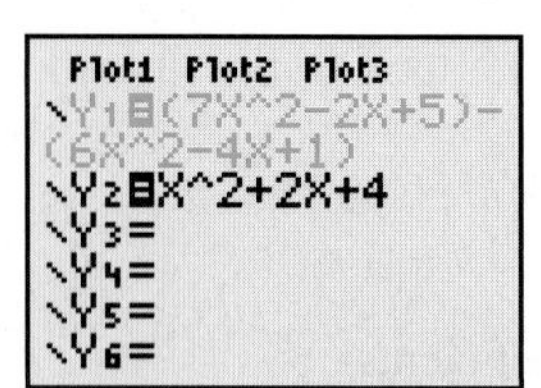

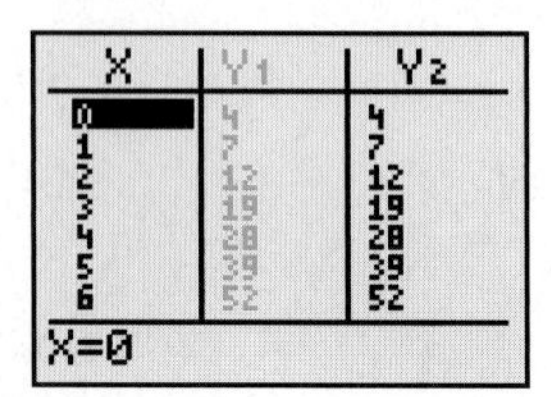

Figure 1 Verify the work

Subtracting Polynomials

To subtract polynomials, first distribute -1; then combine like terms.

Now that we are familiar with polynomial expressions, we turn our attention to *polynomial functions.*

Quadratic Functions

The following are examples of *polynomial functions:*

$$f(x) = 3x^5 - 9x^4 + 8x^2 + 1 \qquad g(x) = 4x^2 - 3x + 5 \qquad h(x) = \frac{2}{3}x$$

A **polynomial function** is a function whose equation can be put into the form $f(x) = P$, where P is a polynomial in terms of the variable x. If P is a quadratic (2nd-degree) polynomial, we call the function a *quadratic function.*

DEFINITION Quadratic function

A **quadratic function** is a function whose equation can be put into the form

$$f(x) = ax^2 + bx + c$$

where $a \neq 0$. This form is called the **standard form.**

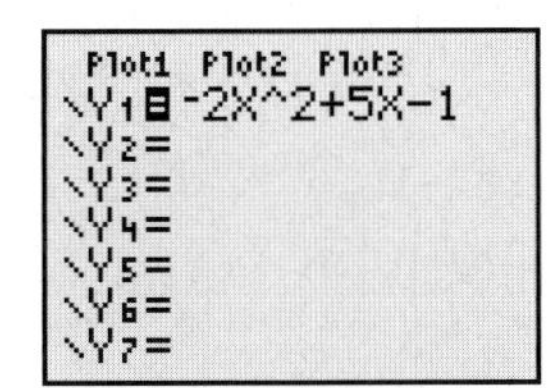

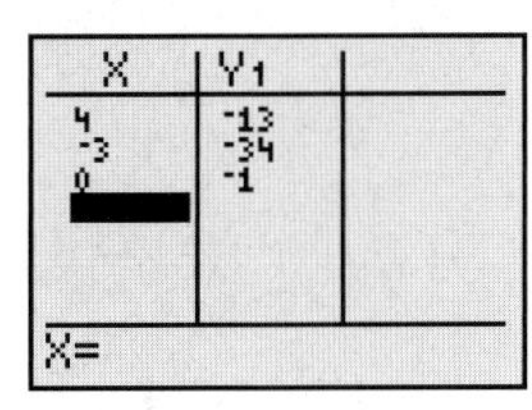

Figure 2 Verify the work

For example, the function $f(x) = 8x^2 - 2x + 7$ is quadratic, because it is in the form $f(x) = ax^2 + bx + c$, with $a = 8$, $b = -2$, and $c = 7$.

Example 5 Evaluating a Quadratic Function

For $f(x) = -2x^2 + 5x - 1$, find the following.

1. $f(4)$ 2. $f(-3)$ 3. $f(0)$

Solution

1. $f(4) = -2(4)^2 + 5(4) - 1 = -2(16) + 5(4) - 1 = -13$
2. $f(-3) = -2(-3)^2 + 5(-3) - 1 = -2(9) + 5(-3) - 1 = -34$
3. $f(0) = -2(0)^2 + 5(0) - 1 = -2(0) + 5(0) - 1 = -1$

We use a graphing calculator table to verify our work (see Fig. 2). ■

The function $f(x) = x^2$ is quadratic, because it is in the form $f(x) = ax^2 + bx + c$, with $a = 1$, $b = 0$, and $c = 0$.

Table 2 Input-Output Pairs of $f(x) = x^2$

x	$f(x)$
-3	$(-3)^2 = 9$
-2	$(-2)^2 = 4$
-1	$(-1)^2 = 1$
0	$0^2 = 0$
1	$1^2 = 1$
2	$2^2 = 4$
3	$3^2 = 9$

Example 6 Graphing a Quadratic Function

Sketch the graph of $f(x) = x^2$.

Solution

First, we list some input–output pairs of $f(x) = x^2$ in Table 2. Then, we plot the corresponding points and sketch a curve that contains the points (see Fig. 3).

We use ZStandard followed by ZSquare on a graphing calculator to verify our graph (see Fig. 4).

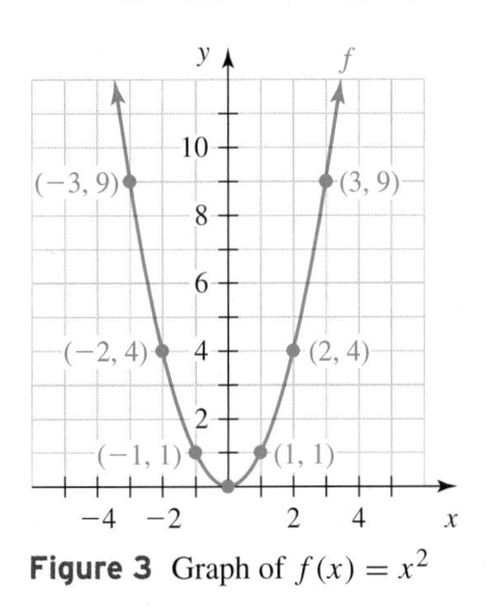

Figure 3 Graph of $f(x) = x^2$

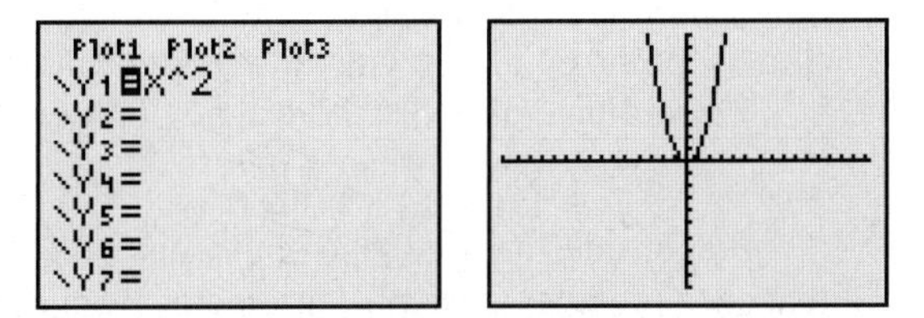

Figure 4 Verify the work ■

The graph of a quadratic function is called a **parabola.** The curve sketched in Fig. 3 is a parabola. Two more examples of parabolas are sketched in Figs. 5 and 6.

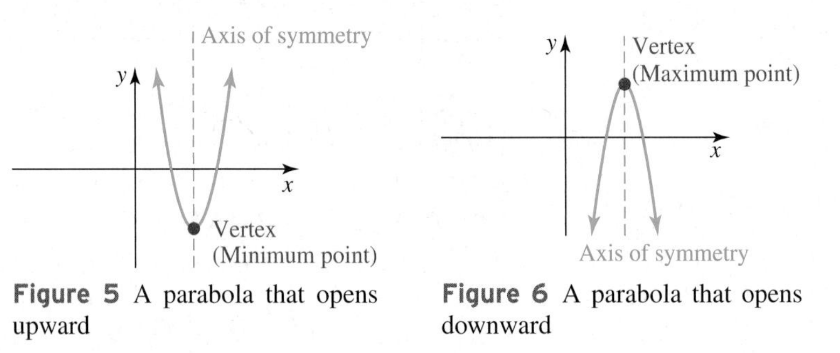

Figure 5 A parabola that opens upward

Figure 6 A parabola that opens downward

There is a difference between parabolas and lines that is worth noting. The lowest point of a parabola that *opens upward* (see Fig. 5) is called the **minimum point.** The highest point of a parabola that *opens downward* (see Fig. 6) is called the **maximum point.** The minimum point or maximum point of a parabola is called the **vertex** of the parabola. In contrast, lines do not have a lowest or highest point.

The vertical line that passes through a parabola's vertex is called the **axis of symmetry** (see Figs. 5 and 6). The part of the parabola that lies to the left of the axis of symmetry is the mirror reflection of the part that lies to the right.

Example 7 Using a Graph to Find Values of a Function

A graph of a quadratic function f is sketched in Fig. 7.

1. Find $f(4)$.
2. Find x when $f(x) = -3$.
3. Find x when $f(x) = 5$.
4. Find x when $f(x) = 6$.

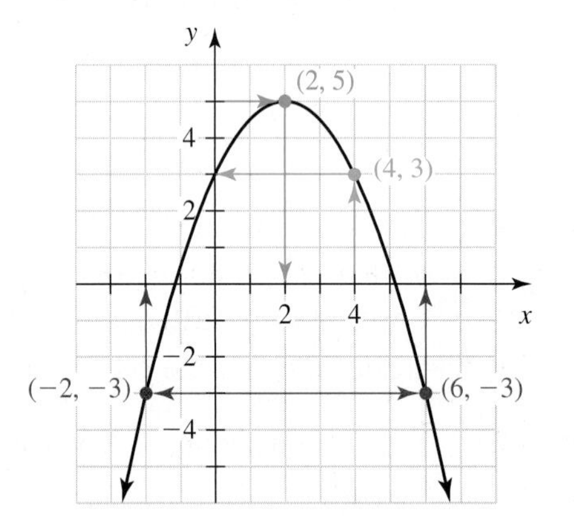

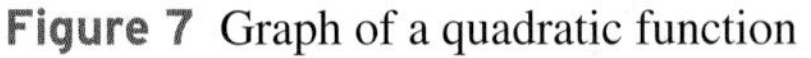

Figure 7 Graph of a quadratic function

Solution

1. The blue arrows in Fig. 7 show that the input $x = 4$ leads to the output $y = 3$. So, $f(4) = 3$.
2. The red arrows in Fig. 7 show that the output $y = -3$ originates from the two inputs $x = -2$ and $x = 6$. That is, $f(-2) = -3$ and $f(6) = -3$. So, the values of x are -2 and 6 when $f(x) = -3$.
3. The green arrows in Fig. 7 show that the output $y = 5$ originates from the single input $x = 2$. That is, $f(2) = 5$. So, the value of x is 2 when $f(x) = 5$. (There is a single input because the vertex (2, 5) is the only point on the parabola that has a y-coordinate equal to 5.)
4. No point of the downward-opening parabola is above the vertex, which has a y-coordinate of 5. So, there is no point on the parabola with $f(x) = 6$. ■

By considering the shape of any parabola that opens upward or downward, we see that each (output) value of y originates from either two, one, or no (input) values of x.

Cubic Function

If a polynomial function can be put into the form $f(x) = P$, where P is a cubic (3rd-degree) polynomial, we call the function a *cubic function.*

> **DEFINITION** Cubic function
>
> A **cubic function** is a function whose equation can be put into the form
>
> $$f(x) = ax^3 + bx^2 + cx + d$$
>
> where $a \neq 0$.

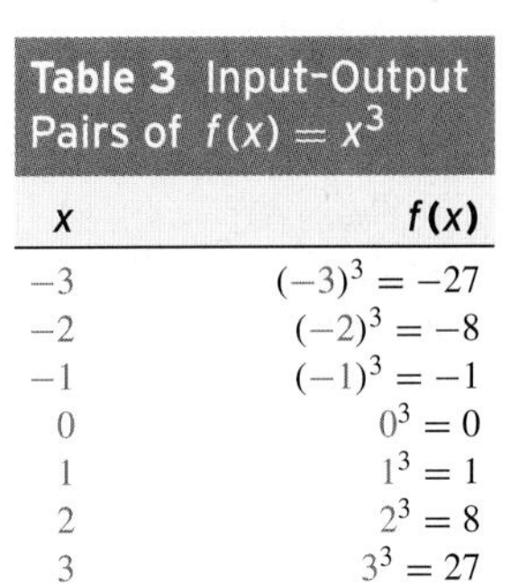

Table 3 Input-Output Pairs of $f(x) = x^3$

x	$f(x)$
-3	$(-3)^3 = -27$
-2	$(-2)^3 = -8$
-1	$(-1)^3 = -1$
0	$0^3 = 0$
1	$1^3 = 1$
2	$2^3 = 8$
3	$3^3 = 27$

Here are some examples of cubic functions:

$$f(x) = -5x^3 + 9x^2 - 7x + 2 \quad g(x) = 8x^3 - 5x + 1 \quad h(x) = x^3 - 4x^2$$

Example 8 Graphing a Cubic Function

Sketch the graph of $f(x) = x^3$.

Solution

First, we list some input–output pairs of $f(x) = x^3$ in Table 3. Then, we plot the corresponding points and sketch a curve that contains the points (see Fig. 8). ■

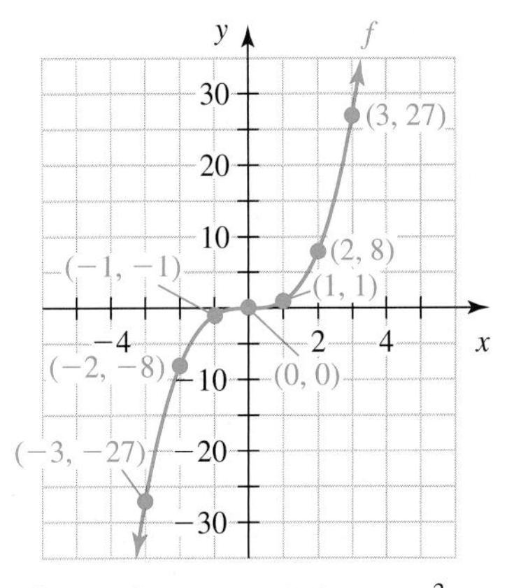

Figure 8 Graph of $f(x) = x^3$

Four graphs of typical cubic functions are shown in Fig. 9.

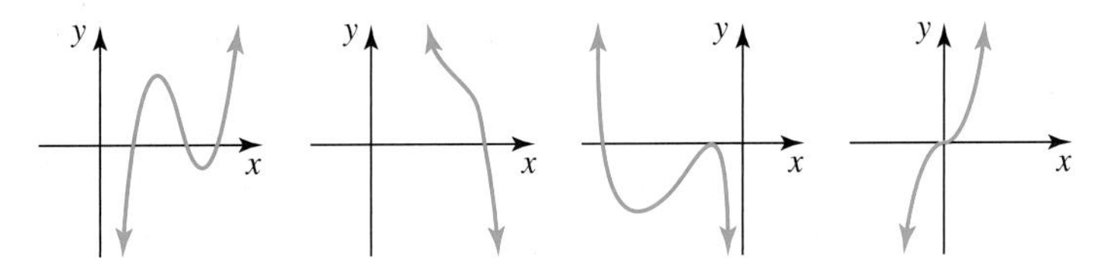

Figure 9 Graphs of typical cubic functions

Sum Function and Difference Function

We can add two functions or subtract two functions to form a "new" function.

> **DEFINITION** Sum function, difference function
>
> If f and g are functions and x is in the domain of both functions, then we can form the following functions:
>
> - **Sum function** $f + g$, where $(f + g)(x) = f(x) + g(x)$
> - **Difference function** $f - g$, where $(f - g)(x) = f(x) - g(x)$

For $f(x) = 5x$ and $g(x) = 3x$, we have $(f+g)(x) = f(x) + g(x) = 5x + 3x = 8x$. So, $(f + g)(x) = 8x$. And $(f - g)(x) = f(x) - g(x) = 5x - 3x = 2x$. So, $(f - g)(x) = 2x$.

Example 9 Finding a Sum Function and a Difference Function

Let $f(x) = 5x^2 - x + 3$ and $g(x) = -2x^2 + 9x - 7$.

1. Find an equation of the sum function $f + g$.
2. Find $(f + g)(2)$.
3. Find an equation of the difference function $f - g$.
4. Find $(f - g)(2)$.

Solution

1. $(f+g)(x) = f(x) + g(x)$ — Definition of sum function

$= (5x^2 - x + 3) + (-2x^2 + 9x - 7)$ — Substitute $5x^2 - x + 3$ for $f(x)$ and $-2x^2 + 9x - 7$ for $g(x)$.

$= 5x^2 - 2x^2 - x + 9x + 3 - 7$ — Rearrange terms.

$= 3x^2 + 8x - 4$ — Combine like terms.

2. $(f+g)(2) = 3(2)^2 + 8(2) - 4 = 24$

3. $(f-g)(x) = f(x) - g(x)$ — Definition of difference function

$= (5x^2 - x + 3) - (-2x^2 + 9x - 7)$ — Substitute $5x^2 - x + 3$ for $f(x)$ and $-2x^2 + 9x - 7$ for $g(x)$.

$= (5x^2 - x + 3) - 1(-2x^2 + 9x - 7)$ — $a - b = a - 1b$

$= 5x^2 - x + 3 + 2x^2 - 9x + 7$ — Distributive law

$= 5x^2 + 2x^2 - x - 9x + 3 + 7$ — Rearrange terms.

$= 7x^2 - 10x + 10$ — Combine like terms.

4. $(f-g)(2) = 7(2)^2 - 10(2) + 10 = 18.$ ■

Modeling with Sum Functions and Difference Functions

Suppose that A and B represent quantities. Then $A + B$ represents the sum of the two quantities.

The difference $A - B$ tells us how much more there is of one quantity than the other quantity. For example, suppose that $A = 8$ and $B = 5$. Then $A - B = 8 - 5 = 3$ tells us that A is 3 more than B. Now suppose that $A = 5$ and $B = 8$. Then $A - B = 5 - 8 = -3$ tells us that A is 3 less than B.

The Meaning of the Sign of a Difference

If a difference $A - B$ is positive, then A is more than B. If a difference $A - B$ is negative, then A is less than B.

In Example 10, we use a sum function and a difference function to describe an authentic situation.

Example 10 Using a Sum Function and a Difference Function to Model a Situation

In Exercise 29 of Homework 3.1, the enrollments (in millions) at U.S. colleges $W(t)$ and $M(t)$ for women and men, respectively, are modeled by the system

$$W(t) = 0.15t + 5.76$$
$$M(t) = 0.072t + 5.43$$

where t is the number of years since 1980 (see Table 4).

Table 4 College Enrollments

	Enrollment (millions)	
Year	**Women**	**Men**
1985	6.4	5.8
1990	7.5	6.3
1995	7.9	6.3
2000	8.6	6.7
2003	9.3	7.3

Source: *U.S. Census Bureau*

1. Find an equation of the sum function $W + M$.
2. Perform a unit analysis of the expression $W(t) + M(t)$.
3. Find $(W + M)(31)$. What does it mean in this situation?
4. Find an equation of the difference function $W - M$.
5. Find $(W - M)(31)$. What does it mean in this situation?

Solution

1. $(W + M)(t) = W(t) + M(t)$ — Definition of sum function

$= (0.15t + 5.76) + (0.072t + 5.43)$ — Substitute $0.15t + 5.76$ for $W(t)$ and $0.072t + 5.43$ for $M(t)$.

$= 0.222t + 11.19$ — Combine like terms.

2. For the expression $W(t) + M(t)$, we have

$$\underbrace{W(t)}_{\text{millions of female students}} \quad + \quad \underbrace{M(t)}_{\text{millions of male students}}$$

The units of the expression are millions of students.

3. $(W + M)(31) = 0.222(31) + 11.19 = 18.072$

This means that the total enrollment for women and men in $1980 + 31 = 2011$ will be 18.072 million students, according to the model.

4. $(W - M)(t) = W(t) - M(t)$ — Definition of difference function

$= (0.15t + 5.76) - 1(0.072t + 5.43)$ — Substitute $0.15t + 5.76$ for $W(t)$ and $0.072t + 5.43$ for $M(t)$; $a - b = a - 1b$

$= 0.15t + 5.76 - 0.072t - 5.43$ — Distributive law

$= 0.078t + 0.33$ — Combine like terms.

5. $(W - M)(31) = 0.078(31) + 0.33 = 2.748$

This means that in 2011 women's enrollment will exceed men's enrollment by 2.748 million students, according to the models. ■

group exploration

Using a difference function to solve a system

1. Let $f(x) = -3x + 7$ and $g(x) = 2x - 3$.
   a. Graph by hand the functions f and g on the same coordinate system.
   b. Find an equation of the difference function $f - g$.
   c. Find $(f - g)(1)$. Refer to your graphs in part (a) to explain why your result is positive.
   d. Find $(f - g)(3)$. Refer to your graphs in part (a) to explain why your result is negative.
   e. Find $(f - g)(2)$. Refer to your graphs in part (a) to explain why your result is 0.
   f. Refer to your graphs in part (a) to solve the system

$$y = -3x + 7$$
$$y = 2x - 3$$

   Explain why your work in part (e) shows that the x-coordinate of the solution of the system is 2.

2. Let $f(x) = -x + 5$ and $g(x) = 2x - 4$.
   a. Find an equation of the difference function $f - g$.
   b. Find x when $(f - g)(x) = 0$.
   c. Without using graphing, substitution, or elimination, state the x-coordinate of the solution of the system that follows. Then use any method to find the y-coordinate of the solution.

$$y = -x + 5$$
$$y = 2x - 4$$

TIPS FOR SUCCESS: Complete the Rest of the Assignment

If you have spent a good amount of time trying to solve an exercise but can't, consider going on to the next exercise in the assignment. The next exercise may involve a different concept or a more familiar situation. After completing the rest of the assignment, you may be able to complete the exercise(s) you skipped. One explanation: You may have learned or remembered some concept in a later exercise that relates to the exercise with which you were struggling.

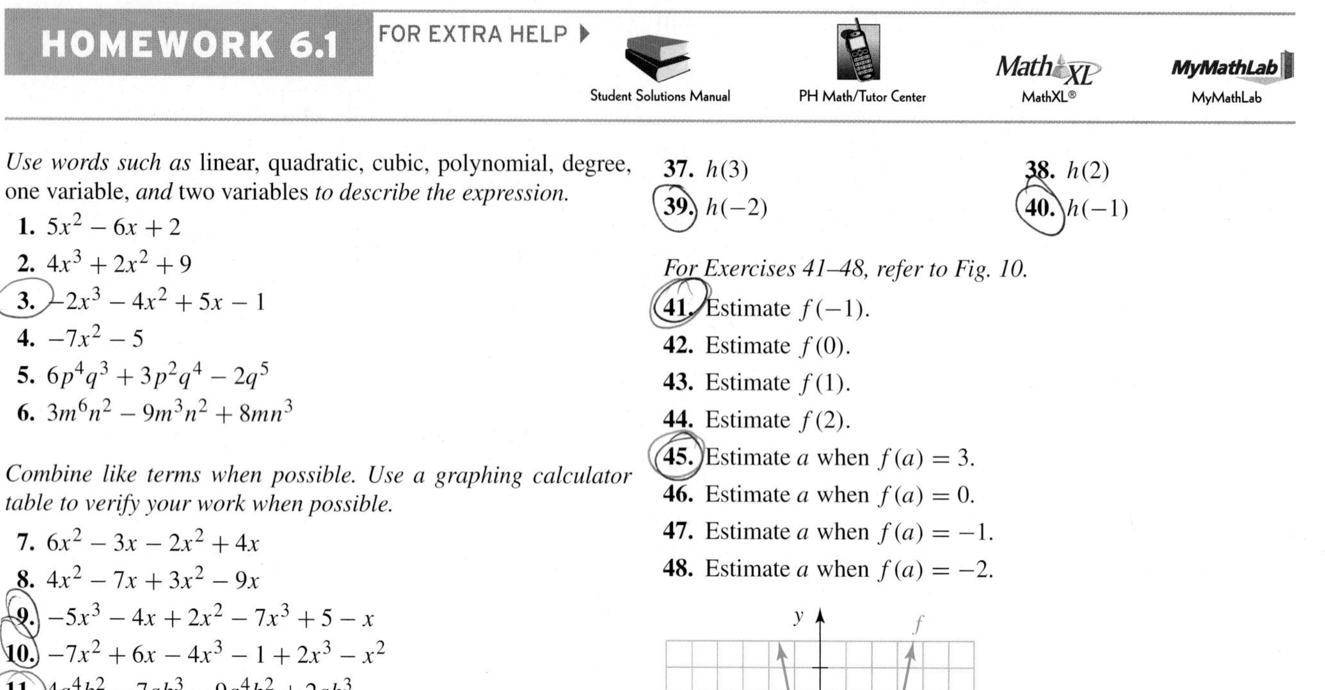

HOMEWORK 6.1

FOR EXTRA HELP ▶ Student Solutions Manual · PH Math/Tutor Center · MathXL® · MyMathLab

Use words such as linear, quadratic, cubic, polynomial, degree, one variable, *and* two variables *to describe the expression.*

1. $5x^2 - 6x + 2$
2. $4x^3 + 2x^2 + 9$
3. $-2x^3 - 4x^2 + 5x - 1$
4. $-7x^2 - 5$
5. $6p^4q^3 + 3p^2q^4 - 2q^5$
6. $3m^6n^2 - 9m^3n^2 + 8mn^3$

Combine like terms when possible. Use a graphing calculator table to verify your work when possible.

7. $6x^2 - 3x - 2x^2 + 4x$
8. $4x^2 - 7x + 3x^2 - 9x$
9. $-5x^3 - 4x + 2x^2 - 7x^3 + 5 - x$
10. $-7x^2 + 6x - 4x^3 - 1 + 2x^3 - x^2$
11. $4a^4b^2 - 7ab^3 - 9a^4b^2 + 2ab^3$
12. $2m^3n + 4mn^2 - 8m^3n - 7mn^2$
13. $2x^4 - 4x^3y + 2x^2y^2 + x^3y - 2x^2y^2 + xy^3$
14. $3r^4 - r^3t - 4r^2t^2 + 6rt^2 + 4r^2t^2 - 4rt^3$

Perform the addition. Use a graphing calculator table to verify your work when possible.

15. $(3x^2 - 5x - 2) + (6x^2 + 2x - 7)$
16. $(5x^2 + 3x - 6) + (-7x^2 - 5x + 1)$
17. $(-2x^3 + 4x - 3) + (5x^3 - 6x^2 + 2)$
18. $(-5x^3 - 8x^2 + 4) + (-4x^3 + x - 9)$
19. $(8a^2 - 7ab + 2b^2) + (3a^2 + 4ab - 7b^2)$
20. $(6t^2 + 2tw - w^2) + (-4t^2 - 9tw - 3w^2)$
21. $(2m^4p + m^3p^2 - 7m^2p^3) + (m^3p^2 + 7m^2p^3 - 8mp^3)$
22. $(x^3y - 5x^2y^2 + 2xy^3) + (5x^2y^2 - 4xy^3 + 7y^4)$

Perform the subtraction. Use a graphing calculator table to verify your work when possible.

23. $(2x^2 + 4x - 7) - (9x^2 - 5x + 4)$
24. $(6x^2 + 3x - 1) - (4x^2 - 8x + 3)$
25. $(6x^3 - 3x^2 + 4) - (-7x^3 + x - 1)$
26. $(5x^3 - x + 6) - (-9x^3 + 4x^2 - 6)$
27. $(8m^2 + 3mp - 5p^2) - (-2m^2 - 7mp - 4p^2)$
28. $(b^2 - 6bc + 4c^2) - (5b^2 + 3bc - 2c^2)$
29. $(a^3b - 5a^2b^2 + ab^3) - (5a^2b^2 - 7ab^3 + b^3)$
30. $(6x^4y + x^3y^2 - 3x^2y^3) - (4x^3y^2 - 3x^2y^3 - 5xy^4)$

For the functions $f(x) = -2x^2 - 5x + 3$, $g(x) = 3x^2 - 8x - 1$, *and* $h(x) = 2x^3 - 4x$, *find the following.*

31. $f(3)$
32. $g(2)$
33. $g(-4)$
34. $f(-4)$
35. $f(0)$
36. $g(0)$
37. $h(3)$
38. $h(2)$
39. $h(-2)$
40. $h(-1)$

For Exercises 41–48, refer to Fig. 10.

41. Estimate $f(-1)$.
42. Estimate $f(0)$.
43. Estimate $f(1)$.
44. Estimate $f(2)$.
45. Estimate a when $f(a) = 3$.
46. Estimate a when $f(a) = 0$.
47. Estimate a when $f(a) = -1$.
48. Estimate a when $f(a) = -2$.

Figure 10 Exercises 41–48

For Exercises 49–56, refer to Table 5, which lists values of a quadratic function f.

49. Find $f(0)$.
50. Find $f(3)$.
51. Find $f(4)$.
52. Find $f(6)$.
53. Find x when $f(x) = 19$.
54. Find x when $f(x) = 3$.
55. Find x when $f(x) = 1$.
56. Find x when $f(x) = 0$.

Table 5 Some Values of a Quadratic Function f (Exercises 49–57)

x	$f(x)$
0	19
1	9
2	3
3	1
4	3
5	9
6	19

57. Consider the quadratic function f described in Table 5.
    a. Find x when $f(x) = 9$.
    b. Explain why f does not have an inverse function.

58. The values of a quadratic function f are listed in Table 6. Estimate the x-intercept(s) and the y-intercept(s).

Table 6 Values of a Function f (Exercise 58)

x	$f(x)$
−3	5.35
−2	0.17
−1	−3.61
0	−5.99
1	−6.97
2	−6.55
3	−4.73
4	−1.51
5	3.11
6	9.13

Graph the function by hand. To begin, substitute the values −2, −1, 0, 1, and 2 for x. Make other substitutions as necessary. Use a graphing calculator to verify your work.

59. $f(x) = 2x^2$

60. $f(x) = 3x^2$

61. $f(x) = -2x^2$

62. $f(x) = -3x^2$

63. $f(x) = 3x^3$

64. $f(x) = 2x^3$

65. $f(x) = -3x^3$

66. $f(x) = -2x^3$

For $f(x) = 4x^2 - 2x + 8$, $g(x) = 7x^2 + 5x - 1$, *and* $h(x) = -3x^2 - 4x - 9$, *find an equation of the given function; then evaluate the function at the indicated value.*

67. $f + g$, $(f + g)(3)$

68. $f + h$, $(f + h)(3)$

69. $f - h$, $(f - h)(4)$

70. $g - h$, $(g - h)(4)$

For $f(x) = 2x^3 - 4x + 1$, $g(x) = -3x^2 + 5x - 3$, *and* $h(x) = x^3 - 3x^2 + 2x$, *find an equation of the given function; then evaluate the function at the indicated value.*

71. $f + g$, $(f + g)(2)$

72. $f + h$, $(f + h)(2)$

73. $f - h$, $(f - h)(-1)$

74. $g - h$, $(g - h)(-1)$

75. In Exercise 5 of Homework 3.3, the annual U.S. consumption (in gallons per person) $M(t)$ and $S(t)$ of milk and soft drinks, respectively, is modeled by the system

$$M(t) = -0.28t + 35.64$$
$$S(t) = 0.86t + 8.3$$

where t is the number of years since 1950 (see Table 7).

Table 7 Annual U.S. Consumption of Milk and Soft Drinks

	Annual Consumption (gallons per person)	
Year	**Milk**	**Soft Drinks**
1950	36.4	10.8
1960	32.6	13.4
1970	29.8	24.3
1980	26.5	35.1
1990	24.3	46.2
2000	22.6	49.3

Source: *USDA/Economic Research Service*

a. Find an equation of the sum function $M + S$.
b. Perform a unit analysis of the expression $M(t) + S(t)$.
c. Find $(M + S)(60)$. What does it mean in this situation?
d. Find an equation of the difference function $M - S$.
e. Find $(M - S)(60)$. What does it mean in this situation?

76. In Exercise 28 of Homework 3.1, the annual U.S. consumption (in pounds per person) $C(t)$ and $R(t)$ of chicken and red meat, respectively, are modeled by the system

$$C(t) = 0.99t - 45.68$$
$$R(t) = -0.62t + 174.40$$

where t is the number of years since 1900 (see Table 8).

Table 8 Annual U.S. Consumptions of Chicken and Red Meat

	Annual Consumption (pounds per person)	
Year	**Chicken**	**Red Meat**
1970	27.4	131.9
1975	26.3	125.8
1980	32.7	126.4
1985	36.4	124.9
1990	42.4	112.2
1995	48.2	113.6
2000	54.2	113.7
2003	57.5	111.9

Source: *U.S. Department of Agriculture*

a. Find an equation of the sum function $C + R$.
b. Perform a unit analysis of the expression $C(t) + R(t)$.
c. Find $(C + R)(130)$. What does it mean in this situation?
d. Find an equation of the difference function $C - R$.
e. Find $(C - R)(130)$. What does it mean in this situation?

77. The average base pay $B(t)$ (in thousands of dollars) of an MBA graduate is modeled by the function

$$B(t) = 2.3t^2 - 12.1t + 98$$

where t is the number of years since 2000 (see Table 9). An MBA graduate's average signing bonus (in thousands of dollars) is modeled by the function $S(t) = 1.57t^2 - 10.9t + 28$, where t is the number of years since 2000.

Table 9 MBA Average Base Pay and Average Signing Bonuses

	Average Pay/Bonus (thousands of dollars)	
Year	**Base Pay**	**Signing Bonus**
2001	88.6	18.9
2002	82.0	13.0
2003	80.9	8.9
2004	88.8	10.6
2005	93.8	12.8

Source: *Graduate Management Admission Council*

a. Find an equation of the sum function $B + S$.
b. Perform a unit analysis of the expression $B(t) + S(t)$.
c. Find $(B + S)(11)$. What does it mean in this situation?
d. Find an equation of the difference function $B - S$.
e. Find $(B - S)(11)$. What does it mean in this situation?

78. Mattel's® sales (in billions of dollars) $N(t)$ of toys other than Barbie® can be modeled by the function $N(t) = 0.17t + 3$, where t is the number of years since 2000 (see Table 10). Mattel's Barbie sales (in billions of dollars) $B(t)$ can be modeled by the function $B(t) = -0.036t^2 + 0.176t + 1.28$, where t is the number of years since 2000.

Table 10 Mattel's Non-Barbie Sales and Barbie Sales

	Sales (billions of dollars)	
Year	**Non-Barbie**	**Barbie**
2001	3.18	1.41
2002	3.35	1.50
2003	3.41	1.45
2004	3.66	1.41
2005	3.88	1.24

Source: *Harris Nesbitt*

a. Find an equation of the sum function $N + B$.
b. Perform a unit analysis of the expression $N(t) + B(t)$.
c. Find $(N + B)(8)$. What does it mean in this situation?
d. Find an equation of the difference function $N - B$.
e. Find $(N - B)(8)$. What does it mean in this situation?
f. Find $(N - B)(11)$. What does it mean in this situation? Explain why model breakdown has likely occurred. [**Hint:** Find $B(11)$.]

79. A student tries to find the difference of the polynomials $6x^2 + 8x + 5$ and $2x^2 + 4x + 3$:

$$\begin{aligned}&(6x^2 + 8x + 5) - (2x^2 + 4x + 3)\\ &= 6x^2 + 8x + 5 - 2x^2 + 4x + 3\\ &= 4x^2 + 12x + 8\end{aligned}$$

Describe any errors. Then find the difference correctly.

80. Let f and g be functions, both with independent variable x. A student says that $(f + g)(x) = f(x) + g(x)$ as a result of the distributive law. Explain why the student is incorrect. Then create two functions f and g so that you can show the meaning of $(f + g)(x) = f(x) + g(x)$.

81. Let $f(x) = 3x + 7$ and $g(x) = 5x + 2$.
a. Find equations of the difference function $f - g$ and the difference function $g - f$.
b. Find $(f - g)(2)$ and $(g - f)(2)$. Compare the results.
c. Find $(f - g)(4)$ and $(g - f)(4)$. Compare the results.
d. Find $(f - g)(7)$ and $(g - f)(7)$. Compare the results.
e. Summarize your findings from parts (b)–(d). Explain why this makes sense.

82. a. Is it possible for the sum of two quadratic polynomials to be the given type of polynomial? If yes, give an example. If no, explain.
i. cubic
ii. quadratic
iii. linear
iv. constant
b. Summarize the possible results for the sum of two quadratic polynomials.

83. Describe how to add two polynomials. Describe how to subtract two polynomials. (See page 4 for guidelines on writing a good response.)

84. Explain how it is sometimes useful to use a sum function or difference function to model an authentic situation. (See page 4 for guidelines on writing a good response.)

Related Review

Match the given type of function to the appropriate graph in Fig. 11.

85. $f(x) = mx + b$, $m < 0$ and $b > 0$

86. $f(x) = mx + b$, $m > 0$ and $b < 0$

87. $f(x) = ax^2 + bx + c$

88. $f(x) = ax^3 + bx^2 + cx + d$

89. $f(x) = ab^x$, $a > 0$ and $b > 1$

90. $f(x) = ab^x$, $a > 0$ and $0 < b < 1$

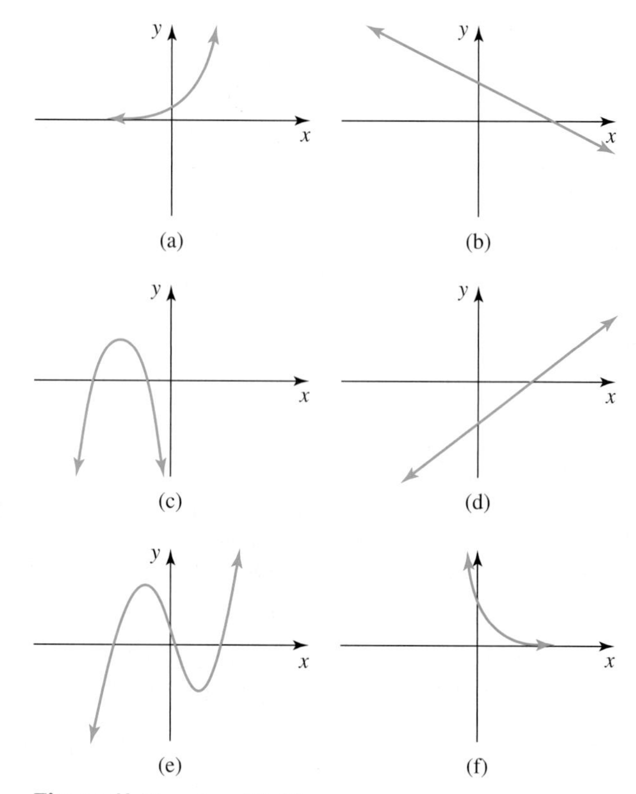

Figure 11 Exercises 85–90

Expressions, Equations, Functions, and Graphs

Perform the indicated instruction. Then use words such as linear, quadratic, cubic, exponential, logarithmic, polynomial, degree, function, one variable, *and* two variables *to describe the expression, equation, or system.*

91. Simplify $\left(\dfrac{25b^5c^{-7}}{4b^{-3}c}\right)^{1/2}$.

92. Solve $7(3)^x - 51 = 83$. Round any solutions to the fourth decimal place.

93. Graph $f(x) = -2(2)^x$ by hand.

94. Solve $7(3x) - 51 = 83$. Round any solutions to the fourth decimal place.

6.2 MULTIPLYING POLYNOMIAL EXPRESSIONS AND FUNCTIONS

Objectives

- Know the meaning of *binomial* and *trinomial.*
- Multiply polynomials.
- *Simplify the square of a binomial.*
- Find the product of two *binomial conjugates.*
- Evaluate quadratic functions, and write them in standard form.
- Find *product functions.*
- Use product functions to describe authentic situations.

In Section 6.1, we added and subtracted polynomials and worked with sum functions and difference functions. In this section, we will multiply polynomials and work with *product functions*.

Monomials, Binomials, and Trinomials

We refer to a polynomial as a monomial, a **binomial,** or a **trinomial,** depending on whether it has one, two, or three nonzero terms, respectively:

Name	Examples	Meaning
monomial	$4x^7$, x^3, $-9x^2y^3$, -3	one nonzero term
binomial	$4x^2-9$, $5x^3+x^2$, $-2y+5$	two nonzero terms
trinomial	$5x^3-3x+6$, $3x^2+7x+2$	three nonzero terms

Multiplication of Monomials

We can use the product property for exponents (Section 4.1) to help us multiply monomials.

Example 1 Finding the Product of Two Monomials

Find the product $4x^3y^2(-2xy^6)$.

Solution

We rearrange factors so that the coefficients are adjacent, the powers of x are adjacent, and the powers of y are adjacent:

$$\begin{aligned} 4x^3y^2(-2xy^6) &= 4(-2)(x^3x^1)(y^2y^6) && \text{Rearrange factors; } x = x^1 \\ &= -8x^4y^8 && \text{Multiply; add exponents: } b^mb^n = b^{m+n} \end{aligned}$$

■

WARNING An expression such as $4x^3y^2(-2xy^6)$ is *not* the same as an expression such as $4x^3y^2 - 2xy^6$: The first expression is a product, whereas the second is a difference. We can tell that $4x^3y^2(-2xy^6)$ is a product because there is no operation symbol between the expressions $4x^3y^2$ and $(-2xy^6)$.

Multiplication of a Monomial and a Polynomial

We can use the distributive law to help us find the product of a monomial and a polynomial.

Example 2 Finding the Product of a Monomial and a Polynomial

Find the product.

1. $-4x(6x+3)$

2. $7p^2t(3p^2t-2pt^2+5t^3)$

Solution

1. $-4x(6x+3) = -4x \cdot 6x - 4x \cdot 3$ Distributive law
$= -24x^2 - 12x$ $x \cdot x = x^2$

2. $7p^2t(3p^2t - 2pt^2 + 5t^3)$
$= 7p^2t \cdot 3p^2t - 7p^2t \cdot 2pt^2 + 7p^2t \cdot 5t^3$ Distributive law
$= 21p^4t^2 - 14p^3t^3 + 35p^2t^4$ Add exponents: $b^m b^n = b^{m+n}$ ■

Multiplication of Two Polynomials

We can also use the distributive law to help us find the product of two binomials. For instance, we can find the product $(a+b)(c+d)$ by using the distributive law three times:

$$(a+b)(c+d) = a(c+d) + b(c+d)$$ Distribute $(c+d)$.
$$= ac + ad + bc + bd$$ Distribute a and distribute b.

By examining the expression $ac + ad + bc + bd$, we can find the product $(a+b)(c+d)$ more directly by adding the four products formed by multiplying each term in the first sum by each term in the second sum:

$$(a+b)(c+d) = ac + ad + bc + bd$$

ad, ac, bc, bd

After using this technique, we combine like terms if possible.

Multiplying Two Polynomials

To multiply two polynomials, multiply each term in the first polynomial by each term in the second polynomial. Then combine like terms if possible.

Example 3 Finding the Product of Two Binomials

Find the product.

1. $(x+4)(x+8)$ **2.** $(5p-6w)(3p+2w)$

Solution

1. $(x+4)(x+8) = x \cdot x + x \cdot 8 + 4 \cdot x + 4 \cdot 8$ Multiply pairs of terms.
$= x^2 + 8x + 4x + 32$ $x \cdot x = x^2$
$= x^2 + 12x + 32$ Combine like terms.

2. $(5p-6w)(3p+2w)$ Multiply pairs of terms.
$= 5p \cdot 3p + 5p \cdot 2w - 6w \cdot 3p - 6w \cdot 2w$
$= 15p^2 + 10pw - 18pw - 12w^2$ $x \cdot x = x^2$
$= 15p^2 - 8pw - 12w^2$ Combine like terms. ■

Example 4 Finding Products

Find the product.

1. $(2a^2 - 5b^2)(4a^2 - 3b^2)$ **2.** $4x(x^2+2)(x-3)$

Solution

1. $(2a^2 - 5b^2)(4a^2 - 3b^2)$
$= 2a^2 \cdot 4a^2 - 2a^2 \cdot 3b^2 - 5b^2 \cdot 4a^2 + 5b^2 \cdot 3b^2$ Multiply pairs of terms.
$= 8a^4 - 6a^2b^2 - 20a^2b^2 + 15b^4$ Add exponents: $b^m b^n = b^{m+n}$
$= 8a^4 - 26a^2b^2 + 15b^4$ Combine like terms.

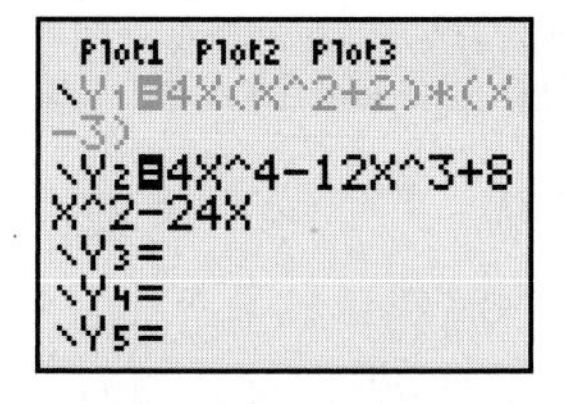

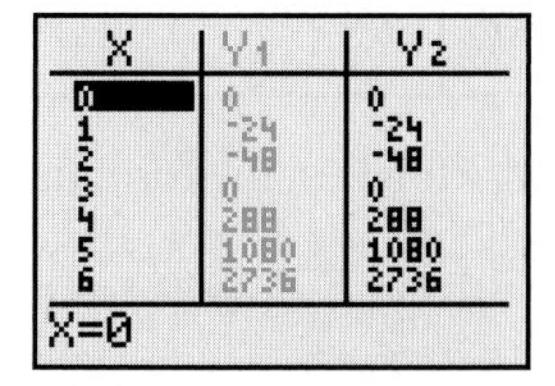

X	Y1	Y2
0	0	0
1	-24	-24
2	-48	-48
3	0	0
4	288	288
5	1080	1080
6	2736	2736

X=0

Figure 12 Verify the work

2. We begin by multiplying $4x$ by both terms in the binomial x^2+2:

$$\begin{aligned} 4x(x^2+2)(x-3) &= (4x\cdot x^2 + 4x\cdot 2)(x-3) && \text{Distributive law} \\ &= (4x^3+8x)(x-3) && \text{Add exponents: } b^m b^n = b^{m+n} \\ &= 4x^3\cdot x - 4x^3\cdot 3 + 8x\cdot x - 8x\cdot 3 && \text{Multiply pairs of terms.} \\ &= 4x^4 - 12x^3 + 8x^2 - 24x && \text{Add exponents: } b^m b^n = b^{m+n} \end{aligned}$$

We use a graphing calculator table to verify our work (see Fig. 12). ■

We can find the product of two polynomials of any degree in a similar fashion. The key is to multiply each term in the first polynomial by each term in the second polynomial. Then combine like terms.

Example 5 Finding the Products of Two Polynomials

Find the product.

1. $(2x+y)(5x^2-3xy+4y^2)$

2. $(x^2-3x+2)(x^2+x-5)$

Solution

1. To begin, we multiply each term in the first polynomial by each term in the second polynomial:

$$\begin{aligned} &(2x+y)(5x^2-3xy+4y^2) && \text{Multiply pairs of terms.} \\ &= 2x\cdot 5x^2 - 2x\cdot 3xy + 2x\cdot 4y^2 + y\cdot 5x^2 - y\cdot 3xy + y\cdot 4y^2 \\ &= 10x^3 - 6x^2y + 8xy^2 + 5x^2y - 3xy^2 + 4y^3 && \text{Add exponents: } b^m b^n = b^{m+n} \\ &= 10x^3 - 6x^2y + 5x^2y + 8xy^2 - 3xy^2 + 4y^3 && \text{Rearrange terms.} \\ &= 10x^3 - x^2y + 5xy^2 + 4y^3 && \text{Combine like terms.} \end{aligned}$$

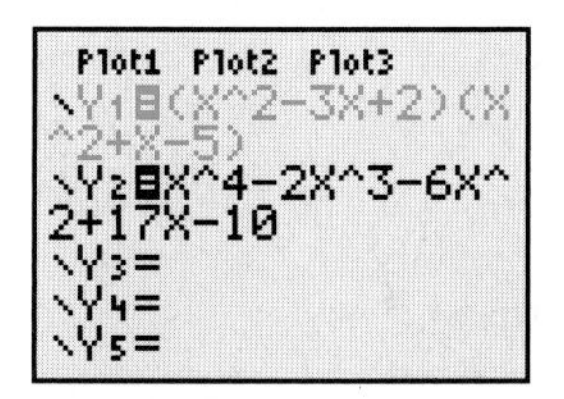

X	Y1	Y2
0	-10	-10
1	0	0
2	0	0
3	14	14
4	90	90
5	300	300
6	740	740

X=0

Figure 13 Verify the work

2. $(x^2-3x+2)(x^2+x-5)$

$$\begin{aligned} &= x^2\cdot x^2 + x^2\cdot x - x^2\cdot 5 - 3x\cdot x^2 - 3x\cdot x + 3x\cdot 5 + 2\cdot x^2 + 2\cdot x - 2\cdot 5 \\ &= x^4 + x^3 - 5x^2 - 3x^3 - 3x^2 + 15x + 2x^2 + 2x - 10 \\ &= x^4 + x^3 - 3x^3 - 5x^2 - 3x^2 + 2x^2 + 15x + 2x - 10 \\ &= x^4 - 2x^3 - 6x^2 + 17x - 10 \end{aligned}$$

We use a graphing calculator table to verify our work (see Fig. 13). ■

Square of a Binomial

How do we square a sum? The square of the binomial $x+3$ is $(x+3)^2$:

$$\begin{aligned} (x+3)^2 &= (x+3)(x+3) && y^2 = yy \\ &= x^2 + 3x + 3x + 9 && \text{Multiply pairs of terms.} \\ &= x^2 + 6x + 9 && \text{Combine like terms.} \end{aligned}$$

We have simplified $(x+3)^2$ by writing it as x^2+6x+9. We **simplify the square of a binomial** by writing it as an expression that does not have parentheses.

Now, we generalize and simplify $(A+B)^2$:

$$\begin{aligned} (A+B)^2 &= (A+B)(A+B) && y^2 = yy \\ &= A^2 + AB + BA + B^2 && \text{Multiply pairs of terms.} \\ &= A^2 + 2AB + B^2 && \text{Combine like terms.} \end{aligned}$$

So, $(A+B)^2 = A^2+2AB+B^2$. We can use similar steps to find the property $(A-B)^2 = A^2-2AB+B^2$.

Squaring a Binomial

$$(A + B)^2 = A^2 + 2AB + B^2 \quad \text{Square of a sum}$$
$$(A - B)^2 = A^2 - 2AB + B^2 \quad \text{Square of a difference}$$

In words: The square of a binomial equals the first term squared, plus (or minus) twice the product of the two terms, plus the second term squared.

We can instead simplify $(x + 3)^2$ by substituting x for A and 3 for B in the formula for the square of sum:

$$\begin{aligned}(A + B)^2 &= A^2 + 2AB + B^2 \\ (x + 3)^2 &= x^2 + 2 \cdot x \cdot 3 + 3^2 && \text{Substitute x for A and 3 for B.} \\ &= x^2 + 6x + 9 && \text{Simplify.}\end{aligned}$$

The result is the same as our result from writing $(x + 3)^2$ as $(x + 3)(x + 3)$ and then multiplying. So, there are two ways to simplify $(x + 3)^2$. Similarly, there are two ways to simplify the square of any binomial. If you experiment with both methods in the homework, you will be able to make an informed choice of method for future problems.

Example 6 Simplifying Squares of Binomials

Simplify.

1. $(x + 7)^2$
2. $(5t - 4w)^2$
3. $(3r^2 + 2y^2)^2$

Solution

1. We substitute x for A and 7 for B:

$$\begin{aligned}(A + B)^2 &= A^2 + 2AB + B^2 \\ (x + 7)^2 &= x^2 + 2 \cdot x \cdot 7 + 7^2 && \text{Substitute.} \\ &= x^2 + 14x + 49 && \text{Simplify.}\end{aligned}$$

Another way to simplify $(x + 7)^2$ is to write it as $(x + 7)(x + 7)$ and then multiply each term in the first binomial by each term in the second binomial:

$$\begin{aligned}(x + 7)^2 &= (x + 7)(x + 7) && b^2 = bb \\ &= x^2 + 7x + 7x + 49 && \text{Multiply pairs of terms.} \\ &= x^2 + 14x + 49 && \text{Combine like terms.}\end{aligned}$$

2. We substitute $5t$ for A and $4w$ for B:

$$\begin{aligned}(A - B)^2 &= A^2 - 2AB + B^2 \\ (5t - 4w)^2 &= (5t)^2 - 2 \cdot 5t \cdot 4w + (4w)^2 && \text{Substitute.} \\ &= 25t^2 - 40tw + 16w^2 && \text{Simplify.}\end{aligned}$$

Another way to simplify $(5t - 4w)^2$ is to write it as $(5t - 4w)(5t - 4w)$ and then multiply each term in the first binomial by each term in the second binomial:

$$\begin{aligned}(5t - 4w)^2 &= (5t - 4w)(5t - 4w) && b^2 = bb \\ &= 25t^2 - 20tw - 20tw + 16w^2 && \text{Multiply pairs of terms.} \\ &= 25t^2 - 40tw + 16w^2 && \text{Combine like terms.}\end{aligned}$$

3. $$\begin{aligned}(3r^2 + 2y^2)^2 &= (3r^2)^2 + 2(3r^2)(2y^2) + (2y^2)^2 && (A+B)^2 = A^2 + 2AB + B^2;\ \text{substitute } 3r^2 \text{ for } A \text{ and } 2y^2 \text{ for } B. \\ &= 9r^4 + 12r^2y^2 + 4y^4 && \text{Simplify.}\end{aligned}$$ ■

In Example 6, we found that both $x^2 + 14x + 49$ and $25t^2 - 40tw + 16w^2$ are *squares* of binomials. Both of these *trinomials* are called perfect-square trinomials. A **perfect-square trinomial** is a trinomial equivalent to the square of a binomial.

Here we compare squaring a product with squaring a sum:

$$(AB)^2 = A^2B^2 \quad \text{Square of a product}$$
$$(A+B)^2 = A^2 + 2AB + B^2 \quad \text{Square of a sum}$$

WARNING It is a common error to omit the middle term in squaring a binomial—for instance,

$$(x+5)^2 = x^2 + 25 \quad \text{Incorrect}$$
$$(x+5)^2 = x^2 + 10x + 25 \quad \text{Correct}$$

When simplifying $(A+B)^2$, don't omit the middle term $2AB$ of $A^2 + 2AB + B^2$. Likewise, when simplifying $(A-B)^2$, don't omit the middle term $-2AB$ of $A^2 - 2AB + B^2$.

Finding the Product of Binomial Conjugates

Binomials such as $2x+5$ and $2x-5$ are said to be binomial conjugates. We say that the sum of two terms and the difference of the same two terms are **binomial conjugates** of each other.

How do we find the product of the binomial conjugates $A+B$ and $A-B$? We multiply as we would for any other two binomials:

$$\begin{aligned}(A+B)(A-B) &= A^2 - AB + AB - B^2 && \text{Multiply pairs of terms.}\\ &= A^2 - B^2 && \text{Combine like terms.}\end{aligned}$$

We see that the product of $A+B$ and $A-B$ is a *difference of two squares,* $A^2 - B^2$.

Product of Binomial Conjugates

$$(A+B)(A-B) = A^2 - B^2$$

In words: The product of two binomial conjugates is the difference of the square of the first term and the square of the second term.

By the commutative law of multiplication, we have

$$(A-B)(A+B) = A^2 - B^2$$

Example 7 Multiplying Binomial Conjugates

Find the product.

1. $(x+6)(x-6)$
2. $(3p-8q)(3p+8q)$
3. $(4m^2 - 7rt)(4m^2 + 7rt)$
4. $(x+3)(x-3)(x^2+9)$

Solution

1. We substitute x for A and 6 for B:

$$\begin{aligned}(A+B)(A-B) &= A^2 - B^2\\ (x+6)(x-6) &= x^2 - 6^2 && \text{Substitute.}\\ &= x^2 - 36 && \text{Simplify.}\end{aligned}$$

2. We substitute $3p$ for A and $8q$ for B:

$$\begin{aligned}(A-B)(A+B) &= A^2 - B^2\\ (3p-8q)(3p+8q) &= (3p)^2 - (8q)^2 && \text{Substitute.}\\ &= 9p^2 - 64q^2 && \text{Simplify.}\end{aligned}$$

3. $(4m^2 - 7rt)(4m^2 + 7rt) = (4m^2)^2 - (7rt)^2$ $(A - B)(A + B) = A^2 - B^2$
$= 16m^4 - 49r^2t^2$ Simplify.

4. $(x + 3)(x - 3)(x^2 + 9) = (x^2 - 9)(x^2 + 9)$ $(A + B)(A - B) = A^2 - B^2$
$= (x^2)^2 - 9^2$ $(A - B)(A + B) = A^2 - B^2$
$= x^4 - 81$ Simplify. ■

Function Notation

So far in this section, we have been working with polynomial expressions. Now we will apply the skills we have learned to polynomial functions.

Example 8 Evaluating a Quadratic Function

For $f(x) = x^2 - 5x$, find the following.

1. $f(a - 3)$
2. $f(a + 2) - f(a)$

Solution

1. $f(a - 3) = (a - 3)^2 - 5(a - 3)$ Substitute $a - 3$ for x.
$= a^2 - 6a + 9 - 5a + 15$ $(A - B)^2 = A^2 - 2AB + B^2$; distributive law
$= a^2 - 11a + 24$ Combine like terms.

2. $f(a + 2) - f(a)$
$= [(a + 2)^2 - 5(a + 2)] - [a^2 - 5a]$ Substitute $a + 2$ for x; substitute a for x.
$= a^2 + 4a + 4 - 5a - 10 - a^2 + 5a$ $(A + B)^2 = A^2 + 2AB + B^2$; distributive law; subtract.
$= 4a - 6$ Combine like terms. ■

Recall from Section 6.1 that a quadratic function in the form $f(x) = ax^2 + bx + c$ is in standard form.

Example 9 Writing a Quadratic Function in Standard Form

Write $f(x) = -3(x - 4)^2 + 8$ in standard form.

Solution

We begin by simplifying $(x - 4)^2$, because we work with exponents before we multiply or add:

$$\begin{aligned} f(x) &= -3(x - 4)^2 + 8 && \text{Original equation} \\ &= -3(x^2 - 8x + 16) + 8 && (A - B)^2 = A^2 - 2AB + B^2 \\ &= -3x^2 + 24x - 48 + 8 && \text{Distributive law} \\ &= -3x^2 + 24x - 40 && \text{Combine like terms.} \end{aligned}$$

We use graphing calculator graphs to verify our work (see Fig. 14). ■

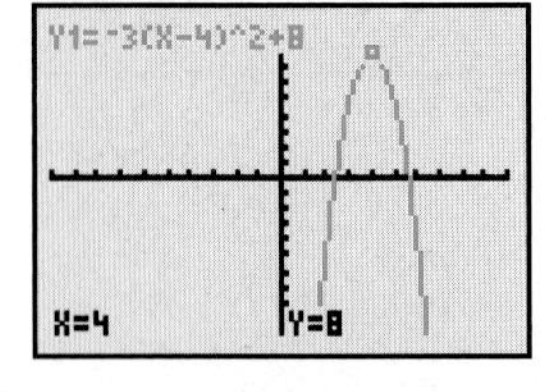

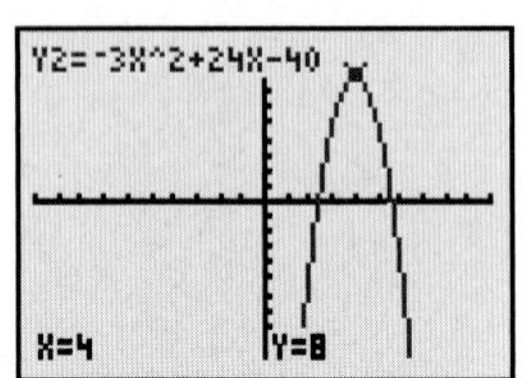

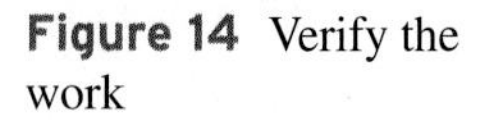
Figure 14 Verify the work

Product Function

In Section 6.1, we worked with sum functions and difference functions. Now we will work with *product functions*. For example, if $f(x) = 2x$ and $g(x) = x + 3$, then it follows that $f(x) \cdot g(x) = 2x(x + 3) = 2x^2 + 6x$. The function $f(x) \cdot g(x) = 2x^2 + 6x$ is called the product function of f and g.

> **DEFINITION** Product function
>
> If f and g are functions and x is in the domain of both functions, then we can form the **product function $f \cdot g$:**
>
> $$(f \cdot g)(x) = f(x) \cdot g(x)$$

Example 10 Finding the Product Function

Let $f(x) = 3x + 7$ and $g(x) = 5x - 2$.

1. Find an equation of the product function $f \cdot g$.
2. Find $(f \cdot g)(2)$.

Solution

1.
$$\begin{aligned}(f \cdot g)(x) &= f(x) \cdot g(x) && \text{Definition of product function}\\ &= (3x + 7)(5x - 2) && \text{Substitute } 3x+7 \text{ for } f(x) \text{ and } 5x-2 \text{ for } g(x).\\ &= 15x^2 - 6x + 35x - 14 && \text{Multiply pairs of terms.}\\ &= 15x^2 + 29x - 14 && \text{Combine like terms.}\end{aligned}$$

2. $(f \cdot g)(2) = 15(2)^2 + 29(2) - 14 = 104$ ■

Sometimes we can find a meaningful model by finding the product of two models.

Example 11 Using a Product Function to Model a Situation

The annual cost of state corrections (prisons and related costs) per person in the United States can be modeled by the function

$$C(t) = 2.75t + 102$$

where $C(t)$ is the cost (in dollars per person) in the year that is t years since 1990 (see Table 11). The U.S. population can be modeled by the function

$$P(t) = 3.3t + 248$$

where $P(t)$ is the population (in millions) at t years since 1990.

Table 11 Costs of State Corrections; U.S. Population

Year	Cost (dollars per person)	Year	U.S. Population (millions)
1995	117	1990	248.8
1996	119	1993	257.8
1997	120	1996	265.2
1998	123	1999	279.0
1999	128	2002	287.9
2000	128	2004	293.7
2001	134		

Source: *U.S. Census Bureau*

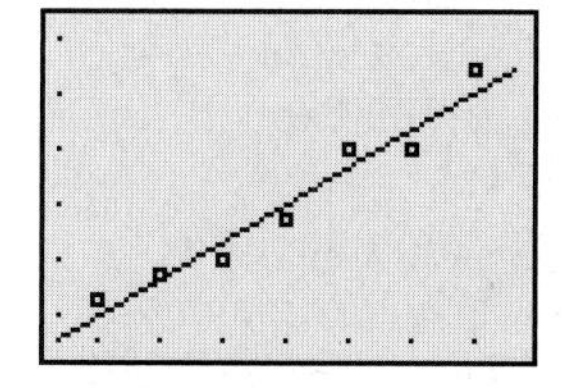

Figure 15 Cost scattergram and model

1. Check that the models fit the data well.
2. Find an equation of the product function $C \cdot P$.
3. Perform a unit analysis of the expression $C(t) \cdot P(t)$.
4. Find $(C \cdot P)(20)$. What does it mean in this situation?
5. Use a graphing calculator graph to determine whether the function $C \cdot P$ is increasing, decreasing, or neither for values of t between 5 and 20. What does your result mean in this situation?

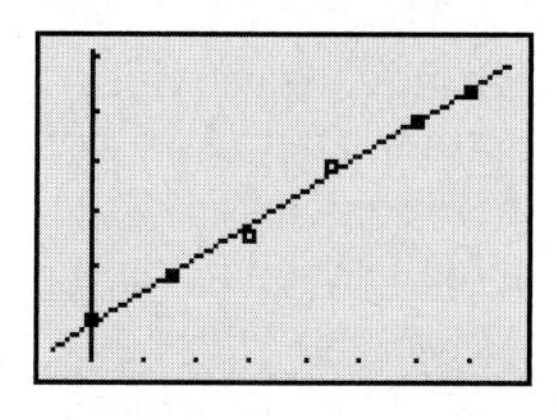

Figure 16 Population scattergram and model

Solution

1. We check the fit of the cost model in Fig. 15 and the fit of the population model in Fig. 16. The models appear to fit the data well.
2.
$$\begin{aligned}(C \cdot P)(t) &= C(t) \cdot P(t) && \text{Definition of product function}\\ &= (2.75t + 102)(3.3t + 248) && \text{Substitute } 2.75t+102 \text{ for } C(t) \text{ and } 3.3t+248 \text{ for } P(t).\\ &= 9.075t^2 + 682t + 336.6t + 25{,}296 && \text{Multiply pairs of terms.}\\ &= 9.075t^2 + 1018.6t + 25{,}296 && \text{Combine like terms.}\end{aligned}$$

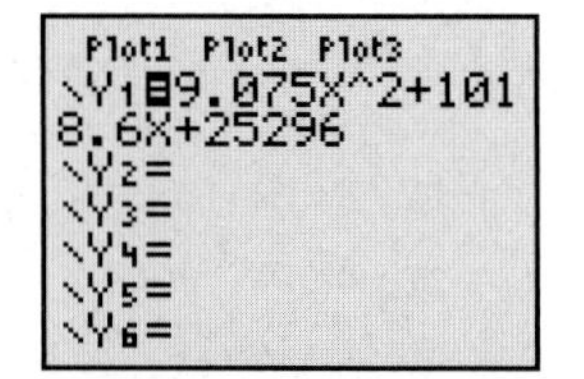

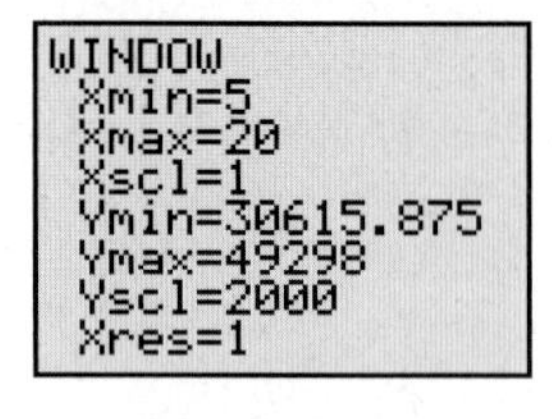

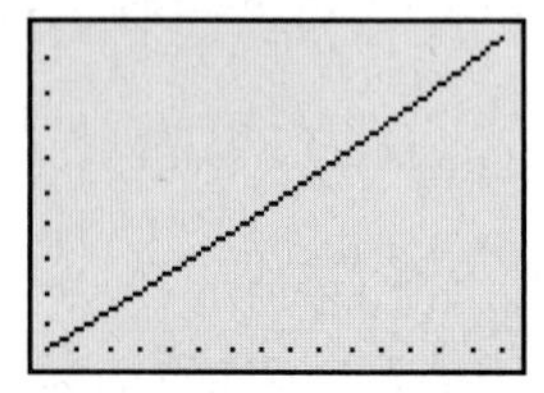

Figure 17 Graph of the function $C \cdot P$

3. For the expression $C(t) \cdot P(t)$, we have

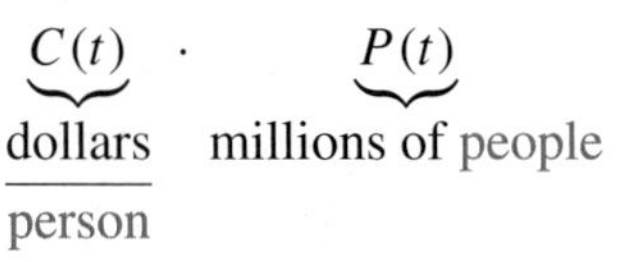

The units of the expression are millions of dollars.

4. $(C \cdot P)(20) = 9.075(20)^2 + 1018.6(20) + 25{,}296 = 49{,}298$. This means that the total cost of state corrections will be \$49,298 million (\$49.298 billion) in 2010, according to the model.

5. To graph the model, we enter the function, then press WINDOW, and set Xmin to be 5 and Xmax to be 20. Then we use ZoomFit to draw the graph (see Fig. 17). For values of t between 5 and 20, the model is increasing. This means that the total cost of state corrections has been increasing since 1995 and will continue to increase until 2010. ■

group exploration

Looking ahead: Factoring trinomials

In Problem 1 of Example 3, we found that the product $(x+4)(x+8)$ is equivalent to $x^2 + 12x + 32$. Working backward, we write $x^2 + 12x + 32 = (x+4)(x+8)$, and we call the process *factoring*. We say that $(x+4)(x+8)$ is a factored polynomial. In general, a *factored polynomial* is a product of two or more polynomials.

1. Complete Table 12.

Table 12 Multiplying Some Quadratic Polynomials

Factored Polynomial	Last Terms	Product of Factored Polynomial	Coefficient of x	Constant Term
$(x+3)(x+4)$	3 and 4	$x^2 + 7x + 12$	7	12
$(x+5)(x-3)$	5 and -3	$x^2 + 2x - 15$	2	-15
$(x-2)(x+6)$				
$(x-4)(x-3)$				

2. For each row of Table 12, what connection do you notice between the last terms of the factored polynomial and the coefficient of x of the product of the factored polynomial? Explain why this happens.

3. For each row of Table 12, what connection do you notice between the last terms of the factored polynomial and the constant term of the product of the factored polynomial? Explain why this happens.

4. **a.** Do the observations that you have made in this exploration apply to the polynomial $(2x+3)(x+4)$? If so, show that they do. If not, explain why not in terms of how you find the product $(2x+3)(x+4)$.

   b. Do the observations that you have made in this exploration apply to the polynomial $(x+5)(3x+4)$? If so, show that they do. If not, explain why not in terms of how you find the product $(x+5)(3x+4)$.

   c. Discuss, in general, when your observations apply and when they do not.

TIPS FOR SUCCESS: Review Your Notes as Soon as Possible

Do you get confused by notes you wrote during class, even though the class activities made sense to you? If this happens a lot, review your notes as soon after class as possible. Even reviewing your notes for just a few minutes between classes will help. This will increase your likelihood of remembering what you learned in class and will give you the opportunity to add more comments to your notes while the class experience is fresh in your mind. Teaming with a classmate to review notes can also be helpful.

HOMEWORK 6.2

FOR EXTRA HELP ▶ Student Solutions Manual · PH Math/Tutor Center · MathXL® · MyMathLab

Find the product. Use a graphing calculator table to verify your work when possible.

1. $3x^2(6x^4)$
2. $5x^3(7x^6)$
3. $2a^3b^5(-4a^2b^3)$
4. $7w^4y^3(-3w^3y^6)$
5. $-6x(5x-2)$
6. $-9x(4x-4)$
7. $5ab^2(4a^2-7ab+3b^2)$
8. $-7w^2y(6w^2+2wy-8y^2)$
9. $(x+3)(x+6)$
10. $(x+9)(x+2)$
11. $(3m-2)(5m+4)$
12. $(4p-7)(2p+3)$
13. $(8x-3)(4x-1)$
14. $(5x-6)(3x-2)$
15. $(1.7x-2.4)(2.3x+1.2)$
16. $(3.5x-1.3)(2.2x-4.7)$
17. $(2a+5b)(3a-7b)$
18. $(8p+t)(7p-4t)$
19. $(4x-9y)(5x-2y)$
20. $(3m-4p)(3m-6p)$
21. $(2a^2-5b^2)(7a^2+3b^2)$
22. $(4a^2-9b^2)(3a^2-2b^2)$
23. $3x^2(2x-5)(4x+1)$
24. $2x^2(3x+2)(2x-5)$
25. $5x(x^2+3)(x-4)$
26. $2x(x^2-7)(x+3)$
27. $(3x+2)(4x^2+5x-3)$
28. $(5x+3)(2x^2-7x-4)$
29. $(a+b)(a^2-ab+b^2)$
30. $(a-b)(a^2+ab+b^2)$
31. $(4x-3y)(2x^2-xy+5y^2)$
32. $(3x-5y)(4x^2+2xy-y^2)$
33. $(x^2+2x-3)(x^2-x+2)$
34. $(x^2-4x+1)(x^2-3x+2)$
35. $(2x^2+xy-3y^2)(x^2-2xy+y^2)$
36. $(x^2+2xy-y^2)(3x^2-xy-2y^2)$

Simplify. Use a graphing calculator table to verify your work when possible.

37. $(x+5)^2$
38. $(x+9)^2$
39. $(x-8)^2$
40. $(x-1)^2$
41. $(3x+5)^2$
42. $(7x+2)^2$
43. $(2.6x-3.2)^2$
44. $(6.7x+1.9)^2$
45. $(4a+3b)^2$
46. $(3p+7t)^2$
47. $(2x^2-6y^2)^2$
48. $(5m^2-4p^2)^2$
49. $-2x(2x+5)^2$
50. $-5x(3x+2)^2$

Find the product. Use a graphing calculator table to verify your work when possible.

51. $(x-4)(x+4)$
52. $(x-7)(x+7)$
53. $(3x+6)(3x-6)$
54. $(5x+9)(5x-9)$
55. $(2r-8t)(2r+8t)$
56. $(7m-4n)(7m+4n)$
57. $(3rt-9w)(3rt+9w)$
58. $(7ab-5c)(7ab+5c)$
59. $(8a^2+3b^2)(8a^2-3b^2)$
60. $(5p^2+7q^2)(5p^2-7q^2)$
61. $(x-2)(x+2)(x^2+4)$
62. $(x-1)(x+1)(x^2+1)$
63. $(3a+2b)(3a-2b)(9a^2+4b^2)$
64. $(2m-5n)(2m+5n)(4m^2+25n^2)$

For $f(x)=x^2-3x$, find the following.

65. $f(5b)$
66. $f(6b)$
67. $f(c+4)$
68. $f(c+1)$
69. $f(b-3)$
70. $(b-2)$
71. $f(a+2)-f(a)$
72. $f(a+3)-f(a)$
73. $f(a+h)-f(a)$
74. $f(a)-f(a-h)$

Write the quadratic function in standard form. Use a graphing calculator table or graph to verify your work.

75. $f(x)=(x+6)^2$
76. $f(x)=(x-5)^2$
77. $f(x)=2(x+3)^2+1$
78. $f(x)=4(x-2)^2-7$
79. $f(x)=-3(x-5)^2-1$
80. $f(x)=-2(x-4)^2+3$

For $f(x)=2x-3$, $g(x)=3x+2$, $h(x)=2x^2-4x+3$, and $k(x)=3x^2+x-5$, find an equation of the given product function; then evaluate the product function at the indicated value.

81. $f\cdot g$, $(f\cdot g)(3)$
82. $g\cdot h$, $(g\cdot h)(3)$
83. $f\cdot h$, $(f\cdot h)(2)$
84. $h\cdot k$, $(h\cdot k)(2)$
85. $f\cdot f$, $(f\cdot f)(4)$
86. $g\cdot g$, $(g\cdot g)(4)$

For $f(x)=4x+1$, $g(x)=5x+3$, $h(x)=3x^2-x-2$, and $k(x)=2x^2-4x+3$, find an equation of the given product function; then evaluate the product function at the indicated value.

87. $f\cdot g$, $(f\cdot g)(-1)$
88. $g\cdot h$, $(g\cdot h)(-1)$
89. $f\cdot h$, $(f\cdot h)(-2)$
90. $h\cdot k$, $(h\cdot k)(-2)$
91. $h\cdot h$, $(h\cdot h)(1)$
92. $k\cdot k$, $(k\cdot k)(1)$

93. The average value per acre of U.S. farmland (in dollars per acre) $V(t)$ is modeled by the function $V(t)=55t+557$, where t is the number of years since 1990 (see Table 13). The amount of U.S. farmland (in millions of acres) $A(t)$ is modeled by the function $A(t)=-2.9t+976$, where t is the number of years since 1990.

Table 13 Values and Acres of U.S. Farmland

Year	Average Value (dollars per acre)	Amount of Farmland (millions of acres)
1993	740	968
1995	840	961
1997	930	956
1999	1030	948
2001	1150	941
2003	1270	939
2004	1360	937

Sources: *National Agriculture Statistics Service; Agricultural Statistics Board*

a. Check that the models fit the data well.
b. Find an equation of the product function $V\cdot A$.
c. Perform a unit analysis of the expression $V(t)\cdot A(t)$.

d. Find $(V \cdot A)(20)$. What does it mean in this situation?

e. Use a graphing calculator graph to determine whether the function $V \cdot A$ is increasing, decreasing, or neither for values of t between 8 and 20. What does your result mean in this situation? Explain how that is possible, given that the amount of farmland has been decreasing.

94. The average hourly costs (in dollars per hour) $C(t)$ of a General Motors® employee is modeled by $C(t) = 6.4t + 52$, where t is the number of years since 2000 (see Table 14). The number of General Motors employees (in millions) $N(t)$ is modeled by $N(t) = -8.4t + 216$, where t is the number of years since 2000.

Table 14 Average Hourly Employee Costs; Numbers of Employees

Year	Average Employee Cost (dollars per hour per employee)	Number of Employees (thousands)
2000	52	215
2001	58	208
2002	63	199
2003	78	192
2004	74	181

Source: *General Motors*

a. Check that the models fit the data well.

b. Find an equation of the product function $C \cdot N$.

c. Perform a unit analysis of the expression $C(t) \cdot N(t)$.

d. Find $(C \cdot N)(10)$. What does it mean in this situation?

e. Predict the total annual cost of all General Motor employees in 2010. Assume that, on average, employees work eight-hour shifts five days a week for 49 weeks each year.

f. Use a graphing calculator graph to determine whether the function $C \cdot N$ is increasing, decreasing, or neither for values of t between 0 and 8. What does your result mean in this situation? Explain how that is possible, given that the number of employees has been decreasing.

95. The average monthly cell phone bill (in dollars per month) $B(t)$ is modeled by the function $B(t) = -0.24t^2+7.2t-3.3$, where t is the number of years since 1990 (see Table 15). The number of cell phone subscribers (in millions) $N(t)$ is modeled by the function $N(t) = 0.7t^2 + 3t + 4$, where t is the number of years since 1990.

Table 15 Average Monthly Cell Phone Bills and Numbers of Subscribers

Year	Average Bill (dollars per month)	Year	Number of Subscribers (millions)
1998	39.43	1990	5.3
1999	41.24	1992	11.0
2000	45.27	1994	24.1
2001	47.37	1996	44.0
2002	48.40	1998	69.2
2003	49.91	2000	109.5
2004	50.64	2002	140.5
		2004	182.1

Source: *Cellular Telecommunications & Internet Association*

a. Check that the models fit the data well.

b. Find an equation of the product function $B \cdot N$.

c. Perform a unit analysis of the expression $B(t) \cdot N(t)$.

d. Find $(B \cdot N)(21)$. What does it mean in this situation?

e. Use a graphing calculator graph to determine whether the function $B \cdot N$ is increasing, decreasing, or neither for values of t between 8 and 20. What does your result mean in this situation?

96. In Example 6 of Section 2.3, we modeled the average salaries of professors at four-year public colleges and universities. A reasonable model is $S(t) = 1.7t + 7$, where $S(t)$ is the average salary (in thousands of dollars) at t years since 1970 (see Table 16). The number of such professors (in thousands) $N(t)$ is modeled by the function $N(t) = 0.43t^2 - 4t + 481$, where t is the number of years since 1970.

Table 16 Average Salaries and Numbers of Professors

Year	Average Salary (thousands of dollars per professor)	Year	Number of Professors (thousands)
1975	16.6	1975	467
1980	22.1	1980	494
1985	31.2	1985	504
1990	41.9	1991	591
1995	49.1	1995	647
2000	57.7	1999	714
2004	65.0	2003	816

Sources: *American Association of University Professors; U.S. National Center for Education Statistics*

a. Check that the models fit the data well.

b. Find an equation of the product function $S \cdot N$.

c. Perform a unit analysis of the expression $S(t) \cdot N(t)$.

d. Find $(S \cdot N)(41)$. What does it mean in this situation?

e. Use a graphing calculator graph to determine whether the function $S \cdot N$ is increasing, decreasing, or neither for values of t between 5 and 40. What does your result mean in this situation?

97. A student tries to simplify $(x + 8)^2$:

$$(x + 8)^2 = x^2 + 64$$

Describe any errors. Then simplify the expression correctly.

98. A student tries to simplify $(2x - 9y)^2$:

$$(2x - 9y)^2 = 4x^2 - 81y^2$$

Describe any errors. Then simplify the expression correctly.

99. **a.** Use a graphing calculator table to show that $(x + 2)^2$ and $x^2 + 2^2$ are not equivalent.

b. Simplify $(x + 2)^2$.

c. Use a graphing calculator table to show that $(x + 2)^2$ and your result in part (b) are equivalent.

100. **a.** Use a graphing calculator table to show that $(x - 4)^2$ and $x^2 - 4^2$ are not equivalent.

b. Simplify $(x - 4)^2$.

c. Use a graphing calculator table to show that $(x - 4)^2$ and your result in part (b) are equivalent.

101. A student tries to find the product $7x(-2x)$:

$$7x(-2x) = 5x$$

Describe any errors. Then find the product correctly.

102. A student tries to simplify $(3x + 5)^2$:

$$(3x + 5)^2 = 9x^2 + 15x + 25$$

Describe any errors. Then simplify the expression correctly.

103. Which expressions are equivalent?

$(2x - 5)(3x + 4)$ $6x^2 + 7x - 20$ $3x(2x - 2) - x - 20$
$(3x - 4)(2x + 5)$ $6x^2 - 7x - 20$ $(3x + 4)(2x - 5)$

104. Which expressions are equivalent?

$(8x + 1)(x - 4)$ $(4x - 2)(2x + 2)$ $8x^2 - 31x - 4$
$8x^2 + 31x - 4$ $(x - 4)(8x + 1)$ $4(2x^2 - 1) - 31x$

105. Show that the property $(A - B)^2 = A^2 - 2AB + B^2$ is correct.

106. Describe the various types of products of polynomials and squares of binomials that we have discussed in this section. Describe how to find such products and simplify such squares.

Related Review

Simplify.

107. $7x^{1/3} - 5x^{1/3}$

108. $\left(x^{1/3}x^{1/4}\right)^2$

109. $\left(7x^{1/3}\right)\left(-5x^{1/3}\right)$

110. $\left(x^{1/3} + x^{1/4}\right)^2$

Find the product or simplify to help you decide whether the function is linear or quadratic. Use a graphing calculator graph to verify your work.

111. $f(x) = (2x - 5)(3x - 1)$

112. $f(x) = (4x - 7)(4x + 7)$

113. $f(x) = x^2 - (x + 1)^2$

114. $f(x) = (x + 6)^2 - (x - 6)^2$

Simplify. Write your result as a single logarithm with a coefficient of 1.

115. $\log_b(x + 5) + \log_b(x - 3)$

116. $\log_b(x - 4) + \log_b(x - 7)$

117. $2\log_b(w - 3) + \log_b(w + 3)$

118. $\log_b(p - 2) + 2\log_b(p + 2)$

Expressions, Equations, Functions, and Graphs

Perform the indicated instruction. Then use words such as linear, quadratic, cubic, exponential, logarithmic, polynomial, degree, function, one variable, *and* two variables *to describe the expression, equation, or system.*

119. Find the product $4x(3x + 5)(2x - 3)$.

120. Graph $f(x) = 9\left(\frac{1}{3}\right)^x$ by hand.

121. Write $f(x) = -3(x - 4)^2 + 5$ in standard form.

122. Simplify $\left(\frac{3b^4c^{-5}}{9b^{-2}c^{-3}}\right)^{-2}$.

6.3 FACTORING TRINOMIALS OF THE FORM $x^2 + bx + c$; FACTORING OUT THE GCF

Objectives

- Know that multiplying and *factoring* are reverse processes.
- Factor a trinomial of the form $x^2 + bx + c$.
- Know the meaning of a *prime* polynomial.
- Factor out the *greatest common factor (GCF)* of a polynomial.
- *Completely factor* polynomials.
- Know when to factor out the opposite of the GCF of a polynomial.

We know how to multiply 2 and 5, as follows: $2 \cdot 5 = 10$. Can we work backward? Here, we *factor* the number 10: $10 = 2 \cdot 5$. Note that factoring 10 is the reverse of multiplying 2 and 5. Next, we will make similar observations about polynomials.

Multiplying Polynomials Versus Factoring Polynomials

In Section 6.2, we found products of polynomials; for example,

$$(x + 2)(x + 3) = x^2 + 3x + 2x + 6 \quad \text{Multiply pairs of terms.}$$
$$= x^2 + 5x + 6 \quad \text{Combine like terms.}$$

In this section, we will learn how to work backward. That is, we will learn how to write $x^2 + 5x + 6$ as a product. This process is called *factoring*.

We **factor** a polynomial by writing it as a product. We say that $(x + 2)(x + 3)$ is a **factored polynomial** and that both $(x + 2)$ and $(x + 3)$ are factors of the polynomial.

Comparing Multiplying with Factoring

Multiplying and factoring are reverse processes. For example,

Multiplying
$$\xrightarrow{\hspace{2cm}}$$
$$(x+2)(x+3) = x^2 + 5x + 6$$
$$\xleftarrow{\hspace{2cm}}$$
Factoring

Factoring a Trinomial of the Form $x^2 + bx + c$

To see how to factor $x^2 + 5x + 6$, let's take another look at how we find the product $(x+2)(x+3)$:

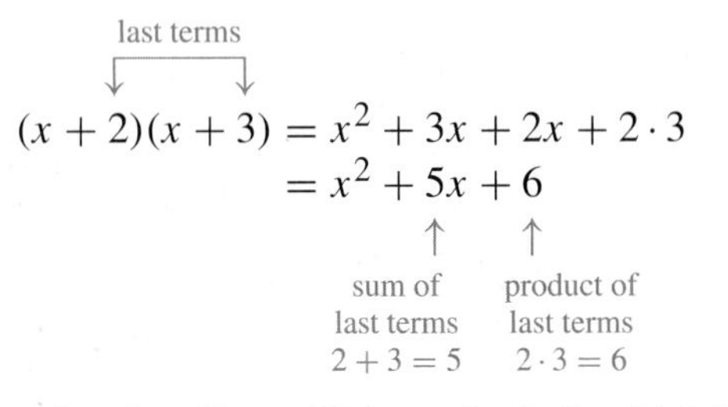

For $x^2 + 5x + 6$, notice that the coefficient of x is 5, which is the sum of 2 and 3, the *last terms* of $(x+2)(x+3)$. Also, the constant term of $x^2 + 5x + 6$ is 6, which is the product of 2 and 3. Now we find the product $(x+p)(x+q)$:

$$\begin{aligned}(x+p)(x+q) &= x^2 + qx + px + pq && \text{Multiply pairs of terms.}\\ &= x^2 + px + qx + pq && \text{Rearrange terms.}\\ &= x^2 + (p+q)x + pq && \text{Distributive law}\end{aligned}$$

In the result, we see that the coefficient of x is the sum of the last terms p and q and that the constant term is the product of the last terms p and q. This observation can help us factor some quadratic trinomials.

TRINOMIALS WITH POSITIVE CONSTANT TERMS

For the trinomial $x^2 + 5x + 6$, the constant term, 6, is positive. In Examples 1 and 2, we will factor two more trinomials whose constant term is positive.

Example 1 Factoring a Trinomial of the Form $x^2 + bx + c$

Factor $x^2 + 11x + 24$.

Solution

To factor $x^2 + 11x + 24$, we need two integers whose product is 24 and whose sum is 11. We try only positive integers, since both their product and their sum must be positive. Here are the possibilities:

Product = 24	**Sum = 11?**
$1(24) = 24$	$1 + 24 = 25$
$2(12) = 24$	$2 + 12 = 14$
$3(8) = 24$	$3 + 8 = 11 \leftarrow$ Success!
$4(6) = 24$	$4 + 6 = 10$

Since $3(8) = 24$ and $3 + 8 = 11$, we conclude that the last terms of the factors are 3 and 8:

$$x^2 + 11x + 24 = (x+3)(x+8)$$

We check the result by finding the product $(x+3)(x+8)$:

$$(x+3)(x+8) = x^2 + 8x + 3x + 24 = x^2 + 11x + 24$$

By the commutative law, $(x+3)(x+8) = (x+8)(x+3)$, so we can write the factors $x + 3$ and $x + 8$ in either order. ■

We now summarize how to factor a trinomial of the form $x^2 + bx + c$.

> **Factoring $x^2 + bx + c$**
>
> To factor $x^2 + bx + c$, look for two integers p and q whose product is c and whose sum is b. That is, $pq = c$ and $p + q = b$. If such integers exist, the factored polynomial is
>
> $$(x + p)(x + q)$$

Example 2 Factoring a Trinomial of the Form $x^2 + bx + c$

Factor $x^2 - 8x + 16$.

Solution

To factor $x^2 - 8x + 16$, we need two integers whose product is 16 and whose sum is -8. Since the product 16 is positive, the two integers must have the same sign. Therefore, both integers must be negative, because the sum -8 is negative. Here are the possibilities:

X	Y1	Y2
0	16	16
1	9	9
2	4	4
3	1	1
4	0	0
5	1	1
6	4	4

X=0

Figure 18 Verify the work

Product = 16	**Sum = −8?**
$-1(-16) = 16$	$-1 + (-16) = -17$
$-2(-8) = 16$	$-2 + (-8) = -10$
$-4(-4) = 16$	$-4 + (-4) = -8 \leftarrow$ Success!

Since $-4(-4) = 16$ and $-4 + (-4) = -8$, we conclude that the last terms of the factors are -4 and -4:

$$x^2 - 8x + 16 = (x - 4)(x - 4) = (x - 4)^2$$

Our result, $(x-4)^2$, is the square of a binomial. So, the original expression, $x^2-8x+16$, is a perfect-square trinomial (Section 6.2).

We use a graphing calculator table to verify our work (see Fig. 18). ■

When the constant term of a trinomial is positive, we need to consider only certain possibilities for the factors of that constant term. In Example 1, we worked with only *positive* factors of the positive constant term, because the coefficient of the middle term was *positive*. In Example 2, we worked with only *negative* factors of the positive constant term, because the coefficient of the middle term was *negative*.

> **Factoring $x^2 + bx + c$ with c Positive**
>
> To factor a trinomial of the form $x^2 + bx + c$ with a positive constant term c,
>
> - If b is positive, look for two *positive* integers whose product is c and whose sum is b. For example,
>
> $$x^2 + 10x + 16 = (x + 2)(x + 8)$$
>
> (10: Positive b; 16: Positive c; 2 and 8: Both last terms are positive.)
>
> - If b is negative, look for two *negative* integers whose product is c and whose sum is b. For example,
>
> $$x^2 - 9x + 18 = (x - 3)(x - 6)$$
>
> (-9: Negative b; 18: Positive c; -3 and -6: Both last terms are negative.)

TRINOMIALS WITH NEGATIVE CONSTANT TERMS

How do we factor a quadratic trinomial for which the constant term is negative?

Example 3 Factoring a Trinomial of the Form $x^2 + bx + c$

Factor $w^2 - 3w - 18$.

Solution

To factor $w^2 - 3w - 18$, we need two integers whose product is -18 and whose sum is -3. Since the product -18 is negative, the two integers must have different signs. Here are the possibilities:

Product = −18	**Sum = −3?**
$1(-18) = -18$	$1 + (-18) = -17$
$2(-9) = -18$	$2 + (-9) = -7$
$3(-6) = -18$	$3 + (-6) = -3 \leftarrow$ Success!
$6(-3) = -18$	$6 + (-3) = 3$
$9(-2) = -18$	$9 + (-2) = 7$
$18(-1) = -18$	$18 + (-1) = 17$

Since $3(-6) = -18$ and $3 + (-6) = -3$, we conclude that the last terms of the factors are 3 and -6:

$$w^2 - 3w - 18 = (w + 3)(w - 6)$$

We check the result by finding the product $(w + 3)(w - 6)$:

$$(w + 3)(w - 6) = w^2 - 6w + 3w - 18 = w^2 - 3w - 18$$ ■

When the constant term of a trinomial is negative, the integers whose product equals that negative constant term have different signs. For instance, in Example 3, we worked with only integers with different signs whose product is -18.

Factoring $x^2 + bx + c$ with c Negative

To factor a trinomial of the form $x^2 + bx + c$ with a negative constant term c, look for two integers with *different* signs whose product is c and whose sum is b. For example,

$$x^2 + 2x - 15 = (x - 3)(x + 5)$$

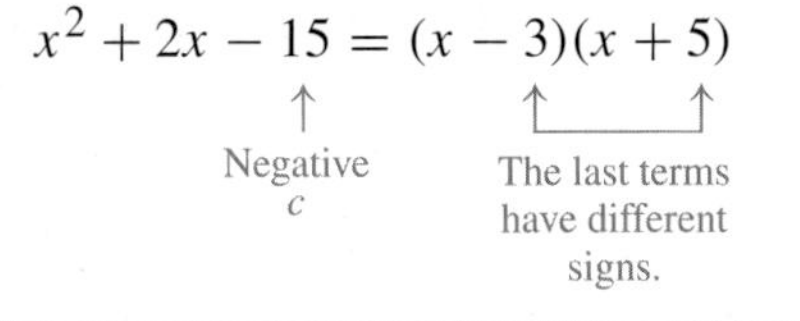

We can use a similar method to factor trinomials that have two variables.

Example 4 Factoring a Trinomial with Two Variables

Factor $a^2 + 6ab + 8b^2$.

Solution

To help us find the last terms, we write the trinomial in the form $a^2 + (6b)a + 8b^2$. We need two monomials whose product is $8b^2$ and whose sum is $6b$. So, the last terms are $2b$ and $4b$:

$$a^2 + 6ab + 8b^2 = (a + 2b)(a + 4b)$$

We check by finding the product $(a + 2b)(a + 4b)$:

$$(a + 2b)(a + 4b) = a^2 + 4ab + 2ab + 8b^2 = a^2 + 6ab + 8b^2$$ ■

Prime Polynomials

Just as a prime number has no positive factors other than itself and 1, a polynomial that cannot be factored is called **prime.**

Consider the polynomial $x^2 + 5x + 10$. To factor this polynomial, we need two integers whose product is 10 and whose sum is 5. We try only positive integers, since both their product and sum must be positive. Here are the possibilities:

Product = 10	**Sum = 5?**
$1(10) = 10$	$1 + 10 = 11$
$2(5) = 10$	$2 + 5 = 7$

Neither of the possible sums equals 5, so we conclude that the trinomial $x^2 + 5x + 10$ is prime.

Example 5 Identifying a Prime Polynomial

Factor $-14 + 6x + x^2$.

Solution

First, we write $-14 + 6x + x^2$ in descending order to avoid confusion about the coefficients:

$$x^2 + 6x - 14$$

To factor $x^2 + 6x - 14$, we need two integers whose product is -14 and whose sum is 6. Since the product -14 is negative, the two integers must have different signs. Here are the possibilities:

Product = −14	**Sum = 6?**
$1(-14) = -14$	$1 + (-14) = -13$
$2(-7) = -14$	$2 + (-7) = -5$
$7(-2) = -14$	$7 + (-2) = 5$
$14(-1) = -14$	$14 + (-1) = 13$

Because none of the sums equal 6, we conclude that the trinomial $x^2 + 6x - 14$ is prime. So, the original trinomial $-14 + 6x + x^2$ is prime. ■

Factoring Out the GCF

Consider the polynomial $3x + 12$. Note that 3 is a common factor of both $3x$ (where $3x = 3 \cdot x$) and 12 (where $12 = 3 \cdot 4$):

$$3x + 12 = 3 \cdot x + 3 \cdot 4$$

We use the distributive law to "factor out" the common factor 3:

$$3x + 12 = 3 \cdot x + 3 \cdot 4 = 3(x + 4)$$

So, we factor $3x + 12$ as $3(x + 4)$. We check the result by finding the product $3(x + 4)$:

$$3(x + 4) = 3x + 12$$

Example 6 Factoring Out a Common Factor

Factor.

1. $10x^2 - 15x$ **2.** $8x^3 + 20x^2$

Solution

1. The expression $5x$ is a common factor of $10x^2 = 5x \cdot 2x$ and $15x = 5x \cdot 3$. So, we use the distributive law to factor out $5x$:

$$10x^2 - 15x = 5x \cdot 2x - 5x \cdot 3 \quad \text{5x is a common factor.}$$
$$= 5x(2x - 3) \quad \text{Factor out 5x.}$$

2. The expression $4x^2$ is a common factor of $8x^3 = 4x^2 \cdot 2x$ and $20x^2 = 4x^2 \cdot 5$. So, we use the distributive law to factor out $4x^2$:

$$8x^3 + 20x^2 = 4x^2 \cdot 2x + 4x^2 \cdot 5 \quad 4x^2 \text{ is a common factor.}$$
$$= 4x^2(2x + 5) \quad \text{Factor out } 4x^2.$$ ■

In Problem 2 of Example 6, notice that $2x$ is also a common factor of $8x^3 = 2x \cdot 4x^2$ and $20x^2 = 2x \cdot 10x$. So, we could have factored $8x^3 + 20x^2$ by factoring out $2x$ rather than $4x^2$:

$$8x^3 + 20x^2 = 2x(4x^2 + 10x)$$

However, this result is not *completely factored;* we can still factor $4x^2 + 10x$ by factoring out $2x$:

$$8x^3 + 20x^2 = 2x(4x^2 + 10x) = 2x \cdot 2x(2x + 5) = 4x^2(2x + 5)$$

Although we have found the same final result, it was more efficient to factor out $4x^2$, which has a larger coefficient and a higher degree than $2x$. We call $4x^2$ the *greatest common factor* of $8x^3$ and $20x^2$.

DEFINITION Greatest common factor

The **greatest common factor (GCF)** of two or more terms is the monomial with the largest coefficient and the highest degree that is a factor of all the terms.

For both polynomials in Example 6, the common factor that we factored out of the polynomial was the GCF. In Example 7, we factor another polynomial by factoring out the GCF.

X	Y1	Y2
0	0	0
1	-12	-12
2	168	168
3	1188	1188
4	4128	4128
5	10500	10500
6	22248	22248

X=0

Figure 19 Verify the work

Example 7 Factoring Out the GCF

Factor $18x^4 - 30x^2$.

Solution

We begin by factoring $18x^4$ and $30x^2$:

$$18x^4 = 2 \cdot 3 \cdot 3 \cdot x \cdot x \cdot x \cdot x$$
$$30x^2 = 2 \cdot 3 \cdot 5 \cdot x \cdot x$$

There are four common factors: 2, 3, x, and x. So, the GCF is $6x^2$:

$$18x^4 - 30x^2 = 6x^2 \cdot 3x^2 - 6x^2 \cdot 5 \quad \text{6x}^2 \text{ is the GCF.}$$
$$= 6x^2(3x^2 - 5) \quad \text{Factor out } 6x^2.$$

We use a graphing calculator table to verify our work (see Fig. 19). ■

After you factor a polynomial, verify your work by finding the product of your result or by using a graphing calculator table.

So far, we have factored out the GCF for some binomials with one variable. We can also factor out the GCF for polynomials with more than two terms and more than one variable.

Example 8 Factoring Out the GCF

Factor $14p^4t + 21p^2t^2 - 70pt$.

Solution

We begin by factoring $14p^4t$, $21p^2t^2$, and $70pt$:

$$14p^4t = 2 \cdot 7 \cdot p \cdot p \cdot p \cdot p \cdot t$$
$$21p^2t^2 = 3 \cdot 7 \cdot p \cdot p \cdot t \cdot t$$
$$70pt = 2 \cdot 5 \cdot 7 \cdot p \cdot t$$

There are three common factors: 7, p, and t. So, the GCF is $7pt$:

$$14p^4t + 21p^2t^2 - 70pt = 7pt \cdot 2p^3 + 7pt \cdot 3pt - 7pt \cdot 10 \quad \text{7pt is the GCF.}$$
$$= 7pt(2p^3 + 3pt - 10) \quad \text{Factor out 7pt.}$$ ■

Completely Factoring Polynomials

After we factor the GCF out of a polynomial, we must check whether the result can be factored further by using factoring techniques discussed earlier in this section. If a result cannot be factored further, it is said to be **completely factored.**

Example 9 Completely Factoring a Polynomial

Factor $3x^3 + 21x^2 + 36x$.

Solution

The GCF of $3x^3$, $21x^2$ and $36x$ is $3x$:

$$3x^3 + 21x^2 + 36x = 3x(x^2 + 7x + 12)$$

To factor $x^2 + 7x + 12$, we need two integers whose product is 12 and whose sum is 7. Since the product and sum must be positive, we try only positive integers:

(We have temporarily put aside the GCF, $3x$.)	**Product = 12**	**Sum = 7?**
	$1(12) = 12$	$1 + 12 = 13$
	$2(6) = 12$	$2 + 6 = 8$
	$3(4) = 12$	$3 + 4 = 7 \leftarrow$ Success!

Because $3(4) = 12$ and $3 + 4 = 7$, we conclude that the last terms of the factors are 3 and 4:

$$x^2 + 7x + 12 = (x + 3)(x + 4)$$

So,

$$3x^3 + 21x^2 + 36x = 3x(x^2 + 7x + 12) = 3x(x + 3)(x + 4)$$ ■

To factor $3x^3 + 21x^2 + 36x$ in Example 9, we first factored out the GCF, $3x$, and then factored the resulting trinomial, $x^2 + 7x + 12$. In general, **when the leading coefficient of a polynomial is positive and the GCF is not 1, first factor out the GCF.** (We will soon discuss what to do when the leading coefficient of a polynomial is negative.)

WARNING When we factor a trinomial by looking for integers whose product and sum are the appropriate values, we can easily forget about the GCF by the time we have found the other factors. If there is more factoring to be done after you have factored out the GCF, write a note several lines down that reminds you to include the GCF in your result.

Example 10 Completely Factoring a Polynomial

Factor $5x^3y^2 - 10x^2y^3 - 40xy^4$.

Solution

The GCF of $5x^3y^2$, $10x^2y^3$, and $40xy^4$ is $5xy^2$:

$$\begin{aligned} 5x^3y^2 - 10x^2y^3 - 40xy^4 &= 5xy^2(x^2 - 2xy - 8y^2) \\ &= 5xy^2[x^2 - (2y)x - 8y^2] \end{aligned}$$

To factor $x^2 - (2y)x - 8y^2$, we need two monomials whose product is $-8y^2$ and whose sum is $-2y$. So, the last terms are $-4y$ and $2y$:

$$x^2 - 2xy - 8y^2 = (x - 4y)(x + 2y)$$

Hence,

$$5x^3y^2 - 10x^2y^3 - 40xy^4 = 5xy^2(x^2 - 2xy - 8y^2) = 5xy^2(x - 4y)(x + 2y)$$ ■

WARNING It is a common error to forget to factor a polynomial *completely*. In Example 10, we factored the GCF, $5xy^2$, out of $5x^3y^2 - 10x^2y^3 - 40xy^4$:

$$5x^3y^2 - 10x^2y^3 - 40xy^4 = 5xy^2(x^2 - 2xy - 8y^2) \quad \text{Not completely factored}$$

We were not done factoring, because we could still factor $x^2 - 2xy - 8y^2$:

$$5xy^2(x^2 - 2xy - 8y^2) = 5xy^2(x - 4y)(x + 2y) \quad \text{Completely factored}$$

When factoring a polynomial, always *completely* factor it.

Factoring Out the Opposite of the GCF of a Polynomial

How do we factor a polynomial in which the leading coefficient is negative? Consider the polynomial $-2r^4 + 18r^3 - 40r^2$, which has a negative leading coefficient: -2.

> **Factoring When the Leading Coefficient Is Negative**
>
> When the leading coefficient of a polynomial is negative, first factor out the opposite of the GCF.

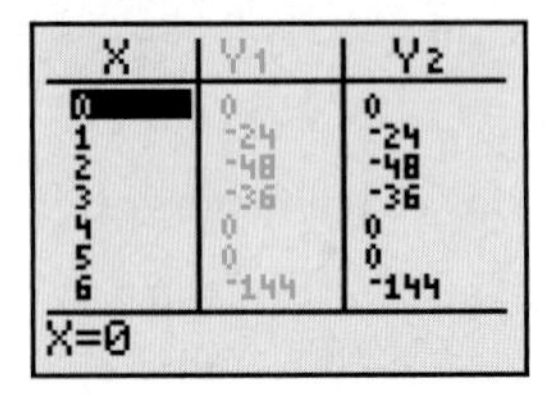

Figure 20 Verify the work

Example 11 Factoring Out the Opposite of the GCF

Factor $-2r^4 + 18r^3 - 40r^2$.

Solution

For $-2r^4 + 18r^3 - 40r^2$, the GCF is $2r^2$. The leading coefficient of $-2r^4 + 18r^3 - 40r^2$ is -2, which is negative. So, we first factor out the opposite of the GCF:

$$
\begin{aligned}
-2r^4 + 18r^3 - 40r^2 &= -2r^2(r^2 - 9r + 20) && \text{Factor out } -2r^2 \text{, opposite of GCF.} \\
&= -2r^2(r-5)(r-4) && \text{Find two integers whose product is 20 and whose sum is } -9.
\end{aligned}
$$

We use a graphing calculator table to verify our work (see Fig. 20). ■

In summary, there are two aspects to factoring that are important to remember:

1. If the leading coefficient of a polynomial is positive and the GCF is not 1, first factor out the GCF. If the leading coefficient is negative, first factor out the opposite of the GCF.
2. Always *completely* factor a polynomial.

group exploration

Factors of an expression and x-intercepts of a function

In this exploration, you will explore a connection between factors of a polynomial of the form $x^2 + bx + c$ and x-intercepts of the polynomial function $f(x) = x^2 + bx + c$.

1. Factor $x^2 + x - 2$.
2. Use ZDecimal to draw a graph of the function $f(x) = x^2 + x - 2$. What are the x-intercepts?
3. What connection do you notice between your result in Problem 1 and the x-intercepts of f? Explain why this connection makes sense.
4. The graph of a function $g(x) = x^2 + bx + c$ is sketched in the indicated figure. Use the graph to help you factor $x^2 + bx + c$. Then find the values of b and c.

a. Fig. 21 **b.** Fig. 22

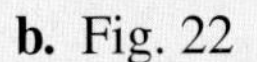

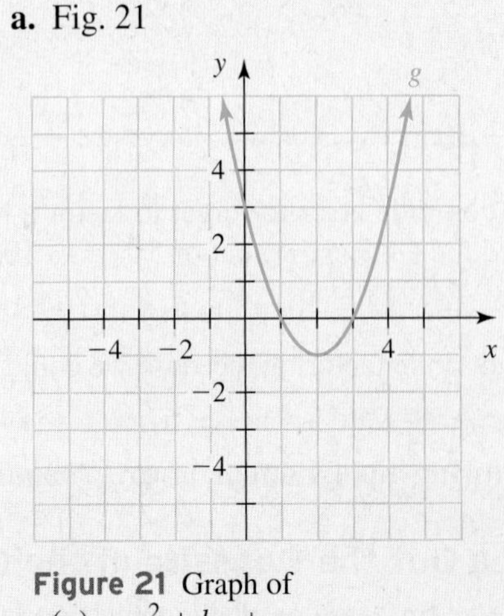

Figure 21 Graph of $g(x) = x^2 + bx + c$

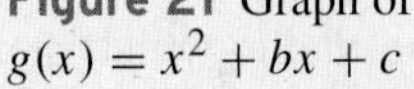

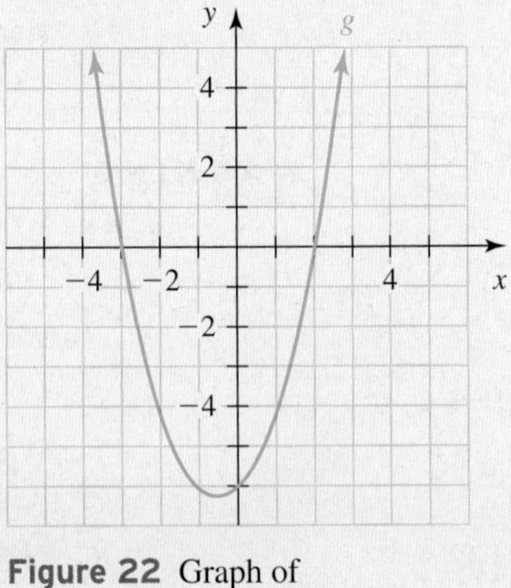

Figure 22 Graph of $g(x) = x^2 + bx + c$

HOMEWORK 6.3

FOR EXTRA HELP ▶ Student Solutions Manual | PH Math/Tutor Center | MathXL® | MyMathLab

Factor when possible. Use a graphing calculator table to verify your work when possible.

1. $x^2 + 11x + 28$
2. $x^2 + 17x + 30$
3. $x^2 - 8x + 12$
4. $x^2 - 8x + 15$
5. $r^2 - 4r - 32$
6. $k^2 - 2k - 24$
7. $x^2 + 5x - 14$
8. $x^2 + 3x - 40$
9. $x^2 - 7x - 12$
10. $x^2 - 11x - 18$
11. $x^2 + 10x + 25$
12. $x^2 + 14x + 49$
13. $t^2 - 18t + 81$
14. $w^2 - 16w + 64$
15. $4x - 5 + x^2$
16. $12 + x^2 - 13x$
17. $a^2 + 12ab + 20b^2$
18. $p^2 + 17pt + 16t^2$
19. $w^2 - 5wy + 4y^2$
20. $x^2 - 10xy + 21y^2$
21. $p^2 + 3pq - 28q^2$
22. $r^2 + 4rt - 12t^2$
23. $b^2 + 4bc - 16c^2$
24. $w^2 + 2wy - 32y^2$
25. $p^2 - 6pq - 16q^2$
26. $m^2 - mr - 42r^2$

Factor when possible. Use a graphing calculator table to verify your work when possible.

27. $3x + 21$
28. $5x + 20$
29. $16x^2 - 12x$
30. $21x^2 - 35x$
31. $9y^5 + 18y^3$
32. $16w^4 + 24w^2$
33. $3ab - 12a^2b$
34. $25p^2q - 45pq^3$
35. $18a^4b^2 + 12a^2b^3$
36. $32x^3y + 40x^2y^4$
37. $-14x^5y + 63x^2y^2$
38. $-22x^3y^3 + 33xy^2$
39. $2x^2 + 12x + 18$
40. $7x^2 + 21x + 14$
41. $3x^2 - 3x - 18$
42. $4x^2 - 4x - 80$
43. $15k - 50 + 5k^2$
44. $-36 + 6t^2 + 6t$
45. $-4x^2 + 24x - 36$
46. $-2x^2 + 24x - 72$
47. $-x^2 + 11x - 10$
48. $-x^2 - 2x + 35$
49. $3w^2 - 27w - 60$
50. $2p^2 - 20p - 32$
51. $4x^3 - 24x^2 + 32x$
52. $2x^3 - 10x^2 + 12x^2$
53. $a^4 - 21a^3 + 20a^2$
54. $t^4 - 5t^3 - 36t^2$
55. $5x^2y + 45xy^2 + 40y^3$
56. $2x^3 + 18x^2y + 28xy^2$
57. $4x^4y - 12x^3y^2 - 40x^2y^3$
58. $3a^3b - 3a^2b^2 - 36ab^3$
59. $-2x^3y^2 + 16x^2y^3 - 32xy^4$
60. $-5x^4y + 20x^3y^2 - 20x^2y^3$
61. A student tries to factor $2x^2 + 16x + 30$:
$$2x^2 + 16x + 30 = 2\left(x^2 + 8x + 15\right)$$
Describe any errors. Then factor the polynomial correctly.
62. A student tries to factor $x^2 + 10x + 24$:
$$\begin{aligned} x^2 + 10x + 24 &= (x+4)(x+6) \\ &= x^2 + 6x + 4x + 24 \\ &= x^2 + 10x + 24 \end{aligned}$$
Describe any errors. Then factor the polynomial correctly.
63. A student tries to factor $12x^3 + 18x^2$:
$$12x^3 + 18x^2 = 2x^2(6x + 9)$$
Describe any errors. Then factor the polynomial correctly.
64. Two students try to factor the polynomial $x^2 + 14x + 48$:

Student A	**Student B**
$x^2 + 14x + 48$	$x^2 + 14x + 48$
$= (x+6)(x+8)$	$= (x+8)(x+6)$

Are both students, one student, or neither student correct? Explain.
65. Which polynomials are equivalent?
$(x-3)(x+6)$ $\quad x^2 - 3x - 18$ $\quad x^2 - 9x - 18$ $\quad x^2 + 3x - 18$ $\quad (x+6)(x-3)$
66. Which polynomials are equivalent?
$x^2 + 4x - 45$ $\quad (x+5)(x-9)$ $\quad x^2 - 4x - 45$ $\quad x^2 - 45$ $\quad (x-9)(x+5)$
67. a. Factor $x^2 - 5x + 4$.
b. Use ZDecimal on a graphing calculator to graph the function $f(x) = x^2 - 5x + 4$. What are the x-intercepts?
c. What connection do you notice between your result in part (a) and the x-intercepts of f? Explain why this connection makes sense.
68. a. Factor $x^2 - 6x + 8$.
b. Use ZDecimal on a graphing calculator to graph the function $f(x) = x^2 - 6x + 8$. What are the x-intercepts?
c. What connection do you notice between your result in part (a) and the x-intercepts of f? Explain why this connection makes sense.
69. Give three examples of a quadratic polynomial in which $x + 4$ is a factor.
70. Give three examples of a cubic polynomial in which $x - 6$ is a factor.
71. Find all possible values of k such that $x^2 + kx + 28$ can be factored.
72. Find all possible values of k such that $x^2 + kx - 32$ can be factored.
73. Compare the process of factoring a polynomial with that of multiplying polynomials.
74. Describe the various factoring techniques addressed in this section. Give an example to illustrate each technique. Finally, explain how to recognize polynomials to which each technique applies.

Related Review

If the polynomial is not factored, then factor it. If it is factored, then find the product.

75. $(x+5)(x-3)$
76. $3(3x-2)(2x-5)$
77. $k^2 - 7k - 30$
78. $6p^2 - 22p + 20$
79. $(7x-5)(7x+5)$
80. $\left(2x^2 - x + 3\right)\left(3x^2 + 4x - 2\right)$
81. $81r^2 - 49$
82. $9y^2 - 42y + 49$

Expressions, Equations, Functions, and Graphs

Perform the indicated instruction. Then use words such as linear, quadratic, cubic, exponential, logarithmic, polynomial, degree, function, one variable, *and* two variables *to describe the expression, equation, or system.*

83. Factor $x^2 + 3x - 28$.

84. Find an approximate equation $y = ab^x$ of an exponential curve that contains the points (3, 98) and (8, 9). Round a and b to the second decimal place.

85. Find the product $(3w - 4)(2w^2 + 3w - 5)$.

86. Solve $3(2)^{2x-1} = 238$. Round any solutions to the fourth decimal place.

6.4 FACTORING POLYNOMIALS

"The great thing about mathematics is that, unlike liberal arts, it is objective rather than subjective. It is great to be able to work on a problem and know that there is a correct and an incorrect answer."

—Colleen S., student

Objectives

- *Factor* a polynomial with four terms *by grouping.*
- *Factor a trinomial by trial and error.*
- Know how to rule out possibilities when factoring by trial and error.
- *Factor a trinomial by grouping.*

In this section, we will first factor some polynomials that have four terms by using a technique called *factoring by grouping*. Then we will factor trinomials of the form ax^2+bx+c where $a \neq 1$. (In Section 6.3, we did so where $a = 1$.) We will discuss two methods: trial and error, and writing such a trinomial as a polynomial with four terms so that we can try to factor by grouping. These two methods give equivalent results.

Factoring a Polynomial with Four Terms by Grouping

In Section 6.3, we factored out a monomial GCF. For example, here we factor out the monomial w from the polynomial $x(w) + 3(w)$:

$$x(w) + 3(w) = (x + 3)(w)$$

We can also factor out a binomial GCF. For example, here we factor out the binomial $y + 5$ from the polynomial $x(y + 5) + 3(y + 5)$:

$$x(y + 5) + 3(y + 5) = (x + 3)(y + 5)$$

We can factor some polynomials that contain four terms by first factoring the first two terms and the last two terms. For example,

$$\underbrace{xy + 5x}_{\text{factor}} + \underbrace{3y + 15}_{\text{factor}} = x(y + 5) + 3(y + 5) \quad \text{Factor both pairs of terms.}$$
$$= (x + 3)(y + 5) \quad \text{Factor out GCF, } y + 5.$$

We call this method *factoring by grouping*.

Example 1 Factoring by Grouping

Factor $3x^3 - 18x^2 - 2x + 12$.

Solution

We begin by factoring the first two terms and the last two:

$$3x^3 - 18x^2 - 2x + 12 = 3x^2(x - 6) - 2(x - 6) \quad \text{Factor both pairs of terms.}$$
$$= (3x^2 - 2)(x - 6) \quad \text{Factor out GCF, } x - 6.$$

We verify the result by finding the product $(3x^2 - 2)(x - 6)$:

$$(3x^2 - 2)(x - 6) = 3x^3 - 18x^2 - 2x + 12$$

■

WARNING It is a common error to think that a polynomial such as $3x^2(x-6)-2(x-6)$ in Example 1 is factored. Even though both terms $3x^2(x-6)$ and $2(x-6)$ are factored, the entire expression $3x^2(x-6)-2(x-6)$ is a difference, not a product. The polynomial $(3x^2-2)(x-6)$ in Example 1 *is* factored, because it is a product.

When trying to factor a polynomial with four terms, consider trying to factor it by grouping.

> **Factoring by Grouping**
>
> For a polynomial with four terms, we **factor by grouping** (if it can be done) by
>
> 1. Factoring the first two terms and the last two terms.
> 2. Factoring out the binomial GCF.

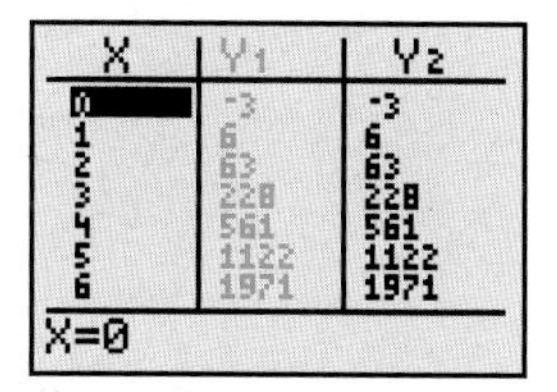

Figure 23 Verify the work

Example 2 Factoring by Grouping

Factor $10x^3-6x^2+5x-3$.

Solution

$$10x^3-6x^2+5x-3 = 2x^2(5x-3)+1(5x-3) \quad \text{Factor both pairs of terms.}$$
$$= (2x^2+1)(5x-3) \quad \text{Factor out GCF, } 5x-3.$$

We use a graphing calculator to verify our work (see Fig. 23). ■

Example 3 Factoring by Grouping

Factor $ax+bx-ay^2-by^2$.

Solution

$$ax+bx-ay^2-by^2 = x(a+b)-y^2(a+b) \quad \text{Factor both pairs of terms.}$$
$$= (x-y^2)(a+b) \quad \text{Factor out GCF, } a+b.$$

We verify the result by finding the product $(x-y^2)(a+b)$:

$$(x-y^2)(a+b) = ax+bx-ay^2-by^2$$ ■

Method 1: Factoring Trinomials by Trial and Error

One way to factor trinomials of the form ax^2+bx+c is to make educated guesses at the factorization and then find the product of these guesses to see if any of them work. This method is called **factoring by trial and error.**

Example 4 Factoring by Trial and Error

Factor $3x^2+14x+8$.

Solution

If we can factor $3x^2+14x+8$, the result will be of the form

$$(3x+?)(x+?)$$

The product of the last terms must be 8, so the last terms must be 1 and 8 or 2 and 4. We can rule out negative last terms in the factors, because the middle term of $3x^2+14x+8$ has the positive coefficient 14. We decide between the two pairs of possible last terms by multiplying:

$$(3x+1)(x+8) = 3x^2+24x+x+8 = 3x^2+25x+8$$
$$(3x+8)(x+1) = 3x^2+3x+8x+8 = 3x^2+11x+8$$
$$(3x+2)(x+4) = 3x^2+12x+2x+8 = 3x^2+14x+8 \leftarrow \text{Success!}$$
$$(3x+4)(x+2) = 3x^2+6x+4x+8 = 3x^2+10x+8$$

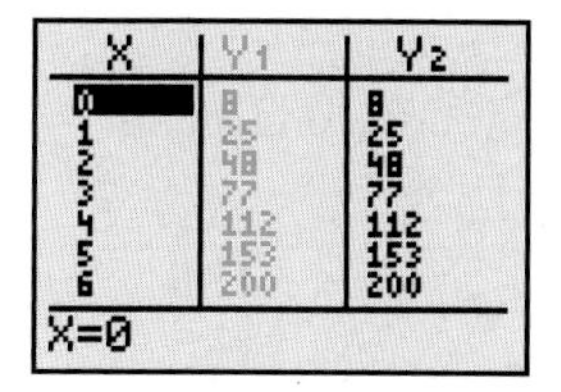

Figure 24 Verify the work

So, $3x^2+14x+8=(3x+2)(x+4)$. We use a graphing calculator table to verify our work (see Fig. 24). ■

In trying to factor a polynomial, once we find the factored polynomial, there is no need to multiply the other possibilities. In Example 4, we multiplied all possible factorizations of $3x^2 + 14x + 8$ only to show how to organize the work in case the last possibility is the correct one.

To use the method shown in Example 4, it is helpful to be able to multiply two binomials in one step. Consider the product of $5x + 2$ and $3x + 2$:

$$\begin{aligned}(5x+2)(3x+2) &= 15x^2 + 10x + 6x + 4\\ &= 15x^2 + 16x + 4\end{aligned}$$

To find the product in one step, we must combine the like terms $10x$ and $6x$ mentally. Note that these like terms come from the product of the two *outer terms* and the product of the two *inner terms* of $(5x + 2)(3x + 2)$:

outer terms; Add these terms mentally.

$$\begin{aligned}(5x+2)(3x+2) &= 15x^2 + 10x + 6x + 4\\ &= 15x^2 + 16x + 4\end{aligned}$$

inner terms

Example 5 Factoring by Trial and Error

Factor $2x^2 - 5x - 25$.

Solution

If we can factor $2x^2 - 5x - 25$, the result will be of the form

$$(2x + ?)(x + ?)$$

The product of the last terms must be -25, so the last terms must be 1 and -25, 5 and -5, or -1 and 25, where we can write each pair in either order. We decide among the three pairs of possible last terms by multiplying:

$$\begin{aligned}(2x+1)(x-25) &= 2x^2 - 49x - 25\\ (2x-25)(x+1) &= 2x^2 - 23x - 25\\ (2x+5)(x-5) &= 2x^2 - 5x - 25 \leftarrow \text{Success!}\\ (2x-5)(x+5) &= 2x^2 + 5x - 25\\ (2x-1)(x+25) &= 2x^2 + 49x - 25\\ (2x+25)(x-1) &= 2x^2 + 23x - 25\end{aligned}$$

Therefore, $2x^2 - 5x - 25 = (2x + 5)(x - 5)$. ■

Factoring $ax^2 + bx + c$ by Trial and Error

To **factor a trinomial** of the form $ax^2 + bx + c$ **by trial and error:** If the trinomial can be factored as a product of two binomials, then the product of the coefficients of the first terms of the binomials is equal to a and the product of the last terms of the binomials is equal to c. For example,

Coefficients of first terms: $5 \cdot 3 = 15 = a$

$$15x^2 + 26x + 8 = (5x + 2)(3x + 4)$$

$a = 15$ $b = 26$ $c = 8$

Last terms: $2 \cdot 4 = 8 = c$

To find the correct factored expression, multiply the possible products and identify those for which the coefficient of x is b.

RULING OUT POSSIBILITIES WHILE FACTORING BY TRIAL AND ERROR

Example 6 shows how to rule out possible factorizations to help speed up the process of factoring.

Example 6 Ruling Out Possibilities

Factor $10x^2 - 19x + 6$.

Solution

If we can factor $10x^2 - 19x + 6$, the result will be in one of these two forms:

$$(10x + ?)(x + ?) \qquad (5x + ?)(2x + ?)$$

The product of the last terms must be 6, so the last terms must be -1 and -6, or -2 and -3, where we can write each pair in either order. We can rule out positive last terms, because the middle term of $10x^2 - 19x + 6$ has a negative coefficient, -19.

Since the terms of $10x^2 - 19x + 6$ do not have a common factor of 2, we can also rule out products that have a factor of 2. For example, we can rule out $(10x - 6)(x - 1)$, because it has a factor of 2:

$$(10x - 6)(x - 1) = 2(5x - 3)(x - 1)$$

We decide among the remaining possible last terms by multiplying:

$$(10x - 1)(x - 6) = 10x^2 - 61x + 6$$

Contains factor of 2, rule out: $(10x - 2)(x - 3)$

$$(10x - 3)(x - 2) = 10x^2 - 23x + 6$$

Contains factor of 2, rule out: $(5x - 1)(2x - 6)$

$$(5x - 6)(2x - 1) = 10x^2 - 17x + 6$$

$$(5x - 2)(2x - 3) = 10x^2 - 19x + 6 \leftarrow \text{Success!}$$

Contains factor of 2, rule out: $(5x - 3)(2x - 2)$

So, $10x^2 - 19x + 6 = (5x - 2)(2x - 3)$. ■

FACTORING OUT THE GCF, THEN FACTORING BY TRIAL AND ERROR

When factoring a polynomial, recall from Section 6.3 that if the GCF is not 1, then we first factor out the GCF and continue factoring if possible. Always completely factor a polynomial.

Example 7 Completely Factoring a Polynomial

Factor $6x^3y^2 + 26x^2y^3 + 24xy^4$.

Solution

To factor $6x^3y^2 + 26x^2y^3 + 24xy^4$, we first factor out the GCF, $2xy^2$:

$$2xy^2(3x^2 + 13xy + 12y^2)$$

If we can factor further, the result will be in the form

$$2xy^2(3x + ?)(x + ?)$$

The product of the last terms must be $12y^2$, so the last terms must be y and $12y$, $2y$ and $6y$, or $3y$ and $4y$, where we can write each pair in either order. We decide by multiplying and rule out any choice that contains a factor the original equation does not:

$$(3x + y)(x + 12y) = 3x^2 + 37xy + 12y^2$$

Contains factor of 3, rule out: $(3x + 12y)(x + y)$

$$(3x + 2y)(x + 6y) = 3x^2 + 20xy + 12y^2$$

Contains factor of 3, rule out: $(3x + 6y)(x + 2y)$

Contains factor of 3, rule out: $(3x + 3y)(x + 4y)$

$$(3x + 4y)(x + 3y) = 3x^2 + 13xy + 12y^2 \leftarrow \text{Success!}$$

(We have temporarily put aside the GCF, $2xy^2$.)

So, $6x^3y^2 + 26x^2y^3 + 24xy^4 = 2xy^2(3x^2 + 13xy + 12y^2) = 2xy^2(3x + 4y)(x + 3y)$. ■

Method 2: Factoring Trinomials by Grouping

Instead of using trial and error to factor a trinomial, we can factor by grouping.

To factor a trinomial of the form $x^2 + bx + c$, recall from Section 6.3 that we look for two integers whose product is c and whose sum is b. To factor a trinomial of the form $ax^2 + bx + c$, we must look for two integers whose product is ac and whose sum is b.

Factoring $ax^2 + bx + c$ by Grouping

To **factor a trinomial** of the form $ax^2 + bx + c$ **by grouping** (if it can be done),

1. Find pairs of numbers whose product is ac.
2. Determine which of the pairs of numbers from step 1 has the sum b. Call this pair of numbers m and n.
3. Write the bx term as $mx + nx$:

$$ax^2 + bx + c = ax^2 + mx + nx + c$$

4. Factor $ax^2 + mx + nx + c$ by grouping.

Another name for this technique is the ***ac*-method.**

Example 8 Factoring a Trinomial by Grouping

Factor $3x^2 + 14x + 8$ by grouping.

Solution

Here, $a = 3$, $b = 14$, and $c = 8$.

Step 1. Find the product ac: $ac = 3(8) = 24$.

Step 2. We want to find two numbers m and n that have the product $ac = 24$ and the sum $b = 14$:

Product = 24	**Sum = 14?**
$1(24) = 24$	$1 + 24 = 25$
$2(12) = 24$	$2 + 12 = 14$ ← Success!
$3(8) = 24$	$3 + 8 = 11$
$4(6) = 24$	$4 + 6 = 10$

Since $2(12) = 24$ and $2 + 12 = 14$, we conclude that the two numbers m and n are 2 and 12.

Step 3. We write the bx term, $14x$, as the sum $mx + nx$:

$$3x^2 + 14x + 8 = 3x^2 + 2x + 12x + 8$$

Step 4. We factor $3x^2 + 2x + 12x + 8$ by grouping:

$$3x^2 + 2x + 12x + 8 = x(3x + 2) + 4(3x + 2) \quad \text{Factor both pairs of terms.}$$
$$= (x + 4)(3x + 2) \quad \text{Factor out GCF, } (3x + 2).$$

■

In step 3 of Example 8, we could switch the mx and nx terms to get $3x^2 + 12x + 2x + 8$ and still be able to factor by grouping in step 4. (Try it.)

In Example 4, we used trial and error to factor $3x^2 + 14x + 8$ as $(3x + 2)(x + 4)$. In Example 8, we factored it as $(x + 4)(3x + 2)$ by using grouping. The two results are equivalent. In general, the results from factoring a trinomial by trial and error and factoring a trinomial by grouping are equivalent.

Example 9 Factoring a Trinomial by Grouping

Factor $6x^2 - 7x + 2$ by grouping.

Solution

Here, $a = 6$, $b = -7$, and $c = 2$.

Step 1. Find the product ac: $ac = 6(2) = 12$.

Step 2. We want to find two numbers m and n that have the product $ac = 12$ and the sum $b = -7$:

Product = 12	Sum = −7?
$-1(-12) = 12$	$-1 + (-12) = -13$
$-2(-6) = 12$	$-2 + (-6) = -8$
$-3(-4) = 12$	$-3 + (-4) = -7$ ← Success!

Since $-3(-4) = 12$ and $-3 + (-4) = -7$, we conclude that the two numbers m and n are -3 and -4.

Step 3. We write $6x^2 - 7x + 2 = 6x^2 - 3x - 4x + 2$.

Step 4. We factor $6x^2 - 3x - 4x + 2$ by grouping:

$$6x^2 - 3x - 4x + 2 = 3x(2x - 1) - 2(2x - 1) \quad \text{Factor both pairs of terms.}$$
$$= (3x - 2)(2x - 1) \quad \text{Factor out GCF, } (2x - 1).$$

■

Example 10 Factoring Out the GCF, Then Factoring by Grouping

Factor $20x^4 - 40x^3 - 25x^2$.

Solution

First, we factor out the GCF, $5x^2$:

$$20x^4 - 40x^3 - 25x^2 = 5x^2(4x^2 - 8x - 5)$$

Next, we use grouping to try to factor $4x^2 - 8x - 5$, where $a = 4$, $b = -8$, and $c = -5$.

Step 1. Find the product ac: $ac = 4(-5) = -20$.

Step 2. We want to find two numbers m and n that have the product $ac = -20$ and the sum $b = -8$:

(We have temporarily put aside the GCF, $5x^2$.)

Product = −20	Sum = −8?
$1(-20) = -20$	$1 + (-20) = -19$
$2(-10) = -20$	$2 + (-10) = -8$ ← Success!
$4(-5) = -20$	$4 + (-5) = -1$
$5(-4) = -20$	$5 + (-4) = 1$
$10(-2) = -20$	$10 + (-2) = 8$
$20(-1) = -20$	$20 + (-1) = 19$

Since $2(-10) = -20$ and $2 + (-10) = -8$, we conclude that the two numbers m and n are 2 and -10.

Step 3. We write $4x^2 - 8x - 5 = 4x^2 + 2x - 10x - 5$.

Step 4. We factor $4x^2 + 2x - 10x - 5$ by grouping:

$$4x^2 + 2x - 10x - 5 = 2x(2x + 1) - 5(2x + 1) \quad \text{Factor both pairs of terms.}$$
$$= (2x - 5)(2x + 1) \quad \text{Factor out GCF, } (2x + 1).$$

So, $20x^4 - 40x^3 - 25x^2 = 5x^2(4x^2 - 8x - 5) = 5x^2(2x - 5)(2x + 1)$. We use a graphing calculator table to verify our work (see Fig. 25). ■

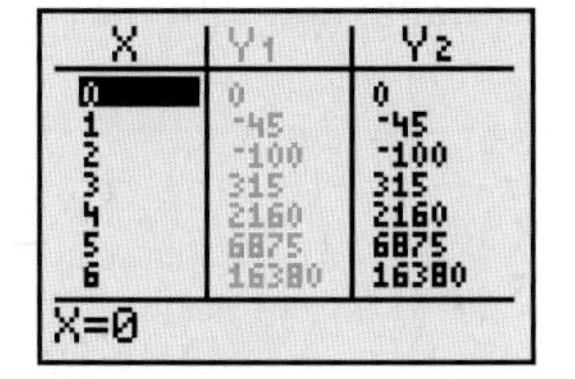

Figure 25 Verify the work

group exploration

Factoring polynomials

1. A student tries to factor $2x^2 - 17x - 30$:
$$2x^2 - 17x - 30 = (2x - 5)(x - 6)$$
Multiply $(2x - 5)(x - 6)$ to show that the work is incorrect. Then factor $2x^2 - 17x - 30$ correctly.
2. A student tries to factor $2x^2 + 10x + 12$:
$$2x^2 + 10x + 12 = (2x + 4)(x + 3)$$
Explain why the work is not correct. Then factor the polynomial correctly.

3. A student tries to factor $x^3 - 3x^2 + 2x - 6$:

$$x^3 - 3x^2 + 2x - 6 = x^2(x - 3) + 2(x - 3)$$

Explain why the student has not succeeded in factoring the given expression. Then factor it correctly.

4. A student tries to factor $2x^2 - x - 6$. Since the product of -3 and 2 is -6 and the sum of -3 and 2 is -1, the student does the following work:

$$2x^2 - x - 6 = (2x - 3)(x + 2)$$

Multiply $(2x - 3)(x + 2)$ to show that the work is incorrect. Explain what is wrong with the student's reasoning. Then factor the polynomial correctly.

group exploration

Looking ahead: Factoring the difference of two squares

1. Find the product $2(x + 4)$. Then factor $2x + 8$. Compare finding the product $2(x + 4)$ with factoring $2x + 8$.
2. Find the product.
   a. $(x - 3)(x + 3)$ **b.** $(x - 4)(x + 4)$
   c. $(x - 5)(x + 5)$ **d.** $(2x - 7)(2x + 7)$
3. Factor the polynomial.
   a. $x^2 - 9$ **b.** $x^2 - 36$
   c. $16x^2 - 25$ **d.** $4x^2 - 9$
4. Describe in general how to factor the difference of two squares.

TIPS FOR SUCCESS: Choose a Good Time and Place to Study

To improve your effectiveness at studying, take stock of when and where you are best able to study. Tracy, a student who lives in a sometimes distracting household, completes her assignments at the campus library just after she attends her classes. Gerome, a morning person, gets up early so that he can study before classes. Being consistent in the time and location for studying can help, too. Research has shown that, after a person repeats a daily activity for about 21 days, the activity becomes habit. Even if it takes willpower to shuffle your schedule so that you can study at your prime time and location, things will start to feel comfortable and familiar within three weeks.

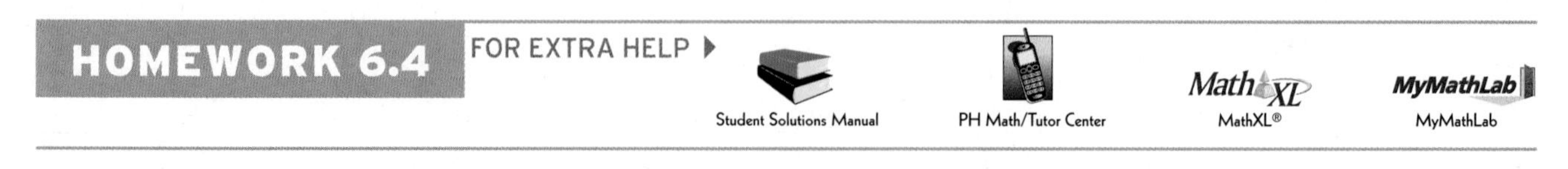

Factor when possible. Use a graphing calculator table to verify your work when possible.

1. $x^3 + 3x^2 + 4x + 12$
2. $x^3 + 2x^2 + 5x + 10$
3. $5x^3 - 20x^2 + 3x - 12$
4. $2x^3 - 4x^2 + 5x - 10$
5. $6m^3 - 15m^2 + 2m - 5$
6. $15k^3 - 10k^2 + 3k - 2$
7. $10x^3 + 25x^2 - 2x - 5$
8. $6x^3 + 9x^2 - 2x - 3$
9. $ax - 3ay - 2bx + 6by$
10. $ax - 5ay - 3bx + 15by$
11. $5a^2x + 2a^2y - 5bx - 2by$
12. $3ax + 7ay - 3b^2x - 7b^2y$

Factor when possible. Use a graphing calculator table to verify your work when possible.

13. $3x^2 + 11x + 10$
14. $5x^2 + 13x + 6$
15. $2x^2 - x - 15$
16. $3x^2 - 19x - 14$
17. $5p^2 - 21p + 4$
18. $7t^2 - 24t + 9$
19. $4x^2 + 16x + 15$
20. $6x^2 + 19x + 8$
21. $9x^2 - 4x - 8$
22. $8x^2 - 5x - 6$
23. $1 + 9w^2 - 6w$
24. $-10m + 1 + 25m^2$

25. $15x^2 + x - 6$

26. $8x^2 + 9x - 14$

27. $6x^2 - 17x + 12$

28. $4x^2 - 21x + 20$

29. $16y^2 - 29y - 6$

30. $12w^2 - 19w - 10$

31. $10a^2 + 21ab + 9b^2$

32. $8m^2 + 26mn + 15n^2$

33. $20x^2 + 17xy - 3y^2$

34. $16p^2 + 31pt - 2t^2$

35. $6w^2 - 3wy - 4y^2$

36. $9m^2 - 5mp - 6p^2$

37. $4r^2 - 20ry + 25y^2$

38. $9b^2 - 24bc + 16c^2$

39. $6x^2 + 26x - 20$

40. $6x^2 + 9x - 60$

41. $-12x^2 + 3x + 9$

42. $-10x^2 - 8x + 24$

43. $12x - 32x^2 + 16x^3$

44. $-65x^2 + 25x + 30x^3$

45. $30x^4 + 4x^3 - 2x^2$

46. $56x^4 - 44x^3 + 8x^2$

47. $36t^3 + 48t^2w + 16tw^2$

48. $8x^2y - 40xy^2 + 50y^3$

49. $20a^3b^2 + 30a^2b^3 - 140ab^4$

50. $15a^3b - 36a^2b^2 + 12ab^3$

Factor when possible. Use a graphing calculator table to verify your work when possible.

51. $x^2 - 6x - 40$

52. $x^2 + 3x - 18$

53. $3w^3 - 6w^2 + 5w - 10$

54. $3t^3 + 15t^2 - 4t - 12$

55. $3x^4 - 21x^3y - 54x^2y^2$

56. $6a^2b^3 - 36ab^4 + 48b^5$

57. $m^2 - 11mp - 30p^2$

58. $r^2 - 11rt - 24t^2$

59. $6x^2 - 19x + 10$

60. $8x^2 + 37x - 15$

61. $x^2 + xy - 30y^2$

62. $x^2 - 3xy - 28y^2$

63. $-6r^3 + 24r^2 - 24r$

64. $-5k^3 + 40k^2 - 80k$

65. $-10 + 12x^2 + 2x$

66. $3x - 6 + 30x^2$

67. $a^2x - 3a^2y - 2bx + 6by$

68. $2ax - 4ay - 3bx + 6by$

69. $x^2 - 10x - 16$

70. $x^2 - 13x - 12$

71. $10p^3t^2 + 22p^2t^3 - 24pt^4$

72. $24w^4y - 44w^3y^2 - 40w^2y^3$

73. A student tries to factor $x^3 + 5x^2 - 3x - 15$:

$$x^3 + 5x^2 - 3x - 15 = x^2(x+5) - 3(x+5)$$

Describe any errors. Then factor the expression correctly.

74. A student tries to factor $4x^2 + 8x + 3$:

$$4x^2 + 8x + 3 = 4x(x+2) + 3$$

Describe any errors. Then factor the expression correctly.

75. A student tries to factor the polynomial $3x^2 - 9x - 30$:

$$3x^2 - 9x - 30 = (3x - 15)(x + 2)$$

Describe any errors. Then factor the polynomial correctly.

76. A student tries to factor the polynomial $4x^2 + 32x + 60$:

$$\begin{aligned} 4x^2 + 32x + 60 &= (2x+6)(2x+10) \\ &= 2(x+3)(2)(x+5) \\ &= 4(x+3)(x+5) \end{aligned}$$

Is the work correct? If no, explain. If yes, is there an easier method?

77. Which of the following expressions are equivalent?

$2(x-2)(x-6)$ $\quad 2(x^2 - 8x + 12)$ $\quad 2x^2 - 16x + 24$

$(x-2)(2x-12)$ $\quad (x-2)(2x-6)$ $\quad (2x-2)(x-6)$

$2(x-4)^2 - 8$ $\quad (2x-4)(x-6)$

78. Describe the various factoring techniques addressed in this section and in Section 6.3. Give an example to illustrate each technique. Explain how to recognize polynomials to which each technique applies.

Related Review

If the polynomial is not factored, then factor it. If the polynomial is factored, then find the product.

79. $12x^3 - 27x$

80. $2x^3 - 20x^2 + 42x$

81. $-2(3p+5)(4p-3)$

82. $5y(2y-3)(2y+3)$

83. $2x^3 - 5x^2 - 18x + 45$

84. $24x^2 - 48x - 3x^3$

85. $(3k+4)(2k^2 - k + 3)$

86. $(w^2 - 5)(w - 7)$

Expressions, Equations, Functions, and Graphs

Perform the indicated instruction. Then use words such as linear, quadratic, cubic, exponential, logarithmic, polynomial, degree, function, one variable, *and* two variables *to describe the expression, equation, or system.*

87. Simplify $-2x(3x-5)^2$.

88. Solve $3(2t-5) - 2(2t+3) = 12$.

89. Factor $8x^3 - 40x^2 + 50x$.

90. Find the product $(5x-2)(3x^2 - 4x + 2)$.

6.5 FACTORING SPECIAL BINOMIALS; A FACTORING STRATEGY

Objectives

- Factor differences of squares.
- Factor sums of cubes and differences of cubes.
- Know a factoring strategy.

In this section, we will first discuss how to factor three special types of binomials: differences of squares, sums of cubes, and differences of cubes. Then we will discuss a factoring strategy to help us sift through the many factoring techniques we have discussed in this chapter and select the best techniques to factor a given polynomial completely.

The Difference of Two Squares

In Section 6.2, we found the product of two binomial conjugates by using the property $(A+B)(A-B) = A^2 - B^2$. The expression $A^2 - B^2$ is the difference of two squares. To factor a difference of two squares, we can use that property in reverse.

Difference of Two Squares

$$A^2 - B^2 = (A+B)(A-B)$$

In words: The difference of the squares of two terms is the product of the sum of the terms and the difference of the terms.

Example 1 Factoring Differences of Two Squares

Factor.

1. $x^2 - 49$
2. $36w^2 - 25y^2$

Solution

1. Since $x^2 - 49 = (x)^2 - (7)^2$, we substitute x for A and 7 for B:

$$\begin{array}{c} A^2 - B^2 = (A+B)(A-B) \\ \downarrow \quad \downarrow \qquad \downarrow \quad \downarrow \quad \downarrow \quad \downarrow \\ x^2 - 49 = x^2 - 7^2 = (x+7)(x-7) \end{array}$$

2. Since $36w^2 - 25y^2 = (6w)^2 - (5y)^2$, we substitute $6w$ for A and $5y$ for B:

$$\begin{array}{c} A^2 - B^2 = (A+B)(A-B) \\ \swarrow \quad \downarrow \qquad \downarrow \quad \downarrow \quad \searrow \quad \searrow \\ 36w^2 - 25y^2 = (6w)^2 - (5y)^2 = (6w+5y)(6w-5y) \end{array}$$

■

WARNING The binomial $x^2 + 25$ is prime. It is a common error to think that this polynomial can be factored as $(x+5)^2$, but if you simplify $(x+5)^2$, you'll see that the result is $x^2 + 10x + 25$, not $x^2 + 25$. In general, **except for factoring out the GCF, a sum of squares cannot be factored.**

Example 2 Factoring Differences of Two Squares

Factor.

1. $4x^3 - 36x$
2. $16p^4 - 1$

Solution

1. To factor $4x^3 - 36x$, we first factor out the GCF, $4x$:

$$\begin{aligned} 4x^3 - 36x &= 4x(x^2 - 9) && \text{Factor out GCF, } 4x. \\ &= 4x(x+3)(x-3) && A^2 - B^2 = (A+B)(A-B) \end{aligned}$$

2. The binomial $16p^4 - 1$ is a difference of squares, since $(4p^2)^2 = 16p^4$ and $1^2 = 1$:

$$\begin{aligned} 16p^4 - 1 &= (4p^2)^2 - 1^2 && \text{Write as difference of squares.} \\ &= (4p^2+1)(4p^2-1) && A^2 - B^2 = (A+B)(A-B) \\ &= (4p^2+1)(2p+1)(2p-1) && 4p^2 + 1 \text{ is prime; } A^2 - B^2 = (A+B)(A-B) \end{aligned}$$

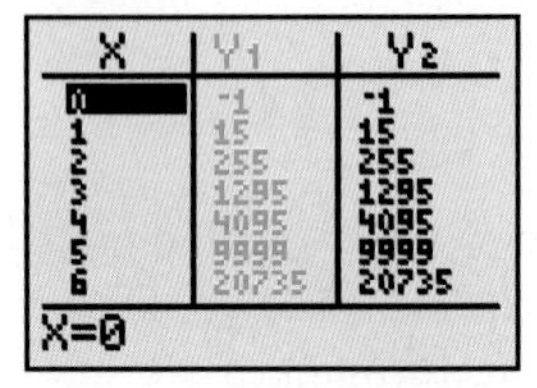

Figure 26 Verify the work

We use a graphing calculator table to verify our work (see Fig. 26). ■

The Sum or Difference of Two Cubes

So far, we have discussed how to factor a difference of squares. We can also factor a sum of cubes and factor a difference of cubes.

To see how to factor the sum of two cubes, we begin by multiplying the expressions $A + B$ and $A^2 - AB + B^2$:

$$\begin{aligned}(A + B)(A^2 - AB + B^2) &= A \cdot A^2 - A \cdot AB + A \cdot B^2 + B \cdot A^2 - B \cdot AB + B \cdot B^2 \\ &= A^3 - A^2B + AB^2 + A^2B - AB^2 + B^3 \\ &= A^3 + B^3\end{aligned}$$

So, $(A+B)(A^2 - AB + B^2) = A^3 + B^3$. The right-hand side of the equation, $A^3 + B^3$, is a sum of two cubes. By similar work, we can also find a property for the difference of two cubes.

Sum or Difference of Two Cubes

$$A^3 + B^3 = (A + B)(A^2 - AB + B^2) \quad \text{Sum of two cubes}$$
$$A^3 - B^3 = (A - B)(A^2 + AB + B^2) \quad \text{Difference of two cubes}$$

We can use these two properties to factor any polynomial that is a sum or difference of two cubes. To use these properties, it will help to memorize the following cubes:

$$2^3 = 8 \qquad 3^3 = 27 \qquad 4^3 = 64 \qquad 5^3 = 125 \qquad 10^3 = 1000$$

Example 3 Factoring a Sum and a Difference of Two Cubes

Factor.

1. $x^3 + 8$ **2.** $x^3 - 125$

Solution

1.

$$A^3 + B^3 = (A + B)(A^2 - AB + B^2)$$
$$\begin{aligned}x^3 + 8 = x^3 + 2^3 &= (x + 2)(x^2 - x \cdot 2 + 2^2) && \text{Factor.} \\ &= (x + 2)(x^2 - 2x + 4) && \text{Simplify.}\end{aligned}$$

The trinomial $x^2 - 2x + 4$ is prime, so we have completely factored $x^3 + 8$.

2.

$$A^3 - B^3 = (A - B)(A^2 + AB + B^2)$$
$$\begin{aligned}x^3 - 125 = x^3 - 5^3 &= (x - 5)(x^2 + x \cdot 5 + 5^2) && \text{Factor.} \\ &= (x - 5)(x^2 + 5x + 25) && \text{Simplify.}\end{aligned}$$

The trinomial $x^2 + 5x + 25$ is prime, so we have completely factored $x^3 - 125$. We use a graphing calculator table to verify our work (see Fig. 27). ■

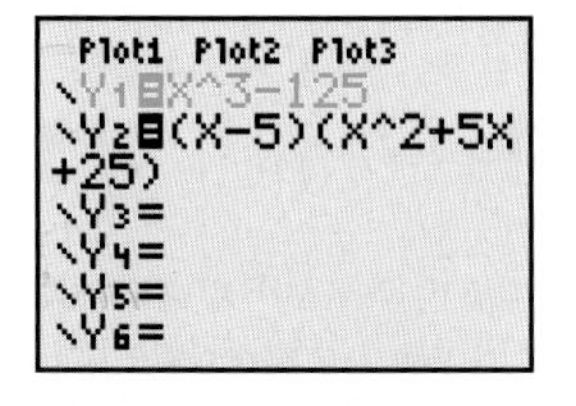

X	Y1	Y2
0	-125	-125
1	-124	-124
2	-117	-117
3	-98	-98
4	-61	-61
5	0	0
6	91	91

X=0

Figure 27 Verify the work

Example 4 Factoring a Sum and a Difference of Two Cubes

Factor.

1. $64t^3 + 27w^3$ **2.** $3x^5 - 24x^2y^3$

Solution

1.

$$\begin{aligned}64t^3 + 27w^3 &= (4t)^3 + (3w)^3 && \text{Write as a sum of cubes.} \\ &= (4t + 3w)((4t)^2 - 4t \cdot 3w + (3w)^2) && A^3 + B^3 = (A + B)(A^2 - AB + B^2) \\ &= (4t + 3w)(16t^2 - 12tw + 9w^2) && \text{Simplify.}\end{aligned}$$

The trinomial $16t^2 - 12tw + 9w^2$ is prime, so we have completely factored $64t^3 + 27w^3$.

2. For $3x^5 - 24x^2y^3$, first we factor out the GCF, $3x^2$:

$$
\begin{aligned}
3x^5 - 24x^2y^3 &= 3x^2(x^3 - 8y^3) && \text{Factor out GCF, } 3x^2. \\
&= 3x^2(x^3 - (2y)^3) && \text{Write as a difference of cubes.} \\
&= 3x^2(x - 2y)(x^2 + x \cdot 2y + (2y)^2) && A^3 - B^3 = (A - B)(A^2 + AB + B^2) \\
&= 3x^2(x - 2y)(x^2 + 2xy + 4y^2) && \text{Simplify.}
\end{aligned}
$$

The trinomial $x^2 + 2xy + 4y^2$ is prime, so we have completely factored $3x^5 - 24x^2y^3$. ■

Consider the properties for the sum of cubes and the difference of cubes:

$$A^3 + B^3 = (A + B)(A^2 - AB + B^2) \qquad A^3 - B^3 = (A - B)(A^2 + AB + B^2)$$

Provided that we have first factored out the GCF (or its opposite), we can assume that the trinomials $A^2 - AB + B^2$ and $A^2 + AB + B^2$ are prime.

Example 5 Factoring the Difference of Two Sixth Powers

Factor $n^6 - p^6$.

Solution

Since $n^6 = (n^3)^2$ and $p^6 = (p^3)^2$, we can begin to factor the binomial by using the property for a difference of two squares:

$$
\begin{aligned}
n^6 - p^6 &= (n^3)^2 - (p^3)^2 && \text{Write as a difference of two squares.} \\
&= (n^3 + p^3)(n^3 - p^3) && A^2 - B^2 = (A + B)(A - B) \\
&= (n + p)(n^2 - np + p^2)(n - p)(n^2 + np + p^2) && A^3 + B^3 = (A + B)(A^2 - AB + B^2); \\
& && A^3 - B^3 = (A - B)(A^2 + AB + B^2)
\end{aligned}
$$

■

Although we could have started to factor $n^6 - p^6$ by writing it as a difference of two cubes, $(n^2)^3 - (p^2)^3$, this would lead to very challenging factoring after we had used the property for a difference of two cubes.

A Factoring Strategy

We will now discuss a five-step factoring strategy that will help us determine the best factoring techniques to use to factor a given polynomial completely.

Five-Step Factoring Strategy

These five steps can be used to factor many polynomials (steps 2–4 can be applied to the entire polynomial or to a factor of the polynomial):

1. If the leading coefficient is positive and the GCF is not 1, factor out the GCF. If the leading coefficient is negative, factor out the opposite of the GCF.
2. For a binomial, try using one of the properties for the difference of two squares, the sum of two cubes, or the difference of two cubes.
3. For a trinomial of the form $ax^2 + bx + c$,
   a. If $a = 1$, try to find two integers whose product is c and whose sum is b.
   b. If $a \neq 1$, try to factor by using trial and error or by grouping.
4. For an expression with four terms, try factoring by grouping.
5. Continue applying steps 2–4 until the polynomial is completely factored.

Example 6 Factoring a Polynomial

Factor $x^4 - 2x^3 + 1000x - 2000$.

Solution

Since $x^4 - 2x^3 + 1000x - 2000$ has four terms, we try to factor it by grouping:

$$\begin{aligned} x^4 - 2x^3 + 1000x - 2000 \\ &= x^3(x-2) + 1000(x-2) \\ &= (x^3 + 1000)(x-2) && \text{Factor by grouping.} \\ &= (x+10)(x^2 - 10x + 100)(x-2) && A^3 + B^3 = (A+B)(A^2 - AB + B^2) \end{aligned}$$

■

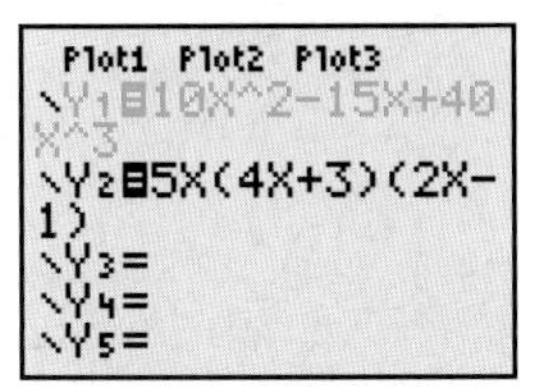

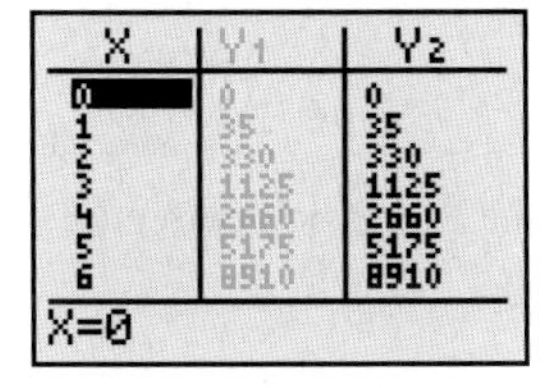

Figure 28 Verify the work

Example 7 Factoring a Polynomial

Factor $10x^2 - 15x + 40x^3$.

Solution

First, we write $10x^2 - 15x + 40x^3$ in descending order:

$$\begin{aligned} 10x^2 - 15x + 40x^3 &= 40x^3 + 10x^2 - 15x && \text{Rearrange terms.} \\ &= 5x(8x^2 + 2x - 3) && \text{Factor out GCF, } 5x. \\ &= 5x(4x+3)(2x-1) && \text{Factor by trial and error.} \end{aligned}$$

We use a graphing calculator table to verify our work (see Fig. 28). ■

Example 8 Factoring a Polynomial

Factor $50t^2w^2 - 8w^4$.

Solution

For $50t^2w^2 - 8w^4$, the GCF is $2w^2$. First, we factor out $2w^2$:

$$50t^2w^2 - 8w^4 = 2w^2(25t^2 - 4w^2)$$

Since the factor $25t^2 - 4w^2$ has two terms, we check to see whether it is the difference of two squares, which it is. So, we have

$$50t^2w^2 - 8w^4 = 2w^2(25t^2 - 4w^2) = 2w^2(5t + 2w)(5t - 2w)$$

■

Example 9 Factoring a Polynomial

Factor $3a^3b - 21a^2b^2 + 18ab^3$.

Solution

For $3a^3b - 21a^2b^2 + 18ab^3$, the GCF is $3ab$. First, we factor out $3ab$:

$$3a^3b - 21a^2b^2 + 18ab^3 = 3ab(a^2 - 7ab + 6b^2)$$

Since the factor $a^2 - 7ab + 6b^2$ is a trinomial with leading coefficient 1, we try to find two monomials whose product is $6b^2$ and whose sum is $-7b$. The monomials are $-b$ and $-6b$, so we have

$$3a^3b - 21a^2b^2 + 18ab^3 = 3ab(a^2 - 7ab + 6b^2) = 3ab(a-b)(a-6b)$$

■

group exploration

Looking ahead: Zero factor property

1. What can you say about A or B if $AB = 0$?
2. What can you say about A or B if $A(B-1) = 0$?
3. What can you say about x if $x(x-1) = 0$?
4. Solve $x^2 - x = 0$. [**Hint:** Does this have something to do with Problem 3?]
5. Solve $2x^2 - 6x = 0$.
6. Solve $x^2 - 8x + 15 = 0$.

HOMEWORK 6.5

FOR EXTRA HELP ▸ Student Solutions Manual | PH Math/Tutor Center | MathXL® | MyMathLab

Factor when possible. Use a graphing calculator table to verify your work when possible.

1. $x^2 - 25$
2. $x^2 - 9$
3. $a^2 - 36$
4. $t^2 - 81$
5. $4x^2 - 49$
6. $25x^2 - 16$
7. $9x^2 + 100$
8. $49x^2 + 4$
9. $16p^2 - 25t^2$
10. $81k^2 - 49r^2$
11. $75x^2 - 12$
12. $45x^2 - 20$
13. $18a^3b - 32ab^3$
14. $50m^3t - 72mt^3$
15. $16x^4 - 81$
16. $81x^4 - 1$
17. $t^4 - w^4$
18. $a^4 - 81b^4$

Factor when possible. Use a graphing calculator table to verify your work when possible.

19. $x^3 + 27$
20. $x^3 + 64$
21. $x^3 - 8$
22. $x^3 - 27$
23. $m^3 + 1$
24. $r^3 - 1$
25. $8x^3 + 27$
26. $27x^3 + 64$
27. $125x^3 - 8$
28. $1000x^3 - 27$
29. $27p^3 + 8t^3$
30. $8w^3 + 125y^3$
31. $27x^3 - 64y^3$
32. $64c^3 - 125d^3$
33. $5x^3 + 40$
34. $10x^3 + 640$
35. $2x^4 - 54xy^3$
36. $4x^5 - 32x^2y^3$
37. $k^6 - 1$
38. $t^6 - 64$
39. $64x^6 - y^6$
40. $x^6 - y^6$

Factor when possible. Use a graphing calculator table to verify your work when possible.

41. $a^2 - 3ab - 28b^2$
42. $p^2 - 9pt + 20t^2$
43. $2x^4 - 16xy^3$
44. $4x^3y - 4y^4$
45. $-7x - 18 + x^2$
46. $2x + x^2 - 80$
47. $4x^3y - 8x^2y^2 - 96xy^3$
48. $2x^4y + 24x^3y^2 + 70x^2y^3$
49. $-k^2 + 12k - 36$
50. $-t^2 + 18t - 81$
51. $4x^2 + 9x + 6$
52. $8x^2 - 13x + 10$
53. $x^3 - 2x^2 - 9x + 18$
54. $4x^3 - x^2 - 16x + 4$
55. $6x^4 - 33x^3 + 45x^2$
56. $15x^4 + 55x^3 - 20x^2$
57. $32m^2 - 98t^2$
58. $75r^2 - 27y^2$
59. $8x^2 + 10x - 3$
60. $6x^2 + 11x - 10$
61. $12x^2y - 26xy^2 - 10y^3$
62. $36x^4 - 21x^3y + 3x^2y^2$
63. $x^2 - 11x - 24$
64. $x^2 - 13x - 36$
65. $125x^3 + 27$
66. $64x^3 + 125$
67. $p^2 + 18p + 81$
68. $r^2 + 20r + 100$
69. $20x^3 - 8x^2 - 5x + 2$
70. $9x^3 + 27x^2 - x - 3$
71. $49x^2 + 14x + 1$
72. $64x^2 - 16x + 1$
73. $25x^2 + 81y^2$
74. $36m^2 + 49n^2$
75. $2w^3y + 250y^4$
76. $5a^3b^2 + 5000b^5$
77. $-3x^3 + 3x^2 + 90x$
78. $-2x^3 + 20x^2 - 32x$
79. $27x^3 - 75x$
80. $98x^3 - 18x$
81. $81p^4 - 16q^4$
82. $16m^4 - n^4$

83. A student tries to factor $x^3 - 8$:

$$x^3 - 8 = (x - 2)(x^2 + 4x + 4)$$

Describe any errors. Then factor the expression correctly.

84. A student tries to factor $x^3 + 27$:

$$x^3 + 27 = (x + 3)(x^2 + 3x + 9)$$

Describe any errors. Then factor the expression correctly.

85. A student tries to factor $4x^2 + 100$:

$$4x^2 + 100 = 4(x^2 + 25) = 4(x + 5)(x + 5) = 4(x + 5)^2$$

Describe any errors. Then factor the expression correctly.

86. A student tries to factor $x^3 + a^3$:

$$x^3 + a^3 = (x + a)(x^2 - 2ax + a^2)$$

Describe any errors. Then factor the expression correctly.

87. a. Factor $x^2 - 4$.

b. Use ZStandard on a graphing calculator to graph the function $f(x) = x^2 - 4$. (For a closer look, try ZDecimal.) What are the x-intercepts?

c. What connection do you notice between your result in part (a) and the x-intercepts of f? Explain why this connection makes sense.

88. a. Factor $x^3 - x^2 - 4x + 4$.

b. Use ZStandard on a graphing calculator to graph the function $f(x) = x^3 - x^2 - 4x + 4$. (For a closer look, try ZDecimal.) What are the x-intercepts?

c. What connection do you notice between your result in part (a) and the x-intercepts of f? Explain why this connection makes sense.

89. Describe how to factor a difference of two squares, a sum of two cubes, and a difference of two cubes.

90. Describe, in your own words, a strategy for factoring polynomials.

Related Review

If the expression is not factored, then factor it. If the expression is factored, then find the product.

91. $(3x - 7)(3x + 7)$
92. $(8x + 1)(8x - 1)$
93. $36p^2 - 49$
94. $81c^2 - 25$
95. $(t - 5)(t^2 + 5t + 25)$
96. $(w + 3)(w^2 - 3w + 9)$
97. $27p^3 + 1$
98. $8m^3 - 125$

Expressions, Equations, Functions, and Graphs

Perform the indicated instruction. Then use words such as linear, quadratic, cubic, exponential, logarithmic, polynomial, degree, function, one variable, *and* two variables *to describe the expression, equation, or system.*

99. Graph $2(3x - 2y) = 3x + 4$ by hand.

100. Solve $\log_3(4w + 2) = 4$.

101. Solve:

$$3x - 5y = 21$$
$$2x + 7y = -17$$

102. Simplify $\left(8b^{-6}c^3\right)^{2/3}\left(4b^4c^{-6}\right)^{1/2}$.

6.6 USING FACTORING TO SOLVE POLYNOMIAL EQUATIONS

Objectives

- Know the *zero factor property.*
- Use factoring to solve *quadratic equations in one variable.*
- Find the x-intercept(s) of polynomial functions.
- Know a connection between x-intercepts of a function and the solutions of a related equation in one variable.
- Use factoring to solve *cubic equations in one variable.*
- Use graphing to solve a polynomial equation in one variable.
- Use a *quadratic model* to make estimates and predictions about an authentic situation.
- Solve area-of-rectangle problems.

In this section, we will discuss how to use factoring to help us solve equations. We will use this skill to help us make estimates and predictions for some authentic situations.

Zero Factor Property

In Section 6.1, we graphed quadratic equations in *two* variables. In this section, we will solve quadratic equations in *one* variable, such as

$$x^2 - 3x - 28 = 0 \qquad 25x^2 - 49 = 0 \qquad 12 = 9x^2 + 4$$

A **quadratic equation in one variable** is an equation that can be put into the form

$$ax^2 + bx + c = 0$$

where a, b, and c are constants and $a \neq 0$. The connection between solving a quadratic equation and factoring an expression lies in the *zero factor property.*

Zero Factor Property

Let A and B be real numbers:

$$\text{If } AB = 0, \text{ then } A = 0 \text{ or } B = 0.$$

In words: If the product of two numbers is zero, then at least one of the numbers must be zero.

Solving Quadratic Equations in One Variable

We can use the zero factor property to help us solve some quadratic equations in one variable.

Example 1 Solving a Quadratic Equation

Solve $(x - 5)(x + 2) = 0$.

Solution

$(x - 5)(x + 2) = 0$	Original equation
$x - 5 = 0$ or $x + 2 = 0$	Zero factor property
$x = 5$ or $x = -2$	Add 5 to both sides./Subtract 2 from both sides.

So, the solutions are 5 and -2. We check that both 5 and -2 satisfy the original equation:

Check $x = 5$	**Check $x = -2$**
$(x-5)(x+2)=0$	$(x-5)(x+2)=0$
$(5-5)(5+2)\stackrel{?}{=}0$	$(-2-5)(-2+2)\stackrel{?}{=}0$
$0(7)\stackrel{?}{=}0$	$-7(0)\stackrel{?}{=}0$
$0\stackrel{?}{=}0$	$0\stackrel{?}{=}0$
true	true

■

Example 2 Solving a Quadratic Equation

Solve $w^2 - 2w - 8 = 0$.

Solution

$$w^2 - 2w - 8 = 0 \quad \text{Original equation}$$
$$(w+2)(w-4) = 0 \quad \text{Factor left side.}$$
$$w + 2 = 0 \quad \text{or} \quad w - 4 = 0 \quad \text{Zero factor property}$$
$$w = -2 \quad \text{or} \quad w = 4 \quad \text{Subtract 2 from both sides./ Add 4 to both sides.}$$

We check that both -2 and 4 satisfy the original equation:

Check $w = -2$	**Check $w = 4$**
$w^2-2w-8=0$	$w^2-2w-8=0$
$(-2)^2-2(-2)-8\stackrel{?}{=}0$	$4^2-2(4)-8\stackrel{?}{=}0$
$4+4-8\stackrel{?}{=}0$	$16-8-8\stackrel{?}{=}0$
$0\stackrel{?}{=}0$	$0\stackrel{?}{=}0$
true	true

So, the solutions are -2 and 4. ■

The key step in solving a quadratic equation of the form $ax^2 + bx + c = 0$ is to factor the left side of the equation so that we can apply the zero factor property.

Connection Between x-Intercepts and Solutions

There is an important connection between x-intercepts of the graph of a function and the solutions of a related equation in one variable. We begin to investigate this connection in Example 3.

Example 3 Finding x-Intercepts of a Quadratic Function

Find the x-intercepts of the graph of $f(x) = x^2 - 7x + 10$.

Solution

To find the x-intercepts, we substitute 0 for $f(x)$ and solve for x:

$$0 = x^2 - 7x + 10 \quad \text{Substitute 0 for } f(x).$$
$$0 = (x-5)(x-2) \quad \text{Factor right-hand side.}$$
$$x - 5 = 0 \quad \text{or} \quad x - 2 = 0 \quad \text{Zero factor property}$$
$$x = 5 \quad \text{or} \quad x = 2$$

So, the x-intercepts are (2, 0) and (5, 0). We use "zero" on a graphing calculator to verify our work (see Fig. 29). ■

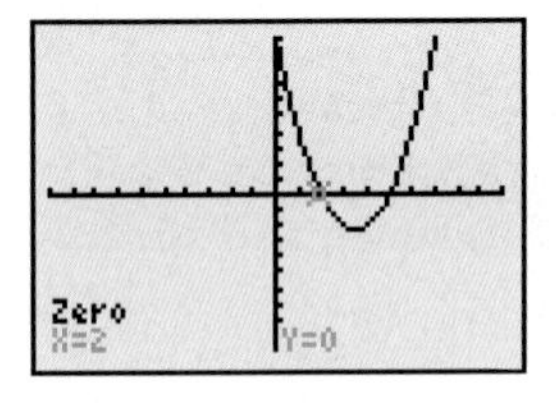

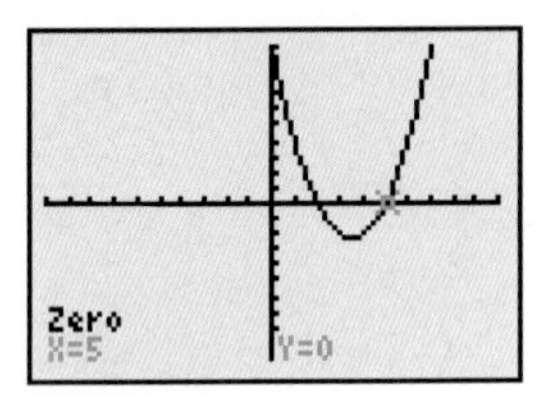

Figure 29 Verify the work

In general, we find the x-intercepts of the graph of $f(x) = ax^2 + bx + c$ by solving the equation $ax^2 + bx + c = 0$.

Connection Between x-Intercepts and Solutions

Let f be a function. If k is a real-number solution of the equation $f(x) = 0$, then $(k, 0)$ is an x-intercept of the graph of the function f. Also, if $(k, 0)$ is an x-intercept of the graph of f, then k is a solution of $f(x) = 0$.

This property suggests that we can verify our solutions of a quadratic equation $ax^2 + bx + c = 0$ by using a graphing calculator to find the x-intercepts of the graph of $f(x) = ax^2 + bx + c$.

Solving More Quadratic Equations

In Example 4, we solve quadratic equations and use a graphing calculator to verify our work.

Example 4 Solving Quadratic Equations

Solve.

1. $4x^2 + 26x + 30 = 0$ **2.** $x^2 - 10x + 25 = 0$

Solution

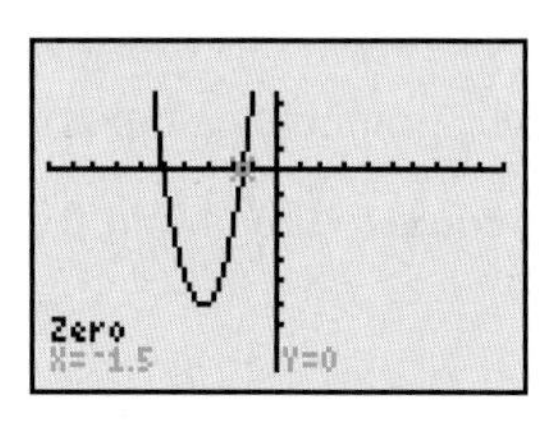

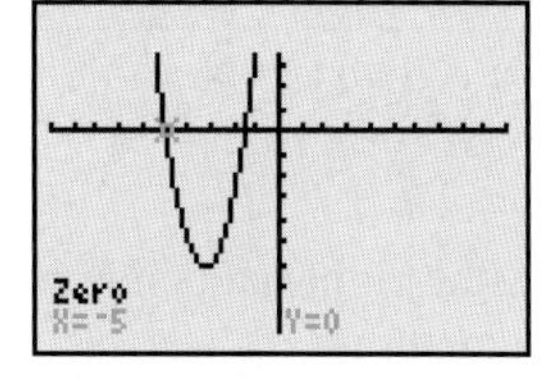

Figure 30 Verify the work

1.

$$\begin{aligned} 4x^2 + 26x + 30 &= 0 && \text{Original equation} \\ 2(2x^2 + 13x + 15) &= 0 && \text{Factor out GCF, 2.} \\ 2(2x + 3)(x + 5) &= 0 && \text{Completely factor left side.} \end{aligned}$$

Now we can apply a variation of the zero factor property: If $2AB = 0$, then $A = 0$ or $B = 0$. Here, we take $(2x + 3)$ to be A and $(x + 5)$ to be B:

$$\begin{aligned} 2x + 3 &= 0 \quad \text{or} \quad x + 5 = 0 \qquad \text{Zero factor property} \\ 2x &= -3 \quad \text{or} \quad x = -5 \\ x &= -\frac{3}{2} \quad \text{or} \quad x = -5 \end{aligned}$$

We use "zero" on a graphing calculator to check that the x-intercepts of the graph of $f(x) = 4x^2 + 26x + 30$ are $\left(-\frac{3}{2}, 0\right)$ and $(-5, 0)$. See Fig. 30.

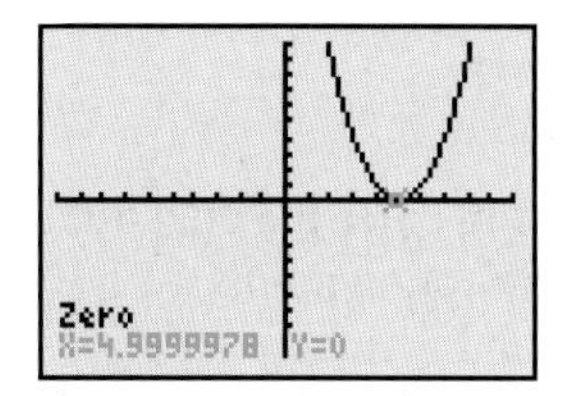

Figure 31 Verify the work

2.

$$\begin{aligned} x^2 - 10x + 25 &= 0 && \text{Original equation} \\ (x - 5)(x - 5) &= 0 && \text{Factor left side.} \\ x - 5 &= 0 && \text{Zero factor property} \\ x &= 5 \end{aligned}$$

Since using the zero factor property yields the *one* equation $x - 5 = 0$, there is one solution: 5. We use "zero" on a graphing calculator to check that the x-intercept of the graph of $f(x) = x^2 - 10x + 25$ is $(5, 0)$. See Fig. 31. ■

In Example 4, we found that the equation $4x^2 + 26x + 30 = 0$ has two solutions, whereas the equation $x^2 - 10x + 25 = 0$ has one solution. What are the possible numbers of solutions of a quadratic equation? To decide, note that a quadratic function can have two x-intercepts, one x-intercept, or no x-intercepts (see Fig. 32).

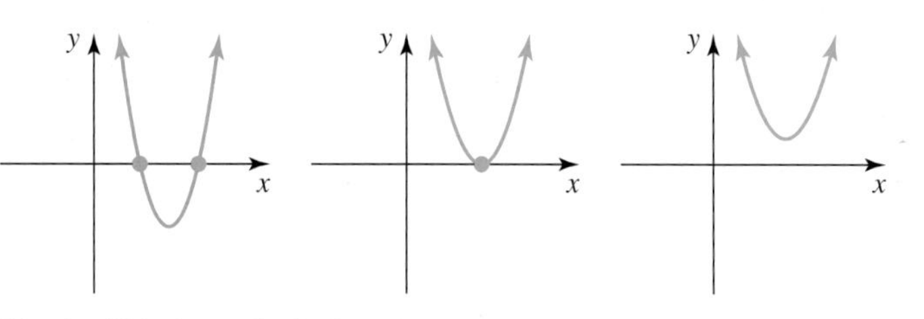

Figure 32 A quadratic function can have two, one, or no x-intercepts

Since the number of real-number solutions of an equation $ax^2 + bx + c = 0$ is equal to the number of x-intercepts of the function $f(x) = ax^2 + bx + c$, we conclude that **the solution set of a quadratic equation in one variable may contain two real numbers, one real number, or no real numbers.** We will solve quadratic equations that have no real-number solutions in Chapter 7.

In Example 4, both original equations had one side that was 0. If neither side of a quadratic equation in one variable is 0, we must first use properties of equality to get one side to be 0 so that we can then apply the zero factor property.

Example 5 Solving a Quadratic Equation

Solve $2x^2 - 8x = 5x - 20$.

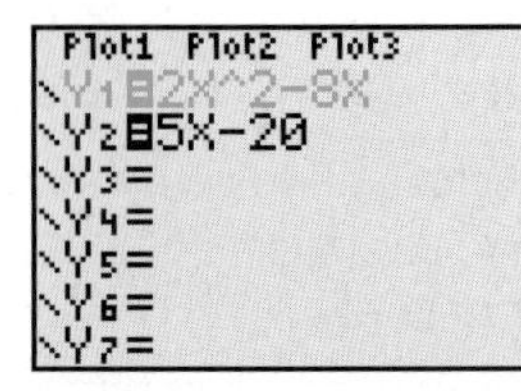

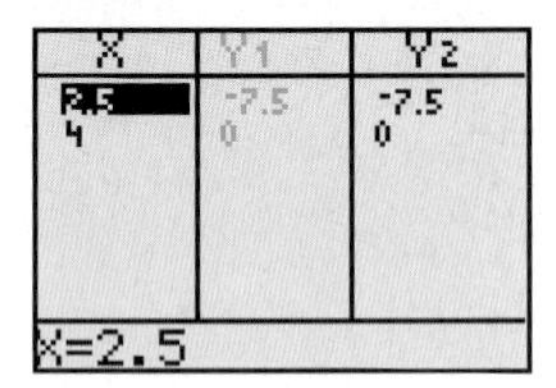

Figure 33 Verify the work

Solution

$$
\begin{aligned}
2x^2 - 8x &= 5x - 20 && \textit{Original equation}\\
2x^2 - 13x + 20 &= 0 && \textit{Write in } ax^2 + bx + c = 0 \textit{ form.}\\
(2x - 5)(x - 4) &= 0 && \textit{Factor left side.}\\
2x - 5 = 0 \quad &\text{or} \quad x - 4 = 0 && \textit{Zero factor property}\\
2x = 5 \quad &\text{or} \quad x = 4 &&\\
x = \frac{5}{2} \quad &\text{or} \quad x = 4 &&
\end{aligned}
$$

To verify that $\frac{5}{2}$ is a solution of $2x^2 - 8x = 5x - 20$, we can use a graphing calculator table to check that, for input $\frac{5}{2}$, the output for $y = 2x^2 - 8x$ is equal to the output for $y = 5x - 20$. We do similarly for input $x = 4$ (see Fig. 33). ■

Example 6 Solving a Quadratic Equation

Solve $(x + 2)(x - 4) = 7$.

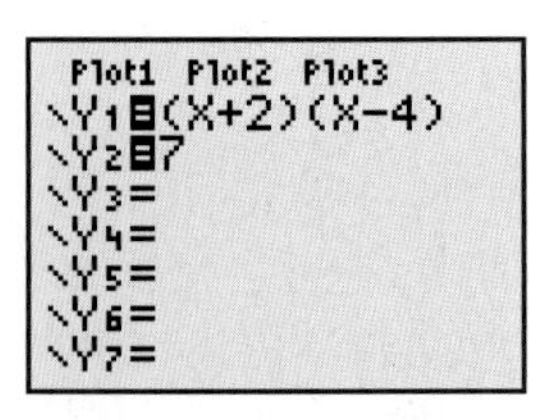

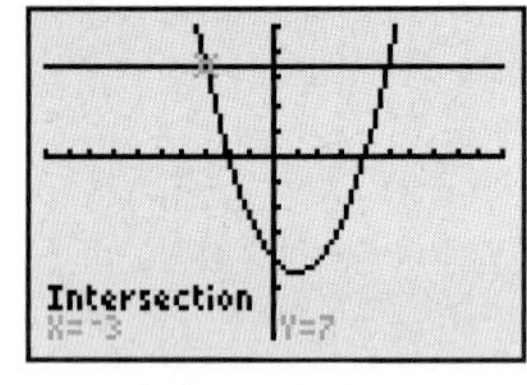

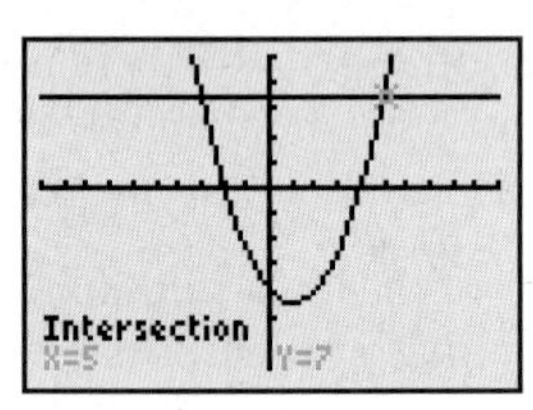

Figure 34 Verify the work

Solution

Although the left-hand side of $(x + 2)(x - 4) = 7$ is factored, the right-hand side is not zero. First, we find the product on the left-hand side of the equation:

$$
\begin{aligned}
(x + 2)(x - 4) &= 7 && \textit{Original equation}\\
x^2 - 2x - 8 &= 7 && \textit{Find product on left side.}\\
x^2 - 2x - 15 &= 0 && \textit{Write in } ax^2 + bx + c = 0 \textit{ form.}\\
(x - 5)(x + 3) &= 0 && \textit{Factor left-hand side.}\\
x - 5 = 0 \quad &\text{or} \quad x + 3 = 0 && \textit{Zero factor property}\\
x = 5 \quad &\text{or} \quad x = -3 &&
\end{aligned}
$$

Therefore, the solutions are -3 and 5. To verify the work, we can enter the equations $y = (x+2)(x-4)$ and $y = 7$ and use "intersect" to find the intersection points $(-3, 7)$ and $(5, 7)$. See Fig. 34. The x-coordinates of these points, -3 and 5, are the solutions of the original equation, which checks. ■

Example 7 Solving a Quadratic Equation That Contains Fractions

Solve $\frac{1}{2}t^2 + \frac{1}{3} = \frac{5}{6}t$.

Solution

To clear the equation of fractions, we multiply both sides by the least common denominator (LCD), 6:

$$
\begin{aligned}
\frac{1}{2}t^2 + \frac{1}{3} &= \frac{5}{6}t && \textit{Original equation}\\
6 \cdot \frac{1}{2}t^2 + 6 \cdot \frac{1}{3} &= 6 \cdot \frac{5}{6}t && \textit{Multiply both sides by LCD, 6.}\\
3t^2 + 2 &= 5t && \textit{Simplify.}\\
3t^2 - 5t + 2 &= 0 && \textit{Write in } ax^2 + bx + c = 0 \textit{ form.}\\
(3t - 2)(t - 1) &= 0 && \textit{Factor left side.}\\
3t - 2 = 0 \quad &\text{or} \quad t - 1 = 0 && \textit{Zero factor property}\\
3t = 2 \quad &\text{or} \quad t = 1 &&\\
t = \frac{2}{3} \quad &\text{or} \quad t = 1 &&
\end{aligned}
$$

■

Now that we can solve quadratic equations in one variable, we can find all inputs of a quadratic function for a given output.

Example 8 Finding an Input and an Output of a Quadratic Function

Let $f(x) = x^2 - 3x - 23$.

1. Find $f(5)$.
2. Find x when $f(x) = 5$.

Solution

1. $f(5) = 5^2 - 3(5) - 23 = 25 - 15 - 23 = -13$
2. We substitute 5 for $f(x)$ in the equation $f(x) = x^2 - 3x - 23$:

$$5 = x^2 - 3x - 23 \quad \text{Substitute 5 for } f(x).$$
$$0 = x^2 - 3x - 28 \quad \text{Write in } 0 = ax^2 + bx + c \text{ form.}$$
$$0 = (x - 7)(x + 4) \quad \text{Factor right-hand side.}$$
$$x - 7 = 0 \quad \text{or} \quad x + 4 = 0 \quad \text{Zero factor property}$$
$$x = 7 \quad \text{or} \quad x = -4$$

Next, we verify that $f(7) = 5$ and $f(-4) = 5$:

Check that $f(7) = 5$

$$f(x) = x^2 - 3x - 23$$
$$f(7) = 7^2 - 3(7) - 23$$
$$= 5$$

Check that $f(-4) = 5$

$$f(x) = x^2 - 3x - 23$$
$$f(-4) = (-4)^2 - 3(-4) - 23$$
$$= 5$$

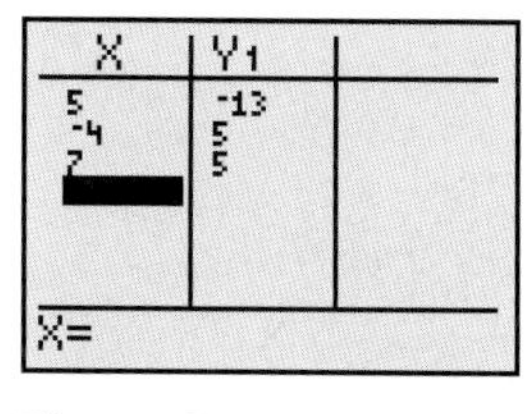

Figure 35 Verify the work

Or we can verify our work in Problems 1 and 2 by using a graphing calculator table (see Fig. 35). ■

Recall from Section 5.1 that the inverse of a function g sends each output of g to its corresponding input(s). From Example 8, we see that the function $f(x) = x^2 - 3x - 23$ is not invertible, since the output 5 corresponds to not one input but two: -4 and 7. It turns out that no quadratic function is invertible.

Cubic Equations in One Variable

So far in this section, we have solved quadratic equations. We will now solve some cubic equations, such as

$$4x^3 - 2x^2 - 36x + 18 = 0 \qquad 2x^3 = 42x + 8x^2$$

A **cubic equation in one variable** is an equation that can be put into the form

$$ax^3 + bx^2 + cx + d = 0$$

where a, b, c, and d are constants and $a \neq 0$. We can solve some cubic equations by applying the zero factor property to three factors:

$$\text{If } ABC = 0, \text{ then } A = 0, B = 0, \text{ or } C = 0.$$

Example 9 Solving a Cubic Equation

Solve $2x^3 = 42x + 8x^2$.

Solution

$$2x^3 = 42x + 8x^2 \quad \text{Original equation}$$
$$2x^3 - 8x^2 - 42x = 0 \quad \text{Write in } ax^3 + bx^2 + cx + d = 0 \text{ form.}$$
$$2x(x^2 - 4x - 21) = 0 \quad \text{Factor out GCF, } 2x.$$
$$2x(x - 7)(x + 3) = 0 \quad \text{Completely factor left side.}$$
$$2x = 0 \quad \text{or} \quad x - 7 = 0 \quad \text{or} \quad x + 3 = 0 \quad \text{Zero factor property}$$
$$x = 0 \quad \text{or} \quad x = 7 \quad \text{or} \quad x = -3$$

So, the solutions are -3, 0, and 7. ■

Note that solving a cubic equation in one variable is similar to solving a quadratic equation in one variable. We now summarize the steps used to solve either type of equation.

Solving Quadratic or Cubic Equations by Factoring

If an equation can be solved by factoring, we solve it by the following steps:

1. Write the equation so that one side of the equation is equal to zero.
2. Factor the nonzero side of the equation.
3. Apply the zero factor property.
4. Solve each equation that results from applying the zero factor property.

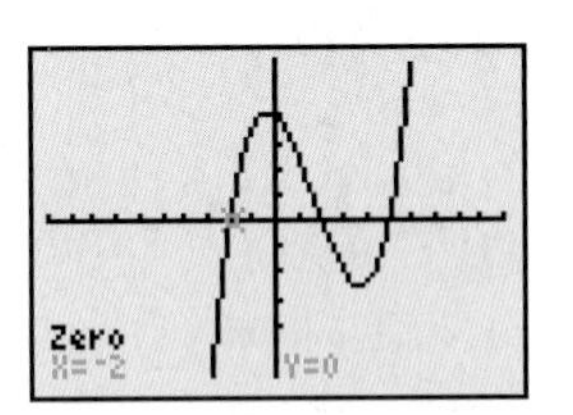

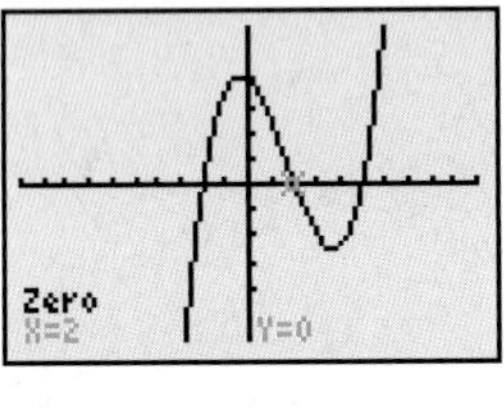

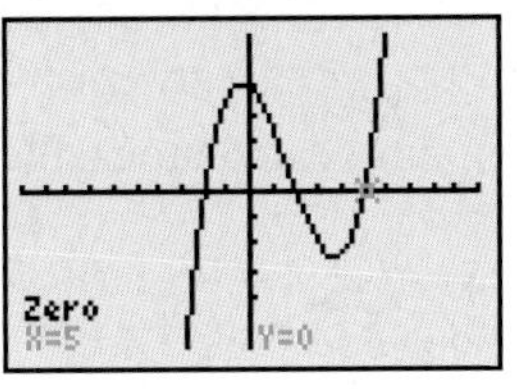

Figure 36 Verify the work

Example 10 Finding x-Intercepts of the Graph of a Cubic Function

Find the x-intercepts of the graph of $f(x) = x^3 - 5x^2 - 4x + 20$.

Solution

We substitute 0 for $f(x)$ and solve for x:

$$x^3 - 5x^2 - 4x + 20 = 0 \qquad \text{Substitute 0 for } f(x).$$

$$\left.\begin{aligned} x^2(x-5) - 4(x-5) &= 0 \\ (x^2-4)(x-5) &= 0 \end{aligned}\right\} \qquad \text{Factor by grouping.}$$

$$(x+2)(x-2)(x-5) = 0 \qquad A^2 - B^2 = (A+B)(A-B)$$

$$x+2=0 \quad \text{or} \quad x-2=0 \quad \text{or} \quad x-5=0 \qquad \text{Zero factor property}$$

$$x=-2 \quad \text{or} \quad x=2 \quad \text{or} \quad x=5$$

So, the x-intercepts are $(-2, 0)$, $(2, 0)$, and $(5, 0)$. We use "zero" on a graphing calculator to verify our work (see Fig. 36). ■

WARNING It is a common error to try to apply the zero factor property to an equation such as $x^2(x-5) - 4(x-5) = 0$ and incorrectly conclude that the solutions are 0 and 5. The expression $x^2(x-5) - 4(x-5)$ is *not* factored, because it is a difference, not a product. Only after we factor the left side of the equation $x^2(x-5) - 4(x-5) = 0$ can we apply the zero factor property.

In Example 10, we worked with the cubic function $f(x) = x^3 - 5x^2 - 4x + 20$. Recall from Section 6.1 that a cubic function is a function whose equation can be put into the form $f(x) = ax^3 + bx^2 + cx + d$, where $a \neq 0$.

The cubic function sketched in graph (a) of Fig. 37 has exactly three x-intercepts; the cubic function sketched in graph (b) has exactly two x-intercepts; and both cubic functions sketched in graphs (c) and (d) have exactly one x-intercept. It turns out that any cubic function has exactly one, two, or three x-intercepts.

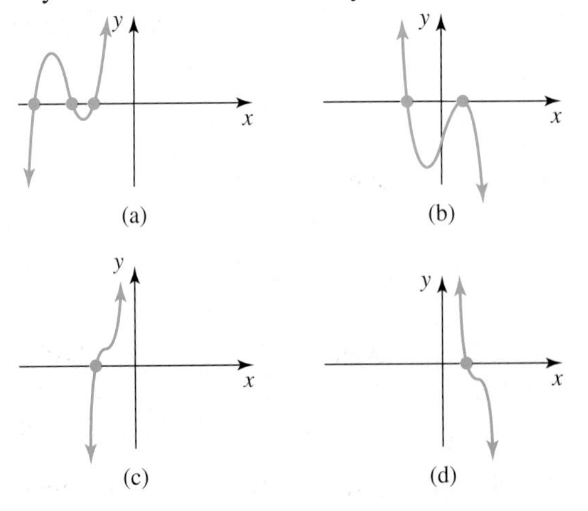

Figure 37 Graphs of typical cubic functions

The number of real-number solutions of a cubic equation $ax^3 + bx^2 + cx + d = 0$ is equal to the number of x-intercepts of the function $f(x) = ax^3 + bx^2 + cx + d$, so we conclude that **the solution set of a cubic equation in one variable may contain one, two, or three real numbers.**

Solving Polynomial Equations in One Variable by Graphing

Consider the equation $x^2 - x - 7 = -x^2$. If we add x^2 to both sides, the result is $2x^2 - x - 7 = 0$. Since the left side of the equation, $2x^2 - x - 7$, is prime (try it), we cannot solve the equation by factoring. However, we will show in Example 11 that we can solve it by graphing.

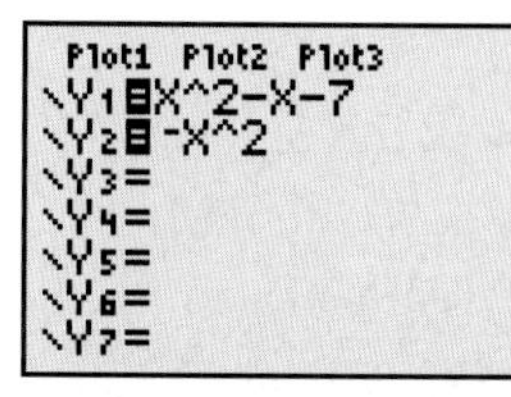

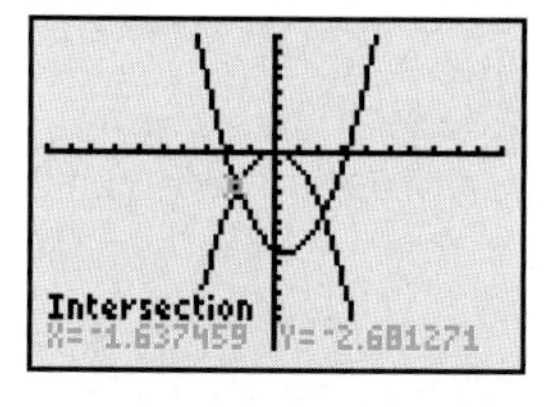

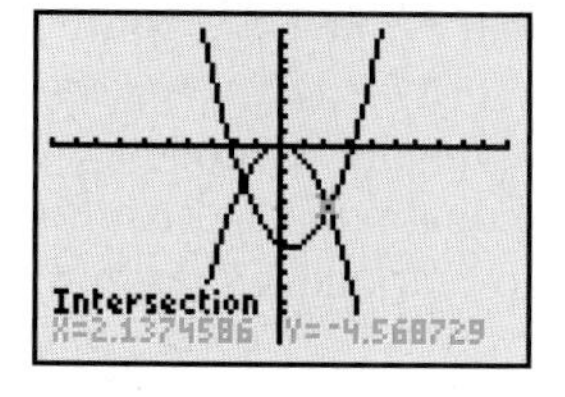

Figure 38 Solve the system

Example 11 Using Graphing to Solve an Equation in One Variable

Use graphing to solve $x^2 - x - 7 = -x^2$.

Solution

We use "intersect" on a graphing calculator to find the solutions of the system

$$y = x^2 - x - 7$$
$$y = -x^2$$

See Fig. 38.

The approximate solutions of the system are $(-1.64, -2.68)$ and $(2.14, -4.57)$. The x-coordinates of these ordered pairs, -1.64 and 2.14, are the approximate solutions of the equation $x^2 - x - 7 = -x^2$. ■

In Sections 7.4 and 7.5, we will discuss two symbolic methods that can be used to solve $x^2 - x - 7 = -x^2$.

Using a Quadratic Function to Model a Situation

In Sections 6.1 and 6.2, we used linear and quadratic functions to model authentic situations. Now we focus on using quadratic functions to perform modeling.

DEFINITION Quadratic model

A **quadratic model** is a quadratic function, or its graph, that describes the relationship between two quantities in an authentic situation.

Table 17 Annual Revenues of American Express

Year	Revenue (billions of dollars)
2000	23.7
2001	22.6
2002	23.8
2003	25.8
2004	29.1

Source: *CSI*

Example 12 Modeling with a Quadratic Function

The annual revenues of American Express® are shown in Table 17 for various years.

Let r be the annual revenue (in billions of dollars) of American Express in the year that is t years since 2000. A model of the situation is $f(t) = \frac{2}{3}t^2 - \frac{4}{3}t + 23.7$.

1. Use a graphing calculator to draw the graph of the model and, in the same viewing window, the scattergram of the data. Does the model fit the data well?
2. Predict the revenue in 2010.
3. Predict when the revenue will be \$39.7 billion.

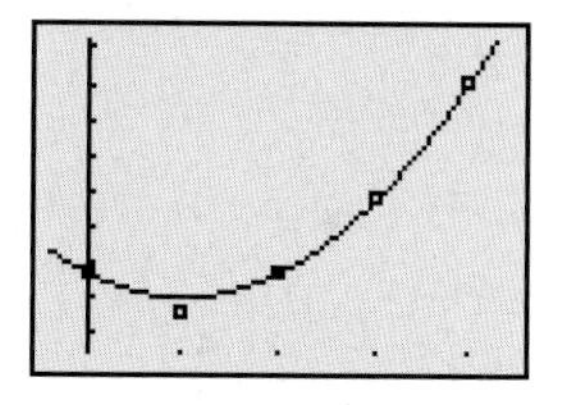

Figure 39 Check the fit

Solution

1. The graph of the model and the scattergram of the data are shown in Fig. 39. The model appears to fit the data fairly well.
2. To predict the revenue in 2010, we find $f(10)$:

$$f(10) = \frac{2}{3}(10)^2 - \frac{4}{3}(10) + 23.7 \approx 77.03$$

The revenue in 2010 will be about \$77.0 billion, according to the model.

3. To predict when the revenue will be \$39.7 billion, we substitute 39.7 for $f(t)$ in $f(t) = \frac{2}{3}t^2 - \frac{4}{3}t + 23.7$ and solve for t:

$$
\begin{aligned}
39.7 &= \frac{2}{3}t^2 - \frac{4}{3}t + 23.7 && \text{Substitute 39.7 for } f(t).\\
0 &= \frac{2}{3}t^2 - \frac{4}{3}t - 16 && \text{Write in } 0 = at^2 + bt + c \text{ form.}\\
0 &= 2t^2 - 4t - 48 && \text{Multiply both sides by LCD, 3.}\\
0 &= 2(t^2 - 2t - 24) && \text{Factor out GCF, 2.}\\
0 &= 2(t-6)(t+4) && \text{Completely factor right-hand side.}\\
t - 6 = 0 \quad &\text{or} \quad t + 4 = 0 && \text{Zero factor property}\\
t = 6 \quad &\text{or} \quad t = -4
\end{aligned}
$$

Figure 40 Verify the work

The revenue was \$39.7 billion in 1996 and will be \$39.7 billion in 2006, according to the model. We use a graphing calculator table to verify our work in Problems 2 and 3 (see Fig. 40). ■

To make a prediction about the dependent variable of a quadratic model, we substitute the chosen value of the independent variable into the equation and solve for the dependent variable.

To make a prediction about the independent variable of a quadratic model, we substitute the chosen value of the dependent variable into the equation and solve for the independent variable. That involves writing the equation so that one side is zero and factoring the nonzero side.

Area of Rectangular Objects

The area A of a rectangle is given by the formula $A = LW$, where L is the length and W is the width.

Example 13 Solving an Area Problem

A person has a rectangular garden with a width of 9 feet and a length of 12 feet. She plans to place mulch outside of the garden to form a border of uniform width. If she has just enough mulch to cover 100 square feet of land, determine the width of the border.

Solution

We use the five-step problem-solving method of Section 3.4, but for step 2 we find an equation in one variable rather than a system of two equations in two variables (and solve the equation in step 3).

Step 1: Define each variable. Let x be the width (in feet) of the mulch border (see Fig. 41).

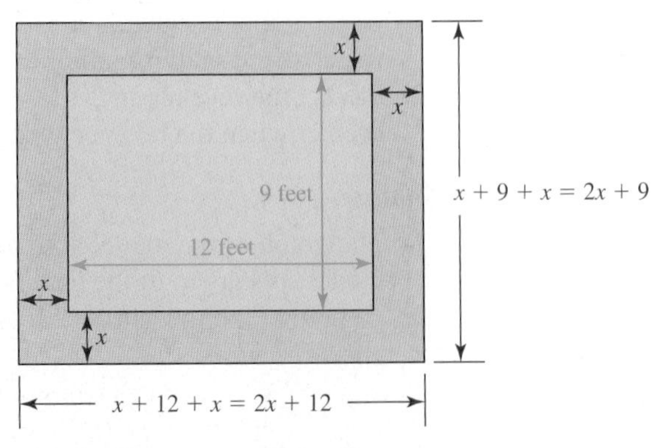

Figure 41 Garden with border

Step 2: Write an equation in one variable. For the outer rectangle (garden and border), the length is $2x + 12$ and the width is $2x + 9$ (see Fig. 41). The area of just the garden is $12 \cdot 9 = 108$ square feet, and the area of the border is given as 100 square feet. So, the area of the outer rectangle is $108 + 100 = 208$ square feet. We use the formula $A = LW$ to write an equation:

$$\overbrace{208}^{\text{area}} = \overbrace{(2x + 12)}^{\text{length}} \cdot \overbrace{(2x + 9)}^{\text{width}}$$

Step 3: Solve the equation.

$208 = (2x + 12)(2x + 9)$	Equation from step 2
$208 = 4x^2 + 18x + 24x + 108$	Find product on right side.
$208 = 4x^2 + 42x + 108$	Combine like terms.
$0 = 4x^2 + 42x - 100$	Write in $0 = ax^2 + bx + c$ form.
$0 = 2(2x^2 + 21x - 50)$	Factor out GCF, 2.
$0 = 2(2x + 25)(x - 2)$	Factor by trial and error.
$2x + 25 = 0 \quad \text{or} \quad x - 2 = 0$	Zero factor property
$2x = -25 \quad \text{or} \quad x = 2$	
$x = -\frac{25}{2} \quad \text{or} \quad x = 2$	

Step 4: Describe each result. For the result $x = -\frac{25}{2}$, the border width is negative. Model breakdown has occurred, because a width must be positive. The border width is 2 feet.

Step 5: Check. If the border width is 2 feet, then the outer rectangle has a width of 13 feet and a length of 16 feet. Therefore, the total area of the outer rectangle is $13 \cdot 16 = 208$ square feet. This checks with our calculation near the beginning of our work. ■

group exploration

Finding equations of quadratic functions

1. Use ZStandard to sketch graphs of the functions $f(x) = (x - 2)(x + 3)$, $g(x) = 2(x-2)(x+3)$, $h(x) = \frac{1}{2}(x-2)(x+3)$, and $k(x) = -(x-2)(x+3)$.
   a. What do you notice about the x-intercepts of f, g, h, and k?
   b. For a function of the form $y = a(x - 2)(x + 3)$, describe the effect that the value of a has on the graph of the function. Sketch more graphs with varying values of a if you are unsure.
2. Find a possible equation of the function sketched in Fig. 42. Use a graphing calculator to verify your work.
3. Find an equation of the function sketched in Fig. 43. Use a graphing calculator to verify your work. [**Hint:** A point on the graph of an equation satisfies the equation.]

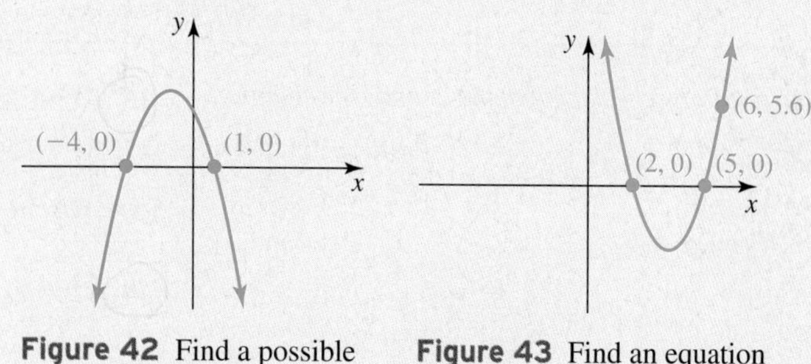

Figure 42 Find a possible equation (Problem 2)

Figure 43 Find an equation (Problem 3)

group exploration

Looking ahead: Significance of a, h, and k for $y = a(x - h)^2 + k$

1. Use ZStandard followed by ZSquare to draw a graph of $y = x^2$.
2. Graph these equations of the form $y = x^2 + k$ in order, and describe in terms of k how you could "move" the graph of $y = x^2$ to get each graph:
$$y = x^2 + 1,\ y = x^2 + 2,\ y = x^2 + 3, \text{ and } y = x^2 + 4$$
Do the same with these equations:
$$y = x^2 - 1,\ y = x^2 - 2,\ y = x^2 - 3, \text{ and } y = x^2 - 4$$
3. Graph these equations of the form $y = (x - h)^2$ in order, and describe in terms of h how you could "move" the graph of $y = x^2$ to get each graph:
$$y = (x - 1)^2,\ y = (x - 2)^2,\ y = (x - 3)^2, \text{ and } y = (x - 4)^2$$
Do the same with these equations:
$$y = (x + 1)^2,\ y = (x + 2)^2,\ y = (x + 3)^2, \text{ and } y = (x + 4)^2$$
4. In this problem, you will explore the graphical significance of the constant a in functions of the form $y = ax^2$. From Problems 2 and 3, you should have an idea of how to go about it. Do this now and describe what you find. Don't forget to try negative values of a as well as values of a between 0 and 1.
5. **a.** Graph these equations in order, and explain how the graphs relate to your observations in Problems 2, 3, and 4:
$$y = x^2,\ y = 0.5x^2,\ y = -0.5x^2,\ y = -0.5(x + 1)^2, \text{ and}$$
$$y = -0.5(x + 1)^2 - 6$$
b. Using your graph of $y = -0.5(x + 1)^2 - 6$, find the coordinates of the vertex. Compare these coordinates with the equation $y = -0.5(x + 1)^2 - 6$. What do you notice?
6. Summarize your findings about a, h, and k in terms of how you could move or adjust the graph of $y = x^2$ to get the graph of $y = a(x - h)^2 + k$. Also, discuss how the coordinates of the vertex are related to a, h, and k. If you are unsure, continue exploring.

TIPS FOR SUCCESS: Scan Test Problems

When you take a test, it is best to scan the test problems quickly, pick the problems with which you feel most comfortable, and complete those problems first. By doing so, you will warm up, gain confidence, and perhaps do better on the rest of the test. Also, you will probably have a better idea of how to allot your time on the remaining problems.

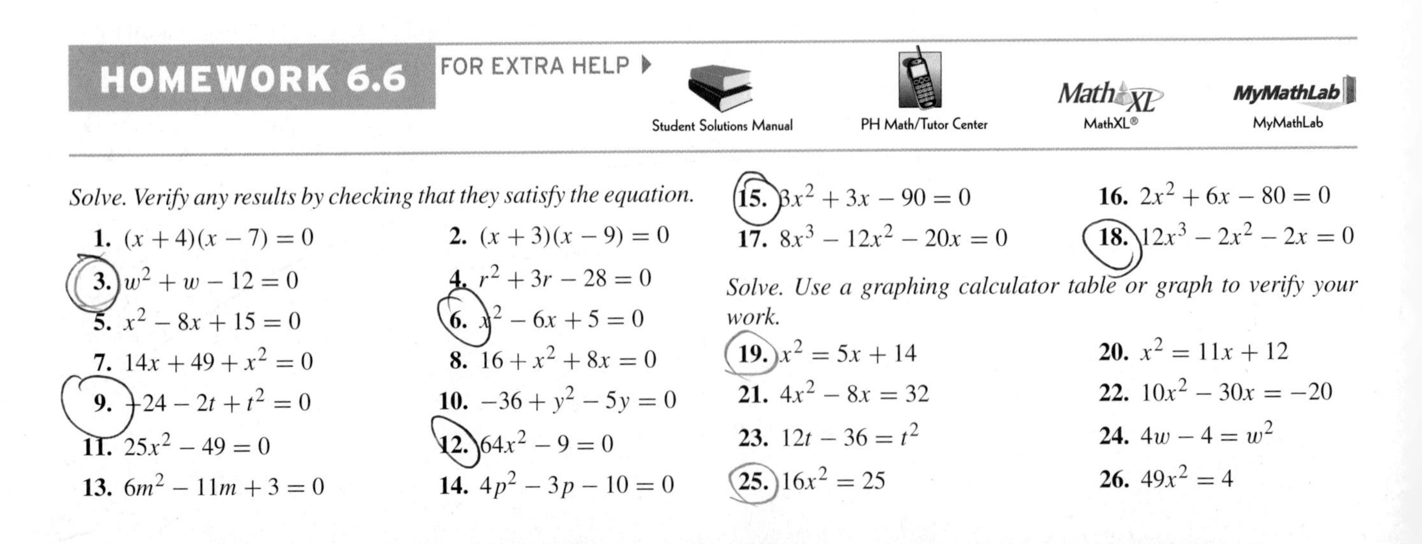

HOMEWORK 6.6

FOR EXTRA HELP ▶ Student Solutions Manual · PH Math/Tutor Center · MathXL® · MyMathLab

Solve. Verify any results by checking that they satisfy the equation.

1. $(x + 4)(x - 7) = 0$
2. $(x + 3)(x - 9) = 0$
3. $w^2 + w - 12 = 0$
4. $r^2 + 3r - 28 = 0$
5. $x^2 - 8x + 15 = 0$
6. $x^2 - 6x + 5 = 0$
7. $14x + 49 + x^2 = 0$
8. $16 + x^2 + 8x = 0$
9. $-24 - 2t + t^2 = 0$
10. $-36 + y^2 - 5y = 0$
11. $25x^2 - 49 = 0$
12. $64x^2 - 9 = 0$
13. $6m^2 - 11m + 3 = 0$
14. $4p^2 - 3p - 10 = 0$
15. $3x^2 + 3x - 90 = 0$
16. $2x^2 + 6x - 80 = 0$
17. $8x^3 - 12x^2 - 20x = 0$
18. $12x^3 - 2x^2 - 2x = 0$

Solve. Use a graphing calculator table or graph to verify your work.

19. $x^2 = 5x + 14$
20. $x^2 = 11x + 12$
21. $4x^2 - 8x = 32$
22. $10x^2 - 30x = -20$
23. $12t - 36 = t^2$
24. $4w - 4 = w^2$
25. $16x^2 = 25$
26. $49x^2 = 4$

27. $6x^3 - 24x = 0$

28. $4x^3 - 100x = 0$

29. $3r^2 = 6r$

30. $5p^2 = 35p$

31. $9x = -2x^2 + 5$

32. $10x = -8x^2 - 3$

33. $2x^3 = 6x^2 + 36x$

34. $36x = 24x^2 - 4x^3$

35. $18y^3 + 3y^2 = 6y$

36. $8x^3 - 14x^2 = 4x$

37. $20x = -4x^2 - 25$

38. $24x = -9x^2 - 16$

39. $\frac{1}{4}x^2 - \frac{1}{2}x = 6$

40. $\frac{1}{8}x^2 - \frac{3}{4}x = 2$

41. $\frac{a^2}{2} - \frac{a}{6} = \frac{1}{3}$

42. $\frac{t^2}{5} - \frac{t}{2} = -\frac{1}{5}$

43. $x^2 - \frac{1}{25} = 0$

44. $x^2 - \frac{1}{49} = 0$

45. $(x + 2)(x + 5) = 40$

46. $(x + 3)(x - 2) = 24$

47. $4r^3 - 2r^2 - 36r + 18 = 0$

48. $y^3 + 3y^2 - 4y - 12 = 0$

49. $9x^3 - 12 = 4x - 27x^2$

50. $3x^2 - 4x = 12 - x^3$

51. $2x(x + 1) = 5x(x - 7)$

52. $x(x - 3) = 3x(x - 4)$

53. $4p(p - 1) - 24 = 3p(p - 2)$

54. $2w^2 + 2(w - 1) = w(1 - w)$

55. $(x^2 + 5x + 6)(x^2 - 5x - 24) = 0$

56. $(x^2 - 7x + 6)(x^2 + 3x - 4) = 0$

Find all x-intercepts. Verify your intercept(s) graphically by using a graphing calculator.

57. $f(x) = x^2 - 9x + 20$

58. $f(x) = x^2 - 4x - 21$

59. $f(x) = 36x^2 - 25$

60. $f(x) = 16x^2 - 81$

61. $f(x) = 24x^3 - 14x^2 - 20x$

62. $f(x) = 6x^3 + 15x^2 + 6x$

63. $f(x) = x^3 + 2x^2 - x - 2$

64. $f(x) = 4x^3 - 12x^2 - 9x + 27$

For Exercises 65–68, let $f(x) = x^2 - x - 6$.

65. Find $f(3)$.

66. Find $f(-4)$.

67. Find x when $f(x) = 14$.

68. Find x when $f(x) = 6$.

For Exercises 69–72, use the graph of $y = \frac{1}{2}x^2 + x - \frac{7}{2}$ shown in Fig. 44 to solve the given equation.

69. $\frac{1}{2}x^2 + x - \frac{7}{2} = 4$

70. $\frac{1}{2}x^2 + x - \frac{7}{2} = -2$

71. $\frac{1}{2}x^2 + x - \frac{7}{2} = -4$

72. $\frac{1}{2}x^2 + x - \frac{7}{2} = -5$

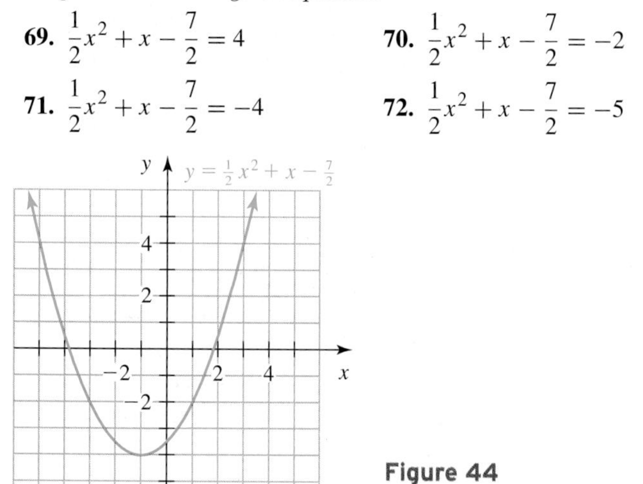

Figure 44 Exercises 69–72

For Exercises 73–76, use the graphs of $y = x^3 - 3x^2 + 1$ and $y = x - 2$ shown in Fig. 45 to solve the given equation or system.

73. $x^3 - 3x^2 + 1 = -3$

74. $x^3 - 3x^2 + 1 = -1$

75. $x^3 - 3x^2 + 1 = x - 2$

76. $y = x^3 - 3x^2 + 1$
$y = x - 2$

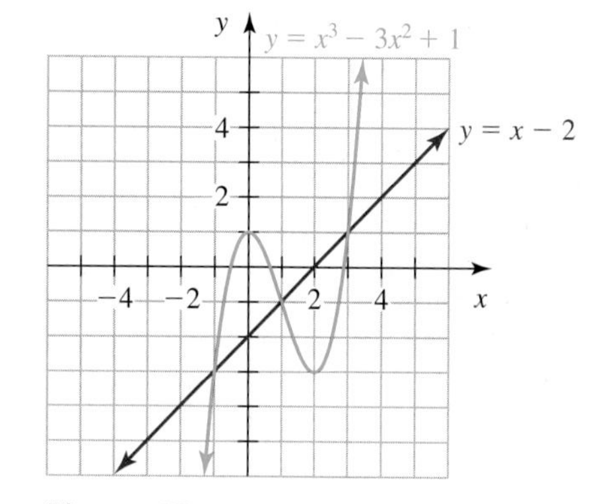

Figure 45 Exercises 73–76

Solve by using "intersect" on a graphing calculator. Round the solution(s) to the second decimal place.

77. $2x^2 - 5x - 3 = -x + 5$

78. $-x^2 + 2x + 4 = x - 1$

79. $-x^3 + 4x^2 - 2x = 4 - x$

80. $x^3 - 4x^2 + 5 = x - 5$

For Exercises 81–84, solve by referring to the values of the function $y = 3x^2 - 6x + 1$ shown in Table 18.

81. $3x^2 - 6x + 1 = 25$

82. $3x^2 - 6x + 1 = 1$

83. $3x^2 - 6x + 1 = -3$

84. $3x^2 - 6x + 1 = -2$

Table 18 Exercises 81-84

x	y
−2	25
−1	10
0	1
1	−2
2	1
3	10
4	25

85. A student tries to solve the equation $x^2 = x$:

$$x^2 = x$$
$$\frac{x^2}{x} = \frac{x}{x}$$
$$x = 1$$

Describe any errors. Then solve the equation correctly.

86. A student tries to solve the equation $(x - 3)(x - 7) = 12$:

$$(x - 3)(x - 7) = 12$$
$$x - 3 = 2 \quad \text{or} \quad x - 7 = 6$$
$$x = 5 \quad \text{or} \quad x = 13$$

Describe any errors. Then solve the equation correctly.

87. A student tries to solve the equation $x^3 + 4x^2 - 9x - 36 = 0$:

$$x^3 + 4x^2 - 9x - 36 = 0$$
$$x^2(x + 4) - 9(x + 4) = 0$$
$$x + 4 = 0$$
$$x = -4$$

Describe any errors. Then solve the equation correctly.

88. A student tries to solve the equation $2x^2 - 18x + 36 = 0$:

$$2x^2 - 18x + 36 = 0$$
$$2(x^2 - 9x + 18) = 0$$
$$2(x-3)(x-6) = 0$$
$$x = 2, x = 3, \text{ or } x = 6$$

Describe any errors. Then solve the equation correctly.

89. The numbers of participants in the Ford Ironman World Championship in Hawaii are shown in Table 19 for various years. Participants attempt to swim 2.4 miles, bike 112 miles, and run 26.2 miles (a marathon).

Table 19 Numbers of Participants in the Ironman World Championship

Year	Number of Participants
1990	1387
1992	1364
1994	1405
1996	1421
1998	1487
2000	1531
2002	1607
2004	1734
2005	1744

Source: *Ford Ironman World Championship*

Let n be the number of participants at t years since 1990. A linear model of the situation is $n = L(t) = 26t + 1314$. An exponential model is $n = E(t) = 1324(1.017)^t$. A quadratic model is $n = Q(t) = 2t^2 - 4t + 1374$.

a. Use a graphing calculator to draw the graphs of all three models and, in the same viewing window, the scattergram of the data. Which model describes the situation best?

b. Use ZOOM OUT on a graphing calculator to help you determine which model predicts the largest participation for years between 2005 and 2020.

c. Find the n-intercept of each model. Which n-intercept describes the situation best?

d. Use Q to estimate when there were 1390 participants.

90. The percentages of Americans who think that the environment should be protected even at the risk of curbing economic growth are shown in Table 20 for various years.

Table 20 Percentages of Americans Who Think That the Environment Should Be Given Priority

Year	Percent
2000	69
2001	57
2002	54
2003	47
2004	49
2005	53

Source: *The Gallup Organization*

Let $p = f(t)$ be the percentage of Americans at t years since 2000 who think that the environment should be protected even at the risk of curbing economic growth.

a. Use a graphing calculator to draw a scattergram to describe the data. Is it best to use a linear, an exponential, or a quadratic function to model the data? Explain.

b. Use a graphing calculator to draw the graph of the model $f(t) = 2t^2 - 13t + 69$ and, in the same viewing window, the scattergram of the data. Does the model fit the data well?

c. What is the p-intercept? What does it mean in this situation?

d. Estimate when 63% of Americans thought that the environment should be given priority.

e. The economy did poorly during most of 2000 through 2003 and has done fairly well from 2003 through 2005. How *might* this fact be related to the data in Table 20?

91. Moody's Investors Service® evaluates the investment quality of companies. Although a B2 rating is six steps above Moody's worst rating, many companies rated B2 eventually default on their bonds (see Table 21).

Table 21 Percentage of Companies with a B2 Rating That Defaulted on Bonds

Years After Being Rated B2	Percent
2	12
4	24
6	32
8	36
10	41

Source: *Moody's Investors Service*

Let $p = f(t)$ be the percentage of companies with a B2 rating that defaulted on their bonds at t years since being rated B2. A model of the situation is $f(t) = -\frac{1}{3}t^2 + \frac{22}{3}t$.

a. Use a graphing calculator to draw the graph of the model and, in the same viewing window, the scattergram of the data. Does the model fit the data well?

b. Find $f(7)$. What does it mean in this situation?

c. Find t when $f(t) = 7$. What does it mean in this situation?

d. Find the t-intercepts. What do they mean in this situation?

92. The total lengths of fiber-optic cable that telecom companies have are shown in Table 22 for various years.

Table 22 Total Lengths of Fiber-Optic Cable That Telecom Companies Have

Year	Total Length of Cable (millions of miles)
1996	21
1997	24
1998	31
1999	42
2000	60
2001	72

Source: *KMI Corporation*

Let $d = Q(t)$ be the total length of cable (in millions of miles) that telecom companies have at t years since 1996. A quadratic model of the situation is $Q(t) = \frac{3}{2}t^2 + 3t + 20$.

a. Use a graphing calculator to draw the graph of the model and, in the same viewing window, the scattergram of the data. Does the model fit the data well?

b. Predict the total length of fiber-optic cable that telecom companies will have in 2010. If all of the cable were placed end to end, how many times could it be wrapped around Earth? Earth's radius at the equator is about 3963 miles. [**Hint:** Use the formula for the circumference C of a circle with radius r: $C = 2\pi r$.]

c. Find t when $Q(t) = 32$. What does it mean in this situation?

d. The situation can be described by the exponential model $E(t) = 19.48(1.3)^t$. Does model Q or model E describe the situation better for years before 1996? Explain.

93. The percentages of Americans who think that labor unions in the United States will become stronger are shown in Table 23 for various years.

Table 23 Percentages of Americans Who Think That Labor Unions Will Become Stronger

Year	Percent
1999	25
2001	24
2002	23
2004	21
2005	19

Source: *The Gallup Organization*

Let $p = f(t)$ be the percentage of Americans at t years since 1999 who think that labor unions in the United States will become stronger. A model of the situation is $f(t) = -\frac{1}{8}t^2 - \frac{1}{4}t + 25$.

a. Use a graphing calculator to draw the graph of the model and, in the same viewing window, the scattergram of the data. Does the model fit the data well?

b. Find the p-intercept. What does it mean in this situation?

c. Predict the percentage of Americans in 2011 who will think that labor unions will become stronger.

d. Estimate when 15% of Americans thought that labor unions will become stronger.

94. First-quarter (January through March) Internet sales of goods and services are shown in Table 24 for various years.

Table 24 First-Quarter Internet Sales

Year	First-Quarter Sales (billions of dollars)
2001	8.3
2002	9.8
2003	12.8
2004	16.0
2005	19.8

Source: *U.S. Census Bureau*

Let $s = f(t)$ be first-quarter Internet sales (in billions of dollars) at t years since 2000. A model of the situation is $f(t) = \frac{1}{3}t^2 + t + 6.5$.

a. Use a graphing calculator to draw the graph of the model and, in the same viewing window, the scattergram of the data. Does the model fit the data well?

b. Find the s-intercept. What does it mean in this situation?

c. Predict the first-quarter Internet sales in 2011.

d. Predict in which year first-quarter Internet sales will be $24.5 billion.

95. The average annual amounts of money each African has received from foreign aid are shown in Table 25 for various years.

Table 25 Average Annual Amounts of Money Each African Has Received from Foreign Aid

Year	Average Amount (dollars)
1994	32
1996	25
1998	22
2000	19
2002	19
2003	23
2004	28

Source: *World Bank*

Let $M = f(t)$ be the average money (in dollars per person) Africans have received from foreign aid in the year that is t years since 1990. A model of the situation is $f(t) = 0.41t^2 - 7.95t + 58$.

a. Use a graphing calculator to draw the graph of the model and, in the same viewing window, the scattergram of the data. Does the model fit the data well?

b. Predict the average money each African will receive from foreign aid in 2010.

c. Predict when Africans will receive an average of $58 per person from foreign aid.

96. The consumer confidence index is a measure of Americans' confidence in the economy. Consumer confidence indexes are shown in Table 26 for January of various years.

Table 26 Consumer Confidence Indexes in January

Year	Consumer Confidence Index (points)
2000	144.7
2001	115.7
2002	97.8
2003	78.8
2004	96.8

Source: *NFO WorldGroup*

Let $I = f(t)$ be the consumer confidence index (in points) at t years since 2000. A quadratic model of the situation is $f(t) = 6.64t^2 - 39.81t + 146.6$.

a. Use a graphing calculator to draw the graph of the model and, in the same viewing window, the scattergram of the data. Does the model fit the data well?

b. Use f to estimate the consumer confidence indexes in 2002 and 2005. According to the model, in which of these two years did Americans have more confidence in the economy?

c. Estimate when the consumer confidence index was 146.6 points.

97. A rectangular rug has an area of 60 square feet. If its length is 2 feet more than twice its width, find the dimensions of the rug.

98. A rectangular garden has an area of 65 square feet. If its length is 3 feet less than twice its width, find the dimensions of the garden.

99. The length of a rectangle is 4 centimeters more than the width. If both the width and length were doubled, the area would be 48 square centimeters. Find the dimensions of the original rectangle.

100. The length of a rectangle is 6 meters more than the width. If both the width and length were doubled, the area would be 108 square meters. Find the dimensions of the original rectangle.

101. A person has a rectangular garden with a width of 6 feet and a length of 10 feet. To form a border of uniform width, he plans to place mulch around the outside of the garden. If he has just enough mulch to cover 80 square feet of land, determine the width of the border.

102. A person has a rectangular garden with a width of 8 feet and a length of 12 feet. To form a border of uniform width, she plans to put sod around the outside of the garden. If she has just enough sod to cover 44 square feet of land, determine the width of the border.

103. A rectangular painting (not including the frame) has a width of 10 inches and a length of 14 inches. If the area of the frame (of uniform width) is 52 square inches, what is the width of the frame?

104. A rectangular painting (not including the frame) has a width of 9 inches and a length of 15 inches. If the area of the frame (of uniform width) is 112 square inches, what is the width of the frame?

105. Give an example of a function whose x-intercepts are $(-5, 0)$ and $(1, 0)$. Verify your work by using a graphing calculator.

106. Give three examples of quadratic functions for which each function's only x-intercept is $(2, 0)$.

107. Give an example of a cubic equation in one variable whose solutions are -4, 0, and 2.

108. Give an example of a quadratic equation in one variable that has 4 as its only solution.

109. The graph of a function h is sketched in Fig. 46. Find a possible equation of h. Verify your equation by using a graphing calculator.

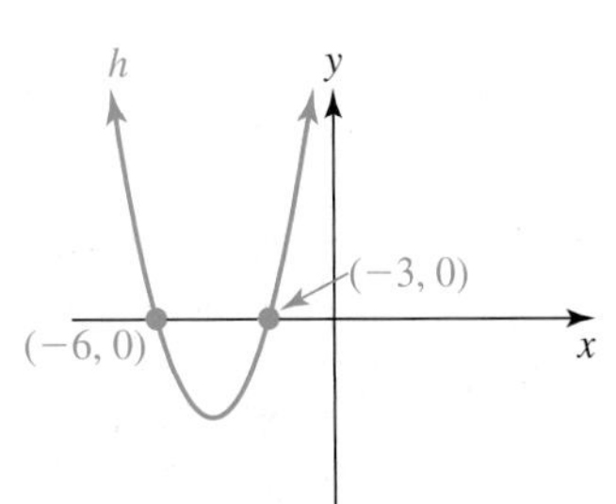

Figure 46 Exercise 109

110. The graph of a function g is sketched in Fig. 47. Find a possible equation of g. Verify your equation by using a graphing calculator.

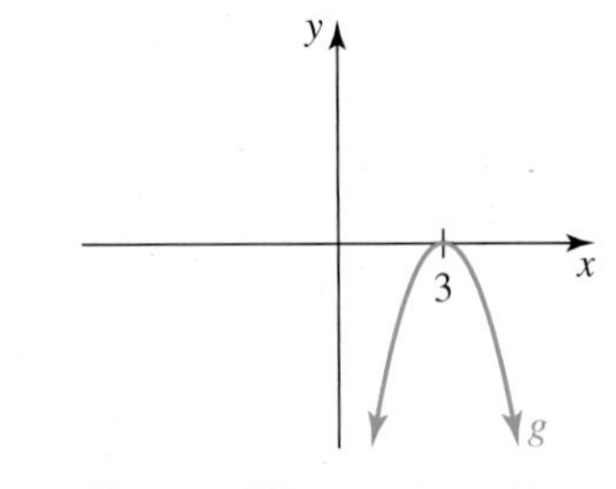

Figure 47 Exercise 110

111. Explain why a quadratic equation in one variable cannot have three solutions.

112. Explain how to solve a quadratic or cubic equation in one variable.

Related Review

Factor or solve, as appropriate.

113. $x^2 + 5x + 6$

114. $25x^2 - 64 = 0$

115. $x^2 + 5x + 6 = 0$

116. $25x^2 - 64$

117. $3p^3 + 8p^2 + 4p = 0$

118. $4w^3 - 20w^2 - 9w + 45$

119. $3p^3 + 8p^2 + 4p$

120. $4w^3 - 20w^2 - 9w + 45 = 0$

Solve. Round approximate solutions to the fourth decimal place.

121. $3x^2(x - 4) = 12x(x - 3)$

122. $4(3x - 5) - 2(5x + 1) = 20$

123. $3b^8 + 39 = 217$

124. $4w^3 - 20w^2 - w = -5$

125. $5(2)^t - 24 = 97$

126. $\log_4(r^3) = 5$

127. $\log(x + 3) + \log(x + 6) = 1$

128. $\log_2(2x + 3) + \log_2(x - 2) = 2$

Expressions, Equations, Functions, and Graphs

Give an example of the following. Then solve, simplify, or graph, as appropriate.

129. quadratic function

130. system of two linear equations in two variables

131. quadratic equation in one variable

132. linear function

133. exponential equation in one variable

134. difference of two logarithmic expressions with equal bases

CHAPTER SUMMARY

Key Points OF CHAPTER 6

Monomial (Section 6.1)
A **monomial** is a constant, a variable, or a product of a constant and one or more variables raised to *counting number* powers.

Polynomial (Section 6.1)
A **polynomial,** or **polynomial expression,** is a monomial or a sum of monomials.

Degree (Section 6.1)
The **degree of a term** in one variable is the exponent on the variable. The degree of a term in two or more variables is the sum of the exponents on the variables. The **degree of a polynomial** is the highest degree of any nonzero term of the polynomial.

Like terms (Section 6.1)
Like terms are either constant terms or variable terms that contain the same variable(s) raised to exactly the same power(s).

Combine like terms (Section 6.1)
To combine like terms, add the coefficients of the terms.

Add polynomials (Section 6.1)
To add polynomials, combine like terms.

Subtract polynomials (Section 6.1)
To subtract polynomials, first distribute -1; then combine like terms.

Quadratic function (Section 6.1)
A **quadratic function** is a function whose equation can be put into the form

$$f(x) = ax^2 + bx + c$$

where $a \neq 0$. This form is called the **standard form.**

Cubic function (Section 6.1)
A **cubic function** is a function whose equation can be put into the form

$$f(x) = ax^3 + bx^2 + cx + d$$

where $a \neq 0$.

Sum function and difference function (Section 6.1)
If f and g are functions and x is in the domain of both functions, then we can form the following functions:

- **Sum function $f + g$,** where $(f + g)(x) = f(x) + g(x)$
- **Difference function $f - g$,** where $(f - g)(x) = f(x) - g(x)$

Meaning of a difference (Section 6.1)
If a difference $A - B$ is positive, then A is more than B. If a difference $A - B$ is negative, then A is less than B.

Multiply polynomials (Section 6.2)
To multiply two polynomials, multiply each term in the first polynomial by each term in the second polynomial. Then combine like terms if possible.

Square of a sum (Section 6.2)
$(A + B)^2 = A^2 + 2AB + B^2$

Square of a difference (Section 6.2)
$(A - B)^2 = A^2 - 2AB + B^2$

Product of binomial conjugates (Section 6.2)
$(A + B)(A - B) = A^2 - B^2$

Product function (Section 6.2)
If f and g are functions and x is in the domain of both functions, then we can form the **product function $f \cdot g$:** $(f \cdot g)(x) = f(x) \cdot g(x)$.

Factor (Section 6.3)
We **factor** a polynomial by writing it as a product.

Multiplying versus factoring (Section 6.3)
Multiplying and factoring are reverse processes.

Factoring $x^2 + bx + c$ (Section 6.3)
To factor $x^2 + bx + c$, look for two integers p and q whose product is c and whose sum is b. That is, $pq = c$ and $p + q = b$. If such integers exist, the factored polynomial is $(x + p)(x + q)$.

Factoring $x^2 + bx + c$ with c positive (Section 6.3)
To factor a trinomial of the form $x^2 + bx + c$ with a positive constant term c,

- If b is positive, look for two *positive* integers whose product is c and whose sum is b.
- If b is negative, look for two *negative* integers whose product is c and whose sum is b.

Factoring $x^2 + bx + c$ with c negative (Section 6.3)
To factor a trinomial of the form $x^2 + bx + c$ with a negative constant term c, look for two integers with *different* signs whose product is c and whose sum is b.

Prime polynomial (Section 6.3)
A polynomial that cannot be factored is called **prime.**

GCF (Section 6.3)	The **greatest common factor (GCF)** of two or more terms is the monomial with the largest coefficient and the highest degree that is a factor of all the terms.
Factor out GCF (Section 6.3)	When the leading coefficient of a polynomial is positive and the GCF is not 1, first factor out the GCF.
Completely factor (Section 6.3)	When factoring a polynomial, always *completely* factor it.
Factoring when the leading coefficient is negative (Section 6.3)	When the leading coefficient of a polynomial is negative, first factor out the opposite of the GCF.
Factoring by grouping (Section 6.4)	For a polynomial with four terms, we **factor by grouping** (if it can be done) by **1.** Factoring the first two terms and the last two terms. **2.** Factoring out the binomial GCF.
Factoring $ax^2 + bx + c$ by trial and error (Section 6.4)	To **factor a trinomial** of the form $ax^2 + bx + c$ **by trial and error:** If the trinomial can be factored as a product of two binomials, then the product of the coefficients of the first terms of the binomials is equal to a and the product of the last terms of the binomials is equal to c. To find the correct factored expression, multiply the possible products and identify those for which the coefficient of x is b.
Factoring $ax^2 + bx + c$ by grouping (ac-method) (Section 6.4)	To **factor a trinomial** of the form $ax^2 + bx + c$ **by grouping** (if it can be done), **1.** Find pairs of numbers whose product is ac. **2.** Determine which of the pairs of numbers from step 1 has the sum b. Call this pair of numbers m and n. **3.** Write the bx term as $mx + nx$: $ax^2 + bx + c = ax^2 + mx + nx + c$ **4.** Factor $ax^2 + mx + nx + c$ by grouping. Another name for this technique is the ***ac*-method.**
Difference of two squares (Section 6.5)	$A^2 - B^2 = (A + B)(A - B)$
Sum of squares (Section 6.5)	Except for factoring out the GCF, a sum of squares cannot be factored.
Sum of two cubes (Section 6.5)	$A^3 + B^3 = (A + B)(A^2 - AB + B^2)$
Difference of two cubes (Section 6.5)	$A^3 - B^3 = (A - B)(A^2 + AB + B^2)$
Factoring strategy (Section 6.5)	These five steps can be used to factor many polynomials (steps 2–4 can be applied to the entire polynomial or to a factor of the polynomial): **1.** If the leading coefficient is positive and the GCF is not 1, factor out the GCF. If the leading coefficient is negative, factor out the opposite of the GCF. **2.** For a binomial, try using one of the properties for the difference of two squares, the sum of two cubes, or the difference of two cubes. **3.** For a trinomial of the form $ax^2 + bx + c$, **a.** If $a = 1$, try to find two integers whose product is c and whose sum is b. **b.** If $a \neq 1$, try to factor by using trial and error or by grouping. **4.** For an expression with four terms, try factoring by grouping. **5.** Continue applying steps 2–4 until the polynomial is completely factored.
Quadratic equation (Section 6.6)	A **quadratic equation in one variable** is an equation that can be put into the form $ax^2 + bx + c = 0$, where a, b, and c are constants and $a \neq 0$.
Zero factor property (Section 6.6)	Let A and B be real numbers: If $AB = 0$, then $A = 0$ or $B = 0$.
Connection between x-intercepts and solutions (Section 6.6)	Let f be a function. If k is a real-number solution of the equation $f(x) = 0$, then $(k, 0)$ is an x-intercept of the graph of the function f. Also, if $(k, 0)$ is an x-intercept of the graph of f, then k is a solution of $f(x) = 0$.
Number of solutions: quadratic equation (Section 6.6)	The solution set of a quadratic equation in one variable may contain two real numbers, one real number, or no real numbers.
Cubic equation (Section 6.6)	A **cubic equation in one variable** is an equation that can be put into the form $ax^3 + bx^2 + cx + d = 0$ where a, b, c, and d are constants and $a \neq 0$.

Solving quadratic or cubic equations by factoring (Section 6.6)	If an equation can be solved by factoring, we solve it by the following steps: **1.** Write the equation so that one side of the equation is equal to zero. **2.** Factor the nonzero side of the equation. **3.** Apply the zero factor property. **4.** Solve each equation that results from applying the zero factor property.
Number of solutions: cubic equation (Section 6.6)	The solution set of a cubic equation in one variable may contain one, two, or three real numbers.
Quadratic model (Section 6.6)	A **quadratic model** is a quadratic function, or its graph, that describes the relationship between two quantities in an authentic situation.
Area of a rectangle (Section 6.6)	The area A of a rectangle is given by the formula $A = LW$, where L is the length and W is the width.

CHAPTER 6 REVIEW EXERCISES

For Exercises 1 and 2, perform the operation.

1. $(-7x^3 + 5x^2 - 9) + (2x^3 - 8x^2 + 3x)$

2. $(5a^3b - 2a^2b^2 + 9ab^3) - (8a^3b + 4a^2b^2 - ab^3)$

3. For $f(x) = 3x^2 - 5x + 2$, find $f(-2)$.

Values of a quadratic function f are listed in Table 27. For Exercises 4 and 5, refer to this table.

4. Find $f(2)$.

5. Find x when $f(x) = 9$.

Table 27 Some Values of a Quadratic Function f (Exercises 4 and 5)

x	y
0	1
1	6
2	9
3	10
4	9
5	6
6	1

For $f(x) = 3x^3 - 7x^2 - 4x + 2$ and $g(x) = -2x^3 + 5x^2 - 3x + 1$, find an equation of the given function; then evaluate the function at the indicated value.

6. $(f + g)(x)$, $(f + g)(2)$

7. $(f - g)(x)$, $(f - g)(-3)$

Find the product or simplify, as appropriate.

8. $(x - 7)(x + 7)$

9. $8a^2b(-5a^3b^5)$

10. $(4p + 9t)(2p - 5t)$

11. $(4x - 3)(5x^2 - 2x + 4)$

12. $(3x + 7y)^2$

13. $(6p^2 - 9t^3)(6p^2 + 9t^3)$

14. $-3rt^3(2r^2 - 5rt + 3t^2)$

15. $-4x(3x - 2)^2$

16. $(3m^2 - mp + 2p^2)(2m^2 + 3mp - 4p^2)$

For Exercises 17 and 18, if $f(x) = x^2 - 2x$, find the following.

17. $f(a - 4)$

18. $f(a + 3) - f(a)$

19. Write the function $f(x) = -2(x - 4)^2 + 3$ in standard form.

20. For $f(x) = 3x - 7$ and $g(x) = 2x^2 - 4x + 3$, find an equation of the product function $f \cdot g$. Then find $(f \cdot g)(3)$.

Factor when possible.

21. $x^2 - 25$

22. $x^2 - 12x + 36$

23. $a^2 + 5ab - 36b^2$

24. $16a^5b^3 - 20a^3b^2$

25. $-11x - 24 + x^2$

26. $3w^2 - 5wy - 8y^2$

27. $81t^4 - 16w^4$

28. $6x^4 + 20x^3 - 16x^2$

29. $36x^2 + 49$

30. $x^2 - 3x - 54$

31. $2y^3 - 54$

32. $5r^2t + 30rt^2 + 45t^3$

33. $2ax - 10ay - 3bx + 15by$

For Exercises 34–40, solve.

34. $x^2 - 2x - 24 = 0$

35. $64t^2 = 9$

36. $3x(x + 10) = 6x^3$

37. $x^3 - 4x = 12 - 3x^2$

38. $\frac{m^2}{2} - \frac{7m}{6} + \frac{1}{3} = 0$

39. $32x^2 = 24x$

40. $4p(5p - 6) = (2p + 3)(2p - 3)$

41. Find all x-intercepts of the graph of $f(x) = 3x^3 + 3x^2 - 18x$.

For Exercises 42 and 43, use the graphs of $y = \frac{1}{2}x^2 + 2x - \frac{1}{2}$ and $y = x + 1$ shown in Fig. 48 to solve the given equation or system.

42. $\frac{1}{2}x^2 + 2x - \frac{1}{2} = 2$

43. $\frac{1}{2}x^2 + 2x - \frac{1}{2} = x + 1$

Figure 48 Exercises 42 and 43

44. The annual revenue of ExxonMobil® (in billions of dollars) $E(t)$ is modeled by the function $E(t) = 12.7t^2 - 35t + 235$, where t is the number of years since 2000 (see Table 28). The annual revenue of Chevron® (in billions of dollars) $C(t)$ is modeled by the function $C(t) = 8t^2 - 24.5t + 125$, where t is the number of years since 2000.

Table 28 Revenues of ExxonMobil and Chevron

Year	Revenue (billions of dollars)	
	ExxonMobil	Chevron
2000	236	126
2001	214	111
2002	207	100
2003	250	129
2004	300	157
2005	375	200

Source: *Bloomberg Financial Markets*

a. Find an equation of the sum function $E + C$.
b. Perform a unit analysis of the expression $E(t) + C(t)$.
c. Find $(E + C)(10)$. What does it mean in this situation?
d. Find an equation of the difference function $E - C$.
e. Find $(E - C)(10)$. What does it mean in this situation?

45. The total numbers of nights that extended-stay hotel rooms are available are shown in Table 29 for various years.

Table 29 Total Numbers of Available Nights for Extended-Stay Hotel Rooms

Year	Total Number of Available Nights (millions)
2000	57
2001	67
2002	72
2003	78
2004	82
2005	86

Source: *Smith Travel Research*

Let $f(t)$ be the total number of nights (in millions) that extended-stay hotel rooms are available in the year that is t years since 2000. A quadratic model of the situation is $f(t) = -\frac{3}{5}t^2 + 9t + 57$.

a. Use a graphing calculator to draw the graph of the model and, in the same viewing window, the scattergram of the data. Does the model fit the data well?
b. Predict the total number of available nights for extended-stay hotel rooms in 2011.
c. Predict in which year there will be a total of 87 million available nights for extended-stay hotel rooms.

46. The length of a rectangle is 8 meters more than the width. If both the width and length were doubled, the area would be 192 square meters. Find the dimensions of the original rectangle.

CHAPTER 6 TEST

1. Find the sum:

$$(4a^3b - 9a^2b^2 - 2ab^3) + (-5a^3b + 4a^2b^2 + 3ab^3)$$

2. For $f(x) = 4x^2 + 5x - 9$ and $g(x) = 6x^2 - 3x + 7$, find an equation of the difference function $f - g$; then find $(f - g)(-2)$.

For Exercises 3–6, refer to Fig. 49.

3. Find $f(-3)$.
4. Find x when $f(x) = 3$.
5. Find x when $f(x) = 1$.
6. Find x when $f(x) = -3$.

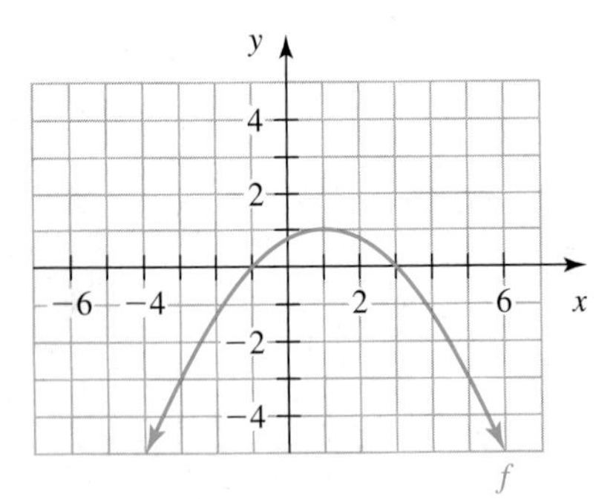

Figure 49 Exercises 3–6

Find the product or simplify, as appropriate.

7. $-2xy^2(7x^2 - 3xy + 6y^2)$
8. $(4x - 7y)(3x + 5y)$
9. $(2w - 5t)(3w^2 - wt + 4t^2)$
10. $3x(2x + 3)^2$
11. $(3x^2 + x - 5)(2x^2 + 4x - 1)$
12. $(4x^2 + 9y^2)(4x^2 - 9y^2)$
13. For $f(x) = x^2 - 3x$, find $f(a - 5)$.
14. Write $f(x) = -3(x + 4)^2 - 7$ in standard form.
15. For $f(x) = 2x^2 - 5x + 4$ and $g(x) = 3x - 2$, find an equation of the product function $f \cdot g$; then find $(f \cdot g)(3)$.

Factor when possible.

16. $x^2 - 10x - 24$
17. $18x + 2x^3 - 12x^2$
18. $-16x^2 - 26x + 12$
19. $9m^2 - 64t^2$
20. $16a^4b - 36a^3b^2 + 18a^2b^3$
21. $54m^3 + 128p^3$

Solve.

22. $25x^2 = 16$
23. $5w^3 - 15w^2 - 50w = 0$
24. $(2x - 7)(x - 3) = 10$
25. $2t^3 + 3t^2 = 18t + 27$
26. $3x(2x - 5) + 4x = 2(x - 3)$
27. Find all x-intercepts of the graph of $f(x) = 10x^2 - 19x + 6$.

28. Use "intersect" on a graphing calculator to solve the equation $\frac{1}{2}x^2 - \frac{2}{3}x - \frac{7}{4} = -\frac{3}{5}x + \frac{1}{2}$. Round your result(s) to the second decimal place.

29. The death rate from heart disease (number of deaths per 100,000 prople) $R(t)$ in the United States is modeled by the function $R(t) = -7.7t + 563$, where t is the number of years since 1960 (see Table 30). The U.S. population (in 100,000s of people) $P(t)$ is modeled by the equation $P(t) = 33t + 1493$, where t is the number of years since 1960.

Table 30 Death Rates from Heart Disease; U.S. Population

Year	Death Rate (number of deaths per 100,000 people)	Year	Population (in 100,000s)
1960	559	1990	2488
1970	493	1993	2578
1980	412	1996	2652
1990	322	1999	2790
2000	258	2002	2879
2003	236	2004	2937

Sources: *U.S. Center for Health Statistics; U.S. Census Bureau*

a. Check that the models fit the data well.
b. Find an equation of the product function $R \cdot P$.
c. Perform a unit analysis of the expression $R(t) \cdot P(t)$.
d. Find $(R \cdot P)(51)$. What does it mean in this situation?
e. Use a graphing calculator graph to determine whether the function $R \cdot P$ is increasing, decreasing, or neither for values of t between 40 and 50. What does your result mean in this situation? Explain how that is possible, given that U.S. population has been increasing.

30. Sales using debit cards (in billions of dollars) are shown in Table 31 for various years.

Table 31 Sales Using Debit Cards

Year	Sales (billions of dollars)
1991	11
1993	27
1995	50
1997	108
1999	222
2001	386

Source: *The Nilson Report*

Let $f(t)$ be the sales using debit cards (in billions of dollars) in the year that is t years since 1990. A model of the situation is $f(t) = 5t^2 - 23t + 38$.

a. Use a graphing calculator to draw the graph of the model and, in the same viewing window, the scattergram of the data. Does the model fit the data well?
b. Find $f(20)$. What does it mean in this situation?
c. Estimate when sales using debit cards were $20 billion.

31. A rectangular painting (not including the frame) has a width of 11 inches and a length of 15 inches. If the area of the frame (of uniform width) is 120 square inches, what is the width of the frame?

Chapter 7

Quadratic Functions

"Like any other art form, math takes practice and a lot of attention. I guess more like a girlfriend."

—*Jose C., student*

Table 1 Capacities at Which U.S. Nuclear Power Plants Are Working

Year	Percent
1975	56
1980	59
1985	58
1990	70
1995	76
2000	88
2004	89

Source: *Energy Information Administration*

Did you know that the 103 nuclear power plants in the United States produce 20% of the energy consumed by this country? The capacities at which U.S. nuclear power plants are working are shown in Table 1 for various years. In Exercise 67 of Homework 7.5, we will predict when U.S. nuclear power plants will be working at full capacity, which might suggest when there will be enough demand for more plants to be built.

In Chapter 6, we worked with polynomial expressions and equations. In this chapter, we will narrow our focus to quadratic expressions and equations. In particular, in Sections 7.1 and 7.2 we will graph quadratic functions. In Sections 7.3–7.5, we will discuss three nonfactoring methods of solving quadratic equations in one variable. In Sections 7.6 and 7.7, we will solve *systems of linear equations in three variables* to help us find equations of parabolas and quadratic models. Finally, in Section 7.8, we will use many of the skills learned earlier in this chapter to help us make estimates and predictions about authentic situations, such as the age of workers who are most likely to use computers on the job.

7.1 GRAPHING QUADRATIC FUNCTIONS IN VERTEX FORM

Objectives

- Know the *vertex form* of a quadratic function.
- Graph a quadratic function in vertex form.
- Find the domain and range of a quadratic function.
- Find a quadratic model in vertex form.

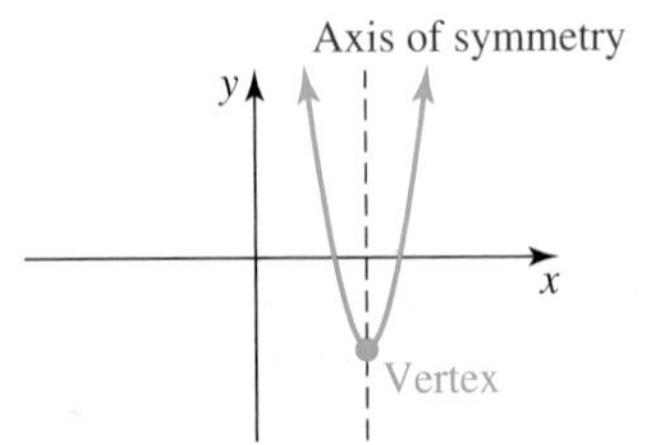

Figure 1 A parabola that opens upward

In this section, we will work with parabolas. As we saw in Section 6.1, each parabola has a vertex and an axis of symmetry (see Figs. 1 and 2). Recall that the part of the parabola that lies to the left of the axis of symmetry is the mirror reflection of the part that lies to the right.

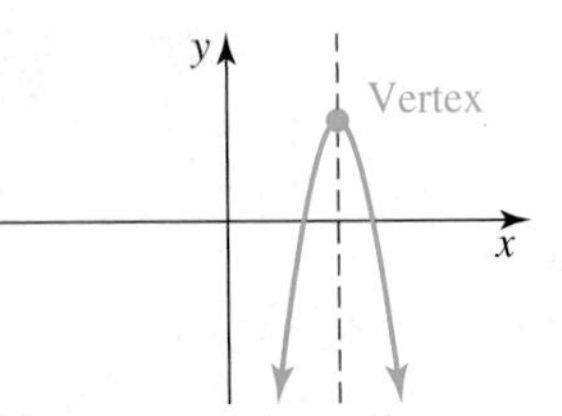

Figure 2 A parabola that opens downward

Vertex Form

Recall from Section 6.1 that a quadratic function is a function whose equation can be put into the standard form $f(x) = ax^2 + bx + c$, where $a \neq 0$, and that the graph of the function is a parabola. Here, we will work with equations in another form, called *vertex form:* $f(x) = a(x-h)^2 + k$, where $a \neq 0$. Any equation in either of these two forms can be written in the other form. So, a function in vertex form is quadratic, and its graph is a parabola.

Vertex Form of a Quadratic Function

Let $f(x) = a(x-h)^2 + k$, where $a \neq 0$. Then f is a quadratic function, and its graph is a parabola. We say that the equation is in **vertex form.**

Here are some quadratic functions in vertex form:

$$f(x) = 3(x-5)^2+4 \qquad g(x) = -4(x+9)^2-1 \qquad h(x) = 6x^2-8 \qquad k(x) = x^2$$

We will see later in this section that for a quadratic function in vertex form, we can find the vertex quickly.

Graphs of Quadratic Functions of the Form $f(x) = ax^2$

To graph quadratic functions in vertex form $f(x) = a(x-h)^2 + k$, we begin by exploring graphs of equations of the form $f(x) = ax^2$ (where $h = 0$ and $k = 0$).

Table 2 Input-Output Pairs of $f(x) = x^2$ and $g(x) = 2x^2$

x	$f(x) = x^2$	$g(x) = 2x^2$
−3	9	18
−2	4	8
−1	1	2
0	0	0
1	1	2
2	4	8
3	9	18

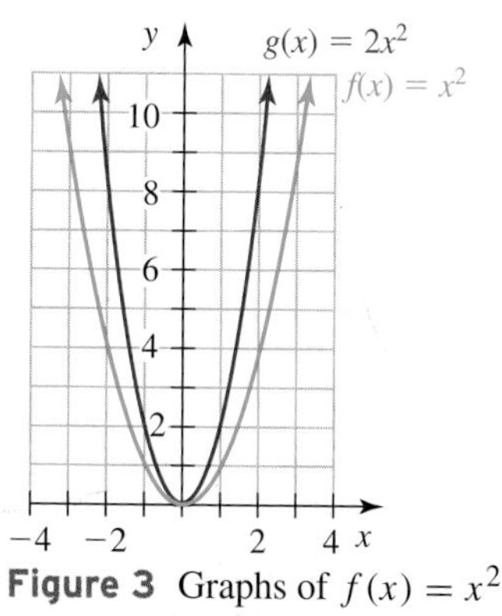

Figure 3 Graphs of $f(x) = x^2$ and $g(x) = 2x^2$

Example 1 Stretching a Graph Vertically

Compare the graph of $g(x) = 2x^2$ with the graph of $f(x) = x^2$.

Solution

We list input–output pairs of f and g in Table 2 and sketch the graphs of f and g in Fig. 3. For example, $g(-2) = 2(-2)^2 = 2(4) = 8$. Therefore, $(-2, 8)$ is a point on the graph of g.

For each value of x, the value of y is twice as large for $g(x) = 2x^2$ as it is for $f(x) = x^2$ (see Table 2). Therefore, the graph of g appears steeper (narrower) than the graph of f. Also, notice that the vertex for both functions is the point $(0, 0)$. ■

Example 2 Reflecting a Graph across the x-Axis

Compare the graph of $f(x) = \frac{1}{2}x^2$ with the graph of $g(x) = -\frac{1}{2}x^2$.

Solution

We list input–output pairs of f and g in Table 3 and sketch the graphs of f and g in Fig. 4.

Table 3 Input-Output Pairs of $f(x) = \frac{1}{2}x^2$ and $g(x) = -\frac{1}{2}x^2$

x	$f(x) = \frac{1}{2}x^2$	$g(x) = -\frac{1}{2}x^2$
−3	4.5	−4.5
−2	2	−2
−1	0.5	−0.5
0	0	0
1	0.5	−0.5
2	2	−2
3	4.5	−4.5

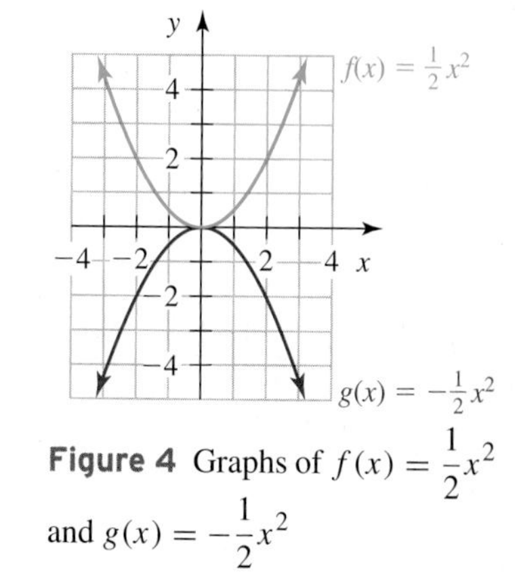

Figure 4 Graphs of $f(x) = \frac{1}{2}x^2$ and $g(x) = -\frac{1}{2}x^2$

The graph of $g(x) = -\frac{1}{2}x^2$ is the reflection across the x-axis of the graph of $f(x) = \frac{1}{2}x^2$. To see why, note that for each value of x, the value of y for $g(x) = -\frac{1}{2}x^2$ is the opposite of the value of y for $f(x) = \frac{1}{2}x^2$ (see Table 3). ■

Our observations made in Examples 1 and 2 suggest the following properties.

Graphs of Quadratic Functions of the Form $f(x) = ax^2$

For a function of the form $f(x) = ax^2$,

- The graph is a parabola with vertex (0, 0).
- If $|a|$ is a large number, then the parabola is steep. (It is narrow.)
- If a is near zero, then the parabola is not steep. (It is wide.)
- If $a > 0$, then the parabola opens upward.
- If $a < 0$, then the parabola opens downward.
- The graph of $y = -ax^2$ is the reflection across the x-axis of the graph of $f(x) = ax^2$.

Translating Graphs

Next, we investigate graphs of functions of the form $f(x) = x^2 + k$.

Example 3 Vertical Translation (Up-Down Shifts)

Compare the graphs of $f(x) = x^2 - 3$, $g(x) = x^2$, and $h(x) = x^2 + 3$.

Solution

We list input–output pairs of f, g, and h in Table 4 and sketch the graphs of the functions in Fig. 5.

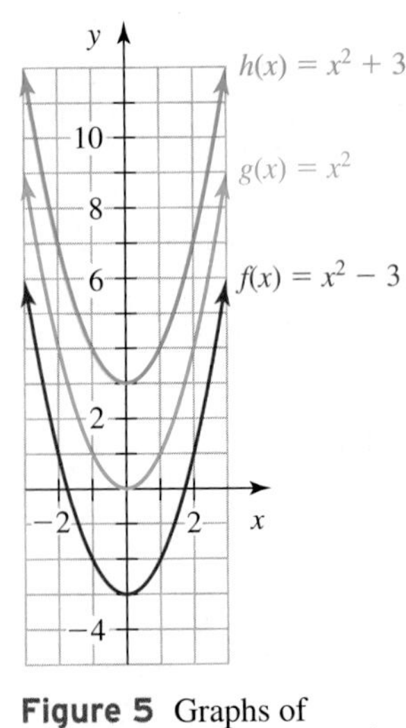

Figure 5 Graphs of $f(x) = x^2 - 3$, $g(x) = x^2$, and $h(x) = x^2 + 3$

Table 4 Input-Output Pairs of $f(x) = x^2 - 3$, $g(x) = x^2$, and $h(x) = x^2 + 3$

x	$f(x) = x^2 - 3$	$g(x) = x^2$	$h(x) = x^2 + 3$
−3	6	9	12
−2	1	4	7
−1	−2	1	4
0	−3	0	3
1	−2	1	4
2	1	4	7
3	6	9	12

For each x value, the y values of $h(x) = x^2 + 3$ are 3 more than the y values of $g(x) = x^2$, which are 3 more than the y values of $f(x) = x^2 - 3$.

To sketch the graph of $h(x) = x^2 + 3$, we *translate* (move) the graph of $g(x) = x^2$ up by 3 units. To sketch the graph of $f(x) = x^2 - 3$, we translate the graph of $g(x) = x^2$ down by 3 units. ■

Example 4 Horizontal Translation (Left-Right Shifts)

Compare the graph of $g(x) = (x - 5)^2$ with the graph of $f(x) = x^2$.

Solution

We list input–output pairs of g in Table 5 and sketch the graphs of f and g in Fig. 6.

Table 5 Input-Output Pairs of $g(x) = (x - 5)^2$

x	$g(x)$
−1	36
0	25
1	16
2	9
3	4
4	1
5	0 ← Vertex
6	1
7	4
8	9

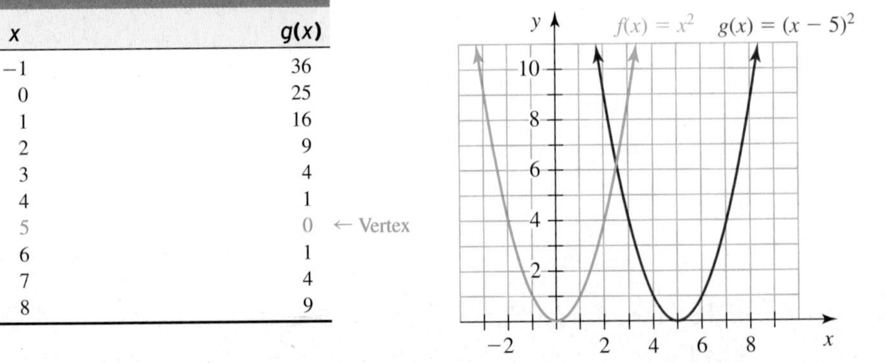

Figure 6 Graphs of $f(x) = x^2$ and $g(x) = (x - 5)^2$

To graph $g(x) = (x - 5)^2$, we translate the graph of $f(x) = x^2$ to the *right* by 5 units. ■

Example 5 Horizontal Translation (Left-Right Shifts)

Compare the graph of $g(x) = (x + 4)^2$ with the graph of $f(x) = x^2$.

Solution

We list input–output pairs of g in Table 6 and sketch the graphs of f and g in Fig. 7.

Table 6 Input-Output Pairs of $g(x) = (x + 4)^2$

x	$g(x)$	
−8	16	
−7	9	
−6	4	
−5	1	
−4	0	← Vertex
−3	1	
−2	4	
−1	9	
0	16	
1	25	

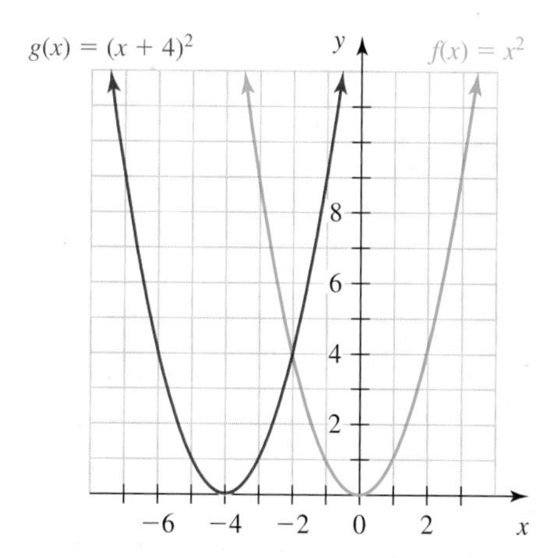

Figure 7 Graphs of $f(x) = x^2$ and $g(x) = (x + 4)^2$

To graph $g(x) = (x+4)^2$, we translate the graph of $f(x) = x^2$ to the *left* by 4 units. ■

Three-Step Method of Graphing a Quadratic Function in Vertex Form

To sketch the graph of $f(x) = a(x - h)^2 + k$, where $a \neq 0$,

1. Sketch the graph of $y = ax^2$.
2. Translate the graph from step 1 to the right by h units if $h > 0$ or to the left by $|h|$ units if $h < 0$.
3. Translate the graph from step 2 up by k units if $k > 0$ or down by $|k|$ units if $k < 0$.

Example 6 Sketching the Graph of a Quadratic Function

Sketch the graph of $f(x) = -(x - 4)^2$.

Solution

The equation of f is already in $f(x) = a(x - h)^2 + k$ form, with $a = -1$, $h = 4$, and $k = 0$. We follow the three-step graphing method:

Step 1. We sketch the graph of $y = -x^2$ in Fig. 8.

Step 2. Since $h = 4$, we translate the graph from step 1 to the right by 4 units.

Step 3. Since $k = 0$, there is no vertical translation.

Table 7 Input-Output Pairs of $f(x) = -(x - 4)^2$

x	$f(x)$	
1	−9	
2	−4	
3	−1	
4	0	← Vertex
5	−1	
6	−4	
7	−9	

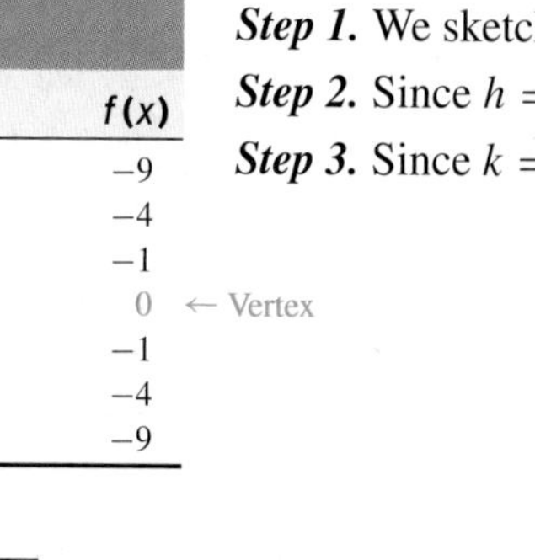

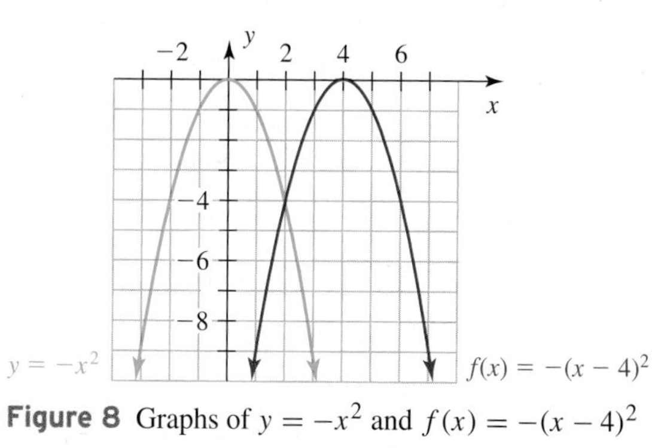

Figure 8 Graphs of $y = -x^2$ and $f(x) = -(x - 4)^2$

We check that the input–output pairs of f listed in Table 7 are points on our sketched parabola. We use integer inputs that are within 3 units of the x-coordinate of the vertex, 4. Also, we use a graphing calculator to verify our sketch (see Fig. 9). ■

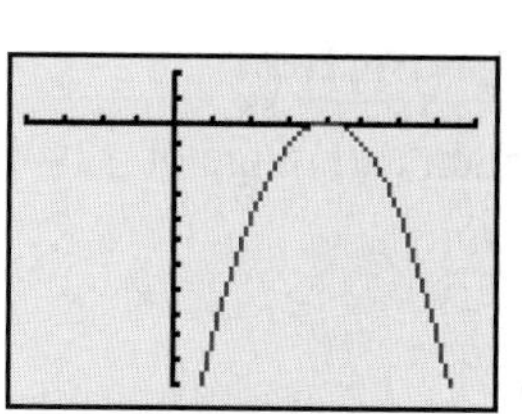

Figure 9 Graph of $f(x) = -(x - 4)^2$

Example 7 Sketching the Graph of a Quadratic Function

Sketch the graph of $f(x) = (x + 3)^2 + 1$.

Solution

First, we write the equation of f in $f(x) = a(x - h)^2 + k$ form:

$$f(x) = [x - (-3)]^2 + 1$$

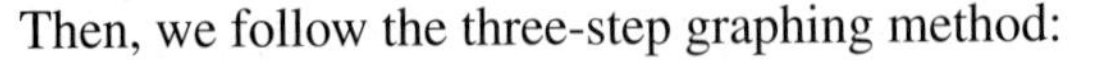

Then, we follow the three-step graphing method:

Step 1. We sketch the graph of $y = x^2$ in Fig. 10.

Step 2. Since $h = -3$, we translate the graph from step 1 to the left by 3 units.

Step 3. Since $k = 1$, we translate the graph from step 2 up by 1 unit.

We check that the input–output pairs of f listed in Table 8 are points on our sketched parabola. We use integer inputs that are within 3 units of the x-coordinate of the vertex, -3. Also, we can use a graphing calculator to verify our work (see Fig. 11).

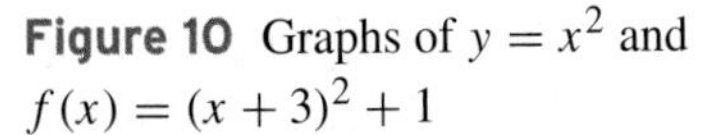

Figure 10 Graphs of $y = x^2$ and $f(x) = (x + 3)^2 + 1$

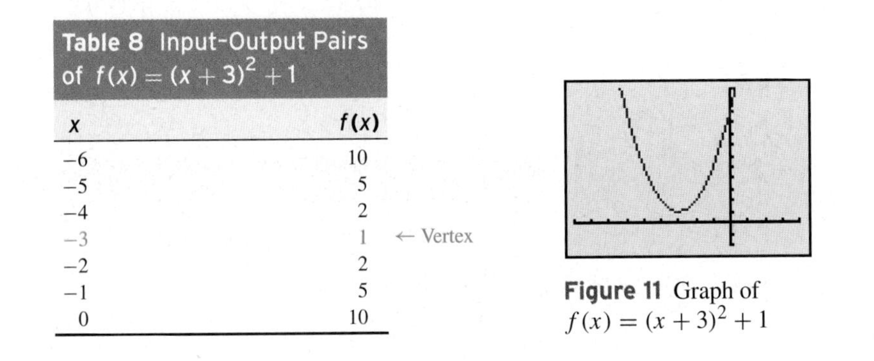

Table 8 Input-Output Pairs of $f(x) = (x + 3)^2 + 1$

x	$f(x)$	
−6	10	
−5	5	
−4	2	
−3	1	← Vertex
−2	2	
−1	5	
0	10	

Figure 11 Graph of $f(x) = (x + 3)^2 + 1$

In Fig. 10, we see that the vertex of $f(x) = (x + 3)^2 + 1$ is the point $(-3, 1)$. This makes sense, since the vertex $(0, 0)$ of $y = x^2$ has been translated to the left by 3 units and up by 1 unit.

Vertex of a Quadratic Function

The vertex of a quadratic function in vertex form, $f(x) = a(x - h)^2 + k$, is the point (h, k).

For example, the vertex of the function $f(x) = -3(x - 1)^2 + 5$ is $(1, 5)$. To find the vertex of the function $g(x) = 6(x + 2)^2 - 7$, we write the equation as $g(x) = 6[x - (-2)]^2 + (-7)$. So, the vertex is $(-2, -7)$.

Domain and Range of a Quadratic Function

Recall from Section 1.6 that the domain of a function is the set of values of x (the independent variable) and that the range of a function is the set of values of y (the dependent variable).

Example 8 Sketching the Graph of a Quadratic Function

Sketch the graph of $f(x) = -2(x + 6)^2 - 2$, and find the domain and range of f.

Solution

First, we write the equation of f in $f(x) = a(x - h)^2 + k$ form:

$$f(x) = -2[x - (-6)]^2 + (-2)$$

Then, we follow the three-step graphing method:

Step 1. We sketch the graph of $y = -2x^2$ in Fig. 12.

Step 2. Since $h = -6$, we translate the graph from step 1 to the left by 6 units.

Step 3. Since $k = -2$, we translate the graph from step 2 down by 2 units.

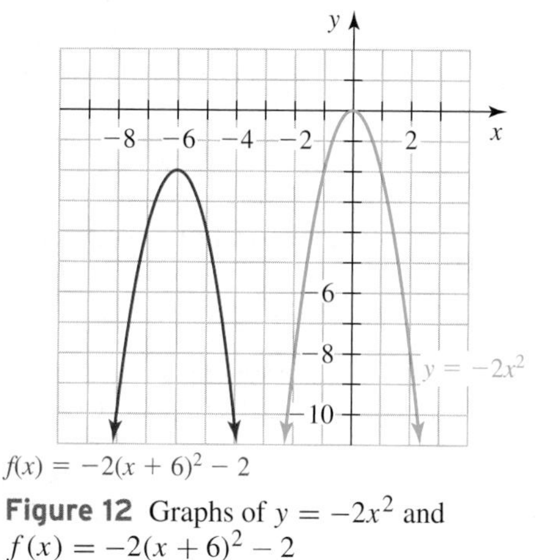

Figure 12 Graphs of $y = -2x^2$ and $f(x) = -2(x + 6)^2 - 2$

Table 9 Input-Output Pairs of $f(x) = -2(x + 6)^2 - 2$

x	$f(x)$	
−8	−10	
−7	−4	
−6	−2	← Vertex
−5	−4	
−4	−10	

We check that the input–output pairs of f listed in Table 9 are points on our sketched parabola. Also, from the equation $f(x) = -2[x - (-6)]^2 + (-2)$, we see that the vertex is $(-6, -2)$, which matches our sketched parabola's vertex. Finally, we use a graphing calculator to verify our sketch (see Fig. 13).

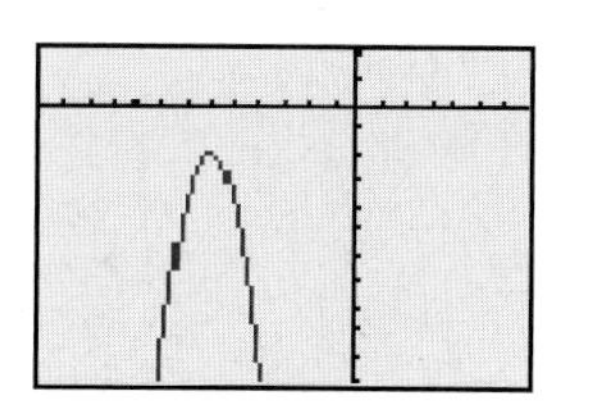

Figure 13 Graph of $f(x) = -2(x + 6)^2 - 2$

Since we can compute a value of $-2(x+6)^2 - 2$ for any real number x, the domain of $f(x) = -2(x + 6)^2 - 2$ is the set of all real numbers. From the graph of f, we see that the vertex $(-6, -2)$ is the maximum point and that values of y are less than or equal to -2. So, the range of f is the set of numbers y where $y \leq -2$. ■

Example 9 Sketching the Graph of a Quadratic Function

Sketch the graph of $h(x) = \frac{1}{3}(x - 5)^2 + 3$, and find the domain and range of h.

Solution

We follow the three-step graphing method:

Step 1. We sketch the graph of $y = \frac{1}{3}x^2$ in Fig. 14.

Step 2. Since $h = 5$, we translate the graph from step 1 to the right by 5 units.

Step 3. Since $k = 3$, we translate the graph from step 2 up by 3 units.

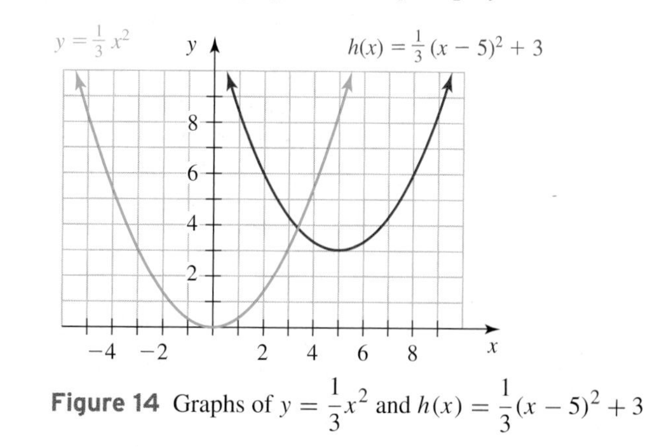

Figure 14 Graphs of $y = \frac{1}{3}x^2$ and $h(x) = \frac{1}{3}(x - 5)^2 + 3$

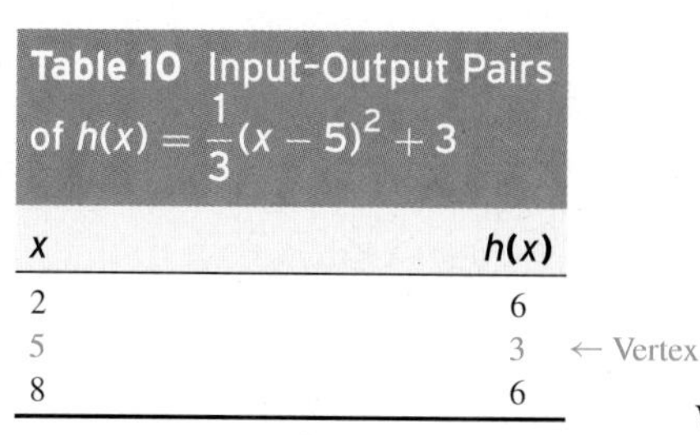

Table 10 Input-Output Pairs of $h(x) = \frac{1}{3}(x - 5)^2 + 3$

x	$h(x)$	
2	6	
5	3	← Vertex
8	6	

We check that the input–output pairs of h listed in Table 10 are points on our sketched parabola. Also, from the equation $h(x) = \frac{1}{3}(x-5)^2+3$, we see that the vertex

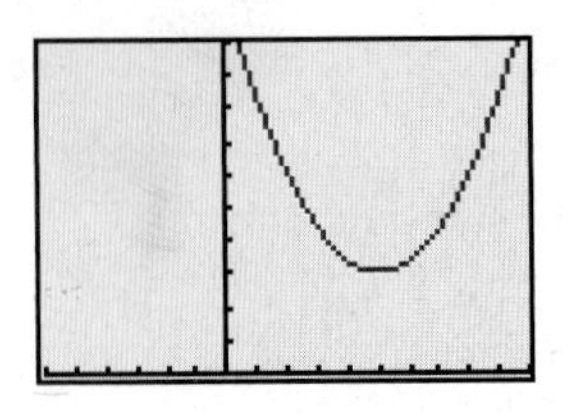

Figure 15 Graph of $h(x) = \frac{1}{3}(x-5)^2 + 3$

is (5, 3), which matches our sketched parabola's vertex. Finally, we use a graphing calculator to verify our work (see Fig. 15).

Since we can compute a value of $\frac{1}{3}(x-5)^2 + 3$ for any real number x, the domain of $h(x) = \frac{1}{3}(x-5)^2 + 3$ is the set of all real numbers. From the graph of h, we see that the vertex (5, 3) is the minimum point and that values of y are greater than or equal to 3. So, the range of f is the set of numbers y where $y \geq 3$. ■

Find a Quadratic Model in Vertex Form

We can use our knowledge of graphing quadratic functions in vertex form to help us find an equation of a quadratic model.

Example 10 Finding an Equation of a Quadratic Model

The numbers of flights with taxi-out times of one hour or more per 1000 flights are shown in Table 11 for various years.

Table 11 Flights with Taxi-Out Times of One Hour or More

Year	Number of Flights That Took One Hour or More to Taxi Out per 1000 Flights
2000	9.5
2001	7.8
2002	6.8
2003	7.5
2004	9.7

Source: *U.S. Bureau of Transportation Statistics*

Let $f(t)$ be the number of flights that took one hour or more to taxi out per 1000 flights at t years since 2000.

1. Find an equation of f.
2. What is the vertex of f? What does it mean in this situation?
3. Use f to estimate the number of flights that took one hour or more to taxi out in 2006.

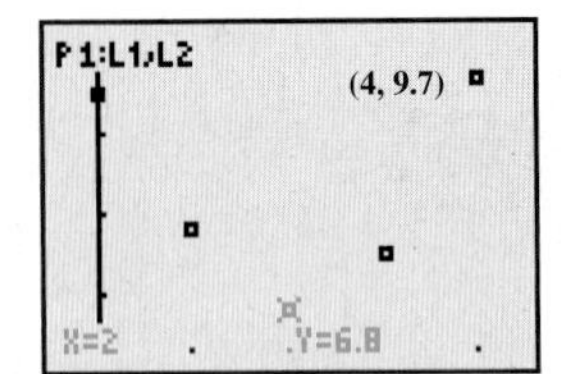

Figure 16 Scattergram of flight data

Solution

1. We use a graphing calculator to draw a scattergram of the data (see Fig. 16). It appears that a quadratic function would describe the situation much better than an exponential function or a linear function. We will find a quadratic function in vertex form, $f(t) = a(t-h)^2 + k$. Although it is not necessary to select the lowest data point, (2, 6.8), to be the vertex (see Fig. 16), it is convenient and satisfactory to do so. This means that $h = 2$ and $k = 6.8$:

$$f(t) = a(t-2)^2 + 6.8$$

Next, we imagine a parabola with vertex (2, 6.8) that comes close to (or contains) the data points. Such a parabola might be the one that contains the data point (4, 9.7). See Fig. 16. To find a, we substitute 4 for t and 9.7 for $f(t)$ in the equation $f(t) = a(t-2)^2 + 6.8$ and solve for a:

$$9.7 = a(4-2)^2 + 6.8 \quad \text{Substitute 4 for } t \text{ and 9.7 for } f(t).$$
$$9.7 = a(2)^2 + 6.8 \quad \text{Subtract.}$$
$$9.7 = 4a + 6.8 \quad \text{Simplify.}$$
$$2.9 = 4a$$
$$0.73 \approx a$$

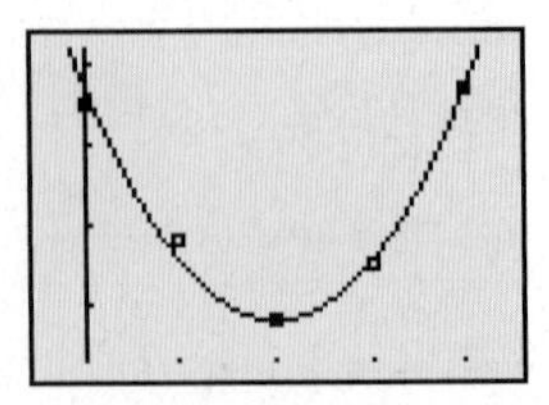

Figure 17 Check the fit

The approximate equation is $f(t) = 0.73(t-2)^2 + 6.8$. We check how well the model fits the data in Fig. 17. Since the graph appears to come close to (or contain) the data points, we conclude that f is a reasonable model of the situation.

2. The vertex of the graph of $f(t) = 0.73(t - 2)^2 + 6.8$ is $(2, 6.8)$, which means that in 2002 there were 6.8 flights that took one hour or more to taxi out per 1000 flights. That is the lowest rate in any year, according to the model.
3. We represent 2006 by $t = 6$. We evaluate f at 6: $f(6) = 0.73(6-2)^2 + 6.8 = 18.48$. About 18.5 flights took one hour or more to taxi out per 1000 flights in 2006, according to the model. ■

Finding a Quadratic Model in Vertex Form

To find a quadratic model in vertex form, given some data,

1. Create a scattergram of the data.
2. Imagine a parabola that comes close to (or contains) the data points, and select a point (h, k) to be the vertex. Although it is not necessary to select a data point, it is often convenient and satisfactory to do so.
3. Select a nonvertex point (not necessarily a data point) of the parabola, substitute the point's coordinates into the equation $f(t) = a(t - h)^2 + k$, and solve for a.
4. Substitute the result you found for a in step 3 into $f(t) = a(t - h)^2 + k$.

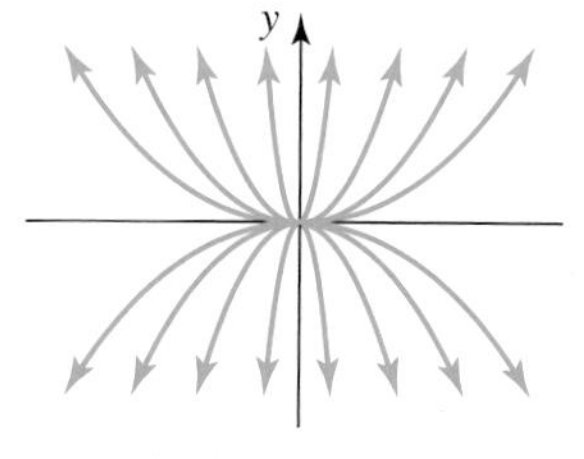

Figure 18 A family of parabolas

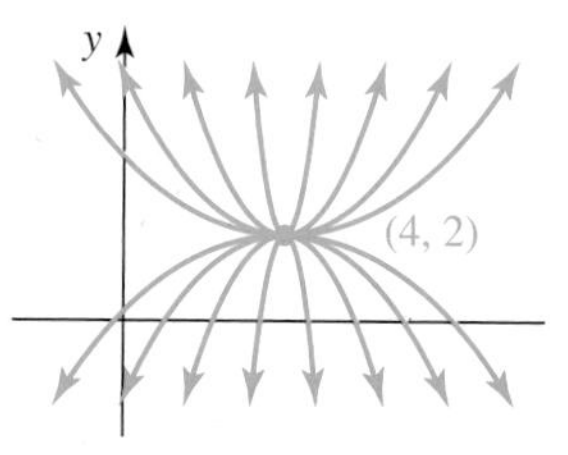

Figure 19 A family of parabolas with vertex (4, 2)

group exploration

Drawing families of parabolas

1. On a graphing calculator, graph eight parabolas to make a design like the one in Fig. 18. List the equations of your parabolas.
2. Now make a design like the one in Fig. 19. List the equations of your parabolas.
3. For a quadratic function of the form $f(x) = a(x - h)^2 + k$, summarize what you have learned about a, h, and k from this exploration and from the rest of this section.

TIPS FOR SUCCESS: Reread a Problem

After you think you have solved a problem, reread it to make sure you have answered its question(s). Also, reread the problem with the solution in mind. If what you read seems to make sense, then you've provided another check of your result(s).

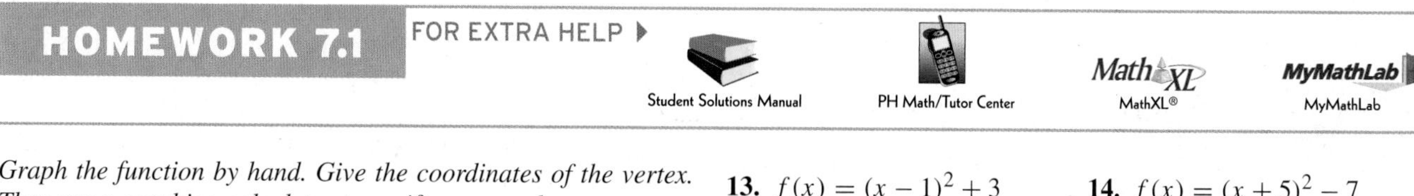

Graph the function by hand. Give the coordinates of the vertex. Then use a graphing calculator to verify your work.

1. $f(x) = 3x^2$
2. $f(x) = 2x^2$
3. $f(x) = -\frac{3}{2}x^2$
4. $f(x) = -\frac{2}{3}x^2$
5. $f(x) = -2x^2 + 5$
6. $f(x) = 2x^2 - 4$
7. $f(x) = (x - 1)^2$
8. $f(x) = (x + 3)^2$
9. $f(x) = -(x + 2)^2$
10. $f(x) = -3(x - 3)^2$
11. $f(x) = (x + 2)^2 - 6$
12. $f(x) = (x - 4)^2 + 2$
13. $f(x) = (x - 1)^2 + 3$
14. $f(x) = (x + 5)^2 - 7$
15. $f(x) = 2(x + 6)^2 - 6$
16. $f(x) = 3(x - 4)^2 + 1$
17. $f(x) = -3(x - 6)^2 - 2$
18. $f(x) = -(x+1)^2 - 2$
19. $f(x) = \frac{1}{3}(x - 2)^2 + 3$
20. $f(x) = -\frac{1}{2}(x + 4)^2 - 2$

Graph the function by hand. Then use a graphing calculator to verify your work. Also, find the domain and range of the function.

21. $f(x) = x^2 - 4$
22. $f(x) = 2x^2 + 1$
23. $f(x) = -x^2 - 3$
24. $f(x) = -2x^2 + 5$

25. $f(x) = (x+4)^2$
26. $f(x) = (x-2)^2$
27. $f(x) = (x+6)^2 + 2$
28. $f(x) = (x-3)^2 + 1$
29. $f(x) = 2(x-1)^2 - 4$
30. $f(x) = 3(x+4)^2 - 7$
31. $f(x) = -(x-5)^2 + 2$
32. $f(x) = -2(x+2)^2 + 5$

33. The numbers of billionaires in the United States are shown in Table 12 for various years.

study example 10!

Table 12 Numbers of Billionaires

Year	Number of Billionaires
2000	298
2001	266
2002	228
2003	262
2004	313

Source: Forbes

Let $n = f(t)$ be the number of billionaires in the United States at t years since 2000.

a. Find a quadratic equation of f in vertex form.
b. What is the vertex of the model? What does it mean in this situation?
c. What is the n-intercept of the model? What does it mean in this situation?
d. Predict the number of billionaires in 2010.

34. The numbers of visitors to Ireland from the United States and Canada are shown in Table 13 for various years.

Table 13 Numbers of Visitors to Ireland from the United States and Canada

Year	Number of Visitors (millions)
2000	1.00
2001	0.91
2002	0.85
2003	0.90
2004	0.98

Source: *Central Statistics Office Ireland*

Let $n = f(t)$ be the number of visitors (in millions) to Ireland from the United States and Canada in the year that is t years since 2000.

a. Find a quadratic equation of f in vertex form.
b. What is the vertex of the model? What does it mean in this situation?
c. What is the n-intercept of the model? What does it mean in this situation?
d. Estimate the number of visitors to Ireland from the United States and Canada in 2006.

35. U.S. Department of Defense spendings are shown in Table 14 for various years. Let $f(t)$ be the U.S. Department of Defense spending (in billions of dollars) in the year that is t years since 1990.

a. Find a quadratic equation of f in vertex form.
b. What is the vertex of the model? What does it mean in this situation?
c. Predict defense spending in 2010.

Table 14 U.S. Department of Defense Spendings

Year	Defense Spending (billions of dollars)
1992	403
1994	358
1996	324
1998	311
2000	323
2002	368
2004	462
2005	454

Sources: *Department of Defense; Government Accountability Office*

36. The numbers of cancer-related charges of discrimination filed with the U.S. Equal Employment Opportunity Commission are shown in Table 15 for various years.

Table 15 Numbers of Cancer-Related Discrimination Charges

Year	Number of Cancer-Related Discrimination Charges
2000	434
2001	454
2002	455
2003	442
2004	427

Source: *U.S. Equal Employment Opportunity Commission*

Let $n = f(t)$ be the number of cancer-related charges of discrimination filed with the U.S. Equal Employment Opportunity Commission in the year that is t years since 2000.

a. Find a quadratic equation of f in vertex form.
b. What is the vertex of the model? What does it mean in this situation?
c. Estimate the number of cancer-related discrimination charges in 2005.

37. Describe the Rule of Four as applied to the function $f(x) = 2(x-1)^2 - 3$:

a. Describe the input–output pairs of f by using a graph.
b. Describe five input–output pairs of f by using a table.
c. Describe the input–output pairs of f by using words.

38. Describe the Rule of Four as applied to the function $g(x) = -3(x+4)^2 + 7$:

a. Describe the input–output pairs of g by using a graph.
b. Describe five input–output pairs of g by using a table.
c. Describe the input–output pairs of g by using words.

39. Let $f(x) = (x-3)^2 + 2$

a. Graph f by hand.
b. Find x when $f(x) = 3$.
c. Find x when $f(x) = 2$.
d. Find x when $f(x) = 1$.

40. Let $f(x) = -(x+2)^2 + 5$.

a. Graph f by hand.
b. Find x when $f(x) = 6$.
c. Find x when $f(x) = 5$.
d. Find x when $f(x) = 4$.

Write an equation of a parabola that meets the given criteria.

41. opens downward and has vertex $(-3, 4)$
42. opens upward and has vertex $(2, -5)$

43. Four functions of the form $y = a(x - h)^2 + k$ are graphed in Fig. 20. Determine whether the constants a, h, and k are positive, negative, or zero for each function.

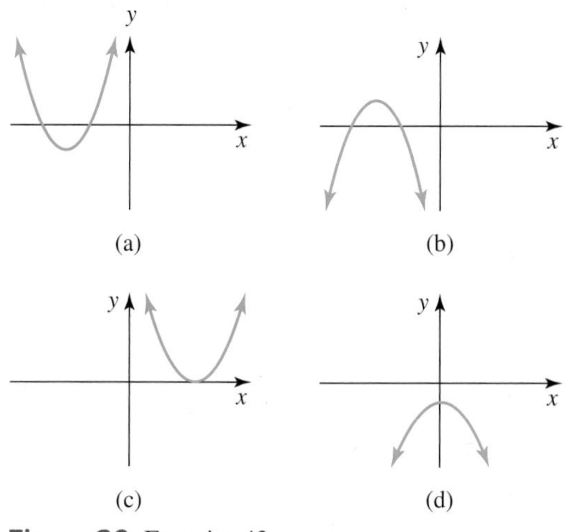

Figure 20 Exercise 43

44. For a quadratic function $f(x) = a(x - h)^2 + k$, for what values of a, h, and k does the graph of f have a maximum point? a minimum point? Describe the maximum or minimum point in terms of a, h, and k.

45. Use a graphing calculator to graph a family of parabolas similar to the family shown in Fig. 21. List the equations of your parabolas.

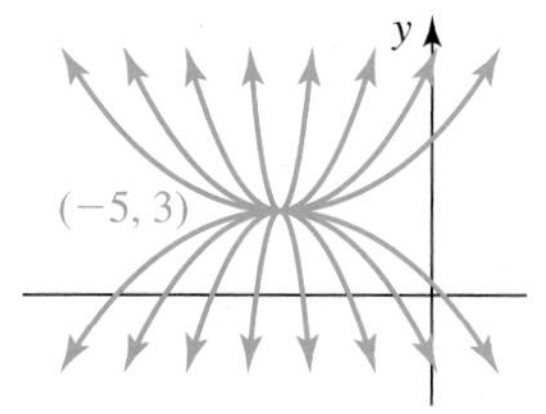

Figure 21 A family of parabolas with vertex $(-5, 3)$

46. Use a graphing calculator to graph a family of parabolas similar to the family shown in Fig. 22. List the equations of your parabolas.

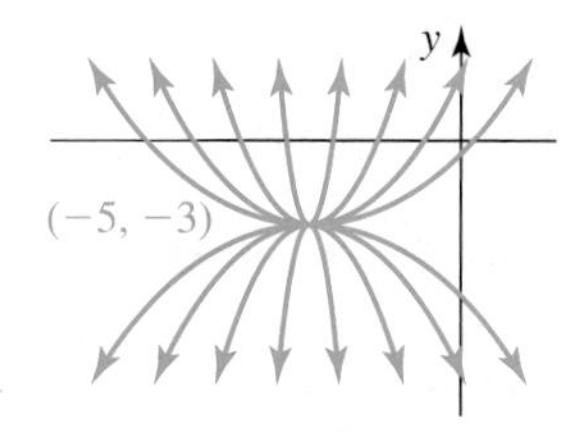

Figure 22 A family of parabolas with vertex $(-5, -3)$

47. Find an equation of the function f graphed in Fig. 23.

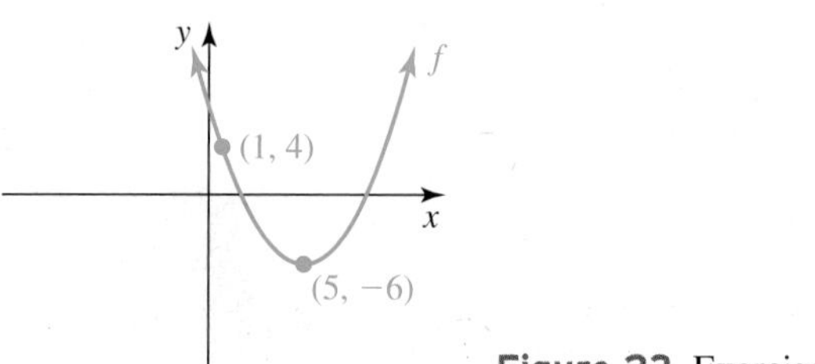

Figure 23 Exercise 47

48. The graph of the function $f(x) = 2.1(x - 2.73)^2 - 3.71$ is shown in Fig. 24. Find an equation of the function g also graphed in Fig. 24.

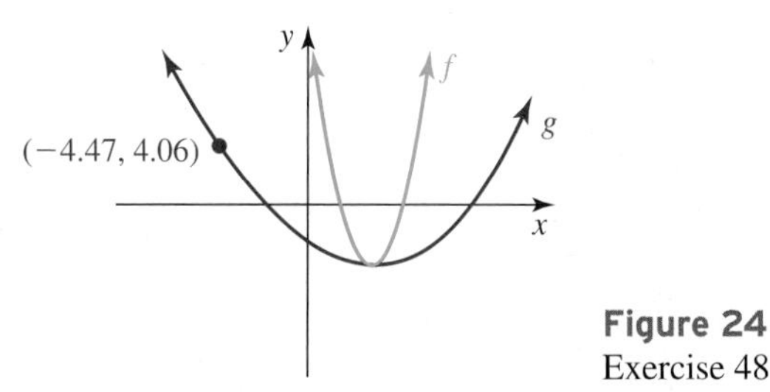

Figure 24 Exercise 48

49. The graph of the function $f(x) = 2.1(x - 2.73)^2 - 3.71$ is shown in Fig. 25. Find an equation of the function g also graphed in Fig. 25. Assume that the graphs of f and g have the same shape.

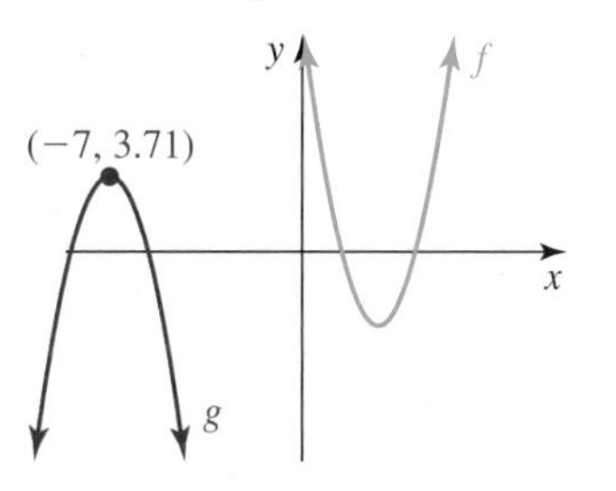

Figure 25 Exercise 49

50. Find equations of the four functions whose graphs produce the design shown in Fig. 26.

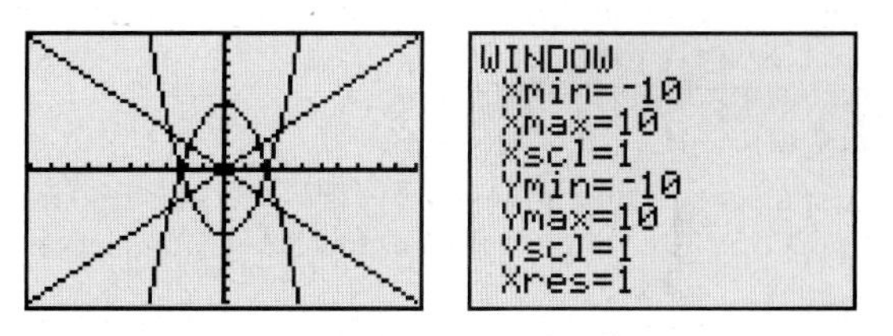

Figure 26 A design created by graphs of functions

For each exercise, decide whether it is possible for a parabola to have the indicated number of x-intercepts. If it is possible, find an equation of such a parabola. If it is not possible, explain why.

51. no x-intercepts

52. one x-intercept

53. two x-intercepts

54. three x-intercepts

55. Solve the system by finding ordered pairs that satisfy both equations. [**Hint:** Graph both equations on the same coordinate system.]

$$y = 2(x - 2)^2 + 5$$
$$y = -3(x - 2)^2 + 5$$

56. Use ZDecimal to draw the graph of $f(x) = 0.7x^2 + 2x - 1$. Then use TRACE to complete Table 16. Verify your table entries by using a graphing calculator table.

Table 16 Values of $f(x) = 0.7x^2 + 2x - 1$

x	$f(x)$
−2	
−1	
0	
1	
2	

57. Use a graphing calculator to graph $y = x^2$.

a. Use the window settings displayed in Fig. 27. (You can get these settings by using ZDecimal.)

b. Use the window settings displayed in Fig. 28. Compare what you see with what you saw in part (a).

c. Use the window settings displayed in Fig. 29. Compare what you see with what you saw in part (a).

d. Explain your results in parts (b) and (c).

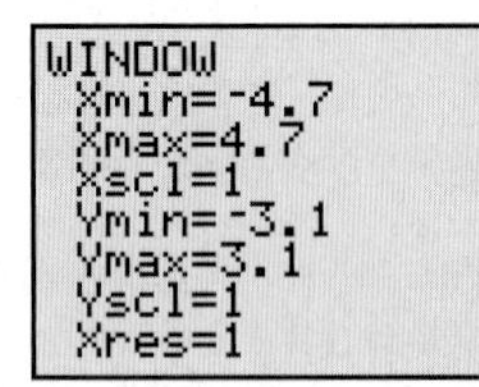

Figure 27 Window for Exercise 57a

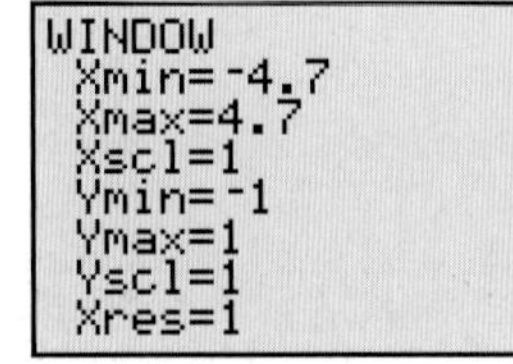

Figure 28 Window for Exercise 57b

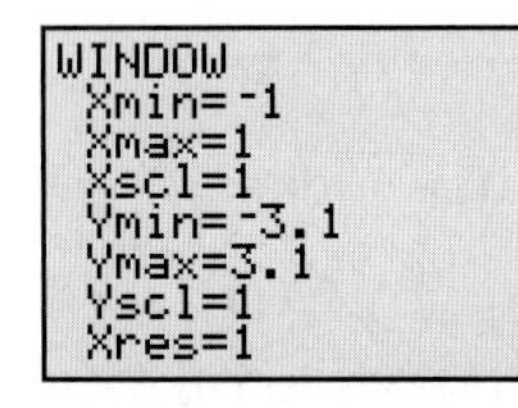

Figure 29 Window for Exercise 57c

58. Use a graphing calculator to graph the function. Record window settings that allow you to see the graph, including the vertex, on the calculator screen.

a. $f(x) = 1000x^2$

b. $g(x) = 1000(x - 1000)^2$

c. $h(x) = (x + 10{,}000)^2 + 10{,}000$

59. A student says that to graph $y = (x - 4)^2$, we translate the graph of $y = x^2$ to the left by 4 units. Is the student correct? Explain.

60. A student says that for the function $y = 2x^2$, the slope is 2. Is the student correct? Explain.

61. A student uses ZStandard on a graphing calculator to graph $f(x) = 0.0001x^2 + 5$. He thought that the graph should be a parabola, but the calculator displays a horizontal line. What would you tell him?

62. The vertex of a parabola is (5, 3), and the parabola passes through the point (8, 10). Find a third point that lies on the parabola.

63. Sketch, on the same coordinate system, the graph of the given function and the graph of $f(x) = a(x - h)^2 + k$ as shown in Fig. 30. Be sure to label which graph is which.

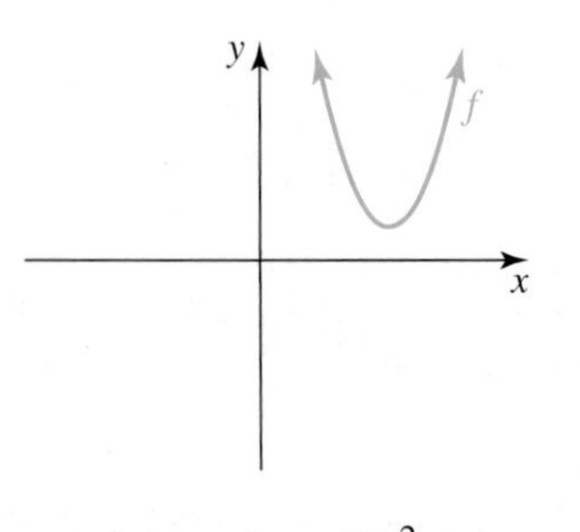

Figure 30 Exercise 63

a. $g(x) = a(x - 2h)^2 + k$

b. $p(x) = a(x - h)^2 + 2k$

c. $q(x) = 2a(x - h)^2 + k$

d. $l(x) = 2a(x - 2h)^2 + 2k$

64. Find two other quadratic functions that have the same domain and range as the function $f(x) = -2(x + 4)^2 + 7$.

65. Describe how to sketch the graph of a function of the form $f(x) = a(x - h)^2 + k, a \neq 0$. Include the effects that the values of a, h, and k have on the graph. (See page 4 for guidelines on writing a good response.)

66. To graph the quadratic function $y = a(x - h)^2 + k$, we translate the graph of $y = a(x - h)^2$ up by k units if $k > 0$ and down by $|k|$ units if $k < 0$. Explain. (See page 4 for guidelines on writing a good response.)

Related Review

67. a. Use a graphing calculator to graph the functions $f(x) = x^3$, $g(x) = (x - 4)^3$, and $h(x) = (x + 4)^3$ in the same viewing screen. Describe how to translate the graph of f to get the graph of g or the graph of h.

b. Use a graphing calculator to graph the functions $f(x) = x^3$, $g(x) = x^3 - 4$, and $h(x) = x^3 + 4$ in the same viewing screen. Describe how to translate the graph of f to get the graph of g or the graph of h.

c. Do the types of translation of $f(x) = x^3$ you performed in parts (a) and (b) match the types of translation we performed with $k(x) = x^2$ in this section? Explain.

d. Graph $f(x) = 2^x$ *by hand*. Then translate this graph to graph the following functions.

i. $g(x) = 2^{x+3}$ **ii.** $h(x) = 2^{x-4}$

iii. $k(x) = 2^x + 2$ **iv.** $p(x) = 2^x - 1$

68. a. Use a graphing calculator to graph the functions $f(x) = 2^x$ and $g(x) = -2^x$ in the same viewing screen. Describe how to reflect the graph of f to get the graph of g.

b. Use a graphing calculator to graph the functions $f(x) = x^4$ and $g(x) = -x^4$ in the same viewing screen. Describe how to reflect the graph of f to get the graph of g.

c. Do the types of reflections you performed in parts (a) and (b) match the type of reflection we performed with $k(x) = x^2$ in this section? Explain.

d. Graph $f(x) = x^3$ *by hand*. Then graph $g(x) = -x^3$ by hand.

Graph $y = 3^x$ by hand. Then use translations and reflections, as appropriate, to graph the given equation. Use a graphing calculator to verify your graph.

69. $f(x) = 3^{x+2} + 1$ **70.** $f(x) = 3^{x-3} - 2$

71. $f(x) = -3^x + 4$ **72.** $f(x) = -3^{x+3}$

Expressions, Equations, Functions, and Graphs

Perform the indicated instruction. Then use words such as linear, quadratic, cubic, exponential, logarithmic, polynomial, degree, function, one variable, *and* two variables *to describe the expression, equation, or system.*

73. Solve $\log_4(3x^2) + 2\log_4(2x^4) = 6$. Round any solutions to the fourth decimal place.

74. Solve:

$$y = -2x + 5$$
$$y = 4x + 23$$

75. Write as a single logarithm: $\log_4(3x^2) + 2\log_4(2x^4)$.

76. Solve $-2x + 5 = 4x + 23$.

77. Write as a single logarithm: $\log_4(3x^2) - 2\log_4(2x^4)$.

78. Graph $y = -2x + 5$ by hand.

7.2 GRAPHING QUADRATIC FUNCTIONS IN STANDARD FORM

Objectives

- Graph a quadratic function in standard form by using two symmetric points to find the vertex.
- Graph a quadratic function in standard form by using the *vertex formula* to find the vertex.
- Find the *minimum value* or *maximum value* of a quantity.

In Section 7.1, we graphed quadratic functions in vertex form, $f(x) = a(x-h)^2 + k$. In this section, we will discuss two methods of graphing quadratic functions in standard form, $f(x) = ax^2 + bx + c$. First, we will use two *symmetric points*. Then we will use the *vertex formula*. After discussing these methods, we will find the *minimum value* or *maximum value* of a quantity.

Method 1: Graphing by Using Two Symmetric Points to Find the Vertex

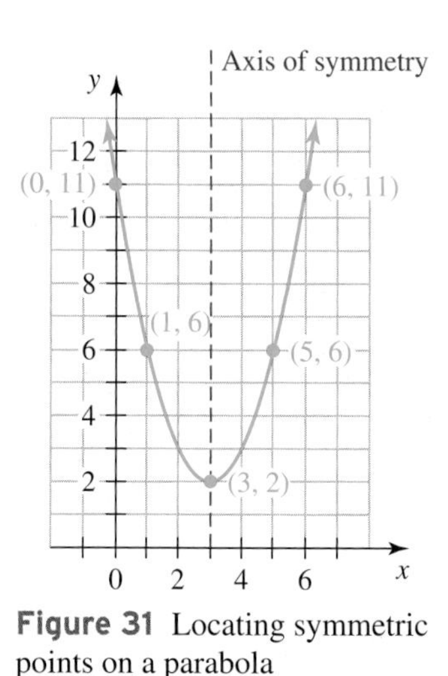

Figure 31 Locating symmetric points on a parabola

Recall from Section 6.1 that the part of the parabola that lies to the left of the axis of symmetry is the mirror reflection of the part that lies to the right (see Fig. 31). For any point on one side of the axis of symmetry, there is a point on the other side that is the "mirror reflection" of the first point. Such a pair of **symmetric points** corresponds to ordered pairs with equal y-coordinates.

For example, in Fig. 31 the point (1, 6) and the point (5, 6) are symmetric points. Since the vertex lies on the axis of symmetry, the x-coordinate of the vertex, 3, is equal to the average of the x-coordinates of the points (1, 6) and (5, 6):

$$\frac{1+5}{2} = 3$$

The average of the x-coordinates of *any* two symmetric points is equal to the x-coordinate of the vertex. As another example, we find the average of the x-coordinates of the y-intercept (0, 11) and its symmetric point (6, 11):

$$\frac{0+6}{2} = 3$$

The only point that does not have a symmetric point is the vertex, which lies on the axis of symmetry.

Using Two Symmetric Points to Find the Vertex

To find the vertex of a parabola in which (p, s) and (q, s) are symmetric points,

1. Find the x-coordinate by using the formula
$$x = \frac{p+q}{2}$$
In words: The x-coordinate is the average of the x-coordinates of any two symmetric points.
2. Find the y-coordinate by evaluating f at the value found in step 1. That is, find
$$f\left(\frac{p+q}{2}\right)$$

In short, the vertex is $\left(\frac{p+q}{2}, f\left(\frac{p+q}{2}\right)\right)$.

Example 1 Find the x-Coordinate of a Vertex

Find the x-coordinate of the vertex of the parabola sketched in the indicated figure.

1. Fig. 32
2. Fig. 33

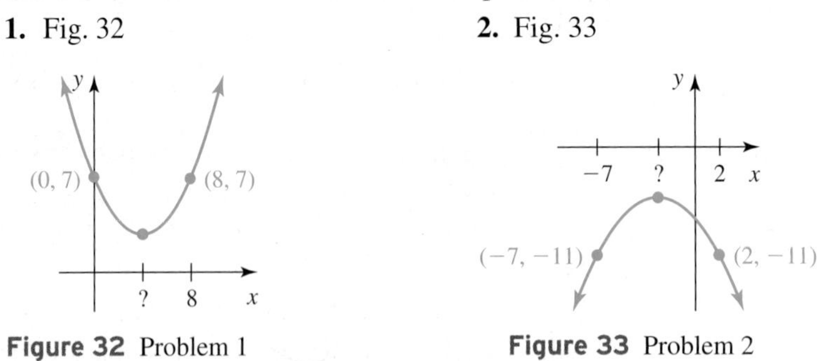

Figure 32 Problem 1 **Figure 33** Problem 2

Solution

1. The x-coordinates of the symmetric points $(0, 7)$ and $(8, 7)$ are 0 and 8, respectively. Since $\frac{0+8}{2} = 4$, the x-coordinate of the vertex is 4.
2. The x-coordinates of the symmetric points $(-7, -11)$ and $(2, -11)$ are -7 and 2, respectively. Since $\frac{-7+2}{2} = -2.5$, the x-coordinate of the vertex is -2.5. ■

Averaging the x-coordinates of the y-intercept and its symmetric point to find the x-coordinate of the vertex is a key step in sketching the graph of a quadratic function in standard form.

Example 2 Sketching the Graph of a Quadratic Function

Sketch the graph of $g(x) = x^2 - 4x + 7$.

Solution

First, we find the y-intercept of $g(x) = x^2 - 4x + 7$:

$$g(0) = 0^2 - 4(0) + 7 = 7$$

The y-intercept is $(0, 7)$. See Fig. 34.

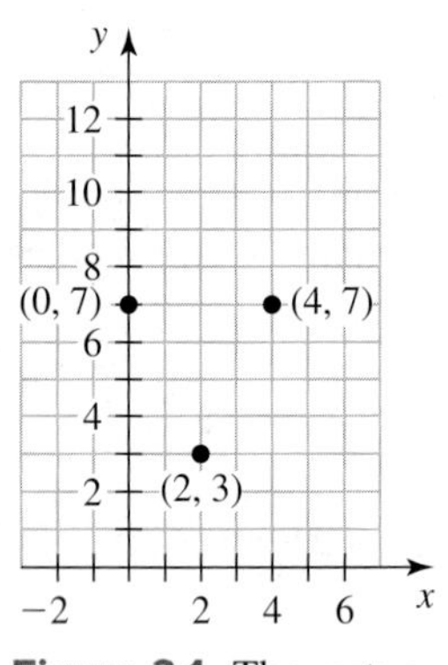

Figure 34 The vertex and two symmetric points

Next, we find the symmetric point of the y-intercept. Since symmetric points have the same height, we know that the y-coordinate of the symmetric point is 7. We find the x-coordinate by substituting 7 for y in the equation $g(x) = x^2 - 4x + 7$ and solving for x:

$$\begin{aligned} 7 &= x^2 - 4x + 7 && \text{Substitute 7 for } y. \\ 0 &= x^2 - 4x && \text{Write in } 0 = ax^2 + bx \text{ form.} \\ 0 &= x(x-4) && \text{Factor right-hand side.} \\ x = 0 \quad &\text{or} \quad x - 4 = 0 && \text{Zero factor property} \\ x = 0 \quad &\text{or} \quad x = 4 \end{aligned}$$

Therefore, the points that have height 7 are $(0, 7)$ and $(4, 7)$; these are symmetric points. To find the x-coordinate of the vertex, we average the x-coordinates of the symmetric points:

$$x\text{-coordinate of vertex} = \frac{0+4}{2} = 2$$

To find the y-coordinate of the vertex, we find $g(2)$:

$$g(2) = 2^2 - 4(2) + 7 = 4 - 8 + 7 = 3$$

So, the vertex is $(2, 3)$.

We can find another pair of symmetric points on the graph by evaluating g at the values of x that are 3 units from $x = 2$ on either side—namely, at $x = -1$ and $x = 5$. At $x = 5$,

$$g(5) = 5^2 - 4(5) + 7 = 25 - 20 + 7 = 12$$

Thus, the graph passes through (5, 12) and its symmetric point, (−1, 12). We plot these points to assist us in sketching a graph of g (see Fig. 35).

We use a graphing calculator to verify our sketch (see Fig. 36). In particular, we check that the vertex (minimum point) is (2, 3). The "minimum" choice on a graphing calculator will closely approximate the vertex (see Section B.19). ■

Figure 35 Graph of g

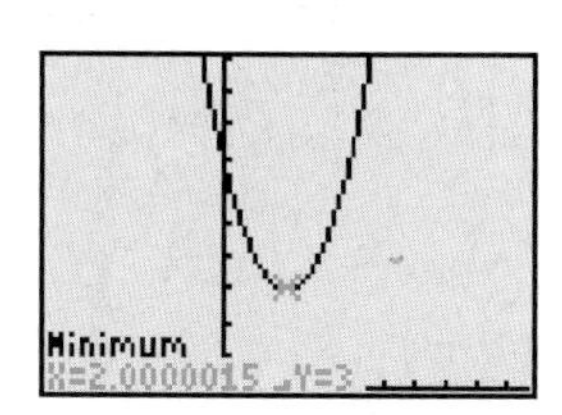

Figure 36 Verify the graph of g

> **Using Symmetric Points to Graph a Quadratic Function in Standard Form (Method 1)**
>
> To sketch a graph of a quadratic function $f(x) = ax^2 + bx + c = 0$, where $b \neq 0$,
>
> 1. Find the y-intercept.
> 2. Find the y-intercept's symmetric point.
> 3. Average the x-coordinates of the two symmetric points to find the x-coordinate of the vertex.
> 4. Find the y-coordinate of the vertex.
> 5. Depending on how accurate your sketch is to be, find additional points on the parabola, as needed.
> 6. Sketch a parabola that contains the points found.

We use these steps to graph a quadratic function $f(x) = ax^2 + bx + c$, where $b \neq 0$. If $b = 0$, the y-intercept is the vertex and, therefore, does not have a symmetric point. In this case, however, $f(x) = ax^2 + 0x + c = ax^2 + c$ is in vertex form, and we can readily use the methods of Section 7.1 to sketch the graph.

Example 3 Sketching the Graph of a Quadratic Function

Sketch the graph of $f(x) = -0.9x^2 - 5.8x - 5.7$.

Solution

We find the y-intercept by finding $f(0)$:

$$f(0) = -0.9(0)^2 - 5.8(0) - 5.7 = -5.7$$

So, the y-intercept is $(0, -5.7)$.

Next, we find the symmetric point of the y-intercept. We substitute -5.7 for $f(x)$ and solve for x:

$$
\begin{aligned}
-5.7 &= -0.9x^2 - 5.8x - 5.7 && \text{Substitute } -5.7 \text{ for } f(x). \\
0 &= -0.9x^2 - 5.8x && \text{Write in } 0 = ax^2 + bx \text{ form.} \\
0 &= -x(0.9x + 5.8) && \text{Factor right-hand side.} \\
-x = 0 \quad &\text{or} \quad 0.9x + 5.8 = 0 && \text{Zero factor property} \\
x = 0 \quad &\text{or} \quad 0.9x = -5.8 \\
x = 0 \quad &\text{or} \quad x = \frac{-5.8}{0.9} \\
&\approx -6.44
\end{aligned}
$$

So, $(0, -5.7)$ and $(-6.44, -5.7)$ are approximate symmetric points. The approximate symmetric point of the y-intercept $(0, -5.7)$ is $(-6.44, -5.7)$.

We find the approximate x-coordinate of the vertex by averaging the x-coordinates of the points $(0, -5.7)$ and $(-6.44, -5.7)$:

$$\frac{0 + (-6.44)}{2} = -3.22$$

We find the approximate y-coordinate of the vertex by computing $f(-3.22)$:

$$f(-3.22) = -0.9(-3.22)^2 - 5.8(-3.22) - 5.7 \approx 3.64$$

So, the vertex is approximately $(-3.22, 3.64)$. See Fig. 37.

Although we could find and plot additional points, we can sketch a fairly accurate graph of f from the three points already found (see Fig. 38). We use a graphing calculator to verify our sketch (see Fig. 39). In particular, we check that the approximate vertex (maximum point) is $(-3.22, 3.64)$.

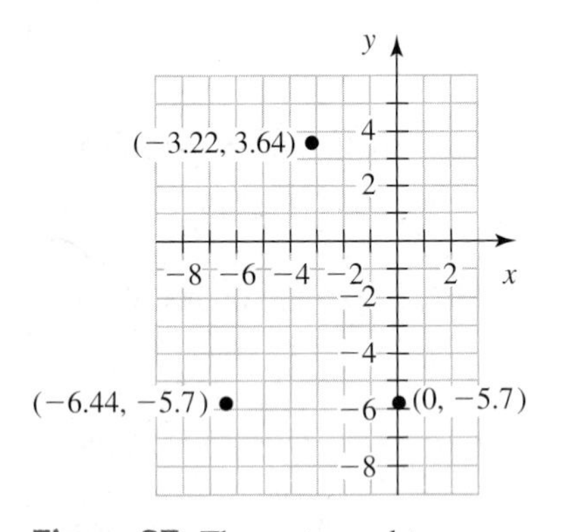

Figure 37 The vertex and two symmetric points

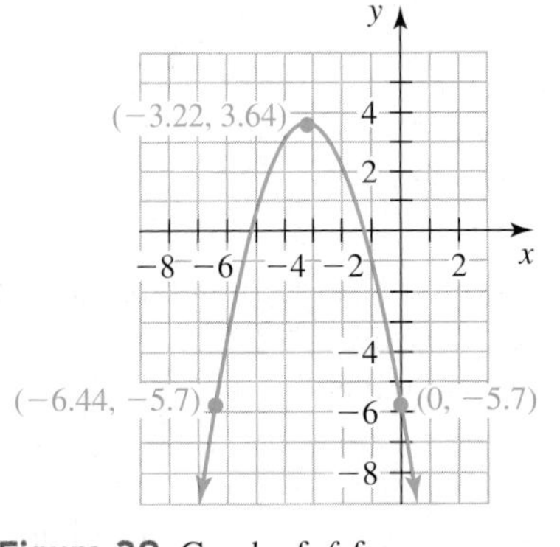

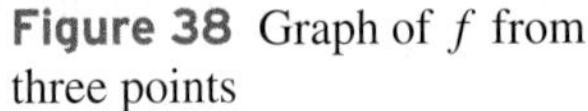
Figure 38 Graph of f from three points

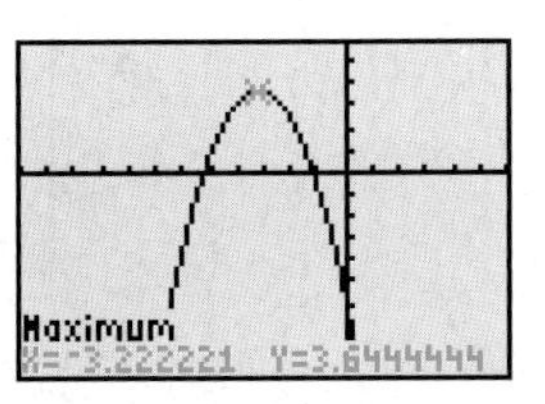

Figure 39 Verify the graph of f

■

Method 2: Graphing by Using the Vertex Formula to Find the Vertex

We can find a formula of the x-coordinate of the vertex of a parabola $f(x) = ax^2 + bx + c$, where $a \neq 0$. If $b = 0$, the vertex is the y-intercept, so the x-coordinate of the vertex is 0.

If $b \neq 0$, then we can find the x-coordinate of the vertex from the first three steps of the graphing method (method 1) we have been using in this section. To begin, we find the y-intercept by finding $f(0)$:

$$f(0) = a(0)^2 + b(0) + c = c$$

The y-intercept is $(0, c)$.

Next, we find the symmetric point of the y-intercept by substituting c for $f(x)$ and solving for x:

$c = ax^2 + bx + c$	Substitute c for $f(x)$.
$0 = ax^2 + bx$	Subtract c from both sides.
$0 = x(ax + b)$	Factor right-hand side.
$x = 0$ or $ax + b = 0$	Zero factor property
$x = 0$ or $ax = -b$	Subtract b from both sides.
$x = 0$ or $x = -\frac{b}{a}$	Divide both sides by a.

The x-coordinate of the symmetric point is $-\frac{b}{a}$.

We find the x-coordinate of the vertex by averaging the x-coordinates of the y-intercept and symmetric point:

$$\begin{aligned} x = \frac{0 + \left(-\frac{b}{a}\right)}{2} &= \frac{-\frac{b}{a}}{2} && 0 + d = d \\ &= -\frac{b}{a} \div 2 && \frac{Q}{R} = Q \div R \\ &= -\frac{b}{a} \cdot \frac{1}{2} && \text{Multiply by reciprocal of 2, which is } \frac{1}{2}. \\ &= -\frac{b}{2a} && \text{Multiply numerators and multiply denominators.} \end{aligned}$$

So, the formula of the x-coordinate of the vertex is $x = -\frac{b}{2a}$. If $b = 0$, this formula gives $x = -\frac{0}{2a} = 0$, which agrees with what we said earlier. So, the formula works for any value of b.

Vertex Formula

To find the vertex of the graph of a quadratic function $f(x) = ax^2 + bx + c$,

1. Find the x-coordinate of the vertex by using the **vertex formula** $x = -\frac{b}{2a}$.
2. Find the y-coordinate of the vertex by evaluating f at the value found in step 1. That is, find $f\left(-\frac{b}{2a}\right)$.

In short, the vertex is $\left(-\frac{b}{2a}, f\left(-\frac{b}{2a}\right)\right)$.

In Example 4, we use the vertex formula to find the vertex of the parabola we sketched in Example 2.

Example 4 Using the Vertex Formula to Find the Vertex

Find the vertex of the graph of $g(x) = x^2 - 4x + 7$.

Solution

Comparing $g(x) = x^2 - 4x + 7$ with $f(x) = ax^2 + bx + c$, we see that $a = 1$ and $b = -4$. We substitute these values of a and b into the formula $x = -\frac{b}{2a}$:

$$x = -\frac{-4}{2(1)} = 2$$

The x-coordinate of the vertex is 2. To find the y-coordinate, we find $f(2)$:

$$f(2) = 2^2 - 4(2) + 7 = 3$$

So, the vertex is (2, 3), as we saw in Example 2. ■

In general, for any quadratic function of the form $f(x) = ax^2 + bx + c$, where $b \neq 0$, both methods we have discussed for finding the vertex give the same result.

Table 17 Input-Output Pairs of $f(x) = 2x^2 + 10x + 7$

x	$f(x)$
−4	−1
−3	−5
−2	−5
−1	−1

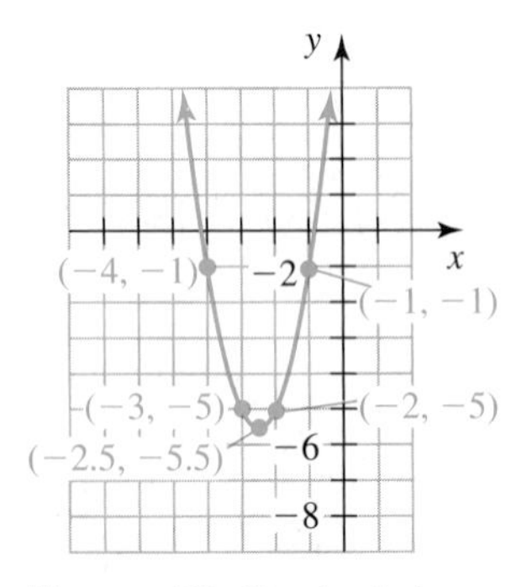

Figure 40 Graph of f from five points

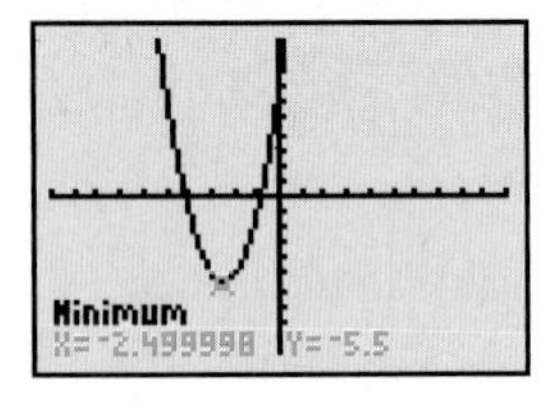

Figure 41 Verify the graph of f

Example 5 Using the Vertex Formula to Graph a Quadratic Function

Sketch the graph of $f(x) = 2x^2 + 10x + 7$.

Solution

To find the x-coordinate of the vertex, we substitute 2 for a and 10 for b in the vertex formula $x = -\dfrac{b}{2a}$:

$$x = -\frac{10}{2(2)} = -2.5$$

The x-coordinate of the vertex is -2.5. To find the y-coordinate, we find $f(-2.5)$:

$$f(-2.5) = 2(-2.5)^2 + 10(-2.5) + 7 = -5.5$$

The vertex is $(-2.5, -5.5)$.

We find additional input–output pairs of f in Table 17. Notice that we select integer input values that are close to the vertex's x-coordinate, -2.5. We plot the found points, then sketch a parabola that contains these points (see Fig. 40).

We use a graphing calculator to verify our sketch (see Fig. 41). In particular, we check that the approximate vertex (minimum point) is $(-2.5, -5.5)$. ■

Example 6 Using the Vertex Formula to Graph a Quadratic Function

Sketch the graph of $f(x) = -2.2x^2 + 6.1x + 1.4$.

Solution

To find the x-coordinate of the vertex, we substitute -2.2 for a and 6.1 for b in the vertex formula $x = -\dfrac{b}{2a}$:

$$x = -\frac{6.1}{2(-2.2)} \approx 1.39$$

The x-coordinate of the vertex is about 1.39. To find the y-coordinate, we find $f(1.39)$:

$$f(1.39) = -2.2(1.39)^2 + 6.1(1.39) + 1.4 \approx 5.63$$

The approximate vertex is $(1.39, 5.63)$.

We find additional input–output pairs of f in Table 18. We plot the found points, then sketch a parabola that contains these points (see Fig. 42). ■

Table 18 Input-Output Pairs of $f(x) = -2.2x^2 + 6.1x + 1.4$

x	$f(x)$
0	1.4
1	5.3
2	4.8
3	−0.1

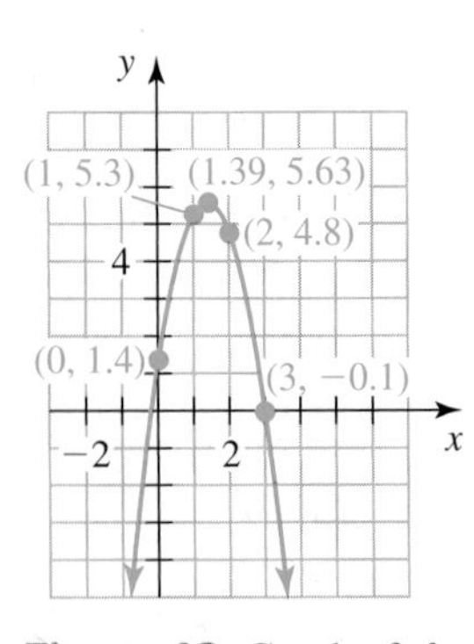

Figure 42 Graph of f using five points

Minimum or Maximum Value

Earlier in this section, we located the vertex to help ourselves graph quadratic functions in standard form. Recall from Section 6.1 that the vertex is also the minimum or maximum point of the parabola. If the vertex is the maximum point, we call the y-coordinate of the vertex the **maximum value** of the function (see Fig. 43). If the vertex is the minimum point, we call the y-coordinate of the vertex the **minimum value** of the function (see Fig. 44).

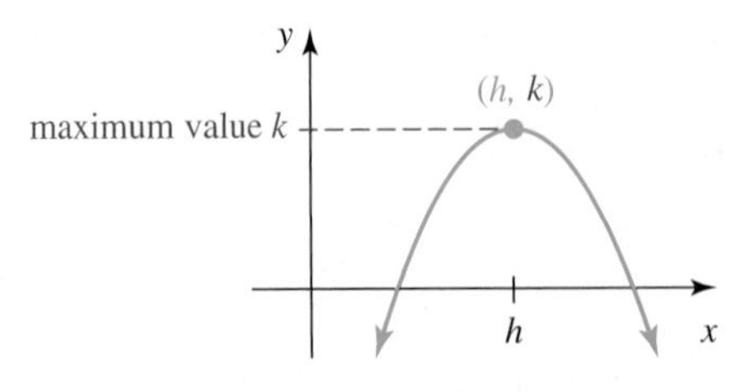

Figure 43 Quadratic function with a maximum value k but no minimum value

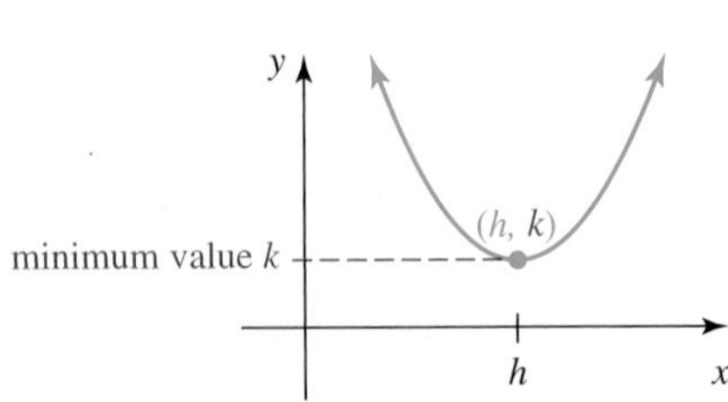

Figure 44 Quadratic function with a minimum value k but no maximum value

If we convert a quadratic function in standard form, $f(x) = ax^2 + bx + c$, into vertex form, $f(x) = a(x - h)^2 + k$, the constant a has the same value for both forms.

So, what we have determined about the graphical significance of a for quadratic functions in vertex form also applies to quadratic functions in standard form.

Maximum or Minimum Value of a Function

For a quadratic function $f(x) = ax^2 + bx + c$ whose graph has vertex (h, k),

- If $a < 0$, then the parabola opens downward and the maximum value of the function is k (see Fig. 43).
- If $a > 0$, then the parabola opens upward and the minimum value of the function is k (see Fig. 44).

In Example 7, we find the vertex of a quadratic model that opens downward to estimate the maximum value of a quantity.

Example 7 Finding the Maximum Value of a Quantity

A fireworks shell is launched into the air. The shell's height (in feet) $h = f(t)$ at time t seconds is modeled well by the function $f(t) = -16t^2 + 200t + 1$. When should the shell explode so that it goes off at the maximum height? What is that height?

Solution

The function f is of the form $f(t) = at^2 + bt + c$, with $a = -16$ (a negative number), so the graph of f is a parabola that opens downward. Therefore, the vertex is the maximum point. We can find the time the shell will reach the maximum height by finding the vertex.

To find the t-coordinate of the vertex, we substitute $a = -16$ and $b = 200$ in the vertex formula $t = -\dfrac{b}{2a}$:

$$t = -\frac{200}{2(-16)} = 6.25$$

The t-coordinate is 6.25. To find the h-coordinate, we evaluate f at 6.25:

$$f(6.25) = -16(6.25)^2 + 200(6.25) + 1 = 626$$

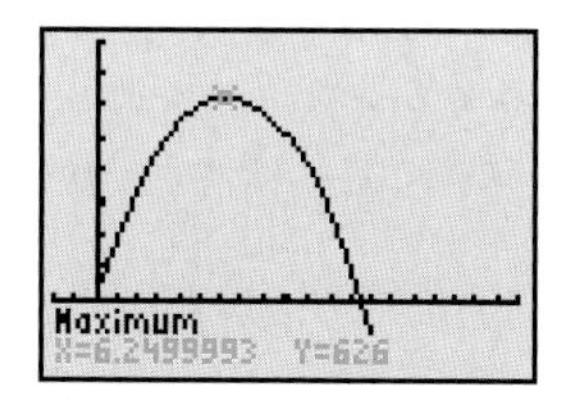

Figure 45 Verify that the vertex is (6.25, 626)

So, the vertex is (6.25, 626). We verify our work by using "maximum" on a graphing calculator (see Fig. 45). For graphing calculator instructions, see Section B.19.

A vertex of (6.25, 626) means that the shell will reach a maximum height of 626 feet at 6.25 seconds and should explode then.

The function $f(t) = -16t^2 + 200t + 1$ describes the shell's height in terms of time, *not* in terms of horizontal distance. So, the graph of f does *not* describe the path of the shell. Its path can be modeled by a quadratic function that describes the height in terms of horizontal distance. ■

Example 8 Finding the Maximum Value of a Quantity

A person plans to use 60 feet of fencing and the side of his house to enclose a rectangular garden (see Fig. 46). What dimensions of the rectangle would give the maximum area? What is that area?

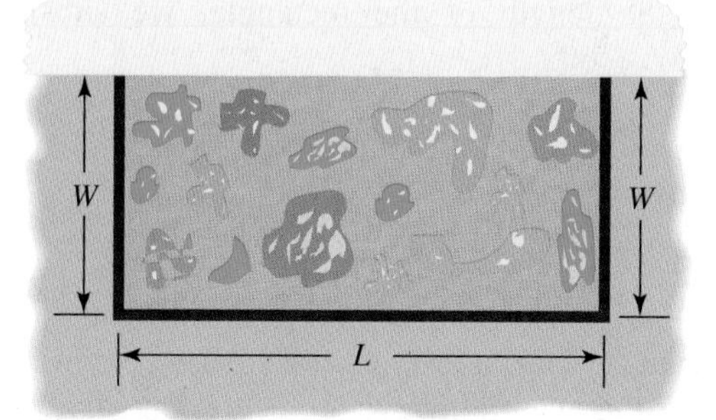

Figure 46 A garden

Solution

We use the five-step problem-solving method of Section 3.4, but in step 2 we use a system of equations to build a function and in step 3 we find the maximum point of the graph of the function.

Step 1: Define each variable. We let W be the width (in feet), L be the length (in feet), and A be the area (in square feet) of the rectangle (see Fig. 46).

Step 2: Use a system of two equations to build a function. No fencing is needed along the house, so the 60 feet of fencing can be used for the three sides of length W, W, and L:

$$W + W + L = 60$$

To obtain our first equation, we combine like terms on the left side:

$$2W + L = 60$$

We use the area formula, $A = LW$, as our second equation. The system is

$$\begin{aligned} 2W + L &= 60 && \text{Equation (1)} \\ A &= LW && \text{Equation (2)} \end{aligned}$$

Next, we will build a function that describes the area of the rectangle. To begin, we solve equation (1) for L:

$$\begin{aligned} 2W + L &= 60 && \text{Equation (1)} \\ L &= -2W + 60 && \text{Subtract } 2W \text{ from both sides.} \end{aligned}$$

Then we substitute $-2W + 60$ for L in equation (2):

$$\begin{aligned} A &= LW && \text{Equation (2)} \\ A &= (-2W + 60)W && \text{Substitute } -2W + 60 \text{ for } L. \\ A &= -2W^2 + 60W && \text{Distributive law} \end{aligned}$$

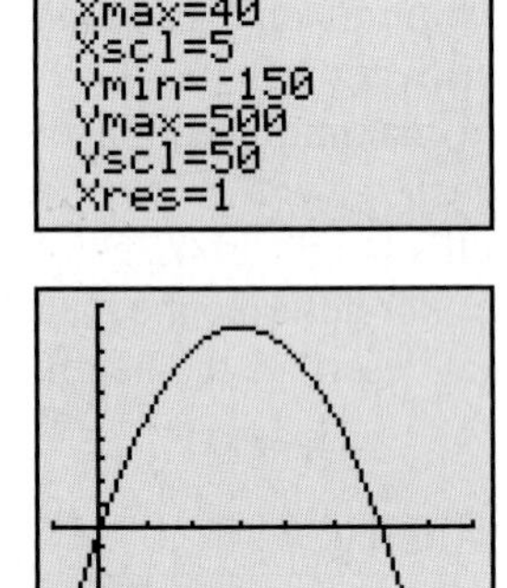

Figure 47 Graph of $A = -2W^2 + 60W$

Step 3: Find the maximum point of the graph of the function. The function $A = -2W^2 + 60W$ is of the form $A = aW^2 + bW + c$, where $a = -2$ (a negative number), so the graph is a parabola that opens downward (see Fig. 47). Therefore, the vertex is the maximum point. We can find the width of the rectangle with maximum area by finding the vertex.

To find the W-coordinate of the vertex, we substitute $a = -2$ and $b = 60$ in the vertex formula $W = -\dfrac{b}{2a}$:

$$W = -\frac{60}{2(-2)} = 15$$

To find the A-coordinate of the vertex, we substitute 15 for W in the equation $A = -2W^2 + 60W$:

$$A = -2(15)^2 + 60(15) = 450$$

To find the length of the rectangle, we substitute 15 for W in the equation $L = -2W + 60$:

$$L = -2(15) + 60 = 30$$

Step 4: Describe each result. The width is 15 feet, the length is 30 feet, and the area is 450 square feet.

Step 5: Check. The amount of fencing is $15 + 15 + 30 = 60$ feet, which checks. To check that the largest area is 450 square feet, we find the area of a couple of other rectangles that would involve 60 feet of fencing (see Fig. 48).

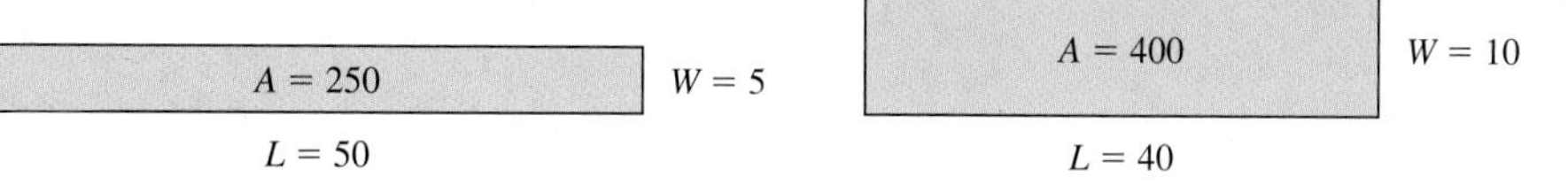

Figure 48 Some other rectangles involving 60 feet of fencing

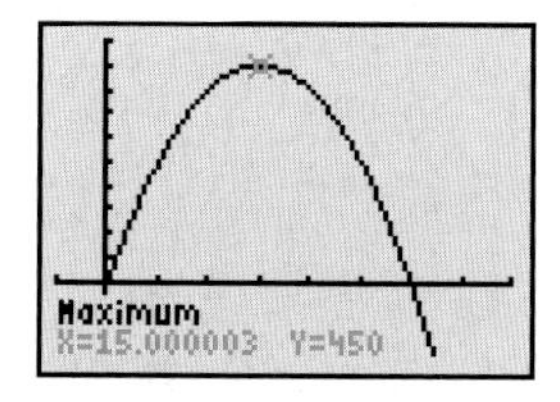

Figure 49 Verify that the maximum point is (15, 450)

Since 450 square feet is larger than the area of either of the two rectangles in Fig. 48, it seems reasonable that our work is correct. Finally, we use "maximum" on a graphing calculator to check that the maximum point of the graph of $A = -2W^2 + 60W$ is (15, 450). See Fig. 49. ■

group exploration

Comparing methods of graphing quadratic functions

1. **a.** Graph the function $f(x) = 2(x-3)^2 - 5$ by using the method discussed in Section 7.1.
   b. Simplify the right-hand side of the equation $f(x) = 2(x-3)^2 - 5$.
   c. Graph the equation you found in part (b) by using two symmetric points to find the vertex (method 1).
   d. Graph the equation you found in part (b) by using the vertex formula to find the vertex (method 2).
   e. Compare your graphs from parts (a), (c), and (d).
2. **a.** Graph the function $g(x) = x^2 + 10x + 25$ by using two symmetric points to find the vertex.
   b. Graph g by using the vertex formula to find the vertex.
   c. Factor the right-hand side of the equation $g(x) = x^2 + 10x + 25$.
   d. Use the method discussed in Section 7.1 to graph the equation you found in part (c).
   e. Compare your graphs from parts (a), (b), and (d).

HOMEWORK 7.2

FOR EXTRA HELP ▶ Student Solutions Manual | PH Math/Tutor Center | MathXL® | MyMathLab

Find the x-coordinate of the vertex of the parabola sketched in the figure.

1. Fig. 50

2. Fig. 51

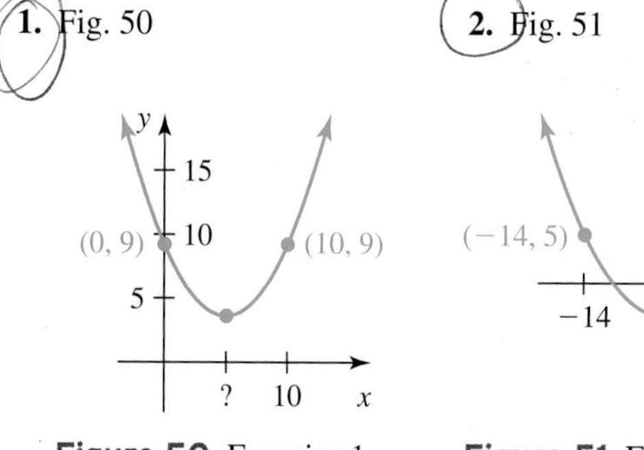

Figure 50 Exercise 1

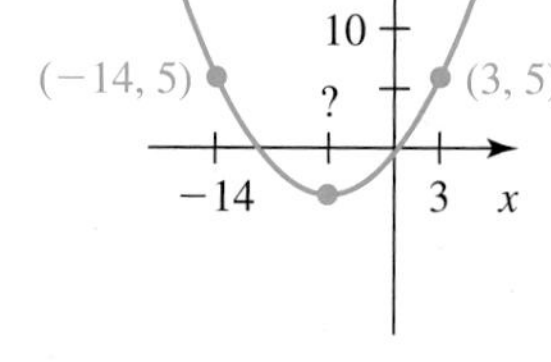

Figure 51 Exercise 2

Find the x-coordinate of the vertex of a parabola that passes through the given points. Round approximate results to the second decimal place.

3. (0, 8) and (6, 8)
4. (0, 5) and (8, 5)
5. (0, −3) and (−7, −3)
6. (0, −6) and (−9, −6)
7. (0, 2) and (7.29, 2)
8. (0, −7) and (15.37, −7)

A parabola has the given vertex and y-intercept. Find another point on the parabola.

9. vertex (2, 5) and y-intercept (0, 9)
10. vertex (−3, −8) and y-intercept (0, 4)

Graph the function by hand. Find the vertex; round approximate coordinates to the second decimal place. Verify your sketch by using a graphing calculator.

11. $y = x^2 - 6x + 7$
12. $y = x^2 - 4x + 5$
13. $y = x^2 + 8x + 9$
14. $y = x^2 + 2x - 7$
15. $y = -x^2 + 8x - 10$
16. $y = -2x^2 + 12x - 9$
17. $y = 3x^2 + 6x - 4$
18. $y = 2x^2 - 4x + 1$
19. $y = -3x^2 + 12x - 5$
20. $y = -2x^2 - 8x - 3$
21. $y = -4x^2 - 9x - 5$
22. $y = -2x^2 + 5x + 3$
23. $y = 2x^2 - 7x + 7$
24. $y = -3x^2 - 2x - 4$
25. $4x^2 - y + 6 = 8x$
26. $6x^2 = 3y - 24x - 15$
27. $y = 2.8x^2 - 8.7x + 4$

28. $y = -1.6x^2 - 4.8x + 3$

29. $y = 3.9x^2 + 6.9x - 3.4$

30. $y = -2.4x^2 + 6.1x - 7.8$

31. $3.6y - 26.3x = 8.3x^2 - 7.1$

32. $5.3 - 2.1y = 9.8x^2 - 3.4x + 8.3$

A parabola has the given x-intercepts. What is the x-coordinate of the vertex? Write your result in decimal form.

33. $(2, 0)$ and $(6, 0)$

34. $(-4, 0)$ and $(3, 0)$

35. $(-9, 0)$ and $(4, 0)$

36. $(-7, 0)$ and $(-3, 0)$

Find the x-intercepts and y-intercept. Next, find the vertex. Write the coordinates in decimal form. Then graph the function by hand. Verify your result by using a graphing calculator.

37. $y = 5x^2 - 10x$

38. $y = 2x^2 - 8x$

39. $y = -2x^2 + 6x$

40. $y = -3x^2 - 6x$

41. $y = x^2 - 10x + 24$

42. $y = x^2 - 4x - 5$

43. $y = x^2 - 8x + 7$

44. $y = x^2 - 10x + 16$

45. $y = x^2 - 9$

46. $y = x^2 - 1$

47. A batter hits a baseball. The ball's height (in feet) $h(t)$ after t seconds is given by $h(t) = -16t^2 + 140t + 3$.
   a. What is the ball's height when the batter makes contact with it?
   b. What is the maximum height of the ball? When does it reach that height?
   c. Graph h by hand.

48. A person on the edge of a cliff throws a stone so that it hits the ground near the base of the cliff. The stone's height (in feet above the base) $h(t)$ after t seconds is given by $h(t) = -16t^2 + 30t + 200$.
   a. Find the vertex. What does it mean in this situation?
   b. Estimate the height of the cliff. State any assumptions that you make.
   c. Did the person throw the stone upward or downward? Explain.

49. The average prices of a gallon of gasoline (in 2006 dollars) are shown in Table 19 for various years.

Table 19 Average Prices of Gasoline

Year	Price (2006 dollars per gallon)
1980	2.59
1985	1.90
1990	1.51
1995	1.28
2001	1.66
2005	2.28
2006*	2.68

Source: *Department of Energy*
*From January to May

Let $f(t)$ be the average price of gasoline (in 2006 dollars per gallon) at t years since 1980.
   a. Use a graphing calculator to draw a scattergram to describe the data. Is it best to use a linear, an exponential, or a quadratic function to model the data? Explain.
   b. Use a graphing calculator to draw the graph of the model $f(t) = 0.0075t^2 - 0.2t + 2.65$ and, in the same viewing window, the scattergram of the data. Does the model fit the data well?
   c. Predict the average price (in 2006 dollars) of gasoline in 2011.
   d. Estimate when the average price (in 2006 dollars) of gasoline was the lowest. What was that average price?

50. Solar energy consumption in the United States is shown in Table 20 for various years.

Table 20 Solar Energy Consumption in the United States

Year	Solar Energy Consumption (trillion Btu)
1990	60
1992	64
1994	69
1996	71
1998	70
2000	66
2002	64

Source: *Energy Information Administration*

Let $f(t)$ be the solar energy consumption (in trillion Btu) in the United States in the year that is t years since 1990.
   a. Use a graphing calculator to draw a scattergram to describe the data. Is it best to use a linear, an exponential, or a quadratic function to model the data? Explain.
   b. Use a graphing calculator to draw the graph of the model $f(t) = -0.24t^2 + 3.2t + 60$ and, in the same viewing window, the scattergram of the data. Does the model fit the data well?
   c. Estimate when solar energy consumption in the United States was 60 trillion Btu.
   d. Estimate when solar energy consumption in the United States was at a maximum. What was that consumption?

51. Americans' average annual expenditures are shown in Table 21 for various age groups.

Table 21 Americans' Average Annual Expenditures

Age Group (years)	Age Used to Represent Age Group (years)	Average Annual Expenditure (thousands of dollars)
under 25	20.0	24.3
25–34	29.5	40.3
35–44	39.5	48.3
45–54	49.5	48.7
55–64	59.5	44.3
65–74	69.5	32.2

Source: *Consumer Expenditure Survey*

Let $f(t)$ be average annual expenditure (in thousands of dollars) of Americans at age t years.
   a. Use a graphing calculator to draw a scattergram to describe the data. Is it best to use a linear, an exponential, or a quadratic function to model the data? Explain.
   b. Use a graphing calculator to draw the graph of the model $f(t) = -0.035t^2 + 3.25t - 26.34$ and, in the same

viewing window, the scattergram of the data. Does the model fit the data well?

c. Estimate the average annual expenditure of 18-year-old Americans.

d. Estimate the age of Americans with the highest average annual expenditure. What is that expenditure?

52. The percentages of Americans who buy newspapers are shown in Table 22 for various age groups.

Table 22 Percentages of Americans Who Buy Newspapers

Age Group (years)	Age Used to Represent Age Group (years)	Percent
25–34	29.5	33
35–44	39.5	43
45–54	49.5	48
55–64	59.5	53
over 65	70	51

Source: *Bureau of Labor Statistics*

Let $f(t)$ be the percentage of Americans at age t years who buy newspapers.

a. Use a graphing calculator to draw a scattergram to describe the data. Is it best to use a linear, an exponential, or a quadratic function to model the data? Explain.

b. Use a graphing calculator to draw the graph of the model $f(t) = -0.017t^2 + 2.15t - 15.57$ and, in the same viewing window, the scattergram of the data. Does the model fit the data well?

c. Estimate the percentage of 21-year-old Americans who buy newspapers.

d. Estimate the age of Americans who are most likely to buy newspapers. What percentage of these Americans buy newspapers?

53. Student-to-faculty ratios at Bates College are shown in Table 23 for various years.

Table 23 Student-to-Faculty Ratios at Bates College

Year	Student-to-Faculty Ratio
1985	13.8
1990	12.2
1995	11.3
2000	10.7
2003	10.8

Source: *Bates College*

Let r be the student-to-faculty ratio at Bates College at t years since 1980. A linear model of the situation is $L(t) = -0.17t + 14.20$. An exponential model of the situation is $E(t) = 14.32(0.986)^t$. A quadratic model is $Q(t) = 0.011t^2 - 0.48t + 15.9$.

a. Use a graphing calculator to draw the graphs of all three models and, in the same viewing window, the scattergram of the data. Which model best describes the situation?

b. Use Zoom Out on a graphing calculator to help you determine which model predicts the lowest student-to-faculty ratios for future years.

c. Use Q to predict when the student-to-faculty ratio will be 15.9.

d. Use Q to estimate when the student-to-faculty ratio was the least. What is that ratio?

54. The numbers of recreational boating fatalities are shown in Table 24 for various years.

Table 24 Numbers of Recreational Boating Fatalities

Year	Number of Fatalities
1980	1392
1985	1104
1990	880
1995	829
2000	701
2002	750

Source: *U.S. Coast Guard*

Let n be the number of recreational boating fatalities in the year that is t years since 1980. A linear model of the situation is $L(t) = -28.64t + 1286.35$. An exponential model is $E(t) = 1295(0.97)^t$. A quadratic model is $Q(t) = 1.54t^2 - 63.35t + 1386.17$.

a. Use a graphing calculator to draw the graphs of all three models and, in the same viewing window, the scattergram of the data. Which model best describes the situation?

b. Use each of the three models to find three estimates of the number of fatalities in 1990. Which is the best estimate? Explain.

c. Use Zoom Out on a graphing calculator to help you determine which model predicts the lowest number of fatalities for any year beyond 2002.

d. Use Q to estimate in which year the number of recreational boating fatalities was the least. What is that number of fatalities?

55. A person plans to use 80 feet of fencing to enclose a rectangular garden. What dimensions of the rectangle would give the maximum area? What is that area?

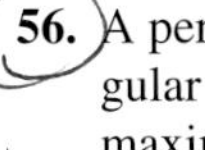

56. A person plans to use 60 feet of fencing to enclose a rectangular patio. What dimensions of the rectangle would give the maximum area? What is that area?

57. A rancher plans to use 400 feet of fencing and a side of his barn to form a rectangular boundary for cattle (see Fig. 52). What dimensions of the rectangle would give the maximum area? What is that area?

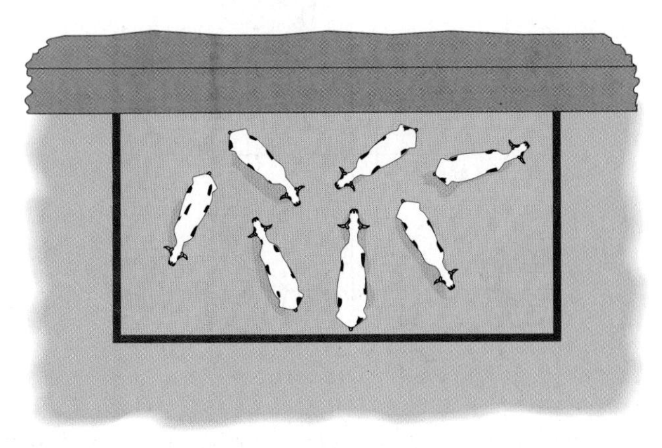

Figure 52 Exercise 57

58. A farmer plans to use 200 feet of fencing and a side of her barn to enclose a rectangular garden (see Fig. 53). What dimensions of the rectangle would give the maximum area? What is that area?

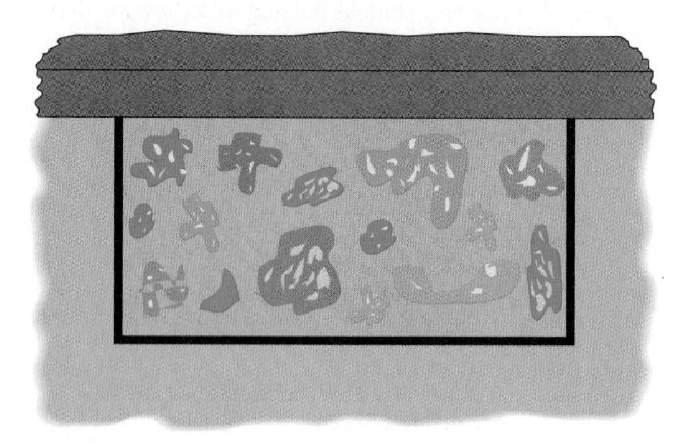

Figure 53 Exercise 58

59. Describe the Rule of Four as applied to the function $f(x) = x^2 - 10x + 18$:
a. Describe the input–output pairs of f by using a graph.
b. Describe five input–output pairs of f by using a table.
c. Describe the input–output pairs of f by using words.

60. Describe the Rule of Four as applied to the function $g(x) = -2x^2 + 4x + 1$:
a. Describe the input–output pairs of g by using a graph.
b. Describe five input–output pairs of g by using a table.
c. Describe the input–output pairs of g by using words.

For Exercises 61–68, refer to the graph sketched in Fig. 54.

61. Find $f(-5)$.

62. Find $f(-3)$.

63. Find x when $f(x) = 3$.

64. Find x when $f(x) = 4$.

65. Find x when $f(x) = 2$.

66. Find x when $f(x) = -1$.

67. Find the maximum value of f.

68. Find the vertex of the graph of f.

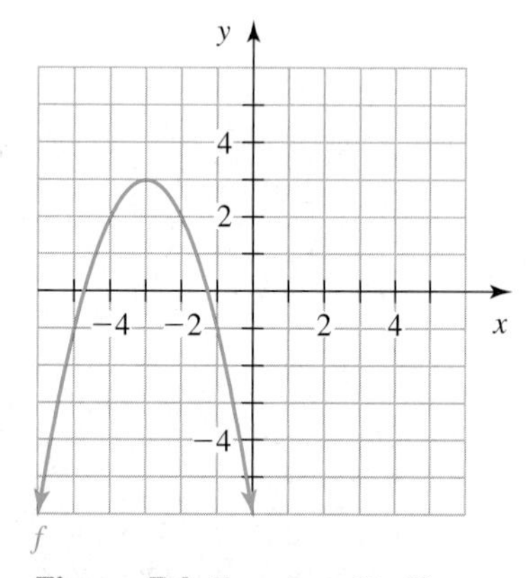

Figure 54 Exercises 61–68

69. a. Find the x-coordinate of the vertex of the parabola $f(x) = x^2 + 4x - 12$ by averaging the x-coordinates of the y-intercept and its symmetric point.
b. Find the x-coordinate of the vertex of the parabola $f(x) = x^2 + 4x - 12$ by averaging the x-coordinates of the x-intercepts.
c. Compare the methods you used in parts (a) and (b). Are your results the same?
d. Which method from parts (a) and (b) is easier to use to find the x-coordinate of the vertex of the parabola $g(x) = 54x^2 - 195x - 216$? Explain.
e. Which method(s) can be used to find the x-coordinate of the vertex of the parabola $h(x) = x^2 + 4x + 6$? Explain.
f. Summarize your findings in this exercise.

70. In this exercise, you will discover how to convert the standard form of a quadratic function to its vertex form.
a. Find the vertex of the parabola $f(x) = 3x^2 - 6x + 7$.
b. Recall that if (h, k) is the vertex of the graph of a quadratic function, then $f(x) = a(x-h)^2 + k$. Use your result from part (a) to determine h and k for $f(x) = 3x^2 - 6x + 7$.
c. Substitute the values of h and k you found from part (b) in $f(x) = a(x - h)^2 + k$.
d. Compare $f(x) = 3x^2 - 6x + 7$ with your result in part (c). Determine the value of a. [**Hint:** You may see this immediately. If not, expand your result in part (c) and compare again with $f(x) = 3x^2 - 6x + 7$.]
e. Substitute your value of a into your result from part (c).
f. Verify your result graphically by comparing the graph of $f(x) = 3x^2 - 6x + 7$ with the graph of $f(x) = a(x-h)^2 + k$ for the values of a, h, and k you found.

71. Input–output pairs for four quadratic functions f, g, h, and k are listed in Table 25. For each function, decide whether $(3, 2)$ is the vertex. If $(3, 2)$ is not the vertex, estimate the coordinates of the vertex.

Table 25 Values of Four Quadratic Functions

x	$f(x)$	$g(x)$	$h(x)$	$k(x)$
1	10	8.2	19.3	11.6
2	4	2.9	7.3	4.4
3	2	2	2	2
4	4	5.5	3.4	4.4
5	10	13.4	11.5	11.6

72. Suppose that $f(x) = ax^2 + bx + c$, where $a > 0$ and g is a linear function. If $f(2) = g(2)$ and $f(5) = g(5)$, which is larger, $f(4)$ or $g(4)$? Explain. [**Hint:** Think graphically.]

73. Explain how to sketch the graph of a quadratic function $f(x) = ax^2 + bx + c$, where $b \neq 0$.

74. The x-coordinate of the vertex of the graph of a quadratic function is equal to the average of the x-coordinates of two symmetric points. Explain. Include a sketch of a graph of a quadratic function.

Related Review

Graph the equation by hand. Use a graphing calculator to verify your work.

75. $y = -3(x + 3)^2 + 5$

76. $y = -3x^2 - 6x + 5$

77. $y = -4\left(\frac{1}{2}\right)^x$

78. $y = -4 + \frac{1}{2}x$

79. We saw in Section 6.1 that the vertex of the graph of $f(x) = a(x-h)^2 + k$, where $a \neq 0$, is (h, k). In this exercise, we will show that that is true.

a. Simplify the right-hand side of $f(x) = a(x - h)^2 + k$ to show that $f(x) = ax^2 - 2ahx + c$, where $c = ah^2 + k$.
b. Use the vertex formula to show that the vertex of f is (h, k).

80. Consider the system

$$y = x^2 - x - 7$$
$$y = x + 1$$

a. Use "intersect" on a graphing calculator to solve the system. [**Hint:** There are two solutions.]
b. Use substitution to solve the system. [**Hint:** After finding the x-coordinates of the two solutions, don't forget to find the y-coordinates.]

Expressions, Equations, Functions, and Graphs

Perform the indicated instruction. Then use words such as linear, quadratic, cubic, exponential, logarithmic, polynomial, degree, function, one variable, *and* two variables *to describe the expression, equation, or system.*

81. Simplify $\dfrac{2b^{-2}c^4(3b^{-5}c^{-1})^2}{8b^{-6}c^{-4}}$.

82. Solve $x(3x - 5) + 2 = x^2 + 5$.

83. Graph $f(x) = -2(3)^x$ by hand.

84. Find the product $-2x(3x - 4)(5x - 2)$.

85. Solve $-2(3)^x = -200$. Round any solutions to the second decimal place.

86. Factor $6x^3 + 3x^2 - 18x$.

7.3 USING THE SQUARE ROOT PROPERTY TO SOLVE QUADRATIC EQUATIONS

Objectives

- Know the *product property* and *quotient property* for square roots.
- Simplify expressions with square roots.
- *Rationalize the denominator* of a fraction.
- Use the *square root property* to solve quadratic equations.
- Make predictions with a quadratic function in vertex form.
- Know the meaning of *pure imaginary number, complex number,* and *imaginary number.*

In this section, we will discuss how to simplify expressions that have square roots. This skill will help us solve some equations that we cannot solve by factoring. We will then be able to make predictions about the dependent variable of a quadratic model in vertex form. Finally, we will define numbers that are not real numbers but can be solutions of quadratic equations.

Product Property for Square Roots

Recall from Section 4.2 that we can represent a principal nth root such as $k^{1/n}$ as $\sqrt[n]{k}$, where $k \geq 0$ if n is even. We call $\sqrt[n]{k}$ a *radical* and refer to k as the *radicand* of $\sqrt[n]{k}$. In this section, we restrict our study to principal square roots $\sqrt{k}$.

The radicand k must be nonnegative for the radical $\sqrt{k}$ to be a real number. For example, $\sqrt{-4}$ is not a real number.

For $k \geq 0$,

$$\sqrt{k} = k^{1/2}$$

We can use properties of exponents to prove the *product property for square roots:* $\sqrt{ab} = \sqrt{a}\sqrt{b}$, where $a \geq 0$ and $b \geq 0$. Here is the proof:

$$\begin{aligned} \sqrt{ab} &= (ab)^{1/2} && \sqrt{k} = k^{1/2} \\ &= a^{1/2}b^{1/2} && \text{Raise factors to } n\text{th power: } (ab)^n = a^n b^n \\ &= \sqrt{a}\sqrt{b} && k^{\frac{1}{2}} = \sqrt{k} \end{aligned}$$

So, $\sqrt{ab} = \sqrt{a}\sqrt{b}$.

Product Property for Square Roots

For $a \geq 0$ and $b \geq 0$,

$$\sqrt{ab} = \sqrt{a}\sqrt{b} \qquad \text{Product property for square roots}$$

In words: The square root of a product is the product of the square roots.

Table 26 Perfect Squares

x	Perfect Square x^2
0	0
1	1
2	4
3	9
4	16
5	25
6	36
7	49
8	64
9	81
10	100
11	121
12	144
13	169
14	196
15	225

Simplifying Expressions with Square Roots

A **perfect square** is a number whose square root is rational. For example, 9 is a perfect square, because $\sqrt{9} = 3 = \frac{3}{1}$ is rational. By squaring the whole numbers from 0 to 15, we can find the integer perfect squares between 0 and 225, inclusive (see Table 26). Memorize the perfect squares shown in this table, because you will work with them again and again.

A radical $\sqrt{k}$ with a counting number k, where $k \geq 2$, is **simplified** if k does not have any perfect square factors other than 1. We can use the product property for square roots to simplify expressions with square roots.

Example 1 Simplifying Radical Expressions

Simplify.

1. $\sqrt{20}$ **2.** $\sqrt{32}$

Solution

1. Note that 4 is the largest perfect-square factor of 20. We write 20 as a product of 4 and 5 and apply the product property for square roots:

$$\begin{aligned} \sqrt{20} &= \sqrt{4 \cdot 5} && \text{4 is largest perfect-square factor.} \\ &= \sqrt{4}\sqrt{5} && \text{Product property: } \sqrt{ab} = \sqrt{a}\sqrt{b} \\ &= 2\sqrt{5} && \sqrt{4} = 2 \end{aligned}$$

```
√(20)
        4.472135955
2√(5)
        4.472135955
```

Figure 55 Verify that $\sqrt{20} = 2\sqrt{5}$

We verify our work by using a graphing calculator (see Fig. 55). To compute $\sqrt{20}$, press [2nd] [x^2] **20** [)] [ENTER].

2. Both 4 and 16 are perfect-square factors of 32. Since 16 is the largest perfect-square factor, we write 32 as the product of 16 and 2 and apply the product property for square roots:

$$\begin{aligned} \sqrt{32} &= \sqrt{16 \cdot 2} && \text{16 is largest perfect-square factor.} \\ &= \sqrt{16}\sqrt{2} && \text{Product property: } \sqrt{ab} = \sqrt{a}\sqrt{b} \\ &= 4\sqrt{2} && \sqrt{16} = 4 \end{aligned}$$

■

In Problem 2 of Example 1, what would happen if we had used the smaller perfect-square factor, 4, rather than the largest perfect-square factor, 16, to simplify $\sqrt{32}$?

$$\sqrt{32} = \sqrt{4 \cdot 8} = \sqrt{4}\sqrt{8} = 2\sqrt{8}$$

The radicand, 8, has a perfect-square factor of 4, so we continue to simplify:

$$\sqrt{32} = 2\sqrt{8} = 2\sqrt{4 \cdot 2} = 2\sqrt{4}\sqrt{2} = 2 \cdot 2\sqrt{2} = 4\sqrt{2}$$

The result is the same as our result in Example 1. This exploration suggests that

- The most efficient way to simplify a square root is to use the *largest* perfect-square factor of the radicand.
- If we don't use the largest perfect-square factor of the radicand, we can simplify the square root by continuing to use perfect-square factors until the radicand has no perfect-square factors other than 1.

Using the Product Property to Simplify Square Roots

To use the product property to simplify a square root,

1. Write the radicand as the product of the *largest* perfect-square factor and another number.
2. Apply the product property for square roots.

Quotient Property for Square Roots

For the square root of a product, we often use the product property for square roots. What property can we use for the square root of a quotient?

Quotient Property for Square Roots

For $a \geq 0$ and $b > 0$,

$$\sqrt{\frac{a}{b}} = \frac{\sqrt{a}}{\sqrt{b}} \quad \text{Quotient property for square roots}$$

In words: The square root of a quotient is the quotient of the square roots.

You will prove the quotient property in Exercise 86.

If a square root has a fractional radicand, we **simplify the radical** by writing it as an expression whose radicand is not a fraction.

Example 2 Simplifying a Radical Expression

Simplify $\sqrt{\frac{7}{64}}$.

Solution

$$\sqrt{\frac{7}{64}} = \frac{\sqrt{7}}{\sqrt{64}} \quad \text{Quotient property: } \sqrt{\frac{a}{b}} = \frac{\sqrt{a}}{\sqrt{b}}$$

$$= \frac{\sqrt{7}}{8} \quad \sqrt{64} = 8$$

■

Rationalizing the Denominator of a Radical Expression

We simplify an expression of the form $\frac{p}{\sqrt{q}}$ by leaving no denominator as a radical expression. We call this process **rationalizing the denominator.**

Example 3 Simplifying Radical Expressions

Simplify.

1. $\frac{2}{\sqrt{3}}$
2. $\sqrt{\frac{5}{18}}$

Solution

1. Since $\sqrt{3}\sqrt{3} = \sqrt{3 \cdot 3} = \sqrt{9} = 3$, we rationalize the denominator of $\frac{2}{\sqrt{3}}$ by multiplying by $\frac{\sqrt{3}}{\sqrt{3}}$:

$$\frac{2}{\sqrt{3}} = \frac{2}{\sqrt{3}} \cdot 1 \quad a = a \cdot 1$$

$$= \frac{2}{\sqrt{3}} \cdot \frac{\sqrt{3}}{\sqrt{3}} \quad \text{Rationalize denominator: } \frac{\sqrt{3}}{\sqrt{3}} = 1$$

$$= \frac{2\sqrt{3}}{\sqrt{9}} \quad \text{Multiply numerators; multiply denominators; product property: } \sqrt{a}\sqrt{b} = \sqrt{ab}$$

$$= \frac{2\sqrt{3}}{3} \quad \sqrt{9} = 3$$

Figure 56 Verify the work

2. $\sqrt{\dfrac{5}{18}} = \dfrac{\sqrt{5}}{\sqrt{18}}$ Quotient property: $\sqrt{\dfrac{a}{b}} = \dfrac{\sqrt{a}}{\sqrt{b}}$

$= \dfrac{\sqrt{5}}{3\sqrt{2}}$ $\sqrt{18} = \sqrt{9 \cdot 2} = \sqrt{9}\sqrt{2} = 3\sqrt{2}$

$= \dfrac{\sqrt{5}}{3\sqrt{2}} \cdot \dfrac{\sqrt{2}}{\sqrt{2}}$ Rationalize denominator: $\dfrac{\sqrt{2}}{\sqrt{2}} = 1$

$= \dfrac{\sqrt{10}}{3\sqrt{4}}$ Multiply numerators; multiply denominators; product property: $\sqrt{a}\sqrt{b} = \sqrt{ab}$

$= \dfrac{\sqrt{10}}{6}$ Simplify.

We verify our work by using a graphing calculator (see Fig. 56). ■

As Problem 1 of Example 3 shows, **to rationalize the denominator of a fraction of the form $\dfrac{p}{\sqrt{q}}$, where q is positive, we multiply the fraction by 1 in the form $\dfrac{\sqrt{q}}{\sqrt{q}}$.**

Summary of Simplifying a Radical Expression

We have discussed various ways to simplify a radical expression. What follows is a summary of those methods.

Simplifying a Radical Expression

To simplify a radical expression,

1. Use the quotient property for square roots so that no radicand is a fraction.
2. Use the product property for square roots so that no radicands have perfect-square factors other than 1.
3. Rationalize the denominators so that no denominator is a radical expression.
4. Continue applying steps 1–3 until the radical expression is completely simplified.

Solving Quadratic Equations of the Form $x^2 = k$, Where $k \geq 0$

We now turn our attention to solving quadratic equations of the form $x^2 = k$, where $k \geq 0$. From Section 4.4, we know that the solutions of the equation $x^2 = k$ are

$$\pm k^{1/2} = \pm\sqrt{k}$$

We call this relationship the **square root property.**

Square Root Property

Let k be a nonnegative constant. Then $x^2 = k$ is equivalent to

$$x = \pm\sqrt{k}$$

We will discuss the case in which k is negative later in this section.

Example 4 Using the Square Root Property to Solve Equations

1. Solve $x^2 = 16$. **2.** Solve $x^2 = 45$.

Solution

1.

$x^2 = 16$ Original equation

$x = \pm\sqrt{16}$ Square root property

$x = \pm 4$ $\sqrt{16} = 4$

2.

$$x^2 = 45 \quad \text{Original equation}$$
$$x = \pm\sqrt{45} \quad \text{Square root property}$$
$$x = \pm\sqrt{9 \cdot 5} \quad \text{9 is a perfect square.}$$
$$x = \pm 3\sqrt{5} \quad \text{Simplify.}$$

We use a graphing calculator to verify that $-3\sqrt{5} \approx -6.71$ and $3\sqrt{5} \approx 6.71$ are solutions (see Fig. 57). ■

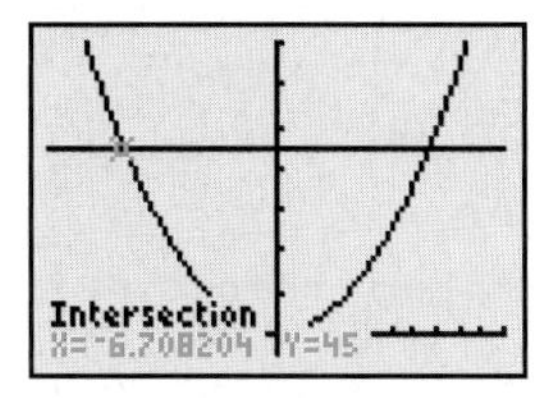

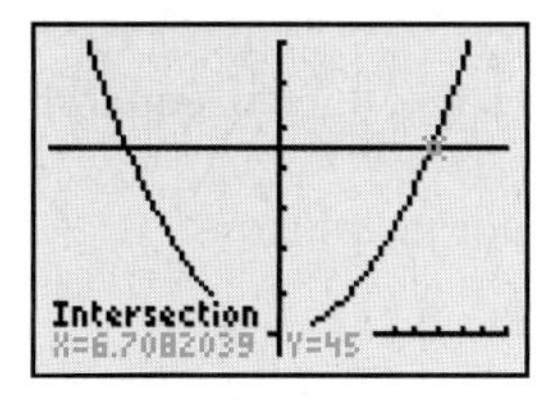

Figure 57 Verify the work

WARNING It is a common error to confuse solving an equation such as $x^2 = 16$ with computing a principal square root such as $\sqrt{16}$. In Problem 1 of Example 4 we found that the solutions of $x^2 = 16$ are the *two* numbers -4 and 4. Recall from Section 4.2 that $\sqrt{16}$ is the *one* number 4.

Example 5 Using the Square Root Property to Solve an Equation

Solve $3x^2 - 5 = 6$.

Solution

$$3x^2 - 5 = 6 \quad \text{Original equation}$$
$$3x^2 = 11 \quad \text{Add 5 to both sides.}$$
$$x^2 = \frac{11}{3} \quad \text{Divide both sides by 3.}$$
$$x = \pm\sqrt{\frac{11}{3}} \quad \text{Square root property}$$
$$= \pm\frac{\sqrt{11}}{\sqrt{3}} \quad \text{Quotient property: } \sqrt{\frac{a}{b}} = \frac{\sqrt{a}}{\sqrt{b}}$$
$$= \pm\frac{\sqrt{11}}{\sqrt{3}} \cdot \frac{\sqrt{3}}{\sqrt{3}} \quad \text{Rationalize denominator.}$$
$$= \pm\frac{\sqrt{33}}{3} \quad \text{Simplify.}$$

We enter $y = 3x^2 - 5$ into a graphing calculator and check that for both inputs $-\frac{\sqrt{33}}{3}$ and $\frac{\sqrt{33}}{3}$, the output is 6 (see Fig. 58). ■

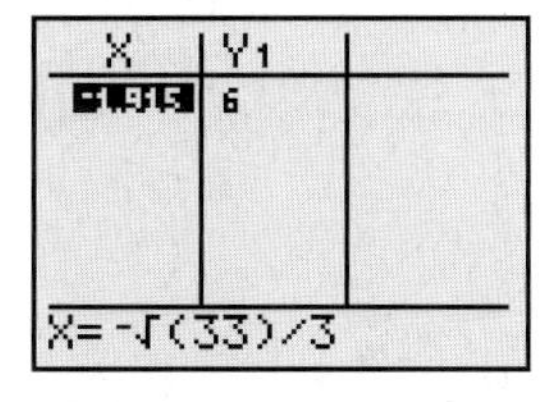

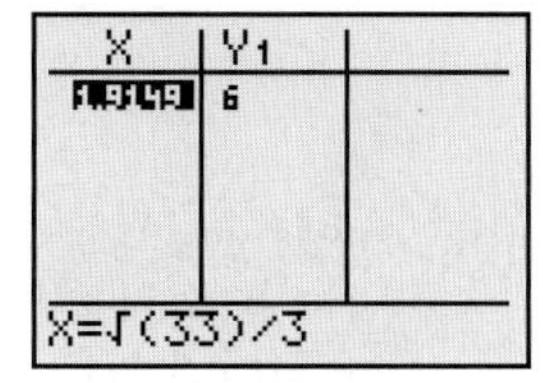

Figure 58 Verify the work

Solving Quadratic Equations of the Form $(px + q)^2 = k$

We can also use the square root property to solve equations that can be put into the form $(px + q)^2 = k$.

Example 6 Using the Square Root Property to Solve an Equation

Solve $(3w - 4)^2 = 36$.

Solution

The base of $(3w - 4)^2$ is $3w - 4$. We can use the square root property to solve the equation $(3w - 4)^2 = 36$:

$$(3w - 4)^2 = 36 \quad \text{Original equation}$$
$$3w - 4 = \pm\sqrt{36} \quad \text{Square root property}$$
$$3w - 4 = \pm 6 \quad \sqrt{36} = 6.$$
$$3w - 4 = -6 \quad \text{or} \quad 3w - 4 = 6 \quad \text{Write as two equations.}$$
$$3w = -2 \quad \text{or} \quad 3w = 10$$
$$w = -\frac{2}{3} \quad \text{or} \quad w = \frac{10}{3}$$

■

Example 7 Using the Square Root Property to Solve an Equation

Solve $\left(x+\frac{5}{2}\right)^2=\frac{31}{4}$.

Solution

$$\left(x+\frac{5}{2}\right)^2=\frac{31}{4} \quad \text{Original equation}$$

$$x+\frac{5}{2}=\pm\sqrt{\frac{31}{4}} \quad \text{Square root property}$$

$$x+\frac{5}{2}=\pm\frac{\sqrt{31}}{\sqrt{4}} \quad \text{Quotient property: } \sqrt{\frac{a}{b}}=\frac{\sqrt{a}}{\sqrt{b}}$$

$$x+\frac{5}{2}=\pm\frac{\sqrt{31}}{2} \quad \sqrt{4}=2$$

$$x=-\frac{5}{2}\pm\frac{\sqrt{31}}{2} \quad \text{Subtract } \frac{5}{2} \text{ from both sides.}$$

$$x=\frac{-5\pm\sqrt{31}}{2} \quad \text{Add/subtract numerators and keep common denominator.}$$

To verify our work, we store each result as x and check that $\left(x+\frac{5}{2}\right)^2$ is equal to $\frac{31}{4}=7.75$ (see Fig. 59). See Section B.20 for graphing calculator instructions.

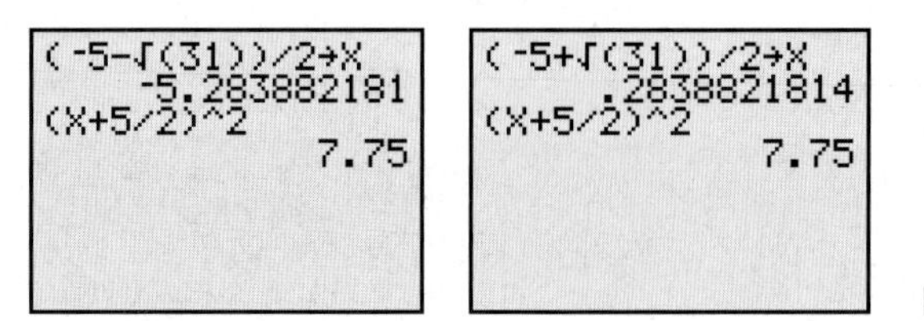

Figure 59 Verify the work ■

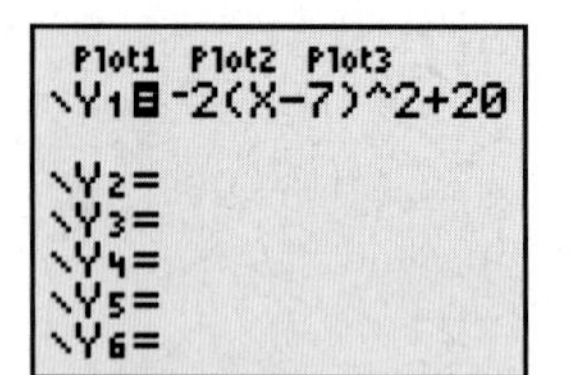

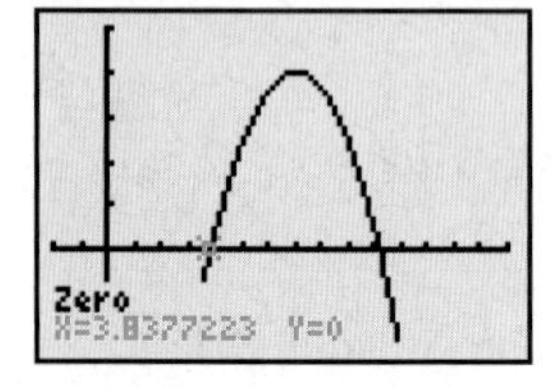

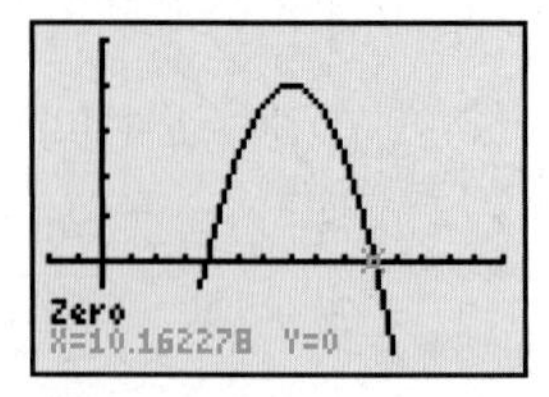

Figure 60 Verify the work

Example 8 Finding x-Intercepts

Find the x-intercepts of the parabola $f(x)=-2(x-7)^2+20$.

Solution

To find the x-intercepts, we substitute 0 for $f(x)$:

$$-2(x-7)^2+20=0 \quad \text{Substitute 0 for } f(x).$$

$$-2(x-7)^2=-20 \quad \text{Subtract 20 from both sides.}$$

$$(x-7)^2=10 \quad \text{Divide both sides by } -2.$$

$$x-7=\pm\sqrt{10} \quad \text{Square root property}$$

$$x=7\pm\sqrt{10} \quad \text{Add 7 to both sides.}$$

The x-intercepts are $(7-\sqrt{10}, 0)$ and $(7+\sqrt{10}, 0)$. In Fig. 60, we use "zero" on a graphing calculator to check that the approximate x-intercepts are (3.84, 0) and (10.16, 0). See Section B.21 for graphing calculator instructions. ■

Making Predictions with a Quadratic Model in Vertex Form

Our work in this section enables us to use a quadratic model in vertex form $f(x)=a(x-h)^2+k$ to make predictions.

Table 27 Average Monthly Participation in Food Stamp Program

Year	Average Monthly Participation (millions)
1996	25.4
1997	22.7
1998	19.8
1999	18.2
2000	17.2
2001	17.3
2002	19.1
2003	21.3
2004	23.9

Source: *U.S. Department of Agriculture*

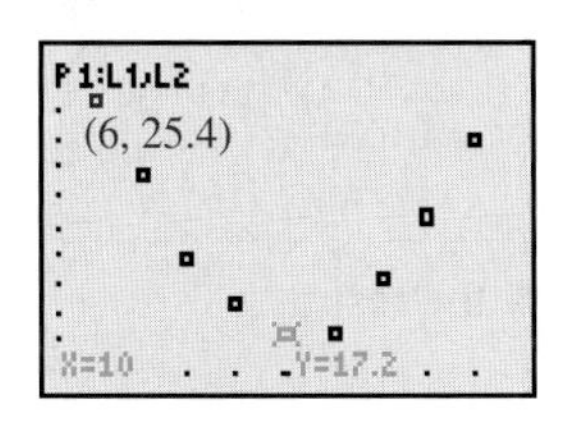

Figure 61 Scattergram of Food Stamp data

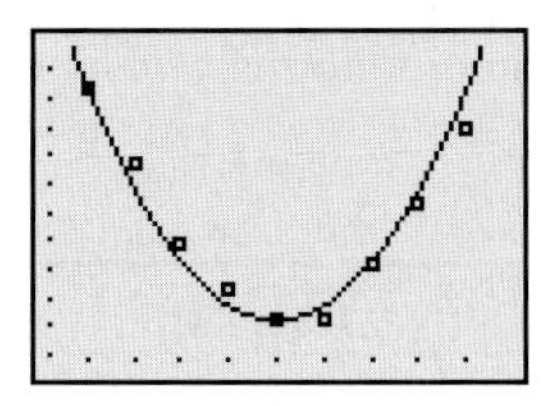

Figure 62 Check the fit

Example 9 Using a Quadratic Model to Make Predictions

Average monthly participation in the Food Stamp Program is shown in Table 27 for various years.

Let $f(t)$ be the average monthly participation (in millions) in the Food Stamp Program in the year that is t years since 1990.

1. Find an equation of f.
2. Estimate when the average monthly participation in the Food Stamp Program was 30 million.

Solution

1. We use a graphing calculator to draw a scattergram of the data (see Fig. 61).

   It appears that a quadratic function will fit the data much better than an exponential model or a linear model. Recall from Section 7.1 that we can find a quadratic function in vertex form $f(t) = a(t - h)^2 + k$. To begin, we select the lowest data point, (10, 17.2), to be the vertex (see Fig. 61). Here, $h = 10$ and $k = 17.2$:

$$f(t) = a(t - 10)^2 + 17.2$$

   Next, we imagine a parabola with vertex (10, 17.2) that comes close to (or contains) the data points. Such a parabola might be the one that contains the data point (6, 25.4). See Fig. 61. To find a, we substitute 6 for t and 25.4 for $f(t)$ in the equation $f(t) = a(t - 10)^2 + 17.2$:

$$\begin{aligned} 25.4 &= a(6 - 10)^2 + 17.2 && \text{Substitute 6 for } t \text{ and 25.4 for } f(t). \\ 25.4 &= a(-4)^2 + 17.2 && \text{Subtract.} \\ 25.4 &= 16a + 17.2 && \text{Simplify.} \\ 8.2 &= 16a && \text{Subtract 17.2 from both sides.} \\ 0.51 &\approx a && \text{Divide both sides by 16.} \end{aligned}$$

   The approximate equation is $f(t) = 0.51(t - 10)^2 + 17.2$. The model fits the data well (see Fig. 62).

2. We substitute 30 for $f(t)$ in the equation $f(t) = 0.51(t - 10)^2 + 17.2$ and solve for t:

$$\begin{aligned} 0.51(t - 10)^2 + 17.2 &= 30 && \text{Substitute 30 for } f(t). \\ 0.51(t - 10)^2 &= 12.8 && \text{Subtract 17.2 from both sides.} \\ (t - 10)^2 &= \frac{12.8}{0.51} && \text{Divide both sides by 0.51.} \\ t - 10 &= \pm\sqrt{\frac{12.8}{0.51}} && \text{Square root property} \\ t &= 10 \pm \sqrt{\frac{12.8}{0.51}} && \text{Add 10 to both sides.} \end{aligned}$$

$$t = 10 - \sqrt{\frac{12.8}{0.51}} \quad \text{or} \quad t = 10 + \sqrt{\frac{12.8}{0.51}} \qquad \text{Write as two equations.}$$

$$t \approx 10 - 5.01 \quad \text{or} \quad t \approx 10 + 5.01 \qquad \text{Approximate square root.}$$

$$t \approx 4.99 \quad \text{or} \quad t \approx 15.01$$

   The model estimates that the average monthly participation in the Food Stamp Program was 30 million people in both 1995 and 2005. ■

Complex Numbers

So far in this section, we have discussed equations of the form $x^2 = k$, where k is nonnegative. What happens when k is a negative number, such as -1?

From Section 4.4, we know that the equation $x^2 = -1$ has no real-number solutions. Now we define a number that is *not* a real number but *is* a solution of $x^2 = -1$.

We define $\sqrt{-1}$ to be a number whose square is -1. We represent this number as i.

DEFINITION Imaginary unit *i*

The **imaginary unit,** written i, is the number whose square is -1. That is,

$$i^2 = -1 \quad \text{and} \quad i = \sqrt{-1}$$

Next, we define the square root of any negative number.

DEFINITION Square root of a negative number

If p is a positive real number, then

$$\sqrt{-p} = i\sqrt{p}$$

If b is a nonzero real number, then we call an expression of the form bi a **pure imaginary number.**

Example 10 Writing Numbers in *bi* Form

Write the number in bi form, where b is a real number. Simplify the result.

1. $\sqrt{-49}$
2. $-\sqrt{-24}$

Solution

1. $\sqrt{-49} = i\sqrt{49} = 7i$
2. $-\sqrt{-24} = -i\sqrt{24} = -i\sqrt{4 \cdot 6} = -2i\sqrt{6}$ ■

We can combine real numbers a and b with the imaginary unit i to form a *complex number* of the form

$$a + bi$$

DEFINITION Complex number

A **complex number** is a number of the form

$$a + bi$$

where a and b are real numbers.

Here are some examples of complex numbers:

$$2 + 5i, \quad 3 - 4i, \quad 5 + 0i = 5, \quad 0 - 7i = -7i$$

Since $a = a + 0i$ and $bi = 0 + bi$, we see that all real numbers and all pure imaginary numbers are complex numbers.

A complex number that is not a real number is called an *imaginary number.*

DEFINITION Imaginary number

An **imaginary number** is a number $a + bi$, where a and b are real numbers and $b \neq 0$.

The complex number $3+7i$ is an imaginary number, since $b = 7 \neq 0$. The complex number $4 = 4 + 0i$ is not an imaginary number, since $b = 0$.

We perform operations with complex numbers in much the same way that we perform operations with polynomials. Here we compare squaring the monomial $5x$ with squaring the pure imaginary number $5i$:

$$(5x)^2 = 5^2x^2 = 25x^2 \qquad (5i)^2 = 5^2i^2 = 25(-1) = -25$$

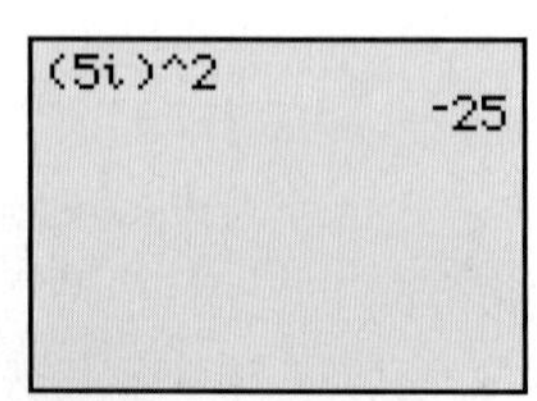

Figure 63 Check the work

We use a graphing calculator to verify our work of squaring $5i$ (see Fig. 63). To compute $(5i)^2$, press [( ] **5** [2nd] [.] [ )] [^] **2** [ENTER].

Solving Quadratic Equations of the Form $x^2 = k$, Where $k < 0$

We can extend the square root property to the case in which k is negative.

Square Root Property

Let k be a real-number constant. Then $x^2 = k$ is equivalent to

$$x = \pm\sqrt{k}$$

Example 11 Using the Square Root Property to Solve Equations

Solve.

1. $y^2 = -9$ **2.** $x^2 = -12$

Solution

1.

$$\begin{aligned} y^2 &= -9 && \text{Original equation} \\ y &= \pm\sqrt{-9} && \text{Square root property} \\ y &= \pm i\sqrt{9} && \sqrt{-p} = i\sqrt{p}, \text{ where } p > 0 \\ y &= \pm 3i && \sqrt{9} = 3 \end{aligned}$$

We check that both $-3i$ and $3i$ satisfy the original equation:

Check $y = -3i$

$$\begin{aligned} y^2 &= -9 \\ (-3i)^2 &\stackrel{?}{=} -9 \\ (-3)^2 i^2 &\stackrel{?}{=} -9 \\ 9(-1) &\stackrel{?}{=} -9 \\ -9 &\stackrel{?}{=} -9 \end{aligned}$$

true

Check $y = 3i$

$$\begin{aligned} y^2 &= -9 \\ (3i)^2 &\stackrel{?}{=} -9 \\ 3^2 i^2 &\stackrel{?}{=} -9 \\ 9(-1) &\stackrel{?}{=} -9 \\ -9 &\stackrel{?}{=} -9 \end{aligned}$$

true

2.

$$\begin{aligned} x^2 &= -12 && \text{Original equation} \\ x &= \pm\sqrt{-12} && \text{Square root property} \\ x &= \pm i\sqrt{12} && \sqrt{-p} = i\sqrt{p}, \text{ where } p > 0 \\ x &= \pm i(2\sqrt{3}) && \sqrt{12} = \sqrt{4 \cdot 3} = 2\sqrt{3} \\ x &= \pm 2i\sqrt{3} && \text{Rearrange factors.} \end{aligned}$$

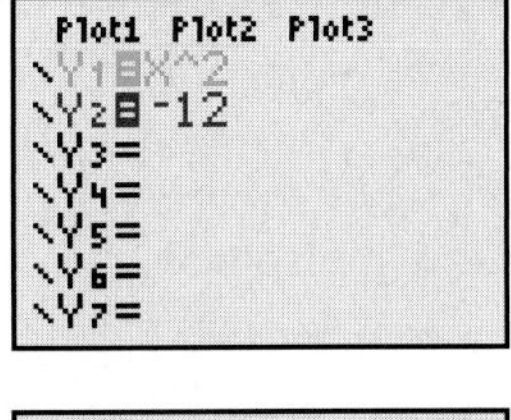

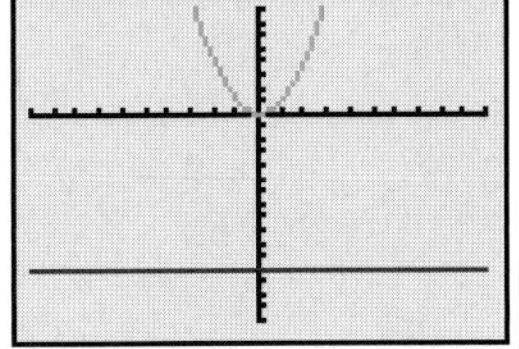

Figure 64 Check the work

The graphical check in Fig. 64 shows that the graphs of $y = x^2$ and $y = -12$ do not intersect. This means that the equation $x^2 = -12$ has no real-number solutions. It *does* have the two imaginary-number solutions: $\pm 2i\sqrt{3}$. ■

Example 12 Using the Square Root Property to Solve an Equation

Solve $(x + 3)^2 = -28$.

Solution

$$\begin{aligned} (x+3)^2 &= -28 && \text{Original equation} \\ x + 3 &= \pm\sqrt{-28} && \text{Square root property} \\ x + 3 &= \pm i\sqrt{28} && \sqrt{-p} = i\sqrt{p}, \text{ where } p > 0 \\ x + 3 &= \pm 2i\sqrt{7} && \sqrt{28} = \sqrt{4 \cdot 7} = 2\sqrt{7} \\ x &= -3 \pm 2i\sqrt{7} && \text{Subtract 3 from both sides.} \end{aligned}$$

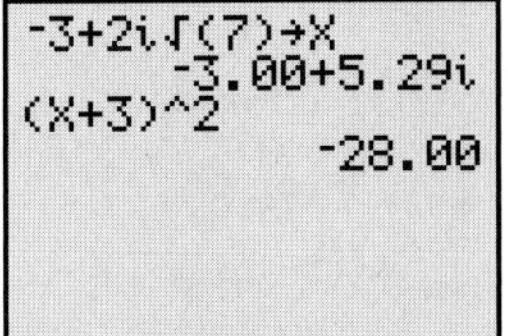

Figure 65 Verify the work

To verify our work, we store each result as x and check that $(x + 3)^2$ is equal to -28 (see Fig. 65). We press MODE and set FLOAT to 2, so the numbers in the result are rounded to the second decimal place. ■

group exploration

Deriving a formula for solving quadratic equations in $a(x-h)^2+k=p$ form

1. Solve $2(x-5)^2+7=10$.
2. Solve the equation $a(x-h)^2+k=p$ for x. Assume that $a \neq 0$ and $\frac{p-k}{a} \geq 0$. [**Hint:** Follow the same steps as in Problem 1.]
3. Use your result from Problem 2 to solve the equation $3(x-4)^2+2=7$. [**Hint:** Substitute the appropriate values for a, h, k, and p into your formula from Problem 2.]

group exploration

Looking ahead: Perfect-square trinomials

1. Consider the true statement
$$(x+k)^2 = x^2+(2k)x+k^2$$
For the trinomial on the right-hand side, the coefficient of x is $2k$ and the constant term is k^2. Describe how to use operations such as adding, subtracting, multiplying, dividing, and/or squaring to change $2k$ into k^2.
2. Show that your description in Problem 1 works for the true statement
$$(x+3)^2 = x^2+6x+9$$
3. Simplify $(x-k)^2$. Make a description similar to the one you made in Problem 1. Show that your description works for the true statement
$$(x-4)^2 = x^2-8x+16$$
4. Find a value of c such that the trinomial can be factored as the square of a binomial $(x+k)^2$. Then factor the trinomial.
   a. x^2+8x+c
   b. $x^2+10x+c$
   c. $x^2-14x+c$
   d. $x^2-18x+c$

HOMEWORK 7.3

FOR EXTRA HELP ▸ Student Solutions Manual | PH Math/Tutor Center | MathXL® | MyMathLab

Simplify.

1. $\sqrt{169}$
2. $\sqrt{196}$
3. $\sqrt{12}$
4. $\sqrt{75}$
5. $\sqrt{\frac{4}{9}}$
6. $\sqrt{\frac{25}{81}}$
7. $\sqrt{\frac{6}{49}}$
8. $\sqrt{\frac{13}{100}}$
9. $\frac{5}{\sqrt{2}}$
10. $\frac{4}{\sqrt{7}}$
11. $\frac{3}{\sqrt{32}}$
12. $\frac{8}{\sqrt{60}}$
13. $\sqrt{\frac{3}{2}}$
14. $\sqrt{\frac{7}{5}}$
15. $\sqrt{\frac{11}{20}}$
16. $\sqrt{\frac{7}{90}}$

Solve. Any solution is a real number.

17. $x^2=25$
18. $x^2=64$
19. $x^2-3=0$
20. $x^2-19=0$
21. $t^2=32$
22. $r^2=24$
23. $5x^2=3$
24. $2x^2=13$
25. $3p^2-11=3$
26. $5w^2-28=5$
27. $(x+4)^2=7$
28. $(x+1)^2=3$
29. $(x-5)^2=27$
30. $(x-9)^2=75$
31. $(8y+3)^2=36$
32. $(4m+7)^2=81$
33. $(9x-5)^2=0$
34. $(5x-4)^2=0$
35. $\left(x+\frac{3}{4}\right)^2=\frac{41}{16}$
36. $\left(x+\frac{5}{6}\right)^2=\frac{51}{36}$

37. $\left(w - \frac{7}{3}\right)^2 = \frac{5}{9}$

38. $\left(p - \frac{2}{5}\right)^2 = \frac{17}{25}$

39. $5(x - 6)^2 + 3 = 33$

40. $2(x - 4)^2 + 5 = 27$

41. $-3(x + 1)^2 + 2 = -5$

42. $-7(x + 8)^2 + 4 = -1$

Find all x-intercepts.

43. $f(x) = x^2 - 17$

44. $f(x) = x^2 - 35$

45. $f(x) = 2(x - 3)^2 - 7$

46. $f(x) = 7(x - 6)^2 - 13$

47. $f(x) = -4(x - 2)^2 - 16$

48. $f(x) = -3(x - 5)^2 - 27$

Simplify.

49. $\sqrt{-36}$

50. $\sqrt{-25}$

51. $-\sqrt{-45}$

52. $-\sqrt{-40}$

53. $\sqrt{-\frac{5}{49}}$

54. $\sqrt{-\frac{7}{9}}$

55. $\sqrt{-\frac{13}{5}}$

56. $\sqrt{-\frac{2}{7}}$

Find all complex-number solutions.

57. $x^2 = -49$

58. $x^2 = -4$

59. $x^2 = -18$

60. $x^2 = -28$

61. $7x^2 + 26 = 5$

62. $4x^2 + 25 = 17$

63. $(m + 4)^2 = -8$

64. $(t + 8)^2 = -45$

65. $\left(x - \frac{5}{4}\right)^2 = -\frac{3}{16}$

66. $\left(x - \frac{1}{2}\right)^2 = -\frac{7}{4}$

67. $-2(y + 3)^2 + 1 = 9$

68. $-2(w + 1)^2 + 7 = 39$

69. The prices of scrap iron and sheet metal are shown in Table 28 for various years.

Table 28 Prices of Scrap Iron and Sheet Metal

Year	Price of Scrap Iron and Sheet Metal (dollars per metric ton)
2000	74
2001	64
2002	94
2003	141
2004	253

Source: *Institute of Scrap Recycling Industries, Inc.*

Let $f(t)$ be the price of scrap iron and sheet metal (in dollars per metric ton) at t years since 2000.

a. Find a quadratic equation of f in vertex form.

b. Predict when the price of scrap iron and sheet metal will be \$1250 per metric ton.

c. Predict the price of scrap iron and sheet metal in 2010.

d. Due to the high price of scrap iron and sheet metal, thieves are stealing manhole covers, light poles, guardrails, and other metallic objects. Predict the scrap-iron value of a 125-pound manhole cover in 2010. (There are 2204.6 pounds in 1 metric ton.)

70. The numbers of poetry books published in Iraq are shown in Table 29 for various years.

Table 29 Numbers of Poetry Books Published in Iraq

Year	Number of Poetry Books
2001	25
2002	14
2003	1
2004	6
2005	16

Source: *Iraq Culture Ministry*

Let $f(t)$ be the number of poetry books published in Iraq in the year that is t years since 2000.

a. Find a quadratic equation of f in vertex form.

b. What is the vertex? What does it mean in this situation?

c. Estimate the number of poetry books that were published in Iraq in 2006.

d. Predict in which year 73 poetry books will be published.

71. Spending by U.S. residents on domestic travel is shown in Table 30 for various years.

Table 30 Spending by U.S. Residents on Domestic Travel

Year	Spending on Domestic Travel (billions of dollars)
2000	499
2001	480
2002	477
2003	494
2004	523

Source: *Travel Industry Association of America*

Let $f(t)$ be the spending (in billions of dollars) by U.S. residents on domestic travel in the year that is t years since 2000.

a. Find a quadratic equation of f in vertex form.

b. What is the vertex of the model? What does it mean in this situation?

c. Predict spending on domestic travel in 2009.

d. Predict when spending on domestic travel will be \$1 trillion. [**Hint:** How many billions are in 1 trillion?]

72. The percentages of the U.S. federal debt owed to foreigners are shown in Table 31 for various years.

Table 31 Percentages of the U.S. Federal Debt Owed to Foreigners

Year	Percent
1980	17
1985	15
1990	18
1995	22
2000	31
2005	45

Source: *White House Office of Management and Budget*

Let $P(t)$ be the percentage of the U.S. federal debt owed to foreigners at t years since 1980.

a. Find a quadratic equation of P in vertex form.

b. What is the vertex of the model? What does it mean in this situation?

c. Use P to predict the percentage of the federal debt owed to foreigners in 2011.
d. In Exercise 38 of Homework 4.5, you found an equation close to $f(t) = 214.76(1.087)^t$, where $f(t)$ is the federal debt (in billions of dollars) at t years since 1960. Use f to predict the federal debt in 2011. Then use your result in part (c) to predict the number of billions of dollars of that debt owed to foreigners in 2011.
e. Use P to predict when all of the debt will be owed to foreigners.

73. A student tries to solve the equation $x^2 - 10x + 25 = 0$:

$$x^2 - 10x + 25 = 0$$
$$x^2 = 10x - 25$$
$$x = \pm\sqrt{10x - 25}$$

Describe any errors. Then solve the equation correctly.

74. A student tries to solve the equation $3(x-5)^2 = 12$:

$$3(x-5)^2 = 12$$
$$(x-5)^2 = 4$$
$$x - 5 = 2$$
$$x = 7$$

Describe any errors. Then solve the equation correctly.

For Exercises 75–80, find approximate solutions of the given equation or system by referring to the graphs shown in Fig. 66. Round results or coordinates of results to the first decimal place.

75. $\frac{1}{2}(x-2)^2 - 1 = -2(x-3)^2 + 4$

76. $-2(x-3)^2 + 4 = \frac{1}{2}x - 4$

77. $-2(x-3)^2 + 4 = 2$

78. $\frac{1}{2}(x-2)^2 - 1 = 0$

79. $y = -2(x-3)^2 + 4$
$y = \frac{1}{2}x - 4$

80. $y = \frac{1}{2}(x-2)^2 - 1$
$y = -2(x-3)^2 + 4$

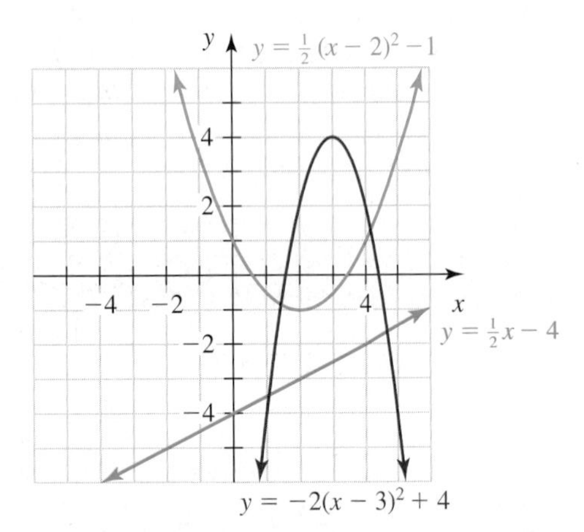

Figure 66 Exercises 75–80

81. Let $f(x) = 2(x-3)^2 + 5$.
a. What is the vertex of the graph of f?
b. Does the graph of f open upward or downward? Explain.
c. How many real-number solutions does each of the following equations have? Explain how you can use your results from parts (a) and (b).
i. $2(x-3)^2 + 5 = 8$
ii. $2(x-3)^2 + 5 = 5$
iii. $2(x-3)^2 + 5 = 1$

82. Let $f(x) = -3(x+2)^2 + 4$.
a. What is the vertex of f?
b. Does the graph of f open upward or downward? Explain.
c. How many real-number solutions does each of the following equations have? Explain how you can use your results from parts (a) and (b).
i. $-3(x+2)^2 + 4 = 6$
ii. $-3(x+2)^2 + 4 = 4$
iii. $-3(x+2)^2 + 4 = 2$

83. a. Use factoring to solve the equation $25x^2 - 49 = 0$.
b. Use the square root property to solve the equation $25x^2 - 49 = 0$.
c. Compare your results for parts (a) and (b).
d. Is it easier to solve $25x^2 - 49 = 0$ by factoring or by using the square root property? Explain.

84. a. Use factoring to solve the equation $(2x-3)^2 = 16$.
b. Use the square root property to solve the equation $(2x-3)^2 = 16$.
c. Compare your results for parts (a) and (b).
d. Is it easier to solve $(2x-3)^2 = 16$ by factoring or by using the square root property? Explain.

85. a. Can the equation $(x+4)^2 = 5$ be solved with the square root property? If yes, do so.
b. Can the equation $(x+4)^2 = 5$ be solved by factoring? If yes, do so.
c. Can all equations that can be solved with the square root property be solved by factoring? Explain.

86. Assume that $a \geq 0$ and $b > 0$. Show that $\sqrt{\frac{a}{b}} = \frac{\sqrt{a}}{\sqrt{b}}$.

For Exercises 87–90, solve for the specified variable. Assume that the constants have values for which the equation has exactly two real-number solutions.

87. $a^2 + b^2 = c^2$, for a

88. $\frac{a^2}{b} + c = d$, for a

89. $(mt+b)^2 = p$, for t

90. $(t+b)^2 + q = p$, for t

91. What forms of quadratic equations can you solve by using the square root property? By using factoring? For example, quadratic equations of the form $x^2 + 2bx + b^2 = 0$ can be solved by factoring. Include examples.

92. Give an example of a quadratic equation in one variable that has the given number and type of solutions. Then solve the equation.
a. two real-number solutions
b. one real-number solution
c. two imaginary-number solutions

Related Review

Solve. Round any results to the fourth decimal place. Verify that they approximately satisfy the equation.

93. $x^6 = 142$

94. $x^4 = 83$

95. $(t-4)^3 = 88$

96. $(p+7)^5 = 61$

97. $3(x+1)^5 - 4 = 44$

98. $4(x-2)^3 + 7 = 79$

Expressions, Equations, Functions, and Graphs

Perform the indicated instruction. Then use words such as linear, quadratic, cubic, exponential, logarithmic, polynomial, degree, function, one variable, *and* two variables *to describe the expression, equation, or system.*

99. Factor $2w^3 + 3w^2 - 18w - 27$.

100. Solve $3(x - 4)^2 - 5 = -7$.

101. Solve $2w^3 + 3w^2 - 18w - 27 = 0$.

102. Simplify $3(x - 4)^2 - 5$.

103. Find the product $\left(5w^2 - 2\right)(4w + 3)$.

104. Graph $f(x) = 3(x - 4)^2 - 5$ by hand.

7.4 SOLVING QUADRATIC EQUATIONS BY COMPLETING THE SQUARE

Objectives

- Know the relationship between b and c in a perfect-square trinomial of the form $x^2 + bx + c$.
- Solve quadratic equations by *completing the square*.
- Know that any quadratic equation can be solved by completing the square.

So far, we have discussed how to solve quadratic equations by factoring and by using the square root property. In this section, we will discuss yet another way, called *completing the square.*

Perfect-Square Trinomials

To begin our study of completing the square, we simplify the square of a sum. For example,

$$(x + 3)^2 = x^2 + 6x + 9$$

So, $x^2 + 6x + 9$ is a perfect-square trinomial. Recall from Section 6.2 that a *perfect-square trinomial* is a trinomial that is equivalent to the square of a binomial.

For $x^2 + 6x + 9$, there is a special connection between the 6 and the 9. If we divide the 6 by 2 and then square the result, we get 9:

$$x^2 + 6x + 9$$

$$\left(\frac{6}{2}\right)^2 = 3^2 = 9$$

This is no coincidence. Consider simplifying $(x - 4)^2$: $(x - 4)^2 = x^2 - 8x + 16$. If we divide -8 by 2 and square the result, we get 16:

$$x^2 - 8x + 16$$

$$\left(\frac{-8}{2}\right)^2 = (-4)^2 = 16$$

For the general case, we simplify $(x + k)^2$, where k is a constant:

$$(x + k)^2 = x^2 + (2k)x + k^2$$

$$\left(\frac{2k}{2}\right)^2 = k^2$$

For $x^2 + (2k)x + k^2$, we see that if we divide the coefficient of x by 2 and square the result, we get the constant term k^2.

Perfect-Square Trinomial Property

For a perfect-square trinomial of the form x^2+bx+c, dividing b by 2 and squaring the result gives c:

$$x^2 + bx + c$$

$$\left(\frac{b}{2}\right)^2 = c$$

Example 1 Factoring Perfect-Square Trinomials

Find the value of c such that the expression is a perfect-square trinomial. Then factor the perfect-square trinomial.

1. $x^2 + 10x + c$ **2.** $x^2 - 9x + c$ **3.** $x^2 + \frac{5}{3}x + c$

Solution

1. We divide 10 by 2 and square the result:

$$\left(\frac{10}{2}\right)^2 = 5^2 = 25 = c$$

The expression is $x^2 + 10x + 25$, with factored form $(x+k)^2$ for some positive integer k. So, we have

$$x^2 + 10x + 25 = (x+k)^2 = x^2 + 2kx + k^2$$

The constant terms 25 and k^2 are equal.

Here, $k^2 = 25$, or $k = 5$ (k is positive). So, the factored form of $x^2 + 10x + 25$ is $(x+5)^2$.

2. We divide -9 by 2 and square the result:

$$\left(\frac{-9}{2}\right)^2 = \frac{81}{4} = c$$

The expression is $x^2 - 9x + \frac{81}{4}$, with factored form $(x-k)^2$ for some positive integer k. So, we have

$$x^2 - 9x + \frac{81}{4} = (x-k)^2 = x^2 - 2kx + k^2$$

The constant terms $\frac{81}{4}$ and k^2 are equal.

Here, $k^2 = \frac{81}{4}$, or $k = \frac{9}{2}$ (k is positive). So, the factored form of $x^2 - 9x + \frac{81}{4}$ is $\left(x - \frac{9}{2}\right)^2$.

3. Dividing by 2 is the same as multiplying by $\frac{1}{2}$. So, we multiply $\frac{5}{3}$ by $\frac{1}{2}$ and square the result:

$$\left(\frac{5}{3}\cdot\frac{1}{2}\right)^2 = \left(\frac{5}{6}\right)^2 = \frac{25}{36} = c$$

The expression is $x^2 + \frac{5}{3}x + \frac{25}{36}$, with factored form $\left(x + \frac{5}{6}\right)^2$. ■

Solving $x^2 + bx + c = 0$ by Completing the Square

Now we can solve a quadratic equation by forming a perfect-square trinomial on one side of the equation.

Example 2 Solving by Completing the Square

Solve $x^2 + 6x = -4$.

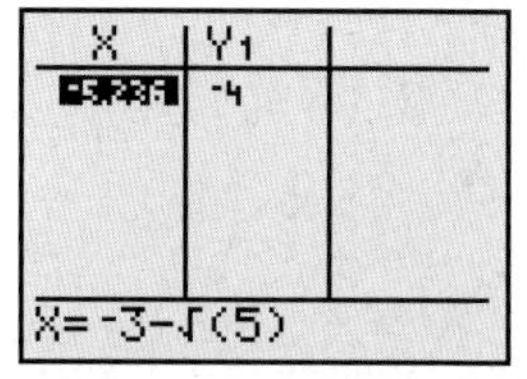

Figure 67 Verify the work

Solution

Since $\left(\frac{6}{2}\right)^2 = 3^2 = 9$, we add 9 to both sides of $x^2 + 6x = -4$ so that the left side will be a perfect-square trinomial:

$$x^2 + 6x = -4 \quad \text{Original equation}$$
$$x^2 + 6x + 9 = -4 + 9 \quad \text{Add 9 to both sides.}$$
$$(x+3)^2 = 5 \quad \text{Factor left side; simplify.}$$
$$x + 3 = \pm\sqrt{5} \quad \text{Square root property}$$
$$x = -3 \pm \sqrt{5} \quad \text{Subtract 3 from both sides.}$$

To check, we enter $y = x^2 + 6x$ in a graphing calculator and find that for both inputs $-3 - \sqrt{5}$ and $-3 + \sqrt{5}$, the output is -4 (see Fig. 67). ■

In Example 2, we added 9 to both sides of $x^2 + 6x = -4$ and factored the left side to get $(x+3)^2$. By adding 9 to both sides of the equation, we say that we *completed the square* for $x^2 + 6x$.

WARNING When solving a quadratic equation such as $x^2 + 6x = -4$ by completing the square, it is a common error to add a number (in this case, 9) to only the left side of the equation. Remember to add the number to *both* sides of the equation.

Example 3 Solving by Completing the Square

Solve $t^2 - 5t + 1 = 0$.

Solution

To solve $t^2 - 5t + 1 = 0$, we first put the equation in the form $t^2 - 5t = k$ before completing the square:

$$t^2 - 5t + 1 = 0 \quad \text{Original equation}$$
$$t^2 - 5t = -1 \quad \text{Subtract 1 from both sides.}$$
$$t^2 - 5t + \frac{25}{4} = -1 + \frac{25}{4} \quad \text{Add } \left(\frac{-5}{2}\right)^2 = \frac{25}{4} \text{ to both sides.}$$
$$\left(t - \frac{5}{2}\right)^2 = -\frac{4}{4} + \frac{25}{4} \quad \text{Factor left side; find LCD.}$$
$$\left(t - \frac{5}{2}\right)^2 = \frac{21}{4} \quad \text{Add numerators and keep common denominator: } \frac{a}{b} + \frac{c}{b} = \frac{a+c}{b}$$
$$t - \frac{5}{2} = \pm\sqrt{\frac{21}{4}} \quad \text{Square root property}$$
$$t - \frac{5}{2} = \pm\frac{\sqrt{21}}{2} \quad \text{Quotient property: } \sqrt{\frac{a}{b}} = \frac{\sqrt{a}}{\sqrt{b}}$$
$$t = \frac{5}{2} \pm \frac{\sqrt{21}}{2} \quad \text{Add } \frac{5}{2} \text{ to both sides.}$$
$$t = \frac{5 \pm \sqrt{21}}{2} \quad \text{Add/subtract numerators and keep common denominator.}$$

Figure 68 Verify the work

We use a graphing calculator to verify our work by storing an approximation of each solution as x and checking that $x^2 - 5x + 1$ is approximately 0 (see Fig. 68). For the calculator entry, the numerator of each fraction must be in parentheses. To store a value, see Section B.20. ■

Solving $ax^2 + bx + c = 0$ by Completing the Square

So far, we have worked with trinomials only of the form $ax^2 + bx + c$, where $a = 1$. What do we do when $a \neq 1$? Consider simplifying $(2x + 5)^2$:

$$(2x + 5)^2 = 4x^2 + 20x + 25$$

Dividing $b = 20$ by 2 and squaring the result does *not* give 25:

$$\left(\frac{20}{2}\right)^2 = 10^2 = 100 \neq 25$$

So, the perfect-square trinomial property given for $ax^2 + bx + c$, where $a = 1$, does not extend to trinomials with $a \neq 1$. However, when solving a quadratic equation of the form $ax^2 + bx + c = 0$ with $a \neq 1$, we can first divide both sides by a to obtain an equation involving "$1x^2$"—an equation to which we *can* apply the property.

Example 4 Solving by Completing the Square

Solve $2x^2 - 16x = -7$.

Solution

$$2x^2 - 16x = -7 \qquad \text{Original equation}$$

$$x^2 - 8x = -\frac{7}{2} \qquad \text{Divide both sides by 2.}$$

$$x^2 - 8x + 16 = -\frac{7}{2} + 16 \qquad \text{Add } \left(\frac{-8}{2}\right)^2 = (-4)^2 = 16 \text{ to both sides.}$$

$$(x - 4)^2 = -\frac{7}{2} + \frac{32}{2} \qquad \text{Factor left side; find LCD.}$$

$$(x - 4)^2 = \frac{25}{2} \qquad \text{Add numerators and keep common denominator: } \frac{a}{b} + \frac{c}{b} = \frac{a + c}{b}$$

$$x - 4 = \pm\sqrt{\frac{25}{2}} \qquad \text{Square root property}$$

$$x - 4 = \pm\frac{5\sqrt{2}}{2} \qquad \sqrt{\frac{25}{2}} = \frac{\sqrt{25}}{\sqrt{2}} = \frac{5}{\sqrt{2}} \cdot \frac{\sqrt{2}}{\sqrt{2}} = \frac{5\sqrt{2}}{2}$$

$$x = 4 \pm \frac{5\sqrt{2}}{2} \qquad \text{Add 4 to both sides.}$$

$$x = \frac{8}{2} \pm \frac{5\sqrt{2}}{2} \qquad \text{Find LCD.}$$

$$x = \frac{8 \pm 5\sqrt{2}}{2} \qquad \text{Add/subtract numerators and keep common denominator.}$$

■

Any quadratic equation can be solved by completing the square. Here is a summary of this method.

Solving a Quadratic Equation by Completing the Square

To solve a quadratic equation by **completing the square:**

1. Write the equation in the form $ax^2 + bx = k$, where a, b, and k are constants.
2. If $a \neq 1$, divide both sides of the equation by a.
3. Complete the square for the expression on the left side of the equation.
4. Solve the equation by using the square root property.

WARNING To solve an equation of the form $ax^2 + bx = k$, where $a \neq 1$, we must divide both sides of the equation by a before completing the square on the left side of the equation.

Example 5 Solving by Completing the Square

Solve $3x^2 + 7x - 5 = 0$.

Solution

$$3x^2 + 7x - 5 = 0 \quad \text{Original equation}$$
$$3x^2 + 7x = 5 \quad \text{Add 5 to both sides.}$$
$$x^2 + \frac{7}{3}x = \frac{5}{3} \quad \text{Divide both sides by 3.}$$
$$x^2 + \frac{7}{3}x + \frac{49}{36} = \frac{5}{3} + \frac{49}{36} \quad \text{Add } \left(\frac{7}{3}\cdot\frac{1}{2}\right)^2 = \left(\frac{7}{6}\right)^2 = \frac{49}{36} \text{ to both sides.}$$
$$\left(x + \frac{7}{6}\right)^2 = \frac{60}{36} + \frac{49}{36} \quad \text{Factor left side; find LCD.}$$
$$\left(x + \frac{7}{6}\right)^2 = \frac{109}{36} \quad \text{Add numerators and keep common denominator: } \frac{a}{b} + \frac{c}{b} = \frac{a+c}{b}$$
$$x + \frac{7}{6} = \pm\sqrt{\frac{109}{36}} \quad \text{Square root property}$$
$$x + \frac{7}{6} = \pm\frac{\sqrt{109}}{6} \quad \text{Quotient property: } \sqrt{\frac{a}{b}} = \frac{\sqrt{a}}{\sqrt{b}}$$
$$x = -\frac{7}{6} \pm \frac{\sqrt{109}}{6} \quad \text{Subtract } \frac{7}{6} \text{ from both sides.}$$
$$x = \frac{-7 \pm \sqrt{109}}{6} \quad \text{Add/subtract numerators and keep common denominator.}$$

■

Solving a Quadratic Equation That Has Imaginary-Number Solutions

In Example 6, we use completing the square to solve a quadratic equation that has imaginary-number solutions.

Example 6 Solving by Completing the Square

Solve $5x^2 - 10x + 45 = 0$.

Solution

$$5x^2 - 10x + 45 = 0 \quad \text{Original equation}$$
$$x^2 - 2x + 9 = 0 \quad \text{Divide both sides by 5.}$$
$$x^2 - 2x = -9 \quad \text{Subtract 9 from both sides.}$$
$$x^2 - 2x + 1 = -9 + 1 \quad \text{Add } \left(\frac{-2}{2}\right)^2 = (-1)^2 = 1 \text{ to both sides.}$$
$$(x - 1)^2 = -8 \quad \text{Factor left side.}$$
$$x - 1 = \pm\sqrt{-8} \quad \text{Square root property}$$
$$x - 1 = \pm i\sqrt{8} \quad \sqrt{-p} = i\sqrt{p}, \text{ where } p > 0$$
$$x - 1 = \pm 2i\sqrt{2} \quad \sqrt{8} = \sqrt{4\cdot 2} = 2\sqrt{2}$$
$$x = 1 \pm 2i\sqrt{2} \quad \text{Add 1 to both sides.}$$

Figure 69 Verify the work

To verify our work, we store each result as x and check that $5x^2 - 10x + 45$ is equal to 0 (see Fig. 69). ■

group exploration

Looking ahead: Deriving a formula for solving quadratic equations of the form $ax^2 + bx + c = 0$

1. Solve the equation $2x^2 + 9x + 3 = 0$.
2. Solve the quadratic equation $ax^2 + bx + c = 0$ for x. [**Hint:** Follow the same steps as in Problem 1.]
3. Use your result from Problem 2 to solve the equation $3x^2 + 11x + 5 = 0$. [**Hint:** Substitute the appropriate values of a, b, and c into your formula from Problem 2.]

TIPS FOR SUCCESS: Solve Problems

Although you can do many things to enhance your learning, there is no substitute for solving problems. Your mathematical ability will respond to solving problems in much the same way that your muscles respond to lifting weights. Muscles greatly increase in strength when you work out intensely, frequently, and consistently.

To learn math, you must be an *active* participant. No amount of watching weight lifters lift, reading about weight-lifting techniques, or conditioning yourself psychologically can replace working out by lifting weights. Similarly, no amount of watching your instructor do problems, reading your text, or listening to a tutor can replace "working out" by solving problems.

HOMEWORK 7.4

FOR EXTRA HELP ▶ Student Solutions Manual · PH Math/Tutor Center · MathXL® · MyMathLab

Find the value of c for which the expression is a perfect-square trinomial. Then factor the perfect-square trinomial.

1. $x^2 + 12x + c$ **2.** $x^2 + 20x + c$
3. $x^2 - 14x + c$ **4.** $x^2 - 18x + c$
5. $x^2 + 7x + c$ **6.** $x^2 + 11x + c$
7. $x^2 - 3x + c$ **8.** $x^2 - 5x + c$
9. $x^2 + \frac{1}{2}x + c$ **10.** $x^2 + \frac{1}{7} + c$
11. $x^2 - \frac{4}{5} + c$ **12.** $x^2 - \frac{3}{4}x + c$

Solve by completing the square. Any solution is a real number.

13. $x^2 + 6x = 1$ **14.** $x^2 + 8x = 3$
15. $p^2 - 2p = 19$ **16.** $r^2 - 10r = 2$
17. $x^2 + 4x - 24 = 0$ **18.** $x^2 + 6x - 9 = 0$
19. $x^2 - 7x = 3$ **20.** $x^2 - 3x = 12$
21. $t^2 + 5t - 4 = 0$ **22.** $w^2 + 9w - 9 = 0$
23. $x^2 - \frac{5}{2}x = \frac{1}{2}$ **24.** $x^2 - \frac{4}{3}x = \frac{5}{3}$
25. $2x^2 + 8x = 3$ **26.** $3x^2 + 12x = 1$
27. $2r^2 - r - 7 = 0$ **28.** $3m^2 - m - 5 = 0$
29. $3x^2 + 4x - 5 = 0$ **30.** $5x^2 + 2x - 2 = 0$
31. $6x^2 - 8x = -1$ **32.** $4x^2 - 6x = 3$
33. $8w^2 + 4w - 3 = 0$ **34.** $6p^2 + 9p + 2 = 0$

Find all complex-number solutions by completing the square.

35. $x^2 + 2x = -7$ **36.** $x^2 + 10x = -28$
37. $x^2 - 6x + 17 = 0$ **38.** $x^2 - 8x + 36 = 0$
39. $k^2 + 3k + 4 = 0$ **40.** $m^2 + 5m + 7 = 0$
41. $x^2 + \frac{2}{3}x + \frac{7}{3} = 0$ **42.** $x^2 + \frac{5}{2}x + \frac{7}{2} = 0$
43. $4r^2 - 3r = -5$ **44.** $3t^2 - 2t = -6$
45. $4p^2 + 6p + 3 = 0$ **46.** $6w^2 + 3w + 3 = 0$

Find all x-intercepts.

47. $f(x) = x^2 - 8x + 3$ **48.** $g(x) = x^2 - 12x + 1$
49. $h(x) = 2x^2 - 5x - 4$ **50.** $f(x) = 3x^2 + 2x - 7$
51. $g(x) = x^2 + 10x + 25$ **52.** $h(x) = x^2 - 8x + 16$

53. A student tries to solve the equation $4x^2 + 6x = 1$ by completing the square:

$$
\begin{aligned}
4x^2 + 6x &= 1 \\
4x^2 + 6x + 9 &= 1 + 9 \\
(2x + 3)^2 &= 10 \\
2x + 3 &= \pm\sqrt{10} \\
2x &= -3 \pm \sqrt{10} \\
x &= \frac{-3 \pm \sqrt{10}}{2}
\end{aligned}
$$

Describe any errors. Then solve the equation correctly.

54. A student tries to solve the equation $x^2 + 8x - 3 = 0$:

$$x^2 - 8x - 3 = 0$$
$$x^2 - 8x = 3$$
$$x^2 - 8x + 16 = 3$$
$$(x-4)^2 = 3$$
$$x - 4 = \pm\sqrt{3}$$
$$x = 4 \pm \sqrt{3}$$

Describe any errors. Then solve the equation correctly.

55. Let $f(x) = x^2 + 6x + 13$.
 a. Find x when $f(x) = 3$.
 b. Find x when $f(x) = 4$.
 c. Find x when $f(x) = 6$.

56. Let $f(x) = 2x^2 - 8x + 9$.
 a. Find x when $f(x) = 2$.
 b. Find x when $f(x) = 1$.
 c. Find x when $f(x) = 0$.

For Exercises 57–62, solve the given equation or system by referring to the solutions of the functions shown in Table 32.

Table 32 Some Solutions of Three Functions (Exercises 57-62)

x	0	1	2	3	4	5	6
$y = x^2 - 5x + 4$	4	0	−2	−2	0	4	10
$y = -\frac{1}{2}x^2 + x - \frac{1}{2}$	−0.5	0	−0.5	−2	−4.5	−8	−12.5
$y = -\frac{1}{4}x^2 - \frac{1}{4}x - \frac{1}{2}$	−0.5	−1	−2	−3.5	−5.5	−8	−11

57. $x^2 - 5x + 4 = -\frac{1}{2}x^2 + x - \frac{1}{2}$

58. $-\frac{1}{2}x^2 + x - \frac{1}{2} = -\frac{1}{4}x^2 - \frac{1}{4}x - \frac{1}{2}$

59. $x^2 - 5x + 4 = -2$

60. $-\frac{1}{2}x^2 + x - \frac{1}{2} = -\frac{1}{2}$

61. $y = -\frac{1}{2}x^2 + x - \frac{1}{2}$
$y = -\frac{1}{4}x^2 - \frac{1}{4}x - \frac{1}{2}$

62. $y = x^2 - 5x + 4$
$y - \frac{1}{2}x^2 + x - \frac{1}{2}$

63. Find nonzero values of a, b, and c such that the equation $ax^2 + bx + c = 0$ has two imaginary-number solutions. Your equation should be different from those in the text. [**Hint:** Begin with an appropriate equation of the form $(x-h)^2 = k$, and simplify the left side of the equation.]

64. Give an example of an equation that can be solved by factoring. Solve the equation by factoring; then solve it by completing the square. Which process was easier? Explain.

65. Compare the methods of solving a quadratic equation by factoring, by using the square root property, and by completing the square. Describe the methods, as well as their advantages and disadvantages.

66. Explain how to solve a quadratic equation by completing the square.

Related Review

67. Factor $w^2 - 10w + 25$.

68. Factor $t^2 + 14t + 49$.

69. Factor $x^2 + \frac{5}{3}x + \frac{25}{36}$.

70. Factor $x^2 - \frac{3}{2}x + \frac{9}{16}$.

Expressions, Equations, Functions, and Graphs

Perform the indicated instruction. Then use words such as linear, quadratic, cubic, exponential, logarithmic, polynomial, degree, function, one variable, *and* two variables *to describe the expression, equation, or system.*

71. Graph $f(x) = 2(3)^x$ by hand.

72. Solve $12x^3 - 27x = 0$.

73. Let $f(x) = 2(3)^x$. Find x when $f(x) = 65$. Round the result to the second decimal place.

74. Factor $12x^3 - 27x$.

75. Solve $\log_3(3x + 2) = 4$. Round the result to the second decimal place.

76. Find the product $(4x - 2)(3x^2 - 2x + 3)$.

7.5 USING THE QUADRATIC FORMULA TO SOLVE QUADRATIC EQUATIONS

Objectives

- Solve quadratic equations by using the *quadratic formula*.
- Use the *discriminant* to determine the number and type of solutions of a quadratic equation.
- Decide which method to use to solve a quadratic equation.
- Use the quadratic formula to make predictions with a quadratic model.

Any quadratic equation can be solved by completing the square. However, this method may be difficult to use on most quadratic equations. An easier option is to use an important equation called the *quadratic formula,* which can also be used to solve *any* quadratic equation.

The Quadratic Formula

To find the quadratic formula, we solve the general quadratic equation

$$ax^2 + bx + c = 0$$

by completing the square. For now, we assume that a is positive:

$ax^2 + bx + c = 0$	General quadratic equation
$x^2 + \frac{b}{a}x + \frac{c}{a} = 0$	Divide both sides by a.
$x^2 + \frac{b}{a}x = -\frac{c}{a}$	Subtract $\frac{c}{a}$ from both sides.
$x^2 + \frac{b}{a}x + \frac{b^2}{4a^2} = -\frac{c}{a} + \frac{b^2}{4a^2}$	Add $\left(\frac{b}{a}\cdot\frac{1}{2}\right)^2 = \left(\frac{b}{2a}\right)^2 = \frac{b^2}{4a^2}$ to both sides.
$\left(x + \frac{b}{2a}\right)^2 = -\frac{c}{a}\cdot\frac{4a}{4a} + \frac{b^2}{4a^2}$	Factor left side; find LCD.
$\left(x + \frac{b}{2a}\right)^2 = \frac{b^2 - 4ac}{4a^2}$	Add numerators and keep common denominator: $\frac{A}{B} + \frac{C}{B} = \frac{A+C}{B}$
$x + \frac{b}{2a} = \pm\sqrt{\frac{b^2 - 4ac}{4a^2}}$	Square root property
$x + \frac{b}{2a} = \pm\frac{\sqrt{b^2 - 4ac}}{2a}$	Quotient property: $\sqrt{\frac{A}{B}} = \frac{\sqrt{A}}{\sqrt{B}}$
$x = -\frac{b}{2a} \pm \frac{\sqrt{b^2 - 4ac}}{2a}$	Subtract $\frac{b}{2a}$ from both sides.
$x = \frac{-b \pm \sqrt{b^2 - 4ac}}{2a}$	Add/subtract numerators and keep common denominator.

We have found a formula (the last line) for the solutions of a quadratic equation $ax^2 + bx + c = 0$, where a is positive. In a similar way, we could derive the same formula for a quadratic equation where a is negative.

Quadratic Formula

The solutions of a quadratic equation $ax^2 + bx + c = 0$ are given by the **quadratic formula:**

$$x = \frac{-b \pm \sqrt{b^2 - 4ac}}{2a}$$

WARNING For the fraction in the quadratic formula, notice that the term $-b$ is part of the numerator:

$$\frac{-b \pm \sqrt{b^2 - 4ac}}{2a} \leftarrow \text{Correct}$$

$$-b \pm \frac{\sqrt{b^2 - 4ac}}{2a} \leftarrow \text{Incorrect}$$

Example 1 Solving by Using the Quadratic Formula

Solve $x^2 - 6x + 8 = 0$.

Solution

Comparing $x^2 - 6x + 8 = 0$ with $ax^2 + bx + c = 0$, we see that $a = 1$, $b = -6$, and $c = 8$. We substitute these values for a, b, and c in the quadratic formula:

$$x = \frac{-(-6) \pm \sqrt{(-6)^2 - 4(1)(8)}}{2(1)} \quad \text{Substitute 1 for } a, -6 \text{ for } b, \text{ and 8 for } c.$$

$$x = \frac{6 \pm \sqrt{4}}{2} \quad \text{Simplify.}$$

$$x = \frac{6 \pm 2}{2} \quad \sqrt{4} = 2$$

$$x = \frac{6 - 2}{2} \quad \text{or} \quad x = \frac{6 + 2}{2} \quad \text{Write as two equations.}$$

$$x = 2 \quad \text{or} \quad x = 4 \quad \text{Simplify./Simplify.}$$

The solutions are 2 and 4. ■

Instead of using the quadratic formula, we could have solved $x^2 - 6x + 8 = 0$ by factoring:

$$x^2 - 6x + 8 = 0 \quad \text{Original equation}$$

$$(x - 2)(x - 4) = 0 \quad \text{Factor left side.}$$

$$x - 2 = 0 \quad \text{or} \quad x - 4 = 0 \quad \text{Zero factor property}$$

$$x = 2 \quad \text{or} \quad x = 4$$

A benefit of the quadratic formula is that we can use it to solve equations that are difficult or even impossible to solve by factoring. In Example 2, we solve an equation that would be impossible to solve by factoring.

Example 2 Solving by Using the Quadratic Formula

Solve $2x^2 = 10x - 3$.

Solution

First, we write $2x^2 = 10x - 3$ in the form $ax^2 + bx + c = 0$:

$$2x^2 - 10x + 3 = 0$$

So, $a = 2$, $b = -10$, and $c = 3$. By the quadratic formula,

$$x = \frac{-(-10) \pm \sqrt{(-10)^2 - 4(2)(3)}}{2(2)} \quad \text{Substitute 2 for } a, -10 \text{ for } b, \text{ and 3 for } c.$$

$$x = \frac{10 \pm \sqrt{76}}{4} \quad \text{Simplify.}$$

$$x = \frac{10 \pm 2\sqrt{19}}{4} \quad \sqrt{76} = \sqrt{4 \cdot 19} = \sqrt{4}\sqrt{19} = 2\sqrt{19}$$

$$x = \frac{10 - 2\sqrt{19}}{4} \quad \text{or} \quad x = \frac{10 + 2\sqrt{19}}{4} \quad \text{Write as two equations.}$$

$$x = \frac{2(5 - \sqrt{19})}{4} \quad \text{or} \quad x = \frac{2(5 + \sqrt{19})}{4} \quad \text{Factor out 2.}$$

$$x = \frac{5 - \sqrt{19}}{2} \quad \text{or} \quad x = \frac{5 + \sqrt{19}}{2} \quad \text{Simplify.}$$

The solutions are $\dfrac{5 \pm \sqrt{19}}{2}$. ■

Example 3 Solving by Using the Quadratic Formula

Solve $\frac{1}{2}x^2 - \frac{5}{4}x = \frac{3}{2}$.

Solution

To begin, we clear the equation of fractions by multiplying both sides by the LCD, 4:

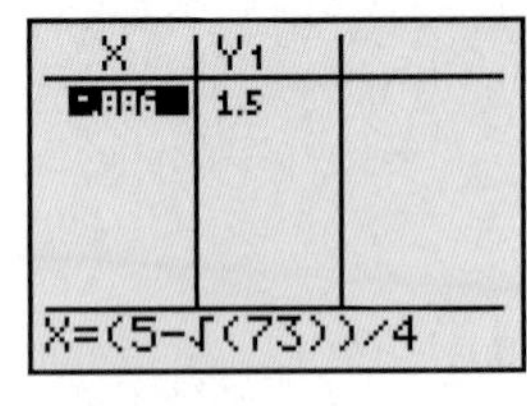

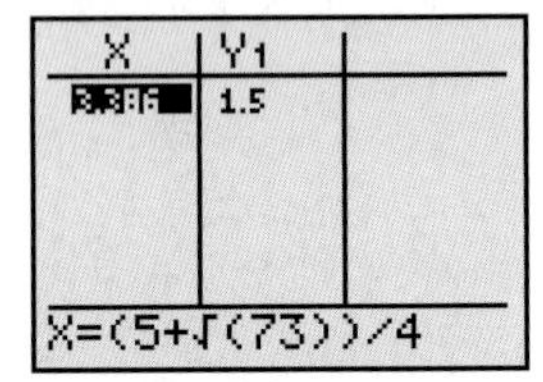

Figure 70 Verify the work

$\frac{1}{2}x^2 - \frac{5}{4}x = \frac{3}{2}$	Original equation
$4 \cdot \frac{1}{2}x^2 - 4 \cdot \frac{5}{4}x = 4 \cdot \frac{3}{2}$	Multiply both sides by LCD, 4.
$2x^2 - 5x = 6$	Simplify.
$2x^2 - 5x - 6 = 0$	Subtract 6 from both sides.
$x = \frac{-(-5) \pm \sqrt{(-5)^2 - 4(2)(-6)}}{2(2)}$	Substitute $a = 2$, $b = -5$, and $c = -6$ in quadratic formula.
$x = \frac{5 \pm \sqrt{73}}{4}$	Simplify.

The solutions are $\frac{5 \pm \sqrt{73}}{4}$. We enter $y = \frac{1}{2}x^2 - \frac{5}{4}x$ in a graphing calculator and check that for both inputs $\frac{5 \pm \sqrt{73}}{4}$, the output is $\frac{3}{2} = 1.5$ (see Fig. 70). ■

Example 4 Finding Approximate Solutions

Find approximate solutions of the equation $2.3x(1.4x - 5.3) = 6.9x - 7.2$.

Solution

First, we write the equation in the form $ax^2 + bx + c = 0$:

Plot1 Plot2 Plot3
\Y1■2.3X(1.4X-5.3)
\Y2■6.9X-7.2
\Y3=
\Y4=
\Y5=
\Y6=

$2.3x(1.4x - 5.3) = 6.9x - 7.2$	Original equation
$3.22x^2 - 12.19x = 6.9x - 7.2$	Distributive law
$3.22x^2 - 19.09x + 7.2 = 0$	Subtract 6.9x from both sides; add 7.2 to both sides.

Then we substitute 3.22 for a, -19.09 for b, and 7.2 for c in the quadratic formula:

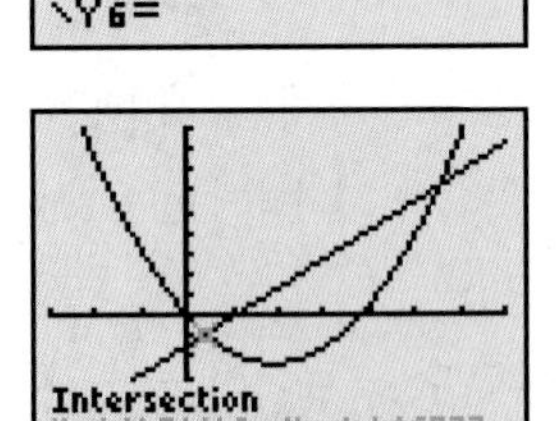

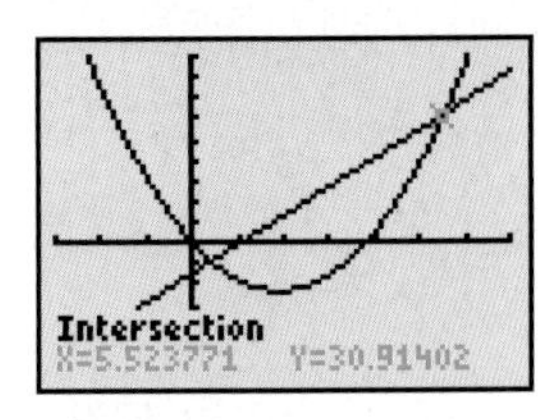

Figure 71 Verify the work

$x = \frac{-(-19.09) \pm \sqrt{(-19.09)^2 - 4(3.22)(7.2)}}{2(3.22)}$	Substitute into quadratic formula.
$x = \frac{19.09 \pm \sqrt{271.6921}}{6.44}$	Simplify.
$x \approx \frac{19.09 \pm 16.48}{6.44}$	Approximate square root.
$x \approx 0.41$ or $x \approx 5.52$	Compute.

We use graphing calculator graphs to verify our work (see Fig. 71). ■

Solving a Quadratic Equation That Has Imaginary-Number Solutions

In Example 5, we use the quadratic formula to solve a quadratic equation that has imaginary-number solutions.

Example 5 Solving by Using the Quadratic Formula

Solve $-3x^2 + 5x - 4 = 0$.

Solution

First, we multiply both sides of the equation by -1 so that we can avoid having a negative denominator after we use the quadratic formula:

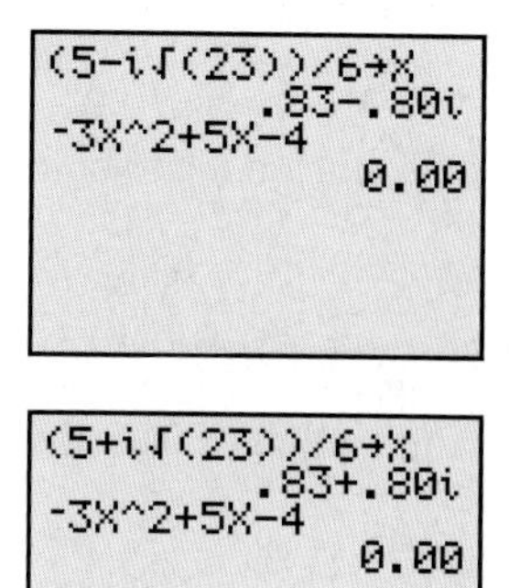

Figure 72 Verify the work

$$-3x^2 + 5x - 4 = 0 \quad \text{Original equation}$$

$$3x^2 - 5x + 4 = 0 \quad \text{Multiply both sides by } -1.$$

(Another benefit is that we have fewer negative numbers to substitute into the quadratic formula.) Then we substitute $a = 3$, $b = -5$, and $c = 4$ in the quadratic formula:

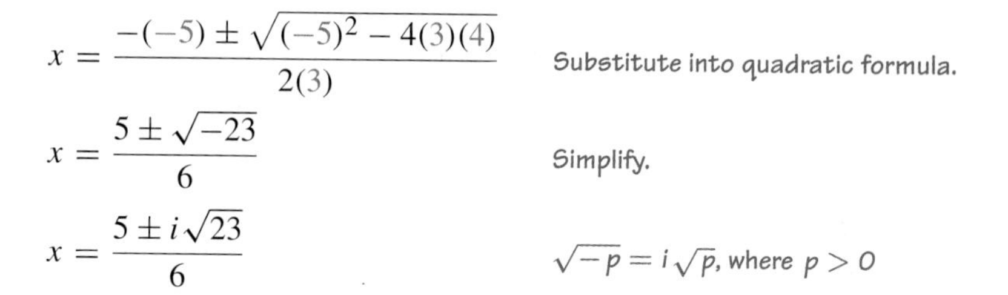

$$x = \frac{-(-5) \pm \sqrt{(-5)^2 - 4(3)(4)}}{2(3)} \quad \text{Substitute into quadratic formula.}$$

$$x = \frac{5 \pm \sqrt{-23}}{6} \quad \text{Simplify.}$$

$$x = \frac{5 \pm i\sqrt{23}}{6} \quad \sqrt{-p} = i\sqrt{p}, \text{ where } p > 0$$

To verify our work, we store each result as x and check that $-3x^2 + 5x - 4$ is equal to 0 (see Fig. 72). We press MODE and set FLOAT to 2 so that the numbers in the result are rounded to the second decimal place.

The graphical check in Fig. 73 shows that the function $y = -3x^2 + 5x - 4$ does not have any x-intercepts. This means that the equation $-3x^2 + 5x - 4 = 0$ has no real-number solutions, which checks. ■

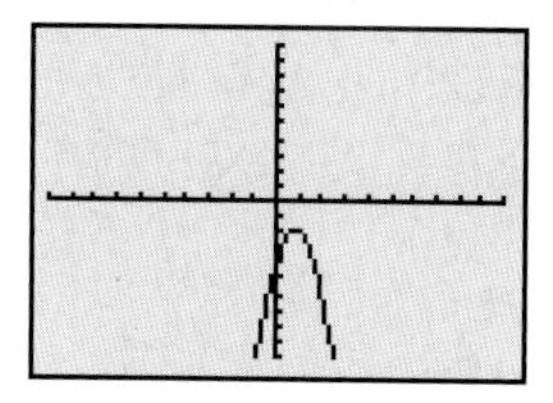
Figure 73 The function $y = -3x^2 + 5x - 4$ has no x-intercepts

Determining the Number of Real-Number Solutions

Recall from Section 6.6 that a quadratic equation can have two, one, or no real-number solutions. How does this fact relate to the quadratic formula?

$$x = \frac{-b \pm \sqrt{b^2 - 4ac}}{2a}$$

The answer lies with the number $b^2 - 4ac$, known as the *discriminant*. If the discriminant is positive, then there are two real-number solutions; see Example 4. If the discriminant is negative, then there are two imaginary-number solutions (and no real-number solutions); see Example 5. Finally, if the discriminant is 0, then the quadratic formula gives

$$x = \frac{-b \pm \sqrt{0}}{2a} = \frac{-b}{2a}$$

and, therefore, there is one real-number solution.

Determining the Number and Type of Solutions

For the quadratic equation $ax^2 + bx + c = 0$, the **discriminant** is $b^2 - 4ac$. Also,

- If $b^2 - 4ac > 0$, there are two real-number solutions.
- If $b^2 - 4ac = 0$, there is one real-number solution.
- If $b^2 - 4ac < 0$, there are two imaginary-number solutions (and no real-number solutions).

Example 6 Determining the Number of Real-Number Solutions

Determine the number and type of solutions of the equation $2x^2 - 3x + 5 = 0$.

Solution

Since $b^2 - 4ac = (-3)^2 - 4(2)(5) = -31 < 0$, we conclude that the quadratic equation $2x^2 - 3x + 5 = 0$ has two imaginary-number solutions (and no real-number solutions). ■

We can also use the discriminant to determine the number of points on a parabola at a given height.

Example 7 Finding the Number of Points at a Given Height

For $f(x) = x^2 - 6x + 12$, find the number of points that lie on the graph of f at the indicated height.

1. $y = 5$ **2.** $y = 3$ **3.** $y = 1$

Solution

1. We substitute 5 for $f(x)$ in the equation $f(x) = x^2 - 6x + 12$:

$$x^2 - 6x + 12 = 5 \quad \text{Substitute 5 for } f(x).$$
$$x^2 - 6x + 7 = 0 \quad \text{Subtract 5 from both sides.}$$

Since $b^2 - 4ac = (-6)^2 - 4(1)(7) = 8 > 0$, we conclude that there are two solutions of the equation $x^2 - 6x + 12 = 5$, which means that two (symmetric) points have height $y = 5$.

2. We substitute 3 for $f(x)$:

$$x^2 - 6x + 12 = 3 \quad \text{Substitute 3 for } f(x).$$
$$x^2 - 6x + 9 = 0 \quad \text{Subtract 3 from both sides.}$$

Since $b^2 - 4ac = (-6)^2 - 4(1)(9) = 0$, we conclude that there is one solution of the equation $x^2 - 6x + 12 = 3$, which means that one point has height $y = 3$. Since the point does not have a symmetric point, it must be the vertex of the parabola.

3. We substitute 1 for $f(x)$:

$$x^2 - 6x + 12 = 1 \quad \text{Substitute 1 for } f(x).$$
$$x^2 - 6x + 11 = 0 \quad \text{Subtract 1 from both sides.}$$

Since $b^2 - 4ac = (-6)^2 - 4(1)(11) = -8 < 0$, we conclude that there are no real-number solutions of the equation $x^2 - 6x + 12 = 1$, which means that no points on the parabola have height $y = 1$. ■

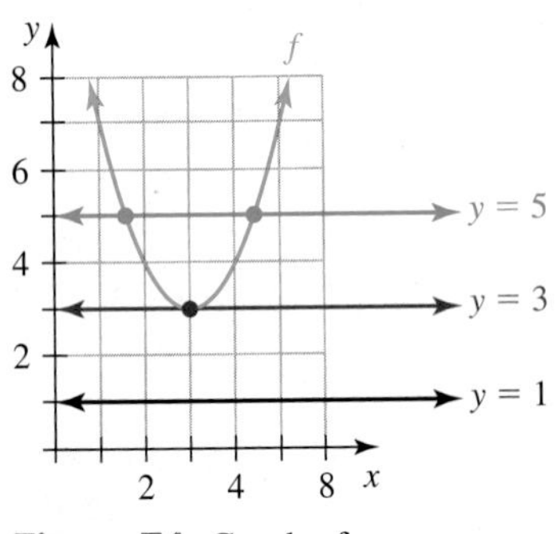

Figure 74 Graph of $f(x) = x^2 - 6x + 12$

In Example 7, we found that the parabola $f(x) = x^2 - 6x + 12$ has exactly two points at height $y = 5$, exactly one point at height $y = 3$, and no points at height $y = 1$. We indicate the three points on the graph of f in Fig. 74.

Deciding Which Method to Use to Solve a Quadratic Equation

In Chapter 6 and in this chapter, we have discussed four ways to solve quadratic equations: factoring, the square root property, completing the square, and the quadratic formula. How do we know which method to use?

Remember that **any quadratic equation can be solved by using the quadratic formula.** Although any equation can also be solved by completing the square, that method is much more difficult to use than the quadratic formula.

A low percentage of quadratic equations can be solved by factoring, because almost all polynomials are prime. However, if an equation *can* be solved by simple factoring techniques, then it is easier to solve by factoring than by any of the other three methods.

Here are some guidelines on deciding which method to use to solve a quadratic equation:

Method:	When to Use:
Factoring	For equations that can easily be put into the form $ax^2 + bx + c = 0$ and where $ax^2 + bx + c$ can easily be factored.
Square root property	For equations that can easily be put into the form $x^2 = k$ or $(x + p)^2 = k$.
Completing the square	When the directions require it.
Quadratic formula	For all equations except those that can easily be solved by factoring or by using the square root property.

Example 8 Deciding Which Method to Use

Solve.

1. $w^2 - 3w - 54 = 0$ **2.** $(x + 3)^2 = 5$

3. $(x + 3)(x - 2) = 2(x + 1)$

Solution

1. The polynomial $w^2 - 3w - 54$ is factorable, so we solve the quadratic equation $w^2 - 3w - 54 = 0$ by factoring:

$$w^2 - 3w - 54 = 0 \quad \text{Original equation}$$
$$(w + 6)(w - 9) = 0 \quad \text{Factor left side.}$$
$$w + 6 = 0 \quad \text{or} \quad w - 9 = 0 \quad \text{Zero factor property}$$
$$w = -6 \quad \text{or} \quad w = 9$$

2. The equation is of the form $(x + p)^2 = k$, so we solve it by using the square root property:

$$(x + 3)^2 = 5 \quad \text{Original equation}$$
$$x + 3 = \pm\sqrt{5} \quad \text{Square root property}$$
$$x = -3 \pm \sqrt{5}$$

3. First, we write the equation in $ax^2 + bx + c = 0$ form:

$$(x + 3)(x - 2) = 2(x + 1) \quad \text{Original equation}$$
$$x^2 + x - 6 = 2x + 2 \quad \text{Multiply; distributive law}$$
$$x^2 - x - 8 = 0 \quad \text{Write in } ax^2 + bx + c = 0 \text{ form.}$$

The polynomial $x^2 - x - 8$ is prime, so we can't solve the equation $x^2 - x - 8 = 0$ by factoring. The equation can't be put into the form $x^2 = k$, and it can't easily be put into the form $(x + p)^2 = k$, so we don't try to solve it by using the square root property. Instead, we substitute $a = 1, b = -1$, and $c = -8$ in the quadratic formula:

$$x = \frac{-(-1) \pm \sqrt{(-1)^2 - 4(1)(-8)}}{2(1)} \quad \text{Substitute into quadratic formula.}$$

$$x = \frac{1 \pm \sqrt{33}}{2} \quad \text{Simplify.}$$

■

Using a Quadratic Function to Model a Situation

We can use the quadratic formula to make a prediction about the independent variable of a quadratic model.

Table 33 Bottled-Water Consumption

Year	Bottled-Water Consumption (billions of gallons)
1990	2.2
1992	2.4
1994	3.0
1996	3.5
1998	4.1
2000	4.9
2002	6.0
2003	6.4

Source: *Beverage Marketing Corporation*

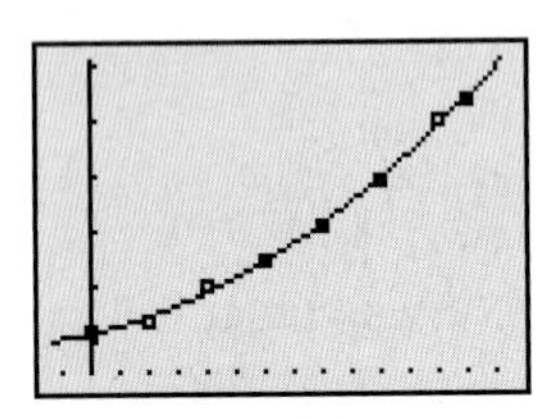

Figure 75 Check the fit

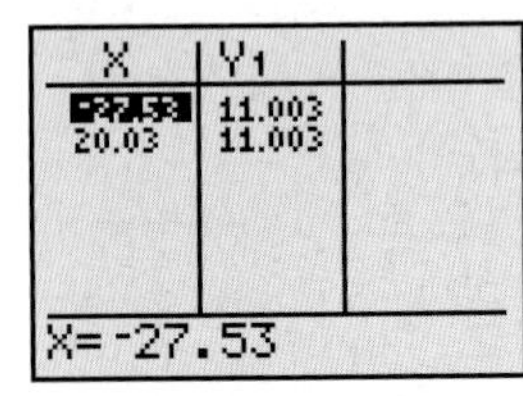

Figure 76 Verify the results

Example 9 Modeling with a Quadratic Function

Annual consumption of bottled water in the United States has been increasing since 1990 (see Table 33). Let $f(t)$ be bottled-water consumption (in billions of gallons) in the year that is t years since 1990. A possible equation of f is

$$f(t) = 0.016t^2 + 0.12t + 2.18$$

1. Verify that f models the data well.
2. Predict when annual consumption will reach 11 billion gallons.

Solution

1. We draw the graph of f and the scattergram of the data in the same viewing window (see Fig. 75). It appears that f is a reasonable model.
2. To predict when annual consumption will be 11 billion gallons, we substitute 11 for $f(t)$ and solve for t:

$$11 = 0.016t^2 + 0.12t + 2.18 \qquad \text{Substitute 11 for } f(t).$$

$$0 = 0.016t^2 + 0.12t - 8.82 \qquad \text{Subtract 11 from both sides.}$$

$$t = \frac{-0.12 \pm \sqrt{0.12^2 - 4(0.016)(-8.82)}}{2(0.016)} \qquad \text{Substitute into quadratic formula.}$$

$$t = \frac{-0.12 \pm \sqrt{0.57888}}{0.032} \qquad \text{Simplify.}$$

$$t \approx \frac{-0.12 \pm 0.7608}{0.032} \qquad \text{Approximate square root.}$$

$$t \approx -27.53 \quad \text{or} \quad t \approx 20.03 \qquad \text{Compute.}$$

We verify the results by entering $y = 0.016t^2 + 0.12t + 2.18$ in a graphing calculator and checking that the inputs -27.53 and 20.03 lead to outputs of about 11 (see Fig. 76).

The inputs -27.53 and 20.03 represent the years 1962 and 2010. The estimate of 1962 is a result of model breakdown, as a little research would show that bottled-water consumption in 1962 was much less than 11 billion gallons. Therefore, we predict that it will be 2010 when bottled-water consumption reaches 11 billion gallons. ■

Recall from Section 6.6 that to make a prediction for the dependent variable of a quadratic model, we substitute a value for the independent variable, then solve for the dependent variable. To make a prediction for the independent variable, we substitute a value for the dependent variable, then solve for the independent variable, usually by using the quadratic formula.

group exploration

Comparing methods of solving quadratic equations

1. Solve the equation $x^2 + 5x + 6 = 0$ by factoring.
2. Solve the equation $x^2 + 5x + 6 = 0$ by completing the square.
3. Solve the equation $x^2 + 5x + 6 = 0$ by using the quadratic formula.
4. Compare your results from Problems 1, 2, and 3. Which method was easiest?
5. Repeat Problems 1–4 for the equation $x^2 + 4x - 7 = 0$.
6. Compare the methods of solving quadratic equations by factoring, by completing the square, and by using the quadratic formula. What are the advantages and disadvantages of each?

group exploration

Looking ahead: Finding an equation of a parabola

In this exploration, you will find an equation of the parabola that passes through the points (1, 8), (2, 15), and (3, 24).

1. Since the point (1, 1) lies on the parabola, the ordered pair (1, 1) should satisfy the equation $y = ax^2 + bx + c$. Find the equation that results from substituting 1 for x and 1 for y. This equation will be in terms of a, b, and c. Find another equation by using the ordered pair (2, 3). Finally, find a third equation by using the ordered pair (3, 9).
2. You should now have three equations, each in terms of a, b, and c. Choose any two of these equations and eliminate c. Then choose another pair of equations and again eliminate c.
3. You should now have two equations, both in terms of a and b, forming a system of two equations in two variables. Solve this system by substitution or elimination.
4. You should now know the values of a and b. Find c by substituting the values of a and b in one of the three equations found in Problem 1.
5. You should now know the values of a, b, and c. Substitute these values into the equation $y = ax^2 + bx + c$ to obtain an equation of the parabola.
6. Verify that the graph of your equation passes through the points (1, 8), (2, 15), and (3, 24) by using a graphing calculator table or graph.

HOMEWORK 7.5

FOR EXTRA HELP ▶ Student Solutions Manual | PH Math/Tutor Center | MathXL® | MyMathLab

Use the quadratic formula to solve the equation. Any solution is a real number.

1. $2x^2 + 5x - 2 = 0$
2. $5x^2 + 5x - 1 = 0$
3. $3x^2 - 6x + 1 = 0$
4. $5x^2 - 4x - 3 = 0$
5. $t^2 = 4t + 3$
6. $w^2 = -2w + 5$
7. $-2x^2 + 5x = 3$
8. $-4x^2 + 7x = -2$
9. $3x^2 - 17 = 0$
10. $2y^2 - 15 = 0$
11. $2y^2 = -5y$
12. $5r^2 = -3r$
13. $\frac{2}{3}x^2 - \frac{5}{6}x = \frac{1}{3}$
14. $\frac{5}{8}x^2 + \frac{3}{4}x = \frac{1}{4}$
15. $(3x + 2)(x - 1) = 1$
16. $(4x - 3)(x - 1) = 4$

Find approximate solutions. Round any results to the second decimal place. Any solution is a real number.

17. $2x^2 = 5x + 4$
18. $3x^2 = 9x + 2$
19. $2.85p^2 - 7.12p = 4.49$
20. $3.98r^2 - 2.17r = 3.68$
21. $-5.4x(x + 9.8) + 4.1 = 3.2 - 6.9x$
22. $7.1x(x - 4.9) - 7.1 = 2.5x + 6.3$

Use the quadratic formula to find all complex-number solutions.

23. $x^2 - 3x + 8 = 0$
24. $x^2 - 3x + 15 = 0$
25. $-w^2 + 2w = 5$
26. $-r^2 + 4r = 7$
27. $\frac{1}{4}x^2 = 2x - \frac{9}{2}$
28. $\frac{5}{6}x^2 = 3x - \frac{7}{2}$
29. $3x(3x - 2) = -2$
30. $2x(5x - 2) = -1$
31. $3k^2 = 4k - 5$
32. $4y^2 = 2y - 1$

Solve by the method of your choice. Any solution is a real number.

33. $4x^2 - 80 = 0$
34. $3x^2 - 36 = 0$
35. $5(w + 3)^2 + 2 = 8$
36. $3(w - 2)^2 - 1 = 6$
37. $m^2 = -12m - 36$
38. $t^2 = 14t - 49$
39. $-24x^2 + 18x = -60$
40. $-16x^2 + 20x = 4$
41. $\frac{1}{3}x^2 - \frac{3}{2}x = \frac{1}{6}$
42. $\frac{1}{8}x^2 - \frac{3}{4}x = \frac{1}{2}$
43. $(x - 5)(x + 2) = 3(x - 1) + 2$
44. $(x + 4)(x - 1) = 5(x + 2) - 1$
45. $25r^2 = 49$
46. $4p^2 = 81$
47. $(x - 1)^2 + (x + 2)^2 = 6$
48. $(x - 3)^2 + (x + 1)^2 = 17$

Use the method of your choice to find all complex-number solutions.

49. $4x^2 = -25$
50. $81x^2 = -49$
51. $-2t^2 + 5t = 6$
52. $-3w^2 + 4w = 2$
53. $(x - 6)^2 + 5 = -43$
54. $(x + 4)^2 - 3 = -66$
55. $(y - 2)(y - 5) = -4$
56. $(k + 3)(k - 2) = -25$

Determine the number and type of solutions.

57. $3x^2 + 4x - 5 = 0$

58. $x^2 - 5x - 8 = 0$

59. $2x^2 - 5x + 7 = 0$

60. $3x^2 - 2x + 5 = 0$

61. $4x^2 = 12x - 9$

62. $9x^2 = 6x - 1$

63. Let $f(x) = x^2 - 4x + 8$. Find the number of points that lie on the graph of f at the indicated height y.

a. $y = 3$ **b.** $y = 4$ **c.** $y = 5$

d. Use a graphing calculator to draw a graph of f, and sketch the graph on paper. Then explain why you found the number of points to be 0, 1, and 2 for parts (a), (b), and (c), respectively.

64. Let $g(x) = -x^2 + 6x - 2$. Find the number of points that lie on the graph of g at the indicated height y.

a. $y = 6$ **b.** $y = 7$ **c.** $y = 8$

d. Use a graphing calculator to draw a graph of g, and sketch the graph on paper. Then explain why you found the number of points to be 2, 1, and 0 for parts (a), (b), and (c), respectively.

65. Let $f(x) = x^2 - 6x + 7$. Find the coordinates of any points on the graph of f at height $y = 2$. Then find the vertex of the graph of f. Finally, sketch a graph of f.

66. Let $g(x) = x^2 + 8x + 6$. Find the approximate coordinates of any points on the graph of g at height $y = -5$. Then find the approximate vertex of the graph of g. Round all coordinates to the second decimal place. Finally, sketch a graph of g.

67. The capacities at which U.S. nuclear power plants are working are shown in Table 34 for various years.

Table 34 Capacities at Which U.S. Nuclear Power Plants Are Working

Year	Percent
1975	56
1980	59
1985	58
1990	70
1995	76
2000	88
2004	89

Source: *Energy Information Administration*

Let $f(t)$ be the capacity (in percent) at which U.S. nuclear power plants are working at t years since 1970. A model of the situation is $f(t) = 0.027t^2 + 0.22t + 53.3$.

a. Use a graphing calculator to draw the graph of the model and, in the same viewing window, the scattergram of the data. Does the model fit the data well?

b. Estimate at what capacity U.S. nuclear power plants were working in 2006.

c. Predict when U.S. nuclear power plants will be working at full (100%) capacity.

d. The last U.S. nuclear power plant to be built was operational in 1996. Explain why it is not surprising that in 2004 applications were filed to build three new nuclear power plants in the United States. (It takes four years to build one.)

68. The numbers of new U.S. hotel openings are shown in Table 35 for various years.

Table 35 New Hotel Openings

Year	Number of Openings
1997	1476
1999	1402
2001	1047
2003	569
2004	131

Source: *Lodging Econometrics*

Let $f(t)$ be the number of new U.S. hotel openings in the year that is t years since 1990. A possible equation of f is $f(t) = -29.59t^2 + 432.99t - 105.53$.

a. Draw the graph of f and the scattergram of the data in the same viewing window. Does f model the data reasonably well?

b. Find t when $f(t) = 800$. What does your result mean in this situation?

c. Find the t-intercepts of the graph of f. What do they mean in this situation?

69. The percentages of Americans who feel good about the morals and values of Americans in general are shown in Table 36 for various years.

Table 36 Percentages of Americans Who Feel Good About the Morals and Values of Americans in General

Year	Percent
1999	36
2000	39
2001	40
2002	44
2003	47
2004	55

Source: *Harris Interactive*

Let $f(t)$ be the percentage of Americans who feel good about the morals and values of Americans in general at t years since 1990. A possible quadratic model of the situation is $f(t) = 0.59t^2 - 10.04t + 79.3$.

a. Use a graphing calculator to draw the graph of the model and, in the same viewing window, the scattergram of the data. Does the model fit the data well?

b. Estimate the percentage of Americans who felt good about the morals and values of Americans in general in 2006.

c. Predict when all Americans will feel good about the morals and values of Americans in general.

70. The number of charter schools—independently run public schools that have their own educational philosophy and curricula—increased greatly in the United States in the 1990s (see Table 37). Let $f(t)$ be the number of charter schools in the United States at t years since 1990. A possible equation of f is $f(t) = 24.27t^2 - 52.76t - 40.68$.

a. Draw the graph of f and the scattergram of the data in the same viewing window. Does f model the data reasonably well?

b. Predict the number of charter schools in 2010. Verify your result by using a graphing calculator graph or table.

c. When will there be an average of 200 charter schools per state? Verify your result by using a graphing calculator graph or table.

Table 37 Charter Schools in the United States

Year	Number of Charter Schools
1992	2
1994	98
1996	425
1998	1092
2000	2000
2001	2372
2002	2700

Source: *U.S. Department of Education*

71. A person throws a stone into the air. The height (in feet) $h(t)$ after t seconds is given by $h(t) = -16t^2 + 52t + 4$.
a. What is the height of the stone after 3 seconds?
b. When is the stone at a height of 30 feet?
c. When does the stone reach the ground?

72. A baseball is hit by a batter. The height (in feet) $h(t)$ of the ball after t seconds is given by $h(t) = -16t^2 + 125t + 4$.
a. What is the height of the ball after 2 seconds?
b. When is the ball at a height of 200 feet?
c. When does the ball reach the ground?

For Exercises 73–78, find approximate solutions of the given equation or system by referring to the graphs shown in Fig. 77. Round results or coordinates of results to the first decimal place.

73. $\frac{1}{2}x^2 + 1 = -x^2 - x + 5$

74. $-x^2 - x + 5 = \frac{2}{5}x - 2$

75. $\frac{1}{2}x^2 + 1 = 4$

76. $-x^2 - x + 5 = 1$

77. $y = \frac{2}{5}x - 2$
$y = -x^2 - x + 5$

78. $y = \frac{1}{2}x^2 + 1$
$y = -x^2 - x + 5$

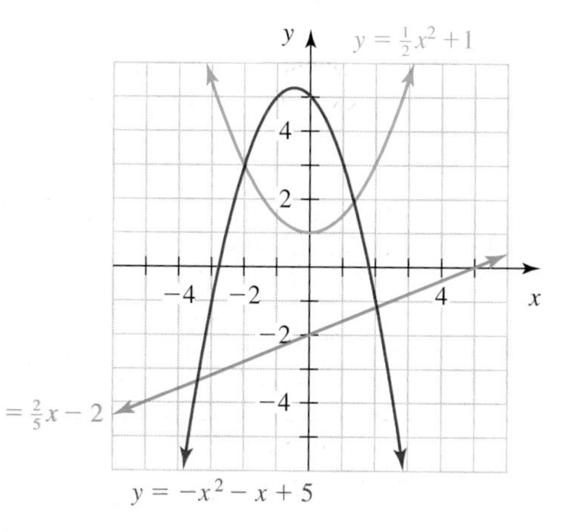

Figure 77 Exercises 73–78

Find all x-intercepts.

79. $f(x) = 2x^2 - x - 7$

80. $f(x) = 3x^2 + 4x - 1$

81. $f(x) = 3x^2 + 2x + 5$

82. $f(x) = -5x^2 + 3x - 2$

83. $f(x) = x^2 + 2x - 5$

84. $f(x) = x^2 - 4x - 3$

85. A student tries to solve $2x^2 + 5x = 1$:

$$x = \frac{-5 \pm \sqrt{5^2 - 4(2)(1)}}{2(2)}$$

$$x = \frac{-5 \pm \sqrt{17}}{4}$$

Describe any errors. Then solve the equation correctly.

86. A student tries to solve $3x^2 + 2x - 4 = 0$:

$$x = \frac{-2 \pm \sqrt{2^2 - 4(3)(-4)}}{2(3)}$$

$$x = \frac{-2 \pm \sqrt{52}}{6}$$

$$x = \frac{-2 \pm 2\sqrt{13}}{6}$$

$$x = \frac{-1 \pm 2\sqrt{13}}{3}$$

Describe any errors. Then solve the equation correctly.

87. The quadratic formula gives the solutions of any equation of the form $ax^2 + bx + c = 0$, where $a \neq 0$.
a. Find a "linear formula" that gives the solution of *any* equation of the form $mx + b = 0$, where $m \neq 0$.
b. Use your linear formula to solve $7x + 21 = 0$. Verify your result by solving $7x + 21 = 0$ in the usual way.

88. a. Use factoring to solve $3x^2 + 2x = 0$.
b. Use factoring to solve $ax^2 + bx = 0$ for x.
c. Use the formula you found in part (b) to solve the equation $3x^2 + 2x = 0$. Compare your results with your results from part (a).
d. Use the quadratic formula to solve $3x^2 + 2x = 0$. Compare your results with your results from part (a).
e. Use the quadratic formula to solve $ax^2 + bx = 0$ for x. Compare your results with your results from part (b).
f. Use the formula you found in part (e) to solve the equation $7x^2 - 3x = 0$.

89. Solve $x^2 - x - 20 = 0$ by factoring, by completing the square, and by using the quadratic formula. Compare your results.

90. Solve $3x^2 = 5x - 2$ by factoring, by completing the square, and by using the quadratic formula. Compare your results.

91. Explain how to determine whether to solve a quadratic equation by factoring, by using the square root property, by completing the square, or by using the quadratic formula. Give examples; for each example, describe the advantages of the method you chose and the disadvantages of the other methods.

92. Describe how to solve a quadratic equation by using the quadratic formula.

Related Review

Perform the indicated operations or solve, as appropriate.

93. $(x + 2)(x - 5)$

94. $2x(3x - 5) = 4$

95. $(x + 2)(x - 5) = 3$

96. $2x(3x - 5)$

97. $-4(x - 2)^2 + 3 = -1$

98. $(x + 3)^2$

99. $-4(x - 2)^2 + 3$

100. $(x + 3)^2 = 7$

Solve. Round approximate solutions to the fourth decimal place.

101. $4b^5 - 12 = 173$

102. $3r^2 = 5r - 2$

103. $4x - (7x - 5) = 3x + 1$

104. $2x^2 - 7x = 3$

105. $7(3)^t + 8 = 271$

106. $\log_3(x + 1) + \log_3(x + 3) = 2$

Expressions, Equations, Functions, and Graphs

Perform the indicated instruction. Then use words such as linear, quadratic, cubic, exponential, logarithmic, polynomial, degree, function, one variable, *and* two variables *to describe the expression, equation, or system.*

107. Factor $8x^2 - 18x + 9$.

108. Simplify $-3(x + 1)^2 + 4$.

109. Find $f(-2)$, where $f(x) = 8x^2 - 18x + 9$.

110. Solve $-3(x + 1)^2 + 4 = -20$.

111. Solve $8x^2 - 18x + 9 = 0$.

112. Graph $f(x) = -3(x + 1)^2 + 4$ by hand.

7.6 SOLVING SYSTEMS OF LINEAR EQUATIONS IN THREE VARIABLES; FINDING QUADRATIC FUNCTIONS

"If I had to select one quality, one personal characteristic that I regard as being most highly correlated with success, whatever the field, I would pick the trait of persistence. Determination. The will to endure to the end, to get knocked down seventy times and get up off the floor saying, 'Here comes number seventy-one!'"

—*Richard M. DeVos*

Objectives

- Know the meaning of *linear equation in three variables.*
- Solve a *system of linear equations in three variables.*
- Find a quadratic equation, in standard form, of a parabola that contains three given points.

In this section, we discuss how to solve a system of three equations. Then we will use this skill to help us find an equation of a parabola that contains three given points. In Section 7.7, we will use the skill to find an equation of a quadratic model.

Linear Equations in Three Variables

In Chapters 1–3, we worked with linear equations in *two* variables. Here, we will work with linear equations in *three* variables.

DEFINITION Linear equation in three variables

A **linear equation in three variables** is an equation that can be put into the form $Ax + By + Cz = D$, where A, B, C, and D are constants and A, B, and C are not all zero.

Here is an example of a linear equation in *three* variables:

$$3x - 5y + 2z = 8$$

An **ordered triple** (x, y, z) represents values of x, y, and z, just as an ordered pair (x, y) represents values of x and y. For example, the ordered triple $(2, -4, 7)$ represents the values $x = 2$, $y = -4$, and $z = 7$. An ordered triple (a, b, c) is a **solution** of an equation in terms of x, y, and z if the equation becomes a true statement when a, b, and c are substituted for x, y, and z, respectively. We say that a solution **satisfies** the equation.

Example 1 Identifying Solutions of a Linear Equation in Three Variables

Decide whether the given ordered triple is a solution of the equation $5x-2y+4z=11$.

1. $(-3, 1, 7)$ **2.** $(2, -9, 6)$

Solution

1. We substitute -3 for x, 1 for y, and 7 for z in $5x-2y+4z=11$:

$$5(-3)-2(1)+4(7) \stackrel{?}{=} 11$$
$$11 \stackrel{?}{=} 11$$
$$\text{true}$$

The ordered triple $(-3, 1, 7)$ is a solution of the equation $5x-2y+4z=11$.

2. We substitute 2 for x, -9 for y, and 6 for z in $5x-2y+4z=11$:

$$5(2)-2(-9)+4(6) \stackrel{?}{=} 11$$
$$52 \stackrel{?}{=} 11$$
$$\text{false}$$

The ordered triple $(2, -9, 6)$ is not a solution of the equation $5x-2y+4z=11$. ■

Solving a System of Linear Equations in Three Variables

A **system of linear equations in three variables** consists of two or more linear equations in three variables. Here is an example of a system of linear equations in three variables:

$$\begin{aligned} 2x-y+3z&=4 \\ x+3y-2z&=-1 \\ 3x-5y+z&=2 \end{aligned}$$

The **solution** of a system of linear equations in three variables is an ordered triple that satisfies *all* of the equations. We can use elimination to solve a system of equations in three variables.

Example 2 Solving a System of Three Equations

Solve the system:

$$\begin{aligned} x+y-z&=-1 && \text{Equation (1)} \\ -4x-y+2z&=-7 && \text{Equation (2)} \\ 2x-2y-5z&=7 && \text{Equation (3)} \end{aligned}$$

Solution

By inspecting the coefficients of the nine variable terms, we see that it is easiest to eliminate y. We add the left sides and add the right sides of equations (1) and (2):

$$\begin{aligned} x+y-z&=-1 && \text{Equation (1)} \\ -4x-y+2z&=-7 && \text{Equation (2)} \\ \hline -3x\qquad+z&=-8 && \text{Equation (4)} \end{aligned}$$

Next, we select equations (1) and (3) and eliminate y again. To do so, we multiply both sides of equation (1) by 2:

$$\begin{aligned} 2x+2y-2z&=-2 && \text{Multiply both sides of equation (1) by 2.} \\ 2x-2y-5z&=7 && \text{Equation (3)} \\ \hline 4x\qquad-7z&=5 && \text{Equation (5)} \end{aligned}$$

Equations (4) and (5) form a system in two variables. To eliminate z, we multiply both sides of equation (4) by 7:

$$\begin{aligned} -21x + 7z &= -56 && \text{Multiply both sides of equation (4) by 7.} \\ 4x - 7z &= 5 && \text{Equation (5)} \\ \hline -17x \quad &= -51 \\ x &= 3 \end{aligned}$$

Next, we substitute 3 for x in equation (4) and solve for z:

$$\begin{aligned} -3(3) + z &= -8 && \text{Substitute 3 for x in equation (4).} \\ z &= 1 \end{aligned}$$

Then, we substitute 3 for x and 1 for z in equation (1) and solve for y:

$$\begin{aligned} 3 + y - 1 &= -1 && \text{Substitute 3 for x and 1 for z in equation (1).} \\ y &= -3 \end{aligned}$$

Therefore, $x = 3$, $y = -3$, and $z = 1$. So, the solution of the system is $(3, -3, 1)$. We check that the ordered triple satisfies *all three* original equations:

$x + y - z = -1$	$-4x - y + 2z = -7$	$2x - 2y - 5z = 7$
$3 + (-3) - 1 \stackrel{?}{=} -1$	$-4(3) - (-3) + 2(1) \stackrel{?}{=} -7$	$2(3) - 2(-3) - 5(1) \stackrel{?}{=} 7$
$-1 \stackrel{?}{=} -1$	$-7 \stackrel{?}{=} -7$	$7 \stackrel{?}{=} 7$
true	true	true

■

Solving a System of Three Linear Equations in Three Variables

To solve a system of three linear equations in three variables,

1. Select a pair of equations and eliminate a variable.
2. Select any other pair of equations and eliminate the *same variable* as in step 1.
3. The equations you found in steps 1 and 2 form a system of linear equations in *two* variables. Use elimination or substitution to solve this system.
4. Substitute the values of two of the variables you found in step 3 into one of the original equations that contains the third variable. Solve for the third variable.
5. Write your solution as an ordered triple.

WARNING Make sure you eliminate the *same* variable in steps 1 and 2. This way, your system in step 3 will be a system of equations in *two* variables (rather than three).

We can plot the point $(3, -3, 1)$ of Example 2 by using an x-axis, a y-axis, and a z-axis as shown in Fig. 78.

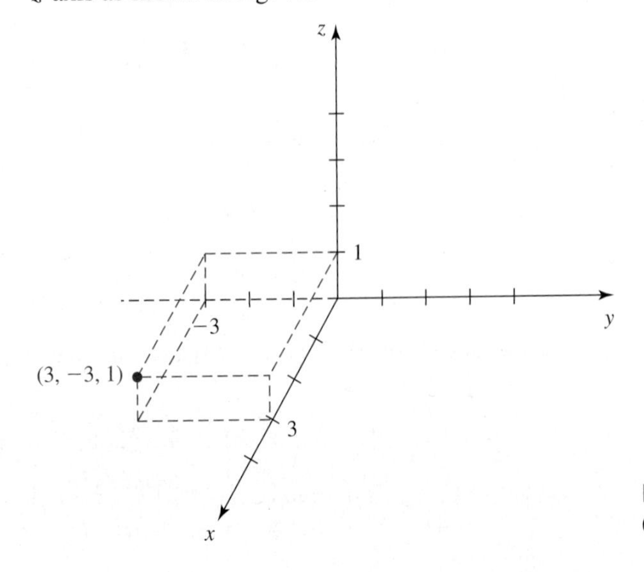

Figure 78 The point $(3, -3, 1)$

The graph of a linear equation in three variables is a plane. So, the graphs of equations (1), (2), and (3) of Example 2 are three planes that intersect at only the point $(3, -3, 1)$.

The intersection of three planes can be one point (see Fig. 79a), a line (see Fig. 79b), a plane (see Fig. 79c), or the empty set (see Fig. 79d). So, the solution set of a system of linear equations in three variables can contain exactly one ordered triple, an infinite number of ordered triples, or no ordered triples (the empty set). We will focus on systems whose solution set contains exactly one ordered triple.

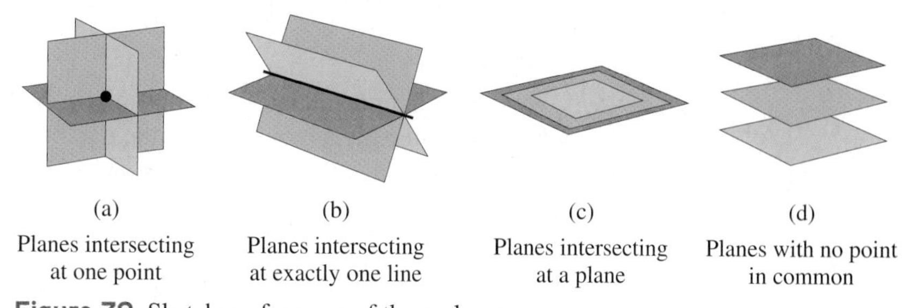

(a) Planes intersecting at one point

(b) Planes intersecting at exactly one line

(c) Planes intersecting at a plane

(d) Planes with no point in common

Figure 79 Sketches of groups of three planes

It is difficult to graph linear equations in three variables, so we will not use graphing to solve systems of such equations.

Example 3 Solving a System of Three Equations

Solve the system

$$\begin{aligned} x + y + z &= 2 && \text{Equation (1)} \\ 10y - z &= 12 && \text{Equation (2)} \\ 2x - 3y &= 3 && \text{Equation (3)} \end{aligned}$$

Solution

Equation (2) does not contain the variable x. To obtain a second equation that does not contain x, we select equations (1) and (3) and eliminate x. We begin by multiplying equation (1) by -2:

$$\begin{aligned} -2x - 2y - 2z &= -4 && \text{Multiply both sides of equation (1) by } -2. \\ 2x - 3y &= 3 && \text{Equation (3)} \\ \hline -5y - 2z &= -1 && \text{Equation (4)} \end{aligned}$$

Equations (2) and (4) form a system in two variables. To eliminate y, we multiply both sides of equation (4) by 2:

$$\begin{aligned} 10y - z &= 12 && \text{Equation (2)} \\ -10y - 4z &= -2 && \text{Multiply both sides of equation (4) by 2.} \\ \hline -5z &= 10 \\ z &= -2 \end{aligned}$$

Next, we substitute -2 for z in equation (2) and solve for y:

$$\begin{aligned} 10y - (-2) &= 12 && \text{Substitute } -2 \text{ for } z \text{ in equation (2).} \\ 10y &= 10 \\ y &= 1 \end{aligned}$$

Then, we substitute 1 for y and -2 for z in equation (1) and solve for x:

$$\begin{aligned} x + 1 + (-2) &= 2 && \text{Substitute 1 for } y \text{ and } -2 \text{ for } z \text{ in equation (1).} \\ x &= 3 \end{aligned}$$

Therefore, $x = 3$, $y = 1$, and $z = -2$. So, the solution of the system is $(3, 1, -2)$. We can check that the ordered triple satisfies all three original equations. ■

Using Points That Are Not y-Intercepts to Find an Equation of a Parabola

Now that we know how to solve a system of linear equations in three variables, we can find an equation of a parabola that contains three given points. In Example 4, we find an equation of a parabola for which none of the given points are y-intercepts.

Example 4 Finding an Equation of a Parabola

Find an equation of a parabola that contains the points (1, 1), (2, 3), and (3, 9).

Solution

Our goal is to find values of the constants a, b, and c for the equation $y = ax^2 + bx + c$. Since the three given points lie on the parabola, each of the ordered pairs (1, 1), (2, 3), and (3, 9) satisfies the equation $y = ax^2 + bx + c$:

$$1 = a(1)^2 + b(1) + c \quad \text{Substitute (1, 1) into } y = ax^2 + bx + c.$$
$$3 = a(2)^2 + b(2) + c \quad \text{Substitute (2, 3) into } y = ax^2 + bx + c.$$
$$9 = a(3)^2 + b(3) + c \quad \text{Substitute (3, 9) into } y = ax^2 + bx + c.$$

We can simplify the right-hand sides of these equations:

$$a + b + c = 1 \quad \text{Equation (1)}$$
$$4a + 2b + c = 3 \quad \text{Equation (2)}$$
$$9a + 3b + c = 9 \quad \text{Equation (3)}$$

We select equations (1) and (2) and eliminate c by multiplying both sides of equation (1) by -1:

$$-a - b - c = -1 \quad \text{Multiply both sides of equation (1) by } -1.$$
$$4a + 2b + c = 3 \quad \text{Equation (2)}$$
$$3a + b = 2 \quad \text{Equation (4)}$$

Next, we select equations (1) and (3) and eliminate c again, once more multiplying both sides of equation (1) by -1:

$$-a - b - c = -1 \quad \text{Multiply both sides of equation (1) by } -1.$$
$$9a + 3b + c = 9 \quad \text{Equation (3)}$$
$$8a + 2b = 8 \quad \text{Equation (5)}$$

Then, we divide both sides of Equation (5) by 2:

$$4a + b = 4 \quad \text{Equation (6)}$$

Equations (4) and (6) form a system in two variables. To eliminate b, we multiply both sides of equation (4) by -1:

$$-3a - b = -2 \quad \text{Multiply both sides of equation (4) by } -1.$$
$$4a + b = 4 \quad \text{Equation (6)}$$
$$a = 2$$

Next, we substitute 2 for a in equation (4) and solve for b:

$$3(2) + b = 2$$
$$b = -4$$

Then, we substitute 2 for a and -4 for b in equation (1) and solve for c:

$$2 + (-4) + c = 1 \quad \text{Substitute 2 for } a \text{ and } -4 \text{ for } b \text{ in equation (1).}$$
$$c = 3$$

Therefore, $a = 2$, $b = -4$, and $c = 3$, and the equation of the parabola is

$$y = 2x^2 - 4x + 3$$

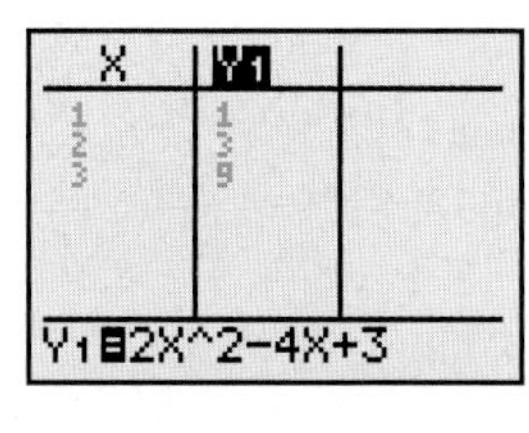

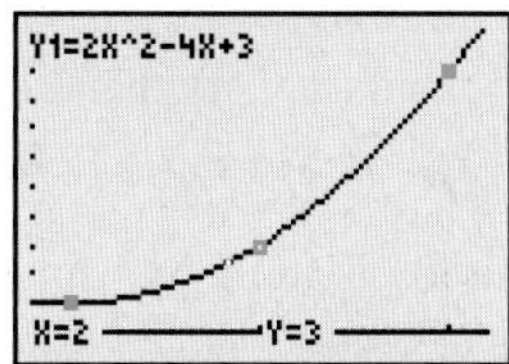

Figure 80 Verify the work

We use a graphing calculator table, as well as a scattergram and graph, to check that the parabola $y = 2x^2 - 4x + 3$ contains the points (1, 1), (2, 3), and (3, 9). See Fig. 80. ■

To find a linear equation $y = mx + b$, we need *two* points to find the *two* constants m and b. To find an exponential equation $y = ab^x$, we need *two* points to find the *two* constants a and b. To find a quadratic equation in standard form, $y = ax^2 + bx + c$, we need *three* points to find the *three* constants a, b, and c.

Finding an Equation of a Parabola

To find an equation of a parabola that contains three given points,

1. Obtain a system of three linear equations in three variables by substituting the coordinates of each of the three given points into the general equation $y = ax^2 + bx + c$.
2. Solve the system you found in step 1.
3. Substitute the values of a, b, and c you found in step 2 into the equation $y = ax^2 + bx + c$.

Using the y-Intercept and Two Other Points

The process of finding an equation of a parabola is considerably easier if one of the three given points is the y-intercept, as we shall see in Example 5.

Example 5 Finding an Equation When One of the Given Points Is the y-Intercept

Find an equation of the parabola that contains the points (0, 1), (3, 7), and (4, 5).

Solution

We begin by substituting the ordered pairs (0, 1), (3, 7), and (4, 5) into $y = ax^2+bx+c$:

$$1 = a(0)^2 + b(0) + c \quad \text{Substitute (0, 1) into } y = ax^2 + bx + c.$$
$$7 = a(3)^2 + b(3) + c \quad \text{Substitute (3, 7) into } y = ax^2 + bx + c.$$
$$5 = a(4)^2 + b(4) + c \quad \text{Substitute (4, 5) into } y = ax^2 + bx + c.$$

Next, we simplify the right-hand sides of these equations:

$$c = 1 \quad \text{Equation (1)}$$
$$9a + 3b + c = 7 \quad \text{Equation (2)}$$
$$16a + 4b + c = 5 \quad \text{Equation (3)}$$

Since $c = 1$, we substitute 1 for c in equations (2) and (3):

$$9a + 3b + 1 = 7 \quad \text{Substitute 1 for } c \text{ in equation (2).}$$
$$16a + 4b + 1 = 5 \quad \text{Substitute 1 for } c \text{ in equation (3).}$$

For each of these two equations, we subtract 1 on both sides:

$$9a + 3b = 6 \quad \text{Equation (4)}$$
$$16a + 4b = 4 \quad \text{Equation (5)}$$

To eliminate b, we multiply both sides of equation (4) by -4 and both sides of equation (5) by 3:

$$-36a - 12b = -24 \quad \text{Multiply both sides of equation (4) by } -4.$$
$$48a + 12b = 12 \quad \text{Multiply both sides of equation (5) by 3.}$$
$$12a = -12$$
$$a = -1$$

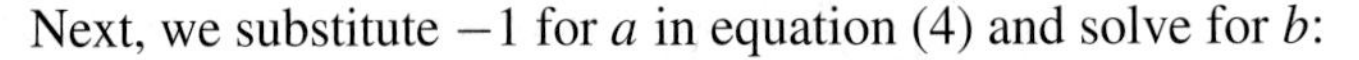

Next, we substitute -1 for a in equation (4) and solve for b:

$$9(-1) + 3b = 6 \qquad \text{Substitute } -1 \text{ for } a \text{ in equation (4).}$$
$$3b = 15$$
$$b = 5$$

Therefore, $a = -1$, $b = 5$, and $c = 1$, and the equation of the parabola is

$$y = -x^2 + 5x + 1$$

We use a graphing calculator table, as well as a scattergram and graph, to verify that the graph of $y = -x^2 + 5x + 1$ contains the points (0, 1), (3, 7), and (4, 5). See Fig. 81. ■

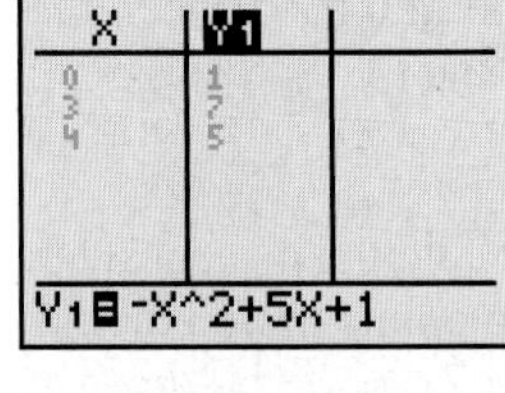

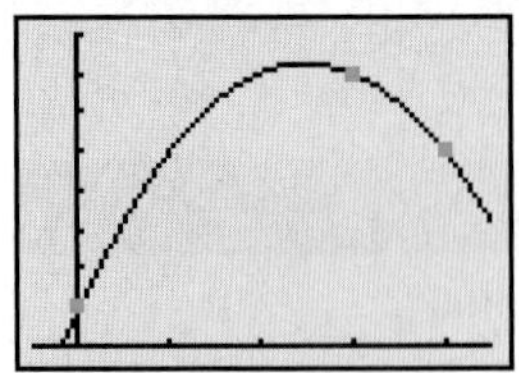

Figure 81 Verify the work

In Example 5, we were able to find an equation of the desired parabola by using elimination once, rather than three times. We need to use elimination only once when one of the three given points is the y-intercept.

group exploration

For any three points, is there a quadratic function that contains them?

1. Find the values of a, b, and c of the function $f(x) = ax^2 + bx + c$, where the graph of f contains the points (0, 1), (1, 4), and (2, 7). What type of function is f? Why did this happen?
2. Do the same for the points (0, 1), (0, 8), and (1, 4). What happens? Is there a function $f(x) = ax^2 + bx + c$ whose graph contains these points? Explain.
3. What must be true of three points so that there is a quadratic function whose graph contains the points? Give an example of three such points, plot them, and sketch the graph of the quadratic function that contains them. Then find the equation and use a graphing calculator to view the graph. Compare the two graphs.

TIPS FOR SUCCESS: Create an Example

When learning a definition or property, try to create an example. While studying the material in this section, you could select three points and determine whether there is a parabola that contains the chosen points, and if so, you could find an equation of the parabola. Creating examples will shed light on many details of a concept and will personalize the information.

HOMEWORK 7.6

FOR EXTRA HELP ▶ Student Solutions Manual · PH Math/Tutor Center · MathXL® · MyMathLab

Solve the system.

1. $\begin{aligned} x + y + z &= 0 \\ x - y + z &= 6 \\ x + 2y - z &= -7 \end{aligned}$

2. $\begin{aligned} x + y + z &= 4 \\ -x + y + 2z &= 1 \\ -x + y - 3z &= -4 \end{aligned}$

3. $\begin{aligned} x + y - z &= -1 \\ 2x - 2y + 3z &= 8 \\ 2x - y + 2z &= 9 \end{aligned}$

4. $\begin{aligned} 2x - 3y + z &= -9 \\ -2x + y - 3z &= 7 \\ x - y + 2z &= -5 \end{aligned}$

5. $\begin{aligned} 3x - y + 2z &= 0 \\ 2x + 3y + 8z &= 8 \\ x + y + 6z &= 0 \end{aligned}$

6. $\begin{aligned} -x + y + z &= 6 \\ x - 2y + 3z &= 5 \\ -2x + y - 2z &= -1 \end{aligned}$

7. $\begin{aligned} 2x + y + z &= 3 \\ 2x - y - z &= 9 \\ x + y - z &= 0 \end{aligned}$

8. $\begin{aligned} x + y + z &= 6 \\ 2x - y + z &= 3 \\ x + 2y - 3z &= -4 \end{aligned}$

9. $\begin{aligned} 2x + 2y + z &= 1 \\ -x + y + 2z &= 3 \\ x + 2y + 4z &= 0 \end{aligned}$

10. $\begin{aligned} 2x - 3y + z &= 2 \\ x - 5y + 5z &= 3 \\ 3x + y - 3z &= 5 \end{aligned}$

11. $\begin{aligned} 2x - y + 2z &= 6 \\ 3x + y - z &= 5 \\ x + 2y + z &= 3 \end{aligned}$

12. $\begin{aligned} 2x + y + z &= -2 \\ 2x - y + 3z &= 6 \\ 3x - 5y + 4z &= 7 \end{aligned}$

13. $\begin{aligned} x - 3z &= 6 \\ y + 2z &= 2 \\ 7x - 3y - 5z &= 14 \end{aligned}$

14. $\begin{aligned} 2x + y &= -2 \\ 3y - z &= -14 \\ x + 2z &= 5 \end{aligned}$

15. $\begin{aligned} 2x - y &= -8 \\ y + 3z &= 22 \\ x - z &= -8 \end{aligned}$

16. $\begin{aligned} x + y &= 1 \\ y - 2z &= 2 \\ x - 3z &= 14 \end{aligned}$

Find an equation of a parabola that contains the given points. Use a graphing calculator to verify that the graph of your equation contains the points.

17. (1, 6), (2, 11), (3, 18)
18. (1, 1), (2, 5), (3, 15)
19. (1, 9), (2, 7), (4, −15)
20. (1, 4), (2, 3), (3, 0)

21. (2, 2), (3, 11), (4, 24)
22. (2, 3), (3, −2), (4, −11)
23. (1, −3), (3, 9), (5, 29)
24. (2, −1), (4, 19), (5, 38)
25. (3, 7), (4, 0), (5, −11)
26. (2, 4), (4, 30), (5, 49)
27. (2, −5), (4, 3), (5, 13)
28. (1, 3), (2, 3), (4, −9)
29. (0, 4), (2, 8), (3, 1)
30. (0, 0), (1, 4), (2, 14)
31. (0, −1), (1, 3), (2, 13)
32. (0, 5), (2, 13), (3, 26)
33. (1, 1), (2, 4), (3, 9)
34. (1, −1), (2, −4), (3, −9)

35. The graph of a quadratic function has y-intercept (0, 4) and x-intercepts (1, 0) and (2, 0). Find an equation of the function.

36. The graph of a quadratic function has y-intercept (0, 8) and x-intercepts (−4, 0) and (2, 0). Find an equation of the function.

37. Find an equation of the parabola sketched in Fig. 82. [**Hint:** Choose three points whose coordinates appear to be integers.]

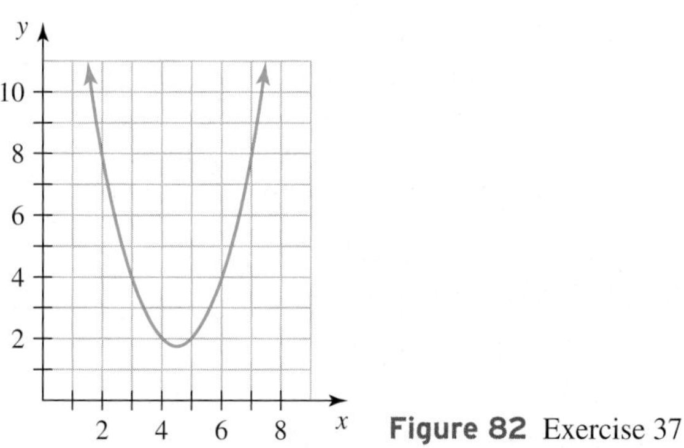

Figure 82 Exercise 37

38. Find an equation of the parabola sketched in Fig. 83. [**Hint:** Choose three points whose coordinates appear to be integers.]

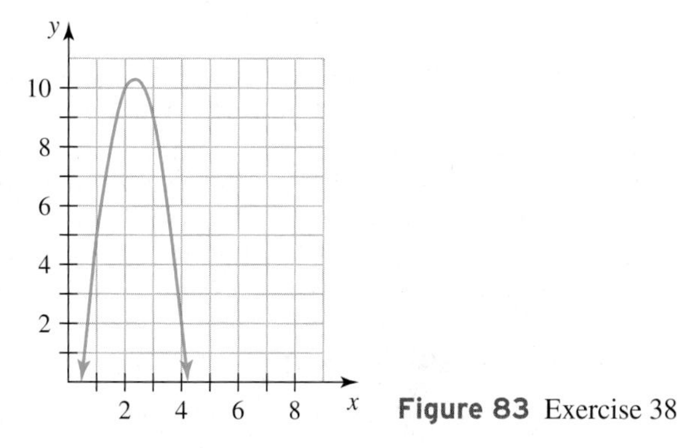

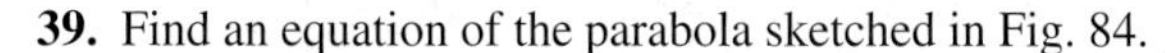

Figure 83 Exercise 38

39. Find an equation of the parabola sketched in Fig. 84.

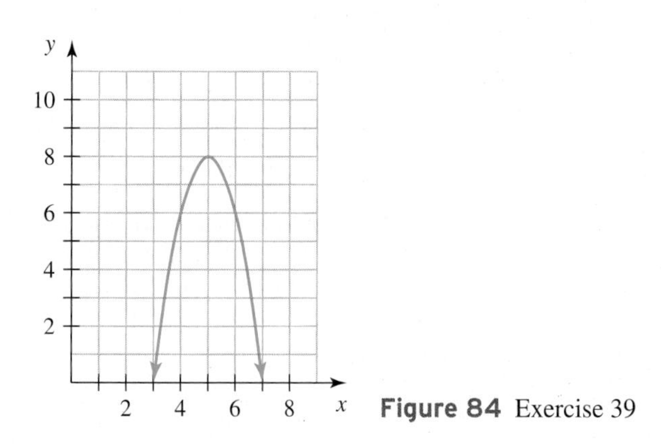

Figure 84 Exercise 39

40. Find an equation of the parabola sketched in Fig. 85.

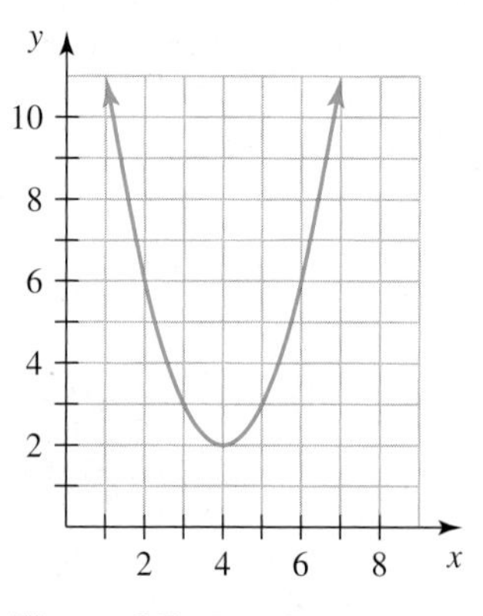

Figure 85 Exercise 40

41. In this exercise, you will show that two points do not determine a parabola.
a. Plot the points (1, 4) and (3, 4) on a coordinate system.
b. Sketch three parabolas that all contain both of the points (1, 4) and (3, 4). Find the vertex of each parabola.
c. Write an equation of each of your three sketched parabolas. [**Hint:** You can use the vertex form $y = a(x - h)^2 + k$.]
d. Explain why two points do not determine a parabola.

42. Find an equation of a parabola that is in quadrants I, III, and IV but not in quadrant II.

43. Choose three points that have different x-coordinates and do not all lie on a line. Find an equation of a parabola that contains the points. [**Hint:** To make the calculations easier, choose one of the points so that it is the y-intercept.]

44. Describe how to find an equation of a parabola that contains three points that have different x-coordinates and do not all lie on a line.

Related Review

Find equations of a linear function and an exponential function such that the graph of each equation contains the given points. Round approximate constants to the second decimal place.

45. (3, 13), (7, 85)

46. (2, 67), (5, 9)

Find equations of a linear function, an exponential function, and a quadratic function such that the graph of each equation contains the given points. Use a graphing calculator to verify your work.

47. (0, 2), (1, 4)

48. (0, 4), (2, 36)

49. Find the values of a, b, and c of $f(x) = ax^2 + bx + c$, where the graph of f contains the points (1, 1), (2, 2), and (3, 3). What type of function is f?

50. Find the values of a, b, and c of $g(x) = ax^2 + bx + c$, where the graph of g contains the points (1, 2), (2, 5), and (3, 8). What type of function is g?

51. Find an equation of a parabola that has vertex (5, −7) and contains the point (8, 11).

52. Find an equation of a parabola that has vertex (−5, 8) and contains the point (−7, −4).

Expressions, Equations, Functions, and Graphs

Perform the indicated instruction. Then use words such as linear, quadratic, cubic, exponential, logarithmic, polynomial, degree, function, one variable, *and* two variables *to describe the expression, equation, or system.*

53. Solve $2x^2 - 10x + 7 = 0$.

54. Write $\log_3(t-4) + \log_3(t+1)$ as a single logarithm.

55. Find $f(2)$, where $f(x) = 2x^2 - 10x + 7$.

56. Solve $\log_3(t-4) + \log_3(t+1) = 2$.

57. Graph $f(x) = 2x^2 - 10x + 7$ by hand.

58. Solve $t^2 - 3t - 13 = 0$.

7.7 FINDING QUADRATIC MODELS

Objectives

- Find an equation of a quadratic model in standard form.
- Determine whether a linear function, an exponential function, a quadratic function, or none of these can be used to model a situation.

In Section 7.6, we used three given points to find an equation of a parabola. In this section, we will use this skill to find an equation of a quadratic model. We will also discuss how to determine whether a linear model, an exponential model, a quadratic model, or none of these can be used to model an authentic situation.

Finding a Quadratic Model in Standard Form

In Example 1, we find an equation of a quadratic model.

Table 38 Numbers of House Calls Paid by Medicare

Year	Number of House Calls (millions)
1996	1.62
1997	1.55
1998	1.48
1999	1.45
2000	1.53
2001	1.60
2002	1.70
2003	1.83
2004	2.06

Source: *CMS Medicare National Procedure Summary*

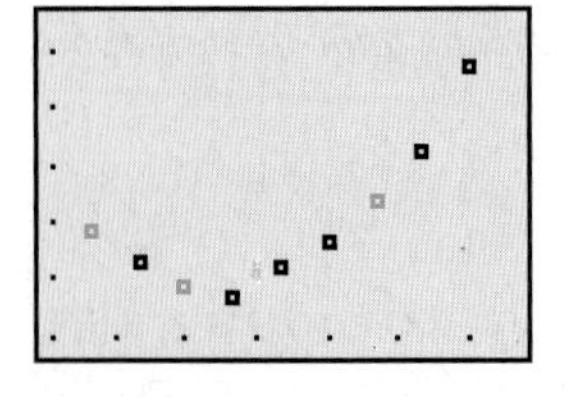

Figure 86 House call scattergram

Example 1 Finding an Equation of a Quadratic Model

The numbers of house calls paid by Medicare are shown in Table 38 for various years. Let $f(t)$ be the number of house calls (in millions) paid by Medicare in the year that is t years since 1995. Find an equation of a model to describe the situation.

Solution

We use a graphing calculator to draw a scattergram to describe the data (see Fig. 86). It looks like a parabola would fit the data well.

Through practice, we envision that the parabola containing the (blue) data points (1, 1.62), (3, 1.48), and (7, 1.70) may be close to the other data points. To find an equation of this parabola, we substitute the three ordered pairs (1, 1.62), (3, 1.48), and (7, 1.70) into the standard form $f(t) = at^2 + bt + c$:

$$1.62 = a(1)^2 + b(1) + c \quad \text{Substitute (1, 1.62) into } f(t) = at^2 + bt + c.$$
$$1.48 = a(3)^2 + b(3) + c \quad \text{Substitute (3, 1.48) into } f(t) = at^2 + bt + c.$$
$$1.70 = a(7)^2 + b(7) + c \quad \text{Substitute (7, 1.70) into } f(t) = at^2 + bt + c.$$

We can simplify the right-hand sides of these equations:

$$a + b + c = 1.62 \quad \text{Equation (1)}$$
$$9a + 3b + c = 1.48 \quad \text{Equation (2)}$$
$$49a + 7b + c = 1.70 \quad \text{Equation (3)}$$

We select equations (1) and (2) and eliminate c by multiplying both sides of equation (1) by -1:

$$-a - b - c = -1.62 \quad \text{Multiply both sides of equation (1) by } -1.$$
$$9a + 3b + c = 1.48 \quad \text{Equation (2)}$$
$$8a + 2b = -0.14 \quad \text{Equation (4)}$$

Next, we select equations (1) and (3) and eliminate c again, once more multiplying both sides of equation (1) by -1:

$$
\begin{array}{rll}
-a - b - c = & -1.62 & \text{Multiply both sides of equation (1) by } -1. \\
49a + 7b + c = & 1.70 & \text{Equation (3)} \\
\hline
48a + 6b \phantom{{}+c} = & 0.08 & \text{Equation (5)}
\end{array}
$$

Equations (4) and (5) form a system in two variables. To eliminate b, we multiply both sides of equation (4) by -3:

$$
\begin{array}{rll}
-24a - 6b = 0.42 & & \text{Multiply both sides of equation (4) by } -3. \\
48a + 6b = 0.08 & & \text{Equation (5)} \\
\hline
24a \phantom{{}+6b} = 0.50 & & \\
a \approx 0.0208 & &
\end{array}
$$

Next, we substitute 0.0208 for a in equation (4) and solve for b:

$$
\begin{aligned}
8(0.0208) + 2b &= -0.14 && \text{Substitute 0.0208 for } a \text{ in equation (4).} \\
0.1664 + 2b &= -0.14 \\
2b &= -0.3064 \\
b &\approx -0.153
\end{aligned}
$$

Then we substitute 0.0208 for a and -0.153 for b in equation (1) and solve for c:

$$
\begin{aligned}
0.0208 + (-0.153) + c &= 1.62 && \text{Substitute 0.0208 for } a \text{ and } -0.153 \text{ for } b \text{ in equation (1).} \\
-0.1322 + c &= 1.62 \\
c &\approx 1.75
\end{aligned}
$$

Finally, we substitute our approximate values of a, b, and c in the general equation $f(t) = at^2 + bt + c$ to obtain our quadratic model:

$$f(t) = 0.0208t^2 - 0.153t + 1.75$$

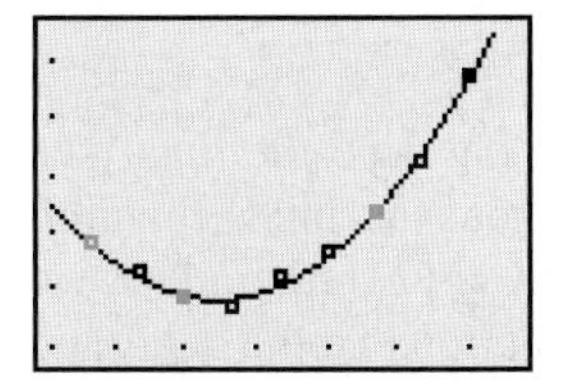

Figure 87 Check the fit

We verify the equation by observing that the graph of f appears to contain the points (1, 1.62), (3, 1.48), and (7, 1.70). See Fig. 87. Since the graph appears to come close to the other data points, we conclude that f is a reasonable model of the situation. ■

Finding an Equation of a Quadratic Model

To find an equation of a quadratic model, given some data,

1. Create a scattergram of the data.
2. Imagine a parabola that comes close to the data points, and choose three points (not necessarily data points) that lie on or close to the parabola.
3. Use the three points to find an equation of the parabola.
4. Use a graphing calculator to verify that the graph of the equation comes close to the points of the scattergram.

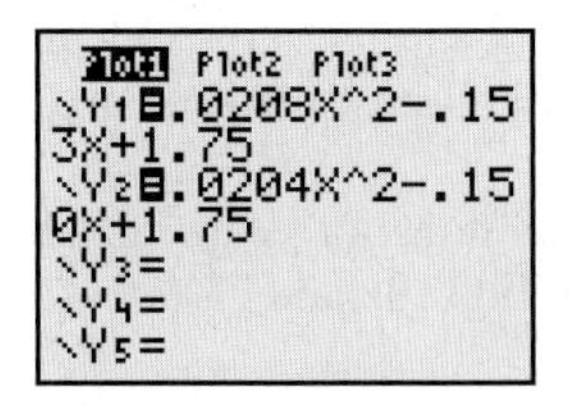

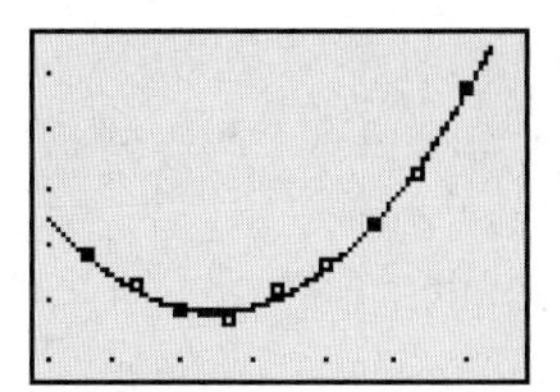

Figure 88 Compare the fit of models f and r

In Example 1, we could have used another procedure to find a quadratic model different from $f(t) = 0.0208t^2 - 0.153t + 1.75$. Instead of choosing three points and solving a system of equations, we could have used a graphing calculator's *quadratic regression*. Quadratic regression gives the model $r(t) = 0.0204t^2 - 0.150t + 1.75$. In Fig. 88, the graphs of the models f and r are so similar that ZoomStat appears to show just one curve. For graphing calculator instructions, see Section B.16.

We call the equation $r(t) = 0.0204t^2 - 0.150t + 1.75$ the **quadratic regression equation** for the given data. We refer to its graph as a **quadratic regression curve.**

Determining Which Type of Model to Use

In Example 2, we will compare how well a linear model, an exponential model, and a quadratic model describe an authentic situation.

Example 2 Determining Which Model to Use

The number of federal documents that are classified has been increasing; so has the cost of keeping these documents secret (see Table 39).

Table 39 Federal Cost of Secrecy

Year	Federal Cost of Secrecy (billions of dollars)
1997	3.4
1998	3.6
1999	3.8
2000	4.3
2001	4.7
2002	5.7
2003	6.5
2004	7.2

Source: *National Security Archive*

Let c be the federal cost of secrecy (in billions of dollars) in the year that is t years since 1990. Find an equation of a model to describe the situation.

Solution

First, we use a graphing calculator to draw a scattergram to describe the data (see Fig. 89). Since the points suggest a curve that "bends," we will not use a linear function to model the data. To decide between an exponential model and a quadratic model, we use a graphing calculator to find the exponential regression equation and the quadratic regression equation:

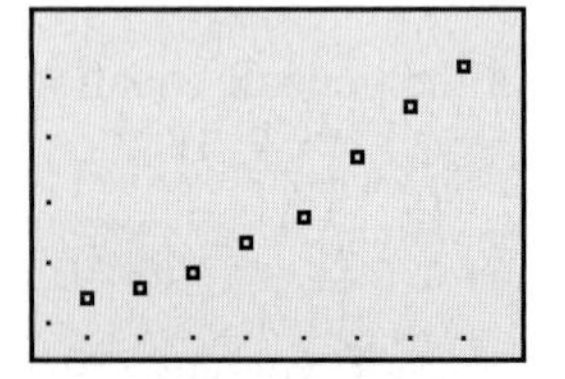

Figure 89 Secrecy scattergram

$$E(t) = 1.44(1.12)^t \quad \text{Exponential regression equation}$$

$$Q(t) = 0.064t^2 - 0.79t + 5.75 \quad \text{Quadratic regression equation}$$

Next, we see how well both regression models fit the data (see Figs. 90 and 91).

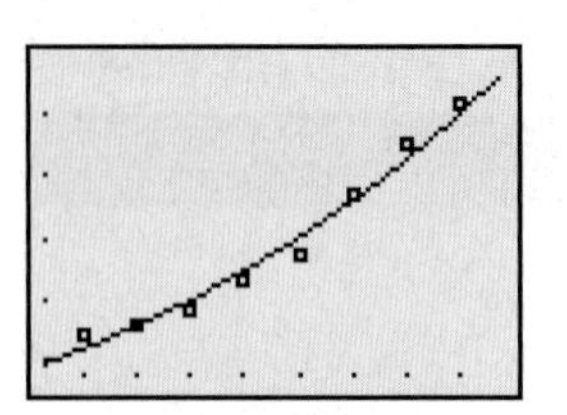

Figure 90 Exponential regression model

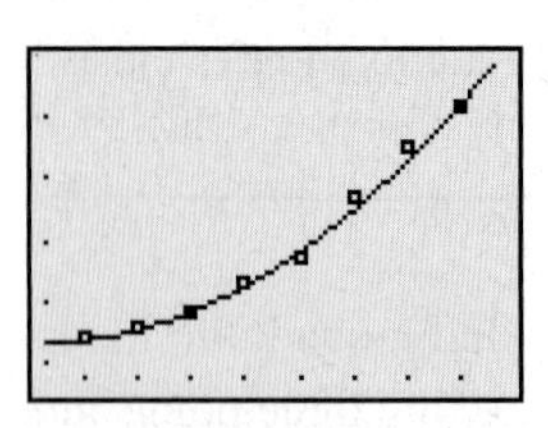

Figure 91 Quadratic regression model

Both models fit the data well, but it appears that Q fits the data a bit better than E does. So, for most years between 1997 and 2004, inclusive, Q estimates the federal cost better than E does. This suggests that Q predicts the federal cost better than E does for at least a few years after 2004.

Does Q estimate the federal cost of secrecy better than E does for years before 1997? To see, we use Zoom Out on a graphing calculator (see Figs. 92 and 93).

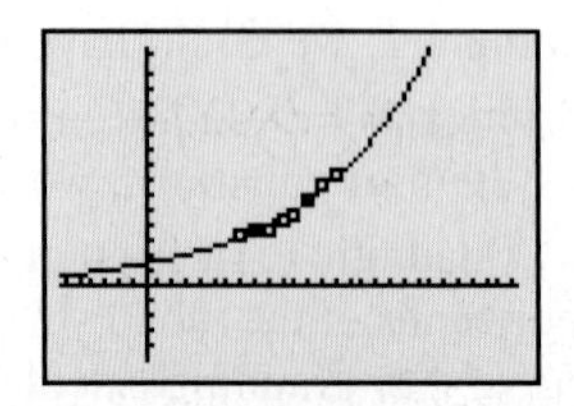

Figure 92 Exponential regression model

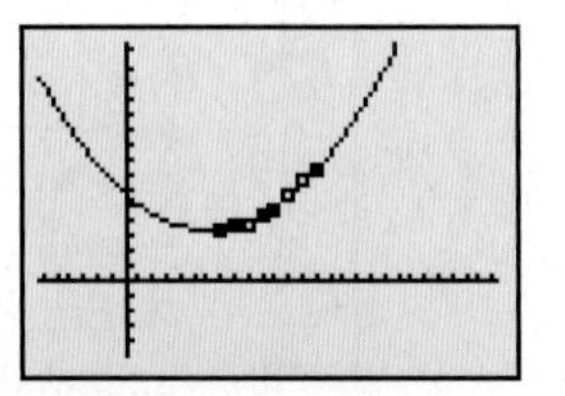

Figure 93 Quadratic regression model

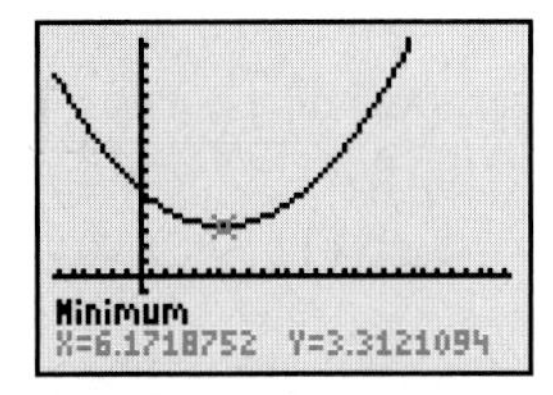

Figure 94 The vertex of Q

It will also help to use "minimum" on a graphing calculator to find the approximate vertex of Q, which is (6.17, 3.31). See Fig. 94. So, Q estimates that the federal cost of secrecy was decreasing for years before 1996, which a little research would show is false. The model E correctly estimates that the federal cost of secrecy was increasing for years before 1997. E is the best model for the years before 1997. ■

Our work in Example 2 suggests two criteria for selecting a model.

Selecting a Model

When performing step 1 of the modeling process, we must decide whether a linear function, an exponential function, a quadratic function, or none of these is suitable for modeling the situation. Here are the criteria for selecting a model:

- The graph of the model should fit the points well.
- The model should make sense within the context of the authentic situation.

In Example 2, we used the criterion of fit to determine that Q is the best model for the years between 1997 and 2004, inclusive. Using this same criterion, we assumed that Q best predicts the federal cost of secrecy for at least a few years after 2004. We used the criterion of context (the cost has generally been increasing over time) to determine that E likely best estimates the cost for at least a few years before 1997.

Here, we outline the four-step modeling process again.

Four-Step Modeling Process

To find a model and make estimates and predictions,

1. Create a scattergram of the data. Decide whether a line, an exponential curve, a parabola, or none of these comes close to the points.
2. Find an equation of your model.
3. Verify that your equation has a graph that comes close to the points in the scattergram. If it doesn't, check for calculation errors or use different points to find the equation. An alternative is to reconsider your choice of model in step 1.
4. Use your equation of the model to draw conclusions, make estimates, and/or make predictions.

We will perform activities from step 4 in Section 7.8.

group exploration

Choosing three "good" points to find a quadratic model

Table 40 lists the average numbers of paid vacation days and holidays for full-time workers at medium-to-large companies for various years of experience. Let D be the average number of paid vacation days and holidays in one year for someone who has worked at a company for t years.

Table 40 Paid Vacation Days and Holidays

Years of Service	Days Off
1	9.4
3	11.2
5	13.6
10	16.6
15	18.8
20	20.4
25	21.6
30	21.9

Source: USA Today

1. Use a graphing calculator to create a scattergram of the vacation data. Does a linear function, an exponential function, or a quadratic function seem to model the data best? Explain.
2. Use the three data points (1, 9.4), (15, 18.8), and (30, 21.9) to find an equation $D = at^2 + bt + c$ of the parabola that comes close to the data points in your scattergram.

3. Draw the graph of your quadratic model and your scattergram in the same viewing window to verify that the parabola passes through the points (1, 9.4), (15, 18.8), and (30, 21.9). Does your quadratic function seem to be a reasonable model of the vacation data?

4. The first three rows in Table 40 give the data points (1, 9.4), (3, 11.2), and (5, 13.6). Had you used these three points, you would have found the equation $D = 0.075t^2 + 0.60t + 8.73$. Compare its graph with the graph you drew in Problem 3. Explain why the graphs look so different.

5. In the future, you will encounter other data sets that can be modeled well by using a quadratic function. Describe a general "game plan" for deciding which three points to use to find a quadratic model.

HOMEWORK 7.7

1. Four scattergrams of data are sketched in Fig. 95. Decide whether a linear function, an exponential function, a quadratic function, or none of these types of functions would be reasonable for modeling the data.

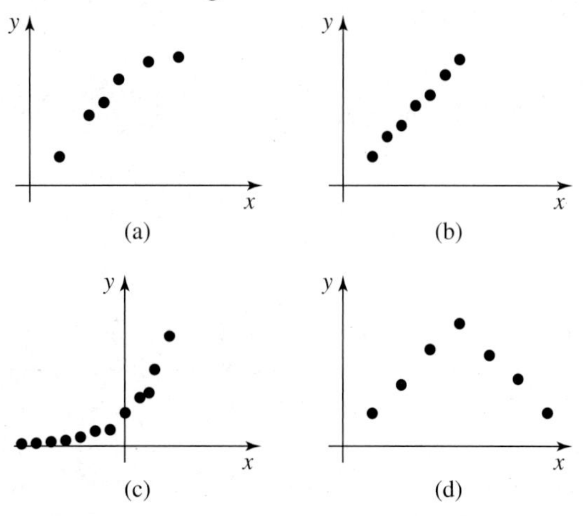

Figure 95 Scattergrams of data, Exercise 1

2. Make a sketch of each scattergram in Fig. 95. Then sketch the graph of the function you would use to model the data.

3. A student believes that the data listed in Table 41 suggest a quadratic relationship, because the values of y increase and then decrease. What would you tell the student?

Table 41 Is There a Quadratic Relationship? (Exercise 3)

x	y
0	3
1	4
2	7
3	12
4	20
5	35
6	20
7	12
8	7
9	4
10	3

4. A student uses the points (2, 2.5), (3, 4.1), and (4, 6.4) to find an equation of a quadratic function to model the data in Table 42. Did the student make a good selection of points? If so, explain; then find the equation of those points. If not, explain; then make a better choice of points and find an equation.

Table 42 A Student Models Some Data (Exercise 4)

x	y
2	2.5
3	4.1
4	6.4
5	7.5
6	8.0
7	7.8
8	7.1
9	5.8
10	3.9
11	1.4
12	−1.7

5. The percentages of teenagers who smoked cigarettes in the past month are shown in Table 43 for various ages.

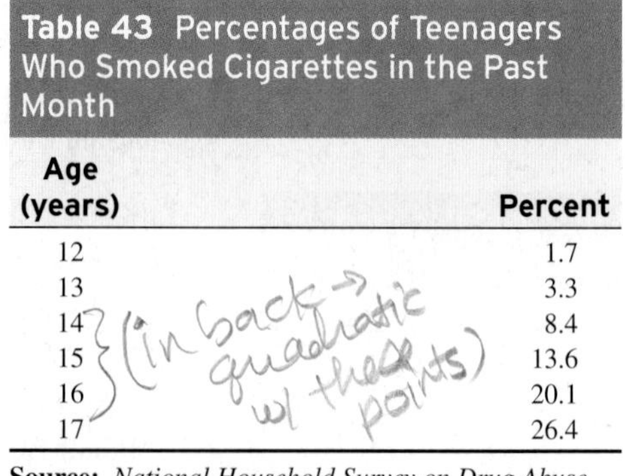

Table 43 Percentages of Teenagers Who Smoked Cigarettes in the Past Month

Age (years)	Percent
12	1.7
13	3.3
14	8.4
15	13.6
16	20.1
17	26.4

Source: *National Household Survey on Drug Abuse*

Let $f(t)$ be the percentage of teenagers at age t years who smoked cigarettes in the past month. Find an equation of f by hand, and find a regression equation of f by using a graphing calculator. Compare the graphs of the two models.

6. Domestic airline fuel prices are shown in Table 44 for various years.

Table 44 Domestic Airline Fuel Prices

Year	Domestic Airline Fuel Price (dollars per gallon)
2001	0.76
2002	0.70
2003	0.84
2004	1.12
2005	1.51

Source: *Bureau of Transportation Statistics*

Let $f(t)$ be the domestic airline fuel price (in dollars per gallon) at t years since 2000. Find an equation of f by hand, and find a regression equation of f by using a graphing calculator. Compare the graphs of the two models.

7. The percentages of major U.S. firms that perform drug tests on employees and/or job applicants are shown in Table 45 for various years.

Table 45 Percentages of Firms That Perform Drug Tests

Year	Percent
1987	22
1990	51
1992	72
1995	78
1998	74
2000	66
2002	62

Source: *American Management Association*

Let $f(t)$ be the percentage of firms that perform drug tests on employees and/or job applicants at t years since 1985.

a. Use the data for the years 1987, 1990, and 2002 to find a quadratic equation of f.

b. The data for the years 1987, 1990, and 1992 lead to the equation $f(t) = 0.17t^2 + 8.5t + 4.3$. The data for the years 1998, 2000, and 2002 lead to the equation $f(t) = 0.5t^2 - 18t + 223.5$. In part (a), you found yet another equation of f. Use a graphing calculator to determine which of these three triples of years gives the best model of the data. Could you have guessed this before you found the equations? If so, explain how.

8. July is the most popular month for Americans to take vacations (see Table 46). Let $f(t)$ be the percentage of Americans who take a vacation in the month that is t months since July.

a. Use the data for August, October, and May to find a quadratic equation of f.

b. The data for July, August, and September lead to the equation $f(t) = -2t^2 - 7t + 43$. The data for November, December, and January lead to the equation $f(t) = -4t^2 + 36t - 69$. The data for December, March, and May lead to $f(t) = 2.07t^2 - 29.20t + 105.33$. In part (a), you found yet another equation of f. Use a graphing calculator to determine which of these four triples of months gives the best model of the data. Could you have guessed this before you found the equations? If so, explain how.

Table 46 Percentages of Americans Who Vacation, by Month

Month	Number of Months Since July t	Percent
July	0	43
August	1	34
September	2	21
October	3	15
November	4	11
December	5	11
January	6	3
February	7	3
March	8	4
April	9	8
May	10	20
June	11	30

Source: *2001 American Express Leisure Travel Index*

9. The U.S. population is shown in Table 47 for various years.

Table 47 U.S. Population

Year	Population (millions)	Year	Population (millions)
1790	3.9	1910	92.4
1800	5.3	1920	106.5
1810	7.2	1930	123.2
1820	9.6	1940	132.1
1830	12.9	1950	151.3
1840	17.1	1960	179.3
1850	23.3	1970	203.3
1860	31.5	1980	226.5
1870	39.9	1990	248.8
1880	50.3	2000	281.4
1890	63.1	2006	299.0
1900	76.1		

Source: *U.S. Census Bureau*

a. Let $f(t)$ be the U.S. population (in millions) at t years since 1790. Find and verify an equation of f.

b. In Example 11 of Section 6.2, we used a linear function to model the U.S. population for the years shown in Table 48. An equation of such a linear model is $L(t) = 3.288t - 409.63$, where $L(t)$ represents the U.S. population (in millions) at t years since 1790.

Table 48 U.S. Population

Year	U.S. Population (millions)
1990	248.8
1993	257.8
1996	265.2
1999	279.0
2002	287.9
2004	293.7

Source: *U.S. Census Bureau*

i. Use a graphing calculator to draw the graphs of f and L and, in the same viewing window, the scattergram of the data *for the years shown in Table 47*. Which function describes the population better for these years? Explain.

ii. Use a graphing calculator to draw the graphs of f and L and, in the same viewing window, the scattergram of the data *for the years shown in Table 48*. Which function describes the population better for these years? Explain.

iii. How can one function be a better model for the time period 1790–2006 but the other function be a better model for part of that period?

iv. Does the portion of the graph of f that you viewed in part (ii) for the years from 1990 to 2004 appear to be linear, quadratic, or exponential? How is this possible?

10. The percentages of Americans who say they do volunteer work are listed in Table 49 for various age groups.

Table 49 Percentages of Americans Who Say They Volunteer

Age Group (years)	Age Used to Represent Age Group (years)	Percent
18–24	21.0	38
25–34	29.5	51
35–54	44.5	55
55–64	59.5	48
65–74	69.5	45
over 74	80.0	34

Source: *The Gallup Organization*

Let $f(A)$ be the percentage of Americans at age A who say they volunteer. Find and verify an equation of f.

11. The annual sales of Miller Lite® beer and Tecate® beer are shown in Table 50 for various years.

Table 50 Annual Sales of Miller Lite Beer and Tecate Beer

	Sales of 2.25-Gallon Cases (millions)	
Year	**Miller Lite**	**Tecate**
2000	3.9	11.0
2001	4.3	12.0
2002	4.9	13.1
2003	6.2	13.5
2004	7.5	14.6

Source: *Adams Beverage Group*

a. Let $f(t)$ and $g(t)$ be the annual sales (in millions of cases) of Miller Lite beer and Tecate beer, respectively, in the year that is t years since 2000. Find an equation of f and an equation of g.

b. Use "intersect" on a graphing calculator to find the intersection points of the graphs of f and g. What do these points mean in this situation?

12. The percentages of California's population who are foreign born and the percentages who were born in other U.S. states are listed in Table 51 for various years.

a. Let $f(t)$ and $g(t)$ be the percentages of California's population that are foreign born and born in other U.S. states, respectively, at t years since 1900. Find and verify regression equations of f and g.

Table 51 Californians Not Originally from California

	Percent	
Year	**Foreign Born**	**Born in Other U.S. States**
1930	18.9	47.0
1940	13.4	50.0
1950	10.0	53.0
1960	8.5	51.0
1970	8.8	47.9
1980	15.1	39.5
1990	21.7	31.8
2000	25.9	23.5

Source: *William Frey analysis of U.S. Census Bureau sources*

b. Use "intersect" on a graphing calculator to find the intersection points of the graphs of f and g. What do these points mean in this situation?

13. Describe how to find an equation of a quadratic function that can be used to model data whose scattergram suggests a quadratic relationship.

14. Describe how to determine whether a linear function, an exponential function, a quadratic function, or none of these can be used to model an authentic situation. Discuss at least two criteria you can use to help you select a type of model.

Related Review

15. The number of people waiting for organ transplants has increased greatly since 1988 (see Table 52).

Table 52 Number of People Waiting for Organ Transplants

Year	Number of People Waiting (thousands)
1988	16
1990	22
1992	30
1994	37
1996	49
1998	62
2000	75
2001	80

Source: *United Network for Organ Sharing*

Let $f(t)$ be the number of people (in thousands) waiting for organ transplants at t years since 1988.

a. Find a linear equation, an exponential equation, and a quadratic equation of f. Compare how well the models fit the data.

b. For the years before 1980, which of the three models likely gives the best estimates of the number of people waiting for organ transplants? [**Hint:** Use a graphing calculator to sketch graphs of the three equations. If you used ZoomStat to form your window, now use Zoom Out.]

16. The number of Home Depot® stores has increased greatly since 1985 (see Table 53). Let $f(t)$ be the number of Home Depot stores at t years since 1985.

Table 53 Numbers of Home Depot Stores

Year	Number of Stores
1985	50
1988	96
1991	174
1994	340
1997	624
2000	1134
2002	1532
2004	1890

Source: The New York Times

a. Find a linear equation, an exponential equation, and a quadratic equation of f. Compare how well the models fit the data.

b. For years before 1987, which of the three models likely gives the best estimates of the number of stores? [**Hint:** Use a graphing calculator to sketch graphs of the three equations. If you used ZoomStat to form your window, now use Zoom Out.]

Expressions, Equations, Functions, and Graphs

Perform the indicated instruction. Then use words such as linear, quadratic, cubic, exponential, logarithmic, polynomial, degree, function, one variable, *and* two variables *to describe the expression, equation, or system.*

17. Solve:

$$\frac{1}{2}x - \frac{2}{3}y = 2$$
$$\frac{4}{3}x + \frac{5}{2}y = 31$$

18. Solve $3x(x-2) = 5 - 2x^2$.

19. Graph $\frac{1}{2}x - \frac{2}{3}y = 2$ by hand.

20. Find the product $(4x-5)(2x^2+x-3)$.

21. Find an equation of a line that contains the points $(-5, -2)$ and $(-2, -7)$.

22. Simplify $\left(\frac{8b^{-2}c^{-4}}{27b^7c^{-10}}\right)^{-1/3}$.

7.8 MODELING WITH QUADRATIC FUNCTIONS

Objectives

- Use a quadratic model to make estimates and predictions.
- Estimate the maximum or minimum value of a quantity.
- Use a system of two quadratic equations to make estimates and predictions.
- Find the maximum revenue of a business venture.

In Section 7.7, we discussed how to find an equation of a quadratic model. In this section, we will use such an equation to make estimates and predictions about an authentic situation. We will also use a system of two quadratic models to make predictions.

Using a Quadratic Function to Make Predictions

In Example 1, we use a quadratic model to make some predictions.

Table 54 Cumulative Unredeemed Miles

Year	Cumulative Unredeemed Miles (trillions of miles)
1986	0.1
1988	0.4
1990	0.9
1992	1.5
1994	2.2
1996	3.2
1998	4.6
2000	6.6
2002	9.1
2004	12.4
2005	14.2

Source: *WebFlyer*

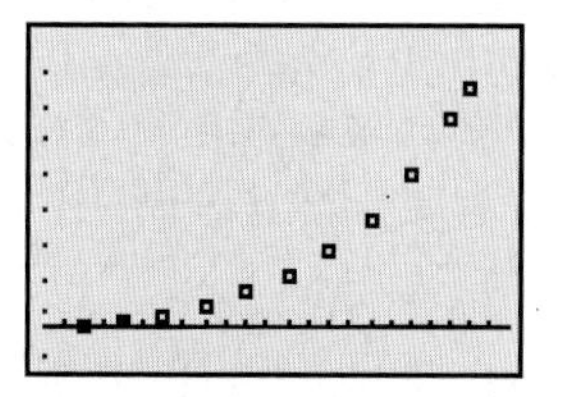

Figure 96 Unredeemed-miles scattergram

Example 1 Using a Quadratic Model to Make Predictions

Airlines originated frequent-flier programs in 1981. The cumulative unredeemed miles are the total number of frequent-flier miles that members have *not* redeemed (spent) from 1981 through a specified year (see Table 54). Let c be the cumulative unredeemed miles (in trillions of miles) at t years since 1980.

1. Find a model to describe the situation.
2. In what years is there model breakdown for certain?
3. Predict the cumulative unredeemed miles in 2010.
4. Predict when the cumulative unredeemed miles will be 25 trillion miles.

Solution

1. First, we use a graphing calculator to draw a scattergram to describe the data (see Fig. 96). Since the points suggest a curve that "bends," we will not use a linear function to model the data. To decide between an exponential model and a quadratic model, we use a graphing calculator to find the exponential regression equation

and the quadratic regression equation:

$$E(t) = 0.064(1.26)^t \qquad \text{Exponential regression equation}$$

$$Q(t) = 0.047t^2 - 0.77t + 3.5 \qquad \text{Quadratic regression equation}$$

Next, we see how well each regression model fits the data (see Figs. 97 and 98).

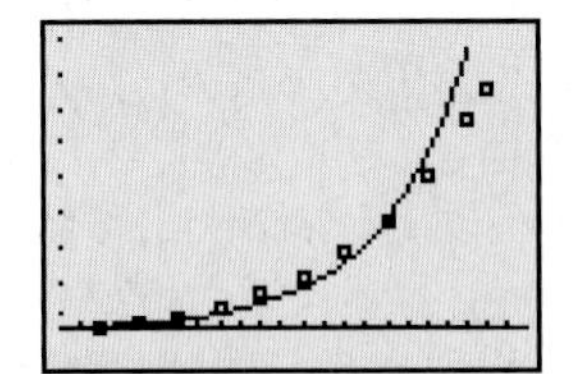

Figure 97 Exponential regression model

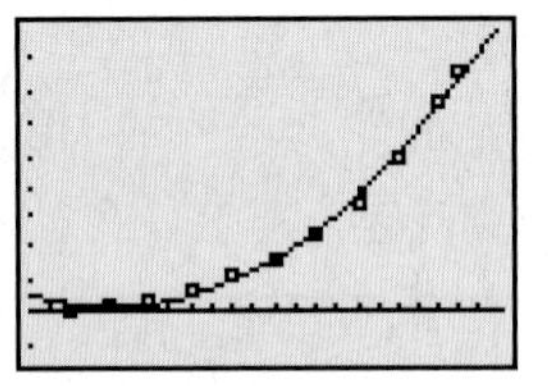

Figure 98 Quadratic regression model

It appears that Q fits the data points better than E does. Since it is likely that Q will give better predictions than E, we will use the quadratic function $Q(t) = 0.047t^2 - 0.77t + 3.5$ as our model.

2. The part of the parabola in Fig. 98 that lies to the left of the vertex suggests that the cumulative unredeemed miles decreased in those years, which is false—cumulative unredeemed miles always increase. So, model breakdown occurs for the part of the model that lies to the left of the vertex. To find the t-coordinate of the vertex, we use the vertex formula $t = -\frac{b}{2a}$. Since our model is $Q(t) = 0.047t^2 - 0.77t + 3.5$, we substitute $a = 0.047$ and $b = -0.77$ into the formula:

$$t = -\frac{-0.77}{2(0.047)} \approx 8.19$$

To find the c-coordinate of the vertex, we evaluate Q at 8.19:

$$Q(8.19) = 0.047(8.19)^2 - 0.77(8.19) + 3.5 \approx 0.35$$

So, the approximate vertex is (8.19, 0.35). To verify our work, we use "minimum" on a graphing calculator (see Fig. 99).

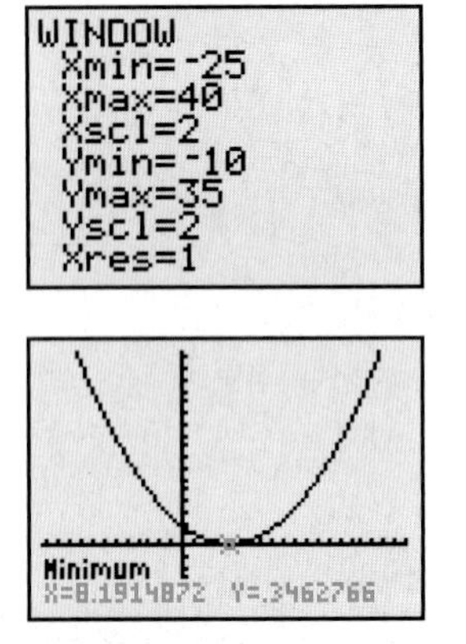

Figure 99 Verify that the approximate vertex is (8.19, 0.35)

We conclude that model breakdown occurs for the years up to 1988.

3. To find the cumulative unredeemed miles in 2010, we evaluate Q at 30:

$$Q(30) = 0.047(30)^2 - 0.77(30) + 3.5 = 22.7$$

So, there will be 22.7 trillion unredeemed miles in 2010, according to the model.

4. To find when there will be 25 trillion unredeemed miles, we substitute 25 for $Q(t)$ and solve for t:

$$0.047t^2 - 0.77t + 3.5 = 25 \qquad \text{Substitute 25 for } Q(t).$$

$$0.047t^2 - 0.77t - 21.5 = 0 \qquad \text{Subtract 25 from both sides.}$$

Next, we apply the quadratic formula:

$$t = \frac{-(-0.77) \pm \sqrt{(-0.77)^2 - 4(0.047)(-21.5)}}{2(0.047)} \qquad \text{Substitute } a = 0.047, b = -0.77, \text{ and } c = -21.5 \text{ in quadratic formula.}$$

$$t \approx -14.71 \quad \text{or} \quad t \approx 31.09 \qquad \text{Compute.}$$

We can use a graphing calculator to verify our work (see Fig. 100). The values of t we found represent the years 1965 and 2011. Model breakdown occurs for 1965, because the frequent-flier programs began in 1981. In Problem 2, we decided that model breakdown occurs for years before 1988, which checks. So, we predict that the cumulative unredeemed miles will be 25 trillion miles in 2011. ■

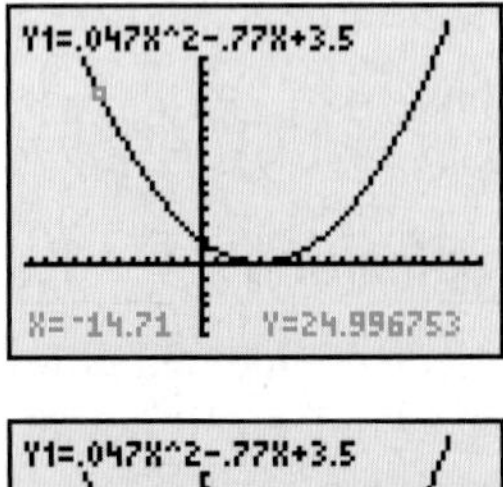

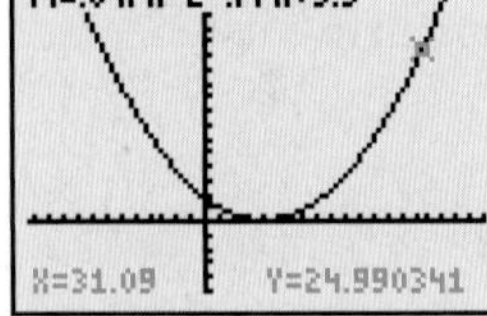

Figure 100 Verify that $t \approx -14.71$ and $t \approx 31.09$

In Example 1, we used a quadratic model to make predictions. Recall from Section 6.6 that to make a prediction about the dependent variable, we substitute a value for the independent variable and solve for the dependent variable. To make a prediction about the independent variable, we substitute a value for the dependent variable and solve for the independent variable, usually by using the quadratic formula.

Finding the Maximum or Minimum Value of a Quantity

In Example 2, we find the vertex of a quadratic function to help us determine the maximum value of the function.

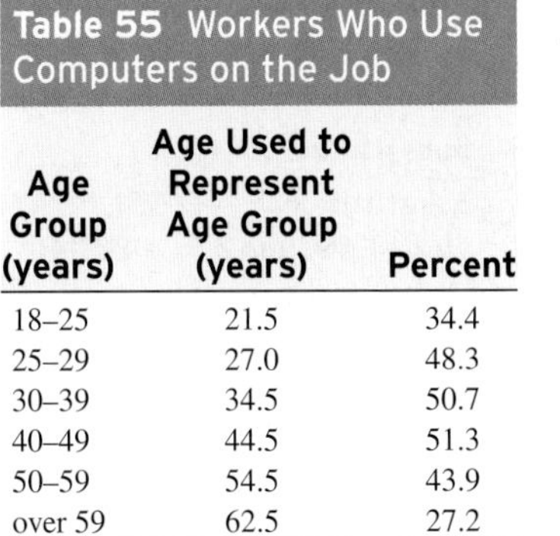

Table 55 Workers Who Use Computers on the Job

Age Group (years)	Age Used to Represent Age Group (years)	Percent
18–25	21.5	34.4
25–29	27.0	48.3
30–39	34.5	50.7
40–49	44.5	51.3
50–59	54.5	43.9
over 59	62.5	27.2

Source: *National Center for Education Statistics*

Example 2 Making Estimates

Table 55 lists the percentages of workers who use computers on the job by age groups. Let $p = f(t)$ be the percentage of workers who use computers at age t years.

1. Find a formula of a function that provides a reasonable model of the computer data.
2. Use f to estimate the percentage of 22-year-old workers who use computers on the job.
3. Estimate the age(s) at which half of workers use computers on the job.
4. Estimate the age of workers who are *most likely* to use computers on the job (maximum percentage). What is that maximum percentage?
5. Find the t-intercepts. What do they mean in this situation?

Solution

1. We begin by drawing a scattergram of the data (see Fig. 101). It looks like a parabola would fit the data well. We can use a graphing calculator to find the quadratic regression model:

$$f(t) = -0.051t^2 + 4.09t - 28.14$$

To verify this result, we draw the scattergram and the graph of f in the same viewing window (see Fig. 102). It appears that f is a reasonable model of the data.

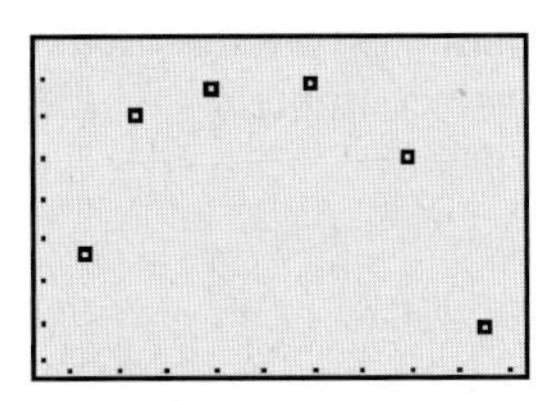

Figure 101 Computer worker scattergram

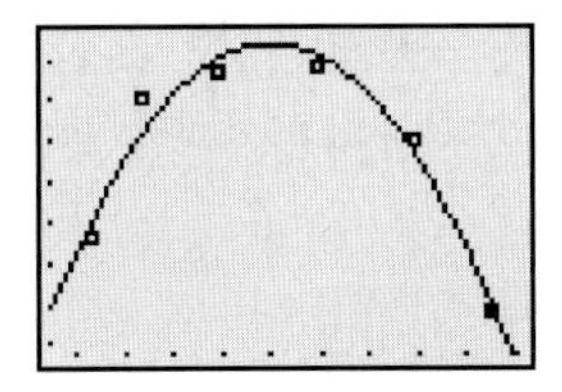

Figure 102 Verify computer worker model

2. Since

$$f(22) = -0.051(22)^2 + 4.09(22) - 28.14 \approx 37.16$$

we estimate that about 37.2% of 22-year-old workers use computers at work. We can verify this computation by using a graphing calculator table or graph.

3. Half of all workers is 50%, so we substitute 50 for $f(t)$ in the equation $f(t) = -0.051t^2 + 4.09t - 28.14$ and solve for t:

$$50 = -0.051t^2 + 4.09t - 28.14$$

Next, we write the equation in $at^2 + bt + c = 0$ form:

$$-0.051t^2 + 4.09t - 78.14 = 0$$

Then we apply the quadratic formula:

$$t = \frac{-4.09 \pm \sqrt{4.09^2 - 4(-0.051)(-78.14)}}{2(-0.051)} \qquad \text{Substitute } a = -0.051,\ b = 4.09, \text{ and } c = -78.14 \text{ in quadratic formula.}$$

$$t \approx 48.80 \quad \text{or} \quad t \approx 31.40 \qquad \text{Compute.}$$

So, according to our model, half of 31-year-old workers and half of 49-year-old workers use computers on the job. We can verify these results by using a graphing calculator table or graph.

4. The function f is of the form $f(t) = at^2 + bt + c$, with $a = -0.051 < 0$, so the graph of f is a parabola that opens downward (see Fig. 102). Therefore, the vertex is the maximum point. We can find the age when workers are most likely to use computers by finding the vertex.

To find the t-coordinate of the vertex, we substitute $a = -0.051$ and $b = 4.09$ in the vertex formula $t = -\dfrac{b}{2a}$:

$$t = -\frac{4.09}{2(-0.051)} \approx 40.10$$

The t-coordinate is about 40. To find the p-coordinate, we evaluate f at 40:

$$f(40) = -0.051(40)^2 + 4.09(40) - 28.14 = 53.86$$

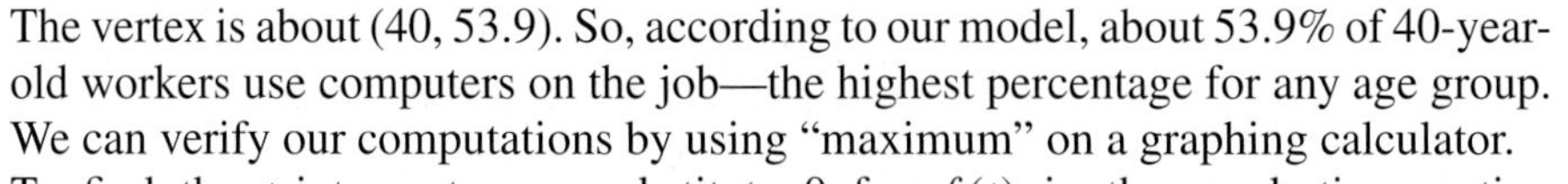

The vertex is about (40, 53.9). So, according to our model, about 53.9% of 40-year-old workers use computers on the job—the highest percentage for any age group. We can verify our computations by using "maximum" on a graphing calculator.

5. To find the t-intercepts, we substitute 0 for $f(t)$ in the quadratic equation $f(t) = -0.051t^2 + 4.09t - 28.14$ and solve for t:

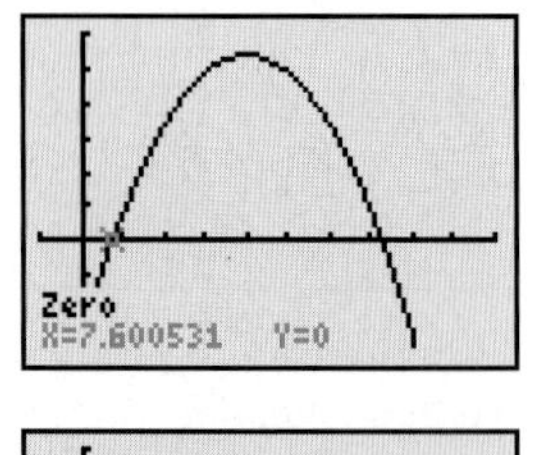

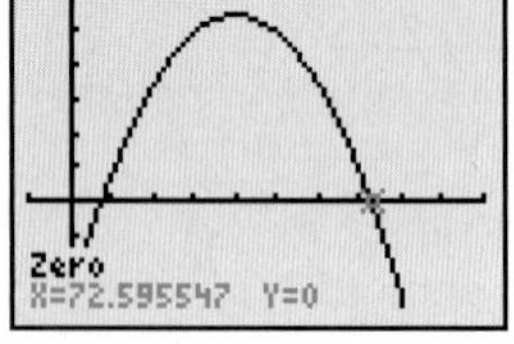

Figure 103 Verify the intercepts

$$0 = -0.051t^2 + 4.09t - 28.14 \qquad \text{Substitute 0 for } f(t).$$

$$t = \frac{-4.09 \pm \sqrt{4.09^2 - 4(-0.051)(-28.14)}}{2(-0.051)} \qquad \text{Substitute } a = -0.051,\ b = 4.09, \text{ and } c = -28.14 \text{ in quadratic formula.}$$

$$t \approx 72.60 \quad \text{or} \quad t \approx 7.60 \qquad \text{Compute.}$$

The t-intercepts are approximately (7.6, 0) and (72.6, 0). We use "zero" on a graphing calculator to verify our work (see Fig. 103). See Section B.21 for graphing calculator instructions.

The model estimates that no 8-year-old workers use computers on the job; this is correct but only because, legally, 8-year-old children cannot work. The model also estimates that no 73-year-old workers use computers on the job. Model breakdown has occurred, because some 73-year-old workers do use computers. ■

Modeling with a System of Quadratic Equations

In Example 3, we make an estimate by working with a system of two quadratic equations in two variables.

Example 3 Modeling with a System of Quadratic Equations

The numbers of videocassettes and DVDs bought by U.S. dealers to sell as rentals are shown in Table 56 for various years.

Table 56 Numbers of Videocassettes and DVDs Bought by U.S. Dealers

Year	Number of Videocassettes (millions of units)	Number of DVDs (millions of units)
1998	57.0	1.6
1999	86.2	8.6
2000	99.4	13.9
2001	86.2	37.1
2002	73.6	79.3
2003	53.2	110.9

Source: *Adams Media Research*

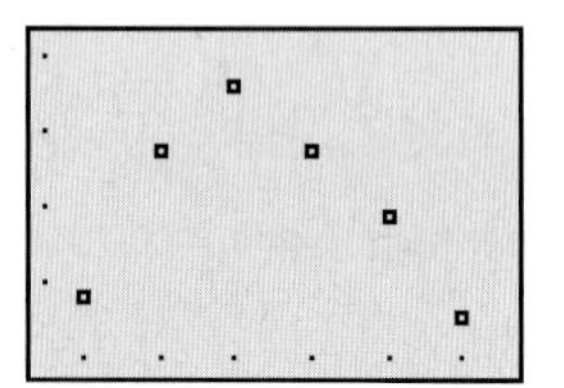

Figure 104 Videocassette scattergram

Estimate when sales of DVDs to dealers overtook sales of videocassettes to dealers.

Solution

Let $v(t)$ and $d(t)$ be the numbers (in millions of units) of videocassettes and DVDs, respectively, bought by U.S. dealers during the year that is t years since 1990. Scattergrams of the data suggest that quadratic functions will model numbers of videocassettes and DVDs bought by dealers (see Figs. 104 and 105).

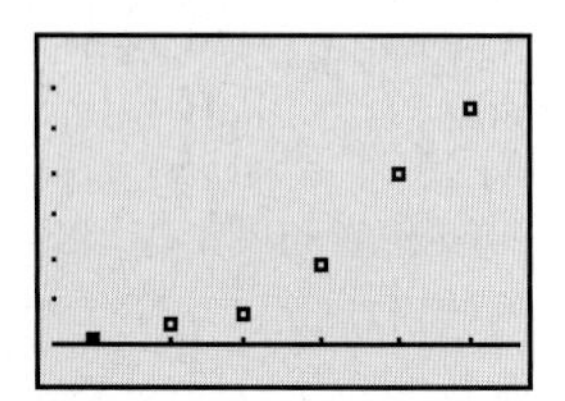

Figure 105 DVD scattergram

Regression equations of v and d are, respectively,

$$S = v(t) = -6.27t^2 + 129.7t - 576.2$$
$$S = d(t) = 4.83t^2 - 79.1t + 326.0$$

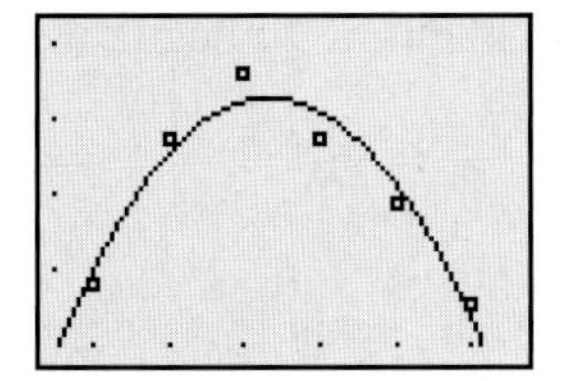

Figure 106 Videocassette scattergram and model

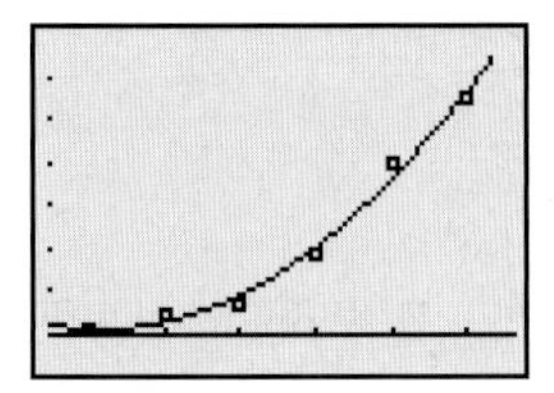

Figure 107 DVD scattergram and model

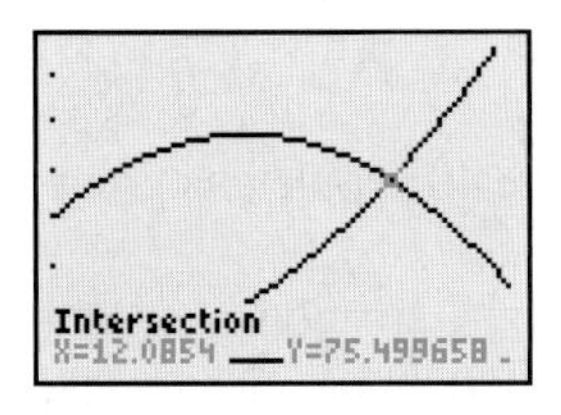

Figure 108 Verify the estimate

The models fit the points in the scattergrams of the two data sets fairly well (see Figs. 106 and 107).

To find when numbers of DVDs bought by dealers equaled numbers of videocassettes bought by dealers, we substitute $-6.27t^2 + 129.7t - 576.2$ for S in the equation $S = 4.83t^2 - 79.1t + 326.0$:

$$-6.27t^2 + 129.7t - 576.2 = 4.83t^2 - 79.1t + 326.0$$
$$-11.1t^2 + 208.8t - 902.2 = 0$$

The quadratic formula gives

$$t = \frac{-208.8 \pm \sqrt{208.8^2 - 4(-11.1)(-902.2)}}{2(-11.1)}$$

$$t \approx 12.09 \quad \text{or} \quad t \approx 6.73$$

The values of t we found represent 1997 and 2002. The year 1997 is not relevant, because we are interested in what happened between 1998 and 2003. So, sales of DVDs to dealers approximately equaled sales of videocassettes to dealers in 2002, according to our models. We use "intersect" on a graphing calculator to verify our work (see Fig. 108). ■

In Example 3, we used substitution with a system of two quadratic equations in two variables. **We can use substitution to solve any system of two quadratic equations if both are in standard form $y = ax^2 + bx + c$.**

Maximum Revenue

Recall from Section 3.4 that if each of n objects has value v, then their total value T is given by $T = vn$. In Example 4, we will find the maximum value of the revenue of a chartered flight.

Example 4 Finding the Maximum Value of a Quantity

A group charters a flight that normally costs \$800 per person. A group discount reduces the fare by \$10 for each ticket sold; the more tickets sold, the lower the per-person fare. There are 60 seats on the plane, including 4 seats for the crew. What size of group would maximize the airline's revenue?

Solution

We use the five-step problem-solving method of Section 3.4, but in step 2 we use a system of equations to build a function and in step 3 we find the maximum point of the graph of the function.

Step 1: Define each variable. We let n be the number of people in the group, p be the price (in dollars per person), and R be the revenue (in dollars).

Step 2: Use a system of two equations to build a function. A group of 3 people would be charged $800 - 3(10) = 770$ dollars per person. A group of 4 people would be charged $800 - 4(10) = 760$ dollars per person. So, a group of n people would be charged $800 - n(10)$ dollars per person. Our first equation is

$$p = -10n + 800$$

For our second equation, the revenue is equal to the price of one ticket times the number of tickets sold:

$$\overbrace{R}^{\text{revenue}} = \overbrace{p}^{\frac{\text{dollars}}{\text{ticket}}} \cdot \overbrace{n}^{\text{tickets}}$$

So, our system is

$$p = -10n + 800 \quad \text{Equation (1)}$$
$$R = pn \quad \text{Equation (2)}$$

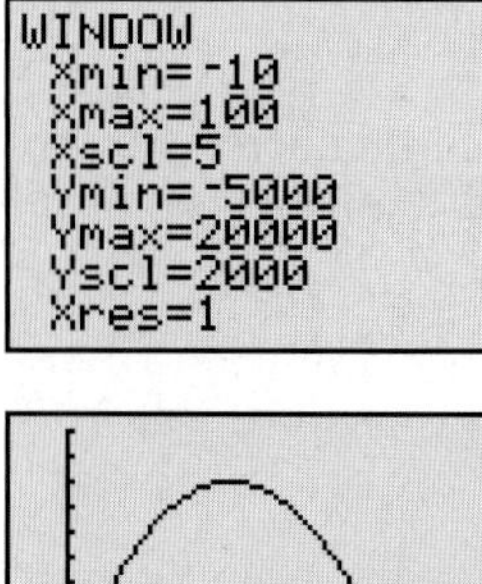

Figure 109 Graph of $R = -10n^2 + 800n$

Next, we will build a revenue function whose independent variable is n. To begin, we substitute $-10n + 800$ for p in equation (2):

$$R = pn \qquad \text{Equation (2)}$$

$$R = (-10n + 800)n \qquad \text{Substitute } -10n + 800 \text{ for } p.$$

$$R = -10n^2 + 800n \qquad \text{Distributive law}$$

Step 3: Find the maximum point of the graph of the function. The graph of the function $R = -10n^2 + 800n$ is a parabola that opens downward (see Fig. 109). So, the parabola has a maximum point (at the vertex).

To find the n-coordinate of the vertex, we use the vertex formula $n = -\frac{b}{2a}$:

$$n = -\frac{800}{2(-10)} = 40$$

To find the R-coordinate of the vertex, we substitute 40 for n in the equation $R = -10n^2 + 800n$:

$$R = -10(40)^2 + 800(40) = 16{,}000$$

To find the price, we substitute 40 for n in the equation $p = -10n + 800$:

$$p = -10(40) + 800 = 400$$

Step 4: Describe each result. The revenue is greatest if there are 40 people in a group. The price is \$400 per person, and the revenue is \$16,000.

Step 5: Check. The revenue from 40 people, each paying \$400, is $40 \cdot 400 = 16{,}000$ dollars, which checks. To check that the largest revenue is \$16,000 dollars, we find the revenue for groups with 1 person fewer or 1 person more than 40 people.

39 people in group

price is $800 - 39(10) = 410$ dollars
revenue is $410 \cdot 39 = 15{,}990$ dollars

41 people in group

price is $800 - 41(10) = 390$ dollars
revenue is $390 \cdot 41 = 15{,}900$ dollars

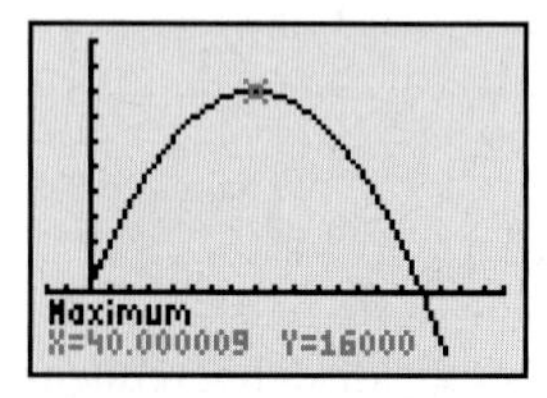

Figure 110 Verify that the maximum point is (40, 16,000)

Since a revenue of \$16,000 is more than either of these (equal) revenues of \$15,900, it seems reasonable that our work is correct. Finally, we use "maximum" on a graphing calculator to check that the maximum point of the parabola $R = -10n^2 + 800n$ is (40, 16,000). See Fig. 110. ■

group exploration

Modeling differences of quantities

Refer to Example 3 for background information on numbers of videocassettes and DVDs bought by U.S. dealers.

1. Complete Table 57.

Table 57 Differences of Numbers of Videocassettes and DVDs Bought by U.S. Dealers

Year	Number of Videocassettes (millions of units)	Number of DVDs (millions of units)	Difference in Numbers of Videocassettes and DVDs (millions of units)
1998	57.0	1.6	$57.0 - 1.6 = 55.4$
1999	86.2	8.6	
2000	99.4	13.9	
2001	86.2	37.1	
2002	73.6	79.3	
2003	53.2	110.9	

2. Let $D(t)$ be the difference in numbers of videocassettes and numbers of DVDs (in millions of units) bought by U.S. dealers during the year that is t years since 1990. Find a regression equation of D. Remember to draw a scattergram of the data first.
3. Which of the following statements is correct? Explain.

$$D(t) = (v + d)(t) \qquad D(t) = (v - d)(t) \qquad D(t) = (d - v)(t)$$

4. In Example 3, we used the following equations for v and d:

$$S = v(t) = -6.27t^2 + 129.7t - 576.2$$
$$S = d(t) = 4.83t^2 - 79.1t + 326.0$$

Substitute $-6.27t^2 + 129.7t - 576.2$ for $v(t)$ and $4.83t^2 - 79.1t + 326.0$ for $d(t)$ in the equation $D(t) = v(t) - d(t)$ to find an equation of D. Simplify your equation and compare it with the equation you found in Problem 2.
5. Find t when $D(t) = 0$. What does your result mean in terms of videocassettes? Compare your result with the result in Example 3.

TIPS FOR SUCCESS: Form a Study Group to Prepare for the Final Exam

To prepare for the final exam, it may be helpful to form a study group. The group could list important concepts of the course and discuss the meaning of these concepts and how to apply them. You could also list important techniques learned in the course and practice these techniques. Set aside some solo study time after the group study session to make sure you can do the mathematics without the help of other members of the study group.

HOMEWORK 7.8

FOR EXTRA HELP ▶

Student Solutions Manual — PH Math/Tutor Center — MathXL® — MyMathLab

1. In Exercise 49 of Homework 7.2, you worked with the model $f(t) = 0.0075t^2 - 0.2t + 2.65$, where $p = f(t)$ is the price (in 2006 dollars per gallon) at t years since 1980 (see Table 58).

Table 58 Average Prices of Gasoline

Year	Price (2006 dollars per gallon)
1980	2.59
1985	1.90
1990	1.51
1995	1.28
2001	1.66
2005	2.28
2006*	2.68

Source: *Department of Energy*
*January to May

a. Find the p-intercept of the model. What does it mean in this situation?
b. Estimate the price of gasoline in 2003. Did you perform interpolation or extrapolation?
c. Predict the price of gasoline in 2010. Did you perform interpolation or extrapolation?
d. Predict when the price of gasoline will be $5 per gallon (in 2006 dollars).

2. In Exercise 10 of Homework 7.7, you found an equation close to $f(A) = -0.0197A^2 + 1.86A + 9.90$, where $f(A)$ is the percentage of Americans at age A who say they do volunteer work (see Table 59).

Table 59 Percentages of Americans Who Say They Volunteer

Age Group (years)	Age Used to Represent Age Group (years)	Percent
18–24	21.0	38
25–34	29.5	51
35–54	44.5	55
55–64	59.5	48
65–74	69.5	45
over 74	80.0	34

Source: *The Gallup Organization*

a. Estimate the percentage of 25-year-old Americans who say they volunteer. Did you perform interpolation or extrapolation?
b. Estimate the percentage of 16-year-old Americans who say they volunteer. Did you perform interpolation or extrapolation?

c. For what age do half of Americans say they volunteer?
d. For what age is the percentage who say they volunteer the greatest? What is that maximum percentage?

3. In Exercise 7 of Homework 7.7, you modeled the percentage of firms $f(t)$ that perform drug tests on employees and/or job applicants at t years since 1985 (see Table 60). The quadratic regression equation is $f(t) = -0.64t^2 + 14.47t - 3.40$.

Table 60 Percentages of Firms That Perform Drug Tests

Year	Percent
1987	22
1990	51
1992	72
1995	78
1998	74
2000	66
2002	62

Source: *American Management Association*

a. Find the t-intercepts. What do they mean in this situation?
b. For what values of t is there model breakdown for certain? Which years are represented by these values?
c. Estimate the maximum percentage of firms that performed drug tests. Estimate the year in which this percentage occurred.
d. Estimate when half of firms performed drug tests.

4. A householder is the person in whose name a house, condominium, or apartment is owned or rented. The percentages of householders who own a home are shown in Table 61 for various age groups.

Table 61 Percentages of Householders Who Own a Home

Age Group (years)	Age Used to Represent Age Group (years)	Percent Who Own a Home
15–24	19.5	18
25–34	29.5	46
35–44	39.5	66
45–54	49.5	75
55–64	59.5	80
65–74	69.5	81
75–84	79.5	77

Source: *U.S. Census Bureau*

Let $f(a)$ be the percentage of householders who own a home at age a years.
a. Find an equation of f.
b. Find the a-intercepts. What do they mean in this situation?
c. For what values of a is there model breakdown for certain? Which years are represented by these values?
d. Estimate the age of householders who are most likely to own a home. What percentage of householders at this age own a home?
e. Estimate all ages at which half of householders own homes.

5. In Exercise 9 of Homework 7.7, you found an equation close to $P = f(t) = 0.0068t^2 - 0.13t + 6.55$, where $f(t)$ is the U.S. population (in millions) at t years since 1790 (see Table 62).

Table 62 U.S. Population

Year	Population (millions)	Year	Population (millions)
1790	3.9	1910	92.4
1800	5.3	1920	106.5
1810	7.2	1930	123.2
1820	9.6	1940	132.1
1830	12.9	1950	151.3
1840	17.1	1960	179.3
1850	23.3	1970	203.3
1860	31.5	1980	226.5
1870	39.9	1990	248.8
1880	50.3	2000	281.4
1890	63.1	2006	299.0
1900	76.1		

Source: *U.S. Census Bureau*

a. Find $f(220)$. What does it mean in this situation?
b. Find t where $f(t) = 315$. What does your result mean in this situation?
c. Graph f by hand.
d. For what values of t is there model breakdown? Which years are represented by these values?
e. Sketch a qualitative graph that describes the relationship between t and P for all time, past and future.

6. The average times it takes for lightweight vehicles to accelerate from 0 mph to 60 mph (average acceleration time) are shown in Table 63 for various years.

Table 63 Average Acceleration Times

Year	Average Acceleration Time (seconds)
1985	13.5
1990	11.9
1995	11.1
2000	10.7
2004	10.4

Source: *Environmental Protection Agency*

Let $A = f(t)$ be the average acceleration time (in seconds) for a lightweight vehicle at t years since 1985.
a. Find an equation of f.
b. Find $f(12)$. What does it mean in this situation?
c. Find t when $f(t) = 11.5$. What does it mean in this situation?
d. Find the vertex of the model. What does it mean in this situation?
e. Assuming that the acceleration time of lightweight vehicles continues to decrease,
   i. Find the values of t for which there would be model breakdown for certain. Which years are represented by those values?
   ii. Sketch a qualitative graph that describes the relationship between t and A for all time, past and future.

7. The total numbers of home runs hit by Barry Bonds are shown in Table 64 for various years.

Table 64 Total Numbers of Home Runs Hit by Barry Bonds

Year	Total Number of Home Runs
1986	16
1990	117
1995	292
2000	494
2004	703

Source: *Baseball-reference.com*

Let $f(t)$ be the total number of home runs hit by Barry Bonds at t years since 1980.

a. Find an equation of f.

b. Use f to estimate when Barry Bonds tied Hank Aaron's record of 755 lifetime home runs.

c. Due to injury, Bonds hit only 5 home runs in 2005. Find $f(25)$ and $f(26)$. Then find the difference $f(26) - f(25)$. Does your result suggest that Bonds tied Aaron's record in 2006? Explain.

d. Bonds turned 42 years old on July 24, 2006. Discuss any assumptions that you made in part (c), and explain why Bonds may not have tied Aaron's record in 2006.

8. The percentages of Americans who say they've forgotten to do something special for their significant other on Valentine's Day are shown in Table 65 for various ages.

Table 65 Americans Who Forgot to Do Something Special on Valentine's Day

Age Group (years)	Age Used to Represent Age Group (years)	Percent
25–34	29.5	11
35–44	39.5	17
45–54	49.5	24
55–64	59.5	27
over 64	70.0	29

Source: *CARAVAN for DHL*

Let $f(a)$ be the percentage of Americans at age a years who say they've forgotten to do something special for their significant other on Valentine's Day.

a. Find an equation of f.

b. Find the a-intercepts. What do they mean in this situation?

c. Find the vertex. What does it mean in this situation?

d. Estimate the percentage of 21-year-old Americans who say they've forgotten to do something special for their significant other on Valentine's Day. The actual percent is 19%. Has model breakdown occurred? Explain.

e. For what values of a is there model breakdown for certain? Which years are represented by these values? Are there other values of a for which it is likely there is model breakdown? If so, find those values of a and the years represented by them.

9. In Exercise 11 of Homework 7.7, beer sales are modeled by the system

$$s = f(t) = 0.18t^2 + 0.2t + 3.9$$
$$s = g(t) = 0.87t + 11.1$$

where $f(t)$ and $g(t)$ represent annual beer sales (in millions of cases) of Miller Lite beer and Tecate beer, respectively, at t years since 2000 (see Table 66).

Table 66 Annual Sales of Miller Lite Beer and Tecate Beer

	Sales of 2.25-Gallon Cases (millions)	
Year	Miller Lite	Tecate
2000	3.9	11.0
2001	4.3	12.0
2002	4.9	13.1
2003	6.2	13.5
2004	7.5	14.6

Source: *Adams Beverage Group*

Use substitution or elimination to predict when the annual sales of the two brands of beer will be equal.

10. Women tend to be younger than men in their first marriage. The median ages at which women and men were first married are shown in Table 67 for various years.

Table 67 Median Ages at First Marriages

	Median Age	
Year	Women (years)	Men (years)
1940	21.5	24.3
1950	20.3	22.8
1960	20.3	22.8
1970	20.8	23.2
1980	22.0	24.5
1990	24.0	26.3
2000	25.1	26.8

Source: *U.S. Census Bureau*

Let $W(t)$ be the median age (in years) at which women were first married and $M(t)$ be the median age (in years) at which men were first married, both at t years since 1900.

a. Find a regression equation of women's median marrying age and another regression equation of men's median marrying age.

b. Predict when the median age or ages at which women and men first marry will be equal, if ever.

11. In the 2000 presidential election, the race between George W. Bush and Al Gore was so close that there was concern about the accuracy of vote-counting systems, including punch cards, lever machines, and paper ballots. Punch cards and lever systems are gradually being replaced with optical scan or other modern electronic vote-counting systems (see Table 68).

a. Let $f(t)$ be the percentage of voters using punch cards or lever machines and $g(t)$ be the percentage of voters using optical scan or other modern electronic systems, both at t years since 1990. Find regression equations of f and g.

b. Use substitution to estimate when the percentage of voters using optical scan or other modern electronic systems equaled the percentage of voters using punch cards or lever machines. Is your result a major (even-numbered) election year (congressional or presidential)? If not, in which even-numbered year were the percentages closest?

Table 68 Vote-Counting Systems

	Percent of Registered Voters	
Year	**Punch Card or Lever Machine**	**Optical Scan or Other Modern Electronic System**
1990	71.7	13.5
1992	69.4	16.5
1994	63.1	25.8
1996	58.0	32.3
1998	52.9	36.4
2002	35.7	54.3
2004	26.0	65.0

Source: *Federal Election Commission*

c. There are three other ways to vote: DataVote (punch holes in ballots that haven't been perforated), paper (hand-marked ballots are counted manually), and mixed systems. Let $h(t)$ be the percentage of voters using any of these three ways to vote at t years since 1990. Which of the statements that follow is correct? Explain.

$$h(t) = (f+g)(t) \qquad h(t) = 100 - (f+g)(t)$$
$$h(t) = (f-g)(t) \qquad h(t) = 100 - (f-g)(t)$$

d. Find an equation of h.
e. Find $h(10)$. What does it mean in this situation?

12. In Exercise 12 of Homework 7.7, the percentages of California's population $f(t)$ and $g(t)$ who are foreign born and born in other U.S. states, respectively, are modeled by the system

$$p = f(t) = 0.0111t^2 - 1.32t + 48.24$$
$$p = g(t) = -0.0117t^2 + 1.17t + 22.79$$

where t is the number of years since 1900 (see Table 69).

Table 69 Californians Not Originally from California

	Percent	
Year	**Foreign Born**	**Born in Other U.S. States**
1930	18.9	47.0
1940	13.4	50.0
1950	10.0	53.0
1960	8.5	51.0
1970	8.8	47.9
1980	15.1	39.5
1990	21.7	31.8
2000	25.9	23.5

Source: *William Frey analysis of U.S. Census Bureau sources*

a. Use substitution to estimate when the percentages of foreign born and those born in other U.S. states were equal. What is that percentage? That year, what percentage of California's population was originally from California?
b. Let $h(t)$ be the percentage of California's population that was originally from California at t years since 1900. Which of the following statements is correct? Explain.

$$h(t) = 100 - (f-g)(t) \qquad h(t) = (f-g)(t)$$
$$h(t) = 100 - (f+g)(t) \qquad h(t) = (f+g)(t)$$

c. Find an equation of h.
d. Find $h(110)$. What does it mean in this situation?
e. Use a graphing calculator to sketch a graph of h. Is the curve increasing or decreasing for $t \geq 30$? What does it mean in this situation?

13. A ski club charters a bus that normally costs \$250 per person. A group discount reduces the fare by \$5 for each ticket sold; the more tickets sold, the lower the per-person fare. There are 31 seats on the bus, including 1 seat for the driver. What size of group would maximize the bus company's revenue?

14. A company charters a party boat that normally costs \$60 per person. A group discount reduces the fare by \$0.50 for each ticket sold; the more tickets sold, the lower the per-person fare. The maximum capacity of the boat is 80 people, including the crew of 10 people. What size of group would maximize the boat company's revenue?

15. Some students reserve the buffet room in a restaurant for a graduation party. The buffet dinner normally costs \$28 per person. A group discount reduces the price by \$0.20 for each student who attends; the more dinners sold, the lower the per-person price. The room's maximum capacity is 80 people. What size of party would maximize the restaurant's revenue?

16. A group charters a flight that normally costs \$900 per person. A group discount reduces the fare by \$15 for each ticket sold; the more tickets sold, the lower the per-person fare. There are 40 seats on the plane, including 4 seats for the crew. What size of group would maximize the airline's revenue?

17. A person plans to use 160 feet of fencing to enclose a rectangular play area. What dimensions of the rectangle would maximize the area? What is that area?

18. A rancher plans to use 1200 feet of fencing and the side of her barn to form a rectangular boundary for cattle (see Fig. 111). What dimensions of the rectangle would maximize the area? What is that area?

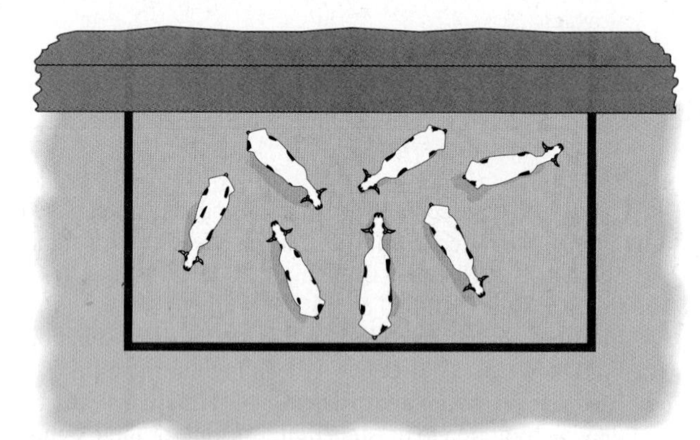

Figure 111 Boundary for cattle

19. For a quadratic model, discuss how to find intercepts, the vertex, and the maximum or minimum value; how to make predictions for the dependent or independent variable; and how to determine values of the independent variable where model breakdown occurs.

20. Describe how you can find a system of quadratic equations in two variables to model an authentic situation. Explain how you can use the system to make an estimate or a prediction about the situation.

Related Review

21. Harnessing wind energy is much more affordable and cleaner than processing oil or coal. Table 70 lists worldwide wind-generating capacities in thousands of megawatts (thousand MW) for various years.

Table 70 Worldwide Wind-Generating Capacities

Year	Wind-Generating Capacity (thousand MW)
1990	1.9
1992	2.5
1994	3.7
1996	6.1
1998	9.7
2000	18.0
2002	31.2
2004	47.6

Source: *World Wind Energy Association*

Let $f(t)$ be the wind-generating capacity (in thousand MW) at t years since 1990.

a. Find a linear regression equation, an exponential regression equation, and a quadratic regression equation of f. Compare how well the models fit the data.

b. For the years before 1990, which equation likely best models the situation?

c. One MW of wind-generating capacity meets the electricity needs of about 1000 people in an industrialized country. Use your exponential model to predict when wind energy could meet the electricity needs of the entire world. Assume that world population is 6.5 billion. [**Hint:** Note that the units of $f(t)$ are *thousands* of megawatts.]

d. Now use your quadratic model to predict when wind energy could meet the electricity needs of the entire world.

e. Explain why the year you predicted in part (d) is so much later than the year you predicted in part (c).

22. The total numbers of time-share weeks owned by vacationers are shown in Table 71 for various years.

Table 71 Total Numbers of Time-share Weeks Owned by Vacationers

Year	Total Number of Time-share Weeks (millions)
1980	0.4
1985	1.0
1990	1.7
1995	2.6
2000	3.8
2003	4.9

Source: *Ragatz Associates*

Let $f(t)$ be the total number of time-share weeks (in millions) owned by vacationers at t years since 1980.

a. Find a linear regression equation, an exponential regression equation, and a quadratic regression equation of f. Compare how well the models fit the data.

b. For the years before 1980, which equation likely best models the situation? Explain.

c. Use the exponential model to predict when vacationers will own a total of 10 million time-share weeks.

d. Use the quadratic model to predict when vacationers will own a total of 10 million time-share weeks.

e. Explain why the year you predicted in part (d) is later than the year you predicted in part (c).

Expressions, Equations, Functions, and Graphs

Give an example of the following. Then solve, simplify, or graph, as appropriate.

23. exponential function
24. linear function
25. quadratic equation in one variable
26. system of two linear equations in two variables
27. quadratic function
28. logarithmic function
29. exponential equation in one variable
30. logarithmic equation in one variable

Chapter Summary

Key Points OF CHAPTER 7

Vertex form of a quadratic function (Section 7.1)

Let $f(x) = a(x - h)^2 + k$, where $a \neq 0$. Then f is a quadratic function, and its graph is a parabola with vertex (h, k). We say that the equation is in **vertex form.**

Graphs of quadratic functions of the form $f(x) = ax^2$ (Section 7.1)

For a function of the form $f(x) = ax^2$,

- The graph is a parabola with vertex (0, 0).
- If $|a|$ is a large number, then the parabola is steep. (It is narrow.)
- If a is near zero, then the parabola is not steep. (It is wide.)
- If $a > 0$, then the parabola opens upward.
- If $a < 0$, then the parabola opens downward.
- The graph of $y = -ax^2$ is the reflection across the x-axis of the graph of $f(x) = ax^2$.

Three-step method of graphing a quadratic function in vertex form (Section 7.1)

To sketch the graph of $f(x) = a(x-h)^2 + k$, where $a \neq 0$,

1. Sketch the graph of $y = ax^2$.
2. Translate the graph from step 1 to the right by h units if $h > 0$ or to the left by $|h|$ units if $h < 0$.
3. Translate the graph from step 2 up by k units if $k > 0$ or down by $|k|$ units if $k < 0$.

Finding a quadratic model in vertex form (Section 7.1)

To find a quadratic model in vertex form, given some data,

1. Create a scattergram of the data.
2. Imagine a parabola that comes close to (or contains) the data points, and select a point (h, k) to be the vertex. Although it is not necessary to select a data point, it is often convenient and satisfactory to do so.
3. Select a nonvertex point (not necessarily a data point) of the parabola, substitute the point's coordinates into the equation $f(t) = a(t-h)^2 + k$, and solve for a.
4. Substitute the result you found for a in step 3 into $f(t) = a(t-h)^2 + k$.

Using two symmetric points to find the vertex (Section 7.2)

To find the vertex of a parabola in which (p, s) and (q, s) are symmetric points,

1. Find the x-coordinate by using the formula $x = \frac{p+q}{2}$. In words: The x-coordinate is the average of the x-coordinates of any two symmetric points.
2. Find the y-coordinate by evaluating f at the value found in step 1. That is, find $f\left(\frac{p+q}{2}\right)$.

Using symmetric points to graph a quadratic function in standard form (method 1) (Section 7.2)

To sketch a graph of a quadratic function $f(x) = ax^2 + bx + c$, where $b \neq 0$,

1. Find the y-intercept.
2. Find the y-intercept's symmetric point.
3. Average the x-coordinates of the two symmetric points to find the x-coordinate of the vertex.
4. Find the y-coordinate of the vertex.
5. Depending on how accurate your sketch is to be, find additional points on the parabola, as needed.
6. Sketch a parabola that contains the points found.

Vertex formula (Section 7.2)

To find the vertex of the graph of a quadratic function $f(x) = ax^2 + bx + c$,

1. Find the x-coordinate of the vertex by using the **vertex formula** $x = -\frac{b}{2a}$.
2. Find the y-coordinate of the vertex by evaluating f at the value found in step 1. That is, find $f\left(-\frac{b}{2a}\right)$.

Maximum or minimum value of a function (Section 7.2)

For a quadratic function $f(x) = ax^2 + bx + c$ whose graph has vertex (h, k),

- If $a < 0$, then the parabola opens downward and the maximum value of the function is k.
- If $a > 0$, then the parabola opens upward and the minimum value of the function is k.

Product property for square roots (Section 7.3)

For $a \geq 0$ and $b \geq 0$, $\sqrt{ab} = \sqrt{a}\sqrt{b}$.

Using the product property to simplify square roots (Section 7.3)

To use the product property to simplify a square root,

1. Write the radicand as the product of the *largest* perfect-square factor and another number.
2. Apply the product property for square roots.

Quotient property for square roots (Section 7.3)

For $a \geq 0$ and $b > 0$, $\sqrt{\frac{a}{b}} = \frac{\sqrt{a}}{\sqrt{b}}$.

Rationalize the denominator (Section 7.3)

To rationalize the denominator of a fraction of the form $\frac{p}{\sqrt{q}}$, where q is positive, multiply the fraction by 1 in the form $\frac{\sqrt{q}}{\sqrt{q}}$.

Simplifying a radical expression (Section 7.3)

To simplify a radical expression,

1. Use the quotient property for square roots so that no radicand is a fraction.
2. Use the product property for square roots so that no radicands have perfect-square factors other than 1.
3. Rationalize the denominators so that no denominator is a radical expression.
4. Continue applying steps 1–3 until the radical expression is completely simplified.

Square root property (Section 7.3)

Let k be a real-number constant. Then $x^2 = k$ is equivalent to $x = \pm\sqrt{k}$.

Imaginary unit i (Section 7.3)

The **imaginary unit,** written i, is the number whose square is -1. That is, $i^2 = -1$ and $i = \sqrt{-1}$.

Square root of a negative number (Section 7.3)

If p is a positive real number, then $\sqrt{-p} = i\sqrt{p}$.

Complex number (Section 7.3)

A **complex number** is a number of the form $a + bi$, where a and b are real numbers.

Imaginary number (Section 7.3)

An **imaginary number** is a number $a + bi$, where a and b are real numbers and $b \neq 0$.

Perfect-square trinomial (Section 7.4)

For a perfect-square trinomial of the form $x^2 + bx + c$, dividing b by 2 and squaring the result gives c:

$$x^2 + bx + c$$

$$\left(\frac{b}{2}\right)^2 = c$$

When completing the square can be used (Section 7.4)

Any quadratic equation can be solved by completing the square.

Solving a quadratic equation by completing the square (Section 7.4)

To solve a quadratic equation by **completing the square,**

1. Write the equation in the form $ax^2 + bx = k$, where a, b, and k are constants.
2. If $a \neq 1$, divide both sides of the equation by a.
3. Complete the square for the expression on the left side of the equation.
4. Solve the equation by using the square root property.

Quadratic formula (Section 7.5)

The solutions of a quadratic equation $ax^2 + bx + c = 0$ are given by the **quadratic formula:**

$$x = \frac{-b \pm \sqrt{b^2 - 4ac}}{2a}$$

Determining the number and type of solutions (Section 7.5)

For the quadratic equation $ax^2 + bx + c = 0$, the **discriminant** is $b^2 - 4ac$. Also,

- If $b^2 - 4ac > 0$, there are two real-number solutions.
- If $b^2 - 4ac = 0$, there is one real-number solution.
- If $b^2 - 4ac < 0$, there are two imaginary-number solutions (and no real-number solutions).

When the quadratic formula can be used (Section 7.5)

Any quadratic equation can be solved by using the quadratic formula.

Guidelines on solving quadratic equations (Section 7.5)

Here are some guidelines on deciding which method to use to solve a quadratic equation:

Method:	When to Use:
Factoring	For equations that can easily be put into the form $ax^2 + bx + c = 0$ and where $ax^2 + bx + c$ can easily be factored.
Square root property	For equations that can easily be put into the form $x^2 = k$ or $(x + p)^2 = k$.
Completing the square	When the directions require it.
Quadratic formula	For all equations except those that can easily be solved by factoring or by using the square root property.

Linear equation in three variables (Section 7.6)

A **linear equation in three variables** is an equation that can be put into the form $Ax + By + Cz = D$, where A, B, C, and D are constants and A, B, and C are not all zero.

System of linear equations in three variables (Section 7.6)

A **system of linear equations in three variables** consists of two or more linear equations in three variables.

Solution (Section 7.6)

The **solution** of a system of linear equations in three variables is an ordered triple that satisfies *all* of the equations.

Solving a system of three linear equations in three variables (Section 7.6)

To solve a system of three linear equations in three variables,

1. Select a pair of equations and eliminate a variable.
2. Select any other pair of equations and eliminate the *same variable* as in step 1.
3. The equations you found in steps 1 and 2 form a system of linear equations in *two* variables. Use elimination or substitution to solve this system.
4. Substitute the values of two of the variables you found in step 3 into one of the original equations that contains the third variable. Solve for the third variable.
5. Write your solution as an ordered triple.

Finding an equation of a parabola (Section 7.6)

To find an equation of a parabola that contains three given points,

1. Obtain a system of three linear equations in three variables by substituting the coordinates of each of the three given points into the general equation $y = ax^2 + bx + c$.
2. Solve the system you found in step 1.
3. Substitute the values of a, b, and c you found in step 2 into the equation $y = ax^2 + bx + c$.

Finding an equation of a quadratic model (Section 7.7)

To find an equation of a quadratic model, given some data,

1. Create a scattergram of the data.
2. Imagine a parabola that comes close to the data points, and choose three points (not necessarily data points) that lie on or close to the parabola.
3. Use the three points to find an equation of the parabola.
4. Use a graphing calculator to verify that the graph of the equation comes close to the points of the scattergram.

Selecting a model (Section 7.7)

When performing step 1 of the modeling process, we must decide whether a linear function, an exponential function, a quadratic function, or none of these is suitable for modeling the situation. Here are the criteria for selecting a model:

- The graph of the model should fit the points well.
- The model should make sense within the context of the authentic situation.

Four-step modeling process (Section 7.7)

To find a model and make estimates and predictions,

1. Create a scattergram of the data. Decide whether a line, an exponential curve, a parabola, or none of these comes close to the points.
2. Find an equation of your model.
3. Verify that your equation has a graph that comes close to the points in the scattergram. If it doesn't, check for calculation errors or use different points to find the equation. An alternative is to reconsider your choice of model in step 1.
4. Use your equation of the model to draw conclusions, make estimates, and/or make predictions.

Solving a system of quadratic equations (Section 7.8)

We can use substitution to solve any system of two quadratic equations if both are in standard form $y = ax^2 + bx + c$.

CHAPTER 7 REVIEW EXERCISES

Sketch by hand the graph of the function.

1. $y = -3x^2$

2. $y = 3(x+1)^2 - 4$

3. $y = -2(x-3)^2 + 5$

4. $y = \frac{1}{2}(x+4)^2 - 2$

5. A function of the form $y = a(x-h)^2 + k$ has been sketched in Fig. 112. Describe the signs of the constants a, h, and k.

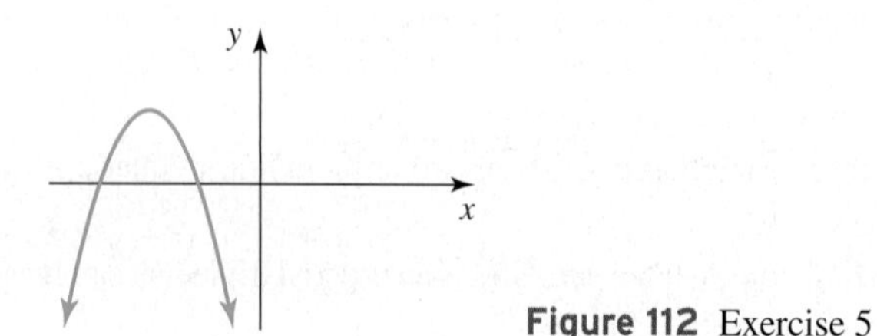

Figure 112 Exercise 5

Graph the function. Find the vertex; round coordinates to the second decimal place.

6. $y = 3x^2 - 12x + 7$

7. $y = -2x^2 + 8x + 5$

8. $1.7x + 2.6x^2 + y = 6.7x^2 - 10x + 2.1$

Simplify.

9. $\sqrt{72}$

10. $\sqrt{\frac{49}{100}}$

Solve. Any solution is a real number.

11. $3x^2 - 2x - 2 = 0$

12. $5x^2 = 7$

13. $5(p-3)^2 + 4 = 7$

14. $(t+1)(t-7) = 4$

15. $2x^2 = 4 - 5x$

16. $4x - x^2 = 1$

17. $5x^2 - 6x = 2$

18. $7x^2 - 20 = 0$

19. $(x + 2)^2 + (x - 3)^2 = 15$

20. $5(5x^2 - 8) = 9$

21. $\frac{3}{2}x^2 - \frac{3}{4}x = \frac{1}{2}$

Find approximate solutions. Round the results to the second decimal place.

22. $2.7x^2 - 5.1x = 9.8$

23. $1.7(x^2 - 2.3) = 3.4 - 2.8x$

Find all complex-number solutions.

24. $-2(x + 4)^2 = 9$

25. $2x^2 = 4x - 7$

Solve by completing the square. Any solution is a real number.

26. $x^2 + 6x - 4 = 0$

27. $2t^2 = -3t + 6$

Find all x-intercepts.

28. $h(x) = 3x^2 + 2x - 2$

29. $k(x) = -5x^2 + 3x - 1$

30. Solve $x^2 - 2x - 8 = 0$ by factoring, completing the square, and using the quadratic formula.

31. Find the number and type of solutions of the equation $3x^2 - 5x + 4 = 0$.

32. Let $f(x) = 3x^2 - 6x + 7$.
   a. Find x when $f(x) = 3$.
   b. Find x when $f(x) = 4$.
   c. Find x when $f(x) = 5$.
   d. Discuss in terms of the graph of f why you found 0, 1, and 2 values of x for parts (a), (b), and (c), respectively.

For Exercises 33–35, find approximate solutions of the equation or system by referring to the graphs shown in Fig. 113. Round results or coordinates of results to the first decimal place.

33. $-\frac{1}{4}(x - 1)^2 + 1 = x^2$

34. $-\frac{1}{4}(x - 1)^2 + 1 = -3$

35. $y = -\frac{1}{4}(x - 1)^2 + 1$
$y = -\frac{1}{3}x - 2$

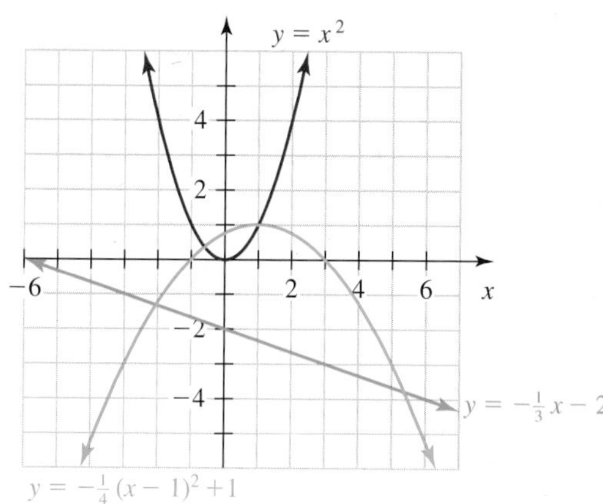

Figure 113 Exercises 33–35

Solve the system.

36. $\begin{aligned} x + 2y - 3z &= -4 \\ 2x - y + z &= 3 \\ 3x + 2y + z &= 10 \end{aligned}$

37. $\begin{aligned} 2x \qquad - 3z &= -4 \\ 3x + y \qquad &= 0 \\ x - 4y + 2z &= 17 \end{aligned}$

Find an equation of the parabola that passes through the three given points.

38. $(2, 9), (3, 18), (5, 48)$

39. $(0, 5), (2, 3), (4, -15)$

40. Find equations of a linear function, an exponential function, and a quadratic function such that the graph of each function contains both of the points $(0, 4)$ and $(1, 2)$.

41. Find an equation of the parabola sketched in Fig. 114.

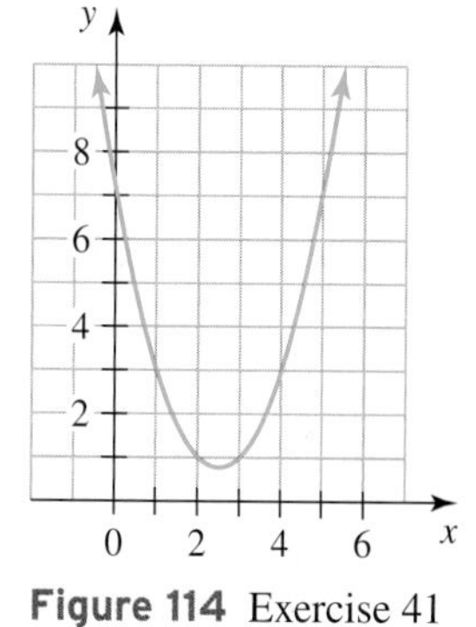

Figure 114 Exercise 41

42. A batter hits a baseball into the air. The height (in feet) $h(t)$ of the baseball after t seconds is given by

$$h(t) = -16t^2 + 100t + 3$$

   a. Find the maximum height of the baseball. When does it reach that height?
   b. A fielder catches the ball at a height of 3 feet. How many seconds after the batter hit the ball did the fielder have to get into position to catch the ball?
   c. Graph h by hand.

43. A farmer plans to use 180 feet of fencing and a side of his barn to enclose a rectangular garden (see Fig. 115). What should the dimensions of the rectangle be so that the area is as large as possible? What is that area?

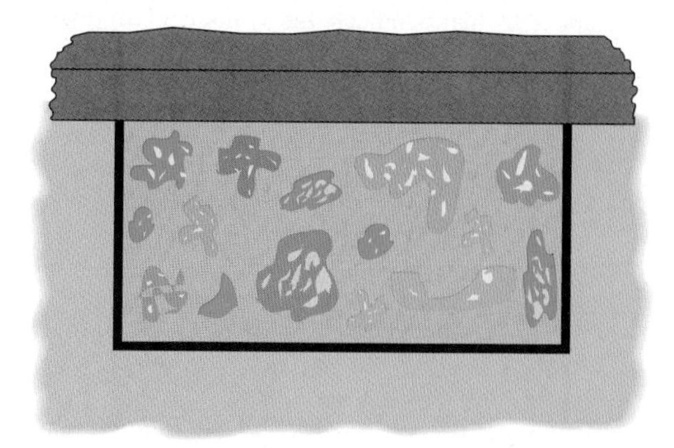

Figure 115 Exercise 43

44. The percentages of television shows that are at least 1 hour long are shown in Table 72 for various years.

Table 72 Television Shows That Are at Least 1 Hour Long

Year*	Percent
1998	48
1999	53
2000	57
2001	59
2002	57

Sources: *American Demographics; Inside.com*
*Each year indicates when the season ended.

Let f be the percentage of television shows that are at least 1 hour long at t years since 1990.

a. Find an equation of f.
b. Find $f(15)$. What does it mean in this situation?
c. Predict when half of all televisions shows will be at least 1 hour long.
d. Find the t-intercepts. What do they mean in this situation?
e. For what years is there model breakdown for certain?

45. The percentages of Americans who get their news every day from nightly network news programs (on ABC, CBS, or NBC) and the percentages of Americans who get their news every day on the Internet are shown in Table 73 for various years.

Table 73 Percentages of Americans Who Get Their News Every Day from Network News Programs and from the Internet

	Percent	
Year	Nightly Network News Programs	Internet
1995	62	3
1998	55	6
1999	52	8
2002	43	15
2004	36	20

Source: *The Gallup Organization*

Let $f(t)$ be the percentage of Americans who get their news every day from nightly network news programs, and let $g(t)$ be the percentage of Americans who get their news on the Internet, both at t years since 1990.

a. Find quadratic regression equations of f and g.
b. Use f and g to predict when the percentage of Americans who get their news every day from the nightly network news programs will equal the percentage of Americans who get their news every day on the Internet.

CHAPTER 7 TEST

1. Graph the function $f(x) = -2(x+5)^2 - 1$ by hand.

2. What can you say about the values of a, h, and k if the equation of the parabola shown in Fig. 116 is $y = a(x-h)^2 + k$?

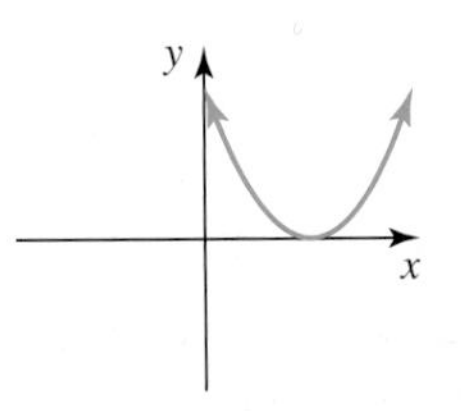

Figure 116 Exercise 2

3. Give an example of a quadratic function that has vertex $(2, -7)$ and no x-intercepts.

4. Graph $f(x) = -2x^2 - 4x + 3$ by hand. Find the vertex.

5. Let $f(x) = x^2 - 2x - 8$.
a. Find the x-intercepts.
b. Find the vertex.
c. Graph f by hand. Indicate the x-intercepts and vertex on the graph.

Simplify.

6. $\sqrt{32}$ **7.** $\sqrt{\dfrac{20}{75}}$

Solve. Any solution is a real number.

8. $x^2 - 3x - 10 = 0$ **9.** $6x^2 = 100$

10. $4(r-3)^2 + 1 = 7$ **11.** $\frac{5}{6}x^2 - \frac{1}{2}x = \frac{2}{3}$

12. $(x-3)(x+5) = 6$ **13.** $2x(x+5) = 4x - 3$

14. $3x^2 - 6x = 1$

15. Find approximate solutions of $3.7x^2 = 2.4 - 5.9x$. Round your results to the second decimal place.

Find all complex-number solutions.

16. $3x^2 - 6x = -5$ **17.** $-2(p+4)^2 = 24$

Solve by completing the square.

18. $x^2 - 8x - 2 = 0$ **19.** $2(x^2 - 4) = -3x$

20. Find the x-intercepts of $f(x) = 3x^2 - 8x + 1$.

21. Find the x-intercepts and vertex of the graph of $f(x) = -2(x-3)^2 + 5$. Round the coordinates to the second decimal place. Then graph f by hand.

22. Find the nonzero value(s) of a such that the equation $ax^2 - 4x + 4a = 0$ has exactly one solution.

23. Find an equation of the parabola that passes through the points $(1, 4)$, $(2, 9)$, and $(3, 16)$.

24. Find an equation of the parabola that has vertex $(5, 3)$ and that contains the point $(3, 11)$.

25. Let $f(x) = x^2 - 6x + 11$. Find the number of points that lie on the graph of f at the indicated height.
a. $y = 1$ **b.** $y = 2$ **c.** $y = 3$

Solve the system.

26.
$$\begin{aligned} x + 4y + 3z &= 2 \\ 2x + y + z &= 10 \\ -x + y + 2z &= 8 \end{aligned}$$

27.
$$\begin{aligned} 2x - 3y &= 4 \\ 3y + 2z &= 2 \\ x - z &= -5 \end{aligned}$$

28. Let $f(t)$ be the height (in feet) of a baseball at t seconds after a batter has hit the ball. An equation of f is

$f(t) = -16t^2 + 80t + 3$. At what time is the ball at its maximum height? What is that height?

29. The percentages of Americans who feel that they are taking a great risk by entering personal information into a pop-up ad are shown in Table 74 for various age groups.

Table 74 Percentages of Americans Who Feel That They Are Taking a Great Risk by Entering Personal Information into a Pop-up Ad

Age Group (years)	Age Used to Represent Age Group (years)	Percent
18–24	21.0	27
25–34	29.5	33
35–44	39.5	37
45–54	49.5	44
55–64	59.5	39
over 64	70.0	23

Source: *Wells Fargo*

Let $f(t)$ be the percentage of Americans at age t years who feel that they are taking a great risk by entering personal information into a pop-up ad.

a. Find an equation of f.

b. Find $f(30)$. What does it mean in this situation?

c. Find t when $f(t) = 30$. What does it mean in this situation?

d. Find the t-intercepts. What do they mean in this situation?

e. Find the vertex. What does it mean in this situation?

30. A company charters a party boat that normally costs $40 per person. A group discount reduces the fare by $0.25 for each ticket sold; the more tickets sold, the lower the per-person fare. The maximum capacity of the boat is 90 people, including the crew of 5 people. What size of group would maximize the boat owner's revenue?

CUMULATIVE REVIEW OF CHAPTERS 1–7

Solve. Any solution is a real number.

1. $81x^2 - 49 = 0$

2. $\log_4(m + 3) = 2$

3. $5x^2 - 2x = 4$

4. $2(3t^2 - 10) = -7t$

5. $(2p - 5)(3p + 4) = 5p - 2$

6. $2(5x + 2) - 1 = 9(x - 3) - (3x - 8)$

7. $2x(3x - 4) + 5 = 4 - x^2$

Solve. Any solution is a real number. Round any results to the fourth decimal place.

8. $3\ln(2w) + 2\ln(3w) = 8$

9. $5b^6 + 4 = 82$

10. $3(2)^{4x-5} = 95$

11. $\log_b(65) = 4$

12. $3e^x - 5 = 49$

13. $3\log_2(x^4) - 2\log_2(4x) = 5$

14. Find all complex-number solutions of $2x^2 - 6x = -5$.

15. Solve $2x^2 + 3x - 6 = 0$ by completing the square.

Solve the system.

16.
$$y = 3x - 1$$
$$2x - 3y = -11$$

17.
$$\frac{1}{2}x - y = \frac{5}{2}$$
$$\frac{2}{5}x - \frac{3}{5}y = \frac{6}{5}$$

18.
$$2x - y + 3z = 1$$
$$3x + 2y - z = -6$$
$$4x - 3y + 2z = -7$$

19. Solve the inequality $2(3x - 4) < 5 - 3(6x + 5)$. Describe the solution set as an inequality, in interval notation, and in a graph.

Simplify.

20. $(2b^4c^{-5})^3(3b^{-1}c^{-2})^4$

21. $\left(\dfrac{6b^8c^{-3}}{8b^{-4}c^{-1}}\right)^3$

Simplify. Write your result as a single logarithm with a coefficient of 1.

22. $3\log_b(4x) - 4\log_b(x^3)$

23. $3\ln(x^7) + 2\ln(x^5)$

Perform the indicated operation.

24. $(3x - 4y)^2$

25. $(5p - 7q)(5p + 7q)$

26. $-3x(x^2 - 5)(x^3 + 8)$

27. $(x^2 - 3x - 4)(x^2 + 4x - 5)$

28. Write $f(x) = -2(x - 5)^2 + 3$ in standard form.

Factor.

29. $m^4 - 16n^4$

30. $x^3 - 13x^2 + 40x$

31. $8p^2 + 22pq - 21q^2$

32. $x^3 + 4x^2 - 9x - 36$

For Exercises 33–37, refer to Table 75.

33. Find a possible equation of f.

34. Find a possible equation of g.

35. Find the slope of k.

36. Find x when $g(x) = 4$.

37. Find $h^{-1}(7)$.

Table 75 Values of Four Functions (Exercises 33–37)

x	f(x)	x	g(x)	x	h(x)	x	k(x)
0	20	1	4	4	7	0	−8
1	17	2	12	5	14	1	−4
2	14	3	36	6	28	2	0
3	11	4	108	7	56	3	4
4	8	5	324	8	112	4	8
5	5	6	972	9	224	5	12

Sketch the graph of the function.

38. $y = -2(x+4)^2 + 3$

39. $y = \log_3(x)$

40. $2x - 5y = 20$

41. $y = 2(3)^x$

42. $y = x^2 + 4x - 5$

43. $y = 8\left(\frac{1}{4}\right)^x$

44. Use the graph of the relation in Fig. 117 to determine the domain and range. Is the relation a function?

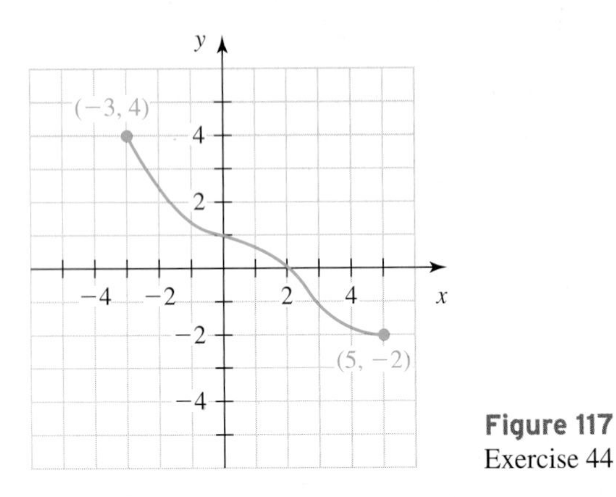

Figure 117 Exercise 44

45. Find the equation of the line that contains the point $(-2, 6)$ and is perpendicular to the line $3x - 4y = 5$.

46. Find an approximate equation $y = ab^x$ of an exponential curve that contains the points $(2, 27)$ and $(5, 83)$. Round a and b to the second decimal place.

47. Find an equation of a parabola that contains the points $(1, -1)$, $(2, 4)$, and $(4, 20)$.

48. Let f be the linear function, g be the exponential function, and h be a quadratic function whose graphs contain the points $(0, 3)$ and $(1, 6)$.

a. Find possible equations of f, g, and h.

b. Use a graphing calculator to draw the graphs of f, g, and h in the same viewing window.

Let $f(x) = -x^2 + 6x - 5$.

49. Find x when $f(x) = 3$.

50. Find the x-intercepts.

51. Graph f by hand.

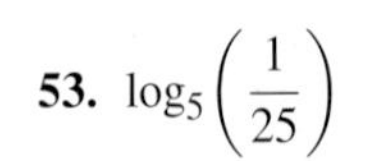

Find the logarithm.

52. $\log_2(16)$

53. $\log_5\left(\frac{1}{25}\right)$

54. The graph of a function f is sketched in Fig. 118. Sketch the graph of f^{-1}.

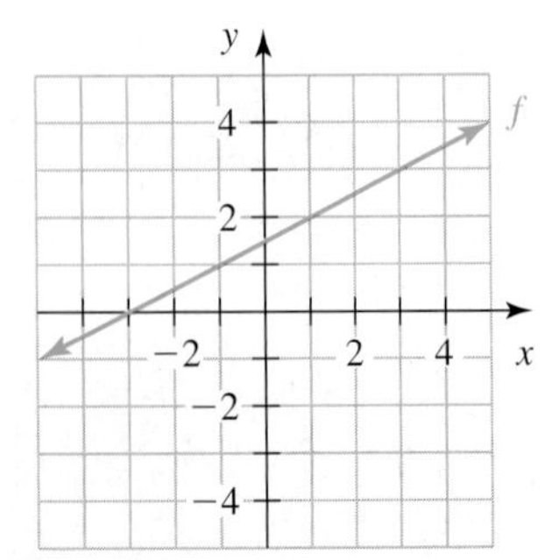

Figure 118 Exercise 54

Find the inverse of the function.

55. $g(x) = 3^x$

56. $f(x) = \frac{2}{5}x + 1$

57. A person plans to invest a total of \$12,000. She will invest in both a First Funds TN Tax-Free I account at 6% annual interest and a W & R International Growth C account at 11% annual interest. How much should she invest in each account to earn \$845 total interest in one year?

58. The numbers of cases of AIDS acquired at birth, by year of diagnosis, in the United States has decayed approximately exponentially from 660 cases in 1995 to 60 cases in 2003 (Source: *Centers for Disease Control and Prevention*). Predict in which year there will be 1 case of AIDS acquired at birth in the United States.

59. The total sales at coffeehouses and doughnut shops are shown in Table 76 for various years.

Table 76 Total Sales at Coffeehouses and Doughnut Shops

Year	Total Sales (billions of dollars)
2000	7.7
2001	8.5
2002	9.5
2003	10.9
2004	12.6
2005	14.4

Source: *Technomic/Mintel*

Let $f(t)$ be the total sales (in billions of dollars) at coffeehouses and doughnut shops in the year that is t years since 2000.

a. Find a linear regression equation, an exponential regression equation, and a quadratic regression equation of f. Compare how well the models fit the data.

b. Find the vertex of the graph of your quadratic model. What does it mean in this situation?

c. For years before 1998, which of the models likely best describes this situation? Explain.

d. Use the exponential model to estimate the percent rate of growth in total sales at coffeehouses and doughnut shops.

e. Use the exponential model to predict when the total sales at coffeehouses and doughnut shops will be \$70 billion.

f. Use the quadratic model to predict when the total sales at coffeehouses and doughnut shops will be \$70 billion.

g. Explain why the year you predicted in part (f) is later than the year you predicted in part (e). [**Hint:** It may help to Zoom Out a couple of times.]

60. As unions have attracted an increasing number of workers in the service industries, the percentage of union members who work in manufacturing has declined (see Table 77).

Table 77 Percentages of Union Members Who Work in Manufacturing

Year	Percent
1985	29
1990	25
1995	21
2000	17
2005	13

Source: *Bureau of Labor Statistics*

Let $p = f(t)$ be the percentage of union members who work in manufacturing at t years since 1980.

a. Find an equation of f.

b. Find $f(30)$. What does it mean in this situation?

c. Find t when $f(t) = 7$. What does it mean in this situation?

d. Find an equation of f^{-1}.

e. Find $f^{-1}(100)$. What does it mean in this situation?

f. Find the t-intercept of the graph of f. What does it mean in this situation?

g. For what values of t is there model breakdown for certain?

61. Although grade school students in Hartford used to have the lowest test scores in Connecticut, their scores have increased due to reform measures. Improvements in their mathematics test scores are shown in Table 78.

Table 78 Hartford and Connecticut Mathematics Test Scores

	Percent of Students Scoring Above Goals	
Year	**Hartford Students**	**Connecticut Average**
1994	20	57
1996	26	59
1998	29	61
2000	35	63

Source: *Connecticut Department of Education*

Let $h(t)$ and $c(t)$ be the percentages of Hartford and all Connecticut grade school students, respectively, scoring above goals on mathematics tests at t years since 1990.

a. Find equations of h and of c.

b. Compare the rate of change of the percentage of the students scoring above goals for students in Hartford versus students in all of Connecticut.

c. Find $h(18)$ and $c(18)$. What do they mean in this situation?

d. Predict when the percentage of Hartford grade school students who score above goals will be equal to the percentage of all Connecticut grade school students who score above goals.

→ give it!

$g(x) = .14x^2 + .65x + 7.69$

Chapter 8

Rational Functions

"Individually [mathematicians] are very different in their mathematical personalities, the kind of mathematics they like, and the way that they do mathematics, but they are alike in one respect. Almost without exception, they love their subject, are happy in their choice of a career, and consider that they are exceptionally lucky in being able to do for a living what they would do for fun."

—*Constance Reid,* Becoming a Mathematician, *1990.*

If you drive a car, how much did you pay for gasoline last year? In Exercise 8 of Homework 8.6, you will find a model that you will use to predict how much an American will pay, on average, for gasoline in 2010 (see Table 1).

Table 1 Average Prices of Gasoline; Vehicle Fuel Consumption

Year	Price of Gasoline (2006 dollars per gallon)	Year	Amount of Fuel Consumed (millions of gallons)
1980	2.59	1975	109,000
1985	1.90	1980	115,000
1990	1.51	1985	121,300
1995	1.28	1990	130,800
2001	1.66	1995	143,800
2005	2.28	2000	162,500
2006	2.68*	2003	169,600

Source: *Federal Highway Administration*
*January–May

In Chapters 6 and 7, we worked with polynomial functions. In this chapter, we will discuss how to divide these functions to form new ones called *rational functions*. We will also work with *rational expressions* and discuss how to simplify them and perform operations with them. We will solve *rational equations,* which will help us use rational functions to make predictions about authentic situations. For example, we will use a rational function to predict the average amount an American will pay for gasoline in 2010.

8.1 FINDING THE DOMAINS OF RATIONAL FUNCTIONS AND SIMPLIFYING RATIONAL EXPRESSIONS

Objectives

- Know the meaning of *rational function* and *vertical asymptote*.
- Find the domain of a rational function.
- Simplify a *rational expression*.
- Know the connection between vertical asymptotes and domains of rational functions.
- Form *quotient functions*.
- Use a *rational model* to describe the percentage of a quantity.

In this section, we will discuss the meaning of a *rational function* and how to find its domain. We will simplify *rational expressions*. We will also discuss how to use a *rational model* to make estimates about authentic situations.

Meaning of a Rational Function

In Chapters 6 and 7, we worked with polynomials. If P and Q are polynomials, where Q is nonzero, we call the ratio $\frac{P}{Q}$ a **rational expression.** The name "rational" refers to "ratio." Here are some examples:

$$\frac{x^3 - 3x + 6}{x^7 - 8}, \quad \frac{-2x^2 + 17}{5x - 1}, \quad -\frac{2}{7x^3}, \quad \frac{3x^2 + 5x - 4}{2x^2 - x + 9}$$

A rational expression is part of an equation of a rational function.

DEFINITION Rational function

A **rational function** is a function whose equation can be put into the form

$$f(x) = \frac{P(x)}{Q(x)}$$

where $P(x)$ and $Q(x)$ are polynomials and $Q(x)$ is nonzero.

In Example 1, we evaluate a rational function at some input values.

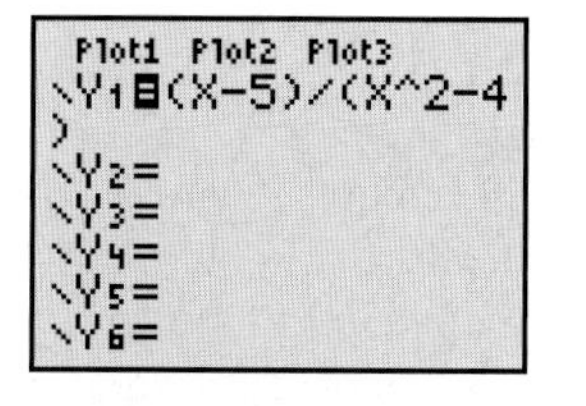

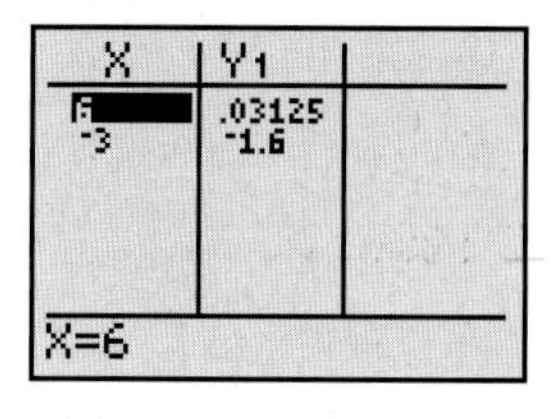

Figure 1 Verify the work

Example 1 Evaluating a Rational Function

Evaluate the function $f(x) = \frac{x - 5}{x^2 - 4}$ at the indicated values.

1. $f(6)$ **2.** $f(-3)$

Solution

1. $f(6) = \frac{6 - 5}{6^2 - 4} = \frac{1}{32}$

2. $f(-3) = \frac{-3 - 5}{(-3)^2 - 4} = \frac{-8}{5} = -\frac{8}{5}$

We use a graphing calculator table to check that $f(6) = \frac{1}{32} = 0.03125$ and $f(-3) = -\frac{8}{5} = -1.6$ (see Fig. 1). ■

WARNING To enter a rational function such as $\frac{x - 5}{x^2 - 4}$ into a graphing calculator, we enclose both the numerator and the denominator in parentheses (see Fig. 1).

Domain of a Rational Function

Our work in Example 1 shows that the numbers 6 and -3 are in the domain of the function $f(x) = \frac{x - 5}{x^2 - 4}$. Are all real numbers in the domain of f? We will explore this issue in Example 2.

To learn about the domain of a rational function, it will help to consider the simpler rational function $f(x) = \frac{5}{x}$ first. Note that $f(0)$ is undefined, since $\frac{5}{0}$ is undefined (we cannot divide by 0). So, 0 is *not* in the domain of f. Since division by nonzero numbers *is* defined, the domain of f is the set of real numbers except 0.

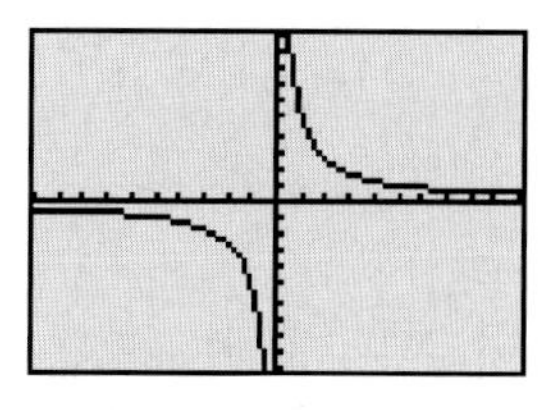

Figure 2 Graph of $f(x) = \frac{5}{x}$

Figure 2 displays a graphing calculator graph of $f(x) = \frac{5}{x}$. The x-axis appears to be a horizontal asymptote of f; that is, in fact, true. The graph of f also appears to get arbitrarily close to, but never intersects, the y-axis; that is also true. We say that the y-axis is a **vertical asymptote** of f. Just as with a horizontal asymptote, a vertical asymptote is *not* part of the graph of a function.

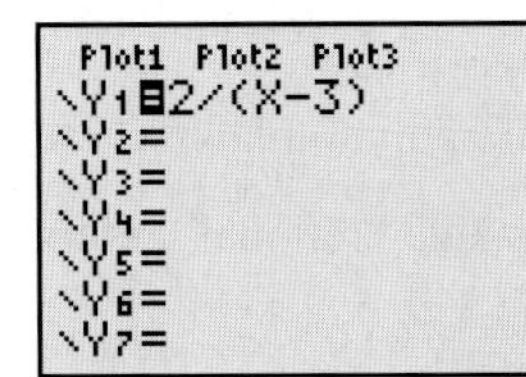

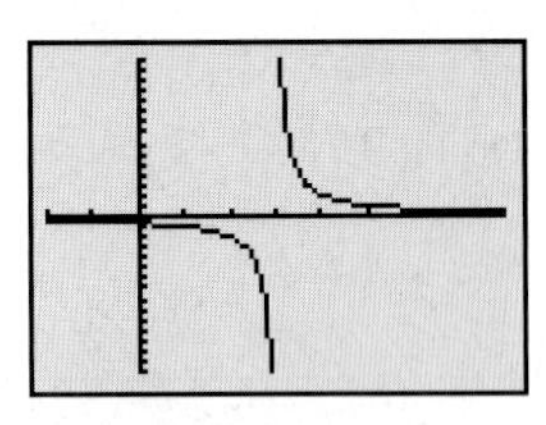

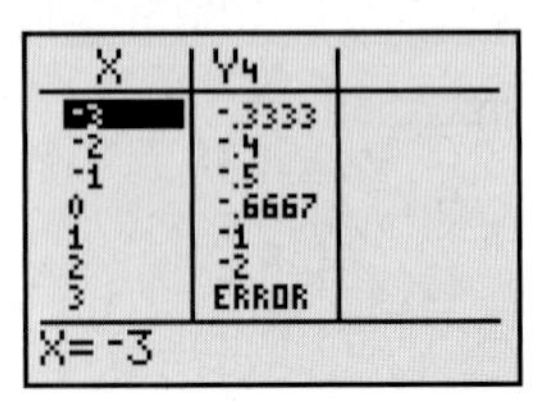

Figure 3 f has a vertical asymptote at $x = 3$, and 3 is not in the domain of f

Example 2 Finding the Domain of a Rational Function

Find the domain of the function.

1. $f(x) = \dfrac{2}{x-3}$

2. $g(x) = \dfrac{x-5}{x^2-4}$

Solution

1. For $f(x) = \dfrac{2}{x-3}$, the number 3 is not in the domain, because $\dfrac{2}{3-3}$ involves a division by zero. No other value of x leads to a division by 0, so the domain is the set of real numbers except 3. In Fig. 3, we draw a graph of f and build a table of input–output pairs of f.

   From the graph of f, it appears that f has a vertical asymptote at $x = 3$; that is true. The "ERROR" message across from $x = 3$ in the table supports the idea that the value 3 for x leads to division by 0. (The TI-83 and TI-84 display either "ERROR" or "ERR:.")

2. To find which values of x lead to a division by 0, we set the denominator of $\dfrac{x-5}{x^2-4}$ equal to 0 and solve for x:

$$x^2 - 4 = 0 \qquad \textit{Set denominator equal to 0.}$$

$$(x+2)(x-2) = 0 \qquad \textit{Factor left side.}$$

$$x + 2 = 0 \quad \text{or} \quad x - 2 = 0 \qquad \textit{Zero factor property}$$

$$x = -2 \quad \text{or} \quad x = 2$$

The numbers -2 and 2 are *not* in the domain, since these values of x lead to divisions by zero. The domain of g is the set of real numbers except -2 and 2. In Fig. 4, we draw a graph of g and build a table of input–output pairs of g. A graphing calculator approximates graphs by plotting many points and connecting the points with curves. The steep lines that are "almost" the lines $x = -2$ and $x = 2$ are *not* part of the graph. Some TI graphing calculators do not show these lines.

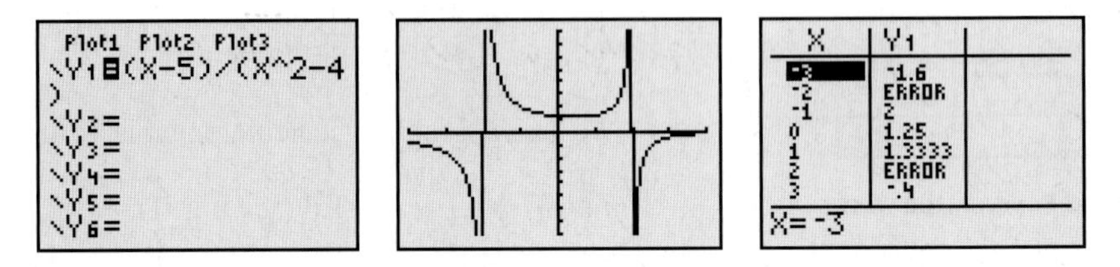

Figure 4 g has vertical asymptotes at $x = -2$ and $x = 2$, and -2 and 2 are not in the domain of g

From the graph of g, it appears that g has vertical asymptotes at $x = -2$ and $x = 2$; that is true. The "ERROR" messages across from $x = -2$ and $x = 2$ in the table support the idea that the values -2 and 2 for x lead to divisions by zero. ■

Our work in Problem 2 of Example 2 suggests the following property.

Domain of a Rational Function

The domain of a rational function $f(x) = \dfrac{P(x)}{Q(x)}$ is the set of real numbers except for those numbers that, when substituted for x, give $Q(x) = 0$.

Example 3 Finding the Domain of a Rational Function

Find the domain of the function.

1. $f(x) = \dfrac{x-4}{x^2+5x+2}$

2. $g(x) = \dfrac{x^2-5x+1}{x^2+10}$

Solution

1. We set the denominator of the right-hand side of $f(x) = \frac{x-4}{x^2+5x+2}$ equal to 0:

$$x^2 + 5x + 2 = 0$$

Then we substitute $a = 1$, $b = 5$, and $c = 2$ in the quadratic formula:

$$x = \frac{-5 \pm \sqrt{5^2 - 4(1)(2)}}{2(1)} \quad \text{Substitute into quadratic formula.}$$

$$= \frac{-5 \pm \sqrt{17}}{2} \quad \text{Simplify.}$$

The domain of f is the set of real numbers except $\frac{-5-\sqrt{17}}{2}$ and $\frac{-5+\sqrt{17}}{2}$.

2. We set the denominator of the right-hand side of $g(x) = \frac{x^2-5x+1}{x^2+10}$ equal to 0 and solve:

$$x^2 + 10 = 0$$
$$x^2 = -10$$

Since a real number squared is nonnegative, we conclude that there are no real-number solutions. For the function g, no values of x lead to division by 0. Therefore, the domain is the set of (all) real numbers. ■

In Problem 1 of Example 2, we found that substituting 3 in the expression $\frac{2}{x-3}$ leads to a division by 0. We call the number 3 an excluded value of the expression.

DEFINITION Excluded value

A number is an **excluded value** of a rational expression if substituting the number into the expression leads to a division by 0.

For instance, 5 is an excluded value of the expression $\frac{7}{x-5}$. This means that 5 is *not* in the domain of the function $f(x) = \frac{7}{x-5}$.

Simplifying Rational Expressions

To simplify a fraction, we factor the numerator and the denominator and use the property $\frac{a}{a} = 1$, where $a \neq 0$. Here, we use these steps to simplify $\frac{8}{12}$:

$$\frac{8}{12} = \frac{4(2)}{4(3)} = \frac{4}{4} \cdot \frac{2}{3} = 1 \cdot \frac{2}{3} = \frac{2}{3}$$

We simplify the rational expression $\frac{2x+6}{x^2-9}$ in a similar manner:

$$\frac{2x+6}{x^2-9} = \frac{2(x+3)}{(x-3)(x+3)} \quad \text{Factor numerator and denominator.}$$

$$= \frac{2}{x-3} \cdot \frac{x+3}{x+3} \quad x+3 \text{ is a factor of both numerator and denominator.}$$

$$= \frac{2}{x-3} \cdot 1 \quad \text{Simplify: } \frac{x+3}{x+3} = 1$$

$$= \frac{2}{x-3} \quad a \cdot 1 = a$$

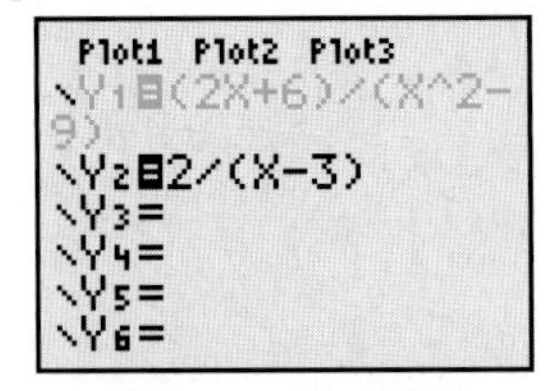

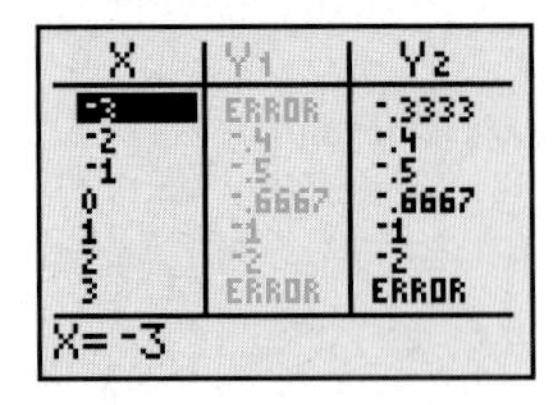

Figure 5 Verify the work

The number 3 is the only excluded value of our result, $\frac{2}{x-3}$. The numbers -3 and 3 are the only excluded values of the original expression, $\frac{2x+6}{x^2-9}$ (try it). When we write

$$\frac{2x+6}{x^2-9} = \frac{2}{x-3}$$

we mean that the two expressions give the same result for each real number substituted for x, except for any excluded values (-3 and 3) of either expression. In Fig. 5, we use a graphing calculator table to verify our work.

A rational expression is in **lowest terms** if the numerator and denominator have no common factors other than 1 or -1. We **simplify a rational expression** by writing it in lowest terms. We also write the numerator and the denominator in factored form.

Simplifying a Rational Expression

To simplify a rational expression,

1. Factor the numerator and the denominator.
2. Use the property

$$\frac{AB}{AC} = \frac{A}{A} \cdot \frac{B}{C} = 1 \cdot \frac{B}{C} = \frac{B}{C}$$

where A and C are nonzero, so that the expression is in lowest terms.

Throughout the rest of this chapter, you may assume that the form $\frac{A}{B}$ represents a rational expression.

In the expression $\frac{AB}{AC}$, note that the polynomial A is a factor of both the numerator and the denominator. The expression $\frac{3x}{7x}$ can be simplified to $\frac{3}{7}$ when $x \neq 0$, because x is a factor of both the numerator and the denominator.

WARNING The expression

$$\frac{3x+2}{7x}$$

is in lowest terms already. Although x is a factor of the term $7x$, it is not a factor of $3x+2$, the (entire) numerator. Likewise, the expression

$$\frac{(x+4)(x+6)+5}{3(x+4)}$$

is in lowest terms already. The expression $x+4$ is not a factor of $(x+4)(x+6)+5$, the (entire) numerator. To see that the rational expression is in lowest terms, we simplify the numerator to get $\frac{x^2+10x+29}{3(x+4)}$ (try it).

Example 4 Simplifying the Right-Hand Side of an Equation

Simplify the right-hand side of the equation $f(x) = \frac{2x^2-6x-20}{2x^2-50}$.

Solution

$$\begin{aligned}
f(x) &= \frac{2x^2-6x-20}{2x^2-50} && \text{Original equation} \\
&= \frac{2(x^2-3x-10)}{2(x^2-25)} && \\
&= \frac{2(x-5)(x+2)}{2(x-5)(x+5)} && \text{Factor numerator and denominator.} \\
&= \frac{x+2}{x+5} && \text{Simplify: } \frac{2(x-5)}{2(x-5)} = 1
\end{aligned}$$

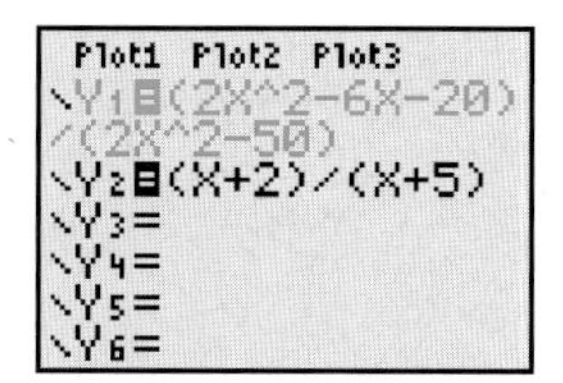

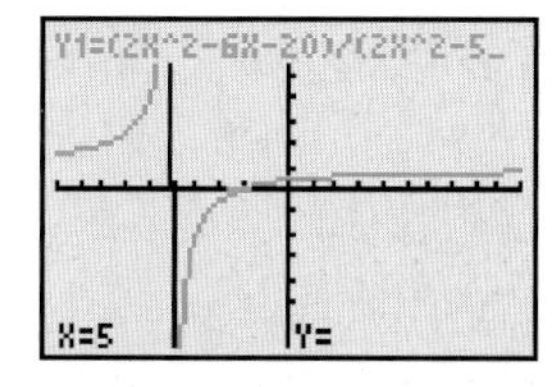

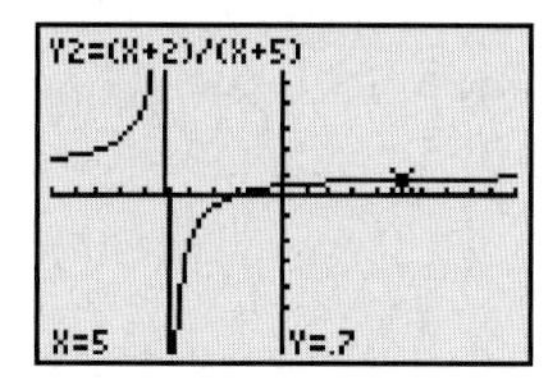

Figure 6 Verify the work

So, we can describe f by the equation $f(x) = \dfrac{x+2}{x+5}$, where the domain is all real numbers except -5 and 5.

To check, we substitute a value—say, 3—for x in both of the expressions $\dfrac{2x^2 - 6x - 20}{2x^2 - 50}$ and $\dfrac{x+2}{x+5}$ and check that the results are equal:

$$\frac{2(3)^2 - 6(3) - 20}{2(3)^2 - 50} = \frac{5}{8}, \qquad \frac{3+2}{3+5} = \frac{5}{8}$$

We perform a more convincing check by comparing graphing calculator graphs for the functions $y = \dfrac{2x^2 - 6x - 20}{2x^2 - 50}$ and $y = \dfrac{x+2}{x+5}$. See Fig. 6. To hand-sketch a graph of $f(x) = \dfrac{2x^2 - 6x - 20}{2x^2 - 50}$, we use an open circle at the point $(5, 0.7)$ to indicate that the point is not part of the graph.

The graphs of the functions are the same for $x \neq -5$ and $x \neq 5$. ■

Connection Between Vertical Asymptotes and Domains

Example 2 suggests that if a rational function has a vertical asymptote at $x = k$, then k is not in the domain of the function, which is true.

In Example 4, we wrote f in the form

$$f(x) = \frac{2(x-5)(x+2)}{2(x-5)(x+5)}$$

by factoring the numerator and the denominator of the right-hand side of the original equation of f. From this form, we see that the domain of f is the set of real numbers except -5 and 5. Yet, $x = -5$ is the only vertical asymptote of f (see Fig. 6). The reason $x = 5$ is not a vertical asymptote has to do with the fact that both the numerator and the denominator of f contain the factor $x - 5$.

Vertical Asymptotes and Domains of Rational Functions

If a rational function g has a vertical asymptote $x = k$, then k is not in the domain of g. If k is not in the domain of a rational function h, then $x = k$ may or may not be a vertical asymptote of h.

Example 5 Simplifying the Right-Hand Side of an Equation

Simplify the right-hand side of the equation $f(x) = \dfrac{x^3 - 5x^2 + 6x}{x^3 - 2x^2 - 9x + 18}$.

Solution

$$f(x) = \frac{x^3 - 5x^2 + 6x}{x^3 - 2x^2 - 9x + 18} \qquad \text{Original equation}$$

$$= \frac{x(x^2 - 5x + 6)}{x^2(x-2) - 9(x-2)}$$

$$= \frac{x(x-3)(x-2)}{(x^2-9)(x-2)} \qquad \text{Factor numerator and denominator.}$$

$$= \frac{x(x-3)(x-2)}{(x+3)(x-3)(x-2)}$$

$$= \frac{x}{x+3} \qquad \text{Simplify: } \frac{(x-3)(x-2)}{(x-3)(x-2)} = 1$$

■

Example 6 Simplifying the Right-Hand Side of an Equation

Simplify the right-hand side of the equation $f(x) = \dfrac{x^2 - x - 12}{4 - x}$.

Solution

$$f(x) = \frac{x^2 - x - 12}{4 - x} \qquad \text{Original equation}$$

$$= \frac{x^2 - x - 12}{-x + 4} \qquad \text{Write denominator in descending order.}$$

$$= \frac{(x - 4)(x + 3)}{-1(x - 4)} \qquad \text{Factor numerator and denominator.}$$

$$= \frac{x + 3}{-1} \qquad \text{Simplify: } \frac{x - 4}{x - 4} = 1$$

$$= -x - 3 \qquad \frac{x + 3}{-1} = -\frac{x + 3}{1} = -(x + 3) = -x - 3$$

■

In Example 7, we simplify a rational expression in two variables.

Example 7 Simplifying a Rational Expression in Two Variables

Simplify $\dfrac{x^2 - 8xy + 16y^2}{x^2 - 16y^2}$.

Solution

$$\frac{x^2 - 8xy + 16y^2}{x^2 - 16y^2} = \frac{(x - 4y)(x - 4y)}{(x + 4y)(x - 4y)} \qquad \text{Factor numerator and denominator.}$$

$$= \frac{x - 4y}{x + 4y} \qquad \text{Simplify: } \frac{x - 4y}{x - 4y} = 1$$

■

Quotient Function

In Sections 6.1 and 6.2, we worked with sum, difference, and product functions. Now we will work with *quotient functions.*

> **DEFINITION** Quotient function
>
> If f and g are functions, x is in the domain of both functions, and $g(x)$ is nonzero, then we can form the **quotient function** $\dfrac{f}{g}$:
>
> $$\left(\frac{f}{g}\right)(x) = \frac{f(x)}{g(x)}$$

For example, if $f(x) = 3x$ and $g(x) = x + 5$, then $\left(\dfrac{f}{g}\right)(x) = \dfrac{f(x)}{g(x)} = \dfrac{3x}{x + 5}$.

Example 8 Quotient Functions

Let $f(x) = 8x^3 - 125$ and $g(x) = 4x^2 - 25$.

1. Find an equation of $\dfrac{f}{g}$. Simplify the right-hand side of the equation.

2. Find $\left(\dfrac{f}{g}\right)(3)$.

Solution

1. $\left(\dfrac{f}{g}\right)(x) = \dfrac{f(x)}{g(x)}$ — Definition of quotient function

$= \dfrac{8x^3 - 125}{4x^2 - 25}$ — Substitute $8x^3 - 125$ for $f(x)$ and $4x^2 - 25$ for $g(x)$.

$= \dfrac{(2x-5)(4x^2+10x+25)}{(2x-5)(2x+5)}$ — Factor numerator and denominator.

$= \dfrac{4x^2+10x+25}{2x+5}$ — Simplify: $\dfrac{2x-5}{2x-5} = 1$

2. $\left(\dfrac{f}{g}\right)(3) = \dfrac{4(3)^2 + 10(3) + 25}{2(3)+5} = \dfrac{91}{11}$ ■

By the definition of a rational function, we see that if P and Q are polynomial functions, then the quotient function $\dfrac{P}{Q}$ is rational.

Use a Rational Model to Describe the Percentage of a Quantity

A **rational model** is a rational function, or its graph, that describes an authentic situation. In this section, we will focus on using a rational model to describe the percentage of a quantity.

Suppose that a student takes a quiz and answers 6 out of 8 questions correctly. To compute the percentage of questions answered correctly, we divide the number of questions answered correctly by the total number of questions and multiply the result by 100:

$$\text{Percentage of correct answers} = \frac{6}{8} \cdot 100\% = 75\%$$

We take similar steps to calculate a percentage in general.

Percentage Formula

If m items out of n items have a certain attribute, then the percentage p (written $p\%$) of the n items that have the attribute is

$$p = \frac{m}{n} \cdot 100$$

We call this equation the **percentage formula.**

We will use the percentage formula to form a rational model in Example 9.

Table 2 Numbers of Internet Users in the United States

Year	Number of Internet Users (millions)
1996	39
1997	60
1998	84
1999	105
2000	122
2001	143
2002	166
2003	183
2004	207

Source: *Jupiter MMXI*

Example 9 Using a Rational Model to Make a Prediction

In Exercise 5 of Homework 2.1, you modeled the number of Internet users in the United States (see Table 2). A reasonable model is $I(t) = 20.72t - 83.94$, where $I(t)$ is the number of Internet users (in millions) at t years since 1990. In Exercise 9 of Homework 7.7, you modeled the U.S. population. A reasonable model is $U(t) = 0.0068t^2 + 2.58t + 251.7$, where $U(t)$ is the U.S. population (in millions) at t years since 1990.

1. Let $P(t)$ be the percentage of Americans who are Internet users at t years since 1990. Find an equation of P.
2. Use P to estimate the percentage of Americans who were Internet users in 2004. Then compute the actual percentage by referring to Table 2 and using the 2004 U.S. population as 293.7 million. Is your result from using the model an underestimate or an overestimate?
3. Find $P(17)$. What does it mean in this situation?

Solution

1. To find the percentage of Americans who are Internet users, we divide the number of Internet users by the number of Americans and multiply the result by 100:

$$P(t) = \frac{I(t)}{U(t)} \cdot 100 \qquad \text{Percentage formula: } p = \frac{m}{n} \cdot 100$$

$$= \frac{20.72t - 83.94}{0.0068t^2 + 2.58t + 251.7} \cdot \frac{100}{1} \qquad \text{Substitute } 20.72t - 83.94 \text{ for } I(t) \text{ and } 0.0068t^2 + 2.58t + 251.7 \text{ for } U(t).$$

$$= \frac{2072t - 8394}{0.0068t^2 + 2.58t + 251.7} \qquad \text{Multiply numerators; multiply denominators.}$$

2. $$P(14) = \frac{2072(14) - 8394}{0.0068(14)^2 + 2.58(14) + 251.7} \approx 71.29$$

The model estimates that about 71.3% of Americans were Internet users in 2004. To find the actual percentage, we divide the number of Internet users in 2004 (from Table 2) by the U.S. population (given) and multiply the result by 100:

$$\frac{207{,}000{,}000}{293{,}700{,}000} \cdot 100\% = 70.48\%$$

Our result from using the model is an overestimate.

3. $$P(17) = \frac{2072(17) - 8394}{0.0068(17)^2 + 2.58(17) + 251.7} \approx 90.18$$

The model predicts that about 90.2% of Americans will be Internet users in 2007. ■

group exploration

Connection between the domain and vertical asymptotes

1. Find the domain of $f(x) = \frac{6}{x}$. Use ZStandard to draw a graph of f. Explain why it makes sense that f does not have a y-intercept. Find any vertical asymptotes of f.
2. Use ZStandard to graph $f(x) = \frac{6}{x}$ and $g(x) = \frac{6}{x-4}$ on the same viewing screen. Describe how you can translate the graph of f to get the graph of g.
3. What is the vertical asymptote of $g(x) = \frac{6}{x-4}$? What do you observe about the connection between the vertical asymptote of g and the domain of g?
4. Use ZStandard to graph $h(x) = \frac{7x-7}{x^2+2x-8}$. What is (are) the vertical asymptote(s)? What is the connection between the vertical asymptote(s) of h and the domain of h?
5. Find the domain of the function $f(x) = \frac{x-2}{2-x}$. Use ZDecimal to graph f. Explain why the graph is so different from the graphs you drew in Problems 1, 2, and 4. [**Hint:** Simplify the right-hand side of the equation of f.] Describe what happens when you use TRACE to try to find the value of y when $x = 2$. Explain.
6. Describe the connection between the vertical asymptote(s) of the function $g(x) = \frac{5x-10}{x^2-7x+10}$ and the domain of g. [**Hint:** Simplify the right-hand side of the equation of g.]

7. True or False? Explain.

a. If a rational function f has a vertical asymptote $x = k$, then k is not in the domain of f.

b. If k is not in the domain of a rational function g, then $x = k$ is a vertical asymptote of g.

TIPS FOR SUCCESS: Stick with It

If you are having difficulty doing an exercise, don't panic! Reread the exercise and reflect on what you have already sorted out about the problem—what you know and where you need to go. Your solution to the problem may be just around the corner.

HOMEWORK 8.1

Evaluate the function at the indicated values if possible. If an indicated value is not in the domain, say so.

1. $f(x) = \frac{x+1}{x^2-9}$; $f(-1)$, $f(2)$, $f(3)$

2. $f(x) = \frac{x-2}{x^2-1}$; $f(-3)$, $f(1)$, $f(2)$

3. $f(x) = \frac{x^3-8}{2x^2+3x-1}$; $f(-1)$, $f(0)$, $f(3)$

4. $f(x) = \frac{x^3+5}{3x^2-x-6}$; $f(-2)$, $f(0)$, $f(1)$

Find the domain of the function. Verify your result with a graphing calculator table or graph.

5. $f(x) = \frac{8}{x}$

6. $f(x) = \frac{9}{x}$

7. $f(x) = \frac{x}{2}$

8. $f(x) = \frac{x}{5}$

9. $f(x) = \frac{x-5}{x+3}$

10. $f(x) = \frac{x+4}{x-9}$

11. $f(x) = \frac{x-3}{2x+1}$

12. $f(x) = \frac{x+4}{5x-7}$

13. $f(x) = \frac{x-9}{x^2-3x-10}$

14. $f(x) = \frac{x-1}{x^2+5x-24}$

15. $f(x) = -\frac{2}{4x^2-25}$

16. $f(x) = -\frac{7}{81x^2-49}$

17. $f(x) = \frac{x+3}{x^2+1}$

18. $f(x) = \frac{x+8}{x^2+4}$

19. $f(x) = \frac{x-10}{2x^2-7x-15}$

20. $f(x) = \frac{x-3}{6x^2+7x-20}$

21. $f(x) = -\frac{x+3}{x^2-3x+6}$

22. $f(x) = -\frac{x+6}{3x^2-x+4}$

23. $f(x) = \frac{x^2+5x-1}{3x^2-2x-7}$

24. $f(x) = \frac{x^2-x+3}{6x^2+4x-1}$

25. $f(x) = \frac{2x-14}{4x^3-8x^2-9x+18}$

26. $f(x) = \frac{5x+10}{x^3+5x^2-4x-20}$

Simplify the right-hand side of the equation. Use a graphing calculator table to verify your work.

27. $f(x) = \frac{20x^7}{15x^4}$

28. $f(x) = \frac{12x^3}{16x^{10}}$

29. $f(x) = \frac{4x-28}{5x-35}$

30. $f(x) = \frac{-4x-20}{6x+30}$

31. $f(x) = \frac{x^2+7x+10}{x^2-7x-18}$

32. $f(x) = \frac{x^2+8x+12}{x^2+9x+18}$

33. $f(x) = \frac{x^2-49}{x^2-14x+49}$

34. $f(x) = \frac{x^2-25}{x^2+10x+25}$

35. $f(x) = \frac{16x^2-25}{8x^2-22x+15}$

36. $f(x) = \frac{9x^2-4}{3x^2-17x+10}$

37. $f(x) = \frac{x-5}{5-x}$

38. $f(x) = \frac{2-x}{x-2}$

39. $f(x) = \frac{4x-12}{18-6x}$

40. $f(x) = \frac{-2x+10}{4x-20}$

41. $f(x) = \frac{6x-18}{9-x^2}$

42. $f(x) = \frac{9-x}{x^2-81}$

43. $f(x) = \frac{x^2+2x-35}{-x^2+3x+10}$

44. $f(x) = \frac{x^2+x-20}{-x^2+11x-28}$

45. $f(x) = \frac{3x^3+21x^2+36x}{x^2-9}$

46. $f(x) = \frac{x^2-49}{2x^3-8x^2-42x}$

47. $f(x) = \frac{x^2-2x-8}{4x^3+8x^2-9x-18}$

48. $f(x) = \frac{25x^3-75x^2-4x+12}{x^2-4x+3}$

49. $f(x) = \frac{x^3+8}{x^2-4}$

50. $f(x) = \frac{x^3+27}{x^2-9}$

51. $f(x) = \frac{3x^2+7x-6}{27x^3-8}$

52. $f(x) = \frac{4x^2-11x+6}{64x^3-27}$

Simplify.

53. $\dfrac{x^2-6xy+9y^2}{x^2-3xy}$

54. $\dfrac{x^2+10xy+25y^2}{xy+5y^2}$

55. $\dfrac{6a^2+ab-2b^2}{3a^2-7ab-6b^2}$

56. $\dfrac{2p^2-9pt+10t^2}{8p^2-18pt-5t^2}$

57. $\dfrac{p^3-q^3}{p^2-q^2}$

58. $\dfrac{mn+n^2}{m^3+n^3}$

For the functions $f(x)=x^2+2x-8$, $g(x)=x^2-8x+12$, $h(x)=3x^2+17x+20$, and $k(x)=3x^2-13x-30$, find an equation of the given quotient function. Evaluate the quotient function at the indicated value.

59. $\dfrac{f}{g}, \left(\dfrac{f}{g}\right)(3)$

60. $\dfrac{g}{k}, \left(\dfrac{g}{k}\right)(2)$

61. $\dfrac{h}{f}, \left(\dfrac{h}{f}\right)(4)$

62. $\dfrac{k}{h}, \left(\dfrac{k}{h}\right)(3)$

For $f(x)=3x^3-x^2$, $g(x)=18x^3+12x^2+2x$, $h(x)=9x^2-1$, and $k(x)=27x^3+1$, find an equation of the given quotient function. Evaluate the quotient function at the indicated value.

63. $\dfrac{f}{h}, \left(\dfrac{f}{h}\right)(-2)$

64. $\dfrac{g}{f}, \left(\dfrac{g}{f}\right)(-1)$

65. $\dfrac{k}{g}, \left(\dfrac{k}{g}\right)(-1)$

66. $\dfrac{h}{k}, \left(\dfrac{h}{k}\right)(0)$

67. In Example 9 of Section 7.3, we modeled the average monthly participation in the Food Stamp Program. A reasonable model is $F(t)=0.465t^2-9.5t+66$, where $F(t)$ is the average monthly participation (in millions) in the year that is t years since 1990 (see Table 3). In Exercise 9 of Homework 7.7, you modeled the U.S. population. A reasonable model is $U(t)=0.0068t^2+2.58t+251.7$, where $U(t)$ is the U.S. population (in millions) at t years since 1990.

Table 3 Average Monthly Participation in Food Stamp Program

Year	Average Monthly Participation (millions)
1996	25.4
1997	22.7
1998	19.8
1999	18.2
2000	17.2
2001	17.3
2002	19.1
2003	21.3
2004	23.9

Source: *U.S. Department of Agriculture*

a. Let $P(t)$ be the percentage of Americans who received food stamps in the year that is t years since 1990. Find an equation of P. Describe any assumptions you have made.

b. Use P to estimate the percentage of Americans who received food stamps in 2004. Then, make the same assumptions that you made in part (a) to compute the actual percentage by referring to Table 3 and using the 2004 U.S. population as 293.7 million. Is your result from using the model an underestimate or an overestimate?

c. Find $P(17)$. What does it mean in this situation?

68. In Exercise 30 of Homework 3.1, you found the system

$$C(t)=4.80t+6.86$$
$$H(t)=1.59t+90.31$$

where $C(t)$ is the number of U.S. households with personal computers and $H(t)$ is the total number of U.S. households, both in millions, at t years since 1990 (see Table 4).

Table 4 Numbers of Households with Personal Computers

Year	Number of Households with Personal Computers (millions)	Total Number of Households (millions)
1995	31.36	99.0
1997	39.56	101.0
1999	50.00	103.6
2001	60.81	108.2
2003	68.78	111.3

Sources: *U.S. Department of Commerce; Media Matrix; U.S. Census Bureau*

a. Let $P(t)$ be the percentage of U.S. households with personal computers at t years since 1990. Find an equation of P.

b. Use P to estimate the percentage of households that had personal computers in 2001. Then compute the actual percentage. Is your result from using the model an underestimate or an overestimate?

c. Use P to predict what percentage of households will have personal computers in 2007. In Example 9, we used an Internet model to predict that about 90.2% of Americans will be Internet users in 2007. Does this mean that there is model breakdown for either the computer model or the Internet model for 2007? Explain.

69. In Example 1 of Section 7.8, we found the function $U(t)=0.047t^2-0.77t+3.5$, where $U(t)$ is the number of unredeemed miles (in trillions of miles) from the frequent-flier programs for airlines at t years since 1980 (see Table 5).

Table 5 Cumulative Unredeemed Miles

Year	Cumulative Unredeemed Miles (trillions of miles)	Cumulative Awarded Miles (trillions of miles)
1986	0.1	0.3
1988	0.4	0.8
1990	0.9	1.5
1992	1.5	2.5
1994	2.2	3.7
1996	3.2	5.2
1998	4.6	7.3
2000	6.6	10.0
2002	9.1	13.3
2004	12.4	17.2
2005	14.2	19.9

Source: *WebFlyer*

a. Let $A(t)$ be the cumulative *awarded* miles (in trillions of miles), which include both redeemed and unredeemed miles, at t years since 1980. Find an equation of A.

b. Let $P(t)$ be the percentage of cumulative awarded miles that are unredeemed at t years since 1980. Find an equation of P.

c. Find $P(30)$. What does it mean in this situation?

d. Use the window settings in Fig. 7 to draw a graph of P. Then use "minimum" to find the minimum point of P for values of t between 0 and 30. What does it mean in this situation?

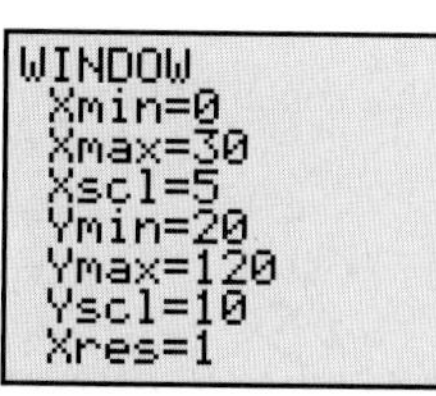

Figure 7 Window settings for Exercise 69d

e. Has the percentage of cumulative unredeemed miles been increasing or decreasing since 1990? Explain.

70. In Example 1 of Section 2.1, we sketched a line to model the number of visitors to the Grand Canyon (see Table 6).

Table 6 Visitors to the Grand Canyon

Year	Average Monthly Participation (millions)
1960	1.2
1970	2.3
1980	2.6
1990	3.8
2000	4.8

Source: *National Park Service*

a. Let $V(t)$ be the number of visitors (in millions) in the year that is t years since 1960. Find an equation of V.

b. In Exercise 9 of Homework 7.7, you modeled the U.S. population. A good model is $U(t) = 0.0068t^2 + 2.17t + 180.41$, where $U(t)$ is the U.S. population (in millions) at t years since 1960. Let $P(t)$ be the percentage of Americans who visit the Grand Canyon in the year that is t years since 1960. Describe any assumptions you have made.

c. Find $P(50)$. What does it mean in this situation?

d. Use the window settings in Fig. 8 to draw a graph of P. Is P increasing, decreasing, or neither for the values of t between 0 and 50? What does that mean in this situation? Can this change be fully explained by the fact that the U.S. population has been increasing for those years? Explain.

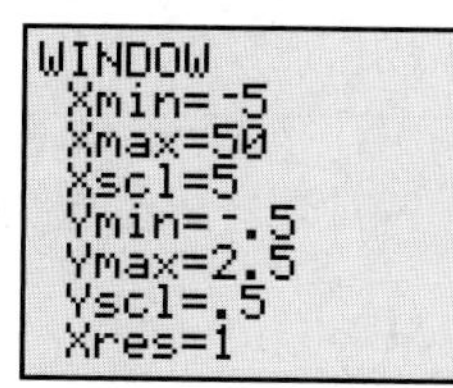

Figure 8 Window settings for Exercise 70d

71. Give examples of three rational functions, each of whose domains are the set of real numbers except -3 and 3.

72. Give examples of three rational functions, each of whose domains are the set of real numbers except $\frac{1}{2}$ and $\frac{1}{3}$.

73. A graphing calculator table for a rational function is shown in Fig. 9. List seven members of the domain. List seven members of the range.

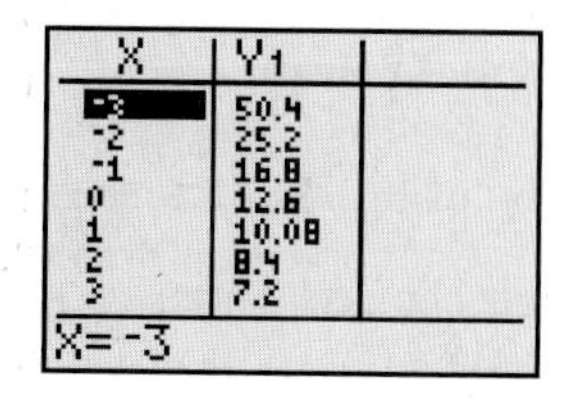

Figure 9 Exercise 73

74. A student decides that the domain of the function

$$f(x) = \frac{6}{x(x-3) + 2(x-3)}$$

is the set of real numbers except 3. After viewing the table shown in Fig. 10, the student believes that his work is correct. What would you tell this student?

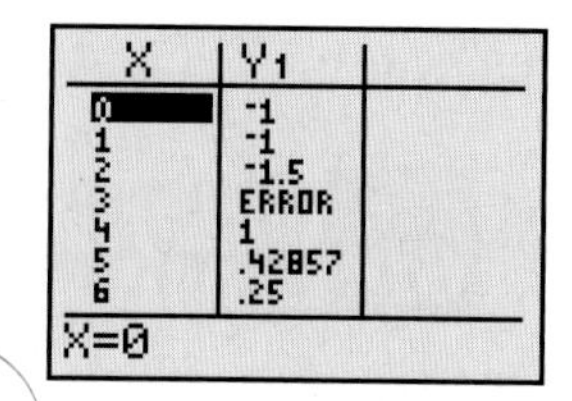

Figure 10 Exercise 74

75. A student states that the domain of the function

$$f(x) = \frac{(x-2)(x-4)}{(x-5)(x-1)}$$

is the set of real numbers except 1, 2, 4, and 5. Is the student correct? Explain.

76. It is a common error to think that 0 is not in the domain of the function $f(x) = \frac{7}{x+4}$. Evaluate f at 0 to show that 0 is in the domain of f.

77. A student tries to simplify the expression $\frac{2(x+4)+3}{(x+4)(x-1)}$:

$$\frac{2(x+4)+3}{(x+4)(x-1)} = \frac{2+3}{x-1} = \frac{5}{x-1}$$

Substitute 3 for x in the original expression and in the student's result. What can you conclude about the student's work, based on the results of your substitutions? Explain.

78. A student tries to simplify the expression $\frac{(x-2)(x-1)}{4(x-2)+5}$:

$$\frac{(x-2)(x-1)}{4(x-2)+5} = \frac{x-1}{4+5} = \frac{x-1}{9}$$

Next, she substitutes 3 for x in the original expression and in her result:

$$\frac{(3-2)(3-1)}{4(3-2)+5} = \frac{2}{9}; \qquad \frac{3-1}{9} = \frac{2}{9}$$

The student concludes that the result $\frac{x-1}{9}$ is correct. What would you tell her?

79. Describe how to find the domain of a rational function. If you believe that a number is not in the domain, describe how you can verify that belief. (See page 4 for guidelines on writing a good response.)

80. Describe how to simplify a rational expression. Explain how you can check that the result and the original expression are equivalent. (See page 4 for guidelines on writing a good response.)

Related Review

Find the domain.

81. $f(x) = 4x - 8$

82. $f(x) = 3(2)^x$

83. $f(x) = 2(x-4)^2 - 5$

84. $f(x) = \frac{x-3}{x^2 - 49}$

Expressions, Equations, Functions, and Graphs

Perform the indicated instruction. Then use words such as linear, quadratic, cubic, exponential, logarithmic, rational, polynomial, degree, function, one variable, *and* two variables *to describe the expression, equation, or system.*

85. Find the domain of $f(x) = \dfrac{x-5}{x^2-4x-21}$.

86. Solve $-2x^2 - 4x + 3 = 0$.

87. Solve $3x^3 + 5x^2 = 12x + 20$.

88. Graph $f(x) = -2x^2 - 4x + 3$ by hand.

89. Factor $8x^3 - 125$.

90. Find an equation of a parabola that contains the points (2, 6), (3, 1), and (4, −8).

8.2 MULTIPLYING AND DIVIDING RATIONAL EXPRESSIONS

Objective

- Multiply and divide rational expressions.

In this section, we will multiply and divide rational expressions.

Multiplication of Rational Expressions

At times in this course, we have multiplied fractions by using the property $\dfrac{a}{b} \cdot \dfrac{c}{d} = \dfrac{ac}{bd}$, where b and d are nonzero. For example,

$$\frac{2}{3} \cdot \frac{5}{7} = \frac{2 \cdot 5}{3 \cdot 7} = \frac{10}{21}$$

We multiply more complicated rational expressions in a similar way.

Multiplying Rational Expressions

If $\dfrac{A}{B}$ and $\dfrac{C}{D}$ are rational expressions and B and D are nonzero, then

$$\frac{A}{B} \cdot \frac{C}{D} = \frac{AC}{BD}$$

In words: To multiply two rational expressions, write the numerators as a product and write the denominators as a product.

Example 1 Multiplying Rational Expressions

Find the product $\dfrac{4x^3}{7x-1} \cdot \dfrac{3x+5}{2x}$. Simplify the result.

Solution

$$\frac{4x^3}{7x-1} \cdot \frac{3x+5}{2x} = \frac{4x^3 \cdot (3x+5)}{(7x-1) \cdot 2x}$$

Multiply numerators and multiply denominators: $\dfrac{A}{B} \cdot \dfrac{C}{D} = \dfrac{AC}{BD}$

$$= \frac{2 \cdot 2 \cdot x \cdot x \cdot x \cdot (3x+5)}{(7x-1) \cdot 2 \cdot x}$$

Factor numerator and denominator.

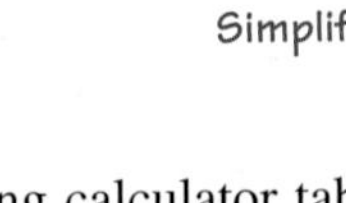

$$= \frac{2x^2(3x+5)}{7x-1}$$

Simplify: $\dfrac{2x}{2x} = 1$

Plot1 Plot2 Plot3
\Y1=(4X^3)/(7X-1)
\Y2=(3X+5)/(2X)
\Y3=Y1*Y2
\Y4=(2X^2(3X+5))/(7X-1)
\Y5=

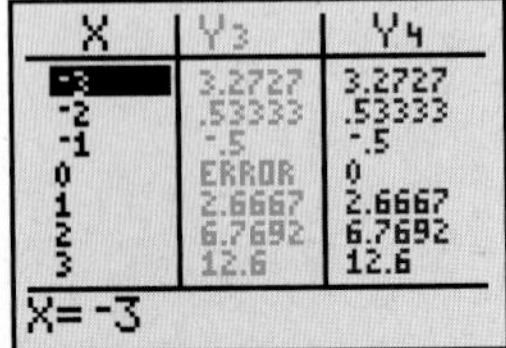

Figure 11 Verify the work

We verify our work by creating a graphing calculator table for $y = \dfrac{4x^3}{7x-1} \cdot \dfrac{3x+5}{2x}$ and $y = \dfrac{2x^2(3x+5)}{7x-1}$. See Fig. 11. (To enter an equation by using Y_n references, see Section B.25.) In the table, $x = 0$ gives an "ERROR" message, because 0 is an excluded value of the original expression. ■

To find the product of rational expressions, we usually begin by factoring the numerators and the denominators if possible. That will put us in a good position to simplify after we have found the product.

Example 2 Multiplying Rational Expressions

Find the product $\frac{k^2-9}{2k^2-k-10} \cdot \frac{4k^2-25}{k^2+4k-21}$. Simplify the result.

Solution

We begin by factoring the numerators and denominators:

$$\frac{k^2-9}{2k^2-k-10} \cdot \frac{4k^2-25}{k^2+4k-21} = \frac{(k-3)(k+3)}{(2k-5)(k+2)} \cdot \frac{(2k-5)(2k+5)}{(k-3)(k+7)}$$

Factor numerators and denominators.

$$= \frac{(k-3)(k+3)(2k-5)(2k+5)}{(2k-5)(k+2)(k-3)(k+7)}$$

Multiply numerators; multiply denominators.

$$= \frac{(k+3)(2k+5)}{(k+2)(k+7)}$$

Simplify: $\frac{(k-3)(2k-5)}{(k-3)(2k-5)} = 1$

■

How to Multiply Rational Expressions

To multiply two rational expressions,

1. Factor the numerators and the denominators.
2. Multiply by using the property $\frac{A}{B} \cdot \frac{C}{D} = \frac{AC}{BD}$, where B and D are nonzero.
3. Simplify the result.

Example 3 Finding a Product Function

Let $f(x) = \frac{35x^2-25x}{x^2-36}$ and $g(x) = \frac{6-x}{15x^4}$.

1. Find an equation of the product function $f \cdot g$.
2. Find $(f \cdot g)(2)$.

Solution

1. $(f \cdot g)(x) = f(x) \cdot g(x)$

Definition of $f \cdot g$

$$= \frac{35x^2-25x}{x^2-36} \cdot \frac{6-x}{15x^4}$$

Substitute $\frac{35x^2-25x}{x^2-36}$ for $f(x)$ and $\frac{6-x}{15x^4}$ for $g(x)$.

$$= \frac{5x(7x-5)}{(x-6)(x+6)} \cdot \frac{-(x-6)}{3 \cdot 5x^4}$$

Factor numerators and denominators; $6-x = -x+6 = -(x-6)$

$$= -\frac{5x(7x-5)(x-6)}{(x-6)(x+6) \cdot 3 \cdot 5 \cdot x^4}$$

Multiply numerators; multiply denominators; $\frac{-a}{b} = -\frac{a}{b}$

$$= -\frac{7x-5}{3x^3(x+6)}$$

Simplify: $\frac{5x(x-6)}{5x(x-6)} = 1$

2. $(f \cdot g)(2) = -\frac{7(2)-5}{3(2)^3(2+6)} = -\frac{9}{192} = -\frac{3}{64}$

■

Division of Rational Expressions

How do we divide two rational expressions? We will need to find the reciprocal of a rational expression to divide rational expressions. The **reciprocal** of $\frac{A}{B}$ is $\frac{B}{A}$. For example, the reciprocal of $\frac{x+2}{9}$ is $\frac{9}{x+2}$.

At times in this course, we have divided fractions by using the property $\frac{a}{b} \div \frac{c}{d} = \frac{a}{b} \cdot \frac{d}{c}$, where b, c, and d are nonzero. For example,

$$\frac{2}{3} \div \frac{5}{7} = \frac{2}{3} \cdot \frac{7}{5} = \frac{14}{15}$$

Dividing by $\frac{5}{7}$ is the same as multiplying by the reciprocal of $\frac{5}{7}$, which is $\frac{7}{5}$. We divide more complicated rational expressions in a similar way.

> **Dividing Rational Expressions**
>
> If $\frac{A}{B}$ and $\frac{C}{D}$ are rational expressions and B, C, and D are nonzero, then
>
> $$\frac{A}{B} \div \frac{C}{D} = \frac{A}{B} \cdot \frac{D}{C}$$
>
> In words: To divide by a rational expression, multiply by its reciprocal.

Example 4 Dividing Two Rational Expressions

Find the quotient $\frac{6x^2}{x-3} \div \frac{4x^7}{x+1}$.

Solution

$$\frac{6x^2}{x-3} \div \frac{4x^7}{x+1} = \frac{6x^2}{x-3} \cdot \frac{x+1}{4x^7} \qquad \text{Multiply by reciprocal of } \frac{4x^7}{x+1}\text{, which is } \frac{x+1}{4x^7}.$$

$$= \frac{2 \cdot 3 \cdot x^2}{x-3} \cdot \frac{x+1}{2 \cdot 2 \cdot x^7} \qquad \text{Factor numerators and denominators.}$$

$$= \frac{2 \cdot 3 \cdot x^2(x+1)}{(x-3) \cdot 2 \cdot 2 \cdot x^7} \qquad \text{Multiply numerators; multiply denominators.}$$

$$= \frac{3(x+1)}{2x^5(x-3)} \qquad \text{Simplify: } \frac{2x^2}{2x^2} = 1$$

X	Y3	Y4
-3	-.0021	-.0021
-2	-.0094	-.0094
-1	ERROR	0
0	ERROR	ERROR
1	-1.5	-1.5
2	-.1406	-.1406
3	ERROR	ERROR

X=-3

Figure 12 Verify the work

We verify our work by creating a graphing calculator table for $y = \frac{6x^2}{x-3} \div \frac{4x^7}{x+1}$ and $y = \frac{3(x+1)}{2x^5(x-3)}$. See Fig. 12. The values $x = -1$, $x = 0$, and $x = 3$ give "ERROR" messages, because each of these values is an excluded value of either the original expression or our result (why?). ■

> **How to Divide Rational Expressions**
>
> To divide two rational expressions,
>
> 1. Write the quotient as a product by using the property $\frac{A}{B} \div \frac{C}{D} = \frac{A}{B} \cdot \frac{D}{C}$, where B, C, and D are nonzero.
> 2. Find the product.
> 3. Simplify.

Example 5 Dividing Two Rational Expressions

Find the quotient $\dfrac{x^3 - y^3}{9x^2 - y^2} \div \dfrac{x^3 + x^2y + xy^2}{6x^2 - xy - y^2}$.

Solution

$$\frac{x^3 - y^3}{9x^2 - y^2} \div \frac{x^3 + x^2y + xy^2}{6x^2 - xy - y^2} = \frac{x^3 - y^3}{9x^2 - y^2} \cdot \frac{6x^2 - xy - y^2}{x^3 + x^2y + xy^2}$$
Multiply by reciprocal of $\dfrac{x^3 + x^2y + xy^2}{6x^2 - xy - y^2}$.

$$= \frac{(x - y)(x^2 + xy + y^2)}{(3x + y)(3x - y)} \cdot \frac{(2x - y)(3x + y)}{x(x^2 + xy + y^2)}$$
Factor numerators and denominators.

$$= \frac{(x - y)(x^2 + xy + y^2)(2x - y)(3x + y)}{(3x + y)(3x - y) \cdot x(x^2 + xy + y^2)}$$
Multiply numerators; multiply denominators.

$$= \frac{(x - y)(2x - y)}{x(3x - y)}$$
Simplify: $\dfrac{(x^2 + xy + y^2)(3x + y)}{(x^2 + xy + y^2)(3x + y)} = 1$

Since our result is in lowest terms, we are done. ■

Example 6 Finding a Quotient Function

Let $f(x) = \dfrac{81x^2 - 49}{3x^2 + 16x + 5}$ and $g(x) = \dfrac{7 - 9x}{18x + 6}$.

1. Find an equation of the quotient function $\dfrac{f}{g}$.

2. Find $\left(\dfrac{f}{g}\right)(3)$.

Solution

1. $$\frac{f}{g}(x) = \frac{f(x)}{g(x)}$$
Definition of $\dfrac{f}{g}$

$$= f(x) \div g(x)$$
$\dfrac{R}{S} = R \div S$

$$= \frac{81x^2 - 49}{3x^2 + 16x + 5} \div \frac{7 - 9x}{18x + 6}$$
Substitute $\dfrac{81x^2 - 49}{3x^2 + 16x + 5}$ for $f(x)$ and $\dfrac{7 - 9x}{18x + 6}$ for $g(x)$.

$$= \frac{81x^2 - 49}{3x^2 + 16x + 5} \cdot \frac{18x + 6}{7 - 9x}$$
Multiply by reciprocal of $\dfrac{7 - 9x}{18x + 6}$.

$$= \frac{(9x - 7)(9x + 7)}{(3x + 1)(x + 5)} \cdot \frac{6(3x + 1)}{-(9x - 7)}$$
Factor numerators and denominators; $7 - 9x = -(9x - 7)$

$$= -\frac{(9x - 7)(9x + 7) \cdot 6(3x + 1)}{(3x + 1)(x + 5)(9x - 7)}$$
Multiply numerators; multiply denominators; $\dfrac{a}{-b} = -\dfrac{a}{b}$

$$= -\frac{6(9x + 7)}{x + 5}$$
Simplify: $\dfrac{(9x - 7)(3x + 1)}{(9x - 7)(3x + 1)} = 1$

2. $$\left(\frac{f}{g}\right)(3) = -\frac{6[9(3) + 7]}{3 + 5} = -\frac{204}{8} = -\frac{51}{2}$$ ■

In Example 7, we combine three rational expressions by using two operations.

Example 7 Combining Three Rational Expressions

Perform the indicated operations:

$$\left(\frac{x^2 - 2x - 48}{x^2 + 8x + 16} \div \frac{3x^2 - 9x}{x^2 - 16}\right) \cdot \frac{6x + 24}{5x + 30}$$

Solution

$$\left(\frac{x^2-2x-48}{x^2+8x+16} \div \frac{3x^2-9x}{x^2-16}\right) \cdot \frac{6x+24}{5x+30} = \left(\frac{x^2-2x-48}{x^2+8x+16} \cdot \frac{x^2-16}{3x^2-9x}\right) \cdot \frac{6x+24}{5x+30}$$

$$= \left(\frac{(x-8)(x+6)}{(x+4)^2} \cdot \frac{(x-4)(x+4)}{3x(x-3)}\right) \cdot \frac{2 \cdot 3 \cdot (x+4)}{5(x+6)}$$

$$= \frac{2 \cdot 3 \cdot (x-8)(x+6)(x-4)(x+4)^2}{3 \cdot 5 \cdot x \cdot (x+4)^2(x-3)(x+6)}$$

$$= \frac{2(x-8)(x-4)}{5x(x-3)}$$

■

group exploration

Looking ahead: Adding and subtracting rational expressions

1. Compare graphing calculator tables for $y = \frac{2}{x} + \frac{3}{x}$ and $y = \frac{5}{x}$. What does this comparison suggest about finding the sum $\frac{A}{B} + \frac{C}{B}$?
2. Compare graphing calculator tables for $y = \frac{5}{x} - \frac{2}{x}$ and $y = \frac{3}{x}$. What does this comparison suggest about finding the difference $\frac{A}{B} - \frac{C}{B}$?
3. Use a graphing calculator table to find outputs of the function $y = \frac{x}{5} - \frac{x+10}{5}$. What do you notice about the outputs? Explain why it makes sense that the outputs are what they are.
4. Compare graphing calculator tables for $y = \frac{2}{x} + \frac{x}{3}$ and $y = \frac{2+x}{x+3}$. Is the following statement, in general, true or false? Explain. [**Hint:** In what way is this sum different from the sum and differences in Problems 1–3?]

$$\frac{A}{B} + \frac{C}{D} = \frac{A+C}{B+D}, \quad \text{where } B, D, \text{ and } B+D \text{ are nonzero}$$

5. State any concepts suggested by this exploration.

HOMEWORK 8.2

FOR EXTRA HELP ▶ Student Solutions Manual | PH Math/Tutor Center | MathXL® | MyMathLab

Perform the indicated operation. Simplify the result.

1. $\frac{5}{x} \cdot \frac{2}{x}$
2. $\frac{x}{9} \cdot \frac{3}{x}$
3. $\frac{7x^5}{2} \div \frac{5x^3}{6}$
4. $\frac{14}{8x^9} \div \frac{21}{6x^5}$
5. $\frac{5p^3}{4p-8} \cdot \frac{3p-6}{10p^8}$
6. $\frac{5m-20}{21m^6} \cdot \frac{7m^2}{3m^2-12m}$
7. $\frac{6x-18}{5x^5} \div \frac{5x-15}{x^3}$
8. $\frac{20x^6}{7x-14} \div \frac{50x^8}{2x-4}$
9. $\frac{4a^3}{9b^2} \cdot \frac{3b^5}{8a}$
10. $\frac{3a^2}{10b^4} \cdot \frac{15b}{7a^5}$
11. $\frac{3y^5}{2x^4} \div \frac{15y^9}{16x^7}$
12. $\frac{7x^4}{8y^2} \div \frac{21x^3}{16y^4}$
13. $\frac{r^2+10r+21}{r-9} \cdot \frac{2r-18}{r^2-9}$
14. $\frac{w^2-3w-28}{w-5} \cdot \frac{5w-25}{w^2-49}$
15. $\frac{x^2+3x+2}{3x-3} \div \frac{x^2-x-6}{6x-6}$

16. $\dfrac{x^2+9x+20}{8x-16} \div \dfrac{x^2-2x-35}{4x-8}$

17. $\dfrac{2x-12}{x+1} \cdot \dfrac{4x+4}{18-3x}$

18. $\dfrac{2x+4}{4x-20} \cdot \dfrac{12-3x}{x+2}$

19. $\dfrac{2k^2-32}{k^2-2k-24} \div \dfrac{k+6}{k^2-7k+6}$

20. $\dfrac{3b^2-27}{b^2-12+35} \div \dfrac{b+3}{b^2-b-20}$

21. $\dfrac{2a^2+3ab}{3a-6b} \cdot \dfrac{a^2-4b^2}{2ab+3b^2}$

22. $\dfrac{2a+10b}{3a^2+ab} \cdot \dfrac{9a^2-b^2}{ab+5b^2}$

23. $\dfrac{4-x}{x^2+10x+25} \div \dfrac{3x^2-9x-12}{25-x^2}$

24. $\dfrac{6-x}{x^2+8x+16} \div \dfrac{4x^2-16x-48}{16-x^2}$

25. $\dfrac{t^2-8t+16}{t^2-2t-3} \cdot \dfrac{3-t}{t^2-16}$

26. $\dfrac{p^2-3p-10}{p^2-4p-12} \cdot \dfrac{6-p}{p^2-25}$

27. $\dfrac{-x^2+7x-10}{2x^2+5x-12} \div \dfrac{-x^2+4}{8x^2-18}$

28. $\dfrac{-x^2-5x+14}{3x^2-5x-12} \div \dfrac{-x^2+4}{5x^2-45}$

29. $\dfrac{-4x-6}{36-x^2} \cdot \dfrac{4x+24}{6x^2+x-12}$

30. $\dfrac{-6x-15}{9-x^2} \cdot \dfrac{6x+18}{4x^2+4x-15}$

31. $\dfrac{9x^2-16}{x+2} \div (3x^2+5x-12)$

32. $\dfrac{16x^2-25}{x+3} \div (4x^2-13x+10)$

33. $\dfrac{6m^2-17m-14}{m^2+6m+9} \cdot \dfrac{9-m^2}{4m^2-49}$

34. $\dfrac{3r^2+7r-20}{r^2+4r+4} \cdot \dfrac{4-r^2}{9r^2-25}$

35. $\dfrac{x^2-4x-32}{x^2+7x+12} \div \dfrac{x^2-2x-48}{x^2+3x-4}$

36. $\dfrac{x^2-6x+8}{x^2-9x+14} \div \dfrac{x^2-x-12}{x^2+x-56}$

37. $\dfrac{p^2+4pt-12t^2}{p^2+pt-12t^2} \cdot \dfrac{p^2+7pt+12t^2}{p^2-7pt+10t^2}$

38. $\dfrac{m^2-6mn+9n^2}{m^2+3mn-10n^2} \cdot \dfrac{m^2+10mn+25n^2}{m^2-4mn+3n^2}$

39. $\dfrac{2x^2-xy-3y^2}{3xy-5y^2} \div \dfrac{4x^2-9y^2}{3x^2-14xy+15y^2}$

40. $\dfrac{4t^2-9tw+2w^2}{9t^2-4w^2} \div \dfrac{4t^2-5tw+w^2}{6tw-4w^2}$

41. $\dfrac{3x^3-15x^2+18x}{x^2+16x+64} \cdot \dfrac{x^2-64}{4x^4-28x^3+40x^2}$

42. $\dfrac{6x^3+6x^2-12x}{x^2-9} \cdot \dfrac{x^2-6x+9}{2x^4+10x^3-12x^2}$

43. $\dfrac{w^2-2w-8}{12w^4+32w^3-12w^2} \div \dfrac{w^2-9w+20}{12w^3+54w^2+54w}$

44. $\dfrac{t^2+9t+18}{6t^4-8t^3-8t^2} \div \dfrac{t^2-5t-24}{12t^3+20t^2+8t}$

45. $\dfrac{x^2+4x-5}{x^3+6x^2-4x-24} \cdot \dfrac{x^2+8x+12}{x^2+10x+25}$

46. $\dfrac{x^2-3x-10}{x^2-12x+36} \cdot \dfrac{x^2-3x-18}{x^3+2x^2-25x-50}$

47. $\dfrac{18x^3+27x^2-8x-12}{3x^2-x-2} \div (6x^2+5x-6)$

48. $\dfrac{4x^3-8x^2-x+2}{2x^2-11x+5} \div (2x^2-3x-2)$

49. $\dfrac{k^3-8}{k^3+27} \cdot \dfrac{k^2-9}{k^2-4}$

50. $\dfrac{y^3-1}{y^2-4} \cdot \dfrac{y^3-8}{y^2-1}$

51. $\dfrac{8x^3-27}{3x^2-6x+12} \div \dfrac{8x^2+12x+18}{6x^3+48}$

52. $\dfrac{64x^3+125}{3x^2-3x+3} \div \dfrac{32x^2-40x+50}{6x^3+6}$

53. $\dfrac{a^2+ab-2b^2}{a^3+b^3} \cdot \dfrac{a^2+2ab+b^2}{a^2-b^2}$

54. $\dfrac{6p+4q}{4p^2+2pq+q^2} \cdot \dfrac{8p^3-q^3}{3p^2+2pq}$

For $f(x)=\dfrac{x^2-6x-16}{x^2+3x-40}$ *and* $g(x)=\dfrac{x^2-64}{x^2-3x-10}$, *find an equation of the given function. Evaluate the function at the indicated value.*

55. $f \cdot g, (f \cdot g)(6)$

56. $\dfrac{f}{g}, \left(\dfrac{f}{g}\right)(9)$

57. $\dfrac{g}{f}, \left(\dfrac{g}{f}\right)(7)$

For $f(x)=\dfrac{1-x^2}{x^2-3x-28}$ *and* $g(x)=\dfrac{x^2-8x+7}{x^2+5x+4}$, *find an equation of the given function. Evaluate the function at the indicated value.*

58. $f \cdot g, (f \cdot g)(2)$

59. $\dfrac{f}{g}, \left(\dfrac{f}{g}\right)(4)$

60. $\dfrac{g}{f}, \left(\dfrac{g}{f}\right)(3)$

Perform the indicated operations. Simplify the result.

61. $\left(\dfrac{20x^7}{x^2-9} \div \dfrac{x^2-14x+24}{5x-15}\right) \cdot \dfrac{x^2+x-6}{8x^{13}}$

62. $\left(\dfrac{8x^2+10x-3}{-2x^2-8x} \cdot \dfrac{x^2+x}{16x^2-1}\right) \div \dfrac{-10x-15}{8x^5}$

63. $\dfrac{12k^3}{k^2-4} \div \left(\dfrac{22k^6}{-6k+12} \cdot \dfrac{k}{11k+22}\right)$

64. $\dfrac{3-p}{20p^5} \div \left(\dfrac{p^2-9}{30p^2} \div \dfrac{8p^2+6p}{3p}\right)$

65. $\left(\left(\dfrac{x-4}{x+5}\right)^2 \cdot \left(\dfrac{x+5}{x-1}\right)^2\right) \div \left(\dfrac{x-4}{x-1}\right)^2$

66. $\dfrac{3x+6}{4x+20} \div \left(\dfrac{x^2-4}{x^2-25} \div \dfrac{x-2}{x-5}\right)^2$

67. A student tries to find the quotient $\dfrac{x-2}{x+8} \div \dfrac{x+8}{x-5}$:

$$\frac{x-2}{x+8} \div \frac{x+8}{x-5} = \frac{x-2}{x-5}$$

Find a value to substitute for x to show that the student's work is incorrect. Then find the quotient correctly. Use a graphing calculator table to verify your result.

68. A student tries to find the product $\dfrac{x+2}{x+5} \cdot \dfrac{x+2}{x-5}$:

$$\frac{x+2}{x+5} \cdot \frac{x+2}{x-5} = \frac{x^2+4}{x^2-25}$$

Find a value to substitute for x to show that the student's work is incorrect. Then find the product correctly. Use a graphing calculator table to verify your result.

69. Perform the indicated operations.

a. $\dfrac{1}{x} \div \dfrac{1}{x}$

b. $\dfrac{1}{x} \div \left(\dfrac{1}{x} \div \dfrac{1}{x}\right)$

c. $\dfrac{1}{x} \div \left(\dfrac{1}{x} \div \left(\dfrac{1}{x} \div \dfrac{1}{x}\right)\right)$

d. $\dfrac{1}{x} \div \left(\dfrac{1}{x} \div \left(\dfrac{1}{x} \div \left(\dfrac{1}{x} \div \dfrac{1}{x}\right)\right)\right)$

e. $\underbrace{\dfrac{1}{x} \div \left(\dfrac{1}{x} \div \left(\dfrac{1}{x} \div \cdots \div \left(\dfrac{1}{x} \div \left(\dfrac{1}{x} \div \dfrac{1}{x}\right)\right)\cdots\right)\right)}_{n \text{ division symbols}}$

70. Describe how to multiply two rational expressions, and describe how to divide two rational expressions.

71. A person plans to drive at a constant speed from San Francisco to Los Angeles. The driving time (in hours) $T(s)$ is given by the equation

$$T(s) = \frac{420}{s}$$

where s is the constant speed (in miles per hour).

a. Find $T(50)$. What does it mean in this situation?

b. Find $T(55)$, $T(60)$, $T(65)$, and $T(70)$.

c. Is T an increasing function or a decreasing function for $s > 0$? Explain why that makes sense in this situation. Use ZStandard and then Zoom Out once or twice to verify your answer.

72. Some students agree to share equally in the expense of renting a beach house for $1200 during spring break.

a. What is the per-student expense if 10 students rent it?

b. What is the per-student expense if 12 students rent it?

c. Let $p(n)$ be the per-student expense (in dollars) for n students to rent the house. Find an equation of p. [**Hint:** Reflect on your work in parts (a) and (b).]

d. Use ZStandard and then Zoom Out twice to draw a graph of p. Is p a decreasing function or an increasing function for $n > 0$? Explain why that makes sense in this situation.

e. Find $p(15)$. What does it mean in this situation?

Related Review

Find equations of $f \cdot g$ *and* $\dfrac{f}{g}$. $\left[\textbf{Hint: } b^n c^n = (bc)^n,\ \dfrac{b^n}{c^n} = \left(\dfrac{b}{c}\right)^n\right]$

73. $f(x) = 8^x$, $g(x) = 2^x$

74. $f(x) = 6^x$, $g(x) = 3^x$

75. $f(x) = 12(6)^x$, $g(x) = 3(2)^x$

76. $f(x) = 10(8)^x$, $g(x) = 2(4)^x$

Expressions, Equations, Functions, and Graphs

Perform the indicated instruction. Then use words such as linear, quadratic, cubic, exponential, logarithmic, rational, polynomial, degree, function, one variable, *and* two variables *to describe the expression, equation, or system.*

77. Write $4\log_b(2x^2) - 2\log_b(3x^3)$ as a single logarithm.

78. Find the product $(5t-3)(4t-6)$.

79. Solve $\log_2(x-3) + \log_2(x-2) = 3$. Round any solutions to the fourth decimal place.

80. Factor $6t^2 - 19t + 10$.

81. Solve $5(4)^x - 23 = 81$. Round any solutions to the fourth decimal place.

82. Solve $6t^2 - 19t + 10 = 0$.

8.3 ADDING AND SUBTRACTING RATIONAL EXPRESSIONS

Objective

- Add and subtract rational expressions.

In Section 8.2, we multiplied and divided rational expressions. How do we add and subtract rational expressions?

Addition of Rational Expressions

At times in this course, we have added fractions by using the property $\frac{a}{b} + \frac{c}{b} = \frac{a+c}{b}$, where b is nonzero. For example,

$$\frac{3}{13} + \frac{5}{13} = \frac{8}{13}$$

We add more complicated rational expressions with a common denominator in a similar way.

Adding Rational Expressions with a Common Denominator

If $\frac{A}{B}$ and $\frac{C}{B}$ are rational expressions and B is nonzero, then

$$\frac{A}{B} + \frac{C}{B} = \frac{A+C}{B}$$

In words: To add two rational expressions with a common denominator, add the numerators and keep the common denominator.

After adding two rational expressions, it may be possible to simplify the result.

Example 1 Adding Two Rational Expressions with a Common Denominator

Find the sum $\frac{x^2+5x}{x^2-9} + \frac{6}{x^2-9}$.

Solution

$$\frac{x^2+5x}{x^2-9} + \frac{6}{x^2-9} = \frac{x^2+5x+6}{x^2-9}$$ Add numerators and keep common denominator: $\frac{A}{B} + \frac{C}{B} = \frac{A+C}{B}$

$$= \frac{(x+2)(x+3)}{(x-3)(x+3)}$$ Factor numerator and denominator.

$$= \frac{x+2}{x-3}$$ Simplify: $\frac{x+3}{x+3} = 1$ ■

Suppose that two brothers, John and Paul, own the following numbers and types of musical instruments:

John	Paul
3 guitars	1 guitar
1 bass guitar	2 bass guitars
1 sitar	1 sitar

The brothers will not give their instruments to each other, but each brother wants to own the same numbers and types of musical instruments as the other brother has.

Since Paul has 1 more bass guitar than John has, John wants 1 more bass guitar. Since John has 2 more guitars than Paul has, Paul wants 2 more guitars. John and Paul have the same number of sitars, so neither brother wants another sitar.

Let's compare this situation with finding the least common denominator (LCD) for the sum $\frac{3}{40} + \frac{7}{50}$. First we find the prime factorization of each denominator:

$$40 = 2 \cdot 2 \cdot 2 \cdot 5$$
$$50 = 2 \cdot 5 \cdot 5$$

What factors do each of the denominators need so that the denominators can become the same? Since 50 has one more 5 factor than 40 has, the denominator 40 needs one

5 factor. Since 40 has two more 2 factors than 50 has, the denominator 50 needs two 2 factors.

We use the fact that $\frac{A}{A} = 1$ when $A \neq 0$ to introduce the one 5 factor for 40 and the two 2 factors for 50:

$$\begin{aligned}
\frac{3}{40}+\frac{7}{50} &= \frac{3}{2\cdot 2\cdot 2\cdot 5}+\frac{7}{2\cdot 5\cdot 5} && \text{Find prime factorization of each denominator.}\\
&= \frac{3}{2\cdot 2\cdot 2\cdot 5}\cdot\frac{5}{5}+\frac{7}{2\cdot 5\cdot 5}\cdot\frac{2\cdot 2}{2\cdot 2} && \text{Introduce missing factors.}\\
&= \frac{3\cdot 5}{2\cdot 2\cdot 2\cdot 5\cdot 5}+\frac{7\cdot 2\cdot 2}{2\cdot 5\cdot 5\cdot 2\cdot 2} && \text{Find products.}\\
&= \frac{15}{200}+\frac{28}{200} && \text{Simplify.}\\
&= \frac{43}{200} && \text{Add numerators and keep common denominator: } \frac{A}{B}+\frac{C}{B}=\frac{A+C}{B}
\end{aligned}$$

This method not only helps us find the LCD but also suggests what forms of $\frac{A}{A}$ to use to introduce missing factors.

Example 2 Adding Two Rational Expressions with Different Denominators

Find the sum $\frac{5}{6x}+\frac{9}{4x^3}$.

Solution

We begin by factoring the denominators:

$$\begin{aligned}
6x &= 2\cdot 3\cdot x\\
4x^3 &= 2\cdot 2\cdot x\cdot x\cdot x
\end{aligned}$$

Since $4x^3$ has one more 2 factor and two more x factors than $6x$ has, the denominator $6x$ needs one more 2 factor and two more x factors. Since $6x$ has a 3 factor and $4x^3$ does not, the denominator $4x^3$ needs a 3 factor.

We use the fact that $\frac{A}{A} = 1$, where A is nonzero, to introduce the missing factors:

X	Y1	Y2
-3	-.3611	-.3611
-2	-.6979	-.6979
-1	-3.083	-3.083
0	ERROR	ERROR
1	3.0833	3.0833
2	.69792	.69792
3	.36111	.36111

X=-3

Figure 13 Verify the work

$$\begin{aligned}
\frac{5}{6x}+\frac{9}{4x^3} &= \frac{5}{2\cdot 3\cdot x}+\frac{9}{2\cdot 2\cdot x\cdot x\cdot x} && \text{Factor denominators.}\\
&= \frac{5}{2\cdot 3\cdot x}\cdot\frac{2\cdot x\cdot x}{2\cdot x\cdot x}+\frac{9}{2\cdot 2\cdot x\cdot x\cdot x}\cdot\frac{3}{3} && \text{Introduce missing factors.}\\
&= \frac{10x^2}{12x^3}+\frac{27}{12x^3} && \text{Find products.}\\
&= \frac{10x^2+27}{12x^3} && \text{Add numerators and keep common denominator.}
\end{aligned}$$

The result is in lowest terms. We use a graphing calculator table to verify our work (see Fig. 13). There are "ERROR" messages across from $x = 0$ in the table, because both the original expression and our result are not defined at 0 (why?). ■

Example 3 Adding Two Rational Expressions with Different Denominators

Find the sum $\frac{2}{p+3}+\frac{4}{p-5}$.

Solution

The denominator $p+3$ needs a $p-5$ factor, and the denominator $p-5$ needs a $p+3$ factor. We use the fact that $\frac{A}{A}=1$, where A is nonzero, to introduce the missing factors:

$$\begin{aligned}
\frac{2}{p+3}+\frac{4}{p-5} &= \frac{2}{p+3}\cdot\frac{p-5}{p-5}+\frac{4}{p-5}\cdot\frac{p+3}{p+3} && \text{Introduce missing factors.}\\
&= \frac{2(p-5)}{(p+3)(p-5)}+\frac{4(p+3)}{(p-5)(p+3)} && \text{Find products.}\\
&= \frac{2(p-5)+4(p+3)}{(p+3)(p-5)} && \text{Add numerators and keep common denominator: } \frac{A}{B}+\frac{C}{B}=\frac{A+C}{B}\\
&= \frac{2p-10+4p+12}{(p+3)(p-5)} && \text{Distributive law}\\
&= \frac{6p+2}{(p+3)(p-5)} && \text{Combine like terms.}\\
&= \frac{2(3p+1)}{(p+3)(p-5)} && \text{Factor numerator.}
\end{aligned}$$

The result is in lowest terms. ■

For the sum of two expressions with different denominators, we first factor the denominators, if possible, to help us find the LCD.

Example 4 Adding Two Rational Expressions with Different Denominators

Find the sum $\frac{3x}{x^2+2xy+y^2}+\frac{2y}{x^2-y^2}$.

Solution

First, we factor the denominators:

$$\begin{aligned}
x^2+2xy+y^2 &= (x+y)(x+y)\\
x^2-y^2 &= (x+y)(x-y)
\end{aligned}$$

Since x^2-y^2 has an $x-y$ factor but $x^2+2xy+y^2$ does not, the denominator $x^2+2xy+y^2$ needs an $x-y$ factor. Since $x^2+2xy+y^2$ has one more $x+y$ factor than x^2-y^2 has, the denominator x^2-y^2 needs an $x+y$ factor:

$$\begin{aligned}
\frac{3x}{x^2+2xy+y^2}+\frac{2y}{x^2-y^2} &= \frac{3x}{(x+y)(x+y)}+\frac{2y}{(x+y)(x-y)} && \text{Factor denominators.}\\
&= \frac{3x}{(x+y)(x+y)}\cdot\frac{x-y}{x-y}+\frac{2y}{(x+y)(x-y)}\cdot\frac{x+y}{x+y} && \text{Introduce missing factors.}\\
&= \frac{3x(x-y)}{(x+y)(x+y)(x-y)}+\frac{2y(x+y)}{(x+y)(x+y)(x-y)} && \text{Find products.}\\
&= \frac{3x(x-y)+2y(x+y)}{(x+y)(x+y)(x-y)} && \text{Add numerators and keep common denominator.}\\
&= \frac{3x^2-3xy+2xy+2y^2}{(x+y)^2(x-y)} && \text{Distributive law}\\
&= \frac{3x^2-xy+2y^2}{(x+y)^2(x-y)} && \text{Combine like terms.}
\end{aligned}$$

The result is in lowest terms. ■

How to Add Two Rational Expressions with Different Denominators

To add two rational expression with different denominators,

1. Factor the denominators of the expressions if possible. Determine which factors are missing.
2. Use the property $\frac{A}{A} = 1$, where A is nonzero, to introduce missing factors.
3. Add the expressions by using the property $\frac{A}{B} + \frac{C}{B} = \frac{A+C}{B}$, where B is nonzero.
4. Simplify.

Example 5 Finding a Sum Function

Let $f(x) = \dfrac{3}{12x^3 - 22x^2 + 6x}$ and $g(x) = \dfrac{x+1}{30x^2 - 10x}$.

1. Find an equation of $f + g$.
2. Find $(f + g)(2)$.

Solution

1. To start, we use the definition of $f + g$:

$$(f+g)(x) = f(x) + g(x)$$
Definition of $f + g$

$$= \frac{3}{12x^3 - 22x^2 + 6x} + \frac{x+1}{30x^2 - 10x}$$
Substitute $\frac{3}{12x^3 - 22x^2 + 6x}$ for $f(x)$ and $\frac{x+1}{30x^2 - 10x}$ for $g(x)$.

Next, we factor the denominators:

$$12x^3 - 22x^2 + 6x = 2x(6x^2 - 11x + 3) = 2 \cdot x \cdot (3x-1)(2x-3)$$
$$30x^2 - 10x = 10x(3x - 1) = 2 \cdot 5 \cdot x \cdot (3x - 1)$$

Since $30x^2 - 10x$ has a 5 factor but $12x^3 - 22x^2 + 6x$ does not, the denominator $12x^3 - 22x^2 + 6x$ needs a 5 factor. Since $12x^3 - 22x^2 + 6x$ has a $2x - 3$ factor but $30x^2 - 10x$ does not, the denominator $30x^2 - 10x$ needs a $2x - 3$ factor:

$$(f+g)(x) = \frac{3}{2 \cdot x \cdot (3x-1)(2x-3)} + \frac{x+1}{2 \cdot 5 \cdot x \cdot (3x-1)}$$
Factor denominators.

$$= \frac{3}{2 \cdot x \cdot (3x-1)(2x-3)} \cdot \frac{5}{5} + \frac{x+1}{2 \cdot 5 \cdot x \cdot (3x-1)} \cdot \frac{2x-3}{2x-3}$$
Introduce missing factors.

$$= \frac{15}{10x(3x-1)(2x-3)} + \frac{(x+1)(2x-3)}{10x(3x-1)(2x-3)}$$
Find products.

$$= \frac{15 + (x+1)(2x-3)}{10x(3x-1)(2x-3)}$$
Add numerators and keep common denominator.

$$= \frac{15 + 2x^2 - x - 3}{10x(3x-1)(2x-3)}$$
Find products.

$$= \frac{2x^2 - x + 12}{10x(3x-1)(2x-3)}$$
Combine like terms.

The result is in lowest terms.

2. $(f+g)(2) = \dfrac{2(2)^2 - (2) + 12}{10(2)[3(2) - 1][2(2) - 3]} = \dfrac{18}{100} = \dfrac{9}{50}$ ■

Subtraction of Rational Expressions

At times in this course, we have subtracted fractions by using the property $\frac{a}{b} - \frac{c}{b} = \frac{a-c}{b}$, where b is nonzero. For example,

$$\frac{6}{11} - \frac{4}{11} = \frac{2}{11}$$

We subtract more complicated rational expressions with a common denominator in a similar way.

Subtracting Rational Expressions with a Common Denominator

If $\frac{A}{B}$ and $\frac{C}{B}$ are rational expressions and B is nonzero, then

$$\frac{A}{B} - \frac{C}{B} = \frac{A-C}{B}$$

In words: To subtract two rational expressions with a common denominator, subtract the numerators and keep the common denominator.

Example 6 Subtracting Two Rational Expressions

Find the difference $\frac{x^2}{x+1} - \frac{x+2}{x+1}$.

Solution

$$\frac{x^2}{x+1} - \frac{x+2}{x+1} = \frac{x^2-(x+2)}{x+1}$$ Subtract numerators and keep common denominator: $\frac{A}{B} - \frac{C}{B} = \frac{A-C}{B}$

$$= \frac{x^2-x-2}{x+1}$$ Simplify.

$$= \frac{(x-2)(x+1)}{x+1}$$ Factor numerator.

$$= x-2$$ Simplify: $\frac{x+1}{x+1} = 1$ ■

WARNING It is a common error to write

$$\frac{x^2}{x+1} - \frac{x+2}{x+1} = \frac{x^2-x+2}{x+1}$$ Incorrect

This work is incorrect. **When subtracting rational expressions, be sure to subtract the *entire* numerator:**

$$\frac{x^2}{x+1} - \frac{x+2}{x+1} = \frac{x^2-(x+2)}{x+1} = \frac{x^2-x-2}{x+1}$$

(See Example 6 for the rest of the work.)

When subtracting rational expressions with different denominators, we use the method discussed earlier in this section to find the LCD.

Example 7 Subtracting Two Rational Expressions with Different Denominators

Find the difference $\frac{5}{4ab^2} - \frac{3}{2a^3b}$.

Solution

We begin by factoring the denominators:

$$4ab^2 = 2 \cdot 2 \cdot a \cdot b \cdot b$$
$$2a^3b = 2 \cdot a \cdot a \cdot a \cdot b$$

Since $2a^3b$ has two more a factors than $4ab^2$ has, the denominator $4ab^2$ needs two more a factors. Since $4ab^2$ has one more 2 factor and one more b factor than $2a^3b$ has, the denominator $2a^3b$ needs one 2 factor and one b factor:

$$\frac{5}{4ab^2} - \frac{3}{2a^3b} = \frac{5}{2\cdot 2\cdot a\cdot b\cdot b} - \frac{3}{2\cdot a\cdot a\cdot a\cdot b} \qquad \text{Factor denominators.}$$

$$= \frac{5}{2\cdot 2\cdot a\cdot b\cdot b}\cdot\frac{a\cdot a}{a\cdot a} - \frac{3}{2\cdot a\cdot a\cdot a\cdot b}\cdot\frac{2\cdot b}{2\cdot b} \qquad \text{Introduce missing factors.}$$

$$= \frac{5a^2}{4a^3b^2} - \frac{6b}{4a^3b^2} \qquad \text{Find products.}$$

$$= \frac{5a^2 - 6b}{4a^3b^2} \qquad \text{Subtract numerators and keep common denominator.}$$

The result is in lowest terms. ■

The steps we take to subtract two rational expressions are similar to the steps we take to add two rational expressions.

How to Subtract Two Rational Expressions with Different Denominators

To subtract two rational expressions with different denominators,

1. Factor the denominators of the expressions if possible. Determine which factors are missing.
2. Use the property $\frac{A}{A} = 1$, where A is nonzero, to introduce missing factors.
3. Subtract the expressions by using the property $\frac{A}{B} - \frac{C}{B} = \frac{A - C}{B}$, where B is nonzero.
4. Simplify.

Example 8 Subtracting Two Rational Expressions with Different Denominators

Find the difference $\dfrac{3x - 1}{2x^2 - 7x - 4} - \dfrac{5}{x^2 - 8x + 16}$.

Solution

$$\frac{3x - 1}{2x^2 - 7x - 4} - \frac{5}{x^2 - 8x + 16} = \frac{3x - 1}{(2x + 1)(x - 4)} - \frac{5}{(x - 4)(x - 4)} \qquad \text{Factor denominators.}$$

$$= \frac{3x - 1}{(2x + 1)(x - 4)}\cdot\frac{x - 4}{x - 4} - \frac{5}{(x - 4)(x - 4)}\cdot\frac{2x + 1}{2x + 1} \qquad \text{Introduce missing factors.}$$

$$= \frac{(3x - 1)(x - 4) - 5(2x + 1)}{(2x + 1)(x - 4)(x - 4)} \qquad \text{Find products; subtract numerators and keep common denominator.}$$

$$= \frac{3x^2 - 13x + 4 - 10x - 5}{(2x + 1)(x - 4)^2} \qquad \text{Find products.}$$

$$= \frac{3x^2 - 23x - 1}{(2x + 1)(x - 4)^2} \qquad \text{Combine like terms.}$$

The result is in lowest terms. ■

Example 9 Subtracting Two Rational Expressions with Different Denominators

Find the difference $\dfrac{y}{y-2} - \dfrac{3}{2-y}$.

Solution

$$\frac{y}{y-2} - \frac{3}{2-y} = \frac{y}{y-2} - \frac{3}{-(y-2)} \qquad 2-y = -y+2 = -(y-2)$$

$$= \frac{y}{y-2} - \left(-\frac{3}{y-2}\right) \qquad \frac{A}{-B} = -\frac{A}{B}$$

$$= \frac{y}{y-2} + \frac{3}{y-2} \qquad A-(-B) = A+B$$

$$= \frac{y+3}{y-2} \qquad \text{Add numerators and keep common denominator: } \frac{A}{B}+\frac{C}{B} = \frac{A+C}{B}$$

■

WARNING The work in Example 9 shows that to find the difference $\dfrac{y}{y-2} - \dfrac{3}{2-y}$, it is *not* necessary to introduce the factor $2-y$ into the denominator $y-2$ or to introduce the factor $y-2$ into the denominator $2-y$. This method applies to any sum or difference of two rational expressions in which the denominators are of the form $A-B$ and $B-A$.

In Example 10, we combine three rational expressions by performing two operations.

Example 10 Performing Operations with Three Rational Expressions

Perform the indicated operations: $\left(\dfrac{x+2}{x^2-x} - \dfrac{6}{x^2-1}\right) + \dfrac{3}{x^2+x}$.

Solution

$$\left(\frac{x+2}{x^2-x} - \frac{6}{x^2-1}\right) + \frac{3}{x^2+x}$$

$$= \left(\frac{x+2}{x(x-1)} - \frac{6}{(x-1)(x+1)}\right) + \frac{3}{x(x+1)} \qquad \text{Factor denominators.}$$

$$= \left(\frac{x+2}{x(x-1)} \cdot \frac{x+1}{x+1} - \frac{6}{(x-1)(x+1)} \cdot \frac{x}{x}\right) + \frac{3}{x(x+1)} \cdot \frac{x-1}{x-1} \qquad \text{Introduce missing factors.}$$

$$= \frac{(x+2)(x+1)-6x}{x(x-1)(x+1)} + \frac{3(x-1)}{x(x-1)(x+1)} \qquad \text{Find products; subtract numerators.}$$

$$= \frac{(x+2)(x+1)-6x+3(x-1)}{x(x-1)(x+1)} \qquad \text{Add numerators.}$$

$$= \frac{x^2+3x+2-6x+3x-3}{x(x-1)(x+1)} \qquad \text{Find products.}$$

$$= \frac{x^2-1}{x(x-1)(x+1)} \qquad \text{Combine like terms.}$$

$$= \frac{(x-1)(x+1)}{x(x-1)(x+1)} \qquad \text{Factor numerator.}$$

$$= \frac{1}{x} \qquad \text{Simplify.}$$

■

Example 11 Finding a Difference Function

Let $f(x) = \dfrac{x-1}{x+1}$ and $g(x) = \dfrac{x+1}{x-1}$.

1. Find an equation of $f - g$.
2. Find $(f - g)(5)$.

Solution

1.
$$(f-g)(x) = f(x) - g(x) \qquad \text{Definition of } f-g$$

$$= \frac{x-1}{x+1} - \frac{x+1}{x-1} \qquad \text{Substitute } \frac{x-1}{x+1} \text{ for } f(x) \text{ and } \frac{x+1}{x-1} \text{ for } g(x).$$

$$= \frac{x-1}{x+1} \cdot \frac{x-1}{x-1} - \frac{x+1}{x-1} \cdot \frac{x+1}{x+1} \qquad \text{Introduce missing factors.}$$

$$= \frac{(x-1)(x-1) - (x+1)(x+1)}{(x-1)(x+1)} \qquad \text{Find products; subtract numerators and keep common denominator: } \frac{A}{B} - \frac{C}{B} = \frac{A-C}{B}$$

$$= \frac{x^2 - 2x + 1 - \left(x^2 + 2x + 1\right)}{(x-1)(x+1)} \qquad \text{Find products.}$$

$$= \frac{x^2 - 2x + 1 - x^2 - 2x - 1}{(x-1)(x+1)} \qquad \text{Subtract trinomial.}$$

$$= \frac{-4x}{(x-1)(x+1)} \qquad \text{Combine like terms.}$$

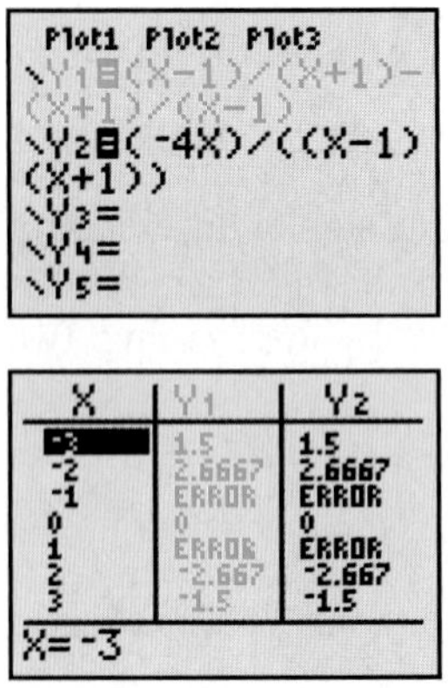

The result is in lowest terms. We use a graphing calculator table to verify our work (see Fig. 14). The values $x = -1$ and $x = 1$ give "ERROR" messages, because -1 and 1 are not in the domain of either the function $y = \dfrac{x-1}{x+1} - \dfrac{x+1}{x-1}$ or the function $y = \dfrac{-4x}{(x-1)(x+1)}$ (why?).

Figure 14 Verify the work

2.
$$(f-g)(5) = \frac{-4(5)}{(5-1)(5+1)} = \frac{-20}{24} = -\frac{5}{6}$$

■

group exploration

Adding and subtracting rational expressions

In Problems 1–3, a student tries to perform an operation and simplify the result. If the work is correct, decide whether there is a more efficient way to do the problem. If the work is incorrect, describe any errors and do the problem correctly.

1. Find the sum $\dfrac{5}{x+1} + \dfrac{2}{x+3}$. Simplify the result.

$$\frac{5}{x+1} + \frac{2}{x+3} = \left(\frac{5}{x+1} + \frac{2}{x+3}\right) \cdot (x+1)(x+3)$$

$$= \frac{5}{x+1} \cdot (x+1)(x+3) + \frac{2}{x+3} \cdot (x+1)(x+3)$$

$$= 5(x+3) + 2(x+1)$$

$$= 7x + 17$$

2. Find the sum $\frac{4}{(x-2)(x+3)} + \frac{1}{x-2}$. Simplify the result.

$$\frac{4}{(x-2)(x+3)} + \frac{1}{x-2} = \frac{4}{(x-2)(x+3)} \cdot \frac{x-2}{x-2} + \frac{1}{x-2} \cdot \frac{(x-2)(x+3)}{(x-2)(x+3)}$$
$$= \frac{4(x-2) + (x-2)(x+3)}{(x-2)(x-2)(x+3)}$$
$$= \frac{4x - 8 + x^2 + x - 6}{(x-2)^2(x+3)}$$
$$= \frac{x^2 + 5x - 14}{(x-2)^2(x+3)}$$
$$= \frac{(x-2)(x+7)}{(x-2)^2(x+3)}$$
$$= \frac{x+7}{(x-2)(x+3)}$$

3. Find the difference $\frac{5x}{x-7} - \frac{3x+4}{x-7}$. Simplify the result.

$$\frac{5x}{x-7} - \frac{3x+4}{x-7} = \frac{5x - 3x + 4}{x-7}$$
$$= \frac{2x+4}{x-7}$$

TIPS FOR SUCCESS: Write a Summary

After each class meeting, consider writing a summary of what you have learned. Your summaries will increase your understanding as well as your memory of concepts and procedures. They will also serve as good references for quizzes and exams.

HOMEWORK 8.3

FOR EXTRA HELP ▶ Student Solutions Manual | PH Math/Tutor Center | MathXL® | MyMathLab

Perform the indicated operation. Simplify the result.

1. $\frac{5}{x} + \frac{2}{x}$

2. $\frac{3}{x^2} + \frac{6}{x^2}$

3. $\frac{x}{x^2-9} + \frac{3}{x^2-9}$

4. $\frac{x}{x^2-4} + \frac{2}{x^2-4}$

5. $\frac{6m^2}{m^2-4m+3} - \frac{4m^2+6m}{m^2-4m+3}$

6. $\frac{7c^2}{c^2+c-6} - \frac{2c^2+10c}{c^2+c-6}$

7. $\frac{3x^2+9x}{x^2+10x+21} - \frac{2x^2+x-15}{x^2+10x+21}$

8. $\frac{2x^2-4x}{3x^2-6x} - \frac{x^2+x-6}{3x^2-6x}$

9. $\frac{2}{x^6} - \frac{4}{x^2}$

10. $\frac{4}{x^5} - \frac{7}{x^3}$

11. $\frac{3}{10x^6} + \frac{5}{12x^4}$

12. $\frac{3}{14x^2} + \frac{4}{21x^9}$

13. $\frac{7}{4a^2b} - \frac{5}{6ab^3}$

14. $\frac{5}{6ab^3} - \frac{2}{9a^4b^2}$

15. $\frac{3}{x+1} + \frac{4}{x-2}$

16. $\frac{3}{x-2} + \frac{2}{x+3}$

17. $\frac{6}{(x+4)(x-6)} - \frac{4}{(x-1)(x+4)}$

18. $\frac{2}{(x+1)(x-2)} - \frac{3}{(x-2)(x+3)}$

19. $\frac{5}{3t-6} - \frac{2}{5t+15}$

20. $\frac{b}{8b-4} - \frac{2}{3b-6}$

21. $\frac{3}{x^2-25}+\frac{5}{x^2-5x}$

22. $\frac{-2}{x^2-2x-3}+\frac{3}{x^2-9}$

23. $\frac{2}{x^2-9}+\frac{3}{x^2-7x+12}$

24. $\frac{4}{x^2-2x}+\frac{1}{x^2-5x+6}$

25. $2+\frac{k-3}{k+1}$

26. $\frac{w+2}{w-5}+3$

27. $2-\frac{2x+4}{x^2+3x+2}$

28. $\frac{6x^2+2x-4}{x^2-1}-5$

29. $\frac{8}{x-6}-\frac{4}{6-x}$

30. $\frac{x}{x-3}-\frac{2}{3-x}$

31. $\frac{2x+1}{x^2-4x-21}+\frac{3}{14-2x}$

32. $\frac{4x+5}{x^2+3x-40}+\frac{2}{15-3x}$

33. $\frac{-2c}{7-2c}-\frac{c+1}{4c^2-49}$

34. $\frac{-5m}{5-3m}-\frac{m+2}{9m^2-25}$

35. $\frac{2a}{a^2-b^2}+\frac{b}{ab-b^2}$

36. $\frac{n}{m^2+3mn}+\frac{3m}{2mn+6n^2}$

37. $\frac{x}{x^2+5x+6}-\frac{3}{x^2+7x+12}$

38. $\frac{x}{x^2+11x+30}-\frac{5}{x^2+9x+20}$

39. $\frac{x-1}{x+2}+\frac{x+2}{x-1}$

40. $\frac{x+4}{x-6}+\frac{x-6}{x+4}$

41. $\frac{y-5}{y-3}-\frac{y+3}{y+5}$

42. $\frac{k+2}{k-1}-\frac{k+1}{k-2}$

43. $\frac{x+4}{x^2-7x+10}-\frac{5}{x^2-25}$

44. $\frac{x-2}{x^2-x-2}-\frac{4}{x^2-1}$

45. $\frac{x+2}{(x-4)(x+3)^2}+\frac{x-1}{(x-4)(x+1)(x+3)}$

46. $\frac{x+1}{(x+2)^2(x+3)}+\frac{x+5}{(x+2)(x+3)(x+4)}$

47. $\frac{c+2}{c^2-4}+\frac{3c}{c^2-2c}$

48. $\frac{w+5}{w^2-25}+\frac{4w}{w^2-5w}$

49. $\frac{x-1}{4x^2+20x+25}-\frac{x+4}{6x^2+17x+5}$

50. $\frac{x-3}{2x^2+11x-6}-\frac{x+2}{2x^2+7x-4}$

51. $\frac{3x-1}{x^2+4x+4}+\frac{2x+1}{3x^2+5x-2}$

52. $\frac{x-3}{3x^2-x-4}+\frac{2x+5}{6x^2+x-12}$

53. $\frac{3p}{p^2-2pq-24q^2}-\frac{2q}{p^2-3pq-18q^2}$

54. $\frac{2x}{x^2-6xy+8y^2}-\frac{5y}{x^2+3xy-10y^2}$

55. $\frac{x-1}{6x^2-24x}+\frac{5}{3x^3-6x^2-24x}$

56. $\frac{3}{2x^3+14x^2+20x}+\frac{x+2}{14x^2+28x}$

Perform the indicated operations. Simplify your result.

57. $\left(\frac{2}{x^2-4}+\frac{3}{x+2}\right)-\frac{1}{2x-4}$

58. $\left(\frac{5}{3x-9}-\frac{2x-1}{x^2-9}\right)+\frac{4}{x+3}$

59. $\frac{3}{t+1}-\left(\frac{2t-3}{t^2+6t+5}+\frac{2}{t+5}\right)$

60. $\frac{2k+11}{k^2+k-6}-\left(\frac{2}{k+3}+\frac{3}{2-k}\right)$

Let $f(x)=\frac{x+3}{x-4}$ and $g(x)=\frac{x+4}{x-3}$. Find an equation of the given function.

61. $f+g$ **62.** $f-g$ **63.** $g-f$

Let $f(x)=\frac{x-2}{x^2-2x-8}$ and $g(x)=\frac{x+1}{3x+6}$. Find an equation of the given function.

64. $f+g$ **65.** $f-g$ **66.** $g-f$

67. A student tries to find the sum $\frac{2}{x+1}+\frac{3}{x+2}$:

$$\frac{2}{x+1}+\frac{3}{x+2}=\frac{2}{x+1}\cdot\frac{1}{x+2}+\frac{3}{x+2}\cdot\frac{1}{x+1}$$
$$=\frac{2}{(x+1)(x+2)}+\frac{3}{(x+2)(x+1)}$$
$$=\frac{5}{(x+1)(x+2)}$$

Describe any errors. Then find the sum correctly.

68. A student tries to find the difference $\frac{6x}{x+4}-\frac{3x+2}{x+4}$:

$$\frac{6x}{x+4}-\frac{3x+2}{x+4}=\frac{6x-(3x+2)}{x+4}$$
$$=\frac{6x-3x+2}{x+4}$$
$$=\frac{3x+2}{x+4}$$

Describe any errors. Then find the difference correctly.

69. A student tries to find the difference $\frac{9x}{x-3}-\frac{5x+1}{x-3}$:

$$\frac{9x}{x-3}-\frac{5x+1}{x-3}=\frac{9x-5x+1}{x-3}$$
$$=\frac{4x+1}{x-3}$$

Find any errors. Then find the difference correctly.

70. Two students try to find the sum $\frac{3}{x-4}+\frac{2}{4-x}$:

Student A's Work

$$\frac{3}{x-4}+\frac{2}{4-x}=\frac{3}{x-4}+\frac{2}{-(x-4)}$$
$$=\frac{3}{x-4}-\frac{2}{x-4}$$
$$=\frac{1}{x-4}$$

Student B's Work

$$\frac{3}{x-4}+\frac{2}{4-x}=\frac{3}{x-4}\cdot\frac{4-x}{4-x}+\frac{2}{4-x}\cdot\frac{x-4}{x-4}$$
$$=\frac{3(4-x)+2(x-4)}{(x-4)(4-x)}$$
$$=\frac{12-3x+2x-8}{(x-4)(4-x)}$$
$$=\frac{4-x}{(x-4)(4-x)}$$
$$=\frac{1}{x-4}$$

Compare the two methods. Are both correct? Explain. Discuss why student A's method is shorter.

71. Describe how to add two rational expressions that have different denominators. Then describe how to subtract two such expressions.

72. When subtracting rational expressions, we subtract the *entire* numerator. Give an example to show how to do this and what can go wrong if we subtract only part of the numerator.

Related Review

Find equations of $f+g$ and $f-g$.

73. $f(x)=6x^2-4x+3$, $g(x)=2x^2-7x-5$

74. $f(x)=3x^2+8x-2$, $g(x)=-5x^2-3x+4$

75. $f(x)=2(5)^x$, $g(x)=-3(5)^x$

76. $f(x)=4(3)^x$, $g(x)=7(3)^x$

Perform the indicated operations. Simplify your result.

77. $\dfrac{4x+5}{x+2}+\left(\dfrac{3x+15}{x^2-4}\cdot\dfrac{x^2-2x}{x^2+7x+10}\right)$

78. $\dfrac{2x-7}{x-5}-\left(\dfrac{2x+10}{x^2+9x+20}\div\dfrac{x^2-25}{3x+12}\right)$

79. $\dfrac{5x+5}{3x+6}\cdot\left(\dfrac{x^2+4x}{x^2+2x+1}+\dfrac{4}{x^2+2x+1}\right)$

80. $\dfrac{x^2-16}{2x-12}\div\left(\dfrac{x^2}{x^2-36}-\dfrac{2x+8}{x^2-36}\right)$

Expressions, Equations, Functions, and Graphs

Perform the indicated instruction. Then use words such as linear, quadratic, cubic, exponential, logarithmic, rational, polynomial, degree, function, one variable, *and* two variables *to describe the expression, equation, or system.*

81. Graph $f(x)=3\left(\frac{1}{3}\right)^x$ by hand.

82. Find the product $(3p-2)(9p^2+6p+4)$.

83. Find all real-number solutions of $5b^4=66$. Round any results to the fourth decimal place.

84. Factor $64p^3-27$.

85. Find an approximate equation $y=ab^x$ of an exponential curve that contains the points (3, 95) and (7, 2). Round the values of a and b to the second decimal place.

86. Solve $6p^3+21p^2=12p$.

8.4 SIMPLIFYING COMPLEX RATIONAL EXPRESSIONS

Objective

- Simplify *complex rational expressions.*

In this section, we will work with complex rational expressions. A **complex rational expression** is a rational expression whose numerator or denominator (or both) is a rational expression. Here are some examples of such expressions:

$$\frac{\frac{x^2}{2}-\frac{3}{x^3}}{\frac{x}{6}+\frac{7}{x^2}},\qquad \frac{\frac{3x}{x-1}}{\frac{x^3}{x-2}},\qquad \frac{5x}{\frac{x^2-2x+1}{x+4}}$$

Here we find the values of two numerical complex rational expressions:

$$\frac{2}{\frac{2}{2}}=\frac{2}{1}=2\qquad\qquad \frac{\frac{2}{2}}{2}=\frac{1}{2}$$

From these two examples, we see that it is important to keep track of the main fraction bar (the longest one, in bold) of the complex fraction.

We will discuss two methods for simplifying complex rational expressions. Ask your instructor whether you are required to know method 1, method 2, or both methods. If the choice is yours, compare the use of method 1 in Examples 1 and 2 with the use of method 2 in Examples 3 and 4. The complex rational expressions in these examples are simplified by both methods so that you can get a sense of the advantages and disadvantages of each.

Method 1: Writing a Complex Rational Expression as a Quotient of Two Rational Expressions

An expression in the form $\frac{R}{S}$, where R and S are themselves expressions, can be written in the form $R \div S$. We use this fact to help simplify a complex rational expression:

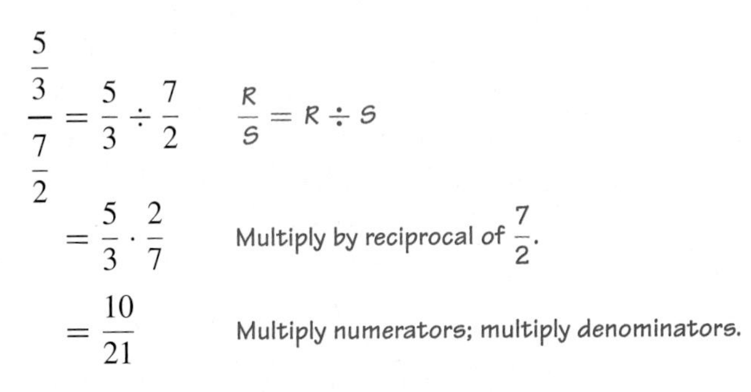

$$\frac{\frac{5}{3}}{\frac{7}{2}} = \frac{5}{3} \div \frac{7}{2} \qquad \frac{R}{S} = R \div S$$

$$= \frac{5}{3} \cdot \frac{2}{7} \qquad \text{Multiply by reciprocal of } \frac{7}{2}.$$

$$= \frac{10}{21} \qquad \text{Multiply numerators; multiply denominators.}$$

We **simplify a complex rational expression** by writing it as a rational expression $\frac{P}{Q}$, with $\frac{P}{Q}$ in lowest terms.

Example 1 Simplifying a Complex Rational Expression by Method 1

Simplify by method 1.

1. $\dfrac{\frac{12}{x}}{\frac{8}{x^3}}$

2. $\dfrac{\frac{x^2 - 9}{x^2 + 2x + 1}}{\frac{2x - 6}{4x + 4}}$

Solution

1. $$\frac{\frac{12}{x}}{\frac{8}{x^3}} = \frac{12}{x} \div \frac{8}{x^3} \qquad \frac{R}{S} = R \div S$$

$$= \frac{12}{x} \cdot \frac{x^3}{8} \qquad \text{Multiply by reciprocal of } \frac{8}{x^3}\text{, which is } \frac{x^3}{8}.$$

$$= \frac{2 \cdot 2 \cdot 3}{x} \cdot \frac{x^3}{2 \cdot 2 \cdot 2} \qquad \text{Factor numerator and denominator.}$$

$$= \frac{2 \cdot 2 \cdot 3 \cdot x^3}{x \cdot 2 \cdot 2 \cdot 2} \qquad \text{Multiply numerators; multiply denominators.}$$

$$= \frac{3x^2}{2} \qquad \text{Simplify: } \frac{2 \cdot 2 \cdot x}{2 \cdot 2 \cdot x} = 1$$

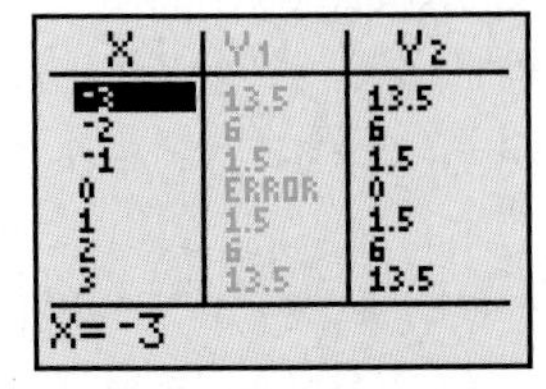

Figure 15 Verify the work

We use a graphing calculator table to verify our work (see Fig. 15). The value $x = 0$ gives an "ERROR" message, because 0 is an excluded value of the numerator (and the denominator) of the original expression.

2. $$\frac{\dfrac{x^2-9}{x^2+2x+1}}{\dfrac{2x-6}{4x+4}} = \frac{x^2-9}{x^2+2x+1} \div \frac{2x-6}{4x+4}$$ $\frac{R}{S} = R \div S$

$$= \frac{x^2-9}{x^2+2x+1} \cdot \frac{4x+4}{2x-6}$$ Multiply by reciprocal of $\frac{2x-6}{4x+4}$.

$$= \frac{(x-3)(x+3)}{(x+1)(x+1)} \cdot \frac{2 \cdot 2(x+1)}{2(x-3)}$$ Factor numerators and denominators.

$$= \frac{(x-3)(x+3) \cdot 2 \cdot 2 \cdot (x+1)}{(x+1)(x+1) \cdot 2(x-3)}$$ Multiply numerators; multiply denominators.

$$= \frac{2(x+3)}{x+1}$$ Simplify: $\frac{2(x-3)(x+1)}{2(x-3)(x+1)} = 1$

■

Next we use method 1 to simplify a complex rational expression that has two rational expressions in the numerator and two rational expressions in the denominator.

Example 2 Simplifying a Complex Rational Expression by Method 1

Simplify $\dfrac{\dfrac{1}{y^2}+\dfrac{3}{2x}}{\dfrac{2}{y}-\dfrac{1}{3x}}$.

Solution

We write both the numerator and the denominator as fractions and simplify as before:

$$\frac{\dfrac{1}{y^2}+\dfrac{3}{2x}}{\dfrac{2}{y}-\dfrac{1}{3x}} = \frac{\dfrac{1}{y^2}\cdot\dfrac{2x}{2x}+\dfrac{3}{2x}\cdot\dfrac{y^2}{y^2}}{\dfrac{2}{y}\cdot\dfrac{3x}{3x}-\dfrac{1}{3x}\cdot\dfrac{y}{y}}$$ Introduce missing factors to get a common denominator, $2xy^2$. Introduce missing factors to get a common denominator, $3xy$.

$$= \frac{\dfrac{2x}{2xy^2}+\dfrac{3y^2}{2xy^2}}{\dfrac{6x}{3xy}-\dfrac{y}{3xy}}$$ Find products.

$$= \frac{\dfrac{2x+3y^2}{2xy^2}}{\dfrac{6x-y}{3xy}}$$ Add numerators and keep common denominator. Subtract numerators and keep common denominator.

$$= \frac{2x+3y^2}{2xy^2} \div \frac{6x-y}{3xy}$$ $\frac{R}{S} = R \div S$

$$= \frac{2x+3y^2}{2xy^2} \cdot \frac{3xy}{6x-y}$$ Multiply by reciprocal of $\frac{6x-y}{3xy}$.

$$= \frac{(2x+3y^2)\cdot 3xy}{2xy^2(6x-y)}$$ Multiply numerators; multiply denominators.

$$= \frac{3(2x+3y^2)}{2y(6x-y)}$$ Simplify: $\frac{xy}{xy} = 1$.

Since our result is in lowest terms, we are done. ■

Using Method 1 to Simplify a Complex Rational Expression

To simplify a complex rational expression by method 1,

1. Write both the numerator and the denominator as fractions.
2. Write the complex rational expression as the quotient of two rational expressions:
$$\frac{\frac{A}{B}}{\frac{C}{D}} = \frac{A}{B} \div \frac{C}{D}, \quad \text{where } B, C, \text{ and } D \text{ are nonzero.}$$
3. Divide the rational expressions.

Method 2: Multiplying by $\frac{\text{LCD}}{\text{LCD}}$

Instead, we can simplify a complex rational expression by first finding the LCD of all of the fractions in the numerator and denominator. Then, we multiply by 1 in the form $\frac{\text{LCD}}{\text{LCD}}$.

In Example 3, we simplify the same complex rational expressions that we simplified in Example 1, but now we use method 2.

Example 3 Simplifying a Complex Rational Expression by Method 2

Simplify by method 2.

1. $\frac{\frac{12}{x}}{\frac{8}{x^3}}$

2. $\frac{\frac{x^2-9}{x^2+2x+1}}{\frac{2x-6}{4x+4}}$

Solution

1. The LCD of $\frac{12}{x}$ and $\frac{8}{x^3}$ is x^3. So, we multiply the complex rational expression by 1 in the form $\frac{x^3}{x^3}$:

$$\frac{\frac{12}{x}}{\frac{8}{x^3}} = \frac{\frac{12}{x} \cdot \frac{x^3}{1}}{\frac{8}{x^3} \cdot \frac{x^3}{1}} \qquad \text{Multiply by } \frac{\text{LCD}}{\text{LCD}}, \frac{x^3}{x^3} = 1.$$

$$= \frac{\frac{12x^3}{x}}{\frac{8x^3}{x^3}} \qquad \text{Simplify.}$$

$$= \frac{12x^2}{8} \qquad \text{Simplify fractions in numerator and in denominator.}$$

$$= \frac{2 \cdot 2 \cdot 3 \cdot x^2}{2 \cdot 2 \cdot 2} \qquad \text{Factor numerator and denominator.}$$

$$= \frac{3x^2}{2} \qquad \text{Simplify: } \frac{2 \cdot 2}{2 \cdot 2} = 1$$

The result is the same as our result in Problem 1 of Example 1.

2. $$\frac{\frac{x^2-9}{x^2+2x+1}}{\frac{2x-6}{4x+4}} = \frac{\frac{(x-3)(x+3)}{(x+1)(x+1)}}{\frac{2(x-3)}{4(x+1)}}$$ Factor numerators and denominators of fractions.

$$= \frac{\frac{(x-3)(x+3)}{(x+1)(x+1)}}{\frac{2(x-3)}{4(x+1)}} \cdot \frac{\frac{4(x+1)(x+1)}{1}}{\frac{4(x+1)(x+1)}{1}}$$ Multiply by $\frac{\text{LCD}}{\text{LCD}}$, $\frac{4(x+1)(x+1)}{4(x+1)(x+1)} = 1$.

$$= \frac{\frac{(x-3)(x+3)\cdot 4(x+1)(x+1)}{(x+1)(x+1)}}{\frac{2(x-3)\cdot 4(x+1)(x+1)}{4(x+1)}}$$ Multiply numerators; multiply denominators.

$$= \frac{2\cdot 2(x-3)(x+3)}{2(x-3)(x+1)}$$ Simplify numerator and denominator.

$$= \frac{2(x+3)}{x+1}$$ Simplify: $\frac{2(x-3)}{2(x-3)} = 1$

The result is the same as our result in Problem 2 of Example 1. ■

In comparing our work in Examples 1 and 3, we see that methods 1 and 2 required about the same number of steps. One advantage that method 1 has over method 2 for *these* complex rational expressions is that method 1 does not require us to find an LCD.

In Example 4, we simplify the same expression as in Example 2, but now we use method 2.

Example 4 Simplifying a Complex Rational Expression by Method 2

Simplify $\dfrac{\frac{1}{y^2}+\frac{3}{2x}}{\frac{2}{y}-\frac{1}{3x}}$ by method 2.

Solution

The LCD of the rational expressions in the numerator and the denominator is $6xy^2$. To simplify, we multiply by $\frac{6xy^2}{6xy^2}$:

$$\frac{\frac{1}{y^2}+\frac{3}{2x}}{\frac{2}{y}-\frac{1}{3x}} = \frac{\frac{1}{y^2}+\frac{3}{2x}}{\frac{2}{y}-\frac{1}{3x}} \cdot \frac{6xy^2}{6xy^2}$$ Multiply by $\frac{\text{LCD}}{\text{LCD}}, \frac{6xy^2}{6xy^2}$.

$$= \frac{\frac{1}{y^2}\cdot\frac{6xy^2}{1}+\frac{3}{2x}\cdot\frac{6xy^2}{1}}{\frac{2}{y}\cdot\frac{6xy^2}{1}-\frac{1}{3x}\cdot\frac{6xy^2}{1}}$$ Distributive law

$$= \frac{6x+9y^2}{12xy-2y^2}$$ Simplify.

$$= \frac{3(2x+3y^2)}{2y(6x-y)}$$ Factor numerator and denominator.

The result is the same as our result in Example 2. ■

WARNING For the first step in Example 4, it would be incorrect to multiply by the fraction $\frac{\text{LCD of the numerator}}{\text{LCD of the denominator}} = \frac{2xy^2}{3xy}$:

$$\frac{\frac{1}{y^2}+\frac{3}{2x}}{\frac{2}{y}-\frac{1}{3x}} = \frac{\frac{1}{y^2}+\frac{3}{2x}}{\frac{2}{y}-\frac{1}{3x}} \cdot \frac{2xy^2}{3xy} \qquad \text{Incorrect}$$

This is incorrect because the expression $\frac{2xy^2}{3xy}$ is not equivalent to 1. It *is* correct to multiply by $\frac{6xy^2}{6xy^2} = 1$.

In comparing our work in Examples 2 and 4, we see that method 2 required fewer steps than method 1 did. In general, when the numerator, denominator, or both contain two rational expressions, method 2 is usually easier to use.

Using Method 2 to Simplify a Complex Rational Expression

To simplify a rational expression by method 2,

1. Find the LCD of all of the fractions in the numerator and denominator.
2. Multiply by 1 in the form $\frac{\text{LCD}}{\text{LCD}}$.
3. Simplify the numerator and the denominator to polynomials.
4. Simplify the rational expression.

In Example 5, we form the quotient function of two rational functions.

Example 5 Finding a Quotient Function by Method 2

Let $f(x) = 2 - \frac{5}{x+2}$ and $g(x) = \frac{x}{x+2} + \frac{x+1}{x^2-4x-12}$.

1. Find an equation of $\frac{f}{g}$.
2. Find $\frac{f}{g}(4)$.

Solution

1.

$$\frac{f}{g}(x) = \frac{f(x)}{g(x)} \qquad \text{Definition of } \frac{f}{g}$$

$$= \frac{2-\frac{5}{x+2}}{\frac{x}{x+2}+\frac{x+1}{x^2-4x-12}} \qquad \text{Substitute for } f(x) \text{ and } g(x).$$

Since the denominator (and the numerator) contains two rational expressions, method 2 should be easier to use than method 1:

$$\frac{f}{g}(x) = \frac{2-\frac{5}{x+2}}{\frac{x}{x+2}+\frac{x+1}{(x-6)(x+2)}} \qquad \text{Factor } x^2-4x-12.$$

$$= \frac{2-\frac{5}{x+2}}{\frac{x}{x+2}+\frac{x+1}{(x-6)(x+2)}} \cdot \frac{(x-6)(x+2)}{(x-6)(x+2)} \qquad \text{Multiply by } \frac{\text{LCD}}{\text{LCD}}.$$

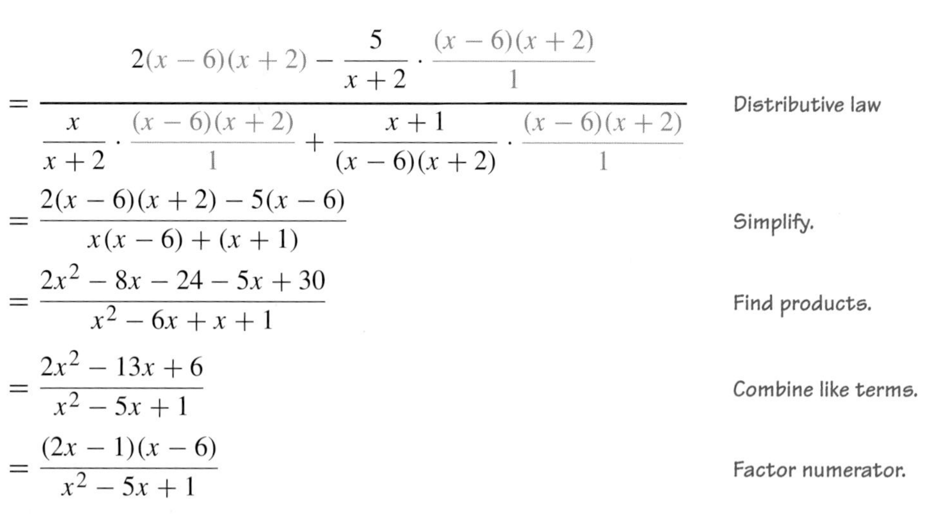

$$= \frac{2(x-6)(x+2) - \frac{5}{x+2} \cdot \frac{(x-6)(x+2)}{1}}{\frac{x}{x+2} \cdot \frac{(x-6)(x+2)}{1} + \frac{x+1}{(x-6)(x+2)} \cdot \frac{(x-6)(x+2)}{1}}$$ Distributive law

$$= \frac{2(x-6)(x+2) - 5(x-6)}{x(x-6) + (x+1)}$$ Simplify.

$$= \frac{2x^2 - 8x - 24 - 5x + 30}{x^2 - 6x + x + 1}$$ Find products.

$$= \frac{2x^2 - 13x + 6}{x^2 - 5x + 1}$$ Combine like terms.

$$= \frac{(2x-1)(x-6)}{x^2 - 5x + 1}$$ Factor numerator.

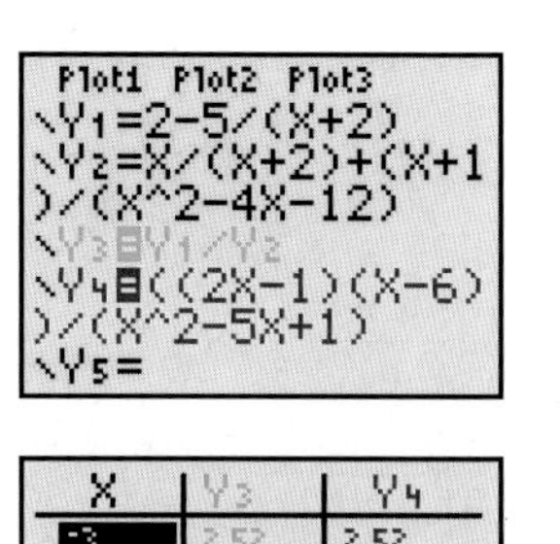

Figure 16 Verify the work

We use a graphing calculator table to verify our work (see Fig. 16). The value $x = -2$ gives an "ERROR" message, because -2 is not in the domain of $\frac{f}{g}$.

2. $\frac{f}{g}(4) = \frac{[2(4) - 1][4 - 6]}{(4)^2 - 5(4) + 1} = \frac{-14}{-3} = \frac{14}{3}$ ■

Our work in Example 5 suggests that if two expressions R and S are rational, then the quotient $\frac{R}{S}$ is rational. This is, indeed, true if the numerator of S is nonzero.

group exploration

Looking ahead: Solving rational equations

1. Solve $\frac{1}{2} + \frac{x}{3} = \frac{5}{6}$.
2. Solve $\frac{1}{x-2} + \frac{2}{3} = \frac{5}{x-2}$. **[Hint:** Multiply both sides by the LCD of all three fractions.]
3. Solve $\frac{x}{x-2} - 5 = \frac{2}{x-2}$. Check whether your result satisfies the equation. What is the solution set? Explain.
4. The equations in Problems 1–3 are called *rational equations in one variable.* What does your work in Problem 3 suggest that you should always do when solving a rational equation in one variable?

HOMEWORK 8.4

FOR EXTRA HELP ▶ Student Solutions Manual | PH Math/Tutor Center | MathXL® | MyMathLab

Simplify.

1. $\dfrac{\frac{2}{x}}{\frac{3}{x}}$ **2.** $\dfrac{\frac{x}{8}}{\frac{x}{6}}$ **3.** $\dfrac{\frac{7}{x^2}}{\frac{21}{x^5}}$ **7.** $\dfrac{\frac{3x-3}{2x+10}}{\frac{6x^2-6}{4x+20}}$ **8.** $\dfrac{\frac{6x-24}{5x-10}}{\frac{3x^2-12}{2x-4}}$

4. $\dfrac{\frac{8}{x^8}}{\frac{24}{x^2}}$ **5.** $\dfrac{\frac{4a^2}{5b}}{\frac{6a}{15b^3}}$ **6.** $\dfrac{\frac{9b^3}{8a}}{\frac{3b}{20a^4}}$ **9.** $\dfrac{\frac{x^2-49}{3x^2-9x}}{\frac{x^2-5x-14}{7x-21}}$ **10.** $\dfrac{\frac{x^2-25}{5x-30}}{\frac{x^2+x-30}{2x^2-12x}}$

11. $\dfrac{\dfrac{25x^2-4}{9x^2-16}}{\dfrac{25x^2-20x+4}{9x^2-24x+16}}$

12. $\dfrac{\dfrac{4x^2+28x+49}{25x^2-30x+9}}{\dfrac{4x^2-49}{25x^2-9}}$

13. $\dfrac{\dfrac{2}{x^3}-\dfrac{3}{x}}{\dfrac{5}{x^3}+\dfrac{4}{x^2}}$

14. $\dfrac{\dfrac{3}{x}-\dfrac{1}{x^2}}{\dfrac{3}{x^2}-\dfrac{1}{x^3}}$

15. $\dfrac{4+\dfrac{3}{x}}{\dfrac{2}{x}-3}$

16. $\dfrac{\dfrac{3}{x}+5}{\dfrac{2}{x}+4}$

17. $\dfrac{\dfrac{5}{2x^3}-4}{\dfrac{1}{6x^3}-3}$

18. $\dfrac{\dfrac{3}{4x^2}-2}{4-\dfrac{5}{8x^2}}$

19. $\dfrac{\dfrac{a^2}{b}-b}{\dfrac{1}{b}-\dfrac{1}{a}}$

20. $\dfrac{\dfrac{m^2}{9n}-n}{\dfrac{1}{n}+\dfrac{3}{m}}$

21. $\dfrac{\dfrac{1}{x}-\dfrac{8}{x^2}+\dfrac{15}{x^3}}{\dfrac{1}{x}-\dfrac{5}{x^2}}$

22. $\dfrac{\dfrac{4}{x}-\dfrac{1}{x^2}+\dfrac{3}{x^3}}{\dfrac{2}{x}-\dfrac{5}{x^2}}$

23. $\dfrac{\dfrac{x}{x-4}-\dfrac{2x}{x+1}}{\dfrac{x}{x+1}-\dfrac{2x}{x-4}}$

24. $\dfrac{\dfrac{2}{x-3}+\dfrac{5}{x-2}}{\dfrac{3}{x-2}-\dfrac{4}{x-3}}$

25. $\dfrac{p+\dfrac{2}{p-4}}{p-\dfrac{3}{p-4}}$

26. $\dfrac{\dfrac{3}{w+1}+2}{4-\dfrac{5}{w-1}}$

27. $\dfrac{\dfrac{1}{x+3}-\dfrac{1}{x}}{3}$

28. $\dfrac{\dfrac{2}{x+4}-\dfrac{2}{x}}{4}$

29. $\dfrac{\dfrac{3}{a+b}-\dfrac{3}{a-b}}{2ab}$

30. $\dfrac{\dfrac{4}{2a-b}-\dfrac{2}{a-3b}}{5ab^2}$

31. $\dfrac{\dfrac{1}{(x+2)^2}-\dfrac{1}{x^2}}{2}$

32. $\dfrac{\dfrac{5}{(x+3)^2}-\dfrac{5}{x^2}}{3}$

33. $\dfrac{\dfrac{6}{2x-8}+\dfrac{10}{x^2-4x}}{\dfrac{1}{x^2-x-12}-\dfrac{2}{x^2-16}}$

34. $\dfrac{\dfrac{4}{3x-9}-\dfrac{1}{x^2-3x}}{\dfrac{2}{x^2-2x-15}+\dfrac{5}{x^2-9}}$

35. $\dfrac{\dfrac{x+7}{x^2+7x+10}-\dfrac{6}{x^2+2x}}{\dfrac{x+1}{x^2+7x+10}+\dfrac{6}{x^2+5x}}$

36. $\dfrac{\dfrac{x-1}{x^2+x-12}+\dfrac{2}{x^2-3x}}{\dfrac{x+2}{x^2+x-12}-\dfrac{3}{x^2+4x}}$

Use the given functions to find an equation of $\dfrac{f}{g}$.

37. $f(x)=\dfrac{5x+10}{x^2-6x+9}$ and $g(x)=\dfrac{4x+8}{x^2-4x+3}$

38. $f(x)=\dfrac{5x-10}{2x^2+x-15}$ and $g(x)=\dfrac{7x-14}{x^2+5x+6}$

39. $f(x)=\dfrac{x}{2}-\dfrac{2}{x}$ and $g(x)=\dfrac{3}{2}+\dfrac{3}{x}$

40. $f(x)=\dfrac{x}{5}-\dfrac{5}{x}$ and $g(x)=\dfrac{8}{5}-\dfrac{8}{x}$

41. $f(x)=\dfrac{2x}{x^2-25}+\dfrac{x+5}{x-5}$ and $g(x)=\dfrac{x-5}{x+5}+\dfrac{3x}{x^2-25}$

42. $f(x)=\dfrac{6x}{x^2-9}+\dfrac{x-3}{x+3}$ and $g(x)=\dfrac{x+3}{x-3}-\dfrac{6x}{x^2-9}$

43. A student tries to simplify the complex rational expression $\dfrac{x}{\dfrac{1}{x}+\dfrac{1}{2}}$:

$$\begin{aligned}\frac{x}{\dfrac{1}{x}+\dfrac{1}{2}} &= x \div \left(\frac{1}{x}+\frac{1}{2}\right)\\ &= x\left(\frac{x}{1}+\frac{2}{1}\right)\\ &= x\cdot(x+2)\\ &= x^2+2x\end{aligned}$$

Describe any errors. Then simplify the expression correctly.

44. A student tries to simplify the complex rational expression $\dfrac{\dfrac{5}{2}+\dfrac{3}{x}}{\dfrac{7}{x}+\dfrac{4}{x^2}}$:

$$\begin{aligned}\frac{\dfrac{5}{2}+\dfrac{3}{x}}{\dfrac{7}{x}+\dfrac{4}{x^2}} &= \frac{\dfrac{5}{2}+\dfrac{3}{x}}{\dfrac{7}{x}+\dfrac{4}{x^2}}\cdot\frac{2x}{x^2}\\ &= \frac{\dfrac{5}{2}\cdot\dfrac{2x}{1}+\dfrac{3}{x}\cdot\dfrac{2x}{1}}{\dfrac{7}{x}\cdot\dfrac{x^2}{1}+\dfrac{4}{x^2}\cdot\dfrac{x^2}{1}}\\ &= \frac{5x+6}{7x+4}\end{aligned}$$

Describe any errors. Then simplify the expression correctly.

45. Simplify the expression by method 1; then simplify it again by method 2. Decide which method you prefer for this expression. Explain.

$$\frac{\dfrac{6}{x^2}-\dfrac{5}{x}}{\dfrac{2}{x^2}+\dfrac{3}{2x}}$$

46. Simplify the expression by method 1; then simplify it again by method 2. Decide which method you prefer for this expression. Explain.

$$\frac{\dfrac{x^2 - 3x - 28}{3x - 6}}{\dfrac{x^2 - 14x + 49}{6x - 12}}$$

47. Describe a complex rational expression. Give an example. Then simplify it.

48. Describe how to simplify a complex rational expression.

Related Review

Simplify.

49. $\dfrac{8x^{-2}y^5}{6x^{-7}y^8}$

50. $\dfrac{6x^4y^{-5}}{15x^{-3}y^{-2}}$

51. $\dfrac{x^{-1} + x^{-2}}{x^{-2} - x^{-1}}$ [**Hint:** First write as a complex rational expression.]

52. $\dfrac{x^{-3} - x^{-1}}{x^{-1} + x^{-2}}$ [**Hint:** First write as a complex rational expression.]

53. $\dfrac{2b^{-2} - 4b}{8b^{-1} - 6b}$

54. $\dfrac{3y + 2y^{-1}}{6y - 4y^{-3}}$

55. When a patient is treated with a radioactive element, the element decreases in amount exponentially from both radioactive (physical) decay and biological means, such as urination. The effective half-life of an element is given by

$$H_e = \frac{1}{\dfrac{1}{H_p} + \dfrac{1}{H_b}}$$

where H_p and H_b are the physical and biological half-lives, respectively.

a. Simplify the right-hand side of the formula.

b. Use the result you found in part (a) to find the effective half-life of sulfur-35, which has a physical half-life of 87.4 days and a biological half-life of 623 days.

56. a. Solve the equation $b^{mx+c} = k$ for x.

b. Students A and B try to solve the equation in part (a). Student A's result is $\dfrac{\log(k) - c\log(b)}{m\log(b)}$. Student B's result is $\dfrac{\dfrac{\log(k)}{\log(b)} - c}{m}$. Simplify student B's result. Are the students' results equivalent?

Expressions, Equations, Functions, and Graphs

Perform the indicated instruction. Then use words such as linear, quadratic, cubic, exponential, logarithmic, rational, polynomial, degree, function, one variable, *and* two variables *to describe the expression, equation, or system.*

57. Find the sum $\dfrac{x - 2}{x^2 - 2x - 24} + \dfrac{x + 4}{x^2 - 8x + 12}$.

58. Solve:
$$y = -\frac{3}{2}x + 4$$
$$\frac{3}{4}x - \frac{1}{2}y = 4$$

59. Find the product $\dfrac{x - 2}{x^2 - 2x - 24} \cdot \dfrac{x + 4}{x^2 - 8x + 12}$.

60. Graph $f(x) = -\dfrac{3}{2}x + 4$ by hand.

61. Find the domain of $f(x) = \dfrac{2x - 1}{x^2 - 2x - 24}$.

62. Find the inverse of $f(x) = -\dfrac{3}{2}x + 4$.

8.5 SOLVING RATIONAL EQUATIONS

"By taking Intermediate Algebra, I realized that I have to be ready mentally to take on this course. I do better when I play the piano before I go to class."

—Ena C., student

Objectives

- Solve *rational equations in one variable.*
- Solve formulas involving rational expressions.
- Compare solving rational equations with simplifying rational expressions.
- Use a rational model to make estimates and predictions for the independent variable.

In Sections 8.1–8.4, we worked with rational expressions. In this section, we will solve rational *equations* in one variable.

Solving Rational Equations in One Variable

A **rational equation in one variable** is an equation in one variable in which both sides can be written as rational expressions. Here are some examples of rational equations in one variable:

$$\frac{3x}{x-4}+\frac{x-2}{5x+3}=2, \qquad \frac{5}{x}=9, \qquad \frac{x}{x-3}+\frac{3}{x^2-6x+9}=\frac{x-5}{x-3}$$

As we have done in earlier sections, we will clear an equation of fractions by multiplying both sides of the equation by the LCD.

With rational equations, it is possible to take the usual steps for solving equations yet arrive at x values that are excluded values for one or more of the fractions in the equation. These values of x are *not* solutions. We call such values **extraneous solutions.**

Example 1 Solving a Rational Equation

Solve the equation $\frac{2}{p}+5=\frac{8}{p}-1$.

Solution

We note that 0 is an excluded value. We clear the equation of fractions by multiplying both sides of the equation by p, which is the LCD of the fractions $\frac{2}{p}$ and $\frac{8}{p}$:

$$\frac{2}{p}+5=\frac{8}{p}-1 \qquad \text{Original equation}$$

$$p\left(\frac{2}{p}+5\right)=p\left(\frac{8}{p}-1\right) \qquad \text{Multiply both sides by LCD, } p.$$

$$p\cdot\frac{2}{p}+p\cdot 5=p\cdot\frac{8}{p}-p\cdot 1 \qquad \text{Distributive law}$$

$$2+5p=8-p \qquad \text{Simplify.}$$

$$6p=6$$

$$p=1$$

Since 1 is not an excluded value, we conclude that 1 is the solution of the equation. We check that 1 satisfies the original equation:

$$\frac{2}{p}+5=\frac{8}{p}-1 \qquad \text{Original equation}$$

$$\frac{2}{1}+5\stackrel{?}{=}\frac{8}{1}-1 \qquad \text{Substitute 1 for } p.$$

$$7\stackrel{?}{=}7 \qquad \text{Simplify.}$$

true

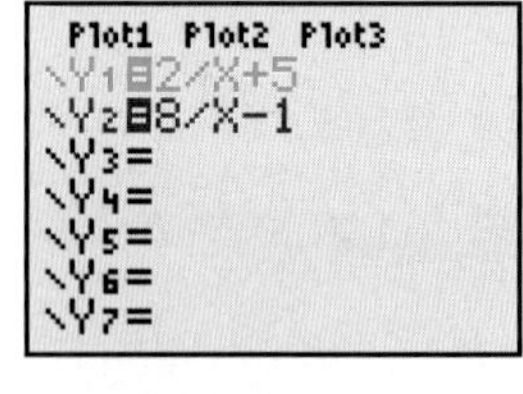

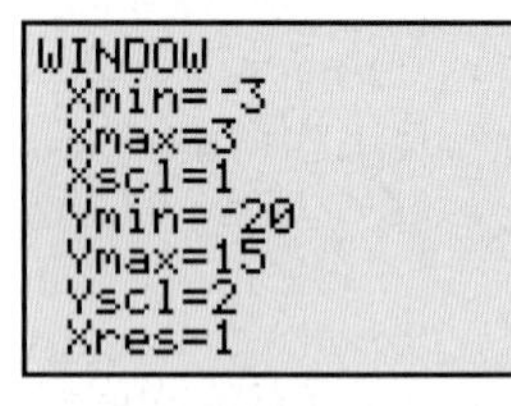

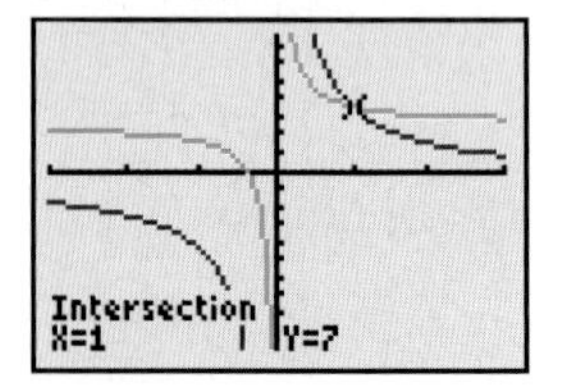

Figure 17 Verify the work

We also check our work by using "intersect" on a graphing calculator (see Fig. 17). ■

Example 2 Solving a Rational Equation

Solve $2-\frac{1}{x-2}=\frac{x-3}{x-2}$.

Solution

We note that 2 is an excluded value, because $x-2$ is in the denominator. We clear the equation of fractions by multiplying both sides of the equation by $x-2$, which is the LCD of the fractions $\frac{1}{x-2}$ and $\frac{x-3}{x-2}$:

$$2 - \frac{1}{x-2} = \frac{x-3}{x-2} \quad \text{Original equation}$$

$$(x-2)\left(2 - \frac{1}{x-2}\right) = (x-2) \cdot \frac{x-3}{x-2} \quad \text{Multiply both sides by LCD, } x-2.$$

$$(x-2) \cdot 2 - (x-2) \cdot \frac{1}{x-2} = (x-2) \cdot \frac{x-3}{x-2} \quad \text{Distributive law}$$

$$(x-2) \cdot 2 - 1 = x - 3 \quad \text{Simplify.}$$

$$2x - 4 - 1 = x - 3 \quad \text{Distributive law}$$

$$x = 2$$

Our result, 2, is *not* a solution, because 2 is an excluded value. Since the only possibility, 2, is not a solution of the original equation, we conclude that no number is a solution. We say that the solution set is the *empty set*. ■

To see where we introduced the extraneous solution 2 in Example 2, notice that 2 does not satisfy the equation

$$(x-2) \cdot 2 - (x-2) \cdot \frac{1}{x-2} = (x-2) \cdot \frac{x-3}{x-2}$$

because 2 is an excluded value of the expression $(x-2) \cdot \frac{x-3}{x-2}$. However, 2 does satisfy the next equation, $(x-2) \cdot 2 - 1 = x - 3$:

$$(x-2) \cdot 2 - 1 = x - 3$$
$$(2-2) \cdot 2 - 1 \stackrel{?}{=} 2 - 3$$
$$-1 \stackrel{?}{=} -1$$
$$\text{true}$$

Since multiplying both sides of a rational equation by the LCD and then simplifying both sides may introduce extraneous solutions, we must always check that any proposed solution is not an excluded value.

To solve a rational equation, we factor the denominators of fractions to help us determine any excluded values, to find the LCD, and, later, to help us simplify rational expressions.

Example 3 Solving a Rational Equation

Solve $\frac{x}{x+2} - \frac{7}{5-x} = \frac{14}{x^2 - 3x - 10}$.

Solution

We begin by factoring the denominators:

$$\frac{x}{x+2} - \frac{7}{5-x} = \frac{14}{x^2 - 3x - 10} \quad \text{Original equation}$$

$$\frac{x}{x+2} - \frac{7}{-(x-5)} = \frac{14}{(x-5)(x+2)} \quad \text{Factor denominators; } 5 - x = -x + 5 = -(x-5)$$

$$\frac{x}{x+2} + \frac{7}{(x-5)} = \frac{14}{(x-5)(x+2)} \quad \frac{a}{-b} = -\frac{a}{b}$$

The excluded values are -2 and 5. Next, we clear the equation of fractions by multiplying both sides by the LCD, $(x-5)(x+2)$:

$$(x-5)(x+2)\left(\frac{x}{x+2}+\frac{7}{x-5}\right)=(x-5)(x+2)\cdot\frac{14}{(x-5)(x+2)} \qquad \text{Multiply both sides by LCD.}$$

On the left-hand side, we use the distributive law. On the right-hand side, we simplify:

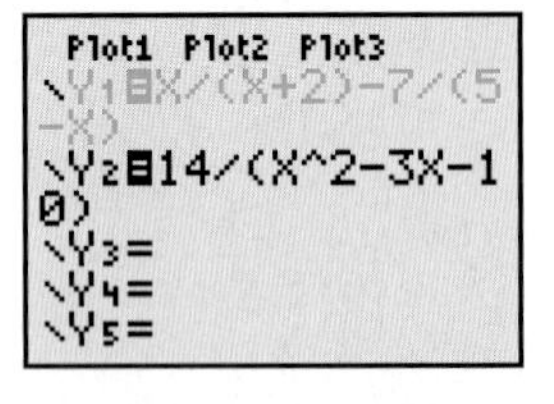

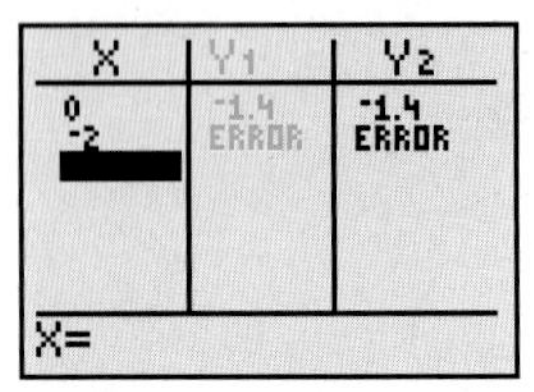

Figure 18 Verify the work

$$\begin{aligned}
(x-5)(x+2)\cdot\frac{x}{x+2}+(x-5)(x+2)\cdot\frac{7}{x-5}&=14 &&\text{Distributive law; simplify.}\\
(x-5)\cdot x+(x+2)\cdot 7&=14 &&\text{Simplify.}\\
x^2-5x+7x+14&=14 &&\text{Distributive law}\\
x^2+2x+14&=14 &&\text{Combine like terms.}\\
x^2+2x&=0 &&\text{Subtract 14 from both sides.}\\
x(x+2)&=0 &&\text{Factor left-hand side.}\\
x=0 \quad\text{or}\quad x+2&=0 &&\text{Zero factor property}\\
x=0 \quad\text{or}\quad x&=-2
\end{aligned}$$

Since -2 is an excluded value, it is *not* a solution. The only solution is 0. We use a graphing calculator table to check our work (see Fig. 18). ■

Solving a Rational Equation in One Variable

To solve a rational equation in one variable,

1. Factor the denominator(s) if possible.
2. Identify any excluded values.
3. Find the LCD of all of the fractions.
4. Multiply both sides of the equation by the LCD, which gives a simpler equation to solve.
5. Solve the simpler equation.
6. Discard any proposed solutions that are excluded values.

Example 4 Finding an Input Value of a Rational Function

Let $f(x)=\frac{x+1}{x-3}-\frac{x-2}{x+3}$. Find x when $f(x)=1$.

Solution

We note that the domain of f is the set of real numbers except -3 and 3. We substitute 1 for $f(x)$ in the equation $f(x)=\frac{x+1}{x-3}-\frac{x-2}{x+3}$ and solve for x:

$$\begin{aligned}
1&=\frac{x+1}{x-3}-\frac{x-2}{x+3}\\
(x-3)(x+3)\cdot 1&=(x-3)(x+3)\left(\frac{x+1}{x-3}-\frac{x-2}{x+3}\right)\\
(x-3)(x+3)&=(x-3)(x+3)\cdot\frac{x+1}{x-3}-(x-3)(x+3)\cdot\frac{x-2}{x+3}
\end{aligned}$$

```
Plot1 Plot2 Plot3
\Y1■(X+1)/(X-3)-
(X-2)/(X+3)
\Y2=
\Y3=
\Y4=
\Y5=
\Y6=
```

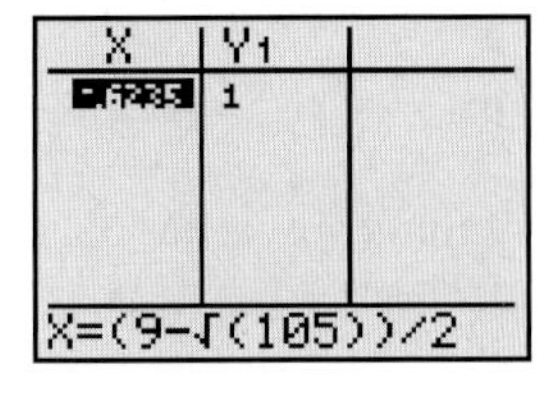

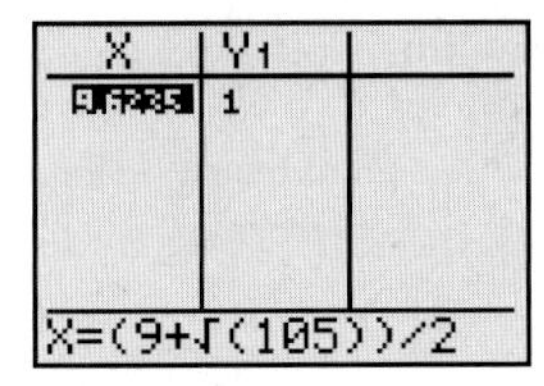

Figure 19 Verify the work

$$
\begin{aligned}
(x-3)(x+3) &= (x+3)(x+1)-(x-3)(x-2) && \text{Simplify.}\\
x^2-9 &= x^2+4x+3-(x^2-5x+6) && \text{Find products.}\\
x^2-9 &= x^2+4x+3-x^2+5x-6 && \text{Subtract trinomial.}\\
x^2-9 &= 9x-3 && \text{Combine like terms.}\\
x^2-9x-6 &= 0 && \text{Write in } ax^2+bx+c=0 \text{ form.}\\
x &= \frac{-(-9)\pm\sqrt{(-9)^2-4(1)(-6)}}{2(1)} && \text{Substitute into quadratic formula.}\\
x &= \frac{9\pm\sqrt{105}}{2} && \text{Simplify.}
\end{aligned}
$$

Since both of our results are in the domain of f, we conclude that if $f(x) = 1$, then $x = \frac{9-\sqrt{105}}{2}$ or $x = \frac{9+\sqrt{105}}{2}$.

We enter $y = \frac{x+1}{x-3} - \frac{x-2}{x+3}$ in a graphing calculator and check that for both inputs $\frac{9-\sqrt{105}}{2}$ and $\frac{9+\sqrt{105}}{2}$, the output is 1 (see Fig. 19). ■

Solving Formulas Involving Rational Expressions

Formulas that contain rational expressions are useful in many fields, such as finance, physics, meteorology, mathematics, electronics, and chemistry. It can be helpful to solve such a formula for one of its variables.

Example 5 Solving a Formula Involving a Rational Expression

The following formula is useful in electronics:

$$I = \frac{\mathcal{E}}{R+r}$$

where I is the current in an electrical circuit, $\mathcal{E}$ is the electromotive force, R is the circuit's resistance, and r is the battery's resistance. Solve the formula for the variable R.

Solution

To begin, we multiply both sides of the equation by the LCD, $R + r$:

$$
\begin{aligned}
I &= \frac{\mathcal{E}}{R+r} && \text{Original formula}\\
(R+r)\cdot I &= (R+r)\cdot\frac{\mathcal{E}}{R+r} && \text{Multiply both sides by LCD, } R+r.\\
RI+rI &= \mathcal{E} && \text{Distributive law; simplify.}\\
RI &= \mathcal{E}-rI && \text{Subtract } rI \text{ from both sides.}\\
\frac{RI}{I} &= \frac{\mathcal{E}-rI}{I} && \text{Divide both sides by } I.\\
R &= \frac{\mathcal{E}-rI}{I} && \text{Simplify.}
\end{aligned}
$$

■

Solving Rational Equations versus Simplifying Rational Expressions

Throughout this course, we have solved equations and simplified expressions. When solving an equation, our objective is to find any *numbers* that satisfy the equation. When simplifying an expression, our objective is to find a simpler, yet equivalent, *expression*.

Solving a Rational Equation versus Simplifying a Rational Expression

To solve a rational equation, clear the fractions in it by multiplying both sides of the equation by the LCD. To simplify a rational expression, do *not* multiply it by the LCD—the only multiplication permissible is multiplication by 1, usually in the form $\frac{A}{A}$, where A is a nonzero polynomial.

Here, we compare solving a rational equation with simplifying a rational expression:

Solving the Equation $\frac{2}{3} = \frac{4}{x}$

The number 0 is an excluded value.

$$\frac{2}{3} = \frac{4}{x} \quad \text{Original equation}$$

$$3x \cdot \frac{2}{3} = 3x \cdot \frac{4}{x} \quad \text{Multiply both sides by LCD, } 3x.$$

$$2x = 12 \quad \text{Simplify.}$$

$$x = 6$$

The result is a number.

Simplifying the Expression $\frac{2}{3} + \frac{4}{x}$

$$\frac{2}{3} + \frac{4}{x} = \frac{2}{3} \cdot \frac{x}{x} + \frac{4}{x} \cdot \frac{3}{3} \quad \text{Introduce missing factors.}$$

$$= \frac{2x}{3x} + \frac{12}{3x} \quad \text{Find products.}$$

$$= \frac{2x + 12}{3x} \quad \frac{A}{B} + \frac{C}{B} = \frac{A+C}{B}$$

The result is an expression.

For the equation $\frac{2}{3} = \frac{4}{x}$, the solution is the *number* 6. We simplify the expression $\frac{2}{3} + \frac{4}{x}$ by writing it as the *expression* $\frac{2x + 12}{3x}$. In general, **the result of solving a rational equation is the empty set or a set of one or more numbers. The result of simplifying a rational expression is an expression.**

Using a Rational Model to Make Predictions About the Independent Variable

In Section 8.1, we modeled the percentage of a quantity by means of a rational model, which we used to make a prediction for the dependent variable. Now that we know how to solve a rational equation in one variable, we can use such a model to make a prediction about the independent variable.

Table 7 Numbers of Internet Users

Year	Number of Internet Users (millions)
1996	39
1997	60
1998	84
1999	105
2000	122
2001	143
2002	166
2003	183
2004	207

Source: *Jupiter MMXI*

Example 6 Using a Rational Model to Make a Prediction

In Example 9 of Section 8.1, we found the model $P(t) = \frac{2072t - 8394}{0.0068t^2 + 2.58t + 251.7}$, where $P(t)$ is the percentage of Americans who are Internet users at t years since 1990 (see Table 7). Predict when 85% of Americans were or will be Internet users.

Solution

We substitute 85 for $P(t)$ in the equation of P and solve for t:

$$85 = \frac{2072t - 8394}{0.0068t^2 + 2.58t + 251.7}$$

$$(0.0068t^2 + 2.58t + 251.7) \cdot 85 = (0.0068t^2 + 2.58t + 251.7) \cdot \frac{2072t - 8394}{0.0068t^2 + 2.58t + 251.7}$$

$$0.578t^2 + 219.3t + 21{,}394.5 = 2072t - 8394$$

$$0.578t^2 - 1852.7 + 29{,}788.5 = 0$$

Figure 20 Verify the work

Now we substitute $a = 0.578$, $b = -1852.7$, and $c = 29{,}788.5$ in the quadratic formula:

$$t = \frac{-(-1852.7) \pm \sqrt{(-1852.7)^2 - 4(0.578)(29{,}788.5)}}{2(0.578)}$$ Substitute into quadratic formula.

$t \approx 16.16 \quad \text{or} \quad t \approx 3189.20$ Compute.

We verify the results by entering $y = \dfrac{2072t - 8394}{0.0068t^2 + 2.58t + 251.7}$ in a graphing calculator and checking that the inputs 16.16 and 3189.20 lead to outputs of about 85 (see Fig. 20).

The inputs 16.16 and 3189.20 represent the years 2006 and 5179. Model breakdown has occurred for the prediction 5179: The year is too far into the future for us to have any faith in this prediction! Therefore, we estimate that 85% of Americans were Internet users in 2006. ■

group exploration

Simplifying versus solving

1. Two students tried to solve $4 = \frac{5}{x} + \frac{3}{x}$. Did one, both, or neither of these students solve the equation correctly? Explain.

Student A

$$\begin{aligned} 4 &= \frac{5}{x} + \frac{3}{x} \\ 4x &= x\left(\frac{5}{x} + \frac{3}{x}\right) \\ 4x &= x \cdot \frac{5}{x} + x \cdot \frac{3}{x} \\ 4x &= 5 + 3 \\ 4x &= 8 \\ x &= 2 \end{aligned}$$

Student B

$$\begin{aligned} 4 &= \frac{5}{x} + \frac{3}{x} \\ &= \frac{5}{x} + \frac{3}{x} - 4 \\ &= \frac{8}{x} - 4 \\ &= \frac{8}{x} - 4 \cdot \frac{x}{x} \\ &= \frac{8}{x} - \frac{4x}{x} \\ &= \frac{-4x + 8}{x} \end{aligned}$$

2. Three students tried to simplify $4 + \frac{5}{x} + \frac{3}{x}$. Which students, if any, simplified the expression correctly? Explain.

Student C

$$\begin{aligned} 4 + \frac{5}{x} + \frac{3}{x} &= x\left(4 + \frac{5}{x} + \frac{3}{x}\right) \\ &= 4x + x \cdot \frac{5}{x} + x \cdot \frac{3}{x} \\ &= 4x + 5 + 3 \\ &= 4x + 8 \end{aligned}$$

Student D

$$\begin{aligned} 4 + \frac{5}{x} + \frac{3}{x} &= 4 \cdot \frac{x}{x} + \frac{5}{x} + \frac{3}{x} \\ &= \frac{4x}{x} + \frac{5}{x} + \frac{3}{x} \\ &= \frac{4x + 8}{x} \end{aligned}$$

Student E

$$\begin{aligned} 4 + \frac{5}{x} + \frac{3}{x} &= 0 \\ x\left(4 + \frac{5}{x} + \frac{3}{x}\right) &= x \cdot 0 \\ 4x + 5 + 3 &= 0 \\ 4x &= -8 \\ x &= -2 \end{aligned}$$

3. **a.** What is the difference in your goals in solving a rational equation versus simplifying a rational expression?
b. Explain how that difference relates to the techniques you use to solve the equation versus simplify the expression.

HOMEWORK 8.5

FOR EXTRA HELP ▶ Student Solutions Manual | PH Math/Tutor Center | MathXL® | MyMathLab

Solve. Any solution is a real number.

1. $\frac{7}{x} = \frac{2}{x} + 1$
2. $\frac{8}{x} = \frac{5}{x} + 3$
3. $\frac{7}{4x} - \frac{5}{6} = \frac{1}{12x}$
4. $\frac{5}{8x} - \frac{7}{4} = \frac{3}{2x}$
5. $\frac{x-2}{x-7} = \frac{5}{x-7}$
6. $\frac{-1}{x-5} = \frac{x-6}{x-5}$
7. $\frac{5}{4p-7} = \frac{2}{2p+3}$
8. $\frac{3}{2y-9} = \frac{4}{3y+5}$
9. $\frac{3}{x+1} + \frac{2}{5} = 1$
10. $\frac{2}{3} = \frac{7}{x-5} + 2$
11. $\frac{1}{x-2} + \frac{1}{x+2} = \frac{4}{x^2-4}$
12. $\frac{3}{x-5} + \frac{2}{x+5} = \frac{30}{x^2-25}$
13. $2 + \frac{4}{k-2} = \frac{8}{k^2-2k}$
14. $1 + \frac{4}{m-5} = \frac{2}{m^2-5m}$
15. $\frac{-48}{x^2-2x-15} - \frac{6}{x+3} = \frac{7}{x-5}$
16. $\frac{-36}{x^2+x-20} - \frac{2}{x-4} = \frac{4}{x+5}$
17. $\frac{x^2-23}{2x^2-5x-3} + \frac{2}{x-3} = \frac{-1}{2x+1}$
18. $\frac{4x^2-24x}{3x^2-x-2} + \frac{3}{3x+2} = \frac{-4}{x-1}$
19. $\frac{w+7}{w^2-9} = \frac{-w+2}{w-3}$
20. $\frac{t-6}{t^2-4} = \frac{-t+1}{t-2}$
21. $3 + \frac{2}{x} = \frac{4}{x^2}$
22. $\frac{5}{x} = \frac{3}{x^2} - 4$
23. $\frac{5}{r^2-3r+2} - \frac{1}{r-2} = \frac{r+6}{3r-3}$
24. $\frac{3}{p^2-6p+9} + \frac{p-2}{3p-9} = \frac{p}{2p-6}$
25. $\frac{2x}{x+1} - \frac{3}{2} = \frac{-2}{x+2}$
26. $-\frac{1}{2} + \frac{x}{x-1} = \frac{-1}{x+3}$
27. $\frac{x-4}{x^2-7x+12} - \frac{x+2}{x-3} = 0$
28. $\frac{x+2}{x^2+x-30} - \frac{x+3}{x-5} = 0$
29. $\frac{t}{t-3} = 2 - \frac{5}{3-t}$
30. $4 - \frac{k}{k-5} = \frac{1}{5-k}$
31. $\frac{12}{9-x^2} + \frac{3}{x+3} = \frac{-2}{x-3}$
32. $\frac{-2}{x+5} + \frac{3}{x-5} = \frac{-20}{25-x^2}$
33. $\frac{x+2}{x-3} - \frac{x-3}{x+2} = \frac{5x}{x^2-x-6}$
34. $\frac{x-4}{x+1} + \frac{x+1}{x-4} = \frac{13x}{x^2-3x-4}$
35. $\frac{2y}{y-2} - \frac{2y-5}{y^2-7y+10} = \frac{-4}{y-5}$
36. $\frac{3p}{p+1} - \frac{4p-1}{p^2-2p-3} = \frac{-2}{p-3}$
37. $\frac{x-2}{x^2-2x-3} + \frac{x+5}{x^2-1} = \frac{x+3}{x^2-4x+3}$
38. $\frac{x+4}{x^2+x-2} + \frac{x+1}{x^2-4} = \frac{x-3}{x^2-3x+2}$

Find all complex-number solutions.

39. $\frac{5}{x} - \frac{2}{x^2} = 4$
40. $\frac{7}{x} + 5 = -\frac{6}{x^2}$
41. $\frac{2}{t-5} - \frac{3t}{t+5} = \frac{35}{t^2-25}$
42. $\frac{3}{k+4} - \frac{2k}{k-4} = \frac{2}{k^2-16}$
43. $\frac{x-1}{3x-12} + \frac{-x+1}{x-5} = \frac{4x}{x^2-9x+20}$
44. $\frac{-x+6}{2x-6} + \frac{x-1}{x+4} = \frac{2x}{x^2+x-12}$

Find x when y is equal to the indicated value.

45. $f(x) = \frac{3}{x-5}$, $y = 4$
46. $g(x) = \frac{7}{x+4}$, $y = 3$
47. $f(x) = \frac{5}{x-1} + \frac{3}{x+1}$, $y = -1$
48. $g(x) = \frac{2}{x+1} - \frac{4}{x-2}$, $y = -1$

Find all x-intercepts.

49. $f(x) = \frac{x-1}{x-5} - \frac{x+2}{x+3}$
50. $g(x) = \frac{x+4}{x-2} - \frac{x-3}{x+6}$

For Exercises 51–56, solve for the specified variable.

51. $F = \dfrac{mv^2}{r}$, for r

52. $P = \dfrac{nrT}{V}$, for V

53. $F = -\dfrac{GMm}{r^2}$, for M

54. $\dfrac{P_1V_1}{T_1} = \dfrac{P_2V_2}{T_2}$, for P_1

55. $P = \dfrac{A}{1+rt}$, for t

56. $P = \dfrac{A}{1+rt}$, for r

57. In Exercise 69 of Homework 8.1, you found the model $P(t) = \dfrac{4.7t^2 - 77t + 350}{0.058t^2 - 0.82t + 3.6}$, where $P(t)$ is the percentage of cumulative awarded miles (from frequent-flier programs) that are unredeemed at t years since 1980 (see Table 8).

Table 8 Cumulative Unredeemed Miles

Year	Cumulative Unredeemed Miles (trillions of miles)	Cumulative Awarded Miles (trillions of miles)
1986	0.1	0.3
1988	0.4	0.8
1990	0.9	1.5
1992	1.5	2.5
1994	2.2	3.7
1996	3.2	5.2
1998	4.6	7.3
2000	6.6	10.0
2002	9.1	13.3
2004	12.4	17.2
2005	14.2	19.9

Source: *WebFlyer*

Use P to predict when 74% of cumulative awarded miles will be unredeemed.

58. In Exercise 68 of Homework 8.1, you found the model $P(t) = \dfrac{480t + 686}{1.59t + 90.31}$, where $P(t)$ is the percentage of U.S. households with personal computers at t years since 1990 (see Table 9).

Table 9 Numbers of Households with Personal Computers

Year	Number of Households with Personal Computers (millions)	Total Number of Households (millions)
1995	31.36	99.0
1997	39.56	101.0
1999	50.00	103.6
2001	60.81	108.2
2003	68.78	111.3

Sources: *U.S. Department of Commerce; Media Matrix; U.S. Census Bureau*

Use P to predict when 85% of households will have personal computers.

59. In Exercise 78 of Homework 6.1, you worked with the system

$$N(t) = 0.17t + 3$$
$$B(t) = -0.036t^2 + 0.176t + 1.28$$
$$(N + B)(t) = -0.036t^2 + 0.346t + 4.28$$

where $N(t)$ is Mattel's annual sales of toys other than Barbie® and $B(t)$ is Mattel's annual Barbie sales, both in billions of dollars, in the year that is t years since 2000 (see Table 10).

Table 10 Mattel's Non-Barbie Sales and Barbie Sales

Year	Sales (billions of dollars): Non-Barbie	Barbie
2001	3.18	1.41
2002	3.35	1.50
2003	3.41	1.45
2004	3.66	1.41
2005	3.88	1.24

Source: *Harris Nesbitt*

a. Let $P(t)$ be the percentage of Mattel's total annual sales that are Barbie sales in the year that is t years since 2000. Find an equation of P.

b. Find $P(8)$. What does it mean in this situation?

c. Find t when $P(t) = 15$. What does it mean in this situation?

d. Use the window settings in Fig. 21 to graph P. Then use "maximum" on a graphing calculator to find the maximum point of the graph of P for values of t between 0 and 10. What does that point mean in this situation?

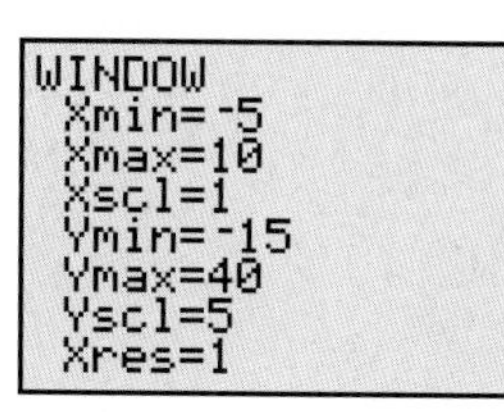

Figure 21 Window settings for Exercise 59d

60. In Exercise 77 of Homework 6.1, you worked with the system

$$B(t) = 2.3t^2 - 12.1t + 98$$
$$S(t) = 1.57t^2 - 10.9t + 28$$
$$(B + S)(t) = 3.87t^2 - 23t + 126$$

where $B(t)$ is the average base pay of an MBA graduate and $S(t)$ is an MBA graduate's average signing bonus, both in thousands of dollars, in the year that is t years since 2000 (see Table 11).

Table 11 MBA Average Base Pays and Average Signing Bonuses

Year	Average Pay/Bonus (thousands of dollars): Base Pay	Signing Bonus
2001	88.6	18.9
2002	82.0	13.0
2003	80.9	8.9
2004	88.8	10.6
2005	93.8	12.8

Source: *Graduate Management Admission Council*

a. Let $P(t)$ be the percentage of an MBA graduate's first year of total earnings (base pay plus signing bonus) that is the signing bonus in the year that is t years since 2000. Find an equation of P.

b. Find $P(6)$. What does it mean in this situation?

c. Find t when $P(t) = 20$. What does it mean in this situation?

d. Use the window settings in Fig. 22 to graph P. Then use "minimum" on a graphing calculator to find the minimum point of the graph of P for values of t between 0 and 10. What does that point mean in this situation?

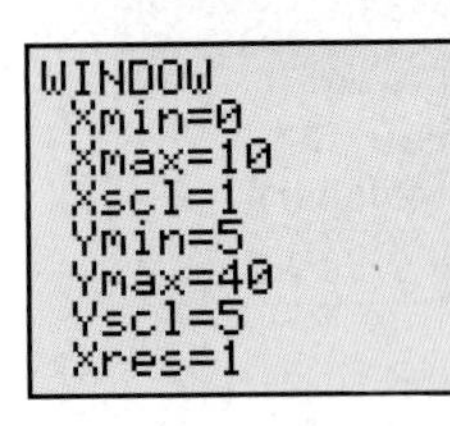

Figure 22 Window settings for Exercise 60d

For Exercises 61–66, find approximate solutions of the given equation or system by referring to the graphs shown in Fig. 23. Round results or coordinates of results to the first decimal place.

61. $\dfrac{5}{x-2} = x^2 - 6x + 10$

62. $\dfrac{5}{x-2} = -x^2 + x - 1$

63. $\dfrac{5}{x-2} = 4$

64. $\dfrac{5}{x-2} = -1$

65. $y = \dfrac{5}{x-2}$
$y = -x^2 + x - 1$

66. $y = \dfrac{5}{x-2}$
$y = x^2 - 6x + 10$

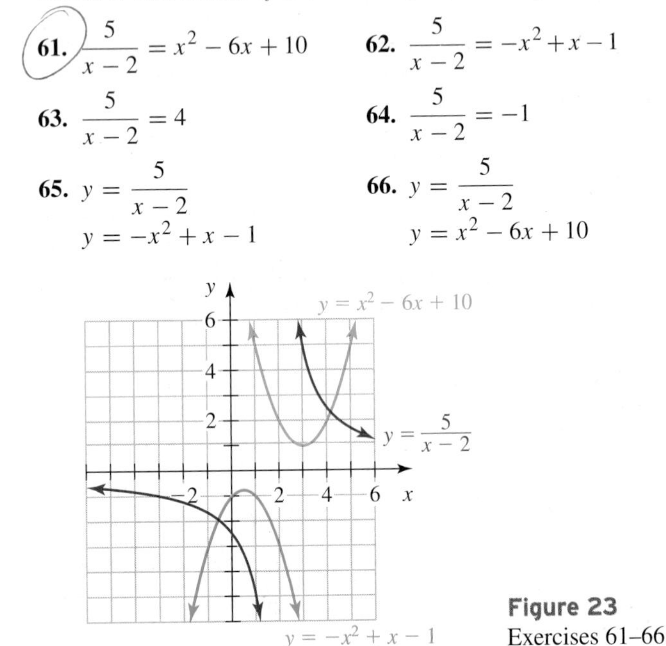

Figure 23 Exercises 61–66

67. Let $f(x) = \dfrac{x+a}{x+b}$. Find values of a and b such that $f(0) = 2$ and $f(1) = \dfrac{5}{2}$.

68. Let $f(x) = \dfrac{x+a}{x+b}$. Find values of a and b such that $f(1) = -2$ and $f(3) = -8$.

69. A student tries to solve the equation $\dfrac{4}{x+2} - \dfrac{2}{x} = \dfrac{1}{x+2}$:

$$\frac{4}{x+2} - \frac{2}{x} = \frac{1}{x+2}$$
$$\frac{4}{x+2}\cdot\frac{x}{x} - \frac{2}{x}\cdot\frac{x+2}{x+2} = \frac{1}{x+2}\cdot\frac{x}{x}$$
$$\frac{4x - 2(x+2)}{x(x+2)} = \frac{x}{x(x+2)}$$
$$\frac{4x - 2x - 4}{x(x+2)} = \frac{x}{x(x+2)}$$
$$\frac{2x-4}{x(x+2)} = \frac{x}{x(x+2)}$$
$$2x - 4 = x$$
$$x = 4$$

Describe a more efficient way to solve the equation.

70. A student tries to simplify the expression $\dfrac{3}{x+1} + \dfrac{3}{x-1}$:

$$\begin{aligned}
&\frac{3}{x+1} + \frac{3}{x-1}\\
&= (x-1)(x+1)\left(\frac{3}{x+1} + \frac{3}{x-1}\right)\\
&= (x-1)(x+1)\cdot\frac{3}{x+1} + (x-1)(x+1)\cdot\frac{3}{x-1}\\
&= 3(x-1) + 3(x+1)\\
&= 3x - 3 + 3x + 3\\
&= 6x
\end{aligned}$$

Describe any errors. Then simplify the expression correctly.

71. A student tries to solve a rational equation in one variable. The result is $\dfrac{x-2}{x^2-4x+1}$. What would you tell the student?

72. Describe how to solve a rational equation in one variable.

Related Review

Solve or simplify, as appropriate. For equations, any solution is a real number.

73. $\dfrac{5}{x} + \dfrac{4}{x+1} - \dfrac{3}{x}$

74. $\dfrac{2}{x+3} + \dfrac{x}{x-3} + \dfrac{-10}{x^2-9}$

75. $\dfrac{5}{t} + \dfrac{4}{t+1} = \dfrac{3}{t}$

76. $\dfrac{2}{k+3} + \dfrac{k}{k-3} = \dfrac{-10}{k^2-9}$

77. $\dfrac{x+2}{x^2-5x+6} - \dfrac{x+1}{x^2-4} = \dfrac{4}{x^2-x-6}$

78. $\dfrac{3}{x^2-4x-12} - \dfrac{x-3}{x^2-7x+6}$

79. $\dfrac{x+2}{x^2-5x+6} - \dfrac{x+1}{x^2-4} + \dfrac{4}{x^2-x-6}$

80. $\dfrac{3}{x^2-4x-12} - \dfrac{x-3}{x^2-7x+6} = \dfrac{x}{x^2+x-2}$

Solve. Any solution is a real number. Round approximate solutions to the fourth decimal place.

81. $2p^3 - p^2 = 8p - 4$

82. $2(2m^2 - m) = 1 - 3m$

83. $2(4)^x + 3 = 106$

84. $\dfrac{3}{x^2-9} - \dfrac{x-2}{x^2-x-12} = \dfrac{-1}{x-4}$

85. $\log_3(5x-4) - \log_3(2x-3) = 2$

86. $2\log_2(x+2) - \log_2(x+3) = 1$

Expressions, Equations, Functions, and Graphs

Perform the indicated instruction. Then use words such as linear, quadratic, cubic, exponential, logarithmic, rational, polynomial, degree, function, one variable, *and* two variables *to describe the expression, equation, or system.*

87. Graph $f(x) = -2(x+2)^2 + 3$ by hand.

88. Solve $\log_5(2x^4) = 3$. Round any solutions to the fourth decimal place.

89. Solve $-2(x+2)^2 + 3 = -15$.

90. Graph $f(x) = \log_2(x)$ by hand.

91. Simplify $-2(x+2)^2 + 3$.

92. Simplify $\log_b(b^3)$.

8.6 MODELING WITH RATIONAL FUNCTIONS

Objectives

- Use a rational function to model the *mean* of a quantity.
- Model the percentage of a quantity.
- Use a rational function to model the time it takes to travel a given distance at a constant speed.

In this section, we will use rational functions to model authentic situations.

Modeling the Mean of a Quantity

To begin, suppose that four students go on a road trip during spring break. The total cost for gas is \$20. Consider the following possibilities:

A. Each person pays \$5 for gas.
B. The amounts contributed for gas are \$4, \$4, \$6, and \$6.
C. One student pays \$20, and the other three students ride for free.

We compute the *mean* amount of money each student spent for gas by dividing the total spent by the number of students:

$$\begin{aligned}\text{mean amount spent per student} &= \frac{\text{total amount spent for gas}}{\text{number of students}} \\ &= \frac{20 \text{ dollars}}{4 \text{ students}} = 5 \text{ dollars per student}\end{aligned}$$

So, the mean amount of money spent per student is \$5. We also say that the *average* amount of money spent per student is \$5.

For scenarios A, B, and C, we make the following observations:

A. The mean gives the (exact) per-person amount if all students pay an equal amount.
B. The mean gives a reasonable estimate of the per-person amount if the students pay nearly the same amount.
C. The mean gives a poor estimate of the per-person amount if the students pay very different amounts.

Computing the Mean

If a quantity Q is divided into n parts, the **mean** amount M of the quantity per part is given by

$$M = \frac{Q}{n}$$

As another example, if a student makes 21 phone calls in 7 days, the mean number of calls he makes per day is $\frac{21}{7} = 3$ calls. The mean is a fairly good estimate of the number of calls on a given day, provided that the student made about the same number of calls each day.

Example 1 Modeling the Mean of a Quantity

The underground band Melted Zipper wants to make and sell a CD of its original songs. It costs about \$1000 to record the music onto a digital audiotape (DAT), \$100 to rearrange the music and improve the sound quality, \$350 for artwork for the cover and inside leaflet, and \$350 to set up production. In addition, it will cost \$2.50 for each CD manufactured.

1. What is the total cost of making 300 CDs?
2. Let $C(n)$ be the total cost (in dollars) of making n CDs. Find an equation of C.
3. Let $P(n)$ be the price the band should set for each CD (in dollars) so that it breaks even by making and selling n CDs. Find an equation of P.
4. Find $P(300)$. What does it mean in this situation?
5. Find n when $P(n) = 10$. What does it mean in this situation?
6. Describe the values of $P(n)$ for large values of n.

Solution

1. First we compute the total *fixed costs*—the costs that do not depend on how many CDs are manufactured:

$$1000 + 100 + 350 + 350 = 1800 \text{ dollars}$$

The band must also pay \$2.50 per CD manufactured. If the band manufactures 300 CDs, this cost, called the *variable cost,* is $2.50(300) = 750$ dollars.

To find the total cost, we add the variable cost and the fixed costs:

$$2.50(300) + 1800 = 2550 \text{ dollars}$$

2. The total cost is equal to \$2.50 times the number of CDs, plus the fixed cost of \$1800:

$$C(n) = 2.50n + 1800$$

3. If the band makes and sells n CDs, it can break even by selling the CD for the amount found by dividing the total cost into n parts. This amount is the mean cost per CD:

$$\begin{aligned} P(n) &= \text{mean cost per CD} \\ &= \frac{\text{total cost}}{\text{number of CDs manufactured}} \\ &= \frac{2.50n + 1800}{n} \end{aligned}$$

4. $P(300) = \dfrac{2.50(300) + 1800}{300} = 8.50$

If the band makes and sells 300 CDs, it must sell each CD for \$8.50 to break even.

5. We substitute 10 for $P(n)$ and solve for n:

$$\begin{aligned} 10 &= \frac{2.50n + 1800}{n} && \text{Substitute 10 for } P(n). \\ n \cdot 10 &= n \cdot \frac{2.50n + 1800}{n} && \text{Multiply both sides by LCD, } n. \\ 10n &= 2.50n + 1800 && \text{Simplify.} \\ 7.5n &= 1800 \\ n &= 240 \end{aligned}$$

If the band can sell each CD for \$10, then it must make 240 CDs to break even.

6. First, we use graphing calculator tables to display input–output pairs of P (see Figs. 24, 25, and 26).

Figure 24 Enter the function

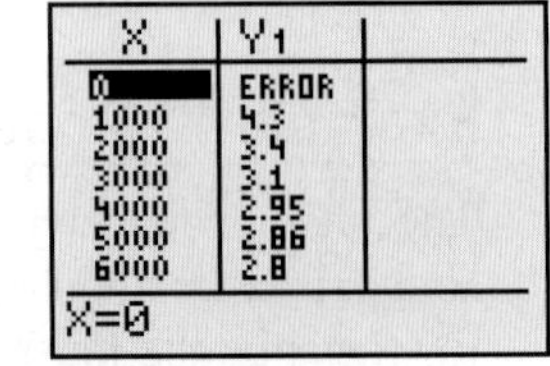

X	Y1
0	ERROR
1000	4.3
2000	3.4
3000	3.1
4000	2.95
5000	2.86
6000	2.8

X=0

Figure 25 Inputs increasing by 1000

X	Y1
0	ERROR
10000	2.68
20000	2.59
30000	2.56
40000	2.545
50000	2.536
60000	2.53

X=0

Figure 26 Inputs increasing by 10,000

From the tables, we see that as n increases, the mean cost (price) decreases (in order to break even). This happens because, as the number of CDs manufactured increases, the more the fixed cost is spread out, so the smaller the fixed cost that each sale has to cover. In fact, if we continue to scroll down a table for larger and larger inputs n, the outputs $P(n)$ approach 2.50, the variable cost in dollars per CD.

We can also study a graphing calculator graph to observe that the break-even CD price approaches \$2.50 for large manufacturing runs (see Figs. 27, 28, and 29). It appears that for large inputs, the height of the graph of P gets close to 2.50.

Figure 27 Enter the functions; Y_2 is the variable cost per disk

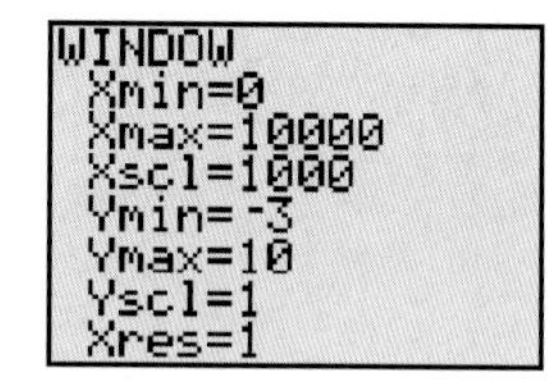

Figure 28 Set up the window to allow for large n

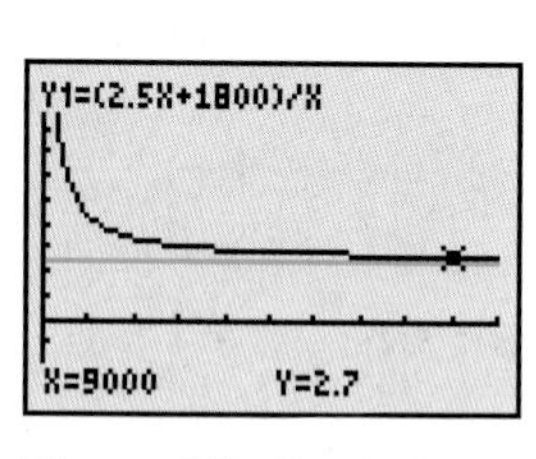

Figure 29 Graph the functions

This means that if the band could sell tens of thousands of CDs, it could price them at a few cents above the \$2.50 cost and still break even, which is probably unrealistic. ■

Example 2 Modeling the Mean of a Quantity

Table 12 Bottled Water Consumption

Year	Bottled Water Consumption (millions of gallons)
1990	2200
1992	2400
1994	3000
1996	3500
1998	4100
2000	4900
2002	6000
2003	6400

Source: *Beverage Marketing Corporation*

In Example 9 of Section 7.5, we modeled the U.S. annual consumption of bottled water. A reasonable model is $B(t) = 16t^2 + 116t + 2185$, where $B(t)$ is U.S. bottled water consumption (in millions of gallons) in the year that is t years since 1990 (see Table 12).

In Exercise 9 of Homework 7.7, you modeled the U.S. population. A reasonable model is $U(t) = 0.0068t^2 + 2.58t + 251.7$, where $U(t)$ is the U.S. population (in millions) at t years since 1990.

1. Let $M(t)$ be the annual mean consumption of bottled water per person (in gallons per person) in the year that is t years since 1990. Find an equation of M.
2. Perform a unit analysis of the equation of M.
3. Find $M(20)$. What does it mean in this situation?
4. Find t when $M(t) = 40$. What does it mean in this situation?

Solution

1. The annual mean consumption of bottled water per person is equal to the total annual consumption divided by the U.S. population:

$$M(t) = \frac{B(t)}{U(t)} = \frac{16t^2 + 116t + 2185}{0.0068t^2 + 2.58t + 251.7}$$

2. Here is a unit analysis of the equation of M:

$$\text{gallons per person} \rightarrow M(t) = \frac{B(t)}{U(t)} = \frac{16t^2 + 116t + 2185}{0.0068t^2 + 2.58t + 251.7} \begin{array}{l} \leftarrow \text{millions of gallons} \\ \leftarrow \text{millions of people} \end{array}$$

The units on both sides of the equation are gallons per person, which suggests that the equation is correct.

3. $$M(20) = \frac{16(20)^2 + 116(20) + 2185}{0.0068(20)^2 + 2.58(20) + 251.7} \approx 35.64$$

The annual mean consumption of bottled water will be 35.64 gallons per person in 2010, according to the model.

4. We substitute 40 for $M(t)$ in the equation of M and solve for t:

$$
\begin{aligned}
40 &= \frac{16t^2 + 116t + 2185}{0.0068t^2 + 2.58t + 251.7} \\
(0.0068t^2 + 2.58t + 251.7) \cdot 40 &= (0.0068t^2 + 2.58t + 251.7) \cdot \frac{16t^2 + 116t + 2185}{0.0068t^2 + 2.58t + 251.7} \\
0.272t^2 + 103.2t + 10{,}068 &= 16t^2 + 116t + 2185 \\
0 &= 15.728t^2 + 12.8t - 7883
\end{aligned}
$$

Now we substitute $a = 15.728, b = 12.8$, and $c = -7883$ in the quadratic formula:

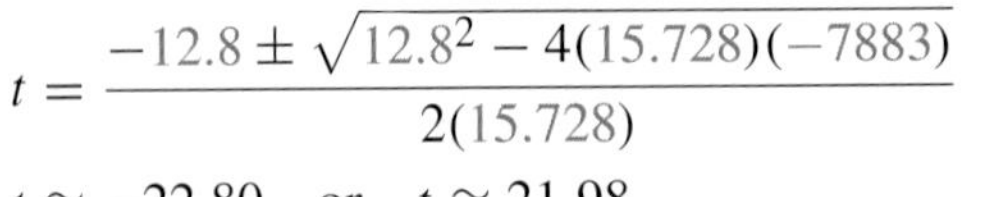

$$t = \frac{-12.8 \pm \sqrt{12.8^2 - 4(15.728)(-7883)}}{2(15.728)} \quad \text{Substitute into quadratic formula.}$$

$$t \approx -22.80 \quad \text{or} \quad t \approx 21.98 \quad \text{Compute.}$$

We verify the results by entering $y = \dfrac{16t^2 + 116t + 2185}{0.0068t^2 + 2.58t + 251.7}$ in a graphing calculator and checking that the inputs -22.80 and 21.98 lead to outputs of about 40 (see Fig. 30).

Figure 30 Verify the work

The inputs -22.80 and 21.98 represent the years 1967 and 2012, respectively. The 1967 estimate indicates model breakdown: Research would show that the annual mean consumption of bottled water in 1967 was much less than 40 gallons per person. Therefore, we predict that it will be 2012 when the annual mean consumption of bottled water reaches 40 gallons per person. ■

Modeling the Percentage of a Quantity

In Sections 8.1 and 8.5, we modeled the percentage of a quantity. In Example 3, we will first find a sum function that will help us find another function that describes the percentage of a quantity.

Example 3 Modeling the Percentage of a Quantity

The numbers of broadband cable subscribers and DSL subscribers in the United States are shown in Table 13 for various years. Let $B(t)$ be the number of broadband cable subscribers and $D(t)$ be the number of DSL subscribers, both in millions, at t years since 1990.

1. Find equations of B and D.
2. Find an equation of the sum function $B + D$. What do the inputs and outputs of $B + D$ mean in this situation?
3. In Exercise 30 of Homework 3.1, you modeled the number of U.S. households. A reasonable model is $H(t) = 1.59t + 90.31$, where $H(t)$ is the total number (in millions) of U.S. households at t years since 1990 (see Table 14). Let $P(t)$ be the percentage of U.S. households that are broadband cable or DSL subscribers. Find an equation of P. Assume that no household subscribes to both services.
4. Predict when 90% of U.S. households will be either broadband cable or DSL subscribers.

Table 13 Numbers of Broadband Cable and DSL Subscribers

	Number of Subscribers (millions)	
Year	**Broadband Cable**	**DSL**
2001	7.0	4.1
2002	11.3	6.4
2003	15.7	9.0
2004	20.0	13.5
2005	24.3	18.6

Source: *Leichtman Research Group*

Table 14 Numbers of Households

Year	Number of Households (millions)
1995	99.0
1997	101.0
1999	103.6
2001	108.2
2003	111.3

Sources: *U.S. Census Bureau*

Solution

1. Scattergrams of the data suggest that we use a linear function to model the number of broadband cable subscribers and a quadratic model to model the number of DSL subscribers (see Figs. 31 and 32). The linear regression equation of B and the quadratic regression equation of D are

$$B(t) = 4.33t - 40.63$$
$$D(t) = 0.536t^2 - 10.32t + 52.85$$

Figure 31 Broadband cable scattergram

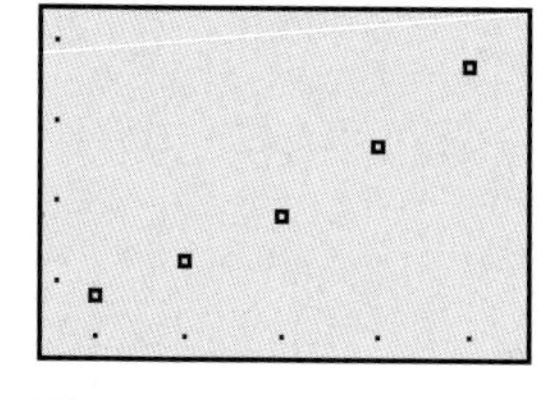

Figure 32 DSL scattergram

The models fit the points in the scattergrams of the two data sets quite well (see Figs. 33 and 34).

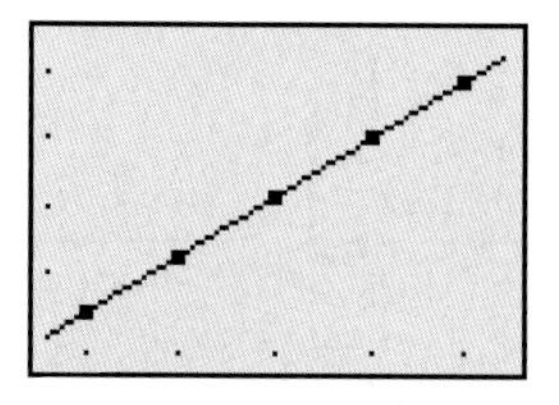

Figure 33 Broadband cable model

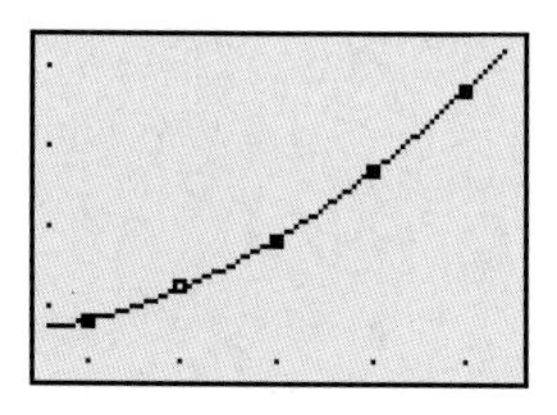

Figure 34 DSL model

2. $(B + D)(t) = B(t) + D(t)$ Definition of sum function

$= (4.33t - 40.63) + \left(0.536t^2 - 10.32t + 52.85\right)$ Substitute.

$= 0.536t^2 - 5.99t + 12.22$ Combine like terms.

The inputs of $B + D$ are the number of years since 1990, and the outputs are the total number of subscribers (in millions).

3. To find the percentage of households that subscribe, we divide the total number of subscribers by the number of households and multiply the result by 100:

$$P(t) = \frac{(B + D)(t)}{H(t)} \cdot 100 \qquad \text{Percent formula: } p = \frac{m}{n} \cdot 100$$

$$= \frac{0.536t^2 - 5.99t + 12.22}{1.59t + 90.31} \cdot \frac{100}{1} \qquad \text{Substitute } 0.536t^2 - 5.99t + 12.22 \text{ for } (B + D)(t) \text{ and } 1.59t + 90.31 \text{ for } H(t).$$

$$= \frac{53.6t^2 - 599t + 1222}{1.59t + 90.31} \qquad \text{Multiply numerators; multiply denominators.}$$

4. We substitute 90 for $P(t)$ in the equation of P:

$$90 = \frac{53.6t^2 - 599t + 1222}{1.59t + 90.31}$$

$$(1.59t + 90.31) \cdot 90 = (1.59t + 90.31) \cdot \frac{53.6t^2 - 599t + 1222}{1.59t + 90.31}$$

$$143.1t + 8127.9 = 53.6t^2 - 599t + 1222$$

$$0 = 53.6t^2 - 742.1t - 6905.9$$

Now we substitute $a = 53.6$, $b = -742.1$, and $c = -6905.9$ in the quadratic formula:

$$t = \frac{-(-742.1) \pm \sqrt{(-742.1)^2 - 4(53.6)(-6905.9)}}{2(53.6)} \qquad \text{Substitute into quadratic formula.}$$

$$t \approx -6.37 \quad \text{or} \quad t \approx 20.22 \qquad \text{Compute.}$$

We verify the results by entering $y = \dfrac{53.6t^2 - 599t + 1222}{1.59t + 90.31}$ in a graphing calculator and checking that the inputs -6.37 and 20.22 lead to outputs of about 90 (see Fig. 35).

The inputs -6.37 and 20.22 represent the years 1984 and 2010, respectively. The 1984 estimate indicates model breakdown: There were no broadband cable or DSL subscribers that year. Therefore, we predict that it will be 2010 when 90% of households will be broadband cable or DSL subscribers. ■

Figure 35 Verify the work

Distance-Speed-Time Applications

How do we model the distance traveled by an object moving at a constant speed? For instance, if a car is driven at 50 mph for 2 hours, it will travel 50 miles in the first hour and 50 miles in the second hour, for a total distance of $50 \cdot 2 = 100$ miles. If a car is driven at 60 miles per hour for 3 hours, it will travel $60 \cdot 3 = 180$ miles.

In general, the (constant) speed of the car multiplied by the amount of time the car is in motion gives the distance traveled.

Distance-Speed-Time Relationship

If an object is moving at a constant speed s for an amount of time t, then the distance traveled d is given by

$$d = st$$

and the time t is given by

$$t = \frac{d}{s}$$

Example 4 Distance-Speed-Time Relationship

A person plans to drive a steady 55 mph on an 80-mile trip. Compute the driving time.

Solution

Since the person is traveling at a constant speed, we use the equation

$$t = \frac{d}{s}$$

We substitute 80 for d and 55 for s in the equation:

$$t = \frac{80}{55} \approx 1.45$$

So, the driving time will be about 1.5 hours. ■

Example 5 Modeling Driving Time

A student at Seattle Central Community College plans to drive from Seattle, Washington, to Eugene, Oregon. The speed limit is 70 mph in Washington and 65 mph in Oregon. She will drive 164 miles in Washington, then 121 miles in Oregon.

1. If the student drives steadily at the speed limits, compute the driving time.
2. If the student exceeds the speed limits, let $T(a)$ be the driving time (in hours) at a mph above the speed limits. Find an equation of T.
3. Find $T(0)$. Compare this result with the result in Problem 1.
4. If the student drives 5 mph over the speed limits, compute the driving time.
5. If the student wants the driving time to be 4 hours, how much over the speed limits would she have to drive?

Solution

1. Since the student drives at a constant speed in Washington, we can use the equation $t = \frac{d}{s}$ to compute the time (in hours) spent driving in Washington:

$$t = \frac{\text{distance in Washington}}{\text{speed in Washington}} = \frac{164}{70}$$

We can also compute the time (in hours) spent driving in Oregon:

$$t = \frac{\text{distance in Oregon}}{\text{speed in Oregon}} = \frac{121}{65}$$

The total driving time is the sum of our two computed times:

$$\frac{164}{70} + \frac{121}{65} \approx 4.2 \text{ hours}$$

2. If the student drives, say, 5 mph over the speed limits, then she will drive $5+70=75$ mph in Washington and $5+65=70$ mph in Oregon. If she drives a miles per hour over the speed limits, she will drive $(a+70)$ mph in Washington and $(a+65)$ mph in Oregon. We use these expressions for speeds to find an equation of T:

$$\begin{aligned} T(a) &= \frac{\text{distance in Washington}}{\text{speed in Washington}} + \frac{\text{distance in Oregon}}{\text{speed in Oregon}} \\ &= \frac{164}{a+70} + \frac{121}{a+65} \end{aligned}$$

3. $T(0) = \dfrac{164}{0+70} + \dfrac{121}{0+65} \approx 4.2$

   The driving time will be about 4.2 hours if the student drives at the speed limits. We found the same result in Problem 1.
4. If the student drives 5 mph over the speed limits, then $a = 5$:

$$T(5) = \frac{164}{5+70} + \frac{121}{5+65} \approx 3.9$$

   The driving time will be about 3.9 hours.
5. If the trip is to take 4 hours, then $T(a) = 4$:

$$\begin{aligned} 4 &= \frac{164}{a+70} + \frac{121}{a+65} \\ (a+65)(a+70)\cdot 4 &= (a+65)(a+70)\cdot\left(\frac{164}{a+70} + \frac{121}{a+65}\right) \\ (a^2+135a+4550)\cdot 4 &= (a+65)(a+70)\cdot\frac{164}{a+70} + (a+65)(a+70)\cdot\frac{121}{a+65} \\ 4a^2+540a+18{,}200 &= (a+65)\cdot 164 + (a+70)\cdot 121 \\ 4a^2+540a+18{,}200 &= 164a + 10{,}660 + 121a + 8470 \\ 4a^2+540a+18{,}200 &= 285a + 19{,}130 \\ 4a^2+255a-930 &= 0 \\ a &= \frac{-255 \pm \sqrt{255^2 - 4(4)(-930)}}{2(4)} \\ a \approx -67.2 \quad &\text{or} \quad a \approx 3.5 \end{aligned}$$

   The value $a = -67.2$ represents driving under the speed limits by 67.2 mph—model breakdown has occurred. So, the student would have to drive 3.5 mph over the speed limits for the driving time to be 4 hours. ■

group exploration

Looking ahead: Inverse variation

Suppose that you intend to drive 100 miles. Let $f(s)$ be the time (in hours) it will take you to drive 100 miles if you drive at a constant speed of s miles per hour.

1. Find an equation of f.
2. Find $f(50)$, $f(55)$, $f(60)$, and $f(70)$. What do your results mean in this situation?

3. Consider completing the 100-mile trip several times, each time at a higher constant speed than the last. What happens to the travel time as the speed gets extremely high? Use a graphing calculator table and graph to verify your answer. (Recall from Section 4.3 that when a function behaves like this, we say that the horizontal axis is a horizontal asymptote of the function.)
4. Consider completing the 100-mile trip several times, each time at a *lower* constant speed than the last. What happens to the travel time as the speed gets extremely close to 0? Use a graphing calculator table and graph to verify your answer.
5. Does the function f have a vertical asymptote? If so, what is it? How does your answer relate to Problem 4?

HOMEWORK 8.6

FOR EXTRA HELP ▶ Student Solutions Manual | PH Math/Tutor Center | MathXL® | MyMathLab

1. The ski club at a community college plans to spend \$1250 to charter a bus for a ski trip. The cost will be split evenly among the students who sign up for the trip. Each student will also pay \$350 for food, lodging, and ski lift tickets.

a. Let $C(n)$ be the total cost (in dollars) for n students going on the trip. Find an equation of C.

b. Let $M(n)$ be the mean cost per student (in dollars per student) if n students go on the trip. Find an equation of M.

c. What is the mean cost per student if 30 students go?

d. The ski trip will be cancelled unless the mean cost per student is \$400 or less. What is the minimum number of students needed to go on the trip?

2. Manhattan is one of the most densely populated regions in the United States. This borough consists of 660,700,000 square feet of land. Let $f(P)$ be the number of square feet of Manhattan each resident would own if Manhattan were divided equally among its P residents.

a. Find an equation of f.

b. Use f to estimate the amount of land per Manhattanite if 1.6 million people live in Manhattan.

c. If the land in the United States were divided equally among U.S. residents, each resident would own about 437,000 square feet of land. How many people could live in Manhattan if each resident owned 437,000 square feet of Manhattan? Compare your answer with the number of people given in part (b).

3. For a 5-year high school reunion, graduates rent out a restaurant that charges a flat fee of \$500 for a band, plus \$50 per person for food and two drinks.

a. Let $T(n)$ be the total cost (in dollars) for n people to attend the reunion. Find an equation of T.

b. Let $M(n)$ be the mean cost per person (in dollars per person) if n people attend the reunion. Find an equation of M.

c. Find $M(270)$. What does it mean in this situation?

d. Find n when $M(n) = 60$. What does it mean in this situation?

e. Complete Table 15 by using a graphing calculator table.

f. Describe $M(n)$ for large n. Explain why your observation makes sense in this situation.

Table 15 Mean Cost per Person for the Reunion

Number of People n	Mean Cost per Person $M(n)$
100	
200	
300	
400	
500	

4. A student agrees to throw a party, provided that each guest shares the expenses equally with him. A four-person local band will play for \$200 and free drinks and snacks. The student estimates that the mean cost of drinks and snacks will be \$3 per person.

a. Suppose that n people (including the host and band members) attend the party. Let $C(n)$ be the party's total cost (in dollars). Find an equation of C.

b. Let $P(n)$ be the equal share of expenses (in dollars) each guest and host contributes. Find an equation of P. [**Hint:** Recall that band members get free drinks and snacks.]

c. If the host and guests are willing to pay at most \$5 each, how many guests must attend to cover expenses?

d. What do the values of $P(n)$ get close to as the values of n get very large? What does the result mean in this situation?

e. If the host and guests are willing to pay only \$2 each, how many guests must attend to cover expenses? Explain why this makes sense.

5. Leasing costs, equipment maintenance, salaries, electricity, and marketing cost a car manufacturer an average of \$90,000 per day to produce a certain type of car. Materials, invoices, and shipping cost the manufacturer an average of \$7000 per car.

a. Suppose that the car manufacturer produces and sells n cars of that type per day. Let $C(n)$ be the total daily cost (in dollars). Find a formula of C.

b. Suppose that the car manufacturer produces and sells n cars of that type per day. Let $B(n)$ be the amount (in dollars) the manufacturer should charge per car to break even by selling n cars. Find an equation of B.

c. Suppose that the car manufacturer produces and sells n cars of that type per day. Let $P(n)$ be the amount (in dollars) the manufacturer should charge per car to make a profit of $2000 per car. Find a formula of P. [**Hint:** Build on your equation from part (b) to find an equation of P.]

d. Find $P(40)$. What does it mean in this situation?

e. What do the values of $P(n)$ get close to as the values of n get very large? What does that mean in this situation?

6. In Example 1 of this section, we found the rational function

$$P = f(n) = \frac{2.50n + 1800}{n}$$

where $f(n)$ models the price (in dollars) the band Melted Zipper should set for its CD to break even by making and selling n CDs. [*Note:* We use different notation than in Example 1. Here, we use f for the name of the function and P for the name of the dependent variable.]

a. If Melted Zipper sets the CD's price at $7, how many CDs must be made and sold for the band to break even?

b. Find an equation of f^{-1}. [**Hint:** Perform steps similar to those in part (a), but do not substitute a value for P. After a couple of steps, factor out n on one side of the equation.]

c. Find $f^{-1}(7)$. Compare this result with that in part (a).

7. The total income of all U.S. households and the numbers of U.S. households are shown in Table 16 for various years.

Table 16 Total Income of All Households; Numbers of Households

Year	Total Income of All Households (billions of dollars)	Number of Households (billions)
1995	3374	0.0990
1997	3738	0.1010
1999	4216	0.1036
2001	4569	0.1082
2003	4821	0.1113

Source: *U.S. Census Bureau*

a. Let $I(t)$ be the total income (in billions of dollars) of all households in the year that is t years since 1990. Find an equation of I.

b. In Exercise 30 of Homework 3.1, you modeled the number of U.S. households. A reasonable model is $H(t) = 0.0016t + 0.09$, where $H(t)$ is the number (in billions) of U.S. households at t years since 1990. Let $M(t)$ be the mean income per household (in dollars) in the year that is t years since 1990. Find an equation of M.

c. Perform a unit analysis of your equation of M.

d. Predict in which year the mean income per household will be $50,000.

e. Use the window settings in Fig. 36 to graph M. Is M increasing, decreasing, or neither for values of t between 5 and 20? What does that mean in this situation?

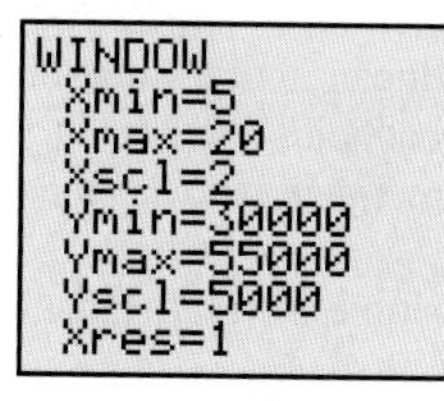

Figure 36 Window settings for Exercise 7e

8. Fuel consumption (in millions of gallons) by vehicles in the United States is listed in Table 17 for various years.

Table 17 Vehicle Fuel Consumption

Year	Amount of Fuel Consumed (millions of gallons)
1975	109,000
1980	115,000
1985	121,300
1990	130,800
1995	143,800
2000	162,500
2003	169,600

Source: *Federal Highway Administration*

a. Let $F(t)$ be the total amount of fuel (in millions of gallons) used during the year that is t years since 1970. Find an equation of F.

b. In Exercise 9 of Homework 7.7, you modeled the U.S. population. A good model is $U(t) = 0.0068t^2 + 2.31t + 202.83$, where $U(t)$ is the U.S. population (in millions) at t years since 1970. Let $M(t)$ be the mean amount of fuel used per person (in gallons per person) during the year that is t years since 1970. Find an equation of M.

c. Find t when $M(t) = 675$. What does it mean in this situation?

d. Use the window settings in Fig. 37 to graph M. From 1975 to the present, when was the mean amount of fuel used per person the lowest during the time period? Has that amount increased or decreased since then?

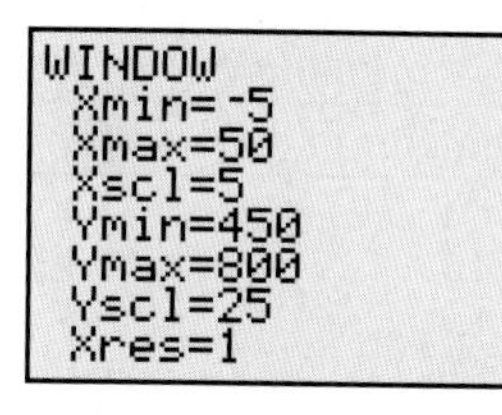

Figure 37 Per-person fuel consumption

e. In Exercise 49 of Homework 7.2, you modeled the average price of gasoline. A reasonable quadratic model is $P(t) = 0.0075t^2 - 0.35t + 5.42$, where $P(t)$ is the price (in 2006 dollars per gallon) at t years since 1970 (see Table 18). Find $(P \cdot M)(40)$. What does it mean in this situation?

Table 18 Average Prices of Gasoline

Year	Price (2006 dollars per gallon)
1980	2.59
1985	1.90
1990	1.51
1995	1.28
2001	1.66
2005	2.28
2006*	2.68

Source: *Department of Energy*
*January–May

f. Is $M(t)$ an underestimate or an overestimate of the mean amount of fuel used per *driver*? Explain.

9. Book sales (including textbooks) at college stores continue to increase (see Table 19).

Table 19 Book Sales at College Stores

Year	Book Sales (millions of dollars)
2000	4265
2001	4571
2002	4899
2003	5086
2004	5479
2005	5703

Source: *Book Industry Study Group, Inc.*

a. Let $B(t)$ be the book sales (in millions of dollars) at college stores during the year that is t years since 1980. Find an equation of B.

b. In Exercise 29 of Homework 3.1, you modeled U.S. college enrollments (in millions) for women and men, $W(t)$ and $M(t)$, respectively, by the system

$$W(t) = 0.15t + 5.76$$
$$M(t) = 0.072t + 5.43$$

where t is the number of years since 1980 (see Table 20). Let $E(t)$ be the total college enrollment (in millions) at t years since 1980. Find an equation of E.

Table 20 College Enrollments

Year	Enrollment (millions) Women	Men
1985	6.4	5.8
1990	7.5	6.3
1995	7.9	6.3
2000	8.6	6.7
2003	9.3	7.3

Source: *U.S. Census Bureau*

c. Let $A(t)$ be the mean amount of money spent on books per student (in dollars per student) at college stores during the year that is t years since 1980. Find an equation of A.

d. Predict the mean amount of money per student that will be spent on books in 2012.

e. During which year will the mean amount from part (d) be $400?

10. a. In Exercise 29 of Homework 3.1, U.S. college enrollments (in thousands) for women and men, $W(t)$ and $M(t)$, respectively, can be modeled by the system

$$W(t) = 149t + 4273$$
$$M(t) = 72t + 4708$$

where t is the number of years since 1970 (see Table 20). Let $E(t)$ be the total college enrollment (in thousands) at t years since 1970. Find an equation of E.

b. In Exercise 20 of the Chapter 3 Test, the numbers (in thousands) of two-year colleges and four-year colleges and universities (postsecondary institutions) $f(t)$ and $g(t)$, respectively, can be modeled by the system

$$f(t) = 0.028t + 0.89$$
$$g(t) = 0.021t + 1.69$$

where t is the number of years since 1970 (see Table 21).

Table 21 Number of Postsecondary Institutions

Year	Number of Postsecondary Institutions (thousands) 2-Year	4-Year
1970	0.886	1.639
1980	1.195	1.957
1990	1.408	2.127
2000	1.721	2.363
2003	1.844	2.324

Source: *National Center for Education Statistics*

Let $N(t)$ be the total number (in thousands) of postsecondary institutions at t years since 1970. Find an equation of N.

c. Let $M(t)$ be the mean enrollment per postsecondary institution at t years since 1970. Find an equation of M.

d. Predict when the mean enrollment will be 3950 students.

e. Use the window settings shown in Fig. 38 to graph M. Has the mean enrollment increased or decreased since 1980?

```
WINDOW
 Xmin=-5
 Xmax=50
 Xscl=5
 Ymin=3300
 Ymax=4100
 Yscl=100
 Xres=1
```

Figure 38 Window settings for Exercise 10e

11. The numbers of U.S. women and men who earned a bachelor's degree are listed in Table 22 for various years.

Table 22 Women and Men Who Earned a Bachelor's Degree

Year	Number of People Who Earned a Bachelor's Degree (thousands) Women	Men
1980	456	474
1985	497	483
1990	560	492
1995	634	526
2000	708	530
2002	742	550

Source: *National Center for Education Statistics*

a. Let $W(t)$ be the number of women and $M(t)$ be the number of men, both in thousands, who earned a bachelor's degree in the year that is t years since 1980. Find equations of W and M.

b. Find an equation of the sum function $W + M$. What do the inputs and outputs of $W + M$ mean in this situation?

c. Among people who earned a bachelor's degree at t years since 1980, let $P(t)$ be the percentage who are men. Find an equation of P.

d. Predict in which year 40% of people who earn a bachelor's degree will be men.

e. Use the window settings shown in Fig. 39 to graph P. Is P increasing, decreasing, or neither for values of t between 0 and 40? How is this possible, given that the number of bachelor's degrees earned by men has been increasing since 1980?

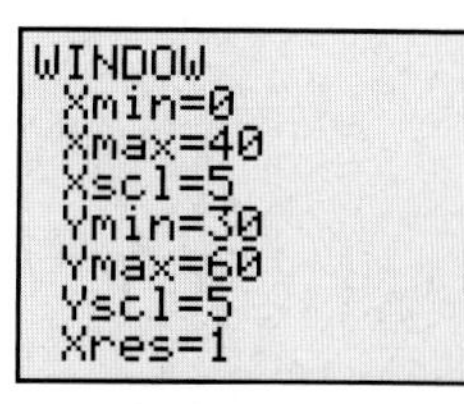

Figure 39 Window settings for Exercise 11e

12. The numbers of women and men who live alone are shown in Table 23 for various years.

Table 23 Numbers of Women and Men Who Live Alone

Year	Number Living Alone (millions)	
	Women	Men
1980	11.3	7.0
1985	12.7	7.9
1990	14.0	9.0
1995	14.6	10.1
2000	15.6	11.2
2004	17.0	12.6

Source: *U.S. Census Bureau*

a. Let $W(t)$ be the number of women who live alone and $M(t)$ be the number of men who live alone, both in millions, at t years since 1980. Find equations of W and M.
b. Find an equation of the sum function $W + M$. What do the inputs and outputs of $W + M$ mean in this situation?
c. Let $P(t)$ be the percentage of people living alone who are women at t years since 1980. Find an equation of P.
d. Predict in which year 57% of people who live alone will be women.
e. Use the window settings shown in Fig. 40 to graph P. Is P increasing, decreasing, or neither for values of t between 0 and 40? How is this possible, given that the number of women who live alone has been increasing since 1980?

WINDOW
Xmin=0
Xmax=40
Xscl=5
Ymin=54
Ymax=64
Yscl=2
Xres=1

Figure 40 Window settings for Exercise 12e

13. The numbers of morning daily newspapers and evening daily newspapers are shown in Table 24 for various years.
a. Let $M(t)$ be the number of morning daily newspapers and $E(t)$ be the number of evening daily newspapers, both at t years since 1980. Find equations of M and E.
b. Find an equation of the sum function $M + E$. What do the inputs and outputs of $M + E$ mean in this situation?
c. Let $P(t)$ be the percentage of newspapers that are morning dailies at t years since 1980. Find an equation of P.
d. Find $P(30)$. What does it mean in this situation?
e. Find t when $P(t) = 75$. What does it mean in this situation?

Table 24 Numbers of Morning Dailies and Evening Dailies

Year	Number of Daily Newspapers	
	Morning	Evening
1980	387	1388
1985	482	1220
1990	559	1084
1995	656	891
2000	766	727
2004	813	653

Source: *Editor & Publisher Co.*

14. In Exercise 59 of the Cumulative Review of Chapters 1–7, you found the model $C(t) = 0.14t^2 + 0.65t + 7.69$, where $C(t)$ is the total sales (in billions of dollars) at coffeehouses and doughnut shops in the year that is t years since 2000. These sales and the total sales at other limited-service restaurants are shown in Table 25 for various years.

Table 25 Total Sales at Coffeehouses and Doughnut Shops and at Other Limited-Service Restaurants

Year	Total Sales (billions of dollars)	
	Coffeehouses and Doughnut Shops	Other Limited-Service Restaurants
2000	7.7	120.4
2001	8.5	124.8
2002	9.5	129.3
2003	10.9	136.8
2004	12.6	150.4
2005	14.4	

Sources: *Technomic/Mintel; U.S. Census Bureau*

a. Let $F(t)$ be the total sales (in billions of dollars) at limited-service restaurants other than coffeehouses and doughnut shops in the year that is t years since 2000. Find an equation of F.
b. Find an equation of the sum function $C + F$. What do the inputs and outputs of $C + F$ mean in this situation?
c. Let $P(t)$ be the percentage of sales at limited-service restaurants that are from coffeehouses or doughnut shops at t years since 1980. Find an equation of P.
d. Predict when 9% of sales at limited-service restaurants will be from coffeehouses and doughnut shops.

15. A person drives 60 mph for 85 miles. Compute the driving time.

16. A person drives at a constant speed for 100 miles in 1.7 hours. What is that speed?

17. A student plans to drive at night from Albuquerque Technical Vocational Institute in Albuquerque, New Mexico, to Dallas, Texas. He will drive 253 miles in New Mexico, then 410 miles in Texas. The speed limit is 75 mph in New Mexico and 65 mph (at night) in Texas.
a. Let $T(a)$ be the driving time (in hours) if the student drives a mph above the speed limits. Find an equation of T.
b. If he drives 3 mph over the speed limits, compute the driving time.
c. By how much would he have to exceed the speed limits for the driving time to be 9 hours? Verify your answer by using a graphing calculator table.

18. A student plans to drive from Oklahoma City, Oklahoma, to Little Rock, Arkansas. She will drive 183 miles in Oklahoma, then 161 miles in Arkansas. The speed limit is 75 mph in Oklahoma and 70 mph in Arkansas.

a. Let $T(a)$ be the driving time (in hours) if the student drives a mph above the speed limits. Find an equation of T.

b. If she drives 5 mph over the speed limits, compute the driving time.

c. By how much would she have to exceed the speed limits for the driving time to be 4 hours? Verify your answer by using a graphing calculator table.

19. A student plans to drive from Ivy Tech Community College in Indianapolis, Indiana, to the University of Illinois in Champaign–Urbana, Illinois. She will drive 83 miles in Indiana, then 37 miles in Illinois. The speed limit is 70 mph in Indiana and 65 mph in Illinois.

a. Let $T(a)$ be the driving time (in hours) if the student drives a mph above the speed limits. Find an equation of T.

b. Find $T(0)$ and $T(10)$. What do they mean in this situation?

c. Find $T(0) - T(10)$. What does it mean in this situation?

d. Find a when $T(a) = 1.6$. What does it mean in this situation?

20. A student plans to drive from North Carolina State University in Raleigh, North Carolina, to the University of Richmond in Richmond, Virginia. He will drive 95 miles in North Carolina, then 78 miles in Virginia. The speed limit is 70 mph in North Carolina and 65 mph in Virginia.

a. Let $T(a)$ be the driving time (in hours) if the student drives a mph above the speed limits. Find an equation of T.

b. Find $T(0)$ and $T(10)$. What do they mean in this situation?

c. Find $T(0) - T(10)$. What does it mean in this situation?

d. Find a when $T(a) = 2.4$. What does it mean in this situation?

21. In Example 5 of this section, we found the equation

$$T(a) = \frac{164}{a+70} + \frac{121}{a+65}$$

where $T(a)$ is the driving time (in hours) for a student to drive from Seattle to Eugene, Oregon, at a mph above the speed limits.

a. Perform the addition on the right side of the equation of T.

b. Use your result in part (a) to find the driving time if the student drives 10 mph over the speed limits.

22. Describe how to use two or more models to form a function that models the mean of a quantity. Compare this process with using two or more models to form a function that models the percentage of a quantity. For both types of modeling, give an example that is different from those in this textbook.

Related Review

23. The risk of having shingles (a painful rash related to chicken pox) increases with age (see Table 26). Let $f(t)$ be the percentage of Americans who have shingles at age t years.

a. Find a linear equation, an exponential equation, and a quadratic equation of f. Compare how well the models fit the data.

Table 26 Percentages of Americans Who Have Shingles

Age (years)	Percent
10	0.5
20	1.3
30	2.7
40	4.8
50	7.5
60	11.9
70	19.7
80	31.8
90	46.1

Source: *J. C. Donahue et al.,* Archives of Internal Medicine, *1995*

b. For ages less than 23 years, explain why there is model breakdown for both the linear model and the quadratic model. How well does the exponential model fit the data for these ages?

c. Use the exponential model to estimate at what age 25% of Americans have shingles.

d. What is the base b of the exponential model $f(t) = ab^t$? What does it mean in this situation?

e. A 2005 study showed that a new vaccine can reduce the number of shingles cases by 51%. If the vaccine is approved and all 85-year-old Americans receive it, use the exponential model to predict the percentage of 85-year-old Americans who will have shingles.

24. World land speed records are shown in Table 27 for various years.

Table 27 Land Speed Records

Year	Record (mph)
1904	91
1914	124
1927	175
1935	301
1947	394
1960	407
1970	622
1983	633
1997	763

Source: *Fédération International de l'Automobile*

Let $r = f(t)$ be the land speed record (in miles per hour) at t years since 1900.

a. Use a graphing calculator to draw a scattergram of the data. Which type of function would best model the data—linear, exponential, or quadratic? Explain.

b. Find an equation of a linear model to describe the data.

c. What is the slope? What does it mean in this situation?

d. The speed of sound is about 748 mph. Use the model you found in part (b) to estimate when a car first broke the sound barrier (that is, exceeded 748 mph). For which year(s) shown in Table 27 did a car actually break the barrier?

e. Find an equation of f^{-1}.

f. Find $f^{-1}(1000)$. What does it mean in this situation?

Expressions, Equations, Functions, and Graphs

Perform the indicated instruction. Then use words such as linear, quadratic, cubic, exponential, logarithmic, rational, polynomial, degree, function, one variable, *and* two variables *to describe the expression, equation, or system.*

25. Factor $75x^3 - 50x^2 - 12x + 8$.

26. Solve $2x^2 + 12x + 13 = 0$.

27. Solve $75x^3 - 50x^2 = 12x - 8$.

28. Find the vertex of $f(x) = 2x^2 + 12x + 13$.

29. Find the product $(x^2 + 2x - 3)(3x^2 - x - 4)$.

30. Graph $f(x) = 2x^2 + 12x + 13$ by hand.

8.7 VARIATION

"Creativity varies inversely with the number of cooks involved in the broth."

—Bernice Fitz-Gibbon

Objectives

- Know the meaning of *direct variation* and *inverse variation.*
- In direct variation and inverse variation, know how a change in the independent variable affects the value of the dependent variable.
- Use a single point to find a *direct variation equation* or an *inverse variation equation.*
- Use direct variation models and inverse variation models to make estimates.
- Use ratios to find a direct variation constant, and use products to find an inverse variation constant.

In this section, we will discuss two simple, yet important, types of equations in two variables: *direct variation equations* and *inverse variation equations.*

Direct Variation

In Example 1, we model a situation by using a direct variation equation.

Example 1 Using a Direct Variation Equation to Model Data

A golf ball is dropped from various heights, and the bounce height is recorded each time (see Table 28). Let B be the bounce height (in inches) of the golf ball after it was dropped from an initial height of d inches. Find an equation of a function that models the situation well.

Table 28 Drop and Bounce Heights of a Golf Ball

Drop Height (inches)	Bounce Height (inches)
6	4.8
12	10.0
18	15.0
24	20.3
30	26.4
36	31.0
42	37.5
48	44.5
54	47.3
60	52.0

Source: *J. Lehmann*

Solution

We begin by drawing a scattergram of the data (see Fig. 41). It appears that the variables d and B are approximately linearly related. The linear regression equation is

$$B = 0.90d - 0.77$$

Although this model fits the data very well (see Fig. 42), it estimates that very small drop heights will have *negative* bounce heights (see Fig. 43). Model breakdown has

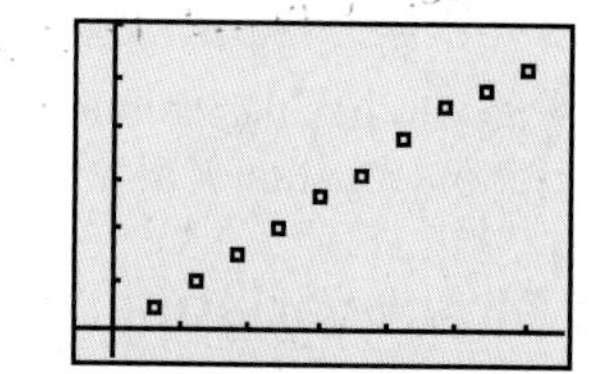

Figure 41 A scattergram of the data

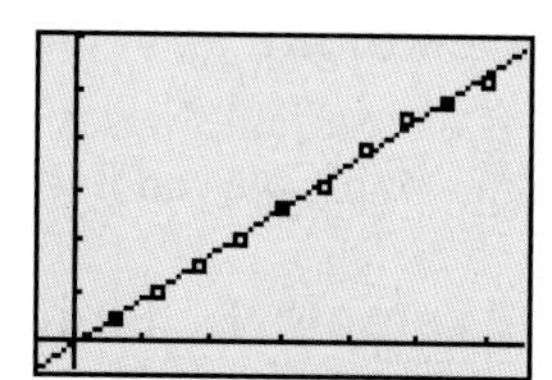

Figure 42 Regression line fits the data well

X	Y1
0	-.77
.1	-.68
.2	-.59
.3	-.5
.4	-.41
.5	-.32
.6	-.23

X=0

Figure 43 Regression line estimates negative bounce heights

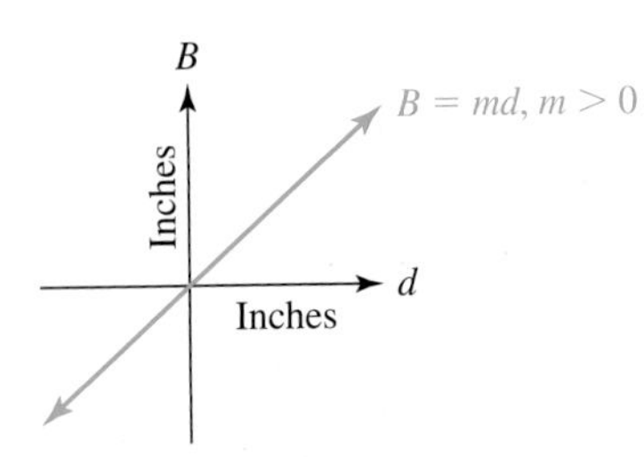

Figure 44 The graph of $B = md$ with $m > 0$

occurred. Also, as the drop heights get closer to 0 inches, the bounce heights should get close to 0 inches. However, the model estimates that as the drop heights get close to 0, the bounce heights will get close to -0.77 inch.

It would be better to find a model of the form $B = md$, where $m > 0$ (see Fig. 44). Such a model estimates a *positive* bounce height for any drop height, including small ones. It also estimates that as drop heights get close to 0 inches, the bounce heights will get close to 0 inches.

If we imagine the line that contains the origin (0, 0) and the data point (30, 26.4), it appears that the line might come close to the other data points (see Fig. 45). To find a slope m for the model $B = md$, we substitute the coordinates of the data point (30, 26.4) into the equation $B = md$:

$$26.4 = m(30) \quad \text{Substitute 30 for } d \text{ and 26.4 for } B.$$

$$\frac{26.4}{30} = m$$

$$m = 0.88$$

The equation is $B = 0.88d$. The model fits the data quite well (see Fig. 45). ■

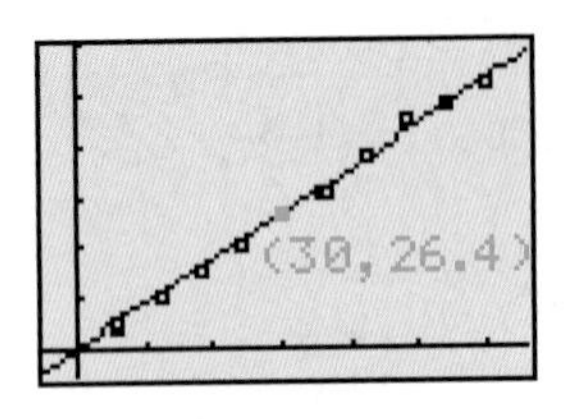

Figure 45 A line that contains the origin and the data point (30, 26.4)

In Example 1, we found the model $B = 0.88d$. This equation is an example of a direct variation equation.

DEFINITION Direct variation

If $y = kx$ for some nonzero constant k, we say that **y varies directly as x** or that **y is proportional to x.** We call k the **variation constant** or the **constant of proportionality.** The equation $y = kx$ is called a **direct variation equation.**

For $y = 5x$, we say that y varies directly as x with variation constant 5. For $p = 2t$, we say that p varies directly as t with variation constant 2.

CHANGES IN VALUES FOR DIRECT VARIATION

A direct variation equation $y = kx$ describes a linear function with slope k and y-intercept (0, 0). In Fig. 46, we sketch three such functions where $k = \frac{1}{2}$, $k = 1$, and $k = 2$.

If k is positive (positive slope), then $y = kx$ is an increasing function. So, if the value of x increases, the value of y increases.

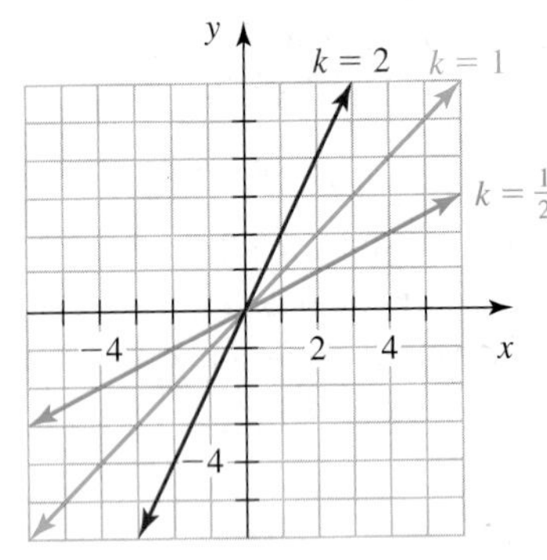

Figure 46 Graphs of $y = kx$, where $k = \frac{1}{2}$, $k = 1$, and $k = 2$

Changes in Values of Variables for Direct Variation

Assume that y varies directly as x with some positive variation constant k:

- If the value of x increases, then the value of y increases.
- If the value of x decreases, then the value of y decreases.

For example, consider the golf ball model $B = 0.88d$. As drop heights increase, so do the bounce heights. As drop heights decrease, so do the bounce heights.

USING ONE POINT TO FIND A DIRECT VARIATION EQUATION

If one variable varies directly as another variable and we know one point that lies on the graph, we can find the variation constant k as well as the direct variation equation.

Example 2 Finding a Direct Variation Equation

The variable y varies directly as x with positive variation constant k.

1. What happens to the value of y as the value of x increases?
2. If $y = 5$ when $x = 3$, find an equation for x and y.

Solution

1. Since y varies directly as x with positive variation constant k, the value of y must increase as the value of x increases.
2. The equation is of the form $y = kx$. We find the constant k by substituting 3 for x and 5 for y:

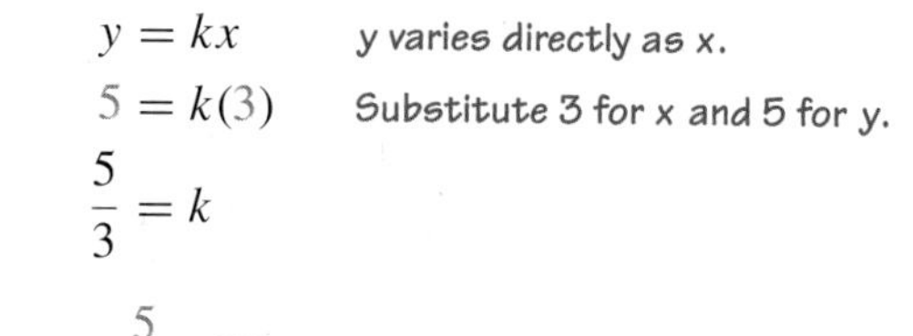

$$y = kx \qquad y \text{ varies directly as } x.$$
$$5 = k(3) \qquad \text{Substitute 3 for } x \text{ and 5 for } y.$$
$$\frac{5}{3} = k$$

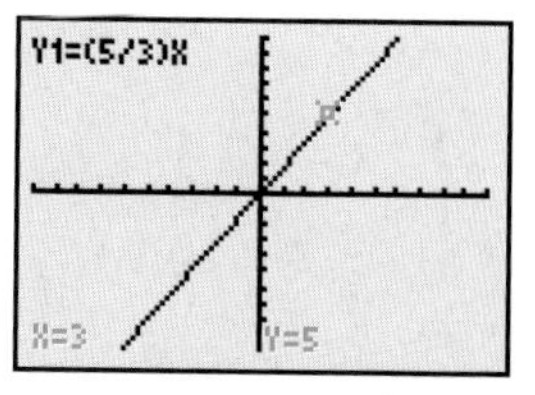

Figure 47 Verify the work

The equation is $y = \frac{5}{3}x$. We use a graphing calculator graph to check that the curve $y = \frac{5}{3}x$ contains the point (3, 5). See Fig. 47. ■

Example 3 Using a Direct Variation Equation

The variable w varies directly as t. If $w = 7$ when $t = 2$, find the value of t when $w = 9$.

Solution

Since w varies directly as t, we can describe the relationship between t and w by the equation $w = kt$. We find the constant k by substituting 2 for t and 7 for w:

$$w = kt \qquad w \text{ varies directly as } t.$$
$$7 = k(2) \qquad \text{Substitute 2 for } t \text{ and 7 for } w.$$
$$\frac{7}{2} = k$$

The equation is $w = \frac{7}{2}t$. We find the value of t when $w = 9$ by substituting 9 for w in the equation $w = \frac{7}{2}t$ and solving for t:

$$9 = \frac{7}{2}t \qquad \text{Substitute 9 for } w.$$
$$18 = 7t \qquad \text{Multiply both sides by LCD, 2.}$$
$$\frac{18}{7} = t$$

So, $t = \frac{18}{7}$ when $w = 9$. ■

In Example 1, we used the function $B = 0.88d$ to model bounce heights of a golf ball. In Exercise 46, you will show that the function $B = 0.80d$ reasonably models bounce heights of a racquetball. Since B varies directly as d for a golf ball and a racquetball, it is a reasonable conjecture that B varies directly as d for another type of ball similar in construction, such as another golf ball, another racquetball, or perhaps even a tennis ball.

Example 4 Finding a Direct Variation Model

Assume that the bounce height B (in inches) of a tennis ball varies directly as the drop height d (in inches). The bounce height of the tennis ball is 20 inches when the ball is dropped from an initial height of 30 inches.

1. Find an equation of B and d.
2. Estimate the bounce height if the drop height is 50 inches.

Solution

1. The equation is of the form $B = kd$. We substitute 30 for d and 20 for B to find the constant k:

$$20 = k(30) \quad \text{Substitute 30 for } d \text{ and 20 for } B.$$
$$\frac{20}{30} = k$$
$$k \approx 0.67$$

The equation is $B = 0.67d$.

2. We substitute 50 for d in the equation $B = 0.67d$:

$$B = 0.67(50) \quad \text{Substitute 50 for } d.$$
$$B = 33.5$$

The bounce height will be 33.5 inches, according to the model. ■

Finding and Using a Direct Variation Model

Assume that a quantity p varies directly as a quantity t. To make estimates about an authentic situation,

1. Substitute the values of a data point into the equation $p = kt$; then solve for k.
2. Substitute the value of k into the equation $p = kt$.
3. Use the equation from step 2 to make estimates of quantity t or p.

In Example 4, we did not run an experiment and perform the usual modeling steps to find the tennis ball model $B = 0.67d$. Instead, we assumed that the model was of the form $B = kd$ because that is the form of our models of the golf ball and the racquetball. The tennis ball model may or may not be accurate. The only way to know for sure is to run an experiment.

Next, we take a closer look at the significance of $k = 0.88$ for the golf ball model $B = 0.88d$. This equation $B = 0.88d$ tells us that the bounce height is equal to 88% of the drop height. We list the values of k and their meanings for the three balls we have investigated:

Type of Ball	*Model*	*Value of k*	*Height of Bounce*
Golf ball	$B = 0.88d$	0.88	88% of drop height
Racquetball	$B = 0.79d$	0.80	80% of drop height
Tennis ball	$B = 0.67d$	0.67	67% of drop height

The value of k takes into account how "bouncy" the ball is.

A VARIABLE VARYING DIRECTLY AS AN EXPRESSION

So far we have described functions in which the dependent variable varies directly as the independent *variable*. Now, for nonzero constant k, we list some examples of functions in which the dependent variable varies directly as an *expression*:

$$y = kx^2 \quad y \text{ varies directly as } x^2.$$
$$p = k\sqrt{t} \quad p \text{ varies directly as } \sqrt{t}.$$
$$F = k\log(r) \quad F \text{ varies directly as } \log(r).$$

So, we use "varies directly" to mean that the dependent variable is equal to a constant times an expression containing the independent variable.

Inverse Variation

In Sections 8.1 and 8.5, we worked with rational equations in two variables. Now we will focus on a very simple type of rational equation in two variables: an inverse variation equation.

> **DEFINITION** Inverse variation
>
> If $y = \frac{k}{x}$ for some nonzero constant k, we say that **y varies inversely as x** or that **y is inversely proportional to x.** We call k the **variation constant** or the **constant of proportionality.** The equation $y = \frac{k}{x}$ is called an **inverse variation equation.**

For $y = \frac{5}{x}$, we say that y varies inversely as x with variation constant 5. For $B = \frac{8}{v}$, we say that B varies inversely as v with variation constant 8.

CHANGES IN VALUES FOR INVERSE VARIATION

In Fig. 48, we graph three equations of the form $y = \frac{k}{x}$ where $k = 1$, $k = 2$, and $k = 4$.

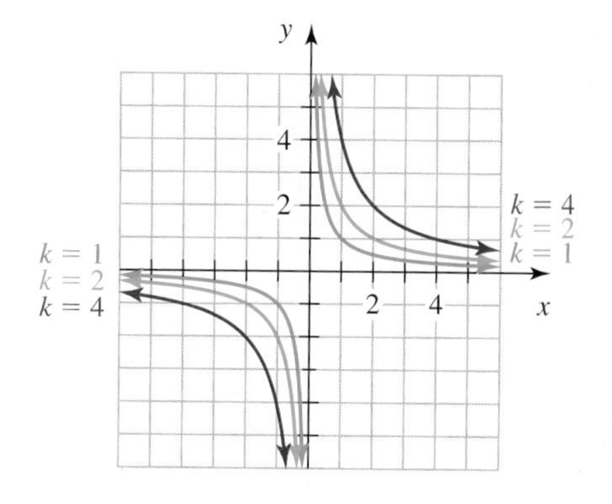

Figure 48 Graphs of $y = \frac{k}{x}$, where $k = 1$, $k = 2$, and $k = 4$

If k is positive, then the function $y = \frac{k}{x}$ is decreasing for positive values of x. So, if the value of x increases, the value of y decreases.

> **Changes in Values of Variables for Inverse Variation**
>
> Assume that y varies inversely as x with some positive variation constant k. For positive values of x:
>
> - If the value of x increases, then the value of y decreases.
> - If the value of x decreases, then the value of y increases.

USE ONE POINT TO FIND AN INVERSE VARIATION EQUATION

If one variable varies inversely as another variable and we know one point that lies on the graph, we can find the variation constant k as well as the inverse variation equation.

Example 5 Finding an Inverse Variation Equation

The variable y varies inversely as x with the positive variation constant k.

1. For positive values of x, what happens to the value of y as the value of x increases?
2. If $y = 4$ when $x = 2$, find an equation for x and y.

Solution

1. Since y varies inversely as x with positive variation constant k and the values of x are positive, the value of y must decrease as the value of x increases.

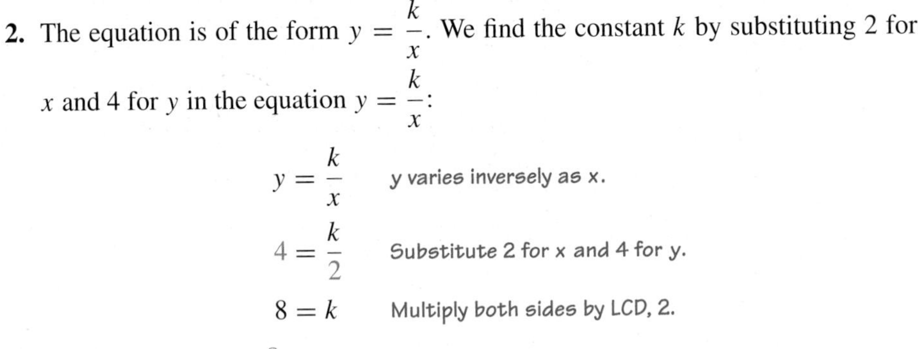

2. The equation is of the form $y = \frac{k}{x}$. We find the constant k by substituting 2 for x and 4 for y in the equation $y = \frac{k}{x}$:

$$y = \frac{k}{x} \quad \text{y varies inversely as x.}$$

$$4 = \frac{k}{2} \quad \text{Substitute 2 for x and 4 for y.}$$

$$8 = k \quad \text{Multiply both sides by LCD, 2.}$$

The equation is $y = \frac{8}{x}$. We use a graphing calculator graph to check that the curve $y = \frac{8}{x}$ contains the point (2, 4). See Fig. 49. ■

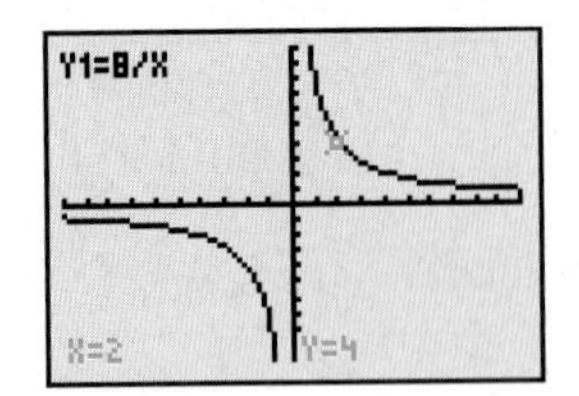

Figure 49 Verify the work

Example 6 Using an Inverse Variation Equation

The variable G varies inversely as r. If $G = 8$ when $r = 3$, find the value of r when $G = 4$.

Solution

Since G varies inversely as r, we can describe the relationship between r and G by the equation $G = \frac{k}{r}$. We find the constant k by substituting 3 for r and 8 for G:

$$G = \frac{k}{r} \quad \text{G varies inversely as r.}$$

$$8 = \frac{k}{3} \quad \text{Substitute 3 for r and 8 for G.}$$

$$24 = k \quad \text{Multiply both sides by LCD, 3.}$$

The equation is $G = \frac{24}{r}$. We find the value of r when $G = 4$ by substituting 4 for G in the equation $G = \frac{24}{r}$ and then solving for r:

$$4 = \frac{24}{r} \quad \text{Substitute 4 for G.}$$

$$4r = 24 \quad \text{Multiply both sides by LCD, r.}$$

$$r = 6$$

So, $r = 6$ when $G = 4$. ■

Table 29 Volumes and Pressures in a Syringe

Volume (cm^3)	Pressure (atm)
3	2.23
4	1.76
5	1.46
6	1.23
7	1.05
8	0.93
9	0.83
10	0.74
11	0.67
12	0.60
13	0.56
14	0.52
15	0.48
16	0.44
17	0.42
18	0.39
19	0.37
20	0.35

Source: *J. Lehmann*

Example 7 Finding an Inverse Variation Model

The more you squeeze a sealed syringe filled with air, the harder it gets to squeeze it further. Some air volumes in cubic centimeters (cm^3) and corresponding pressures in atmospheres (atm) in a sealed syringe are given in Table 29.

Let P be the pressure (in atm) in the syringe at air volume V (in cm^3).

1. Find an equation for V and P.
2. As the value of V increases, what happens to the value of P, according to the model? What does that pattern mean in this situation?
3. Estimate at what volume the pressure will be 5 atm.

Solution

1. We draw a scattergram of the data in Fig. 50. Since the points suggest a curve that "bends," we will not use a linear function to model the data. To decide among an

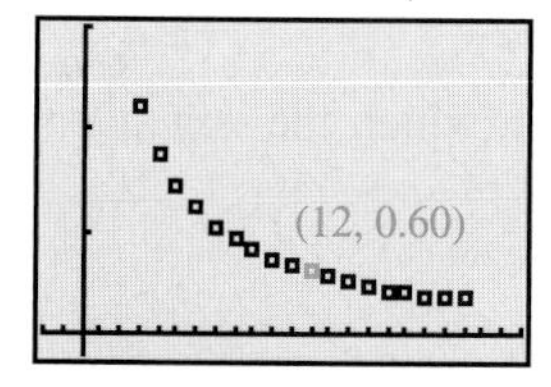

Figure 50 Scattergram of the data

exponential function, a quadratic function, and an inverse variation function, we find equations for each.

First, we find an inverse variation equation. If we imagine an inverse variation curve that contains the data point (12, 0.60), it appears that the curve might come close to the other data points (see Fig. 50). To find its constant k, we substitute the data point (12, 0.60) in the equation $P = \frac{k}{V}$:

$$0.60 = \frac{k}{12} \quad \text{Substitute 12 for } V \text{ and 0.60 for } P.$$

$$7.2 = k \quad \text{Multiply both sides by LCD, 12.}$$

The variation equation is $P = \frac{7.2}{V}$. Next, we use a graphing calculator to find the exponential regression equation and the quadratic regression equation:

$$P = 2.32(0.90)^V \quad \text{Exponential regression equation}$$

$$P = 0.0086V^2 - 0.29V + 2.77 \quad \text{Quadratic regression equation}$$

$$P = \frac{7.2}{V} \quad \text{Inverse variation equation}$$

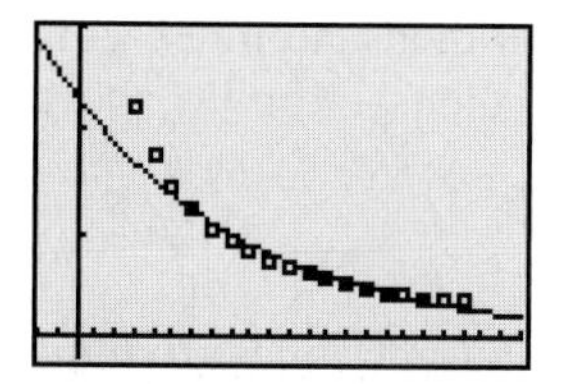

Figure 51 Exponential regression model

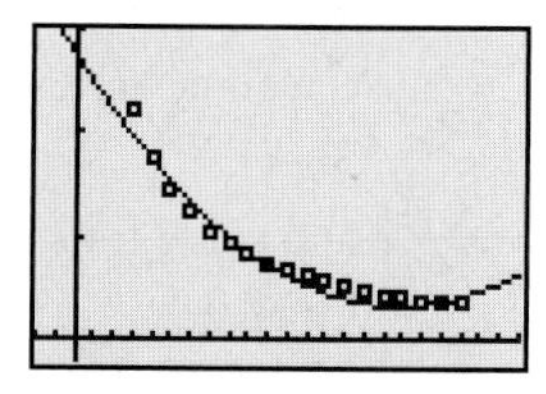

Figure 52 Quadratic regression model

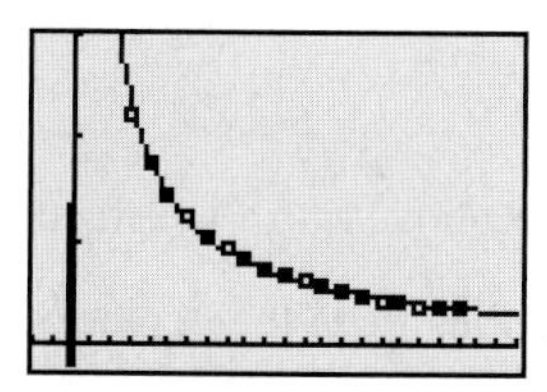

Figure 53 Inverse variation model

Now we see how well each model fits the data (see Figs. 51, 52, and 53). It appears that the inverse variation model $P = \frac{7.2}{V}$ fits the data better than the other two models do.

2. The variable P varies inversely as V, according to our model. So, as the value of V increases, the value of P decreases. The larger the volume of air, the less the pressure will be.
3. We substitute 5 for P in the equation $P = \frac{7.2}{V}$ and solve for V:

$$5 = \frac{7.2}{V} \quad \text{Substitute 5 for } P.$$

$$5V = 7.2 \quad \text{Multiply both sides by LCD, } V.$$

$$V = 1.44 \quad \text{Divide both sides by 5.}$$

The pressure is 5 atm when the air volume is 1.44 cubic centimeters. ■

The equation $P = \frac{k}{V}$ is an excellent model for many situations, as long as the temperature and number of molecules in the container are constant. This equation is a form of **Boyle's law.** The constant k takes into account the temperature and the number of molecules.

Finding and Using an Inverse Variation Model

Assume that a quantity p varies inversely as a quantity t. To make estimates about an authentic situation,

1. Substitute the values of a data point into the equation $p = \frac{k}{t}$; then solve for k.
2. Substitute the value of k into the equation $p = \frac{k}{t}$.
3. Use the equation from step 2 to make estimates of quantity t or p.

A VARIABLE VARYING INVERSELY AS AN EXPRESSION

For nonzero constant k, we list some examples of functions in which the dependent variable varies inversely as an *expression*:

$$y = \frac{k}{x^3} \qquad y \text{ varies inversely as } x^3.$$

$$A = \frac{k}{r-5} \qquad A \text{ varies inversely as } r-5.$$

$$w = \frac{k}{2^t} \qquad w \text{ varies inversely as } 2^t.$$

So, we use "varies inversely" to mean that the dependent variable is equal to a constant divided by an expression containing the independent variable.

Example 8 Finding an Inverse Variation Model

The weight of an object varies inversely as the square of its distance from the center of Earth. An astronaut weighs 180 pounds at sea level (about 4 thousand miles from Earth's center). How much does the astronaut weigh when 2 thousand miles above Earth's surface?

Solution

We let w be the weight (in pounds) of the astronaut when d thousand miles from the center of Earth. Our desired model has the form

$$w = \frac{k}{d^2}$$

The denominator of the fraction on the right-hand side is d^2, because weight varies inversely as the *square* of distance d.

Next, we substitute 4 for d and 180 for w in the equation $w = \frac{k}{d^2}$ and solve for k:

$$180 = \frac{k}{4^2} \qquad \text{Substitute 4 for } d \text{ and 180 for } w.$$

$$180 = \frac{k}{16} \qquad 4^2 = 16$$

$$k = 2880 \qquad \text{Multiply both sides by LCD, 16.}$$

The model is $w = \frac{2880}{d^2}$. When the astronaut is 2 thousand miles above Earth's surface, we see from Fig. 54 that the astronaut is $4 + 2 = 6$ thousand miles from the center of Earth. To find w, we substitute 6 for d in the equation $w = \frac{2880}{d^2}$:

$$w = \frac{2880}{6^2} = 80$$

The astronaut weighs 80 pounds when 2 thousand miles above Earth's surface. ■

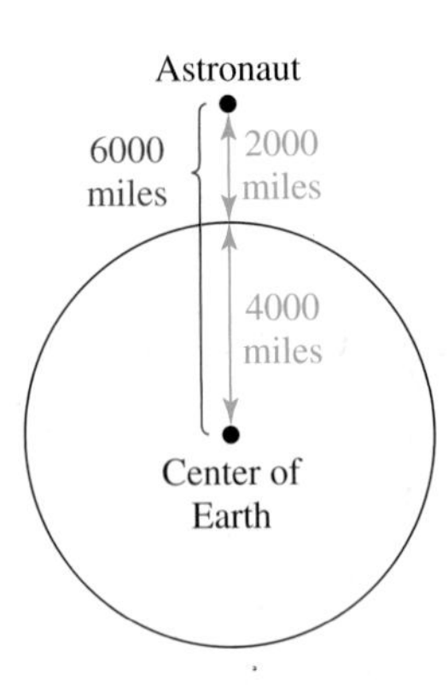

Figure 54 The astronaut is 6000 miles from Earth's center

Using Ratios and Products to Find Variation Constants

First, we will discuss another way to find a direct variation constant. Then, we will discuss another way to find an inverse variation constant.

USING RATIOS TO FIND A DIRECT VARIATION CONSTANT

In Example 9, we use ratios to find a direct variation constant.

Example 9 Finding a Direct Variation Constant with Ratios

Use ratios of bounce heights and drop heights to find a golf ball model of the data shown in Example 1.

Solution

From our work in Example 1, we know that we want an equation of the form $B = kd$. To find k, we begin by solving the equation $B = kd$ for k:

$$k = \frac{B}{d}$$

Next, we use the first two columns of Table 30 to calculate the ratios $\frac{B}{d}$ for the third column.

Table 30 Ratios of Drop and Bounce Heights of a Golf Ball

Drop Height d (inches)	Bounce Height B (inches)	$\frac{B}{d}$
6	4.8	0.80
12	10.0	0.83
18	15.0	0.83
24	20.3	0.85
30	26.4	0.88
36	31.0	0.86
42	37.5	0.89
48	44.5	0.93
54	47.3	0.88
60	52.0	0.87

According to our model $B = kd$, the ratios $\frac{B}{d}$ should be equal to a constant k. The reason the ratios vary a bit is most likely errors in measuring the drop and bounce heights. Or there might be imperfections in the golf ball or the surface of the floor.

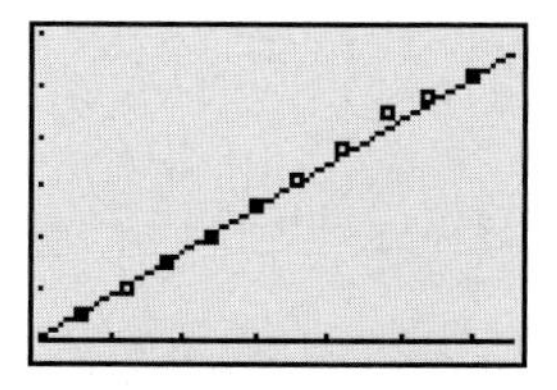

Figure 55 The model $B = 0.86d$

Next, we find the mean of the ratios $\frac{B}{d}$ in the third column:

$$\frac{0.80 + 0.83 + 0.83 + 0.85 + 0.88 + 0.86 + 0.89 + 0.93 + 0.88 + 0.87}{10} \approx 0.86$$

Finally, we substitute 0.86 for k in the equation $B = kd$:

$$B = 0.86d$$

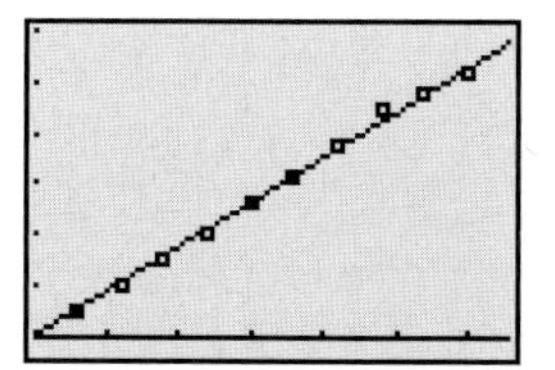

Figure 56 The model $B = 0.88d$

The model fits the data reasonably well (see Fig. 55). However, it appears that the model we found in Example 1 fits the data slightly better (see Fig. 56). ■

We now have two ways to find a direct variation model. The method shown in Example 1 requires few calculations, but finding a good value of k depends on a good selection of an ordered pair. The method shown in Example 9 requires more calculations, but it is not necessary to select an ordered pair to find k.

USING PRODUCTS TO FIND AN INVERSE VARIATION CONSTANT

In Example 10, we use products to find an inverse variation constant.

Example 10 Finding an Inverse Variation Constant with Products

Use products of volumes and pressures to find a syringe model for the data shown in Example 7.

Solution

From our work in Example 7, we know that we want an equation of the form $P = \frac{k}{V}$. To find k, we begin by solving the equation $P = \frac{k}{V}$ for k:

$$k = VP$$

Next, we use the values of V and P in the first two columns and fourth and fifth columns of Table 31 to calculate the products VP for the third and sixth columns, respectively.

Table 31 Products of Volumes and Pressures in a Syringe

Volume V (cm^3)	Pressure P (atm)	VP	Volume V (cm^3)	Pressure P (atm)	VP
3	2.23	6.69	12	0.60	7.20
4	1.76	7.04	13	0.56	7.28
5	1.46	7.30	14	0.52	7.28
6	1.23	7.38	15	0.48	7.20
7	1.05	7.35	16	0.44	7.04
8	0.93	7.44	17	0.42	7.14
9	0.83	7.47	18	0.39	7.02
10	0.74	7.40	19	0.37	7.03
11	0.67	7.37	20	0.35	7.00

According to our model $P = \frac{k}{V}$, the products VP should be equal to a constant k. The reason the products vary a bit is most likely due to errors in measuring the volumes and pressures. Or there might be imperfections in the syringe.

Next, we find the mean of the products PV in the third and sixth columns:

$$\frac{6.69 + 7.04 + 7.30 + 7.38 + 7.35 + \cdots + 7.00}{18} \approx 7.20$$

Finally, we substitute 7.20 for k in the equation $P = \frac{k}{V}$:

$$P = \frac{7.20}{V}$$

This is the same equation we found in Example 7. ■

We now have two ways to find an inverse variation model. The method shown in Example 7 requires few calculations, but finding a good value of k depends on a good selection of an ordered pair. The method shown in Example 10 requires more calculations, but it is not necessary to select an ordered pair to find k.

group exploration

Looking ahead: Simplifying radical expressions

Recall from Section 4.2 that $\sqrt[n]{a} = a^{1/n}$. So,

$$\sqrt[5]{a^3} = \left(a^3\right)^{1/5} = a^{3 \cdot \frac{1}{5}} = a^{3/5}$$

We say that $\sqrt[5]{a^3}$ is in *radical form* and the expression $a^{3/5}$ is in *exponential form*.

1. Write $\sqrt[7]{a^4}$ in exponential form.
2. Write $\sqrt[9]{a^2}$ in exponential form.
3. Write $\sqrt{a^7}$ in exponential form. [**Hint:** $\sqrt{a}$ is shorthand for $\sqrt[2]{a}$.]
4. Write $\sqrt[n]{a^m}$ in exponential form.
5. Write each of the following in exponential form and simplify.
   a. $\sqrt{a^2}$, $\sqrt{a^4}$, $\sqrt{a^6}$, $\sqrt{a^8}$
   b. $\sqrt[3]{a^3}$, $\sqrt[3]{a^6}$, $\sqrt[3]{a^9}$, $\sqrt[3]{a^{12}}$
6. Simplify $\sqrt{x^7}$. [**Hint:** $x^7 = x^6 \cdot x$]
7. Simplify $\sqrt{x^{13}}$.

TIPS FOR SUCCESS: Plan for the Final Exam

Don't wait until the last minute to begin studying for your final exam. Look at your finals schedule and decide how you will allocate your time to prepare for each final.

It is important that you are well rested so that you can fully concentrate during your final exam. Plan to do some fun activities that involve exercise—a great way to neutralize stress.

HOMEWORK 8.7

FOR EXTRA HELP ▶ Student Solutions Manual | PH Math/Tutor Center | MathXL® | MyMathLab

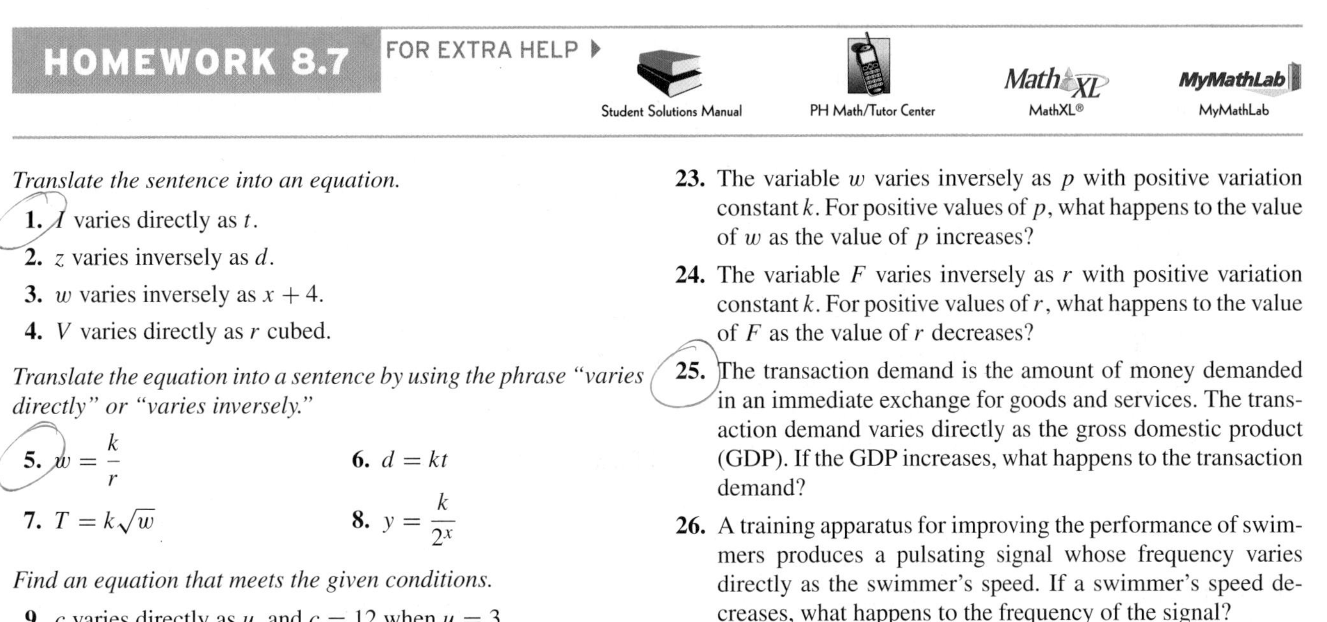

Translate the sentence into an equation.

1. I varies directly as t.
2. z varies inversely as d.
3. w varies inversely as $x + 4$.
4. V varies directly as r cubed.

Translate the equation into a sentence by using the phrase "varies directly" or "varies inversely."

5. $w = \frac{k}{r}$
6. $d = kt$
7. $T = k\sqrt{w}$
8. $y = \frac{k}{2^x}$

Find an equation that meets the given conditions.

9. c varies directly as u, and $c = 12$ when $u = 3$.
10. p varies inversely as d, and $p = 3$ when $d = 5$.
11. w varies inversely as $\sqrt{t}$, and $w = 3$ when $t = 16$.
12. A varies directly as r^2, and $A = 4\pi$ when $r = 2$.

For Exercises 13–20, find the requested value of the variable.

13. If y varies directly as x, and $y = 12$ when $x = 4$, find y when $x = 9$.
14. If p varies directly as t, and $p = 18$ when $t = 3$, find p when $t = 5$.
15. If G varies inversely as r, and $G = 8$ when $r = 3$, find G when $r = 4$.
16. If W varies inversely as u, and $W = 15$ when $u = 2$, find W when $u = 6$.
17. If p varies directly as x^2, and $p = 6$ when $x = 2$, find x when $p = 24$.
18. If y varies directly as w^2, and $y = 6$ when $w = 3$, find w when $y = 54$.
19. If I varies inversely as $r + 2$, and $I = 9$ when $r = 3$, find r when $I = 7$.
20. If D varies inversely as $t + 3$, and $D = 5$ when $t = 4$, find t when $D = 9$.
21. The variable B varies directly as w with positive variation constant k. Describe what happens to the value of B as the value of w increases.
22. The variable y varies directly as t with positive variation constant k. Describe what happens to the value of y as the value of t decreases.
23. The variable w varies inversely as p with positive variation constant k. For positive values of p, what happens to the value of w as the value of p increases?
24. The variable F varies inversely as r with positive variation constant k. For positive values of r, what happens to the value of F as the value of r decreases?
25. The transaction demand is the amount of money demanded in an immediate exchange for goods and services. The transaction demand varies directly as the gross domestic product (GDP). If the GDP increases, what happens to the transaction demand?
26. A training apparatus for improving the performance of swimmers produces a pulsating signal whose frequency varies directly as the swimmer's speed. If a swimmer's speed decreases, what happens to the frequency of the signal?
27. Nerve conduction in muscles varies inversely as a person's height. Will a tall person have more or less nerve conduction than a short person?
28. The frequency of a vibrating guitar string varies inversely with the diameter of the string. Will a large-diameter string have a higher or lower frequency than a small-diameter string?
29. Bernice Fitz-Gibbon said, "Creativity varies inversely with the number of cooks involved in the broth." What did she mean?
30. Craig Bruce said, "Time is a resource whose supply is inversely proportional to its demand." What did he mean?
31. For a Montgomery County (Maryland) resident, the cost of tuition at Montgomery College varies directly as the number of credit hours taken. For academic year 2005–2006, the cost of 15 credit hours was \$1395. What did 12 credit hours cost?
32. For a Jackson County (Michigan) resident, the cost of tuition at Jackson Community College varies directly as the number of billing contact hours taken. For academic year 2006–2007, the cost of 13 billing contact hours was \$1020.50. What did 15 billing contact hours cost?
33. When a stone is tied to a string and whirled in a circle at constant speed, the tension in the string varies inversely as the radius of the circle. If the radius is 60 centimeters, the tension is 80 newtons. Find the tension if the radius is 50 centimeters.
34. The current flowing in an electrical circuit at a constant potential varies inversely as the resistance of the circuit. If the current is 25 amperes when the resistance is 4 ohms, what is the current when the resistance is 5 ohms?

35. The distance that an object falls varies directly as the square of the time the object is in motion. If an object falls for 3 seconds, it will fall 144.9 feet. To estimate the height of a cliff, a person drops a stone at the edge of the cliff and measures how long it takes the stone to reach the base. If it takes 3.4 seconds, what is the height of the cliff?

36. A car is traveling at speed s (in mph) on a dry asphalt road, and the brakes are suddenly applied. The stopping distance d (in feet) varies directly as the square of the speed s. If a car traveling at 60 mph can stop in 120 feet, what is the stopping distance of a car traveling at 70 mph?

37. The intensity of the radiation used to treat a tumor varies inversely as the square of the distance from the machine that emits the radiation. If the intensity is 90 milliroentgens per hour (mr/hr) at a distance of 2.5 meters, at what distance is the intensity 45 mr/hr?

38. The volume of a sphere varies directly as the cube of its radius. A sphere with a 3-foot radius has a volume of 36π cubic feet. How much air is required to fill a beach ball with radius 1.6 feet?

39. The force F (in pounds) required to push a sofa across a floor varies directly as the weight w (in pounds) of the sofa.

a. A person can push a 120-pound sofa across a wood floor by pushing with a force of 50 pounds. Find an equation that describes the relationship between w and F.

b. How much force is required to push a 150-pound sofa across a wood floor?

c. How would an equation of a model for a *carpeted* floor compare with the model you found in part (a)? In particular, how would the variation constants compare?

40. The distance d (in miles) that a student travels by car varies directly as the travel time t (in hours). The student travels 93 miles in 1.5 hours.

a. Is the student traveling at a constant speed? Explain.

b. Find an equation that describes the relationship between time and distance.

c. How far will the student travel in 2 hours?

41. The time T (in seconds) it takes to hear thunder after you see lightning varies directly as the distance d (in feet) from the lightning if the temperature does not vary much during the storm. In a certain storm, it takes 3 seconds to hear thunder when lightning is seen 3313 feet away.

a. Find an equation that describes the relationship between d and T for the storm.

b. If it takes 4 seconds for you to hear thunder after you see lightning, how far away was the lightning?

c. What does the constant k represent in this situation? Explain. [**Hint:** Recall that slope is the rate of change of the *dependent* variable with respect to the *independent* variable.]

d. There is a rule of thumb that the number of seconds it takes you to hear thunder after you see lightning is equal to the number of miles that you are from the lightning. Is this rule of thumb a good approximation? If yes, explain. If no, find a better rule of thumb. [**Hint:** There are 5280 feet in 1 mile.]

42. The force F (in pounds) you must exert on a wrench handle to loosen a bolt on a bike varies inversely as the length L (in inches) of the handle. A force of 40 pounds is needed when the handle is 6 inches long.

a. Find an equation that describes the relationship between handle length and force.

b. If you use an 8-inch-long wrench to loosen the bolt, how much force do you need to apply to the handle?

c. Is it easier to use a wrench with a short or long handle? Explain in terms of the equation you found in part (a).

43. The weight (in pounds) $w = f(d)$ of an object varies inversely as the square of its distance (in thousands of miles) d from the center of Earth.

a. An astronaut weighs 200 pounds at sea level (about 4 thousand miles from Earth's center). Find an equation of f.

b. How much would the astronaut weigh at 1 thousand miles above Earth's surface?

c. At what distance from the center of Earth would the astronaut weigh 1 pound?

d. Estimate how much the astronaut would weigh on the surface of the Moon. The Moon is a mean distance of about 239 thousand miles from Earth. Has model breakdown occurred? Explain.

e. Without finding an equation, discuss how an equation of a model for a 190-pound astronaut would compare with the equation you found in part (a). Discuss how the variation constants would compare.

44. The intensity (in watts per square meter, W/m^2) $I = f(d)$ of a television signal varies inversely as the square of the distance d (in kilometers) from the transmitter. The intensity of a television signal is 30 W/m^2 at a distance of 2.6 km.

a. Find an equation of f.

b. Find $f(1)$, $f(2)$, $f(3)$, and $f(4)$. What do they mean in this situation?

c. Use a graphing calculator to draw a graph of f. What happens to the value of I as the value of d increases, for $d > 0$? What does that mean in this situation?

d. What can you say about the value of $f(d)$ for an extremely large value of d? Explain. What does that mean in this situation?

45. When a guitarist picks the second-thickest string on a six-string guitar (the "open 'A' string"), the string vibrates at a frequency of 110 hertz—that is, 110 times a second. By firmly pressing the "A" string against the fret board, the guitarist shortens the effective length of the string. As a result, the frequency increases (and so does the pitch of the note). See Fig. 57.

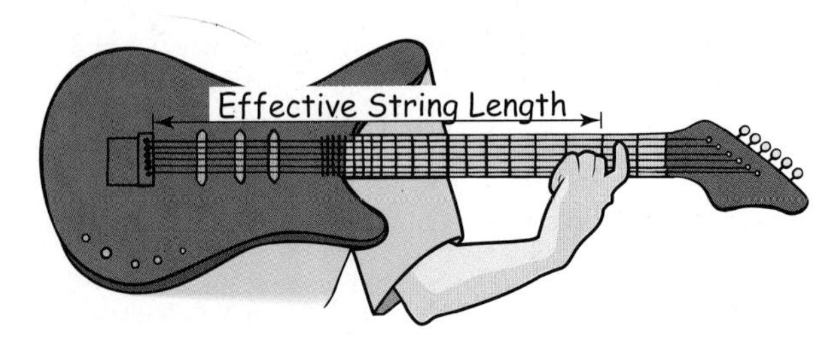

Figure 57 Exercise 45

We use some of the letters of the alphabet, sometimes in conjunction with the "sharp" symbol ♯, to refer to these notes. The frequencies for the open "A" string and the next 12 notes are listed in Table 32.

a. Let F be the frequency (in hertz) of the "A" string when the string's effective length is L inches. Use a graphing calculator to draw a scattergram of the data.

b. Find an equation of a reasonable model to describe the data. Does your model fit the data well?

Table 32 Effective Lengths and Frequencies of 13 Notes on the "A" String

Note	Effective Length of "A" String (inches)	Frequency (hertz)
A	25.50	110.0
A♯	24.07	116.6
B	22.72	123.5
C	21.44	130.8
C♯	20.24	138.6
D	19.10	146.9
D♯	18.03	155.6
E	17.02	164.8
F	16.06	174.6
F♯	15.16	185.0
G	14.31	196.0
G♯	13.51	207.7
A	12.75	220.0

Sources: Math and Music, *Garland and Kahn; J. Lehmann*

c. In a sentence that uses the phrase "varies directly" or "varies inversely," describe how the effective length and frequency of the "A" string are related.

d. When the "A" string is vibrating, what is its frequency if the effective length is 7.58 inches?

e. Use your equation from part (b) to show that if you halve any effective length of the "A" string, the frequency will double. [**Hint:** First, substitute a for L in your model's equation. Then, substitute $\frac{1}{2}a$ for L.]

46. A racquetball is dropped from various heights, and the bounce height is recorded each time (see Table 33).

Table 33 Drop and Bounce Heights of a Racquetball

Drop Height (inches)	Bounce Height (inches)
6	5.0
12	9.3
18	15.0
24	19.6
30	24.0
36	27.6
42	32.8
48	38.0

Source: *J. Lehmann*

Let $B = f(d)$ be the bounce height (in inches) of the racquetball after it is dropped from an initial height of d inches.

a. Use a graphing calculator to draw a scattergram of the data.

b. Find an equation of f.

c. In a sentence using the phrase "varies directly" or "varies inversely," describe how the initial and bounce heights of the racquetball are related.

d. What is the slope of the graph of f? What does it mean in this situation?

e. What is the B-intercept? What does it mean in this situation?

47. As you move away from an object, it appears to decrease in height. To describe this relationship, an algebra professor stood 10 feet from his garage, held a yardstick 1 foot away, and measured the image of his garage. The image of the garage had an *apparent height* of 16.0 inches. He collected apparent heights of the garage at various distances from it (see Table 34).

Table 34 Apparent Heights of a Car Garage

Distance from Garage (feet)	Apparent Height of Garage (inches)
10	16.0
20	7.3
30	4.8
40	3.8
50	3.0
60	2.5
70	2.0

Source: *J. Lehmann*

Let $a = f(d)$ be the apparent height (in inches) of the garage when the professor was d feet from the garage.

a. Use a graphing calculator to draw a scattergram of the data.

b. Find an equation of f.

c. In a sentence using the phrase "varies directly" or "varies inversely," describe how the distance from the garage and the apparent height of the garage are related.

d. Your model f is a decreasing function for $d > 0$. Explain why that makes sense in this situation.

e. Find the apparent height from a distance of 100 feet.

f. Estimate the actual height of the garage. [**Hint:** Think about how the apparent heights were recorded.]

48. A pizza with diameter 12 inches (a "12-inch pizza") and three toppings weighs 36 ounces at Gabriana's Pizza in San Bruno, California. The weight of a three-topping pizza varies directly as the square of the pizza's diameter.

a. Let $W(d)$ be the weight (in ounces) of a three-topping pizza with diameter d inches. Find an equation of W.

b. In a sentence using the phrase "varies directly" or "varies inversely," describe how the diameter and price of the pizza are related.

c. Table 35 lists the prices for various sizes of three-topping pizzas at Gabriana's Pizza. Let $P(d)$ be the price (in dollars) of a three-topping pizza with diameter d inches. Find an equation of P.

Table 35 Prices of Pizzas

Diameter (inches)	Price (dollars)
12	12.20
14	14.80
16	17.40
18	20.00

Source: *Gabriana's Pizza*

d. Let $C(d)$ be the cost per ounce (in dollars per ounce) of a three-topping pizza with diameter d inches. Find an equation of C.

e. Use a graphing calculator to draw a graph of C. Use "maximum" on a graphing calculator to find the highest point of the curve. What does that point represent in this situation?

f. Is C increasing or decreasing for $d > 6$? What does that mean in this situation?

49. A *pendulum* is an object hanging from a lightweight cord. A pendulum can be made by tying one end of some thread to a weight—say, a washer—and attaching the other end of the thread to a surface so that the weight is suspended and can swing freely. The *period* of the pendulum is the amount of time it takes for the weight to swing forward and backward once.

Let T be the period (in seconds) of a pendulum and L be the length (in centimeters) of the thread. Some values of L and T generated from an experiment are shown in Table 36.

Table 36 Periods of a Pendulum

Length L (cm)	Period T (seconds)	$\frac{T}{\sqrt{L}}$
5.0	0.50	
10.0	0.63	
15.0	0.88	
20.0	1.00	
25.0	1.13	
32.5	1.25	
45.0	1.50	
60.0	1.75	
85.0	2.00	
110.0	2.25	

Source: *J. Lehmann*

a. The period of a pendulum varies directly as the square root of the length of the thread. Write an equation involving L, T, and the variation constant k.
b. Solve the equation that you found in part (a) for k.
c. Use the first two columns of Table 36 to complete the third column. What is a reasonable value of k? How did you find this value?
d. Substitute your value of k into your equation from part (a).
e. Draw the graph of your model and a scattergram of the data in the same viewing window. Does your model fit the data well?
f. Estimate the period if the thread is 130 inches long.

50. The illumination from a light bulb decreases as the distance from the bulb increases. Let I be the illumination in milliwatts per square centimeter (mW/cm^2) at d centimeters from a 25-watt light bulb. Some values of d and I generated by experiment are shown in Table 37.

Table 37 Illumination from a Light Bulb

Distance d (cm)	Illumination I (mW/cm^2)	d^2I
70	0.845	
80	0.677	
90	0.546	
100	0.435	
110	0.349	
120	0.293	
130	0.260	
140	0.214	

Source: *J. Lehmann*

a. The illumination from a light source varies inversely as the square of the distance from the source. Write an equation involving d, I, and the variation constant k.
b. Solve the equation that you found in part (a) for k.
c. Use the first two columns of Table 37 to complete the third column. What is a reasonable value of k? How did you find this value?
d. Substitute your value of k into your equation from part (a) to find an equation of a model.
e. Draw the graph of your model and a scattergram of the data in the same viewing window. Does your model fit the data well?
f. Estimate the illumination at 160 centimeters from the bulb.
g. How would an equation of a model for a 50-watt bulb compare with the model you found in part (d)? How would the variation constants compare?

For Exercises 51–56, write a variation equation in the given variables. State the value of the variation constant k.

51. Let $f(L)$ be the area (in square inches) of a rectangle with width 5 inches and length L (in inches).

52. Let $f(s)$ be the area (in square meters) of a square with sides of length s (in meters).

53. A group of n people win a total of \$2 million from a state lottery drawing. Let $f(n)$ be each person's share (in millions of dollars).

54. Let $f(s)$ be the time (in hours) it takes a person to drive 100 miles at a constant speed s (in mph).

55. Let $f(r)$ be the circumference (in inches) of a circle with radius r (in inches).

56. Let $f(r)$ be the area (in square feet) of a circle with radius r (in feet).

57. Suppose that the latest Green Day CD is about to be released. Let $n = f(a)$ be the number of CDs (in millions) that will be sold, where a is the advertising budget (in thousands of dollars). A graph of f is sketched in Fig. 58.

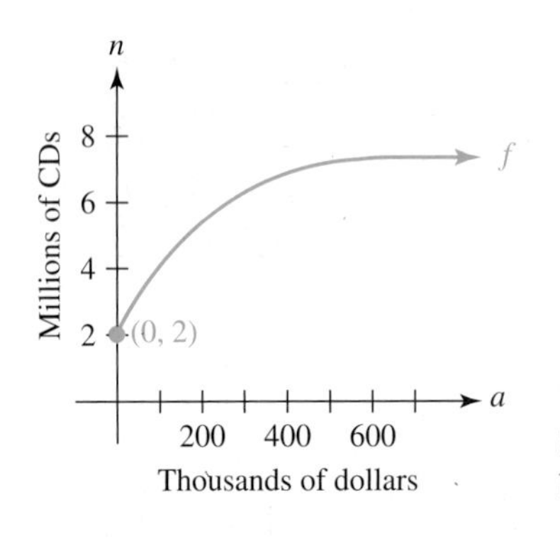

Figure 58 Advertising budget and CD sales

a. Is f an increasing function? Explain.
b. Does the number of CDs sold vary directly as the amount of money spent on advertising? Explain.
c. Compare the meanings of these two statements: "One quantity varies directly as another quantity." "One quantity will increase if the other quantity increases."

58. For a direct variation equation $y = kx$, where k is positive, we know that as the value of x increases, the value of y increases. If k is negative for $y = kx$, what happens to the value of y as the value of x increases?

For Exercises 59–62, decide whether the statement is true or false. Explain.

59. The height of a person varies directly as the person's age.

60. The area of a circle varies directly as the radius of the circle.

61. The temperature of hot coffee varies inversely as the time since it was poured into a cup.

62. The height of a baseball hit straight up varies inversely as the time since it was hit.

63. Assume that y varies directly as x and that the variation constant is k. Does it follow that x varies directly as y? If yes, what is the variation constant? If no, explain. [**Hint:** Solve the equation $y = kx$ for x.]

64. Assume that y varies inversely as x and that the variation constant is k. Does it follow that x varies inversely as y? If yes, what is the variation constant? If no, explain. [**Hint:** Solve the equation $y = \frac{k}{x}$ for x.]

Related Review

65. **a.** If y varies directly as x, are x and y linearly related? Explain.
    b. If w and t are linearly related, does w vary directly as t? Explain.

66. **a.** If a quantity y varies directly as a quantity x, how many data points do you need to find an equation to describe the situation?
    b. If the quantities w and t are linearly related, how many data points do you need to find an equation to describe the situation?

67. The number of words n a typist can type varies directly as the amount of time t (in minutes) he types. He can type 310 words in 5 minutes. What is the slope of the model that describes this situation? What does it mean in this situation?

68. The total cost c (in dollars) of some 5-inch × 7-inch school photos shot by Life-touch® varies directly as the number of photos n that are purchased. It costs $30 for 4 such photos. What is the slope of the model that describes this situation? What does it mean in this situation?

Expressions, Equations, Functions, and Graphs

Give an example of the following. Then solve, simplify, or graph, as appropriate.

69. system of two linear equations in two variables
70. linear equation in one variable
71. difference of two rational expressions
72. quadratic equation in one variable
73. quadratic function
74. rational equation in one variable
75. exponential function
76. the sum of two logarithmic expressions with the same base

CHAPTER SUMMARY

Key Points OF CHAPTER 8

Throughout these key points, assume that A, B, C, and D are polynomials.

Rational function (Section 8.1)
A **rational function** is a function whose equation can be put into the form $f(x) = \frac{P(x)}{Q(x)}$, where $P(x)$ and $Q(x)$ are polynomials and Q is nonzero.

Domain of a rational function (Section 8.1)
The domain of a rational function $f(x) = \frac{P(x)}{Q(x)}$ is the set of all real numbers except for those numbers that, when substituted for x, give $Q(x) = 0$.

Excluded value (Section 8.1)
A number is an **excluded value** of a rational expression if substituting the number into the expression leads to a division by 0.

Simplify a rational expression (Section 8.1)
To simplify a rational expression,
1. Factor the numerator and the denominator.
2. Use the property $\frac{AB}{AC} = \frac{A}{A} \cdot \frac{B}{C} = 1 \cdot \frac{B}{C} = \frac{B}{C}$, where A and C are nonzero, so that the expression is in lowest terms.

Vertical asymptotes and domains of rational functions (Section 8.1)
If a rational function g has a vertical asymptote $x = k$, then k is not in the domain of g. If k is not in the domain of a rational function h, then $x = k$ may or may not be a vertical asymptote of h.

Quotient function (Section 8.1)
If f and g are functions, x is in the domain of both functions, and $g(x)$ is nonzero, then we can form the **quotient function** $\frac{f}{g}$: $\left(\frac{f}{g}\right)(x) = \frac{f(x)}{g(x)}$.

Rational model (Section 8.1)
A **rational model** is a rational function, or its graph, that describes an authentic situation.

Percentage formula (Section 8.1)
If m items out of n items have a certain attribute, then the percentage p (written $p\%$) of the n items that have the attribute is $p = \frac{m}{n} \cdot 100$. We call this equation the **percentage formula.**

Multiplying rational expressions (Section 8.2)	If $\frac{A}{B}$ and $\frac{C}{D}$ are rational expressions and B and D are nonzero, then $\frac{A}{B} \cdot \frac{C}{D} = \frac{AC}{BD}$.
How to multiply rational expressions (Section 8.2)	To multiply two rational expressions, **1.** Factor the numerators and the denominators. **2.** Multiply by using the property $\frac{A}{B} \cdot \frac{C}{D} = \frac{AC}{BD}$, where B and D are nonzero. **3.** Simplify the result.
Dividing rational expressions (Section 8.2)	If $\frac{A}{B}$ and $\frac{C}{D}$ are rational expressions and B, C, and D are nonzero, then $\frac{A}{B} \div \frac{C}{D} = \frac{A}{B} \cdot \frac{D}{C}$.
How to divide rational expressions (Section 8.2)	To divide two rational expressions, **1.** Write the quotient as a product by using the property $\frac{A}{B} \div \frac{C}{D} = \frac{A}{B} \cdot \frac{D}{C}$, where B, C, and D are nonzero. **2.** Find the product. **3.** Simplify.
Adding rational expressions with a common denominator (Section 8.3)	If $\frac{A}{B}$ and $\frac{C}{B}$ are rational expressions and B is nonzero, then $\frac{A}{B} + \frac{C}{B} = \frac{A+C}{B}$.
Subtracting rational expressions with a common denominator (Section 8.3)	If $\frac{A}{B}$ and $\frac{C}{B}$ are rational expressions and B is nonzero, then $\frac{A}{B} - \frac{C}{B} = \frac{A-C}{B}$.
Subtract entire numerator (Section 8.3)	When subtracting rational expressions, be sure to subtract the *entire* numerator.
How to add or subtract two rational expressions with different denominators (Section 8.3)	To add or subtract two rational expressions with different denominators, **1.** Factor the denominators of the expressions if possible. Determine which factors are missing. **2.** Use the property $\frac{A}{A} = 1$, where A is nonzero, to introduce missing factors. **3.** Add the expressions by using the property $\frac{A}{B} + \frac{C}{B} = \frac{A+C}{B}$, where B is nonzero; or subtract the expressions by using the property $\frac{A}{B} - \frac{C}{B} = \frac{A-C}{B}$, where B is nonzero. **4.** Simplify.
Complex rational expression (Section 8.4)	A **complex rational expression** is a rational expression whose numerator or denominator (or both) is a rational expression.
Using method 1 to simplify a complex rational expression (Section 8.4)	To simplify a complex rational expression by method 1, **1.** Write both the numerator and the denominator as fractions. **2.** Write the complex rational expression as the quotient of two rational expressions: $\frac{\frac{A}{B}}{\frac{C}{D}} = \frac{A}{B} \div \frac{C}{D}$, where B, C, and D are nonzero. **3.** Divide the rational expressions.
Using method 2 to simplify a complex rational expression (Section 8.4)	To simplify a complex rational expression by method 2, **1.** Find the LCD of all of the fractions in the numerator and denominator. **2.** Multiply by 1 in the form $\frac{\text{LCD}}{\text{LCD}}$. **3.** Simplify the numerator and the denominator to polynomials. **4.** Simplify the rational expression.
Rational equation in one variable (Section 8.5)	A **rational equation in one variable** is an equation in one variable in which both sides can be written as rational expressions.

Check proposed solutions (Section 8.5)

Since multiplying both sides of a rational equation by the LCD and then simplifying both sides may introduce extraneous solutions, we must always check that any proposed solution is not an excluded value.

Solving a rational equation in one variable (Section 8.5)

To solve a rational equation in one variable,

1. Factor the denominator(s) if possible.
2. Identify any excluded values.
3. Find the LCD of all of the fractions.
4. Multiply both sides of the equation by the LCD, which gives a simpler equation to solve.
5. Solve the simpler equation.
6. Discard any proposed solutions that are excluded values.

Solving a rational equation versus simplifying a rational expression (Section 8.5)

To solve a rational equation, clear the fractions in it by multiplying both sides of the equation by the LCD. To simplify a rational expression, do *not* multiply it by the LCD—the only multiplication permissible is multiplication by 1, usually in the form $\frac{A}{A}$, where A is a nonzero polynomial.

Results of solving rational equations and simplifying rational expressions (Section 8.5)

The result of solving a rational equation is the empty set or a set of one or more numbers. The result of simplifying a rational expression is an expression.

Mean (Section 8.6)

If a quantity Q is divided into n parts, the **mean** amount M of the quantity per part is given by $M = \frac{Q}{n}$.

Distance–speed–time relationship (Section 8.6)

If an object is moving at a constant speed s for an amount of time t, then the distance traveled d is given by $d = st$ and the time t is given by $t = \frac{d}{s}$.

Direct variation (Section 8.7)

If $y = kx$ for some nonzero constant k, we say that **y varies directly as x** or that **y is proportional to x.** We call k the **variation constant** or the **constant of proportionality.** The equation $y = kx$ is called a **direct variation equation.**

Changes in values of variables for direct variation (Section 8.7)

Assume that y varies directly as x with some positive variation constant k:

- If the value of x increases, then the value of y increases.
- If the value of x decreases, then the value of y decreases.

Finding and using a direct variation model (Section 8.7)

Assume that a quantity p varies directly as a quantity t. To make estimates about an authentic situation,

1. Substitute the values of a data point into the equation $p = kt$; then solve for k.
2. Substitute the value of k into the equation $p = kt$.
3. Use the equation from step 2 to make estimates of quantity t or p.

Inverse variation (Section 8.7)

If $y = \frac{k}{x}$ for some nonzero constant k, we say that **y varies inversely as x** or that **y is inversely proportional to x.** We call k the **variation constant** or the **constant of proportionality.** The equation $y = \frac{k}{x}$ is called an **inverse variation equation.**

Changes in values of variables for inverse variation (Section 8.7)

Assume that y varies inversely as x with some positive variation constant k. For positive values of x:

- If the value of x increases, then the value of y decreases.
- If the value of x decreases, then the value of y increases.

Finding and using an inverse variation model (Section 8.7)

Assume that a quantity p varies inversely as a quantity t. To make estimates about an authentic situation,

1. Substitute the values of a data point into the equation $p = \frac{k}{t}$; then solve for k.
2. Substitute the value of k into the equation $p = \frac{k}{t}$.
3. Use the equation from step 2 to make estimates of quantity t or p.

CHAPTER 8 REVIEW EXERCISES

1. For $f(x) = \dfrac{5x - 3}{2x^2 - 3x + 1}$, find $f(0)$ and $f(2)$.

Find the domain of the rational function.

2. $f(x) = \dfrac{5}{4x^2 - 49}$

3. $f(x) = \dfrac{x^2 - 4}{12x^2 + 13x - 35}$

4. $f(x) = \dfrac{3x + 7}{9x^3 + 18x^2 - x - 2}$

Simplify the right-hand side of the equation.

5. $f(x) = \dfrac{3x - 12}{x^2 - 6x + 8}$

6. $f(x) = \dfrac{16 - x^2}{2x^3 - 16x^2 + 32x}$

7. $f(x) = \dfrac{x + 2}{x^3 + 8}$

8. Simplify $\dfrac{6a^2 - 17ab + 5b^2}{3a^2 - 4ab + b^2}$.

9. For $f(x) = x^2 + 3x - 28$ and $g(x) = x^3 - x^2 - 12x$, find an equation of $\dfrac{f}{g}$. Then find $\dfrac{f}{g}(-2)$.

Perform the indicated operation(s).

10. $\dfrac{3x + 6}{2x - 4} \cdot \dfrac{5x - 10}{6x + 12}$

11. $\dfrac{x^2 - 49}{9 - x^2} \cdot \dfrac{2x^3 + 8x^2 - 42x}{5x - 35}$

12. $\dfrac{p^3 - t^3}{p^2 - t^2} \cdot \dfrac{p^2 + 6pt + 5t^2}{p^2t + pt^2 + t^3}$

13. $\dfrac{x^2 - 4}{x^2 + 3x + 2} \div \dfrac{4x^2 - 24x + 32}{x^2 - 5x + 4}$

14. $\dfrac{4 - x}{4x} \div \dfrac{16 - x^2}{16x^2}$

15. $\dfrac{8x^3 + 4x^2 - 18x - 9}{x^2 - 6x + 9} \div \dfrac{4x^2 + 8x + 3}{x^2 - 9}$

16. $\dfrac{x}{x^2 - 5x + 6} + \dfrac{3}{3 - x}$

17. $\dfrac{x}{2x^3 - 3x^2 - 5x} + \dfrac{2}{x^3 - x}$

18. $\dfrac{x - 1}{x^2 - 4} + \dfrac{x + 3}{x^2 - 4x + 4}$

19. $\dfrac{3}{4x - 12} - \dfrac{x}{x^2 - 2x - 3}$

20. $\dfrac{x + 1}{25 - x^2} - \dfrac{x - 4}{2x^2 - 14x + 20}$

21. $\dfrac{2m}{m^2 - 3mn - 10n^2} - \dfrac{4n}{m^2 + 8mn + 12n^2}$

22. $\dfrac{2}{x - 5} - \left(\dfrac{x^2 + 5x + 6}{3x^2 - 75} \div \dfrac{x^2 + 2x}{3x + 15}\right)$

Let $f(x) = \dfrac{x^2 - x - 2}{x^2 + 5x + 6}$ and $g(x) = \dfrac{x + 3}{x + 2}$. Find an equation of the given function.

23. $f \cdot g$ **24.** $f \div g$ **25.** $f + g$ **26.** $f - g$

Simplify.

27. $\dfrac{\dfrac{x - 2}{x^2 - 9}}{\dfrac{x^2 - 4}{x + 3}}$

28. $\dfrac{\dfrac{4}{3x^4} - \dfrac{2}{6x^2}}{\dfrac{1}{2x} + \dfrac{1}{4}}$

Solve. Any solution is a real number.

29. $\dfrac{1}{x + 5} - \dfrac{2}{x - 2} = \dfrac{-14}{x^2 + 3x - 10}$

30. $\dfrac{x}{x + 2} + \dfrac{3}{x + 4} = \dfrac{14}{x^2 + 6x + 8}$

31. $\dfrac{5}{x} + 3 = \dfrac{4}{x^2}$

32. $\dfrac{x - 3}{2x^2 - 7x - 4} - \dfrac{5}{2x^2 + 3x + 1} = \dfrac{x - 1}{x^2 - 3x - 4}$

33. Find all complex-number solutions of the equation $\dfrac{2x}{x + 6} - \dfrac{4}{x - 3} = \dfrac{-37}{x^2 + 3x - 18}$.

34. Find the x-intercept(s) of $f(x) = \dfrac{x - 7}{x + 1} - \dfrac{x + 3}{x - 4}$.

35. Solve the formula $S = \dfrac{a}{1 - r}$ for r.

Translate the equation with positive constant k by using the phrase "varies directly" or "varies inversely" in a sentence.

36. $H = ku^2$

37. $w = \dfrac{k}{\log(t)}$

Find an equation that meets the given conditions.

38. y varies directly as $\sqrt{x}$, and $y = 2$ when $x = 49$.

39. B varies inversely as r^3, and $B = 9$ when $r = 2$.

40. The number of inches of water varies directly as the number of inches of snow. If 20 inches of snow will melt to 2.24 inches of water, to how many inches of water will 37 inches of snow melt?

41. Let m be the mass (in grams) of a ball bearing with radius r (in cm). Table 38 shows values of r and m for ball bearings of various sizes.

Table 38 Radii and Masses of Ball Bearings

Radius r (cm)	Mass m (grams)	$\dfrac{m}{r^3}$
1.0	17.1	
1.2	29.4	
1.4	46.7	
1.6	69.6	
1.8	99.1	
2.0	135.9	

a. The mass of a ball bearing is directly proportional to the radius cubed. Write an equation involving r, m, and k.

b. Solve the equation that you found in part (a) for k.
c. Use the first two columns of Table 38 to complete the third column. What is a reasonable value of k? How did you find this value?
d. Substitute your value of k into your equation from part (a) to find an equation of a model.
e. Draw the graph of your model and a scattergram of the data in the same viewing window. Does your model fit the data well?
f. What is the mass of a ball bearing with radius 2.3 cm?

42. A hotel offers a one-day rental of a conference room (capacity 300) for a flat fee of \$600, plus a per-person charge of \$40 for lunch.
a. Let $C(n)$ be the total cost (in dollars) of renting the room if n people use the room for one day. Find an equation of C.
b. Let $M(n)$ be the mean cost per person (in dollars per person) if n people use the room for one day. Find an equation of M.
c. Find $M(270)$. What does it mean in this situation?
d. Find n when $M(n) = 50$. What does it mean in this situation?

43. In Example 3 of Section 7.8, we found the system

$$S = v(t) = -6.27t^2 + 129.7t - 576.2$$
$$S = d(t) = 4.83t^2 - 79.1t + 326.0$$

where $v(t)$ and $d(t)$ are the numbers (in millions of units) of videocassettes and DVDs, respectively, bought by U.S. dealers to sell as rentals during the year that is t years since 1990 (see Table 39).
a. Find an equation of the sum function $v + d$. What do the inputs and outputs of $v + d$ mean in this situation?

Table 39 Numbers of Videocassettes and DVDs Bought by U.S. Dealers

Year	Number of Videocassettes (millions of units)	Number of DVDs (millions of units)
1998	57.0	1.6
1999	86.2	8.6
2000	99.4	13.9
2001	86.2	37.1
2002	73.6	79.3
2003	53.2	110.9

Source: *Adams Media Research*

b. Let $P(t)$ be the percentage of sales (videocassettes and DVDs) to U.S. dealers that are DVDs at t years since 1990. Find an equation of P.
c. Estimate the percentage of sales (videocassettes and DVDs) to U.S. dealers that were DVDs in 2000. Compare this with the actual percentage.
d. Estimate when all sales (videocassettes and DVDs) to U.S. dealers were DVDs.

44. A student plans to drive 75 miles on an undivided highway and another 40 miles on a divided highway. The speed limits are 50 mph for the undivided highway and 65 mph for the divided highway.
a. Let $T(a)$ be the driving time (in hours) if the student drives a mph above the speed limits. Find an equation of T.
b. Find $T(5)$. What does it mean in this situation?
c. By how much over the speed limits would the student need to drive for the driving time to be 2 hours? Use a graphing calculator table to verify your result.

CHAPTER 8 TEST

Find the domain of the function.

1. $f(x) = \dfrac{5}{6x^2 + 11x - 10}$

2. $g(x) = \dfrac{2}{72 - 2x^2}$

3. $h(x) = \dfrac{x}{3}$

4. Give examples of three functions, each of whose domains are the set of real numbers except -3 and 7.

Simplify the right-hand side of the equation.

5. $f(x) = \dfrac{6 - 3x}{x^2 - 5x + 6}$

6. $f(x) = \dfrac{9x^2 - 1}{18x^3 - 12x^2 + 2x}$

Perform the indicated operation.

7. $\dfrac{5x^4}{3x^2 + 6x + 12} \cdot \dfrac{x^3 - 8}{15x^7}$

8. $\dfrac{p^2 - 4t^2}{p^2 + 6pt + 9t^2} \div \dfrac{p^2 - 3pt + 2t^2}{p^2 + 3pt}$

9. $\dfrac{5x + 12}{-2x^2 - 8x} - \dfrac{2x + 1}{x^2 + 2x - 8}$

10. $\dfrac{x + 2}{x^2 - 9} + \dfrac{3}{x^2 + 11x + 24}$

11. Perform the indicated operations:

$$\frac{3}{x^2 - 2x} \div \left(\frac{x}{5x - 10} - \frac{x - 1}{x^2 - 4}\right)$$

12. Let $f(x) = \dfrac{x + 1}{x - 5}$ and $g(x) = \dfrac{x - 2}{x + 4}$. Find an equation of $f - g$. Then find $(f - g)(0)$.

13. Simplify $\dfrac{5 + \dfrac{2}{x}}{3 - \dfrac{4}{x - 1}}$.

Solve the equation.

14. $\dfrac{2}{x - 1} - \dfrac{5}{x + 1} = \dfrac{4x}{x^2 - 1}$

15. $\dfrac{5}{x - 3} = \dfrac{x}{x - 2} + \dfrac{x}{x^2 - 5x + 6}$

16. Let $f(x) = \dfrac{2}{x - 4} + \dfrac{3}{x + 1}$. Find a when $f(a) = 5$.

Let $f(x) = \dfrac{(x - 5)(x + 2)}{(x - 1)(x + 3)}$.

17. Find $f(-2)$. **18.** Find $f(1)$. **19.** Find x when $y = 0$.

Find an equation that meets the given conditions.

20. W varies directly as t^2, and $W = 3$ when $t = 7$.

21. y varies inversely as $\sqrt{x}$, and $y = 8$ when $x = 25$.

22. It costs a bike manufacturer about \$200 per bike for the materials to manufacture a line of mountain bikes. It also costs \$10,000 each month for the manufacturing plant's lease, electricity, salaries, and so on.

a. Let $C(n)$ be the total monthly cost (in dollars) if n bikes are manufactured that month. Find an equation of C.

b. Let $B(n)$ be the price the manufacturer should set for each bike (in dollars) to break even by making and selling n bikes in a month. Find a formula of B.

c. Let $P(n)$ be the price the manufacturer should set for each bike (in dollars) to earn a profit of \$150 per bike by making and selling n bikes in a month. Find an equation of P.

d. Find $P(100)$. What does it mean in this situation?

23. A student plans to drive from San Francisco, California, to Salt Lake City, Utah. She will drive 400 miles in California, then 920 miles in Nevada and Utah. The speed limit is 70 mph in California and 75 mph in Nevada and Utah.

a. Let $T(a)$ be the driving time (in hours) if the student drives a mph above the speed limits. Find an equation of T.

b. Find $T(5)$. What does it mean in this situation?

c. Find a when $T(a) = 17$. What does your result mean in this situation?

24. The frequency $g(L) = F$ (in hertz) of a tuning fork varies inversely as the square of the length L (in cm) of the prongs. The frequency is 50 hertz when the prong length is 8 cm.

a. Find an equation of g.

b. Find the frequency if the prong length is 6 cm.

c. What is the prong length if the frequency is 200 hertz?

d. Is g increasing or decreasing for $L > 0$? What does that mean in this situation?

Chapter 9

Radical Functions

"Many of life's failures are people who did not realize how close they were to success when they gave up."

—Thomas Edison, 1847–1931

Table 1 Average Numbers of Viewers per Episode of *American Idol*

Year	Average Number of Viewers (millions)
2002	12.0
2003	21.7
2004	25.0
2005	26.6
2006	30.3

Source: *Nielsen Media Research*

Have you ever watched the television show *American Idol*? In 2006, the average number of viewers per episode was 30.3 million viewers. The most recent shows to have that many viewers per episode were *Seinfeld* and *ER* in 1997–1998 (see Table 1). In Exercise 23 of Homework 9.6, you will predict when the average number of viewers per episode will equal the population of California, the most populous U.S. state.

In previous chapters, we worked with linear, exponential, logarithmic, polynomial, and rational functions. In Section 7.3, we worked with radical expressions. In this chapter, we will discuss how to simplify more types of radical expressions, as well as how to perform operations with them. We will also solve *radical equations in one variable*. Finally, we will use *square root functions* to make predictions, such as when the average number of viewers per episode of *American Idol* will equal the population of California.

9.1 SIMPLIFYING RADICAL EXPRESSIONS

Objectives

- Convert expressions in radical form to exponential form and vice versa.
- Know the *product property for radicals* and the *power property for radicals*.
- Simplify radical expressions.
- Know the meaning of *radical function, square root function, radical model,* and *square root model*.
- Use a radical model to make an estimate or prediction.

In this section, we will discuss the meaning of a *radical expression* and how to use properties to simplify such an expression.

Radical Expressions

Recall from Section 4.2 that we can represent a principal nth root such as $x^{1/n}$ as $\sqrt[n]{x}$, where n is a counting number greater than 1. We say that n is the **index** and the symbol $\sqrt{}$ is the **radical sign.** The notation for the principal square root of x, $\sqrt{x}$, is shorthand for $\sqrt[2]{x}$. An expression under the radical sign is called a **radicand.** In $\sqrt[5]{3x+6}$, the radicand is $3x+6$. A radical sign together with an index and a radicand is called a **radical.** Here we label the radical sign, index, and radicand for the radical $\sqrt[4]{2x-7}$:

Index → $\sqrt[4]{2x-7}$ ← Radicand; Radical sign; Radical (entire expression)

Some more radicals are $\sqrt[3]{21}$, $\sqrt{x}$, $\sqrt[7]{x^4}$, $\sqrt{5x-8}$, and $\sqrt[4]{(x+1)^3}$.

An expression that contains a radical is called a **radical expression.** Here are some radical expressions:

$$\sqrt[4]{85}, \quad \sqrt{x}, \quad \sqrt[3]{4x+9}, \quad 2\sqrt[3]{2x}+5\sqrt[4]{7}, \quad (8\sqrt{x}-5)(3\sqrt[3]{x}+4), \quad \frac{2\sqrt{x}-7}{6\sqrt{x}-4}$$

Recall from Section 4.2 that for nonnegative a, $\sqrt[n]{a}$ is the nonnegative number whose nth power is a. If a is negative and n is odd, then $\sqrt[n]{a}$ is the (negative) number whose nth power is a. If a is negative and n is even, then $\sqrt[n]{a}$ is not a real number.

Example 1 Evaluating Radicals

Evaluate the radical.

1. $\sqrt[3]{-8}$
2. $\sqrt[4]{81}$

Figure 1 Verify the work

Solution

1. $\sqrt[3]{-8} = -2$, since $(-2)^3 = -8$.
2. $\sqrt[4]{81} = 3$, since $3^4 = 81$.

We use a graphing calculator to verify our work (see Fig. 1). To enter $\sqrt[3]{-8}$, press MATH **4** (−) **8**) ENTER. To enter $\sqrt[4]{81}$, press **4** MATH **5** (**81**) ENTER. ■

Recall from Section 7.3 that a number whose square root is rational is called a *perfect square*. More generally, a number whose nth root is rational is called a **perfect nth power.** For example, -8 is a perfect 3rd power, or **perfect cube,** because $\sqrt[3]{-8} = -2 = \frac{-2}{1}$ is rational. The number 81 is a perfect 4th power, because $\sqrt[4]{81} = 3 = \frac{3}{1}$ is rational.

Recall from Section 4.2 that if $a^{1/n}$ is defined, then

$$a^{m/n} = \left(a^{1/n}\right)^m = \left(a^m\right)^{1/n}$$

or we can write

$$a^{m/n} = \left(\sqrt[n]{a}\right)^m = \sqrt[n]{a^m}$$

For example, $x^{3/5} = \left(\sqrt[5]{x}\right)^3 = \sqrt[5]{x^3}$. We say that the expressions $\left(\sqrt[n]{x}\right)^m$ and $\sqrt[n]{x^m}$ are in *radical form* and that the expression $x^{m/n}$ is in *exponential form.*

Example 2 Writing Exponential and Radical Forms

If the expression is in exponential form, write it in radical form. If it is in radical form, write it in exponential form.

1. $x^{3/7}$
2. $\sqrt{x^5}$
3. $(3w+1)^{4/5}$
4. $\sqrt[8]{(2x+5)^3}$

Solution

1. $x^{3/7} = \sqrt[7]{x^3}$
2. $\sqrt{x^5}$ is shorthand for $\sqrt[2]{x^5}$. So, $\sqrt{x^5} = x^{5/2}$.
3. $(3w+1)^{4/5} = \sqrt[5]{(3w+1)^4}$
4. $\sqrt[8]{(2x+5)^3} = (2x+5)^{3/8}$ ■

Simplifying Radical Expressions

It will be helpful, at times, to write a radical expression in exponential form and use exponential properties to simplify the result.

Example 3 Simplifying Radical Expressions

Write the expression in exponential form; then simplify the result.

1. $\sqrt[3]{x^3}$
2. $\sqrt[3]{x^{12}}$
3. $\sqrt[3]{x^{21}}$

Solution

1. $\sqrt[3]{x^3} = x^{3/3} = x^1 = x$
2. $\sqrt[3]{x^{12}} = x^{12/3} = x^4$
3. $\sqrt[3]{x^{21}} = x^{21/3} = x^7$ ■

In addition to the numbers that are perfect nth powers, some variable expressions are perfect nth powers. As in Example 3, we can eliminate the radical sign from expressions of the form $\sqrt[3]{x^k}$ when k is a multiple of 3. If $\sqrt[n]{x}$ is defined, we can eliminate the radical sign from $\sqrt[n]{x^k}$ when k is a multiple of the index n. If n is a counting number greater than 1 and k is a multiple of n, we say that x^k is a **perfect *n*th power.**

Here are perfect nth powers for $n = 2$, $n = 3$, and $n = 4$:

Value of n	Name	Examples	Description of Exponents
2	perfect square	$x^2, x^4, x^6, x^8, \ldots$	multiples of 2
3	perfect cube	$x^3, x^6, x^9, x^{12}, \ldots$	multiples of 3
4	perfect 4th power	$x^4, x^8, x^{12}, x^{16}, \ldots$	multiples of 4

The base of a perfect nth power can be any variable expression. For example, $(7x + 2)^{15}$ is a perfect 5th power; furthermore, we can find the 5th root of it:

$$\sqrt[5]{(7x + 2)^{15}} = (7x + 2)^{15/5} = (7x + 2)^3$$

In Section 7.3, we studied the product property for square roots,

$$\sqrt{ab} = \sqrt{a}\sqrt{b}$$

where $a \geq 0$ and $b \geq 0$. The more general product property for radicals

$$\sqrt[n]{ab} = \sqrt[n]{a}\,\sqrt[n]{b}$$

is true for any counting-number index n greater than 1 where $\sqrt[n]{a}$ and $\sqrt[n]{b}$ are defined. Here is the proof:

$$\begin{aligned} \sqrt[n]{ab} &= (ab)^{1/n} && \text{Write in exponential form.} \\ &= a^{1/n}b^{1/n} && (ab)^m = a^m b^m \\ &= \sqrt[n]{a}\,\sqrt[n]{b} && \text{Write in radical form.} \end{aligned}$$

Product Property for Radicals

If $\sqrt[n]{a}$ and $\sqrt[n]{b}$ are defined, then

$$\sqrt[n]{ab} = \sqrt[n]{a}\,\sqrt[n]{b}$$

In words: The nth root of a product is the product of the nth roots.

A radical with index n is **simplified** when the radicand does not have any factors that are perfect nth powers (other than -1 or 1) and the index is as small as possible. We will discuss how to find the smallest possible index later in this section.

Here we use the product property for radicals to simplify $\sqrt{25x^6}$:

$$\begin{aligned} \sqrt{25x^6} &= \sqrt{25}\sqrt{x^6} && \text{Product property: } \sqrt[n]{ab} = \sqrt[n]{a}\,\sqrt[n]{b} \\ &= 5x^3 && \sqrt{25} = 5;\ \sqrt{x^6} = x^3 \text{ for } x \geq 0 \end{aligned}$$

Notice that the radicand $25x^6$ is the product of the perfect squares 25 and x^6. We can use the product property to simplify any radical with index n whose radicand is a product of nth powers.

We can also use the product property to simplify some radicals with index n whose radicand is *not* a product of perfect nth powers only. For those radicals, we write the radicand as a product of one or more perfect nth powers and another expression with no factors that are perfect nth powers. Then we apply the product property.

Example 4 Simplifying Radical Expressions

Simplify. Assume that x is nonnegative.

1. $\sqrt{200}$ **2.** $\sqrt{x^7}$ **3.** $\sqrt{12w^{11}}$

Solution

1. The numbers 4, 25, and 100 are all perfect-square factors of 200. Recall from Section 7.3 that it is most efficient to work with the largest one, which is 100. So, we write 200 as $100 \cdot 2$ and apply the product property for radicals:

$$\begin{aligned} \sqrt{200} &= \sqrt{100 \cdot 2} && \text{100 is largest perfect-square factor.} \\ &= \sqrt{100}\sqrt{2} && \text{Product property: } \sqrt[n]{ab} = \sqrt[n]{a}\sqrt[n]{b} \\ &= 10\sqrt{2} && \sqrt{100} = 10 \end{aligned}$$

2. The powers x^2, x^4, and x^6 are all perfect-square factors of x^7. Since x^6 is the perfect-square factor with the largest exponent, we write x^7 as $x^6 \cdot x$ and apply the product property for radicals:

$$\begin{aligned} \sqrt{x^7} &= \sqrt{x^6 \cdot x} && x^6 \text{ is perfect-square factor with largest exponent.} \\ &= \sqrt{x^6}\sqrt{x} && \text{Product property: } \sqrt[n]{ab} = \sqrt[n]{a}\sqrt[n]{b} \\ &= x^3\sqrt{x} && \sqrt{x^6} = x^3 \text{ for } x \ge 0 \end{aligned}$$

3. The number 4 is the largest perfect-square factor of 12, and the power w^{10} is the perfect-square factor of w^{11} with the largest exponent:

$$\begin{aligned} \sqrt{12w^{11}} &= \sqrt{4 \cdot 3 \cdot w^{10} \cdot w} && \text{4 and } w^{10} \text{ are perfect squares.} \\ &= \sqrt{4 \cdot w^{10} \cdot 3w} && \text{Rearrange factors.} \\ &= \sqrt{4}\sqrt{w^{10}}\sqrt{3w} && \text{Product property: } \sqrt[n]{ab} = \sqrt[n]{a}\sqrt[n]{b} \\ &= 2w^5\sqrt{3w} && \sqrt{4} = 2;\ \sqrt{w^{10}} = w^5 \text{ for } w \ge 0 \end{aligned}$$

■

In Example 4, we had to decide which perfect nth powers to use to factor the radicands. Our work in Example 4 suggests two guidelines:

- **If the radicand has more than one numerical factor that is a perfect nth power, select the largest one.**
- **If the radicand has more than one factor that is a perfect nth power with the same variable base, select the one with the largest exponent.**

Example 5 Simplifying Radical Expressions

Simplify. Assume that all variables are nonnegative.

1. $\sqrt{48x^4y^{13}}$ **2.** $\sqrt{(5x+3)^9}$

Solution

1.

$$\begin{aligned} \sqrt{48x^4y^{13}} &= \sqrt{16 \cdot 3 \cdot x^4 \cdot y^{12} \cdot y} && 16,\ x^4, \text{ and } y^{12} \text{ are perfect squares.} \\ &= \sqrt{16 \cdot x^4 \cdot y^{12} \cdot 3y} && \text{Rearrange factors.} \\ &= \sqrt{16}\sqrt{x^4}\sqrt{y^{12}}\sqrt{3y} && \text{Product property: } \sqrt[n]{ab} = \sqrt[n]{a}\sqrt[n]{b} \\ &= 4x^2y^6\sqrt{3y} && \sqrt{16} = 4;\ \sqrt{x^4} = x^2 \text{ for } x \ge 0;\ \sqrt{y^{12}} = y^6 \text{ for } y \ge 0 \end{aligned}$$

2.

$$\begin{aligned} \sqrt{(5x+3)^9} &= \sqrt{(5x+3)^8(5x+3)} && (5x+3)^8 \text{ is a perfect-square factor.} \\ &= \sqrt{(5x+3)^8}\sqrt{5x+3} && \text{Product property: } \sqrt[n]{ab} = \sqrt[n]{a}\sqrt[n]{b} \\ &= (5x+3)^4\sqrt{5x+3} && \sqrt{(5x+3)^8} = (5x+3)^4 \text{ for } 5x+3 \ge 0 \end{aligned}$$

We use a graphing calculator table to verify our work (see Fig. 2). ■

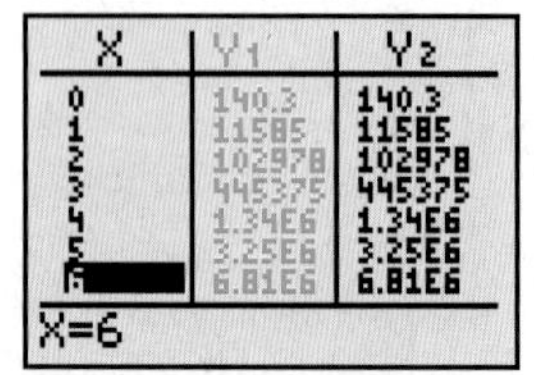

Figure 2 Verify the work

In Example 5, we simplified square root expressions. In Example 6, we simplify radical expressions with index n greater than 2.

Example 6 Simplifying Radical Expressions

Simplify. Assume that all variables are nonnegative.

1. $\sqrt[3]{40x^{17}}$ **2.** $\sqrt[4]{80x^{20}y^{15}}$ **3.** $\sqrt[5]{(2x+7)^{34}}$

Solution

1.
$$\sqrt[3]{40x^{17}} = \sqrt[3]{8 \cdot 5 \cdot x^{15} \cdot x^2} \quad \text{8 and } x^{15} \text{ are perfect 3rd powers (perfect cubes).}$$
$$= \sqrt[3]{8 \cdot x^{15} \cdot 5x^2} \quad \text{Rearrange factors.}$$
$$= \sqrt[3]{8}\sqrt[3]{x^{15}}\sqrt[3]{5x^2} \quad \sqrt[n]{ab} = \sqrt[n]{a}\sqrt[n]{b}$$
$$= 2x^5\sqrt[3]{5x^2} \quad \sqrt[3]{8} = 2;\ \sqrt[3]{x^{15}} = x^5$$

2.
$$\sqrt[4]{80x^{20}y^{15}} = \sqrt[4]{16 \cdot 5 \cdot x^{20} \cdot y^{12} \cdot y^3} \quad 16,\ x^{20}, \text{ and } y^{12} \text{ are perfect 4th powers.}$$
$$= \sqrt[4]{16 \cdot x^{20} \cdot y^{12} \cdot 5y^3} \quad \text{Rearrange factors.}$$
$$= \sqrt[4]{16}\sqrt[4]{x^{20}}\sqrt[4]{y^{12}}\sqrt[4]{5y^3} \quad \sqrt[n]{ab} = \sqrt[n]{a}\sqrt[n]{b}$$
$$= 2x^5y^3\sqrt[4]{5y^3} \quad \sqrt[4]{16} = 2;\ \sqrt[4]{x^{20}} = x^5 \text{ for } x \geq 0;\ \sqrt[4]{y^{12}} = y^3 \text{ for } y \geq 0$$

3.
$$\sqrt[5]{(2x+7)^{34}} = \sqrt[5]{(2x+7)^{30}(2x+7)^4} \quad (2x+7)^{30} \text{ is a perfect 5th power.}$$
$$= \sqrt[5]{(2x+7)^{30}}\sqrt[5]{(2x+7)^4} \quad \sqrt[n]{ab} = \sqrt[n]{a}\sqrt[n]{b}$$
$$= (2x+7)^6\sqrt[5]{(2x+7)^4} \quad \sqrt[5]{(2x+7)^{30}} = (2x+7)^6$$

■

Consider the radical expression $\sqrt[8]{x^6}$, $x \geq 0$. Although the radicand x^6 does not have factors that are perfect 8th powers, we can write the radical with a smaller index:

$$\sqrt[8]{x^6} = x^{6/8} = x^{3/4} = \sqrt[4]{x^3}$$

If $\sqrt[n]{x}$ is defined and the fraction $\frac{m}{n}$ can be simplified, then we can decrease the index of $\sqrt[n]{x^m}$ by writing $\sqrt[n]{x^m} = x^{m/n}$ and simplifying the exponent $\frac{m}{n}$.

Simplifying a radical expression includes writing the result with as small an index as possible.

Example 7 Simplifying Radical Expressions

Simplify. Assume that $x \geq 0$.

1. $\sqrt[12]{(3x+7)^8}$ **2.** $\sqrt[10]{4}$ **3.** $\sqrt[4]{81x^6}$ **4.** $\sqrt[3]{\sqrt{t}}$

Solution

1.
$$\sqrt[12]{(3x+7)^8} = (3x+7)^{8/12} \quad \text{Write in exponential form.}$$
$$= (3x+7)^{2/3} \quad \text{Simplify exponent.}$$
$$= \sqrt[3]{(3x+7)^2} \quad \text{Write in radical form.}$$

2.
$$\sqrt[10]{4} = \sqrt[10]{2^2} \quad 4 = 2^2$$
$$= 2^{2/10} \quad \text{Write in exponential form.}$$
$$= 2^{1/5} \quad \text{Simplify exponent.}$$
$$= \sqrt[5]{2} \quad \text{Write in radical form.}$$

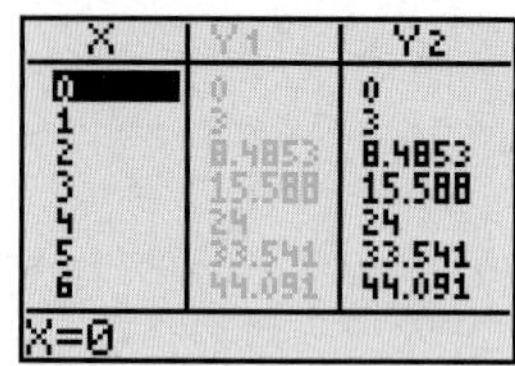

X	Y1	Y2
0	0	0
1	3	3
2	8.4853	8.4853
3	15.588	15.588
4	24	24
5	33.541	33.541
6	44.091	44.091

X=0

Figure 3 Verify the work

3. $\sqrt[4]{81x^6} = \sqrt[4]{81x^4x^2}$ — 81 and x^4 are perfect 4th powers.

$= \sqrt[4]{81}\,\sqrt[4]{x^4}\,\sqrt[4]{x^2}$ — $\sqrt[n]{ab} = \sqrt[n]{a}\,\sqrt[n]{b}$

$= 3x\,\sqrt[4]{x^2}$ — $\sqrt[4]{81} = 3$; $\sqrt[4]{x^4} = x$ for $x \geq 0$

$= 3x \cdot x^{2/4}$ — Write in exponential form.

$= 3x \cdot x^{1/2}$ — Simplify exponent.

$= 3x\sqrt{x}$ — Write in radical form.

We verify our work by comparing graphing calculator tables for $y = \sqrt[4]{81x^6}$ and $y = 3x\sqrt{x}$ for $x \geq 0$ (see Fig. 3).

4. $\sqrt[3]{\sqrt{t}} = \left(t^{1/2}\right)^{1/3}$ — Write in exponential form.

$= t^{\frac{1}{2}\cdot\frac{1}{3}}$ — Multiply exponents: $(b^m)^n = b^{mn}$

$= t^{1/6}$ — Multiply numerators; multiply denominators.

$= \sqrt[6]{t}$ — Write in radical form.

■

Simplifying a Radical Expression

To simplify a radical expression with index n,

1. Find perfect nth-power factors of the radicand.
2. Apply the product property for radicals.
3. Find the nth root of each perfect nth power.
4. Write the radical with as small an index as possible.

Here, we consider the expression $\sqrt[n]{x^n}$, where x is negative. First, we compare $\sqrt{x^2}$ for $x = -4$ and for $x = 4$:

$$\sqrt{(-4)^2} = \sqrt{16} = 4 \qquad \sqrt{4^2} = \sqrt{16} = 4$$

So, when -4 or 4 is substituted for x in the expression $\sqrt{x^2}$, we get the same result: 4. This is precisely what happens if we substitute -4 or 4 in the absolute value (Section A.3) expression $|x|$:

$$|-4| = 4 \qquad |4| = 4$$

These examples suggest that $\sqrt{x^2} = |x|$. It turns out that $\sqrt[n]{x^n} = |x|$ for any even index n.

Here are some examples where n is odd:

$$\sqrt[3]{4^3} = \sqrt[3]{64} = 4 \qquad \sqrt[5]{2^5} = \sqrt[5]{32} = 2$$

$$\sqrt[3]{(-4)^3} = \sqrt[3]{-64} = -4 \qquad \sqrt[5]{(-2)^5} = \sqrt[5]{-32} = -2$$

These examples suggest that $\sqrt[n]{x^n} = x$ if n is odd.

Power Property for Radicals

Let n be a counting number greater than 1:

- If n is even, then $\sqrt[n]{x^n} = |x|$.
- If n is odd, then $\sqrt[n]{x^n} = x$.

For example, $\sqrt[8]{(-5)^8} = |-5| = 5$ and $\sqrt[7]{(-3)^7} = -3$.

Radical Functions

A **radical function** is a function whose equation contains a radical with a variable in the radicand. Here are some examples of radical functions:

$$f(x) = \sqrt[4]{x}, \quad g(x) = 8\sqrt{x} - 3, \quad h(x) = 6\sqrt[3]{(x+7)^2}, \quad k(x) = \frac{\sqrt[3]{x} - 5}{\sqrt{x} + 1}$$

Example 8 Evaluating Radical Functions

For $f(x) = \sqrt[3]{2x - 1}$ and $g(x) = -3\sqrt{x} + 7$, find the following.

1. $f(14)$
2. $f(6)$
3. $g(16)$

Solution

1. $f(14) = \sqrt[3]{2(14) - 1} = \sqrt[3]{27} = 3$
2. $f(6) = \sqrt[3]{2(6) - 1} = \sqrt[3]{11}$
3. $g(16) = -3\sqrt{16} + 7 = -3(4) + 7 = -5$

■

A **square root function** is a radical function in which any radicals are square root radicals. For example, the function $h(x) = 9\sqrt{x - 3} + 4$ is a square root function. In Example 9, we graph the square root function $f(x) = \sqrt{x}$.

Example 9 Graphing a Square Root Function

Sketch the graph of $f(x) = \sqrt{x}$.

Solution

We list some input–output pairs in Table 2. We choose perfect-square inputs because we can find their outputs mentally. Since the radicand of $\sqrt{x}$ must be nonnegative, we cannot choose any negative numbers as inputs. Then we sketch the graph of f (see Fig. 4).

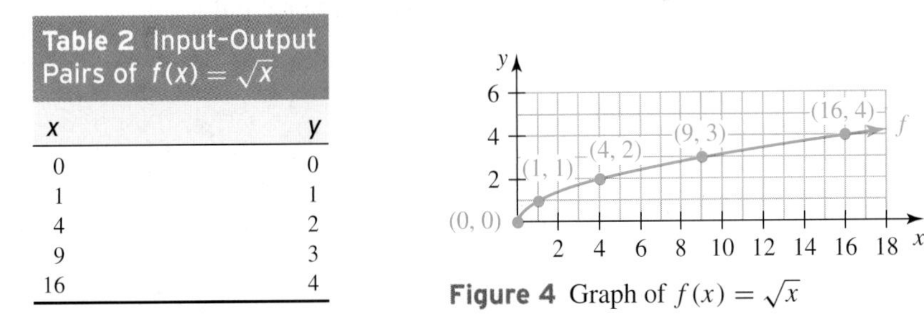

Table 2 Input-Output Pairs of $f(x) = \sqrt{x}$

x	y
0	0
1	1
4	2
9	3
16	4

Figure 4 Graph of $f(x) = \sqrt{x}$

We use a graphing calculator to verify our graph (see Fig. 5). ■

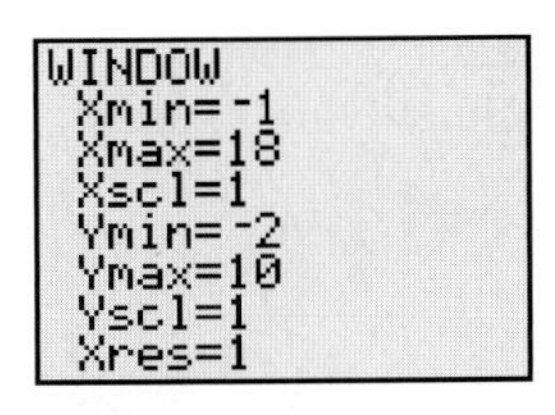

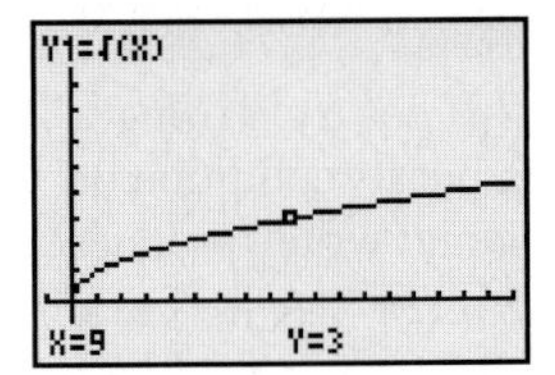

Figure 5 Verify the graph of $f(x) = \sqrt{x}$

We use a graphing calculator to draw the graphs of $y = \sqrt{x}$, $y = \sqrt[3]{x}$, and $y = \sqrt[4]{x}$ (see Fig. 6).

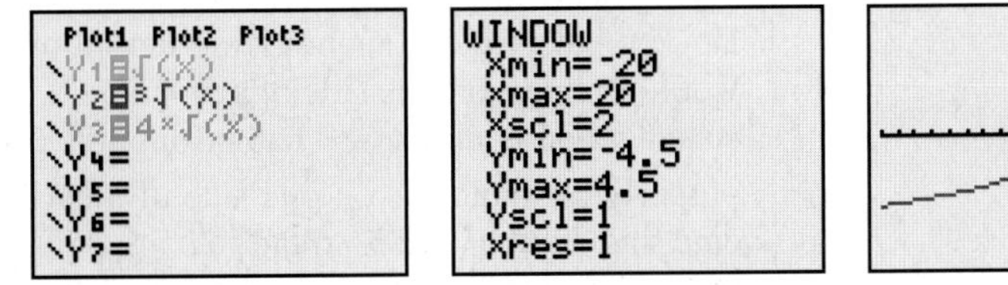

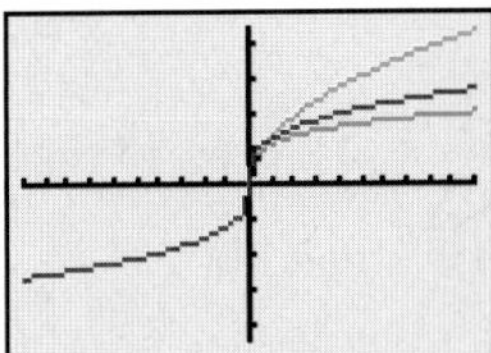

Figure 6 Graphs of $y = \sqrt{x}$, $y = \sqrt[3]{x}$, and $y = \sqrt[4]{x}$

Radical Model

A **radical model** is a radical function, or its graph, that describes an authentic situation. Radical functions can model a variety of situations, including the rise in temperature of an enclosed car, the period of a planet, the length of a braking car's skid marks, and the percentage of e-mails that are spam.

A **square root model** is a square root function, or its graph, that describes an authentic situation. We will work with such a model in Example 10.

Table 3 Percentages of Colleges and Universities Offering Distance Learning

Year	Percent
1997	44
1998	62
1999	69
2002	85
2004	87

Source: *International Data Corporation*

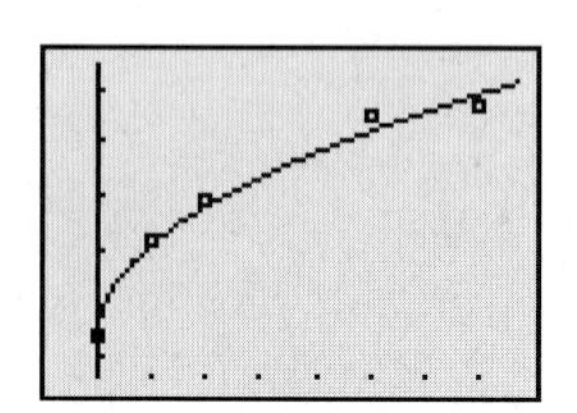

Figure 7 Check the fit

Example 10 Using a Square Root Model to Make a Prediction

The percentages of colleges and universities offering distance learning are shown in Table 3 for various years. Let $f(t)$ be the percentage of colleges and universities that offer distance learning at t years since 1997. A model of the situation is $f(t) = 17\sqrt{t} + 44$.

1. Use a graphing calculator to draw the graph of f and, in the same viewing window, the scattergram of the data. Does the graph of f fit the data well?
2. Predict the percentage of colleges and universities that will offer distance learning in 2007.

Solution

1. The graph of f and the scattergram of the data are shown in Fig. 7. The model appears to fit the data well.
2. We evaluate f at 10:

$$f(10) = 17\sqrt{10} + 44 \approx 97.76$$

About 97.8% of colleges and universities will offer distance learning in 2007, according to the model. ■

group exploration

Index property for radicals

Assume that $x \geq 0$.

1. Write the expression with as small an index as possible. [**Hint:** First write the expression in exponential form.]

   a. $\sqrt[8]{x^6}$ **b.** $\sqrt[6]{x^4}$ **c.** $\sqrt[16]{x^{10}}$
   d. $\sqrt[9]{x^3}$ **e.** $\sqrt[22]{x^4}$ **f.** $\sqrt[30]{x^{25}}$

2. Let k, m, and n be counting numbers, where m and n have no common factors. Write the expression $\sqrt[kn]{x^{km}}$ with as small an index as possible. Include the exponential form of the expression $\sqrt[kn]{x^{km}}$.
3. Describe what your result from Problem 2 tells you about simplifying a radical expression. Use this observation to simplify $\sqrt[20]{x^{16}}$ in one step.

TIPS FOR SUCCESS: Use 3 × 5 Cards

Do you have trouble memorizing definitions and properties? If so, try writing a word or phrase on one side of a 3 × 5 card. On the other side, put its definition or state a property and how it can be applied. For example, you could write "product property for radicals" on one side of a card and "$\sqrt[n]{ab} = \sqrt[n]{a}\sqrt[n]{b}$, where n is a counting number and $\sqrt[n]{a}$ and $\sqrt[n]{b}$ are defined" on the other side. You could also describe, in your own words, the meaning of the property and how you can apply it. Once you have completed a card for each definition and property, shuffle the cards and quiz yourself until you are confident that you know the definitions and properties and how to apply them. Quiz yourself again later to make sure that you have retained the information.

In addition to memorizing definitions and properties, it is important that you continue to strive to understand their meanings and how to apply them.

HOMEWORK 9.1

FOR EXTRA HELP ▶ Student Solutions Manual | PH Math/Tutor Center | MathXL® | MyMathLab

If the expression is in exponential form, write it in radical form. If it is in radical form, write it in exponential form.

1. $x^{2/5}$
2. $x^{3/8}$
3. $\sqrt[4]{x^3}$
4. $\sqrt[9]{x^5}$
5. $\sqrt{w}$
6. $\sqrt[3]{t}$
7. $(2x+9)^{3/7}$
8. $(5x+1)^{5/6}$
9. $\sqrt[7]{(3k+2)^4}$
10. $\sqrt[6]{8m+3}$

Simplify. Assume that each variable is nonnegative.

11. $\sqrt{50}$
12. $\sqrt{20}$
13. $\sqrt{x^8}$
14. $\sqrt{x^{18}}$
15. $\sqrt{36x^6}$
16. $\sqrt{4x^4}$
17. $\sqrt{5a^2b^{12}}$
18. $\sqrt{7a^{10}b^{14}}$
19. $\sqrt{x^9}$
20. $\sqrt{x^{15}}$
21. $\sqrt{24x^5}$
22. $\sqrt{12x^{13}}$
23. $\sqrt{80x^3y^8}$
24. $\sqrt{27x^{10}y^7}$
25. $\sqrt{200a^3b^5}$
26. $\sqrt{75a^{15}b^9}$
27. $\sqrt{(2x+5)^8}$
28. $\sqrt{(3x+4)^2}$
29. $\sqrt{(6t+3)^5}$
30. $\sqrt{(7w+1)^{13}}$
31. $\sqrt[3]{27}$
32. $\sqrt[5]{32}$
33. $\sqrt[6]{x^6}$
34. $\sqrt[9]{x^9}$
35. $\sqrt[3]{8x^3}$
36. $\sqrt[4]{16x^4}$
37. $\sqrt[5]{-32x^{20}}$
38. $\sqrt[3]{-27x^{18}}$
39. $\sqrt[4]{81a^{12}b^{28}}$
40. $\sqrt[5]{32a^{15}b^{30}}$
41. $\sqrt[6]{x^{17}}$
42. $\sqrt[8]{x^{25}}$
43. $\sqrt[3]{-125a^{17}b^{12}}$
44. $\sqrt[3]{-8a^{21}b^{29}}$
45. $\sqrt[5]{64x^{39}y^7}$
46. $\sqrt[4]{32x^{19}y^{13}}$
47. $\sqrt[5]{(6xy)^5}$
48. $\sqrt[7]{(4x^2y)^7}$
49. $\sqrt[4]{(3x+6)^4}$
50. $\sqrt[8]{(5x+2)^8}$
51. $\sqrt[5]{(4p+7)^{20}}$
52. $\sqrt[3]{(3k+5)^{12}}$
53. $\sqrt[6]{(2x+9)^{31}}$
54. $\sqrt[5]{(4x+5)^{43}}$

Simplify. (Write your result with as small an index n as possible.) Assume that each variable is nonnegative.

55. $\sqrt[8]{x^6}$
56. $\sqrt[6]{x^3}$
57. $\sqrt[6]{x^4}$
58. $\sqrt[9]{x^6}$
59. $\sqrt[12]{(2m+7)^{10}}$
60. $\sqrt[21]{(3r+5)^{14}}$
61. $\sqrt[6]{x^{14}}$
62. $\sqrt[8]{x^{22}}$
63. $\sqrt[6]{27}$
64. $\sqrt[8]{25}$
65. $\sqrt[4]{\sqrt[3]{p}}$
66. $\sqrt[5]{\sqrt{t}}$
67. $\sqrt[10]{16x^8}$
68. $\sqrt[12]{125x^9}$
69. $\sqrt[4]{\sqrt{ab}}$
70. $\sqrt[3]{\sqrt[5]{3w}}$

For $f(x)=\sqrt[5]{x}$, $g(x)=\sqrt[3]{3x+2}$, and $h(x)=2\sqrt{x}-5$, find the following.

71. $f(-32)$
72. $f(32)$
73. $g(2)$
74. $g(-22)$
75. $g(-7)$
76. $g(5)$
77. $h(49)$
78. $h(25)$
79. Graph $f(x)=2\sqrt[3]{x}$ by hand. [**Hint:** Evaluate f at some perfect cubes.]
80. Graph $g(x)=3\sqrt[4]{x}$ by hand. [**Hint:** Evaluate g at some perfect 4th powers.]
81. The average temperature rises above the *ambient temperature* (outside temperature) in an enclosed vehicle are shown in Table 4 for various elapsed times.

Table 4 Average Temperature Rises in an Enclosed Vehicle (for Ambient Temperatures between 72°F and 96°F)

Elapsed Time (minutes)	Average Temperature Rise (°F)
0	0
10	19
20	29
30	34
40	38
50	41
60	43

Source: *Jan Null, Department of Geosciences, San Francisco State University*

Let $f(t)$ be the average temperature rise (°F) in a vehicle at t minutes after the vehicle is enclosed. ("Cracking" the windows had little effect on the data.) A model of the situation is $f(t)=8.5\sqrt[5]{t^2}$.

a. Use a graphing calculator to draw the graph of the model and, in the same viewing window, the scattergram of the data. Does the model fit the data well?

b. Estimate the average temperature rise 24 minutes after the vehicle is enclosed if the ambient temperature is 85°F.

c. Estimate the *temperature* inside the car 45 minutes after the vehicle is enclosed if the ambient temperature is 90°F.

d. Since 1998, more than 300 children have died of hyperthermia after being left inside a hot vehicle. A body core temperature of 107°F is usually fatal. Use TRACE to estimate how long it would take for the temperature in an enclosed vehicle to reach 107°F if the ambient temperature is 80°F.

82. The *Beaufort wind scale,* which ranges from 0 to 12, describes wind intensities. Some of the even-numbered *Beaufort numbers* and the corresponding wind speeds and weather conditions are shown in Table 5.

Table 5 Beaufort Numbers, Wind Speeds, and Weather Conditions

Beaufort Number	Wind Speed Group (miles per hour)	Wind Speed Used to Represent Wind Speed Group (miles per hour)	Weather Conditions
0	0	0	calm
2	4–7	5.5	light breeze
4	13–18	15.5	moderate breeze
6	25–31	28.0	strong breeze
8	39–46	42.5	gale
10	55–63	59.0	storm

Source: Stormfax® Weather Almanac

Let $V = f(B)$ be the wind speed (in miles per hour) that corresponds to the Beaufort number B. A model of the situation is $f(B) = 1.87\sqrt{B^3}$.

a. Use a graphing calculator to draw the graph of the model and, in the same viewing window, the scattergram of the data. Does the model fit the data well?

b. The U.S. Coast Guard issues a small-craft advisory when the Beaufort number is 6 or 7. Estimate the wind speed of a near gale, which has Beaufort number 7.

c. Estimate the wind speed in a violent storm, which has Beaufort number 11.

d. The quadratic function $Q(t) = 0.33B^2 + 2.72B - 0.46$ is a model of the situation. Does Q fit the data well? Which model, f or Q, describes the wind speed better when the weather is calm? Explain.

83. A *tsunami* is a fast-moving sea wave typically caused by an underwater earthquake. An earthquake in the Indian Ocean in 2004 caused tsunamis that killed more than 265,000 people. The speed of a tsunami (in meters per second) $f(d)$ can be modeled by the function $f(d) = \sqrt{gd}$, where g is a constant approximately equal to 9.8 and d is the average depth (in meters) of the water.

a. The Indian Ocean has an average depth of 3890 meters. Estimate the speed in meters per second of a tsunami in the Indian Ocean.

b. Find $f(1000)$, $f(2000)$, and $f(3000)$. Is f an increasing or a decreasing function? What does that mean in this situation?

c. As a tsunami slows down, it gets taller. As a tsunami approaches the shore, what happens to the speed and height of the tsunami?

d. Convert your result in part (a) to units of miles per hour. (One mile is approximately 1609 meters.)

84. Suppose that a driver traveling on a dry road slams on the brakes. Let D be the distance (in feet) that the car will skid and S be the speed (in miles per hour) before braking. The relationship between D and S is described by the model $S = \sqrt{30FD}$, where F is a drag factor—a measure of the roughness of the road surface. The drag factor on new concrete is 0.95; on polished concrete or asphalt, 0.75.

a. A motorist involved in an accident claims that he was driving at the speed limit, 50 miles per hour. The car's skid marks on the new concrete road are 120 feet long. Assuming that the motorist applied the brakes suddenly, estimate his speed before braking.

b. If the motorist first applied the brakes lightly, then forcefully applied them after a few seconds, explain why your result in part (a) is an underestimate of his speed before braking.

85. A student says that $\sqrt{x^{16}}$ is equal to x^4, since $\sqrt{16} = 4$. What would you tell this student?

86. Is the statement $\sqrt[n]{a+b} = \sqrt[n]{a} + \sqrt[n]{b}$ true or false, for all values of a and b where $\sqrt[n]{a}$ and $\sqrt[n]{b}$ are defined? [**Hint:** Substitute values of a, b, and n in the two expressions and compare the results.]

87. Write the expression $\sqrt[n]{\sqrt[n]{x}}$ with one radical sign. [**Hint:** Write the expression in exponential form.]

88. a. Use a graphing calculator to draw a graph of the given function. What do you notice about the graph?

i. $y = \sqrt[3]{x^3}$

ii. $y = \sqrt[5]{x^5}$

iii. $y = \sqrt[7]{x^7}$

b. Compare your graphs in part (a) with the graph of $y = x$. Explain how your observation relates to the power property for radicals.

c. Use a graphing calculator to draw a graph of the given function. What do you notice about the graph?

i. $y = \sqrt{x^2}$

ii. $y = \sqrt[4]{x^4}$

iii. $y = \sqrt[6]{x^6}$

d. Compare your graphs in part (c) with the graph of $y = |x|$. Explain how your observation relates to the power property for radicals. [***Graphing Calculator:*** The absolute value choice "abs(" is located in the "NUM" menu. ("NUM" is in the "MATH" menu.)]

89. In this exercise, you will explore another version of the power property. Simplify.

a. $(\sqrt{x})^2$ **b.** $(\sqrt[3]{x})^3$ **c.** $(\sqrt[n]{x})^n$

90. Describe how to simplify a radical expression. (See page 4 for guidelines on writing a good response.)

Related Review

Assume that each variable is nonnegative.

91. a. Simplify $\sqrt{16x^4y^6}$ by using the product property for radicals.

b. Simplify $\sqrt{16x^4y^6}$ first by writing the expression in exponential form, then by using properties of exponents.

c. Compare the results that you found in parts (a) and (b).

92. a. Simplify $\sqrt[3]{8x^6y^9}$ by using the product property for radicals.

b. Simplify $\sqrt[3]{8x^6y^9}$ first by writing the expression in exponential form, then by using properties of exponents.

c. Compare the results you found in parts (a) and (b).

Expressions, Equations, Functions, and Graphs

Perform the indicated instruction. Then use words such as linear, quadratic, cubic, exponential, logarithmic, rational, radical, polynomial, degree, function, one variable, *and* two variables *to describe the expression, equation, or system.*

93. Solve $\dfrac{2x}{x^2+x-6} - \dfrac{3x-1}{x^2+6x+9} = \dfrac{-3}{x+3}$.

94. Solve $2x^2 - 4x - 3 = 0$ by completing the square.

95. Find the difference $\dfrac{2x}{x^2+x-6} - \dfrac{3x-1}{x^2+6x+9}$.

96. Graph $f(x) = 2x^2 - 4x - 3$ by hand.

97. Find the domain of $f(x) = \dfrac{2x}{x^2+x-6}$.

98. Solve $4(2x-5)^2 = 48$.

9.2 ADDING, SUBTRACTING, AND MULTIPLYING RADICAL EXPRESSIONS

Objectives

- Add, subtract, and multiply radical expressions.
- Know another version of the power property for radicals.
- Simplify the square of a radical expression with two terms.

In this section, we add, subtract, and multiply radical expressions. We also simplify the square of a radical with two terms.

Adding and Subtracting Radical Expressions

We can use the distributive law to add like terms, such as $2x$ and $7x$:

$$2x + 7x = (2 + 7)x = 9x$$

How do we add (or subtract) radical expressions? We can again use the distributive law if the radicals are like radicals. We say that $2\sqrt[3]{x}$ and $7\sqrt[3]{x}$ are like radicals, because they have the same index *and* the same radicand. In general, radicals that have the same index and the same radicand are called **like radicals.**

We add the like radicals $2\sqrt[3]{x}$ and $7\sqrt[3]{x}$ as follows:

$$2\sqrt[3]{x} + 7\sqrt[3]{x} = (2 + 7)\sqrt[3]{x} = 9\sqrt[3]{x}$$

To add or subtract like radicals, we use the distributive law. When we add or subtract like radicals, we say that we *combine like radicals.*

Example 1 Combining Like Radicals

Combine like radicals.

1. $3\sqrt{x} + 6\sqrt{x}$
2. $4\sqrt[5]{3xy^2} - 2\sqrt[5]{3xy^2}$
3. $4\sqrt[3]{x} + 5\sqrt[6]{x}$
4. $3\sqrt[4]{x} - 2\sqrt[4]{x+1}$

Solution

1. $3\sqrt{x} + 6\sqrt{x} = (3 + 6)\sqrt{x}$ Distributive law
   $= 9\sqrt{x}$ Add.
2. $4\sqrt[5]{3xy^2} - 2\sqrt[5]{3xy^2} = (4 - 2)\sqrt[5]{3xy^2}$ Distributive law
   $= 2\sqrt[5]{3xy^2}$ Subtract.
3. Since the radicals $4\sqrt[3]{x}$ and $5\sqrt[6]{x}$ have different indexes, we cannot use the distributive law. The expression $4\sqrt[3]{x} + 5\sqrt[6]{x}$ is already in simplified form.
4. Since the radicals $3\sqrt[4]{x}$ and $2\sqrt[4]{x+1}$ have different radicands, we cannot use the distributive law. The expression $3\sqrt[4]{x} - 2\sqrt[4]{x+1}$ is already in simplified form.

■

Example 2 Performing Operations with Radical Expressions

Perform the indicated operations.

1. $3\sqrt[4]{x} + 4\sqrt{x} + 2\sqrt[4]{x} + 7\sqrt{x}$
2. $3(5\sqrt[3]{x+1} - 2) - 4\sqrt[3]{x+1}$

Solution

1. $3\sqrt[4]{x} + 4\sqrt{x} + 2\sqrt[4]{x} + 7\sqrt{x} = (4\sqrt{x} + 7\sqrt{x}) + (3\sqrt[4]{x} + 2\sqrt[4]{x})$ Group like radicals.
   $= (4 + 7)\sqrt{x} + (3 + 2)\sqrt[4]{x}$ Distributive law
   $= 11\sqrt{x} + 5\sqrt[4]{x}$ Add.

2. $3(5\sqrt[3]{x+1}-2)-4\sqrt[3]{x+1} = 3\cdot 5\sqrt[3]{x+1}-3\cdot 2-4\sqrt[3]{x+1}$ Distributive law
$= 15\sqrt[3]{x+1}-4\sqrt[3]{x+1}-6$ Group like radicals.
$= (15-4)\sqrt[3]{x+1}-6$ Distributive law
$= 11\sqrt[3]{x+1}-6$ Subtract. ■

Sometimes, simplifying radicals will allow us to combine like radicals.

Example 3 Adding or Subtracting Radical Expressions

Perform the indicated operation.

1. $\sqrt{45w}+\sqrt{20w}$
2. $5b\sqrt{3a^3}-a\sqrt{12ab^2}$

Solution

1. $\sqrt{45w}+\sqrt{20w} = \sqrt{9\cdot 5w}+\sqrt{4\cdot 5w}$ 9 and 4 are perfect squares.
$= \sqrt{9}\sqrt{5w}+\sqrt{4}\sqrt{5w}$ $\sqrt[n]{ab}=\sqrt[n]{a}\sqrt[n]{b}$
$= 3\sqrt{5w}+2\sqrt{5w}$ $\sqrt{9}=3$; $\sqrt{4}=2$
$= (3+2)\sqrt{5w}$ Distributive law
$= 5\sqrt{5w}$ Add.

2. $5b\sqrt{3a^3}-a\sqrt{12ab^2} = 5b\sqrt{a^2\cdot 3a}-a\sqrt{4\cdot b^2\cdot 3a}$ a^2, 4, and b^2 are perfect squares.
$= 5b\sqrt{a^2}\sqrt{3a}-a\sqrt{4}\sqrt{b^2}\sqrt{3a}$ $\sqrt[n]{ab}=\sqrt[n]{a}\sqrt[n]{b}$
$= 5b\cdot a\cdot\sqrt{3a}-a\cdot 2\cdot b\cdot\sqrt{3a}$ $\sqrt{a^2}=a$ for $a\geq 0$; $\sqrt{4}=2$; $\sqrt{b^2}=b$ for $b\geq 0$
$= 5ab\sqrt{3a}-2ab\sqrt{3a}$ Rearrange factors.
$= (5ab-2ab)\sqrt{3a}$ Distributive law
$= 3ab\sqrt{3a}$ Combine like terms. ■

Example 4 Subtracting Radical Expressions

Find the difference $2\sqrt[3]{16x^4}-4x\sqrt[3]{54x}$.

Solution

$2\sqrt[3]{16x^4}-4x\sqrt[3]{54x} = 2\sqrt[3]{8\cdot x^3\cdot 2x}-4x\sqrt[3]{27\cdot 2x}$ 8, x^3, and 27 are perfect cubes.
$= 2\sqrt[3]{8}\sqrt[3]{x^3}\sqrt[3]{2x}-4x\sqrt[3]{27}\sqrt[3]{2x}$ $\sqrt[n]{ab}=\sqrt[n]{a}\sqrt[n]{b}$
$= 2\cdot 2\cdot x\cdot\sqrt[3]{2x}-4x\cdot 3\cdot\sqrt[3]{2x}$ $\sqrt[3]{8}=2$, $\sqrt[3]{x^3}=x$, $\sqrt[3]{27}=3$
$= 4x\sqrt[3]{2x}-12x\sqrt[3]{2x}$ Multiply.
$= (4x-12x)\sqrt[3]{2x}$ Distributive law
$= -8x\sqrt[3]{2x}$ Combine like terms. ■

Multiplying Radical Expressions

Next, we multiply radical expressions. We will use the product property

$$\sqrt[n]{ab}=\sqrt[n]{a}\sqrt[n]{b}, \qquad \text{where } \sqrt[n]{a} \text{ and } \sqrt[n]{b} \text{ are defined}$$

Here, we multiply $5\sqrt{2x}$ and $4\sqrt{3}$ and simplify the result:

$5\sqrt{2x}\cdot 4\sqrt{3} = 5\cdot 4\cdot\sqrt{2x}\cdot\sqrt{3}$ Rearrange factors.
$= 5\cdot 4\cdot\sqrt{2x\cdot 3}$ Product property
$= 20\sqrt{6x}$ Multiply.

It is good practice to check whether the product of radical expressions can be simplified.

Example 5 Finding Products of Radical Expressions

Find the product.

1. $2\sqrt{6x} \cdot 5\sqrt{2x}$
2. $3\sqrt{5x}(4\sqrt{x} - \sqrt{5})$

Figure 8 Verify the work

Solution

1.
$$
\begin{aligned}
2\sqrt{6x} \cdot 5\sqrt{2x} &= 2 \cdot 5\sqrt{6x} \cdot \sqrt{2x} && \text{Rearrange factors.} \\
&= 2 \cdot 5 \cdot \sqrt{6x \cdot 2x} && \text{Product property} \\
&= 10 \cdot \sqrt{12x^2} && \text{Multiply.} \\
&= 10 \cdot \sqrt{4 \cdot x^2 \cdot 3} && \text{4 and } x^2 \text{ are perfect squares.} \\
&= 10 \cdot 2x\sqrt{3} && \sqrt{4} = 2,\ \sqrt{x^2} = x \text{ for } x \ge 0 \\
&= 20x\sqrt{3} && \text{Multiply.}
\end{aligned}
$$

We verify our work by comparing graphing calculator tables for $y = 2\sqrt{6x} \cdot 5\sqrt{2x}$ and $y = 20x\sqrt{3}$, for $x \ge 0$ (see Fig. 8).

2.
$$
\begin{aligned}
3\sqrt{5x}(4\sqrt{x} - \sqrt{5}) &= 3\sqrt{5x} \cdot 4\sqrt{x} - 3\sqrt{5x} \cdot \sqrt{5} && \text{Distributive law} \\
&= 3 \cdot 4 \cdot \sqrt{5x}\sqrt{x} - 3\sqrt{5x}\sqrt{5} && \text{Rearrange factors.} \\
&= 12\sqrt{5x \cdot x} - 3\sqrt{5x \cdot 5} && \text{Product property} \\
&= 12\sqrt{x^2 \cdot 5} - 3\sqrt{25x} && x^2 \text{ and 25 are perfect squares.} \\
&= 12 \cdot x \cdot \sqrt{5} - 3 \cdot 5 \cdot \sqrt{x} && \sqrt{x^2} = x \text{ for } x \ge 0,\ \sqrt{25} = 5 \\
&= 12x\sqrt{5} - 15\sqrt{x} && \text{Multiply.}
\end{aligned}
$$

■

Note that if $\sqrt[n]{x}$ is defined, then

$$(\sqrt[n]{x})^n = x^{n/n} = x^1 = x$$

This property is helpful in simplifying powers or products of radical expressions.

Another Version of the Power Property for Radicals

If $\sqrt[n]{x}$ is defined, then

$$(\sqrt[n]{x})^n = x$$

In words: The nth power of the nth root of a number is that number.

In particular, we have $(\sqrt{x})^2 = x$ if $x \ge 0$.

Example 6 Simplifying Radical Expressions

Simplify.

1. $(2\sqrt{x} - 7)(3\sqrt{x} + 4)$
2. $(2\sqrt{3x} + 5)(2\sqrt{3x} - 5)$

Solution

1. Multiply each term of the first factor by each term of the second factor, and combine like radicals:

$$
\begin{aligned}
(2\sqrt{x} - 7)(3\sqrt{x} + 4) &= 2\sqrt{x} \cdot 3\sqrt{x} + 2\sqrt{x} \cdot 4 - 7 \cdot 3\sqrt{x} - 7 \cdot 4 && \text{Multiply pairs of terms.} \\
&= 6\sqrt{x^2} + 8\sqrt{x} - 21\sqrt{x} - 28 && \text{Simplify.} \\
&= 6x - 13\sqrt{x} - 28 && (\sqrt[n]{x})^n = x\text{; combine like radicals.}
\end{aligned}
$$

2. We expand $(2\sqrt{3x}+5)(2\sqrt{3x}-5)$ by using $(A+B)(A-B)=A^2-B^2$:

$$
\begin{aligned}
(2\sqrt{3x}+5)(2\sqrt{3x}-5) &= (2\sqrt{3x})^2-5^2 && (A+B)(A-B)=A^2-B^2\\
&= 2^2(\sqrt{3x})^2-5^2 && (AB)^2=A^2B^2\\
&= 4(3x)-25 && \text{Simplify.}\\
&= 12x-25 && \text{Multiply.}
\end{aligned}
$$

We can verify our result by comparing tables for $y=(2\sqrt{3x}+5)(2\sqrt{3x}-5)$ and $y=12x-25$, for $x\geq 0$. ■

Example 7 Simplifying the Square of a Radical Expression with Two Terms

Simplify $(x-\sqrt{3})^2$.

Solution

To begin, we substitute x for A and $\sqrt{3}$ for B in the property for the square of a difference:

$$
\begin{aligned}
(A-B)^2 &= A^2-2AB+B^2\\
(x-\sqrt{3})^2 &= x^2-2(x)\sqrt{3}+(\sqrt{3})^2 && \text{Substitute } x \text{ for } A \text{ and } \sqrt{3} \text{ for } B.\\
&= x^2-2x\sqrt{3}+3 && (\sqrt{x})^2=x \text{ for } x\geq 0
\end{aligned}
$$

Another way to simplify $(x-\sqrt{3})^2$ is to use the fact that $C^2=CC$ and multiply pairs of terms:

$$
\begin{aligned}
(x-\sqrt{3})^2 &= (x-\sqrt{3})(x-\sqrt{3}) && C^2=CC\\
&= x^2-x\sqrt{3}-x\sqrt{3}+\sqrt{3}\sqrt{3} && \text{Multiply pairs of terms.}\\
&= x^2-2x\sqrt{3}+3 && \text{Combine like radicals; } \sqrt{x}\sqrt{x}=x \text{ for } x\geq 0
\end{aligned}
$$

■

WARNING When we simplify $(x+k)^2$, it is important to remember the middle term of $x^2+2kx+k^2$. Likewise, when we simplify $(x-k)^2$, it is important to remember the middle term of $x^2-2kx+k^2$. Do not make the following typical error in simplifying $(\sqrt{x}+\sqrt{5})^2$:

$$
\begin{aligned}
(\sqrt{x}+\sqrt{5})^2 &= (\sqrt{x})^2+(\sqrt{5})^2=x+5 && \text{Incorrect}\\
(\sqrt{x}+\sqrt{5})^2 &= (\sqrt{x})^2+2\sqrt{5}\sqrt{x}+(\sqrt{5})^2=x+2\sqrt{5x}+5 && \text{Correct}
\end{aligned}
$$

Example 8 Simplifying the Square of a Radical Expression with Two Terms

Simplify $(\sqrt{a}+3\sqrt{b})^2$.

Solution

We use the property for the square of a sum:

$$
\begin{aligned}
(\sqrt{a}+3\sqrt{b})^2 &= (\sqrt{a})^2+2\cdot\sqrt{a}\cdot 3\sqrt{b}+(3\sqrt{b})^2 && (A+B)^2=A^2+2AB+B^2\\
&= (\sqrt{a})^2+6\sqrt{ab}+3^2(\sqrt{b})^2 && \sqrt[n]{a}\sqrt[n]{b}=\sqrt[n]{ab};\ (xy)^n=x^ny^n\\
&= a+6\sqrt{ab}+9b && (\sqrt{x})^2=x \text{ for } x\geq 0;\ 3^2=9
\end{aligned}
$$

■

In Example 9, we find products of radical expressions with indexes other than $n=2$.

Example 9 Multiplying Radical Expressions

Find the product.

1. $(2\sqrt[5]{x^2})(7\sqrt[5]{x^4})$

2. $(\sqrt[4]{x^3}+5)(\sqrt[4]{x^3}-6)$

Solution

1. $(2\sqrt[5]{x^2})(7\sqrt[5]{x^4}) = 2\cdot 7\cdot \sqrt[5]{x^2}\sqrt[5]{x^4}$ Rearrange factors.

$= 2\cdot 7\sqrt[5]{x^2\cdot x^4}$ $\sqrt[n]{a}\sqrt[n]{b}=\sqrt[n]{ab}$

$= 14\sqrt[5]{x^6}$ $b^mb^n = b^{m+n}$

$= 14x\sqrt[5]{x}$ $\sqrt[5]{x^6}=\sqrt[5]{x^5x^1}=x\sqrt[5]{x}$

2. $(\sqrt[4]{x^3}+5)(\sqrt[4]{x^3}-6)$

$= \sqrt[4]{x^3}\sqrt[4]{x^3} - \sqrt[4]{x^3}\cdot 6 + 5\cdot\sqrt[4]{x^3} - 5\cdot 6$ Multiply pairs of terms.

$= \sqrt[4]{x^3\cdot x^3} - 6\sqrt[4]{x^3} + 5\sqrt[4]{x^3} - 30$ Product property

$= \sqrt[4]{x^6} - \sqrt[4]{x^3} - 30$ Multiply; combine like radicals.

$= x\sqrt[4]{x^2} - \sqrt[4]{x^3} - 30$ $\sqrt[4]{x^6}=\sqrt[4]{x^4\cdot x^2}=x\sqrt[4]{x^2}$ for $x\ge 0$

$= x\sqrt{x} - \sqrt[4]{x^3} - 30$ $\sqrt[4]{x^2}=x^{2/4}=x^{1/2}=\sqrt{x}$ for $x\ge 0$

We cannot combine the radicals $\sqrt{x}$ and $\sqrt[4]{x^3}$, since the indexes (and the radicands) are different. So, we are done. ■

To multiply two radicals that have the same index, we use the product property. How do we multiply two radicals with *different* indexes? Here, we find the product $\sqrt[3]{x}\cdot\sqrt[4]{x}$:

$\sqrt[3]{x}\cdot\sqrt[4]{x} = x^{\frac{1}{3}}\cdot x^{\frac{1}{4}}$ Write in exponential form.

$= x^{\frac{1}{3}+\frac{1}{4}}$ $a^ma^n = a^{m+n}$

$= x^{\frac{4}{12}+\frac{3}{12}}$ Get a common denominator.

$= x^{\frac{7}{12}}$ Add numerators; keep common denominator.

$= \sqrt[12]{x^7}$ Write in radical form.

Multiplying Two Radicals with Different Indexes but the Same Radicand

To multiply two radicals with different indexes but the same radicand,

1. Write the radicals in exponential form.
2. Use exponential properties to simplify the expression involving exponents.
3. Write the simplified expression in radical form.

Example 10 Simplifying Radical Expressions

Perform the operations. Assume that $x \ge 0$.

1. $2\sqrt{x}(\sqrt[3]{x}-5)$

2. $(\sqrt[3]{x}+3\sqrt[5]{x^2})^2$

Solution

1. $2\sqrt{x}(\sqrt[3]{x}-5) = 2\sqrt{x}\sqrt[3]{x} - 2\sqrt{x}\cdot 5$ Distributive law

$= 2x^{\frac{1}{2}}x^{\frac{1}{3}} - 10\sqrt{x}$ Write in exponential form.

$= 2x^{\frac{1}{2}+\frac{1}{3}} - 10\sqrt{x}$ $a^ma^n = a^{m+n}$

$= 2x^{\frac{3}{6}+\frac{2}{6}} - 10\sqrt{x}$ Get a common denominator.

$= 2x^{\frac{5}{6}} - 10\sqrt{x}$ Add numerators; keep common denominator.

$= 2\sqrt[6]{x^5} - 10\sqrt{x}$ Write in radical form.

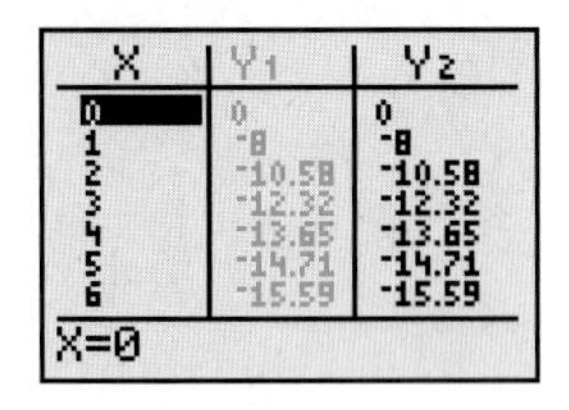

X	Y1	Y2
0	0	0
1	-8	-8
2	-10.58	-10.58
3	-12.32	-12.32
4	-13.65	-13.65
5	-14.71	-14.71
6	-15.59	-15.59

X=0

Figure 9 Verify the work

We can verify our result by comparing tables for $y = 2\sqrt{x}\left(\sqrt[3]{x} - 5\right)$ and $y = 2\sqrt[6]{x^5} - 10\sqrt{x}$, for $x \geq 0$ (see Fig. 9).

2. $\left(\sqrt[3]{x} + 3\sqrt[5]{x^2}\right)^2$

$$= \left(\sqrt[3]{x}\right)^2 + 2\left(\sqrt[3]{x}\right)\left(3\sqrt[5]{x^2}\right) + \left(3\sqrt[5]{x^2}\right)^2 \qquad (A+B)^2 = A^2 + 2AB + B^2$$

$$= \left(\sqrt[3]{x}\right)^2 + 6x^{\frac{1}{3}}x^{\frac{2}{5}} + 3^2\left(\sqrt[5]{x^2}\right)^2 \qquad \text{Write in exponential form.}$$

$$= \sqrt[3]{x^2} + 6x^{\frac{1}{3}+\frac{2}{5}} + 9\sqrt[5]{\left(x^2\right)^2} \qquad \left(\sqrt[n]{x}\right)^m = \sqrt[n]{x^m}$$

$$= \sqrt[3]{x^2} + 6x^{\frac{5}{15}+\frac{6}{15}} + 9\sqrt[5]{x^4} \qquad \text{Get a common denominator.}$$

$$= \sqrt[3]{x^2} + 6x^{\frac{11}{15}} + 9\sqrt[5]{x^4} \qquad \text{Add numerators; keep common denominator.}$$

$$= \sqrt[3]{x^2} + 6\sqrt[15]{x^{11}} + 9\sqrt[5]{x^4} \qquad \text{Write in radical form.}$$

Another way to simplify $\left(\sqrt[3]{x} + 3\sqrt[5]{x^2}\right)^2$ is to use the fact that $C^2 = CC$ and multiply pairs of terms (try it). ■

Simplifying a Radical Expression

To simplify a radical expression,

1. Perform any indicated multiplications.
2. Combine like radicals.
3. For any radical with index n, write the radicand as a product of one or more perfect nth powers and another expression that has no factors that are perfect nth powers. Then apply the product property for radicals.
4. Write any radicals with as small an index as possible.

Depending on the radical expression, we may need to perform these steps in a different order or return to a step at a later stage in the process of simplifying the expression. We will discuss more ways to simplify radical expressions in Section 9.3.

group exploration

Looking ahead: Rationalizing the denominator

In Section 7.3, you "rationalized the denominator" of fractions of the form $\frac{1}{\sqrt{a}}$ by finding an equivalent expression that does not have a radical in any denominator. Here, you will explore how to rationalize the denominator of a fraction with a denominator that is a sum or a difference involving radicals.

1. Perform the indicated multiplication.
   a. $\left(x - \sqrt{2}\right)\left(x + \sqrt{2}\right)$ b. $\left(x + \sqrt{5}\right)\left(x - \sqrt{5}\right)$
   c. $\left(\sqrt{x} - 4\right)\left(\sqrt{x} + 4\right)$ d. $\left(\sqrt{x} + 3\right)\left(\sqrt{x} - 3\right)$
2. What patterns do you notice from your work in Problem 1?
3. Rationalize the denominator of $\frac{1}{\sqrt{x} - 7}$ by performing the multiplication
$$\frac{1}{\sqrt{x} - 7} \cdot \frac{\sqrt{x} + 7}{\sqrt{x} + 7}$$
   Use graphing calculator tables to verify your work.
4. Rationalize the denominator of the expression $\frac{1}{\sqrt{x} + 5}$.
5. Describe how to rationalize the denominator of a radical expression.

HOMEWORK 9.2

Simplify. Use a graphing calculator table to compare your result with the original expression when possible. Assume that each variable is nonnegative.

1. $4\sqrt{x}+5\sqrt{x}$

2. $8\sqrt{x}-4\sqrt{x}$

3. $2\sqrt[3]{5x^2y}-6\sqrt[3]{5x^2y}$

4. $5\sqrt[4]{2xy^3}-7\sqrt[4]{2xy^3}$

5. $3\sqrt{5a}+2\sqrt{3b}-6\sqrt{3b}+7\sqrt{5a}$

6. $4\sqrt{7b}-\sqrt{2a}-4\sqrt{2a}+9\sqrt{7b}$

7. $2\sqrt{x}+5-7\sqrt[3]{x}-9+5\sqrt[3]{x}$

8. $4-6\sqrt[4]{x}+3\sqrt[3]{x}-1-8\sqrt[4]{x}$

9. $6\sqrt[3]{x-1}-3\sqrt[3]{x-1}-2\sqrt{x-1}$

10. $4\sqrt[7]{3x+1}+3\sqrt[7]{3x+1}-5\sqrt[3]{3x+1}$

11. $3.7\sqrt[4]{x}-1.1\sqrt[4]{x}-4.2\sqrt[6]{x}+4.2\sqrt[6]{x}$

12. $4.1\sqrt{x}-2.9\sqrt[3]{x}-5.8\sqrt[3]{x}+2.3\sqrt{x}$

13. $3(7-\sqrt{x}+2)-(\sqrt{x}+2)$

14. $4(1-3\sqrt{x}-8)-(5\sqrt{x}-3)$

15. $7(\sqrt[3]{x}+1)-7(\sqrt[3]{x}-1)$

16. $5(\sqrt[4]{a}-2)+5(2-\sqrt[4]{a})$

17. $\sqrt{12b}+\sqrt{75b}$

18. $\sqrt{8x}-\sqrt{18x}$

19. $\sqrt{18x^5}+2x\sqrt{50x^3}$

20. $\sqrt{27x^7}+2x^2\sqrt{12x^3}$

21. $5\sqrt{4x^3}-x\sqrt{36x}$

22. $2\sqrt{9x^3}-x\sqrt{49x}$

23. $3\sqrt{81x^2}-2\sqrt{100x^2}$

24. $2\sqrt{36x^2}+5\sqrt{16x^2}$

25. $a\sqrt{12b^3}+b\sqrt{75ba^2}$

26. $b\sqrt{20a^5}-a\sqrt{45a^3b^2}$

27. $\sqrt[3]{27x^5}-x\sqrt[3]{8x^2}$

28. $\sqrt[3]{54x^7}+x\sqrt[3]{16x^4}$

29. $y\sqrt[4]{16x^{11}y^4}-3x\sqrt[4]{x^7y^8}$

30. $7x^2\sqrt[5]{x^3y^{10}}+3y\sqrt[5]{32x^{13}y^5}$

Simplify. Use a graphing calculator table to verify your result when possible. Assume that each variable is nonnegative.

31. $3\sqrt{x}\cdot 2\sqrt{x}$

32. $-5\sqrt{x}\cdot 4\sqrt{x}$

33. $-2\sqrt{5x}\cdot 4\sqrt{3x}$

34. $-3\sqrt{10x}\cdot 2\sqrt{5x}$

35. $2\sqrt{7t}(\sqrt{7t}-\sqrt{2t})$

36. $4\sqrt{2k}(\sqrt{8}-3\sqrt{k})$

37. $(2\sqrt{x}+6)(5\sqrt{x}+4)$

38. $(3\sqrt{x}+7)(2\sqrt{x}+5)$

39. $(4\sqrt{x}+\sqrt{3})(2\sqrt{x}-\sqrt{5})$

40. $(5\sqrt{x}-\sqrt{2})(3\sqrt{x}-\sqrt{3})$

41. $(5\sqrt{a}+\sqrt{b})(\sqrt{a}-2\sqrt{b})$

42. $(2\sqrt{a}+\sqrt{b})(\sqrt{a}-\sqrt{b})$

43. $(1-\sqrt{w})(1+\sqrt{w})$

44. $(2+3\sqrt{p})(2-3\sqrt{p})$

45. $(7x+\sqrt{5})(7x-\sqrt{5})$

46. $(4\sqrt{x}+\sqrt{3})(4\sqrt{x}-\sqrt{3})$

47. $(2\sqrt{a}-\sqrt{b})(2\sqrt{a}+\sqrt{b})$

48. $(3\sqrt{a}-\sqrt{b})(3\sqrt{a}+\sqrt{b})$

49. $(5+6\sqrt{x})^2$

50. $(3\sqrt{x}+2)^2$

51. $(4\sqrt{x}-\sqrt{5})^2$

52. $(2\sqrt{x}-\sqrt{7})^2$

53. $(\sqrt{a}+2\sqrt{b})^2$

54. $(3\sqrt{a}-\sqrt{b})^2$

55. $(\sqrt{2x-5}+3)^2$

56. $(\sqrt{3x+4}-5)^2$

57. $\sqrt{x}\sqrt[5]{x}$

58. $\sqrt[4]{x}\sqrt[6]{x}$

59. $\sqrt[5]{x^4}\sqrt[5]{x^3}$

60. $\sqrt[4]{3x^2}\sqrt[4]{3x^2}$

61. $-5\sqrt{m}(\sqrt[4]{2m}-4)$

62. $-4\sqrt[3]{t}(\sqrt{t}+3)$

63. $(\sqrt[3]{x}+1)^2$

64. $(\sqrt[4]{x}-5)^2$

65. $(\sqrt[4]{k}-\sqrt[3]{k})^2$

66. $(2\sqrt[5]{r}+\sqrt{r})^2$

67. $(2\sqrt{x}-6)(3\sqrt[3]{x}+1)$

68. $(4\sqrt[3]{x^2}+1)(5\sqrt[4]{x^2}+2)$

69. $(3\sqrt[4]{x}+5)(3\sqrt[4]{x}-5)$

70. $(2\sqrt[5]{x}+1)(3\sqrt[5]{x}-2)$

71. The flow rate r (in gallons per minute) of water from the nozzle of a firefighter's hose can be modeled by the formula $r=30d^2\sqrt{P}$, where d is the nozzle diameter (in inches) and P is the nozzle pressure (in pounds per square inch). The flow rates of solid bore nozzles are shown in Table 6 for various nozzle pressures and diameters.

Table 6 Flow Rates of Solid Bore Nozzles (gallons per minute)

Nozzle Pressure (pounds per square inch)	Nozzle Diameter (inches)				
	0.5	1.0	1.5	2.0	2.5
40	47	188	423	752	1174
60	58	230	518	921	1438
80	66	266	598	1063	1661
100	74	297	668	1188	1857
120	81	325	732	1302	2034
140	88	352	791	1406	2197
175	98	393	884	1572	2456
200	105	420	945	1681	2626

Source: *Firetactics.com*

a. If the value of P is constant and the value of d is increased, what happens to the value of r? Explain how you can tell this from Table 6, the model's equation, and thinking about the situation.

b. **i.** If the nozzle pressure is 100 pounds per square inch, use the model to estimate the flow rates of water for nozzle diameters of 0.5 inch, 1 inch, 1.5 inches, 2 inches, and 2.5 inches.

ii. Which estimate in part (i) has the largest error? What is that error?

iii. Which estimate in part (i) has the largest percentage error? What is that percentage error? [**Hint:** To find each percentage error, divide the error by the actual amount and multiply the result by 100.]

c. In Virginia the flow rate of water for firefighting must be at least 500 gallons per minute for one- and two-family dwellings that do not exceed 3600 square feet in area. Use the model to determine whether the requirement will be met if a 1.75-inch-diameter nozzle has 45 pounds per square inch of pressure. What is the estimated flow rate?

72. The time it takes for a planet to make one revolution around the Sun is the planet's *period*. The period $f(d)$ (in years) of a planet whose average distance from the Sun is d million kilometers is modeled by the equation

$$f(d) = 0.0005443\sqrt{d^3}$$

a. What is the period of Pluto, whose average distance from the Sun is 5913 million kilometers?

b. Use a graphing calculator table to find Earth's average distance from the Sun.

c. Suppose that in the future we colonize Mars, whose average distance from the Sun is 228 million kilometers. What is the period of Mars? If a person is 20 years old in "Earth years," how old is the person in "Mars years"?

73. A student tries to simplify $(x+\sqrt{7})^2$:

$$(x+\sqrt{7})^2 = x^2 + (\sqrt{7})^2 = x^2 + 7$$

Describe any errors. Then simplify the expression correctly.

74. A student tries to simplify $(x-\sqrt{3})^2$:

$$(x-\sqrt{3})^2 = x^2 - (\sqrt{3})^2 = x^2 - 3$$

Describe any errors. Then simplify the expression correctly.

75. A student tries to find the product $7(2\sqrt{3})$:

$$7(2\sqrt{3}) = 14\sqrt{21}$$

Describe any errors. Then find the product correctly.

76. A student tries to find the product $(3\sqrt{5})(4\sqrt{5})$:

$$(3\sqrt{5})(4\sqrt{5}) = 12\sqrt{5}$$

Describe any errors. Then find the product correctly.

*Write the expression as a single radical. [**Hint:** Write the expression in exponential form.]*

77. 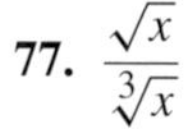 $\dfrac{\sqrt{x}}{\sqrt[3]{x}}$

78. $\sqrt[3]{\sqrt{x}}$

79. **a.** Write the expression $\sqrt[3]{x}\,\sqrt[4]{x}$ as a single radical.

b. Write the expression $\sqrt[k]{x}\,\sqrt[n]{x}$ as a single radical. [**Hint:** Perform steps similar to your work in part (a).]

c. Use your result from part (b) to find the product $\sqrt[3]{x}\,\sqrt[4]{x}$. Compare your result with your result from part (a).

d. Use your result from part (b) to find the product $\sqrt[5]{x}\,\sqrt[7]{x}$.

80. We cannot factor x^2-3 over the integers. We *can* factor x^2-3 over the real numbers:

$$x^2 - 3 = (x-\sqrt{3})(x+\sqrt{3})$$

a. Factor x^2-2 over the real numbers.

b. Factor x^2-5 over the real numbers.

c. Simplify $\dfrac{x^2-2}{x-\sqrt{2}}$.

d. Simplify $\dfrac{x^2-7}{x+\sqrt{7}}$.

81. Describe how to multiply two radical expressions. Include a discussion of various types of formulas, laws, and techniques you can use to find such products.

82. Find two radical expressions whose sum is $9\sqrt{x}+7$ and whose difference is $\sqrt{x}+3$.

Related Review

Simplify. Assume that $x \ge 0$.

83. $3\sqrt{x} - 5\sqrt{x}$

84. $(3\sqrt{x})^2$

85. $(3\sqrt{x})(-5\sqrt{x})$

86. $(3+\sqrt{x})^2$

Expressions, Equations, Functions, and Graphs

Perform the indicated instruction. Then use words such as linear, quadratic, cubic, exponential, logarithmic, rational, radical, polynomial, degree, function, one variable, *and* two variables *to describe the expression, equation, or system.*

87. Write $\log_b(x^2+3x-40) - \log_b(x^2-64)$ as a single logarithm.

88. Factor $2x^2+5x-12$.

89. Solve $\log_2(3x-4) - \log_2(2x-3) = 3$.

90. Solve $2x(2x-3) = 15-2x$.

91. Solve $2(3)^{5x-1} = 35$. Round any solutions to the fourth decimal place.

92. Let $f(x) = 2x^2+5x-12$. Find x when $f(x) = -5$.

9.3 RATIONALIZING DENOMINATORS AND SIMPLIFYING QUOTIENTS OF RADICAL EXPRESSIONS

Objectives

- Rationalize the denominator of a radical expression.
- Simplify a radical expression.
- Know the *quotient property for radicals*.

In Section 9.2, we discussed how to add, subtract, and multiply radical expressions. How do we simplify quotients of radical expressions?

Rationalizing Denominators of Radical Expressions

Recall from Section 7.3 that we simplify an expression of the form $\frac{p}{\sqrt{q}}$ by leaving no denominator as a radical expression and that we call this process *rationalizing the denominator*. For example, to rationalize the denominator of $\frac{2}{\sqrt{5}}$, we multiply by $1 = \frac{\sqrt{5}}{\sqrt{5}}$:

$$\frac{2}{\sqrt{5}} = \frac{2}{\sqrt{5}} \cdot \frac{\sqrt{5}}{\sqrt{5}} = \frac{2\sqrt{5}}{5}$$

In this section, we will rationalize the denominator of each of the following radical expressions:

$$\frac{4}{5\sqrt{3x}} \qquad \frac{3y}{\sqrt[5]{8x^2}} \qquad \frac{5}{3+\sqrt{x}} \qquad \frac{\sqrt{x}+4}{3\sqrt{x}-\sqrt{2}}$$

That is, we will write each expression so that no denominator is a radical expression.

Example 1 Rationalizing a Denominator

Simplify $\frac{4}{5\sqrt{3x}}$.

Solution

Since $\sqrt{3x} \cdot \sqrt{3x} = 3x$ where $x \geq 0$, we rationalize the denominator of $\frac{4}{5\sqrt{3x}}$ by multiplying by $\frac{\sqrt{3x}}{\sqrt{3x}}$:

Plot1 Plot2 Plot3
\Y1=4/(5√(3X))
\Y2=(4√(3X))/(15X)
\Y3=
\Y4=
\Y5=
\Y6=

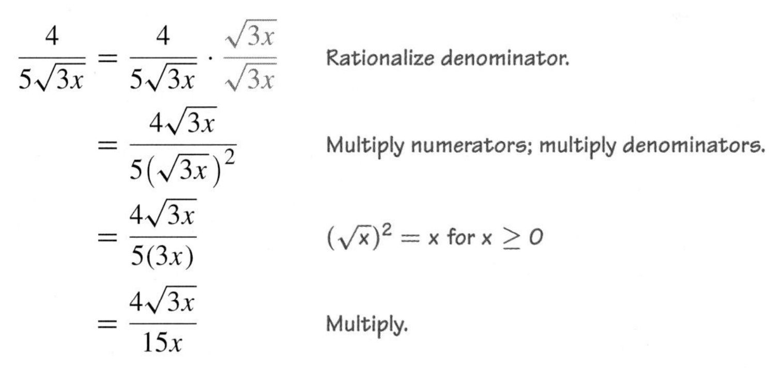

$$\frac{4}{5\sqrt{3x}} = \frac{4}{5\sqrt{3x}} \cdot \frac{\sqrt{3x}}{\sqrt{3x}} \qquad \text{Rationalize denominator.}$$

$$= \frac{4\sqrt{3x}}{5(\sqrt{3x})^2} \qquad \text{Multiply numerators; multiply denominators.}$$

$$= \frac{4\sqrt{3x}}{5(3x)} \qquad (\sqrt{x})^2 = x \text{ for } x \geq 0$$

$$= \frac{4\sqrt{3x}}{15x} \qquad \text{Multiply.}$$

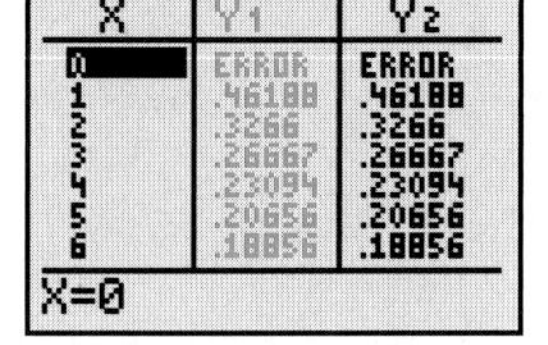

Figure 10 Verify the work

We use a graphing calculator table to verify our work (see Fig. 10). The table has "ERROR" messages across from $x = 0$, because the original expression and our result are not defined at 0 (why?). ■

In Example 2, we will rationalize the denominator for indexes other than 2. With any index n, our intermediate goal is the same: to write the denominator so that the radicand of its radical is a perfect nth power.

Example 2 Rationalizing Denominators

Simplify.

1. $\frac{1}{\sqrt[3]{x}}$ **2.** $\frac{3y}{\sqrt[5]{8x^2}}$

Solution

1. For the radicand to be a perfect cube, x must be multiplied by $x \cdot x = x^2$. So, we multiply $\dfrac{1}{\sqrt[3]{x}}$ by $\dfrac{\sqrt[3]{x^2}}{\sqrt[3]{x^2}}$:

$$\frac{1}{\sqrt[3]{x}} = \frac{1}{\sqrt[3]{x}} \cdot \frac{\sqrt[3]{x^2}}{\sqrt[3]{x^2}} \qquad \text{Rationalize denominator.}$$

$$= \frac{\sqrt[3]{x^2}}{\sqrt[3]{x^3}} \qquad \text{Multiply numerators; multiply denominators.}$$

$$= \frac{\sqrt[3]{x^2}}{x} \qquad \sqrt[3]{x^3} = x$$

2. To become a perfect 5th power, the radicand $8x^2 = 2 \cdot 2 \cdot 2 \cdot x \cdot x$ must be multiplied by $2 \cdot 2 \cdot x \cdot x \cdot x = 4x^3$:

$$\frac{3y}{\sqrt[5]{8x^2}} = \frac{3y}{\sqrt[5]{8x^2}} \cdot \frac{\sqrt[5]{4x^3}}{\sqrt[5]{4x^3}} \qquad \text{Rationalize denominator.}$$

$$= \frac{3y\sqrt[5]{4x^3}}{\sqrt[5]{8x^2 \cdot 4x^3}} \qquad \text{Multiply numerators; multiply denominators.}$$

$$= \frac{3y\sqrt[5]{4x^3}}{\sqrt[5]{32x^5}} \qquad \text{Multiply.}$$

$$= \frac{3y\sqrt[5]{4x^3}}{\sqrt[5]{32}\,\sqrt[5]{x^5}} \qquad \sqrt[n]{ab} = \sqrt[n]{a}\,\sqrt[n]{b}$$

$$= \frac{3y\sqrt[5]{4x^3}}{2x} \qquad \sqrt[5]{32} = 2,\ \sqrt[5]{x^5} = x$$

■

As shown in Example 2, **to rationalize the denominator of a radical expression of the form $\dfrac{A}{\sqrt[n]{x^m}}$, we multiply the expression by a fraction of the form $\dfrac{\sqrt[n]{x^k}}{\sqrt[n]{x^k}}$ so that the radical in the denominator has a perfect *n*th-power radicand.**

Quotient Property

In Section 7.3, we worked with the quotient property for square roots:

$$\sqrt{\frac{a}{b}} = \frac{\sqrt{a}}{\sqrt{b}}, \qquad \text{where } a \ge 0 \text{ and } b > 0$$

Next, we describe the quotient property for any index n.

Quotient Property for Radicals

If $\sqrt[n]{a}$ and $\sqrt[n]{b}$ are defined and b is nonzero, then

$$\sqrt[n]{\frac{a}{b}} = \frac{\sqrt[n]{a}}{\sqrt[n]{b}}$$

In words: The nth root of a quotient is the quotient of the nth roots.

For example, $\sqrt[3]{\dfrac{8}{27}} = \dfrac{\sqrt[3]{8}}{\sqrt[3]{27}} = \dfrac{2}{3}$.

You will prove the quotient property for radicals in Exercise 74.

Using the Quotient Property to Simplify Radical Expressions

If a radical expression has a fractional radicand, we simplify the expression by writing it as an expression in which no radicand is a fraction. We can use the quotient property to help us do this.

Example 3 Simplifying Radical Expressions

Simplify.

1. $\sqrt{\frac{5}{k}}$ **2.** $\sqrt[3]{\frac{7y}{2x^2}}$

Solution

1. $\sqrt{\frac{5}{k}} = \frac{\sqrt{5}}{\sqrt{k}}$ Quotient property

$= \frac{\sqrt{5}}{\sqrt{k}} \cdot \frac{\sqrt{k}}{\sqrt{k}}$ Rationalize denominator.

$= \frac{\sqrt{5k}}{k}$ Multiply numerators; multiply denominators.

2. $\sqrt[3]{\frac{7y}{2x^2}} = \frac{\sqrt[3]{7y}}{\sqrt[3]{2x^2}}$ Quotient property

$= \frac{\sqrt[3]{7y}}{\sqrt[3]{2x^2}} \cdot \frac{\sqrt[3]{4x}}{\sqrt[3]{4x}}$ To become a perfect cube, $2x^2 = 2 \cdot x \cdot x$ must be multiplied by $2 \cdot 2 \cdot x = 4x$.

$= \frac{\sqrt[3]{28xy}}{\sqrt[3]{8x^3}}$ Multiply numerators; multiply denominators.

$= \frac{\sqrt[3]{28xy}}{2x}$ $\sqrt[3]{8} = 2$, $\sqrt[3]{x^3} = x$ ■

Using Conjugates to Rationalize Denominators

Recall from Section 6.2 that we call binomials such as $5x + 2$ and $5x - 2$ binomial conjugates of each other. Similarly, we call the radical expressions $\sqrt{5} + \sqrt{2}$ and $\sqrt{5} - \sqrt{2}$ radical conjugates of each other. We say that the sum of two radicals and the difference of the same radicals are **radical conjugates** of each other.

What happens when we find the product of two radical conjugates? Here we use the property $(A + B)(A - B) = A^2 - B^2$ to find the product $(\sqrt{5} + \sqrt{2})(\sqrt{5} - \sqrt{2})$:

$$(\sqrt{5} + \sqrt{2})(\sqrt{5} - \sqrt{2}) = (\sqrt{5})^2 - (\sqrt{2})^2 \quad (A + B)(A - B) = A^2 - B^2$$
$$= 5 - 2 \quad (\sqrt{x})^2 = x \text{ for } x \geq 0$$
$$= 3 \quad \text{Subtract.}$$

The result contains no radicals. We next list a few expressions, their conjugates, and the products of the expressions and their conjugates:

Expression	Conjugate	Product
$4 - \sqrt{x}$	$4 + \sqrt{x}$	$16 - x$
$7 + 3\sqrt{x}$	$7 - 3\sqrt{x}$	$49 - 9x$
$2\sqrt{x} - 3\sqrt{5}$	$2\sqrt{x} + 3\sqrt{5}$	$4x - 45$

Just as before, we notice that the products of the conjugates contain no radicals. We will use this observation to help us rationalize a denominator in Example 4.

Example 4 Rationalizing a Denominator

Simplify $\dfrac{5}{3+\sqrt{x}}$.

Solution

The conjugate of the denominator is $3-\sqrt{x}$. We can rationalize the denominator of $\dfrac{5}{3+\sqrt{x}}$ by multiplying by $\dfrac{\text{conjugate}}{\text{conjugate}} = \dfrac{3-\sqrt{x}}{3-\sqrt{x}}$:

$$\frac{5}{3+\sqrt{x}} = \frac{5}{3+\sqrt{x}} \cdot \frac{3-\sqrt{x}}{3-\sqrt{x}} \qquad \text{Multiply by } \frac{3-\sqrt{x}}{3-\sqrt{x}}.$$

$$= \frac{5(3-\sqrt{x})}{(3+\sqrt{x})(3-\sqrt{x})} \qquad \text{Multiply numerators; multiply denominators.}$$

$$= \frac{5(3-\sqrt{x})}{3^2-(\sqrt{x})^2} \qquad (A+B)(A-B) = A^2 - B^2$$

$$= \frac{15-5\sqrt{x}}{9-x} \qquad \text{Distributive law; } (\sqrt{x})^2 = x \text{ for } x \geq 0$$

We use a graphing calculator table to verify our work (see Fig. 11). ■

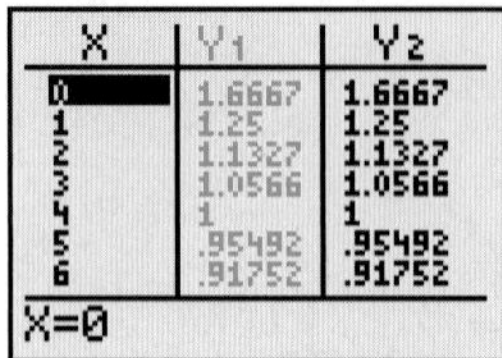

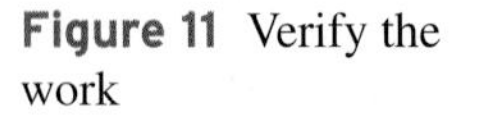

Figure 11 Verify the work

WARNING When we try to rationalize the denominator of $\dfrac{5}{3+\sqrt{x}}$, it is *not* helpful to multiply the fraction by $\dfrac{\sqrt{x}}{\sqrt{x}}$:

$$\frac{5}{3+\sqrt{x}} = \frac{5}{3+\sqrt{x}} \cdot \frac{\sqrt{x}}{\sqrt{x}} = \frac{5\sqrt{x}}{3\sqrt{x}+x}$$

Rather, the conjugate of the denominator is $3-\sqrt{x}$, so we multiply the fraction by $\dfrac{\text{conjugate}}{\text{conjugate}} = \dfrac{3-\sqrt{x}}{3-\sqrt{x}}$, as shown in Example 4.

Rationalizing a Denominator by Using a Radical Conjugate

To rationalize the denominator of a square root expression if the denominator is a sum or difference involving radicals,

1. Determine the radical conjugate of the denominator.
2. Multiply the original fraction by the fraction $\dfrac{\text{conjugate}}{\text{conjugate}}$.
3. Find the product of the denominators by using $(A+B)(A-B) = A^2 - B^2$.

Example 5 Rationalizing a Denominator

Simplify $\dfrac{\sqrt{x}+4}{3\sqrt{x}-\sqrt{2}}$.

Solution

$$\frac{\sqrt{x}+4}{3\sqrt{x}-\sqrt{2}} = \frac{\sqrt{x}+4}{3\sqrt{x}-\sqrt{2}} \cdot \frac{3\sqrt{x}+\sqrt{2}}{3\sqrt{x}+\sqrt{2}} \qquad \text{Conjugate of } 3\sqrt{x}-\sqrt{2} \text{ is } 3\sqrt{x}+\sqrt{2}.$$

$$= \frac{(\sqrt{x}+4)(3\sqrt{x}+\sqrt{2})}{(3\sqrt{x}-\sqrt{2})(3\sqrt{x}+\sqrt{2})} \qquad \text{Multiply numerators; multiply denominators.}$$

$$= \frac{3\sqrt{x}\sqrt{x}+\sqrt{x}\sqrt{2}+4\cdot 3\sqrt{x}+4\sqrt{2}}{(3\sqrt{x})^2-(\sqrt{2})^2} \qquad \text{Multiply pairs of terms; } (A-B)(A+B) = A^2 - B^2$$

$$= \frac{3x+\sqrt{2x}+12\sqrt{x}+4\sqrt{2}}{9x-2} \qquad \text{Simplify.}$$

■

Example 6 Rationalizing a Denominator

Simplify $\dfrac{\sqrt{a}+\sqrt{b}}{\sqrt{a}-\sqrt{b}}$.

Solution

$$\frac{\sqrt{a}+\sqrt{b}}{\sqrt{a}-\sqrt{b}} = \frac{\sqrt{a}+\sqrt{b}}{\sqrt{a}-\sqrt{b}} \cdot \frac{\sqrt{a}+\sqrt{b}}{\sqrt{a}+\sqrt{b}}$$ Conjugate of $\sqrt{a}-\sqrt{b}$ is $\sqrt{a}+\sqrt{b}$.

$$= \frac{(\sqrt{a}+\sqrt{b})(\sqrt{a}+\sqrt{b})}{(\sqrt{a}-\sqrt{b})(\sqrt{a}+\sqrt{b})}$$ Multiply numerators; multiply denominators.

$$= \frac{(\sqrt{a}+\sqrt{b})^2}{(\sqrt{a})^2-(\sqrt{b})^2}$$ $CC = C^2$; $(A-B)(A+B) = A^2 - B^2$

$$= \frac{(\sqrt{a})^2+2\sqrt{a}\sqrt{b}+(\sqrt{b})^2}{(\sqrt{a})^2-(\sqrt{b})^2}$$ $(A+B)^2 = A^2 + 2AB + B^2$

$$= \frac{a+2\sqrt{ab}+b}{a-b}$$ $(\sqrt{x})^2 = x$ for $x \geq 0$; $\sqrt[n]{a}\sqrt[n]{b} = \sqrt[n]{ab}$ ■

group exploration

Looking ahead: Sketching graphs of square root functions

1. Use a graphing calculator to draw a graph of $y = \sqrt{x}$.
2. Use a graphing calculator to compare graphs of $y = 0.5\sqrt{x}$, $y = \sqrt{x}$, $y = 2\sqrt{x}$, and $y = -2\sqrt{x}$. Describe the effect that a has on the graph of $y = a\sqrt{x}$, where $a \neq 0$.
3. Use a graphing calculator to compare graphs of $y = \sqrt{x-2}$, $y = \sqrt{x}$, and $y = \sqrt{x+4}$. Describe the effect that h has on the graph of $y = \sqrt{x-h}$.
4. Use a graphing calculator to compare graphs of $y = \sqrt{x} - 2$, $y = \sqrt{x}$, and $y = \sqrt{x} + 4$. Describe the effect that k has on the graph of $y = \sqrt{x} + k$.
5. Use a graphing calculator to graph

$$y = \sqrt{x}, \quad y = 0.5\sqrt{x}, \quad y = 0.5\sqrt{x+3}, \quad \text{and} \quad y = 0.5\sqrt{x+3} - 2$$

in order, and explain how these graphs relate to your observations in Problems 2, 3, and 4.

6. Sketch the graph of $y = 2\sqrt{x-3} + 1$. Use a graphing calculator to verify your sketch.
7. Describe how a, h, and k affect the graph of $f(x) = a\sqrt{x-h} + k$, where $a \neq 0$. Compare their effects for this function with their effects for the quadratic function $g(x) = a(x-h)^2 + k$.

TIPS FOR SUCCESS: Retake Quizzes and Exams

To study for your final exam, consider retaking your quizzes and other exams. These quizzes and exams can reveal your weak areas. If you have difficulty with a certain concept, you can refer to Homework exercises that address this concept. Reflect on *why* you are having such difficulty, rather than just doing more Homework exercises that address the concept.

HOMEWORK 9.3

FOR EXTRA HELP ▸ Student Solutions Manual · PH Math/Tutor Center · MathXL® · MyMathLab

Simplify. Use a graphing calculator table to verify your result when possible. Assume that each variable is nonnegative.

1. $\dfrac{8}{\sqrt{x}}$
2. $\dfrac{2}{\sqrt{x}}$
3. $\dfrac{3}{\sqrt{5p}}$
4. $\dfrac{2}{\sqrt{7r}}$
5. $\dfrac{4}{3\sqrt{2x}}$
6. $\dfrac{7}{6\sqrt{3x}}$
7. $\dfrac{10}{\sqrt{8k}}$
8. $\dfrac{6}{\sqrt{27t}}$
9. $\sqrt{\dfrac{4}{x}}$
10. $\sqrt{\dfrac{25}{x}}$
11. $\sqrt{\dfrac{7}{2}}$
12. $\sqrt{\dfrac{5}{3}}$
13. $\sqrt{\dfrac{2y}{x}}$
14. $\sqrt{\dfrac{11y}{x}}$
15. $\sqrt{\dfrac{x}{12y}}$
16. $\sqrt{\dfrac{x}{18y}}$
17. $\dfrac{3}{\sqrt{x-4}}$
18. $\dfrac{5}{\sqrt{2x+1}}$
19. $\dfrac{\sqrt{2a^3}}{\sqrt{3b}}$
20. $\dfrac{\sqrt{5b^5}}{\sqrt{7a}}$
21. $\dfrac{2}{\sqrt[3]{5}}$
22. $\dfrac{5}{\sqrt[3]{2}}$
23. $\dfrac{5}{\sqrt[3]{4}}$
24. $\dfrac{1}{\sqrt[3]{25}}$
25. $\dfrac{4}{5\sqrt[3]{x}}$
26. $\dfrac{7}{4\sqrt[3]{x^2}}$
27. $\dfrac{6}{\sqrt[3]{2x^2}}$
28. $\dfrac{1}{\sqrt[3]{9x}}$
29. $\dfrac{7t}{\sqrt[4]{4t^3}}$
30. $\dfrac{2w}{\sqrt[5]{16w^2}}$
31. $\dfrac{\sqrt[3]{x}}{\sqrt{x}}$
32. $\dfrac{\sqrt[5]{x}}{\sqrt[4]{2x}}$
33. $\sqrt[5]{\dfrac{2}{x^3}}$
34. $\sqrt[3]{\dfrac{4}{x^2}}$
35. $\sqrt[4]{\dfrac{4}{9x^2}}$
36. $\sqrt[3]{\dfrac{7}{25x}}$
37. $\sqrt[5]{\dfrac{3w}{4x^4y^2}}$
38. $\sqrt[6]{\dfrac{5w}{8x^2y^3}}$

Simplify. Use a graphing calculator to verify your result when possible.

39. $\dfrac{1}{5+\sqrt{3}}$
40. $\dfrac{1}{1+\sqrt{5}}$
41. $\dfrac{2}{\sqrt{3}+\sqrt{7}}$
42. $\dfrac{4}{\sqrt{2}+\sqrt{5}}$
43. $\dfrac{1}{3\sqrt{r}-7}$
44. $\dfrac{6}{5\sqrt{t}-2}$
45. $\dfrac{\sqrt{x}}{\sqrt{x}-1}$
46. $\dfrac{\sqrt{x}}{\sqrt{x}+1}$
47. $\dfrac{3\sqrt{x}}{4\sqrt{x}-\sqrt{5}}$
48. $\dfrac{4\sqrt{x}}{2\sqrt{x}+\sqrt{6}}$
49. $\dfrac{\sqrt{x}}{\sqrt{x}-y}$
50. $\dfrac{\sqrt{y}}{x-2\sqrt{y}}$
51. $\dfrac{\sqrt{x}-5}{\sqrt{x}+5}$
52. $\dfrac{\sqrt{x}+9}{\sqrt{x}+9}$
53. $\dfrac{2\sqrt{x}+5}{3\sqrt{x}+1}$
54. $\dfrac{4\sqrt{x}-3}{2\sqrt{x}+5}$
55. $\dfrac{6\sqrt{x}+\sqrt{5}}{3\sqrt{x}-\sqrt{7}}$
56. $\dfrac{8\sqrt{x}-\sqrt{3}}{4\sqrt{x}-\sqrt{2}}$
57. $\dfrac{\sqrt{x}-\sqrt{y}}{\sqrt{x}+\sqrt{y}}$
58. $\dfrac{2\sqrt{x}-\sqrt{y}}{3\sqrt{x}-\sqrt{y}}$
59. $\dfrac{1}{\sqrt{x+1}-\sqrt{x}}$
60. $\dfrac{2}{\sqrt{x+3}+\sqrt{x}}$

61. The distance d (in miles) to the horizon at an altitude h feet above sea level is given by the equation

$$d=\sqrt{\frac{3h}{2}}$$

a. Simplify the right side of the horizon–distance equation.
b. The Sears Tower in Chicago is 1450 feet tall. What would be the distance to the horizon from the top of this skyscraper? Assume that the base of the building is at sea level.
c. If an airplane flies at an altitude of 30,000 feet over the skyscraper, what is the distance to the horizon from the airplane?

62. The time (in seconds) $f(d)$ that it takes for an object to fall d feet can be modeled by the equation

$$f(d)=\sqrt{\frac{2d}{g}}$$

where g is the constant 32.2 feet per second squared.
a. Simplify the right-hand side of the model's equation.
b. Find $f(100)$. What does it mean in this situation?
c. In 2002, sky divers jumped off the 1483-foot-tall Petronas Towers (in Malaysia), the tallest habitable buildings in the world. Estimate how long a sky diver was in free fall by finding the time it would take to fall 1483 feet with a closed parachute. Ideally, the parachute opens. If so, is your estimate an underestimate or an overestimate? Explain.
d. Is f an increasing or a decreasing function? Explain why that makes sense in this situation.

63. In the ISO paper-size system, the length-to-width ratio of all pages is $\dfrac{\sqrt{2}}{1}$ (see Fig. 12).

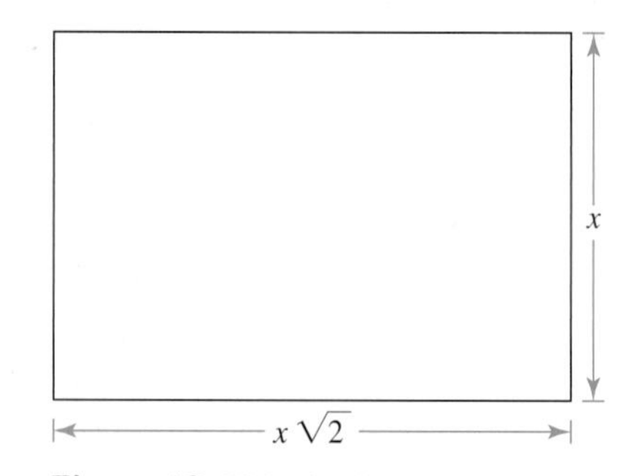

Figure 12 ISO-sized paper

a. Show that if you cut a piece of ISO-sized paper parallel to its shorter side to form two pieces with equal area, each piece will also have a length-to-width ratio of $\dfrac{\sqrt{2}}{1}$ (see Fig. 13). (This property allows two ISO pages of equal size to be photocopied onto one page by setting the magnification factor on a copying machine to $\sqrt{\dfrac{1}{2}}\approx 0.71=71\%$.)

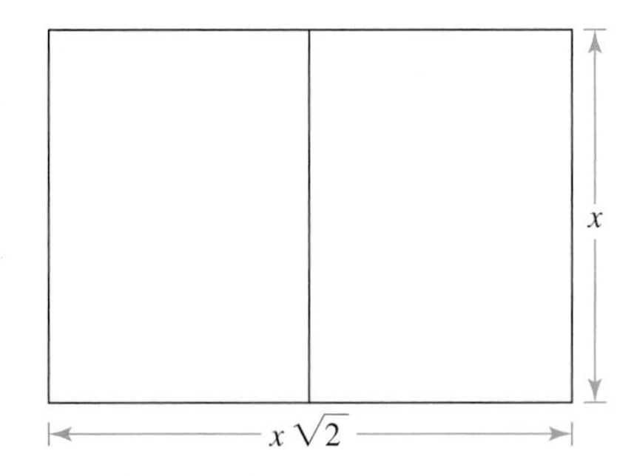

Figure 13 ISO-sized paper cut in two

b. The largest ISO paper size has an area of 1 square meter. Find the (exact) width of a page of this size. Then round your result to the third decimal place.

64. *Escape velocity* is the initial velocity that an object needs to break free of a planet's or moon's gravitational pull. If we ignore the effects of air resistance and the rotation of the planet or moon and assume that there is no continued propulsion, as by a rocket, the escape velocity v (in meters per second) is given by $v = \sqrt{\frac{2GM}{r}}$, where G is a constant equal to 6.67×10^{-11}, M is the mass (in kilograms) of the planet or moon, and r is the object's distance (in meters) from the center of the planet or moon.

a. Simplify the right-hand side of the model's equation.

b. Use Table 7 to find the escape velocity on the planet's or moon's surface at the equator. Round your result to the first decimal place.

i. Earth **ii.** the Moon **iii.** Jupiter

Table 7 Masses and Equatorial Radii of Earth, the Moon, and Jupiter

Planet/Moon	Mass of Planet/Moon (kilograms)	Radius of Planet/Moon (meters)
Earth	5.976×10^{24}	6.378×10^{6}
The Moon	7.349×10^{22}	1.737×10^{6}
Jupiter	1.899×10^{27}	7.149×10^{7}

c. In terms of blastoffs, would a round trip to Jupiter or the Moon require more fuel?

d. Convert your result of part (b, i) to units of miles per hour. (There are approximately 1609.3 meters in 1 mile.)

65. Two students try to rationalize the denominator of the expression $\frac{2}{\sqrt{x}}$:

Student 1's work

$$\frac{2}{\sqrt{x}} = \frac{2}{\sqrt{x}} \cdot \frac{\sqrt{x}}{\sqrt{x}} = \frac{2\sqrt{x}}{\sqrt{x}\sqrt{x}} = \frac{2\sqrt{x}}{x}$$

Student 2's work

$$\frac{2}{\sqrt{x}} = \left(\frac{2}{\sqrt{x}}\right)^2 = \frac{2^2}{\left(\sqrt{x}\right)^2} = \frac{4}{x}$$

Did one, both, or neither of these students rationalize the denominator correctly? Describe any errors and where they occurred.

66. Two students try to rationalize the denominator of $\frac{3}{\sqrt{x^3}}$:

Student 1's work

$$\frac{3}{\sqrt{x^3}} = \frac{3}{\sqrt{x^3}} \cdot \frac{\sqrt{x^3}}{\sqrt{x^3}} = \frac{3\sqrt{x^3}}{\sqrt{x^3}\sqrt{x^3}} = \frac{3\left(x\sqrt{x}\right)}{x^3} = \frac{3\sqrt{x}}{x^2}$$

Student 2's work

$$\frac{3}{\sqrt{x^3}} = \frac{3}{x\sqrt{x}} = \frac{3}{x\sqrt{x}} \cdot \frac{\sqrt{x}}{\sqrt{x}} = \frac{3\sqrt{x}}{x \cdot x} = \frac{3\sqrt{x}}{x^2}$$

Did one, both, or neither of the students rationalize the denominator correctly? Explain.

67. A student tries to rationalize the denominator of $\frac{5}{\sqrt[3]{x}}$:

$$\frac{5}{\sqrt[3]{x}} = \frac{5}{\sqrt[3]{x}} \cdot \frac{\sqrt[3]{x}}{\sqrt[3]{x}} = \frac{5\sqrt[3]{x}}{x}$$

Describe any errors. Then rationalize the denominator correctly.

68. Two students try to rationalize the denominator of the expression $\frac{4}{2+\sqrt{x}}$:

Student 1's work

$$\frac{4}{2+\sqrt{x}} = \frac{4}{2+\sqrt{x}} \cdot \frac{\sqrt{x}}{\sqrt{x}} = \frac{4\sqrt{x}}{2+\sqrt{x}\sqrt{x}} = \frac{4\sqrt{x}}{2+x}$$

Student 2's work

$$\frac{4}{2+\sqrt{x}} = \frac{4}{2+\sqrt{x}} \cdot \frac{2-\sqrt{x}}{2-\sqrt{x}} = \frac{8-4\sqrt{x}}{2^2-\left(\sqrt{x}\right)^2} = \frac{8-4\sqrt{x}}{4-x}$$

Did one, both, or neither of these students rationalize the denominator correctly? Describe any errors and where they occurred.

We rationalize the numerator of a radical expression by finding an equivalent expression whose numerator contains no radicals. Rationalize the numerator of the given expression.

69. $\frac{\sqrt{x}}{3}$

70. $\frac{\sqrt{x}}{\sqrt{2}}$

71. $\frac{\sqrt{x+2}-\sqrt{x}}{2}$

72. $\frac{\sqrt{x+3}-\sqrt{x}}{3}$

73. Simplify the expression.

$$\frac{\frac{1}{\sqrt{x}} - \frac{3}{x}}{\frac{2}{\sqrt{x}} + \frac{1}{x}}$$

74. Prove the quotient property for radicals—that is,

$$\sqrt[n]{\frac{a}{b}} = \frac{\sqrt[n]{a}}{\sqrt[n]{b}}$$

where $\sqrt[n]{a}$ and $\sqrt[n]{b}$ are defined and b is nonzero.

75. Find and simplify the exact solution of $x\sqrt{2} + 3\sqrt{5} = 9\sqrt{5}$.

76. Find the exact solution of $x^2\sqrt{2} + x\sqrt{17} + \sqrt{2} = 0$. Simplify your result. [**Hint:** Use the quadratic formula.]

77. Describe how to rationalize the denominator of a radical expression in which the denominator is a radical.

78. Describe how to rationalize the denominator of a radical expression in which the denominator is a sum or difference involving radicals.

Related Review

79. a. Factor $A^3 + B^3$.
   b. Find the product $(A + B)(A^2 - AB + B^2)$. Explain how your work is related to your work in part (a).
   c. Find the product $(x + 2)(x^2 - 2x + 4)$. Explain how your work is related to the result that you found in part (b).
   d. Find the product $(\sqrt[3]{x} + \sqrt[3]{2})(\sqrt[3]{x^2} - \sqrt[3]{2x} + \sqrt[3]{4})$. Explain how your work is related to the result that you found in part (b).
   e. Rationalize the denominator of $\dfrac{1}{\sqrt[3]{x} + \sqrt[3]{2}}$. [**Hint:** Multiply by 1. See part (d).]

80. a. Factor $A^3 - B^3$.
   b. Find the product $(A - B)(A^2 + AB + B^2)$. Explain how your work is related to your work in part (a).
   c. Find the product $(x - 2)(x^2 + 2x + 4)$. Explain how your work is related to the result that you found in part (b).
   d. Find the product $(\sqrt[3]{x} - \sqrt[3]{2})(\sqrt[3]{x^2} + \sqrt[3]{2x} + \sqrt[3]{4})$. Explain how your work is related to the result that you found in part (b).
   e. Rationalize the denominator of $\dfrac{1}{\sqrt[3]{x} - \sqrt[3]{2}}$. [**Hint:** Multiply by 1. See part (d).]

Expressions, Equations, Functions, and Graphs

Perform the indicated instruction. Then use words such as linear, quadratic, cubic, exponential, logarithmic, rational, radical, polynomial, degree, function, one variable, *and* two variables *to describe the expression, equation, or system.*

81. Find the product $(5x - 4)(3x^2 - 2x - 1)$.

82. Graph $f(x) = -4(3)^x$ by hand.

83. Factor $24x^3 - 3000$.

84. Simplify $\dfrac{(16b^{2/3}c^3)^{1/4}}{(27b^{3/4}c^{-5})^{2/3}}$.

85. Solve $5x^2 - 3 = 4x - 1$.

86. Find any real-number solutions of $5b^4 - 43 = 76$. Round any results to the fourth decimal place.

9.4 GRAPHING AND COMBINING SQUARE ROOT FUNCTIONS

"I love math because I think that it is like magic. Every time I solve a problem, I am always surprised and amazed at what I can do."

—*Ana J., student*

Objectives

- Know the graphical significance of a, h, and k for a square root function of the form $y = a\sqrt{x - h} + k$, where $a \neq 0$.
- Sketch graphs of square root functions.
- Find the domain and range of a square root function.
- Find the sum function, difference function, product function, and quotient function of two square root functions.

In this section, we will graph square root functions and perform operations with square root functions.

Graphing Square Root Functions

Recall from Section 7.1 that to graph a function such as $h(x) = (x - 3)^2$, we translate the graph of $k(x) = x^2$ to the right by 3 units. Can we graph the square root function $g(x) = \sqrt{x - 3}$ by translating the function $f(x) = \sqrt{x}$ in some way? We will explore this question in Example 1.

Example 1 Horizontal Translation

Compare the graph of $g(x) = \sqrt{x - 3}$ with the graph of $f(x) = \sqrt{x}$.

Solution

We list input–output pairs for g in Table 8. We choose inputs that lead to easily found outputs. Then, we sketch graphs of g and f (see Fig. 14).

Table 8 Input-Output Pairs of $g(x) = \sqrt{x-3}$

x	y
3	0
4	1
7	2
12	3
19	4

Figure 14 Graphs of $g(x) = \sqrt{x-3}$ and $f(x) = \sqrt{x}$

The graph of $g(x) = \sqrt{x-3}$ is the translation of the graph of $f(x) = \sqrt{x}$ to the right by 3 units. To see why this makes sense, we solve both equations for x:

$$\begin{aligned} f(x) &= \sqrt{x} \\ y &= \sqrt{x} \\ \sqrt{x} &= y \\ \left(\sqrt{x}\right)^2 &= y^2 \\ x &= y^2 \end{aligned} \qquad \begin{aligned} g(x) &= \sqrt{x-3} \\ y &= \sqrt{x-3} \\ \sqrt{x-3} &= y \\ \left(\sqrt{x-3}\right)^2 &= y^2 \\ x-3 &= y^2 \\ x &= y^2+3 \end{aligned}$$

For each positive value of y, the input value of x for g is 3 more than the input value of x for f. Therefore, the graph of $g(x) = \sqrt{x-3}$ lies 3 units to the right of the graph of $f(x) = \sqrt{x}$. ■

In Example 1, we found that to graph $g(x) = \sqrt{x-3}$, we translate the graph of $f(x) = \sqrt{x}$ to the right by 3 units. This is the same pattern we observed with quadratic functions. To graph the quadratic function $g(x) = (x-3)^2$, we translate the graph of $f(x) = x^2$ to the right by 3 units.

In fact, the values of a, h, and k have similar effects on a square root function $g(x) = a\sqrt{x-h}+k$, where $a \neq 0$, as on a quadratic function $Q(x) = a(x-h)^2+k$. Here are some examples of how we can translate the graph of $y = \sqrt{x}$ to obtain the graph of an equation of the form $y = \sqrt{x-h}+k$:

Function	To graph the function $g(x) = \sqrt{x-h}+k$,
$g(x) = \sqrt{x-2}$	translate the graph of $y = \sqrt{x}$ to the right by 2 units.
$g(x) = \sqrt{x+2}$	translate the graph of $y = \sqrt{x}$ to the left by 2 units.
$g(x) = \sqrt{x}-2$	translate the graph of $y = \sqrt{x}$ down by 2 units.
$g(x) = \sqrt{x}+2$	translate the graph of $y = \sqrt{x}$ up by 2 units.

The graphs of $f(x) = -a\sqrt{x}$ and $g(x) = a\sqrt{x}$ are reflections of each other across the x-axis (see Fig. 15).

If $a > 0$, then $g(x) = a\sqrt{x-h}+k$ is an increasing function and (h, k) is the minimum point (see Fig. 16). **If $a < 0$, then g is a decreasing function and (h, k) is the maximum point** (see Fig. 17).

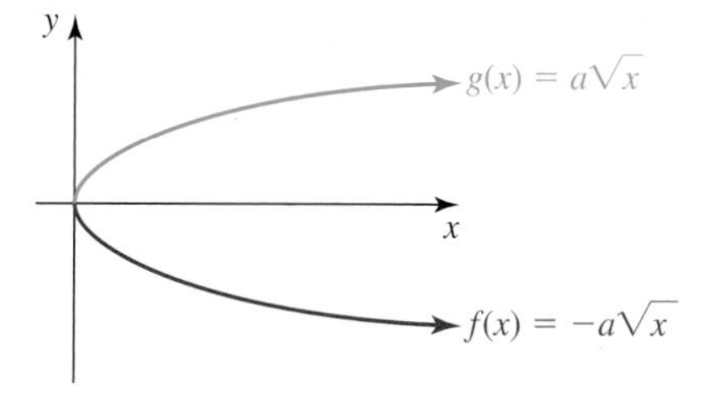

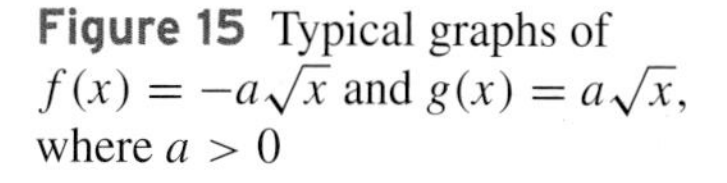

Figure 15 Typical graphs of $f(x) = -a\sqrt{x}$ and $g(x) = a\sqrt{x}$, where $a > 0$

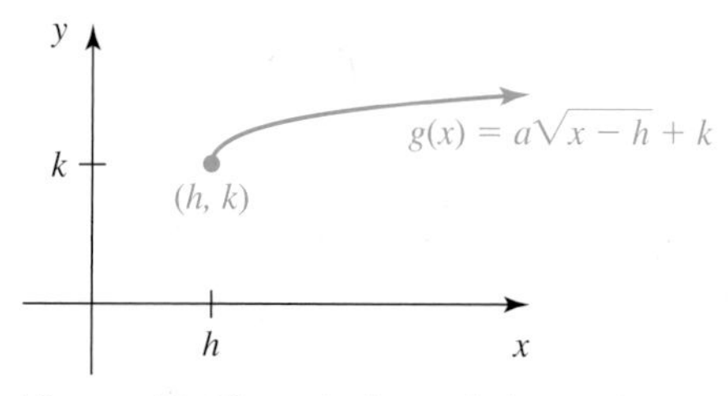

Figure 16 If $a > 0$, then g is increasing with minimum point (h, k)

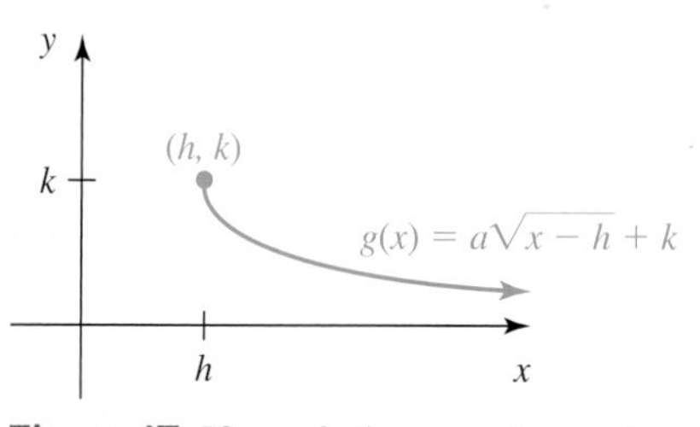

Figure 17 If $a < 0$, then g is decreasing with maximum point (h, k)

If a is a large number, then the graph of $g(x) = a\sqrt{x - h} + k$ rises more quickly than the graph of $y = \sqrt{x - h} + k$. If a is a positive number near zero, then the graph rises more slowly.

Three-Step Method for Graphing a Square Root Function

To sketch the graph of $f(x) = a\sqrt{x - h} + k$, where $a \neq 0$,

1. Sketch the graph of $y = a\sqrt{x}$.
2. Translate the graph sketched in step 1 to the right by h units if $h > 0$ and to the left by $|h|$ units if $h < 0$.
3. Translate the graph sketched in step 2 up by k units if $k > 0$ and down by $|k|$ units if $k < 0$.

The graph of a square root function is called a **square root curve.**

Example 2 Graphing a Square Root Function

Sketch the graph of $g(x) = -2\sqrt{x + 4} - 1$. Find the domain and range of g.

Solution

First, we sketch the graph of $y = -2\sqrt{x}$ in Fig. 18. Next, we translate the graph to the left by 4 units and down by 1 unit.

We check that the solutions of g listed in Table 9 lie on our sketch of the graph of g. Also, we use a graphing calculator to verify our sketch (see Fig. 19).

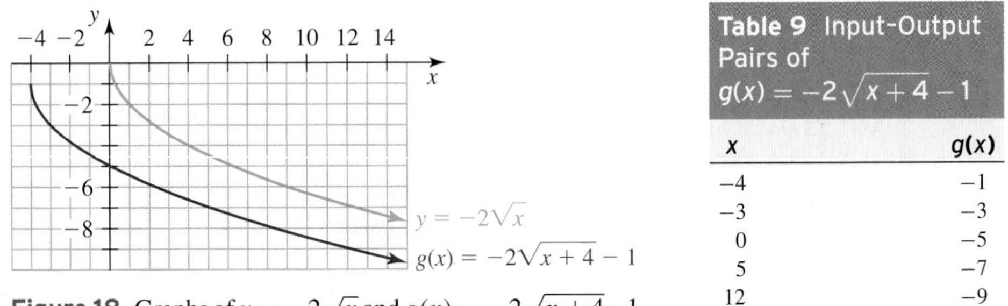

Figure 18 Graphs of $y = -2\sqrt{x}$ and $g(x) = -2\sqrt{x + 4} - 1$

Table 9 Input-Output Pairs of $g(x) = -2\sqrt{x + 4} - 1$

x	$g(x)$
−4	−1
−3	−3
0	−5
5	−7
12	−9

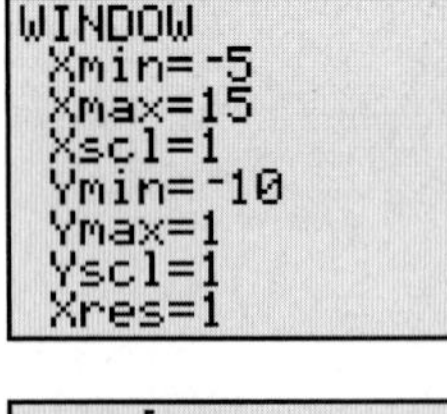

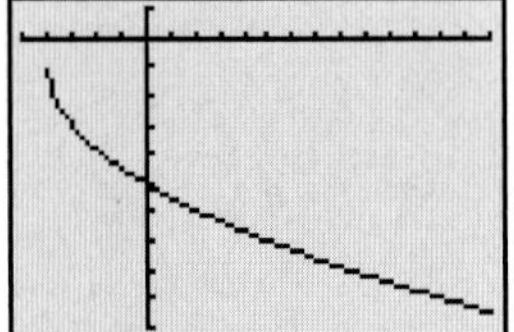

Figure 19 Verify the graph of $g(x) = -2\sqrt{x + 4} - 1$

From the graph of g, we see that the values of x are greater than or equal to -4. So, the domain is the set of numbers x where $x \geq -4$. We can also find the domain by using the fact that the radicand $x + 4$ must be nonnegative:

$$x + 4 \geq 0 \qquad \text{Radicand must be nonnegative.}$$
$$x + 4 - 4 \geq 0 - 4 \qquad \text{Subtract 4 from both sides.}$$
$$x \geq -4$$

From the graph of g, we see that the maximum point is $(-4, -1)$, so the values of y are less than or equal to -1. Therefore, the range is the set of numbers y where $y \leq -1$. ■

Example 3 Graphing a Square Root Function

Sketch the graph of $f(x) = 2\sqrt{x - 3} + 5$. Find the domain and range of f.

Solution

First, we sketch the graph of $y = 2\sqrt{x}$ in Fig. 20. Next, we translate the graph to the right by 3 units and up by 5 units.

We check that the solutions of f listed in Table 10 lie on our sketch of the graph of f. Also, we use a graphing calculator to verify our sketch (see Fig. 21).

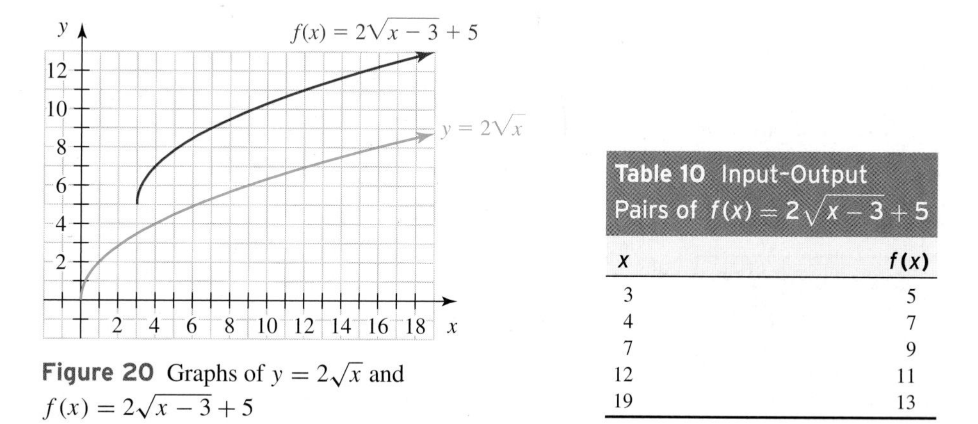

Figure 20 Graphs of $y = 2\sqrt{x}$ and $f(x) = 2\sqrt{x-3} + 5$

Table 10 Input-Output Pairs of $f(x) = 2\sqrt{x-3} + 5$

x	$f(x)$
3	5
4	7
7	9
12	11
19	13

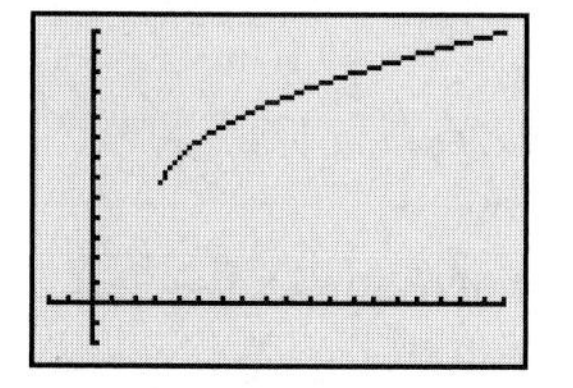

Figure 21 Verify the graph of $f(x) = 2\sqrt{x-3} + 5$

From the graph of f, we see that the values of x are greater than or equal to 3. So, the domain is the set of numbers x where $x \geq 3$. We can also find the domain by using the fact that the radicand $x - 3$ must be nonnegative:

$$x - 3 \geq 0 \qquad \text{Radicand must be nonnegative.}$$
$$x - 3 + 3 \geq 0 + 3 \qquad \text{Add 3 to both sides.}$$
$$x \geq 3$$

From the graph of f, we see that the minimum point is (3, 5), so the values of y are greater than or equal to 5. Therefore, the range is the set of numbers y where $y \geq 5$. ■

Performing Operations with Square Root Functions

In Example 4, we use two square root functions to form a sum function, a difference function, a product function, and a quotient function.

Example 4 Performing Operations with Square Root Functions

Let $f(x) = 2\sqrt{x} - 3$ and $g(x) = 5\sqrt{x} + 4$. Find an equation of the following function.

1. $f + g$ **2.** $f - g$ **3.** $f \cdot g$ **4.** $\dfrac{f}{g}$

Solution

1. $(f + g)(x) = f(x) + g(x) = (2\sqrt{x} - 3) + (5\sqrt{x} + 4)$ Substitute for $f(x)$ and $g(x)$.

$= 7\sqrt{x} + 1$ Combine like radicals.

2. $(f - g)(x) = f(x) - g(x) = (2\sqrt{x} - 3) - (5\sqrt{x} + 4)$ Substitute for $f(x)$ and $g(x)$.

$= 2\sqrt{x} - 3 - 5\sqrt{x} - 4$ Subtract.

$= -3\sqrt{x} - 7$ Combine like radicals.

3. $(f \cdot g)(x) = f(x)g(x) = (2\sqrt{x} - 3)(5\sqrt{x} + 4)$ Substitute for $f(x)$ and $g(x)$.

$= 2\sqrt{x} \cdot 5\sqrt{x} + 2\sqrt{x} \cdot 4 - 3 \cdot 5\sqrt{x} - 12$ Multiply pairs of terms.

$= 10x + 8\sqrt{x} - 15\sqrt{x} - 12$ Simplify.

$= 10x - 7\sqrt{x} - 12$ Combine like radicals.

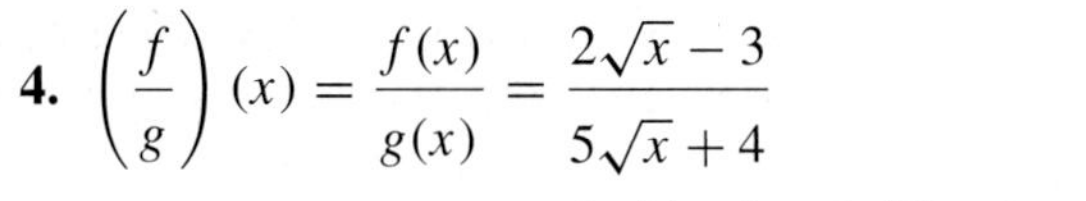

4. $\left(\frac{f}{g}\right)(x) = \frac{f(x)}{g(x)} = \frac{2\sqrt{x}-3}{5\sqrt{x}+4}$ Substitute for $f(x)$ and $g(x)$.

$= \frac{2\sqrt{x}-3}{5\sqrt{x}+4} \cdot \frac{5\sqrt{x}-4}{5\sqrt{x}-4}$ Rationalize denominator.

$= \frac{2\sqrt{x}\cdot 5\sqrt{x} - 2\sqrt{x}\cdot 4 - 3\cdot 5\sqrt{x} + 12}{(5\sqrt{x})^2 - 4^2}$ Multiply numerators; multiply denominators.

$= \frac{10x - 8\sqrt{x} - 15\sqrt{x} + 12}{5^2(\sqrt{x})^2 - 4^2}$ Simplify.

$= \frac{10x - 23\sqrt{x} + 12}{25x - 16}$ Simplify.

We verify our work by comparing graphing calculator tables for $y = \frac{2\sqrt{x}-3}{5\sqrt{x}+4}$ and $y = \frac{10x - 23\sqrt{x} + 12}{25x - 16}$ for $x > 0$ and $x \neq \frac{16}{25}$ (see Fig. 22). ■

Plot1 Plot2 Plot3
\Y1=(2√(X)-3)/(5√(X)+4)
\Y2=(10X-23√(X)+12)/(25X-16)
\Y3=
\Y4=
\Y5=

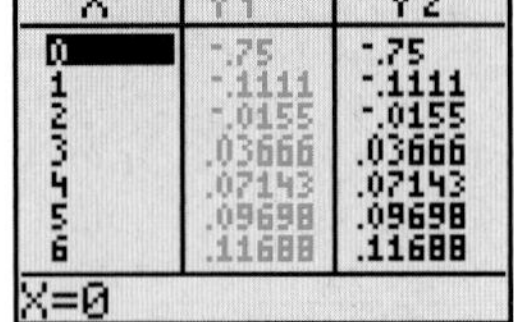

Figure 22 Verify the work

group exploration

Translating and reflecting the absolute value function

1. Complete Table ?? for the absolute value function $y = |x|$. For example, $|-2| = 2$, $|0| = 0$, and $|3| = 3$. To review absolute value, see Section A.3.
2. Sketch a graph of $y = |x|$. Use a graphing calculator to verify your graph. To enter $|x|$, Press MATH ▷ 1 X,T,Θ,n).
3. Translate and/or reflect the graph of $y = |x|$ to sketch the graph of the given function. Use a graphing calculator to verify your sketch.
   a. $y = |x| - 2$
   b. $y = |x - 4|$
   c. $y = -|x + 3|$
   d. $y = -|x - 2| + 5$
4. Sketch the graph of the given function.
   a. $y = 2|x|$
   b. $y = -3|x|$
5. Describe the graphical significance of a, h, and k for a function of the form $y = a|x - h| + k$, where $a \neq 0$.

Table 11 Values of the Function $y = |x|$

x	y
−3	
−2	
−1	
1	
2	
3	

HOMEWORK 9.4

FOR EXTRA HELP ▶ Student Solutions Manual · PH Math/Tutor Center · MathXL® · MyMathLab

Graph the function by hand. Use a graphing calculator to verify your sketch.

1. $y = 2\sqrt{x}$
2. $y = -3\sqrt{x}$
3. $y = -\sqrt{x}$
4. $y = -\frac{1}{2}\sqrt{x}$
5. $y = \sqrt{x} + 3$
6. $y = \sqrt{x} - 5$
7. $y = 2\sqrt{x} - 5$
8. $y = 3\sqrt{x} - 2$
9. $y = -3\sqrt{x} + 4$
10. $y = -2\sqrt{x} + 1$
11. $y = \sqrt{x - 2}$
12. $y = \sqrt{x - 5}$
13. $y = -\sqrt{x + 2}$
14. $y = -2\sqrt{x + 5}$
15. $y = \frac{1}{2}\sqrt{x - 4}$
16. $y = \frac{1}{4}\sqrt{x - 1}$
17. $y = \sqrt{x + 3} + 2$
18. $y = \sqrt{x + 1} + 3$
19. $y = -2\sqrt{x + 3} - 4$
20. $y = -3\sqrt{x + 2} - 1$
21. $y = 4\sqrt{x - 1} - 3$
22. $y = 2\sqrt{x - 6} - 2$
23. $\sqrt{x} + y = 4$
24. $2\sqrt{x} - y = 3$
25. $2y - 6\sqrt{x} = 8$
26. $3y - 6\sqrt{x} = 9$

Graph the function by hand. Use a graphing calculator to verify your sketch. Also, find the domain and range of the function.

27. $y = -2\sqrt{x}$
28. $y = 3\sqrt{x}$
29. $y = \sqrt{x + 2}$
30. $y = \sqrt{x - 6}$
31. $y = \sqrt{x} + 2$
32. $y = \sqrt{x} - 1$
33. $y = \sqrt{x - 5} - 3$
34. $y = \sqrt{x + 1} + 2$

35. $y = 2\sqrt{x+5}+1$

36. $y = 3\sqrt{x-3}-6$

37. $y = -\sqrt{x-2}+4$

38. $y = -2\sqrt{x+4}-2$

39. Describe the Rule of Four as applied to the function $f(x) = 2\sqrt{x} - 3$:

a. Describe five input–output pairs of f by using a table.
b. Describe the input–output pairs of f by using a graph.
c. Describe the input–output pairs of f by using words.

40. Describe the Rule of Four as applied to the function $g(x) = -\sqrt{x} + 4$:

a. Describe five input–output pairs of g by using a table.
b. Describe the input–output pairs of g by using a graph.
c. Describe the input–output pairs of g by using words.

41. Figure 23 shows four functions of the form $y = a\sqrt{x-h}+k$. For each, describe the signs of the constants a, h, and k.

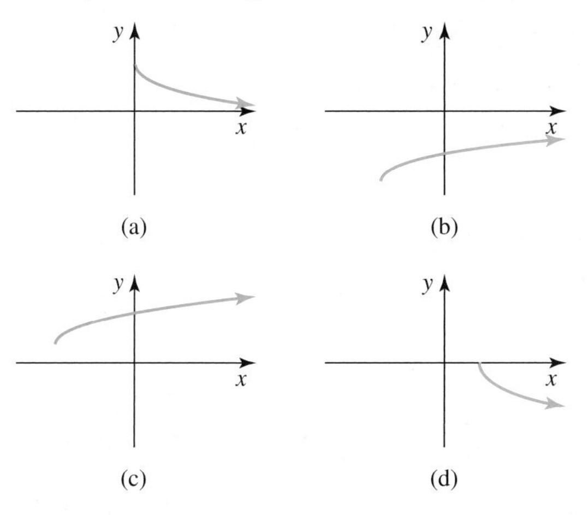

Figure 23 Exercise 41

42. For each part, copy the graph of $f(x) = a\sqrt{x-h}+k$ as shown in Fig. 24. On the same coordinate system, use the graph of f to sketch the graph of the given function. Label each graph.

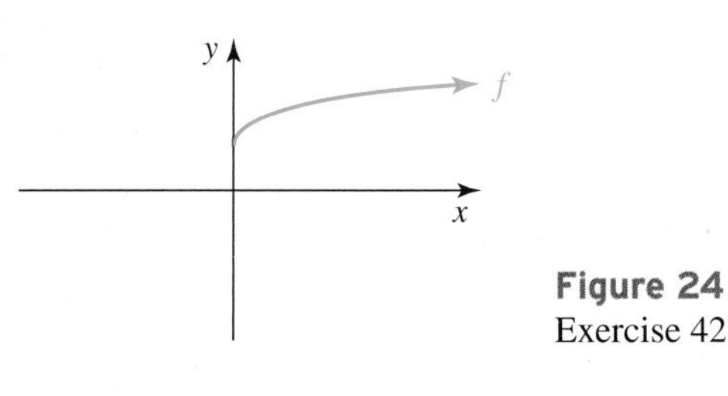

Figure 24 Exercise 42

a. $g(x) = a\sqrt{x-h}+2k$
b. $h(x) = a\sqrt{x-2h}+k$
c. $k(x) = -a\sqrt{x-h}+k$
d. $r(x) = 2a\sqrt{x+h}-k$

43. Use a graphing calculator to graph a family of square root curves similar to the one in Fig. 25. List the equations of your square root curves.

(0, 2)

Figure 25 A family of square root curves—Exercise 43

44. Use a graphing calculator to graph a family of square root curves similar to the one in Fig. 26. List the equations of your square root curves.

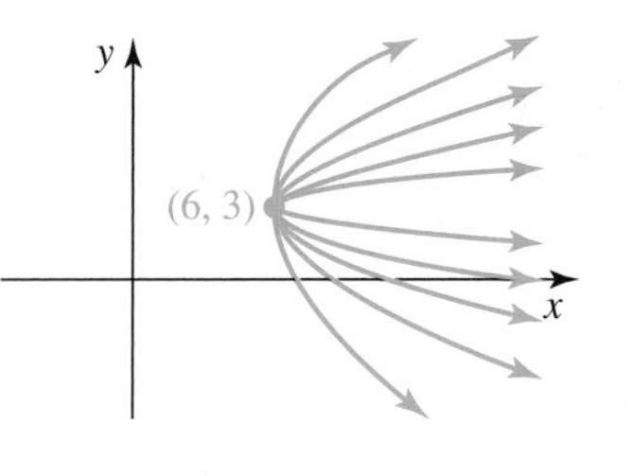

Figure 26 Another family of square root curves—Exercise 44

45. For what values of a, h, and k for the square root function $f(x) = a\sqrt{x-h}+k$, where $a \neq 0$, is there a point on the graph of f that is higher than all other points on the graph? (Recall that this point is the maximum point.) For what values of a, h, and k does f have a minimum point? Describe the maximum or minimum point in terms of a, h, and k.

46. Solve the system.

$$y = 2\sqrt{x-3}+1$$
$$y = -2\sqrt{x-3}+1$$

Evaluate the function $f(x) = 7\sqrt{x} - 3$ at the indicated value of x. Assume that $c \geq 0$.

47. $f(4)$ **48.** $f(0)$ **49.** $f(9c)$ **50.** $f(4c)$

For $f(x) = 5\sqrt{x} - 9$ and $g(x) = 4\sqrt{x} + 1$, find an equation of the given function.

51. $f + g$ **52.** $f - g$ **53.** $f \cdot g$ **54.** $\dfrac{f}{g}$

For $f(x) = 2\sqrt{x} - 3\sqrt{5}$ and $g(x) = 2\sqrt{x} + 3\sqrt{5}$, find an equation of the given function.

55. $f - g$ **56.** $f + g$ **57.** $\dfrac{f}{g}$ **58.** $f \cdot g$

For $f(x) = \sqrt{x+1} - 2$ and $g(x) = \sqrt{x+1} + 2$, find an equation of the given function.

59. $f + g$ **60.** $f - g$ **61.** $f \cdot g$ **62.** $\dfrac{f}{g}$

63. The percentages of e-mails that are spam are shown in Table 12 for various years.

Table 12 Percentages of E-mails That Are Spam

Year	Percent
1999	21
2002	56
2004	68
2005	72
2006	80

Source: *IronPort*

Let $p = f(t)$ be the percentage of e-mails that are spam at t years since 1999. A reasonable model is $f(t) = 21.4\sqrt{t} + 21$.

a. Use a graphing calculator to draw the graph of the model and, in the same viewing window, the scattergram of the data. Does the model fit the data well?
b. Estimate the percentage of e-mails that were spam in 2003. Did you perform interpolation or extrapolation? Explain.
c. Predict the percentage of e-mails that will be spam in 2010. Did you perform interpolation or extrapolation? Explain.

64. The numbers of teenagers ages 11–18 who are adopted from foster care are shown in Table 13 for various years.

Table 13 Numbers of Teenagers Adopted from Foster Care

Year	Number of Teenagers Adopted (thousands)
1998	5.4
1999	7.0
2000	8.0
2001	8.1
2002	9.4
2003	9.0

Source: *U.S. Department of Health and Human Services' Children's Bureau*

Let $f(t)$ be the number (in thousands) of teenagers who are adopted from foster care in the year that is t years since 1998. A reasonable model is $f(t) = 1.75\sqrt{t} + 5.4$.

a. Use a graphing calculator to draw the graph of the model and, in the same viewing window, the scattergram of the data. Does the model fit the data well?

b. Estimate the number of teenagers who were adopted in 2000. Is your result an underestimate or an overestimate? Did you perform interpolation or extrapolation? Explain.

c. Predict the number of teenagers who will be adopted in 2011. Did you perform interpolation or extrapolation? Explain.

For Exercises 65–72 refer to Fig. 27.

65. Estimate $f(-6)$.

66. Estimate $f(-2)$.

67. Estimate $f(0)$.

68. Estimate $f(5)$.

69. Estimate x when $f(x) = 0$.

70. Estimate x when $f(x) = 2$.

71. Estimate x when $f(x) = 3$.

72. Estimate x when $f(x) = 4$.

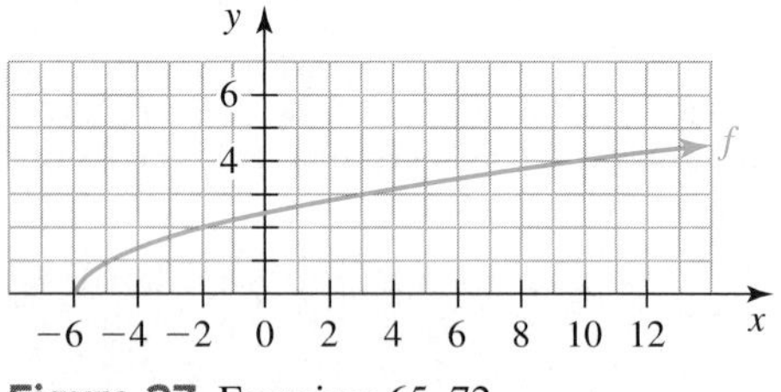

Figure 27 Exercises 65–72

73. Use ZStandard followed by ZSquare to draw the graphs of $f(x) = 2\sqrt{x+3} + 2$ and $g(x) = -2\sqrt{x+3} + 2$ on the same coordinate system. Consider the combined graph as the graph of a single relation. What do you notice about the graph? Is the relation described by this graph a function? Explain.

74. Find two functions that have the same domain and range as the function $f(x) = \sqrt{x+7} - 2$.

75. Describe how to graph the function $g(x) = a\sqrt{x-h} + k$, where $a \neq 0$, given the graph of $f(x) = \sqrt{x}$.

76. Compare the process of graphing a square root function of the form $f(x) = a\sqrt{x-h} + k$, where $a \neq 0$, with that of graphing a quadratic function of the form $g(x) = a(x-h)^2 + k$. How are the processes similar? different?

Related Review

Graph the function by hand.

77. $2x - 5y = 20$

78. $y = -2(x-4)^2 + 3$

79. $y = 2\sqrt{x+3} - 4$

80. $y = -3\sqrt{x-1} + 3$

81. $y = 8\left(\frac{1}{2}\right)^x$

82. $y = 3x^2 - 12x + 9$

Expressions, Equations, Functions, and Graphs

Perform the indicated instruction. Then use words such as linear, quadratic, cubic, exponential, logarithmic, rational, radical, polynomial, degree, function, one variable, *and* two variables *to describe the expression, equation, or system.*

83. Factor $6x^2 - 5x - 6$.

84. Find the product $(3\sqrt{x} - 5)(2\sqrt{x} + 4)$.

85. Let $f(x) = 3x^2 - 2x + 4$. Find x when $f(x) = 6$.

86. Graph $f(x) = -2\sqrt{x-3} + 1$ by hand.

87. Find an equation of a parabola that contains the points (1, 4), (3, 14), and (4, 25).

88. Graph $f(x) = -2(x-3)^2 + 1$.

9.5 SOLVING RADICAL EQUATIONS

Objectives

- Solve radical equations.
- Find the x-intercepts of square root functions.
- Use a radical model to make predictions about the independent variable.

In Sections 9.1–9.3, we worked with radical *expressions*. In this section, we solve radical *equations*. A **radical equation in one variable** is an equation in one variable that contains a radical with the variable in the radicand. Here are some examples of a radical equation in one variable:

$$\sqrt{x} = 5 \quad \sqrt[3]{7x-3} = 2 \quad \sqrt{x-2} = x-4 \quad \sqrt{x+3} - \sqrt{2x-4} = 6$$

We begin by solving square root equations in one variable.

Solving Square Root Equations in One Variable

Consider the square root equation

$$\sqrt{x} = 3$$

If x is nonnegative, then $\left(\sqrt{x}\right)^2 = x$. To get the left side of the equation $\sqrt{x} = 3$ to be x, we square both sides:

$$\left(\sqrt{x}\right)^2 = 3^2$$
$$x = 9$$

So, the solution is 9. This checks out, because $\sqrt{9} = 3$.

Squaring Property of Equality

If A and B are expressions, then all solutions of the equation $A = B$ are *among* the solutions of the equation $A^2 = B^2$. That is, the solutions of an equation are among the solutions of the equation obtained by squaring both sides.

Recall from Section 8.5 that if we clear a rational equation of fractions and arrive at a value of x that is an excluded value, then we call that result an extraneous solution. In general, if a proposed solution of any type of equation is *not* a solution, we call it an **extraneous solution.**

Squaring both sides of an equation can introduce extraneous solutions. Consider the simple equation $x = 5$, whose only solution is 5. Here we square both sides of $x = 5$ and solve by using the square root property (Section 7.3):

$x = 5$	The only solution is 5.
$x^2 = 5^2$	Square both sides.
$x^2 = 25$	$5^2 = 25$
$x = \pm\sqrt{25}$	Square root property
$x = \pm 5$	$\sqrt{25} = 5$

Squaring both sides of the equation $x = 5$ introduced the extraneous solution -5 (which is *not* a solution of the original equation).

Checking Proposed Solutions

Because squaring both sides of a square root equation may introduce extraneous solutions, it is essential to check that each proposed solution satisfies the original equation.

Example 1 Solving a Square Root Equation

Solve the equation $2\sqrt{x} + 5 = 13$.

Solution

First, we isolate $\sqrt{x}$ (get $\sqrt{x}$ alone) on one side of the equation:

$2\sqrt{x} + 5 = 13$	Original equation
$2\sqrt{x} = 8$	Subtract 5 from both sides.
$\sqrt{x} = 4$	Divide both sides by 2.
$\left(\sqrt{x}\right)^2 = 4^2$	Square both sides.
$x = 16$	$\left(\sqrt{x}\right)^2 = x$ for $x \geq 0$; $4^2 = 16$

We check that 16 satisfies the original equation:

$$2\sqrt{x}+5=13 \quad \textit{Original equation}$$
$$2\sqrt{16}+5 \stackrel{?}{=} 13 \quad \textit{Substitute 16 for x.}$$
$$2\cdot 4+5 \stackrel{?}{=} 13 \quad \sqrt{16}=4$$
$$13 \stackrel{?}{=} 13$$
$$\text{true}$$

So, the solution is 16. ■

We see from Example 1 that to solve a square root equation, we isolate a square root term on one side of the equation before we square both sides.

Example 2 Solving a Square Root Equation

Solve $x=\sqrt{x-1}+3$.

Solution

$$x=\sqrt{x-1}+3 \quad \textit{Original equation}$$
$$x-3=\sqrt{x-1} \quad \textit{Isolate radical.}$$
$$(x-3)^2=\left(\sqrt{x-1}\right)^2 \quad \textit{Square both sides.}$$
$$x^2-6x+9=x-1 \quad (A-B)^2=A^2-2AB+B^2;\ (\sqrt{x})^2=x \text{ for } x\geq 0$$
$$x^2-7x+10=0 \quad \textit{Write in } ax^2+bx+c=0 \textit{ form.}$$
$$(x-2)(x-5)=0 \quad \textit{Factor left side.}$$
$$x-2=0 \quad \text{or} \quad x-5=0 \quad \textit{Zero factor property}$$
$$x=2 \quad \text{or} \quad x=5$$

We check that 2 and 5 satisfy the original equation:

Check x = 2	***Check x = 5***
$x=\sqrt{x-1}+3$	$x=\sqrt{x-1}+3$
$2 \stackrel{?}{=} \sqrt{2-1}+3$	$5 \stackrel{?}{=} \sqrt{5-1}+3$
$2 \stackrel{?}{=} 4$	$5 \stackrel{?}{=} 5$
false	true

Since 2 does not satisfy the original equation, it is an extraneous solution. Therefore, the only solution is 5.

We use "intersect" on a graphing calculator to verify our work (see Fig. 28). The point (5, 5) is the only point of intersection. This supports our conclusion that the only solution of the original equation is 5. ■

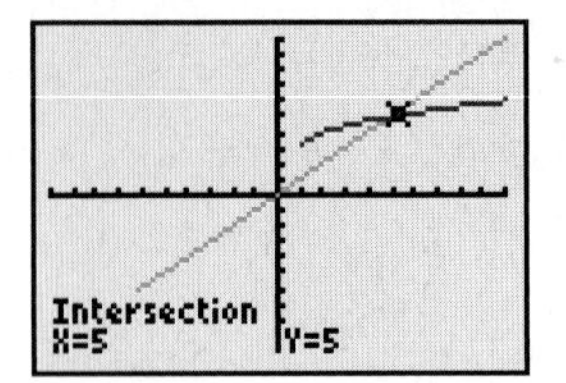

Figure 28 Verify the work

Example 3 Solving a Square Root Equation

Solve $2\sqrt{x-3}-\sqrt{x}=0$.

Solution

First we isolate $2\sqrt{x-3}$; then we square both sides of the equation:

$$2\sqrt{x-3}-\sqrt{x}=0 \quad \textit{Original equation}$$
$$2\sqrt{x-3}=\sqrt{x} \quad \textit{Isolate } 2\sqrt{x-3}.$$
$$\left(2\sqrt{x-3}\right)^2=\left(\sqrt{x}\right)^2 \quad \textit{Square both sides.}$$
$$2^2\left(\sqrt{x-3}\right)^2=\left(\sqrt{x}\right)^2 \quad (AB)^2=A^2B^2$$
$$4(x-3)=x \quad \left(\sqrt{x}\right)^2=x \text{ for } x\geq 0$$
$$4x-12=x$$
$$3x=12$$
$$x=4$$

To verify that 4 is a solution, we enter the function $y = 2\sqrt{x-3} - \sqrt{x}$. Then we check that for the input 4, the output is 0 (see Fig. 29).

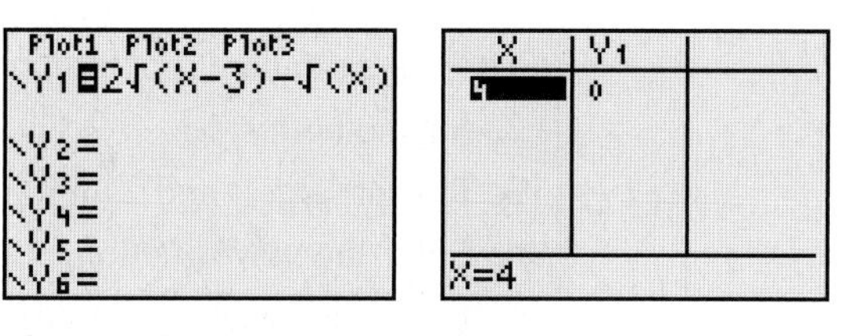

Figure 29 Verify that 4 is a solution ■

In Example 3, after squaring both sides of $2\sqrt{x-3} = \sqrt{x}$, we simplified $\left(2\sqrt{x-3}\right)^2$:

$$\left(2\sqrt{x-3}\right)^2 = 2^2\left(\sqrt{x-3}\right)^2 = 4(x-3) \qquad \text{Correct}$$

WARNING It is a common error to forget to square the coefficient 2 and write

$$\left(2\sqrt{x-3}\right)^2 = 2(x-3) \qquad \text{Incorrect}$$

Recall from Section 4.1 that when we simplify $(AB)^2$, we square both A and B:

$$(AB)^2 = A^2B^2$$

If an equation contains two or more square root terms, we may need to use the squaring property of equality twice.

Example 4 Solving a Square Root Equation

Solve $\sqrt{t-5} - \sqrt{t} = -1$.

Solution

First we isolate the radical $\sqrt{t-5}$; then we square both sides of the equation:

$$\begin{aligned} \sqrt{t-5} - \sqrt{t} &= -1 && \text{Original equation} \\ \sqrt{t-5} &= \sqrt{t} - 1 && \text{Isolate radical } \sqrt{t-5}. \\ \left(\sqrt{t-5}\right)^2 &= \left(\sqrt{t} - 1\right)^2 && \text{Square both sides.} \\ t - 5 &= \left(\sqrt{t}\right)^2 - 2\sqrt{t}\cdot 1 + 1^2 && \left(\sqrt{x}\right)^2 = x \text{ for } x \geq 0; \ (A-B)^2 = A^2 - 2AB + B^2 \\ t - 5 &= t - 2\sqrt{t} + 1 && \left(\sqrt{x}\right)^2 = x \text{ for } x \geq 0 \end{aligned}$$

Next we isolate the radical $\sqrt{t}$ and square both sides:

$$\begin{aligned} 2\sqrt{t} &= t + 1 - t + 5 && \text{Isolate } 2\sqrt{t}. \\ 2\sqrt{t} &= 6 && \text{Combine like terms.} \\ \sqrt{t} &= 3 && \text{Divide both sides by 2.} \\ \left(\sqrt{t}\right)^2 &= 3^2 && \text{Square both sides.} \\ t &= 9 && \left(\sqrt{x}\right)^2 = x \text{ for } x \geq 0 \end{aligned}$$

Now we check that 9 satisfies the original equation:

$$\begin{aligned} \sqrt{t-5} - \sqrt{t} &= -1 && \text{Original equation} \\ \sqrt{9-5} - \sqrt{9} &\stackrel{?}{=} -1 && \text{Substitute 9 for } t. \\ 2 - 3 &\stackrel{?}{=} -1 \\ -1 &\stackrel{?}{=} -1 \\ &\text{true} \end{aligned}$$

So, the solution is 9. ■

Solving a Square Root Equation in One Variable

To solve a square root equation in one variable,

1. Isolate a square root term on one side of the equation.
2. Square both sides.
3. Repeat steps 1 and 2 until no square root terms remain.
4. Solve the new equation.
5. Check that each proposed solution satisfies the original equation.

Example 5 Solving a Square Root Equation

Solve $\sqrt{2x+1} - \sqrt{x+2} = 1$.

Solution

$$\sqrt{2x+1} - \sqrt{x+2} = 1 \quad \text{Original equation}$$

$$\sqrt{2x+1} = \sqrt{x+2} + 1 \quad \text{Isolate radical } \sqrt{2x+1}.$$

$$\left(\sqrt{2x+1}\right)^2 = \left(\sqrt{x+2}+1\right)^2 \quad \text{Square both sides.}$$

$$2x+1 = \left(\sqrt{x+2}\right)^2 + 2\sqrt{x+2}\cdot 1 + 1^2 \quad \left(\sqrt{x}\right)^2 = x \text{ for } x \geq 0,\ (A+B)^2 = A^2 + 2AB + B^2$$

$$2x+1 = x+2+2\sqrt{x+2}+1 \quad \left(\sqrt{x}\right)^2 = x \text{ for } x \geq 0$$

$$x-2 = 2\sqrt{x+2} \quad \text{Isolate radical } 2\sqrt{x+2}.$$

$$(x-2)^2 = \left(2\sqrt{x+2}\right)^2 \quad \text{Square both sides.}$$

$$x^2 - 4x + 4 = 4(x+2) \quad (A-B)^2 = A^2 - 2AB + B^2;\ (bc)^n = b^n c^n$$

$$x^2 - 4x + 4 = 4x + 8 \quad \text{Distributive law}$$

$$x^2 - 8x - 4 = 0 \quad \text{Write in } ax^2 + bx + c = 0 \text{ form.}$$

$$x = \frac{-(-8) \pm \sqrt{(-8)^2 - 4(1)(-4)}}{2(1)} \quad \text{Substitute } a = 1, b = -8, c = -4 \text{ into quadratic formula.}$$

$$= \frac{8 \pm \sqrt{80}}{2} \quad \text{Simplify.}$$

$$= \frac{8 \pm 4\sqrt{5}}{2} \quad \sqrt{80} = \sqrt{16\cdot 5} = 4\sqrt{5}$$

$$= \frac{2(4 \pm 2\sqrt{5})}{2} \quad \text{Factor out 2.}$$

$$= 4 \pm 2\sqrt{5} \quad \text{Simplify.}$$

So, $x = 4 - 2\sqrt{5} \approx -0.47$ or $x = 4 + 2\sqrt{5} \approx 8.47$. We check that the approximations of the solutions approximately satisfy the original equation $\sqrt{2x+1} - \sqrt{x+2} = 1$:

Check $x \approx -0.47$

$\sqrt{2(-0.47)+1} - \sqrt{-0.47+2} \approx -0.9920$

not close to 1

Check $x \approx 8.47$

$\sqrt{2(8.47)+1} - \sqrt{8.47+2} \approx 0.9998$

close to 1

So, the only solution is $4 + 2\sqrt{5}$.

Now we store $4 - 2\sqrt{5}$ for the variable "X" in a graphing calculator and perform a similar, but more precise, check. We do the same for $4 + 2\sqrt{5}$ (see Fig. 30). ■

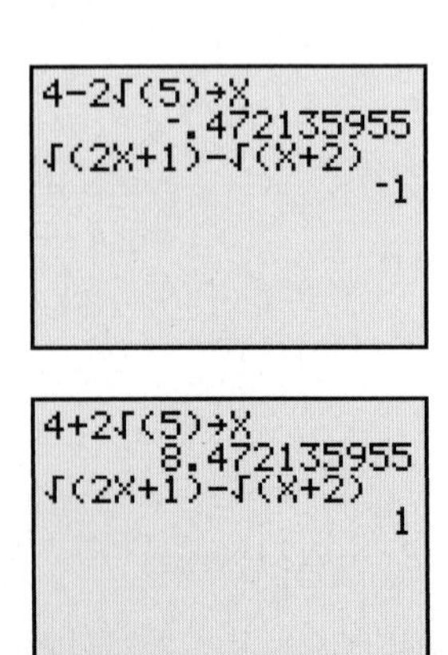

Figure 30 Verify that $4 - 2\sqrt{5}$ is an extraneous solution and $4 + 2\sqrt{5}$ is a solution

In Example 5, we simplified the right-hand side of the equation

$$(\sqrt{2x+1})^2 = (\sqrt{x+2}+1)^2$$

by using the property for the square of a sum, $(A+B)^2 = A^2 + 2AB + B^2$:

$$(\sqrt{x+2}+1)^2 = (\sqrt{x+2})^2 + 2\sqrt{x+2}\cdot 1 + 1^2 \qquad \text{Correct}$$

WARNING Remember that, in general, $(A+B)^2$ is *not* equal to $A^2 + B^2$, so it is incorrect to say that

$$(\sqrt{x+2}+1)^2 = (\sqrt{x+2})^2 + 1^2 \qquad \text{Incorrect}$$

The equation $\sqrt{x+8} = x^2 - 3x - 4$ would be very difficult to solve by using properties such as the squaring property of equality. In Example 6, we use graphing to find approximate solutions of the equation.

Example 6 Using Graphing to Solve a Square Root Equation in One Variable

Use graphing to solve $\sqrt{x+8} = x^2 - 3x - 4$.

Solution

We use "intersect" on a graphing calculator to find the solutions of the system

$$y = \sqrt{x+8}$$
$$y = x^2 - 3x - 4$$

See Fig. 31.

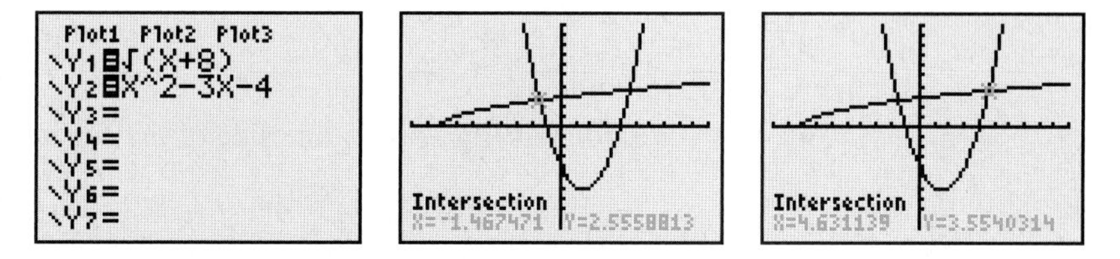

Figure 31 Solve the system

The approximate solutions of the system are $(-1.47,\ 2.56)$ and $(4.63, 3.55)$. The x-coordinates of these ordered pairs, -1.47 and 4.63, are the approximate solutions of the equation $\sqrt{x+8} = x^2 - 3x - 4$. ■

Finding x-Intercepts of a Square Root Function

To find all x-intercepts of the graph of a square root equation in x and y, we substitute 0 for y and solve for x.

Example 7 Finding the x-Intercept of a Square Root Function

Find the x-intercepts of the graph of $f(x) = \sqrt{3x-4} - \sqrt{x+2}$.

Solution

We substitute 0 for $f(x)$ and solve for x:

$$\sqrt{3x-4} - \sqrt{x+2} = 0 \qquad \text{Substitute 0 for } f(x).$$
$$\sqrt{3x-4} = \sqrt{x+2} \qquad \text{Isolate at least one radical.}$$
$$(\sqrt{3x-4})^2 = (\sqrt{x+2})^2 \qquad \text{Square both sides.}$$
$$3x - 4 = x + 2 \qquad (\sqrt{x})^2 = x \text{ for } x \geq 0$$
$$2x = 6$$
$$x = 3$$

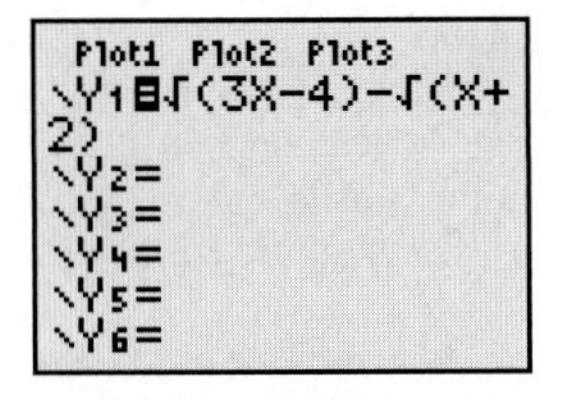

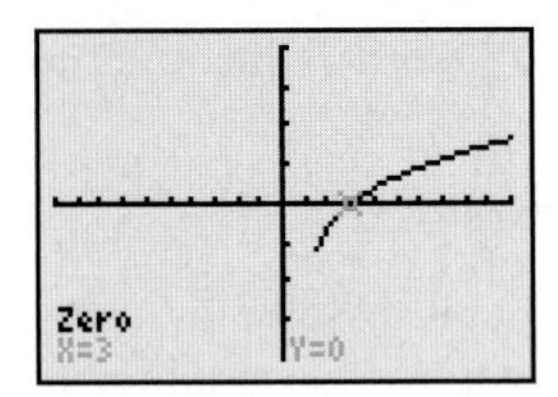

Figure 32 Verify the work

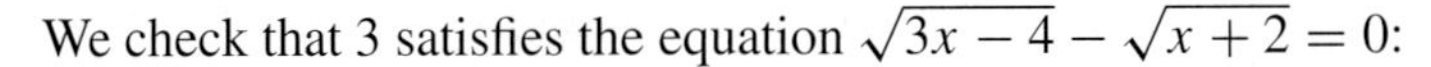

We check that 3 satisfies the equation $\sqrt{3x-4}-\sqrt{x+2}=0$:

$$\sqrt{3x-4}-\sqrt{x+2}=0 \quad \text{Original equation}$$
$$\sqrt{3(3)-4}-\sqrt{3+2}\overset{?}{=}0 \quad \text{Substitute 3 for } x.$$
$$\sqrt{5}-\sqrt{5}\overset{?}{=}0$$
$$0\overset{?}{=}0$$
$$\text{true}$$

The x-intercept is $(3, 0)$. We use "zero" on a graphing calculator to verify our work (see Fig. 32). ■

Solving Radical Equations

So far, we have solved square root equations in one variable. We can solve other types of radical equations in one variable in a similar way.

Power Property of Equality

If A and B are expressions and n is a counting number greater than 1, then all solutions of the equation $A = B$ are *among* the solutions of the equation $A^n = B^n$. That is, the solutions of an equation are among the solutions of the equation obtained by raising both sides to the nth power.

Example 8 Solving a Radical Equation

Solve $\sqrt[3]{2w-4}+7=9$.

Solution

We isolate $\sqrt[3]{2w-4}$ on one side of the equation:

$$\sqrt[3]{2w-4}+7=9 \quad \text{Original equation}$$
$$\sqrt[3]{2w-4}=2 \quad \text{Subtract 7 from both sides.}$$

From the power property for radicals (Section 9.2), $\left(\sqrt[3]{2w-4}\right)^3=2w-4$. To get the left side of the equation $\sqrt[3]{2w-4}=2$ to be $2w-4$, we cube both sides:

$$\left(\sqrt[3]{2w-4}\right)^3=2^3 \quad \text{Cube both sides.}$$
$$2w-4=8 \quad \left(\sqrt[3]{x}\right)^3=x$$
$$2w=12$$
$$w=6$$

We check that 6 satisfies the original equation:

$$\sqrt[3]{2w-4}+7=9 \quad \text{Original equation}$$
$$\sqrt[3]{2(6)-4}+7\overset{?}{=}9 \quad \text{Substitute 6 for } w.$$
$$\sqrt[3]{8}+7\overset{?}{=}9$$
$$2+7\overset{?}{=}9 \quad \sqrt[3]{8}=2$$
$$9\overset{?}{=}9$$
$$\text{true}$$

So, the solution is 6. ■

It is only when we raise both sides of an equation to an *even* power that we might introduce extraneous solutions. However, no matter what type of equation we are solving, it is a good idea to check that any results satisfy the original equation to make sure the work is correct.

Using a Radical Model to Make Predictions About the Independent Variable

Now that we have discussed how to solve radical equations in one variable, we can use radical models to make predictions about the independent variable.

Example 9 Using a Radical Model to Make Predictions

In Example 10 of Section 9.1, we worked with the model $f(t) = 17\sqrt{t} + 44$, where $f(t)$ is the percentage of colleges and universities that offer distance learning at t years since 1997 (see Table 14).

Predict when all colleges and universities will offer distance learning.

Table 14 Percentages of Colleges and Universities Offering Distance Learning

Year	Percent
1997	44
1998	62
1999	69
2002	85
2004	87

Source: *International Data Corporation*

Solution

We substitute 100 for $f(t)$ and solve for t:

$$\begin{aligned} 100 &= 17\sqrt{t} + 44 && \text{Substitute 100 for } f(t). \\ 56 &= 17\sqrt{t} && \text{Subtract 44 from both sides.} \\ \frac{56}{17} &= \sqrt{t} && \text{Divide both sides by 17.} \\ \left(\frac{56}{17}\right)^2 &= \left(\sqrt{t}\right)^2 && \text{Square both sides.} \\ t &\approx 10.85 && \text{Compute.} \end{aligned}$$

All colleges and universities will offer distance learning in 2008, according to the model. Model breakdown seems likely—at least one college or university will probably not offer distance learning. However, it wouldn't be surprising if *nearly* all colleges and universities offered distance learning then. ■

group exploration

Extraneous solutions

1. Solve the equation $\sqrt{3x-2} + 2 = x$. Record each step of your work carefully.
2. In Problem 1, you found that 1 is an extraneous solution and that 6 is the only solution. Now substitute 1 for x in each step you recorded in Problem 1. Which of the equations are satisfied by 1?
3. What does it mean to say that 1 is an extraneous solution? Why do we sometimes get extraneous solutions when we solve square root equations but not when we solve linear, exponential, or quadratic equations?

HOMEWORK 9.5

FOR EXTRA HELP ▶ Student Solutions Manual | PH Math/Tutor Center | MathXL® | MyMathLab

Solve.

1. $\sqrt{x} = 5$

2. $\sqrt{x} = 8$

3. $\sqrt{x} = -2$

4. $\sqrt{x} = -7$

5. $\sqrt[3]{t} = -2$

6. $\sqrt[4]{w} = -3$

7. $3\sqrt{x} - 1 = 5$

8. $5\sqrt{x} + 2 = 37$

9. $\sqrt{x-1} = 2$

10. $\sqrt{x-5} = 3$

11. $\sqrt[4]{r+2} = 2$

12. $\sqrt[3]{b+7} = 3$

13. $\sqrt{5x-7} + 7 = 3$

14. $\sqrt{15+x} + 8 = 2$

15. $\sqrt[3]{2x-5} + 3 = 7$

16. $\sqrt[4]{3x-1} + 5 = 8$

17. $2 - 10\sqrt{6x+3} = -98$

18. $7 - 8\sqrt{2x-1} = -17$

19. $\sqrt{3k+1} = \sqrt{2k+6}$

20. $\sqrt{10m-3} = \sqrt{6m+2}$

21. $\sqrt[4]{6x-3} = \sqrt[4]{2x+17}$

22. $\sqrt[3]{5x-8} = \sqrt[3]{3x+6}$

23. $2\sqrt{1-x} - \sqrt{2x+5} = 0$

24. $2\sqrt{x-1} - \sqrt{3x-1} = 0$

25. $\sqrt{3w+3} = w - 5$

26. $\sqrt{t+10} = t - 2$

27. $\sqrt{12x+13} + 2 = 3x$

28. $\sqrt{x+2} - x = 2$

29. $\sqrt{3x+4} - x = 3$

30. $1 + \sqrt{2x+5} = 2x$

31. $\sqrt{r^2 - 5r + 1} = r - 3$
32. $\sqrt{k^2 + 2k} = k + 5$
33. $\sqrt{x} - 1 = \sqrt{5 - x}$
34. $3 = \sqrt{6 + x} + \sqrt{x}$
35. $\sqrt{x} - \sqrt{2x} = -1$
36. $\sqrt{x} = \sqrt{3x} - 2$
37. $\sqrt{x - 3} + \sqrt{x + 5} = 4$
38. $\sqrt{x + 6} - \sqrt{x - 2} = 2$
39. $\sqrt{2p - 1} + \sqrt{3p - 2} = 2$
40. $\sqrt{3t - 1} - \sqrt{4t + 1} = -1$
41. $\sqrt{\sqrt{x} - 2} = 3$
42. $\sqrt{\sqrt{x} + 1} = 1$
43. $\dfrac{1}{\sqrt{x + 2}} = 3 - \sqrt{x + 2}$
44. $\dfrac{1}{\sqrt{x - 5}} = 2 - \sqrt{x - 5}$

Find an approximate solution of the equation. Round your result to the second decimal place.

45. $5.2\sqrt{x} - 2.8 = 13.9$
46. $4.7\sqrt{x} + 3.1 = 46.9$
47. $1.52 - 4.91\sqrt{3.18x - 7.14} = -0.69$
48. $-7.93 = 5.61 - 3.79\sqrt{4.42 - 9.87x}$

Use "intersect" on a graphing calculator to solve the equation. Round any solutions to the second decimal place.

49. $\sqrt{x + 1} = 4 - \sqrt{x + 3}$
50. $\sqrt{x + 6} = 6 - \sqrt{x + 4}$
51. $\sqrt{x + 3} = x^2 - 4x - 2$
52. $\sqrt{x + 7} = x^2 - 2x - 7$

For Exercises 53–58, estimate all solutions of the equation or system by referring to the graphs shown in Fig. 33. Round results or coordinates of results to the first decimal place.

53. $\sqrt{x + 5} = x^2 - 2x - 4$
54. $\sqrt{x + 5} = -2x - 5$
55. $\sqrt{x + 5} = 1$
56. $\sqrt{x + 5} = 2$
57. $y = \sqrt{x + 5}$
$y = -2x - 5$
58. $y = \sqrt{x + 5}$
$y = x^2 - 2x - 4$

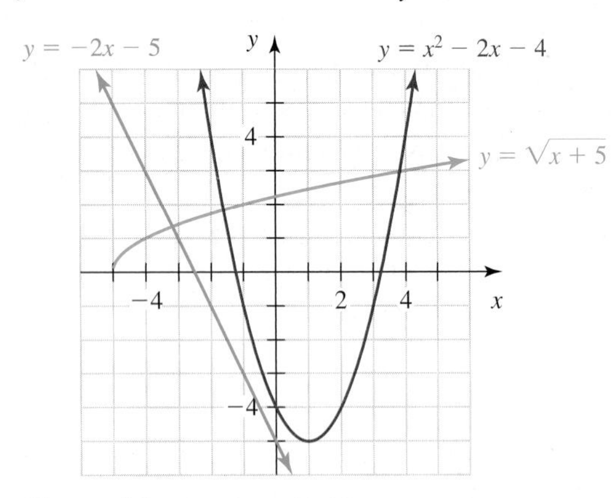

Figure 33 Exercises 53–58

Find all x-intercepts. Use a graphing calculator to verify your work.

59. $h(x) = 3\sqrt{-3x + 4} - 15$
60. $k(x) = 2\sqrt{4x + 1} - 22$
61. $f(x) = \sqrt{3x - 2} - \sqrt{x + 8}$
62. $g(x) = \sqrt{2x + 5} - \sqrt{4x + 1}$
63. $h(x) = 2\sqrt{x + 4} + 3\sqrt{x - 5}$
64. $k(x) = 3\sqrt{x - 1} + 5\sqrt{x - 4}$

Let $f(x) = 3\sqrt{x} - 7$.

65. Find x when $f(x) = -1$.
66. Find x when $f(x) = -8$.

Let $f(x) = -2\sqrt{x - 4} + 5$.

67. Find x when $f(x) = 7$.
68. Find x when $f(x) = 1$.

69. In Exercise 63 of Homework 9.4, you worked with the model $f(t) = 21.4\sqrt{t} + 21$, where $f(t)$ is the percentage of e-mails that are spam at t years since 1999 (see Table 15).

Table 15 Percentages of E-mails That Are Spam

Year	Percent
1999	21
2002	56
2004	68
2005	72
2006	80

Source: *IronPort*

a. Find $f(10)$. What does it mean in this situation?
b. Find t when $f(t) = 95$. What does it mean in this situation?
c. Predict when all e-mails will be spam.

70. In Exercise 64 of Homework 9.4, you worked with the model $f(t) = 1.75\sqrt{t} + 5.4$, where $f(t)$ is the number (in thousands) of teenagers ages 11–18 who are adopted from foster care in the year that is t years since 1998 (see Table 16).

Table 16 Numbers of Teenagers Adopted from Foster Care

Year	Number of Teenagers Adopted (thousands)
1998	5.4
1999	7.0
2000	8.0
2001	8.1
2002	9.4
2003	9.0

Source: *U.S. Department of Health and Human Services' Children's Bureau*

a. Find $f(12)$. What does it mean in this situation?
b. Find t when $f(t) = 12$. What does it mean in this situation?
c. Predict in which year all teenagers awaiting adoption will be adopted. Assume that the number of teenagers awaiting adoption in that year will be the same as in 2000: 38.9 thousand teenagers.

71. For the academic year 2005–2006, tuition at Princeton Day School ranged from $19,200 to $23,600 by grade (kindergarten through 8th grade). In addition, the school charges for supplies and field-trip expenses (see Table 17).

Table 17 Charges for Supplies and Field Trips at Princeton Day School

Grade	Per-Student Charge for Supplies and Field Trips (dollars)
kindergarten	260
1	310
3	365
5	410
6	425
7	425

Source: *Princeton Day School*

Let $f(n)$ be the per-student charge (in dollars) for supplies and field trips for nth grade; kindergarten is represented by $n = 0$. A model of the situation is $f(n) = 257\sqrt[4]{n+1}$.

a. Use a graphing calculator to draw the graph of the model and, in the same viewing window, the scattergram of the data. Does the model fit the data well?

b. Estimate the per-student charge for supplies and field trips for 2nd grade. The actual charge is \$365. Is your result an underestimate or an overestimate?

c. Estimate for what grade the per-student charge for supplies and field trips is \$385.

d. The actual per-student charge for 4th grade is \$450. By referring to the data shown in Table 17, explain why this charge is surprising.

72. The percentages of eligible voters who voted in 2000 are shown in Table 18 for various household income groups.

Table 18 Percentages of Eligible Voters Who Voted in 2000, by Household Income Groups

Income Group (\$1000s)	Income Used to Represent Income Group (\$1000s)	Percent
0–5	2.5	34
5–9.999	7.5	41
10–14.999	12.5	44
15–24.999	20.0	51
25–34.999	30.0	58
35–49.999	42.5	62
50–74.999	62.5	69
over 75	90.0	74

Source: *U.S. Census Bureau*

Let $f(t)$ be the percentage of eligible voters with a household income of d thousand dollars who voted in 2000. A model of the situation is $f(t) = 5.2\sqrt{t} + 27$.

a. Use a graphing calculator to draw the graph of the model and, in the same viewing window, the scattergram of the data. Does the model fit the data well?

b. Estimate the percentage of eligible voters with a household income of \$27 thousand who voted in 2000.

c. Estimate at which household income 65% of eligible voters voted in 2000.

73. A student tries to solve $\sqrt{x^2 + 4x + 5} = x + 3$:

$$\sqrt{x^2 + 4x + 5} = x + 3$$
$$\left(\sqrt{x^2 + 4x + 5}\right)^2 = (x+3)^2$$
$$x^2 + 4x + 5 = x^2 + 9$$
$$4x = 4$$
$$x = 1$$

Describe any errors. Then solve the equation correctly.

74. A student tries to solve $\sqrt{x-2} = x - 4$:

$$\sqrt{x-2} = x - 4$$
$$\left(\sqrt{x-2}\right)^2 = (x-4)^2$$
$$x - 2 = x^2 - 8x + 16$$
$$0 = x^2 - 9x + 18$$
$$0 = (x-3)(x-6)$$
$$x - 3 = 0 \quad \text{or} \quad x - 6 = 0$$
$$x = 3 \quad \text{or} \quad x = 6$$

He says that the solutions are 3 and 6. Is he correct? Explain.

For Exercises 75–80, solve for the specified variable. Assume that the constants have values for which the equation has exactly one real-number solution.

75. $S = \sqrt{gd}$, for d

76. $r = 30d^2\sqrt{P}$, for P

77. $d = \sqrt{\dfrac{3h}{2}}$, for h

78. $T = \sqrt{\dfrac{2d}{g}}$, for d

79. $v = \sqrt{\dfrac{2GM}{R}}$, for R

80. $S = 2\pi\sqrt{\dfrac{L}{32}}$, for L

81. Solve the system; then use "intersect" on a graphing calculator to verify your work:

$$y = 3\sqrt{x} - 4$$
$$y = -2\sqrt{x} + 6$$

82. Create a system of two square root equations that has (4, 5) as its only solution. Verify your system graphically.

83. The first of the following statements is true, yet the last statement is false:

$$2x - x = x$$
$$(2x)^2 - x^2 = x^2$$
$$4x^2 - x^2 = x^2$$
$$3x^2 = x^2$$
$$3 = 1$$

Describe any errors.

84. Describe how to solve square root equations that contain one square root. Also, describe how to solve square root equations that contain two or more square roots.

Related Review

Solve or simplify, as appropriate.

85. $3\sqrt{x} + 4 - 7\sqrt{x} + 1$

86. $2\sqrt{x+1} - 2 + 5\sqrt{x+1} - 9 = 3$

87. $3\sqrt{x} + 4 - 7\sqrt{x} + 1 = -7$

88. $2\sqrt{x+1} - 2 + 5\sqrt{x+1} - 9$

89. $\left(\sqrt{p} + 3\right)\left(\sqrt{p} + 1\right) = 3$

90. $\left(\sqrt{m} - 2\right)\left(\sqrt{m} - 3\right)$

91. $\left(\sqrt{p} + 3\right)\left(\sqrt{p} + 1\right)$

92. $\left(\sqrt{m} - 2\right)\left(\sqrt{m} - 3\right) = 2$

Solve. Round any approximate solutions to the fourth decimal place.

93. $50 - 4(2)^x = -83$

94. $\dfrac{1}{x-2} - \dfrac{2}{x+3} = \dfrac{11}{x^2 + x - 6}$

95. $\sqrt{x+3} - \sqrt{x-2} = 1$

96. $3x^2 + 2x - 4 = 0$

97. $-3(2k - 5) + 1 = 2(4k + 3)$

98. $3b^4 - 29 = 83$

99. $\log_2(5t - 1) = 5$

100. $\log_2(2y + 1) - 2\log_2(y - 1) = 1$

Expressions, Equations, Functions, and Graphs

Perform the indicated instruction. Then use words such as linear, quadratic, cubic, exponential, logarithmic, rational, radical, polynomial, degree, function, one variable, *and* two variables *to describe the expression, equation, or system.*

101. Find the quotient $\dfrac{3x^2 - x - 10}{x^3 - x^2 - x + 1} \div \dfrac{3x^2 - 12}{2x^2 + x - 3}$.

102. Solve $\log_4(7x - 2) = 3$.

103. Find the sum $\dfrac{6}{b-2} + \dfrac{3b}{b^2 - 7b + 10}$.

104. Solve $2(5)^t + 14 = 249$. Round any solutions to the fourth decimal place.

105. Solve $\dfrac{6}{x-2} + \dfrac{3x}{x^2 - 7x + 10} = \dfrac{x}{x-5}$.

106. Write $3\log_b(2x^2) + 2\log_b(3x^7)$ as a single logarithm.

9.6 MODELING WITH SQUARE ROOT FUNCTIONS

Objectives

- Find an equation of a square root curve that contains two given points.
- Find an equation of a square root model.
- Use a square root model to make estimates and predictions.

In this section, we will discuss how to find the equation of a square root function of the form $f(x) = a\sqrt{x} + b$, where $a \neq 0$ and whose graph contains two given points. Then we will find a model of this form that describes an authentic situation and use the model to make estimates and predictions.

Finding a Square Root Function

Recall that the first step in finding an exponential function (Section 4.4) or a quadratic function (Section 7.6) is to substitute any given ordered pairs into a general equation of the function. We will do this same first step to find a square root function.

Example 1 Finding an Equation of a Square Root Function

Find an equation of a square root curve that contains the points (2, 3) and (5, 7).

Solution

We substitute the ordered pairs (2, 3) and (5, 7) into the equation $f(x) = a\sqrt{x} + b$:

$$3 = a\sqrt{2} + b \qquad \text{Substitute 2 for } x \text{ and 3 for } y.$$
$$7 = a\sqrt{5} + b \qquad \text{Substitute 5 for } x \text{ and 7 for } y.$$

We then calculate approximate values for $\sqrt{2}$ and $\sqrt{5}$:

$$1.41a + b = 3 \qquad \text{Equation (1)}$$
$$2.24a + b = 7 \qquad \text{Equation (2)}$$

Next, we multiply both sides of equation (1) by -1:

$$-1.41a - b = -3 \qquad \text{Equation (3)}$$

To eliminate b, we add the left sides and add the right sides of equations (2) and (3) and solve for a:

$$\begin{aligned} 2.24a + b &= 7 \\ -1.41a - b &= -3 \\ \hline 0.83a + 0 &= 4 \\ 0.83a &= 4 \qquad c + 0 = c \\ a &= \frac{4}{0.83} \\ a &\approx 4.82 \end{aligned}$$

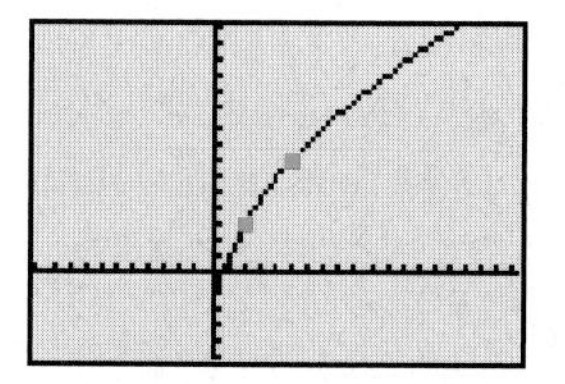

Figure 34 Verify that the graph of $f(x) = 4.82\sqrt{x} - 3.82$ approximately contains the points (2, 3) and (5, 7)

So, our equation has the form $f(x) = 4.82\sqrt{x} + b$. To find b, we substitute the ordered pair (2, 3) into the equation $f(x) = 4.82\sqrt{x} + b$:

$$3 = 4.82\sqrt{2} + b \quad \text{Substitute 2 for } x \text{ and 3 for } y.$$
$$3 - 4.82\sqrt{2} = b \quad \text{Isolate } b.$$
$$b \approx -3.82 \quad \text{Compute.}$$

So, the equation is $f(x) = 4.82\sqrt{x} - 3.82$.

We use a graphing calculator to verify that the graph of f approximately contains the two given points (see Fig. 34). ■

To find an equation of a square root curve that contains two given points, we substitute the points' ordered pairs into the equation $f(x) = a\sqrt{x} + b$ and get a system of two equations. We find values of a and b by solving the system by elimination or substitution.

Finding an Equation of a Square Root Curve

To find an equation of a square root curve that contains two given points,

1. Obtain a system of two linear equations in two variables by substituting the coordinates of both points into the general equation $y = a\sqrt{x} + b$.
2. Solve the system that you found in step 1.
3. Substitute the values of a and b that you found in step 2 into the equation $y = a\sqrt{x} + b$.

Using the y-Intercept and One Other Point to Find an Equation

For an equation of the form $y = a\sqrt{x} + b$, we can find the y-intercept by substituting 0 for x:

$$y = a\sqrt{0} + b = b$$

So, the y-intercept is $(0, b)$.

y-Intercept of the Graph of $y = a\sqrt{x} + b$

The graph of an equation of the form $y = a\sqrt{x} + b$ has y-intercept $(0, b)$.

For example, the graph of the equation $y = 3\sqrt{x} + 7$ has y-intercept (0, 7).

The process of finding an equation of a square root function is easier if one of the two given points is the y-intercept, as we shall see in Example 2.

Example 2 Finding an Equation When One Given Point Is the y-Intercept

Find an equation of a square root curve that contains the points (0, 5) and (6, 19).

Solution

Since the y-intercept of the curve is (0, 5), we can find an equation of the form

$$y = a\sqrt{x} + 5$$

To find a, we substitute the ordered pair (6, 19) in the equation $y = a\sqrt{x} + 5$:

$$19 = a\sqrt{6} + 5 \quad \text{Substitute 6 for } x \text{ and 19 for } y.$$
$$14 = a\sqrt{6} \quad \text{Subtract 5 from both sides.}$$
$$\frac{14}{\sqrt{6}} = a \quad \text{Divide both sides by } \sqrt{6}.$$
$$a \approx 5.72 \quad \text{Compute.}$$

So, the equation is $f(x) = 5.72\sqrt{x} + 5$.

We use a graphing calculator table, as well as a scattergram and graph, to check that the graph of $f(x) = 5.72\sqrt{x} + 5$ approximately contains the points (0, 5) and (6, 19). See Fig. 35.

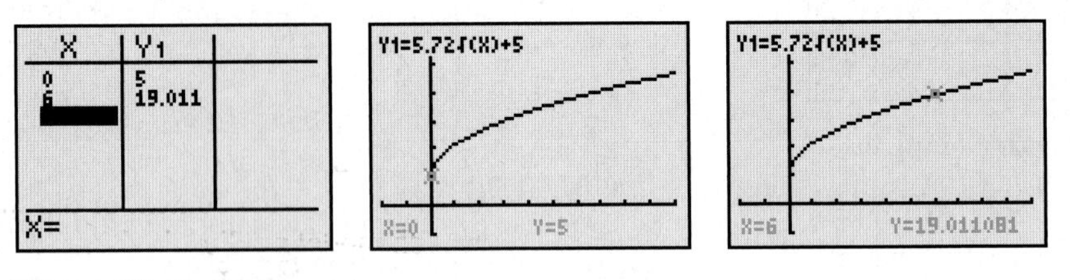

Figure 35 Verify the work

Finding a Square Root Model

Now that we have discussed how to find an equation of a square root function, we can find an equation of a square root model.

Table 19 Numbers of U.S. Coast Guard Drug Arrests

Year	Number of Arrests
2001	114
2002	207
2003	283
2004	326
2005	364

Source: *U.S. Coast Guard*

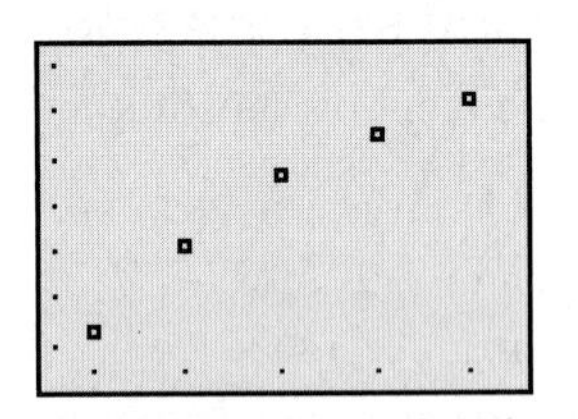

Figure 36 Drug arrests scattergram

Example 3 Finding an Equation of a Square Root Model

The numbers of U.S. Coast Guard drug arrests are shown in Table 19 for various years. Let $f(t)$ be the number of U.S. Coast Guard drug arrests in the year that is t years since 2000. Find an equation of f.

Solution

First we use a graphing calculator to draw a scattergram of the data (see Fig. 36). It appears that the data might be modeled well by a square root function or a quadratic function. To decide which model to use, we find an equation of both types of function and compare the fit of each model.

For the square root model, we use a function of the form

$$f(t) = a\sqrt{t} + b$$

We use the data points (2, 207) and (4, 326) to find values of a and b. To begin, we substitute the ordered pairs (2, 207) and (4, 326) in the equation $f(t) = a\sqrt{t} + b$:

$$207 = a\sqrt{2} + b \quad \text{Substitute (2, 207) in } f(t) = a\sqrt{t} + b.$$
$$326 = a\sqrt{4} + b \quad \text{Substitute (4, 326) in } f(t) = a\sqrt{t} + b.$$

We then calculate that $\sqrt{2} \approx 1.41$ and $\sqrt{4} = 2$:

$$1.41a + b = 207 \quad \text{Equation (1)}$$
$$2a + b = 326 \quad \text{Equation (2)}$$

Next we multiply both sides of equation (1) by -1:

$$-1.41a - b = -207 \quad \text{Equation (3)}$$

To eliminate b, we add the left sides and add the right sides of equations (2) and (3) and solve for a:

$$\begin{aligned} 2a + b &= 326 && \text{Equation (2)} \\ -1.41a - b &= -207 && \text{Equation (3)} \\ \hline 0.59a + 0 &= 119 && \text{Add left sides and add right sides; combine like terms.} \\ a &= \frac{119}{0.59} \\ a &\approx 201.69 \end{aligned}$$

So, our equation has the form $f(t) = 201.69\sqrt{t} + b$. To find b, we substitute the ordered pair (4, 326) in the equation $f(t) = 201.69\sqrt{t} + b$:

$$\begin{aligned} 326 &= 201.69\sqrt{4} + b && \text{Substitute (4, 326) in } f(t) = 201.69\sqrt{t} + b. \\ 326 &= 403.38 + b && \text{Compute.} \\ -77.38 &= b \end{aligned}$$

So, the square root model is $f(t) = 201.69\sqrt{t} - 77.38$.

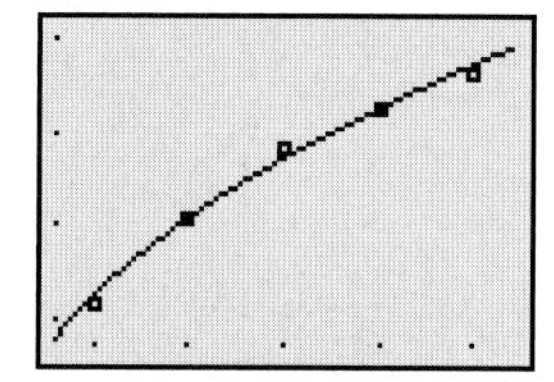

Figure 37 Square root model

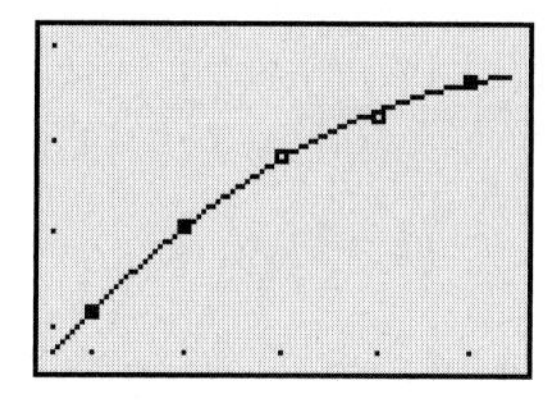

Figure 38 Quadratic model

We use a graphing calculator to find the quadratic regression equation:

$$f(t) = -10.21t^2 + 123.19t + 1.60$$

Next we use a graphing calculator to compare the fits of the two models (see Figs. 37 and 38).

The quadratic model appears to fit the data slightly better than the square root model. However, recall from Section 7.7 that we should also consider whether either model makes sense within the context of the situation. In this situation, it will help to consider what each model predicts in the next several years. So, we use Zoom Out and compare the graphs again (see Figs. 39 and 40).

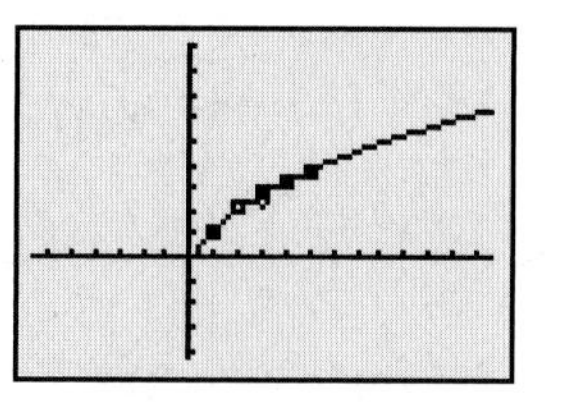

Figure 39 Square root model

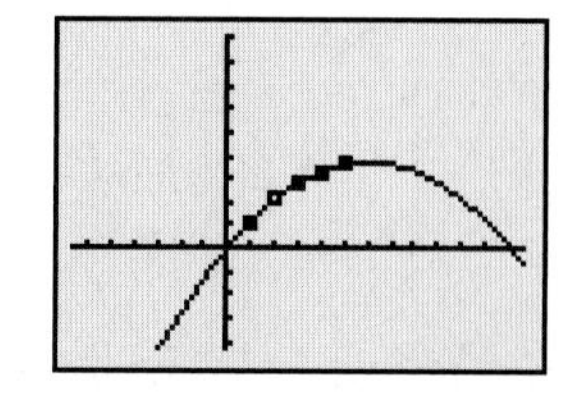

Figure 40 Quadratic model

The square root model predicts that the number of U.S. Coast Guard drug arrests will continue to increase, whereas the quadratic model predicts that the number of arrests will decrease significantly and, in fact, reach 0 arrests (in the year 2012—as we can show, with work). The square root model will probably prove to be the better of the two models. ■

Finding an Equation of a Square Root Model

To find an equation of a square root model, given some data,

1. Create a scattergram of the data.
2. Imagine a square root curve that comes close to the data points, and choose two points (not necessarily data points) that lie on or close to the square root curve.
3. Use the two points to find an equation of the square root curve.
4. Use a graphing calculator to verify that the graph of the equation comes close to the points of the scattergram.

Using a Square Root Model to Make Estimates and Predictions

Once we have found a square root model, we can use it to make estimates and predictions.

Example 4 Using a Square Root Model to Make Predictions

In Example 3, we found the equation $f(t) = 201.69\sqrt{t} - 77.38$, where $f(t)$ is the number of U.S. Coast Guard drug arrests in the year that is t years since 2000. Let $n = f(t)$.

1. Find the n-intercept. What does it mean in this situation?
2. Predict the number of U.S. Coast Guard drug arrests in 2010.
3. Predict in which year there will be 600 U.S. Coast Guard drug arrests.

Solution

1. To find the n-intercept, we find $f(0)$:

$$f(0) = 201.69\sqrt{0} - 77.38 = -77.38$$

The model estimates that a *negative* number of U.S. Coast Guard drug arrests were made in 2000—model breakdown has occurred.

2. We represent 2010 by $t = 10$. We predict the number of drug arrests in 2010 by finding $f(10)$:

$$f(10) = 201.69\sqrt{10} - 77.38 \approx 560.42$$

There will be about 560 U.S. Coast Guard drug arrests in 2010, according to the model.

3. To find in which year there will be 600 drug arrests, we substitute 600 for $f(t)$ and solve for t:

$$600 = 201.69\sqrt{t} - 77.38 \quad \text{Substitute 600 for } f(t).$$

$$677.38 = 201.69\sqrt{t} \quad \text{Add 77.38 to both sides.}$$

$$\frac{677.38}{201.69} = \sqrt{t} \quad \text{Divide both sides by 201.69.}$$

$$\left(\frac{677.38}{201.69}\right)^2 = \left(\sqrt{t}\right)^2 \quad \text{Square both sides.}$$

$$t \approx 11.28 \quad \text{Compute.}$$

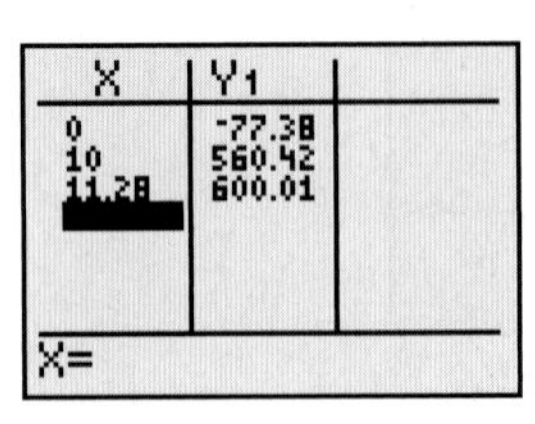

Figure 41 Verify the work

There will be 600 U.S. Coast Guard drug arrests in 2011, according to the model.

We use a graphing calculator table to verify our work of Problems 1–3 (see Fig. 41). ■

We close this section by incorporating square root modeling into the four-step modeling process.

Four-Step Modeling Process

To find a model and make estimates and predictions,

1. Create a scattergram of the data. Decide whether a line, an exponential curve, a parabola, a square root curve, or none of these comes close to the points.
2. Find an equation of your model.
3. Verify that your equation has a graph that comes close to the points in the scattergram. If it doesn't, check for calculation errors or use different points to find the equation. An alternative is to reconsider your choice of model in step 1.
4. Use your equation of the model to draw conclusions, make estimates, and/or make predictions.

group exploration

Looking ahead: Arithmetic sequences

A math tutor charges \$25 per hour to tutor one student, plus \$8 per hour for each additional student.

1. Let $f(n)$ be the amount of money (in dollars) the tutor will charge per hour to work with n students. Find an equation of f.
2. Evaluate f at each given value of n. Explain what each result means in this situation.

   a. $f(3.8)$ **b.** $f(-2)$ **c.** $f(0)$
3. Based on your results in Problem 2, determine a domain of the *model* f. [**Hint:** Which inputs make sense in this situation?]
4. Based on your domain of f, find the range of f. List the values of the range of f in this order: $f(1), f(2), f(3), f(4), \ldots$. What do you notice about these numbers?
5. Sketch a graph of f, but plot only points whose n-coordinates are in the domain. [**Hint:** Your graph will look like a scattergram.]

TIPS FOR SUCCESS: Create a Mind Map for the Final Exam

In preparing for a final exam, consider how all the concepts you have learned are interconnected. One way to help yourself do this is to make a *mind map*. Put the main topic in the middle. Around it attach concepts that relate to it, then concepts that relate to those concepts, and so on.

A portion of a mind map that describes this course is illustrated in Fig. 42. Many more "concept rectangles" could be added to it. You could make one mind map showing an overview of the course and several mind maps for the components of the overview mind map.

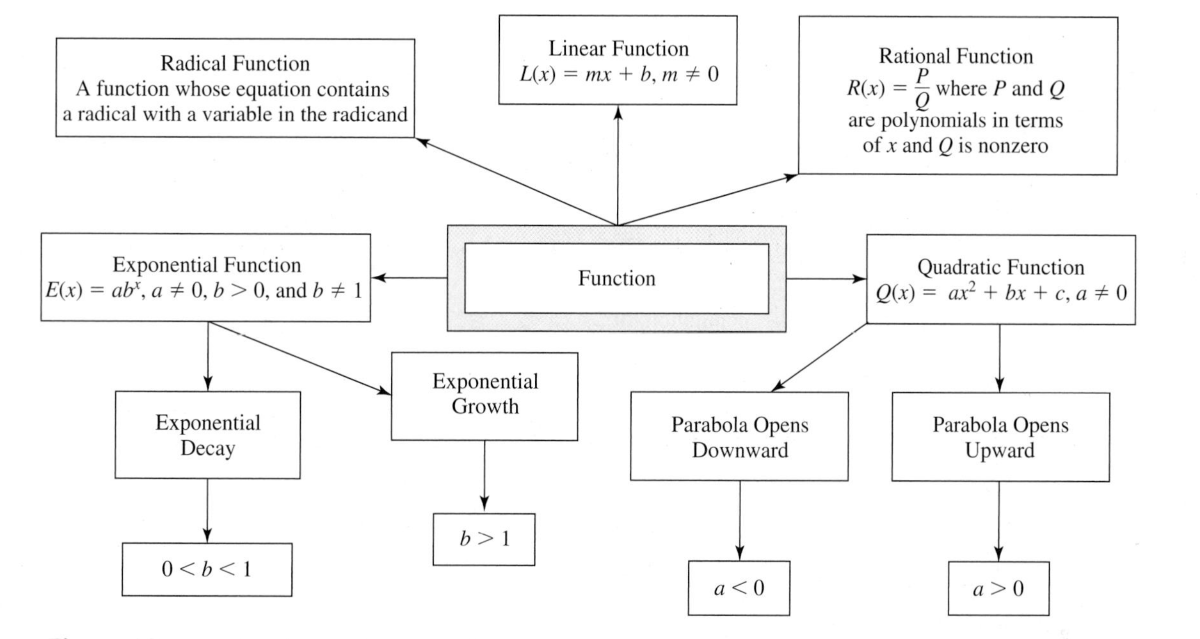

Figure 42 A portion of a mind map describing this course

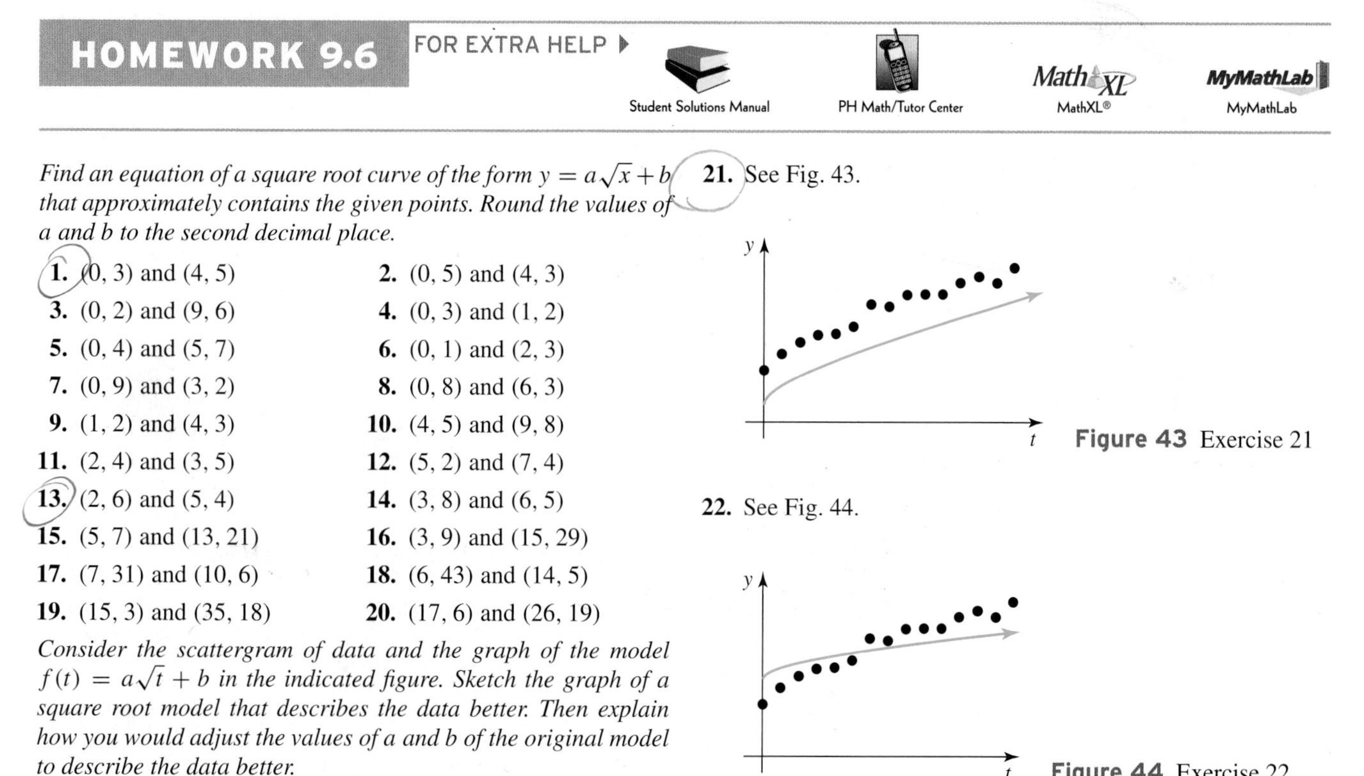

HOMEWORK 9.6

FOR EXTRA HELP ▶ Student Solutions Manual | PH Math/Tutor Center | MathXL® | MyMathLab

Find an equation of a square root curve of the form $y = a\sqrt{x} + b$ that approximately contains the given points. Round the values of a and b to the second decimal place.

1. (0, 3) and (4, 5)
2. (0, 5) and (4, 3)
3. (0, 2) and (9, 6)
4. (0, 3) and (1, 2)
5. (0, 4) and (5, 7)
6. (0, 1) and (2, 3)
7. (0, 9) and (3, 2)
8. (0, 8) and (6, 3)
9. (1, 2) and (4, 3)
10. (4, 5) and (9, 8)
11. (2, 4) and (3, 5)
12. (5, 2) and (7, 4)
13. (2, 6) and (5, 4)
14. (3, 8) and (6, 5)
15. (5, 7) and (13, 21)
16. (3, 9) and (15, 29)
17. (7, 31) and (10, 6)
18. (6, 43) and (14, 5)
19. (15, 3) and (35, 18)
20. (17, 6) and (26, 19)

Consider the scattergram of data and the graph of the model $f(t) = a\sqrt{t} + b$ in the indicated figure. Sketch the graph of a square root model that describes the data better. Then explain how you would adjust the values of a and b of the original model to describe the data better.

21. See Fig. 43.

Figure 43 Exercise 21

22. See Fig. 44.

Figure 44 Exercise 22

23. The average numbers of viewers per episode of the television show *American Idol* are shown in Table 20 for various years.

Table 20 Average Numbers of Viewers per Episode of *American Idol*

Year	Average Number of Viewers (millions)
2002	12.0
2003	21.7
2004	25.0
2005	26.6
2006	30.3

Source: *Nielsen Media Research*

Let $n = f(t)$ be the average number of viewers (in millions) per episode of *American Idol* in the year that is t years since 2002.

a. Find an equation of f.
b. What is the n-intercept? What does it mean in this situation?
c. Predict when the average number of viewers per episode of *American Idol* will be equal to the population of California, the most populous U.S. state. Assume that California's population will be 38 million people at that time.
d. Predict the average number of viewers per episode of *American Idol* in 2011.

24. The percentages of U.S. households owning stocks are shown in Table 21 for various years.

Table 21 Percentages of U.S. Households Owning Stocks

Year	Percent
1983	19
1989	32
1992	37
1995	40
1999	48
2002	50
2005	50

Sources: *ICI/SIA; Federal Reserve Board*

Let $p = f(t)$ be the percentage of U.S. households owning stocks at t years since 1983.

a. Find an equation of f.
b. What is the p-intercept? What does it mean in this situation?
c. Predict the percentage of U.S. households that will own stocks in 2011.
d. Predict when 57% of U.S. households will own stocks.

25. The numbers of U.S. households with webcams are shown in Table 22 for various years. Let $n = f(t)$ be the number of U.S. households (in millions) with webcams at t years since 2000.

a. Find an equation of f.
b. What is the n-intercept? What does it mean in this situation?
c. Find $f(11)$. What does it mean in this situation?
d. Find t when $f(t) = 18$. What does it mean in this situation?

Table 22 Numbers of U.S. Households with Webcams

Year	Number of Households with Webcams (millions)
2000	9.1
2001	11.4
2002	13.1
2003	13.3
2004	13.8
2005	14.4

Source: *IDC*

26. The number of single-parent households headed by fathers has been increasing since 1976. Also rising is the percentage of such fathers who have never been married (see Table 23).

Table 23 Never-Married Single-Parent Fathers

Year	Percent of Single-Parent Fathers Who Have Never Been Married
1982	10.0
1984	17.0
1986	20.0
1988	21.0
1990	24.5
1992	27.0
1994	29.0
1996	29.9
1998	35.0
2000	34.0

Source: *National Center for Children in Poverty*

Let $p = f(t)$ be the percentage of single-parent fathers who have never been married, and let t be the number of years since 1982.

a. Find an equation of f.
b. Find the p-intercept. What does it mean in this situation?
c. Find $f(28)$. What does it mean in this situation?
d. Find t when $f(t) = 40$. What does it mean in this situation?

27. In Exercise 83 of Homework 9.1, you worked with the model $f(d) = \sqrt{9.8d}$, where $f(d)$ is the speed (in meters per second) of a tsunami in which the average water depth is d meters.

a. Before 1856, scientists believed that the average depth of the Pacific Ocean was about 18,000 meters. If that were true, what would be the average speed of a tsunami in the Pacific Ocean?
b. In 1856, scientists used their knowledge of tsunamis to estimate the average depth of the Pacific Ocean. They estimated that the average speed of tsunamis in the Pacific Ocean was between 203 and 210 meters per second. Does this estimate suggest that the average depth of the Pacific Ocean is 18,000 meters? Explain.
c. Use the model to estimate between what two depths is the average depth of the Pacific Ocean.
d. Most current estimates of the average depth of the Pacific Ocean are close to 4280 meters. Is this estimate between the two depths that you found in part (c)?

28. A study of children who were adopted after being in temporary foster care explored the relationship between the children's age when separated from their foster parents and the percentage of children showing problems immediately after the separation (see Table 24).

Table 24 Adopted Children Separated from Foster Parents

Age When Separated from Foster Parents (years)	Middle of Age Group (years)	Percent Showing Problems
<3	1.5	4
3–4	3.5	40
4–5	4.5	70
6–8	7.0	90
9	9.0	100

Source: The Immediate Impact of Separation: Reactions of Infants to a Change in Mother Figures *by Yarrow and Goodwin*

Let $p = f(t)$ be the percentage of adopted children showing problems immediately after separation if they are separated from foster parents at age t years.

a. Find an equation of f.

b. Is f an increasing function, a decreasing function, or neither? What does that mean in this situation?

c. Estimate the percentage of adopted children who show problems if separated at age 10 years. Is this a reasonable estimate? If so, explain why. If not, explain why not and suggest a better value.

d. Sketch a qualitative graph that describes the relationship between t and p for *all* ages of adopted children.

29. Some students ran an experiment to explore the relationship between the time it takes a baseball to fall to the ground and the various heights from which it was dropped (see Table 25). Let $T = S(h)$ be the time (in seconds) it takes a baseball to fall to the ground when dropped from h feet above the ground.

Table 25 Drop Heights and Falling Times of a Baseball*

Drop Height (feet)	Falling Time (seconds)
0.00	0.00
3.28	0.53
13.10	0.96
26.30	1.40
39.40	1.70
52.50	1.94

Source: *J. Lehmann*

*Although a ball cannot be dropped from height zero feet, the data point (0, 0) is included because drop heights near zero will correspond to falling times near zero.

a. Find a square root equation of S.

b. A linear model of the baseball drop data is given by $L(h) = 0.034h + 0.327$. A quadratic model is given by $Q(h) = -0.00058h^2 + 0.064h + 0.165$.

i. Describe how well each of the functions S, L, and Q fits the data points in the scattergram.

ii. Find $S(0)$, $L(0)$, and $Q(0)$. Which function models the baseball situation the best near $h = 0$? Explain.

iii. Use Zoom Out and decide which function could not be a good model for the baseball situation when the drop heights are very large. Explain.

iv. Based on your responses to parts (i)–(iii), determine which of S, L, or Q best models the baseball drop situation. Explain.

v. An equation that is often used to model drop times is $T = \sqrt{\frac{2h}{g}}$, where g is the constant 32.2 feet per second squared. Find an approximate form $T = a\sqrt{h}$ of this equation so that you can compare the model with S.

c. Suppose that you wish to estimate the height of a sheer cliff. You drop a stone from the top of the cliff, and it takes 3 seconds to reach the foot of the cliff. Use your model to estimate the height of the cliff.

d. Use your model to estimate how long it would take for a baseball to reach the ground if it were to be dropped from the top of the Empire State Building, which is 1250 feet tall.

30. The percentages of ex-convicts who have been arrested for a new crime after being out of a state prison for various numbers of years are shown in Table 26.

Table 26 Ex-Convicts Who Have Been Arrested for a New Crime

Number of Years Since Release	Percent
0	0
0.5	30
1	44
2	59
3	68

Source: *U.S. Department of Justice*

Let $S(t)$ be the percentage of convicts released from a state prison who are arrested for a new crime after being out of prison for t years.

a. Find a square root equation of S.

b. A quadratic model of the data is given by

$$Q(t) = -9.05t^2 + 48.09t + 3.47$$

i. Describe how well both of the functions S and Q fit the data points in the scattergram.

ii. Explain in terms of the situation why the model should be increasing for $t \geq 0$. Is function S or Q increasing for $t \geq 0$?

c. Use S to estimate the percentage of ex-convicts who have been arrested for a new crime after being out of prison for 4 years.

d. Use S to estimate after how many years since release all ex-convicts will have been arrested for a new crime.

31. A study of births among 1444 fertile couples recorded the birth order and claims by the parents that conception occurred despite the use of contraception (see Table 27). The percentage, for example, of first births that occurred despite contraception, according to the parents, was 37.6%.

Let $p = f(n)$ be the percentage of births of the nth-born child that occurred despite the use of contraception at the time of conception, as claimed by the parents.

a. Find an equation of f. [**Hint:** Use the points (2, 54.3) and (4, 73.0).]

b. Find $f(7)$. What does it mean in this situation?

c. Find n when $f(n) = 100$. What does your result mean in this situation?

Table 27 Percentages of Births Despite Contraception

Birth Order	Percentage of Conceptions Despite Contraception
1	37.6
2	54.3
3	66.5
4	73.0
5	84.1
6	81.2

Source: Social and Psychological Factors Affecting Fertility *by Whelpton and Kiser*

d. Note that f is an increasing function. What does that mean in this situation? Form a theory to explain why this happens.

32. Describe how to find an equation of a square root function that contains two given points. How is this process different if one point is the y-intercept?

Related Review

33. The numbers of U.S. communities with cameras that photograph drivers who run red lights are shown in Table 28 for various years.

Table 28 Numbers of U.S. Communities with Red-Light Cameras

Year	Number of Communities with Red-Light Cameras
1999	20
2001	28
2003	74
2005	120
2006	160

Source: *Insurance Institute for Highway Safety*

Let $f(t)$ be the number of U.S. communities with red-light cameras at t years since 1990.

a. Use a graphing calculator to draw a scattergram to describe the data. By inspecting the scattergram alone, determine which two of the following functions *might* fit the data well: linear, exponential, quadratic, and square root.

b. Find equations of the two types of functions that you selected in part (a).

c. Compare how well each of your two models fits the data.

d. Which of your two models describes the situation better for years before 1999?

e. Use both of your models to predict when there will be 1200 U.S. communities with red-light cameras—10 times the 2005 level. Refer to the graphs of the two models to explain why your two results are so different.

f. In 2005, Redflex Traffic Systems® sold $26 million worth of red-light cameras. Assuming that Redflex continues to manufacture about 40% of the red-light cameras used in the United States, use your two results in part (e) to help you make two predictions about Redflex's revenue from sales of red-light cameras.

34. The numbers of seizures of methamphetamine (meth) labs in the United States are shown in Table 29 for various years.

Table 29 Numbers of Seizures of Meth Labs

Year	Number of Seizures of Meth Labs (thousands)
2000	9.1
2001	13.4
2002	16.0
2003	17.5
2004	16.0

Source: *Drug Enforcement Administration*

Let $n = f(t)$ be the number (in thousands) of seizures of meth labs in the year that is t years since 2000.

a. Find an equation of f.

b. Estimate the number of meth lab seizures in 2005.

c. Estimate in which years 9 thousand meth labs were seized.

d. What are the t-intercepts? What do they mean in this situation?

e. For what values of t is there model breakdown for certain?

Expressions, Equations, Functions, and Graphs

Give an example of the following. Then solve, simplify, or graph, as appropriate.

35. rational equation in one variable

36. system of two linear equations in two variables

37. difference of two rational expressions

38. quadratic equation in one variable

39. square root equation in one variable

40. expression involving exponents

41. exponential function

42. difference of two logarithmic expressions with the same base

Chapter Summary

Key Points OF CHAPTER 9

Perfect *n*th power (Section 9.1)

If n is a counting number greater than 1, then each of the following is a **perfect *n*th power:**

- A number whose nth root is rational.
- A power x^k, where k is a multiple of n.

Product property for radicals (Section 9.1)

If $\sqrt[n]{a}$ and $\sqrt[n]{b}$ are defined, then $\sqrt[n]{ab} = \sqrt[n]{a}\sqrt[n]{b}$.

Simplified radical (Section 9.1)	A radical with index n is **simplified** when the radicand does not have any factors that are perfect nth powers (other than -1 or 1) and the index is as small as possible.
Selecting perfect nth-power factors (Section 9.1)	To select perfect nth-power factors of radicands, • If the radicand has more than one numerical factor that is a perfect nth power, select the largest one. • If the radicand has more than one factor that is a perfect nth power with the same variable base, select the one with the largest exponent.
Decreasing the index (Section 9.1)	If $\sqrt[n]{x}$ is defined and the fraction $\frac{m}{n}$ can be simplified, then we can decrease the index of $\sqrt[n]{x^m}$ by writing $\sqrt[n]{x^m} = x^{m/n}$ and simplifying the exponent $\frac{m}{n}$.
Simplifying a radical expression (Section 9.1)	To simplify a radical expression with index n, **1.** Find perfect nth-power factors of the radicand. **2.** Apply the product property for radicals. **3.** Find the nth root of each perfect nth power. **4.** Write the radical with as small an index as possible.
Power property for radicals (Section 9.1)	Let n be a counting number greater than 1: • If n is even, then $\sqrt[n]{x^n} = \lvert x \rvert$. • If n is odd, then $\sqrt[n]{x^n} = x$.
Radical function (Section 9.1)	A **radical function** is a function whose equation contains a radical with a variable in the radicand.
Square root function (Section 9.1)	A **square root function** is a radical function in which any radicals are square root radicals.
Radical model (Section 9.1)	A **radical model** is a radical function, or its graph, that describes an authentic situation.
Square root model (Section 9.1)	A **square root model** is a square root function, or its graph, that describes an authentic situation.
Like radicals (Section 9.2)	Radicals that have the same index and the same radicand are called **like radicals.**
Adding or subtracting like radicals (Section 9.2)	To add or subtract like radicals, we use the distributive law.
Power property for radicals (Section 9.2)	If $\sqrt[n]{x}$ is defined, then $\left(\sqrt[n]{x}\right)^n = x$.
Multiplying radicals with the same index (Section 9.2)	To multiply two radicals that have the same index, we use the product property.
Multiplying two radicals with different indexes but the same radicand (Section 9.2)	To multiply two radicals with different indexes but the same radicand, **1.** Write the radicals in exponential form. **2.** Use exponential properties to simplify the expression involving exponents. **3.** Write the simplified expression in radical form.
Simplifying a radical expression (Section 9.2)	To simplify a radical expression, **1.** Perform any indicated multiplications. **2.** Combine like radicals. **3.** For any radical with index n, write the radicand as a product of one or more perfect nth powers and another expression that has no factors that are perfect nth powers. Then apply the product property for radicals. **4.** Write any radicals with as small an index as possible.
Rationalizing the denominator of $\frac{A}{\sqrt[n]{x^m}}$ (Section 9.3)	To rationalize the denominator of a radical expression of the form $\frac{A}{\sqrt[n]{x^m}}$, we multiply the expression by a fraction of the form $\frac{\sqrt[n]{x^k}}{\sqrt[n]{x^k}}$ so that the radical in the denominator has a perfect nth-power radicand.
Quotient property for radicals (Section 9.3)	If $\sqrt[n]{a}$ and $\sqrt[n]{b}$ are defined and b is nonzero, then $\sqrt[n]{\frac{a}{b}} = \frac{\sqrt[n]{a}}{\sqrt[n]{b}}$.

Radical conjugates (Section 9.3)	We say that the sum of two radicals and the difference of the same radicals are **radical conjugates** of each other.
Rationalizing a denominator by using a radical conjugate (Section 9.3)	To rationalize the denominator of a square root expression if the denominator is a sum or difference involving radicals, **1.** Determine the radical conjugate of the denominator. **2.** Multiply the original fraction by the fraction $\frac{\text{conjugate}}{\text{conjugate}}$. **3.** Find the product of the denominators by using $(A+B)(A-B) = A^2 - B^2$.
Reflections across x-axis (Section 9.4)	The graphs of $f(x) = -a\sqrt{x}$ and $g(x) = a\sqrt{x}$ are reflections of each other across the x-axis.
Increasing or decreasing (Section 9.4)	If $a > 0$, then $g(x) = a\sqrt{x-h} + k$ is an increasing function and (h, k) is the minimum point. If $a < 0$, then g is a decreasing function and (h, k) is the maximum point.
Graph rises more quickly or more slowly (Section 9.4)	If a is a large number, then the graph of $g(x) = a\sqrt{x-h} + k$ rises more quickly than the graph of $y = \sqrt{x-h} + k$. If a is a positive number near zero, then the graph rises more slowly.
Three-step method for graphing a square root function (Section 9.4)	To sketch the graph of $f(x) = a\sqrt{x-h} + k$, where $a \neq 0$, **1.** Sketch the graph of $y = a\sqrt{x}$. **2.** Translate the graph sketched in step 1 to the right by h units if $h > 0$ and to the left by $\lvert h \rvert$ units if $h < 0$. **3.** Translate the graph sketched in step 2 up by k units if $k > 0$ and down by $\lvert k \rvert$ units if $k < 0$.
Squaring property of equality (Section 9.5)	If A and B are expressions, then all solutions of the equation $A = B$ are *among* the solutions of the equation $A^2 = B^2$. That is, the solutions of an equation are among the solutions of the equation obtained by squaring both sides.
Checking proposed solutions (Section 9.5)	Because squaring both sides of a square root equation may introduce extraneous solutions, it is essential to check that each proposed solution satisfies the original equation.
Solving a square root equation in one variable (Section 9.5)	To solve a square root equation in one variable, **1.** Isolate a square root term on one side of the equation. **2.** Square both sides. **3.** Repeat steps 1 and 2 until no square root terms remain. **4.** Solve the new equation. **5.** Check that each proposed solution satisfies the original equation.
Power property of equality (Section 9.5)	If A and B are expressions and n is a counting number greater than 1, then all solutions of the equation $A = B$ are *among* the solutions of the equation $A^n = B^n$. That is, the solutions of an equation are among the solutions of the equation obtained by raising both sides to the nth power.
Finding an equation of a square root curve (Section 9.6)	To find an equation of a square root curve that contains two given points, **1.** Obtain a system of two linear equations in two variables by substituting the coordinates of both points into the general equation $y = a\sqrt{x} + b$. **2.** Solve the system that you found in step 1. **3.** Substitute the values of a and b that you found in step 2 into the equation $y = a\sqrt{x} + b$.
y-Intercept of the graph of $y = a\sqrt{x} + b$ (Section 9.6)	The graph of an equation of the form $y = a\sqrt{x} + b$ has y-intercept $(0, b)$.
Finding an equation of a square root model (Section 9.6)	To find an equation of a square root model, given some data, **1.** Create a scattergram of the data. **2.** Imagine a square root curve that comes close to the data points, and choose two points (not necessarily data points) that lie on or close to the square root curve. **3.** Use the two points to find an equation of the square root curve. **4.** Use a graphing calculator to verify that the graph of the equation comes close to the points of the scattergram.

Four-step modeling process (Section 9.6)	To find a model and make estimates and predictions, **1.** Create a scattergram of the data. Decide whether a line, an exponential curve, a parabola, a square root curve, or none of these comes close to the points. **2.** Find an equation of your model. **3.** Verify that your equation has a graph that comes close to the points in the scattergram. If it doesn't, check for calculation errors or use different points to find the equation. An alternative is to reconsider your choice of model in step 1. **4.** Use your equation of the model to draw conclusions, make estimates, and/or make predictions.

CHAPTER 9 REVIEW EXERCISES

If the expression is in exponential form, then write it in radical form. If it is in radical form, then write it in exponential form.

1. $x^{3/7}$ **2.** $\sqrt[5]{(3k+4)^7}$

Simplify. Assume that each variable is nonnegative.

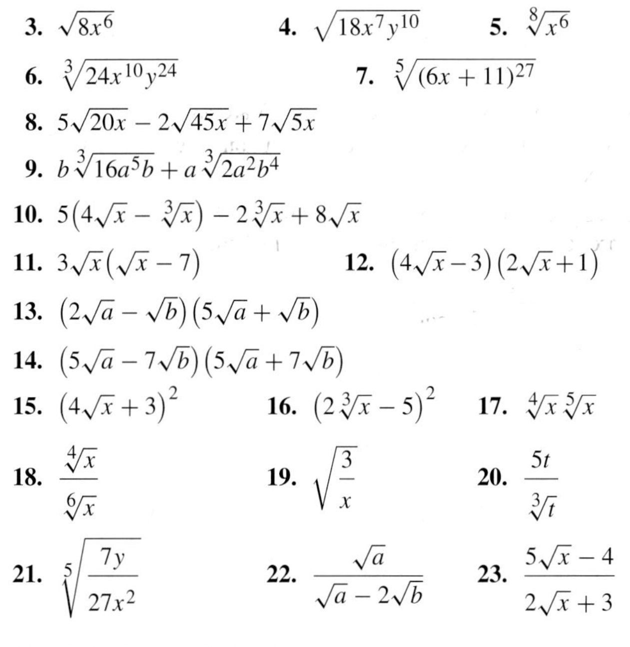

3. $\sqrt{8x^6}$ **4.** $\sqrt{18x^7y^{10}}$ **5.** $\sqrt[8]{x^6}$

6. $\sqrt[3]{24x^{10}y^{24}}$ **7.** $\sqrt[5]{(6x+11)^{27}}$

8. $5\sqrt{20x} - 2\sqrt{45x} + 7\sqrt{5x}$

9. $b\sqrt[3]{16a^5b} + a\sqrt[3]{2a^2b^4}$

10. $5(4\sqrt{x} - \sqrt[3]{x}) - 2\sqrt[3]{x} + 8\sqrt{x}$

11. $3\sqrt{x}(\sqrt{x} - 7)$ **12.** $(4\sqrt{x}-3)(2\sqrt{x}+1)$

13. $(2\sqrt{a} - \sqrt{b})(5\sqrt{a} + \sqrt{b})$

14. $(5\sqrt{a} - 7\sqrt{b})(5\sqrt{a} + 7\sqrt{b})$

15. $(4\sqrt{x}+3)^2$ **16.** $(2\sqrt[3]{x} - 5)^2$ **17.** $\sqrt[4]{x}\sqrt[5]{x}$

18. $\dfrac{\sqrt[4]{x}}{\sqrt[6]{x}}$ **19.** $\sqrt{\dfrac{3}{x}}$ **20.** $\dfrac{5t}{\sqrt[3]{t}}$

21. $\sqrt[5]{\dfrac{7y}{27x^2}}$ **22.** $\dfrac{\sqrt{a}}{\sqrt{a} - 2\sqrt{b}}$ **23.** $\dfrac{5\sqrt{x}-4}{2\sqrt{x}+3}$

Graph the function by hand.

24. $y = -\sqrt{x-5} + 3$ **25.** $y = 2\sqrt{x+4} - 1$

For $f(x) = 3\sqrt{x} + 5$ and $g(x) = 2 - 4\sqrt{x}$, find an equation of the given function.

26. $f + g$ **27.** $f - g$ **28.** $f \cdot g$ **29.** $\dfrac{f}{g}$

Solve.

30. $\sqrt{2x+1} + 4 = 7$ **31.** $\sqrt{2x-4} - x = -2$

32. $\sqrt{x} + 6 = x$ **33.** $\sqrt{13x+4} = \sqrt{5x+20}$

34. $\sqrt{x+2} + \sqrt{x+9} = 7$

35. Solve $3.57 + 2.99\sqrt{8.06x - 6.83} = 14.55$. Round your result to the second decimal place.

36. Solve $\sqrt{x+5} = x^2 - 3x - 4$ by using "intersect" on a graphing calculator. Round any solutions to the second decimal place.

Find all x-intercepts.

37. $f(x) = \sqrt{4x-7} - \sqrt{2x+1}$

38. Consider the scattergram of data and the graph of the model $f(x) = a\sqrt{x} + b$ in Fig. 45. Draw a square root model that fits the data better. Explain how you would adjust the values of a and b of the original model to describe the data better.

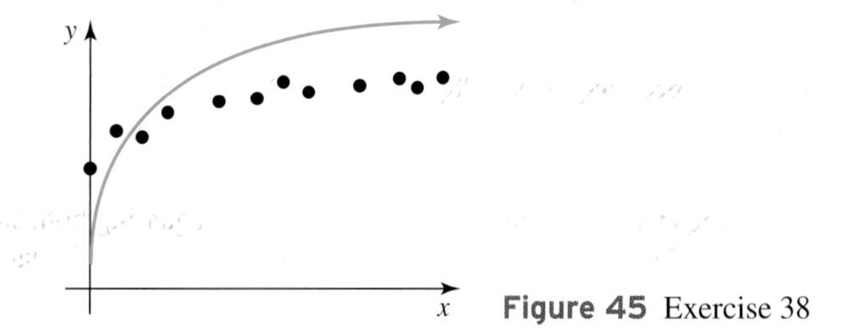

Figure 45 Exercise 38

Find an equation of a square root curve of the form $y = a\sqrt{x} + b$ that approximately contains the given points. Round the values of a and b to the second decimal place.

39. (0, 3) and (4, 8) **40.** (3, 7) and (5, 4)

41. The numbers of households in the United States that have the video game console PlayStation (1 or 2) are shown in Table 30 for various years.

Table 30 Numbers of Households That Have a Sony PlayStation

Year	Number of Households That Have a Sony PlayStation (millions)
1998	16.0
1999	21.9
2000	25.0
2001	27.4
2002	28.8

Source: *International Development Group*

Let $n = f(t)$ be the number of households (in millions) that have a Sony PlayStation (1, 2, or 3) at t years since 1998.*

a. Find an equation of f.

b. Find the n-intercept. What does it mean in this situation?

c. Find $f(14)$. What does it mean in this situation?

d. Find t when $f(t) = 38$. What does it mean in this situation?

*PlayStation 3 came out in November 2006.

CHAPTER 9 TEST

Simplify. Assume that $x \geq 0$.

1. $\sqrt{32x^9y^{12}}$
2. $\sqrt[3]{64x^{22}y^{14}}$
3. $\sqrt[4]{(2x+8)^{27}}$
4. $\dfrac{4\sqrt[3]{x}}{6\sqrt[5]{x}}$
5. $\dfrac{\sqrt{x}+1}{2\sqrt{x}-3}$
6. $4\sqrt{12x^3} - 2x\sqrt{75x} + \sqrt{3x^3}$
7. $3\sqrt{x}\left(6\sqrt{x}-5\right)$
8. $\left(2+4\sqrt{x}\right)\left(3-5\sqrt{x}\right)$
9. $\left(3\sqrt{a}-5\sqrt{b}\right)\left(3\sqrt{a}+5\sqrt{b}\right)$
10. $\left(4\sqrt[5]{x}-3\right)^2$
11. Show that the following statement is true:

$$\frac{\sqrt[n]{x}}{\sqrt[k]{x}} = \sqrt[kn]{x^{k-n}}$$

where n and k are counting numbers greater than 1 and $x > 0$.

12. Graph the function $y = -2\sqrt{x+3}+1$ by hand.
13. Let $f(x) = a\sqrt{x-h}+k$, where $a \neq 0$.
    a. What must be true of a, h, and k for f to have an x-intercept? [**Hint:** Think graphically.]
    b. Now assume that f has an x-intercept. Find the x-intercept in terms of a, h, and k.

For $f(x) = 7 - 3\sqrt{x}$ and $g(x) = 4 + 5\sqrt{x}$, find an equation of the given function.

14. $f+g$
15. $f-g$
16. $f \cdot g$
17. $\dfrac{f}{g}$

Solve.

18. $2\sqrt{x}+3 = 13$
19. $3\sqrt{5x-4} = 27$
20. $3 - 2\sqrt{x} + \sqrt{9-x} = 0$

Let $f(x) = 6 - 4\sqrt{x+1}$.

21. Find $f(8)$.
22. Find a value of x such that $f(x) = -2$.
23. Find all x-intercepts of the graph of

$$f(x) = 3\sqrt{2x-4} - 2\sqrt{2x+1}$$

For Exercises 24 and 25, estimate all solutions of the equation by referring to the graphs shown in Fig. 46. Round results to the first decimal place.

24. $\sqrt{x+4} = x^2 - 4x + 5$
25. $\sqrt{x+4} = 1$

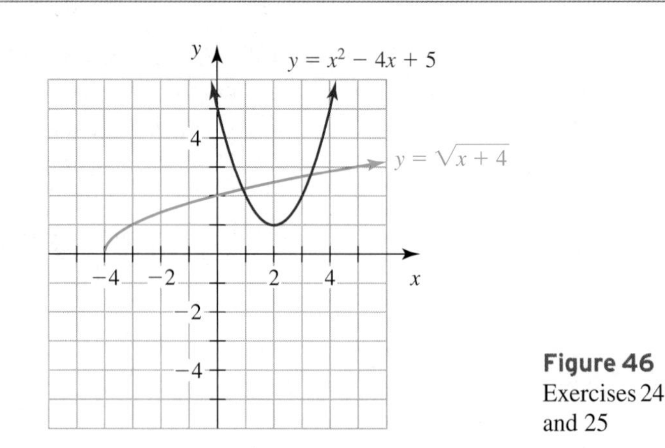

Figure 46 Exercises 24 and 25

26. Consider the scattergram of the data and the graph of the model $y = a\sqrt{x}+b$ sketched in Fig. 47. Draw a square root model that fits the data better; then explain how you would adjust the values of a and b of the original model to describe the data better.

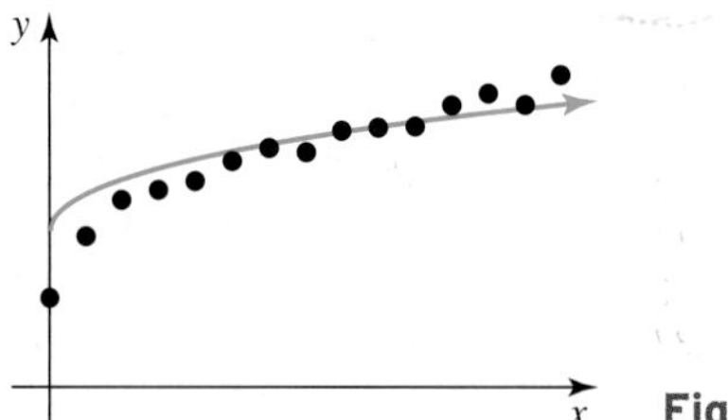

Figure 47 Exercise 26

27. Find an equation of a square root curve of the form $y = a\sqrt{x}+b$ that approximately contains the points $(2, 4)$ and $(5, 6)$.
28. The median heights of boys in the United States are listed in Table 31 for various ages, up to 5 years.

Table 31 Boys' Median Heights

Age (months)	Height (inches)
0	20.5
6	27.0
12	30.8
18	32.9
24	35.0
36	37.5
48	40.8
60	43.4

Source: The Portable Pediatrician for Parents *by Laura Walther Nathanson*

Let $h = f(t)$ be the median height (in inches) of boys who are t months of age.
    a. Find an equation of f.
    b. Estimate the median height of boys who are 6 years old.
    c. Estimate the age at which the median height of boys is 3 feet.
    d. Find the h-intercept. What does it mean in this situation?

Chapter 10

Sequences and Series

"Here, where we reach the sphere of mathematics, we are among processes which seem to some the most inhuman of all human activities and the most remote from poetry. Yet it is here that the artist has the fullest scope of his imagination."

—*Havelock Ellis,* The Dance of Life, *1923*

Table 1 Spending by the Pharmaceutical Industry on Government and Politics

Year	Spending (millions of dollars)
1998	80
1999	97
2000	112
2001	110
2002	137
2003	144
2004	158

Source: *Center for Public Integrity*

Did you know that many organizations spend millions of dollars to try to influence decisions in government and politics? Members of Congress said that the pharmaceutical industry was the fifth most effective special-interest group (Source: *National Journal*). See Table 1. In Exercise 41 of Homework 10.3, you will predict the total spending by the pharmaceutical industry on government and politics from 1998 through 2010.

In this chapter, we reexamine linear and exponential functions from a different perspective—that of lists of numbers called *sequences* and sums of numbers called *series*. We will use these ideas to help us model new situations. We will also discuss how to find sums of quantities efficiently, such as the total spending by the pharmaceutical industry on government and politics.

10.1 ARITHMETIC SEQUENCES

Objectives

- Know the meaning of *sequence, term, term number,* and *arithmetic sequence.*
- Find a formula, term, or term number of an arithmetic sequence.
- Use an arithmetic sequence to make estimates and predictions.

In this section, we use lists of numbers to reexamine linear functions from a different perspective. We will also use these ideas to make estimates and predictions.

Definition of Arithmetic Sequence

Suppose that a math tutor charges \$23 per hour for one student, plus \$6 per hour for each additional student. The tutor will take up to 10 students. We list the total charges (in dollars per hour) for 1 through 10 students in order:

$$23, 29, 35, 41, 47, 53, 59, 65, 71, 77$$

We call this list of numbers a sequence.

> **DEFINITION** Sequence
>
> Any ordered list of numbers is called a **sequence.** Each number is a **term** of the sequence.

A sequence that has a last term, such as 77 in the math tutor sequence, is called a **finite sequence.** A sequence that does not have a last term is called an **infinite sequence.**

For example, the sequence of odd numbers

$$1, 3, 5, 7, 9, \ldots$$

is an infinite sequence. The three dots mean that the pattern of numbers continues without ending.

For the math tutor sequence, notice that the difference between any term and the preceding term is 6:

$$29 - 23 = 6, 35 - 29 = 6, 41 - 35 = 6, \ldots, 77 - 71 = 6$$

We call 6 the common difference of this sequence and say that the sequence is arithmetic.

DEFINITION Arithmetic sequence

If the difference between any term of a sequence and the preceding term is a constant d for every such pair of terms, then the sequence is an **arithmetic sequence.** We call the constant d the **common difference.**

Example 1 Identifying Arithmetic Sequences

Determine whether the sequence is arithmetic. If it is, find the common difference d.

1. $2, 6, 10, 14, 18, \ldots$

2. $80, 77, 74, 71, 68, \ldots$

3. $3, 6, 12, 24, 48, \ldots$

Solution

1. The sequence is arithmetic, because it has a common difference of 4:

$$6 - 2 = 4, 10 - 6 = 4, 14 - 10 = 4, 18 - 14 = 4, \ldots$$

2. The sequence is arithmetic, because it has a common difference of -3:

$$77 - 80 = -3, 74 - 77 = -3, 71 - 74 = -3, 68 - 71 = -3, \ldots$$

3. The sequence is not arithmetic, because we can see from the first two differences that the sequence does not have a common difference:

$$6 - 3 = 3, 12 - 6 = 6$$

■

We use the notation $a_1, a_2, a_3, \ldots$ to denote the terms of a sequence. We say that a_n is the ***n*th term** of the sequence and that its **term number** is n. For the math tutor sequence, we write

$$a_1 = 23, a_2 = 29, a_3 = 35, \ldots, a_{10} = 77$$

where a_n is the charge (in dollars per hour) for n students. For instance, the term 35 is the 3rd term, and its term number is 3.

Finding a Formula of an Arithmetic Sequence

Can we find a formula that describes the terms of an arithmetic sequence? Since the math tutor sequence has a common difference of 6, we add 6 to the first term, 23, to find the second term; we add 6 two times to 23 to find the third term; we add 6 three times to 23 to find the fourth term; and so on:

$$23 + 6 = 29, \ 23 + 6 + 6 = 35, \ 23 + 6 + 6 + 6 = 41, \ldots$$

In general, for an arithmetic sequence with common difference d, we have the terms

$$a_1, \ a_1 + d, \ a_1 + d + d, \ a_1 + d + d + d, \ldots$$

Simplifying, we have

$$a_1, \ a_1 + d, \ a_1 + 2d, \ a_1 + 3d, \ldots$$

We can use a pattern to find a formula that describes any term a_n of an arithmetic sequence:

$a_1 = a_1$

$a_2 = a_1 + d$ — Add d *once* to the first term to get the *second* term.

$a_3 = a_1 + 2d$ — Add d *twice* to the first term to get the *third* term.

$a_4 = a_1 + 3d$ — Add d *three times* to the first term to get the *fourth* term.

$\vdots$

$a_n = a_1 + (n-1)d$ — Add d a total of $(n-1)$ times to the first term to get the nth term.

Formula That Describes the nth Term of an Arithmetic Sequence

If an arithmetic sequence $a_1, a_2, a_3, \ldots, a_n, \ldots$ has the common difference d, then

$$a_n = a_1 + (n-1)d$$

In words: The nth term of an arithmetic sequence is equal to the first term plus $n-1$ times the common difference.

Example 2 Finding a Formula

Find a formula that describes the terms of the math tutor sequence.

Solution

To find a formula that describes the math tutor sequence 23, 29, 35, 41, 47, ..., 77, we substitute $a_1 = 23$ and $d = 6$ in the formula $a_n = a_1 + (n-1)d$:

$$\begin{aligned} a_n &= 23 + (n-1)(6) && \text{Substitute 23 for } a_1 \text{ and 6 for } d. \\ &= 23 + 6n - 6 && \text{Distributive law} \\ &= 6n + 17 \end{aligned}$$

■

The tutor equation $a_n = 6n + 17$ describes a linear function whose inputs are only the numbers of students 1, 2, 3, ..., 10 and whose outputs are only the dollars-per-hour charges 23, 29, 35, ..., 77. For instance,

$$\begin{aligned} a_1 &= 6(1) + 17 = 23 \\ a_2 &= 6(2) + 17 = 29 \\ a_3 &= 6(3) + 17 = 35 \\ a_{10} &= 6(10) + 17 = 77 \end{aligned}$$

We can verify our formula by entering $y = 6x + 17$ in a graphing calculator and checking that the inputs 1, 2, 3, ..., 10 give the outputs 23, 29, 35, ..., 77 (see Fig. 1).

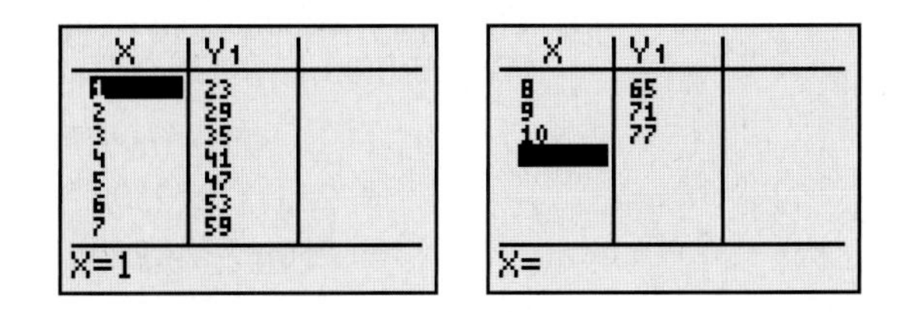

Figure 1 Verify the formula $a_n = 6n + 17$

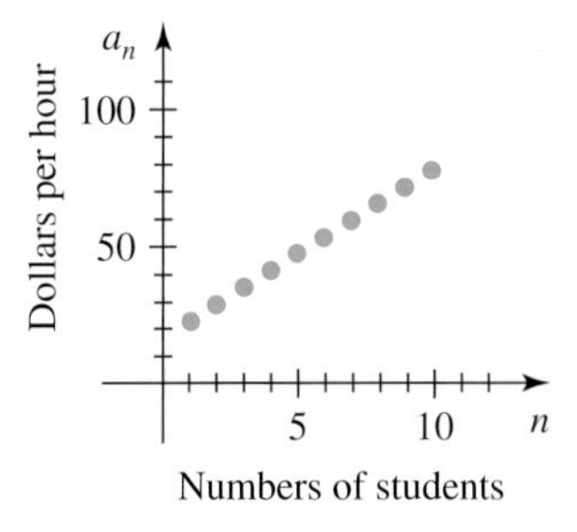

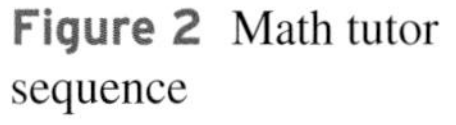

Figure 2 Math tutor sequence

We can also sketch a graph of the math tutor sequence. The graph consists of the 10 input–output pairs (1, 23), (2, 29), (3, 35), ..., (10, 77). For $a_n = 6n + 17$, the coefficient of the independent variable n is 6, so the points lie on an increasing line with slope 6 (see Fig. 2).

Finding a Term or a Term Number of an Arithmetic Sequence

For an arithmetic sequence, if we know the values of three of the four variables a_1, a_n, n, and d, we can find the value of the fourth variable from the formula $a_n = a_1 + (n-1)d$.

As the formula $a_n = a_1 + (n - 1)d$ is valid for arithmetic sequences only, we must first check that a sequence is arithmetic before we use the formula.

Example 3 Finding a Term

Find the 25th term of the sequence 13, 20, 27, 34, 41, ...

Solution

The sequence has a common difference ($d = 7$), so the sequence is arithmetic. We substitute $a_1 = 13$, $n = 25$, and $d = 7$ in the formula $a_n = a_1 + (n - 1)d$:

$$\begin{aligned} a_{25} &= 13 + (25 - 1)(7) && \text{Substitute 13 for } a_1\text{, 25 for } n\text{, and 7 for } d. \\ &= 13 + 24(7) \\ &= 181 \end{aligned}$$

So, the 25th term is $a_{25} = 181$. To verify our work, we enter $y = 13 + (x - 1)(7)$ in a graphing calculator; we check that the first five terms of the sequence are 13, 20, 27, 34, and 41 and that the 25th term is 181 (see Fig. 3). ■

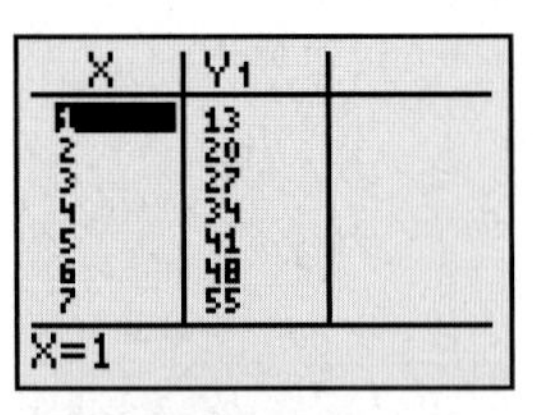

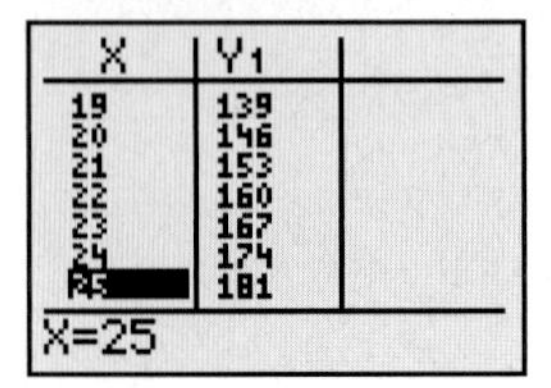

Figure 3 Verify the work

Example 4 Finding a Term Number

The number 23 is a term in the sequence 155, 151, 147, 143, 139, ... What is its term number?

Solution

The sequence has a common difference ($d = -4$), so the sequence is arithmetic. We substitute $a_1 = 155$, $d = -4$, and $a_n = 23$ in the formula $a_n = a_1 + (n-1)d$ and solve for n:

$$\begin{aligned} 23 &= 155 + (n - 1)(-4) && \text{Substitute 155 for } a_1\text{, } -4 \text{ for } d\text{, and 23 for } a_n. \\ 23 &= 155 - 4n + 4 && \text{Distributive law} \\ 4n &= 136 && \text{Isolate } 4n. \\ n &= 34 \end{aligned}$$

So, 23 is the 34th term. In symbols, $a_{34} = 23$. We can use a graphing calculator table to verify our work. ■

Example 5 Modeling with an Arithmetic Sequence

A person's salary is $25,000 in the first year. It will increase by $750 each year. Let a_n be the salary (in dollars) in the nth year.

1. Find a formula that describes a_n.
2. What will be the salary in the 32nd year?
3. In what year will the salary be $40,000?

Solution

1. The salary sequence has a common difference of 750, so the sequence is arithmetic. We substitute $a_1 = 25{,}000$ and $d = 750$ in the formula $a_n = a_1 + (n - 1)d$:

$$\begin{aligned} a_n &= 25{,}000 + (n - 1)(750) && \text{Substitute 25,000 for } a_1 \text{ and 750 for } d. \\ &= 25{,}000 + 750n - 750 && \text{Distributive law} \\ &= 750n + 24{,}250 \end{aligned}$$

The formula is $a_n = 750n + 24{,}250$. We can verify our work with a graphing calculator table by checking that $a_1 = 25{,}000$ and that the common difference is 750.

2. To find the salary in the 32nd year, we substitute $n = 32$ in the formula $a_n = 750n + 24{,}250$:

$$a_{32} = 750(32) + 24{,}250 = 48{,}250$$

The salary will be $48,250 in the 32nd year.

3. To determine when the salary will be \$40,000, we substitute $a_n = 40{,}000$ in the formula $a_n = 750n + 24{,}250$ and solve for n:

$$40{,}000 = 750n + 24{,}250 \quad \text{Substitute 40,000 for } a_n.$$
$$15{,}750 = 750n$$
$$21 = n \quad \text{Divide.}$$

The salary will be \$40,000 in the 21st year. ■

Connection Between a Linear Function and an Arithmetic Sequence

We close this section by noticing a connection between a linear function and an arithmetic sequence. Note that the math tutor formula $a_n = 6n + 17$ describes a linear function whose graph has a slope of 6, which is the common difference of the sequence. If we let $f(n)$ be the tutor's charge (in dollars per hour) for n students, then

$$f(n) = 6n + 17$$

and the sequence 23, 29, 35, 41, 47, ..., 77 is given by

$$f(1), f(2), f(3), \ldots, f(10)$$

Connection Between a Linear Function and an Arithmetic Sequence

If f is a linear function of the form $f(x) = mx + b$, then

$$f(1), f(2), f(3), \ldots$$

is an arithmetic sequence with common difference equal to the slope m.

Example 6 Identifying an Arithmetic Sequence

Let $f(x) = -5x + 38$. Is the sequence $f(1), f(2), f(3), \ldots$ arithmetic? Explain.

Solution

Since f is a linear function whose graph has a slope of -5, the sequence $f(1)$, $f(2)$, $f(3), \ldots$ must be an arithmetic sequence with common difference -5. By computing the outputs, we see that the sequence is 33, 28, 23, 18, 13, ... ■

We know from the slope addition property (Section 1.4) that if the inputs of a linear function increase by 1, the outputs change by the value of the slope. In terms of the slope addition property, it makes sense that the slope of the graph of a linear function and the common difference of the corresponding arithmetic sequence are equal.

HOMEWORK 10.1

Check whether the sequence is arithmetic. If so, find the common difference d.

1. 3, 11, 19, 27, 35, ...

2. 40, 38, 36, 34, 32, ...

3. 1, 5, 7, 11, 13, ...

4. 7, 2, −5, −8, −11, ...

5. −20, −13, −6, 1, 8, ...

6. −2, −5, −8, −11, −14, ...

7. 4, 44, 444, 4444, 44,444, ...

8. 1, 1, 2, 2, 3, 3, ...

Using a_n notation, find a formula of the sequence. Use a graphing calculator table to verify your result.

9. 5, 11, 17, 23, 29, ...

10. 7, 11, 15, 19, 23, ...

11. −4, −15, −26, −37, −48, ...

12. −3, −7, −11, −15, −19, ...

13. 100, 94, 88, 82, 76, ...

14. 72, 69, 66, 63, 60, ...

15. 1, 3, 5, 7, 9, ...

16. 1, 2, 3, 4, 5, ...

Find the indicated term of the sequence. Verify your result with a graphing calculator table.

17. 37th term of 5, 8, 11, 14, 17, ...

18. 52nd term of 4, 19, 34, 49, 64, ...

19. 45th term of 200, 191, 182, 173, 164, ...

20. 21st term of 83, 79, 75, 71, 67, ...

21. a_{96} of 4.1, 5.7, 7.3, 8.9, 10.5, ...

22. a_{31} of 23.8, 21.5, 19.2, 16.9, 14.6, ...

23. a_{400} of 1, 2, 3, 4, 5, ...

24. a_{235} of $-2, -4, -6, -8, -10, \ldots$

Find the term number n of the last term of the finite sequence. Verify your result with a graphing calculator table.

25. 3, 8, 13, 18, 23, ..., 533

26. 4, 10, 16, 22, 28, ..., 1426

27. 7, 15, 23, 31, 39, ..., 695

28. 10, 19, 28, 37, 46, ..., 415

29. $-27, -19, -11, -3, 5, \ldots, 2469$

30. $-11, -5, 1, 7, 13, \ldots, 409$

31. $29, 25, 21, 17, 13, \ldots, -14{,}251$

32. $-27, -39, -51, -63, -75, \ldots, -999$

33. If $a_7 = 24$ and $a_{13} = 66$ are terms of an arithmetic sequence, find a_{40}.

34. If $a_4 = 7$ and $a_{11} = -14$ are terms of an arithmetic sequence, find a_{98}.

35. If $a_{41} = 500$ and $a_{81} = 500$ are terms of an arithmetic sequence, find a_{990}.

36. If $a_{12} = 12$ and $a_{78} = 78$ are terms of an arithmetic sequence, find a_{103}.

37. Let $f(x) = 4x - 2$. Is the sequence $f(1), f(2), f(3), \ldots$ arithmetic? Explain.

38. Let $f(x) = -5x + 1$. Is the sequence $f(1), f(2), f(3), \ldots$ arithmetic? Explain.

39. Let $f(x) = x^2$. Is the sequence $f(1), f(2), f(3), \ldots$ arithmetic? Explain.

40. Let $f(x) = \sqrt{x}$. Is the sequence $f(1), f(2), f(3), \ldots$ arithmetic? Explain.

41. A student tries to find a_{54} of the sequence 2, 7, 11, 16, 20, 25, 29, 34, 38:

$$\begin{aligned} a_{54} &= 2 + (54 - 1)(5) \\ &= 2 + 53(5) \\ &= 267 \end{aligned}$$

Is the student's work correct? Explain.

42. An arithmetic sequence is described by $a_n = 3n + 7$. A student concludes that the first term is 7 and that the common difference is 3. What would you tell the student?

43. Is 2537 a term in the sequence 8, 15, 22, 29, 36, ...? Explain.

44. Is 3901 a term in the sequence 5, 14, 23, 32, 41, ...? Explain.

45. Find an equation of a function f such that $f(1), f(2), f(3), f(4), f(5), \ldots$ is the sequence 8, 17, 26, 35, 44, ...

46. Find an equation of a function g such that $g(1), g(2), g(3), g(4), g(5), \ldots$ is the sequence 75, 65, 55, 45, 35, ...

47. A person's starting salary is \$27,500. Each year, he receives an \$800 raise.

a. Let a_n be his salary (in dollars) for the nth year. Find a formula that describes a_n.

b. What will be his salary for the 22nd year?

c. In what year will his salary first exceed \$50,000?

48. A person's starting salary is \$30,700. At the end of each of the first 9 years, she will get a \$950 raise. After that, she will get an \$1150 raise at the end of each year. What will be her salary for the 17th year?

49. A math instructor estimates that it takes an average of 10 minutes per student to grade students' quizzes and tests each week. This instructor also spends a total of 35 hours per week in classroom activities, holding office hours, planning for classes, and attending committee meetings. Let a_n be the number of hours that the instructor works per week when n students are enrolled in her courses.

a. Find a formula that describes a_n.

b. Find the values of a_1, a_2, a_3, and a_4. What do these four terms mean in this situation?

c. If the instructor has 130 students, how many hours does she work per week?

d. What is the greatest number of students the instructor can have without having to work over 60 hours per week?

50. A full bottle of household glass cleaner holds 22 fluid ounces. It takes about 500 squeezes of the bottle's trigger to use all the cleaner.

a. Let a_n be the number of ounces of cleaner that remains in the bottle after the trigger has been squeezed n times. Find a formula that describes a_n.

b. Find a_1, a_2, a_3, and a_4. What do these four terms mean in this situation?

c. Assume that it takes about 7 squeezes of the trigger to clean one side of a 4×3-foot window. If the bottle starts out full, how much liquid would remain in it after both sides of the 4×3-foot windows have been cleaned in a building that has 32 such windows?

51. The underground rock band Little Muddy spends \$50 to send postcards announcing its latest gig at a club. Three bands are playing that night. Each band gets 30% of the money collected from a cover charge of \$6 per person. Let a_n be Little Muddy's profit (in dollars) if n people pay the cover charge.

a. Find a formula that describes a_n.

b. If the band's profit is \$256, how many people paid the cover charge?

c. The club's maximum capacity is 200 people. Assume that 18 people are on the guest list, 11 people are in the three bands, 6 people work for the club (and that all of these people get in free). What is the greatest profit that Little Muddy can earn, assuming that no one leaves the club until closing time?

d. For what values of n will Little Muddy lose money?

52. The main library in San Francisco has a five-story-high glass sculpture created in 1996 by Nayland Blake. It looks like a star constellation made up of white lights. Each light is actually an illuminated disk displaying the name of an author etched into the glass. The sculpture originally had 160 names, but Blake left room for 200 more names, 5 to be added each year.

Let a_n be the number of names in the sculpture in the nth year, where 1996 is the first year. So, $a_1 = 160$, $a_2 = 165$, $a_3 = 170$, and so on.

a. Find a formula that describes a_n.

b. Predict the number of names there will be in the 33rd year (2028).

c. Graph by hand the library sculpture sequence.

53. The pharmaceutical industry's spending on government and politics is shown in Table 2 for various years.

Table 2 Spending by the Pharmaceutical Industry on Government and Politics

Year	Spending (millions of dollars)
1998	80
1999	97
2000	112
2001	110
2002	137
2003	144
2004	158

Source: *Center for Public Integrity*

Let $f(t)$ be the pharmaceutical industry's spending (in millions of dollars) on government and politics in the year that is t years since 1995.

a. Use a graphing calculator to draw a scattergram to describe the data.

b. Find a linear equation of f.

c. Use a graphing calculator table to find the values of the sequence $f(5)$, $f(6)$, $f(7)$, $f(8)$, $f(9)$. What do they mean in this situation?

d. Predict the pharmaceutical industry's spending on government and politics in 2011.

54. Enrollments in journalism and mass communication programs are shown in Table 3 for various years.

Table 3 Enrollments in Journalism and Mass Communication Programs

Year	Enrollment (thousands)
1996	150
1998	160
2000	179
2002	195
2004	207

Source: Annual Survey of Journalism and Mass Communication Enrollment, *University of Georgia*

Let $f(t)$ be the enrollment (in thousands) in journalism and mass communication programs in the year that is t years since 1995.

a. Use a graphing calculator to draw a scattergram to describe the data.

b. Find a linear equation of f.

c. Use a graphing calculator table to find the values of the sequence $f(1)$, $f(2)$, $f(3)$, $f(4)$, $f(5)$. What do they mean in this situation?

d. Predict the enrollment in journalism and mass communication programs in 2010.

55. Postage for a first-class letter depends on its weight. In November 2006, the cost to mail a letter weighing 1 ounce or less was \$0.39. The postage increased by \$0.24 for each additional ounce through 13 total ounces.

a. Let a_n be first-class postage (in dollars) for a letter that weighs n ounces. Find a formula that describes a_n.

b. Find the postage for a letter that weighs 13 ounces.

c. Postage for a 1-pound large envelope sent by priority mail is \$4.05 (there is no first-class service for this weight). Is this a better deal than the postage that would be required by your formula? (There are 16 ounces in 1 pound.)

d. First-class postage from Boston to Chicago for a 5-pound package is \$9.80. According to your formula, how much would the postage be?

56. Describe an arithmetic sequence. Also, given the first few terms of an arithmetic sequence, explain how to find

- A term with a known term number.
- The term number of a known term.

(See page 4 for guidelines on writing a good response.)

Related Review

57. a. Find the common difference of the arithmetic sequence 5, 7, 9, 11, 13, ...

b. Find the slope of the line that contains the points (1, 5), (2, 7), (3, 9), (4, 11), and (5, 13).

c. Compare your results in parts (a) and (b). Explain.

58. a. Using a_n notation, find a formula of the sequence 20, 17, 14, 11, 8, ...

b. Find an equation of the line that contains the points (2, 17) and (5, 8).

c. Compare your results in parts (a) and (b). Explain.

Expressions, Equations, Functions, and Graphs

Perform the indicated instruction. Then use words such as linear, quadratic, cubic, exponential, logarithmic, rational, radical, polynomial, degree, function, one variable, *and* two variables, *to describe the expression, equation, or system.*

59. Solve $2\sqrt{x+3} - 1 = 5$.

60. Find the product $(4x - 5)(4x + 5)$.

61. Graph $f(x) = 2\sqrt{x+3} - 1$ by hand.

62. Factor $9x^2 - 49$.

63. Find the product $\left(4\sqrt{x} - 5\right)\left(3\sqrt{x} - 2\right)$.

64. Solve $100x^2 + 81 = 0$.

10.2 GEOMETRIC SEQUENCES

Objectives

- Know the meaning of *geometric sequence*.
- Find a formula, term, or term number of a geometric sequence.
- Use a geometric sequence to make estimates and predictions.

In Section 10.1, we worked with arithmetic sequences. In this section, we will study another type of sequence, the geometric sequence.

Definition of Geometric Sequence

Consider the sequence

$$3, 6, 12, 24, 48, \ldots$$

Notice that the *ratio* of any term to its preceding term is 2:

$$\frac{6}{3} = 2, \frac{12}{6} = 2, \frac{24}{12} = 2, \frac{48}{24} = 2, \ldots$$

We call 2 the common ratio, and we call the sequence a geometric sequence.

DEFINITION Geometric sequence

If the ratio of any term of a sequence to the preceding term is a constant r for every such pair of terms, then the sequence is a **geometric sequence.** We call the constant r the **common ratio.**

Example 1 Identifying Geometric Sequences

Determine whether the sequence is geometric. If so, find the common ratio r.

1. 2, 6, 18, 54, 162, ...
2. 4, 8, 24, 48, 240, ...
3. 32, 16, 8, 4, 2, ...

Solution

1. The sequence is geometric, because it has a common ratio of 3:

$$\frac{6}{2} = 3, \frac{18}{6} = 3, \frac{54}{18} = 3, \frac{162}{54} = 3, \ldots$$

2. The sequence is not geometric, because we can see from the first two ratios that it does not have a common ratio:

$$\frac{8}{4} = 2, \frac{24}{8} = 3$$

3. The sequence is geometric, because it has a common ratio of $\frac{1}{2}$:

$$\frac{16}{32} = \frac{1}{2}, \frac{8}{16} = \frac{1}{2}, \frac{4}{8} = \frac{1}{2}, \frac{2}{4} = \frac{1}{2}, \ldots$$

■

Finding a Formula of a Geometric Sequence

Can we find a general formula that describes the terms of a geometric sequence? Earlier, we determined that the geometric sequence 3, 6, 12, 24, 48, ... has the common ratio 2. This means that we multiply the first term, 3, by 2 to find the second term; we multiply 3 by 2 two times to find the third term; we multiply 3 by 2 three times to find the fourth term; and so on:

$$3 \cdot 2 = 6, \ 3 \cdot 2 \cdot 2 = 12, \ 3 \cdot 2 \cdot 2 \cdot 2 = 24, \ldots$$

In general, for a geometric sequence with common ratio r, we have the terms

$$a_1, \ a_1 \cdot r, \ a_1 \cdot r \cdot r, \ a_1 \cdot r \cdot r \cdot r, \ldots$$

Using exponents, we have

$$a_1, \ a_1 r, \ a_1 r^2, \ a_1 r^3, \ldots$$

We can use a pattern to find a formula that describes any term a_n of a geometric sequence:

$a_1 = a_1$

$a_2 = a_1 r$ — Multiply a_1 by r *once* to get the *second* term.

$a_3 = a_1 r^2$ — Multiply a_1 by r *twice* to get the *third* term.

$a_4 = a_1 r^3$ — Multiply a_1 by r *three times* to get the *fourth* term.

$\vdots$

$a_n = a_1 r^{n-1}$ — Multiply a_1 by r a total of $(n-1)$ times to get the nth term.

Formula That Describes the *n*th Term of a Geometric Sequence

If a geometric sequence $a_1, a_2, a_3, \ldots, a_n, \ldots$ has the common ratio r, then

$$a_n = a_1 r^{n-1}$$

In words: The nth term of a geometric sequence is equal to the first term times the $(n-1)$th power of r.

Example 2 Finding a Formula

Find a formula that describes the terms of the sequence 12, 36, 108, 324, 972, ...

Solution

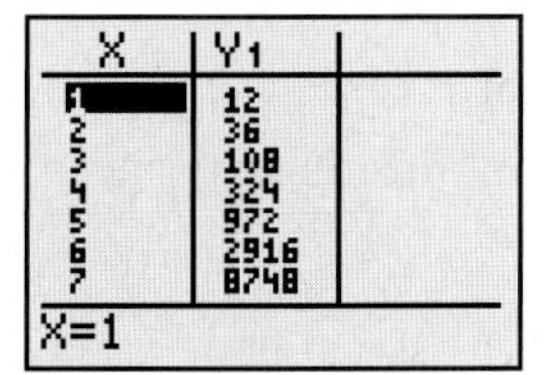

Figure 4 Verify the formula $a_n = 12(3)^{n-1}$

The sequence has a common ratio ($r = 3$), so the sequence is geometric. We substitute $a_1 = 12$ and $r = 3$ in the formula $a_n = a_1 r^{n-1}$:

$$a_n = 12(3)^{n-1}$$

We verify our formula by entering $y = 12(3)^{x-1}$ in a graphing calculator and checking that the first five terms are 12, 36, 108, 324, and 972 (see Fig. 4). ■

Here, we write the formula $a_n = 12(3)^{n-1}$ in the form $a_n = ab^n$:

$$\begin{aligned} a_n &= 12(3)^{n-1} && \text{Original formula} \\ &= 12(3)^n(3)^{-1} && b^{m+n} = b^m b^n \\ &= \frac{12(3)^n}{3} && b^{-n} = \frac{1}{b^n} \\ &= 4(3)^n && \text{Simplify.} \end{aligned}$$

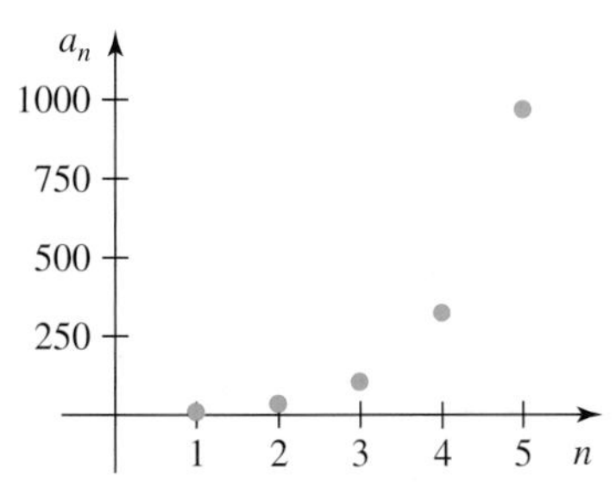

Figure 5 The first five terms of the sequence 12, 36, 108, 324, 972, ...

The formula $a_n = 4(3)^n$ describes an exponential function whose only inputs are the counting numbers 1, 2, 3, 4, 5, ... and whose only outputs are the terms of the sequence 12, 36, 108, 324, 972, ...

In Fig. 5, we sketch a graph of the first five terms of the sequence 12, 36, 108, 324, 972, ... For $a_n = 4(3)^n$, the base 3 is greater than 1, so the points lie on an *increasing* exponential curve.

Finding a Term or a Term Number of a Geometric Sequence

For a geometric sequence, if we know values of three of the four variables a_1, a_n, n, and r, we can find the value of the fourth variable by using the formula $a_n = a_1 r^{n-1}$.

Example 3 Finding a Term

Find the 12th term of the sequence 160, 80, 40, 20, 10, ...

Solution

The sequence has a common ratio $\left(r = \frac{1}{2}\right)$, so the sequence is geometric. We substitute $a_1 = 160$, $n = 12$, and $r = \frac{1}{2}$ in the formula $a_n = a_1 r^{n-1}$:

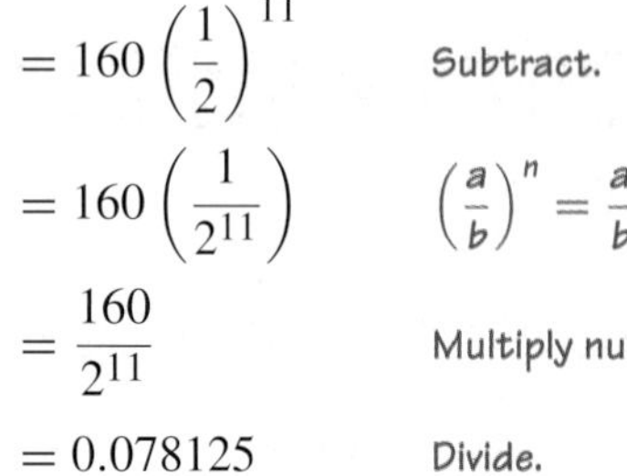

$$
\begin{aligned}
a_{12} &= 160\left(\frac{1}{2}\right)^{12-1} && \text{Substitute 160 for } a_1\text{, 12 for } n\text{, and } \frac{1}{2} \text{ for } r. \\
&= 160\left(\frac{1}{2}\right)^{11} && \text{Subtract.} \\
&= 160\left(\frac{1}{2^{11}}\right) && \left(\frac{a}{b}\right)^n = \frac{a^n}{b^n} \\
&= \frac{160}{2^{11}} && \text{Multiply numerators; multiply denominators.} \\
&= 0.078125 && \text{Divide.}
\end{aligned}
$$

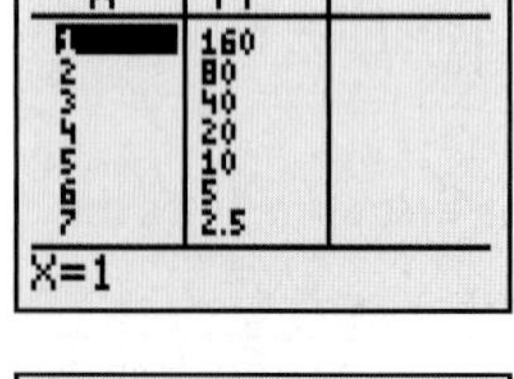

Figure 6 Verify the work

We enter $y = 160\left(\frac{1}{2}\right)^{x-1}$ in a graphing calculator and check that the first five terms are 160, 80, 40, 20, and 10 and that the 12th term is 0.078125 (see Fig. 6). The entry across from $x = 12$ has been rounded. However, the bottom entry of the table displays the exact value $a_{12} = 0.078125$. ■

In working with a sequence, we must first determine whether it is arithmetic, geometric, or neither. An arithmetic sequence has a common difference, whereas a geometric sequence has a common ratio.

Example 4 Finding a Term Number

The number 1,572,864 is a term in the sequence 6, 24, 96, 384, 1536, ... What is its term number?

Solution

The sequence has a common ratio $(r = 4)$, so the sequence is geometric. We substitute $a_1 = 6$, $a_n = 1{,}572{,}864$, and $r = 4$ in the formula $a_n = a_1 r^{n-1}$ and solve for n:

$$
\begin{aligned}
1{,}572{,}864 &= 6(4)^{n-1} && \text{Substitute 6 for } a_1\text{, 1,572,864 for } a_n\text{, and 4 for } r. \\
262{,}144 &= 4^{n-1} && \text{Divide both sides by 6.} \\
\log(262{,}144) &= \log\left(4^{n-1}\right) && \text{Take the logarithm of both sides.} \\
\log(262{,}144) &= (n-1)\log(4) && \text{Power property: } \log_b(x)^p = p\log_b(x) \\
\frac{\log(262{,}144)}{\log(4)} &= n - 1 && \text{Divide both sides by } \log(4). \\
\frac{\log(262{,}144)}{\log(4)} + 1 &= n && \text{Add 1 to both sides.} \\
10 &= n && \text{Compute.}
\end{aligned}
$$

So, 1,572,864 is the 10th term. We can use a graphing calculator table to verify our work. ■

Example 5 Modeling with a Geometric Sequence

A person's salary is \$25,000 in the first year. It will increase by 3% each year. Let a_n be the salary (in dollars) in the nth year.

1. Find a formula that describes a_n.
2. Predict the salary in the 32nd year.

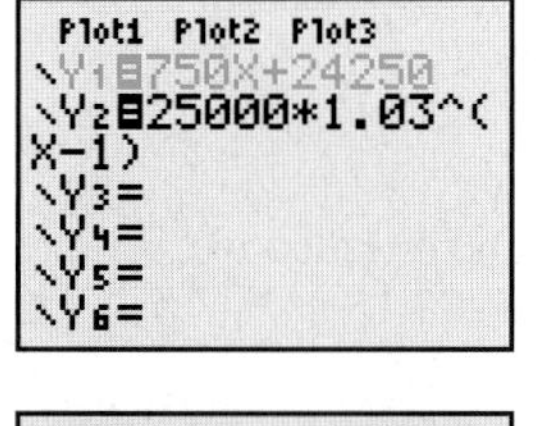

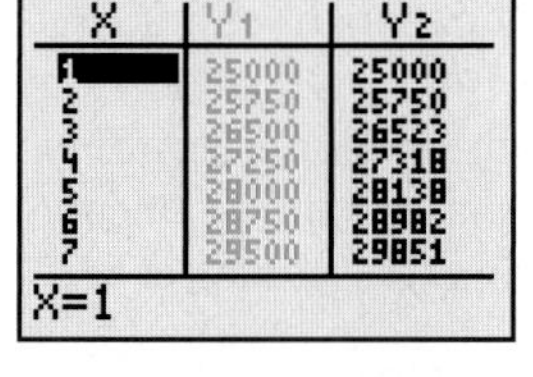

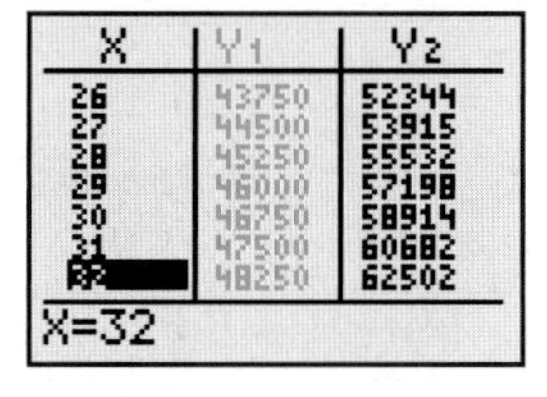

Figure 7 Verify the comparison of the two scenarios

3. Compare the result from Problem 2 with the result from Section 10.1, Example 5, Problem 2, where we assumed that the salary increases by a constant \$750 each year.

Solution

1. The salary in the second year is 103% of \$25,000, or $25{,}000(1.03) = 25{,}750$ dollars. Each year, the salary is equal to 1.03 times the salary in the preceding year. So, a_n is a geometric sequence with the common ratio 1.03. We substitute $a_1 = 25{,}000$ and $r = 1.03$ in the formula $a_n = a_1 r^{n-1}$:

$$a_n = 25{,}000(1.03)^{n-1}$$

2. To find the salary in the 32nd year, we substitute $n = 32$ in the formula $a_n = 25{,}000(1.03)^{n-1}$:

$$a_{32} = 25{,}000(1.03)^{32-1} \approx 62{,}502.01$$

The salary will be \$62,502.01 in the 32nd year.

3. First, we note that 3% of \$25,000 is \$750, so the first raise is the same in both scenarios. In Example 5 of Section 10.1, we found that if the salary increases by \$750 each year, the salary in the 32nd year will be \$48,250, considerably less than the salary of \$62,502 from receiving 3% raises each year.

We can verify our salary comparison by entering the constant-raise formula $a_n = 750n + 24{,}250$ and the percentage-raise formula $a_n = 25{,}000(1.03)^{n-1}$ in a graphing calculator and comparing tables (see Fig. 7). ■

Connection Between an Exponential Function and a Geometric Sequence

Now, we write the geometric sequence $a_n = 25{,}000(1.03)^{n-1}$ of Example 5 in the form $a_n = ab^n$:

$$\begin{aligned} a_n &= 25{,}000(1.03)^{n-1} && \text{Original formula} \\ &= 25{,}000(1.03)^n(1.03)^{-1} && b^{m+n} = b^m b^n \\ &= \frac{25{,}000(1.03)^n}{1.03} && b^{-n} = \frac{1}{b^n} \\ &= \frac{25{,}000}{1.03}(1.03)^n && \text{Write right-hand side in } ab^n \text{ form.} \end{aligned}$$

Notice that the base of the exponential function $a_n = \frac{25{,}000}{1.03}(1.03)^n$ is 1.03, which is also the common ratio of the geometric sequence. If we let $f(n)$ be the person's salary (in dollars) in the nth year, then

$$f(n) = \frac{25{,}000}{1.03}(1.03)^n$$

and the geometric sequence of the salary is

$$f(1), f(2), f(3), \ldots$$

Connection Between an Exponential Function and a Geometric Sequence

If f is an exponential function of the form $f(x) = ab^x$, then

$$f(1), f(2), f(3), \ldots$$

is a geometric sequence with common ratio equal to the base b of f.

Example 6 Identifying a Geometric Sequence

Let $f(x) = 4(3)^x$. Is the sequence $f(1), f(2), f(3), \ldots$ geometric? Explain.

Solution

Since f is an exponential function with base 3, the sequence $f(1), f(2), f(3), \ldots$ must be a geometric sequence with common ratio 3. By computing the outputs, we see that the sequence is $12, 36, 108, 324, 972, \ldots$. ■

Recall from Section 4.3 that the base multiplier property for exponential functions states that if the value of the independent variable increases by 1, then the value of the dependent variable is multiplied by the base of the exponential function. If we think in terms of the base multiplier property, it makes sense that the base of an exponential function is equal to the common ratio of the corresponding geometric sequence.

HOMEWORK 10.2

FOR EXTRA HELP ▶ Student Solutions Manual | PH Math/Tutor Center | MathXL® | MyMathLab

Check whether the sequence is arithmetic, geometric, or neither. If the sequence is geometric, find the common ratio r. If the sequence is arithmetic, find the common difference d.

1. $4, 28, 196, 1372, 9604, \ldots$

2. $0.08, 0.8, 8, 80, 800, \ldots$

3. $13, 6, -1, -8, -15, \ldots$

4. $3, 7, 11, 15, 19, \ldots$

5. $3, 4, 6, 9, 13, \ldots$

6. $62, 57, 54, 42, 39, \ldots$

7. $200, 40, 8, \frac{8}{5}, \frac{8}{25}, \ldots$

8. $96, 48, 24, 12, 6, \ldots$

Find a formula of the sequence. Use a_n notation. Use a graphing calculator table to verify your formula.

9. $3, 6, 12, 24, 48, \ldots$

10. $4, 20, 100, 500, 2500, \ldots$

11. $800, 200, 50, 12.5, 3.125, \ldots$

12. $162, 54, 18, 6, 2, \ldots$

13. $100, 50, 25, 12.5, 6.25, \ldots$

14. $1250, 250, 50, 10, 2, \ldots$

15. $1, 4, 16, 64, 256, \ldots$

16. $5, 15, 45, 135, 405, \ldots$

Find the indicated term of the sequence. Write the result in scientific notation $N \times 10^k$ with N rounded to the fourth decimal place. Use a graphing calculator table to verify your result.

17. 34th term of $4, 20, 100, 500, 2500, \ldots$

18. 103rd term of $2, 8, 32, 128, 512, \ldots$

19. 27th term of $80, 40, 20, 10, 5, \ldots$

20. 25th term of $36, 12, 4, \frac{4}{3}, \frac{4}{9}, \ldots$

21. a_{23} of $8, 16, 32, 64, 128, \ldots$

22. a_{60} of $1, \frac{3}{2}, \frac{9}{4}, \frac{27}{8}, \frac{81}{16}, \ldots$

Find the term number n of the last term of the finite sequence. Verify your result with a graphing calculator.

23. $240, 120, 60, 30, 15, \ldots, 0.46875$

24. $80, 20, 5, 1.25, 0.3125, \ldots, 0.01953125$

25. $0.00224, 0.0112, 0.056, 0.28, 1.4, \ldots, 109{,}375$

26. $0.046875, 0.09375, 0.1875, 0.375, 0.75, \ldots, 192$

The given number is a term in the sequence that follows. Which term is it? Use a graphing calculator table to verify your result.

27. 3,407,872; $13, 26, 52, 104, 208, \ldots$

28. 2,470,629; $3, 21, 147, 1029, 7203, \ldots$

29. 28,697,814; $2, 6, 18, 54, 162, \ldots$

30. 25,165,824; $6, 24, 96, 384, 1536, \ldots$

For f, is the sequence $f(1), f(2), f(3), \ldots$ arithmetic, geometric, or neither? Explain.

31. $f(x) = 2(5)^x$

32. $f(x) = 8\left(\frac{1}{2}\right)^x$

33. $f(x) = 7x - 3$

34. $f(x) = 3x^2$

35. Find an equation of a function f such that $f(1), f(2), f(3), f(4), f(5), \ldots$ is the sequence $8, 24, 72, 216, 648, \ldots$.

36. Find an equation of a function g such that $g(1), g(2), g(3), g(4), g(5), \ldots$ is the sequence $48, 24, 12, 6, 3, \ldots$.

37. Is 9,238,946 a term in the sequence $13, 26, 52, 104, 208, \ldots$? Explain.

38. Is 1,240,029 a term in the sequence $7, 21, 63, 189, 567, \ldots$? Explain.

39. A student tries to find a_{17} for the sequence $2, 6, 10, 14, 18, \ldots$:

$$\begin{aligned} a_n &= a_1 r^{n-1} \\ a_{17} &= 2(3)^{17-1} \\ &= 2(3)^{16} \\ &= 86{,}093{,}442 \end{aligned}$$

Describe any errors. Then find the term correctly.

40. A geometric sequence is described by the formula $a_n = 4(6)^n$. A student concludes that the first term is 4 and the common ratio is 6. What would you tell the student?

41. Assume that a person's salary is \$27,000 in the first year and that the salary increases by 4% each year.

a. Let a_n be the salary (in dollars) in the nth year. Find a formula that describes a_n.

b. What will be the person's salary in the 10th year?

c. In what year will the salary first exceed \$50,000?

42. A person's salary is \$24,000 in the first year.

a. If the salary increases by \$960 each year, calculate the salary in the 2nd year and in the 30th year.

b. If the salary increases by 4% each year, calculate the salary in the 2nd year and in the 30th year.

c. Compare your results for parts (a) and (b). Explain why the salaries are the same in the 2nd year but different in the 30th year.

43. Your ancestors one generation back are your natural parents. Your ancestors two generations back are your natural grandparents. Let a_n be the number of ancestors that you had in the nth generation back.

a. List the first five terms of the sequence a_n.

b. Find a formula that describes a_n.

c. Use your formula to find the number of ancestors in the 8th generation back.

d. Use your formula to find the number of ancestors in the 35th generation back. Explain why model breakdown has occurred. Describe any assumptions you made.

44. A rubber ball is dropped. The height of the ball is measured from the floor to the bottom of the ball. The ball's maximum height after one bounce is 4 feet. The ball's maximum height after the second bounce is 70% of 4 feet, or 2.8 feet. This pattern continues; that is, the maximum height after each bounce is 70% of the maximum height of the preceding bounce. Let a_n be the maximum height (in feet) of the ball after the nth bounce.

a. Find a formula that describes a_n.

b. Predict the ball's maximum height after the 5th bounce.

c. For which bounce does the ball reach at least half a foot for the last time?

d. Graph by hand the first five terms of the bouncing-ball sequence.

e. Does model breakdown occur? Explain.

45. The name Nevaeh has made the fastest climb in U.S. girls' names in more than a century (see Table 4). "Nevaeh" is "Heaven" spelled backward.

Table 4 Numbers of Nevaehs Born

Year	Number of Nevaehs Born
2001	1191
2002	1692
2003	2287
2004	3156
2005	4457

Source: *Social Security Administration*

Let $f(t)$ be the number of Nevaehs born in the year that is t years since 2000.

a. Use a graphing calculator to draw a scattergram to describe the data.

b. Find an exponential equation of f.

c. Use a graphing calculator table to find the values of the sequence $f(1)$, $f(2)$, $f(3)$, $f(4)$, $f(5)$. What do they mean in this situation?

d. Since 2000, Emily has been the most popular girl's name—about 1.25% of girls are named Emily each year. Predict the number of Nevaehs that will be born in 2010. Each year, about 2 million girls are born. What percentage of these girls will be named Nevaeh in 2010, according to the model? Assuming that 1.25% of girls are named Emily in 2010, which of the two names will be more popular?

46. The numbers of subscribers to cellular telephone services in the United States are listed in Table 5 for various years.

Table 5 Cellular Telephone Subscribers

Year	Subscribers (millions)
1990	5.3
1992	11.0
1994	24.1
1996	44.0
1998	69.2
2000	109.5
2001	128.4

Source: *Cellular Telecommunications Industry Association*

Let $f(t)$ be the number of subscribers (in millions) at t years since 1990.

a. Find an exponential equation of f.

b. Use a graphing calculator table to find the values of the sequence $f(0)$, $f(1)$, $f(2)$, $f(3)$, $f(4)$. What do they mean in this situation?

c. Predict when every American will have a cell phone. Assume that the U.S. population is 299 million.

47. Suppose that a rumor is spreading on campus that students will not have to take any final exams this semester. On the first day, 5 students hear the rumor. Each person who hears the rumor tells it, approximately 24 hours later, to exactly 3 students who have not yet heard it. Let a_n be the number of students who hear the rumor on the nth day.

a. Find a formula that describes a_n.

b. How many students will hear the rumor on the 5th day?

c. Use your formula to predict the number of students who will hear the rumor on the 11th day. Has model breakdown occurred? Explain.

d. To find your formula, you made some assumptions about the way the rumor would spread. Describe each assumption, and discuss whether you think it is reasonable.

48. Describe a geometric sequence. Also, given the first few terms of a geometric sequence, explain how to find
- A term with a known term number.
- The term number of a known term.

Related Review

49. a. Find the common ratio of the geometric sequence 7, 14, 28, 56, 112, ...
b. Find the base b of the exponential function $y = ab^x$ that contains the points (1, 7), (2, 14), (3, 28), (4, 56), and (5, 112).
c. Compare your results in parts (a) and (b). Explain why this happened.

50. a. Find a formula of the sequence 486, 162, 54, 18, 6, ... Write your result in the form $a_n = cb^n$. **[Hint:** $b^{n-1} = b^n b^{-1}$]
b. Find an equation of the exponential curve that contains the points (3, 54) and (5, 6).
c. Compare your results in parts (a) and (b). Explain why this happened.

Find a formula of the sequence. Use a_n notation.

51. 14, 19, 24, 29, 34, ...
52. 57, 49, 41, 33, 25, ...
53. 448, 224, 112, 56, 28, ...
54. 4, 12, 36, 108, 324, ...

Find the indicated term of the sequence.

55. a_9 of 2, 10, 50, 250, 1250, ...
56. a_{10} of 3200, 640, 128, 25.6, 5.12, ...
57. a_{99} of 17, 12, 7, 2, −3, ...
58. a_{96} of 9.5, 12.9, 16.3, 19.7, 23.1, ...

Find the term number n of the last term of the finite sequence.

59. 4, 7, 10, 13, 16, ..., 367
60. 88, 81, 74, 67, 60, ..., −801
61. 8192, 2048, 512, 128, 32, ..., 0.0078125
62. 5, 15, 45, 135, 405, ..., 2,657,205

Expressions, Equations, Functions, and Graphs

Perform the indicated instruction. Then use words such as linear, quadratic, cubic, exponential, logarithmic, rational, radical, polynomial, degree, function, one variable, *and* two variables *to describe the expression, equation, or system.*

63. Solve $-3(4)^x = -44$. Round any solutions to the fourth decimal place.

64. Solve: $2x - 7y = 14$
$y = \frac{2}{7}x - 2$

65. Graph $f(x) = -3(4)^x$ by hand.

66. Graph $2x - 7y = 14$ by hand.

67. Write $2\log_b(5x^3) - 3\log_b(2x^7)$ as a single logarithm.

68. Find the inverse of $f(x) = \frac{2}{7}x - 2$.

10.3 ARITHMETIC SERIES

Objectives

- Know the meaning of *arithmetic series.*
- Evaluate the sum of an arithmetic series.
- Use arithmetic series to make estimates and predictions.

So far in this chapter, we have worked with arithmetic sequences and geometric sequences. In the next two sections, we will discuss sums that are related to these sequences.

Definition of Arithmetic Series

Suppose that a person's salary is \$23,000 in the first year and increases by \$2000 each year. Here, we use an arithmetic sequence to describe the salaries (in thousands of dollars) for the first 32 years:

$$23, 25, 27, \ldots, 81, 83, 85$$

To find the *total* earnings (in thousands of dollars) during the first 32 years, we find the sum

$$23 + 25 + 27 + \cdots + 81 + 83 + 85$$

We call this sum an arithmetic series.

DEFINITION Arithmetic series

If the sequence $a_1, a_2, a_3, \ldots, a_n$ is an arithmetic sequence, then the sum $a_1 + a_2 + a_3 + \cdots + a_n$ is an **arithmetic series.** We say that a_i is the ***i*th term** of the series and that its **term number** is i.

For example, for the series $23 + 25 + 27 + \cdots + 81 + 83 + 85$, the number 27 is the third term, and its term number is 3. We use the notation S_n to represent the sum of the first n terms of an arithmetic sequence:

$$S_n = a_1 + a_2 + a_3 + \cdots + a_n$$

Finding the Sum of an Arithmetic Series

Next, we find the total earnings (in thousands of dollars) for 32 years, S_{32}, where

$$S_{32} = 23 + 25 + 27 + \cdots + 81 + 83 + 85$$

We find the sum in a way that suggests a general formula of S_n for any arithmetic series.

To find the sum, we write the equation for S_{32} twice, the second time with the terms in reverse order. Then we add the left-hand sides and add the right-hand sides of the two equations:

$$\begin{array}{rcl} S_{32} &=& 23 + 25 + 27 + \cdots + 81 + 83 + 85 \\ S_{32} &=& 85 + 83 + 81 + \cdots + 27 + 25 + 23 \\ \hline 2S_{32} &=& 108 + 108 + 108 + \cdots + 108 + 108 + 108 \end{array}$$

On the right-hand side of the last equation, the number 108 appears 32 times, so the sum equals $32(108)$:

$$2S_{32} = 32(108) \qquad \text{Equation (1)}$$

$$S_{32} = \frac{32(108)}{2} \qquad \text{Divide both sides by 2: Equation (2)}$$

$$S_{32} = 1728 \qquad \text{Equation (3)}$$

The total earnings for 32 years will be \$1,728,000.

Notice that 108 is the sum of the first and last terms of the series:

$$108 = 23 + 85$$

So, we can write equation (2) as

$$S_{32} = \frac{32(23 + 85)}{2}$$

Our process and result suggest that we can find S_n for any arithmetic series by first multiplying the sum of the first and last terms, $a_1 + a_n$, by the number of terms n and dividing the product by 2.

Formula for Sum of an Arithmetic Series

If $S_n = a_1 + a_2 + a_3 + \cdots + a_n$ is an arithmetic series, then

$$S_n = \frac{n(a_1 + a_n)}{2}$$

We derive the formula $S_n = \dfrac{n(a_1 + a_n)}{2}$ at the end of this section.

To evaluate the sum $S_n = a_1 + a_2 + a_3 + \cdots + a_n$, we must first check that the sequence $a_1, a_2, a_3, \ldots, a_n$ is arithmetic before we use the formula $S_n = \dfrac{n(a_1 + a_n)}{2}$.

Example 1 Evaluating Sums

1. Evaluate S_{50}, where $S_{50} = 3 + 7 + 11 + 15 + 19 + \cdots + 199$.
2. Evaluate S_{80}, where $S_{80} = 60 + 53 + 46 + 39 + 32 + \cdots + (-493)$.

Solution

1. The sequence $3, 7, 11, 15, 19, \ldots, 199$ is arithmetic with common difference $d = 4$. We substitute $n = 50$, $a_1 = 3$, and $a_n = 199$ in the equation $S_n = \dfrac{n(a_1 + a_n)}{2}$:

$$S_{50} = \frac{50(3 + 199)}{2} = 5050$$

2. The sequence $60, 53, 46, 39, 32, \ldots, -493$ is arithmetic with common difference $d = -7$. We substitute $n = 80$, $a_1 = 60$, and $a_n = -493$ in the equation $S_n = \dfrac{n(a_1 + a_n)}{2}$:

$$S_{80} = \frac{80(60 + (-493))}{2} = -17{,}320$$

■

In evaluating S_n for an arithmetic series, we sometimes must use the formula $a_n = a_1 + (n - 1)d$ to find n, a_1, or a_n before we can use the formula $S_n = \dfrac{n(a_1 + a_n)}{2}$.

Example 2 Evaluating a Sum

Evaluate S_{43}, where $S_{43} = 150 + 147 + 144 + 141 + 138 + \cdots + a_{43}$.

Solution

The sequence $150, 147, 144, 141, 138, \ldots, a_{43}$ is arithmetic with common difference $d = -3$. Although we know that $a_1 = 150$ and $n = 43$, we must first find a_{43} before we can use the formula $S_n = \dfrac{n(a_1 + a_n)}{2}$. We find a_{43} by substituting $a_1 = 150$, $n = 43$, and $d = -3$ in the equation $a_n = a_1 + (n - 1)d$:

$$a_{43} = 150 + (43 - 1)(-3) = 24$$

Next, we substitute $n = 43$, $a_1 = 150$, and $a_n = 24$ in the equation $S_n = \dfrac{n(a_1 + a_n)}{2}$:

$$S_{43} = \frac{43(150 + 24)}{2} = 3741$$

■

Example 3 Evaluating a Sum

Evaluate the sum $2 + 8 + 14 + 20 + 26 + \cdots + 338$.

Solution

The sequence $2, 8, 14, 20, 26, \ldots, 338$ is arithmetic with common difference $d = 6$. Although we know that $a_1 = 2$ and $a_n = 338$, we must first find n before we can use the formula $S_n = \dfrac{n(a_1 + a_n)}{2}$. We find n by substituting $a_1 = 2$, $a_n = 338$, and $d = 6$ in the equation $a_n = a_1 + (n - 1)d$:

$$338 = 2 + (n - 1)6 \quad \text{Substitute 2 for } a_1\text{, 338 for } a_n\text{, and 6 for } d.$$
$$336 = (n - 1)6 \quad \text{Subtract 2 from both sides.}$$
$$56 = n - 1 \quad \text{Divide both sides by 6.}$$
$$57 = n$$

Now we substitute $n = 57$, $a_1 = 2$, and $a_n = 338$ in the equation $S_n = \dfrac{n(a_1 + a_n)}{2}$:

$$S_{57} = \frac{57(2 + 338)}{2} = 9690$$

■

Example 4 Modeling with an Arithmetic Series

A person's salary is \$30,000 in the first year and increases by \$1200 each year. Find the person's total earnings for the first 25 years.

Solution

Let a_n be the person's salary (in dollars) in the nth year. The salary sequence $a_1, a_2, a_3, \ldots, a_n$ is arithmetic with common difference 1200. First, we find the salary in the 25th year by substituting $a_1 = 30{,}000$, $n = 25$, and $d = 1200$ in the equation $a_n = a_1 + (n-1)d$:

$$a_{25} = 30{,}000 + (25-1)(1200) = 58{,}800$$

Next, we find S_{25} by substituting $n = 25$, $a_1 = 30{,}000$, and $a_n = 58{,}800$ in the equation $S_n = \dfrac{n(a_1 + a_n)}{2}$:

$$S_{25} = \frac{25(30{,}000 + 58{,}800)}{2} = 1{,}110{,}000$$

The total earnings for 25 years will be \$1,110,000. ■

How do we derive the formula $S_n = \dfrac{n(a_1 + a_n)}{2}$? To begin, consider an arithmetic series:

$$S_n = a_1 + a_2 + a_3 + \cdots + a_n$$

Since the series is arithmetic, each term of the series is found by adding d (the common difference) to the preceding term. This means that a_2 is found by adding d to a_1, so $a_2 = a_1 + d$. Also, a_3 is found by adding d twice to a_1, so $a_3 = a_1 + 2d$. The pattern continues, so the series can be expressed as

$$S_n = a_1 + (a_1 + d) + (a_1 + 2d) + \cdots + a_n$$

If we list backward from a_n, we find the term before a_n by subtracting d, so the term before a_n is $a_n - d$. Likewise, the term before that is $a_n - 2d$. This means that the series can be expressed as

$$S_n = a_1 + (a_1 + d) + (a_1 + 2d) + \cdots + (a_n - 2d) + (a_n - d) + a_n$$

Now, just as we did at the start of the section, we write the equation that describes S_n twice, the second time with the terms in reverse order. Then we add the left-hand sides and add the right-hand sides of the two equations:

$$\begin{array}{rl} S_n = & a_1 + (a_1 + d) + (a_1 + 2d) + \cdots + (a_n - 2d) + (a_n - d) + a_n \\ S_n = & a_n + (a_n - d) + (a_n - 2d) + \cdots + (a_1 + 2d) + (a_1 + d) + a_1 \\ \hline 2S_n = & (a_1 + a_n) + (a_1 + a_n) + (a_1 + a_n) + \cdots + (a_1 + a_n) + (a_1 + a_n) + (a_1 + a_n) \end{array}$$

Notice that the expression $a_1 + a_n$ appears n times on the right-hand side of the last equation. Thus, the sum on the right-hand side is $n(a_1 + a_n)$. We have

$$2S_n = n(a_1 + a_n)$$

$$S_n = \frac{n(a_1 + a_n)}{2} \qquad \text{Divide both sides by 2.}$$

This is the formula we set out to derive, that of the sum of an arithmetic series.

HOMEWORK 10.3

FOR EXTRA HELP ▶ Student Solutions Manual  PH Math/Tutor Center MathXL® MyMathLab

Evaluate the sum of the arithmetic series with the given values of a_1, a_n, and n.

1. $a_1 = 2$, $a_n = 447$, and $n = 90$

2. $a_1 = 7$, $a_n = 187$, and $n = 61$

3. $a_1 = 13$, $a_n = 548$, and $n = 108$

4. $a_1 = 38$, $a_n = 605$, and $n = 82$

5. $a_1 = 37$, $a_n = -1099$, and $n = 72$

6. $a_1 = 208$, $a_n = -386$, and $n = 67$

Evaluate the sum of the series.

7. $S_{74} = 5 + 13 + 21 + 29 + 37 + \cdots + 589$

8. $S_{59} = 14 + 17 + 20 + 23 + 26 + \cdots + 188$

9. $S_{101} = 93 + 89 + 85 + 81 + 77 + \cdots + (-307)$

10. $S_{45} = 131 + 129 + 127 + 125 + 123 + \cdots + 43$

11. $S_{117} = 4 + 4 + 4 + 4 + 4 + \cdots + 4$

12. $S_{46} = -6 + (-6) + (-6) + (-6) + (-6) + \cdots + (-6)$

13. $S_{125} = 3 + 13 + 23 + 33 + 43 + \cdots + a_{125}$

14. $S_{125} = 4 + 14 + 24 + 34 + 44 + \cdots + a_{125}$

15. $S_{81} = 8 + 19 + 30 + 41 + 52 + \cdots + a_{81}$

16. $S_{87} = 11 + 17 + 23 + 29 + 35 + \cdots + a_{87}$

17. $(-15) + (-28) + (-41) + (-54) + (-67) + \cdots + a_{152}$

18. $(-23) + (-26) + (-29) + (-32) + (-35) + \cdots + a_{85}$

19. $(-40) + (-37) + (-34) + (-31) + (-28) + \cdots + a_{137}$

20. $(-29) + (-24) + (-19) + (-14) + (-9) + \cdots + a_{214}$

21. $19 + 25 + 31 + 37 + 43 + \cdots + 247$

22. $14 + 26 + 38 + 50 + 62 + \cdots + 794$

23. $900 + 892 + 884 + 876 + 868 + \cdots + (-900)$

24. $207 + 203 + 199 + 195 + 191 + \cdots + 3$

25. $4 + 7 + 10 + 13 + 16 + \cdots + 340$

26. $1 + 3 + 5 + 7 + 9 + \cdots + 10{,}001$

27. $1 + 2 + 3 + 4 + 5 + \cdots + 10{,}000$

28. $2 + 3 + 4 + 5 + 6 + \cdots + 10{,}001$

*For the given conditions, determine whether the sum of an arithmetic series S_n is positive or negative. Explain why your response makes sense. [**Hint:** Try experimenting with specific values of a_1, d, and n that meet the stated conditions. Then explain why your answer makes sense for any values that meet those conditions.]*

29. $a_1 > 0$, $d > 0$, and n is any counting number

30. $a_1 < 0$, $d < 0$, and n is any counting number

31. $a_1 = -20$, $d = 8$, and n is a very large counting number

32. $a_1 = 10$, $d = -4$, and n is a very large counting number

33. If $f(x) = 7x - 1$, is the series

$$f(1) + f(2) + f(3) + f(4) + \cdots + f(100)$$

arithmetic? Explain.

34. If $g(x) = 4(3)^x$, is the series

$$g(1) + g(2) + g(3) + \cdots + g(50)$$

arithmetic? Explain.

35. A first-year salary is \$28,500. Each year there is a raise of \$1100.
a. Find the salary in the 28th year of work.
b. Find the total earnings for 28 years of work.

36. A first-year salary is \$35,100. Each year there is a raise of \$1400.
a. FInd the salary in the 30th year of work.
b. Find the total earnings for 30 years of work.

37. Two companies have made you job offers. Company A offers a first-year salary of \$35,000 with a \$700 raise at the end of each year. Company B offers a first-year salary of \$27,000 with a \$1500 raise at the end of each year. At which company would your total earnings over 20 years be greater? By how much?

38. Two companies have made you job offers. Company A offers a first-year salary of \$24,500 with a \$1700 raise at the end of each year. Company B offers a first-year salary of \$33,200 with a \$600 raise at the end of each year. At which company would your total earnings over 20 years be greater? By how much?

39. An auditorium has 30 rows of seats. There are 20 seats in the front row, 24 in the second row, 28 in the third row, and so on. In other words, each row has four more seats than the row in front of it.
a. How many seats are in the back row?
b. How many seats are in the auditorium?

40. An auditorium has 50 rows of seats. There are 16 seats in the front row, 18 in the second row, 20 in the third row, and so on. In other words, each row has two more seats than the row in front of it.
a. How many seats are in the auditorium?
b. If a ticket costs \$20 for a seat in the first 10 rows and \$15 for a seat in the remaining rows, what is the revenue for a sellout performance?
c. If 2900 people buy tickets for one performance, describe all possibilities for the revenue from the performance.

41. In Exercise 53 of Homework 10.1, you found the model $f(t) = 12.61t + 44.07$, where $f(t)$ is the pharmaceutical industry's spending (in millions of dollars) on government and politics in the year that is t years since 1995 (see Table 6).

Table 6 Spending by the Pharmaceutical Industry on Government and Politics

Year	Spending (millions of dollars)
1998	80
1999	97
2000	112
2001	110
2002	137
2003	144
2004	158

Source: *Center for Public Integrity*

a. Use f to estimate the pharmaceutical industry's spending on government and politics in 1998.
b. Predict the pharmaceutical industry's spending on government and politics in 2010.
c. Predict the pharmaceutical industry's total spending on government and politics from 1998 through 2010.

42. Sales at food-and-drink places have increased over time as the population has increased (see Table 7).

Table 7 Sales at Food-and-Drink Places

Year	Sales (billions of dollars)
1992	203.4
1994	225.6
1996	242.9
1998	272.6
2000	305.7
2002	332.2
2003	349.4

Source: *U.S. Census Bureau*

Let $f(t)$ be the sales (in billions of dollars) at food-and-drink places at t years since 1990.
a. Use a graphing calculator to draw a scattergram to describe the data.
b. Find a linear equation of f.
c. Find $f(0)$. What does it mean in this situation?
d. Find $f(20)$. What does it mean in this situation?
e. Find $f(0) + f(1) + f(2) + \cdots + f(20)$. What does it mean in this situation? [**Hint:** Think carefully about the value of n.]
f. Many restaurant owners report to the Internal Revenue Service (IRS) that their waitstaffs receive between 8% and 10% of each check as tip income, while in reality most waitstaffs receive between 10% and 20% of each check. Waitstaffs are expected to report to the IRS the portion of their tip income unreported by their restaurant owner, yet most do not. If 60% of food and drink sales in the United States occur in restaurants that have waitstaffs, predict the total amount of unreported waitstaff income from 1990 through 2009.

43. A first-year salary is \$24,800. Each year there is a raise of \$1200.
a. What will be the total amount of money earned in 26 years?
b. What will be the total amount of money earned from raises alone in 26 years?
c. What is the mean amount of money earned per year for the 26 years? For which of the 26 years will this mean be greater than the actual amount of money earned? For which years will this mean be less than the actual amount of money earned?
d. Assume that for each of the 26 years, *taxable income* is equal to salary minus \$4250. Assume also that the federal income tax rate is 15.016% on the first \$25,000 of taxable income and 17.04% on the remaining taxable income. Estimate the total amount paid in federal income tax for the 26 years.

44. Describe an arithmetic series. Also, explain how to evaluate the sum of an arithmetic series $S_n = a_1 + a_2 + a_3 + \cdots + a_n$ if you know a_1, a_n, and the common difference d of the arithmetic sequence $a_1, a_2, a_3, \ldots, a_n$.

Related Review

If the following is a sequence, find the 15th term. If the following is a series, find the sum.

45. $8, 24, 40, 56, 72, \ldots$

46. $8, 24, 72, 216, 648, \ldots$

47. $8 + 24 + 40 + 56 + 72 + \cdots + a_{15}$

48. $8 + (-8) + (-24) + (-40) + (-56) + \cdots + a_{15}$

Expressions, Equations, Functions, and Graphs

Perform the indicated instruction. Then use words such as linear, quadratic, cubic, exponential, logarithmic, rational, radical, polynomial, degree, function, one variable, *and* two variables *to describe the expression, equation, or system.*

49. Find the sum $\dfrac{x-5}{x^2-9} + \dfrac{x+3}{x^2-8x+15}$.

50. Solve $3x(x-2) = 5(x-1)$.

51. Find the product $\dfrac{x-5}{x^2-9} \cdot \dfrac{x+3}{x^2-8x+15}$.

52. Factor $8x^3 + 12x^2 - 2x - 3$.

53. Solve $\dfrac{x-5}{x^2-9} + \dfrac{x+3}{x^2-8x+15} = \dfrac{2}{x-5}$.

54. Find an equation of a parabola that contains the points (2, 7), (4, 15), and (5, 22).

10.4 GEOMETRIC SERIES

Objectives

- Know the meaning of *geometric series*.
- Evaluate the sum of a geometric series.
- Use geometric series to make estimates and predictions.

In Section 10.3, we worked with arithmetic series. In this section, we will discuss another type of series, the geometric series.

Definition of Geometric Series

Consider the geometric sequence

$$3, 6, 12, 24, 48, \ldots, 1536$$

We call the sum

$$3 + 6 + 12 + 24 + 48 + \cdots + 1536$$

a geometric series.

DEFINITION Geometric series

If the sequence $a_1, a_2, a_3, \ldots, a_n$ is a geometric sequence, then the sum $a_1 + a_2 + a_3 + \cdots + a_n$ is a **geometric series.** We say that a_i is the i**th term** of the series and that its **term number** is i.

Finding the Sum of a Geometric Series

Can we derive a general formula that describes the sum of a geometric series $S_n = a_1 + a_2 + a_3 + \cdots + a_n$? In Section 10.2, we described the terms of a geometric sequence $a_1, a_2, a_3, \ldots, a_n$ in terms of a_1 and r:

$$\begin{aligned} a_1 &= a_1 \\ a_2 &= a_1 r \\ a_3 &= a_1 r^2 \\ a_4 &= a_1 r^3 \\ &\vdots \\ a_n &= a_1 r^{n-1} \end{aligned}$$

In each case, the exponent of r is one less than the term number. So, we can express the series $S_n = a_1 + a_2 + a_3 + \cdots + a_n$ as

$$S_n = a_1 + a_1 r + a_1 r^2 + \cdots + a_1 r^{n-1}$$

If we list the terms backward from $a_1 r^{n-1}$, we find that the term before $a_1 r^{n-1}$ will have an exponent of r that is one less than $n - 1$. So, the term before $a_1 r^{n-1}$ is $a_1 r^{n-2}$. Similarly, the term before that is $a_1 r^{n-3}$. This means that the series can be expressed as

$$S_n = a_1 + a_1 r + a_1 r^2 + \cdots + a_1 r^{n-3} + a_1 r^{n-2} + a_1 r^{n-1} \qquad \text{Equation (1)}$$

We multiply both sides of this equation by r to obtain

$$r S_n = a_1 r + a_1 r^2 + a_1 r^3 + \cdots + a_1 r^{n-2} + a_1 r^{n-1} + a_1 r^n \qquad \text{Equation (2)}$$

Next, we multiply both sides of equation (2) by -1, add the left-hand sides, and add the right-hand sides of this new equation and equation (1):

$$\begin{array}{rcccccccccccl} S_n &=& a_1 &+ a_1 r &+ a_1 r^2 &+ \cdots &+ a_1 r^{n-2} &+ a_1 r^{n-1} & \\ -r S_n &=& &- a_1 r &- a_1 r^2 &- \cdots &- a_1 r^{n-2} &- a_1 r^{n-1} &- a_1 r^n \\ \hline S_n - r S_n &=& a_1 &+ 0 &+ 0 &+ \cdots + &0 &+ 0 &- a_1 r^n \end{array}$$

We can simplify both sides of the last equation:

$$S_n - r S_n = a_1 - a_1 r^n$$

We now factor out S_n from the left-hand side and a_1 from the right-hand side:

$$S_n(1 - r) = a_1(1 - r^n) \qquad \text{Factor both sides.}$$

$$S_n = \frac{a_1(1 - r^n)}{1 - r}, \quad r \neq 1 \qquad \text{Divide both sides by } 1 - r.$$

> **Formula for Sum of a Geometric Series**
>
> If $S_n = a_1 + a_2 + a_3 + \cdots + a_n$ is a geometric series with common ratio $r \neq 1$, then
>
> $$S_n = \frac{a_1(1 - r^n)}{1 - r}.$$

To find the sum $S_n = a_1 + a_2 + a_3 + \cdots + a_n$, we must first determine whether the series is arithmetic, geometric, or neither. If the series is arithmetic, we use $S_n = \dfrac{n(a_1 + a_n)}{2}$. If the series is geometric with common ratio $r \neq 1$, we use $S_n = \dfrac{a_1(1 - r^n)}{1 - r}$.

Example 1 Evaluating Sums

Evaluate the sum of the series. Round the result to the fourth decimal place.

1. $S_{15} = 4 + 12 + 36 + 108 + 324 + \cdots + a_{15}$
2. $S_{13} = 486 + 162 + 54 + 18 + 6 + \cdots + a_{13}$

Solution

1. The sequence 4, 12, 36, 108, 324, ..., a_{15} is geometric with common ratio $r = 3$. We substitute $a_1 = 4$, $r = 3$, and $n = 15$ in the equation $S_n = \dfrac{a_1(1 - r^n)}{1 - r}$:

$$S_{15} = \frac{4(1 - 3^{15})}{1 - 3} = 28{,}697{,}812$$

2. The sequence 486, 162, 54, 18, 6, ..., a_{13} is geometric with common ratio $r = \dfrac{1}{3}$. We substitute $a_1 = 486$, $r = \dfrac{1}{3}$, and $n = 13$ in the equation $S_n = \dfrac{a_1(1 - r^n)}{1 - r}$:

$$S_{13} = \frac{486\left(1 - \left(\frac{1}{3}\right)^{13}\right)}{1 - \frac{1}{3}} \approx 728.9995$$

■

In evaluating S_n for a geometric series with common ratio $r \neq 1$, we sometimes must use the formula $a_n = a_1 r^{n-1}$ to find a_1, r, or n before we can use the formula $S_n = \dfrac{a_1(1 - r^n)}{1 - r}$.

Example 2 Evaluating a Sum

Evaluate the sum of the series $24{,}576 + 12{,}288 + 6144 + 3072 + 1536 + \cdots + 3$.

Solution

The sequence 24,576, 12,288, 6144, 3072, 1536, ..., 3 is geometric with common ratio $r = \dfrac{1}{2}$. First, we find the term number n of the last term, 3; then we find S_n. To find n, we substitute $a_1 = 24{,}576$, $a_n = 3$, and $r = \dfrac{1}{2}$ in the equation $a_n = a_1 r^{n-1}$

and solve for n:

$$3 = 24{,}576\left(\frac{1}{2}\right)^{n-1} \quad \text{Substitute 24,576 for } a_1\text{, 3 for } a_n\text{, and } \frac{1}{2} \text{ for } r.$$

$$\frac{3}{24{,}576} = \left(\frac{1}{2}\right)^{n-1} \quad \text{Divide both sides by 24,576.}$$

$$\log\left(\frac{3}{24{,}576}\right) = \log\left(\frac{1}{2}\right)^{n-1} \quad \text{Take the logarithm of both sides.}$$

$$\log\left(\frac{3}{24{,}576}\right) = (n-1)\log\left(\frac{1}{2}\right) \quad \text{Power property: } \log_b(x)^p = p\log_b(x)$$

$$\frac{\log\left(\frac{3}{24{,}576}\right)}{\log\left(\frac{1}{2}\right)} = n - 1 \quad \text{Divide both sides by } \log\left(\frac{1}{2}\right).$$

$$\frac{\log\left(\frac{3}{24{,}576}\right)}{\log\left(\frac{1}{2}\right)} + 1 = n \quad \text{Add 1 to both sides.}$$

$$14 = n \quad \text{Compute.}$$

Next, we substitute $a_1 = 24{,}576$, $r = \frac{1}{2}$, and $n = 14$ in the equation $S_n = \frac{a_1(1 - r^n)}{1 - r}$:

$$S_{14} = \frac{24{,}576\left(1 - \left(\frac{1}{2}\right)^{14}\right)}{1 - \frac{1}{2}} = 49{,}149$$

■

Example 3 Modeling with a Geometric Series

A person's salary is \$30,000 in the first year and increases by 4% at the end of each year.

1. Calculate the person's total earnings for the first 25 years.
2. Compare the result from Problem 1 with the result from Example 4 of Section 10.3, where we assumed that the person's salary increases by a constant \$1200 each year.

Solution

1. Let a_n be the person's salary (in dollars) in the nth year. Since the salary in each year is 104% of the salary in the previous year, the sequence $a_1, a_2, a_3, \ldots, a_n$ is geometric with common ratio 1.04. To find the total earnings, we substitute $a_1 = 30{,}000$, $r = 1.04$, and $n = 25$ in the equation $S_n = \frac{a_1(1 - r^n)}{1 - r}$:

$$S_{25} = \frac{30{,}000(1 - 1.04^{25})}{1 - 1.04} \approx 1{,}249{,}377.25$$

   So, the total earnings will be \$1,249,377.25.
2. First, note that 4% of \$30,000 is \$1200, so the first raise is the same in both scenarios. In Example 4 of Section 10.3, we found that if the person receives constant raises of \$1200, the total earnings will be \$1,110,000 in 25 years, which is about \$140,000 less than the total earnings of \$1,249,377.25 from earning 4% raises each year.

■

HOMEWORK 10.4

Evaluate the sum of the geometric series with the given values of a_1, r, and n. Round any approximate results to the fourth decimal place.

1. $a_1 = 5$, $r = 2$, and $n = 13$

2. $a_1 = 6$, $r = 3$, and $n = 9$

3. $a_1 = 6$, $r = 1.3$, and $n = 12$

4. $a_1 = 10$, $r = 1.5$, and $n = 15$

5. $a_1 = 13$, $r = 0.8$, and $n = 13$

6. $a_1 = 9$, $r = 0.7$, and $n = 12$

7. $a_1 = 2.3$, $r = 0.9$, and $n = 10$

8. $a_1 = 4$, $r = 0.6$, and $n = 11$

Find the sum of the series. Round any approximate results to the fourth decimal place.

9. $2 + 10 + 50 + 250 + 1250 + \cdots + a_{13}$

10. $1 + 2 + 4 + 8 + 16 + \cdots + a_{18}$

11. $600 + 180 + 54 + 16.2 + 4.86 + \cdots + a_{11}$

12. $625 + 500 + 400 + 320 + 256 + \cdots + a_{12}$

13. $3 + 2 + \frac{4}{3} + \frac{8}{9} + \frac{16}{27} + \cdots + a_{10}$

14. $10 + 6 + \frac{18}{5} + \frac{54}{25} + \frac{162}{125} + \cdots + a_{12}$

15. $1 + 4 + 16 + 64 + 256 + \cdots + 67{,}108{,}864$

16. $7 + 21 + 63 + 189 + 567 + \cdots + 33{,}480{,}783$

17. $5 + 6 + 7.2 + 8.64 + 10.368 + \cdots + 21.4990848$

18. $800 + 1120 + 1568 + 2195.2 + 3073.28 + \cdots + 11{,}806.312448$

19. $10{,}000 + 5000 + 2500 + 1250 + 625 + \cdots + 4.8828125$

20. $2500 + 2000 + 1600 + 1280 + 1024 + \cdots + 335.54432$

21. $S_{100} = 1 + 1 + 1 + 1 + 1 + \cdots + 1$

22. $3 + 30 + 300 + 3000 + 30{,}000 + \cdots + 3{,}000{,}000{,}000{,}000$

23. $324 + 108 + 36 + 12 + 4 + \cdots + \frac{4}{729}$

24. $80 + 40 + 20 + 10 + 5 + \cdots + \frac{5}{1024}$

*Let S_n be the sum of a geometric series. For the given conditions, determine whether S_n is positive or negative. Explain. [**Hint:** Try experimenting with specific values of a_1, r, and n that meet the conditions stated. Then explain why your response makes sense for any values that meet those conditions.]*

25. $a_1 > 0$, $r > 0$, and n is a counting number

26. $a_1 < 0$, $r > 0$, and n is a counting number

27. If $f(x) = 7 - x$, is the series

$$f(1) + f(2) + f(3) + \cdots + f(30)$$

arithmetic, geometric, or neither? Explain.

28. If $f(x) = 2(4)^x$, is the series

$$f(1) + f(2) + f(3) + \cdots + f(70)$$

arithmetic, geometric, or neither? Explain.

29. A person's starting salary is \$23,500. Each year, the salary increases by 4%. What will be the person's total earnings after 20 years of work?

30. A person's first-year salary is \$32,000. The salary increases by 3% each year. What will be the person's total earnings after 30 years of work?

31. Two companies make you job offers. Company A offers a first-year salary of \$26,000 and a 5% raise at the end of each year. Company B offers a first-year salary of \$31,000 and a 3% raise at the end of each year. At which company would your total earnings for 30 years be greater? By how much?

32. Two companies are bidding against each other to hire you. Company A offers a first-year salary of \$25,000, a 4% raise at the end of each year, and a \$500 bonus at the end of each year. (The 4% raise is based on the salary, not on the bonus.) Company B offers a first-year salary of \$30,000 and a 3% raise at the end of each year. At which company would your total earnings for 26 years be greater? By how much?

33. In Exercise 43 of Homework 10.2, you found the (greatest possible) number of ancestors a person has in the nth generation back. Find the total number of ancestors a person has through 10 generations back.

34. Suppose that a rumor is spreading in the United States that chlorine in swimming pools causes skin cancer. On the first day, 4 people hear the rumor. Approximately 24 hours after hearing the rumor, each person who hears it tells it to exactly 5 people who have not yet heard it.

a. How many people will have heard the rumor after 10 days?

b. After how many days will everyone in the United States have heard the rumor? Use the 2006 U.S. population of 299 million people.

c. To model the spread of the chlorine-causes-cancer rumor, you made some assumptions about the way the rumor would spread. Describe each assumption, and discuss whether you think it is reasonable.

35. An entrepreneur writes letters to eight people (the first round of letters), explaining that she has found a way for herself and many other people to get rich. On each letter, she has written her name and address. She asks each of the eight people to send her \$5 and to add their name and address below hers, so each letter will now have two names on it. The entrepreneur also instructs each of the eight people to send the list of two names with the instructions to eight more people (the second round). Then all these people should send her \$5, add their names and addresses to the list, send the list of three names to eight more people (the third round), and so on. Each person who receives a letter is instructed to send \$5 to the name at the top of the list. When there are 10 people on the list, the next person should send \$5 to the name at the top of the list, scratch that name off the list, and add his or her name to the bottom of the list. The instructions include a warning that something terrible will happen to those who do not send the money as well as the eight letters. (These letters are called *chain letters* and are illegal.)

Assume that the letters of any one round are received at approximately the same time and that no one receives more than one letter.

a. In which round would the entrepreneur's name be taken off the list? How much money could she receive?

b. By which round would everyone in the world (about 6.5 billion people) have received a letter?

c. How many people will receive money from the chain letters? How much will each of them receive? [**Hint:** With 6.5 billion people, there would be only $32.5 billion to go around.]

36. Suppose you win a contest and choose between two award plans. If you choose plan A, you will receive $100,000 per day for 30 days. If you choose plan B, you will receive 1 cent the first day, 2 cents the second day, 4 cents the third day, and so on (each day you receive twice as much as you did on the preceding day) for 30 days. Which plan would you choose? Explain your reasoning.

37. In Exercise 45 of Homework 10.2, you found a model $f(t) = 865.47(1.39)^t$, where $f(t)$ is the number of girls named Nevaeh in the year that is t years since 2000 (see Table 8).

Table 8 Numbers of Nevaehs Born

Year	Number of Nevaehs Born
2001	1191
2002	1692
2003	2287
2004	3156
2005	4457

Source: *Social Security Administration*

a. Find $f(1)$. What does it mean in this situation?

b. Find $f(11)$. What does it mean in this situation?

c. Find $f(1)+f(2)+f(3)+\cdots+f(11)$. What does it mean in this situation?

38. Prior to cassettes, popular recordings were sold on *eight-track* tape cartridges. In 1980, 89.5 million eight-track cartridges were sold. After that year, sales dropped off sharply due to consumers' preference for cassettes over eight-tracks. (Sound recordings are no longer made on eight-tracks.) The numbers of eight-track cartridges (in millions) sold are listed in Table 9 for various years.

Table 9 Eight-track Cartridge Sales

Year	Sales (millions)
1980	89.5
1981	32.0
1982	20.0
1983	10.0
1984	5.0
1985	1.5

Source: *Recording Industry Association of America*

a. Let $f(t)$ be the number of eight-track cartridges (in millions) sold in the year that is t years since 1980. Find an exponential equation of f.

b. Find $f(0)$. What does it mean in this situation?

c. Use the formula for the sum of a geometric series to find

$$f(0) + f(1) + f(2) + f(3) + f(4) + f(5)$$

[**Hint:** Think carefully about the value of n.]

d. Compare your result from part (c) with the actual total.

e. Predict the total number of eight-track cartridges that would have been sold from 1980 through 2010. Explain why this total is not much more than your total found in part (c).

39. a. Find the sum

$$5 + 10 + 20 + 40 + 80 + \cdots + 2560$$

b. Solve the equation

$$a_n = a_1 r^{n-1}$$

for n.

c. Use the equation you found in part (b) to help you make a substitution for n in the equation

$$S_n = \frac{a_1\left(1 - r^n\right)}{1 - r}$$

d. Use the equation you found in part (c) to find the sum

$$5 + 10 + 20 + 40 + 80 + \cdots + 2560$$

e. Compare the methods you used in parts (a) and (d) to find the sum

$$5 + 10 + 20 + 40 + 80 + \cdots + 2560$$

Which method do you prefer? Explain.

40. Describe a geometric series. Also, explain how to evaluate the sum of a geometric series

$$S_n = a_1 + a_2 + a_3 + \cdots + a_n$$

if you know a_1, a_n, and the common ratio r of the geometric sequence

$$a_1, a_2, a_3, \ldots, a_n$$

Related Review

Find the sum of the series. Round any approximate results to the fourth decimal place.

41. $3 + 9 + 15 + 21 + 27 + \cdots + 351$

42. $351 + 347 + 343 + 339 + 335 + \cdots + 103$

43. $10 + 9 + 8.1 + 7.29 + 6.561 + \cdots + 3.486784401$

44. $7 + 28 + 112 + 448 + 1792 + \cdots + 469{,}762{,}048$

Expressions, Equations, Functions, and Graphs

Give an example of the following. Then solve, simplify, or graph, as appropriate.

45. quadratic function

46. square root function

47. quotient of two rational expressions

48. linear function

49. quadratic equation in one variable

50. product of two radical expressions

51. exponential function

52. rational equation in one variable

53. system of two linear equations in two variables

54. exponential equation in one variable

Chapter Summary

Key Points OF CHAPTER 10

Sequence (Section 10.1)	Any ordered list of numbers is called a **sequence.** Each number is a **term** of the sequence.
Arithmetic sequence (Section 10.1)	If the difference between any term of a sequence and the preceding term is a constant d for every such pair of terms, then the sequence is an **arithmetic sequence.** We call the constant d the **common difference.**
Formula that describes the nth term of an arithmetic sequence (Section 10.1)	If an arithmetic sequence $a_1, a_2, a_3, \ldots, a_n, \ldots$ has the common difference d, then $a_n = a_1 + (n-1)d$.
Check that a sequence is arithmetic (Section 10.1)	As the formula $a_n = a_1 + (n-1)d$ is valid for arithmetic sequences only, we must first check that a sequence is arithmetic before we use the formula.
Connection between a linear function and an arithmetic sequence (Section 10.1)	If f is a linear function of the form $f(x) = mx+b$, then $f(1), f(2), f(3), \ldots$ is an arithmetic sequence with common difference equal to the slope m.
Geometric sequence (Section 10.2)	If the ratio of any term of a sequence to the preceding term is a constant r for every such pair of terms, then the sequence is a **geometric sequence.** We call the constant r the **common ratio.**
Formula that describes the nth term of a geometric sequence (Section 10.2)	If a geometric sequence $a_1, a_2, a_3, \ldots, a_n, \ldots$ has the common ratio r, then $a_n = a_1 r^{n-1}$.
Check that a sequence is geometric (Section 10.2)	In working with a sequence, we must first determine whether it is arithmetic, geometric, or neither. An arithmetic sequence has a common difference, whereas a geometric sequence has a common ratio.
Connection between an exponential function and a geometric sequence (Section 10.2)	If f is an exponential function of the form $f(x) = ab^x$, then $f(1), f(2), f(3), \ldots$ is a geometric sequence with common ratio equal to the base b of f.
Arithmetic series (Section 10.3)	If the sequence $a_1, a_2, a_3, \ldots, a_n$ is an arithmetic sequence, then the sum $a_1 + a_2 + a_3 + \cdots + a_n$ is an **arithmetic series.** We say that a_i is the **ith term** of the series and that its **term number** is i.
Formula for sum of an arithmetic series (Section 10.3)	If $S_n = a_1 + a_2 + a_3 + \cdots + a_n$ is an arithmetic series, then $S_n = \dfrac{n(a_1 + a_n)}{2}$.
Check that a series is arithmetic (Section 10.3)	To evaluate the sum $S_n = a_1 + a_2 + a_3 + \cdots + a_n$, we must first check that the sequence $a_1, a_2, a_3, \ldots, a_n$ is arithmetic before we use the formula $S_n = \dfrac{n(a_1 + a_n)}{2}$.
Using a combination of formulas (Section 10.3)	In evaluating S_n for an arithmetic series, we sometimes must use the formula $a_n = a_1 + (n-1)d$ to find n, a_1, or a_n before we can use the formula $S_n = \dfrac{n(a_1 + a_n)}{2}$.
Geometric series (Section 10.4)	If the sequence $a_1, a_2, a_3, \ldots, a_n$ is a geometric sequence, then the sum $a_1 + a_2 + a_3 + \cdots + a_n$ is a **geometric series.** We say that a_i is the **ith term** of the series and that its **term number** is i.
Formula for sum of a geometric series (Section 10.4)	If $S_n = a_1 + a_2 + a_3 + \cdots + a_n$ is a geometric series with common ratio $r \neq 1$, then $S_n = \dfrac{a_1(1 - r^n)}{1 - r}$.
Check that a series is geometric (Section 10.4)	To find the sum $S_n = a_1 + a_2 + a_3 + \cdots + a_n$, we must first determine whether the series is arithmetic, geometric, or neither. If the series is arithmetic, we use $S_n = \dfrac{n(a_1 + a_n)}{2}$. If the series is geometric with common ratio $r \neq 1$, we use $S_n = \dfrac{a_1(1 - r^n)}{1 - r}$.
Using a combination of formulas (Section 10.4)	In evaluating S_n for a geometric series with common ratio $r \neq 1$, we sometimes must use the formula $a_n = a_1 r^{n-1}$ to find a_1, r, or n before we can use the formula $S_n = \dfrac{a_1(1 - r^n)}{1 - r}$.

CHAPTER 10 REVIEW EXERCISES

Determine whether the following is an arithmetic sequence, an arithmetic series, a geometric sequence, a geometric series, or none of these types of sequences or series.

1. $160, 40, 10, 2.5, 0.625, \ldots$

2. $13+24+35+46+57+\cdots+299$

3. $101, 95, 89, 83, 77, \ldots$

4. $7+\frac{7}{5}+\frac{7}{25}+\frac{7}{125}+\frac{7}{625}+\cdots+\frac{7}{390{,}625}$

Using a_n notation, find a formula of the sequence.

5. $2, 6, 18, 54, 162, \ldots$

6. $9, 4, -1, -6, -11, \ldots$

7. $200, 100, 50, 25, 12.5, \ldots$

8. $3.2, 5.9, 8.6, 11.3, 14, \ldots$

Find the indicated term of the sequence. Find the exact value, or write the result in scientific notation $N \times 10^k$ with N rounded to the fourth decimal place.

9. 47th term of $6, 12, 24, 48, 96, \ldots$

10. 9th term of $768, 192, 48, 12, 3, \ldots$

11. 98th term of $87, 84, 81, 78, 75, \ldots$

12. 87th term of $2.3, 4.9, 7.5, 10.1, 12.7, \ldots$

Find the term number of the last term in the finite sequence.

13. $7, 11, 15, 19, 23, \ldots, 2023$

14. $501, 493, 485, 477, 469, \ldots, -107$

15. The number 470,715,894,135 is a term in the sequence $5, 15, 45, 135, 405, \ldots$. What is its term number?

16. If $a_5 = 52$ and $a_9 = 36$ are terms of an arithmetic sequence, find a_{69}.

17. Find the sum of the first 43 terms of an arithmetic series with $a_1 = 52$ and $a_{43} = -200$.

18. Find the sum of the first 22 terms of a geometric series with $a_1 = 4$, $r = 1.7$, and $n = 22$. Round your result to the fourth decimal place.

Evaluate the sum of the series.

19. $3+6+12+24+48+\cdots+1{,}610{,}612{,}736$

20. $30+36+42+48+54+\cdots+1200$

21. $11+7+3+(-1)+(-5)+\cdots+a_{33}$

22. $531{,}441+177{,}147+59{,}049+19{,}683+6561+\cdots+a_{13}$

23. If $f(x)=4(5)^x$, is $f(1)+f(2)+f(3)+\cdots+f(80)$ an arithmetic sequence, an arithmetic series, a geometric sequence, or a geometric series? Explain.

24. If $f(x) = -9x+40$, is $f(1), f(2), f(3), \ldots, f(80)$ an arithmetic sequence, an arithmetic series, a geometric sequence, or a geometric series? Explain.

25. Two companies have made you job offers. Company A offers a first-year salary of \$28,000 with a 4% raise at the end of each year. Company B offers a first-year salary of \$34,000 with a constant raise of \$1500 each year.

a. What would be the salary in the 25th year at company A? at company B?

b. What would be the total earnings for 25 years of work at company A? at company B?

c. Explain how it is possible for the salary in the 25th year to be greater at company A than at company B, yet the total earnings for 25 years to be greater at company B than at company A.

26. The numbers of traffic deaths at intersections are listed in Table 10 for various years.

Table 10 Deaths at Intersections

Year	Deaths (thousands)
1992	6.2
1994	6.7
1996	7.3
1998	8.0
2000	8.5
2003	9.2

Source: *Federal Highway Administration*

a. Let $f(t)$ be the number of deaths (in thousands) at intersections during the year that is t years since 1990. Assuming that f is a linear function, find an equation of f.

b. Use f to predict the number of deaths at intersections in 2010.

c. Use f to estimate the total number of deaths at intersections from 1990 through 2010.

d. Many communities now use cameras to identify cars that run red lights. If the use of such cameras increases, do you think your results to parts (b) and (c) will be underestimates or overestimates?

CHAPTER 10 TEST

Determine whether the following is an arithmetic sequence, an arithmetic series, a geometric sequence, a geometric series, or none of these types of sequences or series.

1. $3, 6, 12, 24, 48, \ldots$

2. $20, 19, 17, 14, 10, \ldots$

3. $7+35+175+875+4375+\cdots+546{,}875$

4. $69+61+53+45+37+\cdots+5$

Using a_n notation, find a formula of the sequence.

5. $31, 25, 19, 13, 7, \ldots$

6. $6, 24, 96, 384, 1536, \ldots$

7. Find the 87th term of the sequence $4, 7, 10, 13, 16, \ldots$.

8. Find the 16th term of the sequence $6144, 3072, 1536, 768, 384, \ldots$.

Find the term number of the last term of the finite sequence.

9. $-27, -23, -19, -15, -11, \ldots, 1789$

10. $200, 220, 242, 266.2, 292.82, \ldots, 428.717762$

Evaluate the sum of the series. Round any approximate results to the fourth decimal place, or write such results in scientific notation $N \times 10^k$ with N rounded to the fourth decimal place.

11. $27+9+3+1+\frac{1}{3}+\cdots+a_{20}$

12. $4 + 8 + 16 + 32 + 64 + \cdots + 2{,}147{,}483{,}648$

13. $50 + 46 + 42 + 38 + 34 + \cdots + (-78)$

14. $19 + 33 + 47 + 61 + 75 + \cdots + a_{400}$

15. Evaluate the sum of the series. [**Hint:** Begin by writing the series as a sum of two series.]

$$(7 + 2) + (7 \cdot 2 + 2^2) + (7 \cdot 3 + 2^3) + (7 \cdot 4 + 2^4) + (7 \cdot 5 + 2^5) + \cdots + (7 \cdot 20 + 2^{20})$$

16. Let $f(x) = 3x^2 + 1$. Is $f(1) + f(2) + f(3) + \cdots + f(100)$ an arithmetic sequence, an arithmetic series, a geometric sequence, a geometric series, or none of these types of sequences or series? Explain.

17. Let S_n be the sum of an arithmetic series. Determine whether S_n is positive or negative if $a_1 = 10$, $d = -3$, and n is a very large counting number. Explain.

18. Online retail sales are shown in Table 11 for various years. Let $f(t)$ be the online retail sales (in billions of dollars) in the year that is t years since 2000.

a. Use a graphing calculator to draw a scattergram to describe the data.

Table 11 Online Retail Sales

Year	Online Retail Sales (billions of dollars)
2001	40
2002	78
2003	121
2004	136
2005	186

Source: *National Retail Federation*

b. Find a linear equation of f.

c. Find $f(1)$. What does it mean in this situation?

d. Find $f(15)$. What does it mean in this situation?

e. Find $f(1) + f(2) + f(3) + \cdots + f(15)$. What does it mean in this situation?

19. Assume that a person's salary is $32,000 for the first year and increases by 3% each year.

a. Let a_n be the person's salary (in thousands of dollars) in the nth year. Find a formula that describes a_n.

b. When will the salary first exceed $40,000?

c. What will the salary be in the 25th year?

d. What will the total earnings be for the first 25 years?

CUMULATIVE REVIEW OF CHAPTERS 1–10

Solve. Any solution is a real number.

1. $6x^2 + 13x = 5$

2. $\log_3(4x - 7) = 4$

3. $(t + 3)(t - 4) = 5$

4. $\dfrac{1}{w^2 - w - 6} - \dfrac{w}{w + 2} = \dfrac{w - 2}{w - 3}$

5. $5(3x - 2)^2 + 7 = 17$

6. $\log_6(3x) + \log_6(x - 1) = 1$

7. $20 - 4x = 7(2x + 9)$

8. $\sqrt{x + 1} - \sqrt{2x - 5} = 1$

Solve. Any solution is a real number. Round any results to the fourth decimal place.

9. $2b^7 - 3 = 51$

10. $6(3)^x - 5 = 52$

11. $5e^x = 98$

12. Solve $3x^2 - 5x + 1 = 0$ by completing the square.

13. Find all complex-number solutions of $2x^2 = 4x - 3$.

Solve the system.

14. $2x + 4y = 0$
$5x + 3y = 7$

15. $y = 3x + 9$
$4x + 2y = -2$

16. $2x - 3y + 4z = 19$
$5x + y - 5z = -6$
$3x - y + 2z = 13$

17. Solve the inequality $5 - 2(3x - 5) + 1 \geq 2 - 4x$. Describe the solution set as an inequality, in interval notation, and in a graph.

Simplify. Assume that any variable is positive.

18. $(3b^{-2}c^{-3})^4(6b^{-5}c^2)^2$

19. $\dfrac{8b^{1/2}c^{-4/3}}{10b^{3/4}c^{-7/3}}$

20. $3y\sqrt{8x^3} - 2x\sqrt{18xy^2}$

21. $\sqrt{12x^7y^{14}}$

22. $\sqrt[3]{\dfrac{4}{x}}$

23. $\dfrac{3\sqrt{x} - \sqrt{y}}{2\sqrt{x} + \sqrt{y}}$

Simplify. Write your result as a single logarithm with a coefficient of 1.

24. $2\ln(x^4) + 3\ln(x^9)$

25. $4\log_b(x^5) - 5\log_b(2x)$

Perform the indicated operation.

26. $(3a - 5b)^2$

27. $(3\sqrt{k} - 4)(2\sqrt{k} + 7)$

28. $(2x^2 - x + 3)(x^2 + 2x - 1)$

Perform the indicated operation.

29. $\dfrac{x^3 - 27}{2x^2 - 3x + 1} \div \dfrac{2x^3 + 6x^2 + 18x}{4x^2 - 1}$

30. $\dfrac{3x}{x^2 - 10x + 25} - \dfrac{x + 2}{x^2 - 7x + 10}$

31. $\dfrac{4x - x^2}{6x^2 + 10x - 4} \cdot \dfrac{7 - 21x}{x^2 - 8x + 16}$

32. $\dfrac{1}{x^2 + 12x + 27} + \dfrac{x + 2}{x^3 + x^2 - 9x - 9}$

33. Simplify $\dfrac{\dfrac{x + 2}{x^2 - 64}}{\dfrac{x^2 + 4x + 4}{3x + 24}}$.

34. Write $f(x) = -3(x + 3)^2 - 7$ in standard form.

Factor.

35. $4x^3 - 8x^2 - 25x + 50$

36. $2x^3 - 4x^2 - 30x$

37. $6w^2 + 2wy - 20y^2$

38. $100p^2 - 1$

For Exercises 39–43, refer to Fig. 8.

39. Find $f(2)$.

40. Find x when $f(x) = 3$.

41. Find an equation of f.

42. Find the domain of f.

43. Find the range of f.

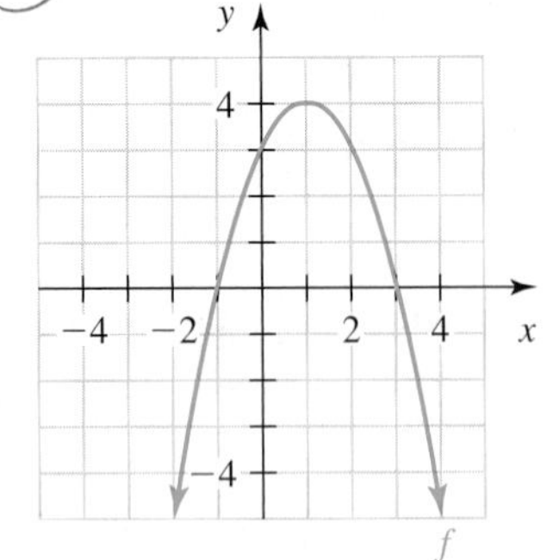

Figure 8 Exercises 39–43

Graph the equation by hand.

44. $y = -3(x - 4)^2 + 3$

45. $y = 2\sqrt{x + 5} - 4$

46. $y = 15\left(\frac{1}{3}\right)^x$

47. $y = 2x^2 + 5x - 1$

48. $2x(x - 3) + y = 5(x + 1)$

49. Find an equation of the line that contains the points $(-3, 2)$ and $(2, -5)$.

50. Find an approximate equation $y = ab^x$ of an exponential curve that contains the points $(3, 95)$ and $(6, 12)$. Round the values of a and b to the second decimal place.

51. Find an equation of a parabola that contains the points $(2, 1)$, $(3, 6)$, and $(4, 15)$.

52. Find an approximate equation of a square root curve $y = a\sqrt{x} + b$ that contains the points $(2, 5)$ and $(6, 17)$. Round the values of a and b to the second decimal place.

53. Let f be the linear function, g be the exponential function, and h be a quadratic function whose graphs contain the points $(0, 2)$ and $(1, 4)$.
a. Find possible equations of f, g, and h.
b. Use a graphing calculator to draw the graphs of f, g, and h in the same viewing window.

Find the logarithm.

54. $\log_3(81)$

55. $\log_b(\sqrt{b})$

Find the inverse of the function.

56. $g(x) = \log_2(x)$

57. $f(x) = -4x - 7$

58. Find the domain of $f(x) = \dfrac{x - 3}{x^2 - 2x - 35}$.

59. Find the 10th term of the sequence $2, 8, 32, 128, 512, \ldots$.

60. Find the term number of the last term of the finite sequence $-86, -82, -78, -74, -70, \ldots, 170$.

Find the sum of the series.

61. $98{,}304 + 49{,}152 + 24{,}576 + 12{,}288 + 6144 + \cdots + 3$

62. $11 + 14 + 17 + 20 + 23 + \cdots + 182$

63. A chemist wants to mix a 15% acid solution and a 30% acid solution to make a 25% acid solution. How many liters of each solution must be mixed to make 6 liters of the 25% solution?

64. The percentages of passenger vehicles sold in the United States that are light trucks are shown in Table 12 for various years.

Table 12 Percentages of Vehicles Sold That Are Light Trucks

Year	Percent
1994	40.3
1996	43.5
1998	47.6
2000	49.0
2002	51.8
2004	55.5

Source: *NADA Industry Analysis Division*

Let $p = f(t)$ be the percentage of passenger vehicles sold that are light trucks in the year that is t years since 1990.
a. Find an equation of f.
b. Find an equation of f^{-1}.
c. Find $f(20)$. What does it mean in this situation?
d. Find $f^{-1}(60)$. What does it mean in this situation?
e. Find the slope of the graph of f. What does it mean in this situation?

65. In 2005, the two most populous nations were China and India, with populations of 1.306 billion and 1.003 billion, respectively. India is expected to surpass China as the most populous nation within the next 50 years. The population of India in 1980 was 0.687 billion.
a. First, assume that India's population is growing linearly. Let $L(t)$ be India's population (in billions) at t years since 1980. Find an equation of L.
b. Now assume that India's population is growing exponentially. Let $E(t)$ be India's population (in billions) at t years since 1980. Find an equation of E.
c. Find $L(70)$ and $E(70)$. What do they mean in terms of India's population?
d. Find $E(70) - L(70)$. What does this result mean in terms of India's population? To get an idea of the size of your result, compare it with 0.420, a prediction of the U.S. population (in billions) in 2050 (Source: *U.S. Census Bureau*).
e. The U.S. Census Bureau predicts that China's population will reach 1.424 billion in 2050. Use first L and then E to predict when India's population will reach that level. Compare your results.

66. The numbers of consumer complaints about improper debt collection practices in the United States are shown in Table 13 for various years. Some collectors have misrepresented the size of debts, made abusive phone calls at all hours, called relatives and employers, or failed to investigate consumers' claims that a debt is paid. Let $f(t)$ be the number (in thousands) of consumer complaints about improper debt collection practices in the year that is t years since 1990.
a. Find an exponential equation of f and a quadratic equation of f. Compare how well the two models fit the data.
b. If the number of complaints has generally increased, which of the two models better describes the situation for years before 1999?
c. Use the exponential model to predict in which year the number of consumer complaints about debt collection will reach 300 thousand complaints.

Table 13 Numbers of Consumer Complaints About Improper Debt Collection Practices

Year	Number of Complaints (thousands)
1999	11
2000	13
2001	16
2002	24
2003	34
2004	59
2005	67

Source: *Federal Trade Commission*

d. Use the quadratic model to predict in which year the number of consumer complaints about debt collection will reach 300 thousand complaints.

e. Explain why the year you predicted in part (d) is later than the year you predicted in part (c).

67. Sales of books and total recreational expenditures in the United States are shown in Table 14 for various years.

a. Let $B(t)$ be the total book sales (in billions of dollars) in the year that is t years since 1980. Find a quadratic equation of B.

b. Let $R(t)$ be the total recreational expenditures (in billions of dollars) in the year that is t years since 1980. Find a linear equation of R.

c. Let $P(t)$ be the percentage of total recreational expenditures that consist of book sales in the year that is t years since 1980. Find an equation of P.

Table 14 Sales of Books and Total Recreational Expenditures

Books		All Forms of Recreation	
Year	Sales (in billions of dollars)	Year	Sales (in billions of dollars)
1982	9.9	1993	340.1
1985	12.6	1996	429.6
1990	19.0	1999	546.1
1995	25.2	2002	628.3
2000	34.5	2003	660.7
2003	39.2		

Sources: *Book Industry Study Group; U.S. Bureau of Economic Analysis*

d. Use the window settings in Fig. 9 to graph P. Then use "minimum" on a graphing calculator to find the minimum point of the graph of P for values of t between 13 and 30. What does that point mean in this situation?

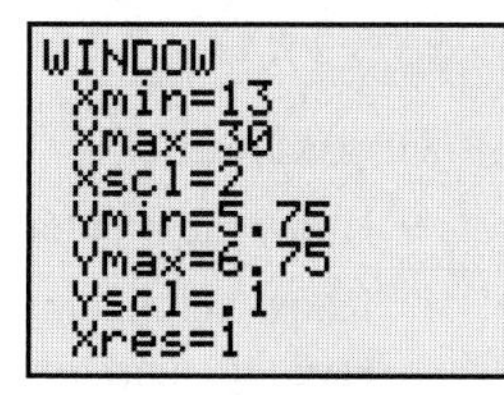

Figure 9 Window settings for Exercise 67d

e. Predict when 6% of recreational expenditures will consist of book sales.

Chapter 11

Additional Topics

"Mathematics abounds in bright ideas. No matter how long and hard one pursues her, mathematics never seems to run out of exciting surprises. And by no means are these gems to be found only in difficult work at an advanced level. All kinds of simple notions are full of ingenuity."

—*Ross Honsberger,* Mathematical Morsels, *Washington, DC: Mathematical Association of America, 1978, p. vii*

11.1 ABSOLUTE VALUE: EQUATIONS AND INEQUALITIES

Objectives

- Know the meaning of *absolute value* in terms of the number line.
- Know the meaning of *absolute value function, absolute value equation in one variable,* and *absolute value inequality in one variable.*
- Solve absolute value equations and absolute value inequalities.

In this section, we will work with *absolute value functions*. We will also solve equations and inequalities that have absolute values.

Absolute Value

We begin by defining the absolute value of a number.

DEFINITION Absolute value

The **absolute value** of a number a, written $|a|$, is the distance from a to 0 on the number line.

For example, $|-6| = 6$, because -6 is a distance of 6 units from 0 (see Fig. 1). Also, $|6| = 6$, because 6 is a distance of 6 units from 0 (see Fig. 1).

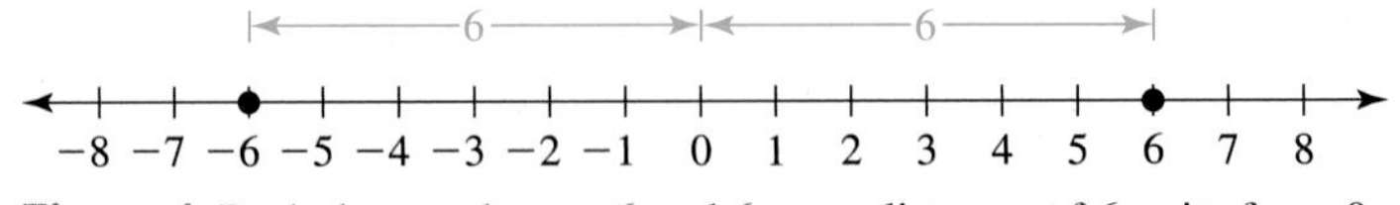

Figure 1 Both the numbers -6 and 6 are a distance of 6 units from 0

Two more examples are $|-3.7| = 3.7$ and $|95| = 95$. Since the absolute value of a number is the *distance* from that number to 0, **the absolute value of any number is nonnegative.**

Absolute Value Function

An **absolute value function** is a function whose equation contains the absolute value of a variable expression. Here are some examples of absolute value functions:

$$f(x) = |x| \qquad g(x) = |5x - 3| + 4 \qquad h(x) = \left|\frac{3x - 7}{2}\right|$$

Table 1 Input-Output Pairs of $f(x) = |x|$

x	$f(x)$
-3	3
-2	2
-1	1
0	0
1	1
2	2
3	3

Example 1 Graphing an Absolute Value Function

Graph $f(x) = |x|$.

Solution

First, we list some input–output pairs of f in Table 1. Then we plot the corresponding points and sketch a "V"-shaped curve through them (see Fig. 2).

We use ZStandard followed by ZSquare to verify our graph (see Fig. 3). To enter $|x|$, press MATH ▷ **1** X,T,Θ,n).

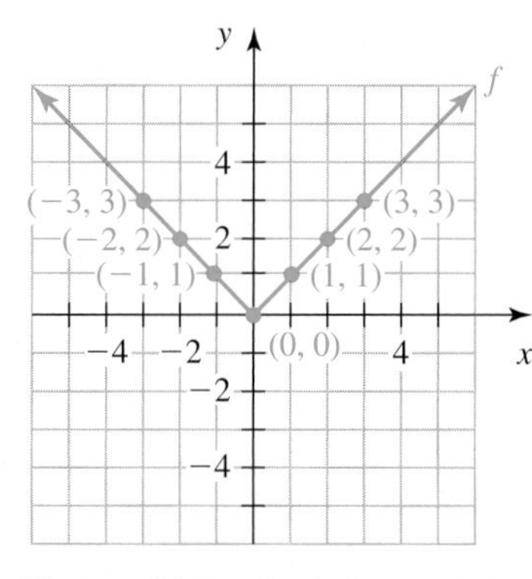

Figure 2 Graph of $f(x) = |x|$

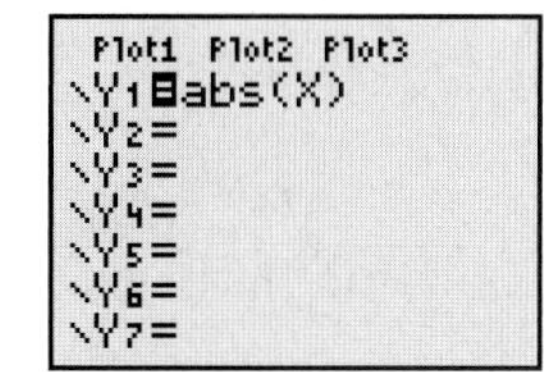

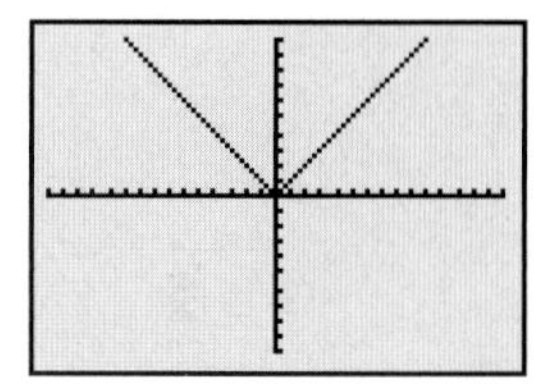

Figure 3 Verify the work

■

Solving Absolute Value Equations in One Variable

We will now solve some equations in *one* variable that contain absolute values. An **absolute value equation in one variable** is an equation in one variable that contains the absolute value of a variable expression. Here are some examples:

$$|x| = 3 \qquad |3x - 7| + 8 = 2 \qquad |4x - 5| = |3 - x| \qquad \left|\frac{2x}{3} - \frac{1}{2}\right| = \frac{5}{6}$$

To solve the absolute value equation $|x| = 4$, we must determine all numbers that are a distance of 4 units from 0. There are two such numbers: -4 and 4. So, the statements $|x| = 4$ and $x = \pm 4$ are equivalent statements.

Absolute Value Property for Equations

For an expression A and a positive constant k, the equation $|A| = k$ is equivalent to

$$A = -k \text{ or } A = k$$

Example 2 Solving an Absolute Value Equation

Solve $|2x + 1| = 11$.

Solution

In $|2x + 1| = 11$, the expression $2x + 1$ represents numbers that are a distance of 11 from 0. These are the numbers -11 and 11:

$$2x + 1 = -11 \quad \text{or} \quad 2x + 1 = 11 \qquad \text{Absolute value property for equations}$$
$$2x = -12 \quad \text{or} \quad 2x = 10$$
$$x = -6 \quad \text{or} \quad x = 5$$

We check that both -6 and 5 satisfy the original equation:

Check $x = -6$	***Check*** $x = 5$
$\lvert 2x + 1\rvert = 11$	$\lvert 2x + 1\rvert = 11$
$\lvert 2(-6) + 1\rvert \stackrel{?}{=} 11$	$\lvert 2(5) + 1\rvert \stackrel{?}{=} 11$
$\lvert -11\rvert \stackrel{?}{=} 11$	$\lvert 11\rvert \stackrel{?}{=} 11$
$11 \stackrel{?}{=} 11$	$11 \stackrel{?}{=} 11$
true	true

■

Example 3 Solving an Absolute Value Equation

Solve $2|x| - 3 = 5$.

Solution

We isolate $|x|$ on one side of the equation; then we use the absolute value property for equations:

$$2|x| - 3 = 5 \quad \text{Original equation}$$
$$2|x| = 8 \quad \text{Add 3 to both sides.}$$
$$|x| = 4 \quad \text{Divide both sides by 2.}$$
$$x = \pm 4 \quad \text{Absolute value property for equations}$$

We use "intersect" on a graphing calculator to verify our work (see Fig. 4).

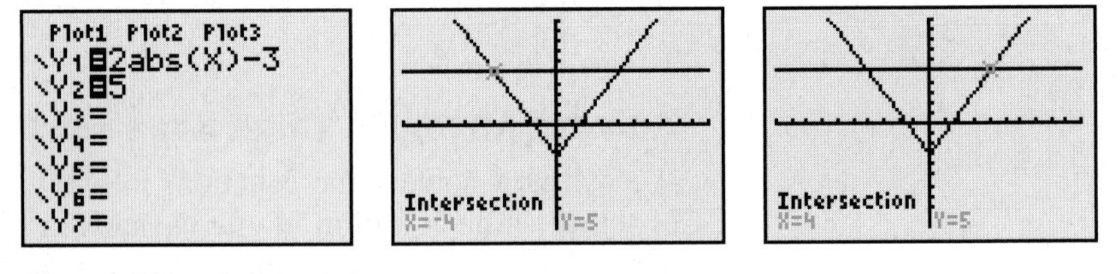

Figure 4 Verify the work

■

Example 4 Solving Absolute Value Equations

Solve.

1. $|4p + 12| = 0$
2. $|2x - 8| = -3$

Solution

1. In $|4p + 12| = 0$, the expression $4p + 12$ represents the one number that is a distance of 0 units from 0. This is the number 0:

$$4p + 12 = 0$$
$$4p = -12$$
$$p = -3$$

2. Since $|2x - 8|$ is nonnegative, the solution set of $|2x - 8| = -3$ is the empty set.

   The graphical check in Fig. 5 shows that the graphs of $y = |2x - 8|$ and $y = -3$ do not intersect. This means that the solution set of the equation $|2x - 8| = -3$ is the empty set, which checks. ■

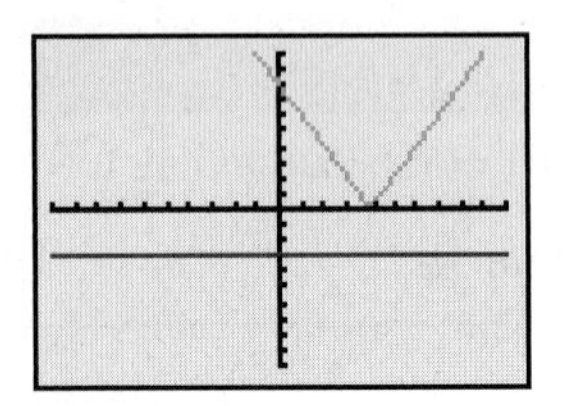

Figure 5 Verify the work

If $|a| = |b|$, what can we say about the numbers a and b? Since the absolute value of a number is its distance from 0, this means that a and b are the same distance from 0 on the number line. So, the numbers must be opposites of each other ($a = -b$) or equal to each other ($a = b$).

Solving an Equation of the Form $|A| = |B|$

For expressions A and B, the equation $|A| = |B|$ is equivalent to

$$A = -B \text{ or } A = B$$

For example, $|x| = |3|$ is equivalent to $x = -3$ or $x = 3$. This makes sense, because the only solutions of $|x| = 3$ are ± 3.

Example 5 Solving an Absolute Value Equation

Solve $|6x - 2| = |4x + 5|$.

Solution

$\|6x - 2\| = \|4x + 5\|$			Original equation
$6x - 2 = -(4x + 5)$	or	$6x - 2 = 4x + 5$	If $\|A\| = \|B\|$, then $A = -B$ or $A = B$.
$6x - 2 = -4x - 5$	or	$6x - 2 = 4x + 5$	Simplify right-hand side of equation./
$10x - 2 = -5$	or	$2x - 2 = 5$	Add 4x to both sides./Subtract 4x from both sides.
$10x = -3$	or	$2x = 7$	
$x = -\dfrac{3}{10}$	or	$x = \dfrac{7}{2}$	

We use a graphing calculator table to verify that $-\dfrac{3}{10} = -0.3$ and $\dfrac{7}{2} = 3.5$ are solutions (see Fig. 6). ■

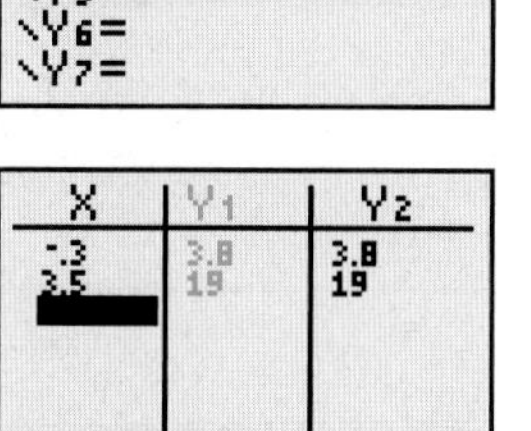
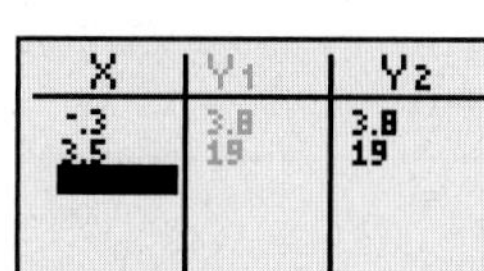

Figure 6 Verify the work

Example 6 Using Graphing to Solve an Equation in One Variable

Use graphing to solve $8 - |x + 3| = 3|x + 2| - 6$.

Solution

We use "intersect" on a graphing calculator to find the solutions of the system

$$y = 8 - |x + 3|$$
$$y = 3|x + 2| - 6$$

See Fig. 7.

The solutions of the system are $(-5.75, 5.25)$ and $(1.25, 3.75)$. The x-coordinates of these ordered pairs, -5.75 and 1.25, are the solutions of $8 - |x + 3| = 3|x + 2| - 6$. ■

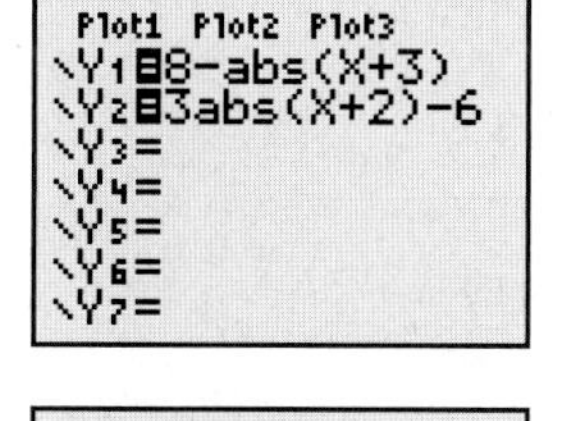
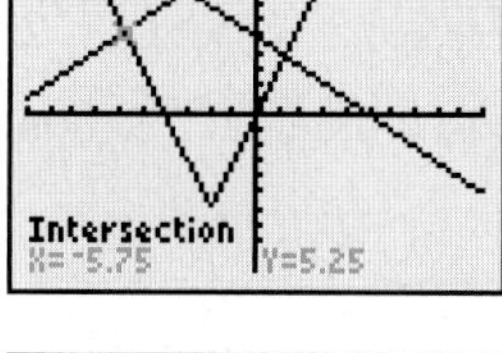

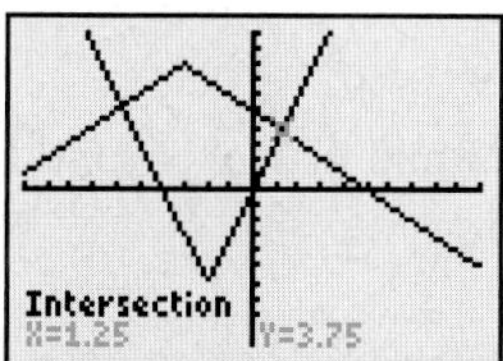

Figure 7 Solve the system

Solving Absolute Value Inequalities in One Variable

We now turn our attention from solving *equations* to solving *inequalities*. An **absolute value inequality in one variable** is an inequality in one variable that contains the absolute value of a variable expression. Here are some absolute value inequalities in one variable:

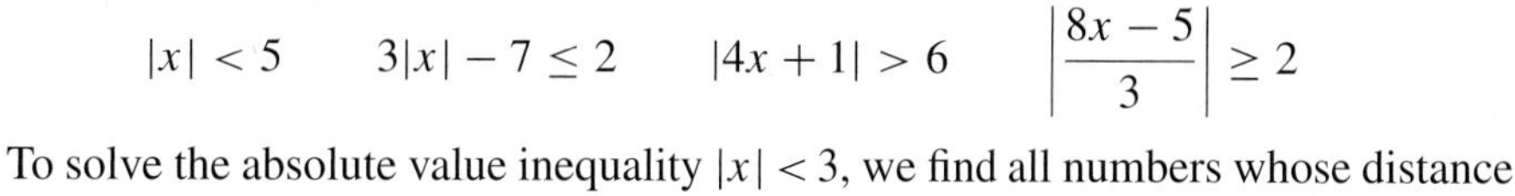

$$|x| < 5 \qquad 3|x| - 7 \le 2 \qquad |4x + 1| > 6 \qquad \left|\frac{8x - 5}{3}\right| \ge 2$$

To solve the absolute value inequality $|x| < 3$, we find all numbers whose distance from 0 is less than 3 units. So, the solutions of $|x| < 3$ are all the numbers between -3 and 3 (see Fig. 8). The solution set is the set of numbers x where $-3 < x < 3$. In interval notation (Section 3.5), the solution set is $(-3, 3)$.

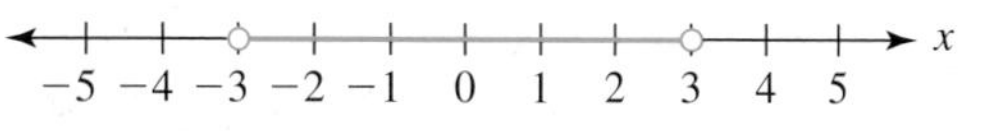

Figure 8 Graph of numbers whose distance from 0 is less than 3 units

The solution set of an absolute value inequality is sometimes best described as the union of two sets. If A and B are sets, then **the union of A and B,** denoted $A \cup B$, is the set of all members of A together with all members of B. So, $(-\infty, 1) \cup (3, \infty)$ is the set of numbers less than 1 together with numbers greater than 3.

To solve $|x| > 3$, we find all numbers whose distance from 0 is more than 3 units. So, the solutions of $|x| > 3$ are all numbers that are either less than -3 *or* greater than 3 (see Fig. 9). The solution set is the set of numbers x where $x < -3$ or $x > 3$, or, in interval notation, $(-\infty, -3) \cup (3, \infty)$.

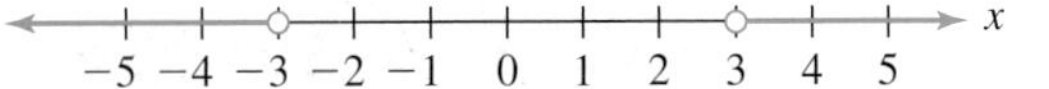

Figure 9 Graph of numbers whose distance from 0 is more than 3 units

Absolute Value Property for Inequalities

For an expression A and a positive constant k,

- The inequality $|A| < k$ is equivalent to $-k < A < k$.
- The inequality $|A| > k$ is equivalent to $A < -k$ or $A > k$.

Example 7 Solving an Absolute Value Inequality

Solve $|2x - 3| \le 9$. Describe the solution set as an inequality, in interval notation, and in a graph.

Solution

In $|2x - 3| \le 9$, the expression $2x - 3$ represents numbers whose distance from 0 is less than or equal to 9. Such a number is between -9 and 9, inclusive:

$$-9 \le 2x - 3 \le 9$$

Next, we solve the inequality $-9 \le 2x - 3 \le 9$:

$$-9 \le 2x - 3 \le 9 \quad \text{Original inequality}$$
$$-9 + 3 \le 2x - 3 + 3 \le 9 + 3 \quad \text{Add 3 to all parts.}$$
$$-6 \le 2x \le 12 \quad \text{Add.}$$
$$-3 \le x \le 6 \quad \text{Divide all parts by 2.}$$

So, the solution set is the set of numbers x where $-3 \le x \le 6$, or, in interval notation, $[-3, 6]$. We graph the solution set in Fig. 10.

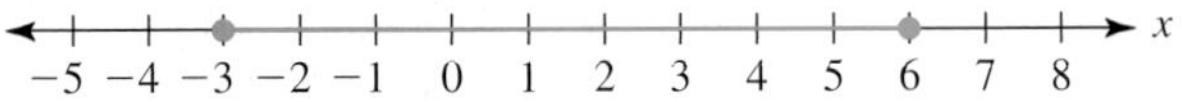

Figure 10 Graph of $-3 \le x \le 6$

To verify our result, we check that for inputs between -3 and 6, inclusive, the outputs of $y = |2x - 3|$ are less than or equal to 9 (see Fig. 11). ■

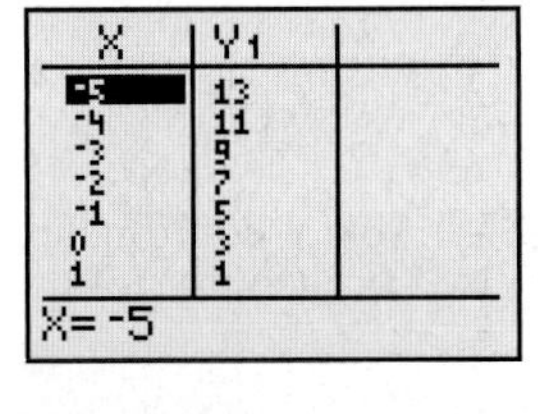

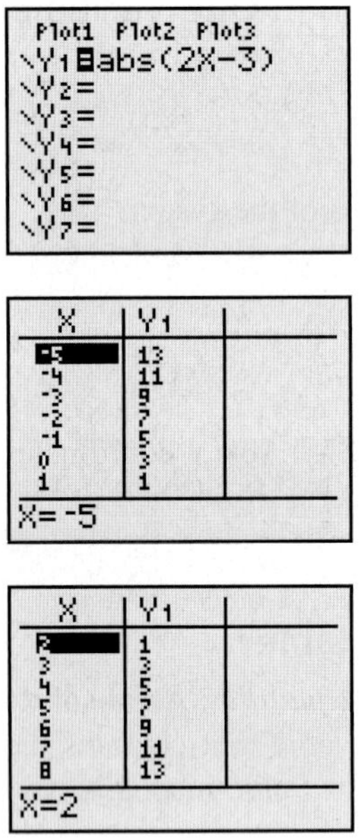

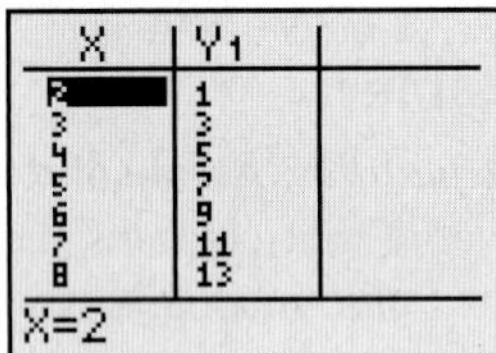

Figure 11 Verify the work

Example 8 Solving an Absolute Value Inequality

Solve $|3t + 4| > 12$. Describe the solution set as an inequality, in a graph, and in interval notation.

Solution

For $|3t + 4| > 12$, the expression $3t + 4$ represents numbers whose distance is more than 12 units from 0. These numbers are less than -12 or greater than 12:

$$3t + 4 < -12 \quad \text{or} \quad 3t + 4 > 12$$
$$3t < -16 \quad \text{or} \quad 3t > 8 \quad \text{Subtract 4 from both sides.}$$
$$t < -\frac{16}{3} \quad \text{or} \quad t > \frac{8}{3} \quad \text{Divide both sides by 3.}$$

We can graph the solution set on a number line (see Fig. 12), or we can describe the solution set in interval notation as $\left(-\infty, -\frac{16}{3}\right) \cup \left(\frac{8}{3}, \infty\right)$.

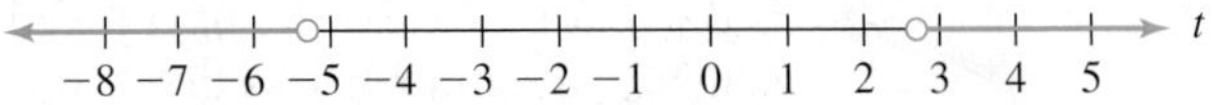

Figure 12 Graph of $t < -\frac{16}{3}$ or $t > \frac{8}{3}$

■

Recall from Section 3.5 that when we multiply or divide both sides of an inequality by a negative number, we reverse the inequality symbol.

Example 9 Solving an Absolute Value Inequality

Solve $7 - |x + 2| > 3$. Describe the solution set as an inequality, in a graph, and in interval notation.

Solution

To begin, we isolate $|x + 2|$ on the left-hand side of the inequality:

$$7 - |x + 2| > 3 \quad \text{Original inequality}$$
$$-|x + 2| > -4 \quad \text{Subtract 7 from both sides.}$$
$$|x + 2| < 4 \quad \text{Multiply both sides by } -1\text{; reverse inequality symbol.}$$

So, $x + 2$ represents numbers whose distance is less than 4 units from 0. These numbers are between -4 and 4:

$$-4 < x + 2 < 4$$
$$-4 - 2 < x + 2 - 2 < 4 - 2 \quad \text{Subtract 2 from all three parts.}$$
$$-6 < x < 2$$

We can graph the solution set on a number line (see Fig. 13), or we can describe the solution set in interval notation as $(-6, 2)$.

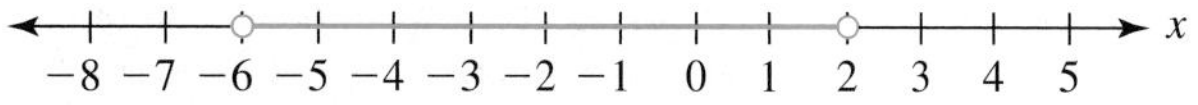

Figure 13 Graph of $-6 < x < 2$

To verify our work, we check that the graph of $y = 7 - |x + 2|$ is above the horizontal line $y = 3$ for values of x between -6 and 2 (see Fig. 14).

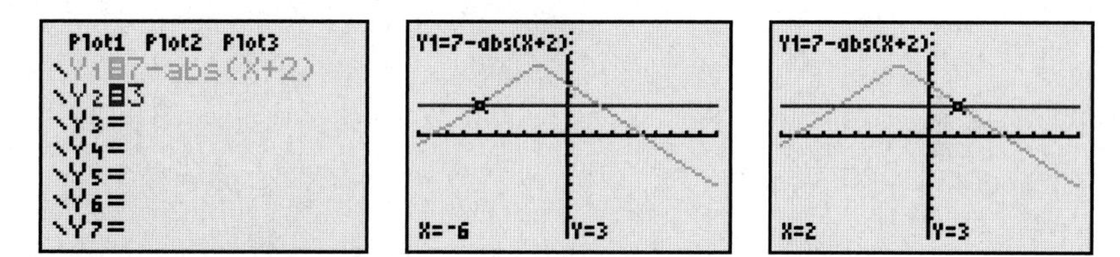

Figure 14 Verify the work ■

Example 10 Solving Absolute Value Inequalities

Solve.

1. $|7x - 10| \leq -2$
2. $|5x - 8| > -1$

Solution

1. Since $|7x - 10|$ is nonnegative, the inequality $|7x - 10| \leq -2$ has an empty-set solution.
2. Since $|5x - 8|$ is nonnegative for *any* real number x, the solution set of $|5x - 8| > -1$ is the set of all real numbers, or, in interval notation, $(-\infty, \infty)$. ■

group exploration

Graphical meaning of $|a - b|$

In this exploration, you will explore the graphical meaning of $|a - b|$.

1. Plot the points 1 and 6 on a number line. What is the distance between 1 and 6? Compare your result with $|1 - 6|$ and with $|6 - 1|$.
2. Plot the points -2 and 3 on a number line. What is the distance between -2 and 3? Compare your result with $|(-2) - 3|$ and with $|3 - (-2)|$.

3. Find the distance between -7 and -3, and compare your result with $|(-7)-(-3)|$ and with $|(-3)-(-7)|$.
4. Describe the graphical meaning of $|a-b|$.
5. Solve the equation. Then find the distance between 5 and each solution. Explain why your result makes sense in terms of the graphical meaning of $|x-5|$.
   a. $|x-5|=1$ **b.** $|x-5|=2$ **c.** $|x-5|=3$
6. Solve the equation or inequality and graph the solutions. Explain why your result makes sense in terms of the graphical meaning of $|x-4|$.
   a. $|x-4|=3$ [**Hint:** For the graph, plot the two solutions.]
   b. $|x-4|<3$
   c. $|x-4|>3$

Key Points OF SECTION 11.1

Absolute value — The **absolute value** of a number a, written $|a|$, is the distance from a to 0 on the number line.

Absolute value is nonnegative — The absolute value of any number is nonnegative.

Absolute value function — An **absolute value function** is a function whose equation contains the absolute value of a variable expression.

Absolute value equation — An **absolute value equation in one variable** is an equation in one variable that contains the absolute value of a variable expression.

Absolute value property for equations — For an expression A and a positive constant k, the equation $|A|=k$ is equivalent to $A=-k$ or $A=k$.

Solving an equation of the form $|A|=|B|$ — For expressions A and B, the equation $|A|=|B|$ is equivalent to $A=-B$ or $A=B$.

Absolute value inequality — An **absolute value inequality in one variable** is an inequality in one variable that contains the absolute value of a variable expression.

Absolute value property for inequalities — For an expression A and a positive constant k,
- The inequality $|A|<k$ is equivalent to $-k<A<k$.
- The inequality $|A|>k$ is equivalent to $A<-k$ or $A>k$.

HOMEWORK 11.1

FOR EXTRA HELP ▶ Student Solutions Manual · PH Math/Tutor Center · MathXL® · MyMathLab

Solve. Use a graphing calculator table or graph to verify your work.

1. $|x|=7$
2. $|x|=4$
3. $|x|=-3$
4. $|x|=-1$
5. $5|p|-3=15$
6. $-7|w|+6=4$
7. $|x+2|=5$
8. $|x-3|=8$
9. $|x-5|=0$
10. $-|x+1|=0$
11. $|3t-1|=11$
12. $|6k+4|=7$
13. $|2x+9|=-6$
14. $|5x-1|=-3$
15. $|4x|+1=9$
16. $|6x|-5=7$
17. $2|a+5|=8$
18. $-3|m-4|=-15$
19. $|2x-5|-4=-3$
20. $|5x+3|-2=5$
21. $|4x-5|=|3x+2|$
22. $|3x+7|=|2x-1|$
23. $|5w+1|=|3-w|$
24. $|2p-4|=|5-p|$
25. $\left|\frac{4x+3}{2}\right|=5$
26. $\left|\frac{3x-5}{6}\right|=2$
27. $\left|\frac{1}{2}x-\frac{5}{3}\right|=\frac{7}{6}$
28. $\left|\frac{3}{4}x+\frac{7}{2}\right|=\frac{1}{3}$
29. $\left|\frac{2}{3}k+\frac{4}{9}\right|=\left|\frac{5}{6}k-\frac{1}{3}\right|$
30. $\left|\frac{5}{6}t-\frac{3}{4}\right|=\left|\frac{1}{4}t+\frac{3}{2}\right|$

Solve. Round any solutions to the second decimal place.

31. $4.7|x|-3.9=8.8$
32. $1.9|x|+4.1=12.8$
33. $|2.1x+5.8|-9.7=10.2$
34. $|3.6x-2.1|+2.8=9.4$

Solve by using "intersect" on a graphing calculator. Round any solutions to the second decimal place.

35. $|x|-3=7-|x+1|$
36. $|x|-1=8-|x-3|$
37. $|x+4|+3=9-2|x+5|$
38. $2|x-2|+1=6-|x-3|$

For Exercises 39–42, use the graphs shown in Fig. 15 to solve the given equation or system.

39. $1 - |x| = -3$

40. $|x + 1| - 4 = 0$

41. $|x + 1| - 4 = 1 - |x|$

42. $y = |x + 1| - 4$
$y = 1 - |x|$

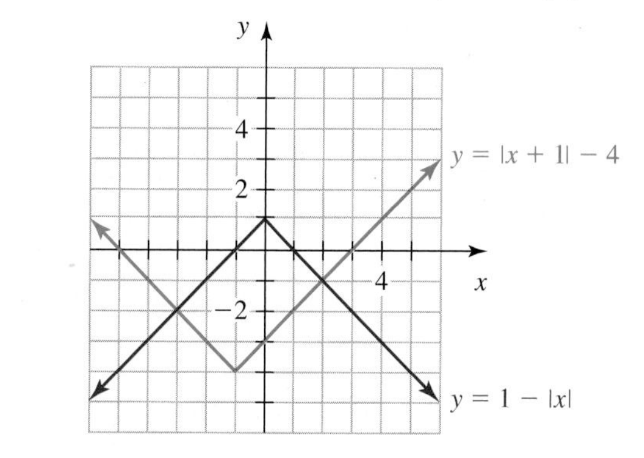

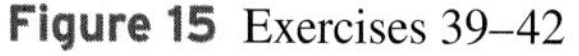

Figure 15 Exercises 39–42

Let $f(x) = 2|x| - 11$.

43. Find $f(-5)$.

44. Find $f(-3)$.

45. Find x where $f(x) = -5$.

46. Find x where $f(x) = -3$.

Let $f(x) = |4x + 7| - 9$.

47. Find $f(-3)$.

48. Find $f(-5)$.

49. Find x where $f(x) = -3$.

50. Find x where $f(x) = -5$.

Solve. Describe the solution set as an inequality, in a graph, and in interval notation.

51. $|x| < 4$

52. $|x| \le 1$

53. $|x| \ge 3$

54. $|x| > 2$

55. $|r| < -3$

56. $|m| > -5$

57. $|x| > 0$

58. $|x| \le 0$

59. $2|x| - 5 > 3$

60. $3|x| - 1 \ge 11$

61. $2 - 5|p| \le -8$

62. $10 - 4|b| < -2$

63. $|x - 6| \ge 7$

64. $|x + 1| > 3$

65. $|2x + 5| < 15$

66. $|3x - 4| \le 25$

67. $|7x + 15| > -4$

68. $|2x - 1| \ge -2$

69. $|0.25t - 1.3| \ge 1.1$

70. $|0.8w - 3.1| < 2.9$

71. $7 - |x + 3| \le 2$

72. $5 - |x - 2| < 4$

73. $\left|\dfrac{x+4}{3}\right| \ge 2$

74. $\left|\dfrac{x-1}{2}\right| > 4$

75. $\left|\dfrac{2x}{5} + \dfrac{3}{2}\right| \le \dfrac{9}{20}$

76. $\left|\dfrac{5x}{3} - \dfrac{1}{4}\right| > \dfrac{7}{12}$

77. Assume that $m \ne 0$ and that the equation $|mx + b| + c = k$ has at least one solution for x. Solve the equation for x.

78. What must be true of the constants m, b, c, and k if the equation $|mx + b| + c = k$ has exactly one solution?

79. A student tries to solve the equation $|x - 5| = 7$:

$$|x - 5| = 7$$
$$x + 5 = 7$$
$$x = 2$$

Describe any errors. Then solve the equation correctly.

80. A student tries to solve the equation $|x + 6| = -3$

$$|x + 6| = -3$$
$$x + 6 = 3 \quad \text{or} \quad x + 6 = -3$$
$$x = -3 \quad \text{or} \quad x = -9$$

Describe any errors. Then solve the equation correctly.

81. A student tries to solve the inequality $|x + 3| < 10$:

$$|x + 3| < 10$$
$$x + 3 < -10 \quad \text{or} \quad x + 3 < 10$$
$$x < -13 \quad \text{or} \quad x < 7$$

Describe any errors. Then solve the inequality correctly.

82. A student tries to solve the inequality $|x + 2| < 7$:

$$|x + 2| < 7$$
$$x + 2 < 7$$
$$x < 5$$

Describe any errors. Then solve the inequality correctly.

83. **a.** Solve $|2x + 3| = 13$.
b. Solve $|2x + 3| < 13$.
c. Solve $|2x + 3| > 13$.
d. Graph the solutions in parts (a), (b), and (c) on the same number line. Use three colors to identify the different solutions. Make some observations about the solutions. Explain these observations.

84. List three numbers that satisfy the inequality $-2|3x - 4| < -4$.

85. Is the statement "$|a + b| = |a| + |b|$ for all real numbers a and b" true or false? Explain. [**Hint:** Try substituting a positive number for a and a negative number for b.]

86. Explain how to solve an inequality of the form $|mx + b| + c < k$, where $m \ne 0$. (See page 4 for guidelines on writing a good response.)

Related Review

Solve. Round any approximate solutions to the fourth decimal place.

87. $\left|x - 5\right| = 4$

88. $\left|x^2 - 5\right| = 4$

89. $\left|2^y - 5\right| = 4$

90. $\left|\sqrt{w} - 5\right| = 4$

91. $\left|\dfrac{2x+3}{x-2} - 5\right| = 4$

92. $|\log_4(x) - 5| = 4$

Solve. Describe the solution set as an inequality, in a graph, and in interval notation.

93. $3(2x) - 5 \le 7$

94. $2(2x - 3) > 6$

95. $3|2x| - 5 \le 7$

96. $2|2x - 3| > 6$

Expressions, Equations, Functions, and Graphs

Perform the indicated instruction. Then use words such as absolute value, linear, quadratic, cubic, exponential, logarithmic, rational, radical, polynomial, degree, function, one variable, *and* two variables *to describe the expression, equation, or system.*

97. Graph $y = 3(x - 2) + 1$ by hand.

98. Factor $4x^3 - 20x^2 + 24x$.

99. Solve $3|x - 2| + 1 = 7$.

100. Solve $4x^3 - 20x^2 + 24x = 0$.

101. Find an equation of a line that contains the points $(-4, 2)$ and $(5, -3)$.

102. Find the product $(2x - 6)\left(3x^2 + 4x - 2\right)$.

Section 11.1 Quiz

Solve.

1. $3|t| - 4 = 11$
2. $5|6r - 5| = 15$
3. $|7x + 1| = -3$
4. $|5x - 2| = |3x + 6|$
5. $\left|\frac{3}{4}x - \frac{1}{2}\right| = \frac{7}{8}$
6. Is the statement "$|a - b| = |a| - |b|$ for all real numbers a and b" true or false? Explain.

Solve. Describe the solution set as an inequality, in a graph, and in interval notation.

7. $3|k| - 4 \geq 2$
8. $|4c - 8| > 12$
9. $7|3x - 2| \leq 42$
10. $|x - 5| < -7$

11.2 LINEAR INEQUALITIES IN TWO VARIABLES; SYSTEMS OF LINEAR INEQUALITIES

Objectives

- Graph a *linear inequality in two variables.*
- Solve a *system of linear inequalities in two variables.*
- Use a system of linear inequalities in two variables to make estimates.

Recall from Section 1.2 that a linear equation in two variables is an equation that can be put into the form $y = mx + b$ or $x = a$, where m, a, and b are constants. A **linear inequality in two variables** is an inequality of the form

$$y < mx + b \quad \text{or} \quad x < a$$

(or with $<$ replaced with $\leq$, $>$, or $\geq$), where m, a, and b are constants. Here are some examples:

$$y < 2x + 9 \qquad y \leq -4x - 7 \qquad 3x - 8y > 24 \qquad x \geq 5$$

First we will work with one linear inequality in two variables. Then we will work with two or more such inequalities.

Linear Inequalities in Two Variables

We use the terms *solution, satisfy,* and *solution set* for an inequality in two variables in much the same way that we have used them for equations in one or two variables and inequalities in one variable.

> **DEFINITION** *Satisfy, solution, solution set,* and *solve* for an inequality in two variables
>
> If an inequality in the two variables x and y becomes a true statement when a is substituted for x and b is substituted for y, we say that the ordered pair (a, b) **satisfies** the inequality and call (a, b) a **solution** of the inequality. The **solution set** of an inequality is the set of all solutions of the inequality. We **solve** the inequality by finding its solution set.

We describe the solution set of an inequality by graphing all of the solutions.

Example 1 Sketching the Graph of an Inequality

Graph $y > \frac{1}{2}x + 1$.

Solution

We begin by sketching a graph of $y = \frac{1}{2}x + 1$ (see Fig. 16).

To investigate how to solve $y > \frac{1}{2}x + 1$, we choose a value of x (say, 4) and find several solutions with an x-coordinate of 4. For $y = \frac{1}{2}x + 1$, if $x = 4$, then $y = \frac{1}{2}(4) + 1 = 3$. So, the point $(4, 3)$ is on the line $y = \frac{1}{2}x + 1$.

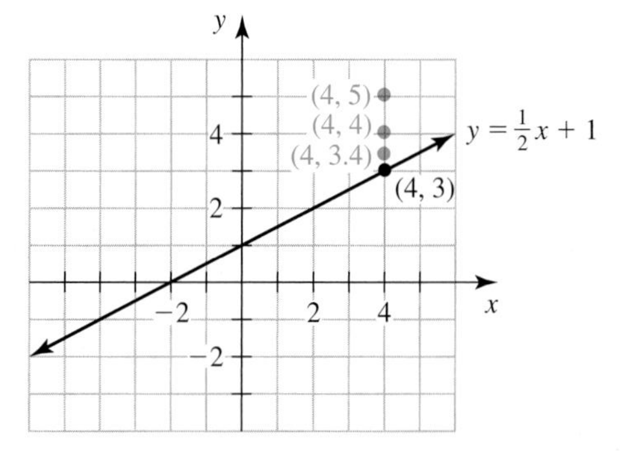

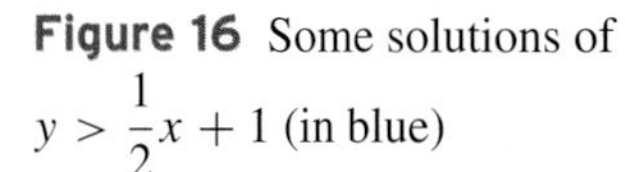

Figure 16 Some solutions of $y > \frac{1}{2}x + 1$ (in blue)

For $y > \frac{1}{2}x + 1$, if $x = 4$, then

$$y > \frac{1}{2}(4) + 1$$
$$y > 3$$

So, if $x = 4$, some possible values of y are $y = 3.4$, $y = 4$, and $y = 5$. The points (4, 3.4), (4, 4), and (4, 5) lie *above* the point (4, 3), which is on the line $y = \frac{1}{2}x + 1$ (see Fig. 16).

We could choose values of x other than 4 and go through a similar argument. These investigations would suggest that the solutions of $y > \frac{1}{2}x + 1$ lie *above* the graph of $y = \frac{1}{2}x + 1$, which is true. In Fig. 17, we shade the region that contains all of the points that represent solutions of $y > \frac{1}{2}x + 1$.

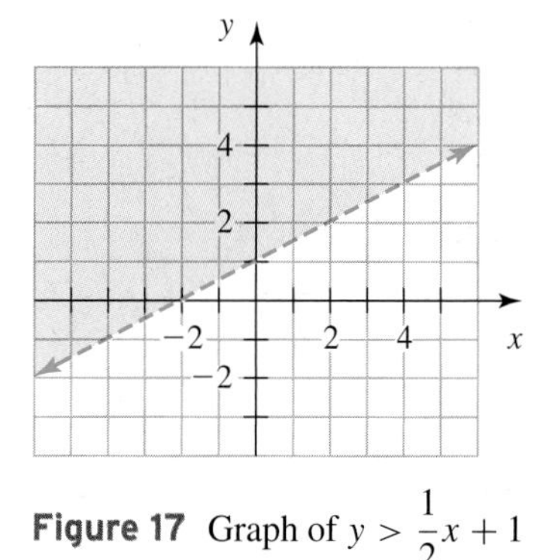

Figure 17 Graph of $y > \frac{1}{2}x + 1$

We *dash* the line $y = \frac{1}{2}x + 1$ in Fig. 17 to indicate that its points are *not* solutions of $y > \frac{1}{2}x + 1$. For example, the point (4, 3) on the line $y = \frac{1}{2}x + 1$ does *not* satisfy the inequality $y > \frac{1}{2}x + 1$:

$y > \frac{1}{2}x + 1$	Original inequality
$3 \overset{?}{>} \frac{1}{2}(4) + 1$	Substitute 4 for x and 3 for y.
$3 \overset{?}{>} 3$	Simplify.
false	

■

We can draw a graph of $y > \frac{1}{2}x + 1$ by using a graphing calculator (see Fig. 18), but we have to imagine that the border $y = \frac{1}{2}x + 1$ is drawn as a dashed line. To shade above a line, press Y= and press ◁ twice. Next, press ENTER as many times as necessary for the triangle shown in Fig. 18 to appear.

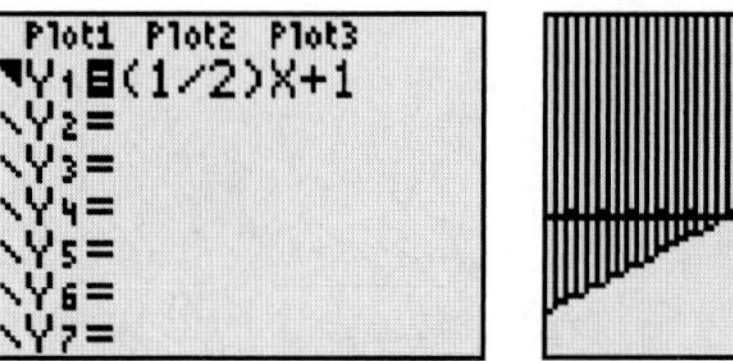

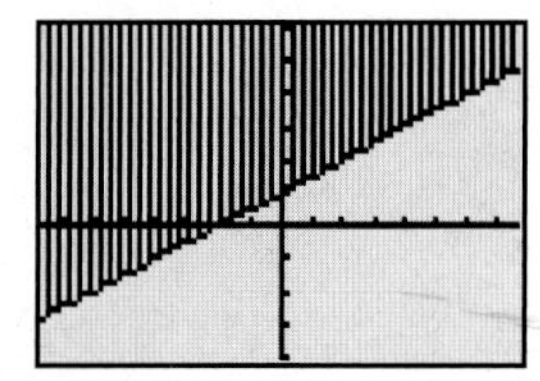

Figure 18 Graph of $y > \frac{1}{2}x + 1$ (imagine that the border line is dashed)

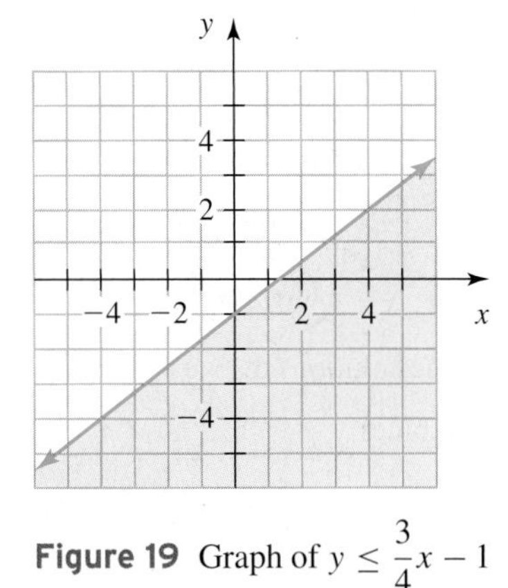

Figure 19 Graph of $y \leq \frac{3}{4}x - 1$

Example 2 Sketching the Graph of an Inequality

Sketch a graph of $y \leq \frac{3}{4}x - 1$.

Solution

The graph of $y \leq \frac{3}{4}x - 1$ is the line $y = \frac{3}{4}x - 1$ and the region below that line (see Fig. 19). We use a *solid* line as the border $y = \frac{3}{4}x - 1$ to indicate that the points on the line $y = \frac{3}{4}x - 1$ are solutions of $y \leq \frac{3}{4}x - 1$. ■

Graph of an Inequality in Two Variables

- The graph of an inequality of the form $y > mx + b$ is the region above the line $y = mx + b$. The graph of an inequality of the form $y < mx + b$ is the region below the line $y = mx + b$. For either inequality, we use a dashed line to show that $y = mx + b$ is not part of the graph.
- The graph of an inequality of the form $y \geq mx + b$ is the line $y = mx + b$ and the region above that line. The graph of an inequality of the form $y \leq mx + b$ is the line $y = mx + b$ and the region below that line.

To sketch a graph of an inequality in two variables, if the variable y is not alone on one side of the inequality, we begin by isolating it. Recall from Section 3.5 that when we multiply or divide both sides of an inequality by a negative number, we must reverse the inequality symbol.

Example 3 Sketching the Graph of an Inequality

Sketch the graph of $-2x - 3y > 6$.

Solution

First, we isolate y on one side of the inequality:

$$-2x - 3y > 6 \quad \text{Original inequality}$$
$$-3y > 2x + 6 \quad \text{Add 2x to both sides.}$$
$$\frac{-3y}{-3} < \frac{2x}{-3} + \frac{6}{-3} \quad \text{Divide both sides by } -3\text{; reverse inequality symbol.}$$
$$y < -\frac{2}{3}x - 2 \quad \text{Simplify.}$$

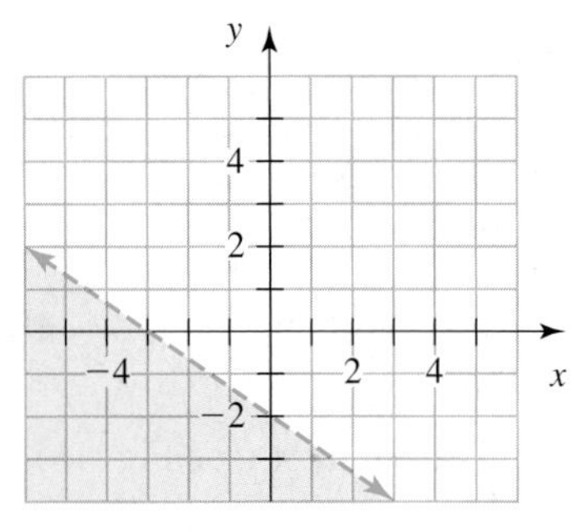

Figure 20 Graph of $-2x - 3y > 6$

The graph of $y < -\frac{2}{3}x - 2$ is the region below the line $y = -\frac{2}{3}x - 2$ (see Fig. 20).

To verify our work, we choose a point on our graph, such as (–3, 1), and check that it satisfies the inequality:

$$-2x - 3y > 6 \quad \text{Original inequality}$$
$$-2(-3) - 3(-1) \overset{?}{>} 6 \quad \text{Substitute } -3 \text{ for x and } -1 \text{ for y.}$$
$$9 \overset{?}{>} 6 \quad \text{Simplify.}$$
$$\text{true}$$

To verify further, we could choose several other points on the graph and check that each point satisfies the inequality.

We could also choose several points that are not on the graph and check that each point does *not* satisfy the inequality. For example, the point (0, 0) is not on the graph, and the ordered pair (0, 0) does not satisfy the inequality $-2x - 3y > 6$:

$-2x - 3y > 6$	Original inequality
$-2(0) - 3(0) \overset{?}{>} 6$	Substitute 0 for x and 0 for y.
$0 \overset{?}{>} 6$	Simplify.
false	

■

WARNING It is a common error to think that the graph of an inequality such as $-2x - 3y > 6$ is the region *above* the line $-2x - 3y = 6$, because the symbol ">" means "is greater than." However, we must first isolate y on the left side of a linear inequality before we can determine whether the graph includes the region that is above or below a line. In fact, in Example 3, we wrote the inequality $-2x - 3y > 6$ as $y < -\frac{2}{3}x - 2$ and concluded that the graph is the region *below* the line $y = -\frac{2}{3}x - 2$.

Example 4 Sketching the Graphs of Inequalities

Sketch the graph of the inequality.

1. $y \le 3$
2. $x > -4$

Solution

1. The graph of $y \le 3$ is the horizontal line $y = 3$ and the region below that line (see Fig. 21).
2. Ordered pairs with x-coordinates *greater than* than -4 are represented by points that lie to the *right* of the vertical line $x = -4$. So, the graph of $x > -4$ is the region to the right of $x = -4$ (see Fig. 22).

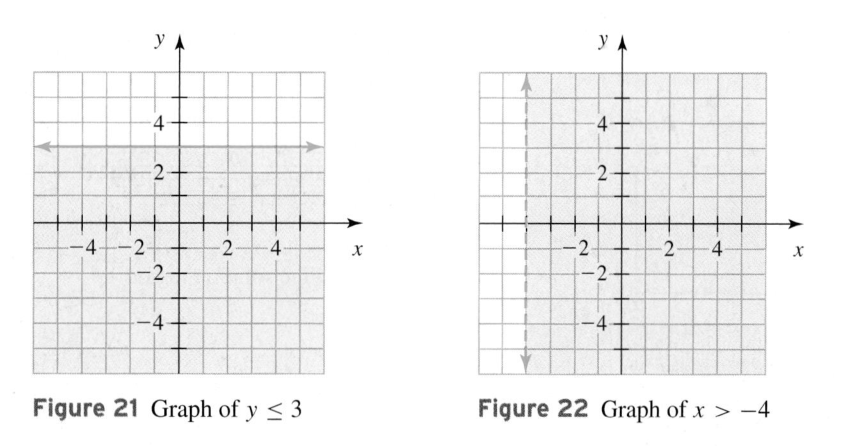

Figure 21 Graph of $y \le 3$

Figure 22 Graph of $x > -4$ ■

Systems of Linear Inequalities in Two Variables

A **system of linear inequalities in two variables** consists of two or more linear inequalities in two variables. Here is an example:

$$y \ge -2x + 1$$
$$y < \frac{1}{2}x - 3$$

> **DEFINITION** *Solution, solution set,* and *solve* for a system of inequalities in two variables
>
> We say that an ordered pair (a, b) is a **solution** of a system of inequalities in two variables if it satisfies all the inequalities in the system. The **solution set** of a system is the set of all solutions of the system. We **solve** a system by finding its solution set.

Recall from Section 3.1 that we can find the solution set of a system of two linear *equations* in two variables by locating any intersection point(s) of the graphs of the two equations. Similarly, **we can find the solution set of a system of inequalities in two variables by locating the intersection of the graphs of all of the inequalities.** This makes sense, because a solution is an ordered pair that satisfies all of the inequalities, meaning that the point that represents the ordered pair lies on the graphs of all of the inequalities.

In this text, we use a graph to describe the solution set of a system of linear inequalities in two variables.

Example 5 Graphing the Solution Set of a System of Inequalities

Solve the system

$$y \geq -2x + 1$$
$$y < \frac{1}{2}x - 3$$

Solution

First, we sketch the graph of $y \geq -2x + 1$ (see Fig. 23, blue region) and the graph of $y < \frac{1}{2}x - 3$ (see Fig. 23, red region). The graph of the solution set of the system is the intersection of the graphs of the inequalities, which is shown in Fig. 24.

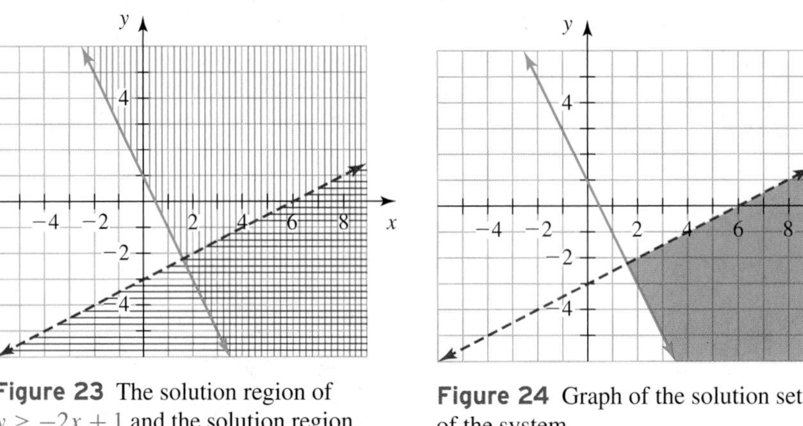

Figure 23 The solution region of $y \geq -2x + 1$ and the solution region of $y < \frac{1}{2}x - 3$

Figure 24 Graph of the solution set of the system

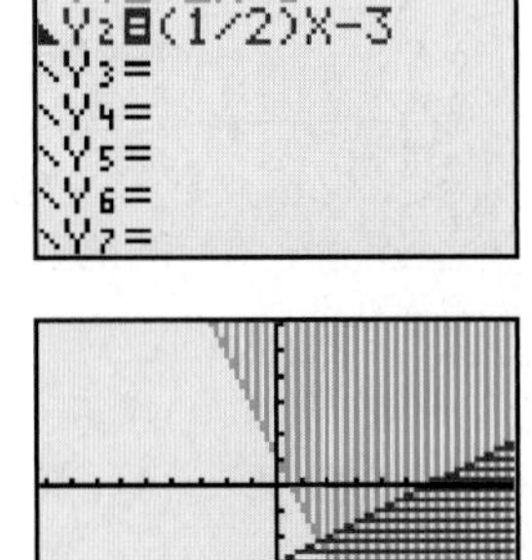

Figure 25 Verify our work (imagine that the border $y = \frac{1}{2}x - 3$ is drawn with a dashed line)

We can use a graphing calculator to draw a graph of the solution set of the system (see Fig. 25), where we imagine that the border $y = \frac{1}{2}x - 3$ is drawn with a dashed line. The graph of the solution set is the region shaded by both vertical lines (in blue) and horizontal lines (in red). ■

Example 6 Graphing the Solution Set of a System

Graph the solution set of the system

$$y \geq 2x - 3$$
$$y \leq -x + 5$$
$$x \geq 0$$
$$y \geq 0$$

Solution

In Fig. 26, we use arrows to indicate the graphs of each of the four inequalities. The solution set of the system is the intersection of the graphs of the four inequalities.

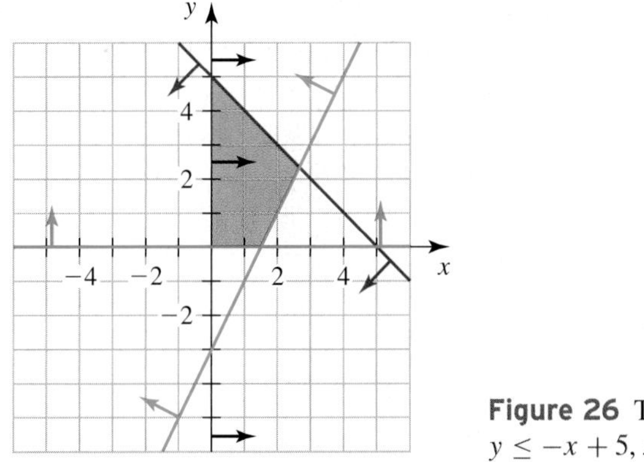

Figure 26 The graphs of $y \geq 2x - 3$, $y \leq -x + 5$, $x \geq 0$, and $y \geq 0$

Modeling with Systems of Inequalities

We can use a system of inequalities to help us make estimates about authentic situations.

Example 7 Using a System of Inequalities to Make Estimates

A person's life expectancy predicts how many *remaining* years the person will live. The life expectancies of U.S. females at birth and at age 20 years are shown in Table 2 for various calendar years.

Table 2 Life Expectancies of U.S. Females at Birth and at Age 20 Years

	Life Expectancy (years)	
Year	**At Birth**	**At Age 20 Years**
1980	77.6	59.0
1985	78.2	59.3
1990	78.8	59.8
1995	78.9	59.9
2000	79.7	60.5
2002	79.9	60.7

Source: *U.S. National Center for Health Statistics*

Let $L = B(t)$ be the life expectancy (in years) at birth and $L = T(t)$ be the life expectancy (in years) at age 20 years, both of U.S. females at t years since 1980. The linear regression models of B and T are, respectively,

$$B(t) = 0.100t + 77.65$$
$$T(t) = 0.076t + 58.96$$

1. Find a system of inequalities that describes the life expectancies of U.S. females from 0 years through 20 years old from 1985 to 2010.
2. Graph the solution set of the system of inequalities that you found in Problem 1.
3. Estimate the life expectancies of U.S. females from 0 years through 20 years old in 2005.

Solution

1. The life expectancies must be less than or equal to the at-birth life expectancies ($L \leq 0.100t + 77.65$) and greater than or equal to the life expectancies at age 20 years ($L \geq 0.076t + 58.96$). We are seeking life expectancies for the calendar years from 1985 through 2010: $t \geq 5$ and $t \leq 30$. So, our system is

$$L \leq 0.100t + 77.65$$
$$L \geq 0.076t + 58.96$$
$$t \geq 5$$
$$t \leq 30$$

2. In Fig. 27, we use arrows to indicate the graphs of each of the four inequalities. The solution set of the system is the intersection of those graphs.

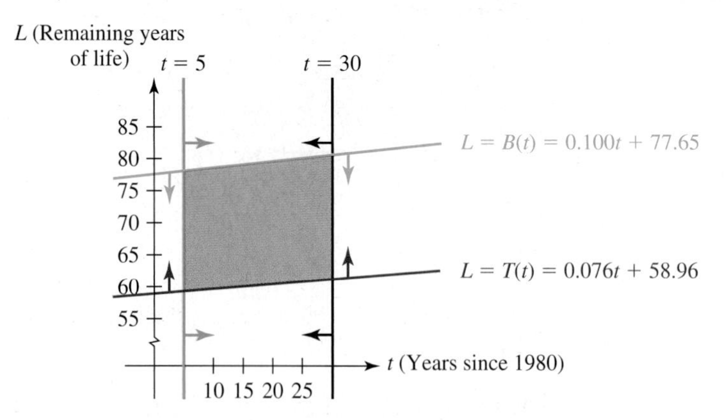

Figure 27 Graph of the solution set of the system

3. The life expectancies of U.S. females from 0 years through 20 years old in the calendar year 2005 are represented by the vertical line segment above $t = 25$ on the t-axis in Fig. 28. From this line segment, we see that the life expectancies are between 60.86 years and 80.15 years, inclusive.

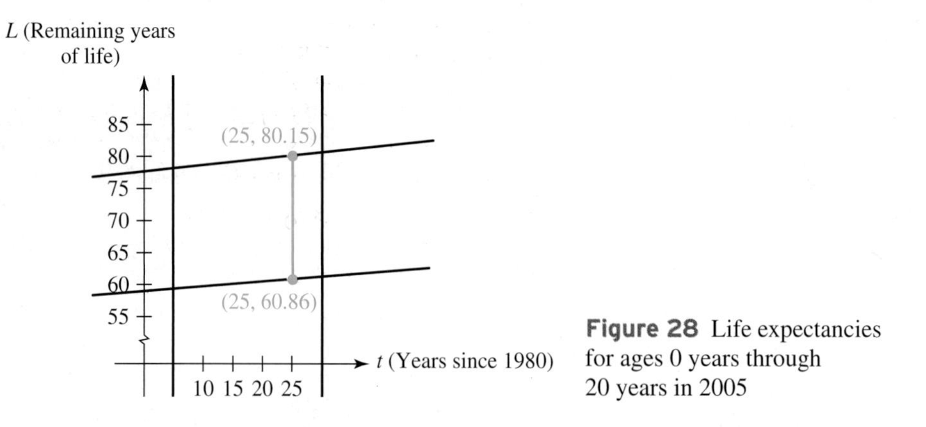

Figure 28 Life expectancies for ages 0 years through 20 years in 2005

We can also use the equations of B and T to find the life expectancies of U.S. females from 0 years through 20 years old in the calendar year 2005. To begin, we evaluate both B and T at 25:

$$B(25) = 0.100(25) + 77.65 = 80.15$$
$$T(25) = 0.076(25) + 58.96 = 60.86$$

This means that the life expectancies are between 60.86 years and 80.15 years, inclusive, which checks. ■

group exploration

Meaning of solution of a system of inequalities in two variables

The graphs of $y = ax + b$ and $y = cx + d$ are sketched in Fig. 29.

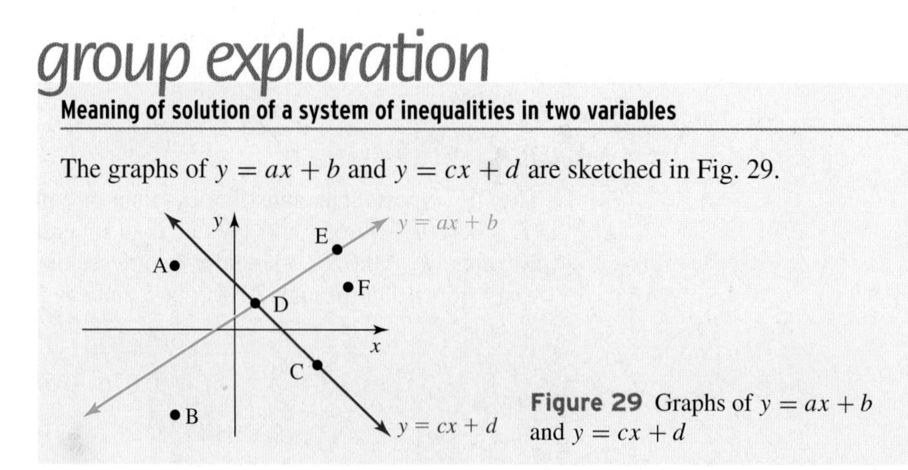

Figure 29 Graphs of $y = ax + b$ and $y = cx + d$

1. For each part, decide which one or more of the points A, B, C, D, E, and F represent ordered pairs that
   a. satisfy the inequality $y < ax + b$
   b. satisfy the inequality $y \geq cx + d$
   c. are solutions of the system of inequalities

$$y < ax + b$$
$$y \geq cx + d$$

2. Write a system of inequalities in terms of a, b, c, d, x, and y such that points B, C, and D are solutions and points A, E, and F are not solutions.
3. Write a system of inequalities in terms of a, b, c, d, x, and y such that points B and C are solutions and points A, D, E, and F are not solutions.

Key Points OF SECTION 11.2

***Satisfy, solution, solution set,* and *solve* for an inequality in two variables**	If an inequality in the two variables x and y becomes a true statement when a is substituted for x and b is substituted for y, we say that the ordered pair (a, b) **satisfies** the inequality and call (a, b) a **solution** of the inequality. The **solution set** of an inequality is the set of all solutions of the inequality. We **solve** the inequality by finding its solution set.
Graph of an inequality in two variables: $y > mx + b$ or $y < mx + b$	The graph of an inequality of the form $y > mx + b$ is the region above the line $y = mx + b$. The graph of an inequality of the form $y < mx + b$ is the region below the line $y = mx + b$. For either inequality, we use a dashed line to show that $y = mx + b$ is not part of the graph.
Graph of an inequality in two variables: $y \geq mx + b$ or $y \leq mx + b$	The graph of an inequality of the form $y \geq mx + b$ is the line $y = mx + b$ and the region above that line. The graph of an inequality of the form $y \leq mx + b$ is the line $y = mx + b$ and the region below that line.
***Solution, solution set,* and *solve* for a system of inequalities in two variables**	We say that an ordered pair (a, b) is a **solution** of a system of inequalities in two variables if it satisfies all the inequalities in the system. The **solution set** of a system is the set of all solutions of the system. We **solve** a system by finding its solution set.
Intersection of graphs	We can find the solution set of a system of inequalities in two variables by locating the intersection of the graphs of all of the inequalities.

HOMEWORK 11.2

FOR EXTRA HELP ▶ Student Solutions Manual | PH Math/Tutor Center | MathXL® | MyMathLab

Graph the inequality by hand.

1. $y \geq 2x - 4$
2. $y > x + 1$
3. $y < -\frac{1}{2}x + 3$
4. $y \leq \frac{2}{3}x - 5$
5. $y \leq -2x + 6$
6. $y \geq -3x + 4$
7. $y > x$
8. $y \leq 2x$
9. $y < -\frac{1}{3}x$
10. $y < -\frac{1}{4}x$
11. $2x + 5y < 10$
12. $3x - 2y \geq 8$
13. $4x - 6y - 6 \geq 0$
14. $2y - x + 1 > 0$
15. $3(x - 2) + y \leq -2$
16. $y > -2(x - 1) + 3$
17. $y \leq 2$
18. $y < -3$
19. $x > -3$
20. $x \geq 1$

Graph by hand the solution set of the system of inequalities.

21. $y \geq \frac{1}{3}x - 2$, $y > -x + 3$
22. $y > \frac{1}{2}x + 3$, $y \leq -2x + 5$
23. $y \leq x - 4$, $y \geq -3x$
24. $y < 2x - 3$, $y < -2x$
25. $y \leq -3x + 9$, $y \geq 2x - 3$, $x \geq 0$, $y \geq 0$
26. $y \leq \frac{1}{3}x + 4$, $y \geq 2x - 5$, $x \geq 0$, $y \geq 0$
27. $y < -x + 5$, $y \leq x + 5$, $y > \frac{1}{2}x + 1$
28. $y > -x - 4$, $y \geq 2x + 6$, $y \leq \frac{1}{2}x + 6$

29. $y \le -3$
$y \ge -5$

30. $y \le 2$
$y > -1$

31. $2x - 4y \le 8$
$3x + 5y \le 10$

32. $3x - 2y < 6$
$4x + 3y > -12$

33. $x - 2y > 6$
$x + 3y \le 3$

34. $x - 4y \ge 8$
$x + 2y < 2$

35. $1 + y \ge \frac{1}{2}(x - 4)$
$3 - y > 2(x - 1)$

36. $2 - y < 3(x + 2)$
$y - 7 \le 2(x - 3)$

37. $5y \le 2x + 20$
$5y \ge 2x + 5$
$x \ge 3$
$x \le 5$

38. $4y < 3x + 20$
$4y > 3x + 8$
$x > 2$
$x < 4$

39. A person's life expectancy predicts how many *remaining* years the person will live. The life expectancies of U.S. males at birth and at age 20 years are shown in Table 3 for various calendar years.

Table 3 Life Expectancies of U.S. Males at Birth and at Age 20 Years

	Life Expectancy (years)	
Year	**At Birth**	**At Age 20 Years**
1980	70.1	51.9
1985	71.1	52.6
1990	71.8	53.3
1995	72.5	53.8
2000	74.3	55.3
2002	74.5	55.6

Source: *U.S. National Center for Health Statistics*

Let $L = B(t)$ be the life expectancy (in years) at birth and $L = T(t)$ be the life expectancy (in years) at age 20 years, both of U.S. males at t years since 1980.

a. Find equations of B and T.
b. Find a system of inequalities that describes the life expectancies of U.S. males from 0 years through 20 years old from 1980 to 2015.
c. Graph the solution set of the system of inequalities that you found in part (b).
d. Predict the life expectancies of U.S. males from 0 years through 20 years old in 2010.

40. Recall from Exercise 12 of Homework 2.1 that the windchill is a measure of how cold you feel when exposed to wind. Table 4 gives the windchills for various temperatures when the wind speed is 10 mph or 20 mph. Let $w = f(t)$ be the windchill when the wind speed is 10 mph and $w = g(t)$ be the windchill when the wind speed is 20 mph, both in degrees Fahrenheit at a temperature t in degrees Fahrenheit.

a. Find equations of f and g.
b. Find a system of inequalities that describes the windchills for temperatures between −20°F and 30°F, inclusive, with wind speeds between 10 mph and 20 mph, inclusive.
c. Graph the solution set of the system of inequalities that you found in part (b).
d. If the temperature is 7°F, what are the windchills for wind speeds between 10 mph and 20 mph, inclusive?

Table 4 Windchills for 10-mph and 20-mph Winds

	Windchill (degrees Fahrenheit)	
Temperature (degrees Fahrenheit)	**10-mph Wind**	**20-mph Wind**
−15	−35	−42
−10	−28	−35
−5	−22	−29
0	−16	−22
5	−10	−15
10	−4	−9
15	3	−2
20	9	4

Source: *National Weather Service*

41. Recommended ski lengths are shown in Table 5 for skiers of various weights and abilities.

Table 5 Recommended Ski Lengths

	Ski Length (centimeters)			
Weight (pounds)	**Beginner**	**Intermediate**	**Advanced**	**Expert**
100.0	130	135	140	145
112.5	135	140	145	150
130.5	140	145	150	155
143.0	145	150	155	160
158.0	150	155	160	165
172.5	155	160	165	170
185.0	165	170	175	180
195.0	175	180	185	190

Source: *backcountry.com*

Let $L = B(w)$ be the recommended ski length for a beginner skier and $L = I(w)$ be the recommended ski length for an intermediate skier, both in centimeters for a skier who weighs w pounds.

a. Find linear equations of B and I.
b. Find a system of inequalities that describes the recommended ski lengths for beginning to intermediate skiers who weigh between 130 and 150 pounds, inclusive.
c. Graph the solution set of the system of inequalities that you found in part (b).
d. Estimate all recommended ski lengths for beginning to intermediate 140-pound skiers.

42. Let $L = A(w)$ be the recommended ski length for an advanced skier and $L = E(w)$ be the recommended ski length for an expert skier, both in centimeters for a skier who weighs w pounds (see Table 5).

a. Find linear equations of A and E.
b. Find a system of inequalities that describes the recommended ski lengths for advanced to expert skiers who weigh between 140 and 160 pounds, inclusive.
c. Graph the solution set of the system of inequalities that you found in part (b).
d. Estimate all recommended ski lengths for advanced to expert 145-pound skiers.

43. A student believes that the graph of $2x - 3y < 6$ is the region below the line $2x - 3y = 6$. What would you tell the student?

44. Find three ordered pairs that are solutions of the inequality $y \le -x - 2$, and find three ordered pairs that are not solutions.

45. Give an example of an inequality in terms of x and y for which $(3, 4)$ is a solution and $(4, 3)$ is not.

46. Give an example of an inequality in terms of x and y for which $(3, 3)$, $(-3, 3)$, and $(-3, -3)$ are solutions and $(3, -3)$ is not.

47. Graph by hand the solution set of the system

$$y \geq 2x + 1$$
$$y \leq 2x + 1$$

48. Describe the solution set of the system

$$y > 2x + 1$$
$$y < 2x + 1$$

49. Explain how to graph an inequality in two variables. When do you reverse the inequality symbol?

50. Explain how to solve a system of inequalities in two variables.

Related Review

Graph by hand.

51. $y = -2x - 3$

52. $5x - 2y = 4$

53. $y < -2x - 3$

54. $5x - 2y \geq 4$

Solve the system if it is a system of linear equations. Graph by hand the solution set of the system if it is a system of inequalities.

55. $2x + y = 8$
$3x - 2y = -2$

56. $2x + y \leq 8$
$3x - 2y \leq -2$

Expressions, Equations, Functions, and Graphs

Perform the indicated instruction. Then use words such as linear, quadratic, cubic, exponential, logarithmic, rational, radical, polynomial, degree, function, one variable, *and* two variables *to describe the expression, equation, or system.*

57. Graph $f(x) = -\frac{3}{5}x + 2$ by hand.

58. Simplify $-2(x - 3)^2 - 1$.

59. Find $f(2)$, where $f(x) = -\frac{3}{5}x + 2$.

60. Solve $-2(x - 3)^2 - 1 = -9$.

61. Let $f(x) = -\frac{3}{5}x + 2$. Find x when $f(x) = 5$.

62. Graph $f(x) = -2(x - 3)^2 - 1$ by hand.

Section 11.2 Quiz

Graph the inequality by hand.

1. $y \leq 2x - 6$

2. $4x - 2y > 8$

3. $-2(y + 3) + 4x \geq -8$

4. $y < -2$

Graph by hand the solution set of the system of inequalities.

5. $y \leq \frac{2}{5}x + 1$
$y < -\frac{1}{4}x + 2$

6. $3x - 4y \geq 12$
$6y - 2x \leq 12$

7. $x - y < 3$
$x + y < 5$
$x > 0$
$y > 0$

8. Find three ordered pairs that are solutions of the inequality $2x - 5y > 10$, and find three ordered pairs that are not.

11.3 PERFORMING OPERATIONS WITH COMPLEX NUMBERS

Objective

- Perform operations with complex numbers.

Recall from Section 7.3 that the *imaginary unit i* is the number whose square is -1. That is,

$$i^2 = -1 \quad \text{and} \quad i = \sqrt{-1}$$

If b is a nonzero real number, then an expression of the form bi is a *pure imaginary number*. A *complex number* is a number of the form $a + bi$, where a and b are real numbers. An *imaginary number* is a number $a + bi$ where a and b are real numbers and $b \neq 0$.

Finally, recall from Section 7.3 that if p is a positive real number, then $\sqrt{-p} = i\sqrt{p}$. For example, $\sqrt{-25} = i\sqrt{25} = 5i$.

In this section, we will perform operations with complex numbers.

Adding, Subtracting, and Multiplying Complex Numbers

Since $i = \sqrt{-1}$, **we perform operations with complex numbers in much the same way as we do with radical expressions.** For example,

$$\left(2 + 3\sqrt{7}\right) + \left(1 + 5\sqrt{7}\right) = 3 + 8\sqrt{7} \qquad \text{Add two radical expressions.}$$
$$(2 + 3i) + (1 + 5i) = 3 + 8i \qquad \text{Add two complex numbers.}$$

If a radicand is a negative real number, we first write the radical in terms of i before performing any operations:

$$\sqrt{-1}\sqrt{-1} = i \cdot i = i^2 = -1 \qquad \text{Correct}$$

$$\sqrt{-1}\sqrt{-1} = \sqrt{-1 \cdot -1} = \sqrt{1} = 1 \qquad \text{Incorrect}$$

So, there is no product property $\sqrt{a}\sqrt{b} = \sqrt{ab}$ when a and b are both negative. To find $\sqrt{-2}\sqrt{-3}$, we first write each radical in terms of i:

$$\sqrt{-2}\sqrt{-3} = i\sqrt{2} \cdot i\sqrt{3} = i^2\sqrt{6} = -\sqrt{6}$$

When we use an operation to combine two complex numbers, we write the result in the form $a + bi$, where a and b are in lowest terms.

Example 1 Performing Operations with Complex Numbers

Perform the indicated operation. Simplify the result.

1. $(5 + 9i) + (3 - 2i)$ **2.** $(3 - \sqrt{-36}) - (2 - \sqrt{-16})$
3. $4i \cdot 6i$ **4.** $\sqrt{-4}\sqrt{-9}$

Solution

1.
$$\begin{aligned} (5 + 9i) + (3 - 2i) &= 5 + 3 + 9i - 2i && \text{Rearrange terms.} \\ &= 8 + 7i && \text{Write in } a + bi \text{ form.} \end{aligned}$$

2.
$$\begin{aligned} (3 - \sqrt{-36}) - (2 - \sqrt{-16}) &= (3 - i\sqrt{36}) - (2 - i\sqrt{16}) && \text{Write in terms of } i. \\ &= (3 - 6i) - (2 - 4i) && \sqrt{36} = 6, \sqrt{16} = 4 \\ &= 3 - 6i - 2 + 4i && \text{Distributive law} \\ &= 3 - 2 - 6i + 4i && \text{Rearrange terms.} \\ &= 1 - 2i && \text{Write in } a + bi \text{ form.} \end{aligned}$$

3.
$$\begin{aligned} 4i \cdot 6i &= 24i^2 && \text{Simplify.} \\ &= 24(-1) && i^2 = -1 \\ &= -24 \end{aligned}$$

4.
$$\begin{aligned} \sqrt{-4}\sqrt{-9} &= i\sqrt{4} \cdot i\sqrt{9} && \text{Write in terms of } i. \\ &= 2i \cdot 3i && \sqrt{4} = 2, \sqrt{9} = 3 \\ &= 6i^2 && \text{Simplify.} \\ &= 6(-1) && i^2 = -1 \\ &= -6 \end{aligned}$$

■

Example 2 Performing Operations with Complex Numbers

Perform the indicated operations. Simplify the result.

1. $9 - 4i(2 - 7i)$ **2.** $(2 + 5i)(3 - 7i)$ **3.** $(3 - 5i)^2$

Solution

1.
$$\begin{aligned} 9 - 4i(2 - 7i) &= 9 - 8i + 28i^2 && \text{Distributive law} \\ &= 9 - 8i + 28(-1) && i^2 = -1 \\ &= 9 - 8i - 28 && \text{Multiply.} \\ &= -19 - 8i \end{aligned}$$

2.
$$\begin{aligned} (2 + 5i)(3 - 7i) &= 2 \cdot 3 - 2 \cdot 7i + 5i \cdot 3 - 5i \cdot 7i && \text{Multiply pairs of terms.} \\ &= 6 - 14i + 15i - 35i^2 && \text{Simplify.} \\ &= 6 + i - 35(-1) && \text{Simplify; } i^2 = -1 \\ &= 6 + i + 35 && \text{Multiply.} \\ &= 41 + i \end{aligned}$$

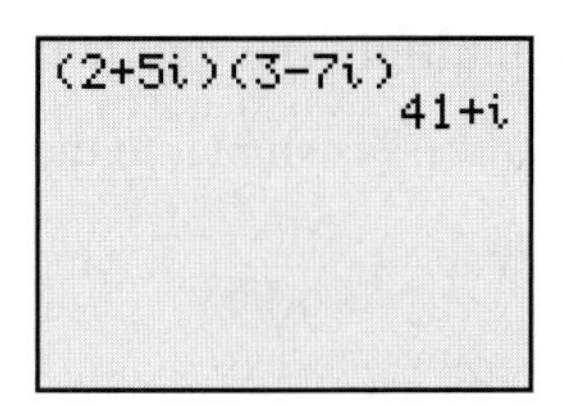

Figure 30 Verify that $(2+5i)(3-7i) = 41+i$

We use a graphing calculator to verify our work (see Fig. 30). Press [2nd] [·] to enter the imaginary unit i.

3.
$$\begin{aligned}(3-5i)^2 &= 3^2 - 2(3)(5i) + (5i)^2 && (A-B)^2 = A^2 - 2AB + B^2\\ &= 9 - 30i + 25i^2 && (bc)^n = b^n c^n\\ &= 9 - 30i + 25(-1) && i^2 = -1\\ &= 9 - 30i - 25 && \text{Multiply.}\\ &= -16 - 30i\end{aligned}$$

■

Recall from Section 9.3 that we call radical expressions such as $7+\sqrt{3}$ and $7-\sqrt{3}$ radical conjugates of each other. Similarly, we call the complex numbers $7+3i$ and $7-3i$ complex conjugates of each other.

DEFINITION Complex conjugate

The complex numbers $a+bi$ and $a-bi$ are called **complex conjugates** of each other.

For example, the complex conjugate of $2-8i$ is $2+8i$. The complex conjugate of $6i$ is $-6i$.

Example 3 Finding the Product of Two Complex Conjugates

Find the product $(4+7i)(4-7i)$.

Solution

$$\begin{aligned}(4+7i)(4-7i) &= 4^2 - (7i)^2 && (A+B)(A-B) = A^2 - B^2\\ &= 16 - 7^2 i^2 && (bc)^n = b^n c^n\\ &= 16 - 49(-1) && i^2 = -1\\ &= 65\end{aligned}$$

■

Dividing Complex Numbers

We **simplify** a quotient of two complex numbers by removing i from the denominator of the quotient. In Example 3, we found that the product of two particular complex conjugates is a real number. In general, **the product of *any* two complex conjugates is a real number.** We can use this generalization to remove i from the denominator of a quotient of two complex numbers.

Example 4 Simplifying the Quotient of Two Complex Numbers

Simplify $\frac{5}{2+3i}$.

Solution

Since the product of the complex conjugates $2+3i$ and $2-3i$ is a real number, we can remove i from the denominator of $\frac{5}{2+3i}$ by multiplying the quotient by $\frac{2-3i}{2-3i}$:

$$\begin{aligned}\frac{5}{2+3i} &= \frac{5}{2+3i}\cdot\frac{2-3i}{2-3i} && \text{The conjugate of } 2+3i \text{ is } 2-3i.\\ &= \frac{10-15i}{4-9i^2} && \text{Distributive law; } (A+B)(A-B) = \left(A^2 - B^2\right)\\ &= \frac{10-15i}{4-9(-1)} && i^2 = -1\\ &= \frac{10-15i}{13} && \text{Simplify.}\\ &= \frac{10}{13} - \frac{15}{13}i && \text{Write in } a+bi \text{ form.}\end{aligned}$$

■

To simplify the quotient of two complex numbers, we multiply the quotient by $\dfrac{\textbf{complex conjugate of the denominator}}{\textbf{complex conjugate of the denominator}}$. Notice that this process is similar to the way we rationalize the denominator of a radical expression such as $\dfrac{5}{2+3\sqrt{7}}$.

Example 5 Simplifying the Quotient of Two Complex Numbers

Simplify $\dfrac{2+4i}{3-5i}$.

Solution

$$\frac{2+4i}{3-5i} = \frac{2+4i}{3-5i}\cdot\frac{3+5i}{3+5i} \qquad \text{The conjugate of } 3-5i \text{ is } 3+5i.$$

$$= \frac{(2+4i)(3+5i)}{(3-5i)(3+5i)} \qquad \text{Multiply numerators; multiply denominators.}$$

$$= \frac{6+10i+12i+20i^2}{9-25i^2} \qquad \text{Multiply pairs of terms; } (A-B)(A+B)=A^2-B^2$$

$$= \frac{6+22i+20(-1)}{9-25(-1)} \qquad \text{Simplify; } i^2=-1$$

$$= \frac{-14+22i}{34} \qquad \text{Simplify.}$$

$$= -\frac{14}{34}+\frac{22}{34}i \qquad \text{Write in } a+bi \text{ form.}$$

$$= -\frac{7}{17}+\frac{11}{17}i \qquad \text{Simplify.}$$

(2+4i)/(3-5i)
-.41+.65i
-(7/17)+(11/17)i
-.41+.65i

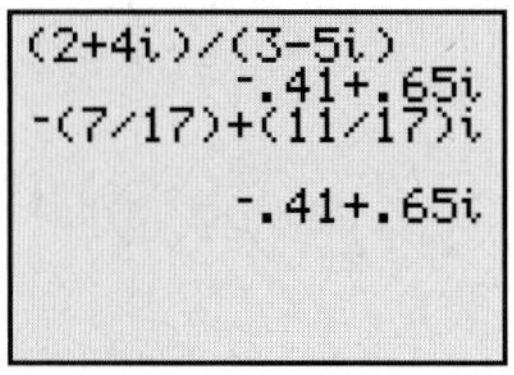

Figure 31 Verify the work

We use a graphing calculator to verify our work (see Fig. 31). Here, we set the float to 2, so the numbers in the result are rounded to the second decimal place. ■

When the denominator of a fraction is a pure imaginary number, the easiest way to remove *i* from the denominator is to multiply the fraction by $\dfrac{i}{i}$.

Example 6 Simplifying the Quotient of Two Complex Numbers

Simplify $\dfrac{2-5i}{3i}$.

Solution

Since the denominator $3i$ is a pure imaginary number, we can remove i from the denominator by multiplying by $\dfrac{i}{i}$:

$$\frac{2-5i}{3i} = \frac{2-5i}{3i}\cdot\frac{i}{i} \qquad \text{Multiply by } \frac{i}{i}.$$

$$= \frac{2i-5i^2}{3i^2} \qquad \text{Multiply numerators; multiply denominators.}$$

$$= \frac{2i-5(-1)}{3(-1)} \qquad i^2=-1$$

$$= \frac{5+2i}{-3} \qquad \text{Multiply.}$$

$$= -\frac{5}{3}-\frac{2}{3}i \qquad \text{Write in } a+bi \text{ form.}$$

■

group exploration

Finding powers of i

1. Find the indicated power of i. Verify your work with a graphing calculator.
   a. i^2
   b. i^3 [**Hint:** $i^3 = i^2 \cdot i$]
   c. i^4 [**Hint:** $i^4 = i^3 \cdot i$]
   d. i^5
2. Continue finding powers of i, such as $i^6, i^7, i^8, \ldots,$ until you see a pattern in your results. Describe the pattern.
3. Find the indicated power of i. Verify your result with a graphing calculator.
   a. i^{23} **b.** i^{41} **c.** i^{102} **d.** i^{400}

Key Points OF SECTION 11.3

Performing operations with complex numbers We perform operations with complex numbers in much the same way as we do with radical expressions.

First write radicals in terms of i If a radicand is a negative real number, we first write the radical in terms of i before performing any operations.

Complex conjugate The complex numbers $a + bi$ and $a - bi$ are called **complex conjugates** of each other.

Simplify a quotient We **simplify** a quotient of two complex numbers by removing i from the denominator of the quotient.

Product of complex conjugates The product of *any* two complex conjugates is a real number.

Simplifying the quotient of two complex numbers To simplify the quotient of two complex numbers, we multiply the quotient by $\frac{\text{complex conjugate of the denominator}}{\text{complex conjugate of the denominator}}$.

Multiplying by $\frac{i}{i}$ When the denominator of a fraction is a pure imaginary number, the easiest way to remove i from the denominator is to multiply the fraction by $\frac{i}{i}$.

HOMEWORK 11.3

FOR EXTRA HELP ▶ Student Solutions Manual | PH Math/Tutor Center | MathXL® | MyMathLab

Perform the indicated operation. If your result is ~~an imaginary~~ a complex number, write it in $a + bi$ form. Use a graphing calculator to verify your work when possible.

1. $(4 - 7i) + (3 + 10i)$
2. $(15 - 2i) + (6 + 17i)$
3. $(5 - \sqrt{-9}) + (2 - \sqrt{-25})$
4. $(7 - \sqrt{-4}) + (9 - \sqrt{-36})$
5. $(6 - 5i) - (2 - 13i)$
6. $(9 - 4i) - (8 - 2i)$
7. $(6 - \sqrt{-49}) - (1 + \sqrt{-81})$
8. $(3 - \sqrt{-1}) - (3 + \sqrt{-64})$
9. $2i \cdot 9i$
10. $4i \cdot 6i$
11. $-10i(-5i)$
12. $-7i(4i)$
13. $\sqrt{-4}\sqrt{-25}$
14. $\sqrt{-49}\sqrt{-16}$
15. $\sqrt{-3}\sqrt{-5}$
16. $\sqrt{-2}\sqrt{-11}$
17. $(8i)^2$
18. $(-4i)^2$
19. $5i(3 - 2i)$
20. $6i(1 + 3i)$
21. $20 - 3i(2 - 7i)$
22. $2 + 4i(3 - 8i)$
23. $(2 + 5i)(3 + 4i)$
24. $(7 + 3i)(10 + 2i)$
25. $(3 - 6i)(5 + 2i)$
26. $(4 - 3i)(2 + 7i)$
27. $(-6 + 4i)(-2 + 7i)$
28. $(-5 + 7i)(-3 + 2i)$
29. $(5 + 4i)(5 - 4i)$
30. $(8 + 3i)(8 - 3i)$
31. $(2 - 9i)(2 + 9i)$
32. $(3 - 7i)(3 + 7i)$
33. $(1 + i)(1 - i)$
34. $(3 + i)(3 - i)$
35. $(2 + 7i)^2$
36. $(6 + 3i)^2$
37. $(4 - 5i)^2$
38. $(7 - 4i)^2$
39. $(-4 + 3i)^2$
40. $(-5 + 2i)^2$
41. $\frac{3}{2 + 5i}$
42. $\frac{4}{4 + 3i}$
43. $\frac{3i}{7 - 2i}$
44. $\frac{2i}{3 - 8i}$
45. $\frac{2 + 3i}{7 + i}$
46. $\frac{5 + 8i}{3 + i}$
47. $\frac{3 + 4i}{3 - 4i}$
48. $\frac{4 - 7i}{4 + 7i}$

49. $\dfrac{3-5i}{2-9i}$ **50.** $\dfrac{4-3i}{1-5i}$ **51.** $\dfrac{5+7i}{4i}$

52. $\dfrac{2+8i}{3i}$ **53.** $\dfrac{7}{5i}$ **54.** $\dfrac{3}{2i}$

55. Two students try to find the product $\sqrt{-2}\sqrt{-8}$:

Student 1's work	***Student 2's work***
$\sqrt{-2}\sqrt{-8} = \sqrt{-2(-8)}$	$\sqrt{-2}\sqrt{-8} = i\sqrt{2}\cdot i\sqrt{8}$
$= \sqrt{16}$	$= i^2\sqrt{16}$
$= 4$	$= -4$

Did one, both, or neither student find the product correctly? Explain.

56. A student tries to find the product $(3+i)(3-i)$:

$$(3+i)(3-i) = 3^2 - i^2 = 9 - 1 = 8$$

Describe any errors. Then find the product correctly.

57. Find nonzero real-number values of a, b, c, and d such that the sum $(a+bi)+(c+di)$
a. is an imaginary number (not pure).
b. is a real number.
c. is a pure imaginary number.

58. Find nonzero real-number values of a, b, c, and d such that the product $(a+bi)(c+di)$
a. is an imaginary number (not pure).
b. is a real number.
c. is a pure imaginary number.

59. The square of a real number is a nonnegative real number. What can you say about the square of a pure imaginary number? Explain.

60. Describe how to multiply two complex numbers. Describe how to simplify the quotient of two complex numbers.

Related Review

Simplify. If your result is an imaginary number, write it in $a+bi$ form.

61. $\dfrac{4}{3+2\sqrt{x}}$ **62.** $\dfrac{7+4\sqrt{x}}{2-5\sqrt{x}}$

63. $\dfrac{4}{3+2i}$ **64.** $\dfrac{7+4i}{2-5i}$

Find all complex-number solutions of the equation.

65. $3x^2 - 2x + 3 = 0$ **66.** $x^2 - 2x + 5 = 0$

67. $5x^2 - 4x = -1$ **68.** $4x^2 - x = -2$

69. $(x-3)(2x+1) = -10$ **70.** $(x+1)(x+2) = x$

71. $x(3x-2) = 2 + 2(x-3)$

72. $5 - 2(x-4) = -2x(x-2)$

73. $(5x+3)^2 = -20$ **74.** $(4x-7)^2 = -18$

Expressions, Equations, Functions, and Graphs

Perform the indicated instruction. Then use words such as linear, quadratic, cubic, exponential, logarithmic, rational, radical, polynomial, degree, function, imaginary number, one variable, *and* two variables *to describe the expression, equation, or system.*

75. Solve $4x^2 - 2x + 3 = 0$.

76. Simplify $2\sqrt{20x^3} - 3x\sqrt{45x} + 4x\sqrt{24}$.

77. Factor $10x^2 - 19x + 6$.

78. Simplify $\dfrac{2\sqrt{x}-5}{3\sqrt{x}+4}$.

79. Find the product $(3i-7)(4i+6)$.

80. Solve $4\sqrt{3x-1} - 3 = 5$.

Section 11.3 Quiz

Perform the indicated operations. If your result is an imaginary number, write it in $a+bi$ form.

1. $(6-2i)+(3-4i)$ **2.** $(3+7i)-(8-2i)$

3. $-4i\cdot 3i$ **4.** $\sqrt{-2}\sqrt{-7}$

5. $(5-3i)(7+i)$ **6.** $(4-3i)^2$

7. $(8+5i)(8-5i)$ **8.** $\dfrac{3+2i}{5-4i}$

9. $\dfrac{5-7i}{6i}$

10. True or false? A complex number times a pure imaginary number must be an imaginary number. Explain.

11.4 PYTHAGOREAN THEOREM, DISTANCE FORMULA, AND CIRCLES

"I'm very well acquainted too with matters mathematical,
I understand equations, both the simple and quadratical,
About the binomial theorem I'm teeming with a lot of news
With many cheerful facts about the square of the hypotenuse."

—Gilbert & Sullivan, *The Pirates of Penzance*

Objectives

- Use the Pythagorean theorem to find the length of a side of a right triangle.
- Use the Pythagorean theorem to make estimates about authentic situations.
- Use the distance formula to find the distance between two points.
- Find or graph the equation of a circle.

In this section, we will discuss a special type of triangle called a *right triangle* and a useful formula relating to right triangles: the Pythagorean theorem. We will also find the distance between two points. Finally, we will find equations of circles and graph equations of circles.

Pythagorean Theorem

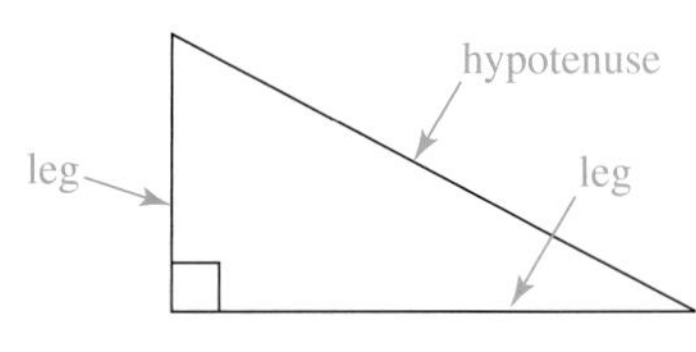

Figure 32 A right triangle

We begin by working with an important type of triangle called a *right triangle*. An angle of 90° is called a *right angle*. If one angle of a triangle measures 90°, the triangle is a **right triangle** (see Fig. 32). The side opposite the right angle is the triangle's longest side. We call that side the **hypotenuse,** and we call the two shorter sides the **legs.**

The **Pythagorean theorem** describes the relationship between the lengths of the legs and hypotenuse of a right triangle.

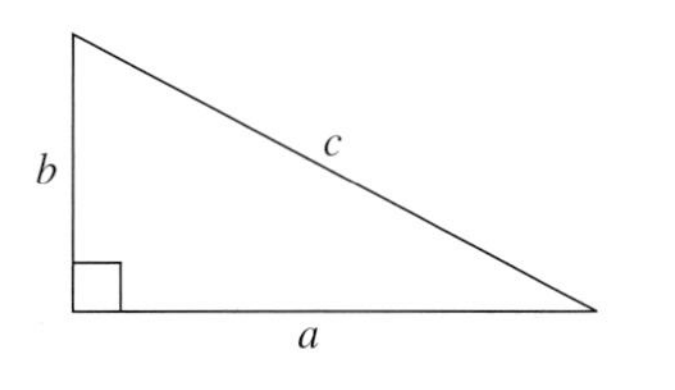

Figure 33 The Pythagorean theorem: $a^2 + b^2 = c^2$

Pythagorean Theorem

If a and b are the lengths of the legs of a right triangle and c is the length of the hypotenuse, then

$$a^2 + b^2 = c^2$$

In words: The sum of the squares of the lengths of the legs is equal to the square of the length of the hypotenuse (see Fig. 33).

If we know the lengths of two of the three sides of a right triangle, how can we use the Pythagorean theorem to find the length of the third side?

Example 1 Finding the Length of a Side of a Right Triangle

The lengths of two sides of a right triangle are given. Find the length of the third side.

1.

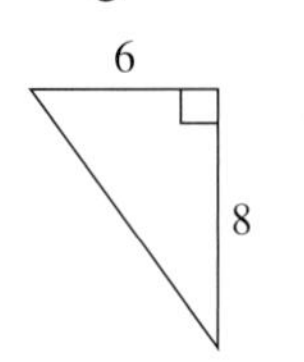

2.

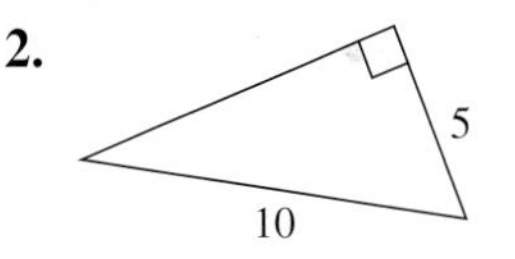

Solution

1. Since the lengths of the legs are given, we are to find the length of the hypotenuse. We substitute $a = 6$ and $b = 8$ into the equation $c^2 = a^2 + b^2$ and solve for c:

$$\begin{aligned} c^2 &= 6^2 + 8^2 && \text{Substitute 6 for } a \text{ and 8 for } b. \\ c^2 &= 36 + 64 && \text{Simplify.} \\ c^2 &= 100 \\ c &= 10 && c \text{ is nonnegative.} \end{aligned}$$

The length of the hypotenuse is 10 units.

2. The length of the hypotenuse is 10 units, and the length of one of the legs is 5 units. We substitute $a = 5$ and $c = 10$ into the equation $a^2 + b^2 = c^2$ and solve for b:

$$\begin{aligned} 5^2 + b^2 &= 10^2 && \text{Substitute 5 for } a \text{ and 10 for } c. \\ 25 + b^2 &= 100 && \text{Simplify.} \\ b^2 &= 75 && \text{Subtract 25 from both sides.} \\ b &= \sqrt{75} && b \text{ is nonnegative.} \\ b &= 5\sqrt{3} && \sqrt{75} = \sqrt{25 \cdot 3} = 5\sqrt{3} \end{aligned}$$

The length of the other leg is $5\sqrt{3}$ units (about 8.66 units). ■

Using the Pythagorean Theorem to Make Estimates About Authentic Situations

The Pythagorean theorem is used in numerous fields, including architecture, physics, engineering, graphic design, mathematics, surveying, chemistry, and aeronautics.

Example 2 Using the Pythagorean Theorem to Find a Distance

A surveyor wants to estimate the distance (in miles) across a lake from point A to point B as shown in Fig. 34. Find that distance.

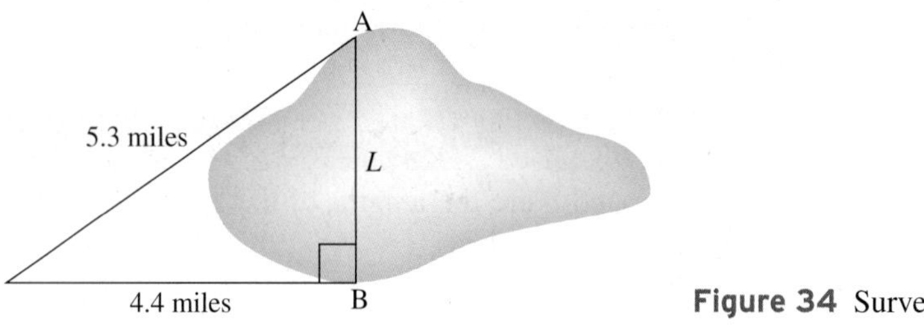

Figure 34 Surveying a lake

Solution

We define L to be the distance (in miles) between points A and B (see Fig. 34). The triangle in Fig. 34 is a right triangle in which the hypotenuse has length 5.3 miles and one of its legs has length 4.4 miles. We use the Pythagorean theorem to find L:

$$4.4^2 + L^2 = 5.3^2 \quad \text{Pythagorean theorem}$$
$$19.36 + L^2 = 28.09 \quad \text{Find the squares.}$$
$$L^2 = 8.73 \quad \text{Subtract 19.36 from both sides.}$$
$$L = \sqrt{8.73} \quad L \text{ is nonnegative.}$$
$$L \approx 3.0 \quad \text{Compute.}$$

So, the distance across the the lake between points A and B is approximately 3.0 miles. ■

Distance Between Two Points

We can use the Pythagorean theorem to find a formula for the distance between two points in the coordinate system. Let (x_1, y_1) and (x_2, y_2) represent two points, where $x_2 > x_1$ and $y_2 > y_1$ (see Fig. 35). We let d be the distance between the two points.

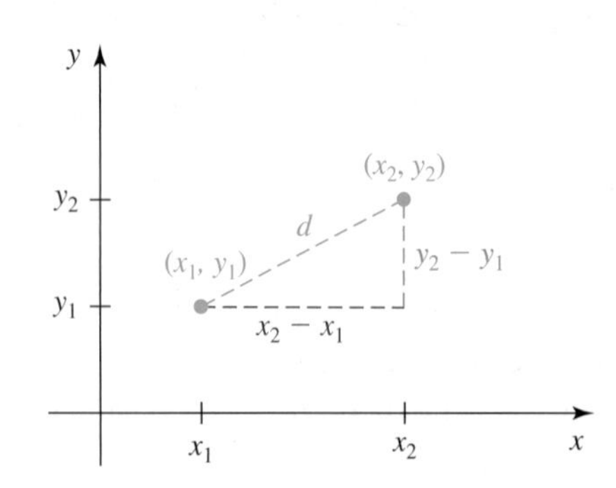

Figure 35 Find the distance between points (x_1, y_1) and (x_2, y_2)

Notice that the triangle shown in Fig. 35 is a right triangle with hypotenuse of length d and legs of lengths $x_2 - x_1$ and $y_2 - y_1$. We apply the Pythagorean theorem:

$$d^2 = (x_2 - x_1)^2 + (y_2 - y_1)^2$$
$$d = \sqrt{(x_2 - x_1)^2 + (y_2 - y_1)^2} \quad d \text{ is nonnegative.}$$

Although we assumed that $x_2 > x_1$ and $y_2 > y_1$ to find the *distance formula,* it can be shown that the formula gives the correct distance between *any* two points.

Distance Formula

The distance d between points (x_1, y_1) and (x_2, y_2) is given by the **distance formula:**

$$d = \sqrt{(x_2 - x_1)^2 + (y_2 - y_1)^2}$$

Example 3 Finding the Distance Between Two Points

Find the distance between $(-2, 5)$ and $(3, -1)$.

Solution

We substitute $x_1 = -2$, $y_1 = 5$, $x_2 = 3$, and $y_2 = -1$ into the distance formula:

$$\begin{aligned} d &= \sqrt{(3-(-2))^2 + (-1-5)^2} && \text{Substitute into distance formula.} \\ &= \sqrt{5^2 + (-6)^2} && \text{Subtract.} \\ &= \sqrt{61} && \text{Simplify.} \end{aligned}$$

The distance between $(-2, 5)$ and $(3, -1)$ is $\sqrt{61}$ units (about 7.81 units). ■

Equation of a Circle

We can use the distance formula to find an equation whose graph is a circle. To see how, we first state the definition of a circle in terms of its *center* and *radius*.

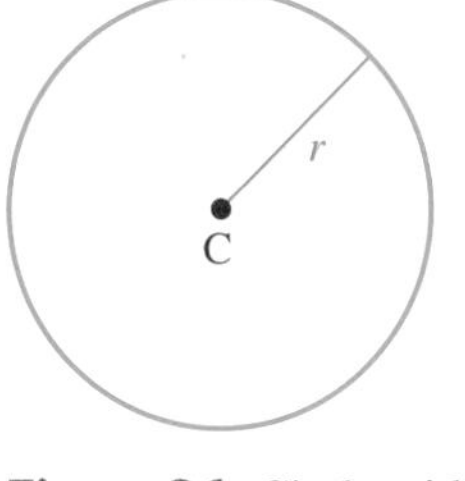

Figure 36 Circle with center C and radius r

DEFINITION Circle

A **circle** with **center** point C and **radius** r, where $r > 0$, is the set of all points in a plane that are r units from point C in that plane. See Fig. 36.

Now we find an equation of a circle with center (h, k) and radius r (see Fig. 37). If (x, y) is a point on the circle, then the distance between (x, y) and (h, k) is the radius r:

$$\sqrt{(x-h)^2 + (y-k)^2} = r \qquad \text{Distance formula}$$

Squaring both sides of the equation gives

$$(x-h)^2 + (y-k)^2 = r^2$$

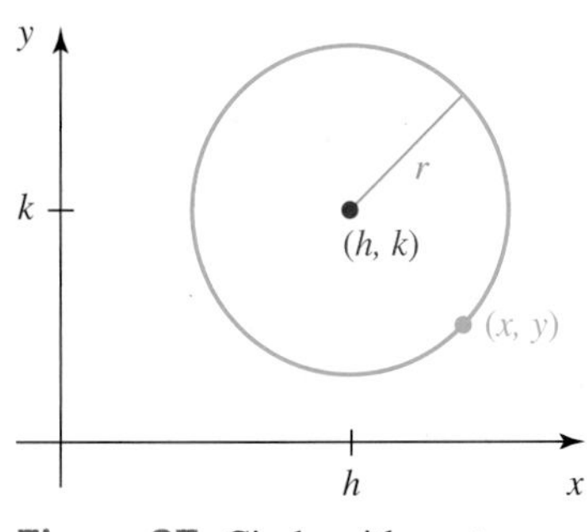

Figure 37 Circle with center (h, k) and radius r

Equation of a Circle

If a circle has center (h, k) and radius r, then an equation of the circle is

$$(x-h)^2 + (y-k)^2 = r^2$$

The graph of an equation of this form with $r > 0$ is a circle with center (h, k) and radius r.

Example 4 Finding an Equation of a Circle

Find an equation of the circle with the given center and radius.

1. Center $(0, 0)$, radius 3
2. Center $(2, -5)$, radius $\sqrt{13}$

Solution

1. We substitute $h = 0$, $k = 0$, and $r = 3$ in $(x-h)^2 + (y-k)^2 = r^2$:

$$\begin{aligned} (x-0)^2 + (y-0)^2 &= 3^2 && \text{Substitute 0 for } h\text{, 0 for } k\text{, and 3 for } r. \\ x^2 + y^2 &= 9 && \text{Simplify.} \end{aligned}$$

2. We substitute $h = 2$, $k = -5$, and $r = \sqrt{13}$ in $(x-h)^2 + (y-k)^2 = r^2$:

$$(x-2)^2 + (y-(-5))^2 = \left(\sqrt{13}\right)^2 \quad \text{Substitute 2 for } h, -5 \text{ for } k, \text{ and } \sqrt{13} \text{ for } r.$$
$$(x-2)^2 + (y+5)^2 = 13 \quad \text{Simplify.}$$

■

The equation of a circle centered at the origin (0, 0) with radius r is

$$(x-0)^2 + (y-0)^2 = r^2$$
$$x^2 + y^2 = r^2$$

Equation of a Circle Centered at the Origin

If a circle has center (0, 0) and radius r, then an equation of the circle is

$$x^2 + y^2 = r^2$$

Example 5 Finding the Center and Radius of a Circle

Determine the center and radius of the circle. Also, sketch the circle.

1. $x^2 + y^2 = 19$
2. $(x+5)^2 + (y-3)^2 = 4$

Solution

1. The equation has the form $x^2 + y^2 = r^2$. Thus, the circle is centered at the origin (0, 0), and

$$r^2 = 19$$
$$r = \sqrt{19} \quad r \text{ is positive.}$$

So, the radius is $\sqrt{19} \approx 4.36$. We sketch the circle in Fig. 38.

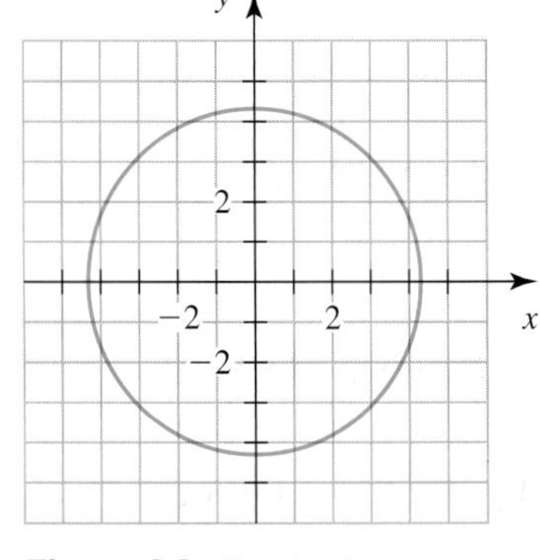

Figure 38 Graph of $x^2 + y^2 = 19$

2. We write $(x+5)^2 + (y-3)^2 = 4$ in the form $(x-h)^2 + (y-k)^2 = r^2$:

$$(x-(-5))^2 + (y-3)^2 = 2^2$$

where $h = -5$, $k = 3$, and $r = 2$. So, the circle has center $(-5, 3)$ and radius 2. We sketch the circle in Fig. 39.

To use a graphing calculator to draw the circle $(x+5)^2 + (y-3)^2 = 4$, we first isolate y:

$$(x+5)^2 + (y-3)^2 = 4 \quad \text{Original equation}$$
$$(y-3)^2 = 4 - (x+5)^2 \quad \text{Subtract } (x+5)^2 \text{ from both sides.}$$
$$y - 3 = \pm\sqrt{4-(x+5)^2} \quad \text{Square root property}$$
$$y = 3 \pm \sqrt{4-(x+5)^2} \quad \text{Add 3 to both sides.}$$

Then we enter $y = 3 - \sqrt{4-(x+5)^2}$ and $y = 3 + \sqrt{4-(x+5)^2}$ and draw the graphs of both functions on the same coordinate system (see Fig. 40). ■

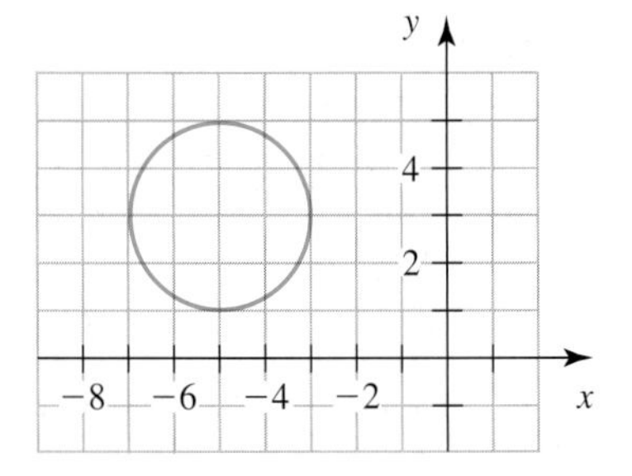

Figure 39 Graph of $(x+5)^2 + (y-3)^2 = 4$

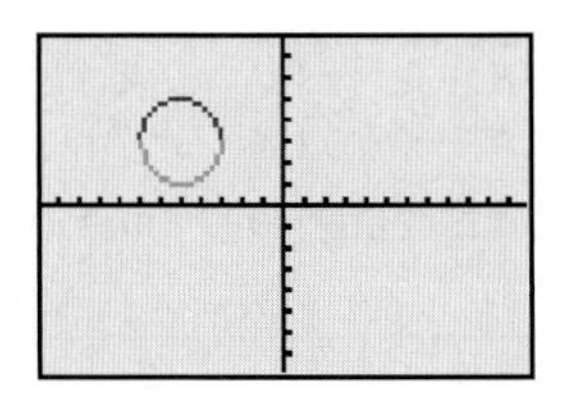

Figure 40 Use ZStandard, then ZInteger, then Zoom In

group exploration

Pythagorean theorem and its converse

For this exploration, you will need a ruler, scissors, and paper. It will help to have a tool for drawing right angles, such as a protractor or graph paper, although a corner of a piece of paper will suffice. For each triangle, assume that c is the length of (one of) the *longest* side(s) and a and b are the lengths of the other sides.

1. Sketch three right triangles of different sizes. Measure the sides and show that for each right triangle, $a^2 + b^2 = c^2$.

2. Now sketch three triangles of different sizes that are *not* right triangles. For these triangles, check whether $a^2 + b^2 = c^2$.
3. Sketch a triangle that has an angle close, but not equal, to 90°. Check whether $a^2 + b^2 \approx c^2$. If you cannot show this for your triangle, repeat the problem with a triangle that has an angle even closer to 90°.
4. If $a = 3$, $b = 5$, and $c = \sqrt{34}$, then $a^2 + b^2 = c^2$. Cut three thin strips of paper that are about 3, 5, and $\sqrt{34}$ inches in length. Form a triangle with the three strips of paper. Is the triangle a right triangle?
5. Find values of a, b, and c, other than the ones in Problem 4, such that $a^2 + b^2 = c^2$. Then repeat Problem 4 with your values.
6. Find three more values of a, b, and c such that $a^2 + b^2 = c^2$. Then repeat Problem 4 with your values.
7. Summarize at least three concepts addressed in this exploration.

Key Points OF SECTION 11.4

Pythagorean theorem	If a and b are the lengths of the legs of a right triangle and c is the length of the hypotenuse, then $a^2 + b^2 = c^2$.
Distance formula	The distance d between points (x_1, y_1) and (x_2, y_2) is given by the **distance formula:** $d = \sqrt{(x_2 - x_1)^2 + (y_2 - y_1)^2}$.
Definition of circle	A **circle** with **center** point C and **radius** r, where $r > 0$, is the set of all points in a plane that are r units from point C in that plane.
Equation of a circle	If a circle has center (h, k) and radius r, then an equation of the circle is $(x - h)^2 + (y - k)^2 = r^2$. The graph of an equation of this form with $r > 0$ is a circle with center (h, k) and radius r.
Equation of a circle centered at the origin	If a circle has center $(0, 0)$ and radius r, then an equation of the circle is $x^2 + y^2 = r^2$.

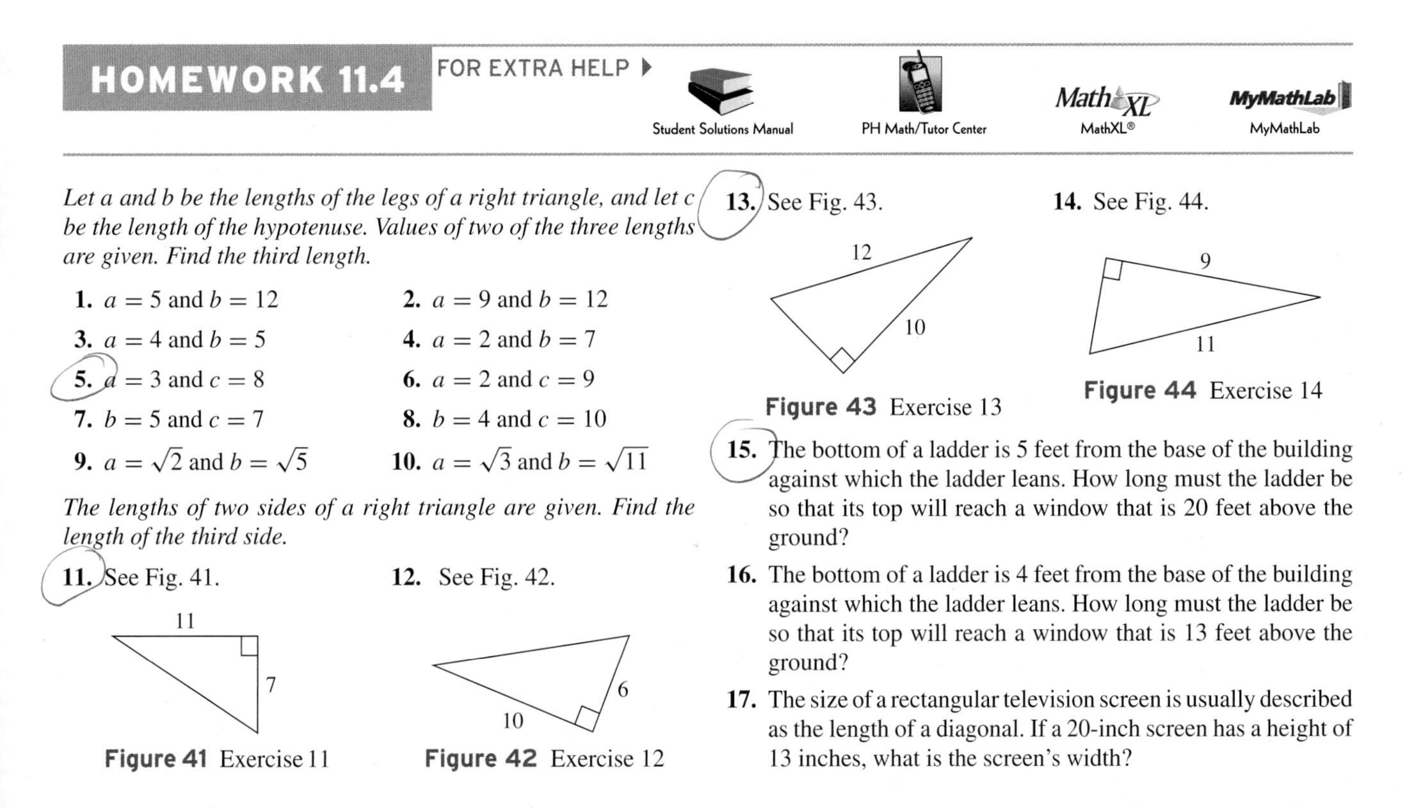

HOMEWORK 11.4

FOR EXTRA HELP ▶ Student Solutions Manual · PH Math/Tutor Center · MathXL® · MyMathLab

Let a and b be the lengths of the legs of a right triangle, and let c be the length of the hypotenuse. Values of two of the three lengths are given. Find the third length.

1. $a = 5$ and $b = 12$
2. $a = 9$ and $b = 12$
3. $a = 4$ and $b = 5$
4. $a = 2$ and $b = 7$
5. $a = 3$ and $c = 8$
6. $a = 2$ and $c = 9$
7. $b = 5$ and $c = 7$
8. $b = 4$ and $c = 10$
9. $a = \sqrt{2}$ and $b = \sqrt{5}$
10. $a = \sqrt{3}$ and $b = \sqrt{11}$

The lengths of two sides of a right triangle are given. Find the length of the third side.

11. See Fig. 41.
12. See Fig. 42.

Figure 41 Exercise 11

Figure 42 Exercise 12

13. See Fig. 43.
14. See Fig. 44.

Figure 43 Exercise 13

Figure 44 Exercise 14

15. The bottom of a ladder is 5 feet from the base of the building against which the ladder leans. How long must the ladder be so that its top will reach a window that is 20 feet above the ground?
16. The bottom of a ladder is 4 feet from the base of the building against which the ladder leans. How long must the ladder be so that its top will reach a window that is 13 feet above the ground?
17. The size of a rectangular television screen is usually described as the length of a diagonal. If a 20-inch screen has a height of 13 inches, what is the screen's width?

18. The size of a rectangular television screen is usually described as the length of a diagonal. If a 25-inch screen has a height of 14 inches, what is the screen's width?

19. A surveyor wants to estimate the distance (in miles) across a lake from point A to point B as shown in Fig. 45. Find that distance.

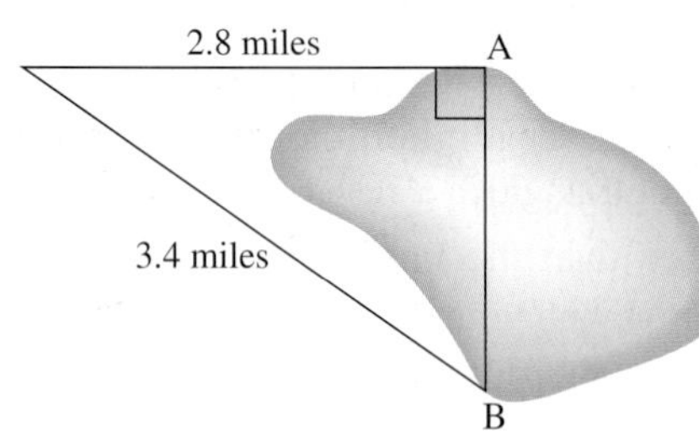

Figure 45 Exercise 19

20. A surveyor wants to estimate the distance (in miles) across a lake from point A to point B as shown in Fig. 46. Find that distance.

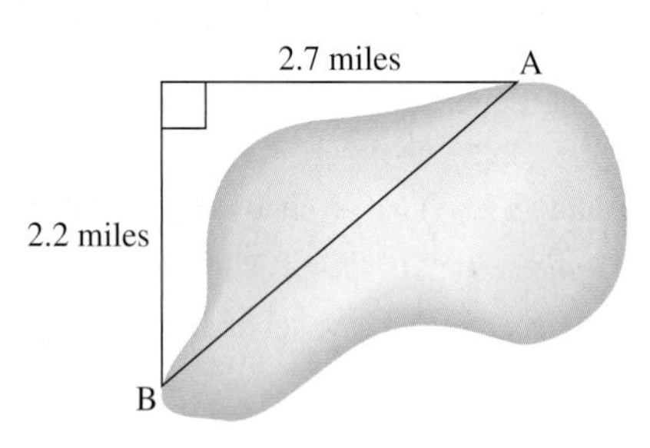

Figure 46 Exercise 20

21. Los Angeles is about 465 miles almost directly south of Reno, Nevada, and is almost directly west of Albuquerque, New Mexico. The distance between Albuquerque and Reno is about 964 miles. What would be the total distance of a trip from Los Angeles to Reno to Albuquerque to Los Angeles?

22. Salt Lake City, Utah, is about 498 miles almost directly south of Helena, Montana, and is almost directly west of Omaha, Nebraska. The distance between Omaha and Helena is about 1144 miles. What would be the total distance of a trip from Salt Lake City to Omaha to Helena to Salt Lake City?

Find the distance between the two given points.

23. $(2, 9)$ and $(8, 1)$ **24.** $(3, 2)$ and $(7, 5)$

25. $(-3, 5)$ and $(4, 2)$ **26.** $(-3, 4)$ and $(2, 7)$

27. $(-6, -3)$ and $(-4, 1)$ **28.** $(-7, 2)$ and $(-5, -6)$

29. $(-4, -5)$ and $(-8, -9)$

30. $(-5, -4)$ and $(-2, -7)$

Find the distance between the two given points. Round your result to the second decimal place.

31. $(2.1, 8.9)$ and $(5.6, 1.7)$ **32.** $(3.2, 7.1)$ and $(6.6, 8.4)$

33. $(-2.18, -5.74)$ and $(3.44, 6.29)$

34. $(-6.41, 1.12)$ and $(2.89, -3.55)$

Find an equation of the circle with the given center and radius.

35. Center $(0, 0)$, radius 7 **36.** Center $(0, 0)$, radius 10

37. Center $(0, 0)$, radius 6.7 **38.** Center $(0, 0)$, radius 2.3

39. Center $(5, 3)$, radius 2 **40.** Center $(4, 7)$, radius 5

41. Center $(-2, 1)$, radius 4

42. Center $(3, -4)$, radius 6

43. Center $(-7, -3)$, radius $\sqrt{3}$

44. Center $(-6, -1)$, radius $\sqrt{2}$

Find the center and radius of the circle. Graph the equation by hand.

45. $x^2 + y^2 = 25$ **46.** $x^2 + y^2 = 9$

47. $x^2 + y^2 = 8$ **48.** $x^2 + y^2 = 17$

49. $(x-3)^2 + (y-5)^2 = 16$ **50.** $(x-2)^2 + (y-4)^2 = 4$

51. $(x+6)^2 + (y-1)^2 = 7$ **52.** $(x-5)^2 + (y+2)^2 = 3$

53. $(x+3)^2 + (y+2)^2 = 1$ **54.** $(x+1)^2 + (y+1)^2 = 1$

55. Find an equation of the circle shown in Fig. 47.

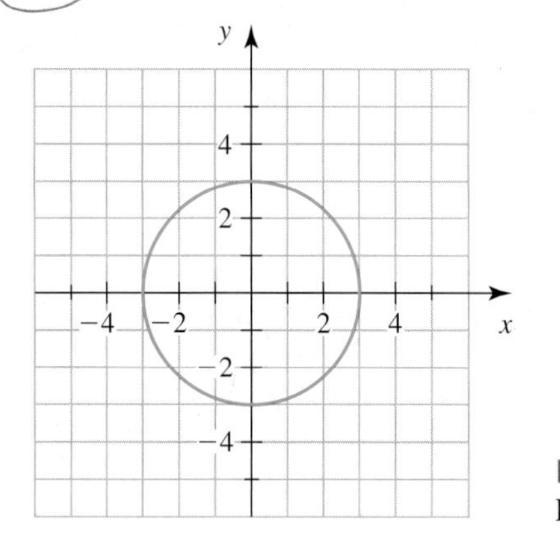

Figure 47 Exercise 55

56. Find an equation of the circle shown in Fig. 48.

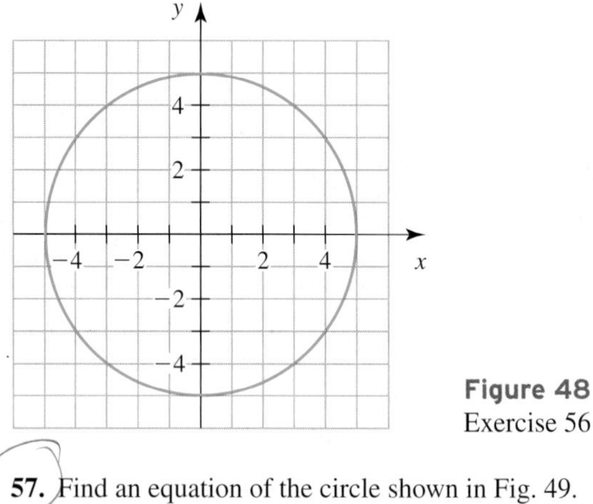

Figure 48 Exercise 56

57. Find an equation of the circle shown in Fig. 49.

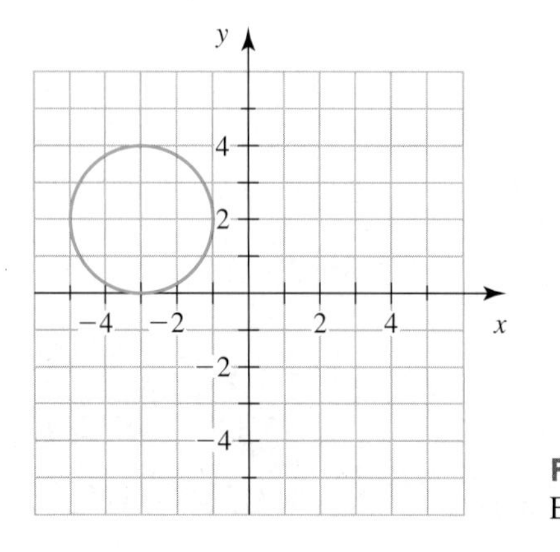

Figure 49 Exercise 57

58. Find an equation of the circle shown in Fig. 50.

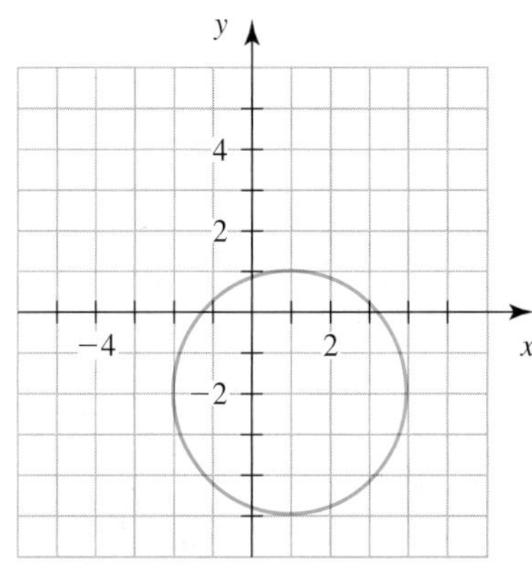

Figure 50 Exercise 58

59. A circle with center (3, 2) contains the point (5, 6). Find an equation of the circle.

60. A circle with center (−4, 3) contains the point (2, −1). Find an equation of the circle.

61. Find equations of two distinct circles that contain the point (5, 3). Sketch by hand the two circles on the same coordinate system.

62. Show with a sketch that there are many circles that contain the points (2, 1) and (4, 6). Which of these circles has the smallest radius? What is an equation of this circle?

63. The Rule of Four can be applied to relations as well as functions. Describe the Rule of Four as applied to the relation $x^2 + y^2 = 16$:
   a. Describe the input–output pairs of the relation by using a graph.
   b. Describe eight input–output pairs by using a table. Round approximate inputs and outputs to the second decimal place.
   c. Describe the input–output pairs by using words.

64. The Rule of Four can be applied to relations as well as functions. Describe the Rule of Four as applied to the relation $(x+1)^2 + (y-3)^2 = 9$:
   a. Describe the input–output pairs of the relation by using a graph.
   b. Describe eight input–output pairs by using a table. Round approximate inputs and outputs to the second decimal place.
   c. Describe the input–output pairs by using words.

65. Give the coordinates of five points that are a distance of 4 units from the point (3, 2).

66. Give the coordinates of five points that are a distance of 1 unit from the point (−4, 1).

67. Is the relation $x^2 + y^2 = 49$ a function? Explain.

68. Is the relation $(x-4)^2 + (y+2)^2 = 16$ a function? Explain.

69. a. For $y = \sqrt{25 - x^2}$, explain why $y \geq 0$ for real-number values of y.
   b. Graph the function $y = \sqrt{25 - x^2}$ by hand. [**Hint:** Square both sides of the equation.] Use a graphing calculator to verify your graph.

70. a. For $y = \sqrt{9 - (x-5)^2}$, explain why $y \geq 0$ for real-number values of y.
   b. Graph the function $y = \sqrt{9 - (x-5)^2}$ by hand. [**Hint:** Square both sides of the equation.] Use a graphing calculator to verify your graph.

71. If the lengths of the legs of a right triangle are equal, we call the triangle an *isosceles right triangle*.
   a. Sketch an example of an isosceles right triangle.
   b. Show that the length of the hypotenuse of an isosceles right triangle is $\sqrt{2}$ times the length of either leg of the triangle. [**Hint:** Let $a = k$ and $b = k$, and apply the Pythagorean theorem.]
   c. If the length of a leg of an isosceles right triangle is 3 units, what is the length of the hypotenuse?
   d. If the length of the hypotenuse of an isosceles right triangle is 5 units, what is the length of each leg?

72. Explain how to graph by hand an equation of the form $(x-h)^2 + (y-k)^2 = r$, where $r > 0$.

Related Review

Sketch the graph.

73. $x + y = 4$

74. $x^2 + y = 4$

75. $x^2 + y^2 = 4$

76. $x^{1/2} + y = 4$

77. $2^x + y = 0$

78. $\left(\frac{1}{2}\right)^x + y = 0$

Expressions, Equations, Functions, and Graphs

Perform the indicated instruction. Then use words such as linear, quadratic, cubic, exponential, logarithmic, rational, radical, polynomial, degree, function, one variable, *and* two variables *to describe the expression, equation, or system.*

79. Graph $f(x) = 2(x-4)^2 - 3$ by hand.

80. Find the domain of $f(x) = \dfrac{x+7}{27x^3 + 18x^2 - 12x - 8}$.

81. Factor $6x^2 - 16x + 8$.

82. Find the difference $\dfrac{3x+5}{x^2+5x-14} - \dfrac{2x}{x^2-4x-21}$.

83. Solve $x(5x-3) = 3(x+1)$.

84. Solve $\dfrac{x+5}{x} - \dfrac{2}{3x^2} = 7$.

Section 11.4 Quiz

1. The length of a leg of a right triangle is 4 inches, and the length of the hypotenuse is 8 inches. Find the length of the other leg.

2. The size of a rectangular television screen is usually described as the length of a diagonal. If a 19-inch screen has a width of 16 inches, what is the screen's height?

Find the distance between the two given points.

3. (−2, −5) and (3, −1)

4. (−3, 2) and (−7, −2)

Find an equation of the circle with given center and radius.

5. Center (−3, 2), radius 6

6. Center (0, 0), radius 2.8

Find the center and radius of the circle. Graph the equation by hand.

7. $x^2 + y^2 = 12$

8. $(x+4)^2 + (y-3)^2 = 25$

9. Find an equation of a circle that has center (2, −1) and contains the point (4, 7).

10. Find equations of two circles that contain the point (0, 0). Sketch the two circles on the same coordinate system.

11.5 ELLIPSES AND HYPERBOLAS

Objective

- Graph equations of ellipses and hyperbolas.

In this text, we work with four types of curves that are cross sections of cones: circles, *ellipses,* parabolas, and *hyperbolas* (see Fig. 51).

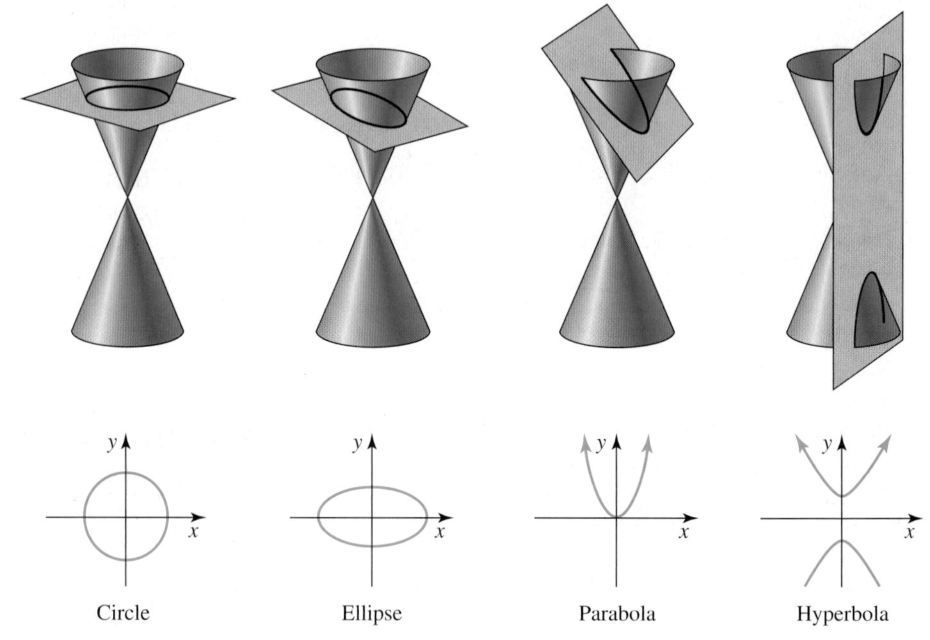

Figure 51 Circles, ellipses, parabolas, and hyperbolas are cross sections of cones

In Chapters 6 and 7, we worked with parabolas. In Section 11.4, we studied circles. In this section, we first discuss ellipses; then we discuss hyperbolas.

Graphing Ellipses

To begin our study of ellipses, we sketch a graph of

$$\frac{x^2}{25} + \frac{y^2}{9} = 1$$

First, we find the y-intercepts by substituting 0 for x and solving for y:

$$\begin{aligned} \frac{0^2}{25} + \frac{y^2}{9} &= 1 && \text{Substitute 0 for } x. \\ \frac{y^2}{9} &= 1 && \text{Simplify.} \\ y^2 &= 9 && \text{Multiply both sides by 9.} \\ y &= \pm 3 && \text{Square root property} \end{aligned}$$

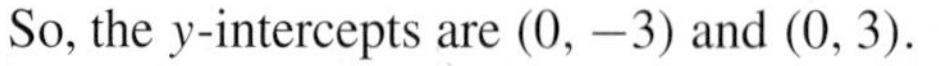

So, the y-intercepts are $(0, -3)$ and $(0, 3)$.

Then, we find the x-intercepts by substituting 0 for y and solving for x:

$$\begin{aligned} \frac{x^2}{25} + \frac{0^2}{9} &= 1 && \text{Substitute 0 for } y. \\ \frac{x^2}{25} &= 1 && \text{Simplify.} \\ x^2 &= 25 && \text{Multiply both sides by 25.} \\ x &= \pm 5 && \text{Square root property} \end{aligned}$$

So, the x-intercepts are $(-5, 0)$ and $(5, 0)$.

All of the points on the graph of the relation have x-coordinates between -5 and 5, inclusive. To see why, we isolate the term $\frac{y^2}{9}$ in the equation $\frac{x^2}{25} + \frac{y^2}{9} = 1$:

$$\frac{y^2}{9} = 1 - \frac{x^2}{25}$$

Note that $\frac{y^2}{9}$ is nonnegative, so

$$1 - \frac{x^2}{25} \geq 0 \qquad 1 - \frac{x^2}{25} \text{ is nonnegative.}$$

$$-\frac{x^2}{25} \geq -1 \qquad \text{Subtract 1 from both sides.}$$

$$x^2 \leq 25 \qquad \text{Multiply both sides by } -25\text{; reverse inequality symbol.}$$

Table 6 Solutions of $\frac{x^2}{25} + \frac{y^2}{9} = 1$

x	y
−4	±1.8
−3	±2.4
−2	±2.75
−1	±2.94
0	±3
1	±2.94
2	±2.75
3	±2.4
4	±1.8

As we set out to show, $x^2 \leq 25$ implies that x is between -5 and 5, inclusive.

Next, we find points on the graph where x is $-4, -3, -2, \ldots, 4$ (see Table 6) and plot these points as well as the intercepts (see Fig. 52).

Finally, we sketch a curve through the points we've plotted (see Fig. 53). The graph is an ellipse.

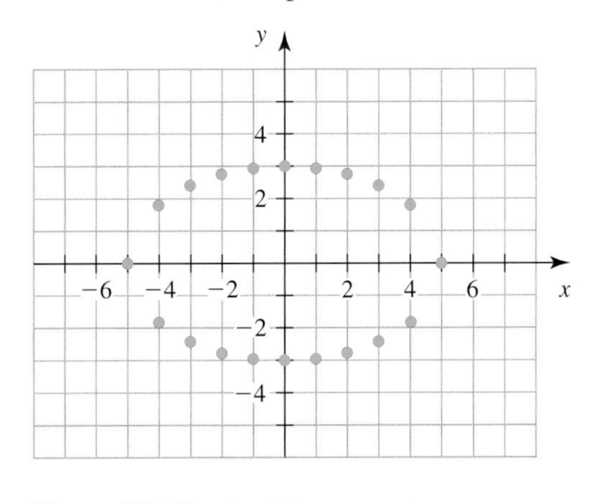

Figure 52 Plot the points

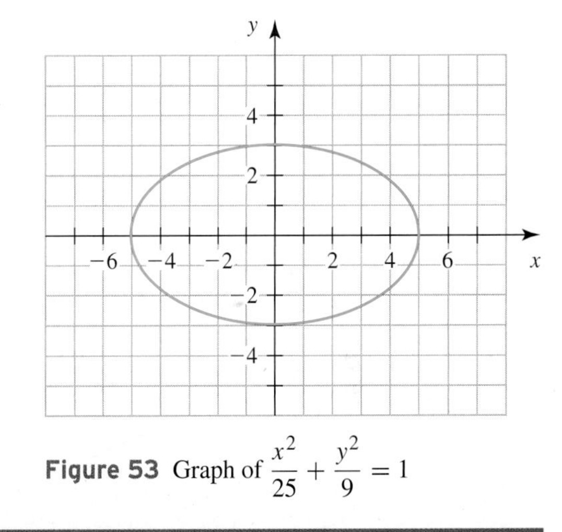

Figure 53 Graph of $\frac{x^2}{25} + \frac{y^2}{9} = 1$

Equation of an Ellipse

An equation that can be put into the form

$$\frac{x^2}{a^2} + \frac{y^2}{b^2} = 1, \qquad \text{where } a > 0 \text{ and } b > 0$$

has an **ellipse** as its graph.

Intercepts of an Ellipse

There is an easier way to sketch the graph of an equation in the form $\frac{x^2}{a^2} + \frac{y^2}{b^2} = 1$, where $a > 0$ and $b > 0$. We begin by finding the x-intercepts:

$$\frac{x^2}{a^2} + \frac{0^2}{b^2} = 1 \qquad \text{Substitute 0 for } y.$$

$$\frac{x^2}{a^2} = 1 \qquad \text{Simplify.}$$

$$x^2 = a^2 \qquad \text{Multiply both sides by } a^2.$$

$$x = \pm\sqrt{a^2} \qquad \text{Square root property}$$

$$x = \pm a \qquad \sqrt{a^2} = a\text{, where } a \geq 0$$

So, the x-intercepts are $(-a, 0)$ and $(a, 0)$.

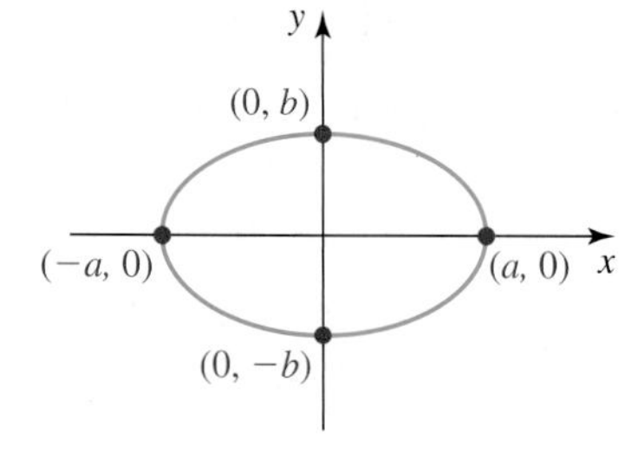

Figure 54 Intercepts of the ellipse $\frac{x^2}{a^2} + \frac{y^2}{b^2} = 1$

By similar steps, we can show that the y-intercepts are $(0, -b)$ and $(0, b)$.

Intercepts of an Ellipse

The ellipse described by

$$\frac{x^2}{a^2} + \frac{y^2}{b^2} = 1$$

has x-intercepts $(-a, 0)$ and $(a, 0)$ and y-intercepts $(0, -b)$ and $(0, b)$. See Fig. 54.

Example 1 Sketching the Graph of an Ellipse

Sketch the graph of $9x^2 + 4y^2 = 36$.

Solution

First, we divide both sides of the equation $9x^2 + 4y^2 = 36$ by 36 so that the right-hand side of the equation is 1:

$$\frac{9x^2}{36} + \frac{4y^2}{36} = \frac{36}{36} \qquad \text{Divide both sides by 36.}$$

$$\frac{x^2}{4} + \frac{y^2}{9} = 1 \qquad \text{Simplify.}$$

The equation is of the form $\frac{x^2}{a^2} + \frac{y^2}{b^2} = 1$, where $a > 0$ and $b > 0$, with $a^2 = 4$ and $b^2 = 9$. Since $a^2 = 4$, we have $a = 2$. So, the x-intercepts are $(-2, 0)$ and $(2, 0)$. Because $b^2 = 9$, we have $b = 3$. So, the y-intercepts are $(0, -3)$ and $(0, 3)$. We plot the intercepts and sketch an ellipse that contains them (see Fig. 55).

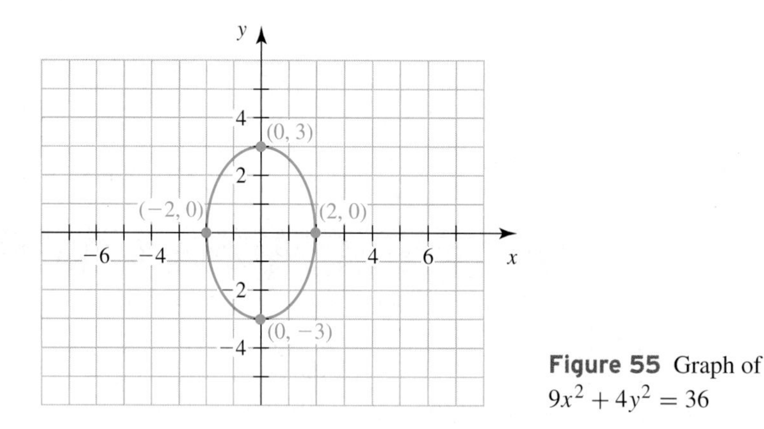

Figure 55 Graph of $9x^2 + 4y^2 = 36$

To use a graphing calculator to draw the ellipse $9x^2 + 4y^2 = 36$, we begin by isolating y:

$$9x^2 + 4y^2 = 36 \qquad \text{Original equation}$$

$$4y^2 = 36 - 9x^2 \qquad \text{Subtract } 9x^2 \text{ from both sides.}$$

$$y^2 = \frac{36 - 9x^2}{4} \qquad \text{Divide both sides by 4.}$$

$$y = \pm\sqrt{\frac{36 - 9x^2}{4}} \qquad \text{Square root property}$$

$$y = \pm\sqrt{\frac{9(4 - x^2)}{4}} \qquad \text{Factor.}$$

$$y = \pm\frac{3}{2}\sqrt{4 - x^2} \qquad \text{Simplify.}$$

Plot1 Plot2 Plot3
\Y1=(3/2)√(4-X^2)
\Y2=-Y1
\Y3=
\Y4=
\Y5=
\Y6=

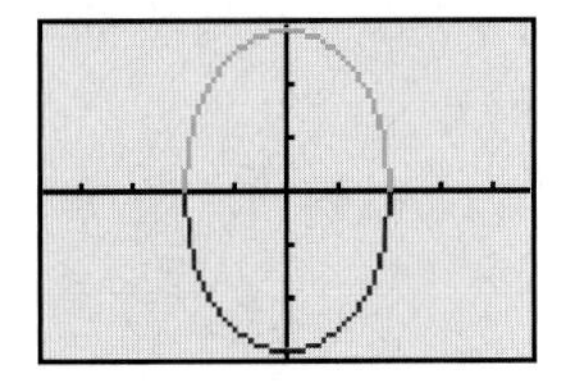

Figure 56 Graph of $9x^2 + 4y^2 = 36$, using ZDecimal

Then, we enter the functions $y = \frac{3}{2}\sqrt{4 - x^2}$ and $y = -\frac{3}{2}\sqrt{4 - x^2}$ (or $Y_2 = -Y_1$) and graph both functions on the same coordinate system (see Fig. 56). ■

Graphing Hyperbolas

We now turn our attention to sketching graphs of hyperbolas. The general equation of a hyperbola is similar to the general equation of an ellipse, except that the left side of the equation is a difference rather than a sum.

Equations of a Hyperbola

An equation that can be put into one of the following forms has a **hyperbola** as its graph:

- An equation that can be put into the form

$$\frac{x^2}{a^2} - \frac{y^2}{b^2} = 1, \qquad \text{where } a > 0 \text{ and } b > 0$$

is a hyperbola with x-intercepts $(-a, 0)$ and $(a, 0)$. See Fig. 57. There are no y-intercepts.

- An equation that can be put into the form

$$\frac{y^2}{b^2} - \frac{x^2}{a^2} = 1, \qquad \text{where } a > 0 \text{ and } b > 0$$

is a hyperbola with y-intercepts $(0, -b)$ and $(0, b)$. See Fig. 58. There are no x-intercepts.

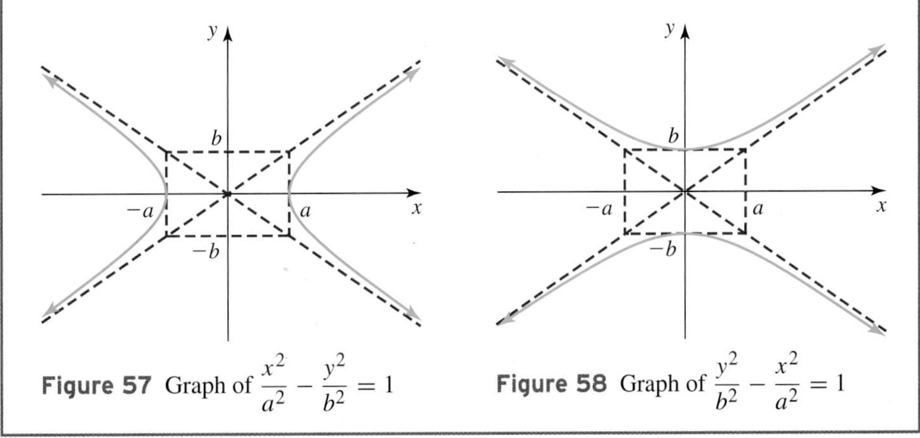

Figure 57 Graph of $\frac{x^2}{a^2} - \frac{y^2}{b^2} = 1$

Figure 58 Graph of $\frac{y^2}{b^2} - \frac{x^2}{a^2} = 1$

Each *pair* of curves in Figs. 57 and 58 is a hyperbola. Each curve is a *branch*. The red dashes are *not* parts of the hyperbola; they are simply tools to guide us in sketching the branches. The dashed rectangles are centered at the origin and stretch a units in both directions horizontally and b units in both directions vertically. Through their opposite corners, we draw the dashed lines that are *inclined asymptotes*. The branches of the hyperbolas approach the inclined asymptotes as $|x|$ gets large.

Graphing Hyperbolas

To sketch a hyperbola on the basis of its equation,

1. Sketch a dashed rectangle whose sides are parallel to the axes and contain the points $(-a, 0)$, $(a, 0)$, $(0, -b)$, and $(0, b)$.
2. Sketch two dashed lines (the inclined asymptotes) that contain the diagonals of the rectangle.
3. Plot the intercepts of the hyperbola.
4. Sketch the branches to contain the intercepts and get closer to the asymptotes as $|x|$ gets large.

Example 2 Sketching the Graph of a Hyperbola

Sketch the graph of $\frac{y^2}{4} - \frac{x^2}{9} = 1$.

Solution

Since the equation is of the form $\frac{y^2}{b^2} - \frac{x^2}{a^2} = 1$, where $a > 0$ and $b > 0$, we have $a^2 = 9$ and $b^2 = 4$. So, $a = 3$ and $b = 2$. We sketch a dashed rectangle that contains the points $(-3, 0)$, $(3, 0)$, $(0, -2)$, and $(0, 2)$; then we sketch the inclined asymptotes (see Fig. 59).

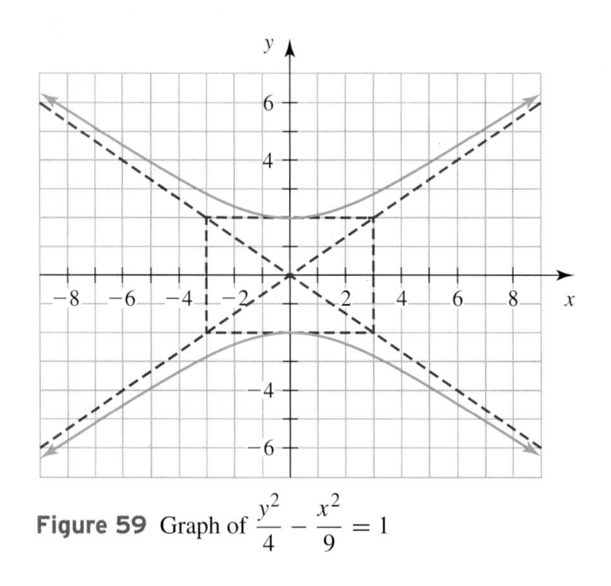

Figure 59 Graph of $\frac{y^2}{4} - \frac{x^2}{9} = 1$

Since the equation is of the form $\frac{y^2}{b^2} - \frac{x^2}{a^2} = 1$ (with $\frac{y^2}{b^2}$ first), the graph has y-intercepts at $(0, -2)$ and $(0, 2)$ and there are no x-intercepts. The branches contain the y-intercepts and approach the inclined asymptotes for large $|x|$. ■

Example 3 Sketching the Graph of a Hyperbola

Sketch the graph of $4x^2 - 25y^2 = 100$.

Solution

We divide both sides of the equation by 100 so that the right-hand side of the equation is equal to 1:

$$4x^2 - 25y^2 = 100 \quad \text{Original equation}$$

$$\frac{4x^2}{100} - \frac{25y^2}{100} = \frac{100}{100} \quad \text{Divide both sides by 100.}$$

$$\frac{x^2}{25} - \frac{y^2}{4} = 1 \quad \text{Simplify.}$$

The equation is of the form $\frac{x^2}{a^2} - \frac{y^2}{b^2} = 1$, where $a > 0$, $b > 0$, $a^2 = 25$, and $b^2 = 4$. So, $a = 5$ and $b = 2$. We sketch the dashed rectangle that contains $(-5, 0)$, $(5, 0)$, $(0, -2)$, and $(0, 2)$ and then sketch the inclined asymptotes (see Fig. 60).

Since the equation is in $\frac{x^2}{a^2} - \frac{y^2}{b^2} = 1$ form (with $\frac{x^2}{a^2}$ first), the graph has x-intercepts at $(-5, 0)$ and $(5, 0)$ and there are no y-intercepts. The branches contain the intercepts and approach the inclined asymptotes for large $|x|$.

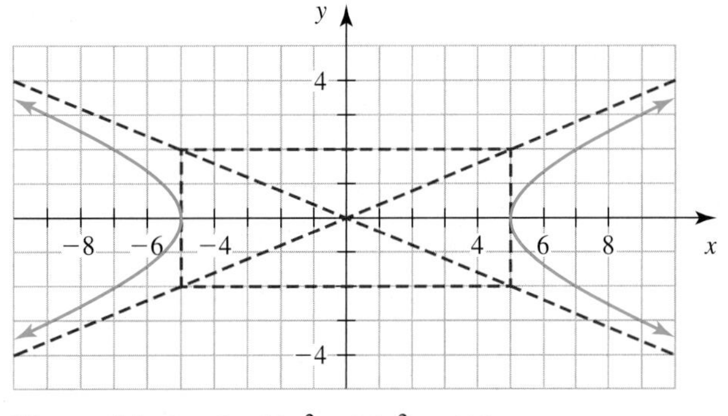

Figure 60 Graph of $4x^2 - 25y^2 = 100$

To use a graphing calculator to draw a graph of $4x^2 - 25y^2 = 100$, we begin by isolating y:

$$4x^2 - 25y^2 = 100 \quad \text{Original equation}$$

$$4x^2 - 100 = 25y^2 \quad \text{Isolate } 25y^2.$$

$$\frac{4x^2 - 100}{25} = y^2 \quad \text{Divide both sides by 25.}$$

$$y = \pm\sqrt{\frac{4x^2 - 100}{25}} \quad \text{Square root property}$$

$$y = \pm\sqrt{\frac{4(x^2 - 25)}{25}} \quad \text{Factor.}$$

$$y = \pm\frac{2}{5}\sqrt{x^2 - 25} \quad \text{Simplify.}$$

Then, we enter the functions $y = \frac{2}{5}\sqrt{x^2 - 25}$ and $y = -\frac{2}{5}\sqrt{x^2 - 25}$ (or $Y_2 = -Y_1$) and draw the graphs on the same coordinate system (see Fig. 61).

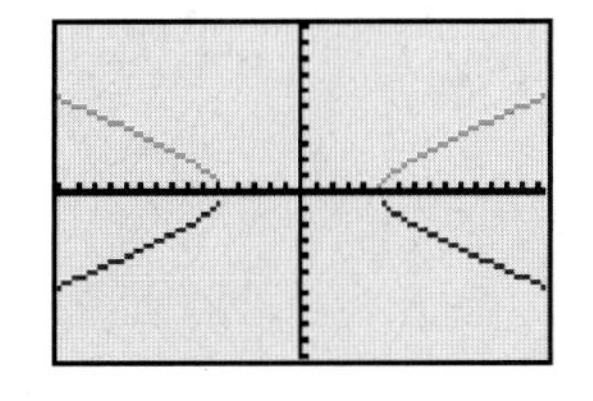

Figure 61 Use ZStandard followed by ZSquare to graph $4x^2 - 25y^2 = 100$

■

group exploration

Graphical significance of *a* and *b* for ellipses and hyperbolas

1. Sketch two ellipses that
   a. intersect in four points. **b.** intersect in two points.
   c. intersect in no points.
2. Write and graph equations to correspond to each of your sketches in Problem 1.
3. Sketch an ellipse and a hyperbola that
   a. intersect in four points. **b.** intersect in two points.
   c. intersect in no points.
4. Write and graph equations to correspond to each of your sketches in Problem 3.
5. Sketch two hyperbolas that
   a. intersect in four points. **b.** intersect in two points.
   c. have no intersection points.
6. Write and graph equations to correspond to each of your sketches in Problem 5.

Key Points OF SECTION 11.5

Throughout these Key Points, assume that a and b are positive.

Equation of an ellipse
An equation that can be put into the form $\frac{x^2}{a^2} + \frac{y^2}{b^2} = 1$ has an **ellipse** as its graph.

Intercepts of an ellipse
The ellipse described by $\frac{x^2}{a^2} + \frac{y^2}{b^2} = 1$ has x-intercepts $(-a, 0)$ and $(a, 0)$ and y-intercepts $(0, -b)$ and $(0, b)$.

An equation that can be put into one of the following forms has a **hyperbola** as its graph:

Equation of a hyperbola with x-intercepts
An equation that can be put into the form $\frac{x^2}{a^2} - \frac{y^2}{b^2} = 1$ is a hyperbola with x-intercepts $(-a, 0)$ and $(a, 0)$. There are no y-intercepts.

Equation of a hyperbola with y-intercepts
An equation that can be put into the form $\frac{y^2}{b^2} - \frac{x^2}{a^2} = 1$ is a hyperbola with y-intercepts $(0, -b)$ and $(0, b)$. There are no x-intercepts.

Graphing hyperbolas
To sketch a hyperbola on the basis of its equation,

1. Sketch a dashed rectangle whose sides are parallel to the axes and contain the points $(-a, 0)$, $(a, 0)$, $(0, -b)$, and $(0, b)$.
2. Sketch two dashed lines (the inclined asymptotes) that contain the diagonals of the rectangle.
3. Plot the intercepts of the hyperbola.
4. Sketch the branches to contain the intercepts and get closer to the asymptotes as $|x|$ gets large.

HOMEWORK 11.5

Graph the equation by hand.

1. $\frac{x^2}{36} + \frac{y^2}{9} = 1$ **2.** $\frac{x^2}{49} + \frac{y^2}{16} = 1$ **3.** $\frac{x^2}{4} + \frac{y^2}{36} = 1$

4. $\frac{x^2}{9} + \frac{y^2}{64} = 1$ **5.** $\frac{x^2}{100} + \frac{y^2}{16} = 1$ **6.** $\frac{x^2}{81} + \frac{y^2}{25} = 1$

7. $25x^2 + 4y^2 = 100$ **8.** $4x^2 + 16y^2 = 64$

9. $9x^2 + 100y^2 = 900$ **10.** $16x^2 + 25y^2 = 400$

11. $x^2 + y^2 = 36$ **12.** $2x^2 + 2y^2 = 50$

13. $x^2 + 25y^2 = 25$ **14.** $64x^2 + y^2 = 64$

15. $5x^2 + 16y^2 = 80$ **16.** $22x^2 + 4y^2 = 88$

17. Find an equation of the ellipse shown in Fig. 62.

Figure 62
Exercise 17

18. Find an equation of the ellipse shown in Fig. 63.

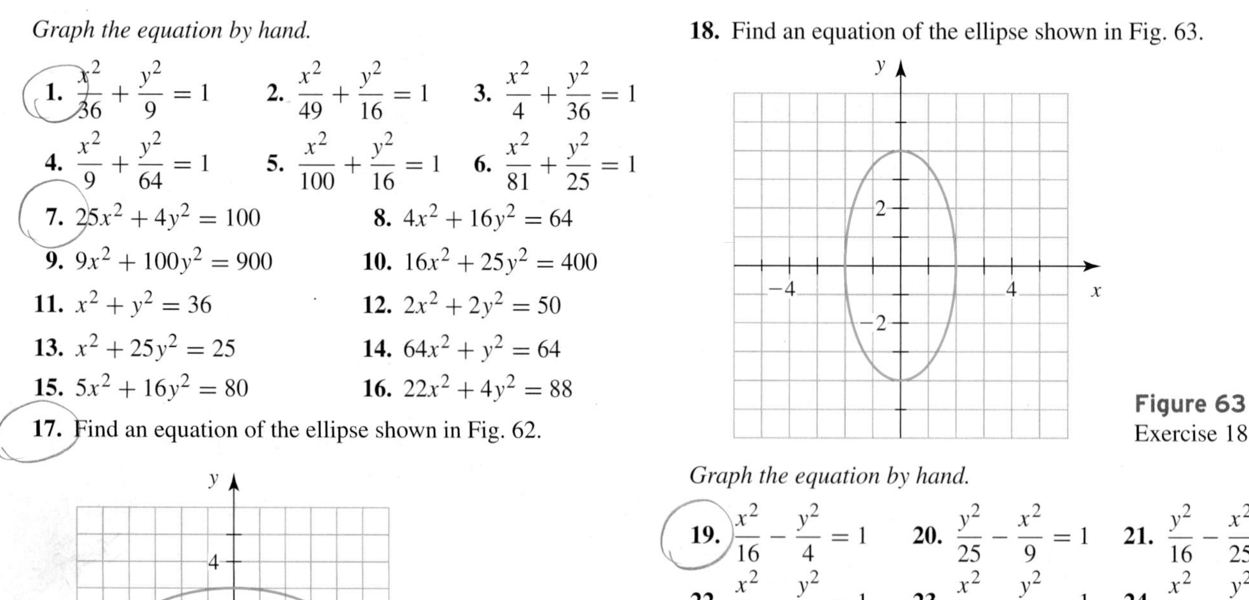

Figure 63
Exercise 18

Graph the equation by hand.

19. $\frac{x^2}{16} - \frac{y^2}{4} = 1$ **20.** $\frac{y^2}{25} - \frac{x^2}{9} = 1$ **21.** $\frac{y^2}{16} - \frac{x^2}{25} = 1$

22. $\frac{x^2}{49} - \frac{y^2}{16} = 1$ **23.** $\frac{x^2}{25} - \frac{y^2}{81} = 1$ **24.** $\frac{x^2}{64} - \frac{y^2}{9} = 1$

25. $16x^2 - 4y^2 = 64$ **26.** $25x^2 - 16y^2 = 400$

27. $x^2 - 9y^2 = 9$ **28.** $y^2 - 4x^2 = 4$

29. $y^2 - x^2 = 4$ **30.** $4x^2 - 4y^2 = 36$

31. $16y^2 - x^2 = 16$ **32.** $25x^2 - y^2 = 25$

33. $25x^2 - 7y^2 = 175$ **34.** $30x^2 - 9y^2 = 270$

Graph the equation by hand.

35. $\frac{x^2}{64} + \frac{y^2}{4} = 1$

36. $\frac{x^2}{49} + \frac{y^2}{100} = 1$

37. $x^2 - y^2 = 1$

38. $x^2 - y^2 = 9$

39. $81x^2 + 49y^2 = 3969$

40. $4x^2 + 36y^2 = 144$

41. $x^2 + y^2 = 1$

42. $x^2 + y^2 = 49$

43. $9y^2 - 4x^2 = 144$

44. $4x^2 - 25y^2 = 100$

45. $\frac{x^2}{25} - \frac{y^2}{25} = 1$

46. $\frac{x^2}{9} - \frac{y^2}{25} = 1$

47. $x^2 + y^2 = 16$

48. $5x^2 + 5y^2 = 45$

49. $9x^2 + 16y^2 = 144$

50. $4x^2 + 9y^2 = 36$

51. $\frac{x^2}{16} + \frac{y^2}{16} = 1$

52. $\frac{x^2}{36} + \frac{y^2}{36} = 1$

53. a. Graph the equation $\frac{x^2}{c} + \frac{y^2}{d} = 1$ by hand for the given values of the constants c and d.

i. $c = 4$ and $d = 16$. **ii.** $c = 4$ and $d = -16$.
iii. $c = -4$ and $d = 16$. **iv.** $c = 4$ and $d = 4$.

b. In terms of the values of c and d, discuss whether the graph of the equation $\frac{x^2}{c} + \frac{y^2}{d} = 1$ is a circle, an ellipse, a hyperbola with x-intercepts, or a hyperbola with y-intercepts.

54. Graph by hand the following equations on the same coordinate system.

a. $\frac{x^2}{36} + \frac{y^2}{9} = 1$ **b.** $\frac{x^2}{36} - \frac{y^2}{9} = 1$ **c.** $\frac{y^2}{9} - \frac{x^2}{36} = 1$

55. The Rule of Four can be applied to relations as well as functions. Describe the Rule of Four as applied to the relation $4x^2 + 25y^2 = 100$:

a. Describe the input–output pairs of the relation by using a graph.
b. Describe eight input–output pairs by using a table. Round approximate inputs and outputs to the second decimal place.
c. Describe the input–output pairs by using words.

56. The Rule of Four can be applied to relations as well as functions. Describe the Rule of Four as applied to the relation $x^2 - 4y^2 = 4$:

a. Describe the input–output pairs of the relation by using a graph.
b. Describe six input–output pairs by using a table. Round approximate inputs and outputs to the second decimal place.
c. Describe the input–output pairs by using words.

57. a. For $y = \frac{5}{2}\sqrt{4 - x^2}$, explain why $y \geq 0$ for real-number values of y.

b. Graph the function $y = \frac{5}{2}\sqrt{4 - x^2}$ by hand. [**Hint:** Square both sides of the equation.] Use a graphing calculator to verify your graph.

58. a. For $y = \frac{2}{3}\sqrt{x^2 - 9}$, explain why $y \geq 0$ for real-number values of y.

b. Graph the function $y = \frac{2}{3}\sqrt{x^2 - 9}$ by hand. Use a graphing calculator to verify your graph. [**Hint:** Square both sides of the equation.]

59. Find equations of five ellipses that do not intersect each other. Sketch the five ellipses on the same coordinate system.

60. Show that the y-intercepts of $\frac{x^2}{a^2} + \frac{y^2}{b^2} = 1$ are the points $(0, -b)$ and $(0, b)$.

61. Is the graph of the equation $x^2 + y^2 = r^2$ with $r > 0$ a circle, an ellipse, both, or neither? Explain. [**Hint:** Is it possible to write the equation in the form $\frac{x^2}{a^2} + \frac{y^2}{b^2} = 1$?]

62. Describe how to graph by hand an equation of the form $\frac{x^2}{a^2} - \frac{y^2}{b^2} = 1$.

Related Review

63. a. Graph by hand the equations $x^2 + y^2 = 1$ and $(x - 3)^2 + (y - 2)^2 = 1$ on the same coordinate system. How can the graph of $x^2 + y^2 = 1$ be translated to get the graph of $(x - 3)^2 + (y - 2)^2 = 1$?

b. Graph by hand the equations $x^2 + y^2 = 1$ and $(x + 3)^2 + (y + 2)^2 = 1$ on the same coordinate system. How can the graph of $x^2 + y^2 = 1$ be translated to get the graph of $(x + 3)^2 + (y + 2)^2 = 1$?

c. Explain how you can translate the graph of an equation of the form $x^2 + y^2 = r^2$ to get the graph of $(x - h)^2 + (y - k)^2 = r^2$, where h, k, and r are constants and $r > 0$.

d. Graph $4x^2 + 25y^2 = 100$ by hand. Then translate your graph to get the graph of $4(x + 2)^2 + 25(y - 5)^2 = 100$.

e. Graph $4x^2 - 25y^2 = 100$ by hand. Then translate your graph to get the graph of $4(x + 2)^2 - 25(y - 5)^2 = 100$.

Graph the equation by hand.

64. $3x - 2y = 8$

65. $y = \log_2(x)$

66. $y = -2(x - 4)^2 + 6$

67. $y = 3\sqrt{x + 5} - 4$

68. $y = -2(3)^x$

69. Is the relation $\frac{x^2}{4} + \frac{y^2}{81} = 1$ a function? Explain.

70. Is the relation $\frac{x^2}{9} - \frac{y^2}{81} = 1$ a function? Explain.

Expressions, Equations, Functions, and Graphs

Perform the indicated instruction. Then use words such as linear, quadratic, cubic, exponential, logarithmic, rational, radical, polynomial, degree, function, one variable, *and* two variables *to describe the expression, equation, or system.*

71. Graph $y = 2x^2 - 8x + 3$ by hand.

72. Solve $\log_3(7x^3) = 5$. Round any solutions to the fourth decimal place.

73. Solve $2x^2 - 8x + 3 = 0$.

74. Graph $f(x) = \log_3(x)$ by hand.

75. Find the product $-5x(2x - 1)(3x - 1)$.

76. Write $4\log_b(2x^2) - 3\log_b(4x^5)$ as a single logarithm.

Section 11.5 Quiz

Graph the equation by hand.

1. $\frac{x^2}{9} + \frac{y^2}{25} = 1$

2. $\frac{y^2}{49} - \frac{x^2}{9} = 1$

3. $4x^2 - y^2 = 16$

4. $16x^2 + 3y^2 = 48$

5. $x^2 - 9y^2 = 81$

6. $4y^2 - 4x^2 = 16$

7. $\frac{x^2}{5} + \frac{y^2}{14} = 1$

8. $\frac{x^2}{8} + \frac{y^2}{3} = 1$

9. Is the relation $\frac{x^2}{9} - \frac{y^2}{4} = 1$ a function? Explain.

10. Find equations of three distinct ellipses that all contain the points $(0, 3)$ and $(0, -3)$.

11.6 SOLVING NONLINEAR SYSTEMS OF EQUATIONS

Objectives

- Know the meaning of a *nonlinear system.*
- Solve nonlinear systems by graphing, substitution, or elimination.

In this section, we solve nonlinear systems. A **nonlinear system of equations** is a system of equations in which *at least* one of the equations is not linear. Here is an example of a nonlinear system:

$$x^2 + y^2 = 4$$
$$y = x^2$$

The graph of $x^2 + y^2 = 4$ is a circle, and the graph of $y = x^2$ is a parabola. Just as with solutions of linear systems, a *solution* of a nonlinear system is an ordered pair that satisfies *all* of the equations in the system. The *solution set* of a nonlinear system is the set of all solutions of the system.

When solving nonlinear systems in this text, we find only solutions that have real-number coordinates.

Solving a Nonlinear System by Graphing and by Substitution

Just as with solving linear systems, we can solve a nonlinear system by graphing. **The solution set of a system of nonlinear equations can be found by locating the intersection of the graphs of *all* of the equations.**

We can also solve some nonlinear systems by substitution.

Example 1 Solving a Nonlinear System

Solve the system

$$y = x^2 - 5$$
$$y = -x + 1$$

by graphing and by substitution.

Solution

The graph of $y = x^2 - 5$ is a parabola, and the graph of $y = -x + 1$ is a line. We sketch graphs of both equations on the same coordinate system (see Fig. 64).

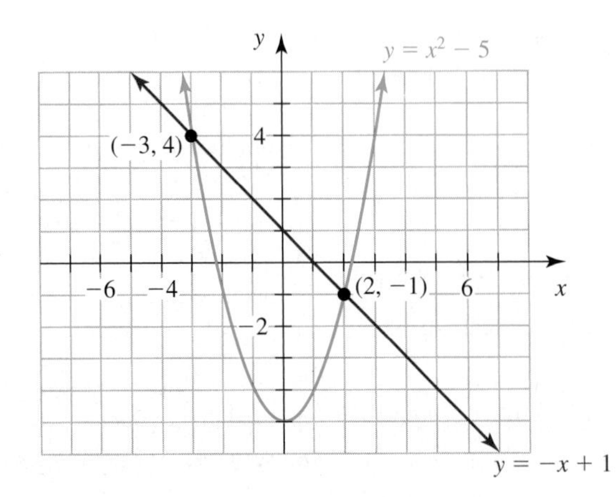

Figure 64 Graphs of $y = x^2 - 5$ and $y = -x + 1$

The graphs appear to intersect at $(-3, 4)$ and $(2, -1)$. The two intersection points are the solutions of the system.

Next, we solve the system

$$y = x^2 - 5$$
$$y = -x + 1$$

by substitution. To begin, we substitute $x^2 - 5$ for y in the equation $y = -x + 1$ and solve for x:

$$\begin{aligned} x^2 - 5 &= -x + 1 && \text{Substitute } x^2 - 5 \text{ for } y. \\ x^2 + x - 6 &= 0 && \text{Write in } ax^2 + bx + c = 0 \text{ form.} \\ (x+3)(x-2) &= 0 && \text{Factor left side.} \\ x + 3 = 0 \quad &\text{or} \quad x - 2 = 0 && \text{Zero factor property} \\ x = -3 \quad &\text{or} \quad x = 2 \end{aligned}$$

We substitute $x = -3$ and $x = 2$ into the equation $y = -x+1$ to find the corresponding values of y:

$$\begin{aligned} y &= -(-3) + 1 \qquad & y &= -2 + 1 \\ &= 4 & &= -1 \end{aligned}$$

So $(-3, 4)$ and $(2, -1)$ are the solutions of the system.

We can check that $(-3, 4)$ and $(2, -1)$ satisfy both equations in the original system. However, the fact that we have solved the system in two different ways and gotten the same result is itself a check of our work. ■

We could also solve the nonlinear system in Example 1 by using "intersect" on a graphing calculator (see Fig. 65).

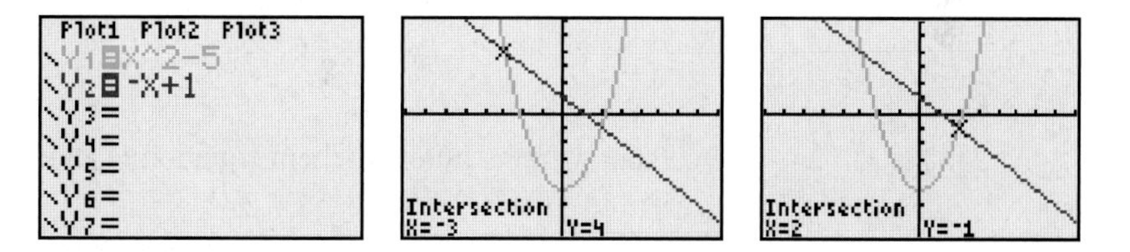

Figure 65 Verify that $(-3, 4)$ and $(2, -1)$ are solutions

Solving a Nonlinear System by Elimination

We can solve some nonlinear systems by elimination.

Example 2 Solving a Nonlinear System

Solve the system

$$x^2 + y^2 = 9$$
$$9x^2 + 4y^2 = 36$$

by graphing and by elimination.

Solution

The graph of $x^2 + y^2 = 9$ is a circle, and the graph of $9x^2 + 4y^2 = 36$ is an ellipse. We sketch the graphs on the same coordinate system (see Fig. 66). (We sketched the ellipse $9x^2 + 4y^2 = 36$ in Example 1 of Section 11.5.)

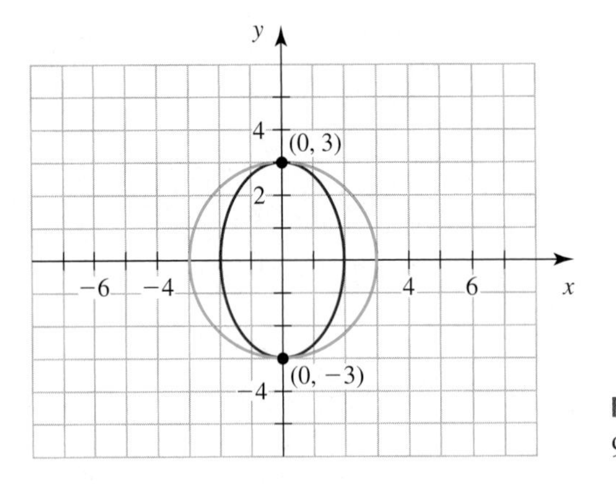

Figure 66 Graphs of $x^2 + y^2 = 9$ and $9x^2 + 4y^2 = 36$

The intersection points appear to be $(0, -3)$ and $(0, 3)$. These points are the solutions of the system.

Now we solve the system

$$x^2 + y^2 = 9 \quad \text{Equation (1)}$$
$$9x^2 + 4y^2 = 36 \quad \text{Equation (2)}$$

by elimination. First, we multiply both sides of equation (1) by -4, yielding the system

$$-4x^2 - 4y^2 = -36 \quad \text{Multiply both sides of equation (1) by } -4.$$
$$9x^2 + 4y^2 = 36 \quad \text{Equation (2)}$$

Next, we add the left-hand sides and add the right-hand sides of the equations and solve for x:

$$\begin{aligned} -4x^2 - 4y^2 &= -36 \\ 9x^2 + 4y^2 &= 36 \\ \hline 5x^2 + \quad 0 &= 0 \\ 5x^2 &= 0 && a + 0 = a \\ x^2 &= 0 && \text{Divide both sides by 5.} \\ x &= 0 && \text{Square root property} \end{aligned}$$

Then, we substitute 0 for x in equation (1) and solve for y:

$$\begin{aligned} 0^2 + y^2 &= 9 && \text{Substitute 0 for } x. \\ y^2 &= 9 && \text{Simplify.} \\ y &= \pm 3 && \text{Square root property} \end{aligned}$$

The solutions are $(0, -3)$ and $(0, 3)$. We got the same result when we solved the system by graphing. ■

If it is reasonable to do so, we first graph the equations of a nonlinear system to determine the number of solutions and to find approximate coordinates of the solutions. Then we solve the system by substitution or elimination.

Example 3 Solving a Nonlinear System

Solve the system

$$x^2 + y^2 = 25$$
$$-4x^2 + 9y^2 = 36$$

by graphing and by elimination. Round coordinates of any solutions to the second decimal place.

Solution

The graph of $x^2 + y^2 = 25$ is a circle, and the graph of $-4x^2 + 9y^2 = 36$ is a hyperbola. We sketch both graphs on the same coordinate system (see Fig. 67). (If we divide both sides of $-4x^2+9y^2=36$ by 36, we have $-\frac{x^2}{9}+\frac{y^2}{4}=1$. We sketched the graph of this equation in Example 2 of Section 11.5.)

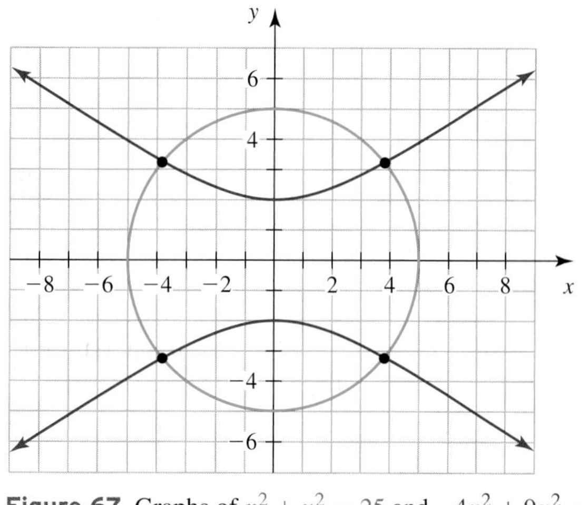

Figure 67 Graphs of $x^2+y^2=25$ and $-4x^2+9y^2=36$

The graphs suggest that there are four solutions: $(-3.8, -3.2)$, $(-3.8, 3.2)$, $(3.8, -3.2)$, and $(3.8, 3.2)$.

Now we solve the system

$$x^2+y^2=25 \quad \text{Equation (1)}$$
$$-4x^2+9y^2=36 \quad \text{Equation (2)}$$

by elimination. First, we multiply both sides of equation (1) by 4, yielding the system

$$4x^2+4y^2=100 \quad \text{Multiply both sides by 4.}$$
$$-4x^2+9y^2=36 \quad \text{Equation (2)}$$

Next, we add the left-hand sides and add the right-hand sides of the equations and solve for y:

$$\begin{aligned} 4x^2+4y^2 &= 100 \\ -4x^2+9y^2 &= 36 \\ \hline 0+13y^2 &= 136 \\ 13y^2 &= 136 && 0+a=a \\ y^2 &= \frac{136}{13} && \text{Divide both sides by 13.} \\ y &= \pm\sqrt{\frac{136}{13}} && \text{Square root property} \\ y &\approx \pm 3.234 && \text{Compute.} \end{aligned}$$

Then, we substitute -3.234 and 3.234 for y in equation (1) and solve for x:

$$\begin{aligned} x^2+(-3.234)^2 &= 25 \\ x^2+3.234^2 &= 25 \\ x^2 &= 25-3.234^2 \\ x &= \pm\sqrt{25-3.234^2} \\ x &\approx \pm 3.81 \end{aligned}$$

$$\begin{aligned} x^2+3.234^2 &= 25 \\ x^2 &= 25-3.234^2 \\ x &= \pm\sqrt{25-3.234^2} \\ x &\approx \pm 3.81 \end{aligned}$$

So, the four approximate solutions are $(-3.81, -3.23)$, $(-3.81, 3.23)$, $(3.81, -3.23)$, and $(3.81, 3.23)$, which agrees with what we found graphically. ■

group exploration

Using graphs to find the number of solutions

For each problem, think graphically. It is not necessary to solve the systems in Problems 2 and 4. Assume that a, b, and r are positive constants.

1. If $r < a$ and $r < b$, explain why there are no solutions with real-number coordinates for the following system:
$$x^2 + y^2 = r^2$$
$$\frac{x^2}{a^2} + \frac{y^2}{b^2} = 1$$
2. If $a < r < b$, explain why the following system has four solutions:
$$x^2 + y^2 = r^2$$
$$\frac{x^2}{a^2} + \frac{y^2}{b^2} = 1$$
3. If $a < r < b$, explain why there are no solutions with real-number coordinates for the following system:
$$x^2 + y^2 = r^2$$
$$\frac{y^2}{b^2} - \frac{x^2}{a^2} = 1$$
4. If $a < r < b$, explain why the following system has four solutions:
$$x^2 + y^2 = r^2$$
$$\frac{x^2}{a^2} - \frac{y^2}{b^2} = 1$$

Key Points OF SECTION 11.6

Nonlinear system of equations A **nonlinear system of equations** is a system of equations in which *at least* one of the equations is not linear.

Solution set of a nonlinear system The solution set of a system of nonlinear equations can be found by locating the intersection of the graphs of *all* of the equations.

First graph a nonlinear system If it is reasonable to do so, we first graph the equations of a nonlinear system to determine the number of solutions and to find approximate coordinates of the solutions. Then we solve the system by substitution or elimination.

HOMEWORK 11.6

FOR EXTRA HELP ▶ Student Solutions Manual | PH Math/Tutor Center | MathXL® | MyMathLab

Solve the system by graphing the equations by hand. Also, solve the system by substitution or elimination.

1. $x^2 + y^2 = 25$
   $4x^2 + 25y^2 = 100$
2. $x^2 + y^2 = 4$
   $4x^2 + 16y^2 = 64$
3. $y = x^2 + 1$
   $y = -x + 3$
4. $y = x^2 - 3$
   $y = 2x$
5. $y = x^2 - 2$
   $y = -x^2 + 6$
6. $y = 2x^2 - 8$
   $y = x^2 + 1$
7. $x^2 + y^2 = 49$
   $x^2 + y^2 = 16$
8. $4x^2 + 9y^2 = 36$
   $9x^2 + 25y^2 = 225$
9. $x^2 + y^2 = 25$
   $y = -x - 1$
10. $x^2 + y^2 = 36$
    $y = x + 6$

11. $y^2 - x^2 = 16$
$y + x^2 = 4$

12. $y^2 - 4x^2 = 4$
$4x^2 + y^2 = 4$

13. $25x^2 - 9y^2 = 225$
$4x^2 + 9y^2 = 36$

14. $y^2 - x^2 = 9$
$4x^2 + 100y^2 = 400$

15. $9x^2 + y^2 = 9$
$y = 3x + 3$

16. $x^2 - y^2 = 16$
$y = -x + 1$

17. $4x^2 + 9y^2 = 36$
$16x^2 + 25y^2 = 225$

18. $16x^2 + 9y^2 = 144$
$x^2 + 4y^2 = 4$

19. $y = \sqrt{x} - 3$
$y = -x - 1$

20. $y = \sqrt{x} + 1$
$y = -\sqrt{x} + 5$

21. $y = 2x^2 - 5$
$y = x^2 - 2$

22. $y = -3x^2 + 7$
$y = -x^2 + 3$

Solve the system by graphing the equations by hand. Also, solve the system by substitution or elimination. Round the coordinates of your solution(s) to the second decimal place.

23. $25y^2 - 4x^2 = 100$
$9x^2 + y^2 = 9$

24. $16x^2 - 4y^2 = 64$
$x^2 + 16y^2 = 16$

25. $25x^2 + 9y^2 = 225$
$x^2 + y^2 = 16$

26. $36x^2 + 4x^2 = 144$
$x^2 + y^2 = 9$

Solve the system by substitution or elimination. Check that any results satisfy both equations.

27. $9x^2 + y^2 = 85$
$2x^2 - 3y^2 = 6$

28. $x^2 - 6y = 34$
$x^2 + y^2 = 25$

29. $4y^2 + x^2 = 25$
$y = -x + 5$

30. $x^2 + 9y^2 = 13$
$y = x - 1$

31. $y = x^2 - 3x + 2$
$y = 2x - 4$

32. $y = 2x^2 - 5x - 11$
$y = x^2 - 3x + 4$

Solve the system of three equations by graphing the equations by hand. Check that any results satisfy each equation.

33. $x^2 + y^2 = 25$
$4x^2 - 25y^2 = 100$
$4x^2 + 25y^2 = 100$

34. $x^2 + y^2 = 1$
$9x^2 + y^2 = 9$
$y = x + 1$

35. Create a nonlinear system of two equations whose solutions are $(-4, 0)$ and $(4, 0)$.

36. Create a nonlinear system of two equations whose solutions are $(0, -3)$ and $(0, 3)$.

37. Find values of c and d such that $(1, 4)$ is a solution of the system

$$2x^2 + cy^2 = 82$$
$$y = x^2 + dx + 5$$

38. Find values of c and d such that $(2, -5)$ is a solution of the system

$$y = cx^2 - 4x^2 - 5$$
$$y = dx - 13$$

39. Explain how to solve a nonlinear system.

40. In your own words, describe a linear system and a nonlinear system. Also, compare the numbers of possible solutions for both types of system.

Related Review

Solve the system by substitution or elimination.

41. $y = 2^x$
$y = 4\left(\frac{1}{2}\right)^x$
[**Hint:** $4 = 2^2$ and $\frac{1}{2} = 2^{-1}$]

42. $y = \log_2(x + 1) + 2$
$y = \log_2(3x + 13) - 1$

Expressions, Equations, Functions, and Graphs

Give an example of the following. Then solve, simplify, or graph, as appropriate.

43. linear function
44. quotient of two radical expressions
45. rational equation in one variable
46. system of two linear equations in two variables
47. exponential function
48. square root function
49. quadratic function
50. difference of two rational expressions
51. quadratic equation in one variable
52. logarithmic equation in one variable

Section 11.6 Quiz

Solve the system.

1. $9x^2 + y^2 = 81$
$x^2 + y^2 = 9$

2. $y = x^2 - 2$
$y = -2x + 1$

3. $y = x^2 + 3$
$y = x^2 - 6x + 9$

4. $25x^2 - 4y^2 = 100$
$9x^2 + y^2 = 9$

5. Solve the system of three equations by graphing the equations by hand.

$$x^2 - y^2 = 16$$
$$x^2 + y^2 = 16$$
$$y = (x + 4)^2$$

6. Create a nonlinear system of two equations whose solution is $(0, 5)$.

Appendix A

Reviewing Prerequisite Material

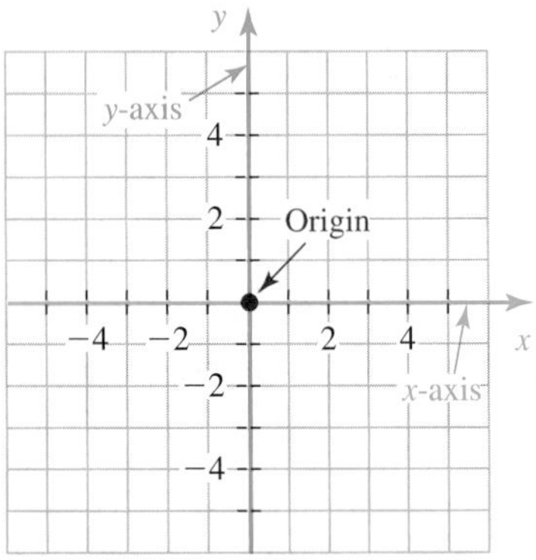

In this appendix, we review skills that you will need for this text. Review these skills before you begin Section 1.2. The answers to exercises in this appendix are located at the end of the Answers to Odd-Numbered Exercises.

Figure 1 Coordinate system

A.1 PLOTTING POINTS

How do we plot points? To start, we draw a horizontal number line called the x-axis and a vertical number line called the y-axis (see Fig. 1). We refer to such a pair of axes as a **coordinate system.** The **origin** is the intersection point of the axes.

Next, we plot the ordered pair (5, 4). To do so, we start at the origin, look 5 units to the right and 4 units up, and draw a dot (see Fig. 2). The points (−4, 2), (−5, −3), and (3, −4) have also been plotted in Fig. 2.

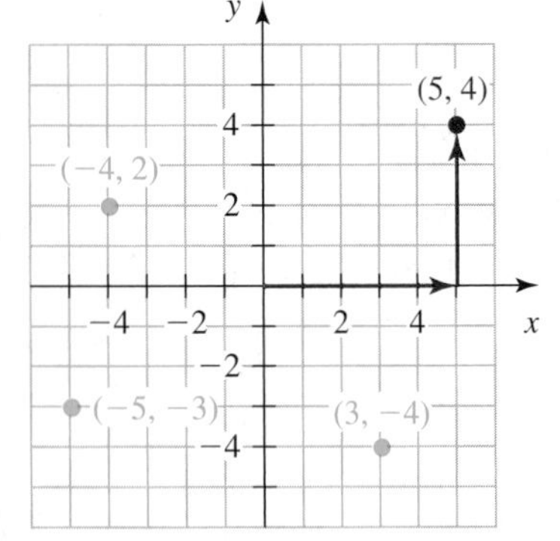

Figure 2 Plotting the points (5, 4), (−4, 2), (−5, −3), and (3, −4)

Plot the given points in a coordinate system.

1. (2, 4) **2.** (−3, 1) **3.** (4, −2) **4.** (−4, −3)
5. (−2, −1) **6.** (0, 1) **7.** (−2, 0) **8.** (0, 0)

A.2 IDENTIFYING TYPES OF NUMBERS

The **counting numbers** are the numbers 1, 2, 3, 4, 5, ... The three dots before the period at the end of the sentence mean that the pattern of the numbers shown continues without ending. In this case, the pattern continues with 6, 7, 8, and so on.

The **integers** are the numbers

$$\ldots, -3, -2, -1, 0, 1, 2, 3, \ldots$$

The three dots on both sides mean that the pattern of the numbers shown continues without ending in both directions. In this case, the pattern continues with −4, −5, −6, and so on, and with 4, 5, 6, and so on.

The **rational numbers** are the numbers that can be written in the form $\frac{n}{d}$, where n and d are integers and d is nonzero. Here are some examples of rational numbers:

$$\frac{2}{3} \qquad \frac{-9}{5} \qquad 6 = \frac{6}{1}$$

A rational number can be written as a decimal number that either terminates or repeats:

$$\frac{5}{8} = \underbrace{0.625}_{\text{terminates}} \qquad \frac{2}{11} = 0.\underbrace{18181818\ldots}_{\text{repeats}}$$

An **irrational** number is a number that can be represented on the number line but is *not* rational. The number $\sqrt{5}$ is the number greater than zero that we multiply by itself to get 5. The number $\sqrt{5}$ is an irrational number. Here are some more examples of irrational numbers:

$$\sqrt{2} \qquad \sqrt{7} \qquad \pi$$

An irrational number can be written as a decimal number that neither terminates nor repeats.

The **real numbers** are all of the numbers represented on the number line. Here are some real numbers:

$$5 \qquad \frac{8}{3} \qquad -7.28 \qquad \pi \qquad 6 \qquad -\frac{9}{7} \qquad 0$$

Consider the following numbers:

$$\frac{2}{9} \qquad 4 \qquad -7.19 \qquad 0 \qquad -2 \qquad \sqrt{17} \qquad 85$$

Which of these numbers are the given type of number?

1. Counting numbers

2. Integers

3. Rational numbers

4. Irrational numbers

5. Real numbers

A.3 ABSOLUTE VALUE

The absolute value of a number a, written $|a|$, is the distance from a to 0 on the number line. So, $|-5| = 5$, because -5 is a distance of 5 units from 0, and $|5| = 5$, because 5 is a distance of 5 units from 0 (see Fig. 3).

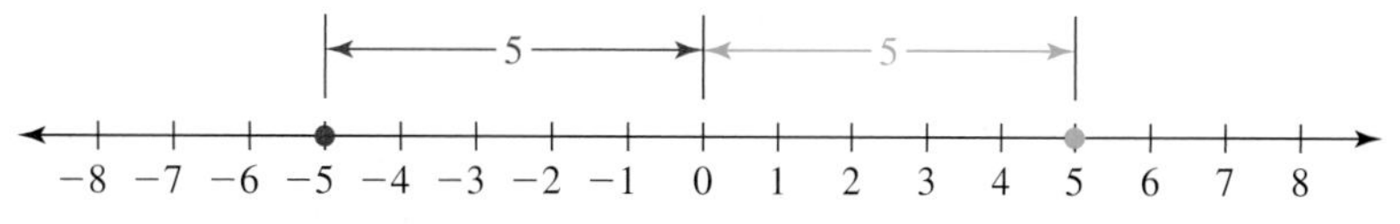

Figure 3 Both numbers −5 and 5 are a distance of 5 units from 0

Here are two more examples:

$$\left|\frac{2}{3}\right| = \frac{2}{3} \qquad \text{and} \qquad |-2.89| = 2.89$$

Compute.

1. $|-3|$

2. $|4.69|$

3. $|0|$

4. $|-\pi|$

A.4 PERFORMING OPERATIONS WITH REAL NUMBERS

In this section, we will review how to multiply, divide, add, and subtract real numbers.

Do you recall how to multiply and divide real numbers? Here are some examples to refresh your memory:

$$2(-5) = -10 \qquad -2(-5) = 10 \qquad \frac{-10}{5} = -2 \qquad \frac{-10}{-5} = 2$$

In general,

- The product or quotient of two numbers with the same sign is positive.
- The product or quotient of two numbers with different signs is negative.

Do you recall how to add real numbers? Here are some examples:

$$2+5=7 \qquad -2+(-5)=-7 \qquad -2+5=3 \qquad 2+(-5)=-3$$

When you add real numbers, it helps to think in terms of the number line. To find $-2+(-5)$, imagine moving 2 units to the left of 0, then 5 more units to the left. Figure 4 illustrates that $-2+(-5)=-7$.

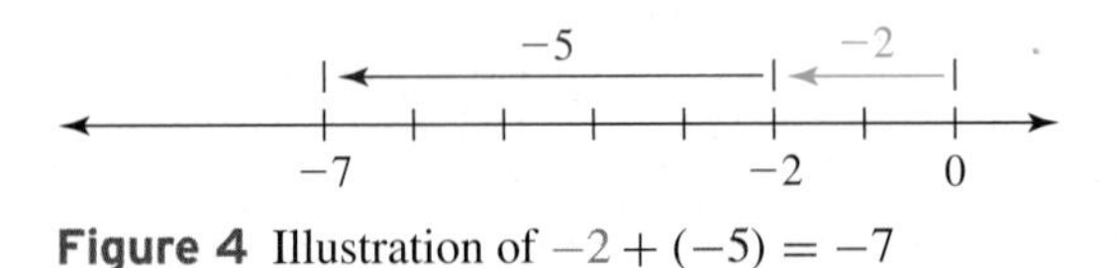

Figure 4 Illustration of $-2+(-5)=-7$

To find $2+(-5)$, imagine moving 2 units to the right of 0, then 5 units to the left of 2. Figure 5 illustrates that $2+(-5)=-3$.

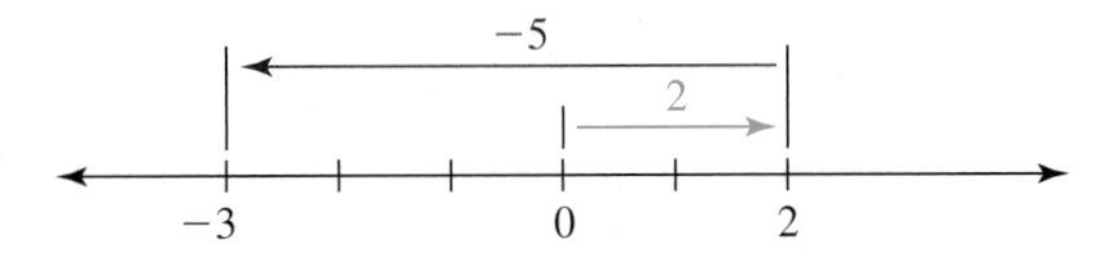

Figure 5 Illustration of $2+(-5)=-3$

Do you recall how to subtract real numbers? Here are some examples:

$$10-2=10+(-2)=8$$

$$10-(-2)=10+[-(-2)]=10+2=12$$

In general, subtracting a number is the same as adding the opposite of that number. In symbols, we write $x-y=x+(-y)$.

Compute.

1. $-3(7)$ **2.** $5(-6)$ **3.** $-9(-4)$ **4.** $-8(-2)$

5. $-(-4)$ [**Hint:** $-(-4)=(-1)(-4)$] **6.** $-(-(-9))$ **7.** $\frac{8}{-2}$ **8.** $\frac{-6}{-2}$

9. $-3+(-5)$ **10.** $-6+(-7)$ **11.** $2+(-8)$ **12.** $-7+3$

13. $-1+6$ **14.** $8+(-3)$ **15.** $-4+(-6)$ **16.** $-2+(-3)$

17. $3-7$ **18.** $2-8$ **19.** $5-(-3)$ **20.** $9-(-4)$

21. $-4-9$ **22.** $-2-4$ **23.** $-1-(-1)$ **24.** $-10-(-6)$

A.5 EXPONENTS

We can use a special notation to describe repeated multiplications. For example, $2^4=2\cdot2\cdot2\cdot2=16$. For any counting number n,

$$b^n = \underbrace{b\cdot b\cdot b\cdot \ldots \cdot b}_{n \text{ factors of } b}$$

We refer to b^n as the **power,** the ***n*th power of *b*,** or ***b* raised to the *n*th power.** We call b the **base** and n the **exponent.** When we calculate a power, we say that we have performed an **exponentiation.** Here we perform some exponentiations:

$$4^3=4\cdot4\cdot4=64 \qquad 2^5=2\cdot2\cdot2\cdot2\cdot2=32$$

$$(-3)^4=(-3)\cdot(-3)\cdot(-3)\cdot(-3)=81 \qquad -3^4=-(3\cdot3\cdot3\cdot3)=-81$$

Perform the indicated exponentiation.

1. 7^2 **2.** 9^2 **3.** 6^3 **4.** 5^4

5. $(-2)^4$ **6.** $(-3)^3$ **7.** -2^4 **8.** -3^3

A.6 ORDER OF OPERATIONS

In this section, we will review the order in which we perform operations:

1. First, perform operations within parentheses or other grouping symbols, starting with the innermost group.
2. Then, perform exponentiations.
3. Next, perform multiplications and divisions, going from left to right.
4. Last, perform additions and subtractions, going from left to right.

For example, here's how to perform the operations in $5 - 8 \div (7-5)^2 + 2 \cdot 3$:

$$
\begin{aligned}
5 - 8 \div (7-5)^2 + 2 \cdot 3 &= 5 - 8 \div (2)^2 + 2 \cdot 3 && \text{Work within parentheses first.} \\
&= 5 - 8 \div 4 + 2 \cdot 3 && \text{Perform exponentiation: } 2^2 = 4. \\
&= 5 - 2 + 2 \cdot 3 && \text{Divide, because the division is to the left of the multiplication.} \\
&= 5 - 2 + 6 && \text{Multiply before adding or subtracting.} \\
&= 3 + 6 && \text{Subtract, because the subtraction is to the left of the addition.} \\
&= 9 && \text{Add.}
\end{aligned}
$$

Perform the operations.

1. $3 + 5 \cdot 2$ **2.** $2(8) - 4$ **3.** $2 + 10 \div (-5)$

4. $14 \div (-2) - 1$ **5.** $2(1-3) + 4 \cdot 2$ **6.** $10(2-7) + 5 \cdot 4$

7. $(5-9)(4+2) \div 8 + 2$ **8.** $(3+5)(2-6) \div 4 + 1$

9. $4(3)^2$ **10.** $-3(2)^3$

11. $-3^2 + (-3)^2 - (-3)^2$ **12.** $2^3 - (-2)^3 + (-2)^3$

13. $5 - 4^2 + (-8) \div (-2)$ **14.** $2^3 - 10 \div (-5) + 1$

15. $6 - (3-1)^3 + 8$ **16.** $10 - (9-6)^3 + 5$

A.7 CONSTANTS, VARIABLES, EXPRESSIONS, AND EQUATIONS

A **variable** is a symbol that represents a quantity that can vary. For example, we can define T to be the temperature (in degrees Fahrenheit) at the top of the Empire State Building. As time passes, the temperature will change. So, T is a variable.

A **constant** is a symbol that represents a specific number (a quantity that does *not* vary). For example, 8, 0, -3.5, and π are constants.

An **expression** is a constant, a variable, or a combination of constants, variables, operation symbols, and grouping symbols, such as parentheses. Here are some examples of expressions:

$$2\pi x - 3y \qquad ax^2 + bx + c \qquad 7 \qquad x^3 - 8 \qquad x \qquad (x-5)^2 + 8x$$

We **evaluate an expression** by substituting a number for each variable in the expression and calculating the result. If a variable appears more than once in the expression, the same number is substituted for that variable each time.

An **equation** consists of an equality sign "=" with expressions on both sides. Here are some examples of equations:

$$2\pi x + 3y = k \qquad ax^2 + bx + c = 0 \qquad 7 = y(y-4) \qquad x^3 - 8 = x^2 + 1 \qquad x = 1$$

Identify each of the following as an expression or an equation.

1. $y = mx + b$

2. $3x^2 - 5x + 4 = 8$

3. $2x - 5\pi + 1$

4. $x^3 - 8$

A.8 DISTRIBUTIVE LAW

Here we review the **distributive law:**

$$a(b + c) = ab + ac$$

For example, $4(x + 5) = 4x + 4 \cdot 5 = 4x + 20$. Also,

$$a(b - c) = ab - ac$$
$$(a + b)c = ac + bc$$
$$(a - b)c = ac - bc$$

For example, $(2x - 7)(3) = 2x \cdot 3 - 7 \cdot 3 = 6x - 21$.

Apply the distributive law.

1. $2(x + 4)$

2. $4(x + 7)$

3. $6(2t - 3)$

4. $5(4w - 6)$

5. $(x + 8)(-3)$

6. $(x + 5)(-4)$

7. $(2x - 9)(-5)$

8. $(3x - 1)(-6)$

9. $2.8(p + 4.1)$

10. $-5.2(b + 3.9)$

A.9 COMBINING LIKE TERMS

A **term** is a constant, a variable, or a product of a constant and one or more variables raised to powers. Terms include 7, y, $4x$, and $9y^2$. **Like terms** are either constant terms or variable terms that contain the same variable(s) raised to the same power(s). Like terms include 3 and 8, $2x$ and $6x$, and $7y^3$ and $4y^3$. (We discuss terms and like terms in Section 6.1.)

We can use the distributive law to add the like terms $3x$ and $5x$:

$$3x + 5x = (3 + 5)x = 8x$$

Notice that we can find $3x + 5x$ in one step by adding the *coefficients,* 3 and 5:

$$\begin{array}{ccccc} 3x & + & 5x & = & 8x \\ \downarrow & & \downarrow & & \uparrow \\ 3 & + & 5 & = & 8 \end{array}$$

We call this process *combining like terms.* Similarly, we can combine like terms to find $9x - 4x$ by adding the coefficients, 9 and -4:

$$\begin{array}{ccccc} 9x & - & 4x & = & 5x \\ \downarrow & & \downarrow & & \uparrow \\ 9 & + & (-4) & = & 5 \end{array}$$

However, we can't add the coefficients in $2x + 7y$. Since $2x$ and $7y$ contain different variables, there is no helpful way to use the distributive law in this case.

Here, we simplify the expression $5x + 9 - 8y + 3x - 2y$ by combining like terms:

$$\begin{aligned} 5x + 9 - 8y + 3x - 2y &= 5x + 3x - 8y - 2y + 9 && \text{Rearrange terms.} \\ &= 8x - 10y + 9 && \text{Combine like terms.} \end{aligned}$$

When possible, apply the distributive law and combine like terms.

1. $4x + 3x$

2. $7x - 2x$

3. $5x - 9y - 3x + 2y$

4. $8x - 4y - 6x + 5y$

5. $7a - 4 + b - 9a - 3b + 2$

6. $4t - 2w + 5 + t - 1 - 8w$

7. $4(2x + 3) + 5(4x - 1)$

8. $5(3x + 2) + 2(3x + 6)$

9. $2(5x - y) - 3(4x + y)$

10. $3(4x - y) - 5(2x + y)$

11. $10 - (3m - 2n) + 4m - 7n$

12. $6 - (6a - 3b) - 5b + 2a$

A.10 SOLVING LINEAR EQUATIONS IN ONE VARIABLE

We say that an equation is an **equation in one variable** if its expressions have exactly one variable. Here are some examples of equations in one variable:

$$3x = 21 \qquad 2x^3 - 1 = 5x^2 + x \qquad 5(2x - 7) + 3 = 18 \qquad x^2 - 2x - 8 = 0$$

A number is a **solution** of an equation in one variable if the equation becomes a true statement when the number is substituted for the variable. We say that the number **satisfies** the equation.

The **solution set** of an equation is the set of all solutions of the equation. We **solve** the equation by finding its solution set.

In this section, we will solve linear equations in one variable. A **linear equation in one variable** is an equation that can be put into the form $mx + b = 0$, where m and b are constants and $m \neq 0$. Here we solve $5(x - 2) = 3x + 4$, a linear equation in one variable:

$$
\begin{aligned}
5(x - 2) &= 3x + 4 && \text{Original equation} \\
5x - 10 &= 3x + 4 && \text{Distributive law} \\
5x - 10 - 3x &= 3x + 4 - 3x && \text{Subtract } 3x \text{ from both sides.} \\
2x - 10 &= 4 && \text{Combine like terms.} \\
2x - 10 + 10 &= 4 + 10 && \text{Add 10 to both sides.} \\
2x &= 14 && a + 0 = a \\
\frac{2x}{2} &= \frac{14}{2} && \text{Divide both sides by 2.} \\
x &= 7 && \text{Simplify.}
\end{aligned}
$$

We check that the original equation becomes a true statement if 7 is substituted for x:

$$
\begin{aligned}
5(x - 2) &= 3x + 4 && \text{Original equation} \\
5(7 - 2) &\stackrel{?}{=} 3(7) + 4 && \text{Substitute 7 for } x. \\
5(5) &\stackrel{?}{=} 21 + 4 && \text{Subtract; multiply.} \\
25 &\stackrel{?}{=} 25 && \text{Multiply; add.}
\end{aligned}
$$

true

So, the solution of the equation is 7.

If an equation contains fractions, it is often helpful to multiply both sides of the equation by the *least common denominator* (*LCD*) of the fractions. To illustrate, we solve $\frac{5}{6}x - \frac{1}{2} = \frac{9}{4}$. To find the LCD of the three fractions in this equation, we list the multiples of 6, the multiples of 2, and the multiples of 4:

Multiples of 6: 6, 12, 18, 24, 30, 36, 42, ...
Multiples of 2: 2, 4, 6, 8, 10, 12, 14, 16, 18, 20, 22, 24, 26, 28, 30, 32, 34, 36, ...
Multiples of 4: 4, 8, 12, 16, 20, 24, 28, 32, 36, 40, ...

Common multiples of 6, 2, and 4 are

$$12, 24, 36, \ldots$$

Notice that 12 is the least (lowest) number in the list. We call it the least common multiple of 6, 2, and 4. The **least common multiple (LCM)** of a group of numbers is the smallest number that is a multiple of *all* of the numbers in the group.

Next, we multiply both sides of $\frac{5}{6}x - \frac{1}{2} = \frac{9}{4}$ by the LCD, 12:

$$\frac{5}{6}x - \frac{1}{2} = \frac{9}{4} \qquad \text{Original equation}$$

$$12\left(\frac{5}{6}x - \frac{1}{2}\right) = 12 \cdot \frac{9}{4} \qquad \text{Multiply both sides by LCD, 12.}$$

$$12 \cdot \frac{5}{6}x - 12 \cdot \frac{1}{2} = 12 \cdot \frac{9}{4} \qquad \text{Distributive law}$$

$$10x - 6 = 27 \qquad \text{Simplify.}$$

$$10x = 33 \qquad \text{Add 6 to both sides.}$$

$$x = \frac{33}{10} \qquad \text{Divide both sides by 10.}$$

Since $\frac{33}{10} = 3.3$, we can use a calculator to check that the original equation becomes a true statement if 3.3 is substituted for x.

Solve.

1. $x + 5 = 9$

2. $x - 3 = 4$

3. $4x = 12$

4. $-3x = 21$

5. $5(w - 3) = 13$

6. $-2(k - 4) = 5$

7. $2x + 5 = 6x - 3$

8. $4x - 7 = 9x + 3$

9. $5 - 4(2x - 3) = 13$

10. $7 - 2(3x + 5) = 19$

11. $\frac{2}{3}t + \frac{1}{4} = \frac{5}{12}$

12. $\frac{5}{9}w + \frac{1}{2} = \frac{7}{6}$

13. $\frac{5}{2}x - \frac{7}{4} = \frac{3}{8}x$

14. $\frac{5}{3}x - \frac{7}{2} = \frac{11}{6}x$

A.11 SOLVING EQUATIONS IN TWO OR MORE VARIABLES

In this section, we will review how to solve *equations in two or more variables*. For example, we solve the equation $5x + 3y = 15$ for y:

$$5x + 3y = 15 \qquad \text{Original equation}$$

$$5x + 3y - 5x = 15 - 5x \qquad \text{Subtract 5x from both sides.}$$

$$3y = 15 - 5x \qquad \text{Combine like terms.}$$

$$3y = -5x + 15 \qquad \text{Rearrange terms on right-hand side.}$$

$$\frac{3y}{3} = \frac{-5x}{3} + \frac{15}{3} \qquad \text{Divide both sides by 3.}$$

$$y = -\frac{5}{3}x + 5 \qquad \text{Simplify.}$$

Here we solve $a(mx + b) = k$ for x:

$a(mx + b) = k$	Original equation
$amx + ab = k$	Distributive law
$amx + ab - ab = k - ab$	Subtract ab from both sides.
$amx = k - ab$	Combine like terms.
$\frac{amx}{am} = \frac{k - ab}{am}$	Divide both sides by am.
$x = \frac{k - ab}{am}$	Simplify left-hand side.

As yet another example, we solve $-\frac{1}{4}x + \frac{7}{10}y = \frac{3}{5}$ for y. To find the LCD of the three fractions in this equation, we list the multiples of 4, the multiples of 10, and the multiples of 5:

Multiples of 4: 4, 8, 12, 16, 20, 24, 28, ...
Multiples of 10: 10, 20, 30, 40, 50, 60, 70, ...
Multiples of 5: 5, 10, 15, 20, 25, 30, 35, ...

The LCD is 20. Next, we multiply both sides of $-\frac{1}{4}x + \frac{7}{10}y = \frac{3}{5}$ by the LCD, 20:

$-\frac{1}{4}x + \frac{7}{10}y = \frac{3}{5}$	Original equation
$20\left(-\frac{1}{4}x + \frac{7}{10}y\right) = 20 \cdot \frac{3}{5}$	Multiply both sides by LCD, 20.
$20\left(-\frac{1}{4}x\right) + 20\left(\frac{7}{10}y\right) = 20 \cdot \frac{3}{5}$	Distributive law
$-5x + 14y = 12$	Simplify.
$-5x + 14y + 5x = 12 + 5x$	Add $5x$ to both sides.
$14y = 5x + 12$	Combine like terms; rearrange terms.
$\frac{14y}{14} = \frac{5x}{14} + \frac{12}{14}$	Divide both sides by 14.
$y = \frac{5}{14}x + \frac{6}{7}$	Simplify.

Solve for the specified variable.

1. $2x + y = 8$, for y

2. $3x - y = 5$, for y

3. $3x - 5y = 15$, for x

4. $3x - 5y = 15$, for y

5. $ax - by = c$, for y

6. $ax - by = c$, for x

7. $-4x + 3y = 2x + 9$, for x

8. $-4x + 3y = 2x + 9$, for y

9. $\frac{1}{2}x - \frac{3}{4}y = \frac{5}{8}$, for y

10. $\frac{3}{4}x - \frac{2}{3}y = \frac{1}{4}$, for y

11. $\frac{x}{a} - \frac{y}{a} = 1$, for x

12. $\frac{x}{a} - \frac{y}{a} = 1$, for y

A.12 EQUIVALENT EXPRESSIONS AND EQUIVALENT EQUATIONS

By the distributive law, we have $2(x + 3) = 2x + 6$. In Table 1, we show that the expressions $2(x + 3)$ and $2x + 6$ attain equal values when we substitute 0, 1, 2, 3, and 4 for x.

Table 1 Substituting Values for x in $2(x + 3)$ and $2x + 6$

x	$2(x + 3)$	$2x + 6$
0	$2(0 + 3) = 6$	$2(0) + 6 = 6$
1	$2(1 + 3) = 8$	$2(1) + 6 = 8$
2	$2(2 + 3) = 10$	$2(2) + 6 = 10$
3	$2(3 + 3) = 12$	$2(3) + 6 = 12$
4	$2(4 + 3) = 14$	$2(4) + 6 = 14$

It turns out that $2(x + 3)$ and $2x + 6$ have equal values when *any* number is substituted for x. We say that $2(x + 3)$ and $2x + 6$ are equivalent expressions. In general, two or more expressions are **equivalent expressions** if, when each variable is evaluated for *any* real number (for which all the expressions are defined), they give equal results.

As another example, consider the expression $2x + 4x + 3$. By combining like terms, we have $2x + 4x + 3 = 6x + 3$. The expressions $2x + 4x + 3$ and $6x + 3$ are equivalent expressions.

We **simplify** an expression such as $2(x - 5) + 4x + 1$ by applying the distributive law to remove parentheses, combining like terms, and/or performing as many operations with numbers as possible. The result is a **simplified expression** that is equivalent to the original expression.

Now we turn our attention to *equivalent equations*. To solve the equation $x + 3 = 7$, we write

$x + 3 = 7$	Original equation
$x + 3 - 3 = 7 - 3$	Subtract 3 from both sides.
$x = 4$	Combine like terms.

Each of the equations $x + 3 = 7$, $x + 3 - 3 = 7 - 3$, and $x = 4$ has 4 as its only solution. So, the three equations have the same solution set. Equations that have the same solution set are called **equivalent equations.**

It is important to know the difference between simplifying an expression and solving an equation. To *simplify an expression,* we find a simpler *equivalent expression.* To *solve an equation,* we find any numbers that *satisfy the equation.*

Determine whether the pair are two equivalent expressions, two equivalent equations, or neither. Explain.

1. $5(x - 4)$ and $5x - 20$

2. $x + 8 = 0$ and $x = -8$

3. $4x - 3x + 8$ and $-12x + 8$

4. $3(x + 1) + 7$ and $3x + 8$

5. $4(x + 2)$ and $4x + 8 = 0$

6. $-3(2x - 5)$ and $-6x + 15$

7. $3x + 1 = 16$ and $3x = 15$

8. $2(x - 3) + 5 = 25$ and $2x = 23$

9. $-3(x - 4) = -18$ and $x = 2$

10. $3x + 4x - 2$ and $2x + x - 2 + 5x$

Appendix B

Using a TI-83 or TI-84 Graphing Calculator

The more you experiment with a graphing calculator, the more comfortable and efficient you will become with it.

A TI graphing calculator can detect several types of errors and display an error message. When this occurs, refer to Section B.26 for explanations of some common error messages and how to fix these types of mistakes. Errors do not hurt the calculator. In fact, you can't hurt the calculator regardless of the order in which you press its keys. So, the more you experiment with the calculator, the better off you will be.

To access a TI-83 command written in yellow above a key, first press 2nd, then the key. Whenever a key must follow the 2nd key, this appendix will use brackets for the key. For example, "Press 2nd [OFF]" means to press 2nd and then press ON (because "OFF" is written in yellow above the ON key). The same applies for TI-84 commands written in blue above a key.

Aside from different-colored keys, the TI-83 and TI-84 are very similar calculators—virtually all of the key combinations for a TI-83 and a TI-84 are the same.

Instructions for using a TI-85 and TI-86 (as well as a TI-82 and TI-83) graphing calculator are available at the website www.prenhall.com/divisions/esm/app/calc_v2/. This site also can serve as a cross reference for TI-83 graphing calculator instructions. Since the TI-83 and TI-84 are so similar, TI-84 users will find this site helpful even though the TI-84 is not mentioned.

B.1 TURNING A GRAPHING CALCULATOR ON OR OFF

To turn a graphing calculator on, press ON. To turn it off, first press 2nd. Then press [OFF].

B.2 MAKING THE SCREEN LIGHTER OR DARKER

To make the screen darker, first press 2nd (then release it); then hold the △ key down for a while. To make the screen lighter, first press 2nd (then release it); then hold the ▽ key down for a while.

B.3 ENTERING AN EQUATION

To enter the equation $y = 2x + 1$,

1. Press Y=.
2. If necessary, press CLEAR to erase a previously entered equation.

When we show two or more buttons in a row, press them one at a time and in order.

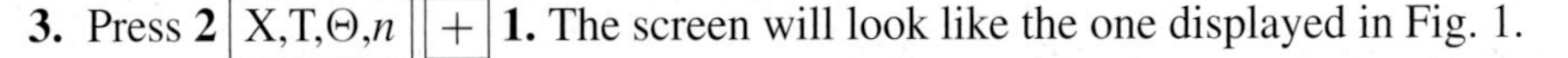
3. **Press 2** X,T,Θ,n + **1.** The screen will look like the one displayed in Fig. 1.
4. If you want to enter another equation, press ENTER. Then type in the next equation.
5. Use the △ or ▽ keys to get from one equation to another.

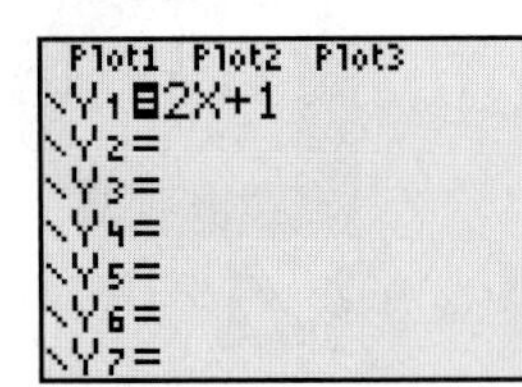

Figure 1 Entering an equation

B.4 GRAPHING AN EQUATION

To graph the equation $y = 2x + 1$,

1. Enter the equation $y = 2x + 1$; see Section B.3.
2. Press ZOOM **6** to draw a graph of your equation between the values of -10 and 10 for both x and y.
3. See Section B.6 if you want to zoom in or zoom out to get another part of the graph to appear on the calculator screen. Or see Section B.7 to change the window format manually; then press GRAPH.

B.5 TRACING A CURVE WITHOUT A SCATTERGRAM

To *trace a curve,* we find coordinates of points on the curve. To trace the line $y = 2x+1$,

1. Graph $y = 2x + 1$ (see Section B.4).
2. Press TRACE.
3. If you see a flashing "×" on the curve, the coordinates of that point will be listed at the bottom of the screen. If you don't see the flashing "×," press ENTER, and your calculator will adjust the viewing window so that you can see it.
4. To find coordinates of points on your curve that are off to the right, press ▷.
5. To find coordinates of points on your curve that are off to the left, press ◁.
6. Find the y-coordinate of a point by entering the x-coordinate. For example, to find the y-coordinate of the point that has x-coordinate 3, press **3** ENTER. The screen will look like the one displayed in Fig. 2. This feature works for values of x between Xmin and Xmax, inclusive (see Section B.7).
7. If more than one equation has been graphed, press ▽ to trace the second equation. Continue pressing ▽ to trace the third equation, and so on. Press △ to return to the previous equation. Notice that the equation of the curve being traced is listed in the upper left corner of the screen.

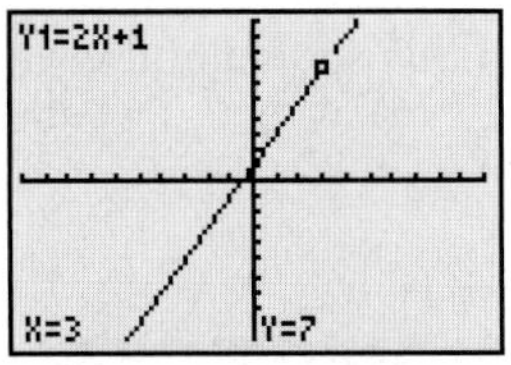

Figure 2 Tracing a curve

B.6 ZOOMING

The ZOOM menu has several features that allow you to adjust the viewing window. Some of the features adjust the values of x that are used when tracing.

- **Zoom In** magnifies the graph around the cursor location. The following instructions are for zooming in on the graph of $y = 2x + 1$:
  1. Graph $y = 2x + 1$ (see Section B.4).
  2. Press ZOOM **2**.
  3. Use ◁, ▷, △, and ▽ to position the cursor on a portion of the line that you want to zoom in on.
  4. To zoom in, press ENTER.

If you lose sight of the line, you can always press TRACE ENTER.

When zooming out, you will return to the original graph only if you did not move the cursor while zooming in.

5. To zoom in on the graph again, you have two options:
   a. To zoom in at the same point, press ENTER.
   b. To zoom in at a new point, move the cursor to the new point; then press ENTER.
6. To return to your original graph, press ZOOM **6**. Or zoom out (see the next instructions) the same number of times you zoomed in.

- **Zoom Out** does the reverse of Zoom In: It allows you to see *more* of a graph. To zoom out, follow the preceding instructions, but press ZOOM **3** instead of ZOOM **2** in step 2.
- **ZStandard** will change your viewing screen so that both x and y will go from -10 to 10. To use ZStandard, press ZOOM **6**.
- **ZDecimal** lets you trace a curve by using the numbers $0, \pm 0.1, \pm 0.2, \pm 0.3, \ldots$ for x. ZDecimal will change your viewing screen so that x will go from -4.7 to 4.7 and y will go from -3.1 to 3.1. To use ZDecimal, press ZOOM **4**.
- **ZInteger** allows you to trace a curve by using the numbers $0, \pm 1, \pm 2, \pm 3, \ldots$ for x. ZInteger can be used for any viewing window, although it will change the view slightly. To use ZInteger, press ZOOM **8** ENTER.
- **ZSquare** will change your viewing window so that the spacing of the tick marks on the x-axis is the same as that on the y-axis. To use ZSquare, press ZOOM **5**.
- **ZoomStat** will change your viewing window so that you can see a scattergram of points that you have entered in the statistics editor. To use ZoomStat, press ZOOM **9**.
- **ZoomFit** will adjust the dimensions of the y-axis to display as much of a curve as possible. The dimensions of the x-axis will remain unchanged. To use ZoomFit, press ZOOM **0**.

B.7 SETTING THE WINDOW FORMAT

To graph the equation $y = 2x + 1$ between the values of -2 and 3 for x and between the values of -5 and 7 for y,

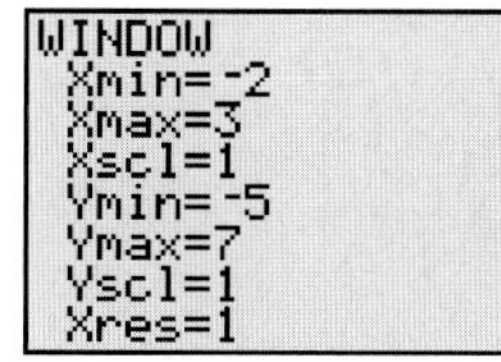

Figure 3 Window settings

1. Enter the equation $y = 2x + 1$ (see Section B.3).
2. Press WINDOW. Then change the window settings so that the window looks like the one displayed in Fig. 3 after you have used steps 3–8.
3. Press (-) **2** ENTER to set the smallest value of x to -2.
4. Press **3** ENTER to set the largest value of x to 3.
5. Press **1** ENTER to set the scaling for the x-axis to increments of 1.
6. Press (-) **5** ENTER to set the smallest value of y to -5.
7. Press **7** ENTER to set the largest value of y to 7.
8. Press **1** ENTER to set the scaling for the y-axis to increments of 1.
9. Press GRAPH to view the graph of $y = 2x + 1$. The screen will look like the graph drawn in Fig. 4.

If you press ZOOM **6** or ZOOM **9** or zoom in or zoom out, your window settings will change accordingly.

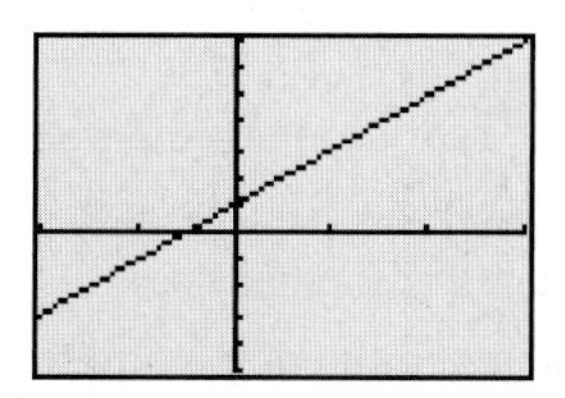

Figure 4 Graph of $y = 2x + 1$

B.8 PLOTTING POINTS IN A SCATTERGRAM

Table 1 Creating a Scattergram

x	y
2	4
3	7
4	10
5	11

To create a scattergram of the data displayed in Table 1,

1. To enter the data, press STAT **1**.
2. If there are numbers listed in the first column (list L_1), clear the column by pressing ◁ as many times as necessary to get to column L_1. Next, presss △ once to get to the top of column L_1. Then press CLEAR ENTER.
3. If there are numbers listed in the second column (list L_2), clear the column by pressing ▷ to move the cursor to column L_2. Then press △ CLEAR ENTER.
4. To return to the first entry position of list L_1, press ◁.
5. Press **2** ENTER **3** ENTER **4** ENTER **5** ENTER to enter the data in column L_1. (If you make a mistake, you can delete an entry by pressing DEL; then insert an entry by pressing 2nd [DEL].)
6. Press ▷ to move to the first entry position of list L_2.
7. Press **4** ENTER **7** ENTER **10** ENTER **11** ENTER to enter the elements of L_2.
8. Press 2nd [STAT PLOT].
9. Press **1** to select Plot 1.
10. Press ENTER to turn Plot 1 on.
11. Press ▽ ENTER to choose the scattergram mode.
12. Press ▽ so that the cursor is at "Xlist." Then press 2nd [L_1].
13. Press ▽ so that the cursor is at "Ylist." Then press 2nd [L_2].
14. Use squares, plus signs, or dots to represent the points plotted on the scattergram. These three symbols are called "Marks." Press ▽ once so that the cursor is on one of the three Mark symbols. Next, press ▷ and/or ◁ to select a symbol. Then press ENTER. The screen will look like the one displayed in Fig. 5.
15. Press ZOOM **9**. The screen will look like the one displayed in Fig. 6.

Make sure that you press CLEAR rather than DEL. If you press DEL, the column will vanish. If you ever do this by mistake, press STAT 5 ENTER to get back the missing column.

If Plot 1 is off, your points will be saved in columns L_1 and L_2, but they will not be plotted.

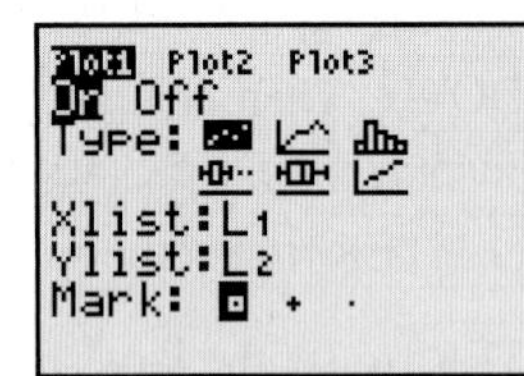

Figure 5 Setting up Plot 1

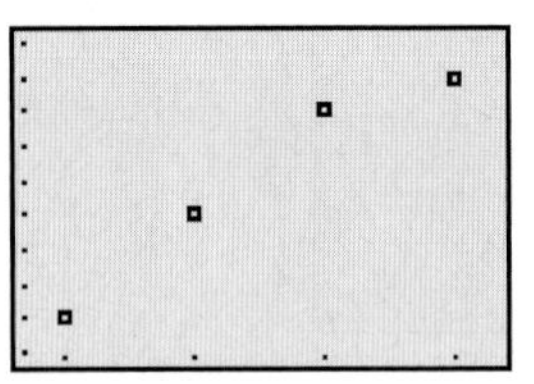
Figure 6 Creating a scattergram

B.9 TRACING A SCATTERGRAM

To see the coordinates of a point in a scattergram,

1. Draw a scattergram (see Section B.8).
2. Press TRACE.
3. Notice the flashing "×" on one of the points of the scattergram. The coordinates of this point are listed at the bottom of the screen.
4. To find the coordinates of the next point to the right, press ▷.
5. To find the coordinates of the next point to the left, press ◁.

Table 2 Creating a Scattergram

x	y
2	4
3	7
4	10
5	11

B.10 GRAPHING EQUATIONS WITH A SCATTERGRAM

To graph the equation $y = 2x + 1$ with a scattergram of the data displayed in Table 2,

1. Enter the equation $y = 2x + 1$ (see Section B.3).
2. Follow the instructions in Section B.8 to draw the scattergram. (The graph of the equation will also be drawn, because you turned the equation on.) The screen will look like the one displayed in Fig. 7.

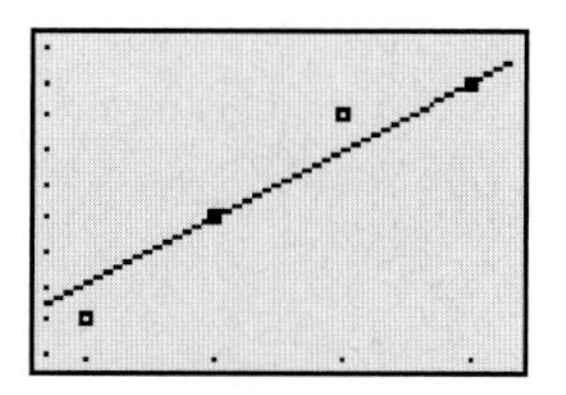

Figure 7 Graphing an equation and a scattergram

B.11 TRACING A CURVE WITH A SCATTERGRAM

To trace a curve with a scattergram,

1. Graph an equation with a scattergram (see Section B.10).
2. Press [TRACE] to trace points that make up the scattergram. Press [TRACE] [▽] to trace points that lie on the curve. If other equations are graphed, continue pressing [▽] to trace the second equation, and so on. Press [△] to begin to return to the scattergram. Notice that the label "P1:L_1, L_2" is in the upper left corner of the screen when Plot 1's points are being traced and that the equation entered in the [Y=] mode is listed in the upper left corner of the screen when the curve is being traced.

Recall from Section B.5 that if you do not see the flashing "×," press [ENTER], and the calculator will adjust the viewing window so that you can see it.

B.12 TURNING A PLOTTER ON OR OFF

To change the on/off status of the plotter,

1. Press [Y=].
2. Press [△]. A flashing rectangle will be on "Plot 1."
3. Press [▷] if necessary to move the flashing rectangle to the plotter you wish to turn on or off.
4. Press [ENTER] to turn your plotter on or off. The plotter is on if the plotter icon is highlighted.

B.13 CREATING A TABLE

To create a table of ordered pairs for the equation $y = 2x + 1$, where the values of x are 3, 4, 5, . . . (see Fig. 8),

1. Enter the equation $y = 2x + 1$ for Y_1 (see Section B.3).
2. Press [2nd] [TBLSET].
3. Press **3** [ENTER] to tell the calculator that the first x value in your table is 3.
4. Press **1** [ENTER] to tell the calculator that the x values in your table increase by 1.
5. Press [ENTER] [▽] [ENTER] to highlight "Auto" for both "Indpnt" and "Depend." The screen will now look like the one displayed in Fig. 9.
6. Press [2nd] [TABLE] to create the table shown in Fig. 8.

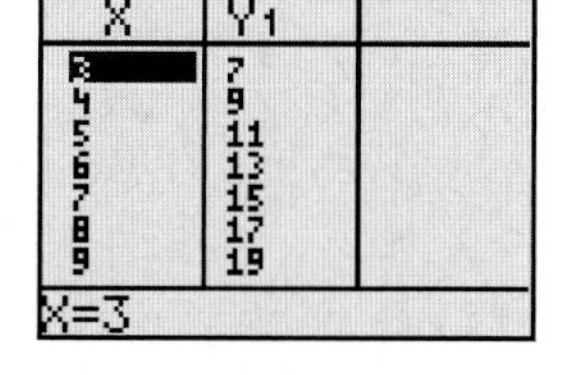

Figure 8 Table of ordered pairs for $y = 2x + 1$

Figure 9 Table setup

B.14 CREATING A TABLE FOR TWO EQUATIONS

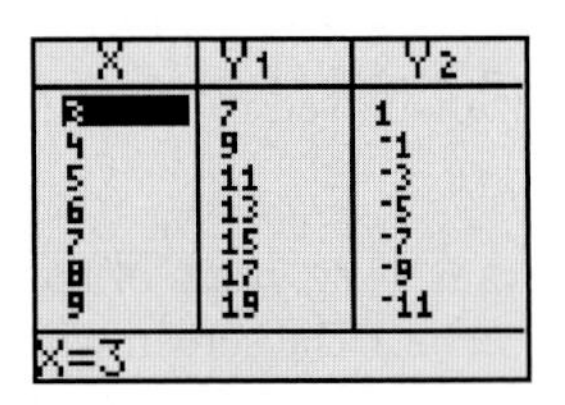

Figure 10 Table for two equations

To create a table of ordered pairs for the equations $y = 2x + 1$ and $y = -2x + 7$, where the values of x are 3, 4, 5, . . . (see Fig. 10),

1. Enter the equation $y = 2x + 1$ for Y_1, and enter the equation $y = -2x + 7$ for Y_2 (see Section B.3).
2. Follow steps 2–5 of Section B.13.

B.15 USING "ASK" IN A TABLE

Table 3 Using "Ask" in a Table with $y = 2x + 1$

x	y
2	
2.9	
5.354	
7	
100	

To use the Ask option in the Table Setup mode to complete Table 3 for $y = 2x + 1$,

1. Enter the equation $y = 2x + 1$ for Y_1 (see Section B.3).
2. Press [2nd] [TBLSET].
3. Press [ENTER] twice. Next, press [▷]. Then press [ENTER]. The Ask option for "Indpnt" will now be highlighted. Make sure that the Auto option for "Depend" is highlighted.
4. Press [2nd] [TABLE].
5. Press **2** [ENTER] **2.9** [ENTER] **5.354** [ENTER] **7** [ENTER] **100** [ENTER]. The screen will now look like the one displayed in Fig. 11.

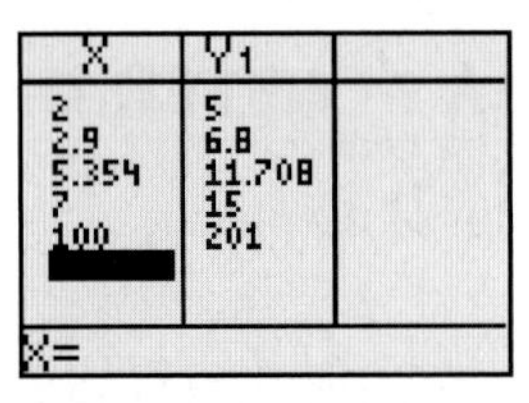

Figure 11 Using "Ask" for a table with $y = 2x + 1$

B.16 FINDING THE REGRESSION CURVE FOR SOME DATA

Table 4 Finding the Regression Line for Data

x	y
2	4
3	7
4	10
5	11

To find the regression line for the data displayed in Table 4,

1. See Section B.8 to create a scattergram of the data in Table 4. Enter your data in the first two columns (L_1 and L_2) of the STAT list editor.
2. Clear the Home Screen by pressing [2nd] [QUIT] [CLEAR].
3. Press [STAT].
4. To choose the CALC menu, press [▷]. The screen should look like the one displayed in Fig. 12.
5. To choose Linear Regression, press **4**. The screen should now look like the one displayed in Fig. 13. (You can choose Quadratic Regression by pressing **5**, Exponential Regression by pressing **0**, or Power Regression by pressing [ALPHA] [A].)
6. Press [ENTER]. The screen should now look like the one displayed in Fig. 14. This means that the equation of the regression line is $y = 2.4x - 0.4$.

You can perform regression on columns other than L_1 and L_2 by listing the two columns, separated by a comma, after the "LinReg(ax+b)" command on the Home screen. For example, "LinReg(ax+b) L_4, L_6" will perform a linear regression on columns 4 and 6 of the STAT list editor.

```
EDIT CALC TESTS
1:1-Var Stats
2:2-Var Stats
3:Med-Med
4:LinReg(ax+b)
5:QuadReg
6:CubicReg
7↓QuartReg
```

Figure 12 CALC menu

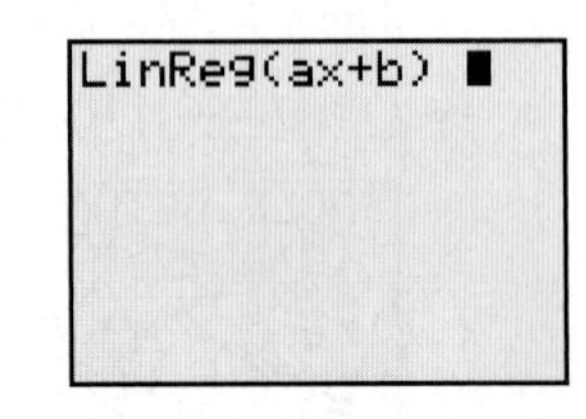

Figure 13 About to find the equation

```
LinReg
 y=ax+b
 a=2.4
 b=-.4
```

Figure 14 The equation

To draw a graph of the regression line, you may either enter the equation manually (see Section B.3) or use the command

LinReg(ax+b) L_1,L_2,Y_1

which saves the equation to Y_1. Here are the keystrokes:

1. Follow the earlier instructions to get "LinReg(ax+b)" on your screen.
2. Press 2nd [L_1] , 2nd [L_2] , .
3. Press VARS ▷ **1** ENTER. The screen should look like the one displayed in Fig. 15.
4. Press ENTER. The screen should now look like the one displayed in Fig. 14. In addition, if you press Y=, the screen will look like the one displayed in Fig. 16.

Figure 15 About to save the equation to Y_1

Plot1 Plot2 Plot3
\Y1=2.4X+-.4
\Y2=
\Y3=
\Y4=
\Y5=
\Y6=
\Y7=

Figure 16 The equation is saved in Y_1

B.17 PLOTTING POINTS IN TWO SCATTERGRAMS

Table 5 Creating the First of Two Scattergrams

x	y
2	4
3	7
4	10
5	11

It is possible to draw two scattergrams on the same calculator screen and use different markings for the two sets of points. To begin, follow the instructions in Section B.8 to create a scattergram of the data values in Table 5.

These data are stored in columns L_1 and L_2. The points are plotted by the plotter called "Plot 1."

You will now create a scattergram of the data values in Table 6.

These data will be stored in columns L_3 and L_4. The points will be plotted by the plotter called "Plot 2." To do this,

1. To enter the data, press STAT **1**.
2. To clear list L_3, press ▷ and/or ◁ to move the cursor to column L_3. Then press △ CLEAR ENTER.
3. To clear list L_4, press ▷ to move the cursor to column L_4. Then press △ CLEAR ENTER.
4. To return to the first entry position of list L_3, press ◁.
5. Press **2** ENTER **2** ENTER **3** ENTER **5** ENTER to enter the elements of L_3.
6. Press ▷ to move to the first entry position of list L_4.
7. Press **11** ENTER **9** ENTER **6** ENTER **4** ENTER to enter the elements of L_4.
8. Press 2nd [STAT PLOT].
9. Press **2** to select "Plot 2."
10. Press ENTER to turn Plot 2 on.
11. Press ▽ twice so that the cursor is at "Xlist." Then press 2nd [L_3].
12. Press ENTER so that the cursor is at "Ylist." Then press 2nd [L_4].

Make sure that you press CLEAR rather than DEL. If you press DEL, the column will vanish. If you ever do this by mistake, press STAT 5 ENTER to get back the missing column.

Table 6 Creating the Second of Two Scattergrams

x	y
2	11
2	9
3	6
5	4

Figure 17 Setting up Plot 2

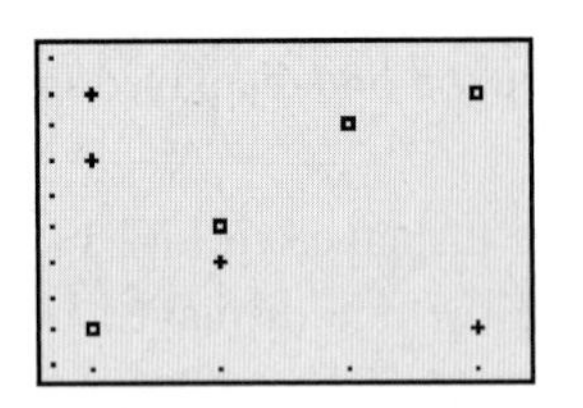

Figure 18 Creating two scattergrams

13. Press ▽ once so that the cursor is on one of the three choices for "Mark." Next, press ▷ and/or ◁ to select a symbol different from the one you used for the first scattergram. Then press ENTER. The screen will look like the one in Fig. 17.
14. Press ZOOM **9** to obtain the two scattergrams with different symbols. The screen will look like the one displayed in Fig. 18.

B.18 FINDING THE INTERSECTION POINT(S) OF TWO CURVES

To find the intersection point of the lines $y = 2x + 1$ and $y = -2x + 7$,

1. Enter the equation $y = 2x + 1$ for Y_1, and enter the equation $y = -2x + 7$ for Y_2 (see Section B.3).
2. By zooming in or out or by changing the window settings, draw a graph of both curves so that you can see an intersection point. For our example, press ZOOM **6**.
3. Press 2nd [CALC]. The screen will look like the one displayed in Fig. 19.
4. Press **5** to select "intersect."
5. You will now see a flashing cursor on your first curve. If there is more than one intersection point on your display screen, move the cursor by pressing ▷ or ◁ so that it is much closer to the intersection point you want to find. The screen will look something like the one displayed in Fig. 20.
6. Press ENTER to put the cursor on the second curve. Press ENTER again to display "Guess?" Press ENTER once more. The screen will look like the one displayed in Fig. 21. The intersection point is (1.5, 4).

Figure 19 Menu of choices

Figure 20 Put cursor near intersection point

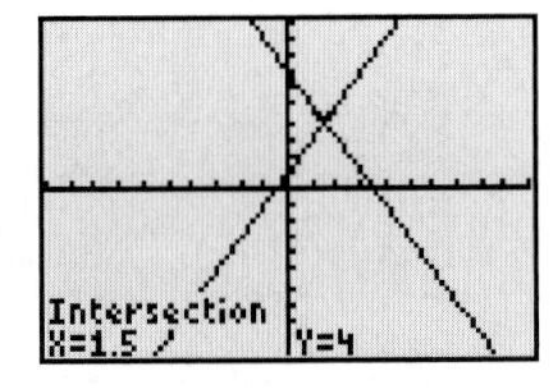

Figure 21 Location of intersection point

B.19 FINDING THE MINIMUM OR MAXIMUM OF A CURVE

To find the minimum point of the curve $y = x^2 - 3x + 1$,

1. Enter the equation $y = x^2 - 3x + 1$ (see Section B.3).
2. Use ZDecimal to draw a graph of the equation (see Section B.6).
3. Press 2nd [CALC].
4. Press **3** to select "minimum."
5. Move the flashing cursor by pressing ◁ or ▷ so that it is to the left of the minimum point, and press ENTER.
6. Move the flashing cursor by pressing ◁ or ▷ so that it is to the right of the minimum point, and press ENTER.
7. Press ENTER. The calculator will display the coordinates of the minimum point—about $(1.50, -1.25)$. See Fig. 22.

   You can find the maximum point of a curve in a similar fashion, but press **4** to select the "maximum" option, rather than the "minimum" option, in step 4.

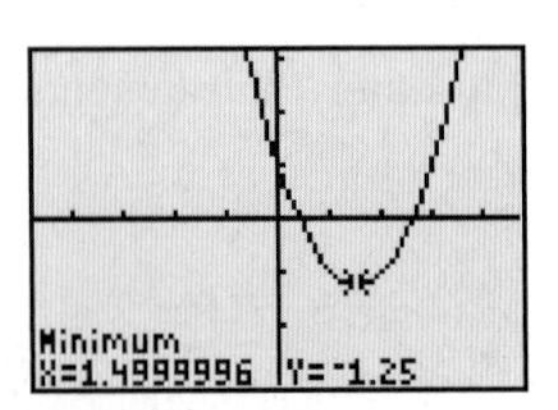

Figure 22 Finding the minimum point of $y = x^2 - 3x + 1$

B.20 STORING A VALUE

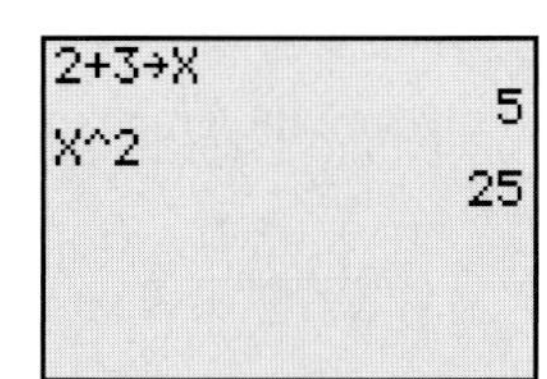

Figure 23 Computing $(2+3)^2 = 25$

It is possible to store a number as x and then perform operations with x. For example, to find $(2+3)^2$,

1. Press **2** [+] **3** [STO▷] [X,T,Θ,n] [ENTER].
2. Press [X,T,Θ,n] [∧] **2** [ENTER]. The screen should now look like the one displayed in Fig. 23.

B.21 FINDING ANY x-INTERCEPTS OF A CURVE

To find the x-intercept of the line $y = x - 2$,

1. Enter the equation $y = x - 2$ (see Section B.3).
2. Use ZDecimal to draw a graph of the equation (see Section B.6).
3. Press [2nd] [CALC].
4. Press **2** to choose the "zero" option.
5. Move the flashing cursor by pressing [◁] or [▷] so that it is to the left of the x-intercept, and press [ENTER]. Or type a number between Xmin and the x-coordinate of the x-intercept, and press [ENTER].
6. Move the flashing cursor by pressing [◁] or [▷] so that it is to the right of the x-intercept, and press [ENTER]. Or type a number between the x-coordinate of the x-intercept and Xmax, and press [ENTER].
7. Press [ENTER]. The screen will look like the one displayed in Fig. 24. The x-intercept is (2, 0).

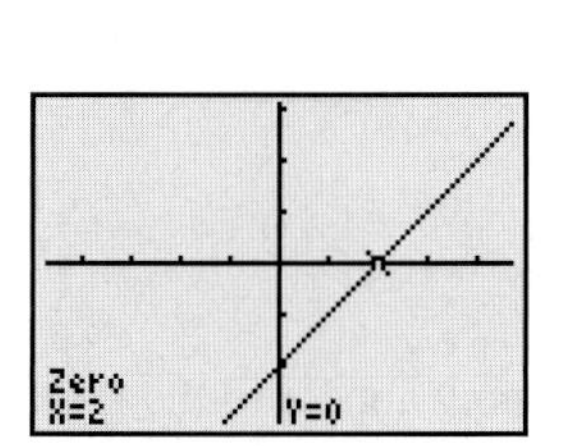

Figure 24 Finding the x-intercept of $y = x - 2$

B.22 TURNING AN EQUATION ON OR OFF

You can graph an equation only if its equals sign is highlighted (the equation is then "on"). Up to 10 equations can be graphed at one time. To change the on–off status of an equation:

1. Press [Y=].
2. Move the cursor to the equation whose status you want to change.
3. Use [◁] to place the cursor over the "=" sign of the equation.
4. Press [ENTER] to change the status.

B.23 FINDING COORDINATES OF POINTS

To find the coordinates of particular points,

1. Press [GRAPH] to get into graphing mode.
2. Press [▷] to get a cursor to appear on the screen. [If you cannot see it, it is probably on one or both of the axes. If it is on an axis, you should still be able to see a small flashing dot.] Notice that the coordinates of the point where the cursor is currently positioned are at the bottom of the screen.
3. Use [◁], [▷], [△], or [▽] to move the cursor left, right, up, or down, respectively.

B.24 GRAPHING EQUATIONS WITH AXES "TURNED OFF"

Suppose that you want to draw a graph of $y = 0$. The axes will obscure the graph of $y = 0$. To graph without the axes appearing on the screen,

Figure 25 Graph of $y = 0$ with axes "turned off"

1. Enter the equation $y = 0$ for Y_1 (see Section B.3).
2. Press [2nd] [FORMAT]. You are now at the FORMAT menu.
3. Press [▽] three times, then press [▷]; then press [ENTER]. "AxesOff" should now be highlighted.
4. Use ZDecimal to have the screen appear like the one displayed in Fig. 25 (see Section B.6).

You can turn the axes back on by highlighting "AxesOn" in the FORMAT menu.

B.25 ENTERING AN EQUATION BY USING Y_n REFERENCES

To enter the complicated equation $y = \frac{x+1}{x-3} \div \frac{x-2}{x+5}$ by using Y_n references,

1. Enter $Y_1 = \frac{x+1}{x-3}$ and $Y_2 = \frac{x-2}{x+5}$ (see Section B.3).
2. Turn both equations off (see Section B.22).
3. Move the flashing cursor to the right of "$Y_3 =$."
4. Press [VARS] [▷] [ENTER].
5. Move the cursor to "1:Y_1" and press [ENTER]. "Y_1" will now appear to the right of "$Y_3 =$" in the [Y=] window.
6. Press [÷].
7. Press [VARS] [▷] [ENTER].
8. Move the cursor to "2:Y_2" and press [ENTER]. "Y_1/Y_2" will now appear to the right of "$Y_3 =$" in the [Y=] window.

B.26 RESPONDING TO ERROR MESSAGES

Here are several common error messages and how to respond to them:

Figure 26 "Syntax" error message

- The **Syntax** error (see Fig. 26) means that you have misplaced one or more parentheses, operations, or commas. The calculator will find this type of error if you choose "Goto" by pressing [▽], then [ENTER]. Your error will be highlighted by a flashing black rectangle.

  The most common "Syntax" error is pressing [(-)] when you should have pressed [−], or vice versa:

  1. Press the [(-)] key when you want to take the opposite of a number or are working with negative numbers. To compute $-5(-2)$, press [(-)] **5** [(] [(-)] **2** [)].
  2. Press the [−] key when you want to subtract two numbers. To compute $5 - 2$, press **5** [−] **2**.

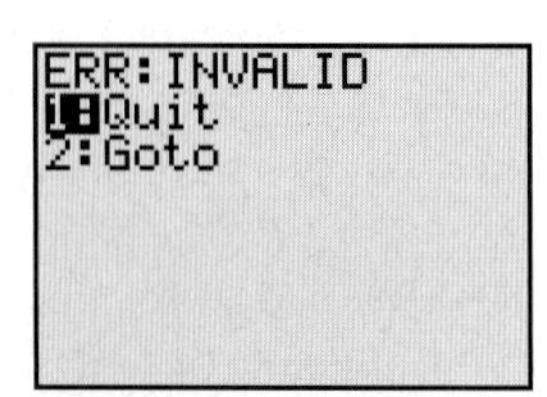

Figure 27 "Invalid" error message

- The **Invalid** error (see Fig. 27) means that you have tried to enter an inappropriate number, expression, or command. The most common "Invalid" error is to try to enter a number that is not between Xmin and Xmax, inclusive, when you use commands such as [TRACE], "minimum," or "maximum."

- The **Invalid dimension** error (see Fig. 28) means that you have the plotter turned on (see Fig. 29) but have not entered any data points in the STAT list editor (see Fig. 30). In this case, first press ENTER to exit the error message display; then either turn the plotter off or enter data in the STAT list editor.

Figure 28 "Invalid dimension" error message

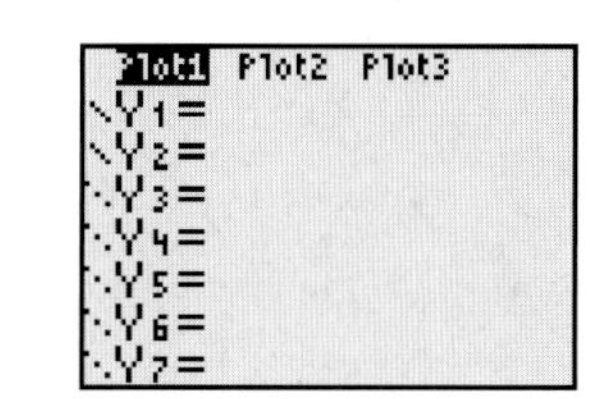

Figure 29 Plotter is on

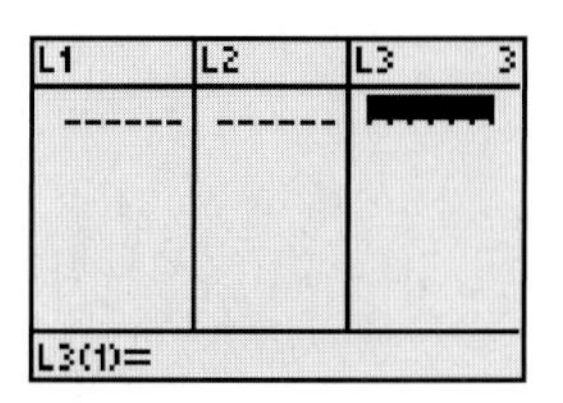

Figure 30 STAT list editor's columns are empty

Figure 31 "Dimension mismatch" error message

- The **Dimension mismatch** error (see Fig. 31) is fixed in two ways:
  1. In the STAT list editor, one column that you are using to plot has more numbers than the other column has (see Fig. 32). In this case, first press ENTER to exit the error message display; then add or delete numbers so that the two columns have the same length.

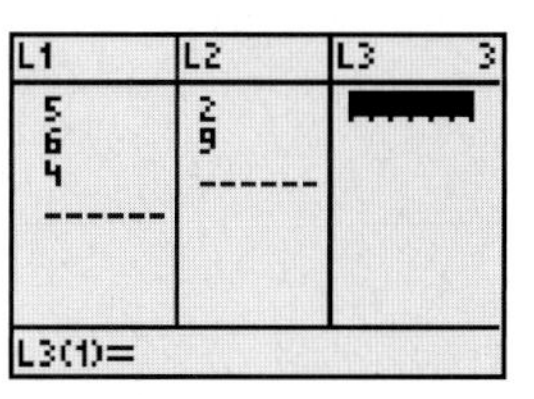

Figure 32 Columns of unequal length in STAT list editor

  2. In the STAT list editor, one column that you are using to plot has more numbers than the other column has, but you didn't notice the difference in length because you deleted one or both of the columns by mistake. You can find the missing column(s) by pressing STAT **5** ENTER.

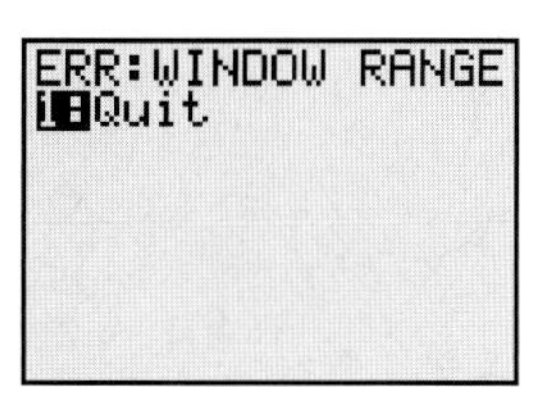

Figure 33 "Window range" error message

- The **Window range** error (see Fig. 33) means one of two things:
  1. You made an error in setting up your window. This usually means that you entered a larger number for *Xmin* than for *Xmax* or that you entered a larger number for *Ymin* than for *Ymax*. In this case, first press ENTER to exit the error message display; then change your window settings accordingly (see Section B.7).
  2. You pressed ZOOM **9** when only one data-point pair was entered in the STAT list editor. (For some TI graphing calculators, the command ZoomStat works only when you have two or more pairs of data points in the STAT list editor.) In this case, first press ENTER to exit the error message display; then either add more points to the STAT list editor or avoid pressing ZOOM **9** and set up your window settings manually (see Section B.7).

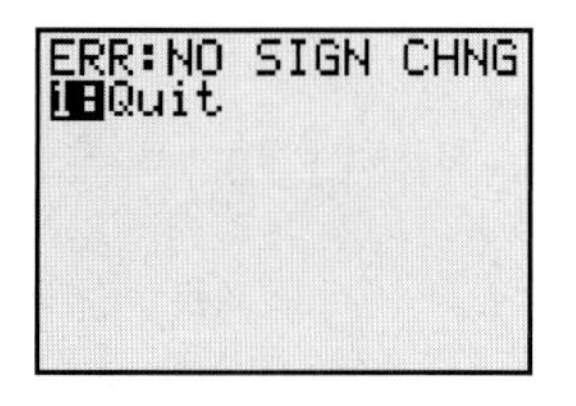

Figure 34 "No sign change" error message

- The **No sign change** error (see Fig. 34) means one of two things:
  1. You are trying to locate a point that does not appear on the screen. For example, you may be trying to find an intersection point of two curves that intersect off screen. Or you may be trying to find a zero of an equation that does not appear on the screen. In this case, press ENTER and change your window settings so that the point you are trying to locate is on the screen.
  2. You are trying to locate a point that does not exist. For example, you may be trying to find an intersection point of two parallel lines. Or you may be trying to find a zero of an equation that does not have one. In this case, press ENTER and stop looking for the point that doesn't exist!

- The **Nonreal answer** error (see Fig. 35) means that your computation did not yield a real number. For example, $\sqrt{-4}$ is not a real number. The calculator will locate this computation if you choose "Goto" by pressing ▽, then ENTER.

ERR:NONREAL ANS
1:Quit
2:Goto

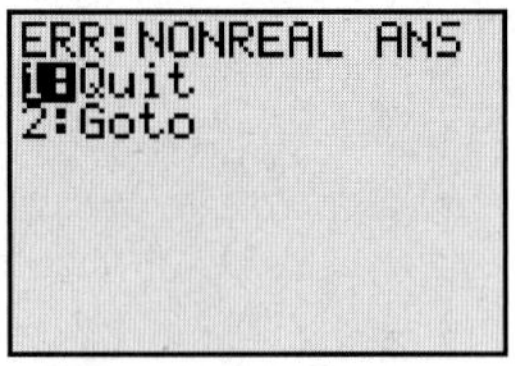

Figure 35 "Nonreal answer" error message

- The **Divide by 0** error (see Fig. 36) means that you asked the calculator to perform a calculation that involves a division by zero. For example, $3 \div (5 - 5)$ will yield the error message shown in Fig. 36.

ERR:DIVIDE BY 0
1:Quit
2:Goto

Figure 36 "Divide by zero" error message

The calculator will locate the division by zero if you choose "Goto" by pressing ▽, then ENTER.

Answers to Odd-Numbered Exercises

Answers to most discussion exercises and to exercises where answers may vary have been omitted.

Chapter 1

Homework 1.1 **1. a.** (d) **b.** (c) **c.** (a) **d.** (b) **3.** N independent, T dependent **5.** F independent, T dependent **7.** L independent, T dependent **9.** I independent, P dependent **11.** r independent, n dependent

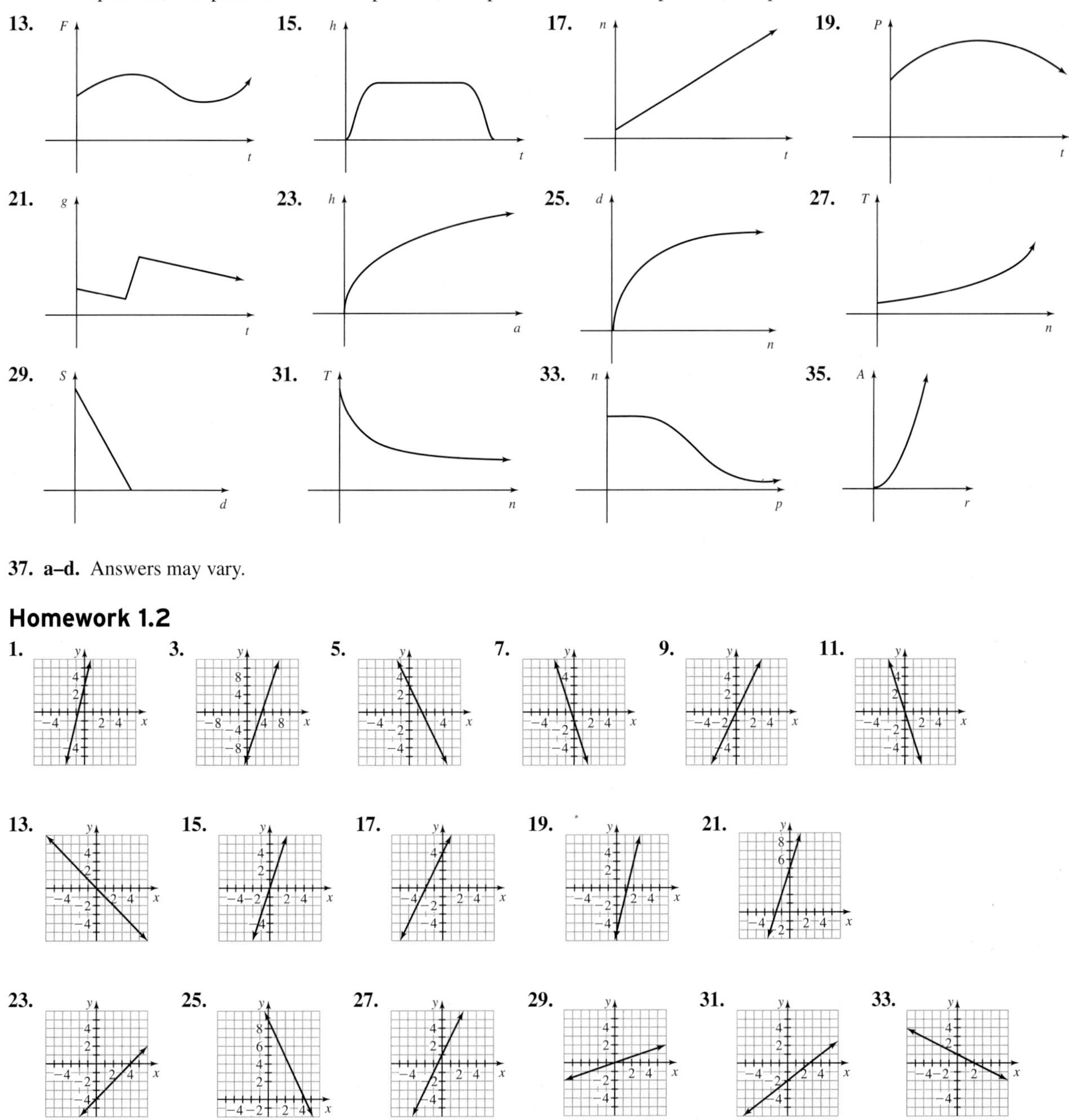

37. a–d. Answers may vary.

Homework 1.2

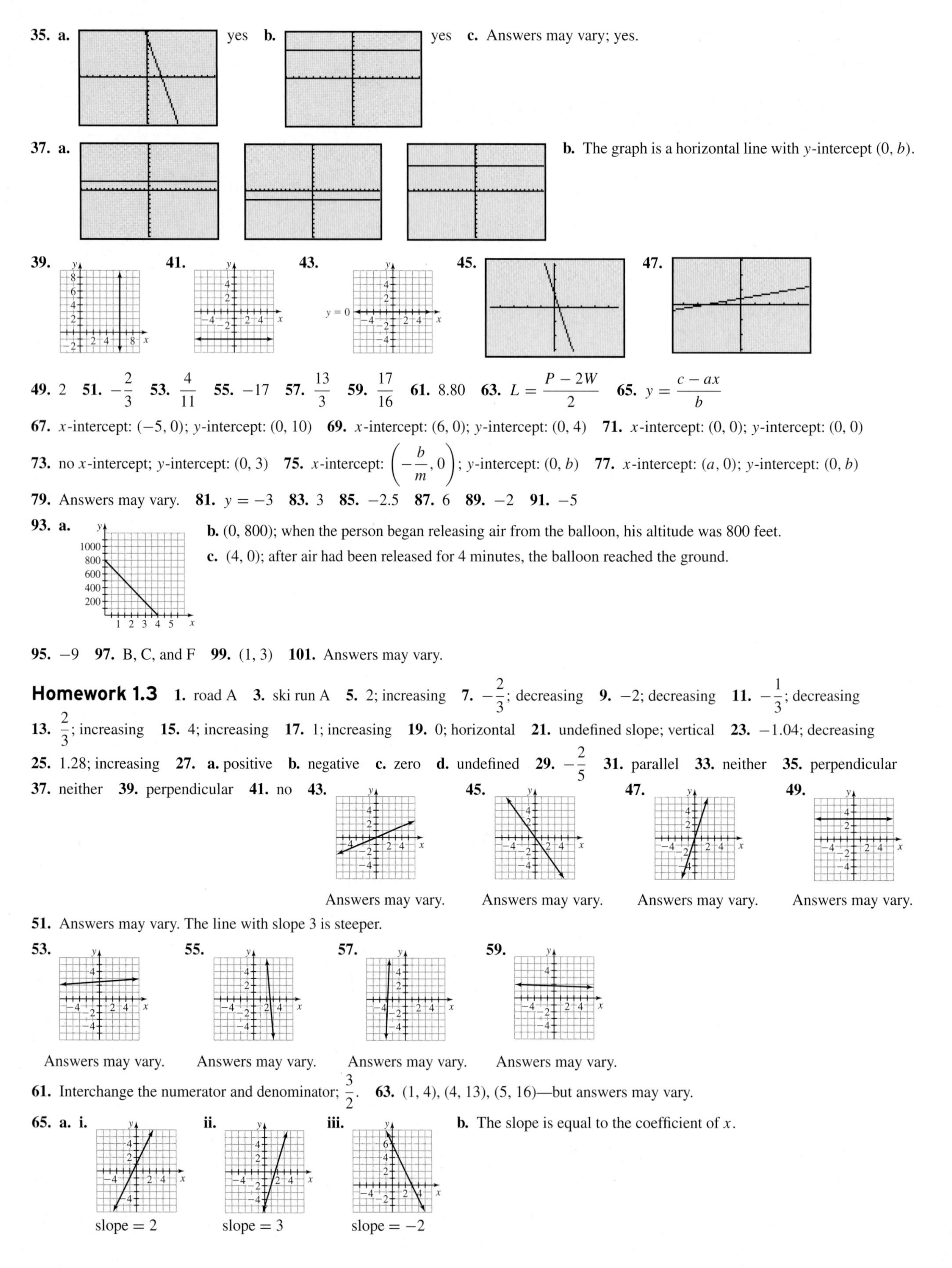

35. a. yes **b.** yes **c.** Answers may vary; yes.

37. a. **b.** The graph is a horizontal line with y-intercept $(0, b)$.

39. **41.** **43.** **45.** **47.**

49. 2 **51.** $-\frac{2}{3}$ **53.** $\frac{4}{11}$ **55.** -17 **57.** $\frac{13}{3}$ **59.** $\frac{17}{16}$ **61.** 8.80 **63.** $L = \frac{P - 2W}{2}$ **65.** $y = \frac{c - ax}{b}$

67. x-intercept: $(-5, 0)$; y-intercept: $(0, 10)$ **69.** x-intercept: $(6, 0)$; y-intercept: $(0, 4)$ **71.** x-intercept: $(0, 0)$; y-intercept: $(0, 0)$

73. no x-intercept; y-intercept: $(0, 3)$ **75.** x-intercept: $\left(-\frac{b}{m}, 0\right)$; y-intercept: $(0, b)$ **77.** x-intercept: $(a, 0)$; y-intercept: $(0, b)$

79. Answers may vary. **81.** $y = -3$ **83.** 3 **85.** -2.5 **87.** 6 **89.** -2 **91.** -5

93. a. **b.** $(0, 800)$; when the person began releasing air from the balloon, his altitude was 800 feet.
c. $(4, 0)$; after air had been released for 4 minutes, the balloon reached the ground.

95. -9 **97.** B, C, and F **99.** $(1, 3)$ **101.** Answers may vary.

Homework 1.3

1. road A **3.** ski run A **5.** 2; increasing **7.** $-\frac{2}{3}$; decreasing **9.** -2; decreasing **11.** $-\frac{1}{3}$; decreasing **13.** $\frac{2}{3}$; increasing **15.** 4; increasing **17.** 1; increasing **19.** 0; horizontal **21.** undefined slope; vertical **23.** -1.04; decreasing **25.** 1.28; increasing **27. a.** positive **b.** negative **c.** zero **d.** undefined **29.** $-\frac{2}{5}$ **31.** parallel **33.** neither **35.** perpendicular **37.** neither **39.** perpendicular **41.** no **43.** Answers may vary. **45.** Answers may vary. **47.** Answers may vary. **49.** Answers may vary.

51. Answers may vary. The line with slope 3 is steeper.

53. Answers may vary. **55.** Answers may vary. **57.** Answers may vary. **59.** Answers may vary.

61. Interchange the numerator and denominator; $\frac{3}{2}$. **63.** $(1, 4)$, $(4, 13)$, $(5, 16)$—but answers may vary.

65. a. i. slope $= 2$ **ii.** slope $= 3$ **iii.** slope $= -2$ **b.** The slope is equal to the coefficient of x.

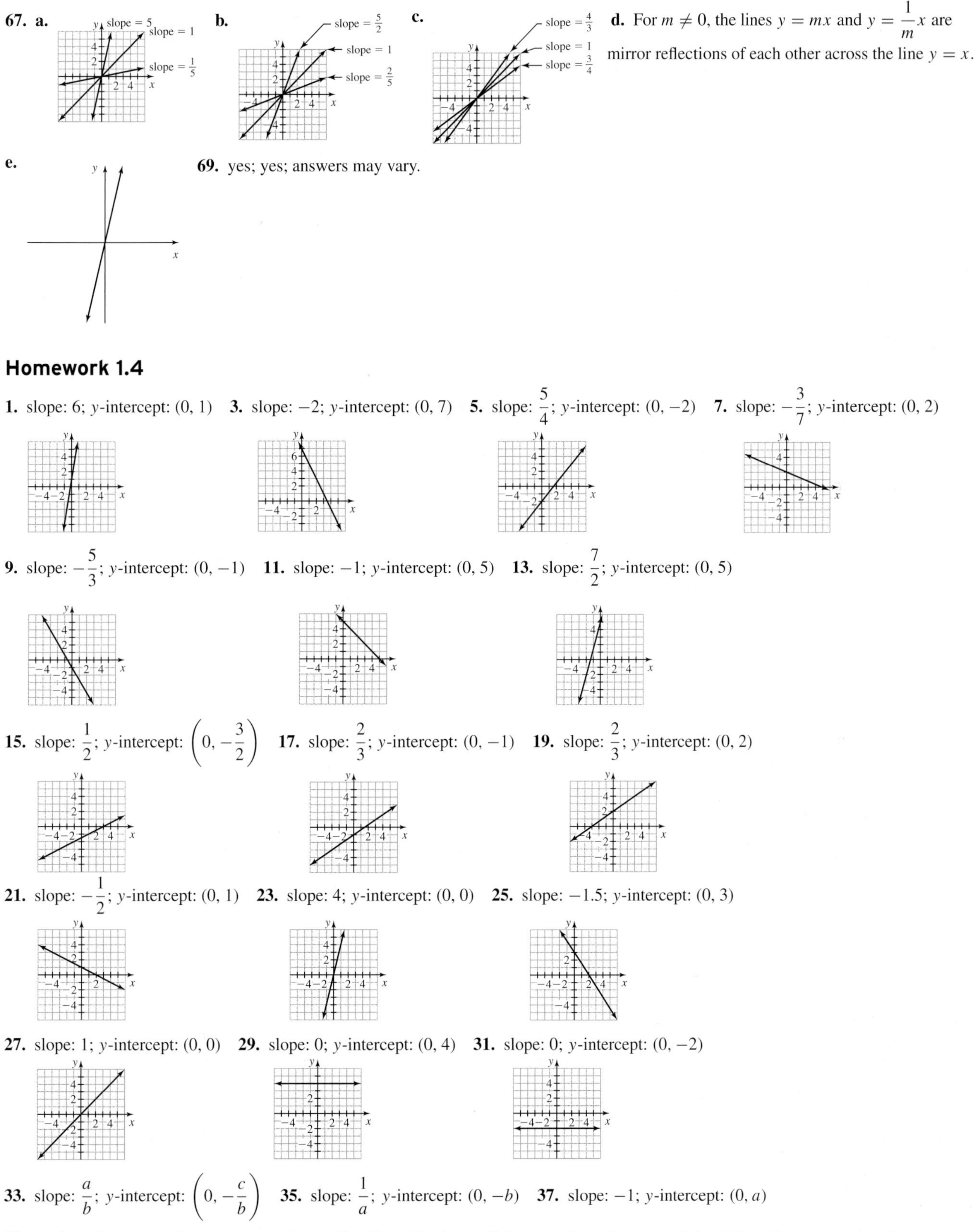

67. a. **b.** **c.** **d.** For $m \neq 0$, the lines $y = mx$ and $y = \frac{1}{m}x$ are mirror reflections of each other across the line $y = x$.

e.

69. yes; yes; answers may vary.

Homework 1.4

1. slope: 6; y-intercept: (0, 1) **3.** slope: -2; y-intercept: (0, 7) **5.** slope: $\frac{5}{4}$; y-intercept: (0, -2) **7.** slope: $-\frac{3}{7}$; y-intercept: (0, 2)

9. slope: $-\frac{5}{3}$; y-intercept: (0, -1) **11.** slope: -1; y-intercept: (0, 5) **13.** slope: $\frac{7}{2}$; y-intercept: (0, 5)

15. slope: $\frac{1}{2}$; y-intercept: $\left(0, -\frac{3}{2}\right)$ **17.** slope: $\frac{2}{3}$; y-intercept: (0, -1) **19.** slope: $\frac{2}{3}$; y-intercept: (0, 2)

21. slope: $-\frac{1}{2}$; y-intercept: (0, 1) **23.** slope: 4; y-intercept: (0, 0) **25.** slope: -1.5; y-intercept: (0, 3)

27. slope: 1; y-intercept: (0, 0) **29.** slope: 0; y-intercept: (0, 4) **31.** slope: 0; y-intercept: (0, -2)

33. slope: $\frac{a}{b}$; y-intercept: $\left(0, -\frac{c}{b}\right)$ **35.** slope: $\frac{1}{a}$; y-intercept: (0, $-b$) **37.** slope: -1; y-intercept: (0, a)

39. set 1: A line comes close to every point; set 2: a line with slope -0.3 passes through every point; set 3: no line comes close to every point; set 4: a line with slope -10 passes through every point.

41.

x	y	x	y	x	y	x	y
1	12	23	69	1	47	30	15
2	15	24	53	2	41	31	24
3	18	25	37	3	35	32	33
4	21	26	21	4	29	33	42
5	24	27	5	5	23	34	51
6	27	28	−11	6	17	35	60

43. a. m: negative; b: positive **b.** m: positive; b: negative **c.** m: 0; b: negative **d.** m: negative; b: 0 **45.** $y = -2x + 3$ **47.** Answers may vary. **49.** Answers may vary. **51.** parallel **53.** neither **55.** parallel **57.** perpendicular **59.** parallel **61.** perpendicular **63. a.** y values, from top: 18, 15, 12, 9, 6, 3, 0 **b.** The amount of gas is decreasing by 3 gallons per hour. The slope is −3. **c.** 20 miles per gallon **65. a.** y values, from top: 26, 28, 30, 32, 34 **b.** Her salary is increasing by $2000 per year. The slope is 2.

67. a. **b.** **c.**

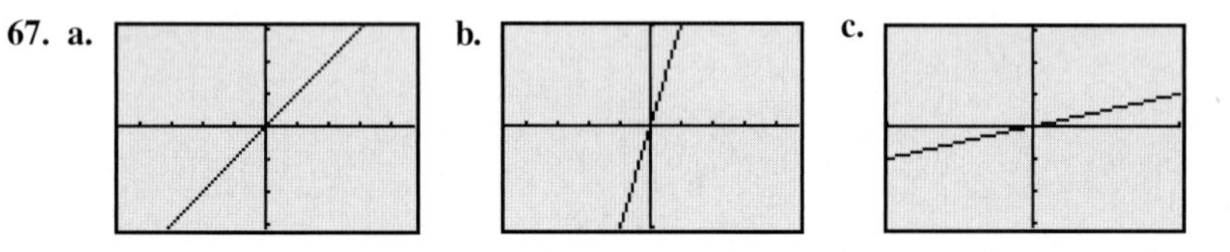

d. The graph appears steeper in part (b) than in part (a). The graph appears less steep in part (c) than in part (a). **e.** no; no; the line will always appear to be increasing and will always pass through the origin and lie in quadrants I and III, but we may make it appear to have any steepness. **69.** $y = 2x - 3$ **71. a.**

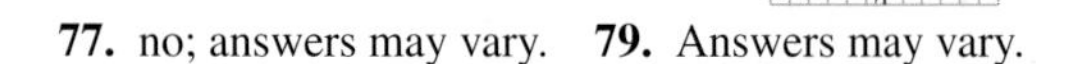

b. $y = 5x - 7$ **73. a.** **b.** $y = -\frac{3}{8}x + \frac{27}{4}$ **75. a.** 0; 0; 0 **b.** 0

77. no; answers may vary. **79.** Answers may vary.

Homework 1.5

1. $y = 3x - 13$; increasing **3.** $y = -2x - 3$; decreasing **5.** $y = \frac{3}{5}x - 5$; increasing **7.** $y = -\frac{1}{6}x - \frac{8}{3}$; decreasing **9.** $y = -\frac{5}{2}x - \frac{23}{2}$; decreasing **11.** $y = 2$; horizontal **13.** $x = 3$; vertical **15.** $y = 1.6x + 0.44$ **17.** $y = -3.24x - 15.18$ **19.** $y = x + 1$ **21.** $y = -2x + 2$ **23.** $y = -2x - 22$ **25.** $y = x$ **27.** $y = \frac{4}{5}x - \frac{3}{5}$ **29.** $y = -\frac{7}{6}x - \frac{8}{3}$ **31.** $y = \frac{5}{2}x + \frac{11}{2}$ **33.** $y = 5$ **35.** $x = -3$ **37.** $y = 0.49x - 1.41$ **39.** $y = 0.46x - 3.47$ **41.** $y = 3x - 7$ **43.** $y = -2x + 2$ **45.** $y = \frac{1}{2}x - 1$ **47.** $y = \frac{3}{4}x + \frac{7}{4}$ **49.** $y = \frac{1}{6}x - \frac{3}{2}$ **51.** $y = 3$ **53.** $x = -5$ **55.** $y = -\frac{1}{2}x + \frac{19}{2}$ **57.** $y = \frac{1}{3}x + \frac{22}{3}$ **59.** $y = \frac{5}{2}x + 2$ **61.** $y = -\frac{5}{4}x + \frac{31}{2}$ **63.** $y = -\frac{1}{2}x - \frac{5}{2}$ **65.** $y = 3$ **67.** $x = 2$ **69.** $y = -2x + 19$ **71.** $y = \frac{1}{3}x - \frac{2}{3}$ **73.** $y = -\frac{3}{2}x + \frac{15}{2}$ **75. a.** yes; answers may vary. **b.** yes; answers may vary. **c.** no; answers may vary. **d.** yes; $y = 0$ **77.** yes; $y = -2x + 7$ **79. a–c.** Answers may vary. **81. a.** Answers may vary. **b.** Answers may vary. **c.** Answers may vary. **d.** no. **83.** Answers may vary. **85.** Answers may vary.

Homework 1.6

1. relations 2 and 3 **3.** no **5.** yes **7.** yes **9.** no **11.** no **13.** yes **15.** yes **17.** yes **19.** yes **21.** no **23.** yes **25.** yes; the graph will pass the vertical line test. **27.** no; the graph will not pass the vertical line test. **29. a.** Answers may vary. **b.** **c.** For each input–output pair, the output is 2 less than 3 times the input.

31. domain: $-4 \le x \le 5$; range: $-2 \le y \le 3$ **33.** domain: $-5 \le x \le 4$; range: $-2 \le y \le 3$ **35.** domain: $-4 \le x \le 4$; range: $-2 \le y \le 2$ **37.** domain: $0 \le x \le 4$; range: $0 \le y \le 2$ **39.** domain: all real numbers; range: $y \le 4$

41. domain: $x \geq 0$; range: $y \geq 0$ **43.** Answers may vary. **45.** Answers may vary. No; an input of $x = 2$ gives two different outputs. **47.** Answers may vary. **49.** Answers may vary. **51.** yes **53.** no **55.** no; no input corresponds to two different outputs, so the definition of a function is not violated.

Chapter 1 Review

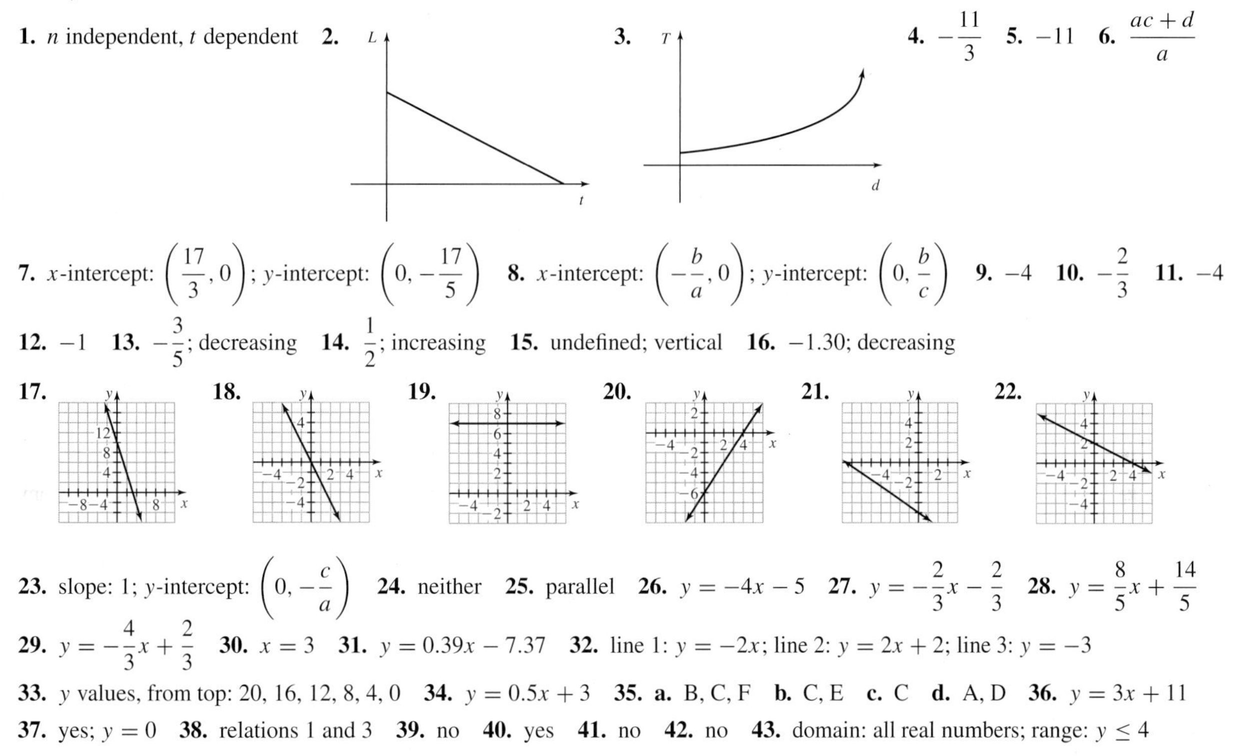

1. n independent, t dependent **2.** **3.** **4.** $-\frac{11}{3}$ **5.** -11 **6.** $\frac{ac+d}{a}$

7. x-intercept: $\left(\frac{17}{3}, 0\right)$; y-intercept: $\left(0, -\frac{17}{5}\right)$ **8.** x-intercept: $\left(-\frac{b}{a}, 0\right)$; y-intercept: $\left(0, \frac{b}{c}\right)$ **9.** -4 **10.** $-\frac{2}{3}$ **11.** -4

12. -1 **13.** $-\frac{3}{5}$; decreasing **14.** $\frac{1}{2}$; increasing **15.** undefined; vertical **16.** -1.30; decreasing

17. **18.** **19.** **20.** **21.** **22.**

23. slope: 1; y-intercept: $\left(0, -\frac{c}{a}\right)$ **24.** neither **25.** parallel **26.** $y = -4x - 5$ **27.** $y = -\frac{2}{3}x - \frac{2}{3}$ **28.** $y = \frac{8}{5}x + \frac{14}{5}$

29. $y = -\frac{4}{3}x + \frac{2}{3}$ **30.** $x = 3$ **31.** $y = 0.39x - 7.37$ **32.** line 1: $y = -2x$; line 2: $y = 2x + 2$; line 3: $y = -3$

33. y values, from top: 20, 16, 12, 8, 4, 0 **34.** $y = 0.5x + 3$ **35. a.** B, C, F **b.** C, E **c.** C **d.** A, D **36.** $y = 3x + 11$

37. yes; $y = 0$ **38.** relations 1 and 3 **39.** no **40.** yes **41.** no **42.** no **43.** domain: all real numbers; range: $y \leq 4$

Chapter 1 Test

1. w independent; v dependent **2.** **3.** Answers may vary. **4.** $\frac{4}{5}$

5. line 1: $y = -\frac{5}{2}x + 10$; line 2: $y = \frac{2}{3}x + 2$; line 3: $x = -3$ **6. a.** k **b.** b **7.** ski run A **8.** y values, from top: 25, 29, 33, 37, 41, 45

9. **10.** **11.** $-\frac{5}{4}$ **12.** $(-1, 10)$, $(8, 4)$, $(11, 2)$; answers may vary. **13.** $y = -\frac{3}{7}x + \frac{29}{7}$

14. $y = -\frac{12}{5}x - \frac{1}{5}$ **15.** no; $y = -2x + 1$ **16.** $y = -\frac{5}{3}x + \frac{17}{3}$ **17.** x-intercept: $\left(\frac{3}{2}, 0\right)$; y-intercept: $(0, -3)$

18. a. Answers may vary. **b.** **c.** For each input–output pair, the output is 4 less than twice the input.

19. Answers may vary. **20.** no **21.** yes **22.** domain: $-3 \leq x \leq 5$; range: $-3 \leq y \leq 4$; yes

Chapter 2

Homework 2.1

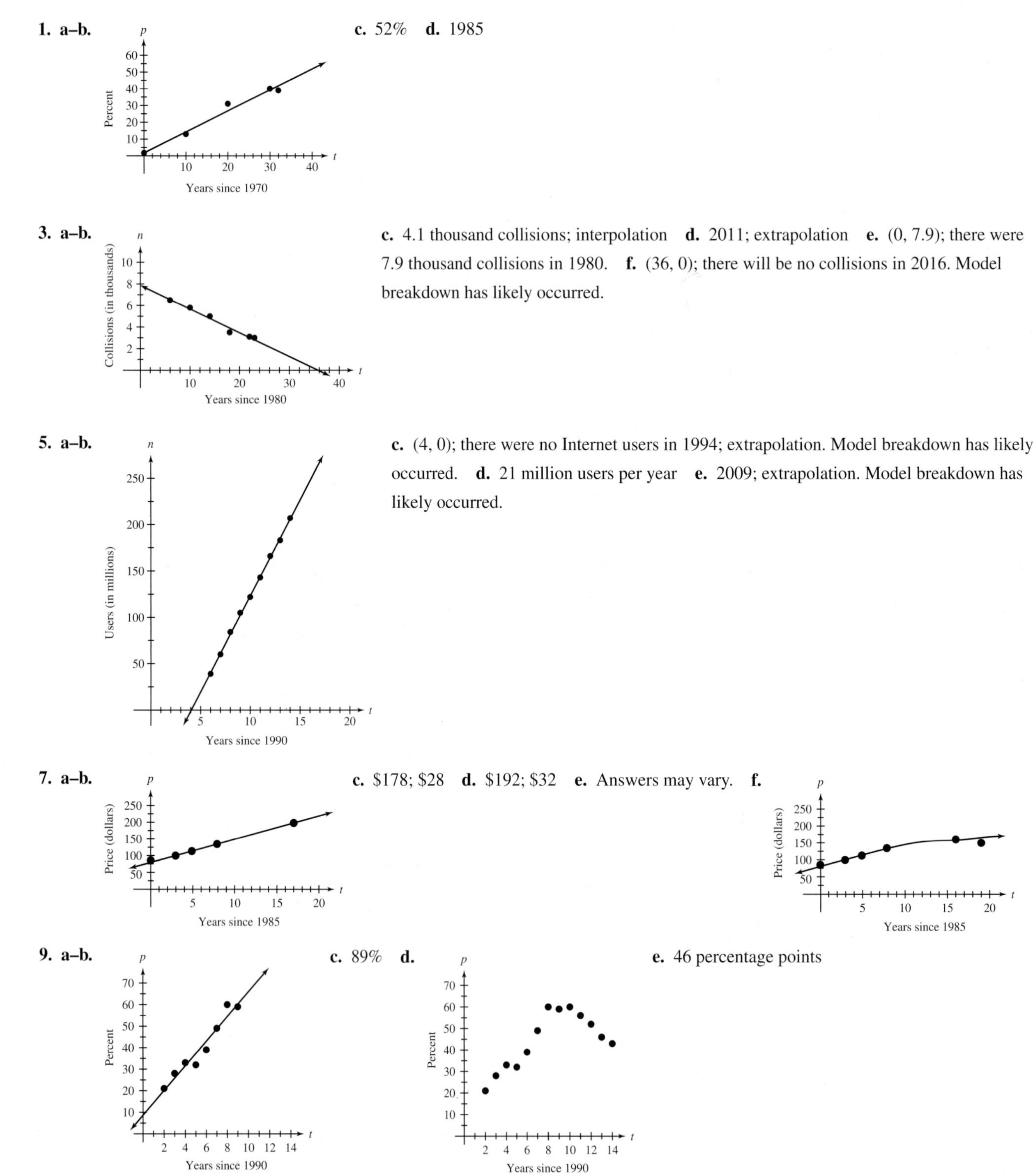

1. a–b.

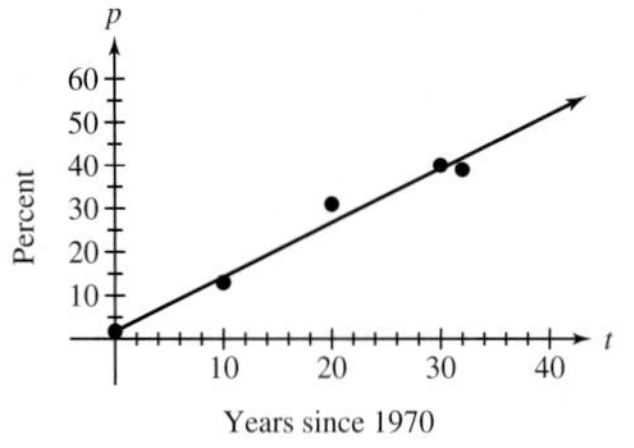

c. 52% **d.** 1985

3. a–b.

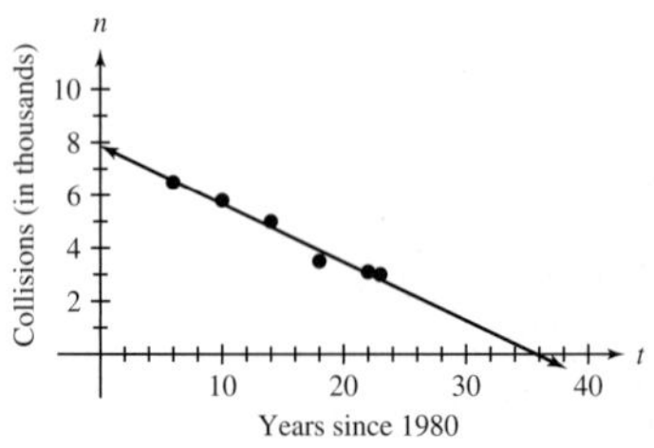

c. 4.1 thousand collisions; interpolation **d.** 2011; extrapolation **e.** (0, 7.9); there were 7.9 thousand collisions in 1980. **f.** (36, 0); there will be no collisions in 2016. Model breakdown has likely occurred.

5. a–b.

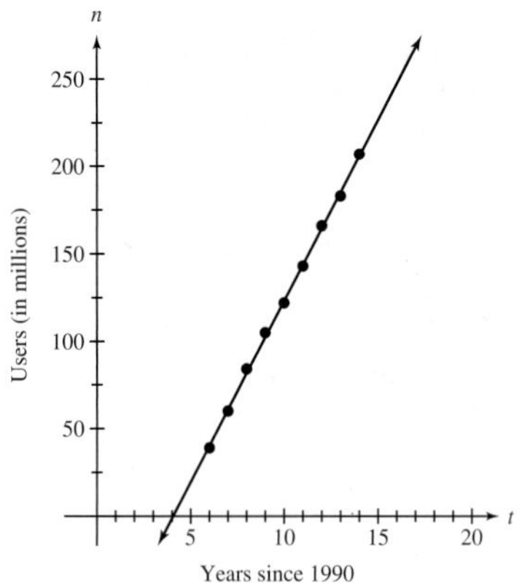

c. (4, 0); there were no Internet users in 1994; extrapolation. Model breakdown has likely occurred. **d.** 21 million users per year **e.** 2009; extrapolation. Model breakdown has likely occurred.

7. a–b.

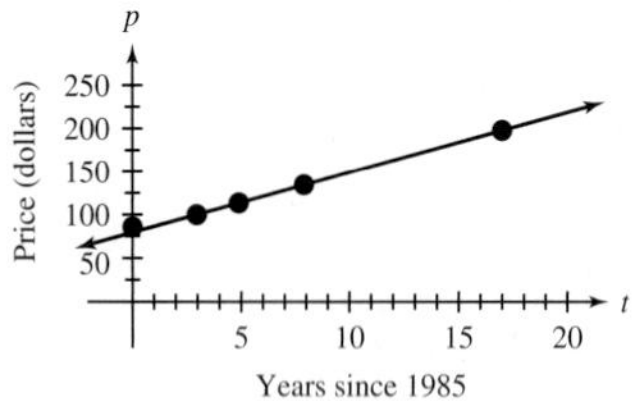

c. \$178; \$28 **d.** \$192; \$32 **e.** Answers may vary. **f.**

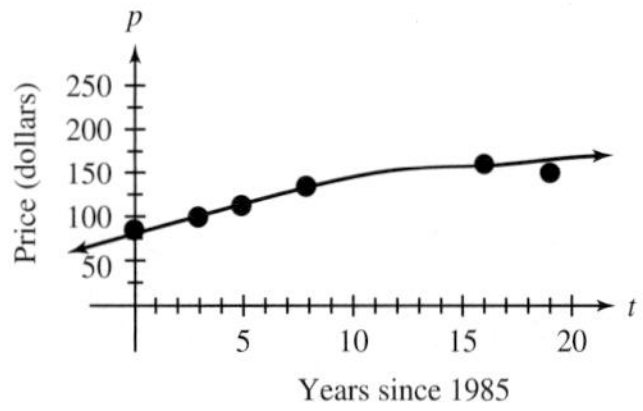

9. a–b.

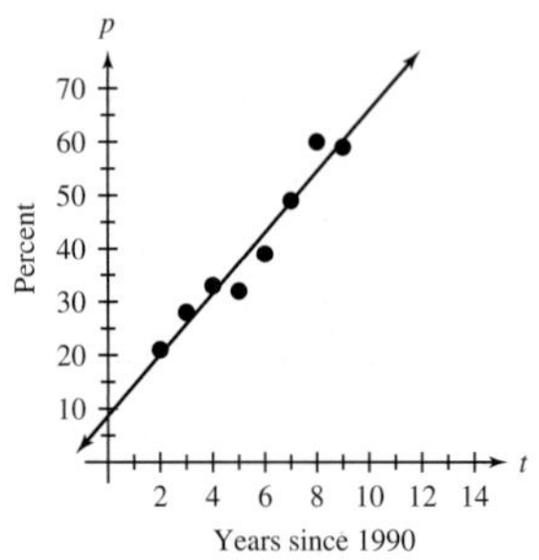

c. 89% **d.**

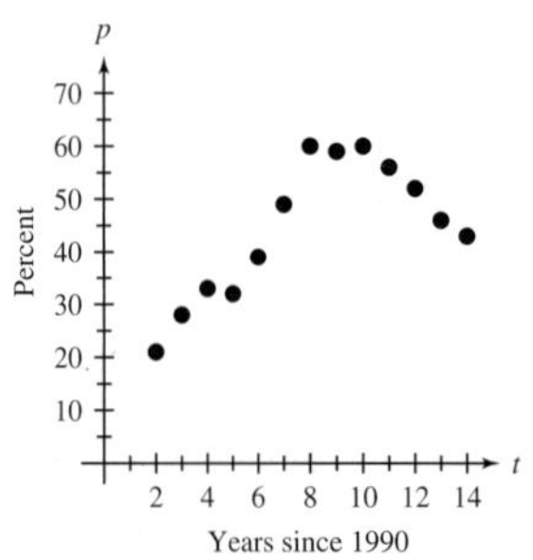

e. 46 percentage points

11. a. (7.5, 0); the difference in percent of sales was 0 percentage points in 1998, according to this model. **b.** (0, 5.1); the difference in percent of sales was 5.1 percentage points in 1990, according to this model. **13.** Answers may vary. **15.** Answers may vary.

Homework 2.2 **1.** $n = 2.3t + 6.4$ **3.** $r = 235t + 705$ **5.** $b = -77.89t + 517.89$ **7.** $L = 1.96a + 15.21$

9. increase b **11.** $y = 2.5x - 2.7$; answers may vary. **13.** student B

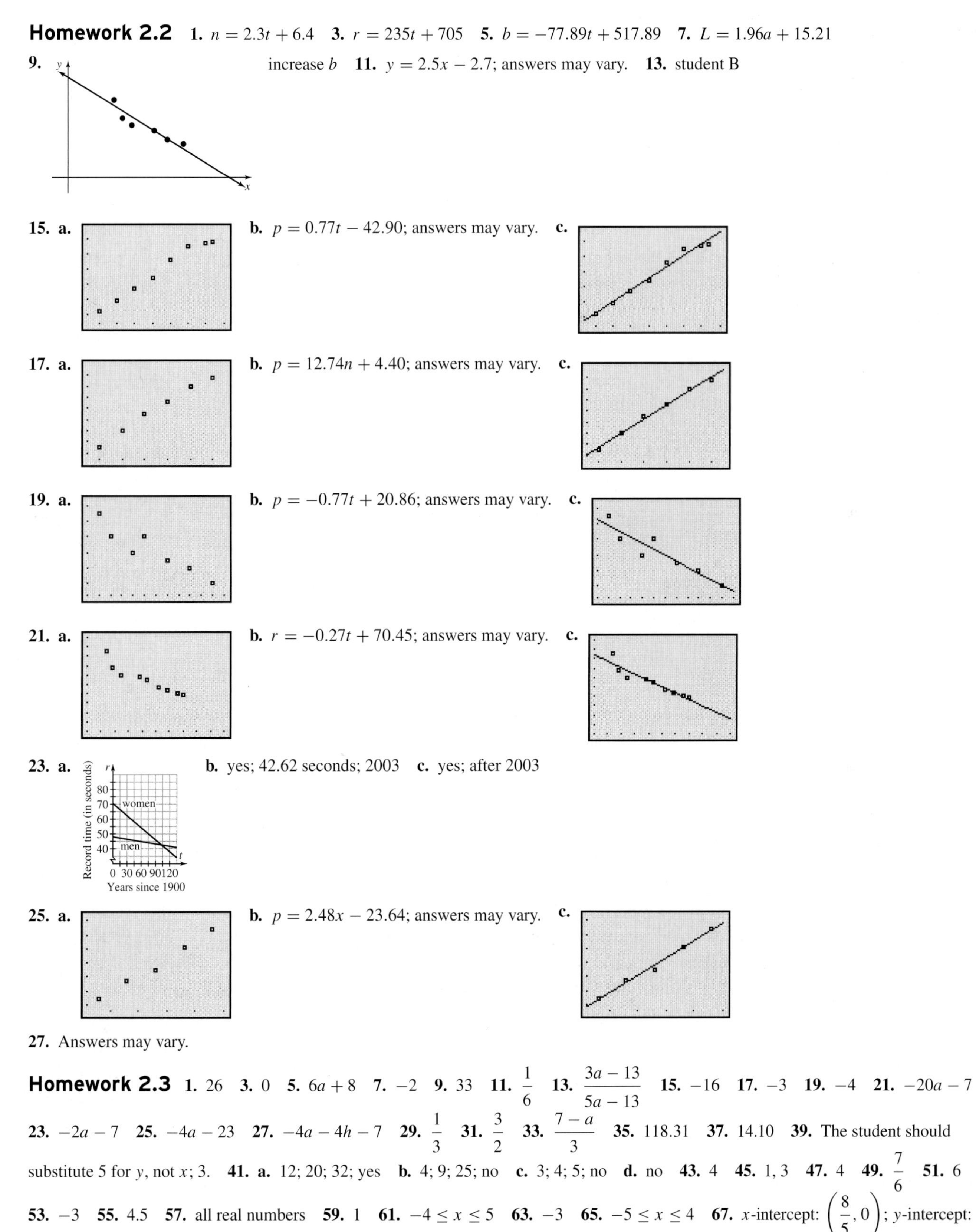

27. Answers may vary.

Homework 2.3 **1.** 26 **3.** 0 **5.** $6a + 8$ **7.** -2 **9.** 33 **11.** $\frac{1}{6}$ **13.** $\frac{3a - 13}{5a - 13}$ **15.** -16 **17.** -3 **19.** -4 **21.** $-20a - 7$ **23.** $-2a - 7$ **25.** $-4a - 23$ **27.** $-4a - 4h - 7$ **29.** $\frac{1}{3}$ **31.** $\frac{3}{2}$ **33.** $\frac{7 - a}{3}$ **35.** 118.31 **37.** 14.10 **39.** The student should substitute 5 for y, not x; 3. **41. a.** 12; 20; 32; yes **b.** 4; 9; 25; no **c.** 3; 4; 5; no **d.** no **43.** 4 **45.** 1, 3 **47.** 4 **49.** $\frac{7}{6}$ **51.** 6 **53.** -3 **55.** 4.5 **57.** all real numbers **59.** 1 **61.** $-4 \le x \le 5$ **63.** -3 **65.** $-5 \le x \le 4$ **67.** x-intercept: $\left(\frac{8}{5}, 0\right)$; y-intercept: $(0, -8)$ **69.** x-intercept: $(0, 0)$; y-intercept: $(0, 0)$ **71.** no x-intercept; y-intercept: $(0, 5)$ **73.** x-intercept: $(6, 0)$; y-intercept: $(0, -3)$ **75.** x-intercept: $(17.52, 0)$; y-intercept: $(0, -45.21)$ **77. a.** Answers may vary. **b.**

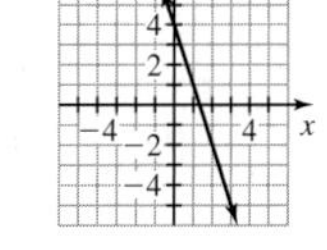

c. The output is 4 more than -3 times the input.

79. a. $f(t) = 0.77t - 42.90$ **b.** 41.8; in 2010, 41.8% of births will be outside marriage. **c.** 112.9; in 2013, 44% of births will be outside marriage. **d.** 2086; model breakdown has likely occurred. **e.** 31.8%; −0.6 percentage points **81. a.** $f(t) = -0.77t + 20.86$ **b.** 16.24; in 1996, baseball was the favorite sport of 16.24% of Americans. **c.** 19.30; in 2009, baseball will be the favorite sport of 6% of Americans. **d.** (0, 20.86); in 1990, baseball was the favorite sport of about 21% of Americans. **e.** (27.09, 0); in 2017, baseball will not be the favorite sport of any Americans. Model breakdown has likely occurred. **83. a.** $f(n) = 12.74n + 4.40$ **b.** (0, 4.40); the price of renting skis for 0 days is $4.40. Model breakdown has occurred. **c.** $93.58

d.

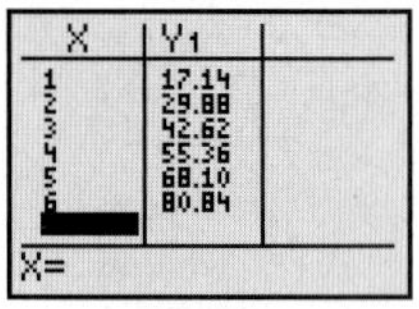

12.74; equal to slope of the graph of f; price increases by $12.74 for each additional day. **e.** 17.14, 14.94, 14.21, 13.84, 13.62, 13.47; 13.47; cost per day is lowest for 6-day package ($13.47 per day). **f.** The cost increase for each additional day is less than the $13 charge per day; the cost per day for the 6-day package is more than the $13 charge per day.

85. a.

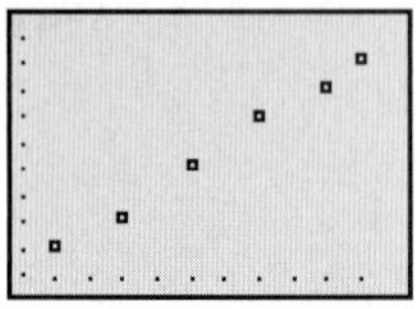

b. $f(t) = 24.07t + 423.92$; yes **c.** 689 million; 67 million boardings **d.** $5.7 billion

87. a. $f(C) = 1.8C + 32$ **b.** 77°F **c.** 4.44°C **89. a.** $f(x) = 2.48x - 23.64$ **b.** 50 points **c.** less than 10 points **d.** 30 students; no **e.** 83 students **91.** $10.6 thousand **93.** 2011 **95. a.** 2009 **b.** 11% **97.** 219 points

99. a. $f(t) = -160t + 640$ **b.**

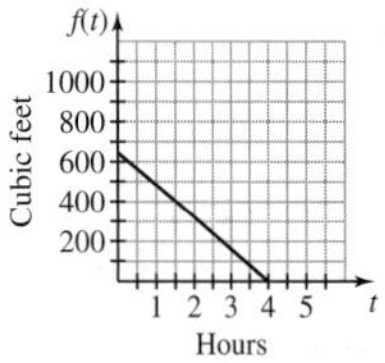

c. domain: $0 \le t \le 4$; range: $0 \le f(t) \le 640$

101. input: 3; output: 5; answers may vary.

Homework 2.4

1. about 1.87 shredder models per year **3.** −50,000 employees per year **5.** −1.8 percentage points per year **7.** $56 per credit hour **9. a.** yes; 70; the student is traveling 70 miles per hour. **b.** $d = 70t$ **11. a.** yes; 6.7; the number of households that pay bills online has increased by 6.7 million per year. **b.** (0, 18.9); in 2003, 18.9 million households paid bills online. **c.** $n = 6.7t + 18.9$ **d.** The units of the expressions on both sides of the equation are millions of households. **e.** 72.5 million households **13. a.** −3.8; the length of a major league baseball game has decreased by 3.8 minutes per year. **b.** $g(t) = -3.8t + 166$ **c.** 2 hours and 19.4 minutes **d.** 2007 **e.** (43.68, 0); in 2047, major league baseball games will, on average, take 0 minutes. Model breakdown has occurred. **15. a.** 25; each textbook costs $25 to rent. **b.** $f(n) = 25n + 1585.50$ **c.** The units of the expressions on both sides of the equation are dollars. **d.** $1685.50; tuition plus rental of 4 textbooks costs $1685.50 per student. **e.** 5; tuition plus rental of 5 textbooks costs $1710.50 per student. **17. a.** −0.05; the car uses 0.05 gallon of gas per mile. **b.** $g(x) = -0.05x + 15.3$ **c.** (306, 0); after she has driven 306 miles, the tank will be empty. **d.** domain: $0 \le x \le 306$; range: $0 \le g(x) \le 15.3$ **e.** 286 miles **19.** 2013
21. 37 minutes **23.** 1.71; the average salary increased by $1710 per year. **25.** −0.27; the record time for the women's 400-meter run decreased by 0.27 second per year. **27.** 1.8; for each 1° temperature increase on the Celsius scale, the Fahrenheit temperature rises by 1.8°. **29. a.** $f(t) = -0.34t + 70.00$ **b.** −0.34; the percentage of the world's population that lives in rural areas has decreased by 0.34 percentage points per year. **c.** 3.4 billion **d.** 2009 **e.** (205.88, 0); in 2156, none of the world's population will live in rural areas. Model breakdown has likely occurred. **31. a.** $f(t) = 129.54t + 1383.04$ **b.** $g(t) = 506.48t + 7180.48$ **c.** 129.54; 506.48; average tuition has increased by $129.54 per year at public colleges and by $506.48 per year at private colleges. **d.** In 2010, the total four-year cost of attending a public college will be about $21,854.20, and that at a private college will be about $92,538.40.
33. a. $f(a) = -0.99a + 29.85$ **b.** −0.99; for each thousand-foot increase in altitude, the pressure decreases by 0.99 inch of mercury. **c. i.** −1.01 inches of mercury per thousand feet; close to −0.99 **ii.** −0.98 inch of mercury per thousand feet; close to −0.99 **iii.** −0.99 inch of mercury per thousand feet; equal to slope **d.** 15.55 inches of mercury

35. a.

t	0	1	2	3	4	5
$f(t)$	0	500	1000	1500	1900	2300

b.

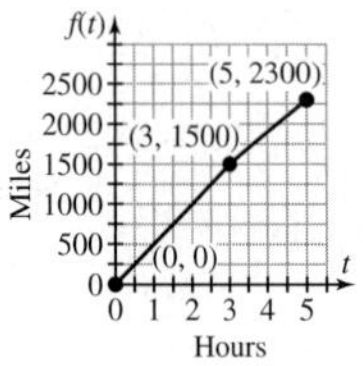

c. no **d.** Answers may vary.

37. **39.**

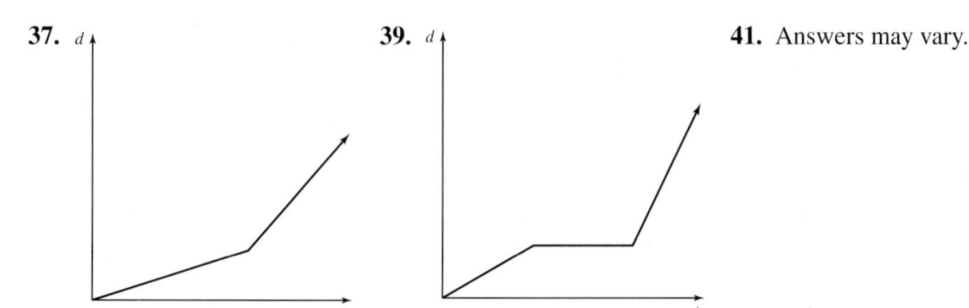

41. Answers may vary.

Chapter 2 Review

1. 20 **2.** 20 **3.** $\frac{3}{4}$ **4.** -9 **5.** $-10a-33$ **6.** $-\frac{9}{2}$ **7.** $-\frac{7}{6}$ **8.** $\frac{a+4}{2}$ **9.** 0 **10.** 1 **11.** 3.6 **12.** -2 **13.** 2 **14.** 4 **15.** $-5 \le x \le 6$ **16.** $-2 \le y \le 4$ **17.** 1 **18.** 4 **19.** 4 **20.** 1 **21.** x-intercept: $\left(\frac{3}{7}, 0\right)$; y-intercept: $(0, 3)$ **22.** x-intercept: none; y-intercept: $(0, 4)$ **23.** x-intercept: $\left(\frac{7}{2}, 0\right)$; y-intercept: $(0, 2)$ **24.** x-intercept: $(30.57, 0)$; y-intercept: $(0, -8.32)$

25. Increase the slope and lower the y-intercept. **26. a.** $f(t) = -1.8t + 13$ **b.** -1.8; the car uses 1.8 gallons of gas per hour. **c.** $(0, 13)$; he started the trip with 13 gallons in the tank. **d.** The units of the expressions on both sides of the equation are gallons. **e.** $(7.2, 0)$; the tank will be empty after 7.2 hours of driving at 65 miles per hour. **f.** domain: $0 \le t \le 7.2$; range: $0 \le A \le 13$

27. a. 962; the average annual personal income has increased by $962 per year. **b.** $g(t) = 962t + 32{,}937$ **c.** 38,709; in 2010, average annual personal income will be $38,709. **d.** 7.34; in 2011, average annual personal income will be $40,000. **28.** 3 times per year **29.** 2012 **30. a.** $f(t) = 1.23t + 20.60$ **b.** 1.23; each year, the IRS standard mileage rate for businesses has increased by 1.23 cents per mile. **c.** $(0, 20.60)$; in 1990, the standard mileage rate was 20.6 cents per mile. **d.** 2010 **e.** $6000 **f.** $(32.5 + 31 + 31 + 31)/4 = 31.375$, which rounds to 31.4. **31. a.** $f(t) = 3.63t - 8.07$ **b.** 3.63; lobbying expenditures have increased by $3.63 million per year. **c.** $(2.22, 0)$; in 1992, lobbying expenditure was 0. Model breakdown has occurred. **d.** 64.53; in 2010, lobbying expenditures will be $64.53 million. **e.** 7.73; in 1998, lobbying expenditures were $20 million. **f.** 2012

Chapter 2 Test

1. -2 **2.** 0 **3.** -1 **4.** -2.7 **5.** -6 **6.** -3 **7.** 3 **8.** 4.5 **9.** $-6 \le x \le 6$ **10.** $-3 \le y \le 1$ **11.** 19 **12.** $-4a + 27$ **13.** $\frac{5}{4}$ **14.** $-\frac{a-7}{4}$ or $\frac{7-a}{4}$ **15.** x-intercept: $\left(\frac{7}{3}, 0\right)$; y-intercept: $(0, -7)$ **16.** x-intercept: $(0, 0)$; y-intercept: $(0, 0)$ **17.** x-intercept: $(24, 0)$; y-intercept: $(0, -8)$ **18. a.** $f(a) = 6.81a - 52.32$ **b.** 6.81; the percentage of teenagers who have driver's licenses has increased by 6.81 percentage points per year. **c.** $(7.68, 0)$; no 8-year-olds have driver's licenses. **d.** 90.7% **e.** 22 years. Model breakdown has likely occurred. **19. a.** $f(t) = 0.38t - 23.19$ **b.** 0.38; the percentage of the U.S. military who are women increases by 0.38 percentage points per year of age. **c.** 14.81; in 2000, 14.8% of the military were women. **d.** 324.18; in 2224, 100% of the military will be women. Model breakdown has likely occurred. **e.** when $t < 61.03$ and when $t > 324.18$ **20.** $663.3 million **21. a.** yes; 6.8; the percentage of community banks that offer Internet banking has increased by 6.8 percentage points per year. **b.** $(0, 41)$; in 2001, 41% of community banks offered Internet banking. **c.** $f(t) = 6.8t + 41$ **d.** 88.6% **e.** 2010 **22.**

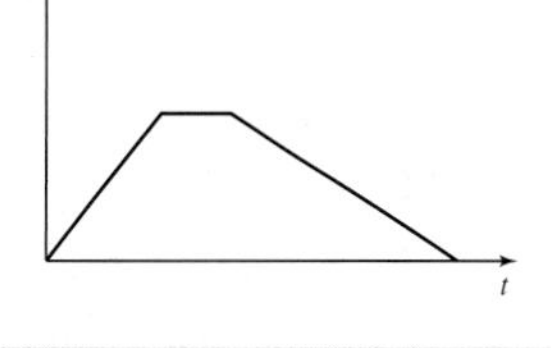

Chapter 3

Homework 3.1 **1.** $(1, 4)$ **3.** $(-6, 6)$ **5.** $(2, 3)$ **7.** $(4, 4)$ **9.** $(0, 5)$ **11.** all points on the line $y = -2x + 3$; dependent system **13.** empty set; inconsistent system **15.** $(4, 2)$ **17.** $(-1.12, -3.69)$ **19.** $(-1.61, 2.04)$ **21.** all points on the line $y = 2x - 1$; dependent system **23.** empty set; inconsistent system **25.** $(3.33, 1.33)$ **27. a.** 37.25 seconds; 34.51 seconds; -1.04 seconds; -0.37 second **b.** The absolute value of the slope of W is greater than the absolute value of the slope of M; the women's winning times are decreasing at a faster rate than the men's winning times. **c.** Answers may vary. **d.** 2116; 18.40 seconds **29. a.** $W = 0.15t + 5.76$; $M = 0.072t + 5.43$ **b.** 1976; 5.1 million students **c.** 18.1 million students

31. a. $L(t) = -0.1t + 4.7$; $W(t) = 0.17t + 1.15$ **b.** 2013; 3.4 billion calls **c.** \$8.38 billion **33. a.** B and E **b.** E and F **c.** E **d.** A, C, and D **35.** $(-1.9, -2.8)$ **37.** $(10, -1)$ **39.** $(3, -4)$ **41.** $(3.5, 19.5)$ **43.** 0 **45.** 5 **47.** -1 **49. a–c.** Answers may vary. **51.** $(2, 5)$ **53.** Answers may vary. **55.** The ordered pair $(1, 2)$ does not satisfy the equation $y = -2x + 9$, so $(1, 2)$ is not a solution of the system; $(2, 5)$ **57.** Answers may vary.

Homework 3.2

1. $(7, 2)$ **3.** $(-5, -3)$ **5.** $(3, -4)$ **7.** $(0, 0)$ **9.** $(-3, -7)$ **11.** $(2, 1)$ **13.** $(-1, 2)$ **15.** $(4, -3)$ **17.** $(4, -7)$ **19.** $(2, 3)$ **21.** $(-1, 4)$ **23.** $(1, -2)$ **25.** $(-4, 3)$ **27.** $(-2, 3)$ **29.** $(-2, 3)$ **31.** $(-3, -4)$ **33.** $(3, -2)$ **35.** $(2, 1)$ **37.** $(5, 4)$ **39.** $(-2, 3)$ **41.** empty set; inconsistent system **43.** $(1, -2)$ **45.** all points on the line $4x - 5y = 3$; dependent system **47.** empty set; inconsistent system **49.** $(3, -2)$ **51.** $(-3, 1)$ **53.** all points on the line $y = \frac{1}{2}x + 3$; dependent system **55.** $(3, 2)$ **57.** $(1, 2)$ **59.** $(2.77, -1.15)$ **61.** $(-4.11, 8.01)$ **63.** $(2, 5)$ **65. a.** $(3, 1)$ **b.** $(3, 1)$ **c.** They are the same. **67.** -3 **69.** 1 **71.** -5 **73.** 4 **75.** 1 **77.** $(-2, 1)$ **79.** 1.57 **81.** -2.42 **83.** -1.94 **85.** 5 **87.** 3 **89.** $(1, 2.8)$ **91.** $(4.7, 21.8)$ **93.** The lines are not parallel; $(200, 403)$ **95.** A: $(0, 0)$; B: $(0, 3)$; C: $(3, 9)$; D: $(6, 8)$; E: $(7.2, 4.4)$; F: $(5, 0)$ **97. a.** $\left(\frac{cp - bd}{ap - bk}, \frac{ad - ck}{ap - bk}\right)$, assuming that $ap - bk \neq 0$ **b.** $\left(\frac{14}{11}, -\frac{4}{11}\right)$ **99.** Answers may vary.

Homework 3.3

1. 2116; 18.40 seconds **3.** 1976; 5.13 million students **5. a.** $M(t) = -0.28t + 35.64$; $S(t) = 0.86t + 8.30$ **b.** 1974; 28.9 gallons per person **c.**

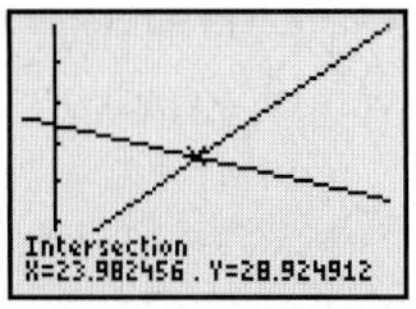

7. a. $K(a) = 0.014a - 0.75$; $M(a) = -0.028a + 1.66$ **b.** 57 years **c.**

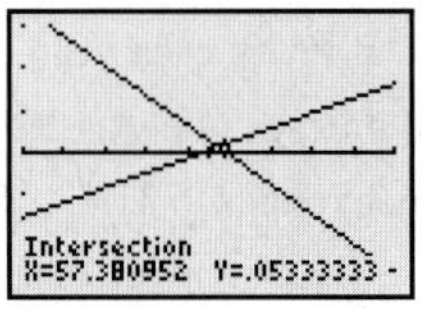

9. a. 1997 **b.** The competition heated up because the two newspapers had approximately equal circulations. **c.** 147 thousand bonus issues **d.** 688 thousand newspapers **e.** overestimate; answers may vary. **11. a.** $T(t) = -1725t + 12{,}281$; $E(t) = -1424t + 10{,}952$ **b.** 2009; \$4665 **c.**

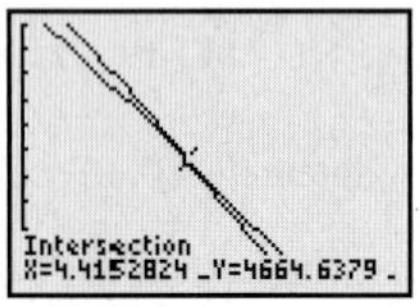

13. a. $J(t) = 72t + 19$; $W(t) = 77t$ **b.** For each equation, the units of the expressions on both sides of the equation are dollars.

d.

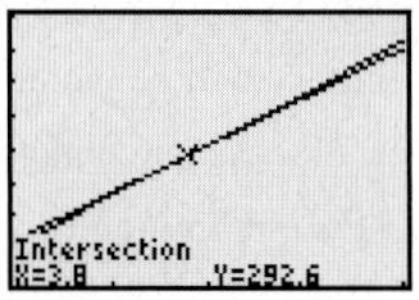

15. 2006 **17. a.** $f(t) = 394.6t + 20{,}656$; $g(t) = 102.6t + 26{,}017$ **b.** 2016

c.

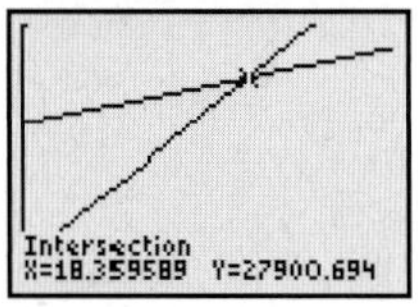

19. 2008; 13.75% **21. a–b.**

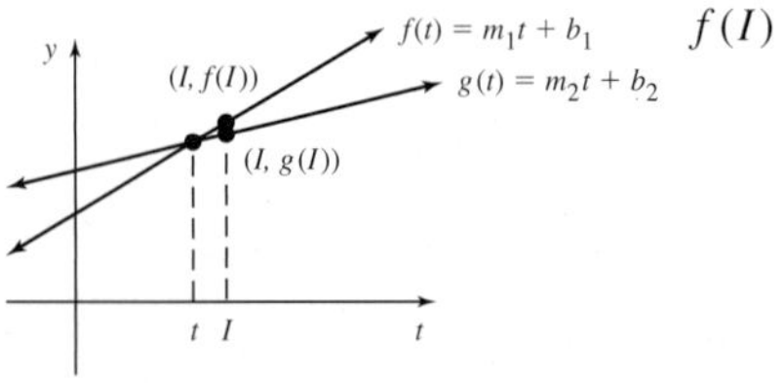

Homework 3.4

1. 3800 tickets at \$27, 1200 tickets at \$40 **3.** 715 of CD *Plans,* 121 of EP *Forbidden Love* **5.** \$25 for balcony seats, \$40 for main-level seats **7.** 1500 part-time students, 4500 full-time students **9. a.** $f(x) = -25x + 1{,}500{,}000$

b.

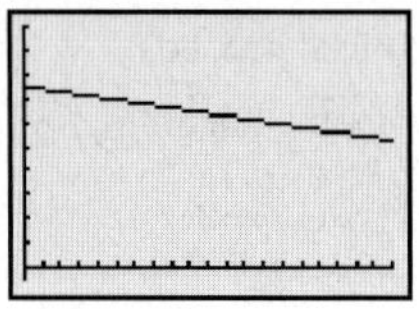

-25; the more \$50 tickets are sold (so, the fewer \$75 tickets are sold), the lower the total revenue (by \$25 for each additional \$50 ticket sold). **c.** 1,100,000; if 16,000 \$50 tickets are sold (and 4000 \$75 tickets are sold), the total revenue will be \$1,100,000. **d.** 17,000 \$50 tickets, 3000 \$75 tickets

11. a. $f(x) = -25x + 840{,}000$ **b.**

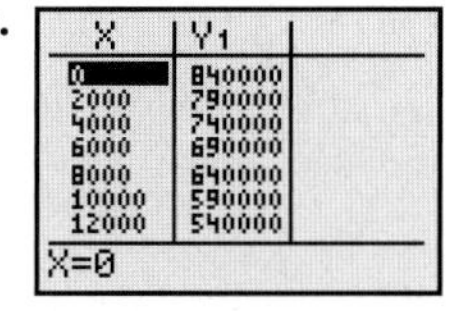

X	Y1
0	840000
2000	790000
4000	740000
6000	690000
8000	640000
10000	590000
12000	540000

X=0

See second column of table. If the number of \$45 tickets sold is 0, 2000, 4000, 6000, 8000, 10,000, or 12,000, the total revenue will be \$840,000, \$790,000, \$740,000, \$690,000, \$640,000, \$590,000, or \$540,000, respectively. **c.** The total revenue will be between \$540,000 and \$840,000, inclusive. **d.** 9500 \$45 tickets, 2500 \$70 tickets

13. a. $f(x) = 134x + 1936$ **b.** 134; if the prices of all tickets are increased by \$1, the total revenue increases by \$134. **c.** coach: \$91; first class: \$333 **15.** \$12,000 in American Funds New Perspective F; \$3000 in Oppenheimer Global Y **17.** \$6000 in GMO Growth III; \$2500 in Gartmore Destinations Mod Agg Svc **19.** \$4000 in Lord Abbett Developing Growth B; \$2000 in Bridgeway Micro-Cap Limited **21.** \$4500 in Dreyfus Premier Worldwide Growth R; \$1500 in Oppenheimer Global Opportunities Y **23. a.** $f(x) = -0.0523x + 810$
b.

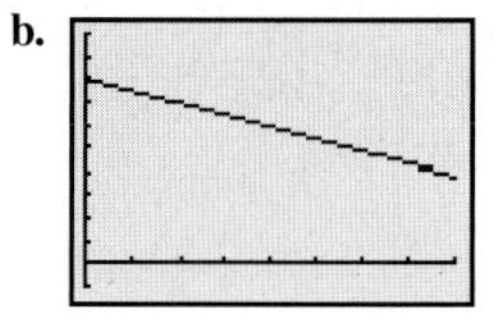

-0.0523; for each additional \$1 invested in the CD (so, \$1 less invested in the mutual fund), the total interest decreases by 5.23 cents. **c.** CD: \$7839.39; mutual fund: \$2160.61

25. a. $f(x) = -0.0695x + 850.5$ **b.** 815.75; if \$500 is invested in the CD (and \$8500 in the mutual fund), the total interest will be \$815.75. **c.** 5043.17; if \$5043.17 is invested in the CD (and \$3956.83 in the mutual fund), the total interest will be \$500. **d.** 155.5; if \$10,000 is invested in the CD, the total interest will be \$155.50. Model breakdown has occurred, because only \$9000 is being invested. **27. a.** $f(x) = -0.101x + 928$ **b.** The total interest will be between \$120 and \$675.50, inclusive. **c.** \$5227.72 should be invested in the CD, and \$2772.28 in the mutual fund. **29. a.** $f(x) = -0.0615x + 540$ **b.** (0, 540); if all of the \$6000 is invested in the mutual fund, the total interest will be \$540. **c.** (8780.49, 0); if \$8780.49 is invested in the CD, the revenue will be \$0. Model breakdown has occurred, because only \$6000 is being invested. **d.** -0.0615; if \$1 more is invested in the CD (and \$1 less in the mutual fund), the total interest will decrease by 6.15 cents. **31.** 4 ounces of 10% solution, 6 ounces of 30% solution **33.** 1 gallon of 5% solution, 2 gallons of 20% solution **35.** 4 cups of 10% solution, 2 cups of 25% solution **37.** 4 liters of 25% solution, 1 liter of water **39.** Answers may vary. **41.** Answers may vary.

Homework 3.5

1.

In Words	Inequality	Graph	Interval Notation
Numbers greater than 3	$x > 3$	0 3	$(3, \infty)$
Numbers less than or equal to -4	$x \le -4$	−4 0	$(-\infty, -4]$
Numbers less than 5	$x < 5$	0 5	$(-\infty, 5)$
Numbers greater than or equal to -1	$x \ge -1$	−1 0	$[-1, \infty)$

3. $x \ge 3$; $[3, \infty)$; **5.** $x \le -3$; $(-\infty, 3]$; **7.** $w < 2$; $(-\infty, 2)$; **9.** $x < 1$; $(-\infty, 1)$; **11.** $x \le 5.6$; $(-\infty, 5.6]$; **13.** $b < -5$; $(-\infty, -5)$; **15.** $x < \frac{5}{3}$; $\left(-\infty, \frac{5}{3}\right)$; **17.** $a < -2.3$; $(-\infty, -2.3)$; **19.** $x \ge 23$; $[23, \infty)$; **21.** $r \le 2$; $(-\infty, 2]$; **23.** $x < -6$; $(-\infty, -6)$; **25.** $t \ge -\frac{22}{9}$; $\left[-\frac{22}{9}, \infty\right)$; **27.** $x \le -\frac{7}{12}$; $\left(-\infty, -\frac{7}{12}\right]$; **29.** $c \ge -31$; $[-31, \infty)$; **31.** $x < -5$; $(-\infty, -5)$; **33.** $1 < x < 5$; $(1, 5)$; **35.** $-5 \le x \le 6$; $[-5, 6]$; **37.** $-3 \le x < 5$; $[-3, 5)$; **39.** $3 < x \le \frac{11}{2}$; $\left(3, \frac{11}{2}\right]$; **41.** years after 2009

43. a. $U(d) = 0.69d + 19.95$; $P(d) = 0.39d + 29.95$ **b.** for miles driven less than $33.\overline{3}$ miles **45.** years after 2005 **47. a.** $W(t) = -2.48t + 50.54$; $M(t) = -4.42t + 77.17$ **b.** before 2004 **49. a.** 4.6 years **b.** for birth years after 2058 **c. i.** younger man **ii.** birth years 1986 and thereafter **51.** When dividing by 3, do not reverse the inequality symbol; $x > -2$

53. a, b. Answers may vary. **55.** Answers may vary. **57. a.** $x = 3$ **b.** $x < 3$ **c.** $x > 3$

d. 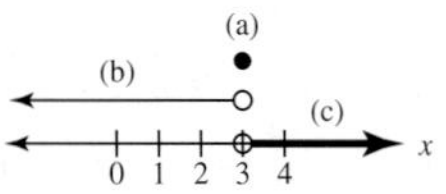

e. 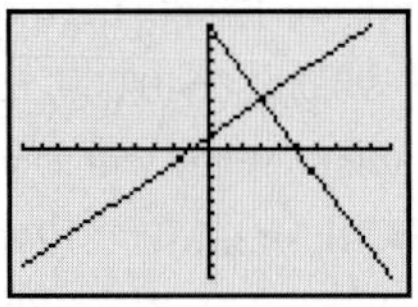

$f(x) < g(x)$ when $x < 3$; $f(x) = g(x)$ when $x = 3$; $f(x) > g(x)$ when $x > 3$.

59. no **61.** no **63.** $x < 2.8$ **65.** Answers may vary.

Chapter 3 Review

1. $(4, -5)$ **2.** $(3, 2)$ **3.** $(-3, 2)$ **4.** empty set; inconsistent system **5.** all points on the line $-4x - 5y = 3$; dependent system **6.** $(1.2, -2.86)$ **7.** $(0, 0)$ **8.** $(2, -3)$ **9.** $(10, 3)$ **10.** $(2, -1)$ **11.** $(5, -1)$ **12. a–c.** Answers may vary. **13.** 1 **14.** -5 **15.** $(0.4, 15.1), (2.6, 21.9)$ **16.** $a = 19, b = 18$ **17.** A: $(0, 0)$; B: $(0, 4)$; C: $(2, 10)$; D: $(5, 8)$; E: $(6, 4)$; F: $\left(\frac{14}{3}, 0\right)$ **18.** empty set

19. $x \le 7$; $(-\infty, 7]$; 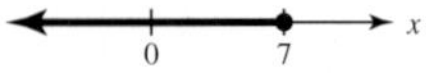

20. $a \le 2.8$; $(-\infty, 2.8]$; 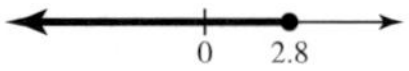

21. $x \le -\frac{7}{16}$; $\left(-\infty, -\frac{7}{16}\right]$; 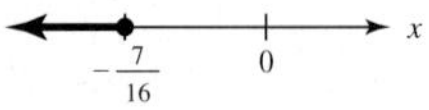

22. $x < -\frac{29}{2}$; $\left(-\infty, -\frac{29}{2}\right)$; 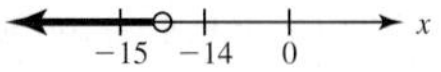

23. $-2 \le x < 3$; $[-2, 3)$; 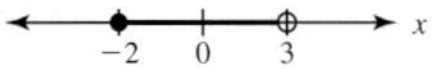

24. a, b. Answers may vary. **25.** When dividing both sides of the inequality by -3, reverse the inequality symbol; $x \ge -2$ **26.** 1 **27.** -5 **28.** -2 **29.** $x > -2$ **30.** Answers may vary. **31. a.** $N(t) = 0.58t + 34.14$; $A(t) = 1.35t + 15.66$ **b.** The slope of N is less than the slope of A. North America's petroleum consumption is increasing by 0.58 quadrillion Btu per year; Asia and Oceania's petroleum consumption is increasing by 1.35 quadrillion Btu per year. **c.** 2004; 48.1 quadrillion Btu **d.** $t > 24$; after 2004, petroleum consumption will be less in North America than in Asia and Oceania. **32. a.** $w(t) = -0.60t + 31.77$; $p(t) = -0.40t + 21.11$ **b.** 2023; -0.2; model breakdown has occurred, because the rating is a negative number. **33. a.** $R(d) = 0.22d + 75$; $U(d) = 0.69d + 29.95$ **b.** 95.85 miles; \$96.09 **c.** for miles driven over 95.85 miles **34.** 2007 **35.** 13,500 tickets at \$55; 6500 tickets at \$70 **36. a.** $f(x) = -0.062x + 1040$ **b.** \$1004.35; if she invests \$575 in Hartford Global Leaders Y (and \$7425 in Mutual Discovery Z), her annual interest will be \$1004.35. **c.** 7500; if she invests \$7500 in Hartford Global Leaders Y (and \$500 in Mutual Discovery Z), her annual interest will be \$575.

Chapter 3 Test

1. $(1, 2)$ **2.** all points on the line $2x - 5y = 3$; dependent **3.** empty set; inconsistent system **4.** $(5, -8)$ **5.** $(-2, 4)$ **6.** Answers may vary. **7.** $(3, 1)$ **8.** $m = 5, b \ne -13$ **9.** $x \le -\frac{12}{13}$; $\left(-\infty, -\frac{12}{13}\right]$; 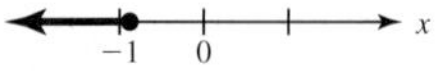

10. $x > \frac{23}{2}$; $\left(\frac{23}{2}, \infty\right)$; 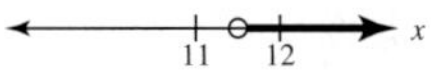

11. $t < 0.8$; $(-\infty, 0.8)$; 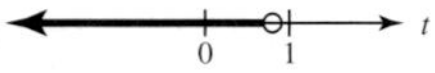

12. $w \ge \frac{2}{41}$; $\left[\frac{2}{41}, \infty\right)$; 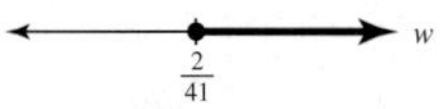

13. 3 **14.** -4 **15.** 2 **16.** $x \le 2$ **17.** -1.35 **18.** $x > 13$

19. a, b. Answers may vary. **20. a.** $f(t) = 28.17t + 886.82$; $g(t) = 21.16t + 1688.39$ **b.** 4598 **c.** 2084; 4108 educational institutions **d.** Answers may vary. **21. a.** $M(t) = 389.6t + 23{,}596$; $G(t) = 161.4t + 25{,}279$ **b.** 2005; \$26,469.34 **c.** years before 2005 **22.** 4 gallons of 10% solution, 6 gallons of 20% solution

Cumulative Review, Chapters 1–3

1. 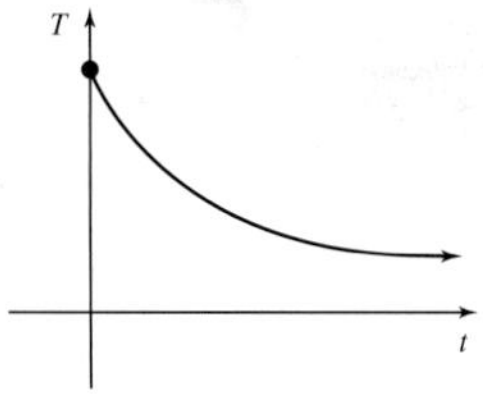

2. 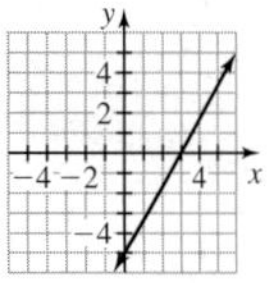

3. 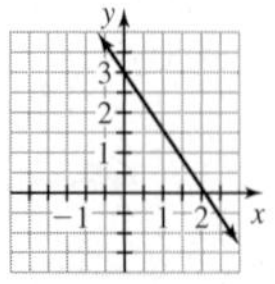

4. $-\frac{3}{7}$ **5.** $y = -\frac{3}{5}x - \frac{9}{5}$ **6.** $y = \frac{5}{3}x + \frac{19}{3}$ **7.** $y = -\frac{5}{2}x - \frac{19}{2}$ **8.** Answers may vary.

9.

x	$f(x)$	x	$g(x)$	x	$h(x)$	x	$k(x)$
0	97	4	−22	1	23	10	−28
1	84	5	−9	2	14	11	−25
2	71	6	4	3	5	12	−22
3	58	7	17	4	−4	13	−19
4	45	8	30	5	−13	14	−16
5	32	9	43	6	−22	15	−13

10. 32 **11.** 8 **12.** 13 **13.** $\frac{32}{9}$ **14.** $\left(\frac{14}{3}, 0\right)$ **15.** (0, 7) **16.** **17.** −3 **18.** 0 **19.** (0, 1)

20. $f(x) = \frac{2}{3}x + 1$ **21.** −3 **22.** $x \le -3$ **23.** $-\frac{20}{9}$ **24.** $\frac{3}{5}$ **25.** $b = cd + k$ **26.** x-intercept: $\left(-\frac{2}{5}, 0\right)$; y-intercept: $\left(0, \frac{2}{3}\right)$

27. $-5 \le x \le 5$ **28.** $-2 \le y \le 3$ **29.** yes **30.** (2, −3) **31.** all points on the line $y = \frac{3}{7}x - 2$ **32.** −2 **33.** 6 **34.** (2, −2)

35. $x \le \frac{10}{11}$; $\left(-\infty, \frac{10}{11}\right]$; **36. a.** 2 **b.** **c.** **d.** (1, 3)

37. a–d. Answers may vary. **38. a.** $f(t) = 0.56t - 0.6$ **b.** 5; in the year 2000, 5000 Chinese children were adopted by American families. **c.** 18.93; in the year 2009, 10,000 Chinese children will be adopted by American families. **d.** 0.56; each year there is an increase of 0.56 thousand, or 560, Chinese children adopted by American families. **e.** −0.04 thousand, or −40 Chinese children (model breakdown has occurred) **39. a.** $f(t) = -17t + 142$ **b.** −17; each year 17 fewer bicyclists younger than 16 are hit and killed by motor vehicles. **c.** (0, 142); in the year 2003, 142 bicyclists younger than 16 were hit and killed by motor vehicles. **d.** (8.35, 0); in 2011, no bicyclists younger than 16 will be hit and killed by motor vehicles (model breakdown has likely occurred). **e.** years after 2011

40. a. $A(t) = 1.3t + 9.5$; $B(t) = 1.8t + 5.2$ **b.**

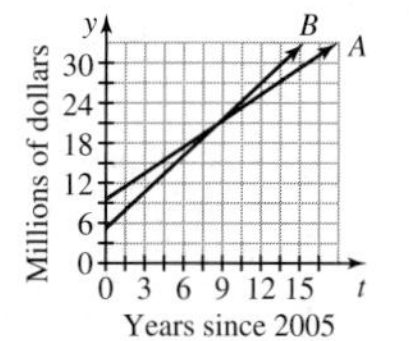

2014; \$20.68 million **c.** 2014; \$20.68 million; they are the same. **d.** $t > 8.6$; company A's sales will be less than company B's sales after 2014.

41. a. $W(t) = -0.064t + 27.00$; $M(t) = -0.029t + 22.10$ **b.** $W(111) = 19.90$; $M(111) = 18.88$; in 2011, the women's record time will be 19.90 seconds and the men's record time will be 18.88 seconds. **c.** The absolute value of W's slope is greater than the absolute value of M's slope. Women's record times are decreasing at a greater rate than men's record times. **d.** Answers may vary. **e.** 2040; 18.04 seconds **f.** $t < 140$; women's record times are greater than men's record times for years before 2040. **g.** W: (421.88, 0); M: (762.07, 0); the women's record time will be 0 seconds in 2322, and the men's record time will be 0 seconds in 2662. Model breakdown has occurred. **h.**

Answers may vary. **42.** \$5000 in UBS Global Equity Y, \$2500 in Fidelity Worldwide

Chapter 4

Homework 4.1 **1.** $\frac{1}{2}$ **3.** 1 **5.** −16 **7.** 16 **9.** 64 **11.** $\frac{5}{6}$ **13.** 49 **15.** 1 **17.** 1 **19.** $\frac{1}{b^2}$ **21.** $-\frac{14}{b^8}$ **23.** $\frac{72}{b}$

25. $108b^{12}c^{22}$ **27.** $\frac{3}{b^{10}c^2}$ **29.** $\frac{32b^{26}}{9c^2}$ **31.** $\frac{1}{b^{25}}$ **33.** $\frac{2}{5b^3}$ **35.** $-\frac{6}{7b^{10}}$ **37.** $-\frac{1}{3bc^5}$ **39.** $-\frac{1}{4b^{10}c^{14}}$ **41.** $\frac{3b^5}{c^{13}}$ **43.** $\frac{54b^{18}}{c^7}$

45. $\frac{b^8c^7}{32}$ **47.** $\frac{36b^6}{49c^{12}}$ **49.** $\frac{81c^8}{b^{24}}$ **51.** 1 **53.** $\frac{1}{bc}$ **55.** $b + c$ **57.** b^{7n} **59.** b^{5n-4} **61.** 54 **63.** $\frac{2}{81}$ **65.** $16(4^a)$ **67.** 16^a

69. a.

x	$f(x)$	x	$f(x)$
-3	$\frac{1}{8}$	1	2
-2	$\frac{1}{4}$	2	4
-1	$\frac{1}{2}$	3	8
0	1	4	16

b. **c.** 1.4 **71.** 396.5 **73.** 0.239 **75.** 520 **77.** 0.00009113

79. -0.000652 **81.** 900,000 **83.** -8 **85.** 5.426×10^7 **87.** 2.3587×10^4 **89.** 9.8×10^{-4} **91.** 3.46×10^{-5} **93.** -4.2215×10^4 **95.** -2.44×10^{-3} **97.** 0.0000063; 0.00013; 3,200,000; 64,000,000 **99.** 3,600,000,000 years **101.** 0.000000063 mole per liter **103.** 1.008×10^7 gallons **105.** 4.7×10^{-7} meter **107. a.**

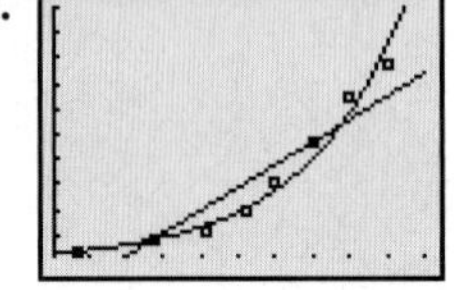

exponential function **b.** 14.4 thousand pairs **c.** 7.7 thousand pairs; answers may vary.

109. Student B was correct; student A should have 5 in the denominator, not -5 in the numerator. **111.** The 3 should stay in the numerator; $\frac{3c^4}{b^2d^7}$ **113.** -2^2, which is -4; $2(-1)$, which is -2; $\left(\frac{1}{2}\right)^2$, which is $\frac{1}{4}$; $2^{-1} = \frac{1}{2}$ (tie); $\left(\frac{1}{2}\right)^{-1}$, which is 2; $(-2)^2 = (2)^2$, which are 4 (tie) **115. a.** 1; 1; 1; 1; 1; 1 **b.** 0; 0; 0; 0; 0; 0 **c.** Answers may vary. **117.** Answers may vary. **119.** 6 **121.** 8 **123.** $(-5, -14)$; a linear system in two variables **125.** -5; a linear equation in one variable

Homework 4.2

1. 4 **3.** 10 **5.** 7 **7.** 5 **9.** 16 **11.** 27 **13.** 4 **15.** 32 **17.** $\frac{1}{3}$ **19.** $-\frac{1}{6}$ **21.** $\frac{1}{32}$ **23.** $\frac{1}{81}$ **25.** 2 **27.** 24 **29.** 49 **31.** 27 **33.** 12 **35.** $\frac{4}{3}$ **37.** -16

39.

x	$f(x)$	x	$f(x)$
$-\frac{3}{4}$	$\frac{1}{8}$	$\frac{1}{4}$	2
$-\frac{1}{2}$	$\frac{1}{4}$	$\frac{1}{2}$	4
$-\frac{1}{4}$	$\frac{1}{2}$	$\frac{3}{4}$	8
0	1	1	16

41. b^2 **43.** $\frac{1}{b^2}$ **45.** $2b^2$ **47.** $\frac{4}{5b^4c^7}$ **49.** $\frac{b}{c^2}$ **51.** $5bcd$ **53.** $3b^6c^2$ **55.** $\frac{c^2}{b^4}$

57. $\frac{5c^3}{3b^4}$ **59.** $2b^{29/35}$ **61.** $b^{7/12}$ **63.** $3b^{29/6}$ **65.** $\frac{8}{b^{1/5}}$ **67.** $\frac{2b^{49/12}}{27c^{5/4}}$ **69.** $b^2 + b$

71. a. exponential function **b.** 89 countries **c.** 1972 **73.** Answers may vary. **75.** Answers may vary.

77. $\frac{8}{3}$ **79.** 2 **81.** $-\frac{8}{3}$ **83.** $\frac{1}{2}$ **85.** a linear function **87.** 6; a linear function

Homework 4.3

1. **3.** **5.** **7.** **9.** **11.**

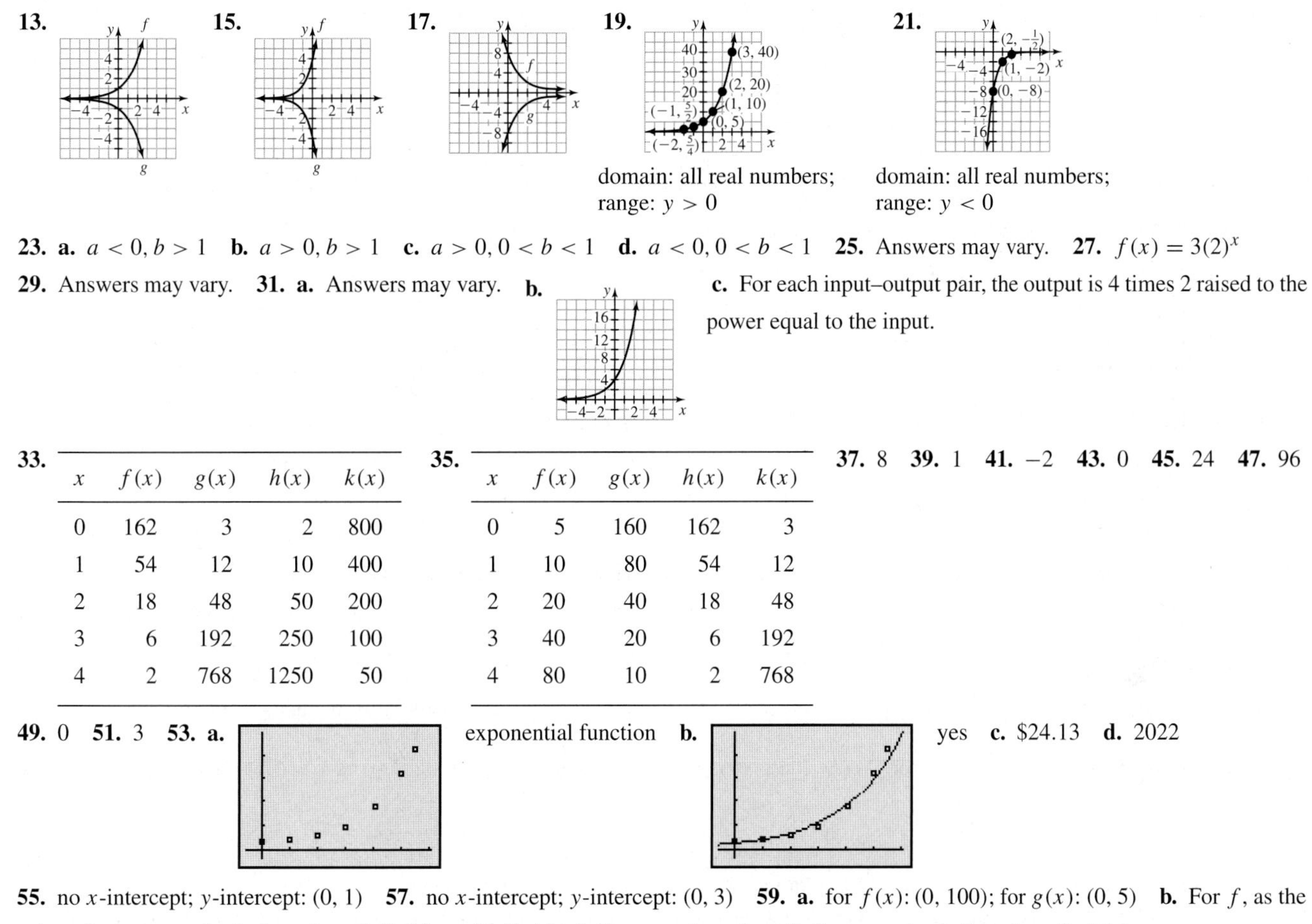

13. **15.** **17.** **19.** domain: all real numbers; range: $y > 0$ **21.** domain: all real numbers; range: $y < 0$

23. a. $a < 0, b > 1$ **b.** $a > 0, b > 1$ **c.** $a > 0, 0 < b < 1$ **d.** $a < 0, 0 < b < 1$ **25.** Answers may vary. **27.** $f(x) = 3(2)^x$
29. Answers may vary. **31. a.** Answers may vary. **b.** **c.** For each input–output pair, the output is 4 times 2 raised to the power equal to the input.

33.

x	$f(x)$	$g(x)$	$h(x)$	$k(x)$
0	162	3	2	800
1	54	12	10	400
2	18	48	50	200
3	6	192	250	100
4	2	768	1250	50

35.

x	$f(x)$	$g(x)$	$h(x)$	$k(x)$
0	5	160	162	3
1	10	80	54	12
2	20	40	18	48
3	40	20	6	192
4	80	10	2	768

37. 8 **39.** 1 **41.** −2 **43.** 0 **45.** 24 **47.** 96

49. 0 **51.** 3 **53. a.** exponential function **b.** yes **c.** \$24.13 **d.** 2022

55. no x-intercept; y-intercept: (0, 1) **57.** no x-intercept; y-intercept: (0, 3) **59. a.** for $f(x)$: (0, 100); for $g(x)$: (0, 5) **b.** For f, as the value of x increases by 1, the value of $f(x)$ is multiplied by 2. For g, as the value of x increases by 1, the value of $g(x)$ is multiplied by 3. **c.** The values of g will eventually be much greater. **d.**

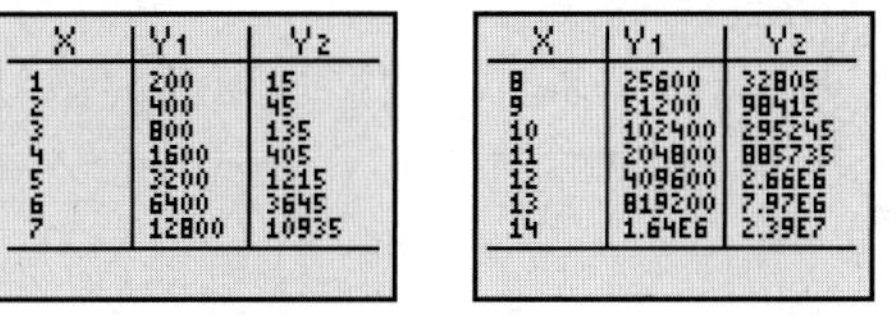

X	Y1	Y2
1	200	15
2	400	45
3	800	135
4	1600	405
5	3200	1215
6	6400	3645
7	12800	10935

X	Y1	Y2
8	25600	32805
9	51200	98415
10	102400	295245
11	204800	885735
12	409600	2.66E6
13	819200	7.97E6
14	1.64E6	2.39E7

61. 13 **63.** $\frac{13}{36}$ **65.** 1 **67.** 0 **69.** $f(x) = g(x)$ **71.** $f(x) = g(x)$ **73.** $f(x) = g(x)$ **75.** $f(x) = g(x)$ **77.** $f(x) = g(x)$
79. $f(x) = g(x)$ **81. a.** no **b.** no **83.** Answers may vary. **85.** Answers may vary. **87.** **89.**

91. x-intercept: (−2, 0); y-intercept: (0, 8) **93.** f might be linear; g might be exponential; h might be exponential; k is neither linear nor exponential. **95.**

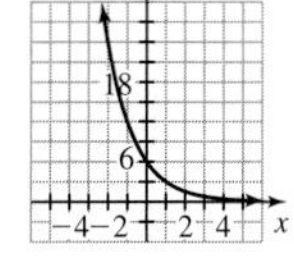

an exponential function **97.** 24; an exponential function

Homework 4.4

1. $f(x) = 4(2)^x$, $g(x) = 36\left(\frac{1}{3}\right)^x$, $h(x) = 5(10)^x$, $k(x) = 250\left(\frac{1}{5}\right)^x$ **3.** $f(x) = 100\left(\frac{1}{2}\right)^x$, $g(x) = -50x + 100$, $h(x) = 4x + 2$, $k(x) = 2(3)^x$ **5.** ±4 **7.** 3 **9.** 2 **11.** ±0.81 **13.** 2.28 **15.** ±1.51 **17.** 2.22 **19.** ±3
21. 1.74 **23.** $(p - k)^{1/n}$ **25.** $(ap - ak)^{1/n}$ **27.** $y = 4(2)^x$ **29.** $y = 3(2.02)^x$ **31.** $y = 87(0.74)^x$ **33.** $y = 7.4(0.56)^x$
35. $y = 5.5(3.67)^x$ **37.** $y = 39.18(0.85)^x$ **39.** $y = 4\left(\frac{1}{2}\right)^x$ **41.** $y = 1.33(3)^x$ **43.** $y = 1.19(1.50)^x$ **45.** $y = 1170.33(0.88)^x$
47. $y = 37.05(0.74)^x$ **49.** $y = 146.91(0.71)^x$ **51.** $y = 0.072(1.57)^x$ **53.** $y = 1.26(1.58)^x$ **55.** (0, 6) **57. a. i.** yes; answers may vary. **ii.** no; answers may vary. **b.** no; answers may vary. **59.** Answers may vary. **61.** b^5 **63.** 2.38 **65.** $\frac{4b^4}{3}$ **67.** ±0.75

69. $L(x) = 4x + 2$; $E(x) = 2(3)^x$

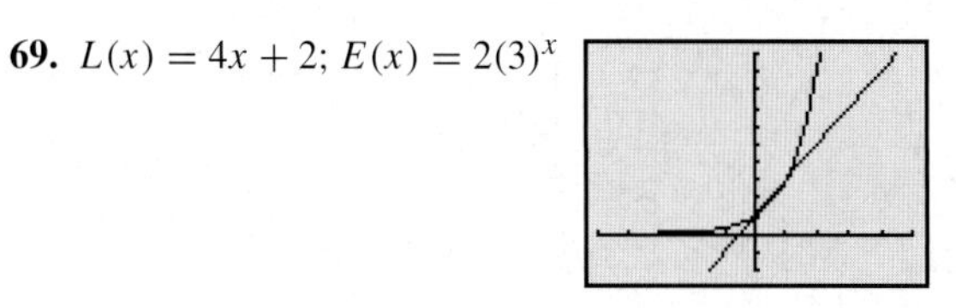

71. could be linear or exponential

73. a. L: (0, 100); E: (0, 3) **b.** $L(x)$ increases by 2; $E(x)$ is multiplied by 2. **c.** $E(x)$ will eventually dominate over $L(x)$.

d.

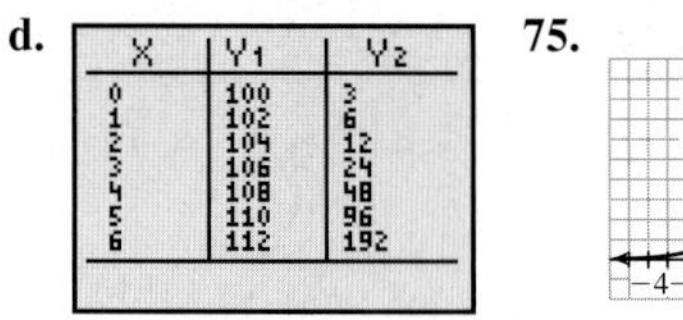

X	Y1	Y2
0	100	3
1	102	6
2	104	12
3	106	24
4	108	48
5	110	96
6	112	192

75. an exponential function **77.** $\frac{3}{8}$; an exponential function

Homework 4.5 **1. a.** $f(t) = 40(3)^t$ **b.** 2,361,960 people **c.** 573,956,280 people. Model breakdown has occurred, because this number exceeds the U.S. population. **3. a.** $f(t) = 8(2)^t$ **b.** 512 billion web pages **c.** 32,323 miles **5. a.** $h(t) = 91(2.14)^t$ **b.** (0, 91); in 2003, the revenue from ring tones was \$91 million. **c.** 2.14; each year, the revenue is 2.14 times that of the previous year. **d.** \$8740 million (\$8.74 billion) **e.** no **7. a.** $D(t) = 194(1.6)^t$ **b.** $S(t) = 15(2.9)^t$ **c.** (4.30, 1466.89); in 2005, the subscribers who got TiVo through DirectTV equaled the number of stand-alone TiVo subscribers, 1467 thousand subscribers (1.467 million subscribers). **9. a.** $f(t) = 3000(1.08)^t$ **b.** 1.08; the account balance increases by 8% per year. **c.** 3000; the initial amount invested was \$3000. **d.** \$9516.51 **11. a.** $f(t) = 4000(2)^{t/6}$ or $f(t) = 4000(1.1225)^t$ **b.** \$40,317.47 **13.** \$8749.97 **15. a.** $g(t) = 984\left(\frac{1}{2}\right)^t$ **b.** (0, 984); in 2003, 984 new copies of the textbook were sold. **c.** 123; in 2006, 123 new copies of the textbook were sold. **d.** 1 year **17. a.** $f(t) = 100\left(\frac{1}{2}\right)^{t/1600}$ or $f(t) = 100(0.999567)^t$ **b.** 95.76% **c.** 25%; 3200 years is two half-lives; so, half of 100% is 50%, and half of that is 25%. **19. a.** $f(t) = 100\left(\frac{1}{2}\right)^{t/7.56}$ or $f(t) = 100(0.9124)^t$ **b.** 75.95% **c.** 32.7 days **21.** 15.87 milligrams **23. a.** 10 years **b.** 6.25% **25.** \$138.7 billion **27.** 26,297 restaurants

29. Decrease b. **31. a.** exponential function **b.** $f(t) = 1.20(1.0162)^t$ **c.** (1.20, 0); the world's population in 1900 was 1.2 billion people. **d.** 7.029; the world's population in 2010 will be 7.029 billion people.

33. a. $f(t) = 100.84(1.41)^t$ **b.** 41% growth per year **c.** 97,278 stores **d.** 2005 **35. a.** $g(t) = 0.66(1.096)^t$ **b.** 0.66; at $t = 0$ (a newborn), the faculty member pays \$0.66. Model breakdown has occurred. **c.** 1.096; the rate increases by 9.6% each year of age. **d.** 49.05; a 47-year-old faculty member must pay \$49.05 per month. **e.** men **37. a.** right-hand column: 1.36, 1.36, 1.33, 1.34, 1.33, 1.36, 1.35 **b.** They are approximately equal. **c.** exponential function **d.** $f(t) = 3.94(1.03)^t$ **e.** right-hand column: 1.27, 1.26, 1.25, 1.21 **f.** no **g.** 2335.3 million (2.3353 billion) people; 2036.8 million (2.0368 billion) people

39. a. by hand, using points (29.5, 34) and (70, 6): $f(t) = 120.28(0.96)^t$; by regression: $r(t) = 113.04(0.96)^t$; **b.** 39.1; 39% of 26-year-olds plan to attend a Halloween party this year. **c.** 0.94 million (940,000) people **d.** (0, 113.04); 113.04% of newborns plan to attend a Halloween party this year. Model breakdown has occurred. **e.** 0.96; for each additional year of age, 4% fewer people will attend a Halloween party this year.

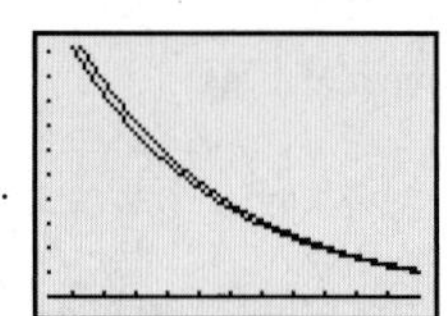

41. Answers may vary. **43. a.** $C(t) = 800(1.03)^t$ **b.** $S(t) = 24t + 800$ **c.** $C(1) = 824$; $C(2) = 848.72$; $S(1) = 824$; $S(2) = 848$ **d.** 1444.89, 1280; compound interest gives a larger balance than simple interest does.

45. a.

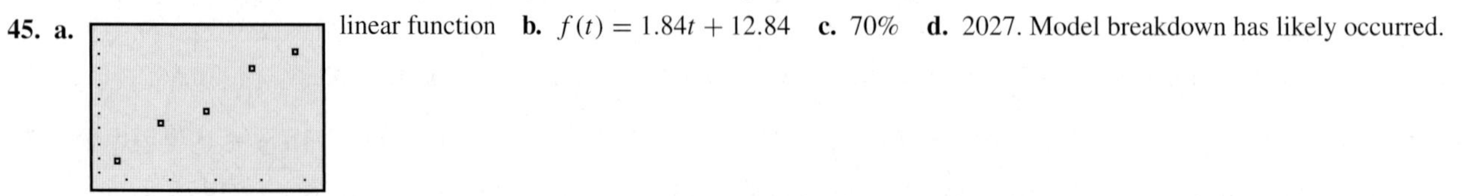

linear function **b.** $f(t) = 1.84t + 12.84$ **c.** 70% **d.** 2027. Model breakdown has likely occurred.

47. Answers may vary. **49.** Answers may vary. **51.** Answers may vary.

Chapter 4 Review

1. 32 **2.** $\frac{bc^9}{4}$ **3.** $\frac{8c^6}{9b^{23}}$ **4.** $\frac{b}{c}$ **5.** 16 **6.** $\frac{1}{8}$ **7.** $\frac{1}{b^{5/3}}$ **8.** $\frac{2b^{11}}{125c^7}$ **9.** $2b^2c$ **10.** $\frac{b^{1/6}}{c^{7/4}}$ **11.** b^{6n+2} **12.** $b^{n/6}$

13. $3^{2x} = (3^2)^x = 9^x$ **14.** $\frac{3}{25}$ **15.** $36(6^a)$ **16.** 7 **17.** $\frac{2}{27}$ **18.** 44,487,000 **19.** 0.0000385 **20.** 5.4×10^7 **21.** -8.97×10^{-3}

22.

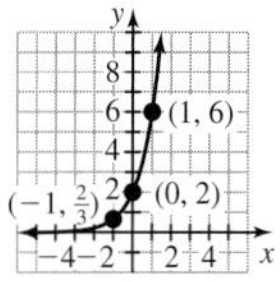

23.

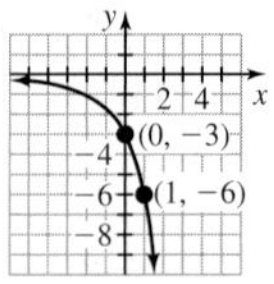

domain: all real numbers; range: $y < 0$ **24.**

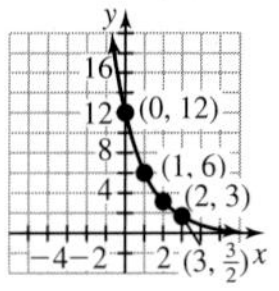

domain: all real numbers; range: $y > 0$

25. 1.84 **26.** ± 2 **27.** ± 1.61 **28.** f is linear, $f(x) = -4x + 34$; g is exponential, $g(x) = \frac{5}{3}(3)^x$; h is neither; k is exponential, $k(x) = 192\left(\frac{1}{2}\right)^x$. **29.** 18 **30.** 5 **31.** $y = 2(1.08)^x$ **32.** $y = 62.11(0.78)^x$ **33.**

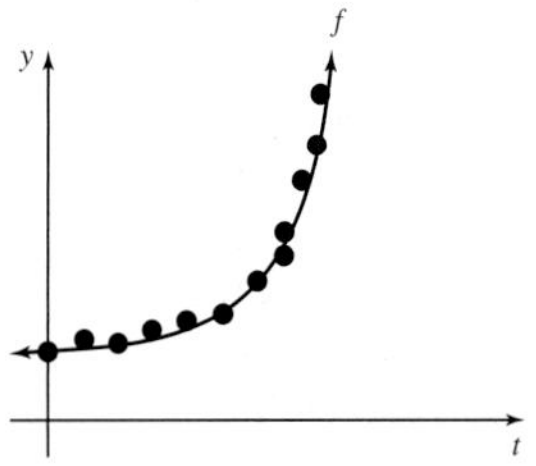

Increase a and decrease b.

34. a. $f(t) = 2000(1.07)^t$ **b.** \$2805.10 **35. a.** $g(t) = 17(2)^t$ **b.** \$1,088,000 **36. a.** $f(t) = 100\left(\frac{1}{2}\right)^{t/5730}$ or $f(t) = 100(0.999879)^t$ **b.** 98.8% **37.** \$2277 million (\$2.277 billion)

38. a.

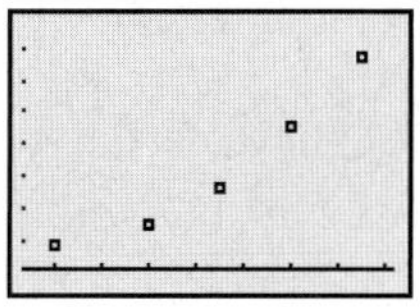

exponential function **b.** $f(t) = 56.22(1.18)^t$ **c.** 56.22; there were 56 Kohl's stores in 1990. **d.** 1.18; the number of stores is growing exponentially by 18% per year. **e.** 409.71; there were 410 Kohl's stores in 2002.

39. a. $f(t) = 44.32(2.01)^t$ **b.** 1.57 million lawsuits **c.** \$785 billion **d.** no

Chapter 4 Test

1. 4 **2.** $-\frac{1}{16}$ **3.** $8b^9c^{24}$ **4.** 1 **5.** $b^{1/6}$ **6.** $\frac{5b}{7c^5}$ **7.** $\frac{4b^4}{c^{14}}$ **8.** $\frac{250b^{14}}{7c^6}$

9. $8^{x/3}2^{x+3} = (2^3)^{x/3}2^{x+3} = 2^x2^{x+3} = 2^{2x+3} = 2^{2x}2^3 = 8(2^2)^x = 8(4)^x$ **10.** $\frac{1}{16}$ **11.** $\frac{1}{8}$

12.

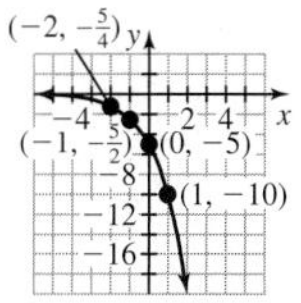

domain: all real numbers; range: $y < 0$ **13.**

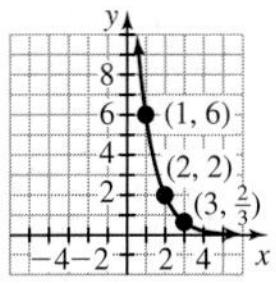

domain: all real numbers; range: $y > 0$

14. Answers may vary. **15.** $f(t) = 160\left(\frac{1}{2}\right)^t$ **16.** ± 1.72 **17.** $y = 70(0.81)^x$ **18.** $y = 0.91(1.77)^x$ **19.** 6 **20.** 1

21. $f(x) = 6\left(\frac{1}{2}\right)^x$ **22. a.** $f(t) = 400(3)^t$ **b.** 291,600; there will be 291,600 leaves on the tree 6 weeks after March 1. Model breakdown has likely occurred. **c.** approximately 2.58×10^{27}; one year after March 1, there will be about 2.58×10^{27} leaves on the tree. Model breakdown has occurred. **23. a.** $f(t) = 773.32(1.073)^t$ **b.** 7.3% per year **c.** 773.32; in 1970, there were 773 multiple births of triplets or more. **d.** 12,952 multiple births of triplets or more

Chapter 5

Homework 5.1

1. 4 **3.** 6 **5.** 5 **7.**

x	$f^{-1}(x)$
2	6
4	5
6	4
8	3
10	2

9. 6 **11.** 1 **13.**

x	$g^{-1}(x)$
2	1
6	2
18	3
54	4
162	5
486	6

15. values of $f^{-1}(x)$: 6, 5, 4, 3, 2, and 1

17. Answers may vary. **19.** 24 **21.** 0

23. **25.** **27.** **29.** **31.** **33.** **35.** **37.** 1 **39.** 6 **41.** 0 **43.** 1 **45.** 3 **47.**

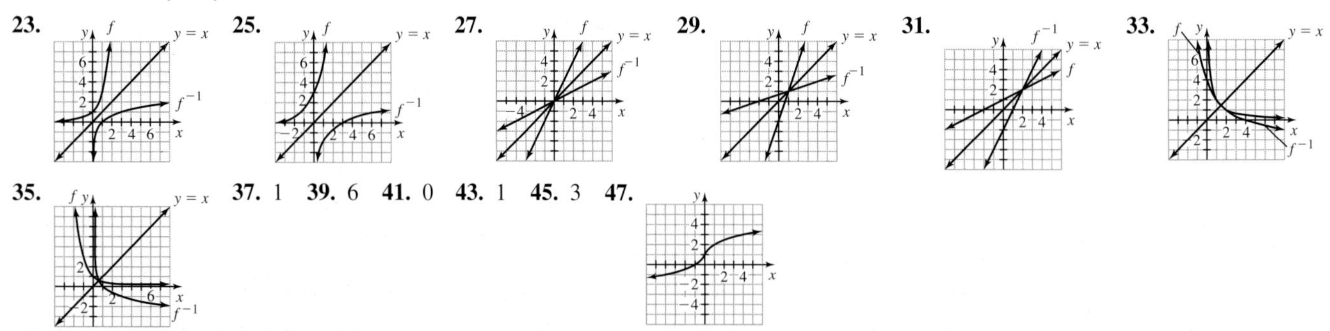

49. a. $f^{-1}(p) = 1.30p + 55.71$ **b.** 34.1; in 2000, 34.1% of births were outside marriage. **c.** 185.71; in 2086, 100% of births will be outside marriage. Model breakdown has likely occurred. **d.** 1.3; the percentage of births outside marriage increases by 1% every 1.3 years.

51. a.

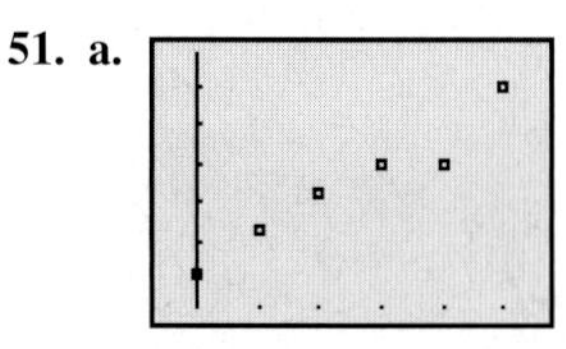

linear function **b.** $f(t) = 6.03t + 51.10$ **c.** $f^{-1}(n) = 0.17n - 8.47$ **d.** 2011 **e.** 2011 **f.** They are the same.

53. a. linear function **b.** $f(a) = 2.17a + 581.49$ **c.** $f^{-1}(c) = 0.46c - 267.97$ **d.** 43 years **e.** 114 years **f.** 0.46; the credit score increases by 1 unit for each age increase of 0.46 year. **55.** $f^{-1}(x) = x - 8$

57. $f^{-1}(x) = -\frac{1}{4}x$ **59.** $f^{-1}(x) = 7x$ **61.** $f^{-1}(x) = -\frac{1}{6}x - \frac{1}{3}$ **63.** $f^{-1}(x) = 2.5x + 19.75$ **65.** $f^{-1}(x) = \frac{3}{7}x - \frac{3}{7}$

67. $f^{-1}(x) = -\frac{6}{5}x - \frac{18}{5}$ **69.** $f^{-1}(x) = \frac{5}{6}x + \frac{1}{3}$ **71.** $f^{-1}(x) = -\frac{1}{8}x - \frac{1}{8}$ **73.** $f^{-1}(x) = x$ **75.** $f^{-1}(x) = x^{\frac{1}{3}}$

77. a. $f^{-1}(x) = \frac{1}{5}x + \frac{9}{5}$ **b.** 11 **c.** $\frac{13}{5}$ **79. a.** $f^{-1}(x) = \frac{1}{3}x + \frac{5}{3}$ **b.** Answers may vary. **c.**

d. For each input–output pair, the output variable is $\frac{5}{3}$ more than $\frac{1}{3}$ the input variable.

81. Subtracting 5 is the inverse operation of adding 5. **83.** no; no

85. a. $f^{-1}(x) = \frac{1}{m}x - \frac{b}{m}$ **b.** Answers may vary. **87. a.** (4, 5) **b.** $f^{-1}(x) = \frac{1}{2}x + \frac{3}{2}$ **c.** $g^{-1}(x) = 2x - 6$ **d.** (5, 4) **e.** The coordinates are interchanged. **89.** (5, 1); a linear system in two variables **91.**

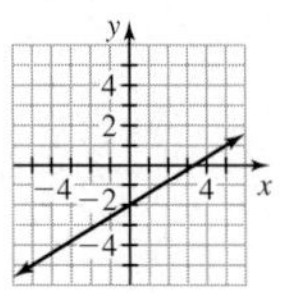

a linear equation in two variables (or a linear function)

Homework 5.2 **1.** 2 **3.** 3 **5.** 4 **7.** 3 **9.** 2 **11.** −1 **13.** −3 **15.** −4 **17.** 0 **19.** 1 **21.** $\frac{1}{2}$ **23.** $\frac{1}{3}$ **25.** $\frac{1}{2}$ **27.** $\frac{1}{4}$ **29.** 2 **31.** 0 **33.** 1 **35.** 4 **37.** −5 **39.** $\frac{1}{2}$ **41.** 0 **43.** $f^{-1}(x) = \log_3(x)$ **45.** $h^{-1}(x) = \log(x)$ **47.** $f^{-1}(x) = 5^x$ **49.** $h^{-1}(x) = 10^x$ **51.** 4 **53.** 1 **55.** 1 **57.** 27 **59.** 3 **61.** 0; $\log_3(1)$

63. **65.** **67.** **69.** (a) **71.** (d) **73. a.** Answers may vary. **b.**

c. For each input–output pair, the output variable is the logarithm, base 5, of the input variable. **75. a.** 2 **b.** 2 **c.** Answers may vary. **77. a.** 9.2 **b.** 7.8 **c.** 25.40 **79.** 0, 20, 40, 60, 80, 100, 120 **81.** Answers may vary.

83. a.

x	f(x)	g(x)	h(x)
1	0	2	2
2	1	4	4
4	2	8	16
8	3	16	256
16	4	32	65,536

b. h; g **85.** $f(x) = 0.58(2.05)^x$; an exponential function

87. $\frac{8b^{7/15}}{c^{1/2}}$; an expression in two variables that involves exponents

Homework 5.3 **1.** $3^5 = 243$ **3.** $10^2 = 100$ **5.** $b^c = a$ **7.** $10^n = m$ **9.** $\log_5(125) = 3$ **11.** $\log(1000) = 3$ **13.** $\log_y(x) = w$ **15.** $\log(q) = p$ **17.** 16 **19.** $\frac{1}{100}$ **21.** 1 **23.** 81 **25.** $\frac{69}{2}$ **27.** 3 **29.** 6561 **31.** 3.3019 **33.** 7 **35.** 2 **37.** 1.7411 **39.** 1.5850 **41.** 2 **43.** 4.8738 **45.** 3.8278 **47.** −0.2281 **49.** 3.4850 **51.** 0.8644 **53.** no real-number solution **55.** 64 **57.** 3.5130 **59.** 2.3587 **61.** 10 **63.** 1.6975 **65.** 1 **67.** 0, 3.7 **69.** 2 **71.** 1.2122 **73.** 1.3618 **75.** 5.4723 **77.** 2 **79.** 5 **81.** (3, 1.5) **83.** $\log[3(8)^x] \neq x\log[3(8)]$; 0.4075

85. $\dfrac{\log\left(\frac{c}{a}\right)}{\log(b)}$ **87.** $\dfrac{\log\left(\frac{c-d}{a}\right)}{k\log(b)}$ **89.** 256 **91.** 0.7925 **93.** 3 **95.** 32 **97. a.** no **b.** no **c.** no **d.** no

99. a. the part of the line $y = 1$ where $x > 0$. **b.** We cannot take the logarithm of a negative number. **c.** $f(x) = \log(x^3) - 3\log(x) + 1 = 3\log(x) - 3\log(x) + 1 = 1$ **101.** Answers may vary; no. **103.** $\frac{29}{5}$ **105.** ±1.7508 **107.** $\frac{16}{11}$ **109.** $\frac{1}{32}$; a logarithmic equation in one variable **111.** a logarithmic function

Homework 5.4 **1. a.** $f(t) = 2000(1.05)^t$ **b.** (0, 2000); the original investment was $2000.00. **c.** $2552.56 **d.** 8.31 years **3.** 6.65 years **5.** 7.27 years; the interest is compounded **7.** 2010 **9. a.** $f(t) = 30(3)^t$ **b.** 196,830 Americans **c.** 15 days **11. a.** $f(d) = 8\left(\frac{1}{2}\right)^{d/5}$ or $f(d) = 8(0.8706)^d$ **b.** 0.29 hour; yes **c.** 97 decibels **13.** 2010 **15.** 2012 **17. a.** 1977 **b.** no **c.** 1.41; the number of Starbucks stores increases by 41% each year. **d.** 2010 **e.** 2043; model breakdown has occurred.

19. a.

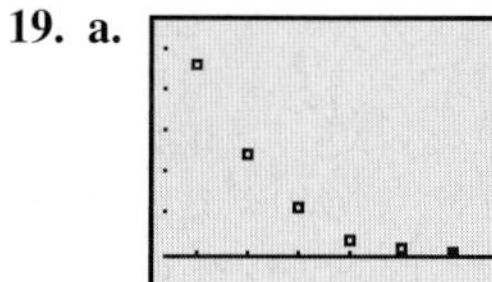

exponential function **b.** $f(t) = 347.97(0.66)^t$ **c.** 0.45 thousand cases, or 450 cases; no **d.** 2021 **e.** 1.67 years

21. a. 2027 **b.** 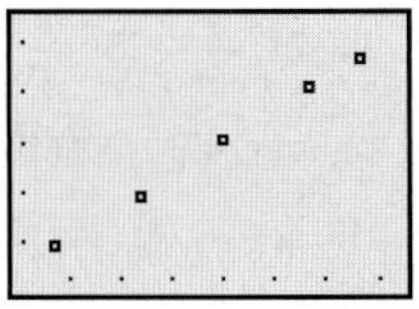 linear function; $g(t) = 0.079t - 1.85$ **c.** 2041; after **d.**

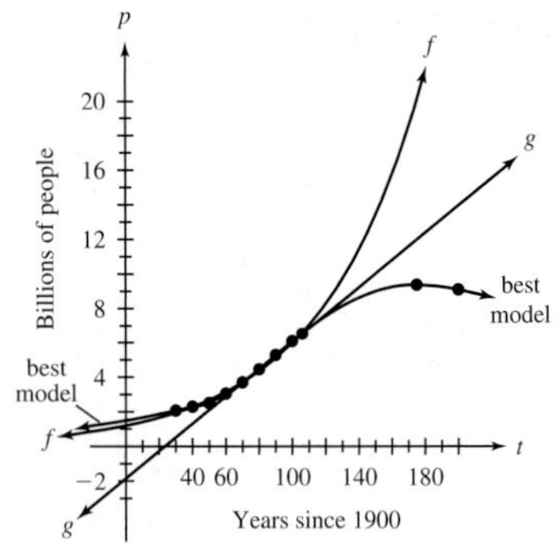

23. a. $f(t) = 0.00067(1.12)^t$ **b.** 1.12; the percentage of seniors with severe memory impairment increases by 12% for each additional year of age. **c.** 1.9% **d.** 85 years **e.** no **25. a.** $E(s) = 0.36(1.0036)^s$; $R(s) = 0.037(1.0049)^s$ **b.** 60%; 39% **c.** 1373 points; 1475 points **d.** 102 points **e.** (1757.58, 199.19); students who score 1758 points have the same chance (199%) of being selected by early decision as by regular decision. Model breakdown has occurred. **27. a.** $f(t) = 100\left(\frac{1}{2}\right)^{t/3.25}$ **b.** 65.28% **c.** after 26.0 days

29. 11,641 years ago **31. a.** 5730 years old **b.** 11,460 years old **c.** 19,035 years old **33.** 1329 years

35. a. $L(t) = 0.32t + 5.34$; $E(t) = 5.34(1.055)^t$ **b.** yes; linear **c.** 0.32; the combined dump capacity increases by 0.32 billion tons per year. **d.** 1.055; the combined dump capacity increases by 5.5% per year. **37.** $-\frac{5}{8b^{1/40}}$; an expression in one variable that involves exponents **39.** 2.2132; an exponential equation in one variable

Homework 5.5

1. $\log_b(3x^2)$ **3.** $\log_b(4x)$ **5.** $\log_b(5t^5)$ **7.** $\log_b\left(\frac{3}{x^3}\right)$ **9.** $\log_b(9x^{11})$ **11.** $\log_b\left(\frac{8m^{12}}{3}\right)$ **13.** 2.0412 **15.** 10.6667 **17.** 1.5275 **19.** 2.1544 **21.** 1.2011 **23.** 3.2702 **25.** 1.7712 **27.** 0.5804 **29.** −2.0431 **31.** 1.6204 **33.** 2.6031 **35.** −2.6876, 1.6964 **37.** $\log_7(x)$ **39.** $\log_s(r)$ **41.** all three students **43.** 1.1402 **45.** 0.8368

47. $\log_b(b^2) = \log_b\left(\frac{b^6}{b^4}\right) = \log_b(b^6) - \log_b(b^4) = 2$ **49.** $\log_b(x) - \log_b(x) = \log_b\left(\frac{x}{x}\right) = \log_b(1)$; hence, $\log_b(1) = 0$

51. a. $\log_2(x^8)$ **b.** 1.8340 **c, d.** Answers may vary. **53.** $\log_2(x^7)$ **55.** 1.4860 **57.** 8.6535 **59.** $\log_9\left(\frac{x^3}{8}\right)$ **61.** $\frac{6b^{13}}{c^{1/12}}$

63. $\log_b(72x^{23})$ **65.** (2, −1) **67.** (8, 5) **69.** all points on the line $y = \frac{2}{3}x - 2$; a dependent linear system of equations in two variables

71. 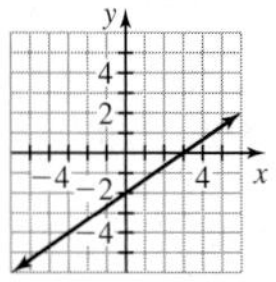 a linear equation in two variables (or a linear function)

Homework 5.6

1. 4.0037 **3.** −0.6931 **5.** 4 **7.** 1 **9.** −1 **11.** 3 **13.** 7.3891 **15.** 15.0855 **17.** 1.8383 **19.** 0.9061 **21.** 1.2777 **23.** 1.9811 **25.** 3.4541 **27.** 1.9458 **29.** 3.3673 **31.** $\ln(12x^5)$ **33.** $\ln(5x)$ **35.** $\ln(8w^{11})$ **37.** $\ln\left(\frac{27}{x}\right)$

39. $\ln(8k^4)$ **41.** 4.2661 **43.** 37.1033 **45.** 2.0654 **47.** 2.1311 **49.** 1.3066 **51.** 1.2377 **53.** −1.6856, 7.0194 **55.** 20

57. 0.1353 **59.** Answers may vary. **61.** $\frac{\ln\left(\frac{c}{a}\right)}{b}$, or $\frac{\ln(c) - \ln(a)}{b}$ **63.** $3\ln(x) = \ln(x^7) - \ln(x^4) = \ln(x^3)$

65. a. 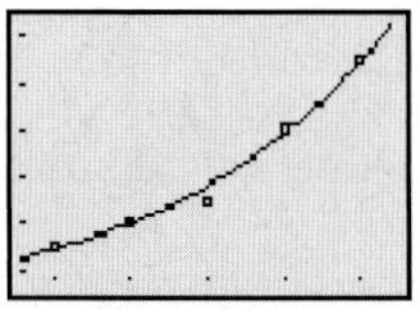 **b.** \$4.07 billion **c.** 2011 **67. a.** 207 degrees Fahrenheit **b.** 3.66 minutes **c.** 70 degrees Fahrenheit

69. a. 20.91 feet **b.** 20.32; the cable's height 4 feet to the left of the right pole is 20.32 feet. **c.** 20 feet **71. a. i.** 3.6960 **ii.** 3.6960 **iii.** They are the same. **b. i.** $\frac{\ln(c)}{\ln(b)}$ **ii.** $\frac{\log(c)}{\log(b)}$ **iii.** They are the same. **c.** Answers may vary. **73.** $\ln(x^5)$ **75.** 2.2255

77. 1.3863 **79.** $\frac{13}{11}$ **81.** ±2 **83.** Answers may vary. **85.** Answers may vary. **87.** Answers may vary.

Chapter 5 Review

1. 4 **2.** 1 **3.** **4.**

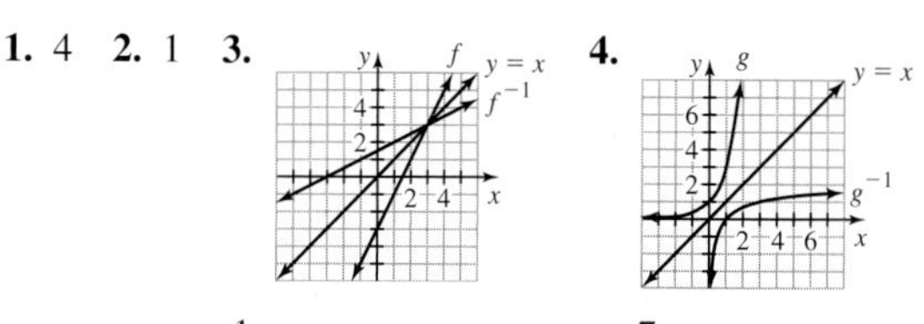

5. a. $f(t) = -16.3t + 507$ **b.** $f^{-1}(n) = -0.061n + 31.10$ **c.** 327.7; in 2011, there will be 328 thousand full-time-equivalent employees for all passenger airlines. **d.** 12.8; in 2013, there will be 300 thousand full-time-equivalent employees for all passenger airlines.

6. $f^{-1}(x) = \frac{1}{3}x$ **7.** $g^{-1}(x) = 2x + \frac{7}{4}$ **8.** 2 **9.** 5 **10.** −2 **11.** −3 **12.** $\frac{1}{3}$ **13.** 1.7712 **14.** 1.6094 **15.** 7 **16.** $h^{-1}(x) = \log_3(x)$ **17.** $h^{-1}(x) = 10^x$ **18.** **19.** $\log_d(k) = t$ **20.** $y^r = w$ **21.** 2.3219 **22.** $\frac{1}{81}$ **23.** 0.4310

24. 2.0886 **25.** 4 **26.** 2.8333 **27.** 1.6507 **28.** 4 **29.** 0 **30.** (4, 2), **31.** 81 **32.** 2.9299 **33.** 1.6309 **34.** 729 **35. a.** $f(t) = 8000(1.05)^t$ **b.** $12,410.63 **c.** 14.2 years **36. a.** $f(t) = 30(4)^t$ **b.** 30,720 leaves **c.** 6 weeks

37. a. exponential function **b.** $f(t) = 0.10(1.0544)^t$ **c.** 19.98; in 2000, 20% of Americans were obese. **d.** 130.40; in 2030, 100% of Americans will be obese. Model breakdown has likely occurred. **e.** 130.40; in 2030, 100% of Americans will be obese. Model breakdown has likely occurred.

38. a. $f(n) = 9.33(1.31)^n$ **b.** 1.31; for each additional cassette, the length increases by 31%. **c.** 9.33; the initial length of the rubber band was 9.33 inches. **d.** 80.92 inches; answers may vary. **e.** 10 cassettes; yes **39.** 14.5 years **40.** $\log_b(3p)$ **41.** $\log_b(72x^5)$ **42.** $\log_b\left(\frac{1}{x^2}\right)$ **43.** $\log_y(w)$ **44.** $\log_b(b^5) - \log_b(b^2) = \log_b(b^3) = \log_b\left(\frac{b^5}{b^2}\right) = 3$ **45.** 3 **46.** 8.4853 **47.** $\ln(256x^5)$ **48.** $\ln(2m)$ **49.** 2.9312 **50.** 7.3891 **51.** 2.8479

Chapter 5 Test

1. **2.** **3. a.** $f(t) = 2.07t + 28.49$ **b.** $f^{-1}(p) = 0.48p - 13.76$ **c.** 2010 **d.** 0.48; the ticket price increases by $1 every 0.48 year.

4. $g^{-1}(x) = \frac{1}{2}x + \frac{9}{2}$ **5.** 4 **6.** −3 **7.** 1.1833 **8.** −1 **9.** $\frac{1}{2}$ **10.** −2 **11.** $h^{-1}(x) = \log_4(x)$ **12.** $f^{-1}(x) = 5^x$ **13.** $\log_s(w) = t$ **14.** $c^d = a$ **15.** 2.6591 **16.** 2.4150 **17.** −0.6964 **18.** 1.67 **19. a.** $f(t) = 643.76(1.074)^t$ **b.** (0, 643.76); the tuition in 1950 was $644. **c.** 7.4% per year **d.** $50,117 **e.** 2014 **20.** 2050 years old **21.** $\log_b(5x^4)$ **22.** $\log_b\left(\frac{2p^{10}}{w^{10}}\right)$ **23.** 11.0227 **24.** 4.4413 **25.** $\ln(25w^{20})$ **26.** 1.4319 **27.** 4.0427

Cumulative Review Chapters 1–5

1. 0.5087 **2.** 86 **3.** 1.3538 **4.** 1.2528 **5.** ±0.7811 **6.** $\frac{7}{10}$ **7.** 1 **8.** 2 **9.** (−1, 2) **10.** (2, 4)

11. $x \geq \frac{9}{10}$; $\left[\frac{9}{10}, \infty\right)$; **12.** $\frac{1600c^4}{b^{23}}$ **13.** $\frac{4b^{5/6}}{3c^{5/4}}$ **14.** $\log_b\left(\frac{x^{26}}{49}\right)$ **15.** $\ln(p^{26})$ **16.** $f(x) = 5(3)^x$

17. $g(x) = 3x + 25$ **18.** −7 **19.** 40 **20.** 0 **21.** **22.** **23.** **24.**

25. $y = -\frac{10}{9}x + \frac{23}{9}$ **26.** $y = 347.56(0.63)^x$ **27.** $\frac{2}{81}$

28. **29.** **30.** 2.6053 **31.** −4 **32.** $\frac{1}{7}$ **33.** 2.0633 **34.** 1 **35.** $f(x) = 2(2)^x$, or $f(x) = (2)^{x+1}$

36. 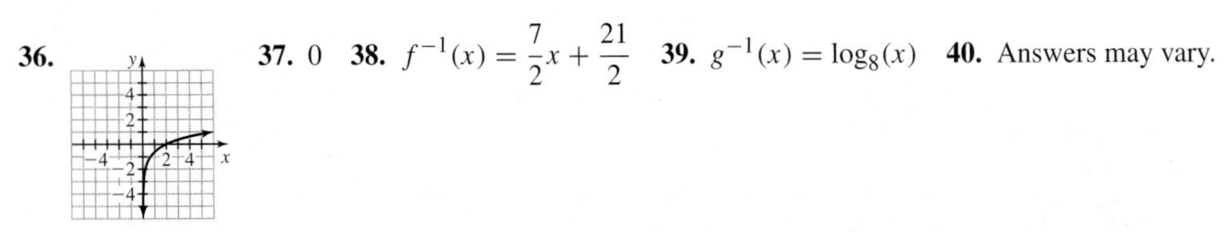 **37.** 0 **38.** $f^{-1}(x) = \frac{7}{2}x + \frac{21}{2}$ **39.** $g^{-1}(x) = \log_8(x)$ **40.** Answers may vary.

41. a. f: (0, 2); g: (0, 2) **b.** As x increases by 1, $f(x)$ increases by 3 and $g(x)$ is multiplied by 3. **c.** g

d. 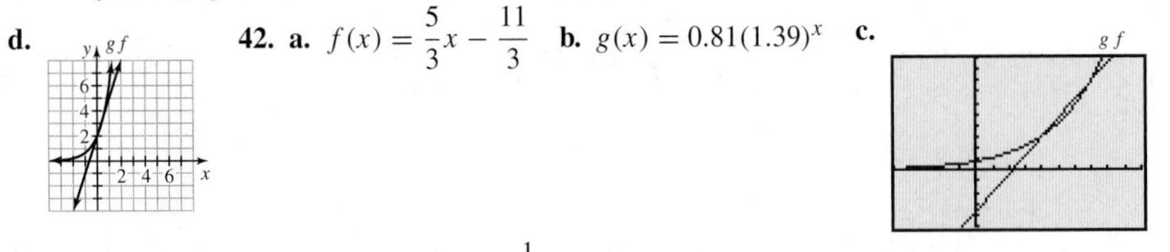 **42. a.** $f(x) = \frac{5}{3}x - \frac{11}{3}$ **b.** $g(x) = 0.81(1.39)^x$ **c.**

43. a. $f(2) = 6$; $g(2) = 9$ **b.** $f^{-1}(x) = \frac{1}{3}x$; $g^{-1}(x) = \log_3(x)$ **c.** $f^{-1}(81) = 27$; $g^{-1}(81) = 4$ **44. a.** $U(x) = 0.69x + 19.95$; $B(x) = 0.45x + 29.95$ **b.** 0.69, 0.45; U-Haul charges \$0.69 per additional mile, and Budget charges \$0.45 per additional mile. **c.** 41.67 miles **d.** $x < 41.67$; below 41.67 miles, U-Haul charges less than Budget. **45.** 10,500 tickets at \$43, 4500 tickets at \$60

46. a. $f(t) = 2.51(1.1)^t$ **b.** (0, 2.51); 2.51 million books were in print in 2003. **c.** 1.1; the number of books in print has increased by 10% per year. **d.** 2010 **47. a.** $f(t) = 20.69(0.78)^t$ **b.** (0, 20.69); in 1990, there were 20.69 cases per 100,000 people. **c.** 22%; the number of cases decreases by 22% each year. **d.** 2002 **e.** 21.46; in 2011, there will be 0.1 case per 100,000 people, or 1 case per 1,000,000 people. **f.** 0.64 case per 100,000 people; yes

48. a. linear function **b.** $f(t) = 1.30t + 102.5$ **c.** slope: 1.3; the average maximum speed increases by 1.3 miles per hour each year. **d.** (−78.85, 0); in 1901, the average maximum speed was 0 miles per hour. Model breakdown has occurred. **e.** 141.5; in 2010, the average maximum speed will be 142 mph. **f.** $f^{-1}(s) = 0.77s - 78.85$ **g.** 32.8; in 2013, the average maximum speed will be 145 miles per hour.

Chapter 6

Homework 6.1

1. quadratic (2nd-degree) polynomial in one variable **3.** cubic (3rd-degree) polynomial in one variable **5.** 7th-degree polynomial in two variables **7.** $4x^2 + x$ **9.** $-12x^3 + 2x^2 - 5x + 5$ **11.** $-5a^4b^2 - 5ab^3$ **13.** $2x^4 - 3x^3y + xy^3$ **15.** $9x^2 - 3x - 9$ **17.** $3x^3 - 6x^2 + 4x - 1$ **19.** $11a^2 - 3ab - 5b^2$ **21.** $2m^4p + 2m^3p^2 - 8mp^3$ **23.** $-7x^2 + 9x - 11$ **25.** $13x^3 - 3x^2 - x + 5$ **27.** $10m^2 + 10mp - p^2$ **29.** $a^3b - 10a^2b^2 + 8ab^3 + b^3$ **31.** −30 **33.** 79 **35.** 3 **37.** 42 **39.** −8 **41.** 3 **43.** −1 **45.** −1, 3 **47.** 1 **49.** 19 **51.** 3 **53.** 0, 6 **55.** 3 **57. a.** 1, 5 **b.** Answers may vary.

59. **61.** **63.** **65.** 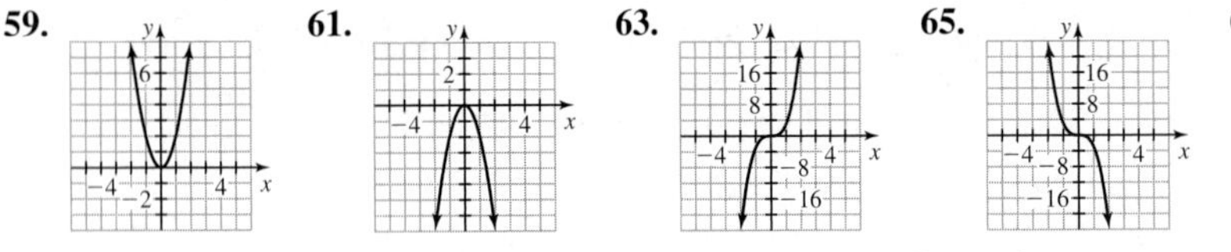 **67.** $(f + g)(x) = 11x^2 + 3x + 7$; 115

69. $(f - h)(x) = 7x^2 + 2x + 17$; 137 **71.** $(f + g)(x) = 2x^3 - 3x^2 + x - 2$; 4 **73.** $(f - h)(x) = x^3 + 3x^2 - 6x + 1$; 9 **75. a.** $(M + S)(t) = 0.58t + 43.94$ **b.** The units of the expressions on both sides of the equation are gallons per person. **c.** 78.74; in 2010, 78.74 gallons of milk and soft drinks (combined) will be consumed per person. **d.** $(M - S)(t) = -1.14t + 27.34$ **e.** −41.06; in 2010, 41.06 fewer gallons of milk than soft drinks will be consumed per person. **77. a.** $(B + S)(t) = 3.87t^2 - 23t + 126$ **b.** The units of the expressions on both sides of the equation are thousands of dollars. **c.** 341.27; in 2011, the average base pay plus average signing bonus for MBA graduates will be \$341,270. **d.** $(B - S)(t) = 0.73t^2 - 1.2t + 70$ **e.** 145.13; in 2011, the average base pay will be \$145,130 more than the average signing bonus. **79.** Answers may vary; $4x^2 + 4x + 2$ **81. a.** $f(x) - g(x) = -2x + 5$; $g(x) - f(x) = 2x - 5$ **b.** 1, −1; the answers are opposites. **c.** −3, 3; the answers are opposites. **d.** −9, 9; the answers are opposites. **e.** In general, $(f - g)(x) = -(g - f)(x)$. **83.** Answers may vary. **85.** (b) **87.** (c) **89.** (a) **91.** $\frac{5b^4}{2c^4}$; an expression in two variables involving exponents **93.** 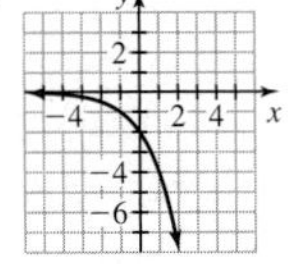 an exponential function

Homework 6.2 **1.** $18x^6$ **3.** $-8a^5b^8$ **5.** $-30x^2+12x$ **7.** $20a^3b^2-35a^2b^3+15ab^4$ **9.** $x^2+9x+18$ **11.** $15m^2+2m-8$ **13.** $32x^2-20x+3$ **15.** $3.91x^2-3.48x-2.88$ **17.** $6a^2+ab-35b^2$ **19.** $20x^2-53xy+18y^2$ **21.** $14a^4-29a^2b^2-15b^4$ **23.** $24x^4-54x^3-15x^2$ **25.** $5x^4-20x^3+15x^2-60x$ **27.** $12x^3+23x^2+x-6$ **29.** a^3+b^3 **31.** $8x^3-10x^2y+23xy^2-15y^3$ **33.** $x^4+x^3-3x^2+7x-6$ **35.** $2x^4-3x^3y-3x^2y^2+7xy^3-3y^4$ **37.** $x^2+10x+25$ **39.** $x^2-16x+64$ **41.** $9x^2+30x+25$ **43.** $6.76x^2-16.64x+10.24$ **45.** $16a^2+24ab+9b^2$ **47.** $4x^4-24x^2y^2+36y^4$ **49.** $-8x^3-40x^2-50x$ **51.** x^2-16 **53.** $9x^2-36$ **55.** $4r^2-64t^2$ **57.** $9r^2t^2-81w^2$ **59.** $64a^4-9b^4$ **61.** x^4-16 **63.** $81a^4-16b^4$ **65.** $25b^2-15b$ **67.** c^2+5c+4 **69.** $b^2-9b+18$ **71.** $4a-2$ **73.** $h^2+2ah-3h$ **75.** $f(x)=x^2+12x+36$ **77.** $f(x)=2x^2+12x+19$ **79.** $f(x)=-3x^2+30x-76$ **81.** $(f\cdot g)(x)=6x^2-5x-6$; 33 **83.** $(f\cdot h)(x)=4x^3-14x^2+18x-9$; 3 **85.** $(f\cdot f)(x)=4x^2-12x+9$; 25 **87.** $(f\cdot g)(x)=20x^2+17x+3$; 6 **89.** $(f\cdot h)(x)=12x^3-x^2-9x-2$; -84 **91.** $(h\cdot h)(x)=9x^4-6x^3-11x^2+4x+4$; 0

93. a.

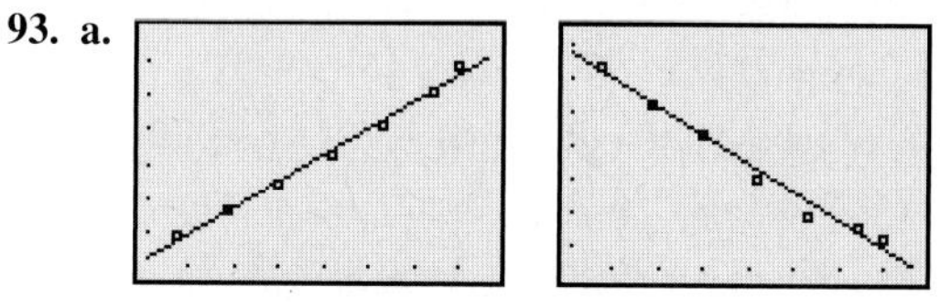

b. $(V\cdot A)(t)=-159.5t^2+52{,}064.7t+543{,}632$ **c.** The units of the expressions on both sides of the equation are millions of dollars. **d.** 1,521,126; in 2010, the total value of U.S. farmland per acre will be \$1,521,126 million, or \$1.52 trillion. **e.** increasing; between 1998 and 2010, the total value of U.S. farmland will increase every year; answers may vary.

95. a.

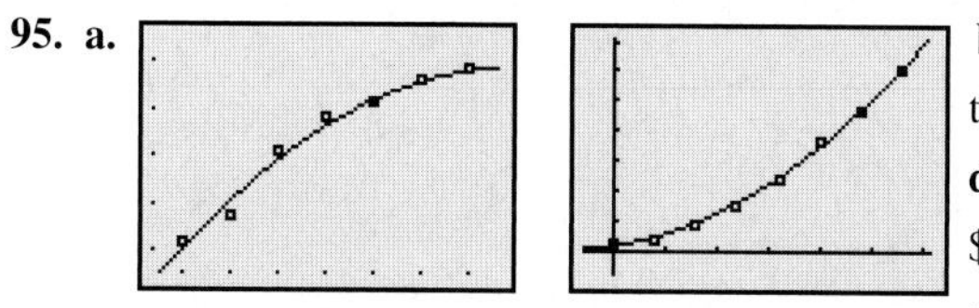

b. $(B\cdot N)(t)=-0.168t^4+4.32t^3+18.33t^2+18.9t-13.2$ **c.** The units of the expressions on both sides of the equation are millions of dollars per month. **d.** 15,801.9; in 2011, the total monthly revenue from cell phones will be \$15,802 million, or \$15.8 billion. **e.** increasing; between 1998 and 2010, the total monthly revenue from cell phones will increase.

97. The middle term, $16x$, is missing; $x^2+16x+64$

99. a.

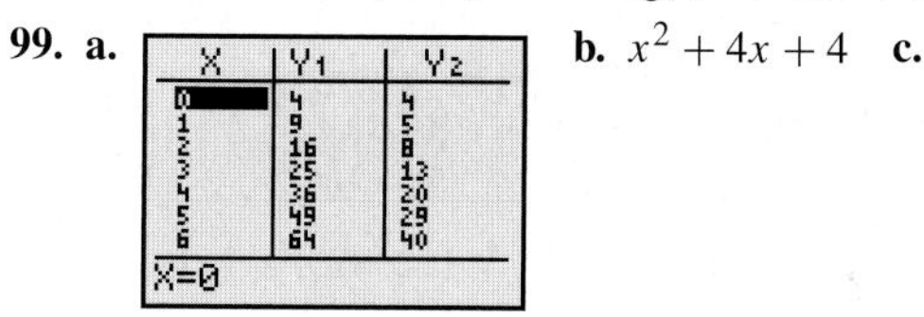

X	Y1	Y2
0	4	4
1	9	5
2	16	8
3	25	13
4	36	20
5	49	29
6	64	40

X=0

b. x^2+4x+4 **c.**

X	Y1	Y2
0	4	4
1	9	9
2	16	16
3	25	25
4	36	36
5	49	49
6	64	64

X=0

101. The student performed subtraction instead of multiplication; $-14x^2$
103. $(2x-5)(3x+4)=3x(2x-2)-x-20=6x^2-7x-20=(3x+4)(2x-5)$; $6x^2+7x-20=(3x-4)(2x+5)$
105. Answers may vary. **107.** $2x^{1/3}$ **109.** $-35x^{2/3}$ **111.** $f(x)=6x^2-17x+5$; quadratic **113.** $f(x)=-2x-1$; linear
115. $\log_b\left(x^2+2x-15\right)$ **117.** $\log_b\left(w^3-3w^2-9w+27\right)$ **119.** $24x^3+4x^2-60x$; a cubic polynomial in one variable
121. $f(x)=-3x^2+24x-43$; a quadratic function

Homework 6.3 **1.** $(x+7)(x+4)$ **3.** $(x-2)(x-6)$ **5.** $(r-8)(r+4)$ **7.** $(x+7)(x-2)$ **9.** prime **11.** $(x+5)^2$ **13.** $(t-9)^2$ **15.** $(x+5)(x-1)$ **17.** $(a+10b)(a+2b)$ **19.** $(w-4y)(w-y)$ **21.** $(p-4q)(p+7q)$ **23.** prime **25.** $(p-8q)(p+2q)$ **27.** $3(x+7)$ **29.** $4x(4x-3)$ **31.** $9y^3\left(y^2+2\right)$ **33.** $3ab(1-4a)$ **35.** $6a^2b^2\left(3a^2+2b\right)$ **37.** $-7x^2y\left(2x^3-9y\right)$ **39.** $2(x+3)^2$ **41.** $3(x-3)(x+2)$ **43.** $5(k+5)(k-2)$ **45.** $-4(x-3)^2$ **47.** $-(x-1)(x-10)$ **49.** $3\left(w^2-9w-20\right)$ **51.** $4x(x-4)(x-2)$ **53.** $a^2(a-1)(a-20)$ **55.** $5y(x+8y)(x+y)$ **57.** $4x^2y(x-5y)(x+2y)$ **59.** $-2xy^2(x-4y)^2$ **61.** The trinomial can be factored further, $2(x+3)(x+5)$ **63.** The GCF is $6x^2$; $6x^2(2x+3)$ **65.** $(x-3)(x+6)=x^2+3x-18=(x+6)(x-3)$ **67. a.** $(x-1)(x-4)$ **b.** $(1,0)$; $(4,0)$ **c.** Answers may vary. **69.** Answers may vary. **71.** $\pm 11, \pm 16, \pm 29$ **73.** Answers may vary. **75.** $x^2+2x-15$ **77.** $(k+3)(k-10)$ **79.** $49x^2-25$ **81.** $(9r+7)(9r-7)$ **83.** $(x+7)(x-4)$; a quadratic polynomial in one variable **85.** $6w^3+w^2-27w+20$; a cubic polynomial in one variable

Homework 6.4 **1.** $(x+3)\left(x^2+4\right)$ **3.** $(x-4)\left(5x^2+3\right)$ **5.** $(2m-5)\left(3m^2+1\right)$ **7.** $(2x+5)\left(5x^2-1\right)$ **9.** $(x-3y)(a-2b)$ **11.** $(5x+2y)\left(a^2-b\right)$ **13.** $(3x+5)(x+2)$ **15.** $(2x+5)(x-3)$ **17.** $(5p-1)(p-4)$ **19.** $(2x+3)(2x+5)$ **21.** prime **23.** $(3w-1)^2$ **25.** $(5x-3)(3x+2)$ **27.** $(3x-4)(2x-3)$ **29.** $(16y+3)(y-2)$ **31.** $(5a+3b)(2a+3b)$ **33.** $(20x-3y)(x+y)$ **35.** prime **37.** $(2r-5y)^2$ **39.** $2(3x-2)(x+5)$ **41.** $-3(4x+3)(x-1)$ **43.** $4x(2x-3)(2x-1)$ **45.** $2x^2(5x-1)(3x+1)$ **47.** $4t(3t+2w)^2$ **49.** $10ab^2(2a+7b)(a-2b)$ **51.** $(x+4)(x-10)$ **53.** $(w-2)\left(3w^2+5\right)$ **55.** $3x^2(x-9y)(x+2y)$ **57.** prime **59.** $(3x-2)(2x-5)$ **61.** $(x-5y)(x+6y)$ **63.** $-6r(r-2)^2$ **65.** $2(6x-5)(x+1)$ **67.** $(x-3y)\left(a^2-2b\right)$ **69.** prime **71.** $2pt^2(5p-4t)(p+3t)$ **73.** The polynomial is not factored; $(x+5)\left(x^2-3\right)$ **75.** The polynomial is not completely factored; $3(x-5)(x+2)$ **77.** $2(x-2)(x-6)=2\left(x^2-8x+12\right)=2x^2-16x+24=(x-2)(2x-12)=2(x-4)^2-8=(2x-4)(x-6)$

79. $3x(2x+3)(2x-3)$ **81.** $-24p^2-22p+30$ **83.** $(2x-5)(x+3)(x-3)$ **85.** $6k^3+5k^2+5k+12$ **87.** $-18x^3+60x^2-50x$; a cubic polynomial in one variable **89.** $2x(2x-5)^2$; a cubic polynomial in one variable

Homework 6.5 **1.** $(x+5)(x-5)$ **3.** $(a+6)(a-6)$ **5.** $(2x+7)(2x-7)$ **7.** prime **9.** $(4p+5t)(4p-5t)$ **11.** $3(5x+2)(5x-2)$ **13.** $2ab(3a+4b)(3a-4b)$ **15.** $(4x^2+9)(2x+3)(2x-3)$ **17.** $(t^2+w^2)(t+w)(t-w)$ **19.** $(x+3)(x^2-3x+9)$ **21.** $(x-2)(x^2+2x+4)$ **23.** $(m+1)(m^2-m+1)$ **25.** $(2x+3)(4x^2-6x+9)$ **27.** $(5x-2)(25x^2+10x+4)$ **29.** $(3p+2t)(9p^2-6pt+4t^2)$ **31.** $(3x-4y)(9x^2+12xy+16y^2)$ **33.** $5(x+2)(x^2-2x+4)$ **35.** $2x(x-3y)(x^2+3xy+9y^2)$ **37.** $(k+1)(k^2-k+1)(k-1)(k^2+k+1)$ **39.** $(2x+y)(4x^2-2xy+y^2)(2x-y)(4x^2+2xy+y^2)$ **41.** $(a-7b)(a+4b)$ **43.** $2x(x-2y)(x^2+2xy+4y^2)$ **45.** $(x-9)(x+2)$ **47.** $4xy(x+4y)(x-6y)$ **49.** $-(k-6)^2$ **51.** prime **53.** $(x-2)(x+3)(x-3)$ **55.** $3x^2(2x-5)(x-3)$ **57.** $2(4m+7t)(4m-7t)$ **59.** $(4x-1)(2x+3)$ **61.** $2y(2x-5y)(3x+y)$ **63.** prime **65.** $(5x+3)(25x^2-15x+9)$ **67.** $(p+9)^2$ **69.** $(5x-2)(2x+1)(2x-1)$ **71.** $(7x+1)^2$ **73.** prime **75.** $2y(w+5y)(w^2-5wy+25y^2)$ **77.** $-3x(x-6)(x+5)$ **79.** $3x(3x+5)(3x-5)$ **81.** $(9p^2+4q^2)(3p+2q)(3p-2q)$ **83.** The difference of cubes is $A^3-B^3=(A-B)(A^2+AB+B^2)$; $(x-2)(x^2+2x+4)$ **85.** The polynomial x^2+4 is prime; $4(x^2+25)$ **87. a.** $(x+2)(x-2)$ **b.** $(-2,0)$, $(2,0)$ **c.** Answers may vary. **89.** Answers may vary. **91.** $9x^2-49$ **93.** $(6p+7)(6p-7)$ **95.** t^3-125 **97.** $(3p+1)(9p^2-3p+1)$

99.

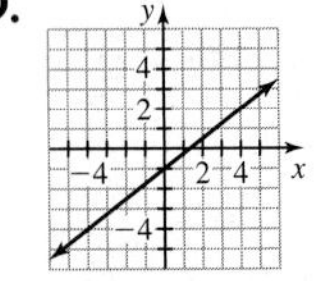

a linear equation in two variables **101.** $(2,-3)$; a system of linear equations in two variables

Homework 6.6 **1.** $-4, 7$ **3.** $-4, 3$ **5.** $3, 5$ **7.** -7 **9.** $-4, 6$ **11.** $\pm\frac{7}{5}$ **13.** $\frac{1}{3}, \frac{3}{2}$ **15.** $-6, 5$ **17.** $-1, 0, \frac{5}{2}$ **19.** $-2, 7$ **21.** $-2, 4$ **23.** 6 **25.** $\pm\frac{5}{4}$ **27.** $-2, 0, 2$ **29.** $0, 2$ **31.** $-5, \frac{1}{2}$ **33.** $-3, 0, 6$ **35.** $-\frac{2}{3}, 0, \frac{1}{2}$ **37.** $-\frac{5}{2}$ **39.** $-4, 6$ **41.** $-\frac{2}{3}, 1$ **43.** $\pm\frac{1}{5}$ **45.** $-10, 3$ **47.** $\pm 3, \frac{1}{2}$ **49.** $-3, \pm\frac{2}{3}$ **51.** $0, \frac{37}{3}$ **53.** $-6, 4$ **55.** $-3, -2, 8$ **57.** $(4,0)$, $(5,0)$ **59.** $\left(-\frac{5}{6}, 0\right)$, $\left(\frac{5}{6}, 0\right)$ **61.** $\left(-\frac{2}{3}, 0\right)$, $(0,0)$, $\left(\frac{5}{4}, 0\right)$ **63.** $(-2,0)$, $(-1,0)$, $(1,0)$ **65.** 0 **67.** $-4, 5$ **69.** $-5, 3$ **71.** -1 **73.** $-1, 2$ **75.** $-1, 1, 3$ **77.** $-1.24, 3.24$ **79.** $-0.81, 1.47, 3.34$ **81.** $-2, 4$ **83.** no real solution **85.** Dividing both sides by x is not allowed if $x=0$; $x=0$ or $x=1$ **87.** The student should have factored the left side of the equation; $x=-4$, $x=-3$, or $x=3$

89. a.

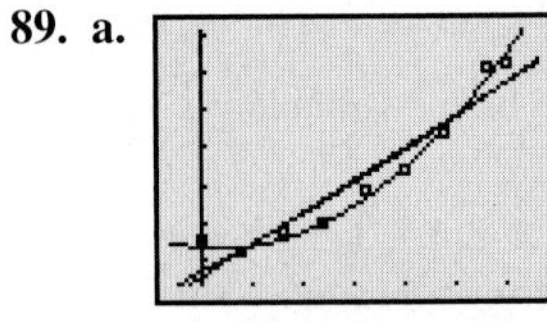

quadratic model **b.** $Q(t)$ **c.** L: $(0, 1314)$; E: $(0, 1324)$; Q: $(0, 1374)$; that of the quadratic model **d.** 1988, 1994

91. a.

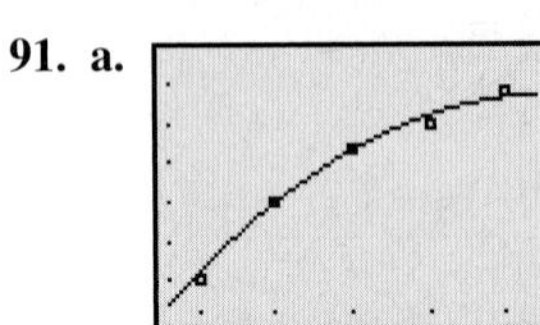

yes **b.** 35; 7 years after being rated B2, 35% of companies default on their bonds. **c.** 1, 21; 1 year and 21 years after being rated B2, 7% of companies default on their bonds. Model breakdown has occurred for the estimate of 21 years. **d.** $(0,0)$, $(22,0)$; no companies default 0 years and 22 years after being rated B2. Model breakdown has occurred for the estimate of 22 years.

93. a.

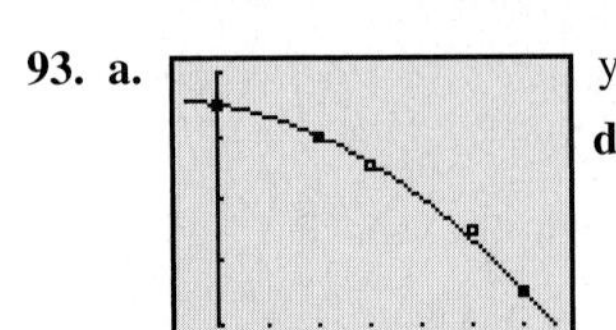

yes **b.** $(0, 25)$; in 1999, 25% of Americans thought that labor unions would become stronger. **c.** 4% **d.** 1989, 2007

95. a. yes **b.** \$63 **c.** 1990, 2009 **97.** width: 5 feet, length: 12 feet

99. width: 2 centimeters, length: 6 centimeters **101.** 2 feet **103.** 1 inch **105.** Answers may vary. **107.** Answers may vary. **109.** Answers may vary. **111.** Answers may vary. **113.** $(x+2)(x+3)$ **115.** $-2, -3$ **117.** $-2, -\frac{2}{3}, 0$ **119.** $p(3p+2)(p+2)$ **121.** $0, 2, 6$ **123.** ± 1.6660 **125.** 4.5969 **127.** -1 **129.** Answers may vary. **131.** Answers may vary. **133.** Answers may vary.

Chapter 6 Review

1. $-5x^3 - 3x^2 + 3x - 9$ **2.** $-3a^3b - 6a^2b^2 + 10ab^3$ **3.** 24 **4.** 9 **5.** 2, 4 **6.** $(f+g)(x) = x^3 - 2x^2 - 7x + 3; -11$ **7.** $(f-g)(x) = 5x^3 - 12x^2 - x + 1; -239$ **8.** $x^2 - 49$ **9.** $-40a^5b^6$ **10.** $8p^2 - 2pt - 45t^2$ **11.** $20x^3 - 23x^2 + 22x - 12$ **12.** $9x^2 + 42xy + 49y^2$ **13.** $36p^4 - 81t^6$ **14.** $-6r^3t^3 + 15r^2t^4 - 9rt^5$ **15.** $-36x^3 + 48x^2 - 16x$ **16.** $6m^4 + 7m^3p - 11m^2p^2 + 10mp^3 - 8p^4$ **17.** $a^2 - 10a + 24$ **18.** $6a + 3$ **19.** $f(x) = -2x^2 + 16x - 29$ **20.** $(f \cdot g)(x) = 6x^3 - 26x^2 + 37x - 21; 18$ **21.** $(x+5)(x-5)$ **22.** $(x-6)^2$ **23.** $(a+9b)(a-4b)$ **24.** $4a^3b^2(4a^2b - 5)$ **25.** prime **26.** $(3w - 8y)(w + y)$ **27.** $(9t^2 + 4w^2)(3t + 2w)(3t - 2w)$ **28.** $2x^2(3x - 2)(x + 4)$ **29.** prime **30.** $(x-9)(x+6)$ **31.** $2(y-3)(y^2 + 3y + 9)$ **32.** $5t(r + 3t)^2$ **33.** $(x - 5y)(2a - 3b)$ **34.** $-4, 6$ **35.** $\pm\frac{3}{8}$ **36.** $-2, 0, \frac{5}{2}$ **37.** $-3, \pm 2$ **38.** $\frac{1}{3}, 2$ **39.** $0, \frac{3}{4}$ **40.** $\frac{3}{4}$ **41.** $(-3, 0), (0, 0), (2, 0)$ **42.** $-5, 1$ **43.** $-3, 1$ **44. a.** $(E + C)(t) = 20.7t^2 - 59.5t + 360$ **b.** The units of the expressions on both sides of the equation are billions of dollars. **c.** 1835; in 2010, the combined revenues of ExxonMobil and Chevron will be \$1835 billion, or \$1.835 trillion. **d.** $(E - C)(t) = 4.7t^2 - 10.5t + 110$ **e.** 475; in 2010, ExxonMobil's revenue will be \$475 billion more than Chevron's revenue. **45. a.**

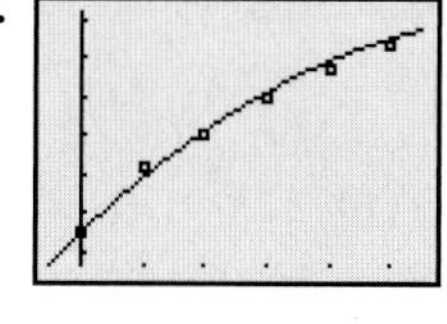

yes **b.** 83 million **c.** 2005, 2010

46. width: 4 meters, length: 12 meters

Chapter 6 Test

1. $-a^3b - 5a^2b^2 + ab^3$ **2.** $-2x^2 + 8x - 16; -40$ **3.** -3 **4.** no such value **5.** 1 **6.** $-3, 5$ **7.** $-14x^3y^2 + 6x^2y^3 - 12xy^4$ **8.** $12x^2 - xy - 35y^2$ **9.** $6w^3 - 17w^2t + 13wt^2 - 20t^3$ **10.** $12x^3 + 36x^2 + 27x$ **11.** $6x^4 + 14x^3 - 9x^2 - 21x + 5$ **12.** $16x^4 - 81y^4$ **13.** $a^2 - 13a + 40$ **14.** $f(x) = -3x^2 - 24x - 55$ **15.** $(f \cdot g)(x) = 6x^3 - 19x^2 + 22x - 8; 49$ **16.** $(x - 12)(x + 2)$ **17.** $2x(x-3)^2$ **18.** $-2(8x - 3)(x + 2)$ **19.** $(3m + 8t)(3m - 8t)$ **20.** $2a^2b(4a - 3b)(2a - 3b)$ **21.** $2(3m + 4p)(9m^2 - 12mp + 16p^2)$ **22.** $\pm\frac{4}{5}$ **23.** $-2, 0, 5$ **24.** $1, \frac{11}{2}$ **25.** $\pm 3, \frac{3}{2}$ **26.** $\frac{2}{3}, \frac{3}{2}$ **27.** $\left(\frac{2}{5}, 0\right), \left(\frac{3}{2}, 0\right)$ **28.** $-2.06, 2.19$ **29. a.** **b.** $(R \cdot P)(t) = -254.1t^2 + 7082.9t + 840{,}559$ **c.** The units of the expressions on both sides of the equation are number of deaths. **d.** 540,872.8; in 2011, there will be 540,873 deaths from heart disease. **e.** decreasing; from 2000 to 2010, the number of deaths from heart disease decreased and will continue decreasing; answers may vary.

30. a.

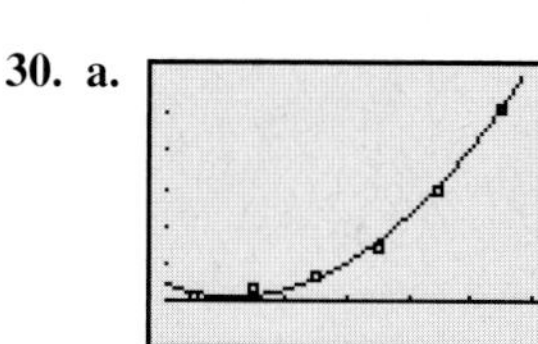

yes **b.** 1578; in 2010, sales using debit cards will be \$1578 billion, or \$1.578 trillion **c.** 1991, 1994

31. 2 inches

Chapter 7

Homework 7.1

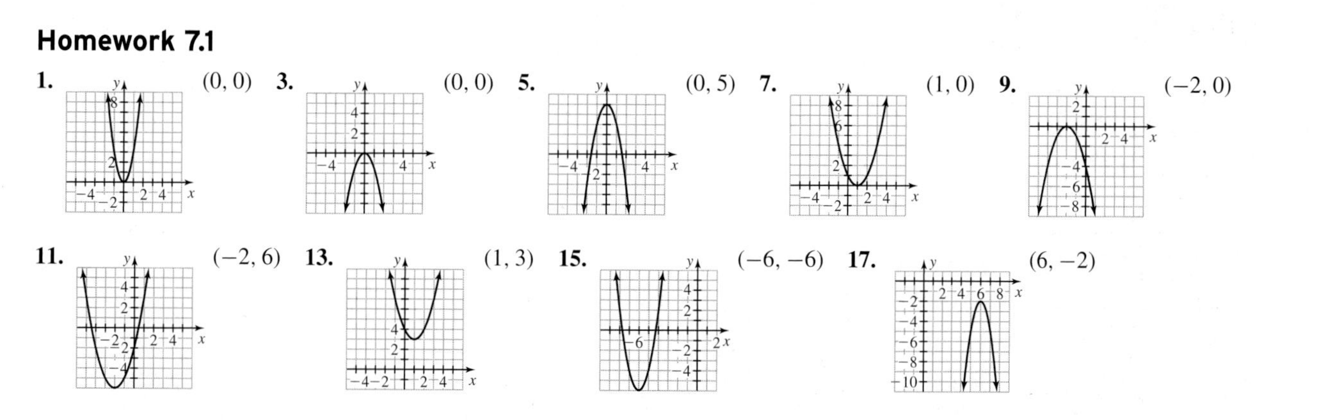

1. $(0, 0)$ **3.** $(0, 0)$ **5.** $(0, 5)$ **7.** $(1, 0)$ **9.** $(-2, 0)$ **11.** $(-2, 6)$ **13.** $(1, 3)$ **15.** $(-6, -6)$ **17.** $(6, -2)$

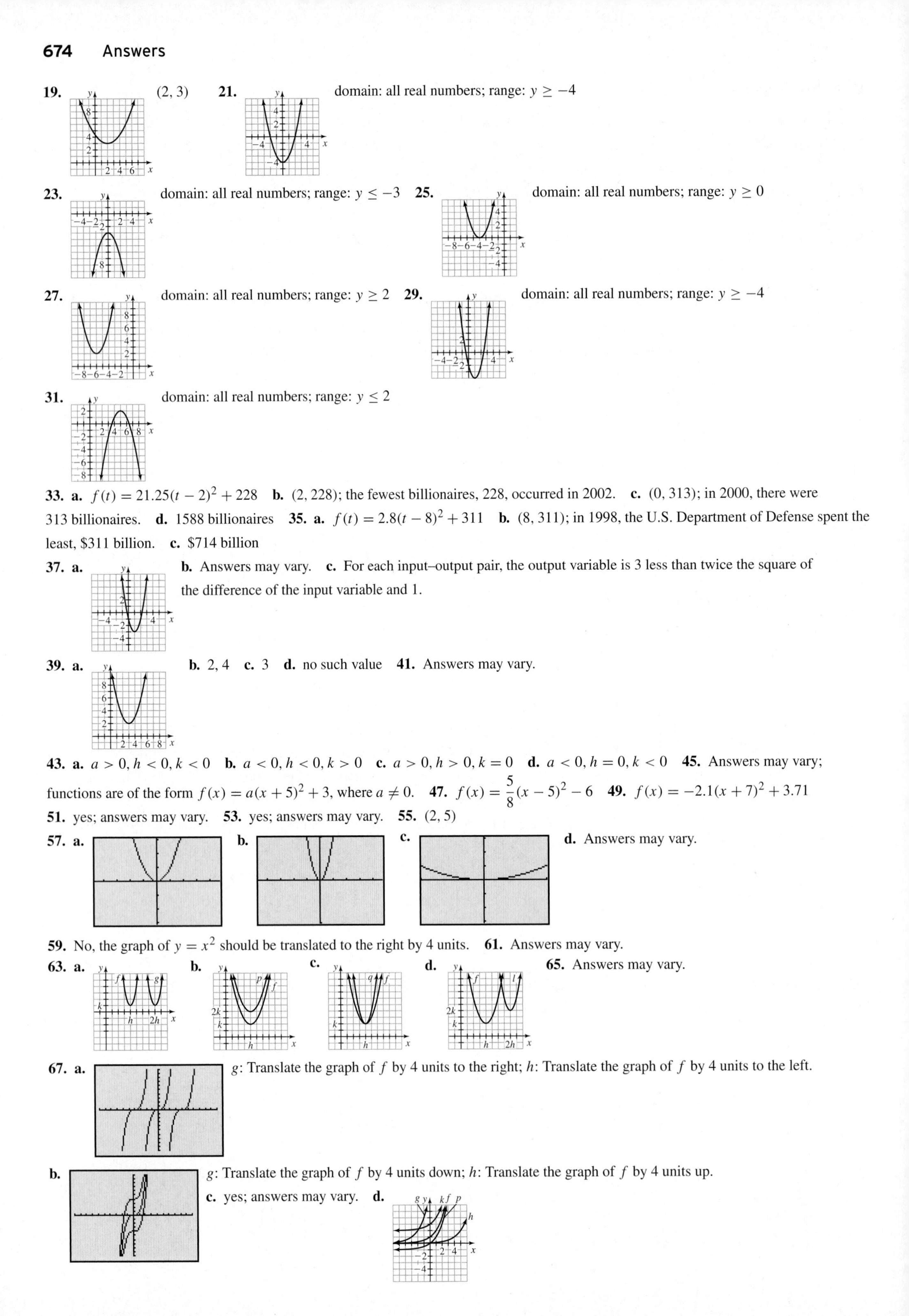

19. (2, 3) **21.** domain: all real numbers; range: $y \geq -4$

23. domain: all real numbers; range: $y \leq -3$ **25.** domain: all real numbers; range: $y \geq 0$

27. domain: all real numbers; range: $y \geq 2$ **29.** domain: all real numbers; range: $y \geq -4$

31. domain: all real numbers; range: $y \leq 2$

33. a. $f(t) = 21.25(t - 2)^2 + 228$ **b.** (2, 228); the fewest billionaires, 228, occurred in 2002. **c.** (0, 313); in 2000, there were 313 billionaires. **d.** 1588 billionaires **35. a.** $f(t) = 2.8(t - 8)^2 + 311$ **b.** (8, 311); in 1998, the U.S. Department of Defense spent the least, \$311 billion. **c.** \$714 billion

37. a. **b.** Answers may vary. **c.** For each input–output pair, the output variable is 3 less than twice the square of the difference of the input variable and 1.

39. a. **b.** 2, 4 **c.** 3 **d.** no such value **41.** Answers may vary.

43. a. $a > 0, h < 0, k < 0$ **b.** $a < 0, h < 0, k > 0$ **c.** $a > 0, h > 0, k = 0$ **d.** $a < 0, h = 0, k < 0$ **45.** Answers may vary; functions are of the form $f(x) = a(x + 5)^2 + 3$, where $a \neq 0$. **47.** $f(x) = \frac{5}{8}(x - 5)^2 - 6$ **49.** $f(x) = -2.1(x + 7)^2 + 3.71$

51. yes; answers may vary. **53.** yes; answers may vary. **55.** (2, 5)

57. a. **b.** **c.** **d.** Answers may vary.

59. No, the graph of $y = x^2$ should be translated to the right by 4 units. **61.** Answers may vary.

63. a. **b.** **c.** **d.** **65.** Answers may vary.

67. a. g: Translate the graph of f by 4 units to the right; h: Translate the graph of f by 4 units to the left.

b. g: Translate the graph of f by 4 units down; h: Translate the graph of f by 4 units up.

c. yes; answers may vary. **d.**

69. **71.**

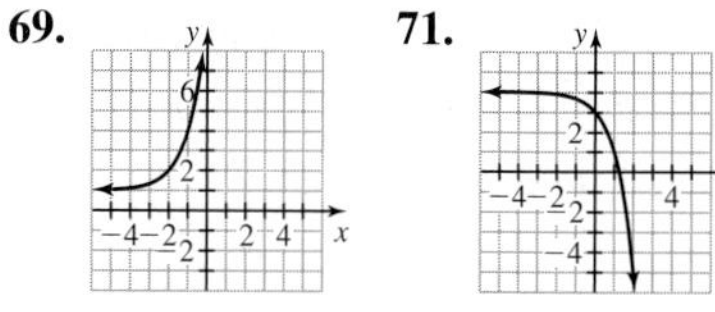

73. ± 1.7919; a logarithmic equation in one variable

75. $\log_4\left(12x^{10}\right)$; a logarithmic expression in one variable **77.** $\log_4\left(\frac{3}{4x^6}\right)$; a logarithmic expression in one variable

Homework 7.2

1. 5 **3.** 3 **5.** -3.5 **7.** 3.65 **9.** (4, 9)

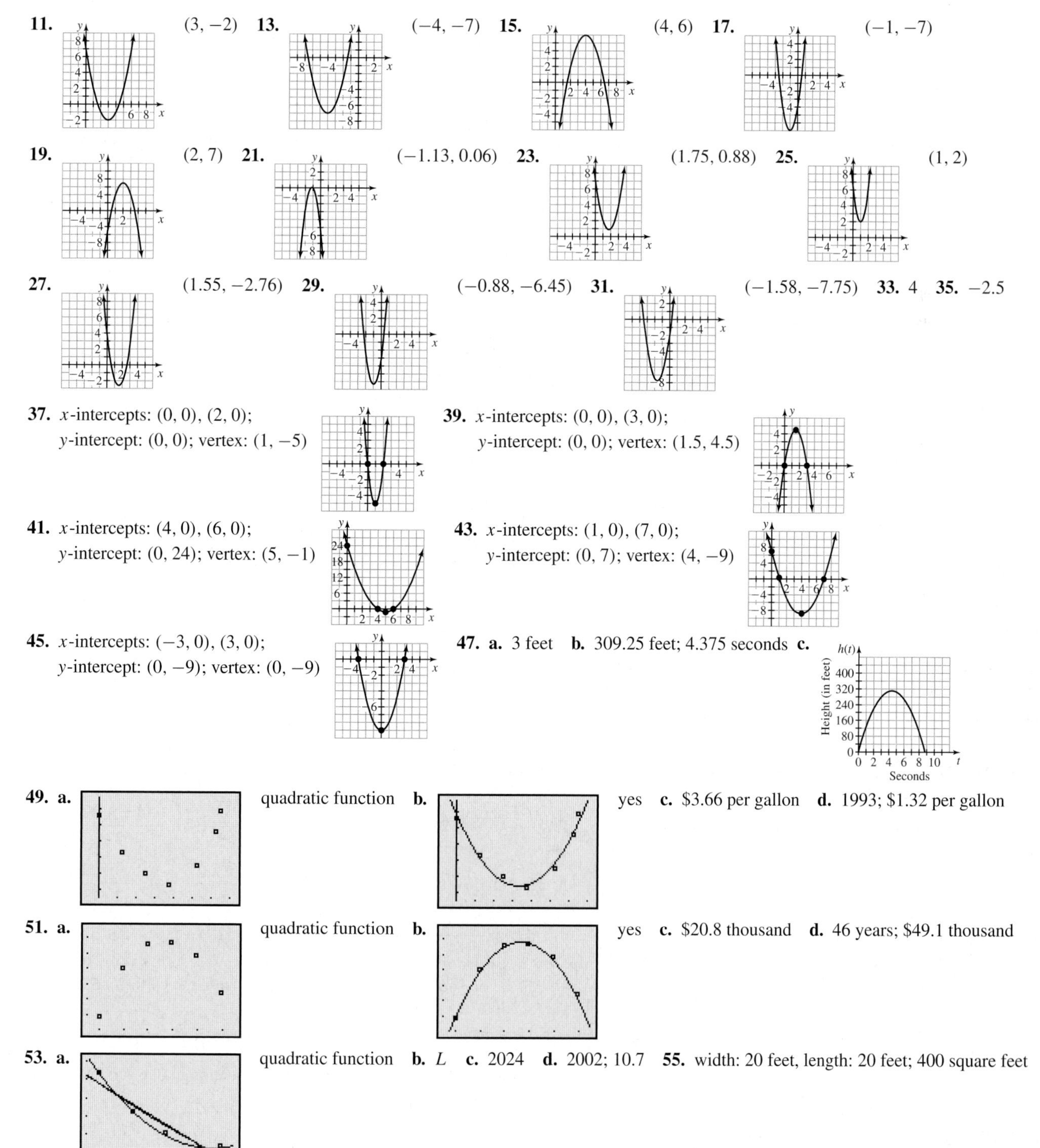

11. $(3, -2)$ **13.** $(-4, -7)$ **15.** (4, 6) **17.** $(-1, -7)$

19. (2, 7) **21.** $(-1.13, 0.06)$ **23.** (1.75, 0.88) **25.** (1, 2)

27. $(1.55, -2.76)$ **29.** $(-0.88, -6.45)$ **31.** $(-1.58, -7.75)$ **33.** 4 **35.** -2.5

37. x-intercepts: (0, 0), (2, 0); y-intercept: (0, 0); vertex: $(1, -5)$

39. x-intercepts: (0, 0), (3, 0); y-intercept: (0, 0); vertex: (1.5, 4.5)

41. x-intercepts: (4, 0), (6, 0); y-intercept: (0, 24); vertex: $(5, -1)$

43. x-intercepts: (1, 0), (7, 0); y-intercept: (0, 7); vertex: $(4, -9)$

45. x-intercepts: $(-3, 0)$, (3, 0); y-intercept: $(0, -9)$; vertex: $(0, -9)$

47. a. 3 feet **b.** 309.25 feet; 4.375 seconds **c.**

49. a. quadratic function **b.** yes **c.** \$3.66 per gallon **d.** 1993; \$1.32 per gallon

51. a. quadratic function **b.** yes **c.** \$20.8 thousand **d.** 46 years; \$49.1 thousand

53. a. quadratic function **b.** L **c.** 2024 **d.** 2002; 10.7 **55.** width: 20 feet, length: 20 feet; 400 square feet

57. width: 100 feet, length: 200 feet; 20,000 square feet **59. a.**

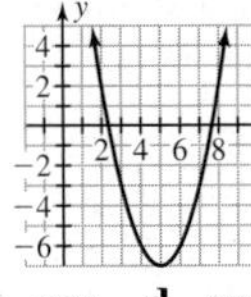

b. Answers may vary. **c.** For each input–output pair, the output is 18 more than the difference between the square of the input and 10 times the input.

61. -1 **63.** -3 **65.** $-2, -4$ **67.** 3 **69. a.** -2 **b.** -2 **c.** yes **d.** averaging x-coordinates of y-intercept and its symmetric point **e.** averaging x-coordinates of y-intercept and its symmetric point **f.** Answers may vary. **71.** for f and k: $(3, 2)$; for g: $(2.7, 1.8)$; for h: $(3.3, 1.7)$ **73.** Answers may vary. **75.** **77.**

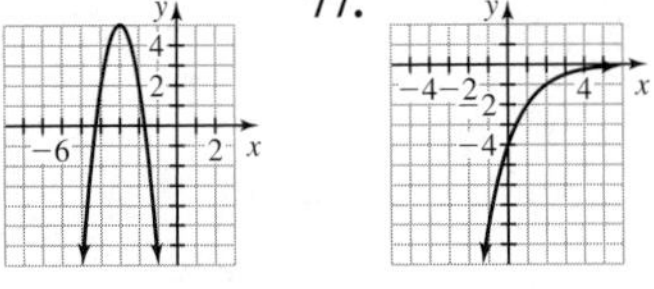

79. a. $a(x-h)^2+k = a\left(x^2-2xh+h^2\right)+k = ax^2-2ahx+ah^2+k = ax^2-2ahx+c$ **b.** x-coordinate of vertex: $-\dfrac{b}{2a} = -\dfrac{-2ah}{2a} = h$; y-coordinate: $f(h) = ah^2-2ah^2+ah^2+k = k$ **81.** $\dfrac{9c^6}{4b^6}$; an expression in two variables involving exponents

83. an exponential function **85.** 4.19; an exponential equation in one variable

Homework 7.3

1. 13 **3.** $2\sqrt{3}$ **5.** $\dfrac{2}{3}$ **7.** $\dfrac{\sqrt{6}}{7}$ **9.** $\dfrac{5\sqrt{2}}{2}$ **11.** $\dfrac{3\sqrt{2}}{8}$ **13.** $\dfrac{\sqrt{6}}{2}$ **15.** $\dfrac{\sqrt{55}}{10}$ **17.** ± 5 **19.** $\pm\sqrt{3}$ **21.** $\pm 4\sqrt{2}$ **23.** $\pm\dfrac{\sqrt{15}}{5}$ **25.** $\pm\dfrac{\sqrt{42}}{3}$ **27.** $-4\pm\sqrt{7}$ **29.** $5\pm 3\sqrt{3}$ **31.** $-\dfrac{9}{8}, \dfrac{3}{8}$ **33.** $\dfrac{5}{9}$ **35.** $\dfrac{-3\pm\sqrt{41}}{4}$ **37.** $\dfrac{7\pm\sqrt{5}}{3}$ **39.** $6\pm\sqrt{6}$ **41.** $\dfrac{-3\pm\sqrt{21}}{3}$ **43.** $\left(-\sqrt{17}, 0\right), \left(\sqrt{17}, 0\right)$ **45.** $\left(\dfrac{6-\sqrt{14}}{2}, 0\right), \left(\dfrac{6+\sqrt{14}}{2}, 0\right)$ **47.** no x-intercepts **49.** $6i$ **51.** $-3i\sqrt{5}$ **53.** $\dfrac{i\sqrt{5}}{7}$ **55.** $\dfrac{i\sqrt{65}}{5}$ **57.** $\pm 7i$ **59.** $\pm 3i\sqrt{2}$ **61.** $\pm i\sqrt{3}$ **63.** $-4\pm 2i\sqrt{2}$ **65.** $\dfrac{5\pm i\sqrt{3}}{4}$ **67.** $-3\pm 2i$ **69. a.** $f(t) = 21(t-1)^2+64$ **b.** 2009 **c.** \$1765 per metric ton **d.** \$100 **71. a.** $f(t) = 8.29(t-1.62)^2+476.88$ **b.** $(1.62, 476.88)$; the minimum spending on domestic travel, \$477 billion, occurred in 2002. **c.** \$927 billion **d.** 2010 **73.** There is still an x on the right-hand side; $x = 5$ **75.** 1.4, 4.2 **77.** 2, 4 **79.** $(1.1, -3.5), (4.7, -1.7)$ **81. a.** $(3, 5)$ **b.** upward **c. i.** 2 **ii.** 1 **iii.** 0 **83. a.** $-\dfrac{7}{5}, \dfrac{7}{5}$ **b.** $-\dfrac{7}{5}, \dfrac{7}{5}$ **c.** They are the same. **d.** Answers may vary. **85. a.** yes; $-4\pm\sqrt{5}$ **b.** no **c.** no; answers may vary. **87.** $\pm\sqrt{c^2-b^2}$ **89.** $\dfrac{-b\pm\sqrt{p}}{m}$ **91.** Answers may vary. **93.** ± 2.2841 **95.** 8.4480 **97.** 0.7411 **99.** $(2w+3)(w+3)(w-3)$; a cubic polynomial in one variable **101.** $\pm 3, -\dfrac{3}{2}$; a cubic equation in one variable **103.** $20w^3+15w^2-8w-6$; a cubic polynomial in one variable

Homework 7.4

1. 36; $(x+6)^2$ **3.** 49; $(x-7)^2$ **5.** $\dfrac{49}{4}$; $\left(x+\dfrac{7}{2}\right)^2$ **7.** $\dfrac{9}{4}$; $\left(x-\dfrac{3}{2}\right)^2$ **9.** $\dfrac{1}{16}$; $\left(x+\dfrac{1}{4}\right)^2$ **11.** $\dfrac{4}{25}$; $\left(x-\dfrac{2}{5}\right)^2$ **13.** $-3\pm\sqrt{10}$ **15.** $1\pm 2\sqrt{5}$ **17.** $-2\pm 2\sqrt{7}$ **19.** $\dfrac{7\pm\sqrt{61}}{2}$ **21.** $\dfrac{-5\pm\sqrt{41}}{2}$ **23.** $\dfrac{5\pm\sqrt{33}}{4}$ **25.** $\dfrac{-4\pm\sqrt{22}}{2}$ **27.** $\dfrac{1\pm\sqrt{57}}{4}$ **29.** $\dfrac{-2\pm\sqrt{19}}{3}$ **31.** $\dfrac{4\pm\sqrt{10}}{6}$ **33.** $\dfrac{-1\pm\sqrt{7}}{4}$ **35.** $-1\pm i\sqrt{6}$ **37.** $3\pm 2i\sqrt{2}$ **39.** $\dfrac{-3\pm i\sqrt{7}}{2}$ **41.** $\dfrac{-1\pm 2i\sqrt{5}}{3}$ **43.** $\dfrac{3\pm i\sqrt{71}}{8}$ **45.** $\dfrac{-3\pm i\sqrt{3}}{4}$ **47.** $(4-\sqrt{13}, 0), (4+\sqrt{13}, 0)$ **49.** $\left(\dfrac{5-\sqrt{57}}{4}, 0\right), \left(\dfrac{5+\sqrt{57}}{4}, 0\right)$ **51.** $(-5, 0)$ **53.** To complete the square, the leading coefficient must be 1; $\dfrac{-3\pm\sqrt{13}}{4}$ **55. a.** no such real-number value **b.** -3 **c.** $-3\pm\sqrt{2}$ **57.** 1, 3 **59.** 2, 3 **61.** $(0, -0.5), (5, -8)$ **63.** Answers may vary. **65.** Answers may vary. **67.** $(w-5)^2$ **69.** $\left(x+\dfrac{5}{6}\right)^2$

71.

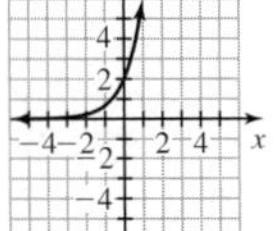

an exponential function **73.** 3.17; an exponential function **75.** 26.33; a logarithmic equation in one variable

Homework 7.5 **1.** $\frac{-5 \pm \sqrt{41}}{4}$ **3.** $\frac{3 \pm \sqrt{6}}{3}$ **5.** $2 \pm \sqrt{7}$ **7.** $1, \frac{3}{2}$ **9.** $\pm\frac{\sqrt{51}}{3}$ **11.** $-\frac{5}{2}, 0$ **13.** $\frac{5 \pm \sqrt{57}}{8}$ **15.** $\frac{1 \pm \sqrt{37}}{6}$ **17.** $-0.64, 3.14$ **19.** $-0.52, 3.02$ **21.** $-8.54, 0.02$ **23.** $\frac{3 \pm i\sqrt{23}}{2}$ **25.** $1 \pm 2i$ **27.** $4 \pm i\sqrt{2}$ **29.** $\frac{1 \pm i}{3}$ **31.** $\frac{2 \pm i\sqrt{11}}{3}$ **33.** $\pm 2\sqrt{5}$ **35.** $\frac{-15 \pm \sqrt{30}}{5}$ **37.** -6 **39.** $-\frac{5}{4}, 2$ **41.** $\frac{9 \pm \sqrt{89}}{4}$ **43.** $3 \pm 3\sqrt{2}$ **45.** $\pm\frac{7}{5}$ **47.** $\frac{-1 \pm \sqrt{3}}{2}$ **49.** $\pm\frac{5}{2}i$ **51.** $\frac{5 \pm i\sqrt{23}}{4}$ **53.** $6 \pm 4i\sqrt{3}$ **55.** $\frac{7 \pm i\sqrt{7}}{2}$ **57.** 2 real solutions **59.** 2 imaginary solutions **61.** 1 real solution

63. a. 0 **b.** 1 **c.** 2 **d.** Answers may vary. **65.** (1, 2), (5, 2); (3, −2);

67. a. yes **b.** 96% **c.** 2008 **d.** Answers may vary. **69. a.** yes **b.** 70% **c.** 2009

71. a. 16 feet **b.** 0.62 second, 2.63 seconds **c.** 3.33 seconds **73.** $-2, 1.3$ **75.** ± 2.4 **77.** $(-3.4, -3.4), (2.0, -1.2)$ **79.** $\left(\frac{1 - \sqrt{57}}{4}, 0\right), \left(\frac{1 + \sqrt{57}}{4}, 0\right)$ **81.** no x-intercepts **83.** $(-1 - \sqrt{6}, 0), (-1 + \sqrt{6}, 0)$ **85.** The equation is not in standard form; $\frac{-5 \pm \sqrt{33}}{4}$ **87. a.** $x = -\frac{b}{m}$ **b.** -3 **89.** The results are the same by all three methods: $-4, 5$ **91.** Answers may vary. **93.** $x^2 - 3x - 10$ **95.** $\frac{3 \pm \sqrt{61}}{2}$ **97.** 1, 3 **99.** $-4x^2 + 16x - 13$ **101.** 2.1529 **103.** $\frac{2}{3}$ **105.** 3.3007 **107.** $(4x - 3)(2x - 3)$; a quadratic polynomial in one variable **109.** 77; a quadratic function **111.** $\frac{3}{4}, \frac{3}{2}$; a quadratic equation in one variable

Homework 7.6 **1.** $(1, -3, 2)$ **3.** $(-2, 15, 14)$ **5.** $(2, 4, -1)$ **7.** $(3, -3, 0)$ **9.** $(-2, 3, -1)$ **11.** $(2, 0, 1)$ **13.** $(3, 4, -1)$ **15.** $(-2, 4, 6)$ **17.** $y = x^2 + 2x + 3$ **19.** $y = -3x^2 + 7x + 5$ **21.** $y = 2x^2 - x - 4$ **23.** $y = x^2 + 2x - 6$ **25.** $y = -2x^2 + 7x + 4$ **27.** $y = 2x^2 - 8x + 3$ **29.** $y = -3x^2 + 8x + 4$ **31.** $y = 3x^2 + x - 1$ **33.** $y = x^2$ **35.** $y = 2x^2 - 6x + 4$ **37.** $y = x^2 - 9x + 22$ **39.** $y = -2x^2 + 20x - 42$ **41. a.** **b–d.** Answers may vary. **43.** Answers may vary.

45. $L(x) = 18x - 41$; $E(x) = 3.18(1.60)^x$ **47.** $L(x) = 2x + 2$; $E(x) = 2(2)^x$; for $Q(x)$, answers may vary. **49.** $f(x) = x$; linear **51.** $y = 2(x - 5)^2 - 7$, or $y = 2x^2 - 20x + 43$ **53.** $x = \frac{5 \pm \sqrt{11}}{2}$; a quadratic equation in one variable **55.** -5; a quadratic function **57.** a quadratic function

Homework 7.7 **1. a.** quadratic **b.** linear **c.** exponential **d.** none **3.** A scattergram of the data does not suggest a quadratic relationship. **5.** by hand, using points (14, 8.4), (15, 13.6), and (16, 20.1): $f(t) = 0.65t^2 - 13.65t + 72.10$; by regression: $f(t) = 0.52t^2 - 9.95t + 45.79$;

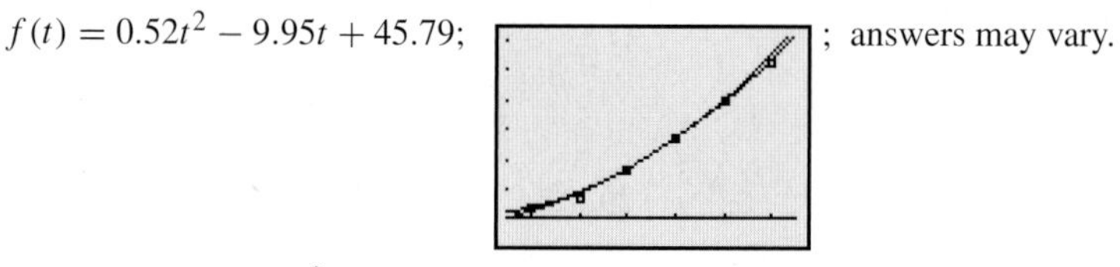

; answers may vary.

7. a. $f(t) = -0.58t^2 + 13.75t - 3.17$ **b.** 1987, 1990, and 2002; yes

9. a. $f(t) = 0.0068t^2 - 0.13t + 6.55$ **b. i.** quadratic function **ii.** linear function

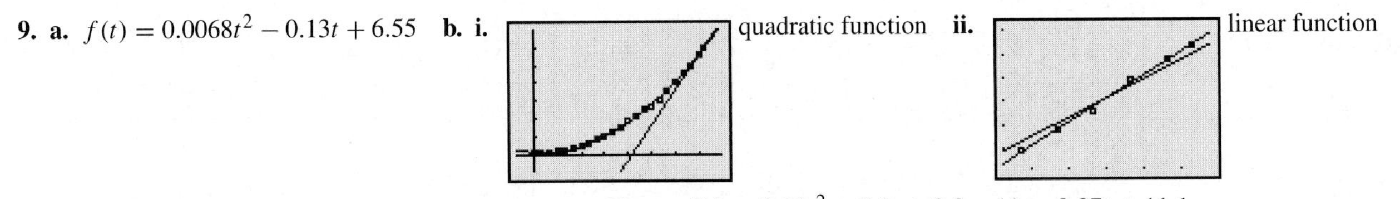

iii. Answers may vary. **iv.** linear; answers may vary. **11. a.** $f(t) = 0.18t^2 + 0.2t + 3.9$; $g(t) = 0.87t + 11.1$ **b.** $(-4.73, 6.98)$, $(8.45, 18.45)$; in 1995, 7.0 million cases of both brands were sold; in 2008, 18.5 million cases of both brands will be sold. **13.** Answers may vary. **15. a.** linear: $f(t) = 5.08t + 11.42$; exponential: $f(t) = 17.26(1.13)^t$; quadratic: $f(t) = 0.18t^2 + 2.71t + 15.85$; all appear to fit the data well. **b.** exponential model **17.** $(12, 6)$; a linear system in two variables

19.

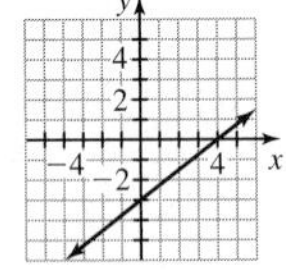

a linear equation in two variables **21.** $y = -\frac{5}{3}x - \frac{31}{3}$; a linear equation in two variables

Homework 7.8

1. a. $(0, 2.65)$; in 1980, the price of gasoline was \$2.65 per gallon (in 2006 dollars per gallon). **b.** \$2.02 per gallon; interpolation. **c.** \$3.40 per gallon; extrapolation **d.** 2015 **3. a.** $(0.24, 0)$, $(22.37, 0)$; in 1985 and 2007, no firms performed drug tests. Model breakdown has likely occurred. **b.** $t < 0.24$ or $t > 22.37$; years before 1985 and years after 2007. **c.** 78%; 1996 **d.** 1990, 2003 **5. a.** 307.07; in 2010, the U.S. population will be 307 million people. **b.** -203.64, 222.75; In 1586, the U.S. population was 315 million (model breakdown has occurred); in 2013, the U.S. population will be 315 million people.

c.

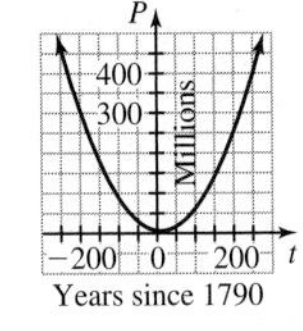

d. $t < 9.56$; years before 1800 **e.**

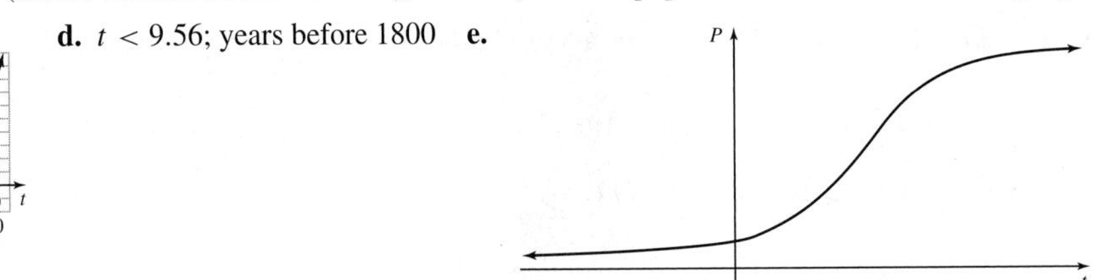

7. a. $f(t) = 0.88t^2 + 11.63t - 85.63$ **b.** 2005 **c.** 755.12, 811.63; 56.51; yes **d.** Answers may vary. **9.** 2008 **11. a.** $f(t) = -0.13t^2 - 1.44t + 71.88$; $g(t) = 3.68t + 10.63$ **b.** 2000; yes **c.** $h(t) = 100 - (f + g)(t)$ **d.** $h(t) = 0.13t^2 - 2.24t + 17.49$ **e.** 8.09; in 2000, 8.1% of registered voters used voting methods other than punch cards, lever machines, or optical scan or other modern electronic systems. **13.** 25 people **15.** 70 people **17.** width: 40 feet, length: 40 feet; 1600 square feet **19.** Answers may vary. **21. a.** $L(t) = 3.04t - 6.16$; $E(t) = 1.58(1.27)^t$; $Q(t) = 0.35t^2 - 1.88t + 3.68$; both the exponential model and the quadratic model fit the data well. The linear model does not. **b.** exponential **c.** 2025 **d.** 2129 **e.** Answers may vary. **23.** Answers may vary. **25.** Answers may vary. **27.** Answers may vary. **29.** Answers may vary.

Chapter 7 Review

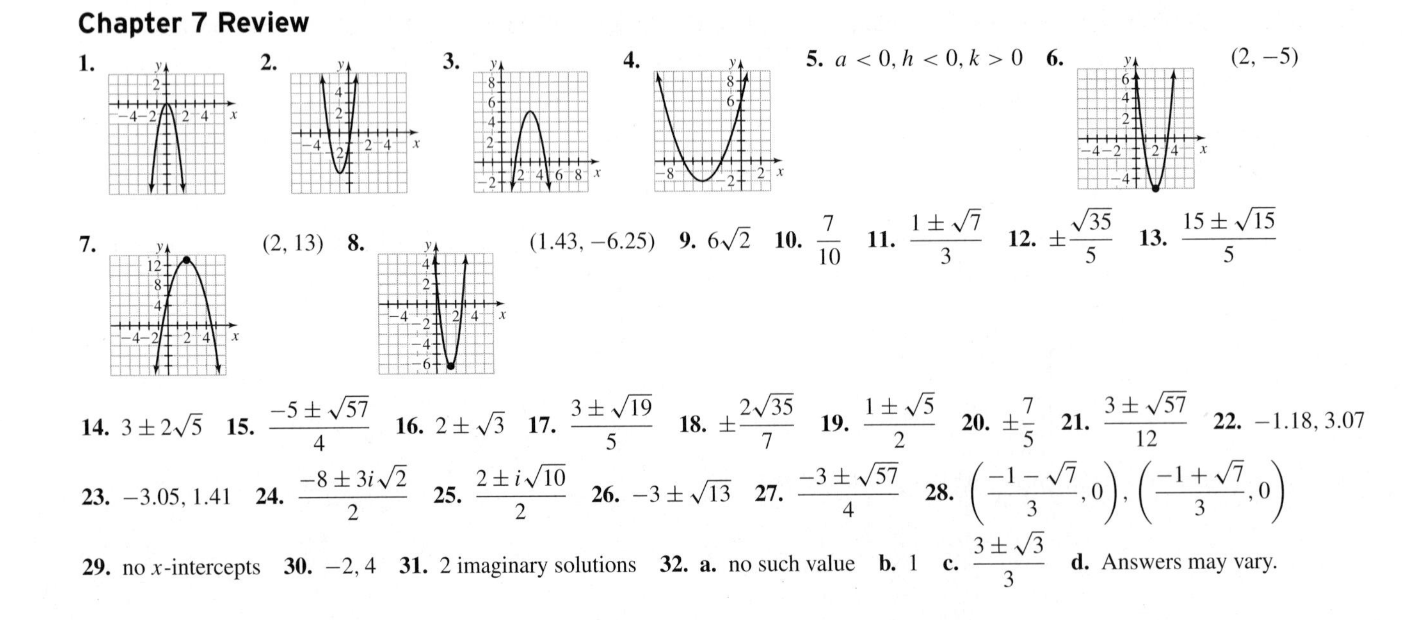

1. **2.** **3.** **4.** **5.** $a < 0, h < 0, k > 0$ **6.** $(2, -5)$

7. $(2, 13)$ **8.** $(1.43, -6.25)$ **9.** $6\sqrt{2}$ **10.** $\frac{7}{10}$ **11.** $\frac{1 \pm \sqrt{7}}{3}$ **12.** $\pm\frac{\sqrt{35}}{5}$ **13.** $\frac{15 \pm \sqrt{15}}{5}$

14. $3 \pm 2\sqrt{5}$ **15.** $\frac{-5 \pm \sqrt{57}}{4}$ **16.** $2 \pm \sqrt{3}$ **17.** $\frac{3 \pm \sqrt{19}}{5}$ **18.** $\pm\frac{2\sqrt{35}}{7}$ **19.** $\frac{1 \pm \sqrt{5}}{2}$ **20.** $\pm\frac{7}{5}$ **21.** $\frac{3 \pm \sqrt{57}}{12}$ **22.** $-1.18, 3.07$ **23.** $-3.05, 1.41$ **24.** $\frac{-8 \pm 3i\sqrt{2}}{2}$ **25.** $\frac{2 \pm i\sqrt{10}}{2}$ **26.** $-3 \pm \sqrt{13}$ **27.** $\frac{-3 \pm \sqrt{57}}{4}$ **28.** $\left(\frac{-1 - \sqrt{7}}{3}, 0\right), \left(\frac{-1 + \sqrt{7}}{3}, 0\right)$

29. no x-intercepts **30.** $-2, 4$ **31.** 2 imaginary solutions **32. a.** no such value **b.** 1 **c.** $\frac{3 \pm \sqrt{3}}{3}$ **d.** Answers may vary.

33. $-0.6, 1$ **34.** $-3, 5$ **35.** $(-2.0, -1.3), (5.4, -3.8)$ **36.** $(1, 2, 3)$ **37.** $(1, -3, 2)$ **38.** $y = 2x^2 - x + 3$
39. $y = -2x^2 + 3x + 5$ **40.** linear: $y = -2x + 4$; exponential: $y = 4\left(\frac{1}{2}\right)^x$; quadratic: answers may vary. **41.** $y = x^2 - 5x + 7$
42. a. 159.25 feet; 3.125 seconds **b.** 6.25 seconds **c.**

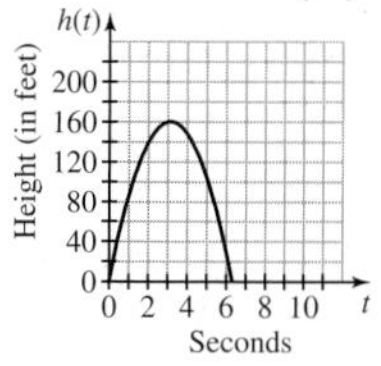

43. width: 45 feet, length: 90 feet; 4050 square feet **44. a.** $f(t) = -1.14t^2 + 25.26t - 81.20$ **b.** 41.2; in 2005, 41% of TV shows were at least 1 hour long. **c.** 2004 **d.** (3.90, 0) and (18.26, 0); in 1994, no TV shows were at least 1 hour long (model breakdown has occurred); in 2008, no TV shows will be at least 1 hour long (model breakdown has likely occurred). **e.** years before 1994 and after 2008
45. a. $f(t) = -0.084t^2 - 1.29t + 70.56$; $g(t) = 0.13t^2 - 0.52t + 2.26$ **b.** 2006

Chapter 7 Test

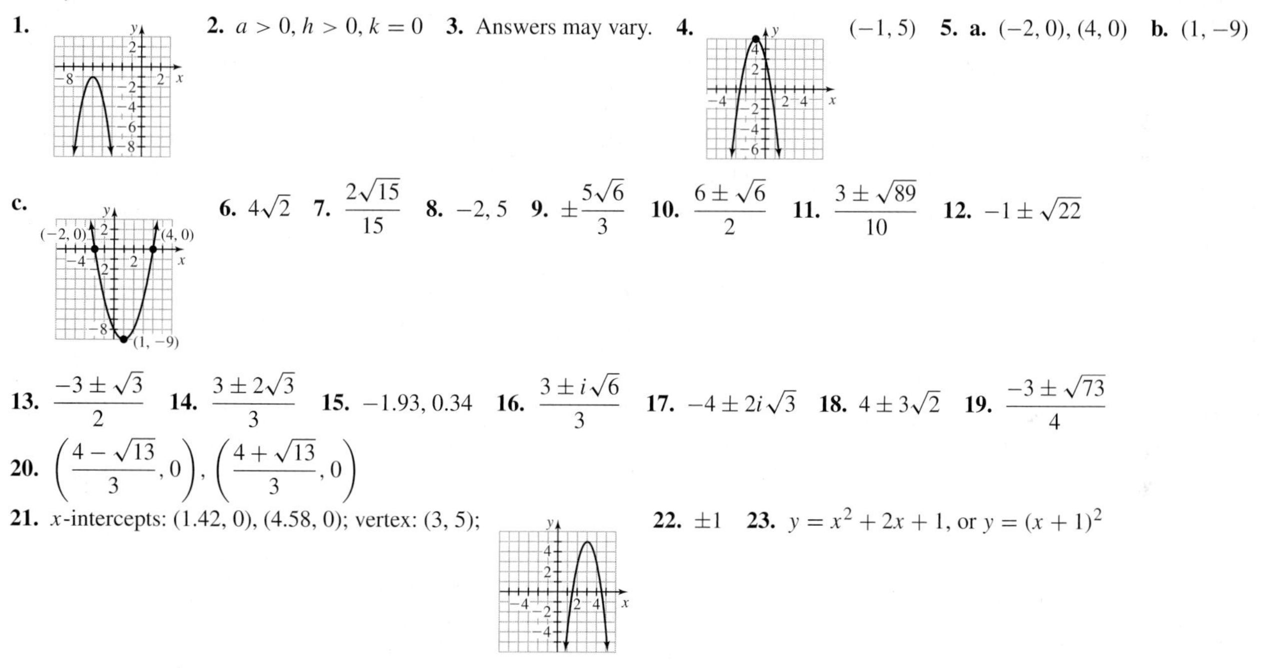

1.

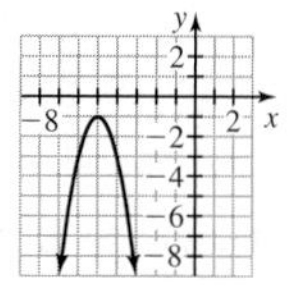

2. $a > 0, h > 0, k = 0$ **3.** Answers may vary. **4.**

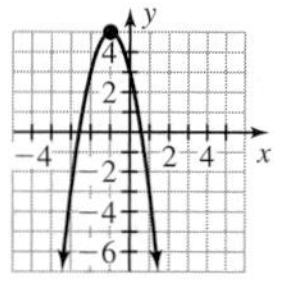

$(-1, 5)$ **5. a.** $(-2, 0), (4, 0)$ **b.** $(1, -9)$

c.

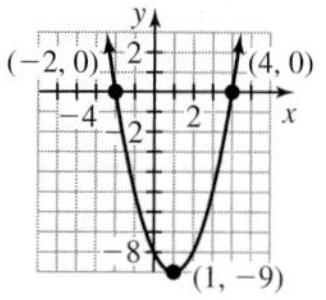

6. $4\sqrt{2}$ **7.** $\frac{2\sqrt{15}}{15}$ **8.** $-2, 5$ **9.** $\pm\frac{5\sqrt{6}}{3}$ **10.** $\frac{6 \pm \sqrt{6}}{2}$ **11.** $\frac{3 \pm \sqrt{89}}{10}$ **12.** $-1 \pm \sqrt{22}$

13. $\frac{-3 \pm \sqrt{3}}{2}$ **14.** $\frac{3 \pm 2\sqrt{3}}{3}$ **15.** $-1.93, 0.34$ **16.** $\frac{3 \pm i\sqrt{6}}{3}$ **17.** $-4 \pm 2i\sqrt{3}$ **18.** $4 \pm 3\sqrt{2}$ **19.** $\frac{-3 \pm \sqrt{73}}{4}$
20. $\left(\frac{4 - \sqrt{13}}{3}, 0\right), \left(\frac{4 + \sqrt{13}}{3}, 0\right)$
21. x-intercepts: (1.42, 0), (4.58, 0); vertex: (3, 5);

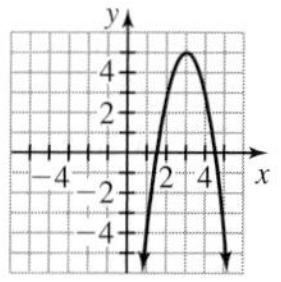

22. ± 1 **23.** $y = x^2 + 2x + 1$, or $y = (x + 1)^2$

24. $y = 2(x - 5)^2 + 3$, or $y = 2x^2 - 20x + 53$ **25. a.** 0 **b.** 1 **c.** 2 **26.** $(4, -8, 10)$ **27.** $(-1, -2, 4)$ **28.** 2.5 seconds; 103 feet
29. a. $f(t) = -0.028t^2 + 2.54t - 15.92$ **b.** 35.08; 35% of 30-year-old Americans feel that they are taking a great risk by entering personal information in a pop-up ad. **c.** 24.93, 65.78; 30% of 25-year-old and 30% of 66-year-old Americans feel that they are taking a great risk. **d.** (6.77, 0), (83.94, 0); no 7-year-old or 84-year-old Americans feel that they are taking a great risk. Model breakdown has likely occurred. **e.** (45.36, 41.68); the age at which the maximum percentage, 42%, of Americans feel that they are taking a great risk is 45 years. **30.** 80 people

Cumulative Review Chapters 1–7

1. $\pm\frac{7}{9}$ **2.** 13 **3.** $\frac{1 \pm \sqrt{21}}{5}$ **4.** $-\frac{5}{2}, \frac{4}{3}$ **5.** $-1, 3$ **6.** $-\frac{11}{2}$ **7.** $\frac{1}{7}, 1$ **8.** 2.1057 **9.** ± 1.5807 **10.** 2.4962 **11.** 2.8394
12. 2.8904 **13.** 1.8661 **14.** $\frac{3 \pm i}{2}$ **15.** $\frac{-3 \pm \sqrt{57}}{4}$ **16.** $(2, 5)$ **17.** $(-3, -4)$ **18.** $(-2, 1, 2)$
19. $x < -\frac{1}{12}$; $\left(-\infty, -\frac{1}{12}\right)$;

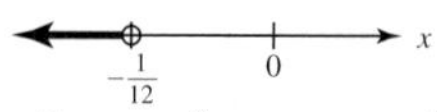

20. $\frac{648b^8}{c^{23}}$ **21.** $\frac{27b^{36}}{64c^6}$ **22.** $\log_b\left(\frac{64}{x^9}\right)$ **23.** $\ln\left(x^{31}\right)$
24. $9x^2 - 24xy + 16y^2$ **25.** $25p^2 - 49q^2$ **26.** $-3x^6 + 15x^4 - 24x^3 + 120x$ **27.** $x^4 + x^3 - 21x^2 - x + 20$
28. $f(x) = -2x^2 + 20x - 47$ **29.** $(m^2 + 4n^2)(m + 2n)(m - 2n)$ **30.** $x(x - 5)(x - 8)$ **31.** $(4p - 3q)(2p + 7q)$
32. $(x + 4)(x + 3)(x - 3)$ **33.** $f(x) = -3x + 20$ **34.** $g(x) = \frac{4}{3}(3)^x$ **35.** 4 **36.** 1 **37.** 4

38. **39.** **40.** **41.** **42.** **43.**

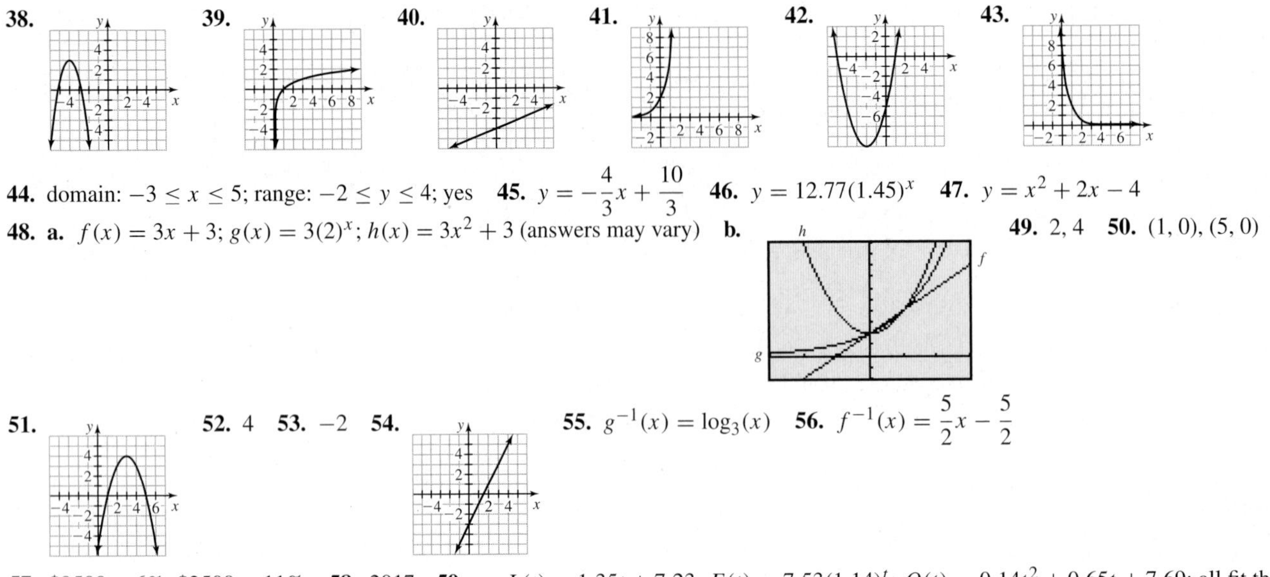

44. domain: $-3 \le x \le 5$; range: $-2 \le y \le 4$; yes **45.** $y = -\frac{4}{3}x + \frac{10}{3}$ **46.** $y = 12.77(1.45)^x$ **47.** $y = x^2 + 2x - 4$
48. a. $f(x) = 3x + 3$; $g(x) = 3(2)^x$; $h(x) = 3x^2 + 3$ (answers may vary) **b.** **49.** 2, 4 **50.** (1, 0), (5, 0)

51. **52.** 4 **53.** -2 **54.** **55.** $g^{-1}(x) = \log_3(x)$ **56.** $f^{-1}(x) = \frac{5}{2}x - \frac{5}{2}$

57. \$9500 at 6%, \$2500 at 11% **58.** 2017 **59. a.** $L(t) = 1.35t + 7.23$; $E(t) = 7.53(1.14)^t$; $Q(t) = 0.14t^2 + 0.65t + 7.69$; all fit the data well. **b.** $(-2.32, 6.94)$; the minimum sales at coffeehouses and doughnut shops, \$6.9 billion, occurred in 1998. **c.** exponential **d.** 14% per year **e.** 2017 **f.** 2019 **g.** Answers may vary. **60. a.** $f(t) = -0.8t + 33$ **b.** 9; in 2010, 9% of union members will work in manufacturing. **c.** 32.5; in 2013, 7% of union members will work in manufacturing. **d.** $f^{-1}(p) = -1.25p + 41.25$ **e.** -83.75; in 1896, 100% of union members worked in manufacturing. Model breakdown has likely occurred. **f.** (41.25, 0); in 2021, no union members will be working in manufacturing. Model breakdown has likely occurred. **g.** $t < -83.75$, $t > 41.25$
61. a. $h(t) = 2.4t + 10.7$, $c(t) = t + 53$ **b.** Hartford: 2.4 percentage points per year; Connecticut: 1 percentage point per year **c.** $h(18) = 53.9$, $c(18) = 71$; in 2008, 53.9% of Hartford students and 71% of Connecticut students will score above goals. **d.** 2020

Chapter 8

Homework 8.1 **1.** $0, -\frac{3}{5}$, undefined **3.** $\frac{9}{2}, 8, \frac{19}{26}$ **5.** all real numbers except 0 **7.** all real numbers **9.** all real numbers except -3 **11.** all real numbers except $-\frac{1}{2}$ **13.** all real numbers except -2 and 5 **15.** all real numbers except $\pm\frac{5}{2}$ **17.** all real numbers **19.** all real numbers except $-\frac{3}{2}$ and 5 **21.** all real numbers **23.** all real numbers except $\frac{1 \pm \sqrt{22}}{3}$ **25.** all real numbers except $\pm\frac{3}{2}$ and 2 **27.** $\frac{4x^3}{3}$ **29.** $\frac{4}{5}$ **31.** $\frac{x+5}{x-9}$ **33.** $\frac{x+7}{x-7}$ **35.** $\frac{4x+5}{2x-3}$ **37.** -1 **39.** $-\frac{2}{3}$ **41.** $-\frac{6}{x+3}$ **43.** $-\frac{x+7}{x+2}$ **45.** $\frac{3x(x+4)}{x-3}$ **47.** $\frac{x-4}{(2x+3)(2x-3)}$ **49.** $\frac{x^2-2x+4}{x-2}$ **51.** $\frac{x+3}{9x^2+6x+4}$ **53.** $\frac{x-3y}{x}$ **55.** $\frac{2a-b}{a-3b}$ **57.** $\frac{p^2+pq+q^2}{p+q}$ **59.** $\frac{f}{g}(x) = \frac{x+4}{x-6}; -\frac{7}{3}$ **61.** $\frac{h}{f}(x) = \frac{3x+5}{x-2}; \frac{17}{2}$ **63.** $\frac{f}{h}(x) = \frac{x^2}{3x+1}; -\frac{4}{5}$ **65.** $\frac{k}{g}(x) = \frac{9x^2-3x+1}{2x(3x+1)}; \frac{13}{4}$ **67. a.** $P(t) = \frac{100F(t)}{U(t)} = \frac{46.5t^2 - 950t + 6600}{0.0068t^2 + 2.58t + 251.7}$; answers may vary. **b.** 8.3%; 8.1%; overestimate **c.** 13.1; in 2007, 13.1% of Americans participated in the Food Stamp Program. **69. a.** $A(t) = 0.058t^2 - 0.82t + 3.60$ **b.** $P(t) = \frac{4.7t^2 - 77t + 350}{0.058t^2 - 0.82t + 3.60}$ **c.** 72.8; in 2010, 73% of cumulative awarded miles will be unredeemed. **d.** (9.3, 40.8); the smallest percentage of unredeemed cumulative miles, 41%, occurred in 1989. **e.** increasing **71.** Answers may vary. **73.** domain: $-3, -2, -1, 0, 1, 2, 3$; range: 50.4, 25.2, 16.8, 12.6, 10.08, 8.4, 7.2 **75.** no; answers may vary. **77.** $\frac{17}{14}, \frac{5}{2}$; the work is incorrect; answers may vary. **79.** Answers may vary. **81.** all real numbers **83.** all real numbers **85.** all real numbers except -3 and 7; a rational function **87.** $-\frac{5}{3}, -2, 2$; a cubic equation in one variable **89.** $(2x-5)\left(4x^2+10x+25\right)$; a cubic expression in one variable

Homework 8.2 **1.** $\frac{10}{x^2}$ **3.** $\frac{21x^2}{5}$ **5.** $\frac{3}{8p^5}$ **7.** $\frac{6}{25x^2}$ **9.** $\frac{a^2b^3}{6}$ **11.** $\frac{8x^3}{5y^4}$ **13.** $\frac{2(r+7)}{r-3}$ **15.** $\frac{2(x+1)}{x-3}$ **17.** $-\frac{8}{3}$ **19.** $\frac{2(k-4)(k-1)}{k+6}$ **21.** $\frac{a(a+2b)}{3b}$ **23.** $\frac{x-5}{3(x+5)(x+1)}$ **25.** $-\frac{t-4}{(t+1)(t+4)}$ **27.** $\frac{2(x-5)(2x+3)}{(x+4)(x+2)}$ **29.** $\frac{8}{(x-6)(3x-4)}$

31. $\frac{3x+4}{(x+2)(x+3)}$ **33.** $-\frac{(3m+2)(m-3)}{(m+3)(2m+7)}$ **35.** $\frac{(x+4)(x-1)}{(x+3)(x+6)}$ **37.** $\frac{(p+6t)(p+3t)}{(p-3t)(p-5t)}$ **39.** $\frac{(x+y)(x-3y)}{y(2x+3y)}$ **41.** $\frac{3(x-3)(x-8)}{4x(x-5)(x+8)}$ **43.** $\frac{3(w+2)(2w+3)}{2w(3w-1)(w-5)}$ **45.** $\frac{x-1}{(x-2)(x+5)}$ **47.** $\frac{1}{x-1}$ **49.** $\frac{(k-3)\left(k^2+2k+4\right)}{(k+2)\left(k^2-3k+9\right)}$ **51.** $(2x-3)(x+2)$ **53.** $\frac{(a+2b)}{\left(a^2-ab+b^2\right)}$ **55.** $\frac{(x-8)^2}{(x-5)^2}$; 4 **57.** $\frac{(x+8)^2}{(x+2)^2}$; $\frac{25}{9}$ **59.** $-\frac{(x+1)^2}{(x-7)^2}$; $-\frac{25}{9}$ **61.** $\frac{25}{2x^6(x-12)}$ **63.** $-\frac{36}{k^4}$ **65.** 1 **67.** Answers may vary; $\frac{(x-2)(x-5)}{(x+8)^2}$ **69. a.** 1 **b.** $\frac{1}{x}$ **c.** 1 **d.** $\frac{1}{x}$ **e.** 1 if n is odd, $\frac{1}{x}$ if n is even **71. a.** 8.4; the drive takes 8.4 hours at a constant speed of 50 mph. **b.** 7.64, 7, 6.46, 6 **c.** decreasing; answers may vary. **73.** $(f \cdot g)(x) = 16^x$; $\left(\frac{f}{g}\right)(x) = 4^x$ **75.** $(f \cdot g)(x) = 36(12)^x$; $\left(\frac{f}{g}\right)(x) = 4(3)^x$ **77.** $\log_b\left(\frac{16x^2}{9}\right)$; a logarithmic expression in one variable **79.** 5.3723; a logarithmic equation in one variable **81.** 2.1893; an exponential equation in one variable

Homework 8.3

1. $\frac{7}{x}$ **3.** $\frac{1}{x-3}$ **5.** $\frac{2m}{m-1}$ **7.** $\frac{x+5}{x+7}$ **9.** $\frac{2-4x^4}{x^6}$ **11.** $\frac{25x^2+18}{60x^6}$ **13.** $\frac{21b^2-10a}{12a^2b^3}$ **15.** $\frac{7x-2}{(x-2)(x+1)}$ **17.** $\frac{2(x+9)}{(x-6)(x-1)(x+4)}$ **19.** $\frac{19t+87}{15(t-2)(t+3)}$ **21.** $\frac{8x+25}{x(x-5)(x+5)}$ **23.** $\frac{5x+1}{(x-4)(x-3)(x+3)}$ **25.** $\frac{3k-1}{k+1}$ **27.** $\frac{2x}{x+1}$ **29.** $\frac{12}{x-6}$ **31.** $\frac{1}{2(x+3)}$ **33.** $\frac{4c^2+13c-1}{(2c-7)(2c+7)}$ **35.** $\frac{3a+b}{(a+b)(a-b)}$ **37.** $\frac{x-2}{(x+2)(x+4)}$ **39.** $\frac{2x^2+2x+5}{(x-1)(x+2)}$ **41.** $-\frac{16}{(y-3)(y+5)}$ **43.** $\frac{x^2+4x+30}{(x-2)(x-5)(x+5)}$ **45.** $\frac{2x^2+5x-1}{(x-4)(x+1)(x+3)^2}$ **47.** $\frac{4}{c-2}$ **49.** $\frac{x^2-15x-21}{(2x+5)^2(3x+1)}$ **51.** $\frac{11x^2-x+3}{(x+2)^2(3x-1)}$ **53.** $\frac{3p^2+7pq-8q^2}{(p-6q)(p+3q)(p+4q)}$ **55.** $\frac{x^2+x+8}{6x(x-4)(x+2)}$ **57.** $\frac{5}{2(x+2)}$ **59.** $\frac{-t+16}{(t+1)(t+5)}$ **61.** $\frac{2x^2-25}{(x-4)(x-3)}$ **63.** $-\frac{7}{(x-4)(x-3)}$ **65.** $\frac{-x^2+6x-2}{3(x-4)(x+2)}$ **67.** First fraction should be multiplied by $\frac{x+2}{x+2}$, second fraction by $\frac{x+1}{x+1}$; $\frac{5x+7}{(x+1)(x+2)}$ **69.** Numerator should be $9x-(5x+1) = 9x-5x-1 = 4x-1$; $\frac{4x-1}{x-3}$ **71.** Answers may vary. **73.** $(f+g)(x) = 8x^2-11x-2$; $(f-g)(x) = 4x^2+3x+8$ **75.** $(f+g)(x) = -5^x$; $(f-g)(x) = (5)^{x+1}$ **77.** $\frac{2\left(2x^2+8x+5\right)}{(x+2)^2}$ **79.** $\frac{5(x+2)}{3(x+1)}$ **81.** [graph] an exponential function **83.** ± 1.9061; a fourth-degree polynomial equation in one variable **85.** $f(x) = 1718.87(0.38)^x$; an exponential function

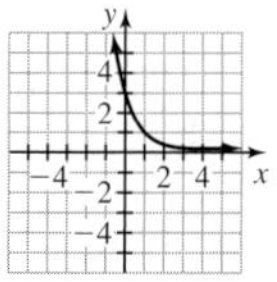

Homework 8.4

1. $\frac{2}{3}$ **3.** $\frac{x^3}{3}$ **5.** $2ab^2$ **7.** $\frac{1}{x+1}$ **9.** $\frac{7(x+7)}{3x(x+2)}$ **11.** $\frac{(5x+2)(3x-4)}{(3x+4)(5x-2)}$ **13.** $\frac{-3x^2+2}{4x+5}$ **15.** $-\frac{4x+3}{3x-2}$ **17.** $\frac{3\left(8x^3-5\right)}{18x^3-1}$ **19.** $a(a+b)$ **21.** $\frac{x-3}{x}$ **23.** $\frac{x-9}{x+6}$ **25.** $\frac{p^2-4p+2}{p^2-4p-3}$ **27.** $-\frac{1}{x(x+3)}$ **29.** $-\frac{3}{a(a+b)(a-b)}$ **31.** $-\frac{2(x+1)}{x^2(x+2)^2}$ **33.** $-\frac{(x+3)(x+4)(3x+10)}{x(x+2)}$ **35.** $\frac{(x+6)(x-5)}{(x+4)(x+3)}$ **37.** $h(x) = \frac{5(x-1)}{4(x-3)}$ **39.** $h(x) = \frac{x-2}{3}$ **41.** $h(x) = \frac{x^2+12x+25}{x^2-7x+25}$ **43.** The reciprocal of $\left(\frac{1}{x}+\frac{1}{2}\right)$ is not $\left(\frac{x}{1}+\frac{2}{1}\right)$; $\frac{2x^2}{x+2}$ **45.** $-\frac{2(5x-6)}{3x+4}$; answers may vary. **47.** Answers may vary. **49.** $\frac{4x^5}{3y^3}$ **51.** $-\frac{x+1}{x-1}$ **53.** $\frac{2b^3-1}{b\left(3b^2-4\right)}$ **55. a.** $\frac{H_pH_b}{H_p+H_b}$ **b.** 76.6 days **57.** $\frac{2\left(x^2+2x+10\right)}{(x-6)(x+4)(x-2)}$; a rational expression in one variable **59.** $\frac{1}{(x-6)^2}$; a rational expression in one variable **61.** all real numbers except -4 and 6; a rational function

Homework 8.5

1. 5 **3.** 2 **5.** empty set **7.** $-\frac{29}{2}$ **9.** 4 **11.** empty set **13.** -2 **15.** empty set **17.** -8 **19.** -1 **21.** $\frac{-1 \pm \sqrt{13}}{3}$ **23.** $-10, 3$ **25.** $\frac{-3 \pm \sqrt{17}}{2}$ **27.** -1 **29.** 1 **31.** empty set **33.** 1 **35.** $\frac{4 \pm \sqrt{22}}{2}$ **37.** $\frac{5 \pm \sqrt{89}}{2}$ **39.** $\frac{5 \pm i\sqrt{7}}{8}$ **41.** $\frac{17 \pm i\sqrt{11}}{6}$ **43.** $\frac{-3 \pm i\sqrt{47}}{4}$ **45.** $\frac{23}{4}$ **47.** $-4 \pm \sqrt{15}$ **49.** $\left(-\frac{7}{5}, 0\right)$ **51.** $r = \frac{mv^2}{F}$ **53.** $M = -\frac{r^2F}{mG}$

55. $t = \frac{A-P}{rP}$ **57.** 2014 **59. a.** $P(t) = \frac{-3.6t^2 + 17.6t + 128}{-0.036t^2 + 0.346t + 4.28}$ **b.** 8.09; in 2008, 8% of Mattel's total sales were Barbie sales. **c.** −2.97, 7.02; in 1997 and 2007, 15% of Mattel's total sales were Barbie sales. **d.** (1.38, 31.02); the maximum percentage of Barbie sales, 31%, occurred in 2001. **61.** 4.2 **63.** 3.3 **65.** (−0.6, −1.9) **67.** $a = -6, b = -3$ **69.** Answers may vary. **71.** Answers may vary. **73.** $\frac{2(3x+1)}{x(x+1)}$ **75.** $-\frac{1}{3}$ **77.** $-\frac{15}{2}$ **79.** $\frac{10x-1}{(x-3)(x+2)(x-2)}$ **81.** $-2, \frac{1}{2}, 2$ **83.** 2.8433 **85.** $\frac{23}{13}$

87.

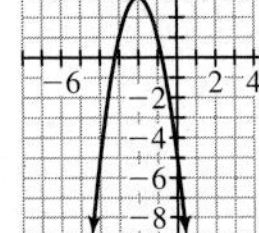

a quadratic function **89.** −5, 1; a quadratic equation in one variable

91. $-2x^2 - 8x - 5$; a quadratic expression in one variable

Homework 8.6

1. a. $C(n) = 350n + 1250$ **b.** $M(n) = \frac{350n + 1250}{n}$ **c.** \$391.67 **d.** 25 students **3. a.** $T(n) = 50n + 500$ **b.** $M(n) = \frac{50n + 500}{n}$ **c.** 51.85; if 270 people attend, the mean cost per person is \$51.85. **d.** 50; if 50 people attend, the mean cost per person is \$60. **e.** second column: 55, 52.5, 51.67, 51.25, 51 **f.** 50; answers may vary. **5. a.** $C(n) = 7000n + 90{,}000$ **b.** $B(n) = \frac{7000n + 90{,}000}{n}$ **c.** $P(n) = \frac{9000n + 90{,}000}{n}$ **d.** 11,250; if 40 cars are produced and sold each day, the price should be \$11,250 per car for the profit to be \$2000 per car. **e.** 9000; if very many cars are produced and sold, a price set a few cents more than \$9000 ensures a profit of \$2000 per car. **7. a.** $I(t) = 186.25t + 2467.35$ **b.** $M(t) = \frac{186.25t + 2467.35}{0.0016t + 0.09}$ **c.** The units of the expressions on both sides of the equation are dollars per household. **d.** 2009 **e.** increasing; between 1995 and 2010, the mean income per household is increasing. **9. a.** $B(t) = 288.6t - 1493$ **b.** $E(t) = 0.222t + 11.19$ **c.** $A(t) = \frac{B(t)}{E(t)} = \frac{288.6t - 1493}{0.222t + 11.19}$ **d.** \$423.21 **e.** 2010 **11. a.** $W(t) = 13.28t + 440.09$; $M(t) = 3.42t + 468.14$ **b.** $(W + M)(t) = 16.70t + 908.23$; at t years since 1970, $(W + M)$ gives the total number (in thousands) of people who have earned a bachelor's degree. **c.** $P(t) = \frac{342t + 46{,}814}{16.70t + 908.23}$ **d.** 2012 **e.** decreasing; the number of bachelor's degrees earned has been increasing at a faster rate for women than for men. **13. a.** $M(t) = 18.13t + 386.86$; $E(t) = -31.52t + 1382.53$ **b.** $(M + E)(t) = -13.39t + 1769.39$; at t years since 1980, $(M + E)$ gives the total number of daily newspapers. **c.** $P(t) = \frac{1813t + 38{,}686}{-13.39t + 1769.39}$ **d.** 68.05; in 2010, 68% of dailies will be morning newspapers. **e.** 33.37; in 2013, 75% of dailies will be morning newspapers. **15.** 1.42 hours **17. a.** $T(a) = \frac{253}{a+75} + \frac{410}{a+65}$ or $T(a) = \frac{663a + 47{,}195}{(a+65)(a+75)}$ **b.** 9.27 hours **c.** 5.16 mph **19. a.** $T(a) = \frac{83}{a+70} + \frac{37}{a+65}$ or $T(a) = \frac{120a + 7985}{(a+65)(a+70)}$ **b.** 1.75, 1.53; the trip will take 1.75 hours at the speed limit and 1.53 hours at 10 mph over the speed limit. **c.** 0.22; the trip will take 0.22 hour less at 10 mph over the speed limit than it would at the speed limit. **d.** −66.61, 6.61; the trip will take 1.6 hours at 66.6 mph below the speed limit (model breakdown) and 1.6 hours at 6.6 mph over the speed limit. **21. a.** $T(a) = \frac{285a + 19{,}130}{(a+70)(a+65)}$ **b.** 3.66 hours **23. a.** $L(t) = 0.53t - 12.22$; $E(t) = 0.43(1.06)^t$; $Q(t) = 0.01t^2 - 0.47t + 6.10$; both the exponential and the quadratic model fit the data well. The linear model does not. **b.** Answers may vary; well. **c.** 70 years **d.** 1.06; the percentage of Americans who have shingles increases by 6% per year. **e.** 29.8%

25. $(3x - 2)(5x + 2)(5x - 2)$; a cubic polynomial in one variable **27.** $-\frac{2}{5}, \frac{2}{5}, \frac{2}{3}$; a cubic equation in one variable

29. $3x^4 + 5x^3 - 15x^2 - 5x + 12$; a fourth-degree polynomial in one variable

Homework 8.7

1. $I = kt$ **3.** $w = \frac{k}{x+4}$ **5.** w varies inversely as r. **7.** T varies directly as the square root of w. **9.** $c = 4u$ **11.** $w = \frac{12}{\sqrt{t}}$ **13.** 27 **15.** 6 **17.** ±4 **19.** $\frac{31}{7}$ **21.** It increases. **23.** It decreases. **25.** It increases. **27.** less **29.** Too many cooks spoil the broth. **31.** \$1116 **33.** 96 newtons **35.** 186.1 feet **37.** 3.5 meters **39. a.** $F = \frac{5}{12}w$ **b.** 62.5 pounds **c.** The constant should increase, because more force would be needed to move the sofa. **41. a.** $T = 0.000906d$ **b.** 4415 feet **c.** The time it takes to hear thunder increases by 0.000906 second for each 1 foot from the lightning strike. **d.** no; the number of seconds divided by 5 is approximately the number of miles to the strike. **43. a.** $f(d) = \frac{3200}{d^2}$ **b.** 128 pounds **c.** 56,569 miles **d.** 0.056 pound; yes; the astronaut would weigh much more than 0.056 pound due to the Moon's gravitational field. **e.** Answers may vary.

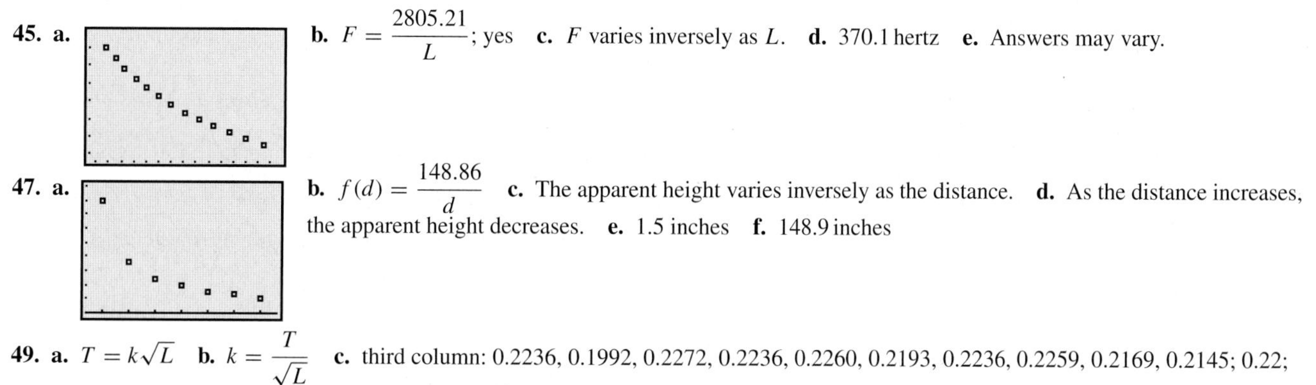

45. a. 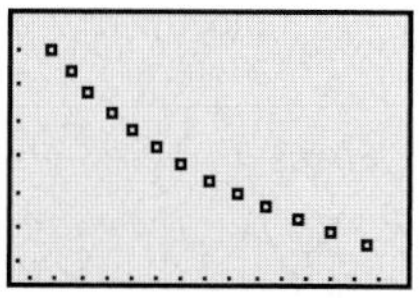 **b.** $F = \frac{2805.21}{L}$; yes **c.** F varies inversely as L. **d.** 370.1 hertz **e.** Answers may vary.

47. a. 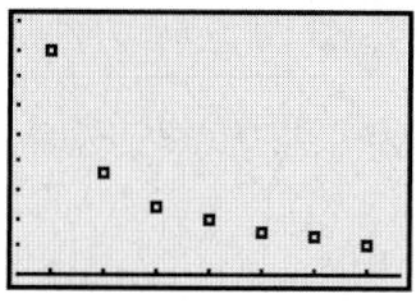 **b.** $f(d) = \frac{148.86}{d}$ **c.** The apparent height varies inversely as the distance. **d.** As the distance increases, the apparent height decreases. **e.** 1.5 inches **f.** 148.9 inches

49. a. $T = k\sqrt{L}$ **b.** $k = \frac{T}{\sqrt{L}}$ **c.** third column: 0.2236, 0.1992, 0.2272, 0.2236, 0.2260, 0.2193, 0.2236, 0.2259, 0.2169, 0.2145; 0.22; average all values in third column. **d.** $T = 0.22\sqrt{L}$ **e.** 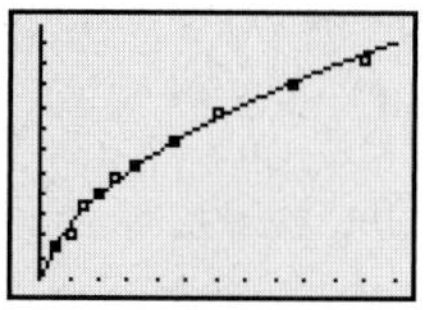 yes **f.** 2.51 seconds **51.** $f(L) = 5L$; 5

53. $f(n) = \frac{2}{n}$; 2 **55.** $f(r) = 2\pi r$; 2π **57. a.** yes **b.** no **c.** Answers may vary. **59.** false **61.** false **63.** yes; $\frac{1}{k}$ **65. a.** yes **b.** not necessarily; answers may vary. **67.** 62; the typist can type 62 words per minute. **69.** Answers may vary. **71.** Answers may vary. **73.** Answers may vary. **75.** Answers may vary.

Chapter 8 Review

1. -3; $\frac{7}{3}$ **2.** all real numbers except $\pm\frac{7}{2}$ **3.** all real numbers except $-\frac{7}{3}$ and $\frac{5}{4}$ **4.** all real numbers except -2 and $\pm\frac{1}{3}$ **5.** $\frac{3}{x-2}$ **6.** $-\frac{x+4}{2x(x-4)}$ **7.** $\frac{1}{x^2-2x+4}$ **8.** $\frac{2a-5b}{a-b}$ **9.** $\frac{f}{g}(x) = \frac{x+7}{x(x+3)}$; $-\frac{5}{2}$ **10.** $\frac{5}{4}$ **11.** $-\frac{2x(x+7)^2}{5(x+3)}$ **12.** $\frac{p+5t}{t}$ **13.** $\frac{x-1}{4(x+1)}$ **14.** $\frac{4x}{x+4}$ **15.** $\frac{(x+3)(2x-3)}{x-3}$ **16.** $-\frac{2}{x-2}$ **17.** $\frac{(x+5)(x-2)}{x(x+1)(x-1)(2x-5)}$ **18.** $\frac{2(x^2+x+4)}{(x-2)^2(x+2)}$ **19.** $-\frac{1}{4(x+1)}$ **20.** $-\frac{(x-3)(3x+8)}{2(x+5)(x-5)(x-2)}$ **21.** $\frac{2(m^2+4mn+10n^2)}{(m+6n)(m+2n)(m-5n)}$ **22.** $\frac{x-3}{x(x-5)}$ **23.** $(f \cdot g)(x) = \frac{(x-2)(x+1)}{(x+2)^2}$ **24.** $\left(\frac{f}{g}\right)(x) = \frac{(x-2)(x+1)}{(x+3)^2}$ **25.** $(f+g)(x) = \frac{2x^2+5x+7}{(x+2)(x+3)}$ **26.** $(f-g)(x) = -\frac{7x+11}{(x+2)(x+3)}$ **27.** $\frac{1}{(x-3)(x+2)}$ **28.** $-\frac{4(x-2)}{3x^3}$ **29.** empty set **30.** $-8, 1$ **31.** $\frac{-5 \pm \sqrt{73}}{6}$ **32.** $-3 \pm 3\sqrt{3}$ **33.** $\frac{5 \pm i}{2}$ **34.** $\left(\frac{5}{3}, 0\right)$ **35.** $r = \frac{S-a}{S}$ **36.** H varies directly as u squared. **37.** w varies inversely as the logarithm of t. **38.** $y = \frac{2}{7}\sqrt{x}$ **39.** $B = \frac{72}{r^3}$ **40.** 4.14 inches **41. a.** $m = kr^3$ **b.** $k = \frac{m}{r^3}$ **c.** third column: 17.1, 17.01, 17.02, 16.99, 16.99, 16.99; 17.02; average all values in third column. **d.** $m = 17.02r^3$ **e.** 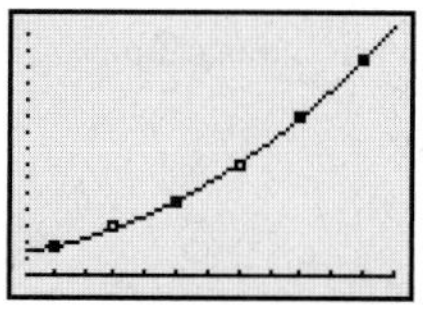 yes **f.** 207.1 grams

42. a. $C(n) = 40n + 600$ **b.** $M(n) = \frac{40n+600}{n}$ **c.** 42.22; when 270 people use the room, the mean cost per person is \$42.22. **d.** 60; when the mean cost per person is \$50, 60 people are using the room. **43. a.** $(v+d)(t) = -1.44t^2 + 50.6t - 250.2$; at t years since 1990, $(v+d)(t)$ gives the sales of videocassettes and DVDs (combined) to U.S. dealers. **b.** $P(t) = \frac{483t^2 - 7910t + 32{,}600}{-1.44t^2 + 50.6t - 250.2}$ **c.** 16.1%; 12.3% **d.** 1996 and 2004; model breakdown has occurred. **44. a.** $T(a) = \frac{75}{a+50} + \frac{40}{a+65}$ **b.** 1.94; when the student drives 5 mph above the speed limits, the driving time is 1.9 hours. **c.** 3.1 mph over the speed limits

Chapter 8 Test

1. all real numbers except $-\frac{5}{2}$ and $\frac{2}{3}$ **2.** all real numbers except ± 6 **3.** all real numbers **4.** Answers may vary. **5.** $f(x) = -\frac{3}{x-3}$ **6.** $f(x) = \frac{3x+1}{2x(3x-1)}$ **7.** $\frac{x-2}{9x^3}$ **8.** $\frac{p(p+2t)}{(p+3t)(p-t)}$ **9.** $\frac{-9x^2-4x+24}{2x(x-2)(x+4)}$ **10.** $\frac{x^2+13x+7}{(x+3)(x-3)(x+8)}$ **11.** $\frac{15(x+2)}{x(x^2-3x+5)}$

12. $(f-g)(x)=\dfrac{6(2x-1)}{(x-5)(x+4)}; \dfrac{3}{10}$ **13.** $\dfrac{(5x+2)(x-1)}{x(3x-7)}$ **14.** empty set **15.** 5 **16.** $2\pm\sqrt{6}$ **17.** 0 **18.** undefined

19. $-2, 5$ **20.** $W=\dfrac{3}{49}t^2$ **21.** $y=\dfrac{40}{\sqrt{x}}$ **22. a.** $C(n)=200n+10{,}000$ **b.** $B(n)=\dfrac{200n+10{,}000}{n}$ **c.** $P(n)=\dfrac{350n+10{,}000}{n}$

d. 450; if the manufacturer makes and sells 100 bikes in a month, the price should be \$450 per bike to make a profit of \$150 per bike.

23. a. $T(a)=\dfrac{400}{a+70}+\dfrac{920}{a+75}$ **b.** 16.83; when she drives 5 mph above the speed limits, the trip takes 16.8 hours. **c.** $-71.58, 4.23$; when the trip takes 17 hours, she is driving 71.58 mph below the speed limits (model breakdown) or 4.2 mph above the speed limits.

24. a. $g(L)=\dfrac{3200}{L^2}$ **b.** 88.89 hertz **c.** 4 cm **d.** decreasing; the longer the prongs, the lower the frequency.

Chapter 9

Homework 9.1

1. $\sqrt[5]{x^2}$ **3.** $x^{3/4}$ **5.** $w^{1/2}$ **7.** $\sqrt[7]{(2x+9)^3}$ **9.** $(3k+2)^{4/7}$ **11.** $5\sqrt{2}$ **13.** x^4 **15.** $6x^3$ **17.** $ab^6\sqrt{5}$ **19.** $x^4\sqrt{x}$ **21.** $2x^2\sqrt{6x}$ **23.** $4xy^4\sqrt{5x}$ **25.** $10ab^2\sqrt{2ab}$ **27.** $(2x+5)^4$ **29.** $(6t+3)^2\sqrt{6t+3}$ **31.** 3 **33.** x **35.** $2x$ **37.** $-2x^4$ **39.** $3a^3b^7$ **41.** $x^2\sqrt[6]{x^5}$ **43.** $-5a^5b^4\sqrt[3]{a^2}$ **45.** $2x^7y\sqrt[5]{2x^4y^2}$ **47.** $6xy$ **49.** $3x+6$ **51.** $(4p+7)^4$ **53.** $(2x+9)^5\sqrt[6]{2x+9}$ **55.** $\sqrt[4]{x^3}$ **57.** $\sqrt[3]{x^2}$ **59.** $\sqrt[6]{(2m+7)^5}$ **61.** $x^2\sqrt[3]{x}$ **63.** $\sqrt{3}$ **65.** $\sqrt[12]{p}$ **67.** $\sqrt[5]{4x^4}$ **69.** $\sqrt[8]{ab}$ **71.** -2 **73.** 2 **75.** $-\sqrt[3]{19}$ **77.** 9 **79.** **81. a.**

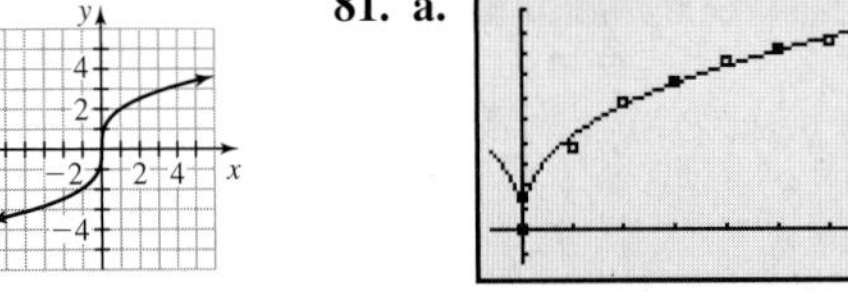

b. 30°F **c.** 129°F **d.** 18 minutes

83. a. 195 meters per second **b.** 99, 140, 171; increasing; the greater the depth, the greater the tsunami's speed **c.** The speed decreases, and the height increases. **d.** 436 miles per hour **85.** Answers may vary. **87.** $\sqrt[n^2]{x}$ **89. a.** x **b.** x **c.** x **91. a.** $4x^2y^3$ **b.** $4x^2y^3$ **c.** They are the same. **93.** $-4\pm\sqrt{26}$; a rational equation in one variable **95.** $\dfrac{-x^2+13x-2}{(x+3)^2(x-2)}$; a rational expression in one variable **97.** all real numbers except -3 and 2; a rational function

Homework 9.2

1. $9\sqrt{x}$ **3.** $-4\sqrt[3]{5x^2y}$ **5.** $10\sqrt{5a}-4\sqrt{3b}$ **7.** $-4-2\sqrt[3]{x}+2\sqrt{x}$ **9.** $3\sqrt[3]{x-1}-2\sqrt{x-1}$ **11.** $2.6\sqrt[4]{x}$ **13.** $25-4\sqrt{x}$ **15.** 14 **17.** $7\sqrt{3b}$ **19.** $13x^2\sqrt{2x}$ **21.** $4x\sqrt{x}$ **23.** $7x$ **25.** $7ab\sqrt{3b}$ **27.** $x\sqrt[3]{x^2}$ **29.** $-x^2y^2\sqrt[4]{x^3}$ **31.** $6x$ **33.** $-8x\sqrt{15}$ **35.** $14t-2t\sqrt{14}$ **37.** $10x+38\sqrt{x}+24$ **39.** $8x-4\sqrt{5x}+2\sqrt{3x}-\sqrt{15}$ **41.** $5a-9\sqrt{ab}-2b$ **43.** $1-w$ **45.** $49x^2-5$ **47.** $4a-b$ **49.** $36x+60\sqrt{x}+25$ **51.** $16x-8\sqrt{5x}+5$ **53.** $a+4\sqrt{ab}+4b$ **55.** $2x+6\sqrt{2x-5}+4$ **57.** $\sqrt[10]{x^7}$ **59.** $x\sqrt[5]{x^2}$ **61.** $-5\sqrt[4]{2m^3}+20\sqrt{m}$ **63.** $\sqrt[3]{x^2}+2\sqrt[3]{x}+1$ **65.** $\sqrt{k}-2\sqrt[12]{k^7}+\sqrt[3]{k^2}$ **67.** $6\sqrt[6]{x^5}+2\sqrt{x}-18\sqrt[3]{x}-6$ **69.** $9\sqrt{x}-25$ **71. a.** increases; answers may vary. **b. i.** 75, 300, 675, 1200, 1875, all in gallons per minute **ii.** 1875 gallons per minute; 18 gallons per minute **iii.** 75 gallons per minute; 1.35% **c.** yes; 616 gallons per minute **73.** The middle term is missing; $x^2+2x\sqrt{7}+7$ **75.** The radicand should not be multiplied by 7; $14\sqrt{3}$ **77.** $\sqrt[6]{x}$ **79. a.** $\sqrt[12]{x^7}$ **b.** $\sqrt[kn]{x^{k+n}}$ **c.** $\sqrt[12]{x^7}$; they are the same. **d.** $\sqrt[35]{x^{12}}$ **81.** Answers may vary. **83.** $-2\sqrt{x}$ **85.** $-15x$ **87.** $\log_b\left(\dfrac{x-5}{x-8}\right)$; a logarithmic expression in one variable **89.** $\dfrac{20}{13}$; a logarithmic equation in one variable **91.** 0.7211; an exponential equation in one variable

Homework 9.3

1. $\dfrac{8\sqrt{x}}{x}$ **3.** $\dfrac{3\sqrt{5p}}{5p}$ **5.** $\dfrac{2\sqrt{2x}}{3x}$ **7.** $\dfrac{5\sqrt{2k}}{2k}$ **9.** $\dfrac{2\sqrt{x}}{x}$ **11.** $\dfrac{\sqrt{14}}{2}$ **13.** $\dfrac{\sqrt{2xy}}{x}$ **15.** $\dfrac{\sqrt{3xy}}{6y}$ **17.** $\dfrac{3\sqrt{x-4}}{x-4}$ **19.** $\dfrac{a\sqrt{6ab}}{3b}$ **21.** $\dfrac{2\sqrt[3]{25}}{5}$ **23.** $\dfrac{5\sqrt[3]{2}}{2}$ **25.** $\dfrac{4\sqrt[3]{x^2}}{5x}$ **27.** $\dfrac{3\sqrt[3]{4x}}{x}$ **29.** $\dfrac{7\sqrt[4]{4t}}{2}$ **31.** $\dfrac{\sqrt[6]{x^5}}{x}$ **33.** $\dfrac{\sqrt[5]{2x^2}}{x}$ **35.** $\dfrac{\sqrt{6x}}{3x}$ **37.** $\dfrac{\sqrt[5]{24wxy^3}}{2xy}$ **39.** $\dfrac{5-\sqrt{3}}{22}$ **41.** $\dfrac{\sqrt{7}-\sqrt{3}}{2}$ **43.** $\dfrac{3\sqrt{r}+7}{9r-49}$ **45.** $\dfrac{x+\sqrt{x}}{x-1}$ **47.** $\dfrac{12x+3\sqrt{5x}}{16x-5}$ **49.** $\dfrac{x+y\sqrt{x}}{x-y^2}$ **51.** $\dfrac{x-10\sqrt{x}+25}{x-25}$ **53.** $\dfrac{6x+13\sqrt{x}-5}{9x-1}$ **55.** $\dfrac{18x+6\sqrt{7x}+3\sqrt{5x}+\sqrt{35}}{9x-7}$ **57.** $\dfrac{x-2\sqrt{xy}+y}{x-y}$ **59.** $\sqrt{x+1}+\sqrt{x}$ **61. a.** $\dfrac{\sqrt{6h}}{2}$ **b.** 47 miles **c.** 212 miles **63. a.** Answers may vary. **b.** $\dfrac{\sqrt[4]{8}}{2}\approx 0.84$ **65.** student 1; answers may vary. **67.** Answers may vary; $\dfrac{5\sqrt[3]{x^2}}{x}$ **69.** $\dfrac{x}{3\sqrt{x}}$ **71.** $\dfrac{1}{\sqrt{x+2}+\sqrt{x}}$ **73.** $\dfrac{2x-7\sqrt{x}+3}{4x-1}$ **75.** $3\sqrt{10}$ **77.** Answers may vary. **79. a.** $(A+B)(A^2-AB+B^2)$ **b.** A^3+B^3; answers may vary. **c.** x^3+8; answers may vary. **d.** $x+2$; answers may vary. **e.** $\dfrac{\sqrt[3]{x^2}-\sqrt[3]{2x}+\sqrt[3]{4}}{x+2}$

81. $15x^3 - 22x^2 + 3x + 4$; a cubic polynomial in one variable **83.** $24(x-5)(x^2+5x+25)$; a cubic polynomial in one variable **85.** $\frac{2 \pm \sqrt{14}}{5}$; a quadratic polynomial in one variable

Homework 9.4

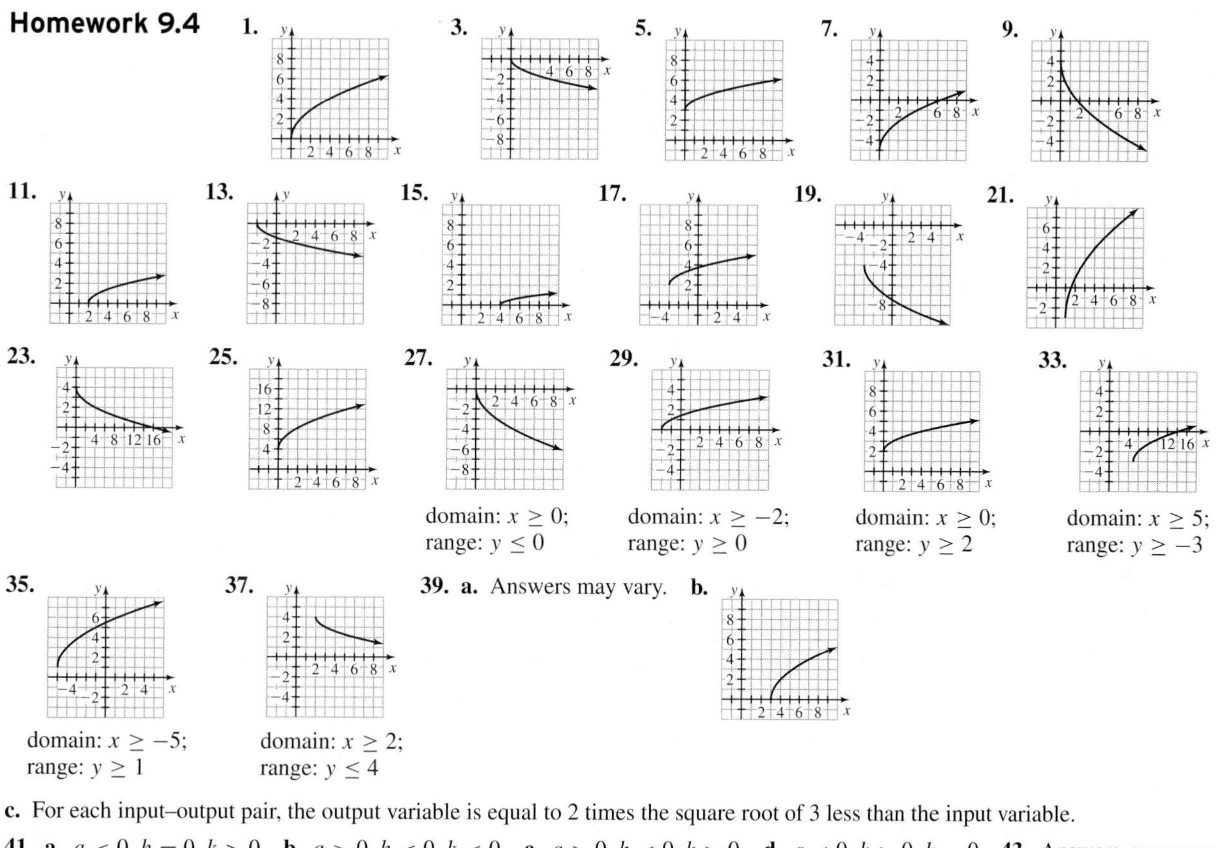

27. domain: $x \geq 0$; range: $y \leq 0$ **29.** domain: $x \geq -2$; range: $y \geq 0$ **31.** domain: $x \geq 0$; range: $y \geq 2$ **33.** domain: $x \geq 5$; range: $y \geq -3$

35. domain: $x \geq -5$; range: $y \geq 1$ **37.** domain: $x \geq 2$; range: $y \leq 4$ **39. a.** Answers may vary. **b.**

c. For each input–output pair, the output variable is equal to 2 times the square root of 3 less than the input variable.

41. a. $a < 0, h = 0, k > 0$ **b.** $a > 0, h < 0, k < 0$ **c.** $a > 0, h < 0, k > 0$ **d.** $a < 0, h > 0, k = 0$ **43.** Answers may vary.

45. If $a < 0$, f has a maximum point at (h, k). If $a > 0$, f has a minimum point at (h, k). **47.** 11 **49.** $21\sqrt{c} - 3$

51. $(f+g)(x) = 9\sqrt{x} - 8$ **53.** $(f \cdot g)(x) = 20x - 31\sqrt{x} - 9$ **55.** $(f-g)(x) = -6\sqrt{5}$ **57.** $\frac{f}{g}(x) = \frac{4x - 12\sqrt{5x} + 45}{4x - 45}$

59. $(f+g)(x) = 2\sqrt{x+1}$ **61.** $(f \cdot g)(x) = x - 3$

63. a.

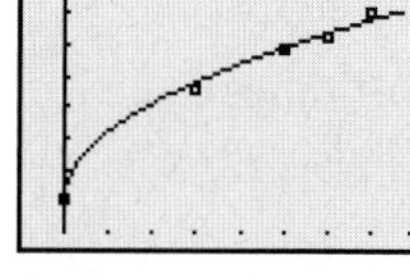

yes **b.** 64%; interpolation **c.** 92%; extrapolation **65.** 0 **67.** 2.4 **69.** −6 **71.** 3

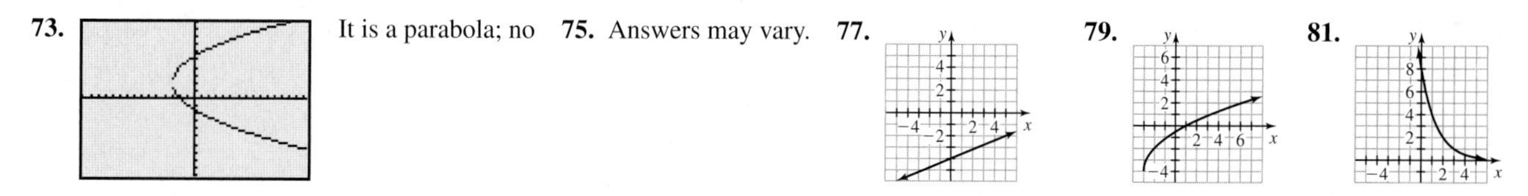

73. It is a parabola; no **75.** Answers may vary. **77.** **79.** **81.**

83. $(3x+2)(2x-3)$; a quadratic polynomial in one variable **85.** $\frac{1 \pm \sqrt{7}}{3}$; a quadratic function **87.** $y = 2x^2 - 3x + 5$; a quadratic function

Homework 9.5

1. 25 **3.** empty set **5.** −8 **7.** 4 **9.** 5 **11.** 14 **13.** empty set **15.** $\frac{69}{2}$ **17.** $\frac{97}{6}$ **19.** 5 **21.** 5 **23.** $-\frac{1}{6}$

25. 11 **27.** 3 **29.** empty set **31.** 8 **33.** 4 **35.** $3 + 2\sqrt{2}$ **37.** 4 **39.** 1 **41.** 121 **43.** $\frac{3 \pm 3\sqrt{5}}{2}$ **45.** 10.31 **47.** 2.31

49. 2.06 **51.** −0.74, 4.97 **53.** −1.6, 3.8 **55.** −4 **57.** (−3.2, 1.4) **59.** (−7, 0) **61.** (5, 0) **63.** no x-intercepts

65. 4 **67.** no such value **69. a.** 88.7; in 2009, 89% of e-mails will be spam. **b.** 11.96; in 2011, 95% of e-mails will be spam.

c. 2013; model breakdown has likely occurred.

71. a. 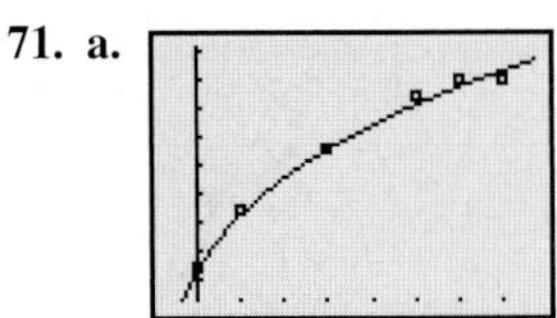 yes **b.** \$338; underestimate **c.** 4th grade **d.** The table seems to indicate that the function is increasing, so we would expect the per-student charge to be between \$365 and \$410.

73. Answers may vary; -2 **75.** $d = \frac{S^2}{g}$ **77.** $h = \frac{2d^2}{3}$ **79.** $R = \frac{2GM}{v^2}$ **81.** (4, 2) **83.** Answers may vary. **85.** $-4\sqrt{x} + 5$ **87.** 9 **89.** 0 **91.** $p + 4\sqrt{p} + 3$ **93.** 5.0553 **95.** 6 **97.** $\frac{5}{7}$ **99.** $\frac{33}{5}$ **101.** $\frac{(3x+5)(2x+3)}{3(x+1)(x-1)(x+2)}$; a rational expression in one variable **103.** $\frac{3(3b-10)}{(b-2)(b-5)}$; a rational expression in one variable **105.** 6; a rational equation in one variable

Homework 9.6

1. $y = \sqrt{x} + 3$ **3.** $y = 1.33\sqrt{x} + 2$ **5.** $y = 1.34\sqrt{x} + 4$ **7.** $y = -4.04\sqrt{x} + 9$ **9.** $y = \sqrt{x} + 1$ **11.** $y = 3.15\sqrt{x} - 0.45$ **13.** $y = -2.43\sqrt{x} + 9.44$ **15.** $y = 10.22\sqrt{x} - 15.86$ **17.** $y = -48.40\sqrt{x} + 159.06$ **19.** $y = 7.34\sqrt{x} - 25.43$ **21.** Increase b. **23. a.** $f(t) = 9.13\sqrt{t} + 12.09$ **b.** (0, 12.09); in 2002, *American Idol* was seen by an average of 12.1 million viewers per episode. **c.** 2010 **d.** 39.5 million viewers **25. a.** $f(t) = 2.13\sqrt{t} + 9.52$ **b.** (0, 9.52); in 2000, 9.5 million households had webcams. **c.** 16.58; in 2011, 16.6 million households will have webcams. **d.** 15.85; in 2016, 18 million households will have webcams. **27. a.** 420 meters per second **b.** no; answers may vary. **c.** between 4205 meters and 4500 meters, inclusive **d.** yes **29. a.** $S(h) = 0.27\sqrt{h}$ **b. i.** Answers may vary. **ii.** 0, 0.327, 0.165; S **iii.** Q; answers may vary. **iv.** S; answers may vary. **v.** $T = 0.25\sqrt{h}$; answers may vary. **c.** 123.46 feet **d.** 9.55 seconds **31. a.** $f(n) = 31.92\sqrt{n} + 9.16$ **b.** 93.61; 93.6% of 7th births occurred despite contraception. **c.** 8.10; all 8th births occurred despite contraception. Model breakdown has likely occurred. **d.** The higher the birth order, the higher the percentage of births that occurred despite contraception; answers may vary.

33. a. 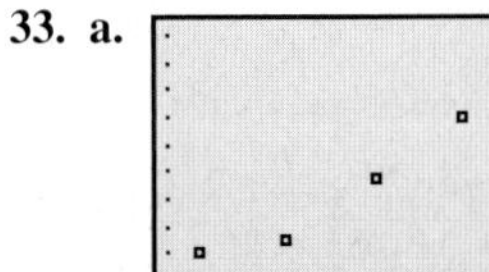 exponential and quadratic **b.** $E(t) = 1.10(1.37)^t$; $Q(t) = 2.56t^2 - 43.86t + 206.06$ **c.** Both fit the data well. **d.** exponential **e.** exponential: 2012; quadratic: 2020; answers may vary. **f.** exponential: Redflex will have a revenue of \$260 million in 2012; quadratic: Redflex will have a revenue of \$260 million in 2020.

35. Answers may vary. **37.** Answers may vary. **39.** Answers may vary. **41.** Answers may vary.

Chapter 9 Review

1. $\sqrt[7]{x^3}$ **2.** $(3k+4)^{7/5}$ **3.** $2x^3\sqrt{2}$ **4.** $3x^3y^5\sqrt{2x}$ **5.** $\sqrt[4]{x^3}$ **6.** $2x^3y^8\sqrt[3]{3x}$ **7.** $(6x+11)^5\sqrt[5]{(6x+11)^2}$ **8.** $11\sqrt{5x}$ **9.** $3ab\sqrt[3]{2a^2b}$ **10.** $28\sqrt{x} - 7\sqrt[3]{x}$ **11.** $3x - 21\sqrt{x}$ **12.** $8x - 2\sqrt{x} - 3$ **13.** $10a - 3\sqrt{ab} - b$ **14.** $25a - 49b$ **15.** $16x + 24\sqrt{x} + 9$ **16.** $4\sqrt[3]{x^2} - 20\sqrt[3]{x} + 25$ **17.** $\sqrt[20]{x^9}$ **18.** $\sqrt[12]{x}$ **19.** $\frac{\sqrt{3x}}{x}$ **20.** $5\sqrt[3]{t^2}$ **21.** $\frac{\sqrt[5]{63x^3y}}{3x}$ **22.** $\frac{a + 2\sqrt{ab}}{a - 4b}$ **23.** $\frac{10x - 23\sqrt{x} + 12}{4x - 9}$ **24.** **25.** **26.** $(f+g)(x) = -\sqrt{x} + 7$ **27.** $(f-g)(x) = 7\sqrt{x} + 3$ **28.** $(f \cdot g)(x) = -12x - 14\sqrt{x} + 10$ **29.** $\frac{6x + 13\sqrt{x} + 5}{2 - 8x}$ **30.** 4 **31.** 2, 4 **32.** 9 **33.** 2 **34.** 7 **35.** 2.52 **36.** -1.36, 4.56 **37.** (4, 0) **38.** Decrease a, increase b. **39.** $y = 2.5\sqrt{x} + 3$ **40.** $y = -5.95\sqrt{x} + 17.31$ **41. a.** $f(t) = 6.4\sqrt{t} + 16$ **b.** (0, 16); in 1998, 16 million households had Sony PlayStations. **c.** 39.95; in 2012, 40 million households will have Sony PlayStations. **d.** 11.82; in 2010, 38 million households will have Sony PlayStations.

Chapter 9 Test

1. $4x^4y^6\sqrt{2x}$ **2.** $4x^7y^4\sqrt[3]{xy^2}$ **3.** $(2x+8)^6\sqrt[4]{(2x+8)^3}$ **4.** $\dfrac{2\sqrt[15]{x^2}}{3}$ **5.** $\dfrac{2x+5\sqrt{x}+3}{4x-9}$ **6.** $-x\sqrt{3x}$ **7.** $18x-15\sqrt{x}$ **8.** $-20x+2\sqrt{x}+6$ **9.** $9a-25b$ **10.** $16\sqrt[5]{x^2}-24\sqrt[5]{x}+9$ **11.** Answers may vary.

12. **13. a.** $a<0$ and $k\geq 0$, or $a>0$ and $k\leq 0$ **b.** $\left(\dfrac{k^2+a^2h}{a^2}, 0\right)$ **14.** $(f+g)(x)=2\sqrt{x}+11$

15. $(f-g)(x)=-8\sqrt{x}+3$ **16.** $(f\cdot g)(x)=-15x+23\sqrt{x}+28$ **17.** $\left(\dfrac{f}{g}\right)(x)=\dfrac{15x-47\sqrt{x}+28}{-25x+16}$ **18.** 25 **19.** 17 **20.** $\dfrac{144}{25}$ **21.** -6 **22.** 3 **23.** $(4, 0)$ **24.** 0.9, 3.3 **25.** -3

26. Increase a, decrease b. **27.** $y=2.43\sqrt{x}+0.56$

28. a. $f(t)=2.90\sqrt{t}+20.5$ **b.** 45.1 inches **c.** 29 months **d.** (0, 20.50); the median height of boys at birth is 20.5 inches.

Chapter 10

Homework 10.1

1. arithmetic; $d=8$ **3.** not arithmetic **5.** arithmetic; $d=7$ **7.** not arithmetic **9.** $a_n=6n-1$ **11.** $a_n=-11n+7$ **13.** $a_n=-6n+106$ **15.** $a_n=2n-1$ **17.** 113 **19.** -196 **21.** 156.1 **23.** 400 **25.** 107 **27.** 87 **29.** 313 **31.** 3571 **33.** 255 **35.** 500 **37.** yes; answers may vary. **39.** no; answers may vary. **41.** no; answers may vary. **43.** no; answers may vary. **45.** $f(n)=9n-1$ **47. a.** $a_n=800n+26{,}700$ **b.** \$44,300 **c.** 30th year **49. a.** $a_n=\dfrac{1}{6}n+35$ **b.** 35.17, 35.33, 35.50, 35.67; they represent the number of hours the instructor would work if she had 1, 2, 3, or 4 students, respectively. **c.** 56.67 hours per week **d.** 150 students **51. a.** $a_n=1.8n-50$ **b.** 170 people **c.** \$247 **d.** integers between 0 and 27, inclusive

53. a. **b.** $f(t)=12.61t+44.07$ **c.** 107.12, 119.73, 132.34, 144.95, 157.56; from 2000 through 2004, the pharmaceutical industry spent (in millions of dollars) 107, 120, 132, 145, 158, respectively, on government and politics. **d.** \$246 million

55. a. $a_n=0.24n+0.15$ **b.** \$3.27 **c.** no **d.** \$19.35 **57. a.** 2 **b.** 2 **c.** They are the same; answers may vary.
59. 6; a radical equation in one variable

61. a radical function **63.** $12x-23\sqrt{x}+10$; a radical expression in one variable

Homework 10.2

1. geometric; $r=7$ **3.** arithmetic; $d=-7$ **5.** neither **7.** geometric; $r=\dfrac{1}{5}$ **9.** $a_n=3(2)^{n-1}$ **11.** $a_n=800\left(\dfrac{1}{4}\right)^{n-1}$ **13.** $a_n=100\left(\dfrac{1}{2}\right)^{n-1}$ **15.** $a_n=4^{n-1}$ **17.** 4.6566×10^{23} **19.** 1.1921×10^{-6} **21.** 3.3554×10^{7} **23.** 10 **25.** 12 **27.** 19 **29.** 16 **31.** geometric **33.** arithmetic **35.** $f(n)=8(3)^{n-1}$ **37.** no; answers may vary. **39.** The sequence is arithmetic, not geometric; 66 **41. a.** $a_n=27{,}000(1.04)^{n-1}$ **b.** \$38,429.42 **c.** 17th year **43. a.** 2, 4, 8, 16, 32 **b.** $a_n=2^n$ **c.** 256 ancestors **d.** 34.36 billion ancestors; answers may vary.

45. a.

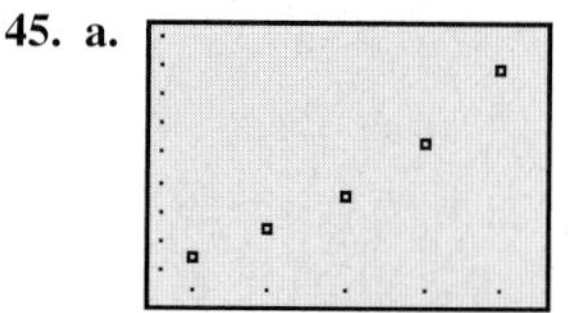

b. $f(t)=865.47(1.39)^t$ **c.** 1203, 1672, 2324, 3231, 4491; from 2001 through 2005, the numbers of Nevaehs born were 1203, 1672, 2324, 3231, and 4491, respectively. **d.** 23,302; 1.17%; Emily

47. a. $a_n = 5(3)^{n-1}$ **b.** 405 students **c.** 295,245 students; yes; answers may vary. **d.** Answers may vary. **49. a.** 2 **b.** 2 **c.** They are the same; answers may vary. **51.** $a_n = 5n + 9$ **53.** $a_n = 448\left(\frac{1}{2}\right)^{n-1}$ **55.** 781,250 **57.** −473 **59.** 122 **61.** 11 **63.** 1.9372; an exponential equation in one variable **65.** an exponential function **67.** $\log_b\left(\frac{25}{8x^{15}}\right)$; a logarithmic expression in one variable

Homework 10.3

1. 20,205 **3.** 30,294 **5.** −38,232 **7.** 21,978 **9.** −10,807 **11.** 468 **13.** 77,875 **15.** 36,288 **17.** −151,468 **19.** 22,468 **21.** 5187 **23.** 0 **25.** 19,436 **27.** 50,005,000 **29.** positive **31.** positive **33.** yes; answers may vary. **35. a.** \$58,200 **b.** \$1,213,800 **37.** A; \$8000 **39. a.** 136 seats **b.** 2340 seats **41. a.** \$82 million **b.** \$233 million **c.** \$2048 million, or \$2.05 billion **43. a.** \$1,034,800 **b.** \$30,000 **c.** \$39,800; years 1 through 13; years 14 through 26 **d.** \$144,559 **45.** 232 **47.** 1800 **49.** $\frac{2(x^2 - 2x + 17)}{(x+3)(x-3)(x-5)}$; a rational expression in one variable **51.** $\frac{1}{(x-3)^2}$; a rational expression in one variable **53.** 13; a rational equation in one variable

Homework 10.4

1. 40,955 **3.** 445.9617 **5.** 61.4266 **7.** 14.9804 **9.** 610,351,562 **11.** 857.1413 **13.** 8.8439 **15.** 89,478,485 **17.** 103.9945 **19.** 19,995.1172 **21.** 100 **23.** 485.9973 **25.** positive **27.** arithmetic **29.** \$699,784.85 **31.** A; \$252,572.15 **33.** 2046 ancestors **35. a.** 11th round; approximately \$6.14 billion **b.** There will be 10 full rounds and part of an 11th round. **c.** nine people; the entrepreneur will get approximately \$6.14 billion; the other eight people will get an average of \$3.30 billion. **37. a.** 1203.00; 1203 Nevaehs were born in 2001. **b.** 32,390.29; 32,390 Nevaehs will be born in 2011. **c.** 112,357; 112,357 Nevaehs will be born from 2001 to 2011, inclusive. **39. a.** 5115 **b.** $n = \frac{\log\left(\frac{a_n r}{a_1}\right)}{\log(r)}$ **c.** $S_n = \frac{a_1\left(1 - r^{\log(a_n r/a_1)/\log(r)}\right)}{1 - r}$ **d.** 5115 **e.** Answers may vary. **41.** 10,443 **43.** 68.6189 **45.** Answers may vary. **47.** Answers may vary. **49.** Answers may vary. **51.** Answers may vary. **53.** Answers may vary.

Chapter 10 Review

1. geometric sequence **2.** arithmetic series **3.** arithmetic sequence **4.** geometric series **5.** $a_n = 2(3)^{n-1}$ **6.** $a_n = -5n + 14$ **7.** $a_n = 200\left(\frac{1}{2}\right)^{n-1}$ **8.** $a_n = 2.7n + 0.5$ **9.** 4.2221×10^{14} **10.** 0.01172 **11.** −204 **12.** 225.9 **13.** 505 **14.** 77 **15.** 24 **16.** −204 **17.** −3182 **18.** 671,173.0723 **19.** 3,221,225,469 **20.** 120,540 **21.** −1749 **22.** 797,161 **23.** geometric series **24.** arithmetic sequence **25. a.** \$71,772.52; \$70,000 **b.** \$1,166,085.43; \$1,300,000 **c.** Answers may vary. **26. a.** $f(t) = 0.28t + 5.64$ **b.** 11.2 thousand deaths **c.** 177.2 thousand deaths **d.** overestimates

Chapter 10 Test

1. geometric sequence **2.** none **3.** geometric series **4.** arithmetic series **5.** $a_n = -6n + 37$ **6.** $a_n = 6(4)^{n-1}$ **7.** 262 **8.** 0.1875 **9.** 455 **10.** 9 **11.** 40.5000 **12.** 4.2950×10^9 **13.** −462 **14.** 1.1248×10^6 **15.** 2,098,620 **16.** none **17.** negative **18. a.**

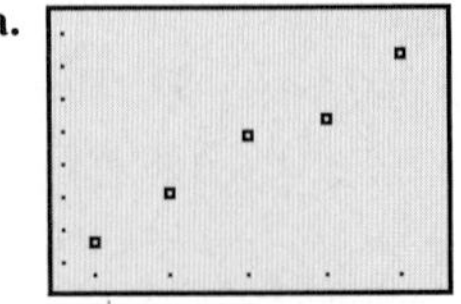

b. $f(t) = 35t + 7.2$ **c.** 42.2; in 2001, online retail sales were \$42 billion. **d.** 532.2; in 2015, online retail sales will be \$532 billion. **e.** 4308; total online retail sales from 2001 through 2015 will be \$4308 billion, or \$4.3 trillion.

19. a. $a_n = 32(1.03)^{n-1}$ **b.** 9th year **c.** \$65,049.41 **d.** \$1,166,696.46

Cumulative Review Chapters 1-10

1. $-\frac{5}{2}, \frac{1}{3}$ **2.** 22 **3.** $\frac{1 \pm \sqrt{69}}{2}$ **4.** $-1, \frac{5}{2}$ **5.** $\frac{2 \pm \sqrt{2}}{3}$ **6.** 2 **7.** $-\frac{43}{18}$ **8.** 3 **9.** 1.6013 **10.** 2.0492 **11.** 2.9755 **12.** $\frac{5 \pm \sqrt{13}}{6}$ **13.** $\frac{2 \pm i\sqrt{2}}{2}$ **14.** (2, −1) **15.** (−2, 3) **16.** (2, −1, 3) **17.** $x \le 7$; $(-\infty, 7]$;

18. $\frac{2916}{b^{18}c^8}$ **19.** $\frac{4c}{5b^{1/4}}$ **20.** 0 **21.** $2x^3y^7\sqrt{3x}$ **22.** $\frac{\sqrt[3]{4x^2}}{x}$ **23.** $\frac{6x - 5\sqrt{xy} + y}{4x - y}$ **24.** $\ln(x^{35})$ **25.** $\log_b\left(\frac{x^{15}}{32}\right)$

26. $9a^2 - 30ab + 25b^2$ **27.** $6k + 13\sqrt{k} - 28$ **28.** $2x^4 + 3x^3 - x^2 + 7x - 3$ **29.** $\frac{(2x+1)(x-3)}{2x(x-1)}$ **30.** $\frac{2x^2 - 3x + 10}{(x-5)^2(x-2)}$

31. $\frac{7x}{2(x-4)(x+2)}$ **32.** $\frac{2x^2 + 9x + 15}{(x-3)(x+1)(x+3)(x+9)}$ **33.** $\frac{3}{(x-8)(x+2)}$ **34.** $f(x) = -3x^2 - 18x - 34$

35. $(x-2)(2x-5)(2x+5)$ **36.** $2x(x-5)(x+3)$ **37.** $2(3w - 5y)(w + 2y)$ **38.** $(10p+1)(10p-1)$ **39.** 3 **40.** 0, 2

41. $f(x) = -x^2 + 2x + 3$ **42.** all real numbers **43.** $y \le 4$

44. **45.** **46.** **47.** **48.**

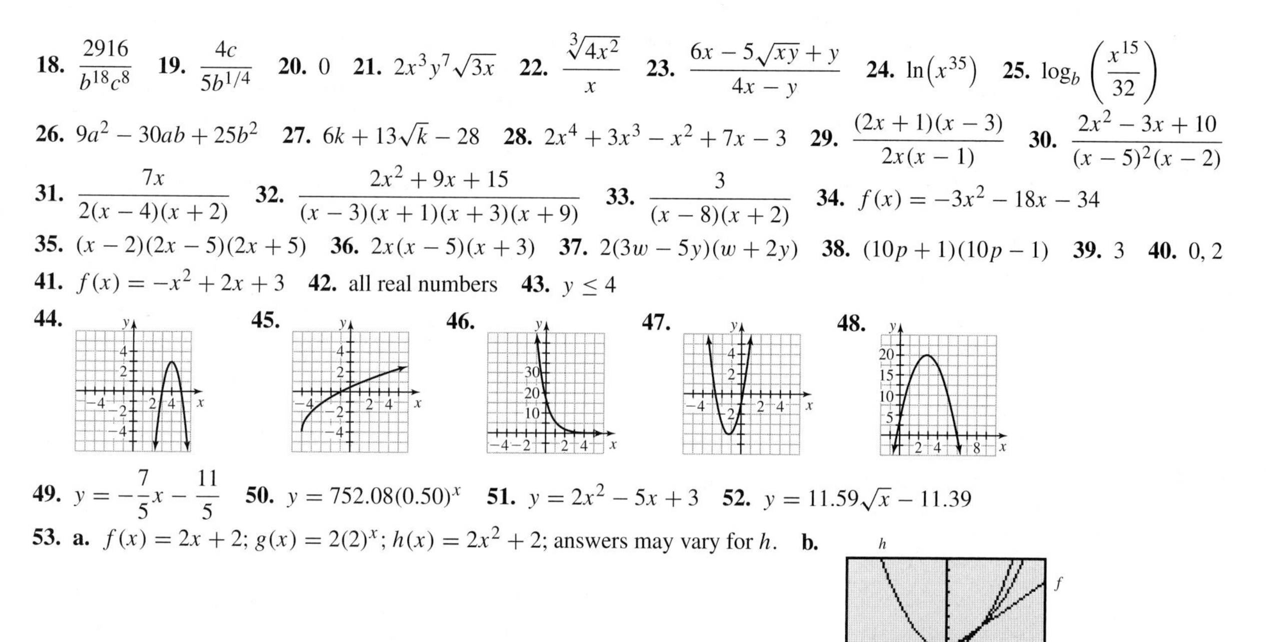

49. $y = -\frac{7}{5}x - \frac{11}{5}$ **50.** $y = 752.08(0.50)^x$ **51.** $y = 2x^2 - 5x + 3$ **52.** $y = 11.59\sqrt{x} - 11.39$

53. a. $f(x) = 2x + 2$; $g(x) = 2(2)^x$; $h(x) = 2x^2 + 2$; answers may vary for h. **b.**

54. 4 **55.** $\frac{1}{2}$ **56.** $g^{-1}(x) = 2^x$ **57.** $f^{-1}(x) = -\frac{1}{4}x - \frac{7}{4}$ **58.** all real numbers except -5 and 7 **59.** 524,288 **60.** 65 **61.** 196,605 **62.** 5597 **63.** 2 liters of the 15% acid solution, 4 liters of the 30% acid solution **64. a.** $f(t) = 1.46t + 34.80$ **b.** $f^{-1}(p) = 0.68p - 23.84$ **c.** 64; in 2010, 64% of passenger vehicles sold will be light trucks. **d.** 16.96; in 2007, 60% of passenger vehicles sold will be light trucks. **e.** 1.46; each year, the percentage of vehicles sold that are light trucks increases by 1.46 percentage points. **65. a.** $L(t) = 0.0126t + 0.687$ **b.** $E(t) = 0.687(1.0153)^t$ **c.** 1.569; 1.989; in 2050, India's population will be 1.569 billion according to the linear model and 1.989 billion according to the exponential model. **d.** 0.42; the difference in India's population in 2050 between the models is 420 million, which is equal to the predicted U.S. population of 420 million for 2050. **e.** linear: 2038; exponential: 2028; answers may vary. **66. a.** exponential: $f(t) = 0.50(1.39)^t$; quadratic: $f(t) = 1.71t^2 - 31.21t + 152.86$; both models fit the data well. **b.** exponential **c.** 2009 **d.** 2012 **e.** Answers may vary. **67. a.** $B(t) = 0.023t^2 + 0.82t + 8.08$ **b.** $R(t) = 32.37t - 81.18$ **c.** $P(t) = \frac{2.3t^2 + 82t + 808}{32.37t - 81.18}$ **d.**

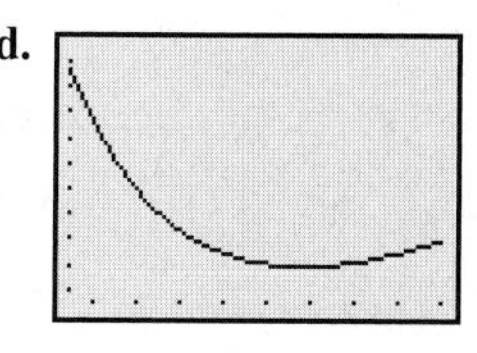

(23.7, 5.9); the minimum percentage of total recreational expenditures that was for book sales, 5.9%, occurred in 2004. **e.** 2010

Chapter 11

Homework 11.1 **1.** ± 7 **3.** empty set **5.** $\pm\frac{18}{5}$ **7.** $-7, 3$ **9.** 5 **11.** $-\frac{10}{3}, 4$ **13.** empty set **15.** ± 2 **17.** $-1, -9$ **19.** 2, 3 **21.** $\frac{3}{7}, 7$ **23.** $-1, \frac{1}{3}$ **25.** $-\frac{13}{4}, \frac{7}{4}$ **27.** $1, \frac{17}{3}$ **29.** $-\frac{2}{27}, \frac{14}{3}$ **31.** ± 2.70 **33.** $-12.24, 6.71$ **35.** $-5.5, 4.5$ **37.** $-6.67, -2.67$ **39.** ± 4 **41.** $-3, 2$ **43.** -1 **45.** ± 3 **47.** -4 **49.** $-\frac{13}{4}, -\frac{1}{4}$ **51.** $-4 < x < 4$; x; $(-4, 4)$

53. $x \le -3$ or $x \ge 3$; x; $(-\infty, -3] \cup [3, \infty)$ **55.** empty set

57. $x < 0$ or $x > 0$; x; $(-\infty, 0) \cup (0, \infty)$ **59.** $x < -4$ or $x > 4$; x; $(-\infty, -4) \cup (4, \infty)$

61. $p \le -2$ or $p \ge 2$; p; $(-\infty, -2] \cup [2, \infty)$

63. $x \le -1$ or $x \ge 13$; x; $(-\infty, -1] \cup [13, \infty)$ **65.** $-10 < x < 5$; x; $(-10, 5)$

67. The set of real numbers; x; $(-\infty, \infty)$ **69.** $t \le 0.8$ or $t \ge 9.6$; t; $(-\infty, 0.8] \cup [9.6, \infty)$

71. $x \le -8$ or $x \ge 2$; x; $(-\infty, -8] \cup [2, \infty)$ **73.** $x \le -10$ or $x \ge 2$; x; $(-\infty, -10] \cup [2, \infty)$

75. $-\frac{39}{8} \le x \le -\frac{21}{8}$; x; $\left[-\frac{39}{8}, -\frac{21}{8}\right]$ **77.** $x = \frac{-b \pm (k - c)}{m}$ **79.** Answers may vary; $-2, 12$

81. Answers may vary; $-13 < x < 7$ **83. a.** $-8, 5$ **b.** $-8 < x < 5$ **c.** $x < -8$ or $x > 5$ **d.** **85.** false

87. 1, 9 **89.** 0, 3.1699 **91.** $-5, 3$ **93.** $x \le 2$; x; $(-\infty, 2]$ **95.** $-2 \le x \le 2$; x; $[-2, 2]$

97. ; a linear function **99.** 0, 4; an absolute value equation in one variable

101. $y = -\frac{5}{9}x - \frac{2}{9}$; a linear equation in two variables

11.1 Quiz **1.** ± 5 **2.** $\frac{1}{3}, \frac{4}{3}$ **3.** empty set **4.** $-\frac{1}{2}, 4$ **5.** $-\frac{1}{2}, \frac{11}{6}$ **6.** false

7. $k \le -2$ or $k \ge 2$; k; $(-\infty, -2] \cup [2, \infty)$ **8.** $x < -1$ or $x > 5$; x; $(-\infty, -1) \cup (5, \infty)$ **9.** $-\frac{4}{3} \le x \le \frac{8}{3}$; x; $\left[-\frac{4}{3}, \frac{8}{3}\right]$ **10.** empty set

Homework 11.2

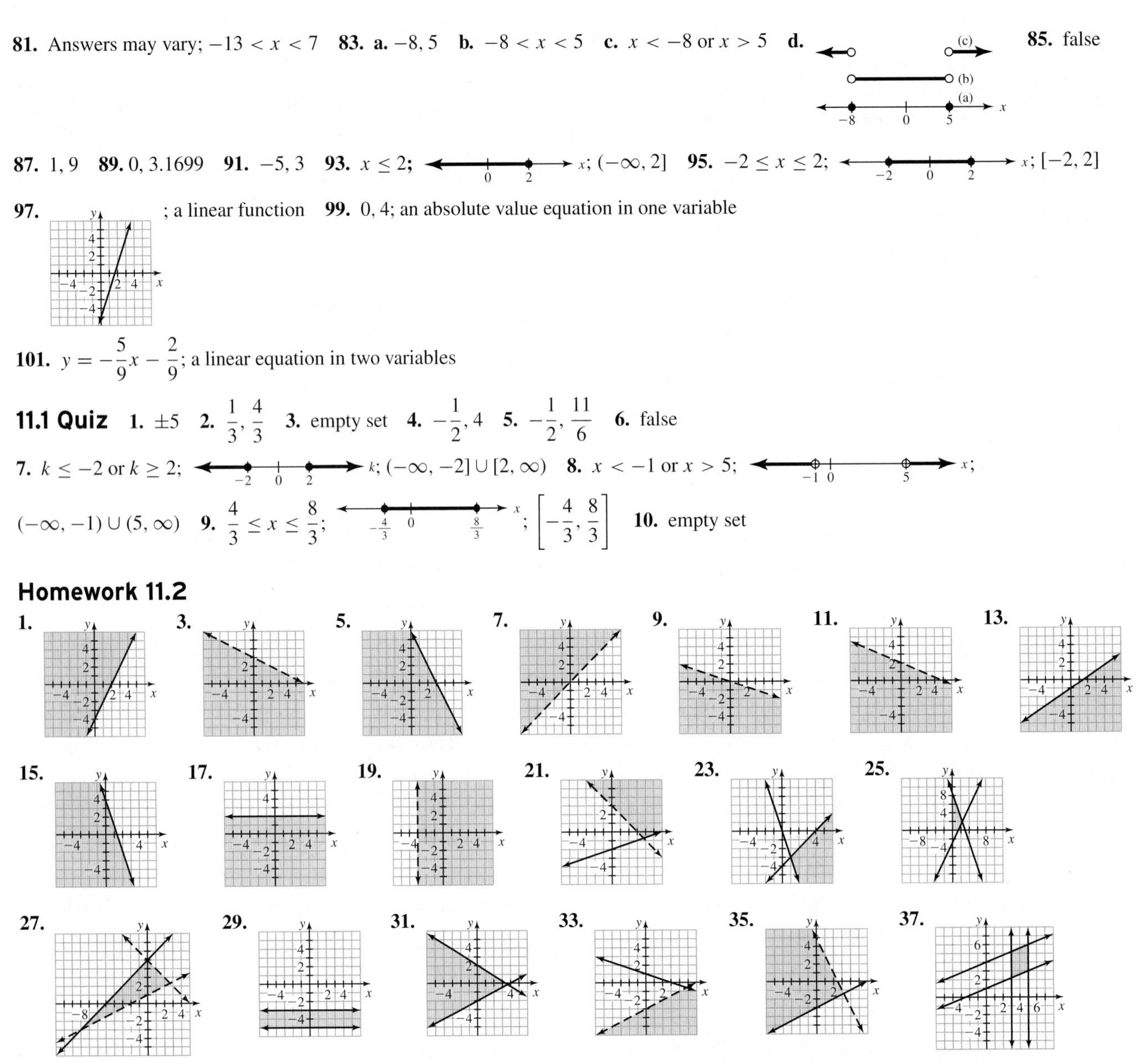

39. a. $B(t) = 0.20t + 69.97$; $T(t) = 0.17t + 51.73$ **b.** The system consists of the inequalities $L \le 0.20t + 69.97$, $L \ge 0.17t + 51.73$, $t \ge 0$, and $t \le 35$. **c.** **d.** between 56.83 years and 75.97 years, inclusive

41. a. $B(w) = 0.44w + 84.21$; $I(w) = 0.44w + 89.21$ **b.** The system consists of the inequalities $L \ge 0.44w + 84.21$, $L \le 0.44w + 89.21$, $w \ge 130$, and $w \le 150$. **c.**

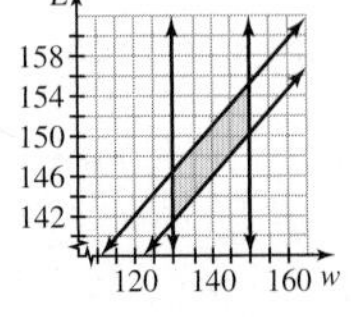

d. from 145.81 centimeters to 150.81 centimeters

43. The graph is the region above the line $2x - 3y = 6$. **45.** Answers may vary.

47. **49.** Answers may vary. **51.** **53.** **55.** (2, 4) **57.** a linear function

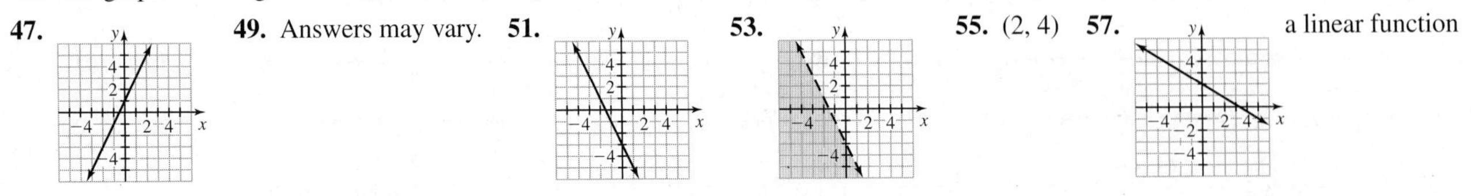

59. $\frac{4}{5}$; a linear function **61.** -5; a linear function

11.2 Quiz

1. 2. 3. 4. 5. 6. 7.

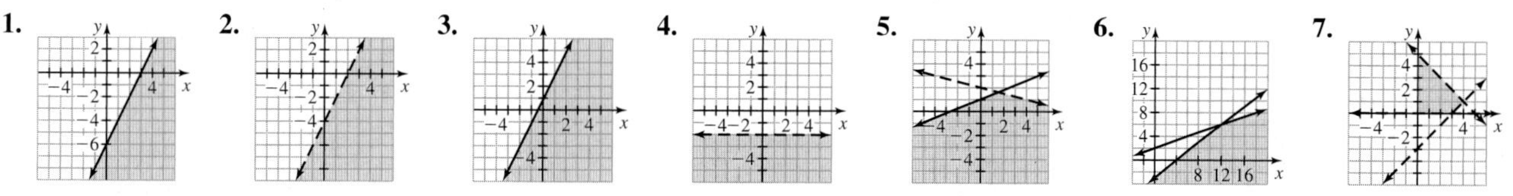

8. Answers may vary.

Homework 11.3 1. $7+3i$ 3. $7-8i$ 5. $4+8i$ 7. $5-16i$ 9. -18 11. -50 13. -10 15. $-\sqrt{15}$ 17. -64 19. $10+15i$ 21. $-1-6i$ 23. $-14+23i$ 25. $27-24i$ 27. $-16-50i$ 29. 41 31. 85 33. 2 35. $-45+28i$ 37. $-9-40i$ 39. $7-24i$ 41. $\frac{6}{29}-\frac{15}{29}i$ 43. $\frac{-6}{53}+\frac{21}{53}i$ 45. $\frac{17}{50}+\frac{19}{50}i$ 47. $-\frac{7}{25}+\frac{24}{25}i$ 49. $\frac{3}{5}+\frac{1}{5}i$ 51. $\frac{7}{4}-\frac{5}{4}i$ 53. $-\frac{7}{5}i$ 55. Student 2's work is correct; answers may vary. 57. a–c. Answers may vary. 59. The result will be a negative real number; answers may vary. 61. $\frac{12-8\sqrt{x}}{9-4x}$ 63. $\frac{12}{13}-\frac{8}{13}i$ 65. $\frac{1\pm 2i\sqrt{2}}{3}$ 67. $\frac{2\pm i}{5}$ 69. $\frac{5\pm i\sqrt{31}}{4}$ 71. $\frac{2\pm 2i\sqrt{2}}{3}$ 73. $\frac{-3\pm 2i\sqrt{5}}{5}$ 75. $\frac{1\pm i\sqrt{11}}{4}$; a quadratic equation in one variable 77. $(5x-2)(2x-3)$; a quadratic expression in one variable 79. $-54-10i$; an imaginary number

11.3 Quiz 1. $9-6i$ 2. $-5+9i$ 3. 12 4. $-\sqrt{14}$ 5. $38-16i$ 6. $7-24i$ 7. 89 8. $\frac{7}{41}+\frac{22}{41}i$ 9. $-\frac{7}{6}-\frac{5}{6}i$ 10. false; answers may vary.

Homework 11.4 1. $c=13$ 3. $c=\sqrt{41}$ 5. $b=\sqrt{55}$ 7. $a=2\sqrt{6}$ 9. $c=\sqrt{7}$ 11. $\sqrt{170}$ 13. $2\sqrt{11}$ 15. $5\sqrt{17}$ feet ≈ 20.6 feet 17. $\sqrt{231}$ inches ≈ 15.2 inches 19. 1.9 miles 21. 2273.4 miles 23. 10 25. $\sqrt{58}$ 27. $2\sqrt{5}$ 29. $4\sqrt{2}$ 31. 8.01 33. 13.28 35. $x^2+y^2=49$ 37. $x^2+y^2=44.89$ 39. $(x-5)^2+(y-3)^2=4$ 41. $(x+2)^2+(y-1)^2=16$ 43. $(x+7)^2+(y+3)^2=3$ 45. (0, 0), 5 47. (0, 0), $2\sqrt{2}$

49. (3, 5), 4 51. $(-6, 1)$, $\sqrt{7}$ 53. $(-3, -2)$, 1 55. $x^2+y^2=9$

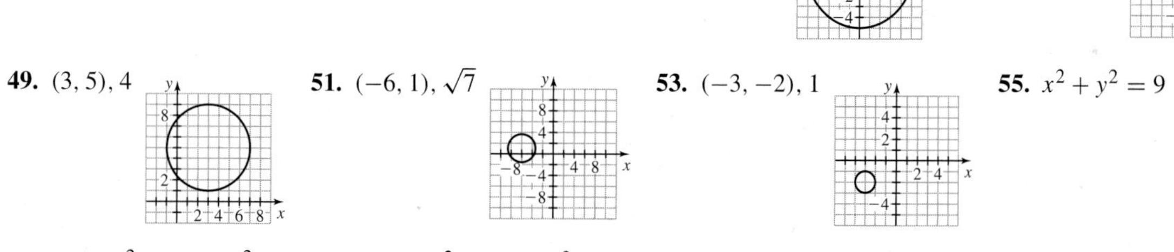

57. $(x+3)^2+(y-2)^2=4$ 59. $(x-3)^2+(y-2)^2=20$ 61. Answers may vary.

63. a. b. Answers may vary. c. The sum of the squares of both variables equals 16.

65. Answers may vary. 67. no; answers may vary. 69. a. Answers may vary. b.

71. a. Answers may vary. b. Answers may vary. c. $3\sqrt{2}$ d. $\frac{5\sqrt{2}}{2}$

73. 75. 77. 79. a quadratic function

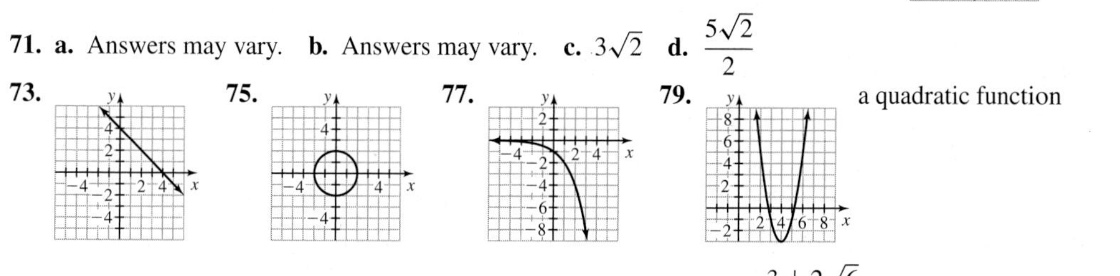

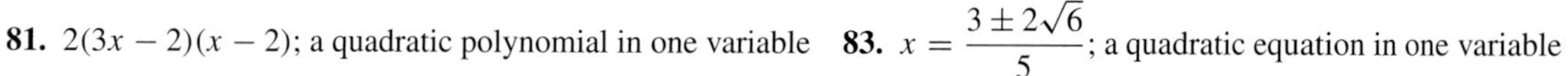

81. $2(3x-2)(x-2)$; a quadratic polynomial in one variable 83. $x=\frac{3\pm 2\sqrt{6}}{5}$; a quadratic equation in one variable

11.4 Quiz **1.** $4\sqrt{3}$ inches $\approx$ 6.9 inches **2.** $\sqrt{105}$ inches $\approx$ 10.2 inches **3.** $\sqrt{41}$ **4.** $4\sqrt{2}$ **5.** $(x+3)^2+(y-2)^2=36$
6. $x^2+y^2=7.84$ **7.** $(0, 0), 2\sqrt{3}$ **8.** $(-4, 3), 5$ **9.** $(x-2)^2+(y+1)^2=68$ **10.** Answers may vary.

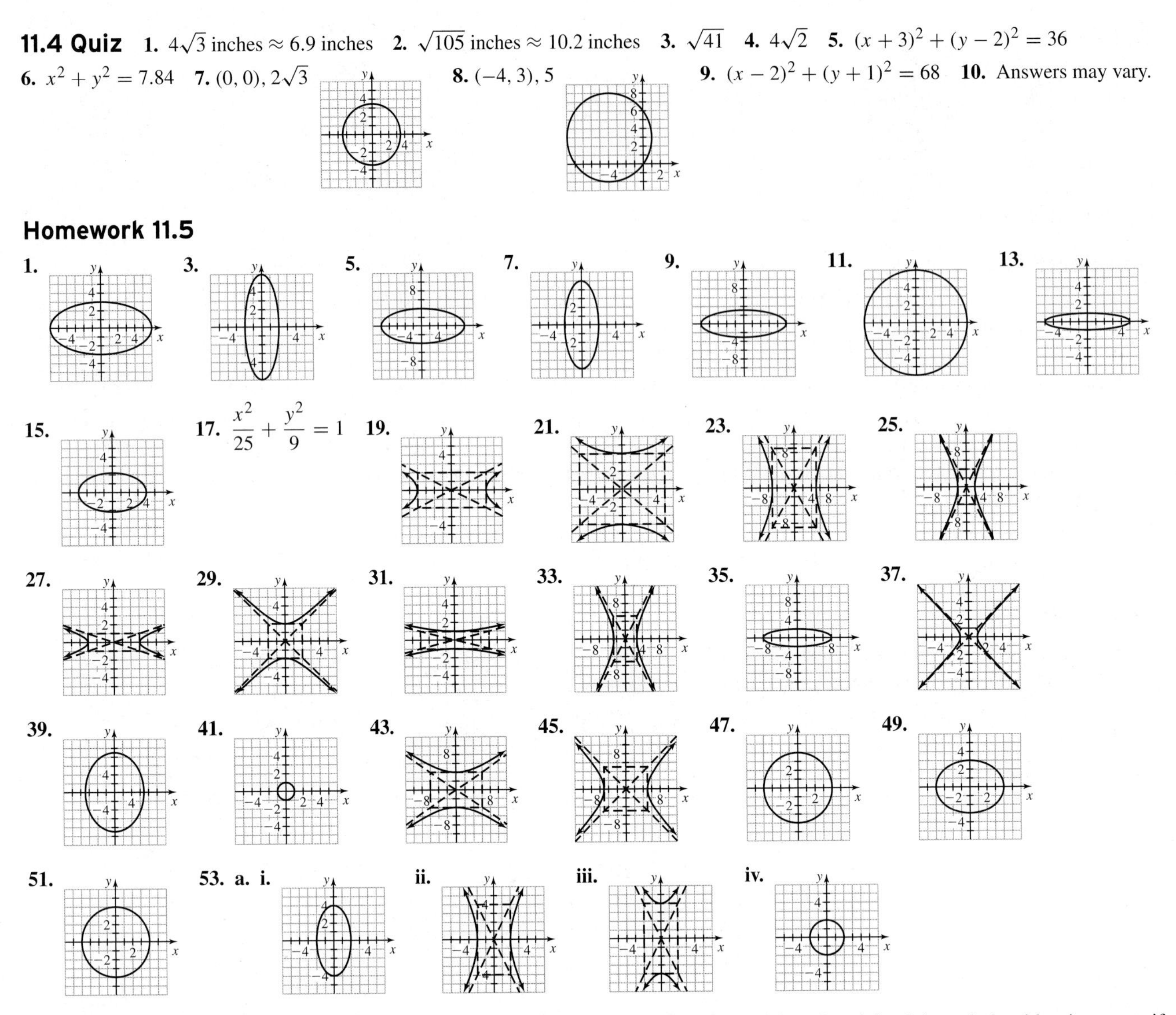

Homework 11.5

17. $\frac{x^2}{25}+\frac{y^2}{9}=1$

53. b. circle if $c=d$, both positive; ellipse if $c \neq d$, both positive; hyperbola with x-intercepts if $c>0$ and $d<0$; hyperbola with y-intercepts if $c<0$ and $d>0$

55. a. **b.** Answers may vary. **c.** Four times the square of x plus 25 times the square of y equals 100.

57. a. Answers may vary. **b.** **59.** Answers may vary. **61.** both a circle and an ellipse

63. a. Translate the graph of $x^2+y^2=1$ by 3 units to the right and 2 units up.

b.

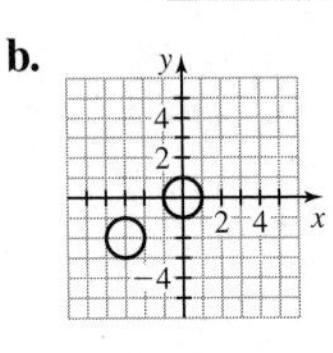

Translate the graph of $x^2+y^2=1$ by 3 units to the left and 2 units down. **c.** Translate the graph of $x^2+y^2=r^2$ by h units to the right if $h>0$ or by $|h|$ units to the left if $h<0$, then by k units up if $k>0$ or by $|k|$ units down if $k<0$.

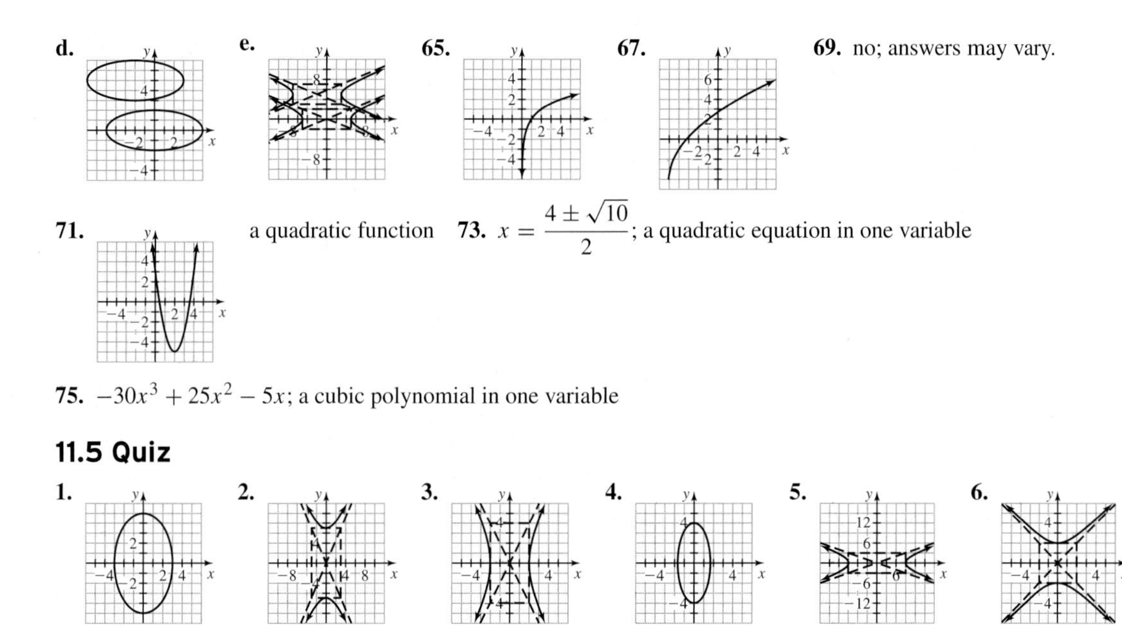

d. **e.** **65.** **67.** **69.** no; answers may vary.

71. a quadratic function **73.** $x = \dfrac{4 \pm \sqrt{10}}{2}$; a quadratic equation in one variable

75. $-30x^3 + 25x^2 - 5x$; a cubic polynomial in one variable

11.5 Quiz

1. **2.** **3.** **4.** **5.** **6.** **7.**

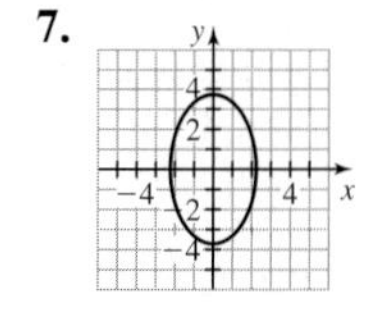

8. **9.** no; answers may vary. **10.** Answers may vary.

Homework 11.6 **1.** $(-5, 0), (5, 0)$ **3.** $(-2, 5), (1, 2)$ **5.** $(-2, 2), (2, 2)$ **7.** empty set **9.** $(-4, 3), (3, -4)$ **11.** $(-3, -5), (0, 4), (3, -5)$ **13.** $(-3, 0), (3, 0)$ **15.** $(-1, 0), (0, 3)$ **17.** empty set **19.** $(1, -2)$ **21.** $\left(-\sqrt{3}, 1\right), \left(\sqrt{3}, 1\right)$ **23.** $(-0.74, -2.02), (-0.74, 2.02), (0.74, -2.02), (0.74, 2.02)$ **25.** $(-2.25, -3.31), (-2.25, 3.31), (2.25, -3.31), (2.25, 3.31)$ **27.** $(-3, -2), (-3, 2), (3, -2), (3, 2)$ **29.** $(3, 2), (5, 0)$ **31.** $(3, 2), (2, 0)$ **33.** $(-5, 0), (5, 0)$ **35.** Answers may vary. **37.** $c = 5, d = -2$ **39.** Answers may vary. **41.** $(1, 2)$ **43.** Answers may vary. **45.** Answers may vary. **47.** Answers may vary. **49.** Answers may vary. **51.** Answers may vary.

11.6 Quiz **1.** $(-3, 0), (3, 0)$ **2.** $(-3, 7), (1, -1)$ **3.** $(1, 4)$ **4.** empty set **5.** $(-4, 0)$ **6.** Answers may vary.

Appendix A

Section A.1 **1–8.**

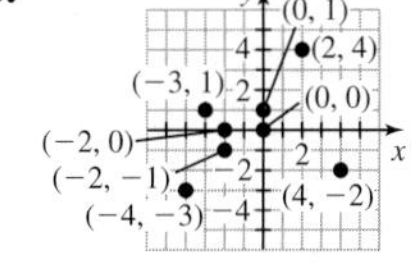

Section A.2 **1.** 4, 85 **2.** 4, 0, −2, 85 **3.** $\frac{2}{9}$, 4, −7.19, 0, −2, 85 **4.** $\sqrt{17}$ **5.** $\frac{2}{9}$, 4, −7.19, 0, −2, $\sqrt{17}$, 85

Section A.3 **1.** 3 **2.** 4.69 **3.** 0 **4.** π

Section A.4 **1.** −21 **2.** −30 **3.** 36 **4.** 16 **5.** 4 **6.** −9 **7.** −4 **8.** 3 **9.** −8 **10.** −13 **11.** −6 **12.** −4 **13.** 5 **14.** 5 **15.** −10 **16.** −5 **17.** −4 **18.** −6 **19.** 8 **20.** 13 **21.** −13 **22.** −6 **23.** 0 **24.** −4

Section A.5 **1.** 49 **2.** 81 **3.** 216 **4.** 625 **5.** 16 **6.** −27 **7.** −16 **8.** −27

Section A.6 **1.** 13 **2.** 12 **3.** 0 **4.** −8 **5.** 4 **6.** −30 **7.** −1 **8.** −7 **9.** 36 **10.** −24 **11.** −9 **12.** 9 **13.** −7 **14.** 11 **15.** 6 **16.** −12

Section A.7 **1.** equation **2.** equation **3.** expression **4.** expression

Section A.8 **1.** $2x + 8$ **2.** $4x + 28$ **3.** $12t - 18$ **4.** $20w - 30$ **5.** $-3x - 24$ **6.** $-4x - 20$ **7.** $-10x + 45$ **8.** $-18x + 6$ **9.** $2.8p + 11.48$ **10.** $-5.2b - 20.28$

Section A.9 **1.** $7x$ **2.** $5x$ **3.** $2x - 7y$ **4.** $2x + y$ **5.** $-2a - 2b - 2$ **6.** $5t - 10w + 4$ **7.** $28x + 7$ **8.** $21x + 22$ **9.** $-2x - 5y$ **10.** $2x - 8y$ **11.** $m - 5n + 10$ **12.** $-4a - 2b + 6$

Section A.10 **1.** 4 **2.** 7 **3.** 3 **4.** -7 **5.** $\frac{28}{5}$ **6.** $\frac{3}{2}$ **7.** 2 **8.** -2 **9.** $\frac{1}{2}$ **10.** $-\frac{11}{3}$ **11.** $\frac{1}{4}$ **12.** $\frac{6}{5}$ **13.** $\frac{14}{17}$ **14.** -21

Section A.11 **1.** $y = -2x + 8$ **2.** $y = 3x - 5$ **3.** $x = \frac{5}{3}y + 5$ **4.** $y = \frac{3}{5}x - 3$ **5.** $y = \frac{ax - c}{b}$ **6.** $x = \frac{by + c}{a}$ **7.** $x = \frac{1}{2}y - \frac{3}{2}$ **8.** $y = 2x + 3$ **9.** $y = \frac{2}{3}x - \frac{5}{6}$ **10.** $y = \frac{9}{8}x - \frac{3}{8}$ **11.** $x = y + a$ **12.** $y = x - a$

Section A.12 **1.** equivalent expressions; answers may vary. **2.** equivalent equations; answers may vary. **3.** neither; answers may vary. **4.** neither; answers may vary. **5.** neither; answers may vary. **6.** equivalent expressions; answers may vary. **7.** equivalent equations; answers may vary. **8.** neither; answers may vary. **9.** neither; answers may vary. **10.** neither; answers may vary.

Index

A

absolute value, 584–92, 629
  defined, 584
absolute value equations in one variable, 585–87
absolute value function(s), 584–85
  translating and reflecting, 530
absolute value inequalities in one variable, 587–90
absolute value property
  for equations, 585
  for inequalities, 588
ac-method, 308
addition
  of complex numbers, 601–3
  order of operations and, 631
  of polynomials, 276–77
  of radical expressions, 511–12
  of rational expressions, 438, 441–44, 448–49
    with a common denominator, 441–42
    with different denominators, 442–44
  of real numbers, 630
addition property of inequalities, 144
affirmations, 39
amplitude, reference, 233
angle, right, 607
annual simple interest rate, 135
approximately exponentially related variables, 200
approximately linearly related variables, 57
area of rectangular objects, 324–25
arithmetic sequence(s), 546, 555–61, 564
  common difference in, 556, 564
  defined, 556
  finding a term or a term number of, 558
  formula of, 556–57
  identifying, 556, 559
  linear function and, 559
  modeling with, 558–59
  term and term number of, 556
arithmetic series, 568–73, 575
  defined, 568–69
  modeling with, 571
  sum of, 569–70
  term and term number of, 568
Ask option, 642
asymptote(s)
  horizontal, 186
  inclined, 617
    defined, 622
    solution of, 622, 626
    solution set of, 622
    solving, 622–25
  vertical, 423
    domains of rational functions and, 427–28, 430–31
average rate of change, 88
axis of symmetry of parabola, 278, 336

B

b^0, 165
$b^{1/n}$, 176
b^n, 162
b^{-n}, 165–66
base
  of exponent, 162, 630
  of exponential model, 203
  of logarithm, 230, 237–38
  of perfect *n*th power, 503
base multiplier property
  to find equations of exponential functions, 193–94
  to find exponential model, 200–201
  graphs of exponential functions and, 184–85
bi form, 366
binomial(s), 285
  factoring, 311–17
    difference of two squares, 310, 312
    strategy for, 314–15
    sum or difference of two cubes, 313–14
  special, 311–17
  square of, 287–89
binomial conjugates, 289–90
boiling point, 63
Boyle's law, 487
branch of hyperbola, 617
breakdown, model, 60

C

Caddy, Eileen, 188
calculators, scientific notation on, 171. *See also* graphing calculator(s)
carbon dating, 246
Celsius scale, 219–20, 222
center of circle, 609
change-of-base property, 256–57
circle(s), 614
  equation of, 609–10
coefficient(s), 632
  leading, 275
combining like terms, 275–76, 632–33
commands. *See* graphing calculator(s)
common difference, 556, 564
common logarithm, 231
common ratio, 562, 564
completely factoring polynomials, 300–301
completing the square, quadratic equations solved by, 371–77, 382, 383
  of form $ax^2 + bx + c = 0$, 374–75
  of form $x^2 + bx + c = 0$, 373–74
  imaginary-number solutions, 375
  perfect-square trinomials, 371–72
complex conjugate, 603
complex number(s), 365–66, 601–6
  adding, subtracting, and multiplying, 601–3
  defined, 366, 601
  dividing, 603–4
complex rational expressions
  defined, 451
  simplifying, 451–59
    by multiplying by LCD ÷ LCD, 454–57
    by writing as a quotient of two rational expressions, 452–54
concepts
  describing, 4
  responding to a general question about, 4–5
conjugates
  binomial, 289–90
  complex, 603
  to rationalize denominators, 521–23
constant(s), 631
constant of proportionality (variation constant), 482, 485
constant rate of change, 91
conversion
  to decimal notation, 169–70
  to scientific notation, 170–71
coordinates, 7
  finding on graphing calculator, 645
coordinate system, 7, 628
counting numbers, 628
Creative Visualization (Gawain), 39
cross-checks, 121
cube(s), 163
  perfect, 502
  sum or difference of two, 313–14
cube root, 176
cubic equations in one variable, 321–22
cubic function(s), 279
  graph of, *x*-intercepts of, 322–23
curve(s)
  decreasing, 3
  exponential, 3, 184
  increasing, 3
  intercept of, 3
  intersection of, 596, 644
  linear, 2
  on TI-83 or TI-84 graphing calculator, 638
    intersection point(s) of two curves, 644
    minimum or maximum values, 644
    x-intercepts, 645

D

D'Angelo, Anthony J., 193
decimal notation, 169–70
decreasing curve, 3
decreasing function, 69
decreasing or increasing property, 185
degree
  of polynomial, 275
  of term, 275
DEL key, 640, 643
denominator
  negative exponent in, 166
  rationalizing the, 361–62, 516
    of radical expressions, 518–20, 521–23
dependent linear system, 107–8
  substitution to solve, 118–19

dependent variable(s), 2–3
  of function, 77
  in linear models, 78
descending order, 275
DeVos, Richard M., 388
difference
  common, 556, 564
  sign of, 280
  square of, 287–89
  of two cubes, 313–14
  of two squares, 289
difference function, 279–81, 448
Dimension mismatch error, 647
direct variation, 481–84
  changes in values for, 482
  defined, 482
  model for, 483–84
  using one point to find equation for, 482–84
  variable varying directly as an expression, 484
discriminant, 381–82
distance between two points, 608–9
distance formula, 608–9
distance–speed–time applications, 473–76
distributive law, 10, 632, 636
  to add or subtract like radicals, 511
Divide by 0 error, 648
division
  of complex numbers, 603–4
  order of operations and, 631
  of rational expressions, 436–38
  of real numbers, 629
division property of inequalities, 145
domain(s)
  of function, 45–46
  of linear model, 81
  of logarithmic functions, 231
  of quadratic function, 340–42
  of rational function, 423–25
    vertical asymptotes and, 427–28, 430–31
  of relation, 42
doubling time, power property to estimate, 246

E

e, exponential models with base, 263
Edison, Thomas, 501
effective half-life, 209
Einstein, Albert, 219
elimination
  to find intersection point, 125
  systems of linear equations solved by, 115–18
    equations in three variables, 389–90
    linear systems, 116–17
    system with fractions, 117–18
  systems of nonlinear equations solved by, 623–25
ellipse(s), 614–17, 619
  equation of, 615
  graphing, 614–15, 616–17
  intercepts of, 615–16
Ellis, Havelock, 555
empty set, 461
  as solution set of linear system, 107
entering equations on graphing calculator, 637–38, 646
equality
  logarithm property of, 238–40
  power property of, 538
  squaring property of, 533
equation(s). *See also specific classes of equations*
  absolute value property for, 585
  adding left sides and adding right sides of two, 115
  of circle, 609–10
  defined, 631
  of ellipse, 615
  entering, on graphing calculator, 637–38, 646
  equivalent, 636
  of exponential functions, 193–99
    base multiplier property to find, 193–94
    comparing methods of finding, 197
    dividing left sides and right sides of, 195–96
    of form $ab^n = k$ for b, 194
    of form $b^n = k$ for b, 194
    two points to find, 195–96
  exponential regression, 205
  fractions in, 633
  graphing, on graphing calculator, 638, 646
  of hyperbola, 617
  linear function as, 42–43, 44–45
  in one variable, 633
    graphing to solve, 119–20
    tables to solve, 120, 121
  satisfying, 7, 592, 633
  solution of, 7, 633
  solution set of, 7, 633
  solving, simplifying an expression vs., 636
  on TI-83 or TI-84 graphing calculator
    entering, 637–38, 646
    graphing, 638, 646
    turning on or off, 645
    Y_n references, 646
  in two or more variables, 634–35
  of vertical and horizontal lines, 12
equivalent equations, 636
equivalent expressions, 636
equivalent statements, 220–21
error messages, on graphing calculator, 637, 646–48
  Dimension mismatch error, 647
  Divide by 0 error, 648
  Invalid dimension error, 647
  Invalid error, 646
  Nonreal answer error, 648
  No sign change error, 647
  Syntax error, 646
  Window range error, 647
errors, 64
  in linear models, 61–62
escape velocity, 525
estimate(s)
  of doubling time, 246–47
  finding model to make, 399
  with linear inequalities, 150–51
  with linear models, 57–58, 77–78
  of maximum or minimum value of quantity, 405–6
  with Pythagorean theorem, 608
  with rational models, 430
  with square root model, 545–46
  with system of inequalities, 597–98
evaluate an expression, 631
excluded value of rational expression, 425
exponent(s), 230, 630–31
  b^0, 165
  $b^{1/n}$, 176
  b^n, 162
  b^{-n}, 165–66
  base of, 162
  defined, 162–63
  expressions involving, 164–65, 166–68
  fraction in, 172
  integer, properties of, 167
  logarithm as, 226
  negative, 165–66
  one-variable equations involving, 194–95
  properties of, 162–64, 167, 178
  rational, 175–83
    defined, 176, 177
    properties of, 177–78
    simplifying expressions involving, 176–77, 178–79
  in scientific notation, 169–71
  simplifying expressions involving, 164–65, 166–68
  zero as, 165
exponential curve(s), 3, 184. *See also* exponential function(s)
  families of, 188
exponential equation(s), 193–99
  dividing left sides and right sides of, 195–96
  finding
    base multiplier property for, 193–94
    comparing methods of, 197
    using two points, 195–96
  of form $ab^n = k$ for b, 194
  of form $b^n = k$ for b, 194
  in one variable, 238, 241
  solving, 239–40, 261–62
    by power property for logarithms, 238, 240
exponential form
  equation in, 236–37
  of expression, 502
exponential function(s), 168–69
  defined, 168
  evaluating, 168–69, 177
  of form $f(x) = ab^x$, 180–81, 183
  graphs of, 183–92
    with $0 < b < 1$, 184
    base multiplier property, 184–85
    as exponential curve, 184
    of form $y = ab^x$ and $y = -ab^x$, 187
    increasing or decreasing property, 185
    intercepts and, 185–86
    reflection property, 186–87
    values from, 187
  inverse of. *See* inverse function(s)
  as inverse of logarithmic functions, 232
  linear functions vs., 169, 193–94
  to model data. *See* exponential model(s)
exponential/logarithmic forms property, 236–37

exponentially related variables, 200
exponential model(s), 200–214
base multiplier property to find, 200–201
base of, 203
e, 263
defined, 200
half-life applications, 201–3
inverse for, 247
linear model vs., 206
power property with, to make predictions, 244–48
using data described in words, 203–4
using data displayed in table, 204–6
exponential properties to simplify radical expressions, 502–3
exponential regression, 205
exponential regression curve, 205
exponentiation, 163, 630–31
expression(s), 631
complex rational. *See* complex rational expressions
equivalent, 636
evaluate an, 631
involving exponents, 164–65, 166–68
radical. *See* radical expression(s)
rational. *See* rational expression(s)
simplified, 636
extraneous solutions, 460–61, 533, 539
extrapolation with linear models, 59–60

F

factoring
perfect-square trinomials, 372
polynomial equations solved by, 317–30
area of rectangular objects, 324–25
cubic equations in one variable, 321–22
in one variable, 323
quadratic equations in one variable, 317–18, 319–22, 382–83
x-intercepts and solutions, 318–19
zero factor property, 317
polynomials, 292, 295–311
completely, 300–301
of form $x^2 + bx + c$, 296–98, 305–6, 308
greatest common factor, 299–300, 307, 309
by grouping, 304–5, 308–9
multiplying vs., 295–96
special binomials, 311–17
strategy for, 314–15
by trial and error, 305–7
with two variables, 298
when leading coefficient is negative, 301–2
quadratic equations solved by, 382–83
factoring by grouping, 304–5, 308–9
Fahrenheit scale, 219–20, 222
family of lines, 30
finite sequence, 555
five-step factoring strategy, 314–15
five-step problem-solving method for systems of linear equations, 131–32
four-step modeling process, 399
formula(s). *See also specific formulas*
defined, 18
involving rational expressions, 463
fraction(s)
in equations, 633
in exponents, 172
graph of linear equations containing, 10
Framingham point scores, 211
frequency, 191
function(s). *See also* exponential function(s); inverse function(s); linear function(s); logarithmic function(s); polynomial function(s); quadratic function(s); radical function(s); rational function(s)
decreasing, 69
defined, 42
dependent variable of, 77
difference, 448
evaluating, 75
increasing, 69
independent variable of, 77
interest problems modeled by, 136–37
inverse, 206–7
invertible, 220–21
linear regression, 68
one-to-one, 225–26
product, 285, 435
quotient, 428–29, 437, 456–57
of sum, 258
sum, 444
table to find output and input of, 75–76
value problems modeled by, 134–35
x-intercepts of, 302
function notation, 74–80, 290
defined, 74, 77
finding intercepts by using, 79–80
with models, 77

G

Gardner, Martin, 162
Gawain, Shakti, 39
GCF (greatest common factor)
defined, 300
factoring out, 299–300, 307
geometric sequence(s), 561–68
common ratio of, 562, 564
defined, 562
exponential function and, 565
finding a term or a term number of, 563–64
formula of, 562–63
identifying, 562, 566
modeling with, 564–65
geometric series, 573–78
defined, 574
modeling with, 576
sum of, 574–76
term and term number of, 574
grade, 17
graph(s)/graphing. *See also* graphing calculator(s); solution(s)
absolute value equation in one variable solved by, 587
of absolute value function, 585
of cubic function, 279
x-intercepts of, 322
defined, 8
to describe authentic situation, 55–56
of ellipses, 614–15, 616–17
equation in one variable solved by, 119–20
equations on graphing calculator, 638, 646
exponential equations in one variable solved by, 241
of exponential functions, 183–92
with $0 < b < 1$, 184
base multiplier property, 184–85
as exponential curve, 184
of form $y = ab^x$ and $y = -ab^x$, 187
increasing or decreasing property, 185
intercepts and, 185–86
reflection property, 186–87
values from, 187
of hyperbola, 617–19
of inverse functions, 222–23
of linear equations, 7–16
containing fractions, 10
distributive law in, 10
horizontal line, 12
intercepts of, 10–11
in three variables, 391
using slope, 27–28
vertical lines, 11–12
in $y = mx + b$ (slope–intercept) form, 9, 13
of linear functions, 43–44, 45–46, 55–65. *See also* linear models
of linear inequalities in two variables, 592–94
logarithmic equation in one variable solved by, 257
of logarithmic functions, 232–33
of nonlinear systems of equations, 622–23, 624
polynomial equations in one variable solved by, 323
of quadratic functions, 277–78, 336–46
comparing methods of, 355
of form $f(x) = ax^2$, 337–38
minimum or maximum value from, 352–55
reflecting across the x-axis, 337
sketching, 340–42, 348–50
in standard form, 347–59
stretching, 337
translating, 338–39
in vertex form, 336–46
qualitative, 1–6
defined, 1
independent and dependent variables in, 2–3
reading, 1–3
sketching, 3–5
of solution set of system of inequalities, 596–97
square root equation solved by, 537
of square root function, 507, 523, 526–29
of system of two linear equations, 105–7
systems of linear equations solved by, 103–13
intersection point of graphs of two models, 104, 105
prediction by using two linear models, 103–4
on TI-83 or TI-84 graphing calculator, 638
with axes turned off, 646
with a scattergram, 641

graphing calculator(s)
command access in, 637
commands
Ask, 642
intersect, 104, 119, 644
OFF, 637
ON, 637
TRACE, 11, 34, 638
ZDecimal, 8, 639
ZInteger, 639
ZoomFit, 639
Zoom In, 638–39
Zoom menu, 638–39
Zoom Out, 639
ZoomStat, 639
ZSquare, 9, 10, 639
ZStandard, 9, 10, 639
coordinates of points on, 645
curve tracing on, 638
intersection point(s) of two curves, 644
minimum or maximum values, 644
x-intercepts, 645
DEL key, 640, 643
equations on
entering, 637–38, 646
graphing, 638, 646
turning on or off, 645
Y_n references, 646
error messages, 637, 646–48
Dimension mismatch error, 647
Divide by 0 error, 648
Invalid dimension error, 647
Invalid error, 646
Nonreal answer error, 648
No sign change error, 647
Syntax error, 646
Window range error, 647
exponential regression on, 205
graphing equations with scattergram, 641
for imaginary unit i, 366
intersect feature on, 104, 119, 644
linear regression feature, 68
maximum of a curve, 644
minimum of a curve, 644
plotter, turning on/off, 641
plotting points in scattergram, 640, 643–44
quadratic regression on, 397
quotient of logarithms on, 239–40
regression curves on, 642–43
scattergrams on
graphing equations with, 641
plotting points, 640, 643–44
tracing, 640
tracing a curve with, 641
screen brightness control, 637
STAT list editor, 642, 647
storing a value in, 645
tables on, 641–42
Ask option, 642
for two equations, 642
TI-82 graphing calculator, 637
TI-83 or TI-84 graphing calculator, 637–48
TI-85 and TI-86 graphing calculators, 637
tracing a curve
with a scattergram, 641
without a scattergram, 638
turning equation on or off, 645
turning on or off, 637
window settings, 639
x-intercepts, finding, 645
zooming in and out on, 638–39
greatest common factor (GCF)
defined, 300
factoring out, 299–300, 307
"green" buildings, 219
gross national product (GNP), 212–13
grouping, factoring by, 304–5, 308–9

H

half-life, 201–3, 246
defined, 201
effective, 209
Herbert, Frank, 244
Honsberger, Ross, 584
horizontal asymptote, 186
horizontal lines, 12
horizontal translation (left–right shifts), 338–39, 526–27
hyperbola(s), 619
equations of, 617
graphing, 617–19
hypotenuse, 607

I

i (imaginary unit), 366, 601
powers of, 605
imaginary number, 601
defined, 366
pure, 366, 601
inclined asymptotes, 617
inconsistent linear system, 107, 108
substitution to solve, 118
increasing curve, 3
increasing function, 69
increasing or decreasing property, 185
independent variable(s), 2–3. *See also* predictions, making
of function, 77
in linear models, 61, 78
index of radical expressions, 501
index property for radicals, 508
inequality(ies)
absolute value, 587–90
absolute value property of, 588
addition property of, 144
division property of, 145
linear
systems of, 595–99
in two variables, 592–95
multiplication property of, 144–45
in one variable
estimates and predictions by using, 150–51
three-part, 149–50
satisfying, 7, 145–49, 592
subtraction property of, 144
inequality symbols, 45
infinite sequence, 555–56
input, 42
integers, 628
intercept(s), 10–11
defined, 3
of ellipse, 615–16
in graphs of exponential functions, 185–86
of linear models, 59–60, 79–80
significance of, 82
interest
compounded annually, 201
defined, 135
problems involving, 135–37
functions to model, 136–37
simple, 135, 213
interpolation with linear models, 59–60
intersect command, 104, 119, 644
intersection
of graphs of inequalities, 596
point, of graphs of two models, 104
and solution set, 105
using substitution and elimination to find intersection point, 124–27
interval, defined, 147
interval notation, 147
Invalid dimension error, 647
Invalid error, 646
inverse for exponential model, 247
inverse of a function, 220
inverse function(s), 206–7, 219–29
defined, 219–20
equation of, 223–25
evaluating, 221
graphing, 222–23
input–output values of, 221–22
one-to-one functions and, 225–26
properties of, 221, 222
inverse variation, 475–76, 484–90
changes in values for, 485
defined, 485
equation for, 485
one point to find, 485–87
model for, 486–87, 488
products to find constants of, 489–90
variable varying inversely as an expression, 488
invertible, 220
irrational numbers, 629
isosceles right triangle, 613

J

Jordan, Michael, 1

L

LCD (least common denominator), 633
LCM (least common multiple), 634
Leadership in Energy and Environmental Design (LEED) certification, 219
leading coefficient, 275
least common denominator (LCD), 633
least common multiple (LCM), 634
left–right shifts (horizontal translation), 338–39, 526–27
legs of triangle, 607
Levitt, Steven D., 74
like radicals, 511
like terms, 275–76, 632
combining, 632–33
line(s). *See also* slope(s)
family of, 30
horizontal, 12

to model data, 55–65. *See also* linear models
vertical, 12
linear equation(s)
approximate, 36
finding, 34–39
given two points, 35
parallel to a given line, 36
perpendicular to a given line, 36–37
selecting points to use for, 38–39
using point–slope form, 37–38
using slope–intercept form, 34–37
graphs of, 7–16
containing fractions, 10
distributive law in, 10
horizontal line, 12
intercepts of, 10–11
using slope, 27–28
vertical lines, 11–12
in $y = mx + b$ (slope–intercept) form, 9, 13
identifying, 29–30
in one variable
defined, 633
solving, 633–34
slope addition property of, 29–30
systems of. *See* systems of linear equations
in three variables, 388–89
defined, 388
graph of, 391
solution satisfying, 388
in two variables, 13
vertical change property of, 27
y-intercept of, 27
linear function(s), 41–50. *See also* linear models; sequence(s); series
defined, 41, 42
domain of, 45–46
equation as, 42–43, 44–45
exponential functions vs., 169, 193–94
graph of, 43–44, 45–46
range of, 45–46
relation and, 42, 46
Rule of Four for, 45
table as, 43
linear inequality(ies)
in one variable, 139–40, 143–54
defined, 146
examples of, 145
models to compare quantities, 143–44
properties of inequalities, 144–45
satisfying, 146
solution of, 146
solution set of, 146, 147–48, 151
solving, 145–49
three-part inequalities, 149–51
in two variables, 592–95
systems of, 595–99
linear models, 46–47, 57–62
analyzing, 93
approximately linearly related variables in, 57
breakdown of, 60
choosing "good points" to find, 69–70
defined, 57
dependent and independent variables of, 61, 78
domain and range of, 81
equations of, 65–74
linear regression, 68
predictions by using, 77–78
rate of change, 91
steps for finding, 68
by using data described in words, 66
by using data displayed in table, 66–68
errors in, 61–62
estimates and predictions, 57–58, 77–78
using data, 80–81
exponential model vs., 206
intercepts of, 59–60, 79–80
significance of, 82
interpolation and extrapolation with, 59–60
modifying, 60–62
scattergrams and, 58, 68
slope of, 82
for approximately linearly related variables, 92–94
steps in making, 78
unit analysis of, 91–92
using, 58
linear regression, 68
linear regression function, 68
linear system, 104–8
dependent system, 107–8
elimination to solve, 116–17
inconsistent system, 107, 108
one-solution system, 105–7, 108
solution set of, 105–7
substitution to solve, 114–15
table of solutions of equations for, 108
logarithm(s), 226
base of, 230, 237–38
common, 231
defined, 230
finding, 230–31
to model authentic situations, 233
properties of, 231, 236–41, 262
change-of-base property, 256–57, 258
comparing, 258
exponential/logarithmic forms property, 236–37
power property, 234, 238, 240, 241, 244–48
product property, 247–48, 253–54, 255
quotient property, 247–48, 254–55, 258
quotient of, 239–40
logarithmic equation(s)
in one variable, 237, 257
solving, 255–56, 261, 263
logarithmic function(s), 230–73
defined, 231
domain of, 231
graphing, 232–33
as inverse of exponential functions, 232
logarithm property of equality, 238–39
lowest terms, rational expression in, 426

M

maximum point, 278
maximum value
finding on graphing calculator, 644
of quadratic functions, 352–55
quadratic functions to find, 405–6
mean, 469–72
minimum point, 278
minimum value
finding on graphing calculator, 644
of quadratic functions, 352–55
quadratic functions to find, 405–6
mixture problems, 137–39
model(s)/modeling
comparing two quantities by using, 150–51
defined, 57
differences of quantities, 408–9
direct variation, 483–84
direct variation equation in, 481–82
exponential, 244–48
four-step process in, 399
inverse of, 223–24
inverse variation, 486–87, 488
linear, 46–47, 57–62
logarithms in, 233
product function in, 291
quadratic, 323–24
in vertex form, 364–65
with quadratic functions, 383–84, 403–13
to find maximum or minimum value of quantity, 405–6
to make predictions, 403–5
radical, 507–8
rational, 429–30
with rational functions, 469–81
distance–speed–time applications, 473–76
mean of quantity, 469–72
percentage of quantity, 472–73
selecting, 398–99
square root, 508
with square root functions, 542–50
with sum functions and difference functions, 280–81
with system of equations, 94
quadratic, 406–7
with systems of inequalities, 597–98
of systems of linear equations, 124–31
rate of change to find, 126–27
table of data to find, 125–26
trial and error to find, 188
model breakdown, 60
monomials, 274, 285
multiplication of polynomial and, 285–86
multiplication of two, 285
Moore, Gordon, 246*n*
Moore's law, 246*n*
multiplication
of complex numbers, 601–3
order of operations and, 631
of polynomials, 285–95
binomial conjugates, 289–90
factoring vs., 295–96
monomial and a polynomial, 285–86
product function, 290–92
quadratic polynomials, 292
square of binomial, 287–89
two monomials, 285
two polynomials, 286–87
of radical expressions, 512–16
of rational expressions, 434–35
of real numbers, 629
repeated, 630
multiplication property of inequalities, 144–45

N

natural logarithm(s), 260–66
  defined, 260
  finding, 261
  properties of, 262
    exponential/natural logarithmic forms property, 261
    power and quotient properties, 263
negative exponents, 165–66
negative self-talk, 39
Newton's law of cooling, 264
nonlinear systems of equations, 622–26
  defined, 622
  solution of, 622, 626
  solution set of, 622
  solving, 622–25
    by elimination, 623–25
    by graphing and by substitution, 622–23, 624
Nonreal answer error, 648
*n*th power, 162, 630
*n*th root, 176
numbers, types of, 628–29

O

OFF command, 637
ON command, 637
one-to-one functions, 225–26
ordered pair, 7
  satisfying both of two given equations, 105
  as solution of system, 105
ordered triple, 388
order of operations, 631
origin, 628
  equation of circle centered at, 610
output, 42

P

parabola(s), 278, 336, 614
  axis of symmetry of, 278, 336
  defined, 3
  equation of, 385
    using points that are not *y*-intercepts, 392–93
    using the *y*-intercept and two other points, 393–94
  families of, 343
  finding *x*-intercepts of, 364
  vertex of, 278
parallel lines, 21, 22, 26
parentheses, order of operations and, 631
percentage, 472–73
percentage formula, 429
perfect cube, 502
perfect *n*th power, 502, 503
perfect square, 360, 502
perfect-square trinomials, 288–89, 297, 368, 371–72, 374
  factoring, 372
perpendicular lines, slope of, 21–22, 26
Péter, Rósza, 274
plotter, turning on/off, 641
plotting points, 628, 640, 643–44
point(s)
  coordinates of, 645
  distance between two, 608–9
  maximum, 278
  minimum, 278
  plotting, 628
    in scattergram, 640, 643–44
  symmetric, 347–49
point–slope form of equation of line, 37–38
polynomial equations, 317–30
  cubic equations in one variable, 321–22
  factoring to solve, 317–30
  in one variable, graphing to solve, 323
  quadratic equations in one variable, 317–22, 382–83
    containing fractions, 320
  of quadratic functions, 325
polynomial function(s), 274–335
  cubic functions, 279
    x-intercepts of graph of, 322–23
  difference function, 279–81, 448
  function notation, 74–80, 290
  product function, 285, 290–92, 435
  quadratic functions. *See* quadratic function(s)
  sum function, 279–81, 444
polynomial(s)/polynomial expression(s), 274–77
  adding and subtracting, 276–77
  binomials, 285
    factoring, 311–17
    special, 311–17
    square of, 287–89
  defined, 277
  degree of, 275
  describing, 275
  factoring, 292, 295–311
    completely, 300–301
    of form $x^2 + bx + c$, 296–98, 305–6, 308
    greatest common factor, 299–300, 307, 309
    by grouping, 304–5, 308–9
    multiplying vs., 295–96
    to solve polynomial equations, 317–30
    special binomials, 311–17
    strategy for, 314–15
    by trial and error, 305–7
    with two variables, 298
    when leading coefficient is negative, 301–2
  monomials, 274, 285
    multiplication of polynomial and, 285–86
    multiplication of two, 285
  multiplying, 285–95
    binomial conjugates, 289–90
    factoring vs., 295–96
    monomial and a polynomial, 285–86
    product function, 290–92
    quadratic polynomials, 292
    square of binomial, 287–89
    two monomials, 285
    two polynomials, 286–87
  in one variable, 275
  prime, 298–99
  terms of, 274
    combining, 275–76
  trinomials, 285
    factoring, 292, 295–304
    with negative constant terms, 298
    perfect-square, 288–89, 297, 368, 371–72, 374
    with positive constant terms, 296–97
power(s), 162–63, 502, 503, 630. *See also* exponent(s)
  of *i*, 605
  power raised to, 163, 167, 178
  product raised to a, 163, 167, 178
  quotient raised to a, 163, 167, 178
  real-number, 184
power property
  of equality, 538
  of logarithms, 234, 238, 240, 241, 244–48
    to make predictions, 244–48
  of natural logarithms, 263
  of radicals, 506, 513
practice exams, 47
predictions, making
  with data described in words, 80–81
  finding models for, 399
  with linear inequalities, 150–51
  with linear inequalities in one variable, 143–54
    models to compare quantities, 143–44
  with linear models, 57–58, 77–78
    equations of, 57–58, 77–78
    using data, 80–81
  power property with exponential models, 244–48
  with quadratic formula, 383–84
  with quadratic function, 403–5
  with quadratic model, 324
    in vertex form, 364–65
  with radical model, 539
  with rational model, 429–30, 464–65
  solving a system for, 125–26
  with square root model, 508, 545–46
  with two linear models, 103–4
prime polynomials, 298–99
principal, 135
principal square root, 176
procedure, describing, 4
product(s). *See also* multiplication
  inverse variation constant from, 489–90
  raised to power, 163, 178
product functions, 285, 290–92, 435
product property
  for exponents, 163, 167, 178
  for logarithms, 247–48, 253–54, 255
  to multiply radical expressions, 512–13
  for radicals, 503
  simplifying square roots by using, 361
  for square roots, 359–60
proportionality, constant of, 482, 485
pure imaginary number, 366, 601
Pythagorean theorem, 606–8
  converse of, 610–11

Q

quadratic equation(s)
  choosing method of solving, 382–83
  comparing methods of solving, 384
  completing the square to solve, 371–77, 382, 383
    of form $ax^2 + bx + c = 0$, 374–75

of form $x^2 + bx + c = 0$, 373–74
imaginary-number solutions, 375
perfect-square trinomials, 371–72
containing fractions, 320–21
factoring to solve, 317–22, 382–83
of form $ax^2 + bx + c = 0$, 376
input and an output of, 321
quadratic formula to solve, 377–88
approximate solutions, 380
imaginary-number solutions, 380–81
for modeling a situation, 383–84
number of real-number solutions, 381–82
statement of, 378
square root property to solve, 359–71, 382, 383
of form $a(x - h)^2 + k = p$, 368
of form $(px + q)^2 = k$, 363–64
of form $x^2 = k, k \geq 0$, 362–63
of form $x^2 = k, k < 0$, 367
system of, modeling with, 406–7
quadratic function(s), 277–78, 292, 336–421
for any three points, 394
domain and range of, 340–42
equations of, 325
evaluating, 290
finding equations of, 325
graphing, 336–46
comparing methods of, 355
of form $f(x) = ax^2$, 337–38
minimum or maximum value from, 352–55
reflecting across the x-axis, 337
sketching, 340–42, 348–50
in standard form, 347–59
stretching, 337
translating, 338–39
in vertex form, 336–46
input and output of, 321
modeling with, 323–25, 383–84, 403–13
to find maximum or minimum value of quantity, 405–6
to make predictions, 403–5
in standard form, 290
vertex of, 340
symmetric points to find, 347–49
vertex formula to find, 350–52
quadratic model(s), 396–403
choosing "good points" to find, 399–400
in standard form, 396–97
in vertex form, 342–43
quadratic regression, 397
quadratic regression curve, 397
qualitative graphs, 1–6
defined, 1
independent and dependent variables in, 2–3
reading, 1–3
sketching, 3–5
intercepts for, 11
quotient
of logarithms, 239–40
raised to power, 163, 167, 178
of two complex numbers, simplifying, 603–4
quotient function, 428–29, 437, 456–57
quotient property
for exponents, 163, 167, 178
for logarithms, 247–48, 254–55, 258
for natural logarithms, 263
for radicals, 520
to simplify radical expressions, 521
for square roots, 361

R

radical(s), 359
defined, 501
evaluating, 502
index property for, 508
like, 511
power property for, 506, 513
product property for, 503
quotient property for, 520
radical conjugates, 521–23
radical equation(s)
in one variable, 532–37
solving, 532–42
power property of equality in, 538
square root equation, 535–37
radical expression(s)
adding and subtracting, 511–12
defined, 502
index of, 501
multiplying, 512–16
rationalizing denominators of, 361–62, 518–20
using conjugates, 521–23
simplifying, 359–62, 425–28, 490, 501–6, 515–16
exponential properties for, 502–3
with index n greater than 2, 505
perfect nth powers, 503
with product property, 361, 503–4
quotient property for, 361–62, 521
radical with smallest index possible, 505–6
rationalizing the denominator, 361–62
solving rational equations vs., 463–64, 465–66
square of, with two terms, 514
in two variables, 428
radical form of expression, 502
radical function(s), 501–54
defined, 507
evaluating, 507
square root function, 507
radical model(s), 507–8
predictions by using, 539
radical sign, 501
radicand, 359, 501
radius, 609, 610
range
of function, 45–46
of linear model, 81
of quadratic function, 340–42
of relation, 42
rate of change, 88–98
average, 88, 93
calculating, 88–91
constant, 91
increasing and decreasing, 89
systems of linear equations by using, 126–27
ratio
common, 562, 564
defined, 88
unit, 88
rational equation(s), 422
solving, 457, 459–68
in one variable, 460–62
simplifying rational expressions vs., 463–64, 465–66
rational exponents, 175–83
defined, 176, 177
properties of, 177–78
simplifying expressions involving, 176–77, 178–79
rational expression(s), 422
addition of, 438, 441–44, 448–49
with a common denominator, 441–42
with different denominators, 442–44
combining three, 437–38
complex
defined, 451
simplifying, 451–59
defined, 423
division of, 436–38
excluded value of, 425
formulas involving, 463
in lowest terms, 426
multiplication of, 434–35
performing operations with three, 447
simplifying, 425–28
solving rational equations vs., 463–64, 465–66
in two variables, 428
subtraction of, 438, 445–49
with a common denominator, 445
with different denominators, 445–47
rational function(s), 422–500
difference function, 448
domain of, 423–25
evaluating, 423
input value of, 462–63
meaning of, 423
modeling with, 469–81
distance–speed–time applications, 473–76
mean of quantity, 469–72
percentage of quantity, 472–73
product function, 435
quotient function, 428–29, 437, 456–57
rational model, 429–30
sum function, 444
rationalizing denominators, 361–62, 516
of radical expressions, 518–20
using conjugates, 521–23
rational model, predictions by using, 464–65
rational numbers, 628
ratios, direct variation constant from, 488–89
real-number powers, 184
real numbers, 629
operations with, 629–30
reciprocal, 436
rectangular objects, area of, 324–25
reference amplitude, 233
reflection property, 186–87
of inverse functions, 222

reflections, 337
of absolute value function, 530
across the x-axis, 527
regression
exponential, 205
linear, 68
quadratic, 397
regression curves, on TI-83 or TI-84 graphing calculator, 642–43
regression line, 68
Reid, Constance, 422
relation, 42, 46
repeating number, 628
revenue, 132
Richter number, 233
Richter scale, 233
right angle, 607
right triangle, 606, 607. *See also* Pythagorean theorem
isosceles, 613
rise, 17–18
Rule of Four, 45
run, 17–18
Russell, Bertrand, 103

S

Sarton, May, 1
satisfy, defined, 7, 105, 145–49, 388, 592, 633
satisfying equations and inequalities, 7, 592, 633
linear inequality in one variable, 145–49
scattergram(s), 55–56
defined, 56
linear modeling and, 58
on TI-83 or TI-84 graphing calculator
graphing equations with, 641
plotting points in, 640, 643–44
tracing, 640
tracing a curve with, 641
scientific notation, 169–71
on calculators, 171
converting from standard decimal notation to, 170–71
converting to standard decimal notation, 169–70
defined, 169
screen brightness control, graphing calculator, 637
sequence(s). *See also* series
arithmetic, 546, 555–61, 564
common difference in, 556, 564
defined, 556
finding a term or a term number of, 558
formula of, 556–57
identifying, 556, 559
linear function and, 559
modeling with, 558–59
term and term number of, 556
defined, 555–56
finite, 555
geometric, 561–68
common ratio of, 562, 564
defined, 562
exponential function and, 565
finding a term or a term number of, 563–64
formula of, 562–63
identifying, 562, 566
modeling with, 564–65
infinite, 555–56
terms of, 555
series
arithmetic, 568–73, 575
defined, 568–69
modeling with, 571
sum of, 569–70
term and term number of, 568
geometric, 573–78
defined, 574
modeling with, 576
sum of, 574–76
term and term number of, 574
set(s)
defined, 7
empty, 107, 461
solution. *See* solution set
shift (left–right or up–down). *See* translation
sign of difference, 280
simple interest, 135, 213
simplified expression, 636
simplifying expressions, 636
complex rational expressions, 451–59
by multiplying by LCD ÷ LCD, 454–57
by writing as a quotient of two rational expressions, 452–54
involving exponents, 164–65, 166–68
rational exponents, 176–77, 178–79
quotient of two complex numbers, 603–4
radical expressions, 359–62, 425–28, 490, 501–6, 515–16
exponential properties for, 502–3
with index n greater than 2, 505
perfect nth powers, 503
with product property, 361, 503–4
quotient property for, 361–62, 521
radical with smallest index possible, 505–6
rationalizing the denominator, 361–62
solving rational equations vs., 463–64, 465–66
square of, with two terms, 514
in two variables, 428
square of binomial, 287
with square roots, 360–61
using the product property, 361
slope(s), 16–23
comparing steepness of two objects, 16–17
defined, 18
finding, 17–19
formula for, 18, 82
graph of linear equations by using, 27–28
of horizontal and vertical lines, 20
of increasing and decreasing lines, 19–20
of linear equation of form $y = mx + b$, 25–26
of linear model, 82
of parallel lines, 21, 22, 26
of perpendicular lines, 21–22, 26
as rate of change, 88–98
average, 88, 93
calculating, 88–91
slope addition property for, 22–23, 29–30
vertical change property for, 27
slope–intercept form of equation of line, 9, 13, 27, 34–37
solution(s). *See also* graph(s)/graphing
checking proposed, 533
defined, 7
of equation in one variable, 633
extraneous, 460–61, 533, 539
of inequality, 592
of systems of inequalities in two variables, 595, 598
of systems of linear equations in three variables, 389
of systems of linear inequality in one variable, 146
of systems of nonlinear equations, 622, 626
solution set, 7, 633
defined, 7
of inequality, 592
of linear inequality in one variable, 146, 147–48, 151
of linear system, 105–7
of nonlinear systems of equations, 622
of system of inequalities in two variables, 595–96
speed–distance–time applications, 473–76
square(s), 163
of binomial, 287–89
difference of two, 289
factoring, 310, 312–13
perfect, 360, 502
square root
of negative number, 366
principal, 176
product property for, 359–61
quotient property for, 361–62
simplifying, 360–62
square root curve, 528
equation of, 543–44
square root equation, 535–37
square root function(s), 507
finding equation of, 542–44
graphs of, 523, 526–29
modeling with, 542–50
operations with, 529–30
x-intercepts of, 537–38
square root model, 508, 544–46
equation of, 545
estimates and predictions by using, 545–46
square root property
defined, 362
quadratic equations solved with, 359–71, 382, 383
of form $a(x - h)^2 + k = p$, 368
of form $(px + q)^2 = k$, 363–64
of form $x^2 = k, k \geq 0$, 362–63
of form $x^2 = k, k < 0$, 367
squaring property of equality, 533
standard form
quadratic function in, 277, 290
graph of, 347–59
quadratic model in, 396–97
STAT list editor, 642, 647
steepness. *See* slope(s)
storing a value, on graphing calculator, 645
Stravinsky, Igor, 55

stretching graphs, 337
stride rate, 73
substitution
to find intersection point, 125–26
linear systems of equations solved by, 113–15
dependent system, 118–19
inconsistent systems, 118
linear systems, 114–15
nonlinear systems of equations solved by, 622–23, 624
subtraction
of complex numbers, 601–3
order of operations and, 631
of polynomials, 276–77
of radical expressions, 511–12
of rational expressions, 438, 445–49
with a common denominator, 445
with different denominators, 445–47
of real numbers, 630
subtraction property of inequalities, 144
sum function, 258, 279–81, 444
sum of two cubes, 313–14. *See also* addition
symmetric points, 347–49
Syntax error, 646
system(s) of equations. *See also* systems of linear equations
modeling with, 94
quadratic equations, 406–7
nonlinear, 622–26
systems of linear equations, 103–61
comparing techniques for solving, 121
elimination to solve, 115–18
equations in three variables, 389–90
linear systems, 116–17
system with fractions, 117–18
five-step problem-solving method for, 131–32
graphs and tables to solve, 103–13
intersection point of graphs of two models, 104, 105
predictions by using two linear models, 103–4
interest problems, 135–37
functions to model, 136–37
mixture problems, 137–39
to model data, 124–31
rate of change to find, 126–27
table of data to find, 125–26
substitution to solve, 113–15
dependent system, 118–19
inconsistent systems, 118
linear systems, 114–15
three-equation systems, 388–96
in three variables, 388–96
solution of, 389
solving, 389–91
two linear equations. *See* linear system
value problems, 132–35
functions to model, 134–35
systems of linear inequalities in one variable, 139–40, 143–54
defined, 146
examples of, 145
models to compare quantities, 143–44
properties of inequalities, 144–45
satisfying, 146
solution of, 146
solution set of, 146, 147–48, 151
solving, 145–49
three-part inequalities, 149–51
systems of linear inequalities in two variables, 595–99
solution and solution set of, 595–97, 598
graphing, 596–97
modeling with, 597–98

T

table(s)
equation in one variable solved by using, 120, 121
equations of linear models by using data displayed in, 66–68
exponential model from, 204–6
to find output and input of function, 75–76
linear function as, 43
of solutions of systems of linear equations in two variables, 108
systems of linear equations solved with, 103–13
intersection point of graphs of two models, 104, 105
predictions by using two linear models, 103–4
systems of linear equations by using, 125–26
on TI-83 or TI-84 graphing calculator, 641–42
Ask option, 642
for two equations, 642
term(s)
of polynomials, 274, 275–76
coefficient of, 275
combining like, 275–76, 632–33
terminating number, 628
term number(s)
of sequence, 555
arithmetic, 556
geometric, 563–64
of series
arithmetic, 568
geometric, 574
three-equation systems, 388–94
three-part inequalities in one variable, 149–50
three-step method of graphing a quadratic function in vertex form, 339–40
three-step process for finding the inverse of a model, 223–24
TI-82 graphing calculator, 637
TI-83 or TI-84 graphing calculator, 637–48. *See also* graphing calculator(s)
TI-85 and TI-86 graphing calculators, 637
time–distance–speed applications, 473–76
Tips for Success
affirmations, 39
asking questions, 197
assignment completion, 281
changing study methods, 172
classmates, getting in touch with, 62
class time, 13
creating examples to learn a definition or property, 394
cross-checking, 121
desire for and faith in success, 188
exercises completed without help, 181
instructor's office hours, 23
math journal, 82
note taking, 140
persistence, 431
problem solving, 376
rereading a problem, 343
reviewing material, 226
reviewing notes, 292
studying with a classmate, 127
study teams, 151
study time and place, 13, 310
taking a break, 241
tests
calmness during, 264
mind maps for final exam, 547
partial credit by showing what you know, 207
planning for final exam, 491
practice exams, 47
retaking quizzes and exams, 523
scanning problems, 326
study group to prepare for final exam, 409
studying in test environment, 94–95
3 × 5 cards as memorizing aids, 508
verifying work, 70
visualization, 109
writing summaries, 449
total-value formula, 132
TRACE command, 11, 34, 638, 640
tracing a curve without a scattergram, 638
tracing a scattergram, 640
translations
of absolute value function, 530
horizontal, 526–27
of quadratic function, 338–39
triangle, right, 606, 607. *See also* Pythagorean theorem
isosceles, 613
trinomial(s), 285
factoring, 292, 295–304
completely, 300–301
of form $x^2 + bx + c$, 296–98
greatest common factor, 299–300, 307
by grouping, 308–9
multiplying vs., 295–96
by trial and error, 305–7
with two variables, 298
when leading coefficient is negative, 301–2
with negative constant terms, 298
perfect-square, 288–89, 297, 368, 371–72, 374
factoring, 372
with positive constant terms, 296–97
tsunami, 510
two-variable linear inequalities, 592–95
systems of, 595–99

U

unit analysis, 91–92
defined, 92
unit *i* (imaginary), 366, 601
unit ratio, 88

unlike terms, 275
up–down shifts (vertical translation), 338

V

value problems, 132–35
  functions to model, 134–35
variable(s)
  approximately exponentially related, 200
  approximately linearly related, 57
    slope of, 92–94
  defined, 631
  dependent, 2–3, 77, 78
  exponentially related, 200
  independent, 2–3, 61, 77, 78
variation, 481–95
  direct, 481–84
    changes in values for, 482
    defined, 482
    model for, 483–84
    ratios to find constant of, 488–89
    using one point to find equation for, 482–84
    variable varying directly as an expression, 484
  inverse, 475–76, 484–90
    changes in values for, 485
    defined, 485
    equation for, 485
    model for, 486–87, 488
    one point to find equation for, 485–87
    products to find constant of, 489–90
    variable varying inversely as an expression, 488
variation constant (constant of proportionality), 482, 485
velocity, escape, 525
vertex
  of parabola, 278
  of quadratic function, 340
    symmetric points to find, 347–49
    vertex formula to find, 350–52
vertex form, quadratic model in, 342–43
  predictions with, 364–65
vertex form of quadratic function, 336–46
vertical asymptotes, 423
  domains of rational functions and, 427–28, 430–31
vertical change property, 27
vertical lines, 12
vertical line test, 43–44, 46
vertical translation (up–down shifts), 338
visualization, 109

W

windchill factor, 65
Window range error, 647
window settings, graphing calculator, 639

X

x-axis, 628
  reflections across, 527
x-coordinate, 7
  of vertex, 348
x-intercept
  of curve, 645
  of function, 302
  of parabola, 364
  solutions to polynomial equations and, 318–19
  of square root function, 537–38

Y

y-axis, 628
y-coordinate, 7
y-intercept
  equation of parabola by using, and two other points, 393–94
  equation of square root curve by using, 543–44
  of exponential function, 185
  of linear equation, 27
Y_n references, 646

Z

ZDecimal command, 8, 639
zero as exponent, 165
zero factor property, 315, 317
ZInteger command, 639
ZoomFit command, 639
Zoom In command, 638–39
Zoom menu, 638–39
Zoom Out command, 639
ZoomStat command, 639
ZSquare command, 9, 10, 639
ZStandard command, 9, 10, 639